Mathematics Without Boundaries

Themistocles M. Rassias • Panos M. Pardalos
Editors

Mathematics Without Boundaries

Surveys in Pure Mathematics

Springer

Editors
Themistocles M. Rassias
Department of Mathematics
National Technical University of Athens
Athens, Greece

Panos M. Pardalos
Dept. Industrial & Systems Engineering
University of Florida
Gainesville, FL, USA

ISBN 978-1-4939-4312-8 ISBN 978-1-4939-1106-6 (eBook)
DOI 10.1007/978-1-4939-1106-6
Springer New York Heidelberg Dordrecht London

Mathematics Subject Classification (2010): 05A16, 05A, 15, 05A18, 11B75, 30C45, 26D15, 31A10, 45P05, 47G10, 47A07, 46B04, 46B20, 46A16, 46E40

Softcover reprint of the hardcover 1st edition 2014

Printed on acid-free paper

Springer is part of Springer Science+Business Media (www.springer.com)

Preface

"Mathematics Without Boundaries: Surveys in Pure Mathematics" consists of chapters written by eminent scientists from the international mathematical community, who present significant advances in several theories, methods and problems of mathematical analysis, discrete mathematics, geometry and their applications.

These contributions focus on both old and recent developments in functional analysis, harmonic analysis, complex analysis, operator theory, combinatorics, functional equations, differential equations as well as a variety of applications.

Furthermore some review works are published in this book which could prove to be particularly useful for a broader audience of readers in mathematical sciences and especially to graduate students who search for the latest information.

It is our pleasure to express our thanks to all the contributors of chapters in this book.

We would like to thank Dr. Michael Batsyn for his invaluable help during the preparation of this publication. Last but not least, we would like to acknowledge the superb assistance that the staff of Springer has provided for the publication of this book.

Athens, Greece, Themistocles M. Rassias
Gainesville, FL, USA Panos M. Pardalos

Contents

Contributors

Dorin Andrica Faculty of Mathematics and Computer Science, Babeş-Bolyai University, Cluj-Napoca, Romania

Javier Fernández de Bobadilla Instituto de Ciencias Matemáticas CSIC-UAM-UC3M-UCM, Consejo Superior de Investigaciones Cientìficas, Madrid, Spain

Daniel Breaz University of Alba Iulia, Alba Iulia, Romania

L.P. Castro CIDMA-Center for Research and Development in Mathematics and Applications, Department of Mathematics, University of Aveiro, Aveiro, Portugal

O. Chau University of La Réunion, PIMENT EA4518, La Réunion, France

Wing-Sum Cheung Department of Mathematics, The University of Hong Kong, Hong Kong, China

Stanisław Domoradzki Faculty of Mathematics and Natural Sciences, University of Rzeszów, Rzeszów, Poland

H. G. Feichtinger Faculty of Mathematics, University Vienna, Wien, Austria

Javier Fresán LAGA, UMR 7539, Institut Galilée, Universitè Paris 13, Villetaneuse, France

H. Fujiwara Graduate School of Informatics, Kyoto University, Kyoto, Japan

Paolo Giordano Fakultät für Mathematik, University of Vienna, Wien, Austria

D. Goeleven University of La Réunion, PIMENT EA4518, La Réunion, France

N. K. Govil Department of Mathematics and Statistics, Auburn University, Auburn, USA

Vijay Gupta Department of Mathematics, Netaji Subhas Institute of Technology, Dwarka, New Delhi, India

Mehdi Hassani University of Zanjan, University Blvd., Zanjan, Iran

Rabha W. Ibrahim Institute of Mathematical Sciences, University Malaya, Kuala Lumpur, Malaysia

Eugen J. Ionascu Math Department, Columbus State University, Columbus, GA, USA

Stamatis Koumandos Department of Mathematics and Statistics, University of Cyprus, Nicosia, Cyprus

Jung Rye Lee Department of Mathematics, Daejin University, Pocheon, South Korea

Fridoun Moradlou Department of Mathematics, Sahand University of Technology, Tabriz, Iran

Dumitru Motreanu Département de Mathématiques, Université de Perpignan, Perpignan, France

Vicente Muñoz Facultad de Ciencias Matemàticas, Universidad Complutense de Madrid, Madrid, Spain

Aniceto Murillo Departamento de Álgebra, Geometría y Topología, Universidad de Málaga, Málaga, Spain

A. B. Németh Faculty of Mathematics and Computer Science, Babeş Bolyai University, Cluj-Napoca, Romania

S. Z. Németh School of Mathematics, University of Birmingham, Birmingham, UK

K. Nowak Wayne, PA, USA

Department of Computer Science, Drexel University, Philadelphia, PA, USA

R. Oujja University of La Réunion, PIMENT EA4518, La Réunion, France

Nihal Yılmaz Özgür Faculty of Science and Literature, Department of Mathematics, Balıkesir University, Balıkesir, Turkey

M. Pap Faculty of Sciences, University of Pécs, Pécs, Hungary

Choonkil Park Research Institute for Natural Sciences, Hanyang University, Seoul, South Korea

Agostino Prástaro Department SBAI-Mathematics, University of Roma La Sapienza, Roma, Italy

Themistocles M. Rassias Department of Mathematics, National Technical University of Athens, Athens, Greece

M.M. Rodrigues CIDMA-Center for Research and Development in Mathematics and Applications, Department of Mathematics, University of Aveiro, Aveiro, Portugal

S. Saitoh Department of Mathematics, Institute of Reproducing Kernels, Kiryu, Japan

Małgorzata Stawiska Mathematical Reviews, Ann Arbor, MI, USA

László Tóth Institute of Mathematics, Universität für Bodenkultur, Vienna, Austria

Department of Mathematics, University of Pécs, Pécs, Hungary

V.K. Tuan Department of Mathematics, University of West Georgia, Carrollton, GA, USA

Mihai Turinici "A. Myller" Mathematical Seminar, "A. I. Cuza" University, Iaşi, Romania

Nicoleta Ularu University of Pitesti, Piteşti, Romania

Abraham Albert Ungar Department of Mathematics, North Dakota State University, Fargo, ND, USA

Patrick Winkert Technische Universität Berlin, Institut für Mathematik, Berlin, Germany

Bicheng Yang Department of Mathematics, Guangdong University of Education, Guangzhou, P. R. China

Xiuzhong Yang College of Mathematics and Information Science, Hebei Normal University, and Hebei Key Laboratory of Computational Mathematics and Applications, Shijiazhuang, Hebei, P. R. China

Chang-Jian Zhao Department of Mathematics, China Jiliang University, Hangzhou, People's Republic of China

Xiaopeng Zhao Department of Mathematics, Zhejiang University, Hangzhou, Zhejiang, P. R. China

Some Unexpected Connections Between Analysis and Combinatorics

Dorin Andrica and Eugen J. Ionascu

2000 AMS Subject Classification: 05A16 (05A15; 05A18; 11B75)

Abstract We go through a series of results related to the k-signum equation $\pm 1^k \pm 2^k \pm \cdots \pm n^k = 0$. We are investigating the number $S_k(n)$ of possible writings and the asymptotic behavior of these numbers, as k is fixed and $n \to \infty$. The results are presented in connections with the Erdös–Surányi sequences. Analytic methods and algebraic ones are employed in order to predict the asymptotic behavior in general and to study in detail various situations for small values of k. Some simplifications and further ramifications are discussed in the end about the recent proof of Andrica–Tomescu conjecture.

Keywords Derivative · Partition of integers · Asymptotic formula · Integral representation · Random variable · Erdös–Surányi sequence

1 Introduction

One of the simplest differential equations, $f' = g$, for a given function g, leads to a classic theory in analysis, part of which, the usual calculus is just the first approximation. If the derivative f' is understood in the usual sense, it turns out that two necessary conditions on g are well known: it must be of first class Baire and it needs to have the Darboux property. This is almost a characterization since, by G. Choquet's result, if a function g is of the first Baire class and it has the Darboux property, there exists a differentiable function f and a homeomorphism h such that $g = f' \circ h$.

D. Andrica (✉)
Faculty of Mathematics and Computer Science, Babeş-Bolyai University,
Str. Kogalniceanu 1, 400084, Cluj-Napoca, Romania
e-mail: andrica@math.ubbcluj.ro

E. J. Ionascu
Math Department, Columbus State University, 4225 University Avenue,
31907, Columbus, GA, USA
e-mail: ionascu@columbusstate.edu; math@ejionascu.ro

T. M. Rassias, P. M. Pardalos (eds.), *Mathematics Without Boundaries*,
DOI 10.1007/978-1-4939-1106-6_1, © Springer Science+Business Media, LLC 2014

We begin Sect. 2 with some results in this area about some elementary functions having discontinuities of the second kind and characterize when these are derivatives, using mostly elementary methods. In the end, we arrive at an analytical interpretation of the number of vectors $\underbrace{[\pm 1, \pm 1, \ldots, \pm 1]}_{n}$ whose dot product with a fixed vector $[a_1, a_2, \cdots, a_n] \in \mathbb{N}^n$ gives a desired output. Of course, there is room for asking questions like, "what is the range of outputs?" or "what is the most favorable output?". Before reading further, we challenge the reader to find the derivative at zero for the function

$$f(x) = \int_0^x \cos\left(\frac{1}{t}\right)\cos\left(\frac{3}{t}\right)\cos\left(\frac{5}{t}\right)\cos\left(\frac{7}{t}\right) dt, \ x \in \mathbb{R}.$$

In Sect. 3, we look at classical sequences that have been studied in connection with questions mentioned earlier. An integral formula of the main combinatorial numbers at interest is derived here. An asymptotic conjecture is formulated as a generalization of Andrica–Tomescu [6] using an argument due to S. Finch [18]. Certainly, probabilistic methods can be employed here and we give in detail this argument based on a variant of central limit theorem due to A. C. Berry [9].

There are famous problems that appeared over the years, like in the Putnam competition or some other mathematics olympiads, which relate intimately to this subject, in Sect. 4.

Finally, in the last section, we go over Sullivan's proof [31] of Andrica–Tomescu conjecture and point out what can be shown in general and what are the difficulties in the general case.

2 Derivatives

In this section, we introduce the class $\mathcal{D}$ consisting all derivatives defined on the real axis, and in Theorem 1, we derive a result about the product of some special derivatives, which opens the door to some difficult problems involving partitions of integers.

Definition 1 A function $f : \mathbb{R} \to \mathbb{R}$ is called a *derivative* if there exists a differentiable function $g : \mathbb{R} \to \mathbb{R}$ such that $g' = f$. We denote by $\mathcal{D}$ the class of all derivatives.

From examples and properties, we noticed that any continuous function is in $\mathcal{D}$. It is known that every function in $\mathcal{D}$ has the Darboux property, that is, it transforms any open interval into an interval (see [27]). It is then clear that $\mathcal{D}$ is a linear space under usual addition of functions and multiplication with real scalars. Because of the Darboux property any function of the form

$$\chi_{a,b}(x) = \begin{cases} a & \text{if } x \neq 0 \\ b & \text{if } x = 0, \end{cases} \tag{1}$$

is in $\mathcal{D}$ if and only if $a = b$.

Lemma 1 *For $a \in \mathbb{R} \setminus \{0\}$ and $b \in \mathbb{R}$, the function*

$$f_{a,b}(x) = \begin{cases} \cos \frac{a}{x} & \text{if } x \neq 0 \\ b & \text{if } x = 0, \end{cases}$$

is in $\mathcal{D}$ if and only if $b = 0$.

Proof We consider the function

$$h(x) = \begin{cases} x^2 \sin \frac{a}{x} & \text{if } x \neq 0 \\ 0 & \text{if } x = 0, \end{cases}$$

which is easy to see that it is differentiable with

$$h'(x) = \begin{cases} 2x \sin \frac{a}{x} - a \cos \frac{a}{x} & \text{if } x \neq 0 \\ 0 & \text{if } x = 0, \end{cases} .$$

Hence, we see that we can write

$$f_{a,b}(x) = \frac{2}{a} \begin{cases} x \sin \frac{a}{x} & \text{if } x \neq 0 \\ 0 & \text{if } x = 0 \end{cases} - \frac{1}{a} h'(x) + \begin{cases} 0 & \text{if } x \neq 0 \\ b & \text{if } x = 0 \end{cases}$$

$$= \frac{2}{a} k(x) - \frac{1}{a} h'(x) + \chi_{0,b}(x).$$

Since k is continuous, $\frac{2}{a}k - \frac{1}{a}h' \in \mathcal{D}$, which implies that $f_{a,b} \in \mathcal{D}$ if and only if $\chi_{0,b} \in \mathcal{D}$, where $\chi_{0,b}$ is a function defined in (1). The previous observation then gives us the claim. □

Lemma 1 allows us to show that $\mathcal{D}$ is not closed under usual multiplication of functions. Indeed, if we take $f_{a,0}^2$ with $f_{a,0}$ as in Lemma 1, we observe that

$$f_{a,0}^2 = \frac{1}{2} \left(f_{2a,0} + \chi_{1,0} \right)$$

and so, if $f_{a,0}^2 \in \mathcal{D}$ that would lead to the contradiction $\chi_{1,0} \in \mathcal{D}$. For other remarks involving such examples we refer to the paper of D. Andrica and S. Buzeţeanu ('85, [4]).

M. Iosifescu ('58, [21]) showed that if f_1 and f_2 are in $\mathcal{D}$, then the product $f_1 f_2$ may not even have the Darboux property. We can recapture this fact by taking

$\phi = f_{1,0} \in \mathcal{D}$ and

$$\psi(x) = \begin{cases} \cos^2 \frac{1}{x} & \text{if } x \neq 0 \\ 1/2 & \text{if } x = 0, \end{cases} = \chi_{1/2,1/2}(x) + \frac{1}{2} f_{2,0}(x)$$

which shows that $\psi \in \mathcal{D}$. Hence, the functions $f_1 = \phi + \psi$ and $f_2 = \phi - \psi$ are in $\mathcal{D}$. Then, we calculate $f_1 f_2 = \phi^2 - \psi^2$ that can be simplified to

$$(f_1 f_2)(x) = \frac{1}{4} \begin{cases} \sin^2 \frac{2}{x} & \text{if } x \neq 0 \\ -1 & \text{if } x = 0. \end{cases}$$

It is clear that this function transforms every interval containing 0 into $[0, 1] \cup \{-1\}$, hence, it does not have the Darboux property.

In comparison to the previous example, a totally surprising result of V. Hruska ('46, [19]) says that for two functions $f, g \in \mathcal{D}$, f/g does have the Darboux property ($g \neq 0$). Other properties of functions in $\mathcal{D}$ are given in the paper of A. M. Bruckner and J. L. Leonard ('66, [12]).

The following lemma is an easy exercise of mathematical induction but it is very useful when we want to get the linear form of a product of cosines.

Lemma 2 *For $x_1, x_2, \ldots, x_m$ arbitrary real numbers, we have*

$$\cos x_1 \cos x_2 \cdots \cos x_m = \frac{1}{2^m} \sum \cos(\pm x_1 \pm x_2 \pm \cdots \pm x_m),$$

where the sum is over all possible choices of signs + and −.

Using this identity one can prove the following generalization of Lemma 1.

Theorem 1 *(D. Andrica, '00, [3]) Let $n_1, \cdots, n_k$ be positive integers such that $n_1 \cdots + n_k \geq 1$, $\alpha_1, \ldots, \alpha_k$ mutually distinct nonzero real numbers, and the function $g_{n_1,\ldots,n_k}^{\alpha_1,\ldots,\alpha_k} : \mathbb{R} \to \mathbb{R}$ be defined by*

$$g_{n_1,\ldots,n_k}^{\alpha_1,\ldots,\alpha_k}(x) = \begin{cases} \cos^{n_1} \frac{\alpha_1}{x} \cdots \cos^{n_k} \frac{\alpha_k}{x} & \text{if } x \neq 0 \\ \alpha & \text{if } x = 0. \end{cases}$$

Then, $g_{n_1,\ldots,n_k}^{\alpha_1,\ldots,\alpha_k} \in \mathcal{D}$ if and only if $\alpha = \frac{1}{2^{n_1+\cdots+n_k}} S(n_1, \ldots, n_k; \alpha_1, \ldots, \alpha_k)$, where the integer $S(n_1, \ldots, n_k; \alpha_1, \ldots, \alpha_k)$ is given by the number of choices of signs + and − such that

$$\underbrace{\pm\alpha_1 \pm \cdots \pm \alpha_1}_{n_1 \text{ times}} \pm \cdots \underbrace{\pm\alpha_k \pm \cdots \pm \alpha_k}_{n_k \text{ times}} = 0.$$

Proof Using Lemma 2, we can write

$$g_{n_1,\ldots,n_k}^{\alpha_1,\ldots,\alpha_k}(x) =$$

$$\begin{cases} \frac{1}{2^{n_1+\cdots+n_k}} \sum_{+,-} \cos\big(\underbrace{\pm\frac{\alpha_1}{x} \pm \cdots \pm \frac{\alpha_1}{x}}_{n_1\ times} \pm \cdots \pm \underbrace{\frac{\alpha_k}{x} \pm \cdots \pm \frac{\alpha_k}{x}}_{n_k\ times}\big) & \text{if } x \neq 0 \\ \alpha & \text{if } x = 0, \end{cases} =$$

$$\frac{1}{2^{n_1+\cdots+n_k}} \sum_{+,-\neq 0} f_{(\underbrace{\pm\alpha_1 \pm \cdots \pm \alpha_1}_{n_1\ times} \pm\cdots \underbrace{\pm\alpha_k \pm \cdots \pm \alpha_k}_{n_k\ times},0)} +$$

$$\begin{cases} \frac{1}{2^{n_1+\cdots+n_k}} S(n_1, \ldots, n_k; \alpha_1, \ldots, \alpha_k) & \text{if } x \neq 0 \\ \alpha & \text{if } x = 0 \end{cases},$$

where the sum $\sum_{+,-\neq 0} f_{(\underbrace{\pm\alpha_1 \pm \cdots \pm \alpha_1}_{n_1\ times} \pm\cdots \underbrace{\pm\alpha_k \pm \cdots \pm \alpha_k}_{n_k\ times},0)}$ is taken over all combinations $\underbrace{\pm\alpha_1 \pm \cdots \pm \alpha_1}_{n_1\ times} \pm \cdots \underbrace{\pm\alpha_k \pm \cdots \pm \alpha_k}_{n_k\ times}$ different from 0. According to Lemma 1, each function $f_{(\underbrace{\pm\alpha_1 \pm \cdots \pm \alpha_1}_{n_1\ times} \pm\cdots \underbrace{\pm\alpha_k \pm \cdots \pm \alpha_k}_{n_k\ times},0)}$, corresponding to such a combination, belongs to $\mathcal{D}$. Finally, it follows that $g_{n_1,\cdots,n_k}^{\alpha_1,\cdots,\alpha_k} \in \mathcal{D}$ if and only if

$$\chi_{\frac{1}{2^{n_1+\cdots+n_k}} S(n_1,\cdots,n_k;\alpha_1,\cdots,\alpha_k),\alpha} \in \mathcal{D},$$

and this is equivalent to the relation $\alpha = \frac{1}{2^{n_1+\cdots+n_k}} S(n_1, \cdots, n_k; \alpha_1, \cdots, \alpha_k)$. □

As an immediate application to Theorem 1, we obtained that the constant α for which the function $g_n^1 : \mathbb{R} \to \mathbb{R}$, defined by

$$g_n^1(x) = \begin{cases} \cos^n \frac{1}{x} & \text{if } x \neq 0 \\ \alpha & \text{if } x = 0, \end{cases}$$

is a derivative, is given by $\frac{1}{2^n} S(n; 1)$, where $S(n; 1)$ is the number of choices of signs $+$ and $-$ such that $\underbrace{\pm 1 \cdots \pm 1}_{n\ times} = 0$. If n is odd, clearly we have $S(n; 1) = 0$. If n is even, we have $S(n; 1) = \binom{n}{n/2}$, hence

$$\alpha = \begin{cases} 0 & \text{if } n \text{ is odd} \\ \frac{1}{2^n}\binom{n}{n/2} & \text{if } n \text{ is even.} \end{cases}$$

This result was obtained with a different method by the first author in [2].

Another application may be obtained when the distinct nonzero real numbers $\alpha_1, \cdots, \alpha_k$ are linear independent over $\mathbb{Q}$. In this case, we can write the relation $\underbrace{\pm\alpha_1 \pm \cdots \pm \alpha_1}_{n_1\ times} \pm \cdots \underbrace{\pm\alpha_k \pm \cdots \pm \alpha_k}_{n_k\ times} = 0$ as $\underbrace{(\pm 1 \pm \cdots \pm 1)}_{n_1\ times}\alpha_1 \pm \cdots \underbrace{(\pm 1 \pm \cdots \pm 1)}_{n_k\ times}\alpha_k = 0$, and from the linear independence, we get $\underbrace{\pm 1 \pm \cdots \pm 1}_{n_1\ times} = \cdots = \underbrace{\pm 1 \pm \cdots \pm 1}_{n_k\ times} = 0.$

Therefore, the function $g_{n_1,\ldots,n_k}^{\alpha_1,\ldots,\alpha_k}$ in Theorem 1, is a derivative if and only if

$$\alpha = \begin{cases} \frac{1}{2^{n_1+\cdots+n_k}}\binom{n_1}{n_1/2}\cdots\binom{n_k}{n_k/2} & \text{if } n_1,\cdots,n_k \quad \text{are even} \\ 0 \quad \text{otherwise.} \end{cases}$$

3 A Problem About the Number of Partitions

This section is a part of the work done by the first author and I. Tomescu in [6] about the interpretation of the number $S(n_1,\ldots,n_k;\alpha_1,\ldots,\alpha_k)$. Let $n_1,\ldots,n_k$ be natural numbers and let M be the multiset

$$M = \{\underbrace{\alpha_1,\cdots,\alpha_1}_{n_1\ times},\cdots,\underbrace{\alpha_k,\cdots,\alpha_k}_{n_k\ times}\}.$$

In [6], it is shown that $S(n_1,\ldots,n_k;\alpha_1,\ldots,\alpha_k)$ equals the number of ordered partitions of M having equal sums, i.e., the number of pairs (C_1, C_2) such that

(i) $C_1 \cup C_2 = M$ and $C_1 \cap C_2 = \emptyset$
(ii) $\sum\limits_{x\in C_1} x = \sum\limits_{x\in C_2} x = \frac{1}{2}\sum_{i=1}^{k} n_i\alpha_i.$

If we assume that M is a set of k positive integers then $S(n_1,\ldots,n_k;\alpha_1,\ldots,\alpha_k)$ is the term not depending on z in the expansion

$$F(z) = \left(z^{\alpha_1} + \frac{1}{z^{\alpha_1}}\right)^{n_1}\left(z^{\alpha_2} + \frac{1}{z^{\alpha_2}}\right)^{n_2}\cdots\left(z^{\alpha_k} + \frac{1}{z^{\alpha_k}}\right)^{n_1}.$$

Let us observe that we can write $F(z) = S(n_1,\ldots,n_k;\alpha_1,\ldots,\alpha_k) + \sum\limits_{j\in\mathbb{Z}\setminus\{0\}} c_j z^j$, for some $c_j \in \mathbb{Z}$. If we set $z = \cos t + i \sin t$, then we get the equivalent form

$$2^{n_1+\cdots+n_k}\prod_{j=1}^{k}(\cos\alpha_j t)^{n_j} = S(n_1,\ldots,n_k;\alpha_1,\ldots,\alpha_k) + \sum_{j\in\mathbb{Z}\setminus\{0\}} c_j(\cos jt + i\sin jt).$$

Integrating this last identity over $[0, 2\pi]$, we obtain the following integral formula for the number $S(n_1,\ldots,n_k;\alpha_1,\ldots,\alpha_k)$:

$$S(n_1,\ldots,n_k;\alpha_1,\ldots,\alpha_k) = \frac{2^{n_1+\cdots+n_k}}{2\pi}\int_0^{2\pi}\prod_{j=1}^{k}(\cos\alpha_j t)^{n_j}\,dt. \tag{2}$$

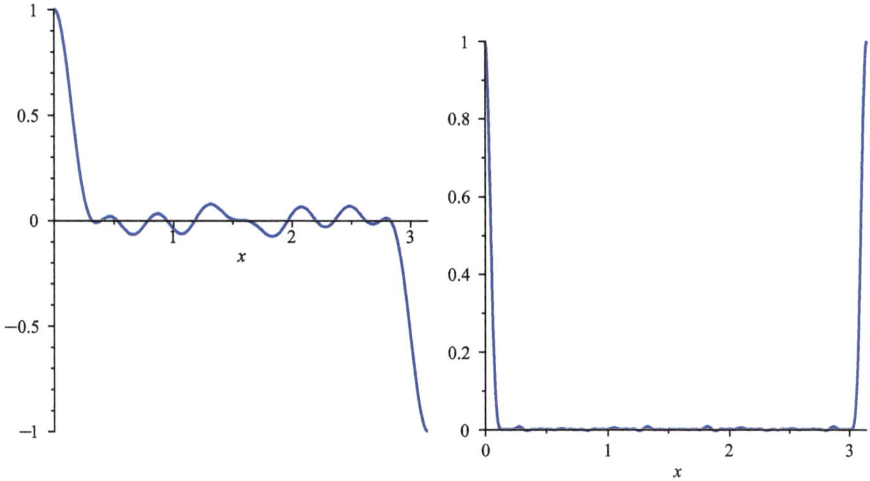

Fig. 1 Graph of f_5 and f_{12} on $[0, \pi]$

One particular case here, which brings into our attention an old combinatorial problem (see [17, 18, 23] [1]), is if $n_1 = n_2 = \cdots = n_k = 1$ and $\alpha_j = j$, $j = 1, 2, \cdots, k$.

Let us shorten the notation by setting $S(n) := S(1, 1, \cdots, 1; 1, 2, 3, \cdots, n)$, which is the number of ways of choosing $+$ and $-$ such that $\pm 1 \pm 2 \pm 3 \pm \cdots \pm n = 0$. The above equation is called by S. Finch [18] the *signum equation*. This is the sequence A063865 in the The On-Line Encyclopedia of Integer Sequences. As an example, we have $S(40) = 5830034720$.

In the paper [6], it is shown that:

(i) $S(n) = 0$ if and only if $n \equiv 1$ or 2 (mod 4)

(ii) for $n \geq 8$ we have $S(n) \geq 6S(n-4)$

(iii) $\lim_{n\to\infty} S(4n)^{\frac{1}{4n}} = \lim_{n\to\infty} S(4n+3)^{\frac{1}{4n+3}} = 2$,

and it was conjectured the following asymptotic formula

$$\lim_{\substack{n\to\infty \\ n\equiv 0 \text{ or } 3 (mod 4)}} \frac{S(n)}{\frac{2^n}{n\sqrt{n}}} = \sqrt{\frac{6}{\pi}},$$

which was recently proved using analytic methods by B.D. Sullivan (2013) (see [31]).

From (2), we easily get

$$S(n) = \frac{2^n}{\pi} \int_0^{\pi} \cos t \cos 2t \cdots \cos nt dt. \tag{3}$$

[1] The problem A5 on the William Putnam Competition (1985) asked to find all all positive integers $m \leq 10$ such that $\int_0^{2\pi} \prod_{j=1}^{m} \cos jx dx$ is not zero.

Using this integral representation one can check for instance, that $S(n) = 0$ if $n \equiv 1$ or $2 \pmod 4$, by showing that the integrant in (3), $f_n(t) := \prod_{j=1}^{n} \cos jt$ is a function symmetric with respect to the point $(\pi/2, 0)$. For clarity, we include the graphs of f_5 and f_{12} in Fig. 1.

For small n, with no difficulty, integrals as in (3) can be computed using, for instance, Mathematica or Maple. It takes only a few seconds to get $S(24) = 187692$.

Another particular case is if $n_1 = n_2 = \cdots = n_k = 1$ and $\alpha_j = j^2$, $j = 1, 2, \cdots, k$. In this case, $S_2(n) := S\left(1, 1, \cdots, 1; 1^2, 2^2, 3^2, \cdots, n^2\right)$ is the sequence A158092 or the number of ways to choose $+$ and $-$ such that $\pm 1 \pm 2^2 \pm 3^2 \pm \cdots \pm n^2 = 0$. In this case, it seems that a similar asymptotic expression yields

$$S_2(n) = 2^n \sqrt{\frac{10}{n^5 \pi}}(1 + o(1)),$$

and Pietro Majer has one more on OEIS (see A158118)

$$S_3(n) := S(1, 1, \cdots, 1; 1^3, 2^3, 3^3, \cdots, n^3) = 2^n \sqrt{\frac{14}{n^7 \pi}}(1 + o(1))$$

as $n \to \infty, n \equiv 0$ or $3 \pmod 4$.

S. Finch [18] has a very interesting probabilistic but heuristic approach to arrive at above two formulas, which allows one to find similar such asymptotic results. The idea of proof suggested by H.-K. Hwang [20] does not seem to generalize to higher powers.

For the sake of completeness, let us include next, Finch's heuristic argument about the asymptotic behavior in general. For $k > 0$, we consider the random variable E_j taking two values: j^k or $-j^k$ with probability 1/2 each. For $i \neq j$, each E_i is independent of E_j. We then let the random variable

$$S_n := E_1 + E_2 + \cdots + E_n. \tag{4}$$

Since each of the E_j has expectation zero, we have $E(S_n) = \sum_{j=1}^{n} E(E_j) = 0$. Because each of the E_j has variation $\sigma_j^2 = Var(E_j) = E\left(|E_j - E(E_j)|^2\right) = j^{2k}$, and the random variables are independent, we get

$$s_n^2 = Var(S_n) = \sum_{j=1}^{n} Var\left(E_j\right) = \sum_{j=1}^{n} j^{2k}.$$

We would like to use the so called Berry–Esseen theorem (see for example [9]), that is a better version of central limit theorem in this case.

Theorem 2 *Let $X_1, X_2, \ldots,$ be independent random variables with $E(X_j) = 0$, $E\left(X_j^2\right) = \sigma_j^2 > 0$ and $\rho_j = E\left(|X_j|^3\right) < \infty$. If $G_n = (X_1 + X_2 + \cdots + X_n)/\sqrt{\sum_{j=1}^{n} \sigma_j^2}$, then*

$$\sup_{x \in \mathbb{R}} |P(G_n \leq x) - \frac{1}{\sqrt{2\pi}} \int_{-\infty}^{x} e^{-\frac{t^2}{2}} dt| \leq C_1 (\sum_{j=1}^{n} \sigma_j^2)^{-1/2} \max_{1 \leq j \leq n} \frac{\rho_j}{\sigma_j^2}, \tag{5}$$

for some universal constant C_1.

In our case, if we take $X_j = E_j$, we get $\rho_j = j^{3k}$ and so $\max\limits_{1\le j\le n}\frac{\rho_j}{\sigma_j^2} = n^k$. The inequality (5) then becomes

$$\sup_{x\in\mathbb{R}}|P(S_n \le xs_n) - \frac{1}{\sqrt{2\pi}}\int_{-\infty}^{x} e^{-\frac{t^2}{2}}dt| \le C_1\frac{n^k}{s_n}.$$

In particular for $y < x$, we have

$$|P(ys_n < S_n \le xs_n) - \frac{1}{\sqrt{2\pi}}\int_{y}^{x} e^{-\frac{t^2}{2}}dt| \le 2C_1\frac{n^k}{s_n}.$$

Let us denote by $S_k(n)$ the sequence $S(1,1,\cdots,1;1^k,2^k,3^k,\cdots,n^k)$. For an arbitrary $\epsilon \in (0,1)$ and $x = \frac{1-\epsilon}{s_n}$, $y = -\frac{1}{s_n}$, the inequality above can be written as

$$\left|\frac{1}{2^n}S_k(n) - \frac{1}{\sqrt{2\pi}}\int_{-\frac{1}{s_n}}^{\frac{1-\epsilon}{s_n}} e^{-\frac{t^2}{2}}dt\right| \le 2C_1\frac{n^k}{s_n}.$$

It is easy to show that for $\alpha > 0$ we have

$$\Sigma_\alpha(n) := \sum_{j=1}^{n} j^\alpha = \frac{n^{\alpha+1}}{\alpha+1} + O(n^\alpha), \tag{6}$$

which implies $s_n = \frac{n^{k+1/2}}{\sqrt{2k+1}}(1+o(1))$. Putting these together we see that there exist an universal constant C such that

$$\left|\frac{1}{2^n}S_k(n) - \sqrt{\frac{2(2k+1)}{\pi n^{2k+1}}}\right| \le \frac{C}{\sqrt{n}}.$$

This inequality suggests that we may conjecture in general that for every $k \in \mathbb{N}$

$$\lim_{\substack{n\to\infty \\ n\equiv 0 \text{ or } 3 (mod 4)}} \frac{S_k(n)}{\frac{2^n}{n^k\sqrt{n}}} = \sqrt{\frac{2(2k+1)}{\pi}}, \tag{7}$$

but at this moment we do not have a proof for $k \ge 2$.

4 Connections to Erdös–Surányi Type Problems

In this section, we present a special class of sequences of distinct positive integers, which give special representations of the integers. Some general results are presented in Theorems 3 and 4, and then we concentrate on specific examples.

Definition 2 We say that a sequence of distinct positive integers $\{a_m\}_{m\geq 1}$ is a *Erdös–Surányi sequence* if every integer may be written in the form

$$\pm a_1 \pm a_2 \pm \cdots \pm a_n$$

in infinitely many ways.

As an example of a Erdös–Surányi sequence, for every $k \in \mathbb{N}$, $a_n = n^k$ (see J.Mitek [25]). For instance, for $k = 1$ we have

$$m = (-1+2) + (-3+4) + \cdots + (-(2m-1) + 2m) + \cdots + \underbrace{[(n+1) - (n+2) - (n+3) + (n+4)]}_{=0}$$

For $k = 2$, we have the original result of Erdös and Surányi mentioned as a problem in the book [17], which is based on the identity $4 = (m+1)^2 - (m+2)^2 - (m+3)^2 + (m+4)^2$ and on the basis cases

$$1 = 1^2,\ 2 = -1^2 - 2^2 - 3^2 + 4^2$$
$$3 = -1^2 + 2^2,\ 4 = 1^2 - 2^2 - 3^2 + 4^2.$$

For $k = 3$, one may use the identity

$$-(m+1)^3 + (m+2)^3 + (m+3)^3 - (m+4)^3$$
$$+(m+5)^3 - (m+6)^3 - (m+7)^3 + (m+8)^3 = 48$$

and induction with a basis step for the first 48 positive integers.

An important result concerning the Erdös–Surányi sequences is the following:

Theorem 3 *(M.O. Drimbe, '83, [15]) Let $\{a_m\}_{m\geq 1}$ be a sequence of distinct positive integers such that $a_1 = 1$ and for every $n \geq 1$, $a_{n+1} \leq a_1 + \cdots + a_n + 1$. If the sequence contains infinitely many odd integers, then it is a Erdös–Surányi sequence.*

Proof Recall that a sequence of distinct positive integers is *complete* if every positive integer can be written as a sum of some distinct of its terms. From the first hypothesis, it follows that the sequence $\{a_m\}_{m\geq 1}$ is complete. For details, we refer to the papers of J. L. Brown Jr. ([10, 11]) and that of M. O. Drimbe ([15]).

Following the recent papers of M. Tetiva ([32, 33]), let us denote by u_n the partial sum $a_1 + \cdots + a_n$, $n \geq 1$. Because the sequence $\{a_m\}_{m\geq 1}$ has infinitely many odd terms, it follows that u_p is even and odd for infinitely many p's. Let m be an arbitrary positive integer, and let p be sufficiently large such that $u_p > m$, and u_p and m are of the same parity. The integer $s = (u_p - m)/2$ is less than u_p. Hence, it is the sum of some distinct terms of the sequence $\{a_m\}_{m\geq 1}$ with indices $\leq p$, that is

$$\frac{1}{2}\left(a_1 + \cdots + a_p - m\right) = \varepsilon_1 a_1 + \cdots + \varepsilon_p a_p,$$

for some $\varepsilon_1, \cdots, \varepsilon_p \in \{0, 1\}$. The last relation is equivalent to $m = (1 - 2\varepsilon_1)a_1 + \cdots + \left(1 - 2\varepsilon_p\right)a_p$, with $1 - 2\varepsilon_1, \cdots, 1 - 2\varepsilon_p \in \{-1, 1\}$, and the proof is finished since p can be selected in infinitely many ways. □

Unfortunately, the sequences $a_n = n^k$ do not satisfy the condition $a_{n+1} \leq a_1 + \cdots + a_n + 1, n \geq 1$, in Theorem 3, we cannot use this result to prove they are Erdös–Surányi sequences.These examples of Erdös–Surányi sequences are particular cases of the following result.

Theorem 4 *(M.O. Drimbe, '88,* [16]*) Let $f \in \mathbb{Q}[X]$ be a polynomial such that for any $n \in \mathbb{Z}$, $f(n)$ is an integer. If the greatest common factor of the terms of the sequence $\{f(n)\}_{n\geq 1}$ is equal to* 1*, then $\{f(n)\}_{n\geq 1}$ is an Erdös–Surányi sequence.*

Using this result, we can obtain other examples of Erdös–Surányi sequences: $a_n = (an-1)^k$ for any $a \geq 2$, and $a_n = \binom{n+s}{s}$ for any $s \geq 2$. Note that it is difficult to obtain a proof by induction for these sequences, similar to those given for the previous examples. Other discussions are given in the recent paper of D. Andrica and E. J. Ionascu ('13, [5]).

In the same spirit with formula (2), we obtain the following.

Theorem 5 *(D. Andrica and D. Văcăreţu, '06, [7]) Given a Erdös–Surányi sequence $\{a_m\}_{m\geq 1}$, then the number of representations of $k \in [-u_n, u_n]$, where $u_n = a_1 + \cdots + a_n$, in the form $\pm a_1 \pm a_2 \pm \cdots \pm a_n$, denoted here by $A_n(k)$, is given by*

$$A_n(k) = \frac{2^n}{\pi} \int_0^{\pi} \cos(kt) \prod_{j=1}^{k} \cos(a_j t) dt. \tag{8}$$

Proof Consider the function F_n is associated to the first n terms of the sequence $\{a_m\}_{m\geq 1}$

$$F_n(z) = (z^{a_1} + \frac{1}{z^{a_1}}) \cdots (z^{a_n} + \frac{1}{z^{a_n}}).$$

We may write

$$F_n(z) = \Sigma_{j=-u_n}^{u_n} A_n(j) z^j,$$

and observe that $A_n(k)$ is the term not containing z in $z^{-k} F_n(z)$. For $z = \cos t + i \sin t$, we have

$$z^{-k} F_n(z) = 2^n(\cos kt - i \sin kt) \cos a_1 t \cdots \cos a_n t =$$

$$A_n(k) + \Sigma_{j\neq k} A_n(j)(\cos(j-k)t + i \sin(j-k)t).$$

Integrating the last identity over the interval $[0, 2\pi]$, we get the result in formula (8).

□

Let us look at the range of the variable S_n defined earlier in (4). For a fixed k, we are going to denote the range of S_n by $\mathcal{R}_k(n)$. In other words, the set $\mathcal{R}_k(n)$ consists of all numbers

$$\pm 1^k \pm 2^k \pm \cdots \pm n^k. \tag{9}$$

To determine $\mathcal{R}_1(n)$ was a 2011 Romanian Olympiad problem, let us include an answer to this problem. The greatest element of the set $\mathcal{R}_1(n)$ is the triangular number $T_n := 1+2+\cdots+n = \frac{n(n+1)}{2}$, and the smallest element of $\mathcal{R}_1(n)$ is clearly $-T_n$. Also, the difference of any two elements of $\mathcal{R}_1(n)$ is an even number. Hence, all elements of $\mathcal{R}_1(n)$ are of the same parity

We claim that

$$\mathcal{R}_1(n) = \{-T_n, -T_n+2, \cdots, T_n-2, T_n\}. \tag{10}$$

Let us define a map on the elements of $\mathcal{R}_1(n) \setminus \{T_n\}$ having values in $\mathcal{R}_1(n)$. First, if $x \in \mathcal{R}_1(n) \setminus \{T_n\}$ is an element for which the writing begins with -1, then by changing -1 by $+1$, we get $x+2 \in \mathcal{R}_1(n)$. If the writing of x begins with $+1$, then consider the first term in the sum with sign $-$. Such a term exists unless $x = T_n$. In this case, we have

$$x = 1+2+\cdots+(j-1)-j\pm\cdots\pm n.$$

By changing the signs of terms $j-1$ and j, it follows that $x+2 \in \mathcal{R}_1(n)$. This shows the claim in (10).

One can wonder what happens if we work within classes modulo m with representations of the form (9). Of course, taking all the numbers in (10) modulo a small m, the chances are that all classes are covered. There is an interesting and a more precise result related to this question which appeared as a problem in the monthly [24].

Let p be an odd prime. The $2^{\frac{p-1}{2}}$ numbers $\pm1\pm2\pm\cdots\pm\frac{p-1}{2}$ represent each nonzero residue class mod p the same number of times.

Indeed, let $z = e^{\frac{2\pi i}{p}}$ be the standard primitive root of order p of the unity, and denote by

$$S = \sum_{\varepsilon_i\in\{-1,1\}} z^{\varepsilon_1+2\varepsilon_2+\cdots+\frac{p-1}{2}\varepsilon_{\frac{p-1}{2}}} = a_0 + a_1 z + \cdots + a_{p-1}z^{p-1},$$

for some integers a_j.

It is clear that a_j is exactly the number of ways the residue j is represented by the numbers $\pm1\pm2\pm\cdots\pm\frac{p-1}{2}$. We need to show that $a_1 = a_2 = \cdots = a_{p-1}$. Our analysis is going to also give the common value. As we have seen several times before, we have $S = \prod_{j=1}^{\frac{p-1}{2}}\left(z^j + z^{-j}\right)$. We observe that

$$\prod_{j=1}^{p-1}\left(z^j+z^{-j}\right) = S\cdot\prod_{j=\frac{p+1}{2}}^{p-1}\left(z^j+z^{-j}\right) = S\cdot\prod_{j=1}^{\frac{p-1}{2}}\left(z^{p-j}+z^{j-p}\right) = S^2.$$

On the other hand, we can easily compute

$$\prod_{j=1}^{p-1}\left(z^j+z^{-j}\right) = \frac{1}{z^{\frac{p(p-1)}{2}}}\cdot\prod_{j=1}^{p-1}\left(1+z^{2j}\right) = \prod_{j=1}^{p-1}\left(1+z^j\right) = 1,$$

where we used the fact that $j \to 2j$ is a bijection of the nonzero remainders modulo p, and that $\prod_{j=1}^{p-1}(1+z^j) = 1$. This last relation immediately follows from the polynomial identity $\prod_{j=1}^{p-1}(X-z^j) = \frac{X^p-1}{X-1}$ by taking $X = -1$.

Putting together these facts, we conclude that $S^2 = 1$ and so $S = \pm 1$. Then, the relation $a_0 \mp 1 + a_1 z + \cdots + a_{p-1} z^{p-1} = 0$, implies $a_0 - S = a_1 = a_2 = \cdots = a_{p-1}$ ([1], Proposition 4, p. 47). Since we must have $a_0 + a_1 + \cdots + a_{p-1} = 2^{\frac{p-1}{2}}$, we get $S + pa_1 = 2^{\frac{p-1}{2}}$, which implies $a_1 = \frac{2^{\frac{p-1}{2}} - S}{p}$ and also

$$S \equiv 2^{\frac{p-1}{2}} \equiv \left(\frac{2}{p}\right) \equiv (-1)^{\frac{p^2-1}{8}} (mod p).$$

Here, we used Euler's Criterion ([26], p. 101) and a standard result on quadratic residues. Hence, $S = (-1)^{\frac{p^2-1}{8}}$ and

$$a_1 = a_2 = \cdots = a_{p-1} = \frac{2^{\frac{p-1}{2}} - (-1)^{\frac{p^2-1}{8}}}{p}.$$

For $k = 2$, the situation with $\mathcal{R}_2(n)$ is almost the same as in case $k = 1$ but there is an interesting new phenomenon, although expected since $\{m^2\}_{m \geq 1}$ is a Erdös–Surányi sequence as we have seen. Let us define the set

$$R_2(n) = \{-\sum\nolimits_2(n), -\sum\nolimits_2(n) + 2, \cdots, \sum\nolimits_2(n) - 2, \sum\nolimits_2(n)\},$$

where $\sum_2(n) = 1^2 + 2^2 + \cdots + n^2 = \frac{n(n+1)(2n+1)}{6}$.

Theorem 6 *For $n \in \mathbb{N}$, $\mathcal{R}_2(n) = R_2(n) \setminus \mathcal{E}_2(n)$, where*

$$\mathcal{E}_2(n) = \{\pm \left(\sum\nolimits_2 (n) - 2j\right) : j \in E\} \textit{ and}$$

$$E := \{2, 3, 6, 7, 8, 11, 12, 15, 18, 19, 22, 23, 24, 27, 28, 31, 32, 33, \tag{11}$$

$$43, 44, 47, 48, 60, 67, 72, 76, 92, 96, 108, 112, 128\}.$$

Proof The set E is known (see A001422) as being the set of all positive integers which cannot be written as a sum of distinct squares. This is connected to an old result of E. M. Wright ([35]) and to the papers of R. Sprague, P. T. Bateman, A. J. Hildebrand, and G. B. Purdy (see [29] and [8]). Also, one can find a simple proof for this fact that was given by P. Jain ([22]). Let us observe that the equation

$$\pm 1^2 \pm 2^2 \pm 3^2 \cdots \pm n^2 = \sum\nolimits_2(n) - 2j$$

can be written equivalently as

$$2j = (1 \mp 1)1^2 + (1 \mp 1)2^2 + (1 \mp 1)3^2 + \cdots (1 \mp 1)n^2 \text{ or}$$

$$j = c_1 1^2 + c_2 2^2 + c_3 3^2 + \cdots + c_n n^2, \text{ with } c_i \in \{0, 1\}.$$

So, a representation is possible if and only if j can be written as a sum of distinct perfect squares. It is known that the only positive integers that cannot be written as a sum of distinct squares are the ones in E. □

For $k = 3$, a similar result can be stated. The list of the numbers (A001476) which cannot be represented as a sum of distinct cubes has 2788 terms. This was obtained by R. E. Dressler and T. Parker in [14]. For $k = 4$, the exceptional set of numbers (A046039) has 889576 elements.

In [30], $P_n(k)$ denotes the number of partitions of k into distinct parts from $1, 2^n, 3^n, \ldots,$ and it is proved that for each n, there are only a finite number of integers which are not the sums of distinct nth powers. That is, there is a positive integer N_n depending only on n such that $P_n(k) > 0$ for all $k > N_n$. This result was extended by H. E. Richert ([28]) to a more general class of sequences.

For a sequence of distinct positive integers $\mathbf{a} = \{a_m\}_{m\geq 1}$, define the *exceptional set* of $\mathbf{a}$ to be the set $E(\mathbf{a})$ constituting all positive integers that cannot be represented as a sum of distinct terms of $\mathbf{a}$. Also, define the range $\mathcal{R}_{\mathbf{a}}(n)$ of the random variable $S_n = E_1 + E_2 + \cdots + E_n$, where E_j takes two values: a_j or $-a_j$ with the probability $1/2$ each. If $\mathbf{a}$ is a Erdös–Surányi sequence, then using a similar argument as in Theorem 6, we get $\mathcal{R}_{\mathbf{a}}(n) = R_{\mathbf{a}}(n) \setminus \mathcal{E}(\mathbf{a})$, where $\mathcal{E}(\mathbf{a}) = \{\pm(u_n - 2j) : j \in E(\mathbf{a})\}$, $u_n = a_1 + \cdots + a_n$, and $R_{\mathbf{a}}(n) = \{-u_n, -u_n + 2, \ldots, u_n - 2, u_n\}$.

As we have seen for the sequence of squares and of cubes, it is challenging to determine the exceptional set for a given Erdös–Surányi sequence. For a better understanding of the difficulty of this problem, we mention here three more examples.

The sequence $\mathbf{t}$ of triangular numbers $T_n = \frac{n(n+1)}{2}$, $(n \geq 1)$, satisfies for every $m \geq 1$ the relation $T_{m+3} - T_{m+3} - T_{m+3} + T_m = 2$. Since we may write $1 = T_1$ and $2 = -T_1 + T_2$, it follows by induction that $\mathbf{t}$ is an Erdös–Surányi sequence. On the other hand, according to the result of H. E. Richert [28], its exceptional set is $E(\mathbf{t}) = \{2, 5, 8, 12, 23, 33\}$, hence the range $\mathcal{R}_{\mathbf{t}}(n) = R_{\mathbf{t}}(n) \setminus \{\pm(u_n - 2j) : \text{ with } j \in E(\mathbf{t})\}$, where $u_n = T_1 + \cdots + T_n = \frac{n(n+1)(n+2)}{6}$.

It is also known that the sequence of primes is an Erdös–Surányi sequence. A nice proof based on Theorem 3 combined with Bertrand's postulate is given by M.O. Drimbe ([15], Proposition 4). According to the result of R. E. Dressler [13], every positive integer, except 1, 2, and 6, can be written as the sum of distinct primes, that is the exceptional set of the sequence $\mathbf{p}$ of primes is $E(\mathbf{p}) = \{1, 4, 6\}$. Therefore, for a given n, the range $\mathcal{R}_{\mathbf{p}}(n) = R_{\mathbf{p}}(n) \setminus \{-u_n + 2, -u_n + 8, -u_n + 12, u_n - 12, u_n - 8, u_n - 2\}$, where $u_n = p_1 + \cdots + p_n$.

Finally, it is not difficult to check that the hypotheses in Theorem 3 are satisfied for the Fibonacci sequence $F_0 = 0$, $F_1 = 1$, $F_{n+2} = F_{n+1} + F_n$, $n \geq 0$. Thus, $\{F_n\}$ is an Erdös–Surányi sequence. On the other hand, it is more or less known Zeckendorf's theorem in [34], which states that every positive integer can be represented uniquely as the sum of one or more distinct Fibonacci numbers in such a way that the sum does not include two consecutive Fibonacci numbers. Such a sum is called *Zeckendorf representation* and it is related to the Fibonacci coding of a positive integer. In this case, the exceptional set $E(\mathbf{f})$ of the Fibonacci sequence, say $\mathbf{f}$, is the empty set. Then, for a given positive integer n, the range $\mathcal{R}_{\mathbf{f}}(n)$ consists of all positive integers between $-(F_{n+2} - 1)$ and $F_{n+2} - 1$ of the same parity as $F_{n+2} - 1 = F_1 + F_2 + \cdots + F_n$.

5 A Proof of Andrica–Tomescu Conjecture

Let us observe that, in general for $k \in \mathbb{N}$, as we have seen in the previous sections

$$S_k(n) = \frac{2^{n-1}}{\pi} \int_0^{2\pi} \cos t \cos 2^k t \cdots \cos n^k t dt.$$

We let $f_{k,n}(x) = \cos t \cos 2^k t \cdots \cos n^k t$, $t \in \mathbb{R}$, which is a 2π-periodic and an even function. Hence, we can write the above as

$$S_k(n) = \frac{2^n}{\pi} \int_0^{\pi} \cos t \cos 2^k t \cdots \cos n^k t dt.$$

It is easy to see that each $x \to \cos\left(j^k x\right)$ is symmetric with respect to the point $\left(\frac{\pi}{2}, 0\right)$ if j is an odd integer and symmetric with respect to the vertical line $x = \pi/2$ if j is even. This means $S_k(n) = 0$ if $n \equiv 1$ or 2 (mod 4). So, we may assume that $n \equiv 0$ or 3 (mod 4) in what follows, but we will see that after the next reduction this makes no difference. The reduction is to shrink the domain of $f_{k,n}$ even further and write

$$S_k(n) = \frac{2^{n+1}}{\pi} \int_0^{\pi/2} \cos t \cos 2^k t \cdots \cos n^k t dt. \tag{12}$$

Let us emphasize that, the asymptotic behavior is the same regardless of what n is, modulo 4, as long as $S_k(n)$ is newly defined as in (12) (the new values of $S_k(n)$ for $n \equiv 1$ or 2 (mod 4) are not integers anymore, in general, but the asymptotic behavior for these values of n is the same, but probably no combinatorial interpretation exists for these values).

Sullivan's idea [31] is to split the integral into two parts, let us say, one over $[0, \frac{\pi}{4n^k}]$ and the other over the interval $\left[\frac{\pi}{4n^k}, \frac{\pi}{2}\right]$. Then show that the first one is $(1 + o(1)))\frac{\sqrt{2(2k+1)\pi}}{2n^{k+\frac{1}{2}}}$ and the second is $o\left(\frac{1}{n^{k+\frac{1}{2}}}\right)$. This is clearly enough to conclude that (7) is true. We will follow this argument and see what parts can be obtained in general.

First, let us observe that each function $x \to \cos\left(j^k x\right)$ $(j = 1 \ldots n)$ on the interval $[0, \frac{\pi}{4n^k}]$ is positive and in fact bounded below by $\frac{1}{\sqrt{2}}$. Hence, we can write $f_{k,n}(x)$ as an exponential:

$$f_{k,n}(x) = \exp\left[\sum_{j=1}^{n} \ln\left(\cos\left(j^k x\right)\right)\right] = \exp\left[\sum_{j=1}^{n} g_j(x)\right],$$

with $g_{j,k}(x) = \ln\left(\cos\left(j^k x\right)\right)$, $j = 1 \ldots n$. We need to compute some derivatives of $g_{j,k}$ and use the Taylor formula to estimate $f_{k,n}(x)$ for $x \in [0, \frac{\pi}{4n^k}]$. First, we have $g'_{j,k}(x) = -j^k \tan j^k x$ and so $g'_{j,k}(0) = 0$. Then, $g''_{j,k}(x) =$

$-j^{2k}\sec^2 j^k x = -j^{2k}(1+\tan^2 j^k x)$ which gives $g''_{j,k}(0) = -j^{2k}$. Finally $g'''_{j,k}(x) = -2j^{3k}\tan j^k x \sec^2 j^k x$. Then, we have

$$f_{k,n}(x) = \exp[-\frac{1}{2}(\sum_{j=1}^{n} j^{2k})x^2 - \frac{x^3}{3}\sum_{j=1}^{n} j^{3k}\tan(j^k c_x)\sec^2(j^k c_x)],\ x, c_x \in [0, \frac{\pi}{4n^k}].$$

Since $0 \le \tan t \sec^2 t \le 2$ for $t \in [0, \pi/4]$, this implies that

$$\exp(-\frac{x^2}{2}\Sigma_{2k}(n) - \frac{2x^3}{3}\Sigma_{3k}(n)) \le f_{k,n}(x) \le \exp(-\frac{x^2}{2}\Sigma_{2k}(n)),\ x \in [0, \frac{\pi}{4n^k}], \tag{13}$$

where as before $\Sigma_\alpha(n) = \sum_{j=1}^{n} j^\alpha$. We need then the following result.

Lemma 3 *For $k \ge 1$, we have $L_1 = L_2 = \sqrt{\frac{(2k+1)\pi}{2}}$, where*

$$L_1 := \lim_{n\to\infty} n^{k+1/2}\int_0^{\frac{\pi}{4n^k}} \exp\left(-\frac{x^2}{2}\Sigma_{2k}(n)\right)dx = \sqrt{\frac{(2k+1)\pi}{2}},\ \textit{and}$$

$$L_2 := \lim_{n\to\infty} n^{k+1/2}\int_0^{\frac{\pi}{4n^k}} \exp\left(-\frac{x^2}{2}\Sigma_{2k}(n) - \frac{2x^3}{3}\Sigma_{3k}(n)\right)dx.$$

Proof For the first limit, let us just change the variable: $x\sqrt{\Sigma_{2k}(n)} = t$. Then, the limit becomes

$$L_1 = \lim_{n\to\infty}\frac{n^{k+1/2}}{\sqrt{\Sigma_{2k}(n)}}\int_0^{\frac{\pi\sqrt{\Sigma_{2k}(n)}}{4n^k}} e^{-t^2/2}dt = \sqrt{\frac{(2k+1)\pi}{2}},$$

since we have seen that $\lim_{n\to\infty}\frac{\sqrt{\Sigma_{2k}(n)}}{n^{k+1/2}} = \frac{1}{\sqrt{2k+1}}$ and $\int_0^\infty e^{-t^2/2}dt = \sqrt{\frac{\pi}{2}}$. For the second limit, we do the same change of variable and obtain

$$L_2 = \lim_{n\to\infty}\frac{n^{k+1/2}}{\sqrt{\Sigma_{2k}(n)}}\int_0^{\frac{\pi\sqrt{\Sigma_{2k}(n)}}{4n^k}} e^{-t^2/2 - \frac{2t^3\Sigma_{3k}(n)}{3\Sigma_{2k}(n)^{3/2}}}dt,$$

which is clearly satisfying $L_2 \le L_1$. Since $e^x \ge 1 + x$ for all x, and $\int_0^a e^{-t^2/2}t^3 dt = 2 - 2exp\left(-\frac{a^2}{2}\right) - a^2 exp\left(-\frac{a^2}{2}\right) \to 2$ as $a \to \infty$, we have

$$L_2 \ge L_1 - \lim_{n\to\infty}\frac{n^{k+1/2}}{\sqrt{\Sigma_{2k}(n)}}\frac{2\Sigma_{3k}(n)}{3\Sigma_{2k}(n)^{3/2}}\int_0^{\frac{\pi\sqrt{\Sigma_{2k}(n)}}{4n^k}} e^{-t^2/2}t^3 dt = L_1.$$

This shows that $L_2 = L_1$. □

From (13) and Lemma 3, we can conclude that

$$\lim_{n\to\infty} n^{k+1/2} \int_0^{\frac{\pi}{4n^k}} f_{k,n}(x)dx = \sqrt{\frac{(2k+1)\pi}{2}}.$$

For $k = 1$, in order to finish the proof, we simply need to show that

$$\lim_{n\to\infty} n^{3/2} \int_{\frac{\pi}{4n}}^{\pi/2} f_{1,n}(x)dx = 0.$$

Sullivan's approach here is to use the Arithmetic–Geometric Mean inequality and observe that

$$[f_{1,n}(x)^2]^{1/n} \le \frac{1}{n}\left(\sum_{j=1}^{n} \cos^2 jx\right) = \frac{1}{2n}\left(n + \sum_{j=1}^{n} \cos 2jx\right) = \frac{1}{2} + \frac{\cos((n+1)x)\sin nx}{2n \sin x}.$$

Hence, for $x \in [\frac{\pi}{2n}, \frac{\pi}{2}]$ we get

$$\left[f_{1,n}(x)^2\right]^{1/n} \le \frac{1}{2} + \frac{1}{\frac{\sin \pi/2n}{\pi/2n}} \frac{1}{\pi} \to c := \frac{\pi+2}{2\pi} < 1.$$

This is enough to conclude that

$$0 \le \lim_{n\to\infty} n^{3/2} \int_{\frac{\pi}{2n}}^{\pi/2} |f_{1,n}(x)|dx \le \lim_{n\to\infty} n^{3/2} \left(\frac{\pi+2}{2\pi}\right)^{n/2} = 0.$$

We observe that there is still a missing interval left unaccounted for, but both limits can be extended a little more so that the two new intervals will cover $[0, \frac{\pi}{2}]$. Indeed, let us consider an $\epsilon \in (0, 1)$ that is going to be determined later. The first limit can be shown to be true for the interval $\left[0, \frac{\pi}{2n}(1-\epsilon)\right]$ since we only need to have the functions $x \to \cos jx$ bounded from below by a positive constant. The constant c above becomes $\frac{\pi(1-\epsilon)+2}{2\pi(1-\epsilon)}$ which is still less than 1 if ϵ is small enough ($\epsilon = 1/2$ is just not good enough).

Let us close by formulating the equivalence of the conjecture (7) for $k \ge 2$:

$$\lim_{n\to\infty} n^{k+1/2} \int_{\frac{\pi(1-\epsilon)}{2n^k}}^{\pi/2} f_{k,n}(x)dx = 0 \text{ for some } \epsilon \in (0, 1).$$

This seems to require a very fine analysis especially in the absence a good upper bound for the sums $\sum_{j=1}^{n} \cos(2j^k x)$ for $x \in \left[\frac{\pi(1-\epsilon)}{2n^k}, \frac{\pi}{2}\right]$.

Acknowledgement The present material was elaborated in the period of the Fall Semester 2013 when the first author was a Mildred Miller Fort Foundation Visiting Scholar at Columbus State University, Georgia, USA. He takes this opportunity to express his gratitude for the nice friendship showed and for the excellent facilities offered during his staying.

References

1. Andreescu, T., Andrica, D.: Complex Numbers from A to $\cdots$ Z, Birkauser, (2005)
2. Andrica, D.: On some trigonometric identities and applications (Romanian), G.M.-B, **11** (1998)
3. Andrica, D.: A combinatorial result concerning the product of two or more derivatives. Bull. Cal. Math. Soc. **92**(4), 299–304 (2000)
4. Andrica, D., Buzeţeanu, S.: On the product of two or more derivatives. Rev. Romaine Math. Pures Appl. **30**, 703–710 (1985)
5. Andrica, D., Ionascu, E.J.: Variations on a result of Erdös and Surányi, INTEGERS 2013: The Erdös Centennial Conference in Combinatorial Number Theory, University of West Georgia, Carrollton, GA, October 24–27, 2013, http:dorinandrica.ro/presentation-INTEGERS-2013.pdf
6. Andrica, D., Tomescu, I.: On an integer sequence related to a product of trigonometric fuctions, and its combinatorial relevance. J. Integer Seq. **5**, (2002), Article 02.2.4
7. Andrica, D., Vacaretu, D.: Representation theorems and almost unimodal sequences. Studia Univ. Babes-Bolyai, Mathematica, Volume LI. Number. **4**, 23–33 (2006)
8. Bateman, P.T., Hildebrand, A.J., Purdy, G.B.: Sums of distinct squares, Acta Arithmetica, LXVII. 4, 349–380 (1994)
9. Berry, A.C.: The accuracy of the Gaussian approximation to the sum of independent variates. Trans. Am. Math. Soc. **49**(1), 122–136 (1941)
10. Brown, J.L., Jr.: Note on complete sequences of integers. Am. Math. Mon. **68**(6), 557–560 (1961)
11. Brown, J.L., Jr.: Generalization of Richert's theorem. Am. Math. Mon. **83**(8) 631–634 (1976)
12. Bruckner, A.M., Leonard, J.L.: Derivatives. Am. Math. Mon. **73**(4), 24–56 (April 1966). (Part 2; Papers in Analysis)
13. Dressler, R.E.: A stronger Bertrand's postulate with an application to partitions. Proc. Am. Math. Soc. **33**(2), 226–228 (1972)
14. Dressler, R.E., Parker, T.: 12, 758, Mathematics of Computation. **28**(125), 313–314 (January 1974)
15. Drimbe, M.O.: A problem of representation of integers (Romanian), G.M.-B, 10–11, 382–383 (1983)
16. Drimbe, M.O.: Generalization of representation theorem of Erdős and Surányi. Ann. Soc. Math. Pol. Series I Comment. Math. XXVII(2), 233–235 (1988)
17. Erdős, P., Surányi, J.: Topics in the Theory of Numbers. Springer, (2003)
18. Finch, S.R.: Signum equations and extremal coefficients. people.fas.harvard.edu/ sfinch/
19. Hruska, V.: Une note sur les fonctions aux valeurs intermédiaires. Časopis pro pestování matematiky a fysiky 71(1–4), 67–69 (1946)
20. Hwang, H.-K.: Review of "On an integer sequence related to a product of trigonometric fuctions, and its combinatorial relevance", MR1938223
21. Iosifescu, M.: Conditions that the product of two derivatives to be a derivative (Russian). Rev. Math. Pures Appl. **4**, 641–649 (1959)
22. Jain, P. A proof of the sum of squares conjecture.
23. Kedlaya, K., Poonen, B., Vakil, R.: The William Lowell Putnam Mathematical Competition 1985–2000: Problems, Solutions, and Commentary (2002)
24. McFarland, R.L.: Problem 6457, American Mathematical Monthly.
25. Mitek, J.: Generalization of a theorem of Erdős and Surányi. Ann. Soc. Math. Pol. Series I Comment. Math. **XXI**, 173–175 (1979)
26. Niven, I., Zuckerman, H., Montgomery, H.: An introduction to the Theory of Numbers. 5th edn. Wiley, New York (1991)
27. Olsen, L.: A New Proof of Darboux's Theorem. Am. Math. Mon. **8**, 713–715 (2004)
28. Richert, H.E.: Über Zerlegungen in paarweise verschiedene Zahlen. Nord. Mat. Tidiskr. **31**, 120–122 (1949)
29. Sprague, R.: Über Zerlegungen in ungleiche Quadratzahlen. Math. Z. **51**, 289–290 (1948)
30. Sprague, R.: Über Zerlegungen in n-te Potenzen mit lauter verschiedenen Grundzahlen. Math. Z. **51**, 466–468 (1948)

31. Sullivan, B.D.: On a Conjecture of Andrica and Tomescu. J. Integer Seq. **16** (2013), Article 13.3.1.
32. Tetiva, M.: A representation problem (Romanian), Recreaţii Matematice. No. 2, 123–127 (2010)
33. Tetiva, M.: A representation theorem II (Romanian), Recreaţii Matematice. No. 1, 5–10 (2012)
34. Zeckendorf, E.: Représentation des nombres naturels par une somme de nombres de Fibonacci ou des nombres de Lucas. Bull. Soc. R. Sci. Liège **41**, 179–182 (1972)
35. Wright, E.M.: The representation of a number as a sum of five or more squares. Quart. J. Math. Oxf. Ser. **4**, 37–51 (1933)

The Hilali Conjecture for Hyperelliptic Spaces

Javier Fernández de Bobadilla, Javier Fresán, Vicente Muñoz and Aniceto Murillo

Abstract The Hilali conjecture predicts that for a simply connected elliptic space, the total dimension of the rational homotopy does not exceed that of the rational homology. Here, we give a proof of this conjecture for a class of elliptic spaces known as hyperelliptic.

Keywords Rational homotopy · Sullivan models · Elliptic spaces · Tor functors

1 Introduction

Let X be a simply connected CW-complex. Then, X is said to be of elliptic type if both $\dim H^*(X, \mathbb{Q}) < \infty$ and $\dim \pi_*(X) \otimes \mathbb{Q} < \infty$. For these spaces, Hilali conjectured in [5] the following:

Conjecture 1 If X is a simply connected CW-complex of elliptic type, then

$$\dim \pi_*(X) \otimes \mathbb{Q} \leq \dim H^*(X, \mathbb{Q}) .$$

J. Fernández de Bobadilla (✉)
Instituto de Ciencias Matemáticas CSIC-UAM-UC3M-UCM, Consejo Superior de Investigaciones Cientìficas, C/Nicolás Cabrera, n° 13–15, Campus Cantoblanco UAM, 28049 Madrid, Spain
e-mail: javier@icmat.es

J. Fresán
LAGA, UMR 7539, Institut Galilée, Universitè Paris 13, 99, Avenue Jean-Baptiste Clément, 93430 Villetaneuse, France
e-mail: fresan@math.univ-paris13.fr

V. Muñoz
Facultad de Ciencias Matemàticas, Universidad Complutense de Madrid, Plaza de Ciencias 3, 28040 Madrid, Spain
e-mail: vicente.munoz@mat.ucm.es

A. Murillo
Departamento de Álgebra, Geometría y Topología, Universidad de Málaga, Ap. 59, 29080 Málaga, Spain
e-mail: aniceto@uma.es

T. M. Rassias, P. M. Pardalos (eds.), *Mathematics Without Boundaries*,
DOI 10.1007/978-1-4939-1106-6_2,

By the theory of minimal models of Sullivan [3], the rational homotopy type of X is encoded in a differential algebra (A, d) called *the minimal model* of X. This is a free graded algebra $A = \Lambda V$, generated by a graded vector space $V = \bigoplus_{k\geq 2} V^k$, and with decomposable differential, i.e., $d : V^k \to \left(\Lambda^{\geq 2} V\right)^{k+1}$. It satisfies that:

$$V^k = (\pi_k(X) \otimes \mathbb{Q})^*,$$
$$H^k(\Lambda V, d) = H^k(X, \mathbb{Q}).$$

Therefore, the Hilali conjecture can be rewritten as follows: for a finite-dimensional graded vector space V (in degrees bigger or equal than two), we have

$$\dim V \leq \dim H^*(\Lambda V, d)$$

for any decomposable differential d on ΛV.

An elliptic space X is called of pure type if its minimal model $(\Lambda V, d)$ satisfies that $V = V^{even} \oplus V^{odd}$, $d(V^{even}) = 0$ and $d(V^{odd}) \subset \Lambda V^{even}$. Also X is called hyperelliptic if $d(V^{even}) = 0$ and $d(V^{odd}) \subset \Lambda^+ V^{even} \otimes \Lambda V^{odd}$.

In his thesis [5] in 1990, Hilali proved Conjecture 1 for elliptic spaces of pure type. The conjecture is known to hold [6, 7] also in several cases: H-spaces, nilmanifolds, symplectic, and cosymplectic manifolds, coformal spaces with only odd-degree generators, and formal spaces. Hilali and Mamouni [6, 7] have also proved Conjecture 1 for hyperelliptic spaces under various conditions in the homotopical and homological Euler characteristics.

The main result of this paper is the following:

Theorem 1 *Conjecture 1 holds for hyperelliptic spaces.*

We shall start by proving it for elliptic spaces of pure type in Sect. 3. This requires reducing the question to a problem about Tor functors of certain modules of finite length over a polynomial ring. We solve it by using a semicontinuity result for the Tor functor. Then, in Sect. 4 we prove Theorem 1 for hyperelliptic spaces. For this we have to prove a semicontinuity result for the homology of elliptic spaces, and apply it to reduce the general case to the case in which the minimal model only has generators of odd degree and zero differential. We give two different proofs of an inequality from which the result follows.

2 Minimal Models

We recall some definitions and results about minimal models [2]. Let (A, d) be a *differential algebra*, that is, A is a (positively) graded commutative algebra over the rational numbers, with a differential d which is a derivation, i.e., $d(a \cdot b) = (da) \cdot b + (-1)^{\deg(a)} a \cdot (db)$, where $\deg(a)$ is the degree of a. We say that A is connected if $A^0 = \mathbb{Q}$, and simply connected if moreover $A^1 = 0$.

A simply connected differential algebra (A, d) is said to be *minimal* if:

1. A is free as an algebra, that is, A is the free algebra ΛV over a graded vector space $V = \oplus_{k\geq 2} V^k$, and
2. For $x \in V^k$, $dx \in (\Lambda V)^{k+1}$ has no linear term, i.e., it lives in $\Lambda V^{>0} \cdot \Lambda V^{>0} \subset \Lambda V$.

Let (A, d) be a simply connected differential algebra. A minimal model for (A, d) is a minimal algebra $(\Lambda V, d)$ together with a quasi-isomorphism $\rho : (\Lambda V, d) \to (A, d)$ (that is, a map of differential algebras such that $\rho_* : H^*(\Lambda V, d) \to H^*(A, d)$ is an isomorphism). A minimal model for (A, d) exists and it is unique up to isomorphism.

Now consider a simply connected CW-complex X. There is an algebra of piecewise polynomial rational differential forms $\left(\Omega^*_{PL}(X), d\right)$ defined in [3, Chap. VIII]. A minimal model of X is a minimal model $(\Lambda V, d)$ for $\left(\Omega^*_{PL}(X), d\right)$. We have that

$$
\begin{aligned}
V^k &= (\pi_k(X) \otimes \mathbb{Q})^* , \\
H^k(\Lambda V, d) &= H^k(X, \mathbb{Q}) .
\end{aligned}
$$

A space X is elliptic [1] if both $\sum \dim \pi_k(X) \otimes \mathbb{Q} < \infty$ and $\sum \dim H^k(X, \mathbb{Q}) < \infty$. Equivalently, if $(\Lambda V, d)$ is the minimal model, we require that both V and $H^*(\Lambda V, d)$ are finite dimensional. For elliptic spaces, the Euler–Poincaré and the homotopic characteristics are well defined:

$$
\begin{aligned}
\chi &= \sum_{i\geq 0} (-1)^i \dim H^i(\Lambda V, \mathbb{Q}), \\
\chi_\pi &= \sum_{i\geq 0} (-1)^i \dim \pi_i(X) \otimes \mathbb{Q} = \dim V^{even} - \dim V^{odd} .
\end{aligned}
$$

We refer the reader to [2, Theorem 32.10] for the proof of the following:

Proposition 1 *Let $(\Lambda V, d)$ be an elliptic minimal model. Then, $\chi \geq 0$ and $\chi_\pi \leq 0$. Moreover, $\chi_\pi < 0$ if and only if $\chi = 0$.*

In his thesis [5], M. Hilali conjectured that for elliptic spaces:

$$
\dim \pi_*(X) \otimes \mathbb{Q} \leq \dim H^*(X, \mathbb{Q}) .
$$

In algebraic terms, this is equivalent to

$$
\dim V \leq \dim H^*(\Lambda V, d) ,
$$

whenever $(\Lambda V, d)$ is a minimal model with $\dim V < \infty$. Note that finiteness of both $\dim H^*(X, \mathbb{Q})$ and $\dim \pi_*(X) \otimes \mathbb{Q}$ is necessary. Otherwise, one can easily construct counterexamples such as $X = S^3 \vee S^3$.

3 Proof of the Hilali Conjecture for Elliptic Spaces of Pure Type

A minimal model $(\Lambda V, d)$ is of *pure type* if $V = V^{even} \oplus V^{odd}$, with

$$d\left(V^{even}\right) = 0, \quad d\left(V^{odd}\right) \subset \Lambda V^{even}.$$

An elliptic space is of pure type if its minimal model is so. These spaces are widely studied in [2, § 32]. By Proposition 1, we have that $\dim V^{even} - \dim V^{odd} \leq 0$. Let $n = \dim V^{even}$ and $n + r = \dim V^{odd}$, where $r \geq 0$. Write $x_1, \ldots, x_n$ for the generators of even degree, and $y_1, \ldots, y_{n+r}$ for the generators of odd degree. Then, $dx_i = 0$, and $dy_j = P_j(x_1, \ldots, x_n)$, where P_j are polynomials without linear terms.

In this section, we prove the following:

Theorem 2 *The Hilali conjecture holds for elliptic spaces of pure type.*

3.1 Expressing the Homology as a Tor Functor

To work over nice modules, we would like to reorder the generators $y_1, \ldots, y_{n+r}$, so that $P_1, \ldots, P_n$ form a regular sequence in $\Lambda(x_1, \ldots, x_n)$. Recall that this means that the image of P_i in $\Lambda(x_1, \ldots, x_n)/(P_1, \ldots, P_{i-1})$ is not a zero divisor, for any $i = 1, \ldots, n$. But this is not possible in general, as shown by the following example.

Example 1 Let $V = \mathbb{Q}\langle x_1, x_2, y_1, y_2, y_3\rangle$, where $\deg(x_1) = 2$ and $\deg(x_2) = 6$. Define a differential d on ΛV by

$$dy_1 = x_1^6 + x_2^2, \quad dy_2 = x_1^9 + x_2^3, \quad dy_3 = x_1^4 x_2 + x_1 x_2^2.$$

Then, $(\Lambda V, d)$ is a pure minimal model. It can be proved that it is elliptic if and only if there exist exact powers of x_1 and x_2. This is the case, since $2x_1^{10} = d\left(x_1^4 y_1 + x_1 y_2 - x_2 y_3\right)$ and $2x_2^4 = d\left(x_2^2 y_1 + x_2 y_2 - x_1^5 y_3\right)$. But for the same reason, models $\left(\Lambda(x_1, x_2, y_i, y_j), d\right)$ are not elliptic for any choice of indices i, j. This amounts to say that dy_i, dy_j are not a regular sequence in $\Lambda(x_1, x_2)$.

However, Halperin showed in [4, Lemma 8] that pure models always admit a basis $z_1, \ldots, z_{n+r}$ of V^{odd} such that $dz_1, \ldots, dz_n$ is a regular sequence in $\Lambda(x_1, \ldots, x_n)$. This basis is not necessarily homogeneous but it is possible to preserve the lower grading induced by the number of odd elements, that is

$$(\Lambda V)_q^p = \left(\Lambda V^{even} \otimes \Lambda^q V^{odd}\right)^p.$$

This grading passes to cohomology and by taking into account the quasi-isomorphisms

$$(\Lambda(x_1, \ldots, x_n, y_1, \ldots, y_{n+r}),\ d) \xrightarrow{\sim} (\Lambda(x_1, \ldots, x_n, z_1, \ldots, z_{n+r}),\ d)$$

$$(\Lambda(x_1, \ldots, x_n, z_1, \ldots, z_n),\ d) \xrightarrow{\sim} (\Lambda(x_1, \ldots, x_n)/(dz_1, \ldots, dz_n),\ d)$$

with respect to the lower grading, one deduces that:

$$H_*(\Lambda V, d) \cong H_*(\Lambda(x_1, \dots, x_n)/(dz_1, \dots, dz_n) \otimes \Lambda(z_{n+1}, \dots, z_{n+r}), d).$$

So let $z_1, \dots, z_{n+r}$ be a basis such that $dz_1, \dots, dz_n$ form a regular sequence. Put $P_j = dz_j$ for $j = 1, \dots, n+r$ and consider the module

$$M = \mathbb{Q}[x_1, \dots, x_n]/(P_1, \dots, P_n)$$

over the ring

$$R = \mathbb{Q}[x_1, \dots, x_n]\,.$$

Consider the ring

$$S = \mathbb{Q}[\lambda_1, \dots, \lambda_r]$$

and the map $f : S \to R$, $\lambda_i \mapsto P_{n+i}$. Then, M becomes an S-module.

Consider also the S-module

$$\mathbb{Q}_0 = S/(\lambda_1, \dots, \lambda_r).$$

Then, we have the following:

Proposition 2 $H_*(\Lambda V, d) \cong \mathrm{Tor}^*_S(M, \mathbb{Q}_0)$.

Proof Let $U = \langle z_1, \dots, z_n \rangle$, $W = \langle z_{n+1}, \dots, z_{n+r} \rangle$ so that $V^{odd} = U \oplus W$. Then, the map $(\Lambda V^{even} \oplus U, d) \to (M, 0)$ is a quasi-isomorphism. Actually, the Koszul complex

$$R \otimes \Lambda^n U \to R \otimes \Lambda^{n-1} U \to \dots \to R \otimes \Lambda^1 U \to R \to M$$

is exact, which means that $(R \otimes \Lambda U, d) \xrightarrow{\sim} (M, 0)$.

Therefore,

$$(\Lambda V, d) = (R \otimes \Lambda U \otimes \Lambda W, d) \xrightarrow{\sim} (M \otimes \Lambda W, d')\,, \qquad (1)$$

is a quasi-isomorphism, where the differential d' is defined as zero on M, and $d' z_{n+i} = \bar{P}_{n+i} \in M$. This can be seen as follows: the map (1) is a map of differential algebras. Grading both algebras in such a way that $\Lambda^k W$ has degree k, we get two spectral sequences. The map between their E_1-terms is

$$H^*(R \otimes \Lambda U, d) \otimes \Lambda W \to M \otimes \Lambda W\,.$$

As this is a quasi-isomorphism, it follows that the map in the E_∞-terms is also an isomorphism. The E_∞-terms are the homology of both algebras in (1). So the map (1) is a quasi-isomorphism.

Finally, we have to identify $H^*(M \otimes \Lambda W, d') \cong \mathrm{Tor}_S^*(M, \mathbb{Q}_0)$. Note that the homology of $(M \otimes \Lambda W, d')$ is computed as follows: Take the Koszul complex

$$S \otimes \Lambda^r W \to S \otimes \Lambda^{r-1} W \to \ldots \to S \otimes \Lambda^1 W \to S \to \mathbb{Q}_0 ,$$

and tensor it with M over S (with the S-module structure given above), to get

$$\left(M \otimes_S (S \otimes \Lambda W), d'\right) = \left(M \otimes \Lambda W, d'\right).$$

The homology of this computes $\mathrm{Tor}_S^*(M, \mathbb{Q}_0)$.

Lemma 1 *Under our assumptions,*

$$\dim \mathrm{Tor}_S^0(M, \mathbb{Q}_0) \geq n+1 \quad \textit{and} \quad \dim \mathrm{Tor}_S^r(M, \mathbb{Q}_0) \geq n+1.$$

Proof Clearly,

$$\mathrm{Tor}_S^0(M, \mathbb{Q}_0) = M \otimes_S \mathbb{Q}_0 = M/\left(\bar{P}_{n+1}, \ldots, \bar{P}_{n+r}\right) = R/\left(P_1, \ldots, P_{n+r}\right).$$

As all the polynomials $P_1, \ldots, P_{n+r}$ have no linear part, this module contains the constant and linear monomials at least, so $\dim \mathrm{Tor}_S^0(M, \mathbb{Q}_0) \geq n+1$.

For the other inequality, note that $\mathrm{Tor}_S^r(M, \mathbb{Q}_0)$ is the kernel of $M \otimes \Lambda^r W \to M \otimes \Lambda^{r-1} W$, i.e., the kernel of

$$(P_{n+1}, \ldots, P_{n+r}) : M \to M \oplus \overset{(r)}{\ldots} \oplus M . \tag{2}$$

Now we use the following fact: As M is a complete intersection R-module (it is the quotient of R by a regular sequence), it has Poincaré duality in the sense that there is a map $M \to \mathbb{Q}$ such that $\Gamma : M \otimes M \overset{\text{mult}}{\longrightarrow} M \to \mathbb{Q}$ is a perfect pairing. Take elements $\nu, \mu_j \in M$, $j = 1, \ldots, n$, such that

$$\Gamma\left(\nu, x_j\right) = 0, \ j = 1, \ldots, n, \qquad \Gamma(\nu, 1) = 1,$$
$$\Gamma\left(\mu_j, x_k\right) = \delta_{jk}, \ j,k = 1, \ldots, n, \qquad \Gamma\left(\mu_j, 1\right) = 0,$$
$$\Gamma(\nu, Q) = \Gamma\left(\mu_j, Q\right) = 0, \ \text{for any quadratic } Q \in R.$$

Since the elements ν, μ_j are in the kernel of (2) and they are linearly independent, we get $\dim \mathrm{Tor}_S^r(M, \mathbb{Q}_0) \geq n+1$.

3.2 Semicontinuity Theorem

We are going to prove a semicontinuity theorem for the Tor functors $\mathrm{Tor}_S^k(M, \mathbb{Q}_0)$ for flat families of modules M of finite length (i.e., finite dimensional as $\mathbb{Q}$-vector spaces).

Consider a variable t. A family of S-modules is a module $\mathcal{M}$ over $S[t]$ such that for each t_0, the S-module

$$M_{t_0} = \mathcal{M}/(t - t_0)$$

is of finite length. We say that $\mathcal{M}$ is flat over $\mathbb{Q}[t]$ if it is a flat $\mathbb{Q}[t]$-module, under the inclusion $\mathbb{Q}[t] \hookrightarrow S[t]$. Consider $\mathcal{M}$ as a $\mathbb{Q}[t]$-module. Then

$$\mathcal{M} \cong \mathbb{Q}[t]^N \oplus \frac{\mathbb{Q}[t]}{(t - t_1)^{b_1}} \oplus \ldots \oplus \frac{\mathbb{Q}[t]}{(t - t_l)^{b_l}},$$

for some $N \geq 0$, $l \geq 0$, $1 \leq b_1 \leq \ldots \leq b_l$. The module is flat if and only if there is no torsion part, i.e., $l = 0$ (to see this, tensor the exact sequence $0 \to \mathbb{Q}[t] \xrightarrow{t-t_i} \mathbb{Q}[t] \to \mathbb{Q}[t]/(t - t_i) \to 0$ with $\mathcal{M}$). Note that for generic ξ, $\text{length}(M_\xi) = N$. Therefore, the flatness is equivalent to $\mathcal{M}/(t - t_i)$ being of length N, i.e.,

$$\mathcal{M} \text{ is flat} \iff \text{length}(M_t) = N, \ \forall t\,.$$

Lemma 2 *For any flat family* $\mathcal{M}$,

$$\dim \text{Tor}_S^k(M_0, \mathbb{Q}_0) \geq \dim \text{Tor}_S^k(M_\xi, \mathbb{Q}_0),$$

for generic $\xi \in \mathbb{Q}$.

Proof Let us resolve $\mathcal{M}$ as an $S[t]$-module:

$$0 \to S[t]^{a_r} \to \ldots \to S[t]^{a_0} \to \mathcal{M} \to 0\,. \quad (3)$$

As $\mathcal{M}$ is flat as $\mathbb{Q}[t]$-module, if we tensor the inclusion $\mathbb{Q}[t] \overset{t}{\hookrightarrow} \mathbb{Q}[t]$ by $\mathcal{M}$ over $\mathbb{Q}[t]$, we have that $\mathcal{M} \overset{t}{\hookrightarrow} \mathcal{M}$ is an inclusion. Hence, the sequence

$$0 \to \mathcal{M} \overset{t}{\hookrightarrow} \mathcal{M} \to \mathcal{M}/(t) \to 0$$

is exact. But this sequence is the sequence $0 \to S[t] \to S[t] \to S[t]/(t) \to 0$ tensored by $\mathcal{M}$ over $S[t]$. Hence, $\text{Tor}^1_{S[t]}(\mathcal{M}, S[t]/(t)) = 0$. Obviously $\text{Tor}^j_{S[t]}(\mathcal{M}, S[t]/(t)) = 0$ for $j \geq 2$ (since the resolution $S[t]/(t)$ has two terms).

Using the above, we can tensor $(3) \otimes_{S[t]} S[t]/(t)$ to get an exact sequence:

$$0 \to S^{a_r} \to \ldots \to S^{a_0} \to M_0 \to 0\,. \quad (4)$$

Now we tensor (4) by $\otimes_S \mathbb{Q}_0$ and take homology to obtain $\text{Tor}_S^*(M_0, \mathbb{Q}_0)$. But

$$(4) \otimes_S \mathbb{Q}_0 = (3) \otimes_{S[t]} \mathbb{Q}_0 = ((3) \otimes_{S[t]} \mathbb{Q}[t]) \otimes_{\mathbb{Q}[t]} \mathbb{Q}[t]/(t) = (5) \otimes_{\mathbb{Q}[t]} \mathbb{Q}[t]/(t)\,,$$

where $\mathbb{Q}_0 = S[t]/(\lambda_1, \ldots, \lambda_r, t)$, and

$$0 \to \mathbb{Q}[t]^{a_r} \to \ldots \to \mathbb{Q}[t]^{a_0} \to \mathcal{F} = \mathcal{M}/(\lambda_1, \ldots, \lambda_r) \to 0. \quad (5)$$

(This is just a complex, maybe not exact.) Analogously,

$$\mathrm{Tor}^*_S(M_0, \mathbb{Q}_\xi) = H^*((5) \otimes_{\mathbb{Q}[t]} \mathbb{Q}[t]/(t-\xi)).$$

So it remains to see that for a complex $L_\bullet$ of free $\mathbb{Q}[t]$-modules like (5), it holds that

$$\dim H^k(L_\bullet \otimes \mathbb{Q}[t]/(t-\xi)) \le \dim H^k(L_\bullet \otimes \mathbb{Q}[t]/(t)),$$

for generic ξ. (Tensor products are over $\mathbb{Q}[t]$, which we omit in the notation henceforth.) For proving this, just split (5) as short exact sequences

$$0 \to Z_i \to L_i \to B_{i-1} \to 0, \tag{6}$$

and note that Z_i, B_i are free $\mathbb{Q}[t]$-modules, being submodules of free modules. So $Z_i = \mathbb{Q}[t]^{z_i}$ and $B_i = \mathbb{Q}[t]^{b_i}$. Now $0 \to B_i \to Z_i \to H^i(L_\bullet) \to 0$ gives that

$$H^i(L_\bullet) = \mathbb{Q}[t]^{z_i - b_i} \oplus \text{torsion}.$$

For generic ξ, we have $\dim H^i(L_\bullet \otimes \mathbb{Q}[t]/(t-\xi)) = z_i - b_i$. Hence,

$$\begin{array}{cccccccc}
0 & \to Z_i \otimes \mathbb{Q}[t]/(t) & \to & L_i \otimes \mathbb{Q}[t]/(t) & \to & B_{i-1} \otimes \mathbb{Q}[t]/(t) & \to 0 \\
 & \downarrow & & \| & & \downarrow & \\
0 & \to Z_i(L_\bullet \otimes \mathbb{Q}[t]/(t)) & \to & L_i \otimes \mathbb{Q}[t]/(t) & \to & B_{i-1}(L_\bullet \otimes \mathbb{Q}[t]/(t)) & \to 0.
\end{array}$$

The first sequence is (6) tensored by $\mathbb{Q}[t]/(t)$. Thus, the last vertical map is surjective, and the first vertical map is injective.

Therefore, we get

$$\begin{aligned}
\dim H^i(L_\bullet \otimes \mathbb{Q}[t]/(t)) &= \dim Z_i(L_\bullet \otimes \mathbb{Q}[t]/(t)) - \dim B_i(L_\bullet \otimes \mathbb{Q}[t]/(t)) \\
&\ge \dim Z_i \otimes \mathbb{Q}[t]/(t) - \dim B_i \otimes \mathbb{Q}[t]/(t) \\
&= \dim H^i(L_\bullet) \otimes \mathbb{Q}[t]/(t) - \dim \mathrm{Tor}_1^{\mathbb{Q}[t]}(H^i(L_\bullet), \mathbb{Q}[t]/(t)) \\
&= z_i - b_i,
\end{aligned}$$

where we have used in the third line that there is an exact sequence

$$\begin{aligned}
0 \to \mathrm{Tor}_1^{\mathbb{Q}[t]}(H^i(L_\bullet), \mathbb{Q}[t]/(t)) &\to B_i \otimes \mathbb{Q}[t]/(t) \\
\to Z_i \otimes \mathbb{Q}[t]/(t) &\to H^i(L_\bullet) \otimes \mathbb{Q}[t]/(t) \to 0,
\end{aligned}$$

and in the fourth line that $\dim(N \otimes \mathbb{Q}[t]/(t)) = \dim \mathrm{Tor}_1^{\mathbb{Q}[t]}(N, \mathbb{Q}[t]/(t))$ for a torsion $\mathbb{Q}[t]$-module N.

3.3 *Proof of Theorem 2*

We proceed to the proof of the Hilali conjecture for elliptic spaces of pure type. We have to prove that

$$\dim H^*(\Lambda V, d) \geq 2n + r.$$

By Proposition 2, we need to prove that $\dim \operatorname{Tor}_S^*(M, \mathbb{Q}_0) \geq 2n + r$. Consider the family

$$\mathcal{M} = \frac{\mathbb{Q}[t, x_1, \ldots, x_n]}{(P_1 + tx_1, \ldots, P_n + tx_n)}.$$

For small t, the hypersurfaces $P_1 + tx_1, \ldots, P_n + tx_n$ intersect in N points near the origin accounted with multiplicity, where $N = \text{length}(M)$. Therefore, $\mathcal{M}$ is a flat family. By Lemma 2, it is enough to bound below $\dim \operatorname{Tor}_S^*\left(M_\xi, \mathbb{Q}_0\right)$. But for generic t, the hypersurfaces $P_1 + tx_1, \ldots, P_n + tx_n$ intersect in N distinct points (at least, it is clear that they intersect in several points and the origin is isolated of multiplicity one). Therefore,

$$\operatorname{Tor}_S^k\left(M_\xi, \mathbb{Q}_0\right) = \operatorname{Tor}_S^k(\mathbb{Q}_0, \mathbb{Q}_0).$$

This is easily computed to have dimension $\binom{r}{k}$ (using the Koszul complex). Therefore , using Lemma 1,

$$\begin{aligned}
\dim\ \operatorname{Tor}_S^*(M, \mathbb{Q}_0) &\geq (n+1) + \sum_{k=1}^{r-1} \dim\ \operatorname{Tor}_S^k(M, \mathbb{Q}_0) + (n+1) \\
&\geq 2n + 2 + \sum_{k=1}^{r-1} \dim\ \operatorname{Tor}_S^k\left(M_\xi, \mathbb{Q}_0\right) \\
&= 2n + 2 + \sum_{k=1}^{r-1} \binom{r}{k} = 2n + 2^r \geq 2n + r.
\end{aligned}$$

Remark 1 The above computation works for $r \geq 1$. If $r = 0$ then we have to prove that $\text{length}(M) \geq 2n$. But then computing the degree 2 nonzero elements in M, we have that they are at least $\binom{n+1}{2} - n$. So for any n,

$$\text{length}(M) \geq 1 + n + \binom{n+1}{2} - n = \frac{1}{2}(n+1)n + 1 \geq 2n.$$

4 The Hyperelliptic Case

A minimal model $(\Lambda V, d)$ of elliptic type is *hyperelliptic* if $V = V^{even} \oplus V^{odd}$, and

$$d\left(V^{even}\right) = 0, \quad d\left(V^{odd}\right) \subset \Lambda^+ V^{even} \otimes \Lambda V^{odd}. \tag{7}$$

An elliptic space is hyperelliptic if its minimal model is so. Note that elliptic spaces of pure type are in particular hyperelliptic.

By Proposition 1 we have that $\dim V^{even} - \dim V^{odd} \leq 0$. Let $n = \dim V^{even}$ and $n + r = \dim V^{odd}$, where $r \geq 0$. Write $x_1, \dots, x_n$ for the generators of even degree, and $y_1, \dots, y_{n+r}$ for the generators of odd degree. Then, $dx_i = 0$, and $dy_j = P_j\left(x_1, \dots, x_n, y_1, \dots, y_{j-1}\right)$, where P_j do not have linear terms.

In this section, we prove the following:

Theorem 3 *The Hilali conjecture holds for hyperelliptic spaces.*

4.1 Semicontinuity for Elliptic Minimal Models

Lemma 3 *Let V be a graded rational finite-dimensional vector space, and let d be a differential for $\Lambda V \otimes \mathbb{Q}[t]$ such that $dt = 0$, where t has degree 0. Take a non-countable field $\mathbf{k} \supset \mathbb{Q}$, $V_\mathbf{k} = V \otimes \mathbf{k}$. We denote by d_ξ the differential induced on $\Lambda V_\mathbf{k} = \Lambda V \otimes \mathbf{k}[t]/(t - \xi)$, for $\xi \in \mathbf{k}$. Then*

$$\dim H\left(\Lambda V_\mathbf{k}, d_\xi\right) \leq \dim H(\Lambda V, d_0),$$

for generic $\xi \in \mathbf{k}$.

Proof Write

$$0 \to \tilde{K} \to \Lambda V \otimes \mathbf{k}[t] \to \tilde{I} \to 0,$$

where $\tilde{K}$ and $\tilde{I}$ are the kernel and image of d, respectively. Note that both $\tilde{K}$ and $\tilde{I}$ are free $\mathbf{k}[t]$-modules, being submodules of $\Lambda V \otimes \mathbf{k}[t]$.

Denote by $\mathbf{k}_\xi = \mathbf{k}[t]/(t - \xi)$. Then, we have a diagram

$$\begin{array}{ccccccccc} 0 & \to & \tilde{K} \otimes \mathbf{k}_\xi & \to & (\Lambda V \otimes \mathbf{k}[t]) \otimes \mathbf{k}_\xi & \to & \tilde{I} \otimes \mathbf{k}_\xi & \to & 0 \\ & & \downarrow & & || & & \downarrow & & \\ 0 & \to & K & \to & \Lambda V_\mathbf{k} & \to & I & \to & 0. \end{array} \tag{8}$$

(Here, the tensor products of all $\mathbf{k}[t]$-modules are over $\mathbf{k}[t]$, and the tensor product $\Lambda V \otimes \mathbf{k}[t]$ is over the rationals.) Therefore, the last vertical map is a surjection, and the first map is an injection.

We have

$$0 \to \tilde{I} \to \tilde{K} \to H(\Lambda V \otimes \mathbf{k}[t], d) \to 0,$$

which is an exact sequence of $\mathbf{k}[t]$-modules. Then, $H(\Lambda V \otimes \mathbf{k}[t], d)$ contains a free part and a torsion part. The torsion is supported at some points, which are at most countably many. Therefore, for generic $\xi \in \mathbf{k}$,

$$0 \to \tilde{I} \otimes \mathbf{k}_\xi \to \tilde{K} \otimes \mathbf{k}_\xi \to H(\Lambda V \otimes \mathbf{k}[t], d) \otimes \mathbf{k}_\xi \to 0$$

is exact. As $\tilde{I} \otimes \mathbf{k}_\xi \twoheadrightarrow I \subset K$ and $\tilde{I} \otimes \mathbf{k}_\xi \subset \tilde{K} \otimes \mathbf{k}_\xi \subset K$, we have that the last map in (8) is an injection, therefore an isomorphism, and thus the first map is also an isomorphism by the snake lemma.

Note that also, when tensoring with $\mathbf{k}(t)$, we have an exact sequence

$$0 \to \tilde{I} \otimes \mathbf{k}(t) \to \tilde{K} \otimes \mathbf{k}(t) \to H(\Lambda V \otimes \mathbf{k}[t], d) \otimes \mathbf{k}(t) \to 0 .$$

Also $H(\Lambda V \otimes \mathbf{k}[t], d) \otimes \mathbf{k}(t) = H(\Lambda V \otimes \mathbf{k}(t), d)$, since $\mathbf{k}(t)$ is a flat $\mathbf{k}[t]$-module. Hence,

$$\begin{aligned} \dim H(\Lambda V_{\mathbf{k}}, d_\xi) &= \dim K - \dim I \\ &= \dim \tilde{K} \otimes \mathbf{k}_\xi - \dim \tilde{I} \otimes \mathbf{k}_\xi \\ &= \dim H(\Lambda V \otimes \mathbf{k}(t), d) . \end{aligned}$$

In the first line, we mean $\dim K - \dim I = \sum_{d \geq 0} \left(\dim K^d - \dim I^d\right)$.

Take now $\xi = 0$. The map $\tilde{K} \to K \to K/I$ factors as $\tilde{K}/\tilde{I} \to K/I$. Tensor this map by $\mathbf{k}_0$ to get $(\tilde{K}/\tilde{I}) \otimes \mathbf{k}_0 \to K/I$. Note that there is an exact sequence

$$\tilde{I} \otimes \mathbf{k}_0 \to \tilde{K} \otimes \mathbf{k}_0 \to (\tilde{K}/\tilde{I}) \otimes \mathbf{k}_0 \to 0,$$

but the first map may not be injective. Then, there is a map

$$\frac{\tilde{K} \otimes \mathbf{k}_0}{\operatorname{Im}(\tilde{I} \otimes \mathbf{k}_0)} = (\tilde{K}/\tilde{I}) \otimes \mathbf{k}_0 \to K/I .$$

By (8), this is an inclusion. Now we have:

$$\begin{aligned} \dim H(\Lambda V, d_\xi) &= \dim H(\Lambda V \otimes \mathbf{k}(t), d) \\ &= \dim (\tilde{K}/\tilde{I}) \otimes \mathbf{k}(t) \\ &\leq \dim (\tilde{K}/\tilde{I}) \otimes \mathbf{k}_0 \\ &= \dim \frac{\tilde{K} \otimes \mathbf{k}_0}{\operatorname{Im}(\tilde{I} \otimes \mathbf{k}_0)} \\ &\leq \dim K/I \\ &= \dim H(\Lambda V_{\mathbf{k}}, d_0) \\ &= \dim_{\mathbb{Q}} H(\Lambda V, d_0) . \end{aligned}$$

4.2 Perturbing the Minimal Model

Let $x_1, \dots, x_n$ denote generators for V^{even}, and $y_1, \dots, y_{n+r}$ generators for V^{odd}. Here, $dx_i = 0$ and $dy_j = P_j\left(x_1, \dots, x_n, y_1, \dots, y_{j-1}\right)$.

We consider the algebra

$$(\Lambda W, d) = (\Lambda V, d) \otimes (\Lambda \bar{y}_1, 0),$$

where $\deg(\bar{y}_1) = \deg(x_1) - 1$. Then

$$\dim H(\Lambda W, d) = 2 \dim H(\Lambda V, d).$$

Consider now the differential δ on ΛW such that $\delta x_j = 0$, $\delta y_j = 0$ and $\delta \bar{y}_1 = x_1$. Hence, $\delta^2 = 0$ and $d\delta = \delta d = 0$. So,

$$d_t = d + t\delta$$

is a differential on $\Lambda W \otimes \mathbf{k}[t]$.

For generic $\xi \in \mathbf{k}$, $(\Lambda W_{\mathbf{k}}, d_\xi)$ verifies that $d_\xi \bar{y}_1 = \xi x_1$. So, for nonzero ξ, there is a KS-extension [8, § 1.4]

$$\left(\Lambda(x_1, \bar{y}_1), d_\xi\right) \longrightarrow \left(\Lambda W_{\mathbf{k}}, d_\xi\right) \longrightarrow (\Lambda(x_2, \dots, x_n, y_1, \dots y_{n+r}), d).$$

As $H(\Lambda(x_1, \bar{y}_1), d_\xi) = \mathbf{k}$, we have that

$$H(\Lambda W_{\mathbf{k}}, d_\xi) \cong H(\Lambda(x_2, \dots, x_n, y_1, \dots y_{n+r}), d).$$

Now we apply Lemma 3 to this to obtain that

$$\dim H(\Lambda(x_2, \dots, x_n, y_1, \dots y_{n+r}), d) \leq \dim H(\Lambda W, d) = 2 \dim H(\Lambda V, d).$$

Repeating the argument n times, we get that

$$\dim H\ (\Lambda(y_1, \dots y_{n+r}), d) \leq 2^n \dim H(\Lambda V, d).$$

But the hyperelliptic condition says that $d = 0$ for the first space, so

$$2^n \dim H(\Lambda V, d) \geq \dim H(\Lambda(y_1, \dots y_{n+r}), d) = 2^{n+r}.$$

This gives

$$\dim H(\Lambda V, d) \geq 2^r. \tag{9}$$

4.3 Another Proof of Inequality (9)

In this section, we present a different proof of the inequality $\dim H(\Lambda V, d) \geq 2^r$ for hyperelliptic spaces. Recall that if A is a commutative graded differential algebra, and if M, N are differential graded A-modules, the differential Tor is defined as

$$\mathrm{Tor}^*(M, N) = H^*(P \otimes_A N),$$

where $P \xrightarrow{\sim} M$ is a semifree resolution, i.e., a quasi-isomorphism from a semifree A-module P to M (see [2, § 6]).

Lemma 4 *Let* $C \xleftarrow{\varphi} A \xrightarrow{\psi} B$ *be morphisms of commutative differential graded algebras. There exists a convergent spectral sequence*

$$E_2^{p,q} = H^p(B) \otimes \mathrm{Tor}_A^q(\mathbb{Q}, C) \Rightarrow \mathrm{Tor}_A^{p+q}(B, C).$$

Proof Decompose φ and ψ as

Then, $\alpha : A \otimes \Lambda W \xrightarrow{\sim} B$ is a semifree resolution of B regarded as A-module, so

$$\mathrm{Tor}_A^*(B, C) = H^*((A \otimes \Lambda W) \otimes_A C).$$

Moreover, $\mathrm{Id} \otimes \beta : (A \otimes \Lambda W) \otimes_A A \otimes \Lambda U \xrightarrow{\sim} (A \otimes \Lambda W) \otimes_A C$ is a quasi-isomorphism and $(A \otimes \Lambda W) \otimes_A (A \otimes \Lambda U) \cong A \otimes \Lambda W \otimes \Lambda U$. Therefore, one gets a rational fibration

$$A \otimes \Lambda W \to A \otimes \Lambda W \otimes \Lambda U \to \Lambda U,$$

whose associated Serre spectral sequence has the form

$$E_2^{p,q} = H^p(A \otimes \Lambda W) \otimes H^q(\Lambda U) \Rightarrow H^{p+q}(A \otimes \Lambda W \otimes \Lambda U).$$

On the one hand, $H^*(A \otimes \Lambda W) = H^*(B)$. On the other hand, since β is a semifree resolution of C, we have that

$$H^*(\Lambda U) = H^*((A \otimes \Lambda U) \otimes_A \mathbb{Q}) = \mathrm{Tor}_A^*(\mathbb{Q}, C).$$

Putting all pieces together, we get

$$E_2^{p,q} = H^p(B) \otimes \mathrm{Tor}_A^q(\mathbb{Q}, C) \Rightarrow \mathrm{Tor}_A^{p+q}(B, C).$$

Theorem 4 *Let* $(\Lambda V, d)$ *be a hyperelliptic minimal model. Then*

$$\dim H(\Lambda V, d) \geq 2^r.$$

Proof Write as usual $x_1, \dots, x_n$ for generators of $X = V^{even}$ and $y_1, \dots, y_{n+r}$ for generators of $Y = V^{odd}$. When we apply the previous lemma to morphisms $\mathbb{Q} \longleftarrow \Lambda X \hookrightarrow \Lambda V$ we get a spectral sequence

$$E_2 = H(\Lambda V, d) \otimes \mathrm{Tor}^*_{\Lambda X}(\mathbb{Q}, \mathbb{Q}) \Rightarrow \mathrm{Tor}^*_{\Lambda X}(\Lambda V, \mathbb{Q}).$$

On the one hand,

$$\mathrm{Tor}^*_{\Lambda X}(\mathbb{Q}, \mathbb{Q}) = H^*(\Lambda(\overline{x}_1, \dots, \overline{x}_n), 0) = \Lambda(\overline{x}_1, \dots, \overline{x}_n),$$

where $\Lambda(x_1, \dots, x_n, \overline{x}_1, \dots, \overline{x}_n) \xrightarrow{\sim} \mathbb{Q}$ is a semifree resolution of $\mathbb{Q}$ regarded as ΛX-module. Hence, $\overline{x}_i$ are all of odd degree.

On the other hand, ΛV is already ΛX-semifree, so

$$\mathrm{Tor}^*_{\Lambda X}(\Lambda V, \mathbb{Q}) = H(\Lambda V \otimes_{\Lambda X} \mathbb{Q}) = H^*(\Lambda(y_1, \ldots, y_{n+k}), 0) = \Lambda(y_1, \ldots, y_{n+k}).$$

Then, the inequality

$$\dim H^*(\Lambda V, d) \cdot \dim \mathrm{Tor}^*_{\Lambda X}(\mathbb{Q}, \mathbb{Q}) \geq \dim \mathrm{Tor}^*_{\Lambda X}(\Lambda V, \mathbb{Q})$$

coming from the spectral sequence translates into

$$2^n \dim H^*(\Lambda V, d) \geq 2^{n+r},$$

so the result follows.

4.4 Proof of Theorem 3

Now we prove the inequality $\dim H(\Lambda V, d) \geq 2n + r$, for the hyperelliptic minimal model.

If $r = 0$, then $\chi_\pi = 0$. So [2, Prop. 32.10] says that the model is pure, and this case is already covered by Remark 1.

If $r > 0$, then $\chi_\pi < 0$. So by Proposition 1, $\chi = 0$, and hence, it is enough to prove that

$$\dim H^{even}(\Lambda V, d) \geq n + \tfrac{r}{2}.$$

Suppose that $r = 1, 2$. As the degree 0 and degree 1 elements give always nontrivial homology classes, then $\dim H^{even}(\Lambda V, d) \geq n + 1$, and we are done.

Therefore, we can assume $r \geq 3$. We use the following fact: if $P(x)$ is a quadratic polynomial on the x, and $P(x) = d\alpha$, $\alpha \in \Lambda V$, then α must be linear, $\alpha \in V^{odd}$ and denoting by d_o the composition

$$V^{odd} \longrightarrow \Lambda^+ V^{even} \otimes \Lambda V^{odd} \twoheadrightarrow \Lambda^+ V^{even},$$

we have $P(x) = d_o\alpha$. So there are at least $\binom{n+1}{2} - (n + r)$ quadratic terms in the homology. Conjecture 1 is proved if

$$\begin{cases} \text{either} & 1 + n + \binom{n+1}{2} - (n + r) \geq n + \frac{r}{2}, \\ \quad\text{or} & 2^r \geq 2n + r. \end{cases} \tag{10}$$

So now assume that (10) does not hold. Then

$$2^r - r \leq 2n - 1, \tag{11}$$

and $1 + \binom{n+1}{2} - n < \frac{3}{2}r$, i.e.,

$$(2n - 1)^2 \leq 12r - 11. \tag{12}$$

Putting together (11) and (12), we get $2^r - r \leq \sqrt{12r-11}$, i.e., $2^r \leq r + \sqrt{12r-11}$. This is easily seen to imply that $r \leq 3$. So, $r = 3$ and $n = 3$.

It remains to deal with the case $n = 3$, $r = 3$, and d_o is an isomorphism of the odd degree elements onto $\Lambda^2 V^{even}$. Let x_1, x_2, x_3 be the even-degree generators, of degrees $d_1 \leq d_2 \leq d_3$, respectively. The degrees of $x_1^2, x_1x_2, x_2^2, x_1x_3, x_2x_3, x_3^2$ are the six numbers

$$2d_1 \leq d_1 + d_2 \leq 2d_2, \quad d_1 + d_3 \leq d_2 + d_3 \leq 2d_3.$$

We have two cases:

- Case $2d_2 \leq d_1 + d_3$. We can arrange the odd generators $y_1, \dots, y_6$ with increasing degree and so that $d_o y_1 = x_1^2, d_o y_2 = x_1x_2, d_o y_3 = x_2^2, d_o y_4 = x_1x_3, d_o y_5 = x_2x_3, d_o y_6 = x_3^2$. Clearly, $dy_1 = x_1^2$. Then, $dy_2 = x_1x_2 + P(x_1)$, where $P(x_1)$ is a polynomial on x_1, i.e., of the form $c\,x_1^n$, $n \geq 2$. But this can be absorbed by a change of variables $y_2 \mapsto y_2 - c\,x_1^{n-2} y_1$. So, we can write $dy_2 = x_1x_2$. Now the even-degree closed elements in $\Lambda(x_1, x_2, x_3, y_1, y_2)$ are again polynomials on x_1, x_2, x_3. So, we can assume $dy_3 = x_2^2$ as before. Continuing the computation, the even-degree closed elements in $\Lambda(x_1, x_2, x_3, y_1, y_2, y_3)$ are either polynomials on the x_i's or a multiple of the element $x_2^2 y_1 y_2 - x_1x_2 y_1 y_3 + x_1^2 y_2 y_3 = d(y_1 y_2 y_3)$, which is exact. Therefore, we can again manage to arrange that $dy_4 = x_1x_3$.
- Case $2d_2 > d_1 + d_3$. Then, we have that $d_o y_3 = x_1x_3$ and $d_o y_4 = x_2^2$. As before, we can arrange $dy_3 = x_1x_3$. Now the even-degree closed elements in $\Lambda(x_1, x_2, x_3, y_1, y_2, y_3)$ are polynomials on the x_i's or a multiple of $x_3 y_1 y_2 - x_2 y_1 y_3 + x_1 y_2 y_3$. But this element has degree $3d_1 + d_2 + d_3 - 2 > 2d_2$, so it must be $dy_4 = x_2^2$.

In either case, dy_1, dy_2, dy_3, dy_4 are x_1^2, x_1x_2, x_2^2 and x_1x_3. Let us assume that we are in the first case to carry over the notation.

Now we compute the even-degree closed elements in $\Lambda(x_1, x_2, x_3, y_1, y_2, y_3, y_4)$. These are polynomials on x_i's or combinations of

$$\begin{aligned}
&x_2^2 y_1 y_2 - x_1x_2 y_1 y_3 + x_1^2 y_2 y_3 = d(y_1 y_2 y_3), \\
&x_3 y_1 y_2 - x_2 y_1 y_4 + x_1 y_2 y_4, \\
&x_1x_3 y_2 y_3 - x_2^2 y_2 y_4 + x_1x_2 y_3 y_4 = d(y_2 y_3 y_4), \text{ and} \\
&x_1x_3 y_1 y_3 + x_1^2 y_3 y_4 - x_2^2 y_1 y_4 = d(y_1 y_3 y_4).
\end{aligned}$$

Only the second one is nonexact, but its degree is strictly bigger than $d_2 + d_3$. So again we can arrange that $dy_5 = x_2x_3$.

Finally, the minimal model is:

$$\begin{cases}
dy_1 = x_1^2, \\
dy_2 = x_1x_2, \\
dy_3 = x_2^2, \\
dy_4 = x_1x_3, \\
dy_5 = x_2x_3, \\
dy_6 = x_3^2 + P(x_i, y_j).
\end{cases}$$

The even-degree closed elements in $\Lambda(x_1, x_2, x_3, y_1, y_2, y_3, y_4, y_5)$ contain at least

$$\alpha_1 = x_3 y_2 y_3 + x_1 y_3 y_5 - x_2 y_2 y_5 ,$$
$$\alpha_2 = x_3 y_1 y_2 - x_2 y_1 y_4 + x_1 y_2 y_4 .$$

At most, one of them does not survive in $H(\Lambda V, d)$, so proving the existence of at least another even-degree cohomology class. Hence, dim $H(\Lambda V, d) \geq 10 \geq 9$, as required.

References

1. Félix, Y.: La Dichotomie Elliptique-Hyperbollique en Homotopie Rationnelle, Astérisque 176. Socièté mathèmatique de France Paris (1989)
2. Félix, Y., Halperin, S., Thomas, J.-C.: Rational Homotopy Theory. Graduate Texts in Mathematics, vol. 205. Springer, Berlin (2000)
3. Griffiths, P.A., Morgan, J.W.: Rational Homotopy Theory and Differential Forms. Progress in Mathematics, vol. 16. Birkhäuser, Basel (1981)
4. Halperin, S.: Finiteness in the minimal models of Sullivan. Trans. Am. Math. Soc. **230**, 173–199 (1977)
5. Hilali, M.R.: Action du tore $\mathbb{T}^n$ sur les espaces simplement connexes. Ph.D. thesis, Université catholique de Louvain, Belgique (1990)
6. Hilali, M.R., Mamouni, M.I.: A conjectured lower bound for the cohomological dimension of elliptic spaces. J. Homotopy Relat. Struct. **3**, 379–384 (2008)
7. Hilali, M.R., Mamouni, M.I.: A lower bound of cohomologic dimension for an elliptic space. Topol. Appl. **156**, 274–283 (2008)
8. Oprea, J., Tralle, A.: Symplectic Manifolds with no Kähler Structure. Lecture Notes in Mathematics, vol. 1661. Springer, Berlin (1997)

Aveiro Discretization Method in Mathematics: A New Discretization Principle

L.P. Castro, H. Fujiwara, M.M. Rodrigues, S. Saitoh and V.K. Tuan

Abstract We found a very general discretization method for solving wide classes of mathematical problems by applying the theory of reproducing kernels. An illustration of the generality of the method is here performed by considering several distinct classes of problems to which the method is applied. In fact, one of the advantages of the present method—in comparison to other well-known and well-established methods—is its global nature and no need of special or very particular data conditions. Numerical experiments have been made, and consequent results are here exhibited. Due to the powerful results which arise from the application of the present method, we consider that this method has everything to become one of the next-generation methods of solving general analytical problems by using computers. In particular, we would like to point out that we will be able to solve very global linear partial differential equations satisfying very general boundary conditions or initial values (and in a somehow independent way of the boundary and domain). Furthermore, we will be able to give an ultimate sampling theory and an ultimate realization of the consequent general reproducing kernel Hilbert spaces. The general theory is here presented in a constructive way, and contains some related historical and concrete examples.

S. Saitoh (✉)
Department of Mathematics, Institute of Reproducing Kernels,
Kawauchi-cho, 5-1648-16, Kiryu 376-0041, Japan
e-mail: saburou.saitoh@gmail.com

L.P. Castro · M.M. Rodrigues
CIDMA–Center for Research and Development in Mathematics and Applications,
Department of Mathematics, University of Aveiro, 3810–193 Aveiro, Portugal
e-mail: castro@ua.pt

M.M. Rodrigues
e-mail: mrodrigues@ua.pt

H. Fujiwara
Graduate School of Informatics, Kyoto University, Kyoto, 606-8501 Japan
e-mail: fujiwara@acs.i.kyoto-u.ac.jp

V.K. Tuan
Department of Mathematics, University of West Georgia,
1601 Maple Street, Carrollton, GA 30118, USA
e-mail: vu@westga.edu

T. M. Rassias, P. M. Pardalos (eds.), *Mathematics Without Boundaries*,
DOI 10.1007/978-1-4939-1106-6_3,

Keywords Reproducing kernel · Discretization · Computer · Numerical · PDE · ODE · Integral equation · Numerical experiment · Generalized inverse · Tikhonov regularization · Real inversion of the Laplace transform · Matrix · Convolution · Singular integral equation · Sampling theory · Analyticity · Smoothness of function

1 Introduction

Nowadays, the importance and influence of reproduction kernel Hilbert spaces is identified in several different subareas of mathematics and their applications. In some cases, such importance is increased when convenient modifications and applications of different techniques are additionally implemented. A major ingredient which gave rise to important and global results was simply the consideration of linear mappings within the framework of Hilbert spaces in a reproduction kernels identification process. This gave rise to a very powerful and also natural use of reproducing kernels in the framework of Hilbert spaces (as it may be seen in [35]) which, in turn, had several very significant consequences during the last decades.

As a somehow natural consequence of the research project proposed by S. Saitoh in the University of Aveiro some years ago, and being developed until now, it was introduced as a new line of research within the Functional Analysis and Applications Group of this university (involving not only researchers from this university but also collaborators from abroad). Among some other fruitful consequences of that project within reproduction kernel Hilbert spaces, in here we would like to present the so-called *Aveiro Discretization Method in Mathematics*. Interestingly enough, this name was identified by S. Saitoh in the night just before one of his talks presented in the "9th International Conference on Mathematical Problems in Engineering, Aerospace and Science" (Vienna, Austria, 10–14 July 2012). In such a talk, it was presented for the first time (in an oral and nondetailed way) the main ideas of the method that we are now presenting in here in a complete form.

As main ingredients for the starting points of the method, we may identify the general theory of reproducing kernels, the Tikhonov regularization procedure, consequent approximate solutions for bounded linear operator equations on Hilbert spaces, and optimal numerical solutions.

In particular, as examples of possible applications, we will be able to solve very general linear partial differential equations (PDEs) satisfying very general boundary conditions and initial values—independently of the boundary and domains. Furthermore, we will be able to give clearly an ultimate sampling theory and ultimate realizations of general reproducing kernel Hilbert spaces. We would like to develop the general theory in a self-contained manner with some related history and many concrete examples.

The remaining part of this work is organized in the following way. In Sect. 2, we establish a general formula giving an optimal inverse by using a finite number of output data in the framework of Hilbert spaces. In Sect. 3, we shall establish the convergence property of our approximate inverses in Sect. 2. In order to show our history for some fundamental inverses with the typical example of the Laplace transform, we shall refer to several typical inverses in Sects. 3–9 and in Sect. 10, we shall give our final and new approximate inversion of the Laplace transform. In Sect. 11, we shall introduce the Paley–Wiener spaces and Sobolev spaces (due to the necessity of these spaces in our new discretization principle). In Sect. 12, we shall derive results for general linear ordinary differential equations, and in Sect. 13, we derive concrete results for various typical partial differential equations. In Sects. 14 and 15, we will introduce results for convolution and singular integral equations, respectively. In Sect. 16, we shall give several numerical experiments, and we end up with a general discussion in Sect. 17 (mostly from the viewpoint of numerical analysis).

2 Principal Formulation of the Inverse by Using a Finite Number of Data

We shall first establish our fundamental inversion formula by using a finite number of data in the framework of Hilbert spaces. In view of this, following [35, 38], we shall start by introducing a general theory for linear mappings in the framework of Hilbert spaces.

Let $\mathcal{H}$ be a Hilbert (possibly finite-dimensional) space, and consider E to be an abstract set and $\mathbf{h}$ a Hilbert $\mathcal{H}$-valued function on E. Then, we are able to consider the linear transform

$$f(p) = (\mathbf{f}, \mathbf{h}(p))_{\mathcal{H}}, \quad \mathbf{f} \in \mathcal{H}, \tag{1}$$

from $\mathcal{H}$ into the linear space $\mathcal{F}(E)$ comprising all the complex valued functions on E. In order to investigate the linear mapping (1), we form a positive definite quadratic form function $K(p,q)$ on $E \times E$ defined by

$$K(p,q) = (\mathbf{h}(q), \mathbf{h}(p))_{\mathcal{H}} \quad \text{on} \quad E \times E. \tag{2}$$

Then, we obtain the following fundamental result.

Proposition 1

(I) *The range of the linear mapping (1) by $\mathcal{H}$ is characterized as the reproducing kernel Hilbert space $H_K(E)$ admitting the reproducing kernel $K(p,q)$ whose characterization is given by the two properties: (i) $K(\cdot,q) \in H_K(E)$ for any $q \in E$ and, (ii) for any $f \in H_K(E)$ and for any $p \in E$, $(f(\cdot), K(\cdot .p))_{H_K(E)} = f(p)$.*

(II) *In general, we have the inequality*

$$\|f\|_{H_K(E)} \leq \|\mathbf{f}\|_{\mathcal{H}}.$$

Here, for any member f of $H_K(E)$, there exists a uniquely determined $\mathbf{f}^* \in \mathcal{H}$ satisfying

$$f(p) = (\mathbf{f}^*, \mathbf{h}(p))_{\mathcal{H}} \quad \text{on} \quad E$$

and

$$\|f\|_{H_K(E)} = \|\mathbf{f}^*\|_{\mathcal{H}}. \tag{3}$$

(III) *In general, we have the inversion formula in (1) in the form*

$$f \mapsto \mathbf{f}^* \tag{4}$$

in (II) by using the reproducing kernel Hilbert space $H_K(E)$.

The typical ill-posed problem (1) will become a well-posed problem, because the image space of (1) is characterized as the reproducing kernel Hilbert space $H_K(E)$ with the isometric identity (3), which may be considered as a generalization of the Pythagorean theorem.

However, this viewpoint is a mathematical one and is not a numerical one, as we shall refer to it later.

We shall consider the inversion in (4) formally. However, this idea will be very important for the general inversions and for our new discretization method.

Following the above general situation, we shall recall the general theory in the following general situation (cf. also [34, 35]): Let $\{\mathbf{v}_j\}$ be a complete orthonormal basis for $\mathcal{H}$. Then, for

$$v_j(p) = \left(\mathbf{v}_j, \mathbf{h}(p)\right)_{\mathcal{H}},$$

we have

$$\mathbf{h}(p) = \sum_j \left(\mathbf{h}(p), \mathbf{v}_j\right)_{\mathcal{H}} \mathbf{v}_j = \sum_j \overline{v_j(p)} \mathbf{v}_j.$$

Hence, by setting

$$\overline{\mathbf{h}}(p) = \sum_j v_j(p) \mathbf{v}_j,$$

we may write

$$\overline{\mathbf{h}}(\,\cdot\,) = \sum_j v_j(\,\cdot\,) \mathbf{v}_j.$$

Thus, we shall define

$$\left(f, \overline{\mathbf{h}}(p)\right)_{H_K} = \sum_j \left(f, v_j\right)_{H_K} \mathbf{v}_j.$$

For simplicity, we will write as follows:

$$H_K = H_K(E),$$

here and in the sequel. Then, we have the following result.

Proposition 2 *We assume that for $f \in H_K$, it holds*

$$\left(f, \overline{\mathbf{h}}\right)_{H_K} \in \mathcal{H}$$

and for all $p \in E$,

$$(f, (\mathbf{h}(p), \mathbf{h}(\cdot))_{\mathcal{H}})_{H_K} = \left((f, \overline{\mathbf{h}})_{H_K}, \mathbf{h}(p)\right)_{\mathcal{H}}.$$

Then, we have

$$\|f\|_{H_K} \leqq \|(f, \overline{\mathbf{h}})_{H_K}\|_{\mathcal{H}}. \tag{5}$$

If $\{\mathbf{h}(p); p \in E\}$ is complete in $\mathcal{H}$, then the equality takes place in (5).
Furthermore, we assume:

$$\left(\mathbf{f}_0, (f, \overline{\mathbf{h}})_{H_K}\right)_{\mathcal{H}} = ((\mathbf{f}_0, \mathbf{h})_{\mathcal{H}}, f)_{H_K} \quad for \quad \mathbf{f}_0 \in N(L).$$

Then, we have, for $\mathbf{f}^*$ in (II) and (III),

$$\mathbf{f}^* = \left(f, \overline{\mathbf{h}}\right)_{H_K}.$$

In particular, note that the basic assumption $\left(f, \overline{\mathbf{h}}\right)_{H_K} \in \mathcal{H}$ in Proposition 2, is, in general, not valid for many analytical problems and we need some delicate treatment for the inversion (see [35]). In view of this, let us analyze again the possibilities for the linear mapping defined by (1). In order to derive a general inversion formula for (1) that is widely applicable in analysis, we shall assume that both Hilbert spaces $\mathcal{H}$ and H_K are given as $\mathcal{H} = L_2(T, dm)$, $H_K \subset L_2(E, d\mu)$, on the sets T and E, respectively. Note that for $dm, d\mu$ measurable sets T, E we assume that they are the Hilbert spaces comprising $dm, d\mu - L_2$ integrable complex-valued functions, respectively. Therefore, we shall consider the integral transform

$$f(p) = \int_T F(t)\overline{h(t,p)}dm(t). \tag{6}$$

Here, $h(t,p)$ is a function on $T \times E$, $h(\cdot, p) \in L_2(T, dm)$, and $F \in L_2(T, dm)$. The corresponding reproducing kernel for (2) is given by

$$K(p,q) = \int_T h(t,q)\overline{h(t,p)}dm(t) \quad \text{on} \quad E \times E.$$

The norm of the reproducing kernel Hilbert space H_K is represented as $L_2(E, d\mu)$. Under these situations, we obtain the following result.

Proposition 3 *We assume that an approximating sequence $\{E_N\}_{N=1}^{\infty}$ of E satisfies*

(a)$E_1 \subset E_2 \subset \cdots \subset E_n \subset \cdots$,
(b)$\bigcup_{N=1}^{\infty} E_N = E$,
(c)$\int_{E_N} K(p,p)d\mu(p) < \infty, \quad N = 1, 2, \ldots.$

Then, for $f \in H_K$ *satisfying* $\int_{E_N} f(p)h(t,p)d\mu(p) \in L_2(T,dm)$ *for any* N*, the sequence*

$$\left\{\int_{E_N} f(p)h(t,p)d\mu(p)\right\}_{N=1}^{\infty} \tag{7}$$

converges to F^* *in* (4) *in Proposition 1 in the sense of the* $L_2(T,dm)$ *norm.*

Practically, for many cases, the assumptions in Proposition 3 will be satisfied automatically, and so Proposition 3 may be applied in all those cases. To construct the inversion formula is, in general, difficult. However, the formulation in Proposition 3 may be considered as a natural one. Indeed, this may be realized if we recognize that it was given as the strong convergence in the Hilbert space $L_2(T,dm)$.

Proposition 3 will give a new viewpoint and method for the fundamental Fredholm integral equation of the first kind (6). Our method and solution will have the following properties:

(P1) We use the naturally determined reproducing kernel Hilbert space H_K which is determined by the integral kernel.
(P2) The solution is given in the sense of $\mathcal{H}$ norm convergence.
(P3) Our solution (inverse) is given by f^* in Proposition 1.
(P4) For the ill-posed problem in (1), our method gives a well-posed solution.

In general, in a linear problem, when we cannot guarantee the eventual uniqueness of a solution continuously dependent on the data, then the linear problem is called an *ill-posed problem* in the Hadamard sense. The linear problem (1) or (6) has been considered a typical ill-posed problem. Our method gives a characterization of the image, that is, the solution space. For the inverse or solution, we consider the minimum norm one (among many solutions) and then we have the isometric relation in Proposition 1; that is, our solution has uniqueness and continuity properties. However, this is a mathematical and theoretical viewpoint. In practical and physical linear systems, the observation data will be a finite number of data containing error or noises, and so we will meet various delicate numerical problems and there exists a field of ill-posed problems. In order to overcome such cases, we examined several methods, as we shall state in the sequel.

Our new idea is based on the approximate realization of the abstract Hilbert space H_K by taking a finite number of points of E, because, in general, the reproducing kernel Hilbert space H_K has a complicated structure. Therefore, we will implement the following procedure: By taking a finite number of points $\{p_j\}_{j=1}^{n}$, we set

$$K\left(p_j, p_{j'}\right) := a_{jj'}. \tag{8}$$

Consequently, if the matrix $A := \| a_{jj'} \|$ is positive definite, then, the corresponding norm in H_A comprising the vectors $\mathbf{x} = (x_1, x_2, \ldots, x_n)$ is determined by

$$\|\mathbf{x}\|_{H_A}^2 = \mathbf{x}^* \widetilde{A}\mathbf{x},$$

where $\widetilde{A} = \overline{A^{-1}} = \|\widetilde{a_{jj'}}\|$ (see [35], p. 250). This property may be confirmed directly in a nondifficult way.

When we approximate the reproducing kernel Hilbert space H_K by the vector space H_A, then from Proposition 2, the following theorem is directly derived.

Theorem 1 *In the linear mapping*

$$f(p) = (\mathbf{f}, \mathbf{h}(p))_{\mathcal{H}}, \quad \mathbf{f} \in \mathcal{H} \tag{9}$$

for

$$\{p_1, p_2, \ldots, p_n\}$$

the minimum norm inverse $\mathbf{f}^*_{A_n}$ *satisfying*

$$f(p_j) = \left(\mathbf{f}, \mathbf{h}(p_j)\right)_{\mathcal{H}}, \quad \mathbf{f} \in \mathcal{H} \tag{10}$$

is given by

$$\mathbf{f}^*_{A_n} = \sum_j \sum_{j'} f\left(p_j\right) \widetilde{a_{jj'}} \mathbf{h}\left(p_{j'}\right), \tag{11}$$

where $\widetilde{a_{jj'}}$ *are assumed the elements of the complex conjugate inverse of the positive definite Hermitian matrix* A_n *constituted by the elements*

$$a_{jj'} = \left(\mathbf{h}(p_{j'}), \mathbf{h}(p_j)\right)_{\mathcal{H}}.$$

Here, the positive definiteness of A_n *is a basic assumption.*

3 Convergence of the Approximate Inverses

The following theorem deals with the convergence of our approximate inverses in Theorem 1. The case of the Laplace transform was proved in [43].

Theorem 2 *Let* $\{p_j\}_{j=1}^{\infty}$ *be a sequence of distinct points on* E*, that is the positive definiteness in Theorem 1 for any* n *and a uniqueness set for the reproducing kernel Hilbert space* H_K*; that is, for any* $f \in H_K$*, if all* $f(p_j) = 0$*, then* $f \equiv 0$*. Then, in the space* $\mathcal{H}$*, it holds*

$$\lim_{n\to\infty} \mathbf{f}^*_{A_n} = \mathbf{f}^*. \tag{12}$$

Proof By Theorem 1, we obtain the monotonicity

$$\|\mathbf{f}^*_{A_n}\|_{\mathcal{H}} \leq \|\mathbf{f}^*_{A_m}\|_{\mathcal{H}}, \quad \text{if} \quad 0 < n < m,$$

and

$$\|\mathbf{f}^*_{A_n}\|_{\mathcal{H}} \leq \|\mathbf{f}^*\|_{\mathcal{H}}, \quad \text{if} \quad n > 0.$$

The sequence $\{\mathbf{f}^*_{A_n}\}$ is uniformly bounded in norm by $\|\mathbf{f}^*\|_{\mathcal{H}}$ in the Hilbert space $\mathcal{H}$, therefore, there exists a subsequence $\{\mathbf{f}^*_{A_{n_j}}\}$ that converges weakly to a member $\mathbf{f}^{**} \in \mathcal{H}$ and $\|\mathbf{f}^{**}\|_{\mathcal{H}} \le \liminf_{j\to\infty} \|\mathbf{f}^*_{A_{n_j}}\|_{\mathcal{H}}$. Then, we have

$$\lim_{j\to\infty} f^*_{A_{n_j}}(p) := \lim_{j\to\infty} \left(\mathbf{f}^*_{A_{n_j}}, \mathbf{h}(p)\right)_{\mathcal{H}} =: f^*(p) \in H_K.$$

In particular, since $f^*_{A_{n_j}}(p_k) = f(p_k)$ when $n_j \ge k$, we have

$$f^*(p_k) = \lim_{j\to\infty} f^*_{A_{n_j}}(p_k) = f(p_k), \quad k > 0.$$

From the uniqueness assumption, $f^*(z) = f(z)$. Therefore, $\mathbf{f}^{**} = \mathbf{f}^*$.

The sequence $\left\{\|\mathbf{f}^*_{A_{n_j}}\|_{\mathcal{H}}\right\}$ is monotone increasing, and bounded above by $\|\mathbf{f}^*\|_{\mathcal{H}}$, hence,

$$\|\mathbf{f}^*\|_{\mathcal{H}} \le \liminf_{j\to\infty} \|\mathbf{f}^*_{A_{n_j}}\|_{\mathcal{H}} \le \|\mathbf{f}^*\|_{\mathcal{H}}.$$

Consequently,

$$\lim_{j\to\infty} \|\mathbf{f}^*_{A_{n_j}}\|_{\mathcal{H}} = \|\mathbf{f}^*\|_{\mathcal{H}}.$$

The subsequence $\{\mathbf{f}^*_{A_{n_j}}\}$ converges weakly to $\mathbf{f}^*$, and the sequence of their norms $\{\|\mathbf{f}^*_{A_{n_j}}\|_{\mathcal{H}}\}$ converges to $\|\mathbf{f}^*\|_{\mathcal{H}}$. Therefore, the subsequence $\{\mathbf{f}^*_{A_{n_j}}\}$ converges strongly to $\mathbf{f}^*$. We claim now that the whole sequence $\{\mathbf{f}^*_{A_n}\}$ converges strongly to $\mathbf{f}^*$. If not, there would exist an $\epsilon > 0$ and a subsequence $\{\mathbf{f}^*_{A_{n_l}}\}$ such that

$$\|\mathbf{f}^* - \mathbf{f}^*_{A_{n_l}}\|_{\mathcal{H}} \ge \epsilon, \quad l > 0.$$

By the same techniques as above, from the sequence $\{\mathbf{f}^*_{A_{n_l}}\}$ one can choose a subsequence convergent strongly to $\mathbf{f}^*$, that is a contradiction. Hence, $\mathbf{f}^*_{A_n}$ converges in norm to $\mathbf{f}^*$.

The result is a surprisingly simple and pleasant one. Indeed, we can obtain directly the ultimate realization of the reproducing kernel Hilbert spaces and the ultimate sampling theory that are much more simpler than the known derivations (cf. [35], pp. 92–96).

Theorem 3 (Ultimate realization of reproducing kernel Hilbert spaces) *In our general situation and for a uniqueness set $\{p_j\}$ of the set E satisfying the linearly independence in Theorem 1, we obtain*

$$\|f\|^2_{H_K} = \|\mathbf{f}^*\|^2_{\mathcal{H}} = \lim_{n\to\infty} \sum_j \sum_{j'} f(p_j)\, \widetilde{a_{jj'}}\, \overline{f(p_{j'})}. \tag{13}$$

Theorem 4 (Ultimate sampling theory) *In our general situation and for a uniqueness set $\{p_j\}$ of the set E satisfying the linearly independence in Theorem 1, we obtain*

$$f(p) = \lim_{n\to\infty} \left(\mathbf{f}^*_{A_n}, \mathbf{h}(p)\right)_{\mathcal{H}}$$

$$= \lim_{n\to\infty}\left(\sum_j\sum_{j'} f(p_j)\widetilde{a_{jj'}}\mathbf{h}(p_{j'}), \mathbf{h}(p)\right)_{\mathcal{H}}$$

$$= \lim_{n\to\infty}\sum_j\sum_{j'} f\left(p_j\right)\widetilde{a_{jj'}} K\left(p, p_{j'}\right). \tag{14}$$

In Theorems 1–4, there are several relevant observations. Namely, for any given finite number of $f(p_j)$, $j = 1, 2, \ldots, n$, the result in Theorem 1 is valid. Meanwhile, Theorems 2 and 3 are valid when we consider the sequence $f(p_j)$, $j = 1, 2, \ldots,$ for any member f of H_K. The sequence $f(p_j)$, $j = 1, 2, \ldots,$ for any member f of H_K is characterized by the convergence of (13) in Theorem 3. Then, any member f of H_K is represented by (14) in terms of the sequence $f(p_j)$, $j = 1, 2, \ldots,$ in Theorem 4.

Here, we shall consider the following problem: In our general setting in Proposition 1, we assume that we observed the values $f(p_j) = \alpha_j, j = 1, 2, \ldots, n$, for a finite number of points $\{p_j\}$. Then, for the whole value $f(p)$ of the set E, how can we consider it?

One idea is to consider the function $f_1(p)$: among the functions satisfying the conditions $f(p_j) = \alpha_j, j = 1, 2, \ldots, n$, we consider the minimum norm member $f_1(p)$ in $H_K(E)$. This function $f_1(p)$ is determined by the following formula (as we can see easily)

$$f_1(p) = \sum_{j=1}^{n} C_j K\left(p, p_j\right)$$

where, the constants $\{C_j\}$ are determined by the formula

$$\sum_{j=1}^{n} C_j K\left(p_{j'}, p_j\right) = \alpha_{j'}, \quad j' = 1, 2, \ldots, n$$

(being clear that we assume that $\| K\left(p_{j'}, p_j\right) \|$ is positive definite).

Meanwhile, by Theorem 1, we can consider the function $f_2(p)$ defined by

$$f_2(p) = \left(\mathbf{f}^*_{A_n}, \mathbf{h}(p)\right)_{\mathcal{H}}$$

in terms of $\mathbf{f}^*_{A_n}$ in Theorem 1. This interpolation formula is dependent on the linear system.

We shall show our short history for our various inverse formulas by the typical real inversion of the Laplace transform

$$(Lf)(p) = \int_0^\infty e^{-tp} F(t)\,dt, \tag{15}$$

and our first main aim is to obtain a consequent *real inversion formula* following the new idea.

4 On the Known Real Inversion Formulas of the Laplace Transform

In order to see what is known about the situation on the real inversion formula of the Laplace transform, we shall first recall the well-known inversion formulas. Here, we denote by f the Laplace transform of F; $f = LF$.

The most popular formulas may be the following:

$$\lim_{n\to\infty} \frac{(-1)^n}{n!} \left(\frac{n}{t}\right)^{n+1} f^{(n)}\left(\frac{n}{t}\right) = F(t), \tag{16}$$

$$\lim_{n\to\infty} \prod_{k=1}^{n} \left(1 + \frac{t}{k}\frac{d}{dt}\right) \frac{n}{t} f\left(\frac{n}{t}\right) = F(t). \tag{17}$$

For (16), we refer to the books by Post [31] and Widder [45], and for (17) we refer to the book by Widder [45].

We obtained the analytical real inversion formula for the Laplace transform after the long steps within the method and reasoning of Sect. 2 (see also [35]):

$$f(p) = \int_0^\infty e^{-pt} F(t) dt \tag{18}$$

for $p > 0$, where $F : \mathbb{R}^+ \to \mathbb{C}$ is a measurable function satisfying

$$\int_0^\infty |F(t)|^2 dt < \infty. \tag{19}$$

For the polynomial of degree $2\,N + 2$,

$$P_N(\xi) := \sum_{0\le\nu\le n\le N} \frac{(-1)^{\nu+1}(2n)!\xi^{n+\nu}}{(n+1)!(n-\nu)!(n+\nu)!\nu!}$$
$$\cdot \left\{ \frac{(2n+1)\xi^2}{n+\nu+1} - \left(\frac{2n+1}{n+\nu+1} + 3n + 1\right)\xi + n(n+\nu+1) \right\},$$

we set

$$F_N(t) = \int_0^\infty f(p) e^{-pt} P_N(pt) dp. \tag{20}$$

Then, we have

$$\lim_{N\to\infty} \int_0^\infty |F(t) - F_N(t)|^2 dt = 0. \tag{21}$$

Furthermore, the estimation of the truncation error of $F_N(t)$ was also given in [1, 40].

We recall also the formula by [32, p. 221]: For the Laplace transform

$$\int_0^b e^{-pt} F(t) dt = f(p), \tag{22}$$

we have

$$F(t) = \frac{2tb^{-1}}{\pi} \frac{d}{du} \int_0^u \frac{G(v)}{(u-v)^{\frac{1}{2}}} dv \Big|_{u=t^2b^{-2}};$$

$$G(v) = v^{-\frac{1}{2}} \frac{2}{\pi} \int_0^\infty dy \cos\left(y \cosh^{-1} v^{-1}\right) \cosh \pi y$$

$$\cdot \int_0^\infty dz \cos(zy)(\cosh z)^{-\frac{1}{2}} \int_0^\infty dp f(p) J_0 \left(p \frac{b}{(\cosh z)^{\frac{1}{2}}}\right),$$

where J_0 denotes the Bessel function. Unfortunately, in this very complicated formula, the characteristic properties on both functions F and f such that the inversion formula holds true were not given.

Peng and Chung (cf. [29]) gave the formula

$$F(t) = \lim_{\sigma \to \infty} \overline{F}_\sigma(t), \qquad \overline{F}_\sigma(t) = \sum_{j=1}^{\infty} \frac{(-1)^{j-1}}{(j-1)!} e^{j\sigma} \frac{\sigma}{t} f\left(\frac{j\sigma}{t}\right).$$

Additionally, Tuan and Duc ([44]) gave

$$F(t) = \lim_{n \to \infty} n \sum_{j=1}^{\infty} \frac{(-1)^{j-1}}{(j-1)!} e^{njt} f(nj).$$

Furthermore, we also refer to the related articles [35, 39].

The problem may also be related to analytic extension problems (cf. [24, 25, 39]). In addition, in direct connection with Laplace transform inversion problems, we also would like to refer to [3, 4, 14, 20, 23, 26, 30, 46].

For the above typical difficult problem, we proceeded in the following ways.

5 Best Approximations

Let L be any bounded linear operator from a reproducing kernel Hilbert space H_K into a Hilbert space $\mathcal{H}$. Then, the following problem is a classical and fundamental problem known as the best approximate mean square norm problem. For any member $\mathbf{d}$ of $\mathcal{H}$, we would like to find

$$\inf_{f \in H_K} \|Lf - \mathbf{d}\|_{\mathcal{H}}.$$

It is clear that we are considering operator equations, generalized solutions, and corresponding generalized inverses within the framework of $f \in H_K$ and $\mathbf{d} \in \mathcal{H}$, having in mind

$$Lf = \mathbf{d}. \tag{23}$$

However, this problem has a complicated structure, specially in the infinite dimension Hilbert spaces case, leading in fact to the consideration of generalized inverses (in the Moore–Penrose sense). Following our theory, we can realize its complicated structure. Anyway, the problem turns out to be well posed within the reproducing kernels theory framework in the following proposition.

Proposition 4 *For any member* $\mathbf{d}$ *of* $\mathcal{H}$*, there exists a function* $\tilde{f}$ *in* H_K *satisfying*

$$\inf_{f \in H_K} \|Lf - \mathbf{d}\|_{\mathcal{H}} = \|L\tilde{f} - \mathbf{d}\|_{\mathcal{H}} \tag{24}$$

if and only if, for the reproducing kernel Hilbert space H_k *admitting the kernel defined by* $k(p,q) = (L^*LK(\cdot,q), L^*LK(\cdot,p))_{H_K}$,

$$L^*\mathbf{d} \in H_k. \tag{25}$$

Furthermore, when there exists a function $\tilde{f}$ *satisfying* (24)*, there exists a uniquely determined function that minimizes the norms in* H_K *among the functions satisfying the equality, and its function* $f_{\mathbf{d}}$ *is represented as follows:*

$$f_{\mathbf{d}}(p) = \left(L^*\mathbf{d}, L^*LK(\cdot,p)\right)_{H_k} \quad on \quad E. \tag{26}$$

Here, the adjoint operator L^* of L, as we see, from

$$(L^*\mathbf{d})(p) = \left(L^*\mathbf{d}, K(\cdot,p)\right)_{H_K} = (\mathbf{d}, LK(\cdot,p))_{\mathcal{H}}$$

is represented by known $\mathbf{d}$, L, $K(p,q)$, and $\mathcal{H}$. From this proposition, we see that the problem is well formulated by the theory of reproducing kernels (i.e., the existence, uniqueness, and representation of the solutions are well formulated). In particular, note that the adjoint operator is represented in a good way; this fact will be very important in our framework. The extremal function $f_{\mathbf{d}}$ is the *Moore–Penrose generalized inverse* $L^{\dagger}\mathbf{d}$ of the equation $Lf = \mathbf{d}$. The criteria (25) is involved and so the Moore–Penrose generalized inverse $f_{\mathbf{d}}$ is not good, when the data contain error or noises in some practical cases. We shall see this complicated structure by using the Laplace transform in the next section.

6 The Laplace Transform by Moore–Penrose Generalized Inverses

Here, we will consider the (modified) Laplace transform

$$(Lf)(p) := p(\mathcal{L}f)(p) = p\int_0^{\infty} e^{-tp} f(t)\,dt. \tag{27}$$

Let us fist fix some notations. The $\mathrm{ABS}[0,\infty)$ denotes the set of all real-valued absolutely continuous functions on $[0,\infty)$ (and we recall that we are using $\mathbb{R}^+$ for denoting the half-line $(0, \infty)$). By a "weight" we simply mean a continuous function $w : \mathbb{R}^+ \to \mathbb{R}^+$. Given a weight w, we define

$$L_2(w) := \{f : \mathbb{R}^+ \to \mathbb{R} \, : \, f \text{is measurable and} \|f\|_{L_2(w)} < \infty\},$$
$$H_K(w) := \{f \in \mathrm{ABS}[0, \infty) \, : \, f(0) = 0, \, \|f\|_{H_K(w)} < \infty\},$$

where the norms are given by

$$\|f\|_{L_2(w)} := \left(\int_0^\infty |f(p)|^2 \, w(p) \, dp\right)^{\frac{1}{2}}, \quad \|f\|_{H_K(w)} := \left(\int_0^\infty |f'(t)|^2 \, w(t) \, dt\right)^{\frac{1}{2}},$$

respectively.

Let us now recall some fundamental results.

Proposition 5 [41] *Assume that for a given pair w and u of weights,*

$$M := \iint_{\mathbb{R}^+ \times \mathbb{R}^+} e^{-2t\,p} \, w(t)^{-1} u(p) \, dt \, dp < \infty. \tag{28}$$

Then, L is compact from $H_K(w)$ into $L_2(u)$ with norm less than or equal to $\sqrt{M}$.

Corollary 1 *For $u, w \in L_1(\mathbb{R}^+)$, then, L is compact from $H_K(w^{-1})$ into $L_2(u)$.*

In the paper [41], there were given many concrete examples satisfying (28) and the conditions in Corollary 1, and it also derived real inversion formulas of the Laplace transform but in terms of the singular values decomposition and the Tikhonov regularization. This allowed a general numerical and real inversion formula of the Laplace transform.

Recall that the reproducing kernel Hilbert space $H_K(w)$ admits the concrete representation of the reproducing kernel

$$K_w(s, t) = \int_0^{\min(s,t)} w(\xi)^{-1} \, d\xi. \tag{29}$$

In order that (29) is finite, we need to assume that

$$\int_0^1 \frac{d\xi}{w(\xi)} < \infty. \tag{30}$$

Under the situation, we obtained:

Theorem 5 [13] *For positive continuous functions $u \in L_1(\mathbb{R})$ and w on $[0, \infty)$ satisfying the conditions (28) and (30), the Moore–Penrose generalized inverse $g^\dagger$ of the (modified) Laplace transform L from $H_K(w)$ into $L_2(u)$ exists uniquely if and only if it is the image of the derivative of a function in the space $H_K(w)$. Then, when we take a finite number of points $\{p_j\}_{j=1}^n$, an approximate inverse depending on the points $\{p_j\}_{j=1}^n$ is represented as follows:*

$$g_A^\dagger(x) = \sum_{j,j'=1}^{n} \left(L^* g\right)(p_j) \widetilde{a_{jj'}} L^* L K_w\left(p_{j'}, x\right), \tag{31}$$

where

$$(L^*g)(p_j) = \int_0^{p_j} \mathcal{L}(gu)(t)\frac{dt}{w(t)},$$

$$\left(L^*LK_w(p_j,x)\right) = \int_0^x \int_0^{p_j} \frac{dt}{w(t)}\frac{d\tau}{w(\tau)}(\mathcal{L}u)(t+\tau),$$

and $\widetilde{a_{jj'}}$ *are the elements of the inverse of the positive definite symmetric matrix constituted by the elements*

$$a_{jj'} = \int_0^{p_j}\int_0^{p_{j'}}\int_0^{\infty} \frac{(\mathcal{L}u)(t+\xi)(\mathcal{L}u)(\tau+\xi)}{w(\xi)}\frac{dt}{w(t)}\frac{d\tau}{w(\tau)}d\xi.$$

A pointwise estimation of the inverse is given in terms of the points $\{p_j\}_{j=1}^n$ *by*

$$\left|g_A^{\dagger}(x)\right| \leq \left(\sum_{j,j'=1}^{n} (L^*g)(p_j)\widetilde{a_{jj'}}(L^*g)(p_{j'})\right)^{1/2} \cdot \left(\sum_{j,j'=1}^{n} L^*LK_w\left(p_j,x\right)\widetilde{a_{jj'}}L^*LK_w\left(p_{j'},x\right)\right)^{1/2}.$$

As a more powerful and practical method, we introduced the idea of the Tikhonov regularization.

7 The Tikhonov Regularization

We shall consider some practical and more concrete representation in the extremal functions involved in the Tikhonov regularization by using the theory of reproducing kernels.

Furthermore, we shall need error estimates when **d** contains error or noises. For this fundamental problem, we obtain the following conclusion.

Theorem 6 *Let* $L : H_K \to \mathcal{H}$ *be a bounded linear operator, and define the inner product*

$$\langle f_1, f_2\rangle_{H_{K_\lambda}} = \alpha\,\langle f_1, f_2\rangle_{H_K} + \langle Lf_1, Lf_2\rangle_{\mathcal{H}}$$

for $f_1, f_2 \in H_K$. *Then* $\left(H_K, \langle\cdot,\cdot\rangle_{H_{K_\alpha}}\right)$ *is a reproducing kernel Hilbert space whose reproducing kernel is given by*

$$K_\alpha(p,q) = \left[(\alpha + L^*L)^{-1}K_q\right](p).$$

Here, $K_\alpha(p,q)$ *is the solution* $\widetilde{K}_\alpha(p,q)$ *of the functional equation*

$$\widetilde{K}_\alpha(p,q) + \frac{1}{\alpha}\left(L\widetilde{K}_q, LK_p\right)_{\mathcal{H}} = \frac{1}{\alpha}K(p,q) \tag{32}$$

that is corresponding to the Fredholm integral equation of the second kind for many concrete cases. Moreover, we are using

$$\widetilde{K}_q = \widetilde{K}_\alpha(\cdot, q) \in H_K \quad \textit{for} \quad q \in E, \qquad K_p = K(\cdot, p) \quad \textit{for} \quad p \in E.$$

Theorem 7 *The map*

$$f \in H_K \mapsto \left\{\alpha \, \|f \, : \, H_K\|^2 + \|Lf - \mathbf{d} \, : \, \mathcal{H}\|^2\right\} \in \mathbb{R}$$

attains the minimum and the minimum is attained only at $f_{\mathbf{d},\alpha} \in H_K$ such that

$$(f_{\mathbf{d},\alpha})(p) = \langle \mathbf{d}, LK_\alpha(\cdot, p)\rangle_{\mathcal{H}}.$$

Furthermore, $(f_{\mathbf{d},\alpha})(p)$ satisfies

$$|(f_{\mathbf{d},\alpha})(p)| \leq \sqrt{\frac{K(p,p)}{2\alpha}} \|\mathbf{d}\|_{\mathcal{H}}. \tag{33}$$

This theorem means that in order to obtain good approximate solutions, we must take a sufficiently small α. However, here we have restrictions for them, as we see, when $\mathbf{d}$ moves to $\mathbf{d}'$, by considering $f_{\mathbf{d},\alpha}(p) - f_{\mathbf{d}',\alpha}(p)$ in connection with the relation of the difference $\|\mathbf{d} - \mathbf{d}'\|_{\mathcal{H}}$. This fact is a very natural one, because we cannot obtain good solutions from the data containing errors. Here, we wish to know how to take a small α a prior and what is the bound for it.

The basic idea may be given as follows. We examine for various α tending to zero, the corresponding extremal functions. By examining the sequence of the extremal functions, when it converges to some function numerically and after that when the sequence diverges numerically, it will give the bound for α numerically (see [18, 19]).

For up-to-date versions of the Tikhonov regularization using the theory of reproducing kernels, see [9, 10].

8 Real and Numerical Inversion Formulas of the Laplace Transform by the Tikhonov Regularization

We shall consider the inversion formula of the Laplace transform

$$(\mathcal{L}F)(p) = f(p) = \int_0^\infty e^{-pt} F(t) dt, \quad p > 0$$

for some natural function spaces. For more general functions, we shall apply their transforms suitably in order to apply the results. In order to consider a bounded linear operator by the Laplace transform, we shall recall a natural function space from [40].

On the positive real line $\mathbb{R}^+$, we shall consider the norm

$$\left\{\int_0^\infty |F'(t)|^2 \frac{1}{t} e^t dt\right\}^{1/2}$$

for absolutely continuous functions F satisfying $F(0) = 0$. This space H_K admits the reproducing kernel

$$K(t,t') = \int_0^{\min(t,t')} \xi e^{-\xi} d\xi. \tag{34}$$

Then, we have

$$\int_0^\infty |(\mathcal{L}F)(p)p|^2 dp \leq \frac{1}{2}\|F\|_{H_K}^2. \tag{35}$$

That is, $(\mathcal{L}F)(p)p$ is a bounded linear operator from H_K into $L_2(\mathbb{R}^+, dp) = L_2(\mathbb{R}^+)$. By using this reproducing kernel Hilbert space, we obtain the next significative result, following our general method.

Proposition 6 *For any $g \in L_2(\mathbb{R}^+)$ and for any $\alpha > 0$, in the sense*

$$\inf_{F \in H_K} \left\{ \alpha \int_0^\infty |F'(t)|^2 \frac{1}{t} e^t dt + \|(\mathcal{L}F)(p)p - g\|_{L_2(\mathbb{R}^+)}^2 \right\}$$
$$= \alpha \int_0^\infty |F_{\alpha,g}^{*\prime}(t)|^2 \frac{1}{t} e^t dt + \|(\mathcal{L}F_{\alpha,g}^*)(p)p - g\|_{L_2(\mathbb{R}^+)}^2 \tag{36}$$

there exists a uniquely determined best approximate function $F_{\alpha,g}^*$ and it is represented by

$$F_{\alpha,g}^*(t) = \int_0^\infty g(\xi)\,(\mathcal{L}K_\alpha(\cdot,t))\,(\xi)\xi d\xi. \tag{37}$$

Here, $K_\alpha(\cdot,t)$ is determined by the functional equation for $K_{\alpha,t'} = K_\alpha(\cdot,t')$, $K_t = K(\cdot,t)$,

$$K_\alpha(t,t') = \frac{1}{\alpha}K(t,t') - \frac{1}{\alpha}((\mathcal{L}K_{\alpha,t'})(p)p, (\mathcal{L}K_t)(p)p)_{L_2(\mathbb{R}^+)}. \tag{38}$$

We calculate the approximate inverse $F_{\alpha,g}^*(t)$ by using (37). By taking the Laplace transform of (38) with respect to t, by changing the variables t and t'

$$(\mathcal{L}K_\alpha(\cdot,t))(\xi) = \frac{1}{\alpha}(\mathcal{L}K(\cdot,t))(\xi) - \frac{1}{\alpha}((\mathcal{L}K_{\alpha,t})(p)p, (\mathcal{L}(\mathcal{L}K_\cdot)(p)p))(\xi))_{L_2(\mathbb{R}^+)}. \tag{39}$$

Here,

$$K(t,t') = \begin{cases} -te^{-t} - e^{-t} + 1 & \text{for} \quad t \leq t' \\ -t'e^{-t'} - e^{-t'} + 1 & \text{for} \quad t \geq t', \end{cases}$$

$$(\mathcal{L}K(\cdot,t'))(p) = e^{-t'p}e^{-t'}\left[\frac{-t'}{p(p+1)} + \frac{-1}{p(p+1)^2}\right] + \frac{1}{p(p+1)^2},$$

$$\int_0^\infty e^{-qt'}(\mathcal{L}K(\cdot,t'))(p)dt' = \frac{1}{pq(p+q+1)^2}.$$

Therefore, by setting as $(\mathcal{L}K_\alpha(\cdot,t))(\xi)\xi = H_\alpha(\xi,t)$, we obtain the *Fredholm integral equation of the second kind*:

$$\alpha H_\alpha(\xi,t) + \int_0^\infty \frac{H_\alpha(p,t)}{(p+\xi+1)^2}dp = -\frac{e^{-t\xi}e^{-t}}{\xi+1}\left(t+\frac{1}{\xi+1}\right)+\frac{1}{(\xi+1)^2}, \quad (40)$$

which is corresponding to (32). By solving this integral equation, H. Fujiwara derived a very reasonable numerical inversion formula for the integral transform and he expanded very good algorithms for numerical and real inversion formulas of the Laplace transform. Consequently, international patents by its software are being forwarded to Japan, USA, Canada, and the EU (for more detailed references and comments for this equation, see [38]).

Figure 1 is an example for $\mathcal{L}F(p) = \exp(-p)$ for which $F(t) = \delta_1(t)$ in the distribution sense, and Fig. 2 is for

$$\mathcal{L}F(p) = \frac{e^{-p}}{p(1+e^{-p})}$$

in which $F(t)$ is a square wave function.

In both figures, **a** is computed with large regularization parameters $\alpha \geq 10^{-12}$, and **b** is computed with small regularization parameters $\alpha = 10^{-100}, 10^{-400}$.

At this moment, theoretically we shall use the whole data of the output, in fact, 6000 data. Surprisingly enough, Fujiwara gave the solutions with $\alpha = 10^{-400}$ and *600 digits precision*. The core of the above mentioned and corresponding patent is *10 GB* data for the solutions. See [17, 18, 36, 38, 41] for the basic references.

This great numerical power of Fujiwara has been giving a great impact within their associated numerical problems.

In [8], we gave an exact real inversion formula such that for any given finite number of points $\{p_j\}$ and for any given values $\{d_j\}$, we construct an optimal inversion F satisfying $f(p_j) = d_j$ whose results are derived from the general concept of the next section. Moreover, it is interesting to realize that this concept was derived from a generalization of the Pythagorean theorem.

Proposition 7 *For any given n values* $\mathbf{d} = \{d_j\}_{j=1}^n$, *among the Laplace transforms taking the values*

$$f(p_j) = d_j, \quad j = 1,2,\ldots,n,$$

and among their inverses, the uniquely determined function $F_{\mathbf{d}}^*(t)$ *with the minimum norm* $||F||_{H_K}$ *is uniquely determined and it is represented by*

$$F_{\mathbf{d}}^*(t) = \sum_{j,j'=1}^{n} \left(d_j p_j\right) \widetilde{a}_{jj'} \left(\mathbf{L}^*\mathbf{e}_{j'}\right)(t).$$

Here, $\widetilde{a}_{jj'}$ *is the element of the positive definite Hermitian matrix formed by the element*

$$a_{jj'} = \frac{1}{(p_j + p_{j'} + 1)^2}$$

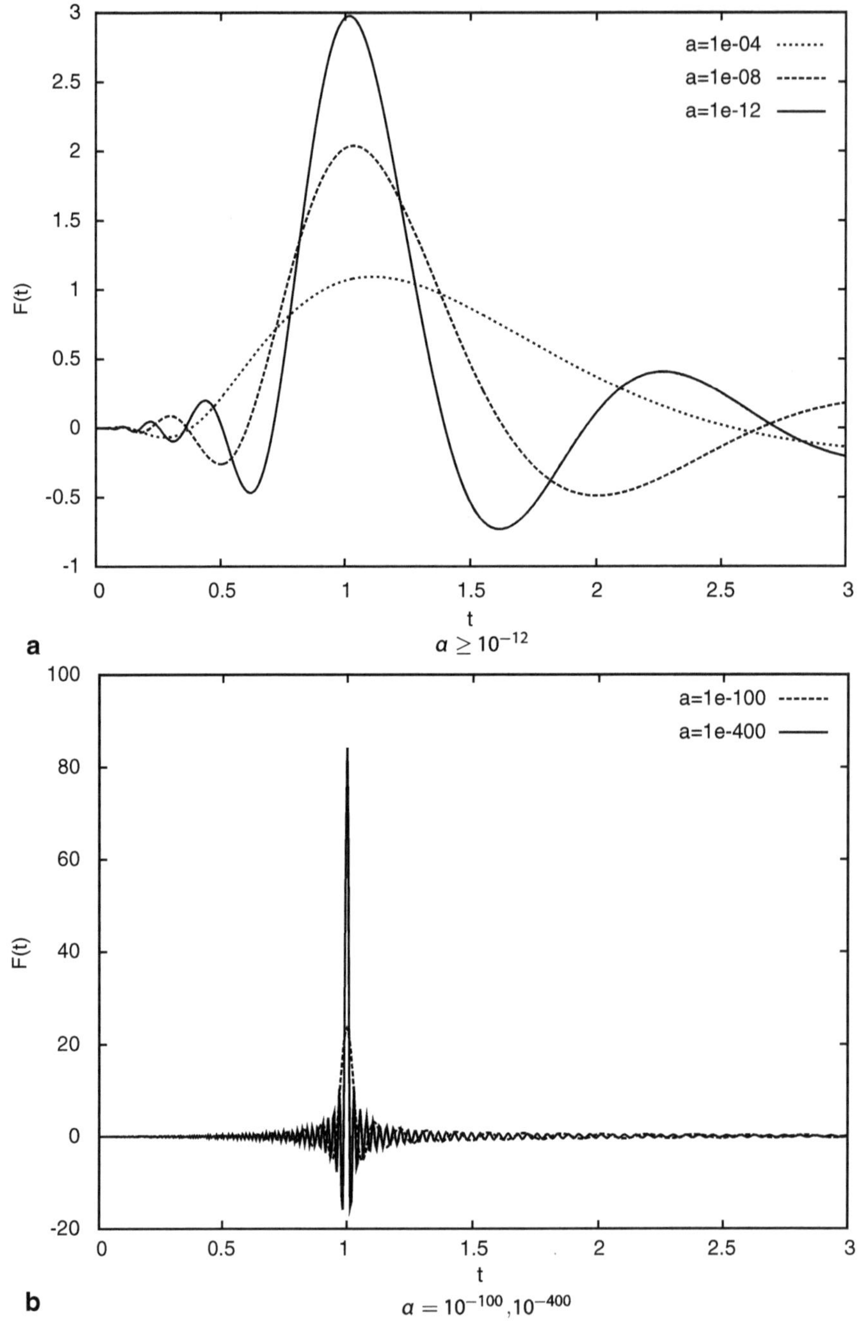

Fig. 1 Numerical results for the delta function δ_1

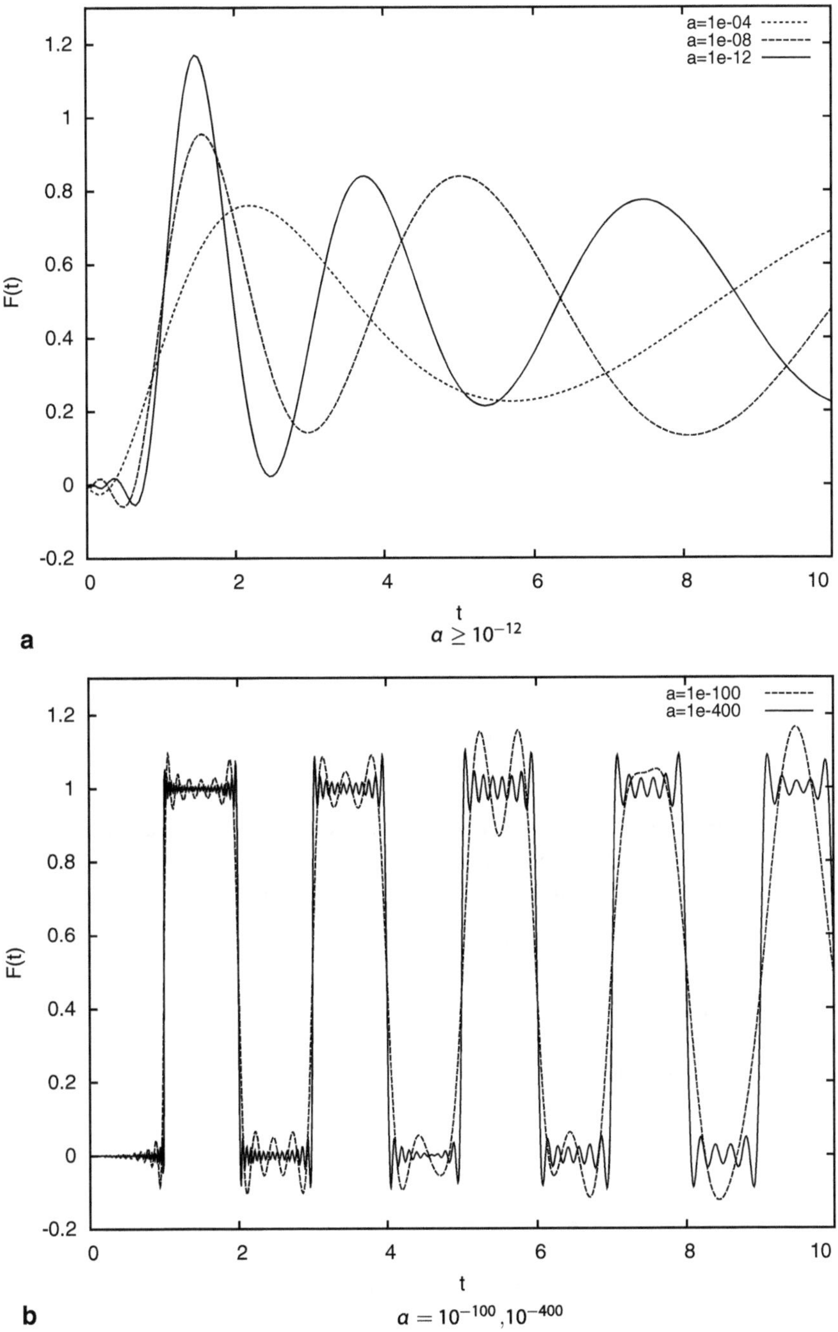

Fig. 2 Numerical results for a square wave function

and

$$
\begin{aligned}
(\mathbf{L}^*\mathbf{e}_j)(t) &= ((\mathbf{L}^*\mathbf{e}_j)(\,\cdot\,), K(\cdot,t))_{H_K}\\
&= (\mathbf{e}_j, \mathbf{L}K(\cdot,t))_{\mathbb{R}^n}\\
&= e^{-tp_j}e^{-t}\left[\frac{-t}{(p_j+1)} + \frac{-1}{(p_j+1)^2}\right] + \frac{1}{(p_j+1)^2}.
\end{aligned}
$$

9 Inversion Mapping from Many Kinds of Information Data

We shall give a fundamental concept to inversions from many kinds of information data. For an abstract set Λ, we shall consider an operator-valued function L_λ on Λ,

$$\Lambda \longrightarrow L_\lambda \tag{41}$$

where L_λ are bounded linear operators from a Hilbert space $\mathcal{H}$ into various Hilbert spaces $\mathcal{H}_\lambda$,

$$L_\lambda : \mathcal{H} \mapsto \mathcal{H}_\lambda. \tag{42}$$

In particular, we are interested in the inversion formula

$$L_\lambda x \mapsto x, \quad x \in \mathcal{H}. \tag{43}$$

Here, we consider $\{L_\lambda x; \lambda \in \Lambda\}$ as informations obtained from x and we wish to determine x from the informations. However, the informations $L_\lambda x$ belong to various Hilbert spaces $\mathcal{H}_\lambda$, and so, in order to unify the informations in a sense, we shall take fixed elements $\mathbf{b}_{\lambda,\omega} \in \mathcal{H}_\lambda$ and consider the linear mapping from $\mathcal{H}$,

$$X_{\mathbf{b}}(\lambda,\omega) = \left(L_\lambda x, \mathbf{b}_{\lambda,\omega}\right)_{\mathcal{H}_\lambda} = \left(x, L_\lambda^*\mathbf{b}_{\lambda,\omega}\right)_{\mathcal{H}}, \quad x \in \mathcal{H}, \tag{44}$$

into a linear space comprising functions on $\Lambda \times \Omega$. For the informations $L_\lambda x$, we shall consider $X_{\mathbf{b}}(\lambda,\omega)$ as observations (measurements, in fact) for x depending on λ and ω. For this linear mapping (44), we form the positive definite quadratic function $K_{\mathbf{b}}(\lambda,\omega;\lambda',\omega')$ on $\Lambda \times \Omega$ defined by

$$K_{\mathbf{b}}\left(\lambda,\omega;\lambda',\omega'\right) = \left(L_{\lambda'}^*\mathbf{b}_{\lambda',\omega'}, L_\lambda^*\mathbf{b}_{\lambda,\omega}\right)_{\mathcal{H}} = \left(L_\lambda L_{\lambda'}^*\mathbf{b}_{\lambda',\omega'}, \mathbf{b}_{\lambda,\omega}\right)_{\mathcal{H}_\lambda} \quad \text{on} \quad \Lambda \times \Omega. \tag{45}$$

Then, we can apply our theory. The concept was derived from the generalization of the Pythagorean theorem in the following way (cf. [2, 33, 37]): Let $x \in \mathbb{R}^n$ and $\{\mathbf{e}_j\}_{j=1}^n$ be linearly independent unit vectors. We consider the linear mappings

$$L : x \mapsto \left\{x - \left(x, \mathbf{e}_j\right)\mathbf{e}_j\right\}_{j=1}^n \tag{46}$$

from $\mathbb{R}^n$ into $\mathbb{R}^n$. Then, we wish to establish an isometric identity and inversion formula for the operators. Recall the Pythagorean theorem for $n = 2$. By our operator versions, we can establish the desired results.

Note that in (46), for $n \geq 3$, if we consider

$$\left\{ \|x - (x, \mathbf{e}_j)\mathbf{e}_j\| \right\}_{j=1}^{n} \tag{47}$$

as scalar valued mappings, then the mappings are, in general, no longer linear. So, we must consider the operator valued mappings (46) in order to obtain isometric mappings in the framework of Hilbert spaces.

We see that some related equations were considered as follows [28, p. 128–157]: Let H, H_j, $j = 1, 2, \ldots, p$, be Hilbert spaces and let

$$R_j : H \mapsto H_j, \quad j = 1, 2, \ldots, p, \tag{48}$$

be linear continuous maps from H onto H_j. Let $g_j \in H_j$ be given. Then, consider the problem to compute $f \in H$ such that

$$R_j f = g_j, \quad j = 1, 2, \ldots, p. \tag{49}$$

These equations are very important in the theory of computerized tomography by discretization. The typical method is Kaczmarz's method based on an iterative method by using the orthogonal projections P_j in H onto the affine subspaces $R_j f = g_j$. See [28] for the details.

As for our direct solutions for (49) it seems that the result is stable for the sake of (49) as data.

10 New Real Inversion Formulas of the Laplace Transform

We shall state the new formulas by our new discretization method. The following theorem was derived directly from our general Theorem 1:

Theorem 8 *For any fixed positive continuous function w on $[0, \infty)$ satisfying the condition*

$$\int_0^\infty e^{-xt} \frac{dt}{w(t)} < \infty, \quad x > 0, \tag{50}$$

the best approximate inverse $F_A^*(t)$ of the Laplace transform

$$(LF)(p) = \int_0^\infty e^{-tp} F(t)\, dt,$$

for any function $F \in L_2(w)$ satisfying

$$\int_0^\infty |F(t)|^2 \, w(t) dt < \infty$$

is given by the formula by taking a finite number of points $\{p_j\}_{j=1}^n$

$$F_A^*(t) = \sum_{j,j'=1}^{n} (LF)(p_j)\, \widetilde{a_{jj'}} \frac{e^{-p_{j'}t}}{w(t)} \tag{51}$$

where $\widetilde{a_{jj'}}$ *are the elements of the inverse of the positive definite symmetric matrix A constituted by the elements*

$$a_{jj'} = \int_0^\infty e^{-(p_j+p_{j'})\xi} \frac{d\xi}{w(\xi)}.$$

This function $F_A^*(t)$ *is the optimal inverse in the sense that among the functions of the space* $L_2(w)$ *whose Laplace transforms are* $(LF)(p_j)$ *it minimizes the norms in* $L_2(w)$.

In our approach of the numerical inversion formula of the Laplace transform, the effectiveness of the inversion may be reduced to some good selection of the weights w.

We shall list up several examples in this section from [11, 21]:

1. For $w(t) = t^{1-\mu}$,

$$L\left(\frac{1}{w}\right)(p) = \frac{1}{p^\mu}\Gamma(\mu), \quad \mu > 0.$$

2. For $w(t) = \frac{e^t}{t}$,

$$L\left(\frac{1}{w}\right)(p) = \frac{1}{(p+1)^2}.$$

3. For $w(t) = (t+1)^{-2n}$,

$$L\left(\frac{1}{w}\right)(p) = \frac{(2n)!}{(p+1)^{2n+1}}\, e_{2n}(p)$$

   for

$$e_k(p) = 1 + p + \frac{p^2}{2!} + \ldots + \frac{p^k}{k!}; \quad n = 1, 2, \ldots.$$

4. For $w(t) = \frac{1+e^t}{t^{n-1}}$,

$$L\left(\frac{1}{w}\right)(p) = (n-1)! \sum_{k=1}^{\infty} \frac{(-1)^{k-1}}{(p+k)^n}, \quad p > -1, n = 1, 2, \ldots.$$

5. For $w(t) = \frac{t}{1-\frac{t+2}{2t}(1-e^{-t})}$,

$$L\left(\frac{1}{w}\right)(p) = -1 + \left(p + \frac{1}{2}\right)\log\frac{p+1}{p}.$$

6. For $w(t) = \frac{t^2}{(e^{-t}-1)^n}$,

$$L\left(\frac{1}{w}\right)(p) = \sum_{k=0}^{n} \binom{n}{k} (p+n-k)\log(p+n-k)(2q+p)\log(2q+p),$$

$$n = 2, 4.$$

7. For $w(t) = \frac{t^3}{(1-e^{-qt})^n}$,

$$L\left(\frac{1}{w}\right)(p) = \frac{1}{2}\sum_{k=2}^{n} (-1)^{k-1} \binom{n}{k} (p+kq)^2 \log(p+kq), \quad n > 2,$$

$$p > 0, \ qn + p > 0.$$

8. For $w(t) = \frac{t^2}{(1-e^{-qt})^2}$,

$$L\left(\frac{1}{w}\right)(p) = (2q+p)\log(2q+p) - 2(p+q)\log(p+q) + p\log p,$$

$$p > 0, \ 2q > -p.$$

We shall now state very concrete examples from [43].

Example 1 As a first example take $p_k = k - \frac{1}{2}$. Then $a_{kl} = \frac{1}{k+l-1}$, and the matrix A_n is the Hilbert matrix

$$A_n = \left(\frac{1}{k+l-1}\right)_{n\times n}.$$

The inverse of A_n is well known (see [5, 15]). In fact,

$$\widetilde{a}_{lk} = (-1)^{k+l}(k+l-1)\binom{n+k-1}{n-l}\binom{n+l-1}{n-k}\binom{k+l-2}{k-1}^2.$$

Consequently,

$$F^*_{A_n}(z) = \sum_{k,l=1}^{n} (-1)^{k+l}(k+l-1)$$
$$\cdot \binom{n+k-1}{n-l}\binom{n+l-1}{n-k}\binom{k+l-2}{k-1}^2 \frac{F\left(k-\frac{1}{2}\right)}{z+l-\frac{1}{2}}. \quad (52)$$

Moreover,

$$\lim_{n\to\infty} \sum_{k,l=1}^{n} (-1)^{k+l}(k+l-1)$$

$$\cdot \binom{n+k-1}{n-l}\binom{n+l-1}{n-k}\binom{k+l-2}{k-1}^2 \frac{F\left(k-\frac{1}{2}\right)}{z+l-\frac{1}{2}} = F(z), \quad \Re(z) > 0,$$

where the convergence is both pointwise and in $\mathcal{H}^2(\mathbb{C}_+)$ norm, and the real variable inverse Laplace transform has the form

$$f(t) = \lim_{n\to\infty} \sum_{k,l=1}^{n} (-1)^{k+l}(k+l-1)$$
$$\cdot \binom{n+k-1}{n-l}\binom{n+l-1}{n-k}\binom{k+l-2}{k-1}^2 F\left(k-\frac{1}{2}\right) e^{-(l-\frac{1}{2})t}.$$

Example 2

Take now $p_k = k + \alpha/2$ with $\alpha > -2$. Then, $a_{kl} = \frac{1}{k+l+\alpha}$, and the matrix A_n is a Hankel matrix

$$A_n = \left(\frac{1}{k+l+\alpha}\right)_{n\times n}.$$

The inverse of A_n is less known in this case (see [5]). We have

$$\begin{aligned}\widetilde{a}_{lk} =& (-1)^{k+l}(k+l+\alpha)\binom{n+k+\alpha}{n-l}\binom{n+l+\alpha}{n-k}\\ &\cdot \binom{k+l+\alpha-1}{k-1}\binom{k+l+\alpha-1}{l-1}.\end{aligned} \tag{53}$$

Consequently,

$$\begin{aligned}F^*_{A_n}(z) =& \sum_{k,l=1}^{n} (-1)^{k+l}(k+l+\alpha)\binom{n+k+\alpha}{n-l}\binom{n+l+\alpha}{n-k}\\ &\cdot \binom{k+l+\alpha-1}{k-1}\binom{k+l+\alpha-1}{l-1}\frac{F(k+\alpha/2)}{z+l+\alpha/2}.\end{aligned} \tag{54}$$

Moreover,

$$\lim_{n\to\infty} \sum_{k,l=1}^{n} (-1)^{k+l}(k+l+\alpha)\binom{n+k+\alpha}{n-l}\binom{n+l+\alpha}{n-k}$$
$$\cdot \binom{k+l+\alpha-1}{k-1}\binom{k+l+\alpha-1}{l-1}\frac{F(k+\alpha/2)}{z+l+\alpha/2} = F(z), \quad \Re(z) > 0,$$

where the convergence is both pointwise and in $\mathcal{H}^2(\mathbb{C}_+)$ norm, and the real variable inverse Laplace transform has the form

$$f(t) = \lim_{n\to\infty} \sum_{k,l=1}^{n} (-1)^{k+l}(k+l+\alpha)\binom{n+k+\alpha}{n-l}\binom{n+l+\alpha}{n-k}$$

$$\cdot \binom{k+l+\alpha-1}{k-1}\binom{k+l+\alpha-1}{l-1} F\left(k+\alpha/2\right) e^{-(l+\alpha/2)t}.$$

Example 3

Choose now $p_k = \dfrac{\alpha}{2} + \dfrac{1}{k}$. Then,

$$A_n = \left(\frac{1}{\alpha + \frac{1}{k} + \frac{1}{l}}\right)_{n\times n} = \left(\frac{kl}{\alpha kl + k + l}\right)_{n\times n},$$

and we have

$$\tilde{a}_{lk}^{n} = (-1)^{k+l} \frac{\prod_{j=1}^{n}\left[\left(\alpha+\frac{1}{j}+\frac{1}{k}\right)\left(\alpha+\frac{1}{j}+\frac{1}{l}\right)\right]}{(\alpha+\frac{1}{k}+\frac{1}{l})\prod_{j=1}^{k-1}(\frac{1}{k}-\frac{1}{j})\prod_{j=k+1}^{n}(\frac{1}{j}-\frac{1}{k})\prod_{j=1}^{l-1}(\frac{1}{l}-\frac{1}{j})\prod_{j=l+1}^{n}(\frac{1}{j}-\frac{1}{l})}$$

$$= (-1)^{k+l} \frac{\dfrac{\prod_{j=1}^{n}(\alpha jk+j+k)(\alpha jl+j+l)}{k^n l^n (n!)^2}}{\dfrac{\alpha kl+k+l}{kl}\cdot\dfrac{(-1)^{k-1}}{k^{k-1}}\cdot\dfrac{(-1)^{n-k}(n-k)!}{k^{n-k}\dfrac{n!}{k!}}\cdot\dfrac{(-1)^{l-1}}{l^{l-1}}}$$

$$\cdot \frac{(-1)^{n-l}(n-l)!}{l^{n-l}\dfrac{n!}{l!}}.$$

Finally, we obtain

$$\tilde{a}_{lk}^{n} = (-1)^{k+l} \frac{\prod_{j=1}^{n}\left[(\alpha jk+j+k)(\alpha jl+j+l)\right]}{(\alpha kl+k+l)\, k!\,(n-k)!l!(n-l)!}, \tag{55}$$

and

$$F^{*}_{A_n}(z) = \sum_{k,l=1}^{n}(-1)^{k+l}\frac{\prod_{j=1}^{n}\left[(\alpha jk+j+k)(\alpha jl+j+l)\right]}{(\alpha kl+k+l)\, k!\,(n-k)!l!(n-l)!}\frac{F\left(\frac{1}{k}+\frac{\alpha}{2}\right)}{z+\frac{1}{l}+\frac{\alpha}{2}}.$$

Moreover,

$$\lim_{n\to\infty}\sum_{k,l=1}^{n}(-1)^{k+l}\frac{\prod_{j=1}^{n}\left[(\alpha jk+j+k)(\alpha jl+j+l)\right]}{(\alpha kl+k+l)\, k!\,(n-k)!l!\,(n-l)!}\frac{F\left(\frac{1}{k}+\frac{\alpha}{2}\right)}{z+\frac{1}{l}+\frac{\alpha}{2}} = F(z), \Re(z)>0,$$

where the convergence is both pointwise and in the $\mathcal{H}^2(\mathbb{C}_+)$ norm, and the real variable inverse Laplace transform has the form

$$f(t) = \lim_{n\to\infty} \sum_{k,l=1}^{n} (-1)^{k+l} \frac{\prod_{j=1}^{n} [(\alpha jk + j + k)(\alpha jl + j + l)]}{(\alpha kl + k + l)k!(n-k)!l!(n-l)!} F\left(\frac{1}{k} + \frac{\alpha}{2}\right) e^{-(\frac{1}{l}+\frac{\alpha}{2})t}.$$

Therefore, the last formula recovers the Laplace inverse from data on a finite interval $\left(\frac{\alpha}{2}, \frac{\alpha}{2}+1\right]$.

11 Paley–Wiener Spaces and Sobolev Spaces

The Aveiro discretization method in mathematics is essentially the application of Theorem 1 by using the backward transform method (stated in detail below) to analytical problems like general linear differential equations and integral equations.

Then, we shall use basically the Paley–Wiener and the Sobolev reproducing kernel Hilbert spaces. So, we need to introduce these spaces and we will do it in a simpler and natural way.

We refer to J. R. Higgins [22] and F. Stenger [42] for the details in connection with the sampling theory. The method which is using the Paley–Wiener spaces is called the *sinc* method. At first we shall fix notations and basic results in the sampling theory following the book by F. Stenger [42], and at the same time we shall show the basic relation of the sampling theory and the theory of reproducing kernels. Let us consider the following domain

$$D_h = \left(-\frac{\pi}{h}, \frac{\pi}{h}\right)^n, \qquad h > 0.$$

We shall consider the Fourier transform for functions F in $L_2(\mathbb{R}^n, D_h)$ by

$$f(z) = \frac{1}{(2\pi)^n} \int_{\mathbb{R}^n} \chi_{D_h}(\xi) F(\xi) e^{-iz\cdot\xi} d\xi. \tag{56}$$

Here, $\xi = (\xi_1, \xi_2, \dots, \xi_n)$, $d\xi = d\xi_1 d\xi_2 \dots d\xi_n$, $z \cdot \xi = z_1\xi_1 + z_2\xi_2 \dots z_n\xi_n$ and $\chi_{D_h}(\xi)$ is the characteristic function of the domain D_h.

In order to identify the image space following the theory of reproducing kernels, we form the reproducing kernel

$$\begin{aligned} K_h(z, \overline{u}) &= \frac{1}{(2\pi)^n} \int_{\mathbb{R}^n} \chi_{D_h}(\xi) e^{-iz\cdot\xi} \overline{e^{-iu\cdot\xi}} d\xi \\ &= \prod_{\nu}^{n} \frac{1}{\pi(z_\nu - \bar{u}_\nu)} \sin \frac{\pi}{h} \frac{1}{\pi(z_\nu - \bar{u}_\nu)} \end{aligned}$$

and it forms the Paley–Wiener space, i.e., the image space of (56) is called the Paley–Wiener space $W\left(\frac{\pi}{h}\right)$ comprising all analytic functions of exponential type satisfying for each ν, with some constant C_ν, $z_\nu \to \infty$

$$|f(z_1, \dots, z_\nu, z_{\nu+1}, \dots, z_n)| \leq C_\nu \exp\left(-\frac{\pi |z_\nu|}{h}\right)$$

and that

$$\int_{\mathbb{R}^n} |f(x)|^2 dx < \infty.$$

Then, for $j = (j_1, j_2, \dots, j_n) \in \mathbb{Z}^n$, we have the identity

$$\frac{1}{(2\pi)^n} \int_{\mathbb{R}^n} |F(\xi)|^2 d\xi = h^n \sum_j |f(jh)|^2 = \int_{\mathbb{R}^n} |f(x)|^2 dx.$$

This means that

$$f(x) = (f(\cdot), K_h(\cdot, x))_{H_{K_h}} = h^n \sum_j f(jh) K_h(jh, x) = \int_{\mathbb{R}^n} f(\xi) K_h(\xi, x) d\xi$$

and the whole data $f(x)$ is represented by the discrete data $\{f(jh)\}_j$ (sampling theorem).

For a general theory for the sampling theory and error estimates for some finite points $\{hj\}$, see [35].

Next, we shall introduce the Sobolev Hilbert spaces which we will be using. For this purpose, let $m > \frac{n}{2}$ be an integer, and denote by ${}_N C_K$ the binomial coefficient. Then, we have

$$W^{m,2}(\mathbb{R}^n) = H_K(\mathbb{R}^n), \tag{57}$$

where $W^{m,2}(\mathbb{R}^n)$ denotes the Sobolev space whose norm is given by

$$\|F\|_{W^{m,2}(\mathbb{R}^n)} = \sqrt{\sum_{\nu=0}^{m} {}_m C_\nu \left(\sum_{\alpha \in \mathbb{Z}_+^n, |\alpha| \leq \nu}^{\nu} \frac{\nu!}{\alpha!} \int_{\mathbb{R}^n} \left| \frac{\partial^\nu F(x)}{\partial x^\nu} \right|^2 dx \right)} \tag{58}$$

and

$$K(x, y) = \frac{1}{(2\pi)^n} \int_{\mathbb{R}^n} \frac{\exp(i(x-y)\cdot\xi)}{(1+|\xi|^2)^m} d\xi. \tag{59}$$

In particular, note that if $m > \frac{n}{2}$ then $W^{m,2}(\mathbb{R}^n)$ is embedded into $\mathrm{BC}(\mathbb{R}^n)$.

A generalization of the above spaces is given in the following way when considering $s > \frac{n}{2}$. Define

$$K(x, y) := \frac{1}{(2\pi)^n} \int_{\mathbb{R}^n} \left(1+|\xi|^2\right)^{-s} \exp(i(x-y)\cdot\xi)\, d\xi. \tag{60}$$

Then, we have

$$H_K(\mathbb{R}^n) = H^s(\mathbb{R}^n), \tag{61}$$

where the norm is given by

$$\|f\|_{H^s(\mathbb{R}^n)} = \left(\int_{\mathbb{R}^n} (1 + |\xi|^2)^s |\mathcal{F}f(\xi)|^2 \, d\xi \right)^{\frac{1}{2}}. \tag{62}$$

We can now point out a very simple example. The space $H_S(\mathbb{R})$ is made up of absolutely continuous functions F on $\mathbb{R}$ with the norm

$$\|F\|_{H_S(\mathbb{R})} := \sqrt{\int_{\mathbb{R}} (F(x)^2 + F'(x)^2) dx}. \tag{63}$$

The Hilbert space $H_S(\mathbb{R})$ admits the reproducing kernel

$$K(x, y) := \frac{1}{2\pi} \int_{\mathbb{R}} \frac{1}{1+\xi^2} \exp\left(i(x-y)\xi\right) d\xi = \frac{1}{2} e^{-|x-y|}. \tag{64}$$

Note, in particular, when constructing the reproducing kernel associated with the Paley–Wiener spaces and the Sobolev spaces the only significant difference appear in the change of the characteristic functions by the factors or weight $\left(1 + |\xi|^2\right)^{-s}$.

In the sequel, we shall introduce typical results following our new discretization method.

12 Ordinary Linear Differential Equations

Ordinary differential equations is certainly a subarea of mathematics having a very significant number applications to the real world. There are plenty of examples of their use in physics, but also in chemistry, biology, economics, etc. Unfortunately, despite the great amount of applications, only the simplest ordinary differential equations admit solutions given by explicit formulas. However, some properties of the solutions of a given differential equation may be determined without finding their exact form. If a self-contained formula for the solution is not available, the solution may be numerically approximated by using computers if a suitable numerical method is implemented.

The argument presented here will show that our results will be valid in a very general situation. However, to state the results and methods in a simpler manner, we consider the following prototype differential operator:

$$Ly := \alpha y'' + \beta y' + \gamma y. \tag{65}$$

Here, we shall consider a very general situation in which the coefficients are *arbitrary* functions on their nature and on a general interval I.

We wish to construct some natural solution of

$$Ly = g \tag{66}$$

for a very general function g on a general interval I.

For a practical construction of the solution, we obtain the following formula.

Theorem 9 *Let us fix a positive number h and take a finite number of points $\{t_j\}_{j=1}^n$ of I such that*

$$(\alpha(t_j), \beta(t_j), \gamma(t_j)) \neq 0$$

for each j. Then, the optimal solution y_h^A of the Eq. (66) *is given by*

$$y_h^A(t) = \frac{1}{2\pi}\int_{-\pi/h}^{\pi/h} F_h^A(\xi)e^{-it\xi}d\xi$$

in terms of the function $F_h^A \in L_2(-\pi/h, +\pi/h)$ in the sense that F_h^A has the minimum norm in $L_2(-\pi/h, +\pi/h)$ among the functions $F \in L_2(-\pi/h, +\pi/h)$ satisfying, for the characteristic function $\chi_h(t)$ of the interval $(-\pi/h, +\pi/h)$:

$$\frac{1}{2\pi}\int_{\mathbb{R}} F(\xi)\left[\alpha(t)(-\xi^2) + \beta(t)(-i\xi) + \gamma(t)\right]\chi_h(\xi)\exp(-it\xi)d\xi = g(t) \tag{67}$$

for all $t = t_j$ and for the function space $L_2(-\pi/h, +\pi/h)$.

The best extremal function F_h^A is given by

$$F_h^A(\xi) = \sum_{j,j'=1}^{n} g(t_j)\widetilde{a_{jj'}}\overline{(\alpha(t_{j'})(-\xi^2) + \beta(t_{j'})(-i\xi) + \gamma(t_{j'}))}\exp(it_{j'}\xi). \tag{68}$$

Here, we are indirectly using the matrix $A = \{a_{jj'}\}_{j,j'=1}^n$ formed by the elements

$$a_{jj'} = K_{hh}\left(t_j, t_{j'}\right)$$

with

$$\begin{aligned} K_{hh}(t,t') = \frac{1}{2\pi}\int_{\mathbb{R}} & [\alpha(t)(-\xi^2) + \beta(t)(-i\xi) + \gamma(t)] \\ & \overline{[\alpha(t')(-\xi^2) + \beta(t')(-i\xi) + \gamma(t')]} \\ & \cdot \chi_h(\xi)\exp(-i(t-t')\xi)d\xi \end{aligned} \tag{69}$$

which is positive definite and the elements $\widetilde{a_{jj'}}$ directly used in (68) *are the elements of the inverse of $\overline{A}$ (the complex conjugate of A).*

Therefore, the optimal solution y_h^A of the Eq. (66) *is given by*

$$y_h^A(t) = \sum_{j,j'=1}^{n} g(t_j)\, \widetilde{a_{jj'}}\, \frac{1}{2\pi} \left[-\overline{\alpha(t_{j'})} \int_{-\frac{\pi}{h}}^{\frac{\pi}{h}} \xi^2 e^{-i(t-t_{j'})\xi} d\xi \right.$$

$$\left. +i\overline{\beta(t_{j'})} \int_{-\frac{\pi}{h}}^{\frac{\pi}{h}} \xi e^{-i(t-t_{j'})\xi} d\xi + \overline{\gamma(t_{j'})} \int_{-\frac{\pi}{h}}^{\frac{\pi}{h}} e^{-i(t-t_{j'})\xi} d\xi \right].$$

We shall refer to the situation of the theorem. At first, we are considering approximate solutions of the differential equation (66) and, at this point, we are considering the Paley–Wiener function spaces with parameter h as the approximating function spaces. The function spaces are formed by analytic functions of the entire functions of exponential type that are decreasing to zero within exponential orders. Next, by using the Fourier inversion, the differential equation (66) may be transformed into (67). However, to solve the integral equation (67) is very difficult for the generality of the coefficient functions. So, we assume that (67) is valid on some finite number of points t_j. This assumption will be very reasonable for the discretization of the integral equation. By this assumption we can obtain an optimal approximate solution in a very simple way, by the fundamental Theorem 1.

Here, we assume that Eq. (66) is valid on I and so, as some practical case we would like to consider the equation in (66) on I satisfying some boundary conditions. In the present case, the boundary conditions are given as zero at infinity for $I = \mathbb{R}$.

However, our result gives the approximate general solutions satisfying boundary values. For example, for a finite interval (a,b), we consider $t_1 = a$ and $t_n = b$ and $\alpha(t_1) = \beta(t_1) = \alpha(t_n) = \beta(t_n) = 0$. Then, we can obtain the approximate solution having the arbitrary given boundary values $y_h^A(t_1)$ and $y_h^A(t_n)$. In addition, by a simple modification we may give the general approximate solutions satisfying the corresponding boundary values.

For a finite interval case I, following the boundary conditions, we can consider the corresponding reproducing kernels by the Sobolev Hilbert spaces. However, the concrete representations of the reproducing kernels are involved and depending on the boundary conditions. However, despite this great generality of the problem, we can still consider them and we can use them, as we shall see.

As an easily understandable circunstancie, for a smaller h we can obtain better approximate solutions. in addition, for the representation (69) of the reproducing kernel $K_{hh}(t,t')$, we can calculate it easily.

As a very surprising fact, which in a sense exhibits the power of the method, we would like to point out that for variable coefficients linear differential equations, we can represent their approximate solutions satisfying their boundary conditions without using integrals.

13 Linear Partial Differential Equations

Partial differential equations (PDEs) arise frequently in the formulation of fundamental laws of nature and in the mathematical analysis of a wide variety of problems in applied mathematics, mathematical physics, and engineering science. This subject plays a central role in a modern mathematical sciences, especially in physics, geometry, and analysis. Many problems of physical interest are described by PDE with appropriate initial and/or boundary conditions.

Typically, a given PDE will only be accessible by means of numerical solutions, and analytic solutions in a practical or research scenario are often impossible. However, it is vital to understand the general theory in order to conduct a sensible investigation. For example, we may need to understand what type of PDE we have to ensure the numerical solution is valid. Indeed, certain types of equations need appropriate boundary conditions; without a knowledge of the general theory it is possible that the problem may be ill-posed and the solution will be erroneous.

13.1 Second-Order Differential Equations

The argument presented here will show that our results will valid in a very general situation. However, in order to state the results and methods in a simpler manner, we shall start by considering the following prototype of partial differential operator of (hyperbolic) type:

$$Lu := \vartheta \frac{\partial^2 u}{\partial t^2} - \alpha \frac{\partial^2 u}{\partial x^2} - \gamma u. \tag{70}$$

Here, we shall consider a very general situation that the coefficients are *arbitrary* functions in time t and space x on a general domain $D_{T,I} = (0,T) \times I$, where I is any interval. We wish, for example, to construct some natural solution of

$$Lu(t,x) = g(t,x), \quad (t,x) \in (0,T) \times I, \tag{71}$$

satisfying the boundary condition

$$u(t,x) = 0, \quad (t,x) \in (0,T) \times \partial I,$$

and the initial conditions

$$u(0,x) = u_0(x), \quad x \in I,$$

and

$$u_t(0,x) = v_0(x), \quad x \in I,$$

for very general functions g, u_0 and v_0 on a general domain $D_{T,I}$.

For a practical construction of the solution, we obtain the following formula.

Theorem 10 *Let us fix a positive number h and take a finite number of points* $\{(t_\iota, x_j)\}_{\iota,j=1}^{n}$ *of* $D_{T,I}$ *such that*

$$\left(\vartheta(t_\iota, x_j), \alpha(t_\iota, x_j), \gamma(t_\iota, x_j)\right) \neq 0$$

as vectors, for each (ι, j)*. Then, the optimal solution* u_h^A *of the Eq.* (71) *is given by (for* $X = (t,x)$*):*

$$u_h^A(t,x) = \frac{1}{(2\pi)^2} \int_{D_h} F_h^A(\xi) e^{-iX\cdot\xi} d\xi$$

in terms of the function $F_h^A \in L_2(D_h)$, $D_h = ((-\pi/h, +\pi/h) \times (-\pi/h, +\pi/h))$ in the sense that F_h^A has the minimum norm in $L_2(D_h)$ among the functions $F \in L_2(D_h)$ satisfying, for the characteristic function $\chi_{D_h}(\xi)$ of D_h:

$$\frac{1}{(2\pi)^2}\int_{\mathbb{R}^2} F(\xi)[\vartheta(t,x)(-\xi_1^2) - \alpha(t,x)(-\xi_2^2) - \gamma(t,x)]\chi_{D_h}(\xi)\exp(-iX\cdot\xi)d\xi = g(t,x) \tag{72}$$

for all $X = X_{\iota j}$, i.e., $(t,x) = (t_\iota, x_j)$ and for the function space $L_2(D_h)$.

The best extremal function F_h^A is given by

$$F_h^A(\xi) = \sum_{\iota,j,\iota',j'=1}^{n} g(t_\iota, x_j)\widetilde{a_{\iota j,\iota' j'}} \cdot \overline{(\vartheta(t_{\iota'},x_{j'})(-\xi_1^2) - \alpha(t_{\iota'},x_{j'})(-\xi_2^2) - \gamma(t_{\iota'},x_{j'}))}\exp(iX_{\iota' j'}\cdot\xi). \tag{73}$$

Here, we are using the matrix $A = \{a_{\iota j,\iota' j'}\}$ $\left(\iota,\iota',j,j' = 1,\ldots,n\right)$ formed by the elements

$$a_{\iota j,\iota' j'} = K_{hh}\left(X_{\iota j}, X_{\iota' j'}\right)$$

with

$$K_{hh}(X,X') = \frac{1}{(2\pi)^2}\int_{\mathbb{R}^2}\left[\vartheta(t,x)(-\xi_1^2) - \alpha(t,x)(-\xi_2^2) - \gamma(t,x)\right]$$

$$\cdot\overline{[\vartheta(t',x')(-\xi_1^2) - \alpha(t',x')(-\xi_2^2) - \gamma(t',x')]}\chi_{D_h}(\xi)\exp(-i(X-X')\cdot\xi))d\xi$$

which is positive definite and the $\widetilde{a_{\iota j,\iota' j'}}$ are the elements of the inverse of $\overline{A}$ (the complex conjugate of A).

Therefore, the optimal solution u_h^A of the Eq. (71) is given by

$$u_h^A(t,x) = \sum_{\iota,j,\iota',j'=1}^{n} g\left(t_\iota, x_j\right)\widetilde{a_{\iota j,\iota' j'}}$$

$$\frac{1}{(2\pi)^2}\left[-\overline{\vartheta(t_{\iota'},x_{j'})}\int_{-\frac{\pi}{h}}^{\frac{\pi}{h}}\xi_1^2 e^{-i(t-t_{\iota'})\xi_1}\int_{-\frac{\pi}{h}}^{\frac{\pi}{h}} e^{-i(x-x_{j'})\xi_2}d\xi_2 d\xi_1\right.$$

$$+\overline{\alpha(t_{\iota'},x_{j'})}\int_{-\frac{\pi}{h}}^{\frac{\pi}{h}} e^{-i(t-t_{\iota'})\xi_1}\int_{-\frac{\pi}{h}}^{\frac{\pi}{h}}\xi_2^2 e^{-i(x-x_{j'})\xi_2}d\xi_2 d\xi_1$$

$$\left.-\overline{\gamma(t_{\iota'},x_{j'})}\int_{-\frac{\pi}{h}}^{\frac{\pi}{h}} e^{-i(t-t_{\iota'})\xi_1}\int_{-\frac{\pi}{h}}^{\frac{\pi}{h}} e^{-i(x-x_{j'})\xi_2}d\xi_2 d\xi_1\right].$$

In the next result, we present the optimal solution for a particular typical case of Eq. (71).

Corollary 2 *If* $\vartheta(t,x) \equiv 1, \alpha(t,x) \equiv 1, \gamma \equiv 0$, *then Eq.* (71) *reduces to an inhomogeneous wave equation, and then the optimal solution of* (71) *is given as*

$$u_h^A(t,x) = \sum_{\iota,j,\iota',j'=1}^{n} g(t_\iota, x_j)\widetilde{a_{\iota j,\iota' j'}} \frac{1}{(2\pi)^2}\left[-\int_{-\frac{\pi}{h}}^{\frac{\pi}{h}} \xi_1^2 e^{-i(t-t_{\iota'})\xi_1} \int_{-\frac{\pi}{h}}^{\frac{\pi}{h}} e^{-i(x-x_{j'})\xi_2} d\xi_2 d\xi \right.$$
$$\left. + \int_{-\frac{\pi}{h}}^{\frac{\pi}{h}} e^{-i(t-t_{\iota'})\xi_1} \int_{-\frac{\pi}{h}}^{\frac{\pi}{h}} \xi_2^2 e^{-i(x-x_{j'})\xi_2} d\xi_2 d\xi_1 \right].$$

We shall refer to the situation of Theorem 10. At first, we are considering approximate solutions of the differential equation (71) and at this point, we are considering the Paley–Wiener function spaces with parameter h as approximating function spaces. So, the differential operator (70) for these function spaces is well defined. Next, by using the Fourier inversion, the differential equation (71) may be transformed into (72). However, to solve the integral equation (72) is very difficult for the generality of the coefficient functions. Therefore, as previously exposed, we assume that (72) is valid on some finite number of points (t_ι, x_j). This assumption will be very reasonable for the discretization of the integral equation. By this assumption we can obtain an optimal approximate solutions in a very simple way.

Here, we assume that Eq. (71) is valid on $D_{T,I}$ and so, as some practical case we would like to consider the equation in (71) on $D_{T,I}$ satisfying some initial and boundary conditions. In the present case, the boundary conditions are given as zero at infinity for $I = \mathbb{R}$ by the property of the Paley–Wiener spaces.

Since the functions in the Paley–Wiener spaces are decreasing exponentially, for the zero boundary condition solutions, for a large interval I, Theorem 10 will give good approximate solutions. If h is smaller, then we can consider the interval I as a smaller one.

However, our result gives the approximate general solutions satisfying initial and boundary conditions. For example, for a finite interval $I = (a,b)$, we consider $t_1 = 0$ and

$$\vartheta(0,x_j) = \alpha(0,x_j) = 0.$$

Then, $\gamma(0,x_j) \neq 0$ and we have

$$\gamma(0,x_j)u(0,x_j) = g(0,x_j) \Rightarrow u_0(x) = \frac{g(0,x_j)}{\gamma(0,x_j)}.$$

Thus, we see that we can consider the initial condition to be given in a very general situation. Additionally, we realize that we can obtain the approximate solution having the boundary and initial conditions from the above formulations.

However, in the above formula, we cannot give the initial velocity $v_0(x_j)$ at the same point x_j but we will further refer to this problem.

13.1.1 When We Give Initial Values and Initial Velocities

In Theorem 10, we cannot give initial values and initial velocities at the same points x_j and so, we shall now consider precisely this situation and will give a formula giving the initial values and initial velocities at the same points x_j.

We are considering the integral representation, in (72),

$$u(t,x) = \frac{1}{(2\pi)^2}\int_{\mathbb{R}^2} F(\xi)\left[\vartheta(t,x)(-\xi_1^2) - \alpha(t,x)(-\xi_2^2) - \gamma(t,x)\right] \chi_{D_h}(\xi)\exp(-iX\cdot\xi)d\xi. \quad (74)$$

In order to consider the derivative $u_t(t,x)$, we shall assume the time-independent coefficients case (however, by assuming the time differentiability for the coefficients, we can derive a similar, but a more complicated formula). Then, we have

$$u(t,x) = \frac{1}{(2\pi)^2}\int_{\mathbb{R}^2} F(\xi)\left[\vartheta(x)(-\xi_1^2) - \alpha(x)(-\xi_2^2) - \gamma(x)\right]\chi_{D_h}(\xi) \exp(-iX\cdot\xi)d\xi \quad (75)$$

and

$$u_t(t,x) = \frac{1}{(2\pi)^2}\int_{\mathbb{R}^2} F(\xi)\left[\vartheta(x)(-\xi_1^2) - \alpha(x)(-\xi_2^2) - \gamma(x)\right] (-i\xi_1)\chi_{D_h}(\xi)\exp(-iX\cdot\xi)d\xi. \quad (76)$$

Therefore, by considering the linear mappings (72) and (76) from $L_2(D_h)$ into

$$\{u(t_\iota, x_j), \iota, j = 1,2,\ldots,n, \text{ and } u_t(0,x_j), j = 1,2,\ldots,n\}, \quad (77)$$

we can obtain the desired formula, similarly.

Furthermore, note that we are giving the approximate solution for the homogeneous equation ($g \equiv 0$) satisfying boundary and initial conditions.

The surprising fact in here is that for variable coefficients linear PDEs, we can represent their approximate solutions satisfying their initial and boundary conditions.

13.1.2 Construction of Approximate Solutions and Proof of Theorem 10

Since the structure of the Hilbert space $H_{K_{hh}}$ in Theorem 10 is, in general, very complicated and abstract, we shall approximate the space by taking a finite number of points and by using matrix theory. In order to realize approximately the inner product in the space $H_{K_{hh}}$ with some practical sense, we shall consider the natural and general approximate realization of the space $H_{K_{hh}}$.

By taking a finite number of points $\left\{(t_\iota, x_j)\right\}_{\iota,j=1}^{n}$, we set

$$K_{hh}\left(X_{\iota j}, X_{\iota' j'}\right) := a_{\iota j,\iota' j'},$$

and we will define $A := \left[a_{\imath j,\imath' j'}\right]_{\imath,j=1}^{n}$. Then, if the matrix A is positive definite, the corresponding norm in H_A comprising the vectors

$$\mathbf{y} = (y_{11}, y_{12}, \ldots, y_{1n}, y_{21}, y_{22}, \ldots, y_{2n}, \ldots, y_{n1}, y_{n2}, \ldots, y_{nn})$$

is determined by

$$\|\mathbf{y}\|_{H_A}^2 = \mathbf{y}^* \widetilde{A} \mathbf{y},$$

where $\widetilde{A} := [\widetilde{a_{\imath j,\imath' j'}}]$ (see [35, p. 250]).

Now, from (71), we set

$$a_{\imath j,\imath' j'} = K_{hh}\left(X_{\imath j}, X_{\imath' j'}\right), \quad \imath, \imath', j, j' = 1, \ldots, n.$$

If A is positive definite, then the minimum norm solution $F_h^A(\xi)$ of the integral transform (72),

$$F \in L_2(D_h) \longrightarrow H_{K_{hh}},$$

is given by the following formula (having in mind our fundamental Theorem 1):

$$F_h^A(\xi) = \sum_{\imath,j,\imath',j'=1}^{n} g(t_\imath, x_j)\widetilde{a_{\imath j,\imath' j'}}$$
$$\overline{(\vartheta(t_{\imath'}, x_{j'})(-\xi_1^2) - \alpha(t_{\imath'}, x_{j'})(-\xi_2^2) - \gamma(t_{\imath'}, x_{j'}))} \exp(i X_{\imath' j'} \cdot \xi).$$

Associated with (72), it remains to observe that

$$[\vartheta(t_\imath, x_j)(-\xi_1^2) - \alpha(t_\imath, x_j)(-\xi_2^2) - \gamma(t_\imath, x_j)] \exp(-iX_{\imath j} \cdot \xi), \quad \imath, j = 1, 2, \ldots, n,$$

are linearly independent in the space $L_2(D_h)$ if and only if

$$(\vartheta(t_\imath, x_j), \alpha(t_\imath, x_j), \gamma(t_\imath, x_j)) \neq 0$$

as vectors, for each $\imath, j$.

We can directly prove this proposition by mathematical induction but, instead, in here, we shall prove it by a different method using associated differential operators, cf. [12]. In that paper, Y. Sawano gave the noble proof in the following reasoning. Assume that

$$\sum_{\imath,j=1}^{n} P_{\imath,j}\{\vartheta(t_\imath, x_j)(-\xi_1^2) - \alpha(t_\imath, x_j)(-\xi_2^2) - \gamma(t_\imath, x_j)\} \exp(-iX_{\imath j} \cdot \xi) \equiv 0.$$

For each $\imath, j$, consider the differential operator

$$P_{\imath,j}(D) = \left(\frac{d}{d\xi_1} + it_\imath\right)^3 \left(\frac{d}{d\xi_2} + ix_j\right)^3.$$

Then, since a well-known theorem about ordinary differential equations shows that

$$\text{Ker}\left(\Pi_{\imath,j}^{n} P_{\imath,j}(D)\right) = \oplus_{\imath,j=1}^{n} \text{Ker}(P_{\imath,j}(D))$$

in the sense of direct sum, we have

$$P_{\imath,j}\left\{\vartheta(t_\imath,x_j)(-\xi_1^2) - \alpha(t_\imath,x_j)(-\xi_2^2) - \gamma(t_\imath,x_j)\right\}\exp\left(-iX_{\imath j}\cdot\xi\right) \equiv 0$$

for each $\imath$, j and hence

$$P_{\imath,j}\left\{\vartheta(t_\imath,x_j)(-\xi_1^2) - \alpha(t_\imath,x_j)(-\xi_2^2) - \gamma(t_\imath,x_j)\right\} \equiv 0,$$

This shows the desired result that $P_{\imath,j} = 0$ for $\imath, j = 1, 2, \ldots, n$.

Hence, we obtain the main Theorem 10.

The argument presented in here will show that our results will be valid in a very general situation. However, in order to state the results and methods in a simpler manner, we shall consider a prototype partial differential operator of parabolic nature

$$Lu := \vartheta\frac{\partial u}{\partial t} - \alpha\frac{\partial^2 u}{\partial x^2} - \beta\frac{\partial u}{\partial x} - \gamma u. \tag{78}$$

We shall consider a very general situation that the coefficients are *arbitrary* functions in time t and space x on a general domain $D_{T,I} = (0,T)\times I$, I is any interval.

We wish, for example, to construct some natural solution of

$$\begin{aligned} Lu(t,x) &= g(t,x), \quad (t,x)\in(0,T)\times I, \\ u(t,x) &= 0, \quad (t,x)\in(0,T)\times\partial I, \\ u(0,x) &= u_0(x), \quad x\in I, \end{aligned} \tag{79}$$

for very general functions g and u_0 on a general domain $D_{T,I}$.

For a practical construction of the solution, we obtain the following formula (upon a finite number of data as it may be recognized in the statement).

Theorem 11 *Let us fix a positive number h and take a finite number of points $\{(t_\imath,x_j)\}_{\imath,j=1}^n$ of $D_{T,I}$ such that*

$$\left(\vartheta(t_\imath,x_j), \alpha(t_\imath,x_j), \beta(t_\imath,x_j), \gamma(t_\imath,x_j)\right) \neq 0$$

as vectors, for each $(\imath, j)$. Then, the optimal solution u_h^A of the Eq. (79) *is given by*

$$u_h^A(t,x) = \frac{1}{(2\pi)^2}\int_{D_h} F_h^A(\xi)e^{-iX\cdot\xi}d\xi \qquad \textit{for } X = (t,x)$$

in terms of the function $F_h^A \in L_2(D_h)$, with $D_h = ((-\pi/h, +\pi/h)\times(-\pi/h, +\pi/h))$ in the sense that F_h^A has the minimum norm in $L_2(D_h)$ among the functions $F \in L_2(D_h)$ satisfying

$$\frac{1}{(2\pi)^2}\int_{\mathbb{R}^2} F(\xi)[\vartheta(t,x)(-i\xi_1) - \alpha(t,x)(-\xi_2^2) - \beta(t,x)(-i\xi_2) - \gamma(t,x)]$$

$$\cdot\chi_{D_h}(\xi)\exp(-iX\cdot\xi)d\xi = g(t,x), \quad (80)$$

where $\chi_{D_h}(\xi)$ *denotes the characteristic function of* D_h, *and for all* $X = X_{\imath j}$, *i.e.,* $(t,x) = (t_\imath, x_j)$ *and for the function space* $L_2(D_h)$.

The best extremal function F_h^A *is given by*

$$\begin{aligned} F_h^A(\xi) = &\sum_{\imath,j,\imath',j'=1}^{n} g(t_\imath,x_j)\widetilde{a_{\imath j,\imath' j'}} \\ &\overline{\cdot(\vartheta(t_{\imath'},x_{j'})(-i\xi_1) - \alpha(t_{\imath'},x_{j'})(-\xi_2^2) - \beta(t_{\imath'},x_{j'})(-i\xi_2) - \gamma(t_{\imath'},x_{j'}))} \\ &\cdot\exp(iX_{\imath' j'}\cdot\xi). \end{aligned} \quad (81)$$

We are considering in here the matrix $A = \{a_{\imath j,\imath' j'}\}$ $(\imath,\imath',j,j' = 1,\ldots,n)$ *formed by the elements*

$$a_{\imath j,\imath' j'} = K_{hh}(X_{\imath j}, X_{\imath' j'})$$

with

$$\begin{aligned} K_{hh}(X,X') = &\frac{1}{(2\pi)^2}\int_{\mathbb{R}^2}\left[\vartheta(t,x)(-i\xi_1) - \alpha(t,x)(-\xi_2^2) - \beta(t,x)(-i\xi_2) - \gamma(t,x)\right] \\ &\overline{\cdot\left[\vartheta(t',x')(-i\xi_1) - \alpha(t',x')(-\xi_2^2) - \beta(t',x')(-i\xi_2) - \gamma(t',x')\right]} \\ &\cdot\chi_{D_h}(\xi)\exp(-i(X-X')\cdot\xi))d\xi \end{aligned} \quad (82)$$

which is positive definite and the $\widetilde{a_{\imath j,\imath' j'}}$ *are the elements of the inverse of* $\overline{A}$ *(the complex conjugate of* A*).*

Therefore, the optimal solution u_h^A *of the Eq.* (79) *is given by*

$$\begin{aligned} u_h^A(t,x) = &\sum_{\imath,j,\imath',j'=1}^{n} g(t_\imath,x_j)\,\widetilde{a_{\imath j,\imath' j'}} \\ &\frac{1}{(2\pi)^2}\Bigg[i\overline{\vartheta(t_{\imath'},x_{j'})}\int_{-\frac{\pi}{h}}^{\frac{\pi}{h}}\xi_1 e^{-i(t-t_{\imath'})\xi_1}\int_{-\frac{\pi}{h}}^{\frac{\pi}{h}}e^{-i(x-x_{j'})\xi_2}d\xi_2 d\xi_1 \\ &+\overline{\alpha(t_{\imath'},x_{j'})}\int_{-\frac{\pi}{h}}^{\frac{\pi}{h}}e^{-i(t-t_{\imath'})\xi_1}\int_{-\frac{\pi}{h}}^{\frac{\pi}{h}}\xi_2^2 e^{-i(x-x_{j'})\xi_2}d\xi_2 d\xi_1 \\ &-i\overline{\beta(t_{\imath'},x_{j'})}\int_{-\frac{\pi}{h}}^{\frac{\pi}{h}}e^{-i(t-t_{\imath'})\xi_1}\int_{-\frac{\pi}{h}}^{\frac{\pi}{h}}\xi_2 e^{-i(x-x_{j'})\xi_2}d\xi_2 d\xi_1 \\ &-\overline{\gamma(t_{\imath'},x_{j'})}\int_{-\frac{\pi}{h}}^{\frac{\pi}{h}}e^{-i(t-t_{\imath'})\xi_1}\int_{-\frac{\pi}{h}}^{\frac{\pi}{h}}e^{-i(x-x_{j'})\xi_2}d\xi_2 d\xi_1\Bigg]. \end{aligned}$$

In the next two results, we present the optimal solution for particular typical cases of Eq. (79).

Corollary 3 *If* $\vartheta(t,x) \equiv 1, \alpha(t,x) \equiv 1, \beta(t,x) \equiv 0$ *and* $\gamma(t,x) \equiv 0$, *then Eq.* (79) *reduces to an inhomogeneous heat equation and then the optimal solution of* (79) *is given as*

$$u_h^A(t,x) = \sum_{\iota,j,\iota',j'=1}^{n} g(t_\iota, x_j)\widetilde{a_{\iota j,\iota' j'}} \frac{1}{(2\pi)^2} \left[i \int_{-\frac{\pi}{h}}^{\frac{\pi}{h}} \xi_1 e^{-i(t-t_{\iota'})\xi_1} \int_{-\frac{\pi}{h}}^{\frac{\pi}{h}} e^{-i(x-x_{j'})\xi_2} d\xi_2 d\xi_1 \right.$$
$$\left. + \int_{-\frac{\pi}{h}}^{\frac{\pi}{h}} e^{-i(t-t_{\iota'})\xi_1} \int_{-\frac{\pi}{h}}^{\frac{\pi}{h}} \xi_2^2 e^{-i(x-x_{j'})\xi_2} d\xi_2 d\xi_1 \right].$$

Corollary 4 *If* $\vartheta(t,x) \equiv 1, \alpha(t,x) \equiv \frac{i\hbar}{2m}, \beta(t,x) \equiv 0$ *and* $\gamma(t,x) \equiv 0$, *then Eq.* (79) *reduces to an inhomogeneous Schrödinger equation and then the optimal solution of* (79) *is given as*

$$u_h^A(t,x) = \sum_{\iota,j,\iota',j'=1}^{n} g(t_\iota, x_j)\widetilde{a_{\iota j,\iota' j'}} \frac{1}{(2\pi)^2} \left[i \int_{-\frac{\pi}{h}}^{\frac{\pi}{h}} \xi_1 e^{-i(t-t_{\iota'})\xi_1} \int_{-\frac{\pi}{h}}^{\frac{\pi}{h}} e^{-i(x-x_{j'})\xi_2} d\xi_2 d\xi_1 \right.$$
$$\left. - \frac{i\hbar}{2\,m} \int_{-\frac{\pi}{h}}^{\frac{\pi}{h}} e^{-i(t-t_{\iota'})\xi_1} \int_{-\frac{\pi}{h}}^{\frac{\pi}{h}} \xi_2^2 e^{-i(x-x_{j'})\xi_2} d\xi_2 d\xi_1 \right].$$

Here, we assume that Eq. (79) is valid on $D_{T,I}$ and so, as some practical case we would like to consider the equation in (79) on $D_{T,I}$ satisfying some initial and boundary conditions. In the present case, the boundary conditions are given as zero at infinity for $I = \mathbb{R}$ by the property of the Paley–Wiener spaces.

Since the functions in the Paley–Wiener spaces are decreasing exponentially, for the zero boundary condition solutions, for a large interval I, Theorem 11 will give good approximate solutions. If h is smaller, then we can consider the interval I as a more smaller one.

However, our result gives the approximate general solutions satisfying initial and boundary conditions. For example, for a finite interval $I = (a,b)$, we consider $t_1 = 0$ and

$$\vartheta(0,x_j) = \alpha(0,x_j) = \beta(0,x_j) = 0,$$

then $\gamma(0,x_j) \neq 0$ and we have

$$\gamma(0,x_j)u(0,x_j) = g(0,x_j) \Rightarrow u_0(x_j) = \frac{g(0,x_j)}{\gamma(0,x_j)}.$$

Then, we see that we can consider the initial condition which may be given, in a very general situation.

We will now turn to a different type of equations. Anyway, once again, the argument presented in here will show that our results will continue valid in a very general situation. However, in order to state the results and methods in a simpler manner, we shall consider the following two-dimensional second-order differential operator of Laplace type:

$$Lu := \alpha \frac{\partial^2 u}{\partial x^2} + \beta \frac{\partial^2 u}{\partial y^2}, \tag{83}$$

where I is any open interval.

The main goal is to construct some natural solution of

$$Lu(x, y) = g(x, y), \quad (x, y) \in I \times I, \tag{84}$$

satisfying the boundary condition

$$u(x, y) = u_f(x, y), \qquad (x, y) \in \partial I \times \partial I,$$

where the function u_f is defined on the boundary of $I \times I$ and such that

$$u_f \to 0 \quad \text{as} \quad |X| \to \infty,$$

with $X = (x, y)$.

For a practical construction of the solution, we obtain the following formula (once again by considering the technique of taking a finite number of data as it may be recognized in the statement).

Theorem 12 *Let us fix a positive number h and take a finite number of points* $\{(x_\iota, y_j)\}_{\iota,j=1}^n$ *of* $D_{I,I}$ *such that*

$$\big(\alpha(x_\iota, y_j), \beta(x_\iota, y_j)\big) \neq 0$$

as vectors, for each (x_ι, y_j)*. Then, the optimal solution* u_h^A *of the Eq. (84) is given by, for* $X = (x, y)$,

$$u_h^A(x, y) = \frac{1}{(2\pi)^2} \int_{D_h} F_h^A(\xi) e^{-iX \cdot \xi} d\xi$$

in terms of the function $F_h^A \in L_2(D_h)$, $D_h = ((-\pi/h, +\pi/h) \times (-\pi/h, +\pi/h))$ *in the sense that* F_h^A *has the minimum norm in* $L_2(D_h)$ *among the functions* $F \in L_2(D_h)$ *satisfying*

$$\frac{1}{(2\pi)^2} \int_{\mathbb{R}^2} F(\xi)[\alpha(x, y)\xi_1^2 + \beta(x, y)\xi_2^2)]\chi_{D_h}(\xi) \exp(-iX \cdot \xi) d\xi + g(x, y) = 0 \tag{85}$$

for all $X = X_{\iota j}$*, i.e.,* $(x, y) = (x_\iota, y_j)$ *and for the function space* $L_2(D_h)$.

The best extremal function F_h^A *is given by*

$$F_h^A(\xi) = - \sum_{\iota,j,\iota',j'=1}^{n} g(t_\iota, x_j)\, \widetilde{a_{\iota j,\iota' j'}}\, \overline{\big(\alpha(x_{\iota'}, y_{j'})\xi_1^2 + \beta(x_{\iota'}, y_{j'})\xi_2^2\big)} \exp(iX_{\iota' j'} \cdot \xi).$$

Moreover, the matrix $A = \{a_{\iota j,\iota' j'}\}$ $(\iota, \iota', j, j' = 1, \dots, n)$ *formed by the elements*

$$a_{\iota j,\iota' j'} = K_{hh}\left(X_{\iota j}, X_{\iota' j'}\right)$$

with

$$K_{hh}(X, X') = \frac{1}{(2\pi)^2}\int_{\mathbb{R}^2} (\alpha(x,y)\xi_1^2 + \beta(x,y)\xi_2^2)\overline{(\alpha(x',y')\xi_1^2 + \beta(x',y')\xi_2^2)}$$
$$\cdot\, \chi_{D_h}(\xi)\exp(-i(X - X')\cdot\xi))d\xi \tag{86}$$

which is positive definite and we are using above elements $\widetilde{a_{\iota j,\iota' j'}}$*, which are the entries of the inverse of* $\overline{A}$ *(the complex conjugate of A).*

Therefore, the optimal solution u_h^A *of the Eq.* (84) *is given by*

$$u_h^A(x,y) = -\sum_{\iota,j,\iota',j'=1}^{n} g\left(x_\iota, y_j\right)\widetilde{a_{\iota j,\iota' j'}}\frac{1}{(2\pi)^2}$$
$$\cdot\,[\overline{\alpha(x_{\iota'}, y_{j'})}\int_{-\frac{\pi}{h}}^{\frac{\pi}{h}} \xi_1^2 e^{-i(x-x_{\iota'})\xi_1}\int_{-\frac{\pi}{h}}^{\frac{\pi}{h}} e^{-i(y-y_{j'})\xi_2}d\xi_2 d\xi_1$$
$$+\,\overline{\beta(x_{\iota'}, y_{j'})}\int_{-\frac{\pi}{h}}^{\frac{\pi}{h}} e^{-i(x-x_{\iota'})\xi_1}\int_{-\frac{\pi}{h}}^{\frac{\pi}{h}} \xi_2^2 e^{-i(y-y_{j'})\xi_2}d\xi_2 d\xi_1].$$

In the next two results, we present the optimal solution for the particular typical case of Eq. (84).

Corollary 5 *If* $\alpha(t,x) \equiv 1$ *and* $\beta(t,x) \equiv 1$*, then Eq.* (84) *reduces to an inhomogeneous Poisson equation and then the optimal solution of* (84) *is given as*

$$u_h^A(x,y) = -\sum_{\iota,j,\iota',j'=1}^{n} g\left(x_\iota, y_j\right)\widetilde{a_{\iota j,\iota' j'}}\frac{1}{(2\pi)^2}$$
$$\left[\int_{-\frac{\pi}{h}}^{\frac{\pi}{h}} \xi_1^2 e^{-i(x-x_{\iota'})\xi_1}\int_{-\frac{\pi}{h}}^{\frac{\pi}{h}} e^{-i(y-y_{j'})\xi_2}d\xi_2 d\xi_1\right.$$
$$\left.+\int_{-\frac{\pi}{h}}^{\frac{\pi}{h}} e^{-i(x-x_{\iota'})\xi_1}\int_{-\frac{\pi}{h}}^{\frac{\pi}{h}} \xi_2^2 e^{-i(y-y_{j'})\xi_2}d\xi_2 d\xi_1\right].$$

Corollary 6 *If* $\alpha(t,x) \equiv y$ *and* $\beta(t,x) \equiv 1$*, then Eq.* (84) *reduces to an inhomogeneous Tricomi equation and then the optimal solution of* (84) *is given as*

$$u_h^A(x,y) = -\sum_{\iota,j,\iota',j'=1}^{n} g(x_\iota, y_j)\widetilde{a_{\iota j,\iota' j'}}\frac{1}{(2\pi)^2}$$

$$\left[\overline{y_{j'}}\int_{-\frac{\pi}{h}}^{\frac{\pi}{h}} \xi_1^2 e^{-i(x-x_{\iota'})\xi_1}\int_{-\frac{\pi}{h}}^{\frac{\pi}{h}} e^{-i(y-y_{j'})\xi_2}d\xi_2 d\xi_1 \right.$$

$$\left. + \int_{-\frac{\pi}{h}}^{\frac{\pi}{h}} e^{-i(x-x_{\iota'})\xi_1}\int_{-\frac{\pi}{h}}^{\frac{\pi}{h}} \xi_2^2 e^{-i(y-y_{j'})\xi_2}d\xi_2 d\xi_1\right].$$

Note that, the Tricomi equation is of mixed type and has been associated with transonic flow, where u may be thought of as a stream function or a perturbation velocity potential, when considering a source g and boundary data u_f to be specified. The equation is elliptic for $y > 0$, hyperbolic for $y < 0$, and parabolic for $y = 0$, an interface called the sonic line as it separates regions of subsonic and supersonic flow. It is well known that the Eq. (84) yields a well-posed problem if u_f is prescribed only on suitable portions of the boundary, a study initiated by Tricomi.

13.2 First-Order Partial Differential Equations

In this section, we will consider a general two-dimensional first-order (hypoelliptic) differential operator,

$$Lu := \alpha(x,y)\frac{\partial u}{\partial x} + \beta(x,y)\frac{\partial u}{\partial y}. \tag{87}$$

The idea is to implement our method so that some natural solution of

$$Lu(x,y) = g(x,y), \quad \text{on } \mathbb{R}^2. \tag{88}$$

will be constructed.

Theorem 13 *Let us fix a positive number h and take a finite number of points $\{(x_\iota, y_j)\}_{\iota,j=1}^n$ of $\mathbb{R}^2$ such that*

$$\big(\alpha(x_\iota, y_j), \beta(x_\iota, y_j)\big) \neq 0$$

as vectors. Then, the optimal solution u_h^A of the Eq. (88) is given by

$$u_h^A(x,y) = \frac{1}{(2\pi)^2}\int_{D_h} F_h^A(\xi)e^{-iX\cdot\xi}d\xi \qquad \text{for } X = (x,y)$$

in terms of the function $F_h^A \in L_2(D_h)$, $D_h = ((-\pi/h, +\pi/h) \times (-\pi/h, +\pi/h))$ in the sense that F_h^A has the minimum norm in $L_2(D_h)$ among all the functions $F \in L_2(D_h)$ satisfying

$$\frac{1}{(2\pi)^2}\int_{\mathbb{R}^2} F(\xi)[-i\alpha(x,y)\xi_1 - i\beta(x,y)\xi_2)]\chi_{D_h}(\xi)\exp(-iX\cdot\xi)d\xi = g(x,y) \tag{89}$$

for all $X = X_{\iota j}$*, i.e.,* $(x, y) = (x_\iota, y_j)$ *and for the function space* $L_2(D_h)$*.*

The best extremal function F_h^A *is given by*

$$F_h^A(\xi) = \sum_{\iota,j,\iota',j'=1}^{n} g(t_\iota, x_j)\widetilde{a_{\iota j,\iota' j'}}\left[\overline{-i\alpha(x_{\iota'}, y_{j'})\xi_1 - i\beta(x_{\iota'}, y_{j'})\xi_2)}\right]\exp(iX_{\iota' j'}\cdot\xi),$$

where $\widetilde{a_{\iota j,\iota' j'}}$ *are the entries of the inverse of* $\overline{A}$*, the complex conjugate of the matrix* $A = \{a_{\iota j,\iota' j'}\}\ \left(\iota,\iota', j, j' = 1, \ldots, n\right)$ *formed by the elements*

$$a_{\iota j,\iota' j'} = K_{hh}\left(X_{\iota j}, X_{\iota' j'}\right)$$

with

$$\begin{aligned} K_{hh}(X, X') = \frac{1}{(2\pi)^2}\int_{\mathbb{R}^2} &[-i\alpha(x, y)\xi_1 - i\beta(x, y)\xi_2)] \\ &\left[\overline{(-i\alpha(x', y')\xi_1 - i\beta(x', y')\xi_2)}\right]\chi_{D_h}(\xi)\exp(-i(X - X')\cdot\xi))d\xi \end{aligned} \tag{90}$$

which is a positive definite.

Therefore, the optimal solution u_h^A *of the Eq.* (88) *is given by*

$$\begin{aligned} u_h^A(x, y) = \sum_{\iota,j,\iota',j'=1}^{n} & g\left(x_\iota, y_j\right)\widetilde{a_{\iota j,\iota' j'}}\frac{1}{(2\pi)^2} \\ &\cdot [i\overline{\alpha(x_{\iota'}, y_{j'})}\int_{-\frac{\pi}{h}}^{\frac{\pi}{h}} \xi_1 e^{-i(x-x_{\iota'})\xi_1}\int_{-\frac{\pi}{h}}^{\frac{\pi}{h}} e^{-i(y-y_{j'})\xi_2}d\xi_2 d\xi_1 \\ &+ i\overline{\beta(x_{\iota'}, y_{j'})}\int_{-\frac{\pi}{h}}^{\frac{\pi}{h}} e^{-i(x-x_{\iota'})\xi_1}\int_{-\frac{\pi}{h}}^{\frac{\pi}{h}} \xi_2 e^{-i(y-y_{j'})\xi_2}d\xi_2 d\xi_1]. \end{aligned}$$

In the next result, we present the optimal solution for a particular case of Eq. (88).

Corollary 7 *If* $\alpha(x, y) = -1 + 2i(x + y)$ *and* $\beta(x, y) = -i + 2(x + y)$*, then Eq.* (88) *reduces to an inhomogeneous, i.e.,*

$$-\frac{\partial u}{\partial x} - i\frac{\partial u}{\partial y} + 2i(x + y)\frac{\partial u}{\partial z} = g(x, y),$$

with $\frac{\partial}{\partial z}$ *a complex differential operator and then the optimal solution is given as*

$$\begin{aligned} u_h^A(x, y) = \sum_{\iota,j,\iota',j'=1}^{n} & g(x_\iota, y_j)\widetilde{a_{\iota j,\iota' j'}}\frac{1}{(2\pi)^2} \\ &\cdot\left[\left(-i + 2x_{\iota'} + 2y_{j'}\right)\int_{-\frac{\pi}{h}}^{\frac{\pi}{h}} \xi_1 e^{-i(x-x_{\iota'})\xi_1}\int_{-\frac{\pi}{h}}^{\frac{\pi}{h}} e^{-i(y-y_{j'})\xi_2}d\xi_2 d\xi_1\right. \\ &\left.+\left(-1 + 2ix_{\iota'} + 2iy_{j'}\right)\int_{-\frac{\pi}{h}}^{\frac{\pi}{h}} e^{-i(x-x_{\iota'})\xi_1}\int_{-\frac{\pi}{h}}^{\frac{\pi}{h}} \xi_2 e^{-i(y-y_{j'})\xi_2}d\xi_2 d\xi_1\right]. \end{aligned}$$

13.3 Existence Problem of the Differential Equation

Following (I) and (II) in Proposition 1, we shall first consider the existence of the solution of the previous differential equations. For the image identification of the integral transform ((72), (80), (85), (89)), from the corresponding representation ((74), (82), (86), (90)), we must realize the reproducing kernel Hilbert space $H_{K_{hh}}$ admitting the reproducing kernel $K_{hh}(X, X')$. However, we know that if ((74), (82), (86), (90)) are not calculated in a good way, in general, its realization is very complicated and the inversion of the integral transform ((72), (80), (85), (89)) are involved. So, we shall consider the existence problem with a numerical treatment.

Following a general approach of the theory of reproducing kernels, we shall consider what in our opinion is a very natural and general method.

First, let us recall several facts and some notation. For two positive semidefinite matrices $K^{(1)}(p,q)$ and $K^{(2)}(p,q)$ on E, if $K^{(2)}(p,q) - K^{(1)}(p,q)$ is a positive semidefinite matrix on E, then we shall write it as follows:

$$K^{(1)} \ll K^{(2)}.$$

We see that the symbol $\ll$ is a partial ordering in the class of all positive semidefinite matrices on E and, moreover, the following result holds.

Proposition 8 *For two positive semidefinite matrices $K^{(1)}(p,q)$ and $K^{(2)}(p,q)$ on E,*

$$H_{K^{(1)}} \subseteq H_{K^{(2)}} \quad (\text{as} \quad \text{members})$$

if and only if there exists a positive constant Γ such that

$$K^{(1)}(p,q) \ll \Gamma^2 \, K^{(2)}(p,q).$$

Here, the minimum of such constants Γ coincides with the norm of the inclusion map J from $H_{K^{(1)}}$ into $H_{K^{(2)}}$.

By using this Proposition 8, we can give completely the existence condition of the solutions as follows:

Theorem 14 *For any function g, there exists a solution $u(x, y) \in W\left(\frac{\pi}{h}\right)$ satisfying* (71), (79), (84), (88) *if and only if for some constant Γ*

$$g(x, y)\overline{g(x', y')} \ll \Gamma^2 K_{hh}(X, X'); \tag{91}$$

that is, for any finite number of points $\{(t_\iota, x_j)\}$ and for any constants $\{C_{\iota j}\}$,

$$\left| \sum_{\iota,j=1}^{n} C_{\iota j} g(t_\iota, x_j) \right|^2 \leq \Gamma^2 \sum_{\iota,j,\iota',j'=1}^{n} C_{\iota j}\overline{C_{\iota' j'}} K_{hh}\left(X_{\iota j}, X_{\iota' j'}\right) \tag{92}$$

holds with $X_{\iota j} = (x_\iota, y_j)$, $X_{\iota' j'} = \left(x_{\iota'}, y_{j'}\right)$. If this is the case, then the norm of g in $H_{K_{hh}}$ is given by the infimum Γ satisfying those inequalities.

At this moment, we cannot provide any further information about the uniqueness of the solutions.

14 Convolution Integral Equations

Let us now introduce the function spaces $\mathcal{F}(\rho_j)$ $(j = 1, 2, 3)$, where ρ_j are nonnegative and integrable functions on $\mathbb{R}$. Each of these spaces collects all elements F_j which satisfy

$$\int \frac{|F_j(t)|^2}{\rho_j(t)} dt < \infty,$$

on the support of ρ_j, and are simply $F_j = 0$ on the outside of the support of ρ_j. Additionally, we also set

$$\Omega(t; \rho_1, \rho_2, \rho_3) := \rho_1 * (2\pi + \rho_2 + \rho_3)$$

(for the usual convolution $*$).

In particular, note that all the functions $F_j \in \mathcal{F}(\rho_j)$ $(j = 1, 2, 3)$ belong to $L_1(\mathbb{R})$, as we see from the Schwarz inequality.

We set

$$f_j(x) = (L_j F_j)(x) = \frac{1}{2\pi} \int_{\mathbb{R}} F_j(t) \exp(it \cdot x)\, dt \tag{93}$$

and

$$g(x) = \frac{1}{(2\pi)^2} \int_{\mathbb{R}} G(t) \exp(ix \cdot t)\, dt. \tag{94}$$

Theorem 15 *Let ρ_j $(j = 1, 2, 3)$ be nonnegative and integrable functions on $\mathbb{R}$ which allow us to consider the just introduced spaces $\mathcal{F}(\rho_j)$.*

For any fixed $F_j \in \mathcal{F}(\rho_j)$, $j = 2, 3$ (with $\alpha + F_2 + F_3$ not zero on a positive measure set), and for a function G satisfying

$$\int_{\mathbb{R}} |G(t)|^2\, \Omega(t; \rho_1, \rho_2, \rho_3)^{-1} dt < \infty,$$

we consider the integral equation

$$2\pi\alpha F_1(t) + \int_{\mathbb{R}} F_1(\xi) F_2(t - \xi) d\xi + \int_{\mathbb{R}} F_1(\xi) F_3(t - \xi) d\xi = G(t). \tag{95}$$

When the function $\alpha + f_2(x) + f_3(x)$ is not vanishing, for any function g satisfying the condition

$$\int_{\mathbb{R}} \left| \int_{\mathbb{R}} \frac{g(x)}{\alpha + f_2(x) + f_3(x)} \right|^2 \frac{dt}{\rho_1(t)} < \infty, \tag{96}$$

there exists a uniquely determined solution F_1 belonging to $\mathcal{F}(\rho_1)$ and the solution F_1 is represented by

$$F_1(t) = \frac{1}{2\pi} \int_{\mathbb{R}} \frac{g(x)}{\alpha + f_2(x) + f_3(x)} \exp(-it \cdot x) dx. \tag{97}$$

In this Theorem 15, we can see that the function space where g belongs to is a very natural one for considering in the integral Eq. (97) together with the other function spaces for the integral kernels and solutions, from the theory of reproducing kernels.

The criteria (96) will be, in general, involved and so, we, in general, proposed numerical calculation algorithms of some general fractional functions by using the Tikhonov regularization and the theory of reproducing kernels, cf. [6].

For some practical construction of the solution, we obtain the following formula.

Theorem 16 *Assume that* $\alpha + f_2(x) + f_3(x)$ *is not zero on a positive measure set. We take a finite number of points* $\{p_j\}_{j=1}^n$ *satisfying that* $\alpha + f_2(p_j) + f_3(p_j)$ *is not zero, for any* j.

Then, the optimal solution F_1^A *of the Eq.* (95) *satisfying that, for the equation*

$$\alpha f_1(x) + f_1(x)f_2(x) + f_1(x)f_3(x) = g(x), \tag{98}$$

which is transformed to the integral Eq. (95) *by the Fourier integrals, and for all* j,

$$\alpha f_1(p_j) + f_1(p_j)f_2(p_j) + f_1(p_j)f_3(p_j) = g(p_j) \tag{99}$$

is given by

$$F_1^A(t) = \sum_{j,j'=1}^{n} g(p_j)\, \widetilde{a_{jj'}}\, \aleph(p_{j'}, t). \tag{100}$$

Here, the matrix A *formed by the elements*

$$a_{jj'} = K_\varphi\left(p_j, p_{j'}\right)$$

with

$$K_\varphi(x, y) = K_1(x, y)(\alpha + f_2(x) + f_3(x))\overline{(\alpha + f_2(y) + f_3(y))},$$

and where

$$K_1(x, y) = \frac{1}{2\pi}\int_{\mathbb{R}} \exp\left(i(x - y)\cdot t\right)\rho_1(t)\, dt$$

is positive definite and $\widetilde{a_{jj'}}$ *are the elements of the inverse of* $\overline{A}$ *(the complex conjugate of* A*). Further,* $\aleph(p_{j'}, t)$ *are given by*

$$\exp\left(-ip_{j'}\cdot t\right)\overline{(\alpha + f_2(p_{j'}) + f_3(p_{j'}))}.$$

The optimal solution means that it minimizes the norm squares

$$\frac{1}{2\pi}\int_{\mathbb{R}} |F_1(t)|^2 \rho_1(t)^{-1} dt$$

among the functions F_1 *satisfying the above conditions.*

The point value estimation of the approximate solution $F_1^A(t)$ is given in terms of the points $\{p_j\}_{j=1}^n$ by

$$|F_1^A(t)| \leq \left(\sum_{j,j'=1}^{n} g(p_j)\, \widetilde{a_{jj'}}\, \overline{g(p_{j'})} \right)^{1/2} \cdot \left(\sum_{j,j'=1}^{n} \aleph(p_j,t)\, \widetilde{a_{jj'}}\, \overline{\alpha(p_{j'},t)} \right)^{1/2} .$$

In particular, note that for the approximate solution (100), we do not use the values $\{f_1(p_j)\}$; this means that we do not solve the algebraic equations (99). This point of view will be very interesting and important for the formula (100).

15 Singular Integral Equations

Singular integral equations arise in a great variety of problems which are useful for modeling a wide range of applied problems (e.g., in acoustics, elasticity, fluid dynamics, radiative equilibrium, and fracture mechanics—just to cite a few subareas of applied sciences). Despite a great variety of existing methods, singular integral equations are typically difficult to solve analytically, and therefore, additional knowledge about their approximate solutions is still needed. Anyway, many numerical methods have been developed already for such type of equations. Some of those are known as the method of interpolatory quadrature rules based on piecewise polynomials, Legendre polynomial method, Bernstein polynomial method, Taylor series expansion method, Galerkin method, and collocation methods. However, these methods typically require a lot of computational effort in several steps of their approximation techniques. See the original paper [7] for the detailed references and survey for the singular integral equations.

In here, we will apply our technique to the following Carleman singular integral equation over a general interval, and having very general real or complex valued functions a and b as coefficients:

$$(Ly)(t) := a(t)y(t) + \frac{b(t)}{\pi i}\text{p.v.}\int_{\mathbb{R}} \frac{y(\zeta)}{\zeta - t}\, d\zeta = g(t), \qquad t \in \mathcal{I}, \tag{101}$$

where g is given a very general function, and $\mathcal{I}$ is a general interval on the real line. As mentioned, we shall call (101) an equation of Carleman type because it was discussed by T. Carleman, who first gave an explicit solution of this equation in closed form (when considering some conditions on the coefficients).

In case of $a^2(t) - b^2(t) \neq 0$, on $\mathcal{I}$, the operator L is called a regular type operator (see [27]). An explicit solution of the Eq. (101) for a regular type operator (in convenient spaces of functions) is well known. However, when $a^2(t) - b^2(t) = 0$ (for t in the integrability domain), there are solutions if and only if g satisfies a special condition. The analysis of this case is important, e.g., for the kinked crack problem (see [16]). Accordingly, we shall exhibit here the application of our method to this equation which will give simple and natural approximate solutions for linear singular integral equations including the case where the condition of a regular type operator is violated. We will deal with a surprisingly general linear singular integral equation by our new discretization method.

Theorem 17 *We fix a positive number h and take a finite number of points* $\{t_j\}_{j=1}^n$ *of* $\mathcal{I}$ *satisfying that*

$$\{[a(t_j) + b(t_j)\,\mathrm{sgn}(\xi)]\exp(-it_j\xi)\}$$

are linearly independent in the space $L_2(-\pi/h, +\pi/h)$.

Then, the optimal solution y_h^A *of the Eq.* (101) *has the form*

$$y_h^A(t) = \frac{1}{2\pi}\int_{-\pi/h}^{\pi/h} F_h^A(\xi)e^{-it\xi}\,d\xi$$

satisfying $F_h^A \in L_2(-\pi/h, +\pi/h)$ *in the sense that* F_h^A *has the minimum norm in* $L_2(-\pi/h, +\pi/h)$ *among the functions* $F \in L_2(-\pi/h, +\pi/h)$ *satisfying, for the characteristic function* $\chi_h(t)$ *of the interval* $(-\pi/h, +\pi/h)$*:*

$$a(t)(\mathcal{F}^{-1}(F\chi_h))(t) + \frac{b(t)}{i}(-1)(\mathcal{H}\mathcal{F}^{-1}(F\chi_h))(t) = 2\pi g(t) \tag{102}$$

for all $t = t_j$. *Here, we are using the Fourier transform* $\mathcal{F}$ *given by*

$$\mathcal{F}[f](\xi) = \frac{1}{\sqrt{2\pi}}\int_{-\infty}^{\infty} f(t)e^{i\xi t}\,\mathrm{d}t,$$

and the Hilbert transform $\mathcal{H}$,

$$[\mathcal{H}y](t) = \frac{1}{\pi}p.v.\int_{-\infty}^{\infty}\frac{y(\zeta)}{t-\zeta}\,\mathrm{d}\zeta.$$

The best extremal function F_h^A *is given by*

$$F_h^A(\xi) = \sum_{j,j'=1}^{n} g(t_j)\,\widetilde{a_{jj'}}\,\overline{(a(t_{j'}) + b(t_{j'})\,sgn(\xi))}\,\exp(it_{j'}\cdot\xi).$$

Here, the matrix A formed by the elements

$$a_{jj'} = K_{hh}(t_j, t_{j'})$$

with

$$K_{hh}(t,t') = \frac{1}{2\pi}\int_{\mathbb{R}}[a(t) + b(t)\,\mathrm{sgn}(\xi)]\,\overline{[a(t') + b(t')\,sgn(\xi)]}$$
$$\chi_h(\xi)\exp(-i(t-t')\xi)d\xi$$

which is positive definite and $\widetilde{a_{jj'}}$ *are the elements of the inverse of* $\overline{A}$ *(the complex conjugate of A).*

Therefore, the optimal solution y_h^A of the Eq. (101) *is given by*

$$y_h^A(t) = \sum_{j,j'=1}^{n} g(t_j)\widetilde{a_{jj'}} \left(\overline{(a(t_{j'})} \frac{1}{\pi(t-t_{j'})} \sin \frac{\pi}{h}(t-t_{j'}) \right.$$

$$\left. + \overline{b(t_{j'})} \frac{1}{\pi i(t-t_{j'})} \left(1 - \exp\left(-\frac{\pi}{h}(t-t_{j'})i\right)\right)\right).$$

In particular, note that for the approximate solution of y_h^A, we do not use the values $(\mathcal{F}^{-1}(F\chi_h))(t_j)$. This means that we do not need to solve the algebraic equations (102). This point of view will be very interesting and important for obtaining the optimal and approximate solution of (101).

Furthermore, for the coefficients a and b, they are arbitrary functions in their nature. Note that we do not assume even their measurable property.

16 Numerical Examples Associated with Theorem 9

We shall show some numerical experiments of the proposed method for ordinary differential equations. In the following, we set $h = 1$; we seek our solution in the Paley–Wiener space $W(\pi)$ with equispaced collocation points.

Example 4 We consider an ordinary differential equation

$$y''(t) = 4, \quad -1 \le t \le 1,$$

without any boundary conditions. Equispaced points $\{t_j\}$ in the interval $[-1, 1]$ are employed as the collocation points; more precisely, $t_j = -1 + 2(j-1)/(n-1)$, $j = 1, 2, \cdots, n$ for some positive number n.

Figure 3 shows numerical results with 100 digits precision computation by multiple-precision arithmetic environment *exflib* [17], which provides arbitrarily accurate approximation of real numbers and their arithmetic on digital computers.

Here, we give a remark that *optimality* of the solution given in Theorem 9 is not L_2-norm minimality. The general solution is $y(t) = 2t^2 + at + b$, $a, b \in \mathbb{C}$, and the minimum norm solution in $L_2(-1, 1)$ is $y_0(t) = 2t^2 - 2/3$. On the other hand, numerical solutions shown in Fig. 3 depend on n, and none of them approximate the minimum norm solution $y_0(t)$. However, note that they are different only by a shift because in here we do not impose boundary conditions.

Example 5 We consider a boundary value problem

$$y''(t) = 4 \qquad (-1 < t < 1), \qquad y(-1) = 1, \quad y(1) = 3,$$

whose solution is $y(t) = 2t^2 + t$.

The numerical results with $n = 20$ are shown in Fig. 4a, and maximum errors are shown in Fig. 4b.

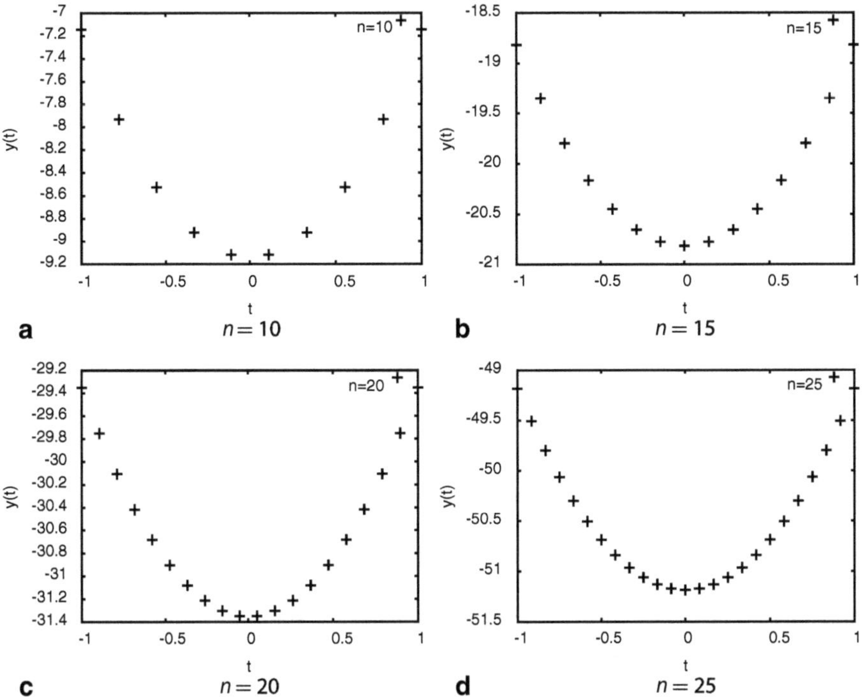

Fig. 3 Numerical results for Example 4 by 100 decimal digits precision

The former implies that the proposed method gives the classical solution under appropriate settings, and the latter shows that computational errors decay exponentially with respect to n.

Example 6 We consider an initial value problem

$$t^3 y''(t) + t y'(t) = -25t^3 \sin(5t) + 5t \cos(5t) \qquad (-1 < t \le 1),$$
$$y(-1) = \sin(5), \quad y'(-1) = 5\cos(5),$$

and we set collocation points to $t_j = -1 + 2j/(n-2)$, $j = 1, 2, \cdots, n-2$.

Numerical results shown in Fig. 5 have a good coincidence with the exact solution $y(t) = \sin(5t)$.

Example 7 Let us consider an initial value problem

$$y''(t) = g(t) \qquad (-1 < t \le 1), \quad y(-1) = y'(-1) = 0,$$

where

$$g(t) = \begin{cases} 0, & t < 0; \\ t, & t \ge 0. \end{cases}$$

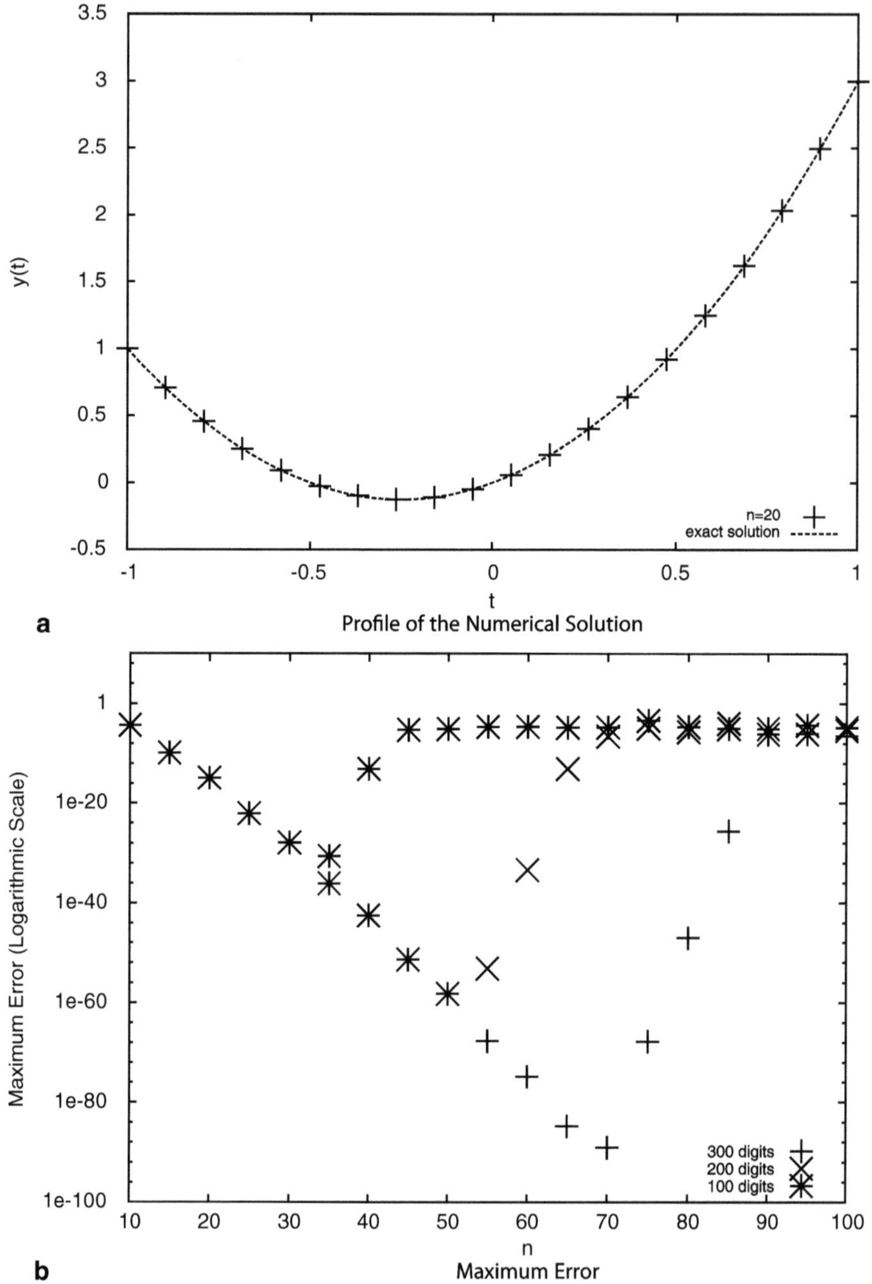

Fig. 4 Numerical results for the boundary value problem (Example 5) by multiple-precision arithmetic

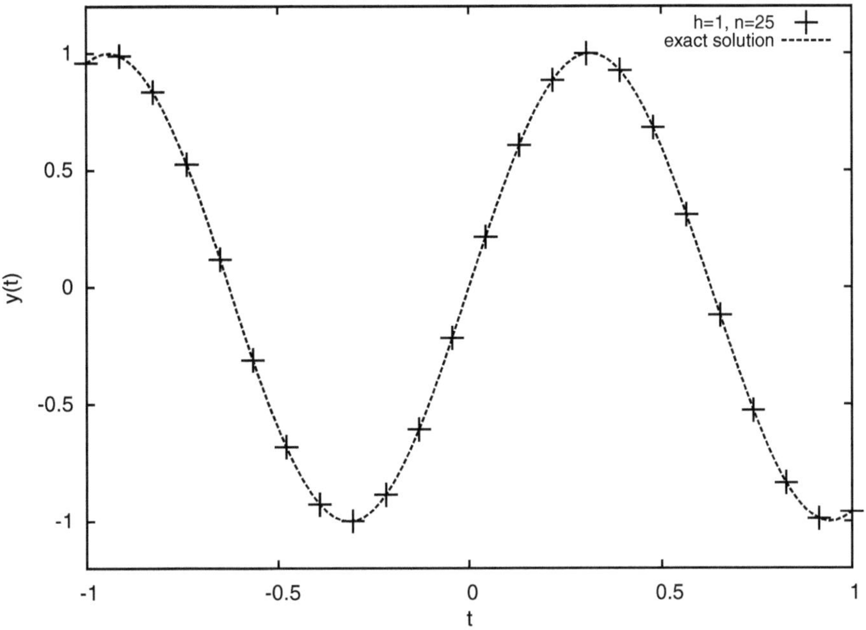

Fig. 5 Numerical results for the initial value problem (Example 7) for $n = 25$ with 100 decimal digits. Maximum error is approximately 10^{-13}

For the problem, there exists a unique solution

$$y(t) = \begin{cases} 0, & t < 0; \\ \dfrac{t^3}{6}, & t \geq 0, \end{cases}$$

while our numerical results shown in Fig. 6 do not approximate it.

We give a remark that the exact solution is not an analytic function but a C^2-class function, though the proposed method assumes that the solution belongs to the Paley–Wiener space $W(\frac{\pi}{h})$, where h is an approximation parameter. We employ $h = 1$ in Fig. 6, and more small h in Fig. 7. These results imply that the space does not have enough approximation ability if the solution is not an analytic function.

This numerical experiment seems to show the need of the application of suitable Sobolev spaces as basic approximate function spaces. Of course, we are, at this moment, interested in the classical solutions for the corresponding differential equations.

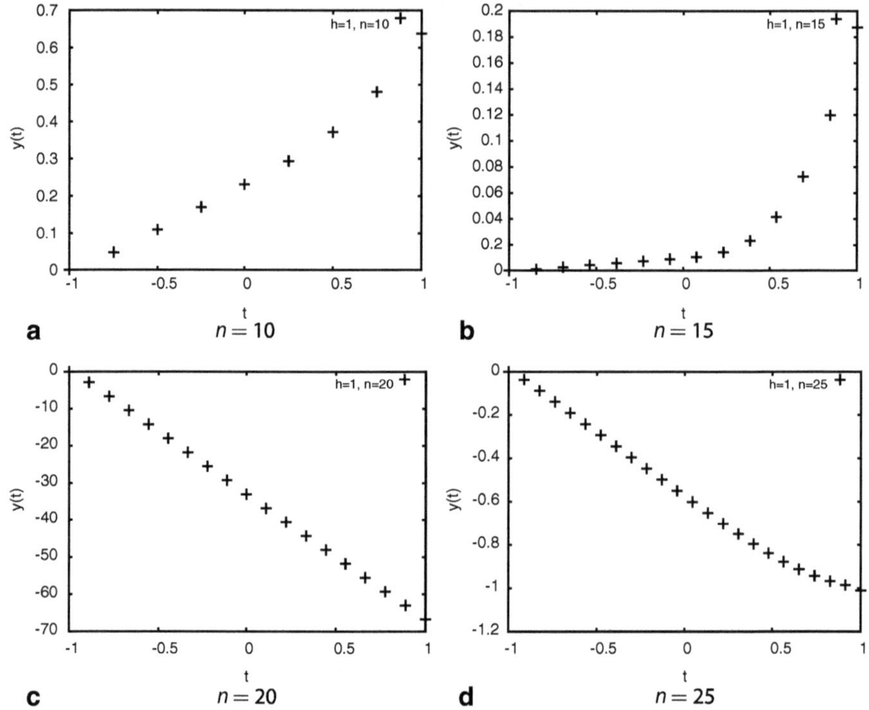

Fig. 6 Numerical results for Example 7 by 500 decimal digits precision, $h = 1$

17 Final Discussion

In our new discretization method, we will need the precision in some deep way and huge computer resources. However, these both requirements were prepared by H. Fujiwara already (e.g., recall the case of the Laplace transform). Meanwhile, our method is relying on very general linear problems which may be dealt with as shown, for arbitrary linear PDEs with very general domains. Furthermore, boundary values and initial values can also be given with wide freedom. Of course, the essential new restriction of the method is that it can be considered for only a finite number of the points of time and space for the equations.

We are looking for some optimal solutions satisfying the differential equations at the given discrete points and so, we are free from important restrictions on the domains which occur on ordinary methods. For instance, this is not the case of the *finite element method* and the *difference method*, which are depending seriously on the domains. In our case, we can consider the problems on any domains. Furthermore, we can consider the error estimates for the true solutions and our approximate solutions over the global space and time spaces depending on the numbers of the points. However, these error estimates are depending on the specific situation, case

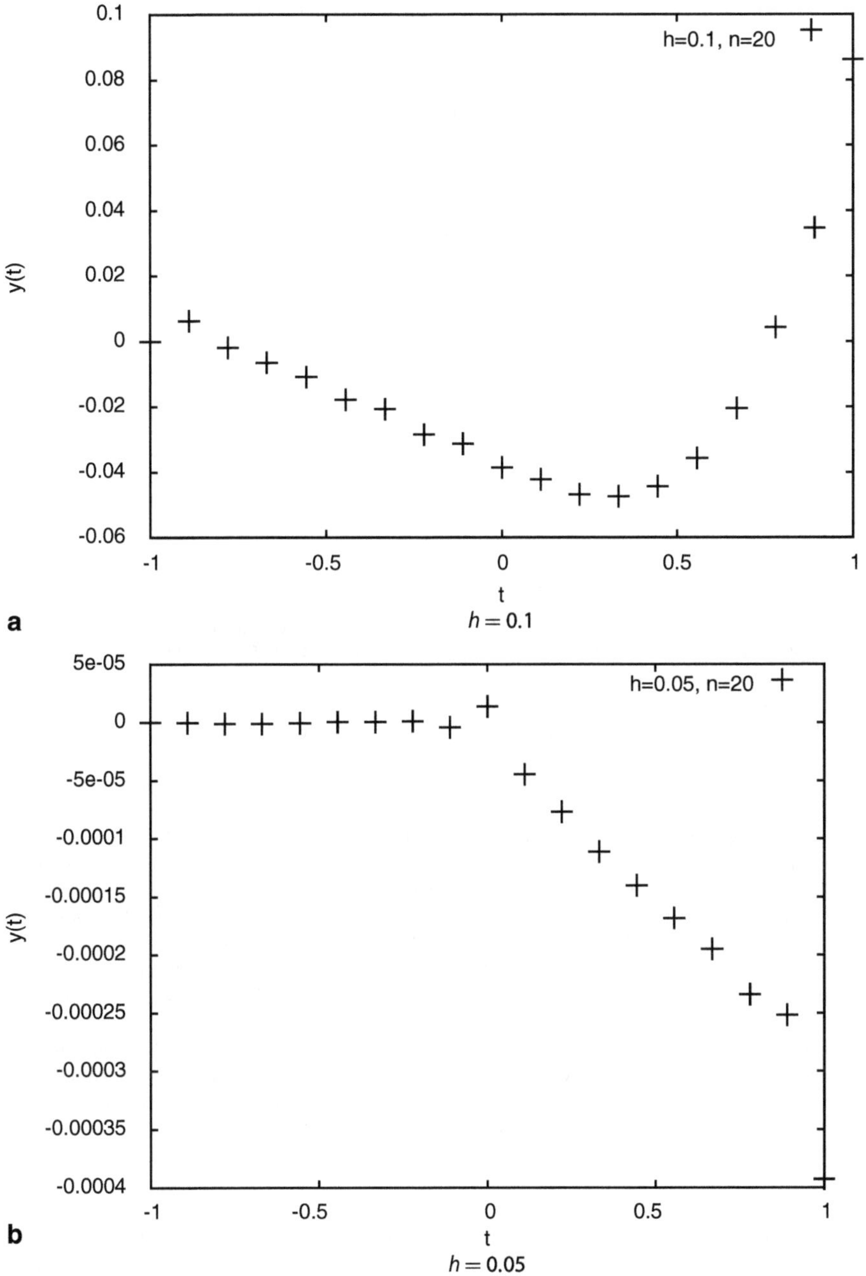

Fig. 7 Numerical results for Example 7 by 500 decimal digits precision, $n = 20$

by case. Here, the error estimates mean that our approximate solutions of the differential equations satisfying the equations for some discrete number of points and the solutions of the equations are near in some sense. For the unidimensional case, we will be easily able to image its meaning. However, on some higher-dimensional cases, its detail and exact meaning are delicate.

Anyhow, error estimates for our approximate solutions are entirely new open problems.

Finally, as a lateral mathematical consequence, we would like to point out that these problems will generate new interesting inequalities.

Acknowledgment This work was supported in part by Portuguese funds through the CIDMA–Center for Research and Development in Mathematics and Applications, and the Portuguese Foundation for Science and Technology (FCT), within project PEst–OE/MAT/UI4106/2014. The second author is supported by Grant-in-Aid for Young Scientists (B) (No.23740075). The fourth author is supported in part by the Grant-in-Aid for the Scientific Research (C)(2)(No. 24540113).

References

1. Amano, K., Saitoh, S., Yamamoto, M.: Error estimates of the real inversion formulas of the Laplace transform. Integral Transform. Spec. Funct. **10**(3–4), 165–178 (2000)
2. Asaduzzaman, M., Saitoh, S.: Inverses of a family of matrices and generalizations of Pythagorean theorem. Panam. Math. J. **13**(4), 45–53 (2003)
3. Benedetto, J.: The Laplace transform of generalized functions. Can. J. Math. **18**, 357–374 (1966)
4. Boas, R.P.: Inversion of a generalized Laplace integral. Proc. Nat. Acad. Sci. U. S. A. **28**, 21–24 (1942)
5. Collar, A.R.: On the reciprocation of certain matrices. Proc. R. Soc. Edinb. **59**, 195–206 (1939)
6. Castro, L.P., Saitoh, S.: Fractional functions and their representations. Complex Anal. Oper. Theory **7**(4), 1049–1063. doi:10.1007/s11785-011-0154-1
7. Castro, L.P., Saitoh, S.: Optimal and approximate solutions of singular integral equations by means of reproducing kernels. Complex Anal. Oper. Theory **7**(6), 1839–1851 (2013)
8. Castro, L.P., Rojas, E.M., Saitoh, S.: Inversion from different kinds of information and real inversion formulas of the Laplace transform from a finite number of data. Math. Eng. Sci. Aerosp. MESA **1**(2), 181–190 (2010)
9. Castro, L.P., Fujiwara, H., Saitoh, S., Sawano, Y., Yamada, A., Yamada, M.: Fundamental error estimates inequalities for the Tikhonov regularization using reproducing kernels. In: Bandle, C., et al. (eds.) Inequalities and Applications 2010. International Series of Numerical Mathematics, vol. 161, pp. 87–101. Springer, Basel (2012)
10. Castro, L.P., Saitoh, S., Sawano, Y., Silva, A.S.: Discrete linear differential equations. Analysis. **32**, 181–198 (2012)
11. L.H., Son, W., Tutschke, edts, *Interactions between Real and Complex Analysis*, Science and Technics Publishing House, Ministry for Science and Technology of Vietnam, Ha Noi, 185–223 (2012)
12. Castro, L.P., Saitoh, S., Sawano, Y., Tuan, N.M.: Approximate solutions of arbitrary linear ordinary differential equations (manuscript)
13. Castro, L.P., Saitoh, S., Sawano, Y., Yamada, A.: Real inversion of the Laplace transform by a Moore–Penrose generalized inverse construction (manuscript)
14. Cohen, A.M.: Numerical Methods for Laplace Transform Inversion. Numerical Methods and Algorithms, vol. 5. Springer, New York (2007)

15. Choi, M.D.: Tricks or treats with the Hilbert matrix. Am. Math. Mon. **90**, 301–312 (1983)
16. Cotterell, B., Rice, J.R.: Slightly curved or kinked cracks. Int. J. Fract. **16**, 155–169 (1980)
17. Fujiwara, H.: Applications of reproducing kernel spaces to real inversions of the Laplace transform. RIMS Koukyuuroku **1618**, 188–209 (2008)
18. Fujiwara, H.: Numerical real inversion of the Laplace transform by reproducing kernel and multiple-precision arithmetic. In: Progress in Analysis and its Applications. Proceedings of the 7th International ISAAC Congress, pp. 289–295. World Scientific, Singapore (2010)
19. Fujiwara, H., Matsuura, T., Saitoh, S., Sawano, Y.: Numerical real inversion of the Laplace transform by using a high-accuracy numerical method. In: Further Progress in Analysis, pp. 574–583. World Scientific, Hackensack (2009)
20. González, B.J., Negrin, E.R.: A distributional inversion formula for a generalization of the Stieltjes and Poisson transforms. Integral Transform. Spec. Funct. **20**(12), 897–903 (2009)
21. Gradshteyn, I.S., Ryzhik, I.M.: Table of Integrals Series, and Products. Academic, New York (1980)
22. Higgins, J.R.: Sampling Theory in Fourier and Signal Analysis: Foundations. Clarendon, Oxford (1996)
23. Indratno, S.W., Ramm, A.G.: Inversion of the Laplace transform from the real axis using an adaptive iterative method. Int. J. Math. Math. Sci. **38**, Art. ID 898195 (2009). doi:10.1155/2009/898195
24. Kryzhniy, V.V.: Regularized inversion of integral transformations of Mellin convolution type. Inverse Probl. **19**, 1227–1240 (2003)
25. Kryzhniy, V.V.: Numerical inversion of the Laplace transform: Analysis via regularized analytic continuation. Inverse Probl. **22**, 579–597 (2006)
26. Manandhar, R.P., Debnath, L.: Representation of functions as the Post–Widder inversion operator of generalized functions. Int. J. Math. Math. Sci. **7**(2), 371–396 (1984)
27. Muskhelishvili, N.I.: Singular Integral Equations. Noordhoff, Groningen (1972)
28. Natterer, F.: The Mathematics of Computerized Tomography. Classics in Applied Mathematics, vol. 32. Society for Industrial and Applied Mathematics (SIAM), Philadelphia (2001)
29. Peng, J., Chung, S.-K.: Laplace transforms and renerators of semigroups of operators. Proc. Am. Math. Soc. **126**(8), 2407–2416 (1998)
30. Phillips, R.S.: An inversion formula for Laplace transforms and semi-groups of linear operators. Ann. Math. **59**, 325–356 (1954)
31. Post, E.L.: Generalized differentiation. Trans. Am. Math. Soc. **32**(4), 723–781 (1930)
32. Ramm, A.G.: Multidimensional Inverse Scattering Problems. Monographs and Surveys in Pure and Applied Mathematics, vol. 51. Longman Scientific and Technical, Harlow (1992)
33. Rassias, Th.M., Saitoh, S.: The Pythagorean theorem and linear mappings. Panam. Math. J. **12**, 1–10 (2002)
34. Saitoh, S.: Hilbert spaces induced by Hilbert space valued functions. Proc. Am. Math. Soc. **89**, 74–78 (1983)
35. Saitoh, S.: Integral Transforms, Reproducing Kernels and their Applications. Pitman Research Notes in Mathematics Series, vol. 369. Addison-Wesley/Longman, Harlow/Boston (1997)
36. Saitoh, S.: Approximate real inversion formulas of the Laplace transform. Far East J. Math. Sci. **11**, 53–64 (2003)
37. Saitoh, S.: Inverses of a family of bounded linear operators, generalized Pythagorean theorems and reproducing kernels. In: Lavrent'ev, M.M., et al. (eds.) Ill-posed and Non-classical Problems of Mathematical Physics and Analysis. Inverse Ill-posed Problems Series, vol. 41, pp. 125–141. VSP, Utrecht (2003)
38. Saitoh, S.: Theory of reproducing kernels: Applications to approximate solutions of bounded linear operator functions on Hilbert spaces. Am. Math. Soc. Transl. Ser. 2. **230** (2010). (Amer. Math. Soc., Providence, RI)
39. Saitoh, S., Hayashi, N., Yamamoto, M. (eds.): Analytic Extension Formulas and their Applications. Kluwer Academic, Dordrecht (2001)

40. Saitoh, S., Tuan, V.K., Yamamoto, M.: Conditional stability of a real inverse formula for the Laplace transform. Z. Anal. Anwend. **20**(1), 193–202 (2001)
41. Sawano, Y., Fujiwara, H., Saitoh, S.: Real inversion formulas of the Laplace transform on weighted function spaces. Complex Anal. Oper. Theory. **2**, 511–521 (2008)
42. Stenger, F.: Numerical Methods Based on Sinc and Analytic Functions. Springer Series in Computational Mathematics, vol. 20. Springer, New York (1993)
43. Tuan, V.K., Hong, N.T.: Interpolation Formula in Hardy space, Integral Transforms and Special Functions, **24**(8), 664–671, (2013), http://dx.doi.org/10.1080/10652469.2012.749874
44. Tuan, V.K., Duc, D.T.: A new real inversion formula of the Laplace transform and its convergence rate. Fract. Calc. Appl. Anal. **5**(4), 387–394 (2002)
45. Widder, D.V.: The Laplace Transform. Princeton Mathematical Series, vol. 6. Princeton University Press, Princeton (1941)
46. Yakubovich, S.: A real inversion formula for the bilateral Laplace transform (Russian). Izv. Nats. Akad. Nauk Armenii Mat. **40**(3), 67–79 (2005). (English translation in J. Contemp. Math. Anal. **40**(3), 66–77 (2005))

Variational Inequality Models Arising in the Study of Viscoelastic Materials

O. Chau, D. Goeleven and R. Oujja

Abstract In this paper, we present a synthesis of different results obtained recently in the papers (Chau and Awbi, Appl Anal 83:635–648, 2004; Addi et al. Discret Contin Dyn Syst 31:1039–1051, 2011; Adly et al. Numer Algebra Control Optim 2:89–101, 2012; Adly and chau, to appear in Mathematical Programming; Chau et al., Int J Appl Math Mech, 2012). It concerns the study of contact problems for viscoelastic materials with possible thermal effects. We first describe a general thermo-viscoelastic model involving a thermo-viscoelastic Kelvin–Voigt constitutive law, a temperature field governed by the heat equation and a subdifferential surface contact condition. Then, we study a model which describes the frictional contact between a short memory thermo-viscoelastic body and a given rigid foundation. The free boundary contact problem for a long memory viscoelastic material is also considered. Finally, we provide numerical simulations for different fundamental examples of thermal contact problems.

Keywords Variational inequality · Evolution inequality · Fixed point · Weak solutions · Dynamic process · Friction · Thermo-viscoelasticity · Subdifferential contact condition · Nonclamped condition · Numerical analysis · Numerical simulations

1 Thermo-Viscoelastic Models

Because of their considerable impact in everyday life and their multiple open problems, contact mechanics still remain a rich and fascinating domain of challenge. The literature devoted to various aspects of the subject is considerable, it concerns the

O. Chau (✉) · D. Goeleven · R. Oujja
University of La Réunion, PIMENT EA4518, Saint-Denis Messag cedex 9, 97715 La Réunion, France
e-mail: oanh.chau@univ-reunion.fr

D. Goeleven
e-mail: daniel.goeleven@univ-reunion.fr

R. Oujja
e-mail: rachid.oujja@univ-reunion.fr

T. M. Rassias, P. M. Pardalos (eds.), *Mathematics Without Boundaries*,
DOI 10.1007/978-1-4939-1106-6_4,

modelling, the mathematical analysis as well as the numerical approximation of the related problems.

For example, many food materials used in process engineering are viscoelastic [18] and consequently, mathematical models can be very helpful in understanding various problems related to the product development, packing, transport, shelf life testing, thermal effects, and heat transfer. It is thus important to study mathematical models that can be used to describe the dynamical behavior of a given viscoelastic material subjected to various highly nonlinear and even nonsmooth phenomena like contact, friction, and thermal effects.

A panoply of tools and approaches are needed to face the multiple difficulties of the problems. An abstract functional nonlinear analysis could be found in [5, 7, 14, 20, 26]. An early attempt at the study of contact problems for elastic and viscoelastic materials within the applied mathematical analysis framework was introduced in the pioneering reference works [12, 13, 19, 22]. For the error estimates analysis and numerical approximation, the reader can refer to [11, 15, 17]. Further extensions to nonconvex contact conditions with nonmonotone and possible multi-valued constitutive laws led to the active domain of nonsmooth mechanics within the framework of the so called hemivariational inequalities, for a mathematical as well as a mechanical treatment, we refer to [16, 23]. Finally, the mathematical, mechanical, and numerical state of the art can be found in the proceedings [25].

The basic mechanical contact problem is the following.

We consider a deformable body which occupies a bounded domain $\Omega \subset \mathbb{R}^d$ ($d = 1$ or 2 or 3), with a Lipschitz boundary Γ and let ν denote the unit outer normal on Γ. The body is acted upon by given forces and tractions. As a result, its mechanical state evolves over the time interval $[0, T]$, $T > 0$. We assume that the boundary Γ of Ω is partitioned into three disjoint measurable parts Γ_1, Γ_2, and Γ_3. The body is clamped on $\Gamma_1 \times (0, T)$. Here, we are interested in various natures of Γ_1. In case of classical fixed condition, the property $\text{meas}(\Gamma_1) > 0$ holds (see [1, 3, 8, 10]), which allows to use the well known Korn's inequality. In case of a free boundary, we consider $\text{meas}(\Gamma_1) = 0$, where Γ_1 is reduced to one point or eventually may be an empty set. This last case presents a source of additional difficulties and new approach is necessary (see [2, 10]). We suppose also that surface tractions of density f_2 act on $\Gamma_2 \times (0, T)$. The solid is in frictional contact with a rigid obstacle on $\Gamma_3 \times (0, T)$, where various contact conditions may be considered. Moreover, a volume force of density f_0 acts on the body in $\Omega \times (0, T)$ (see figure).

In this paper, $u = (u_i)$ denotes the displacement field, $\sigma = \big(\sigma_{ij}\big)$ is the stress field and $\varepsilon(u) = \big(\varepsilon_{ij}(u)\big)$ denotes the linearized strain tensor.

In what follows, for simplification, we do not indicate explicitly the dependence of functions with respect to $x \in \Omega \cup \Gamma$ and $t \in [0, T]$. Everywhere in the sequel, the indexes i and j run from 1 to d, summation over repeated indices is implied and the index that follows a comma represents the partial derivative with respect to the corresponding component of the independent variable. Moreover the dot above represents the time derivative, i.e.,

$$\dot{u} = \frac{du}{dt}, \qquad \ddot{u} = \frac{d^2u}{dt^2}.$$

Let us denote the mass density by $\rho : \Omega \longrightarrow \mathbb{R}_+$. The dynamical evolution of the body is described by the following equation of motion,

$$\rho\, \ddot{u} = \text{Div}\ \sigma + f_0 \quad \text{in} \quad \Omega \times (0, T).$$

Here, $\ddot{u}$ represents the acceleration of the dynamical process.

1.1 Thermo-Viscoelastic Constitutive Law

For viscoelastic materials, the body follows a constitutive law of Kelvin–Voigt's type in the form

$$\sigma(t) = \mathcal{A}\varepsilon(\dot{u}(t)) + \mathcal{G}\varepsilon(u(t)),$$

where $\mathcal{A}$ et $\mathcal{G}$ are generally non linear functions, $\mathcal{A}$ represents the viscosity operator, and $\mathcal{G}$ the elasticity operator.

In the case of linear Kelvin–Voigt constitutive law, we have

$$\sigma_{ij} = a_{ijkl}\, \varepsilon_{kl}(\dot{u}) + g_{ijkl}\, \varepsilon_{kl}(u),$$

where $\mathcal{A} = (a_{ijkl})$ is the viscosity tensor and $\mathcal{G} = (g_{ijkl})$ the elasticity tensor.

This last law is qualified as of short memory, for it is instantaneous and takes place at each time t.

The long memory viscoelastic constitutive law is defined by

$$\sigma(t) = \mathcal{A}\varepsilon(\dot{u}(t)) + \mathcal{G}\varepsilon(u(t)) + \int_0^t \mathcal{B}(t-s)\, \varepsilon(u(s))\, ds \quad \text{in} \quad \Omega.$$

Here, $\mathcal{B}$ is the so called tensor of relaxation which defines the long memory behavior of the material. The above convolution term represents a kind of sum of all the elasticity of the body through the past, from the initial time to the present time. Of course, as a particular case, when $\mathcal{B} \equiv 0$, we recover the usual viscoelasticity of short memory.

In order to complete the last law with some additional thermal effects, we consider the following Kelving–Voigt's long memory thermo-viscoelastic constitutive law

$$\sigma(t) = \mathcal{A}\varepsilon(\dot{u}(t)) + \mathcal{G}\varepsilon(u(t)) + \int_0^t \mathcal{B}(t-s)\, \varepsilon(u(s))\, ds - \theta(t)\, C_e \quad \text{in} \quad \Omega,$$

where $C_e := (c_{ij})$ represents the thermal expansion tensor and θ is the temperature field.

1.2 The Temperature Field

We suppose that the evolution of the temperature field θ is governed by the heat equation (see [7, 8]) obtained from the conservation of energy and defined by the following differential equation

$$\dot{\theta} - \mathrm{div}(K\,\nabla\theta) = r(\dot{u}(t)) + q(t),$$

where $K = (k_{ij})$ represents the thermal conductivity tensor, $\mathrm{div}(K\,\nabla\theta) = (k_{ij}\theta_{,j})_{,i}$, $q(t)$ the density of volume heat sources, and $r(\dot{u}(t))$ a nonlinear function of the velocity. Usually, the following linear function is used

$$r(\dot{u}(t)) = -c_{ij}\dot{u}_{i,j}(t).$$

The associated temperature boundary condition on Γ_3 is described by

$$k_{ij}\,\theta_{,i}\,n_j = -k_e\,(\theta - \theta_R) \quad \text{on } \Gamma_3 \times (0,T),$$

where θ_R is the temperature of the foundation and k_e is the heat exchange coefficient between the body and the obstacle.

1.3 Subdifferential Contact Condition

Let us here describe the surface contact condition on Γ_3. We model the frictional contact with a general subdifferential boundary condition of the form

$$u \in U, \quad \varphi(v) - \varphi(\dot{u}) \geq -\sigma\nu\,(v - \dot{u}) \qquad \forall v \in U. \tag{1}$$

In this condition, U represents the set of contact admissible test functions, $\sigma\nu$ denotes the Cauchy stress vector on the contact boundary, and $\varphi : \Gamma_3 \times \mathbb{R}^d \longrightarrow \mathbb{R}$ is a given convex function. The inequality in (1) holds almost everywhere on the contact surface. Various situations may be modeled by such a condition. Examples and detailed explanations of inequality problems in contact mechanics which lead to boundary conditions of this form can be found in [9, 23].

Here, we present some examples of contact and dry friction laws which lead to such subdifferential inequality.

Example 1 Bilateral contact with Tresca's friction law. This contact condition can be found in [13, 23]. It is written in the form of the following boundary condition:

$$\begin{cases} u_\nu = 0, \quad |\sigma_\tau| \leq g, \\ |\sigma_\tau| < g \implies \dot{u}_\tau = 0, \\ |\sigma_\tau| = g \implies \dot{u}_\tau = -\lambda\sigma_\tau, \ \ \lambda \geq 0 \end{cases} \qquad \text{on } \Gamma_3 \times (0,T). \tag{2}$$

Here, $g \geq 0$ represents the friction bound, i.e., the magnitude of the limiting friction traction at which slip begins. The contact is assumed to be bilateral, i.e., there is no loss of contact during the process.

The set of admissible test functions U consists of those elements of H_1 whose normal component vanishes on Γ_3.

Moreover, it is straightforward to show that if $\{u, \sigma\}$ is a pair of regular functions satisfying (1) then

$$\sigma\, \nu(v - \dot{u}) \geq g|\dot{u}_\tau| - g|v_\tau| \qquad \forall v \in U,$$

a.e. on $\Gamma_3 \times (0, T)$. We get the following contact functional

$$\varphi(v) = g|v_\tau|.$$

Example 2 Viscoelastic contact with Tresca's friction law. We consider the contact problem with the boundary conditions

$$\begin{cases} -\sigma_\nu = k|\dot{u}_\nu|^{r-1}\dot{u}_\nu, \ |\sigma_\tau| \leq g, \\ |\sigma_\tau| < g \implies \dot{u}_\tau = 0, \text{ on } \Gamma_3 \times (0, T). \\ |\sigma_\tau| = g \implies \dot{u}_\tau = -\lambda\sigma_\tau, \ \lambda \geq 0 \end{cases} \tag{3}$$

Here, $g, k \geq 0$ and the normal contact stress depends on a power of the normal speed (this condition may describe the motion of a body, say a wheel, on a fine granular material, say the sand on a beach. We have $U = H_1$, $0 < r \leq 1$, and

$$\varphi(v) = \frac{k}{r+1}|v_\nu|^{r+1} + g|v_\tau|.$$

Example 3 Viscoelastic contact with friction. Here, the body is moving on sand or a granular material and the normal stress is proportional to a power of the normal speed, while the tangential shear is proportional to a power of the tangential speed. We choose the following boundary conditions:

$$-\sigma_\nu = k|\dot{u}_\nu|^{r-1}\dot{u}_\nu, \ \sigma_\tau = -\mu|\dot{u}_\tau|^{p-1}\dot{u}_\tau \qquad \text{on } \Gamma_3 \times (0, T). \tag{4}$$

Here, $\mu \in L^\infty(\Gamma_3)$ and $k \in L^\infty(\Gamma_3)$ are positive functions and $0 < p, r \leq 1$. We choose $U = H_1$, $V = \{\ v \in H_1 \mid v = 0 \text{ on } \Gamma_1\ \}$ and

$$\varphi(v) = \frac{k}{r+1}|v_\nu|^{r+1} + \frac{\mu}{p+1}|v_\tau|^{p+1}.$$

Remark 1 In the examples above, the normal pressure as well as the tangential stress are related to the powers of the normal and tangential speeds. This is dictated by the structure of the functional φ, which depends only on the surface velocity.

1.4 Notation and Functional Spaces

In this short section, we present the notations we shall use and some preliminary materials for functional spaces. For further details, we refer the reader to [13].

We denote by S_d the space of second order symmetric tensors on $\mathbb{R}^d$ $(d = 2, 3)$, while " $\cdot$ " and $|\cdot|$ will represent the inner product and the Euclidean norm on S_d and $\mathbb{R}^d$. Let $\Omega \subset \mathbb{R}^d$ be a bounded domain with a Lipschitz boundary Γ and let ν denote the unit outer normal on Γ. We also use the following notation:

$$H = \left\{ u = (u_i) \mid u_i \in L^2(\Omega) \right\}, \qquad \mathcal{H} = \left\{ \sigma = \left(\sigma_{ij}\right) \mid \sigma_{ij} = \sigma_{ji} \in L^2(\Omega) \right\},$$
$$H_1 = \{ u \in H \mid \varepsilon(u) \in \mathcal{H} \}, \qquad \mathcal{H}_1 = \{ \sigma \in \mathcal{H} \mid \text{Div } \sigma \in H \}.$$

Here, $\varepsilon : H_1 \longrightarrow \mathcal{H}$ and $\text{Div} : \mathcal{H}_1 \longrightarrow H$ are the deformation and the divergence operators, respectively, defined by:

$$\varepsilon(u) = \left(\varepsilon_{ij}(u)\right), \quad \varepsilon_{ij}(u) = \frac{1}{2}\left(u_{i,j} + u_{j,i}\right), \quad \text{Div } \sigma = \left(\sigma_{ij,j}\right).$$

The spaces H, $\mathcal{H}$, H_1, and $\mathcal{H}_1$ are real Hilbert spaces endowed with the canonical inner products given by:

$$\langle u, v\rangle_H = \int_\Omega u_i v_i \, dx, \qquad \langle \sigma, \tau\rangle_{\mathcal{H}} = \int_\Omega \sigma_{ij}\tau_{ij}\, dx,$$
$$\langle u, v\rangle_{H_1} = \langle u, v\rangle_H + \langle \varepsilon(u), \varepsilon(v)\rangle_{\mathcal{H}}, \qquad \langle \sigma, \tau\rangle_{\mathcal{H}_1} = \langle \sigma, \tau\rangle_{\mathcal{H}} + \langle \text{Div } \sigma, \text{Div } \tau\rangle_H.$$

The associated norms on the spaces H, $\mathcal{H}$, H_1, and $\mathcal{H}_1$ are denoted by $|\cdot|_H$, $|\cdot|_{\mathcal{H}}$, $|\cdot|_{H_1}$, and $|\cdot|_{\mathcal{H}_1}$, respectively.

Let $H_\Gamma = H^{\frac{1}{2}}(\Gamma)^d$ and let $\gamma : H_1 \longrightarrow H_\Gamma$ be the trace map. For every element $u \in H_1$, we also use the notation u to denote the trace γu of u on Γ and we denote by u_ν and u_τ the normal and the tangential components of u on Γ given by:

$$u_\nu = u \cdot \nu, \qquad u_\tau = u - u_\nu \nu. \tag{5}$$

Let H'_Γ be the dual of H_Γ and let $< \cdot, \cdot >$ denote the duality pairing between H'_Γ and H_Γ. For every $\sigma \in \mathcal{H}_1$, $\sigma\nu$ can be defined as the element in H'_Γ which satisfies:

$$\langle \sigma\nu, \gamma u\rangle = \langle \sigma, \varepsilon(u)\rangle_{\mathcal{H}} + \langle \text{Div } \sigma, u\rangle_H \qquad \forall u \in H_1. \tag{6}$$

Let also σ_ν and σ_τ denote the normal and tangential traces of σ, respectively. If σ is a smooth function, e.g., $\sigma \in C^1$, then

$$\langle \sigma\nu, \gamma u\rangle = \int_\Gamma \sigma\nu \cdot u\, da \qquad \forall u \in H_1 \tag{7}$$

where da is the surface measure element and

$$\sigma_\nu = (\sigma\nu) \cdot \nu, \qquad \sigma_\tau = \sigma\nu - \sigma_\nu \nu. \tag{8}$$

Finally, we recall that $C([0,T];X)$ is the space of continuous functions from $[0,T]$ to X; while $C^m([0,T];X)$ $(m \in \mathbb{N}^*)$ is the set of m times differentiable functions. Then, $\mathcal{D}(\Omega)$ denotes the set of infinitely differentiable real functions with compact support in Ω. We will also use the Lesbesgue spaces $L^p(0,T;X)$; and the Sobolev spaces:

$$W^{m,p}(0,T;X), \quad H_0^m(\Omega) := \{w \in W^{m,2}(\Omega),\ w = 0 \text{ on } \Gamma\},$$

where $m \geq 1$ and $1 \leq p \leq +\infty$.

2 Dynamic Contact Problems with Clamped Condition

In this section, we present the results obtained in [1, 3, 8, 10] with the usual fixed condition. In [8], O. Chau and B. Awbi analyze a problem which describes the frictional contact between a short memory thermo-viscoelastic body and a rigid foundation. The process is assumed to be quasistatic and the contact is modeled by a normal damped response condition with friction law. Moreover, heat exchange condition has been taken into account on the contact surface. The mechanical model is described as a coupled system of a variational elliptic equality for the displacements and a differential heat equation for the temperature. Then, the authors present a variational formulation of the problem and establish the existence and uniqueness of weak solution in using general results on evolution equations with monotone operators and fixed point arguments. In [1], the constitutive law has been extended to a long memory viscoelastic type and the contact has been modeled by a general subdifferential condition on the velocity. K. Addi et al. derived weak formulations for the models and established existence and uniqueness results. The proofs are based on evolution variational inequalities, in the framework of monotone operators and fixed point methods. The quasistatic evolution in these two latter works have been then extended to dynamic process for long memory thermo-viscoelastic materials in [3]. Finally, O. Chau et al. complete the study by numerical approximations in [10], where analysis of error order estimate and various simulations have been provided. The dynamic mechanical problem for long memory thermo-viscoelastic materials subjected to subdifferential contact condition and to clamped condition is then formulated as follows.

Problem Q Find a displacement field $u : \Omega \times [0,T] \longrightarrow \mathbb{R}^d$, a stress field $\sigma : \Omega \times [0,T] \longrightarrow S_d$, and a temperature field $\theta : \Omega \times [0,T] \longrightarrow \mathbb{R}_+$ such that for a.e. $t \in (0,T)$:

$$\sigma(t) = \mathcal{A}\varepsilon(\dot{u}(t)) + \mathcal{G}\varepsilon(u(t)) + \int_0^t \mathcal{B}(t-s)\,\varepsilon(u(s))\,ds - \theta(t)\,C_e \quad \text{in} \quad \Omega, \tag{9}$$

$$\ddot{u}(t) = \operatorname{Div}\sigma(t) + f_0(t) \quad \text{in} \quad \Omega, \tag{10}$$

$$u(t) = 0 \quad \text{on} \quad \Gamma_1, \tag{11}$$

$$\sigma(t)\nu = f_2(t) \quad \text{on} \quad \Gamma_2, \tag{12}$$

$$u(t) \in U, \quad \varphi(w) - \varphi(\dot{u}(t)) \geq -\sigma(t)\nu \cdot (w - \dot{u}(t)) \quad \forall w \in U \quad \text{on } \Gamma_3, \tag{13}$$

$$\dot{\theta}(t) - \operatorname{div}(K_c \, \nabla\theta(t)) = -c_{ij} \, \frac{\partial \dot{u}_i}{\partial x_j}(t) + q(t) \quad \text{on} \quad \Omega, \tag{14}$$

$$-k_{ij} \, \frac{\partial \theta}{\partial x_j}(t) \, n_i = k_e \, (\theta(t) - \theta_R) \quad \text{on} \quad \Gamma_3, \tag{15}$$

$$\theta(t) = 0 \quad \text{on} \quad \Gamma_1 \cup \Gamma_2, \tag{16}$$

$$\theta(0) = \theta_0 \quad \text{in} \quad \Omega, \tag{17}$$

$$u(0) = u_0, \quad \dot{u}(0) = v_0 \quad \text{in} \quad \Omega. \tag{18}$$

Here, we suppose that $\text{meas}(\Gamma_1) > 0$ and the mass density $\rho \equiv 1$. We also suppose that the set of contact admissible test functions verifies

$$\mathcal{D}(\Omega)^d \subset U \subset H_1.$$

Finally, u_0, v_0, and θ_0 represent the initial displacement, velocity, and temperature, respectively.

To obtain the variational formulation of the mechanical problems (41)–(49) we need additional notations. Thus, let V denote the closed subspace of H_1 defined by

$$\mathcal{D}(\Omega)^d \subset V = \{ v \in H_1 \mid v = 0 \quad \text{on} \quad \Gamma_1 \} \cap U.$$

We set

$$E = \left\{ \eta \in H^1(\Omega), \, \eta = 0 \quad \text{on} \quad \Gamma_1 \cup \Gamma_2 \right\}, \quad F = L^2(\Omega).$$

Since meas $\Gamma_1 > 0$, Korn's inequality holds, i.e., there exists $C_K > 0$, which depends only on Ω and Γ_1 such that

$$\|\varepsilon(v)\|_{\mathcal{H}} \geq C_K \, \|v\|_{H_1} \qquad \forall v \in V.$$

A proof of Korn's inequality may be found in [21, p. 79].

On V we consider the inner product given by

$$(u, v)_V = (\varepsilon(u), \varepsilon(v))_{\mathcal{H}} \qquad \forall \, u, \, v \, \in V,$$

and let $\| \cdot \|_V$ be the associated norm, i.e.,

$$\|v\|_V = \|\varepsilon(v)\|_{\mathcal{H}} \qquad \forall \, v \in V.$$

It follows that $\| \cdot \|_{H_1}$ and $\| \cdot \|_V$ are equivalent norms on V and therefore ,$(V, \| \cdot \|_V)$ is a real Hilbert space. Moreover, by the Sobolev's trace theorem, we have a constant $C_0 > 0$ depending only on Ω, Γ_1, and Γ_3 such that

$$\|v\|_{L^2(\Gamma_3)} \leq C_0 \, \|v\|_V \qquad \forall \, v \in V.$$

In the study of the mechanical problem (41)–(49), we assume the following conditions (see e.g., [3, 20]).

The viscosity operator $\mathcal{A} : \Omega \times S_d \longrightarrow S_d$ satisfies:

$$\begin{cases} (a) & \text{there exists } L_{\mathcal{A}} > 0 \text{ such that} \\ & |\mathcal{A}(x,\varepsilon_1) - \mathcal{A}(x,\varepsilon_2)| \leq L_{\mathcal{A}} |\varepsilon_1 - \varepsilon_2| \\ & \forall \varepsilon_1,\, \varepsilon_2 \in S_d,\ a.e.\ x\ \in \Omega\,, \\ (b) & \text{there exists } m_{\mathcal{A}} > 0 \text{ such that} \\ & (\mathcal{A}(x,\varepsilon_1) - \mathcal{A}(x,\varepsilon_2)) \cdot (\varepsilon_1 - \varepsilon_2) \geq m_{\mathcal{A}}\, |\varepsilon_1 - \varepsilon_2|^2 \\ & \forall \varepsilon_1,\, \varepsilon_2 \in S_d,\ a.e.\ x \in \Omega\,, \\ (c) & x \longmapsto \mathcal{A}(x,\varepsilon) \text{ is Lebesgue measurable on } \Omega, \forall \varepsilon \in S_d\,, \\ (d) & \text{the mapping } x \longmapsto \mathcal{A}(x,0) \in \mathcal{H}. \end{cases} \tag{19}$$

The elasticity operator $\mathcal{G} : \Omega \times S_d \longrightarrow S_d$ satisfies:

$$\begin{cases} (a) & \text{there exists } L_{\mathcal{G}} > 0 \text{ such that} \\ & |\mathcal{G}(x,\varepsilon_1) - \mathcal{G}(x,\varepsilon_2)| \leq L_{\mathcal{G}} |\varepsilon_1 - \varepsilon_2| \\ & \forall \varepsilon_1,\, \varepsilon_2 \in S_d,\ a.e.\ x \in \Omega\,, \\ (b) & x \longmapsto \mathcal{G}(x,\varepsilon) \text{ is Lebesgue measurable on } \Omega, \forall \varepsilon \in S_d\,, \\ (c) & \text{the mapping } x \longmapsto \mathcal{G}(x,0) \in \mathcal{H}. \end{cases} \tag{20}$$

The relaxation tensor $\mathcal{B} \,:\, [0,T] \times \Omega \times S_d \longrightarrow S_d,\ (t,x,\tau) \longmapsto \left(B_{ijkh}(t,x)\,\tau_{kh}\right)$ satisfies

$$\begin{cases} \text{(i) } B_{ijkh} \in W^{1,\infty}(0,T;L^{\infty}(\Omega)), \\ \text{(ii) } \mathcal{B}(t)\sigma \cdot \tau = \sigma \cdot \mathcal{B}(t)\tau \\ \qquad \forall \sigma,\tau \in S_d,\ \text{a.e. } t \in (0,T),\ \text{a.e. in } \Omega. \end{cases} \tag{21}$$

We suppose that the body forces and surface tractions satisfy

$$f_0 \in W^{1,2}(0,T;H), \qquad f_2 \in W^{1,2}\left(0,T;L^2(\Gamma_2)^d\right). \tag{22}$$

We assume that the thermal tensor and the heat source density satisfy the conditions:

$$C_e = \left(c_{ij}\right), \quad c_{ij} = c_{ji} \in L^{\infty}(\Omega), \quad q \in W^{1,2}\left(0,T;L^2(\Omega)\right). \tag{23}$$

The boundary thermic data are supposed to satisfy the regularity condition:

$$k_e \in L^{\infty}\left(\Omega;\mathbb{R}^+\right), \quad \theta_R \in W^{1,2}\left(0,T;L^2(\Gamma_3)\right). \tag{24}$$

We suppose that the thermal conductivity tensor verifies the usual symmetry and ellipticity properties, i.e., for some $c_k > 0$ and for all $(\xi_i) \in \mathbb{R}^d$:

$$K_c = \left(k_{ij}\right), \quad k_{ij} = k_{ji} \in L^{\infty}(\Omega), \quad k_{ij}\,\xi_i\xi_j \geq c_k\,\xi_i\xi_i. \tag{25}$$

We assume that the initial data satisfy the conditions

$$u_0\ \in\ V, \quad v_0\ \in\ V, \quad \theta_0 \in E. \tag{26}$$

On the contact surface, the following frictional contact function

$$\psi(w) := \int_{\Gamma_3} \varphi(w)\, da$$

is assumed to verify

$$\begin{cases} \text{(i) } \psi : V \longrightarrow \mathbb{R} \quad \text{is well defined, continuous and convex,} \\ \text{(ii) there exists a sequence of differentiable convex functions} \\ \qquad (\psi_n) : V \longrightarrow \mathbb{R} \quad \text{such that} \quad \forall w \in L^2(0,T;V), \\ \qquad \int_0^T \psi_n(w(t))\, dt \longrightarrow \int_0^T \psi(w(t))\, dt, \ n \longrightarrow +\infty, \\ \text{(iii) for all sequence } (w_n) \text{ and } w \text{ in } W^{1,2}(0,T;V) \text{ such that} \\ \qquad w_n \rightharpoonup w, \ w_n' \rightharpoonup w' \text{ weakly in } L^2(0,T;V), \\ \qquad \text{then } \liminf_{n \longrightarrow +\infty} \int_0^T \psi_n(w_n(t))\, dt \geq \int_0^T \psi(w(t))\, dt, \\ \text{(iv) if } w \in V, \ w = 0 \ \text{ on } \Gamma_3, \ \text{then } \forall n \in \mathbb{N}, \ \psi_n'(w) = 0_{V'}. \end{cases} \tag{27}$$

Here $\psi_n'(\mathrm{v})$ denotes the Fréchet derivative of ψ_n at v.

Using Green's formula, we obtain the variational formulation of the mechanical problem Q as follows.

Problem QV Find $u : [0,T] \to V, \theta : [0,T] \to E$ satisfying a.e. $t \in (0,T)$:

$$\begin{cases} \langle \ddot{u}(t) + A\,\dot{u}(t) + B\,u(t) + C\,\theta(t),\ w - \dot{u}(t) \rangle_{V' \times V} \\ + \left(\int_0^t B(t-s)\,\varepsilon(u(s)) \right) ds, \varepsilon(w) - \varepsilon(\dot{u}(t)))_{\mathcal{H}} + \psi(w) - \psi(\dot{u}(t)), \\ \qquad \geq \langle f(t),\ w - \dot{u}(t) \rangle_{V' \times V} \quad \forall w \in V, \\ \dot{\theta}(t) + K\,\theta(t) = R\dot{u}(t) + Q(t) \qquad \text{in} \quad E', \\ u(0) = u_0, \quad \dot{u}(0) = v_0, \quad \theta(0) = \theta_0. \end{cases}$$

Here, the operators and functions $A,\ B : V \longrightarrow V', C : E \longrightarrow V', \psi : V \longrightarrow \mathbb{R}$, $K : E \longrightarrow E'$, $R : V \longrightarrow E'$, $f : [0,T] \longrightarrow V'$, and $Q : [0,T] \longrightarrow E'$ are defined by $\forall v \in V, \forall w \in V, \forall \tau \in E, \forall \eta \in E$:

$$\langle A\,v, w \rangle_{V' \times V} = (\mathcal{A}(\varepsilon v), \varepsilon w)_{\mathcal{H}},$$

$$\langle B\,v, w \rangle_{V' \times V} = (\mathcal{G}(\varepsilon v), \varepsilon w)_{\mathcal{H}},$$

$$\langle C\tau, w \rangle_{V' \times V} = -(\tau\, C_e,\ \varepsilon w)_{\mathcal{H}},$$

$$\langle f(t), w \rangle_{V' \times V} = (f_0(t), w)_H \ + (f_2(t), w)_{(L^2(\Gamma_2))^d}\,,$$

$$\langle Q(t), \eta \rangle_{E' \times E} = \int_{\Gamma_3} k_e\, \theta_R(t)\, \eta\, dx + \int_{\Omega} q(t)\, \eta\, dx,$$

$$\langle K\,\tau, \eta \rangle_{E' \times E} = \sum_{i,j=1}^{d} \int_{\Omega} k_{ij}\, \frac{\partial \tau}{\partial x_j}\, \frac{\partial \eta}{\partial x_i}\, dx + \int_{\Gamma_3} k_e\, \tau \cdot \eta\, da,$$

$$\langle R\,v, \eta \rangle_{E' \times E} = -\int_{\Omega} c_{ij}\, \frac{\partial v_i}{\partial x_j}\, \eta\, dx.$$

Let us recall now the following main mathematical result (see for details [3]):

Theorem 1 *Assume that* (19)–(27) *hold. Then, there exists a unique solution* $\{u, \theta\}$ *to problem* QV *with the regularity:*

$$\begin{cases} u \in W^{2,2}(0,T;V) \cap W^{2,\infty}(0,T;H), \\ \theta \in W^{1,2}(0,T;E) \cap W^{1,\infty}(0,T;F). \end{cases}$$

2.1 Analysis of a Numerical Scheme

In this section, we study a fully-discrete numerical approximation scheme of the variational problem QV (see [10]). For this purpose, we suppose in the following that the conditions on the data (19)–(27) of Theorem 1 are satisfied. In particular, we have

$$f \in C\left([0,T]; V'\right), \quad Q \in C\left([0,T]; E'\right).$$

Let $\{u, \theta\}$ be the unique solution of the problem QV and let us introduce the velocity variable

$$v(t) = \dot{u}(t), \quad \forall t \in [0,T].$$

Then

$$u(t) = u_0 + \int_0^t v(s)\, ds, \quad \forall t \in [0,T].$$

From Theorem 1, we see that $\{v, \theta\}$ verify for all $t \in [0,T]$:

$$\begin{cases} \langle \dot{v}(t) + A\, v(t) + B\, u(t) + C\, \theta(t), w - v(t) \rangle_{V' \times V} \\ + \left(\int_0^t B(t-s)\, \varepsilon(u(s))\, ds, \varepsilon(w) - \varepsilon(\dot{u}(t)) \right)_{\mathcal{H}} + \psi(w) - \psi(v(t)) \\ \quad \geq \langle f(t), w - v(t) \rangle_{V' \times V}, \quad \forall w \in V. \end{cases} \tag{28}$$

$$\langle \dot{\theta}(t), \eta \rangle_F + \langle K\, \theta(t), \eta \rangle_{E' \times E} = \langle R v(t), \eta \rangle_{E' \times E} + \langle Q(t), \eta \rangle_{E' \times E}, \quad \forall \eta \in E. \tag{29}$$

$$u(0) = u_0, \quad v(0) = v_0, \quad \theta(0) = \theta_0, \tag{30}$$

with the regularity:

$$\begin{cases} v \in W^{1,2}(0,T;V) \cap W^{1,\infty}(0,T;H), \\ \theta \in W^{1,2}(0,T;E) \cap W^{1,\infty}(0,T;F). \end{cases} \tag{31}$$

In this section, we make the following additional assumptions on the solution and contact function:

$$v \in W^{2,2}(0,T;H), \tag{32}$$

$$\theta \in W^{2,2}(0,T;F), \tag{33}$$

$$\psi \text{ is Lipschitz continuous on } V. \tag{34}$$

Let $V^h \subset V$ and $E^h \subset E$ be a family of finite dimensional subspaces, with $h > 0$, a discretization parameter. We divide the time interval $[0,T]$ into N equal parts: $t_n = n\,k$, $n = 0, 1, \ldots, N$, with the time step $k = T/N$. For a continuous function $w \in C([0,T];X)$ with values in a space X, we use the notation $w_n = w(t_n) \in X$. Then, from (28) and (29), we introduce the following fully discrete scheme.

Problem P^{hk}**.** Find $v^{hk} = \left\{v_n^{hk}\right\}_{n=0}^{N} \subset V^h, \theta^{hk} = \left\{\theta_n^{hk}\right\}_{n=0}^{N} \subset E^h$ such that

$$v_0^{hk} = v_0^h, \quad \theta_0^{hk} = \theta_0^h \tag{35}$$

and for $n = 1, \cdots, N$,

$$\begin{aligned}
&\left(\frac{v_n^{hk} - v_{n-1}^{hk}}{k}, w^h - v_n^{hk}\right)_H + \langle A\, v_n^{hk}, w^h - v_n^{hk}\rangle_{V'\times V} + \langle B\, u_{n-1}^{hk}, w^h - v_n^{hk}\rangle_{V'\times V} \\
&+ \langle C\,\theta_{n-1}^{hk}, w^h - v_n^{hk}\rangle_{V'\times V} + \psi\left(w^h\right) - \psi\left(v_n^{hk}\right) \\
&+ \left(k\sum_{m=0}^{n-1} \mathcal{B}\,(t_n - t_m)\; \varepsilon\left(u_m^{hk}\right), \varepsilon\left(w^h\right) - \varepsilon\left(v_n^{hk}\right)\right)_{\mathcal{H}} \\
&\quad \geq \langle f_n, w^h - v_n^{hk}\rangle_{V'\times V}, \quad \forall\, w^h \in V^h,
\end{aligned} \tag{36}$$

$$\begin{aligned}
&\left(\frac{\theta_n^{hk} - \theta_{n-1}^{hk}}{k}, \eta^h\right)_F + \langle K\,\theta_n^{hk}, \eta^h\rangle_{E'\times E} \\
&= \langle R\, v_n^{hk}, \eta^h\rangle_{E'\times E} + \langle Q_n, \eta^h\rangle_{E'\times E}, \quad \forall\,\eta^h \in E^h,
\end{aligned} \tag{37}$$

where

$$u_n^{hk} = u_{n-1}^{hk} + k\, v_n^{hk}, \quad u_0^{hk} = u_0^h. \tag{38}$$

Here, $u_0^h \in V^h, v_0^h \in V^h, \theta_0^h \in E^h$ are suitable approximations of the initial values u_0, v_0, θ_0.

For $n = 1, \ldots, N$, suppose that $u_{n-1}^{hk}, v_{n-1}^{hk}, \theta_{n-1}^{hk}$ are known. We may then calculate v_n^{hk} by (36), θ_n^{hk} by (37) and u_n^{hk} by (38). Hence, the discrete solution $v^{hk} \subset V^h$, $\theta^{hk} \subset E^h$ exists and is unique.

We now turn to an error analysis of the numerical solution. The main result of this section is the following one (see for details [10]).

Theorem 2 *We keep the assumptions of Theorem 1. Under the additional assumptions (32)–(33), then for the unique solution $v^{hk} \subset V^h$, $\theta^{hk} \subset E^h$ of the discrete problem P^{hk}, we have the following error estimate*

$$
\begin{aligned}
&\max_{1\le n\le N}\|v_n - v_n^{hk}\|_H^2 + \left(k\sum_{n=1}^{N}\|v_n - v_n^{hk}\|_V^2\right)\\
&+\max_{1\le n\le N}\|\theta_n - \theta_n^{hk}\|_F^2 + \left(k\sum_{n=1}^{N}\|\theta_n - \theta_n^{hk}\|_E^2\right)\\
&\le c\,\|u_0 - u_0^h\|_V^2 + c\,\|v_0 - v_0^{hk}\|_H^2 + c\,\|\theta_0 - \theta_0^h\|_F^2\\
&+ c\max_{1\le n\le N}\|v_n - w_n^h\|_H + c\max_{1\le n\le N}\|\theta_n - \eta_n^h\|_F^2\\
&+ c\,k\sum_{j=1}^{N}\|v_j - w_j^h\|_V^2 + c\,k\sum_{j=1}^{N}\|\theta_j - \eta_j^h\|_E^2 \qquad (39)\\
&+ c\left(\sum_{j=1}^{N-1}\|\left(v_j - w_j^h\right) - \left(v_{j+1} - w_{j+1}^h\right)\|_H\right)^2\\
&+ c\left(\sum_{j=1}^{N-1}\|\theta_j - \eta_j^h - \left(\theta_{j+1} - \eta_{j+1}^h\right)\|_F\right)^2\\
&+ c\,k^2 + c\,k\sum_{j=1}^{N}\|v_j - w_j^h\|_V,
\end{aligned}
$$

where for $j = 1,\ldots,N$, $w_j^h \in V^h$, $\eta_j^h \in E^h$ are arbitrary.

The inequality (39) is a basis for error estimates for particular choice of the finite-dimensional subspace V^h and under additional data and solution regularities.

As a typical example, let us consider $\Omega \subset \mathbb{R}^d$, $d \in \mathbb{N}^*$, a polygonal domain. Let $\mathcal{T}^h$ be a regular finite element partition of Ω. Let $V^h \subset V$ and $E^h \subset E$ be the finite element space consisting of piecewise polynomials of degree $\le m-1$, with $m \ge 2$, according to the partition $\mathcal{T}^h$. Denote by $\Pi_V^h : H^m(\Omega)^d \to V^h$ and $\Pi_E^h : H^m(\Omega) \to E^h$ the corresponding finite element interpolation operator. Recall (see e.g., [11]) that:

$$
\begin{cases}
\|w - \Pi_V^h w\|_{H^l(\Omega)^d} \le c h^{m-l}\,|w|_{H^m(\Omega)^d}, & \forall w \in H^m(\Omega)^d,\\
\|\eta - \Pi_E^h \eta\|_{H^l(\Omega)} \le c h^{m-l}\,|\eta|_{H^m(\Omega)}, & \forall \eta \in H^m(\Omega),
\end{cases}
$$

where $l = 0$ (for which $H^0 = L^2$) or $l = 1$.

In the following, we assume the additional data and solution regularities

$$
\begin{cases}
u_0 \in H^{\alpha+1}(\Omega)^d,\\
v \in C\left([0,T]; H^{2\alpha+1}(\Omega)^d\right), \quad \dot{v} \in L^1\left(0,T; H^{\alpha}(\Omega)^d\right),\\
\theta \in C\left([0,T]; H^{\alpha+1}(\Omega)\right), \quad \dot{\theta} \in L^1\left(0,T; H^{\alpha}(\Omega)\right).
\end{cases}
\qquad (40)
$$

Here

$$\alpha = m - 1 \geq 1.$$

We remark that the previous properties already hold for $\alpha = 1$, except for

$$v \in C\left([0,T]; H^3(\Omega)^d\right) \quad \text{and} \quad \theta \in C\left([0,T]; H^2(\Omega)\right).$$

Then, we choose in (39) the elements

$$u_0^h = \Pi_V^h\, u_0, \quad v_0^h = \Pi_V^h\, v_0, \quad \theta_0^h = \Pi_E^h\, \theta_0,$$

and

$$w_j^h = \Pi_V^h\, v_j, \quad \eta_j^h = \Pi_E^h\, \theta_j, \quad j = 1 \cdots N.$$

From the assumptions (40), we have:

$$\begin{aligned} \|u_0 - u_0^h\|_V \leq ch^\alpha, \quad \|e_0\|_H \leq ch^\alpha, \quad \|\varepsilon_0\|_F \leq ch^\alpha, \\ A_0 \leq ch^\alpha, \quad B_0 \leq ch^{2\alpha}, \\ A_3 \leq ch^\alpha, \quad B_3 \leq ch^\alpha, \\ k\, A_2 \leq ch^{2\alpha}, \quad k\, B_2 \leq ch^{2\alpha}, \quad k\, \hat{B}_2 \leq ch^{2\alpha}. \end{aligned}$$

Using these estimates in (39), we conclude to the following error estimate result.

Theorem 3 *We keep the assumptions of Theorem 2. Under the additional assumptions* (40), *we obtain the error estimate for the corresponding discrete solution* v_n^{hk}, θ_n^{hk}, $n = 1, \ldots, N$:

$$\max_{0 \leq n \leq N} \|v_n - v_n^{hk}\|_H + \left(k \sum_{n=0}^{N} \|v_n - v_n^{hk}\|_V^2\right)^{1/2}$$
$$+ \max_{0 \leq n \leq N} \|\theta_n - \theta_n^{hk}\|_F + \left(k \sum_{n=0}^{N} \|\theta_n - \theta_n^{hk}\|_E^2\right)^{1/2} < c\,(h^\alpha + k).$$

In particular, for $\alpha = 1$, *we have*

$$\max_{0 \leq n \leq N} \|v_n - v_n^{hk}\|_H + \left(k \sum_{n=0}^{N} \|v_n - v_n^{hk}\|_V^2\right)^{1/2}$$
$$+ \max_{0 \leq n \leq N} \|\theta_n - \theta_n^{hk}\|_F + \left(k \sum_{n=0}^{N} \|\theta_n - \theta_n^{hk}\|_E^2\right)^{1/2} < c\,(h + k).$$

2.2 Numerical Computations

Here, we consider two typical examples of thermal contact problems with Tresca's friction law, the first one is bilateral and the second one obeys a normal damped response condition (see [3, 10]). We provide numerical simulations for the discrete schemes in Sect. 3 in using MATLAB computation codes.

Example 4 Thermal bilateral contact problem with Tresca's friction law.
The contact condition on Γ_3 is bilateral, and satisfies (see e.g., [13, 22]):

$$\begin{cases} u_\nu = 0, \quad |\sigma_\tau| \leq g, \\ |\sigma_\tau| < g \implies \dot{u}_\tau = 0, \\ |\sigma_\tau| = g \implies \dot{u}_\tau = -\lambda\sigma_\tau, \text{for some } \lambda \geq 0, \end{cases} \qquad \text{on } \Gamma_3 \times (0, T).$$

Here, g represents the friction bound, i.e., the magnitude of the limiting friction traction at which slip begins, with $g \in L^\infty(\Gamma_3)$, $g \geq 0$ a.e. on Γ_3. The corresponding admissible displacement space is:

$$V := \{w \in H_1, \text{ with } w = 0 \text{ on } \Gamma_1,\ w_\nu = 0 \text{ on } \Gamma_3\},$$

and the subdifferential contact function is given by:

$$\varphi(x, y) = g(x)|y_{\tau(x)}| \quad \forall x \in \Gamma_3,\ y \in \mathbb{R}^d,$$

where $y_{\tau(x)} := y - y_{\nu(x)}\nu(x)$, $y_{\nu(x)} := y \cdot \nu(x)$, with $\nu(x)$ denoting the unit normal at $x \in \Gamma_3$. Then, the function

$$\psi(v) := \int_{\Gamma_3} g\,|v_\tau|\,da, \quad \forall v \in V$$

is well defined on V and is Lipschitz continuous on V (see [3, 10]).

For our computations, we consider a rectangular open set, linear elastic and long memory viscoelastic operators. We set:

$$\Omega = (0, L_1) \times (0, L_2),$$

$$\Gamma_1 = (\{0\} \times [0, L_2]), \Gamma_2 = [0, L_1] \times \{L_2\} \cup (\{L_1\} \times [0, L_2]), \Gamma_3 = [0, L_1] \times \{0\},$$

$$(\mathcal{G}\,\tau)_{ij} = \frac{E\kappa}{1-\kappa^2}\,(\tau_{11} + \tau_{22})\,\delta_{ij} + \frac{E}{1+\kappa}\tau_{ij}, \quad 1 \leq i, j \leq 2,\ \tau \in S_2,$$

$$(\mathcal{A}\,\tau)_{ij} = \mu\,(\tau_{11} + \tau_{22})\,\delta_{ij} + \eta\,\tau_{ij}, \quad 1 \leq i, j \leq 2,\ \tau \in S_2,$$

$$(\mathcal{B}(t)\,\tau)_{ij} = B_1(t)\,(\tau_{11} + \tau_{22})\,\delta_{ij} + B_2(t)\,\tau_{ij}, \quad 1 \leq i, j \leq 2,\ \tau \in S_2,\ t \in [0, T].$$

Here, E is the Young's modulus, κ is the Poisson's ratio of the material, δ_{ij} denotes the Kronecker symbol, and μ, η are viscosity constants.

We use spaces of continuous piecewise affine functions $V^h \subset V$ and $E^h \subset E$ as families of approximating subskpaces. For our computations, we considered the following data (IS unity):

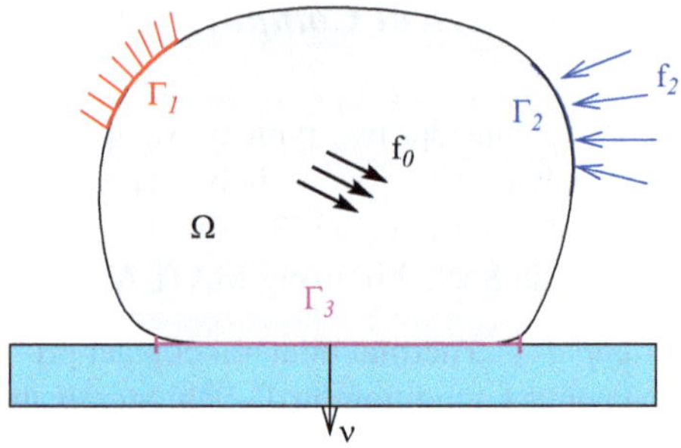

Fig. 1 The mechanical contact problem

$$L_1 = L_2 = 1, \quad T = 1,$$
$$\mu = 10, \quad \eta = 10, \quad E = 2, \quad \kappa = 0.1,$$
$$f_0(x,t) = (0,\ -t), \quad f_2(x,t) = (1,\ 0), \quad \forall t \in [0,T],$$
$$c_{ij} = k_{ij} = k_e = 1,\ 1 \le i,\ j \le 2, \quad q = 1,$$
$$B_1(t) = B_2(t) = 10^{-2}\, e^{-t}, \quad \forall t \in [0,T],$$
$$u_0 = (0,0), \quad v_0 = (0,0), \quad \theta_0 = 0.$$

The initial configuration is represented in Fig. 1.

Then, we show in Fig. 2 the deformed configurations at final time, where the relaxation coefficients are positive and decreasing for the two different types of Tresca's friction bounds. For small friction bound, where $g(x,0) = \frac{x}{2}$, $0 \le x \le 1$, we observe on the contact surface a slip phenomena in the direction of the surface fraction on Γ_2. This means that the friction bound has been obtained in the zone of the values of x near to 1. Whereas for large friction bound, e.g., for $g(x,0) = 10\,x$, $0 \le x \le 1$, then slip in the direction of the traction could not be realized. In Fig. 3, we compute the Von Mise norm, which gives a global measure of the stress in the body. The maxima of the norm could be seen in the neighborhood of the point (0, 1) for small friction bounds and in the neighborhood of the point $(1,0)$ for large friction bounds. In Fig. 4, we show the influence of the different temperatures of the foundation on the temperature field of the body. We observe larger deformations of the body for greater temperature of the foundation.

Example 5 Thermal contact problem with normal damped response and Tresca's friction law.

The normal damped response contact condition with Tresca's friction law is defined by:

$$\begin{cases} -\sigma_\nu = k_0\, |\dot{u}_\nu|^{r-1}\dot{u}_\nu, \quad |\sigma_\tau| \le g, \\ |\sigma_\tau| < g \implies \dot{u}_\tau = 0, \quad \text{on } \Gamma_3 \times (0,T). \\ |\sigma_\tau| = g \implies \dot{u}_\tau = -\lambda\sigma_\tau, \quad \text{for some } \lambda \ge 0, \end{cases}$$

Here, $0 < r < 1$ and $g,\ k_0 \in L^\infty(\Gamma_3)$, $g \ge 0$, $k_0 \ge 0$. The coefficient k_0 represents the hardness of the foundation and g is the friction threshold.

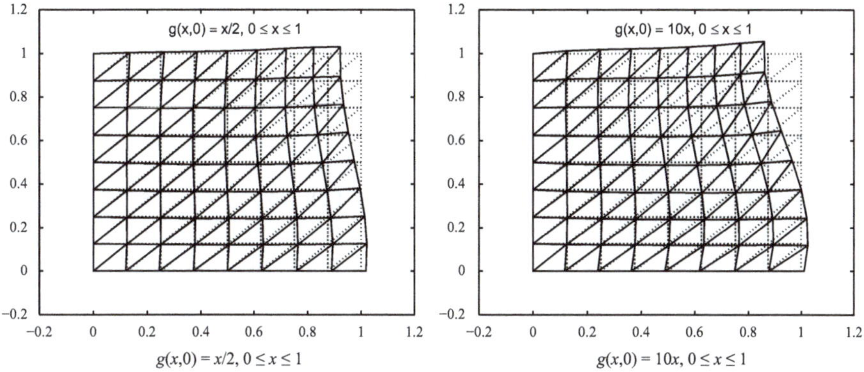

Fig. 2 Deformed configurations at final time, $\theta_R(t) = 1,\ 0 \le t \le 1$

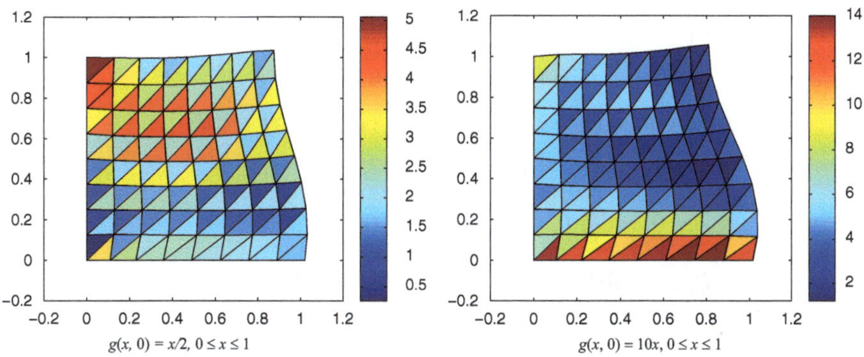

Fig. 3 Von Mise norm in deformed configurations, $\theta_R(t) = 1,\ 0 \le t \le 1$

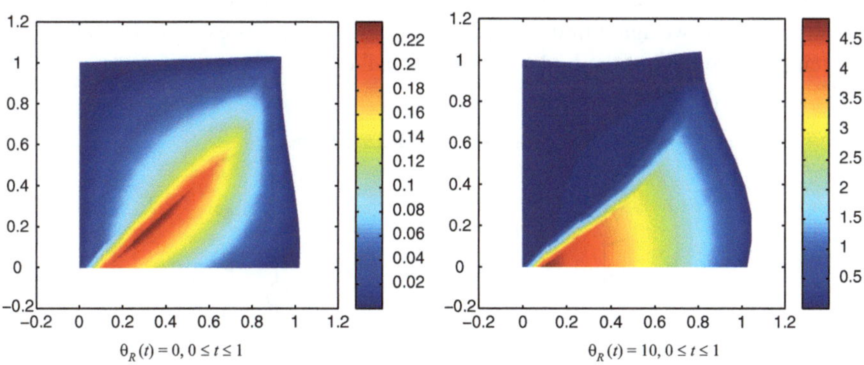

Fig. 4 Temperature field at final time, $g(x, 0) = \frac{x}{2},\ 0 \le x \le 1$

The admissible displacement space is given by:

$$V := \{w \in H_1, \text{ with } w = 0 \text{ on } \Gamma_1\}$$

and the subdifferential contact function is:

$$\varphi(x, y) = \frac{1}{r+1} k_0(x) \, |y_{\nu(x)}|^{r+1} + g(x)|y_\tau(x)| \quad \forall x \in \Gamma_3, \; y \in \mathbb{R}^d.$$

Then, setting $p := r + 1$, we have the contact function well defined on V by

$$\psi(v) := \int_{\Gamma_3} \frac{k_0}{p} \, |v_\nu|^p \, da, + \int_{\Gamma_3} g \, |v_\tau| \, da, \quad \forall v \in V.$$

We also verify that ψ is Lipschitz continuous on V (see [3, 10]).

For our computations, we again take the previous rectangular open set, with linear elasticity and viscoelasticity, and use the following data (IS unity):

$$\begin{aligned}
&L_1 = L_2 = 1, \quad T = 1,\\
&\mu = 10, \quad \eta = 10, \quad E = 2, \quad \kappa = 0.1,\\
&f_0(x,t) = (0, -t), \quad f_2(x,t) = (1, 0), \quad \forall t \in [0, T],\\
&c_{ij} = k_{ij} = k_e = 1, \; 1 \le i, \, j \le 2, \quad q = 1,\\
&g(x, 0) = \frac{x}{2}, \; 0 \le x \le 1, \quad r = 0.5,\\
&B_1(t) = B_2(t) = 10^{-2} \, e^{-t}, \quad \forall t \in [0, T],\\
&u_0 = (0, 0), \quad v_0 = (0, 0), \quad \theta_0 = 0.
\end{aligned}$$

We show in Fig. 5, the deformed configurations at final time, through the body for different normal damped response coefficients k_0, we verify that the penetrability of the foundation depends on its coefficient of hardness. In Fig. 6, we compute the Von Mise norm. Larger stress near the contact surface is then observed for hard obstacle. Finally in Fig. 7, we again find the influence of the temperature of the foundation on the temperature field of the body and on the final deformed configurations.

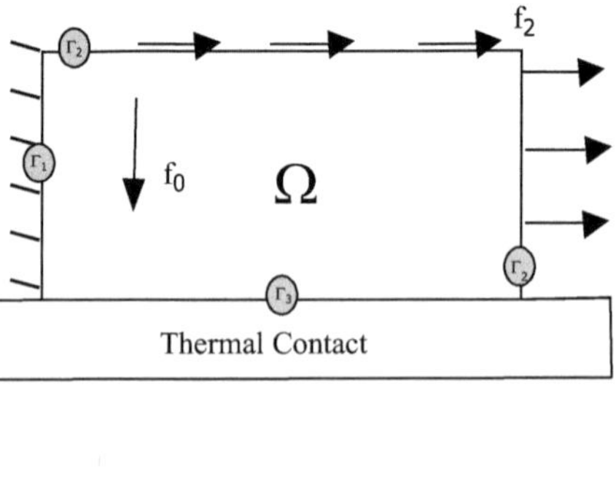

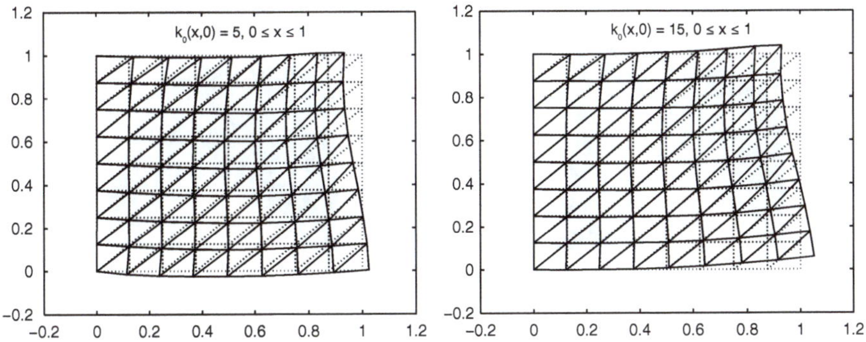

Fig. 5 Deformed configurations at final time, $\theta_R(t) = 1,\ 0 \leq t \leq 1$

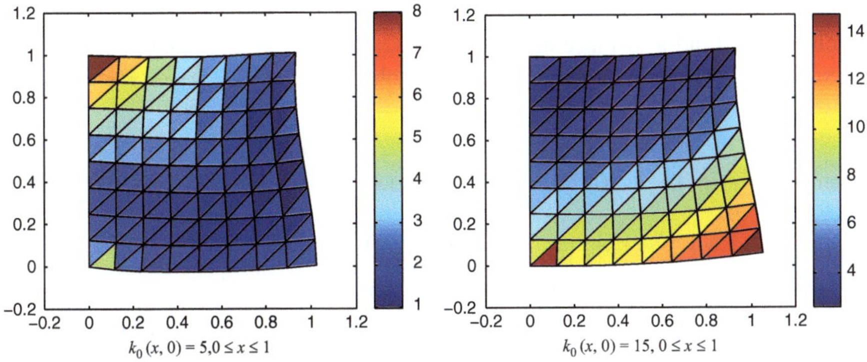

Fig. 6 Von Mise norm in deformed configurations, $\theta_R(t) = 1,\ 0 \leq t \leq 1$

3 Dynamic Contact Problems with Free Boundary Condition

We present here a class of dynamic thermal subdifferential contact problems with friction for long memory viscoelastic materials and without the clamped condition. The boundary Γ of the body Ω is partitioned into three disjoint measurable parts Γ_1, Γ_2, and Γ_3, with meas$(\Gamma_1) = 0$.

The model leads to a system defined by a second order evolution inequality, coupled with a first order evolution equation. We establish an existence and uniqueness result. Finally, a fully discrete scheme for numerical approximations is provided and corresponding various numerical computations in dimension two will be given for the cases where Γ_1 is reduced to one point or is an empty set (see [2, 10]).

The dynamic mechanical problem for long memory thermo-viscoelastic materials subjected to subdifferential contact condition and to non clamped condition is then formulated as follows.

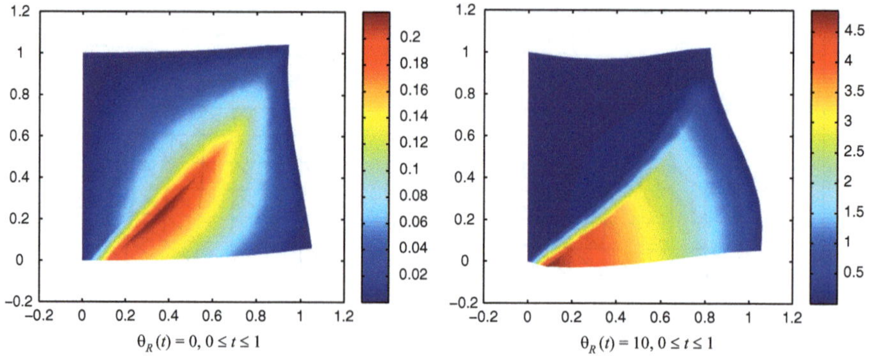

Fig. 7 Temperature field at final time, $k_0(x,0) = 15,\ 0 \le x \le 1$

Problem Q Find a displacement field $u : \Omega \times [0,T] \longrightarrow \mathbb{R}^d$ and a stress field $\sigma : \Omega \times [0,T] \longrightarrow S_d$ and a temperature field $\theta : \Omega \times [0,T] \longrightarrow \mathbb{R}_+$ such that for a.e. $t \in (0,T)$:

$$\sigma(t) = \mathcal{A}\varepsilon(\dot{u}(t)) + \mathcal{G}\varepsilon(u(t)) + \int_0^t \mathcal{B}(t-s)\,\varepsilon(u(s))\,ds - \theta(t)\,C_e \quad \text{in} \quad \Omega, \tag{41}$$

$$\ddot{u}(t) = \operatorname{Div}\sigma(t) + f_0(t) \quad \text{in} \quad \Omega, \tag{42}$$

$$\sigma(t)\nu = f_2(t) \quad \text{on} \quad \Gamma_2, \tag{43}$$

$$u(t) \in U, \quad \varphi(w) - \varphi(\dot{u}(t)) \ge -\sigma(t)\nu \cdot (w - \dot{u}(t)) \quad \forall w \in U \quad \text{on}\ \Gamma_3, \tag{44}$$

$$\dot{\theta}(t) - \operatorname{div}(K_c\,\nabla\theta(t)) = -c_{ij}\,\frac{\partial\,\dot{u}_i}{\partial\,x_j}(t) + q(t) \quad \text{on} \quad \Omega, \tag{45}$$

$$-k_{ij}\,\frac{\partial\,\theta}{\partial\,x_j}(t)\,n_i = k_e\,(\theta(t) - \theta_R) \quad \text{on} \quad \Gamma_3, \tag{46}$$

$$\theta(t) = 0 \quad \text{on} \quad \Gamma_2, \tag{47}$$

$$\theta(0) = \theta_0 \quad \text{in} \quad \Omega, \tag{48}$$

$$u(0) = u_0, \quad \dot{u}(0) = v_0 \quad \text{in} \quad \Omega. \tag{49}$$

It is worth to notice that the new feature here is due to the absence of the usual claimed condition. However, there is coerciveness with regard to the temperature by (15).

To derive the variational formulation of the mechanical problems (5)–(18), let us introduce the spaces V and E defined by

$$\mathcal{D}(\Omega)^d \subset V = H_1 \cap U,$$
$$E = \left\{\eta \in H^1(\Omega),\ \eta = 0 \quad \text{on} \quad \Gamma_2\right\}, \quad F = L^2(\Omega).$$

On V, we consider the inner product given by

$$(u, v)_V = (\varepsilon(u), \varepsilon(v))_{\mathcal{H}} + (u, v)_H \qquad \forall\, u,\, v \in V,$$

and the associated norm

$$\|v\|_V^2 = \|\varepsilon(v)\|_{\mathcal{H}}^2 + \|v\|_H^2 \qquad \forall\, v \in V.$$

It follows that $\|\cdot\|_{H_1}$ and $\|\cdot\|_V$ are equivalent norms on V and therefore $(V, \|\cdot\|_V)$ is a real Hilbert space.

In the study of the mechanical problems (5)–(18), we put again the analogous assumptions as in Chap. 2 on the different operators and data.

The viscosity operator $\mathcal{A} : \Omega \times S_d \longrightarrow S_d$, $(x, \tau) \longmapsto \big(a_{ijkh}(x)\, \tau_{kh}\big)$ is linear on the second variable and satisfies the usual properties of ellipticity and symmetry, i.e.,

$$\begin{cases} \text{(i)}\ \ a_{ijkh} \in W^{1,\infty}(\Omega), \\ \text{(ii)}\ \ \mathcal{A}\sigma \cdot \tau = \sigma \cdot \mathcal{A}\tau \ \ \forall \sigma, \tau \in S_d, \text{ a.e. in } \Omega, \\ \text{(iii) there exists } m_{\mathcal{A}} > 0 \text{ such that} \\ \qquad \mathcal{A}\tau \cdot \tau \geq m_{\mathcal{A}}\, |\tau|^2 \quad \forall \tau \in S_d, \text{ a.e. in } \Omega. \end{cases} \tag{50}$$

The elasticity operator $\mathcal{G} : \Omega \times S_d \longrightarrow S_d$ satisfies:

$$\begin{cases} \text{(i)} \quad \text{there exists } L_{\mathcal{G}} > 0 \text{ such that} \\ \qquad |\mathcal{G}(x, \varepsilon_1) - \mathcal{G}(x, \varepsilon_2)| \leq L_{\mathcal{G}} |\varepsilon_1 - \varepsilon_2| \\ \qquad \forall \varepsilon_1,\, \varepsilon_2 \in S_d,\ a.e.\ x \in \Omega, \\ \text{(ii)} \quad x \longmapsto \mathcal{G}(x, \varepsilon) \text{ is Lebesgue measurable on } \Omega, \forall \varepsilon \in S_d, \\ \text{(iii)} \ \text{the mapping } x \longmapsto \mathcal{G}(x, 0) \in \mathcal{H}. \end{cases} \tag{51}$$

The relaxation tensor $\mathcal{B} : [0, T] \times \Omega \times S_d \longrightarrow S_d$, $(t, x, \tau) \longmapsto \big(B_{ijkh}(t, x)\, \tau_{kh}\big)$ satisfies

$$\begin{cases} \text{(i) } B_{ijkh} \in W^{1,\infty}(0, T; L^\infty(\Omega)), \\ \text{(ii) } \mathcal{B}(t)\sigma \cdot \tau = \sigma \cdot \mathcal{B}(t)\tau \\ \qquad \forall \sigma, \tau \in S_d, \text{ a.e. } t \in (0, T), \text{ a.e. in } \Omega. \end{cases} \tag{52}$$

We suppose the body forces and surface tractions satisfy

$$f_0 \in W^{1,2}(0, T; H), \qquad f_2 \in W^{1,2}\left(0, T; L^2(\Gamma_F)^d\right). \tag{53}$$

For the thermal tensors and the heat sources density, we suppose that

$$C_e = (c_{ij}), \quad c_{ij} = c_{ji} \in L^\infty(\Omega), \quad q \in W^{1,2}\left(0, T; L^2(\Omega)\right). \tag{54}$$

The boundary thermal data satisfy

$$k_e \in L^\infty(\Omega; \mathbb{R}^+), \quad \theta_R \in W^{1,2}\left(0, T; L^2(\Gamma_3)\right). \tag{55}$$

The thermal conductivity tensor verifies the usual symmetry end ellipticity: for some $c_k > 0$ and for all $(\xi_i) \in \mathbb{R}^d$,

$$K_c = (k_{ij}), \quad k_{ij} = k_{ji} \in L^\infty(\Omega), \quad k_{ij}\,\xi_i\xi_j \geq c_k\,\xi_i\xi_i. \tag{56}$$

Finally, we have to put technical assumptions on the initial data and the subdifferential condition on the contact surface as to use classical results on first order set valued evolution equations. Here, we use a general theorem taken in ([5], p. 46), in a simplified case, which is enough for our proposal and applications.
We assume that the initial data satisfy the conditions

$$u_0 \in V, \quad v_0 \in V \cap H_0^2(\Omega)^d, \quad \theta_0 \in E \cap H_0^2(\Omega). \tag{57}$$

On the contact surface, the following frictional contact function

$$\psi(w) := \int_{\Gamma_3} \varphi(w)\,da$$

verifies

$$\begin{cases} \text{(i) } \psi : V \longrightarrow \mathbb{R} \quad \text{is well defined, continuous, and convex,} \\ \text{(ii) there exists a sequence of differentiable convex functions} \\ \quad (\psi_n) : V \longrightarrow \mathbb{R} \quad \text{such that} \quad \forall w \in L^2(0,T;V), \\ \quad \int_0^T \psi_n(w(t))\,dt \longrightarrow \int_0^T \psi(w(t))\,dt,\ n \longrightarrow +\infty, \\ \text{(iii) for all sequence } (w_n) \text{ and } w \text{ in } W^{1,2}(0,T;V) \text{ such that} \\ \quad w_n \rightharpoonup w,\ w_n' \rightharpoonup w' \text{ weakly in } L^2(0,T;V), \\ \quad \text{then } \liminf\limits_{n \longrightarrow +\infty} \int_0^T \psi_n(w_n(t))\,dt \geq \int_0^T \psi(w(t))\,dt, \\ \text{(iv) if } w \in V \text{ and } w = 0 \text{ on } \Gamma_3, \text{ then } \forall n \in \mathbb{N},\ \psi_n'(w) = 0_{V'}. \end{cases} \tag{58}$$

The weak formulation of the mechanical problem Q is then formulated as follows.

Problem QV Find $u : [0,T] \to V, \theta : [0,T] \to E$ satisfying a.e. $t \in (0,T)$:

$$\begin{cases} \langle \ddot{u}(t) + A\,\dot{u}(t) + B\,u(t) + C\,\theta(t),\ w - \dot{u}(t)\rangle_{V'\times V} \\ + \left(\int_0^t B(t-s)\,\varepsilon(u(s))\,ds, \varepsilon(w) - \varepsilon(\dot{u}(t))\right)_{\mathcal{H}} + \psi(w) - \psi(\dot{u}(t)) \\ \qquad \geq \langle f(t),\ w - \dot{u}(t)\rangle_{V'\times V} \quad \forall w \in V, \\ \dot{\theta}(t) + K\,\theta(t) = R\dot{u}(t) + Q(t) \qquad \text{in} \quad E', \\ u(0) = u_0, \quad \dot{u}(0) = v_0, \quad \theta(0) = \theta_0. \end{cases}$$

The different operators are here defined as in Chap. 2. Then, we obtain our main existence and uniqueness result stated as below (see for details [2]):

Theorem 4 *Assume that* (19)–(58) *hold, then there exists a unique solution* $\{u, \theta\}$ *to problem* QV *with the regularity:*

$$\begin{cases} u \in W^{2,2}(0,T;V) \cap W^{2,\infty}(0,T;H), \\ \theta \in W^{1,2}(0,T;E) \cap W^{1,\infty}(0,T;F). \end{cases} \tag{59}$$

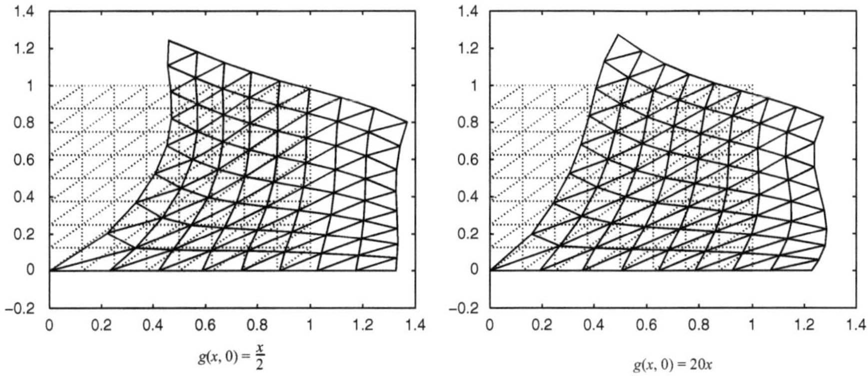

Fig. 8 Deformed configurations at final time, $\theta_R(t) = 0,\ 0 \le t \le 1$

3.1 *Numerical Simulations A*

Here, Γ_1 is reduced to one point (see [10]). We take again the two typical examples and the analogous data as in Chap. 2, except the following:

$$\Omega = (0, L_1) \times (0, L_2),$$
$$\Gamma_1 = \{(0,0)\}; \quad \Gamma_2 = (\{0\} \times]0, L_2]) \cup (]0, L_1] \times \{L_2\}) \cup (\{L_1\} \times [0, L_2]),$$
$$\Gamma_3 =]0, L_1[\times \{0\},$$
$$f_2(x,t) = (0,\ 0), \quad \forall x \in (\{0\} \times]0, L_2]), \quad \forall t \in [0, T],$$
$$f_2(x,t) = (1,\ 0), \quad \forall x \in (]0, L_1] \times \{L_2\}) \cup (\{L_1\} \times [0, L_2]), \quad \forall t \in [0, T].$$

Similar conclusions as in Chap. 2 can be stated. See Fig.8 to Fig.13.

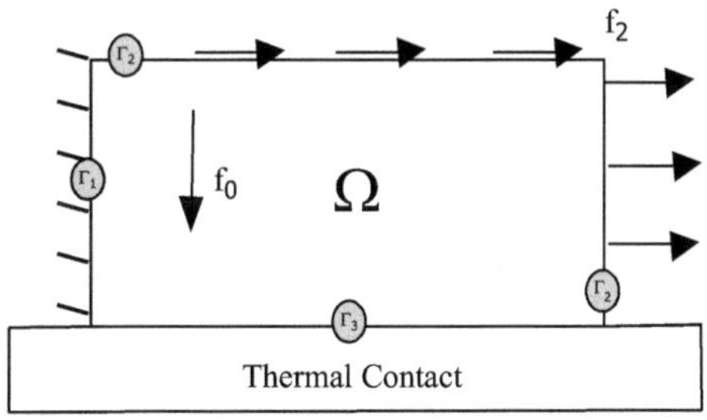

3.2 *Numerical Simulations B*

Here, $\Gamma_1 = \emptyset$, $\Gamma_F = \Gamma_2$, $f_F = f_2$, $\Gamma_u = \Gamma_3$ (see [2]). We take again the two typical examples as in Chap. 2 with the following data:

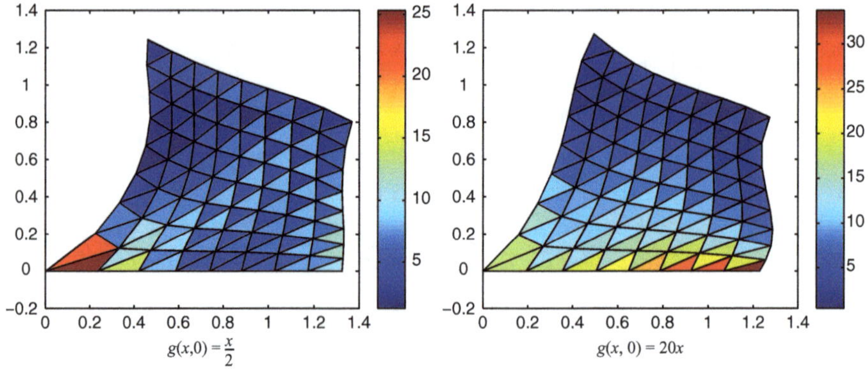

Fig. 9 Von Mises' norm in deformed configurations, $\theta_R(t) = 0,\ 0 \le t \le 1$

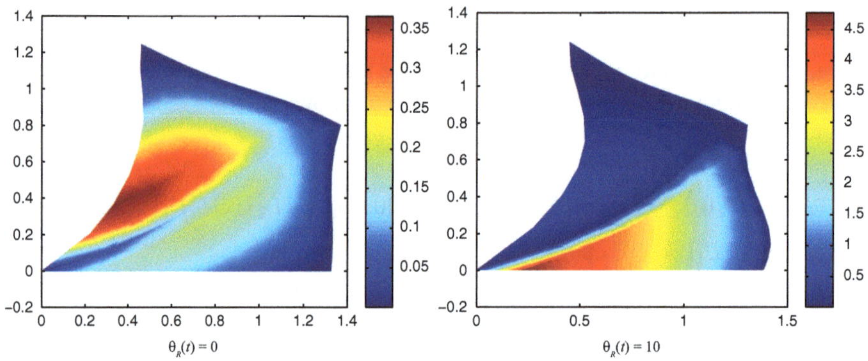

Fig. 10 Temperature field at final time, $g(x,0) = \frac{x}{2},\ 0 \le x \le 1$

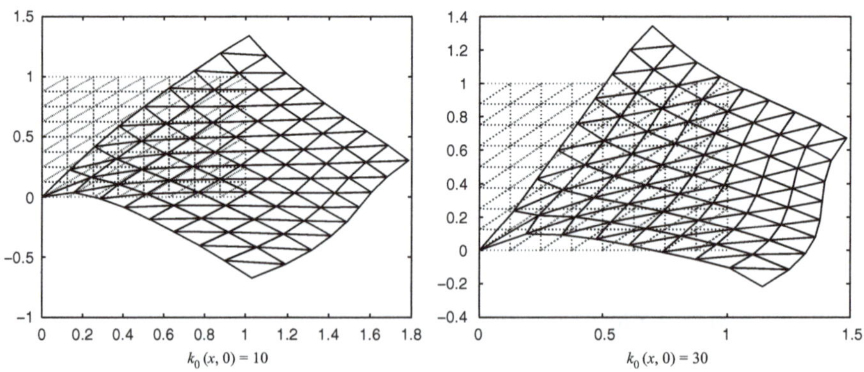

Fig. 11 Deformed configurations at final time, $\theta_R(t) = 0,\ 0 \le t \le 1$

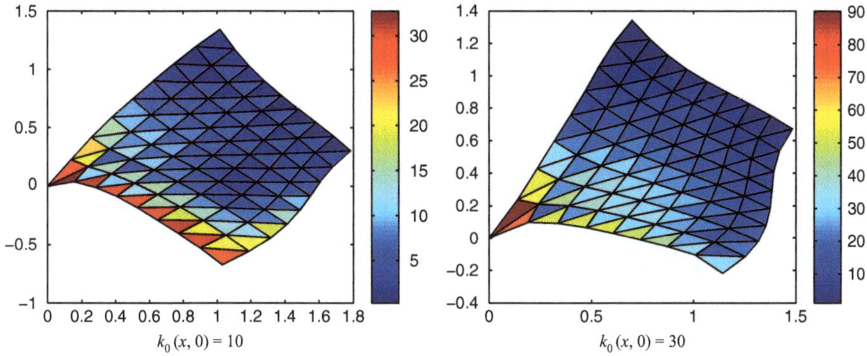

Fig. 12 Von Mises' norm in deformed configurations, $\theta_R(t) = 0,\ 0 \leq t \leq 1$

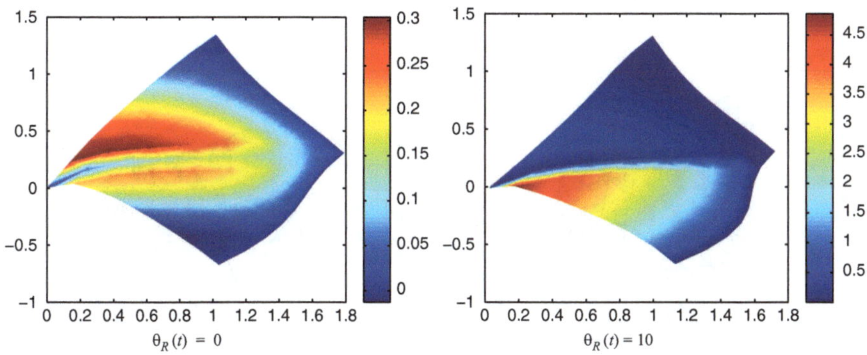

Fig. 13 Temperature field at final time, $k_0(x, 0) = 10,\ 0 \leq x \leq 1$

$$\Omega = (0, L_1) \times (0, L_2),$$

$$\Gamma_F = (\{0\} \times [0, L_2]) \cup ([0, L_1] \times \{L_2\}) \cup (\{L_1\} \times [0, L_2]), \quad \Gamma_u =]0, L_1[\times \{0\},$$

$$f_2(x,t) = (0,\ 0), \quad \forall x \in \{0\} \times [0, L_2], \quad \forall t \in [0, T],$$

$$f_2(x,t) = (1,\ 0), \quad \forall x \in (]0, L_1] \times \{L_2\}) \cup (\{L_1\} \times [0, L_2]), \quad \forall t \in [0, T].$$

Analogous conclusions as in Sect. 2 can be stated. See Fig.14 to Fig.19.

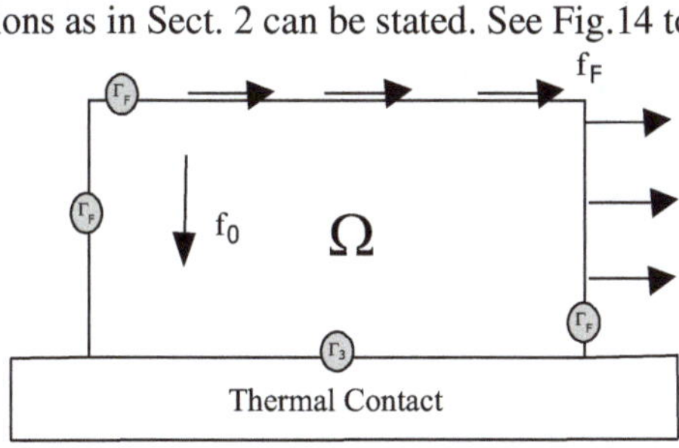

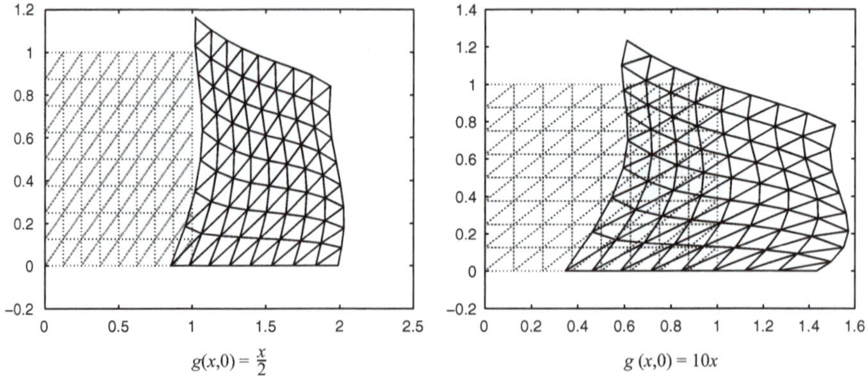

Fig. 14 Deformed configurations at final time, $\theta_R(t) = 10,\ 0 \le t \le 1$

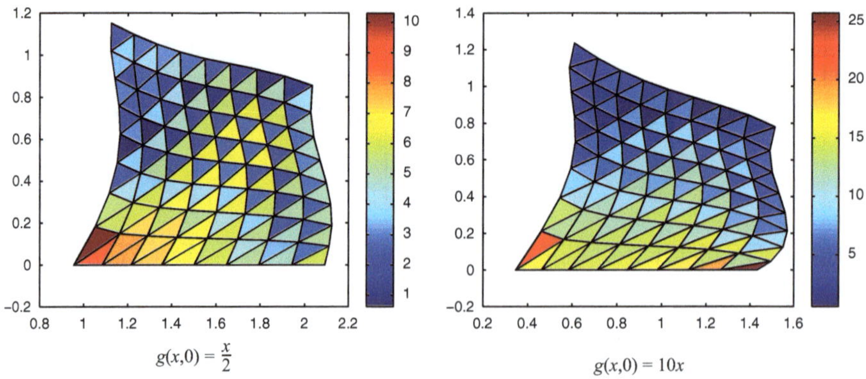

Fig. 15 Von Mises' norm in deformed configurations, $\theta_R(t) = 10,\ 0 \le t \le 1$

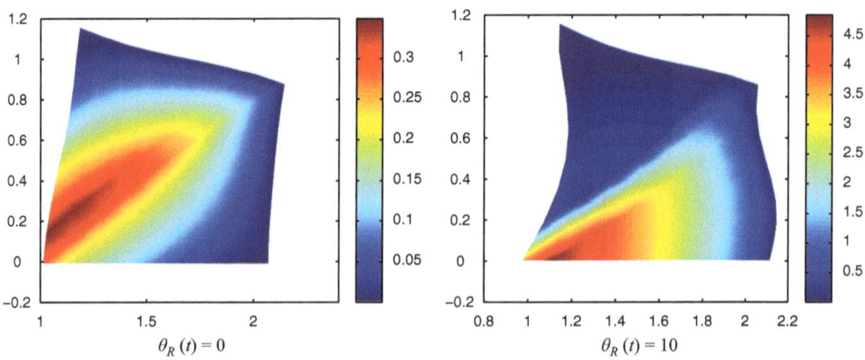

Fig. 16 Temperature field at final time, $g(x, 0) = \frac{x}{2},\ 0 \le x \le 1$

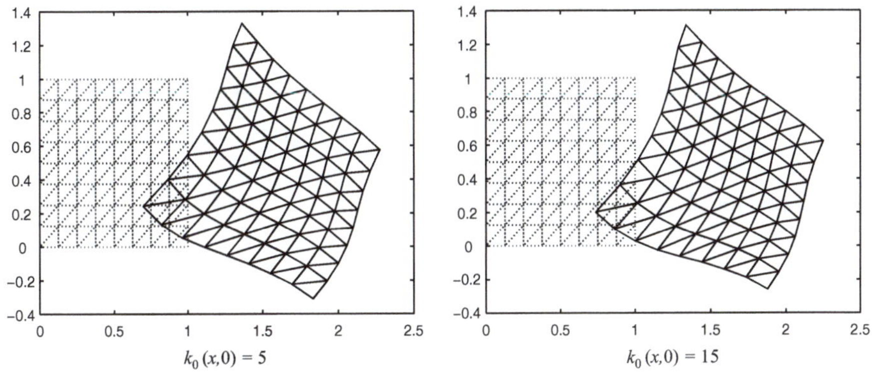

Fig. 17 Deformed configurations at final time, $\theta_R(t) = 10,\ 0 \le t \le 1$

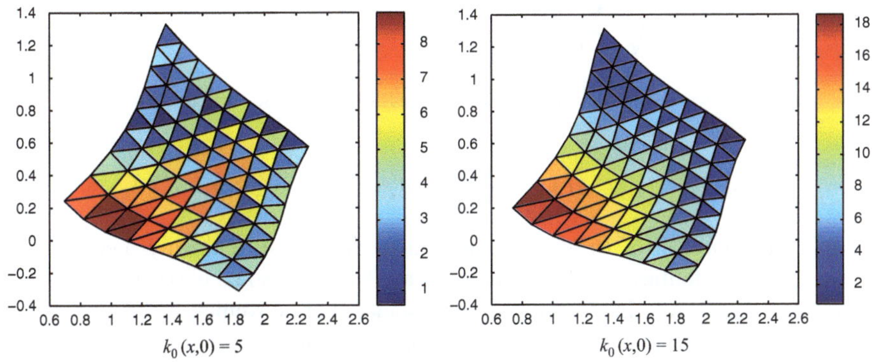

Fig. 18 Von Mises' norm in deformed configurations, $\theta_R(t) = 10,\ 0 \le t \le 1$

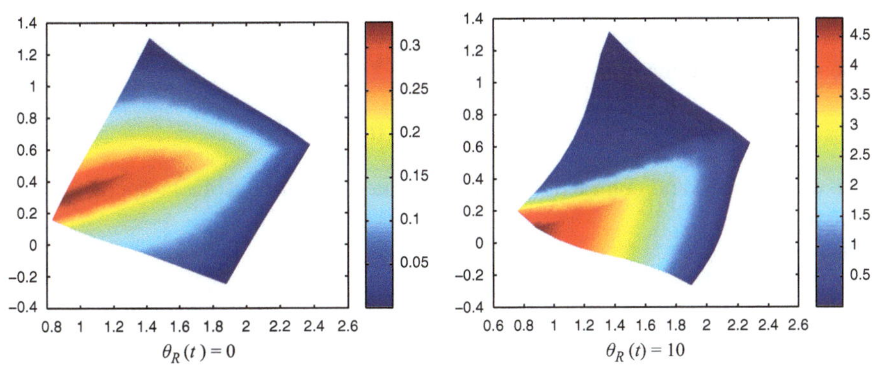

Fig. 19 Temperature field at final time, $k_0(x,0) = 15,\ 0 \le x \le 1$

4 A Duality Numerical Method

The fully discrete scheme (36) is equivalent to the variational inequality

$$\left\langle \mathcal{A} v_n^{hk}, w^h - v_n^{hk} \right\rangle + \psi\left(w^h\right) - \psi\left(v_n^{hk}\right) \geq \left\langle \mathcal{L}, w^h - v_n^{hk} \right\rangle_{V' \times V}, \forall w^h \in V^h \tag{60}$$

where operator $\mathcal{A} : V \to V'$ is defined by

$$\langle \mathcal{A} v, w \rangle = \left(\frac{v}{k}, w \right)_H + \langle A\, v, w \rangle_{V' \times V}, \quad \forall w \in V^h, \tag{61}$$

and $\mathcal{L} : V \to \mathbb{R}$ is defined by

$$\begin{aligned} \langle \mathcal{L}, w \rangle = & \left\langle \frac{v_{n-1}^{hk}}{k} - B\, u_{n-1}^{hk} - C\, \theta_{n-1}^{hk} + f(t_n), w \right\rangle_{V' \times V} \\ & - \left(k \sum_{m=0}^{n-1} \mathcal{B}\,(t_n - t_m)\, \varepsilon\left(u_m^{hk}\right), \varepsilon(w) \right)_{\mathcal{H}}, \quad \forall w \in V^h. \end{aligned} \tag{62}$$

For clearness, we drop the indexes and consider in the sequel the problem: Find $v \in V^h$ such that

$$\langle \mathcal{A} v, w - v \rangle + \psi(w) - \psi(v) \geq \langle \mathcal{L}, w - v \rangle_{V' \times V}, \quad \forall w \in V^h. \tag{63}$$

Numerical approach of (63) can be placed in the frame of duality methods for variational inequalities [4]. These methods are based on some classical results for monotone maps given in [6] and [24] for instance. For convenience, we give first a brief introduction to the monotonous maximal operator theory.

Let G be a maximal monotone multivalued map on a Hilbert space H, and let λ be a nonnegative parameter. It can be proved that for all $f \in H$ there exists a unique $y \in H$ such that

$$f \in (I + \lambda G)(y).$$

The single-valued map

$$J_\lambda^G = (I + \lambda G)^{-1}$$

is a well-defined and contraction map on H, its called the resolvent operator of G (see [6]).

The map

$$G_\lambda = \frac{I - J_\lambda^G}{\lambda}$$

is called the Moreau–Yosida approximation of G. It is a maximal monotone, single-valued, and $\frac{1}{\lambda}$-Lipschitz continuous map.

Moreover, G_λ satisfies the following important property on which is based our method.

Lemma 1 *Let be G a maximal monotone map on a Hilbert space H and G_λ, with $\lambda > 0$, its Yosida approximation. Then, for all y and u in H, we have*

$$u \in G(y) \Longleftrightarrow u = G_\lambda(y + \lambda u). \tag{64}$$

Proof Let be $u = G_\lambda(x)$. Then

$$\begin{aligned} u = \frac{I - J_\lambda^G}{\lambda} x &\Longleftrightarrow \lambda u = x - J_\lambda x \\ &\Longleftrightarrow J_\lambda x = x - \lambda u \\ &\Longleftrightarrow x \in (I + \lambda G)(x - \lambda u) = x - \lambda u + \lambda G(x - \lambda u) \\ &\Longleftrightarrow \lambda u \in \lambda G(x - \lambda u) \\ &\Longleftrightarrow u \in G(x - \lambda u) \end{aligned}$$

and by taking $x - \lambda u = y$ we get: $u \in G(y) \Longleftrightarrow u = G_\lambda(y + \lambda u)$. □

Now returning to problem (63), the frictional contact function ψ is continuous and convex, therefore, its subdifferential $\partial\psi$ is a maximal monotone operator in V_h, and (63) can be written using the subdifferential operator:
Find $v \in V^h$ such that

$$\begin{cases} \langle \mathcal{A}v, w\rangle + (\gamma, v)_{L^2(\Gamma_3)} = \langle \mathcal{L}, w\rangle_{V'\times V}, \ \forall w \in V^h, \\ \gamma \in \partial\psi(v). \end{cases} \tag{65}$$

Using relation (64), we obtain equivalently:
Find $v \in V^h$ such that

$$\begin{cases} \langle \mathcal{A}v, w\rangle + (\gamma, w)_{L^2(\Gamma_3)} = \langle \mathcal{L}, w\rangle_{V'\times V}, \ \forall w \in V^h, \\ \gamma = (\partial\psi)_\lambda(v + \lambda\gamma), \end{cases} \tag{66}$$

where $\lambda > 0$ and $(\partial\psi)_\lambda$ is the Yosida approximation of $\partial\psi$.

Thereby, we apply the following algorithm to solve (66).

(0) Start with some arbitrary value of the multiplier γ^0.
(1) For γ^j known, compute v_j solution to

$$\langle \mathcal{A}v^j, w\rangle + \left(\gamma^j, w\right)_{L^2(\Gamma_3)} = \langle \mathcal{L}, w\rangle_{V'\times V}, \ \forall w \in V^h. \tag{67}$$

(2) Update multiplier γ^j as

$$\gamma^{j+1} = (\partial\psi)_\lambda \left(v^j + \lambda\gamma^j\right). \tag{68}$$

(3) Go to (1) until stop criterion is reached.

Theorem 5 *If $\mathcal{A}$ is an elliptic operator and $\lambda > \frac{1}{2\alpha}$, where α is the elliptic constant of $\mathcal{A}$, we have*

$$\lim_{j \to \infty} \|v^j - v\| = 0.$$

Proof The mapping $(\partial\psi)_\lambda$ is $\dfrac{1}{\lambda}$-Lipschitz and thus

$$\begin{aligned}
\|\gamma - \gamma^{j+1}\|^2 &= \|(\partial\psi)_\lambda(v + \lambda\gamma) - (\partial\psi)_\lambda \left(v^j + \lambda\gamma^j\right)\|^2 \\
&\leq \frac{1}{\lambda^2}\|(v + \lambda\gamma) - \left(v^j + \lambda\gamma^j\right)\|^2 \\
&= \frac{1}{\lambda^2}\| \left(v - v^j\right) + \lambda\left(\gamma - \gamma^j\right)\|^2 \\
&= \frac{1}{\lambda^2}\|v - v^j\|^2 + \frac{2}{\lambda}\left((v - v^j), (\gamma - \gamma^j)\right) + \|\gamma - \gamma^j\|^2.
\end{aligned}$$

Therefore

$$\|\gamma - \gamma^j\|^2 - \|\gamma - \gamma^{j+1}\|^2 \geq -\frac{1}{\lambda^2}\|v - v^j\|^2 - \frac{2}{\lambda}\left((v - v^j), (\gamma - \gamma^j)\right). \tag{69}$$

Using now (66) and (67), we obtain

$$\langle \mathcal{A}(v - v^j),\, w\rangle + \left(\gamma - \gamma^j, w\right) = 0, \ \forall w \in V^h.$$

Thus

$$\begin{aligned}
\alpha\|v - v^j\|^2 &\leq \left\langle \mathcal{A}\left(v - v^j\right), \left(v - v^j\right)\right\rangle \\
&= -\left(\gamma - \gamma^j, v - v^j\right).
\end{aligned}$$

Substituting it in (69), we get

$$\begin{aligned}
\|\gamma - \gamma^j\|^2 - \|\gamma - \gamma^{j+1}\|^2 &\geq -\frac{1}{\lambda^2}\|v - v^j\|^2 + \frac{2\alpha}{\lambda}\|v - v^j\|^2 \\
&= \frac{1}{\lambda}\left(2\alpha - \frac{1}{\lambda}\right)\|v - v^j\|^2.
\end{aligned}$$

Recalling that $\lambda > \dfrac{1}{2\alpha}$, we obtain

$$\|\gamma - \gamma^j\|^2 - \|\gamma - \gamma^{j+1}\|^2 \geq \|v - v^j\|^2 \geq 0.$$

The sequence $\left(\|\gamma - \gamma^j\|^2\right)_{j \geq 0}$ is decreasing and positive, therefore

$$\lim_{j \to \infty} \|\gamma^j - \gamma\|^2 = 0$$

and finally

$$\lim_{j\to\infty} \|v^j - v\|^2 = 0. \qquad \square$$

Remark 2 Under symmetric property of operator $\mathcal{A}$ and if ψ is differentiable this algorithm is the Uzawa one to reach the saddle point of the Lagrangien:

$$L(v,q) = \frac{1}{2}\langle \mathcal{A}v,\, v\rangle - \langle \mathcal{L},\, w\rangle + (q, \psi(v))$$

$$L(u,q) \le L(u,p) \le L(v,p).$$

Now we turn out to determine the Yosida approximation $(\partial\psi)_\lambda$. Note that

$$(\partial\psi)_\lambda = \left((\partial\psi)^{-1} + \lambda I\right)^{-1}. \tag{70}$$

Indeed, $\forall u, x \in V$ we have from (64):

$$\begin{aligned} u = (\partial\psi)_\lambda(x) &\Longleftrightarrow u \in (\partial\psi)(x - \lambda u) \\ &\Longleftrightarrow (x - \lambda u) \in (\partial\psi)^{-1}(u) \\ &\Longleftrightarrow x \in \left(\lambda I + (\partial\psi)^{-1}\right)(u) \\ &\Longleftrightarrow u = \left(\lambda I + \left(\partial\psi\right)^{-1}\right)^{-1}(x). \end{aligned}$$

and from (70), we note that the Yosida approximation of $(\partial\psi)$ and the resolvent of $(\partial\psi)^{-1}$ are linked by

$$(\partial\psi)_\lambda(x) = J_{1/\lambda}^{\partial\psi^{-1}}\left(\frac{x}{\lambda}\right). \tag{71}$$

Let be $u, x \in V$, we have

$$\begin{aligned} u = (\partial\psi)_\lambda(x) &\Longleftrightarrow u = ((\partial\psi)^{-1} + \lambda I)^{-1}(x) \\ &\Longleftrightarrow x \in \left((\partial\psi)^{-1} + \lambda I\right)(u) \\ &\Longleftrightarrow \frac{x}{\lambda} \in \left(\frac{1}{\lambda}(\partial\psi)^{-1} + I\right)(u) \\ &\Longleftrightarrow u = \left(\frac{1}{\lambda}(\partial\psi)^{-1} + I\right)^{-1}(x/\lambda) = J_{1/\lambda}^{\partial\psi^{-1}}(x/\lambda). \end{aligned}$$

Definition 1 The map

$$\begin{aligned} \psi^* : V &\longrightarrow \mathbb{R} \\ x &\longmapsto \sup_{y\in V}\{(x,y)_{L^2(\Gamma_3)} - \psi(y)\} \end{aligned}$$

is the Fenchel conjugate of ψ.

Theorem 6 *The Fenchel conjugate ψ^* is well defined, continuous, and convex. Its subdifferential is the inverse subdifferential $(\partial\psi)^{-1}$ of ψ, and we have*

$$y \in (\partial\psi)(x) \Longleftrightarrow x \in (\partial\psi^*)(y). \tag{72}$$

From (71) and (72), we get the equality linking the resolvent of $\partial\psi$ and the Yosida approximation of $(\partial\psi^*)$,

$$J_\lambda^{\partial\psi}(x) = (\partial\psi^*)_{\frac{1}{\lambda}}(\lambda x). \tag{73}$$

Let us set:

$$K = \left\{ f \in L^2(\Gamma_3) : (f, w) - \psi(w) \le 0,\ \forall w \in L^2(\Gamma_3) \right\}.$$

Theorem 7 *The Fenchel conjugate ψ^* satisfies*

$$\psi^* = I_K, on L^2(\Gamma_3), \tag{74}$$

where I_K is the indicator function of K.

Proof Let $f \in L^2(\Gamma_3)$ be given. There are two possibilities. If there exists $w \in L^2(\Gamma_3)$ such that

$$(f, w) - \psi(w) > 0,$$

then for $r > 0$:

$$(f, rw) - \psi(rw) = r((f, w) - \psi(w)),$$

and

$$\psi^*(f) = \sup_{w \in L^2(\Gamma_3)} \{(f, y) - \psi(w)\} = +\infty.$$

If such a w does not exist then

$$(f, w) - \psi(w) \le 0,\ \forall w \in L^2(\Gamma_3),$$

but this quantity vanishes for $w = 0$, so that:

$$\psi^*(f) = 0$$

□

Consequently, we can compute $(\partial\psi^*)$ as (∂I_K) and we obtain

$$(\partial\psi^*)(y) = (\partial I_K)(y) = N_K(y),\ \forall y \in K, \tag{75}$$

where

$$N_K(y) = \left\{ f \in L^2(\Gamma_3) : (f, w - y) \le 0,\ \forall w \in K \right\}.$$

is the normal cone of K in y.

On the other hand, we can easily prove that the Yosida approximation of (∂I_K) is

$$(\partial I_K)_\lambda = \frac{I - P_K}{\lambda}, \tag{76}$$

where P_K is the projection operator on K.
Now, taking into account (73), (75), and (76), we have for all $x \in V$

$$\begin{aligned} J_\lambda^{\partial\psi}(x) &= \left(\partial\psi^*\right)_{\frac{1}{\lambda}}(\lambda x) \\ &= (\partial I_K)_{\frac{1}{\lambda}}(\lambda x) \\ &= \lambda(I - P_K)(\lambda x) \\ &= \lambda^2 x - \lambda P_K(\lambda x). \end{aligned}$$

We get finally

$$\begin{aligned} (\partial\psi)_\lambda(x) &= \frac{I - J_\lambda^{\partial\psi}}{\lambda}(x) \\ &= \frac{x - \lambda^2 x + \lambda P_K(\lambda x)}{\lambda} \\ &= \frac{1-\lambda^2}{\lambda} x + P_K(\lambda x). \end{aligned}$$

Therefore, the multiplier γ^j in (68) is updated by the formula

$$\gamma^{j+1} = \frac{1-\lambda^2}{\lambda}\left(v^j + \lambda\gamma^j\right) + P_K\left(\lambda v^j + \lambda^2\gamma^j\right). \tag{77}$$

References

1. Addi, K., Chau, O., Goeleven, D.: On some frictional contact problems with velocity condition for elastic and visco-elastic materials. Discret. Contin. Dyn. Syst. **31**(4), 1039–1051 (2011)
2. Adly, S., Chau, O.: On some dynamical thermal non clamped contact problems. To appear in Mathematical Programming, series B
3. Adly, S., Chau, O., Rochdi, M.: Solvability of a class of thermal dynamical contact problems with subdifferential conditions. Numer. Algebra Control Optim. (NACO) **2**(1), 89–101 (2012)
4. Bermudez, A., Moreno, C.: Duality methods for solving variational ineq ualities. Comput. Math. Appl. **7**, 43–58 (1981)
5. Brézis, H.: Problèmes unilatéraux. J. Math. Pure Appl. **51**, 1–168 (1972)
6. Brézis, H.: Operateurs maximaux monotones et semigroups de contractions dans les espaces de Hilbert. North-Holland, Amterdam (1973)
7. Brézis, H.: Analyse fonctionnelle, Théorie et Application. Masson, Paris (1987)
8. Chau, O., Awbi, B.: Quasistatic thermovisoelastic frictional contact problem with damped response. Appl. Anal. **83**(6), 635–648 (2004)

9. Chau, O., Motreanu, D., Sofonea, M.: Quasistatic frictional problems for elastic and viscoelastic materials. Appl. Math. **47**(4), 341–360 (2002)
10. Chau, O., Goeleven, D., Oujja, R.: A numerical treatment of a class variational inequalities arising in the study of viscoelastic materials. Submitted to Int. J. Appl. Math. Mech. (May 2012)
11. Ciarlet, P.G.: The Finite Element Method for Elliptic Problems. North Holland, Amsterdam (1978)
12. Ciarlet, P.G.: Mathematical Elasticity, Vol. I : Three-Dimensional Elasticity. North Holland, Amsterdam (1988)
13. Duvaut, G., Lions, J.L.: Les Inéquations en Mécanique et en Physique. Dunod, Paris (1972)
14. Ekeland, I., Teman, R.: Analyse convexe et problèmes variationelles. Gauthier-Villars, Paris (1984)
15. Glowinski, R.: Numerical Methods for Nonlinear Variational problems. Springer-Verlag, New York (1984)
16. Goeleven, D., Motreanu, D., Dumont, Y., Rochdi, M.: Variational and Hemivariational Inequalities, Theory, Methods and Applications, Volume I: Unilateral Analysis and Unilateral Mechanics. Kluwer Academic, Dordrecht (2003)
17. Han, W., Reddy, B.D.: Plasticity: Mathematical Theory and Numerical Analysis. Springer-Verlag, New York (1999)
18. Jeantet, R., Croguennec, T., Schuck, P., Brulé, G.: Science des Aliments. Lavoisier, Paris (2006)
19. Kikuchi, N.N., Oden, J.T.: Contact Problems in Elasticity : A Study of Variational Inequalities and Finite Element Methods. SIAM, Philadelphia (1988)
20. Lions, J.L.: Quelques méthodes de résolution des problèmes aux limites non linéaires. Dunod et Gauthier-Villars, Paris (1969)
21. Nečas, J., Hlavaček, I.: Mathematical Theory of Elastic and Elastoplastic Bodies: An Introduction. Elsevier, Amsterdam (1981)
22. Panagiotopoulos, P.D.: Inequality Problems in Mechanical and Applications. Birkhauser, Basel (1985)
23. Panagiotopoulos, P.D.: Hemivariational Inequalities, Applications in Mechanics and Engineering. Springer-Verlag, Berlin (1993)
24. Pazy, A.: Semigroups of non-linear contractions in Hilbert Spaces. In: Prodi, G. (ed.) Problems in Nonlinear Analysis (C.I.M.E. Ed.). Cremonese, Roma (1971)
25. Raous, M., Jean, M., Moreau, J.J. (eds.): Contact Mechanics. Plenum Press, New York (1995)
26. Zeidler, E.: Nonlinear Functional Analysis and its Applications. Springer, New York (1997)

Lucjan Emil Böttcher and his Mathematical Legacy

Stanisław Domoradzki and Małgorzata Stawiska

Abstract This contribution concerns the life and work of Lucjan (Łucjan) Emil Böttcher (1872–1937), a Polish mathematician. Besides biographical and bibliographical information, it contains a survey of his mathematical achievements in the theory of iteration and holomorphic dynamics. Some documents are presented for the first time.

Keywords Nineteenth and twentieth century mathematics · Iteration · Holomorphic dynamics · Warsaw · Leipzig · Lvov

1 Introduction

The name of Lucjan Emil Böttcher is familiar to mathematicians interested in functional equations, theory of iterations or dynamics of holomorphic functions. It is associated with Böttcher's equation $F(f(z)) = [F(z)]^n$, where $f(z) = z^n + a_{n+1}z^{n+1} + \ldots,\quad n \geq 2$ is a known function which is analytic in some neighborhood of the point 0 in the complex plane) [38], Böttcher's coordinate (the unknown function in Böttcher's equation, also called Böttcher's function or Böttcher's map) and Böttcher's theorem (establishing existence of Böttcher's map under certain conditions). But overall, he remains a relatively obscure figure and his works are scarcely known.

Here, we try to present Lucjan Emil Böttcher as a mathematician truly "without borders." Born in Warsaw (then under Russian rule) in a Polish family of Evangelical Lutheran denomination (where Roman Catholics constituted a majority), he studied mathematics in Warsaw and engineering in Lvov (then in Austro-Hungarian monarchy), received his doctorate in Leipzig, and worked as an academic teacher in Lvov (which after 1918 became a part of newly independent Poland). We write in detail

S. Domoradzki (✉)
Faculty of Mathematics and Natural Sciences, University of Rzeszów, ul. prof. S. Pigonia 1, 35–959 Rzeszów, Poland
e-mail: domoradz@ur.edu.pl

M. Stawiska
Mathematical Reviews, 416 Fourth St., Ann Arbor, MI 48103, USA
e-mail: stawiska@umich.edu

T. M. Rassias, P. M. Pardalos (eds.), *Mathematics Without Boundaries*,
DOI 10.1007/978-1-4939-1106-6_5,

about his education, people who influenced him, his career and obstacles he faced in its advancement, and his activities in scientific societies. We add some newly found biographical details to what the first author of this article has already written about Böttcher's life and activities [14, 15]. In particular, we compile a complete (to the best of our knowledge) bibliography of Böttcher's publications and, on its basis, discuss his mathematical ideas and results, along with their later development and impact. We highlight the places where particular concepts appear in his works. For example, all authors quoting Böttcher's theorem (as well as the function and equation named after him) refer to his paper published in 1904 in Russian, while we find out that the result already appeared in one of his papers, written in Polish, in 1898.

Quite recently, another contribution by Böttcher got recognition. Namely, it was pointed out in [3], [18], and [35] that Böttcher gave examples of rational maps for which the whole sphere is the "chaotic" set, 20 years before Samuel Lattès independently came up with maps with the same property, which are now known as "Lattès examples." Although one such example appeared even earlier in the work of Ernst Schröder, Böttcher was the first to consider these examples from the dynamical point of view, and (according to Milnor [35]) he was the first to use the term "chaotic" in reference to their behavior. Moreover, Alexander et al. [3] acknowledged,

> (...) he seems to have made a conceptual leap that would not be seen in print in the French study until Fatou's 1906 *Comptes Rendus* notice: given a particular function f, he viewed the sphere as partitioned into convergence regions by boundary curves. (p. 177)

Such a partition arose from the study of the convergence of Newton's method for a quadratic equation undertaken by Arthur Cayley; other mathematicians, including Schröder, attempted to describe convergence regions in other cases, without much progress. It was Böttcher who formulated several general properties of boundary curves of the regions of convergence (the boundaries are now known to be contained in the Julia set) and described them explicitly in the simplest cases (of monomials and Chebyshev polynomials). He also stated an upper bound for the number of "regions of convergence" (properly speaking, of nonrepelling cycles) of a rational function in terms of the number of critical points. The bound was later conjectured by Fatou (who himself proved a weaker estimate) and was proved to be sharp, by M. Shishikura, only in 1980s. However, it should also be noted that Böttcher presented very few proofs of his statements and his account is mostly schematic, sometimes hypothetic, plainly speculative, or even mistaken. This was the main reason for his being underappreciated during his lifetime. His forerunning insights had to wait for full development by other mathematicians who rediscovered them independently. These significant ideas, while few in number, not only have their place in history but are very much a part of today's mathematics and deserve to be widely known. So does their author.

2 The Life of Lucjan Emil Böttcher

Lucjan Emil Böttcher was born on January 7 (21),[1] 1872, in Warsaw, in a Lutheran family. His father was Piotr, and his mother was Anna née Kraus. In Warsaw,[2] in the years 1881–1885, he attended Herman Benni's[3] four-class real school; later, in 1886–1891, he attended Pankiewicz's[4] six-class real school. Having completed the latter he passed an exam on the material of six classes in a state real school in Warsaw. His intention was to study mathematics at the university in Warsaw, for which he needed education in classics and the maturity exam (matura). He completed his education in the classical gymnasium in Łomża (then in the Polish Kingdom under the Russian rule, which since 1883 was also known as Vistula Country), where he passed his maturity exam in 1893. Then he began studying mathematics at the Imperial University of Warsaw. He was a student in the academic year 1893–1894. As he noted in his CV, he had to leave the university because of his participation in a demonstration in the honor of Colonel Kiliński.[5] In Warsaw, he attended lectures in mathematical analysis (by Nikolay Yakovlevich Sonin), analytic geometry (by Vassily Afanasyevich Anisimov), descriptive astronomy (by Ehrenfeucht), as well as in general chemistry and physics. Later he moved to Lvov and enrolled as a student in the Division of Machine Construction of the Polytechnic School, where he studied in the years 1894–1897. He obtained a so called half diploma and passed the first state exam. As a student in Lvov, he was active in the students' engineering circle. In the c.k. (imperial and royal) Polytechnic School in Lvov, a student had to pass the following mathematical subjects in the first 2 years of study: mathematics-Course I (6 h of lectures in the winter semester and 6 h of lectures in the summer semester), descriptive geometry (5 h of lectures and 10 h of exercises, called repetitory, each semester, respectively), repetitory in elementary mathematics (2 h each, respectively), repetitory in higher mathematics (2 h each).

---

[1] 21 is listed in the official documents of the Lvov Polytechnic related to Böttcher's retirement; in parentheses there is February 2. In other documents, e.g., Böttcher's CV, January 7 is listed. These discrepancies are due to the differences between the Julian and Gregorian calendars. See personal file, L. Böttcher, fond 26, op. 5–58, Lvov District Archive.

[2] There was no Polish state at that time. Poland was partitioned among three occupants: Russia, Austria, and Prussia. Warsaw was under Russian occupation; Lvov belonged to the Austro-Hungarian monarchy.

[3] Herman Benni (1834–1900), an Evangelical Lutheran pastor, a graduate of theological studies in Dorpat. In 1880 he opened a private men's School, which was closed after 5 years at the excuse of lectures being conducted in Polish. This was a time of intensified russification. See p. 239, [41].

[4] Jan Pankiewicz (1816–1899), a graduate of the St. Petersburg university, where he obtained the degree of candidate in philosophy. In 1841, he taught mathematics in the Real Gymnasium in Warsaw and descriptive geometry in the School of Fine Arts. A school director and inspector, a translator of works in mathematics and chemistry, e.g., A. M. Legendre's *Beginnings of Geometry* (1844), an author of entries in mathematics in Orgelbrand's *Encyklopedia Powszechna*. The school which Böttcher attended was founded in 1876. Pankiewicz managed it until 1894.

[5] Jan Kiliński (1760–1819), a shoemaker by trade, a leader of the burghers of Warsaw during the Kościuszko insurection in 1794, promoted to the rank of colonel. Imprisoned in St. Petersburg in 1794–1796.

It should be added that for the first state exam in the Division of Machine Construction, the following material in mathematics was compulsory: mathematics, course I and II, and descriptive geometry. It is noteworthy that already in 1895, Böttcher published in Lvov lithographed materials for students in differential and integral calculus.[6]

At the beginning of 1897, he interrupted his course of technical studies in Lvov and moved to Leipzig in order to study mathematics. There, he attended the lectures of the following professors: Sophus Lie (*Theory of differential invariants, Theory of differential equations, Theory of continuous transformation groups*; seminars *Theory of integral invariants* and *Differential equations*), Adolph Mayer (*Higher analytical mechanics*), Friedrich Engel (*Differential equations, Algebraic equations, Non-Euclidean geometry*) and Felix Hausdorff (*Similarity transformations*). He finished his studies in 1898, obtaining the Doctor of Philosophy degree on the basis of the dissertation "Beiträge zu der Theorie der Iterationsrechnung" (published by Oswald Schmidt, Leipzig) and complying with other procedures. He was extremely industrious; in a very short time, he managed to prepare the doctoral dissertation. Formally, his work was supervised by Sophus Lie, an outstanding mathematician, known, among other things, for his study of continuous transformation groups (now called Lie groups).[7] Böttcher's intention was to treat in his thesis the theory of iteration from the point of view of Lie groups. Even though this ambitious attempt remained largely unsuccessful, he managed to outline several deep ideas. There was some disagreement among the committee members regarding evaluation of his thesis; in the official application to grant Böttcher the doctoral degree, the signature of Wilhelm Scheibner, who did not want to supply an official report, is struck out. Lie then engaged in correspondence with the university's officials in support of Böttcher, explaining importance of Böttcher's investigations and their relation with his own research:

> As both the author and Mr. Scheibner indicate the relationship of the submitted work to my concept of one-parameter groups of transformations, I agree in part with these comments. The relationship, however, lies a little deeper. In 1874, I thought that every finite transformation of a finite continuous group is contained in a one-parameter subgroup. In 1883, I formulated the question whether this fundamental theorem also applies to infinite continuous groups. However, since this issue exceeds not only my strength, but also the strength of the current analysis, I restricted myself mainly to showing only for specific examples that this question can be answered in the affirmative.
> Various authors, including Mr. Böttcher, considered the same issue for a particularly important group, namely the group of all point transformations. However, I cannot admit that the author has managed to definitively substantiate significant new results. Despite all of this, his considerations, which testify to the diligence and talent, have their value. (…). In any case, I (as well as Mr. Scheibner) agree that this attempt be accepted as a thesis and we also agree regarding the grade II. I am choosing such a good grade because Mr Böttcher himself chose the topic and developed it independently [15].

[6] Information from Böttcher's CV. We were unable to find these lectures.
[7] Among Lie's students, there were Georg Scheffers and Kazimierz Żorawski.

Ultimately, Böttcher's thesis was evaluated as IIa (admodum laudabilis, the second highest grade) by Lie, who wrote that

> [t]he candidate is an intelligent mathematician, possessing good and solid knowledge.

Some materials from the University Archive in Leipzig (Universitätsarchiv Leipzig, Phil. Fak., Prom., 714, Bl 7) are presented in the Appendix.

The Polish version of Böttcher's thesis, modified and expanded, was published in three parts in the years 1899–1902 in "Prace Matematyczno-Fizyczne," practically the first Polish journal devoted solely to mathematical and natural sciences, which was privately financed by a distinguished publisher and historian of mathematics, Samuel Dickstein (1851–1939).

After returning from Leipzig to Lvov, Böttcher was appointed to the post of an assistant in c.k. Polytechnic School in Lvov. He worked there from October 1, 1898 to September 30, 1910, initially as the chair of mechanical technology and later (since 1899) as the chair of mathematics. He had his Ph.D. diploma from Leipzig nostrified in Lvov in 1901. Since that year, he made multiple attempts to obtain a license to lecture at the Lvov University. We present the copies of the documents in the Appendix. Let us outline his application for the license to lecture (*venia legendi* and habilitation). He first applied for admission to habilitation on October 2, 1901. The committee appointed for his habilitation proceeding met on on January 13, 1902, and the unanimously negative motion by the committee "not to admit to the further stages of habilitation" was accepted by the Faculty Council on March 14, 1902. In 1911, Böttcher obtained *veniam legendi* in mathematics in the c.k. Polytechnic School in Lvov. It was possible to obtain habilitation in the Polytechnic School in Lvov during the late 1890s. This possibility was open also for practicing professionals without Ph.D. Some of them complemented this procedure later with Ph.D. degrees from other universities and polytechnic schools of the Austro-Hungarian monarchy. We would like to point out different roles played by habilitation at universities and at polytechnic schools.

Since 1910, Böttcher had a position of an *adiunkt*. In 1911, he was also a privat-dozent of mathematics (additionally, he could lecture during the years 1912–1918). In the years 1920–1935, in the period of the Second Republic, he was a docent in the chair of mathematics, still employed as an *adiunkt*. Before obtaining habilitation at the Polytechnic School, he was an assistant to P. Dziwiński. They jointly ran recitations in elementary mathematics and in mathematics course II. Böttcher lectured on elements of higher mathematics in the Division of Architecture from 1910 to 1927. Between two wars, he lectured in the Division of Civil and Hydraulic Engineering on applied mathematics, theory of vectors, difference equations, notions and methods of elementary mathematics, methods of computation, and, in the Division of Mechanics, on theoretical mechanics and calculus of variations.

In 1911, Böttcher requested at the Faculty of Philosophy of the Lvov University that his license to lecture at the c.k. Polytechnic School be also recognized at the university. His request was denied. Another time he applied for habilitation in 1918. The committee's opinion was that

> Dr. Böttcher's works do not yield any positive scientific results. There are many formal manipulations and computations in them; essential difficulties are usually dismissed with a few words without deeper treatment. The content and character diverges significantly from current research.

Once again, already in free Poland, Böttcher requested recognition of his license to lecture on May 1, 1919, taking into account his research work (he was an author of about 20 works in mathematics; see the list and the analysis in further parts of the article). This time, too, the decision was unfavorable.
Böttcher took part in scientific meetings of mathematicians and philosophers, also in Conventions of Polish Naturalists and Physicians. In many of those conventions, there were mathematical and physical sections. He presented results of his research on functional equations. He also participated in Polish Philosophical Convention I in Lvov in 1923. His talk at this convention, "On Russell's antinomy" (published in 1927), was cited in W. Tatarkiewicz's "History of Philosophy," used by many generations of Polish students.

Böttcher belonged to Polish Mathematical Society in Lvov, which was created in 1917 at the initiative of Lvov mathematicians J. Puzyna (1856–1919), Z. Janiszewski (1888–1920), H. Steinhaus (1887–1972), A. Łomnicki (1881–1941), P. Dziwiński (1851–1936) and the philosopher T. Czeżowski (1889–1981).[8] Böttcher is not mentioned among speakers, but he is listed as an active member. Participation in conventions, the society's work, and activities in an engineering circle as a student allow us to view him as an extremely engaged and open scientist, devoted not only to mathematics.

Böttcher published about 20 papers in significant Polish, Russian, and French journals. Their topics and results are presented in a subsequent part of the article. He also dealt with mechanics: he published a lithographed "Lecture on general mechanics" and the article "Some remarks on the principle of inertia." In the (contemporaneous) article [31], Böttcher's lecture is characterized as follows:

> [it is] very carefully developed, including kinematics, statics and dynamics, with examples taking into account the needs of technologists. One can regret that this course was not prepared by the author to be printed as a textbook.

As noticed in Chap. 3 of [36], Böttcher also made an original attempt to express the foundations of mechanics through some psychological concepts. Additionally, he was concerned with the problems of mathematical education in high school and he wrote high school textbooks.[9] In 1911, he published in Warsaw "Principles of elementary algebra, adapted to the curriculum in the Polish Kingdom." This textbook was written in the spirit of the so-called Meran program, which recommended

---

[8] In 1919, Mathematical Society in Kraków was created, which in 1920 was transformed into Polish Mathematical Society. The operations of the Polish–Bolshevik war interrupted the activities of Polish Mathematical Society in Lvov in the second half of 1920. At the proposal of Polish Mathematical Society in Kraków, the Society in Lvov dissolved and then reconstituted as the Lvov Branch of Polish Mathematical Society.

[9] He was also involved in proofreading and editing other authors' works; [45] mentions his work on a text by Placyd Dziwiński and [11] lists him as an editor of Władysław Zajączkowski's lectures.

initiating students into thinking in terms of functions, and was meant for students interested in mathematics. It is vast, with 704 pages. Böttcher was also a wholehearted advocate of introducing differential and integral calculus at school. He considered the process of differentiation to be easier than division or taking roots of real numbers, or computing compound interest.[10] He also wrote a geometry textbook for high schools, "Principles of geometry with numerous exercises" (Warsaw 1908). The text contains material in planimetry and has 322 pages. In the reviewer's opinion,[11] the author was going beyond the curriculum, too often relating to the material in elementary physics, but his drawings were prepared very meticulously, which could facilitate understanding of the material by students. Let us note that Böttcher published texts for Polish schools under the Russian rule (where most instruction at that time was conducted in Russian), despite well-developed market for publications in Lvov. Therefore, the reviewer wrote:

> It could also be advantageously used in our high schools, despite being somewhat too vast, which however can be amended; it could be particularly recommended to more diligent students for studying and working out numerous exercises.

After 1912, Böttcher developed interest in spiritualism and metapsychology.[12] He wrote books on these topics, e.g., *"Table-turning"* (1912)[13] which is "a practical outline containing instructions for conducting seances with table-turning."
He retired on August 31, 1935.

Böttcher married Maria Wolle (in 1900) and had four children: Władysław (born 1901), a Defender of Lvov in 1919 [14]; Marian (1911); and two other, whose names are unknown. Here is how Paweł Böttcher, Lucjan's grandson, remembers his family:

> Both my maternal and paternal grandparents came from Lvov. After the war they were repatriated to Bielsko-Biała. They hoped it was only temporary. They did not want to take root there. They did not talk a lot about what they lost. This is why I know so little today. My paternal grandfather, Lucjan Böttcher, taught mathematics at the Lvov Polytechnic, where also my maternal grandfather was a student . . .

Paweł's father was Marian. Mr. Paweł Böttcher, whom the first author contacted, did not have any more information related to his family, as he himself noted in his memoirs from the Eastern Borderlands.

Lucjan Böttcher died on May 29, 1937, in Lvov.

---

[10] The idea of introducing differential and integral calculus occurred at Convention X of Polish Naturalists and Physicians, after meetings of the section of mathematics and physics and the section of scientific education. Böttcher noticed in talks about education in physics that the ideas of modern physics were contained in school curricula.

[11] The review by A. Wilk, [46].

[12] The interest might have been influenced by Wilhelm Wundt, whose lectures in psychology Böttcher attended as a student in Leipzig. The authors thank Prof. Cezary Domański for pointing out this possible connection. One of Wundt's students in 1870s was Julian Ochorowicz (1850–1917), a Polish scientist, publicist and inventor, who studied hypnosis, occultism, spiritualism, and telepathy. It should be noted, however, that Wundt was an experimentalist and considered the notion of soul to be irrelevant for explaining psychological phenomena.

[13] In the series: Biblioteka Wiedzy Ogólnej, published by Kultura i Sztuka, Lwów, 1912. There is a copy in the special collection of the Jagiellonian Library in Kraków.

[14] cf. Z. Popławski, [37].

3 The Bibliography of L. Böttcher

The bibliography was compiled as a result of search in: the journals *Jahrbuch Fortschritte der Mathematik, Prace Matematyczno-Fizyczne, Wiadomości Matematyczne, Muzeum*, the database Zentralblatt MATH, the bibliographic guide by Byelous [11] (items 2–8 and 11 in the list of other publications), "The Polish Bibliography" by K. Estreicher [19] and Böttcher's personal file in the District Archive in Lviv.[15] The list may be incomplete. The author self-published some of his works.

3.1 Mathematical Publications

1. Zasadnicze podstawy teoryi iteracyi [Essential foundations of the theory of iterations], pp. 8, *Pamiętnik Towarzystwa Politechnicznego we Lwowie*, issue 1, s. 126–133. a supplement to *Czasopismo Techniczne*, Lwów, 1897.
2. Beiträge zu der Theorie der Iterationsrechnung, published by Oswald Schmidt, Leipzig, pp. 78, 1898 (doctoral dissertation).
3. Przyczynki do teoryi rachunku iteracyjnego [Contributions to the theory of iterational calculus], *Wiadomości Matematyczne*, vol. II(1898), pp. 224–229 (the author's discussion of his doctoral dissertation).
4. Kilka słów z dziedziny rachunku iteracyjnego [A few words in the subject of iterational calculus], an offprint from *Czasopismo Techniczne*, vol. 17 (1899), pp. 56–57, Lwów. Also as a separate offprint, Lwów 1899, 5 p.
5. Zasady rachunku iteracyjnego (część pierwsza i część druga) [Principles of iterational calculus (part one and two)], *Prace Matematyczno - Fizyczne*, vol. X (1899–1900), pp. 65–86, 86–101.
6. Równania funkcyjne podstawnicze [Substitutional functional equations], *Wiadomości Matematyczne*, vol. IV (1900), pp. 233–235
7. O własnościach wyznaczników funkcyjnych [On properties of functional determinants], *Rozprawy Wydziału Matematyczno - Przyrodniczego Akademii Umiejętności w Krakowie*, vol. 38 (general volume) (1901); series II, vol. 18 (1901), pp. 382–389. Also as a separate offprint, Kraków 1901, 10 p.
8. Zasady rachunku iteracyjnego (część III) [Principles of iterational calculus (part III)], *Prace Matematyczno - Fizyczne*, vol. XII (1901), pp. 95–111
9. Zasady rachunku iteracyjnego (część III, dokończenie) [Principles of iterational calculus (part III, completion)], *Prace Matematyczno - Fizyczne*, vol. XIII (1902), pp. 353–371
10. Rozwijanie na szeregi potegowe funkcyi, określonej równaniem algebraicznym nieprzywiedlnem f(x,y) = 0 [Expansion into power series of functions defined by an algebraic irreducible equation f(x,y)=0], *Wiadomości Matematyczne*, vol. VII (1903), pp. 1–21. Also as a separate offprint, Warszawa 1903.

[15] Bïtcher is not mentioned in any of the editions of J. C. Poggendorff's biographical dictionary.

11. Glavn"yshiye zakony skhodimosti iteratsiy i ikh prilozheniya k" analizu [The principal laws of convergence of iterates and their application to analysis], *Bulletin de la Societe Physico-Mathematique de Kasan*, vol. XIII (1, 1903), pp. 1–37, XIV (2, 1904), pp. 155–200, XIV (3, 1904), pp. 201–234.
12. Iteracye funkcyi liniowej [Iterations of a linear function], *Wiadomości Matematyczne*, vol. VIII (1904), pp. 291–307. Also as a separate offprint, Warszawa 1904.
13. Iteracye funkcyi liniowej (ciag dalszy i dokończenie) [Iterations of a linear function (continuation and completion)], *Wiadomości Matematyczne*, vol. IX (1905), pp. 77–86. Also as a separate offprint, Warszawa 1905.
14. Nowa metoda całkowania układu n równań funkcyjnych, liniowych pierwszego rzędu postaci: $U_i(z) = \sum_{j=1}^{j=n} A_{i,j}(z)U_j F(z)$ [New method of integration of a system of n functional linear equations of the first order, of the form $U_i(z) = \sum_{j=1}^{j=n} A_{i,j}(z)U_j F(z)$], (a separate offprint), Lwów 1905, Druk. Związkowa, 16 p.
15. Notatka o rozwiązaniu równania funkcyjnego $\Psi f(z) - \Psi(z) = F(z)$ [A note on solution of the functional equation $\Psi f(z) - \Psi(z) = F(z)$], *Wiadomości Matematyczne*, vol. XIII (1909), pp. 17–21. Also as a separate offprint, Warszawa 1909.
16. Nouvelle méthode d'intégration d'un système de n équations fonctionnelles linéaires du premier ordre de la forme $U_i(z) = \sum_{j=1}^{j=n} A_{i,j}(z)U_j F(z)$, *Annales scientifiques de l'Ecole Normale Supérieure*, vol. 26 (1909), pp. 519–543.
17. Zasady rachunku iteracyjnego [Principles of iterational calculus], *Wektor* 9 (1912), pp. 501–513.
18. Iteratsiya $f_x(z)$ algebraicheskoy funktsii $f(z)$ kak" metatranstsendental'naya funktsiya ukazatelya x [Iteration $f_x(z)$ of an algebraic function as a metatranscendental function of the exponent x], *Bulletin de la Societe Physico-Mathematique de Kasan*, tome XVIII (3, 1912), pp. 106–125. Also as a separate offprint, Kasan 1913.
19. Przyczynek do rachunku iteracyj funkcyi algebraicznej wymiernej całkowitej [A contribution to the calculus of iterations of an algebraic rational entire function], *Wiadomości Matematyczne* XVI (1912), pp. 201–206.
20. Iteracye funkcyi kwadratowej [Iterations of a quadratic function], *Wiadomości Matematyczne*, v. XVIII (1914), pp. 83–132. Also as a separate offprint, Warszawa 1914.

3.2 *Other Publications of L. E. Böttcher*

Textbooks in mathematics and mechanics, remarks on mathematics education, mechanics, philosophy, logic, and occult/spiritualism:

1. Repetytoryum Wyższej Matematyki, Rachunek różniczkowy (55 ss.), Rachunek całkowy (47 ss.), (litografowany podręcznik), [A review course in higher mathematics, Differential calculus (55 pp.), Integral calculus (48 pp.), (a lithographed text)], Lwów, 1895.
2. O podziale kąta na trzy części równe [On dividing an angle into three equal parts], *Czasopismo Techniczne*, vol. 13(1895), p. 106.
3. Znużenie w szkole [Weariness at school], *Czasopismo Techniczne*, vol. 17(1899), pp. 4–5.
4. Teorya wyznaczników, cz. 1 [Theory of determinants, part 1], Lwów, 1899, pp. 118.
5. Z teoryi równan liczebnych [From the theory of numerical equations], *Czasopismo Techniczne*, 18 (1900), pp. 306–307.
6. Praktyczne rozwiązywanie liczebnych algebraicznych równan stopni wyższych [Practical solutions of numerical algebraic equations of higher degrees], *Czasopismo Techniczne* 19 (1901), 7–8, s. 15–17.
7. Przybliżony rachunek rzeczywistych pierwiastków [Approximate computation of real roots], *Czasopismo Techniczne* 19 (1901), pp. 91–92, s. 114.
8. Kilka uwag z powodu artykułu p. Bronisława Biegeleisena "U podstaw mechaniki", [Some remarks on the occasion of the article by Mr. Bronisław Biegeleisen, "At the foundations of mechanics"], *Czasopismo Techniczne* 20 (1902), pp. 147–148.
9. Obliczanie funkcyj trygonometrycznych kątów bardzo małych [Computing trigonometric functions of very small angles], *Czasopismo Techniczne* 20 (1902), pp. 255–256.
10. Wykłady mechaniki. Wyd. Kazimierz Bartel, rps. powiel. [Lectures in mechanics. Publ. by K. Bartel [16], mimeographed manuscript] Lwów 1904, cm 25, s. 308 (in the collection of the National Library in Warsaw; no title page, table of contents or bibliography).
11. Wykład mechaniki ogólnej, (litografowany podręcznik) [A lecture in general mechanics (a lithographed text)], published by Wacław Kutyłowski-Sokół in quarto, pp. 738 (Lwów 1905) (information after [31]).
12. Śp. Władysław Folkierski [Late Władysław Folkierski], *Czasopismo Techniczne* 22(1904), pp. 217–221.
13. Kilka uwag o zasadzie bezwładności [Some remarks on the principle of inertia], *Czasopismo Techniczne*, vol. 23(1905), pp. 237–240, 253–255, 269–271.
14. Kilka uwag w sprawie reformy nauczania matematyki w szkołach średnich [Some remarks on the reform of mathematics education in high school], *Muzeum*, vol. 23(1906), pp. 163–168.
15. Nauka matematyki w szkole średniej [The study of mathematics in high school], *Wszechświat*, vol. 26 (1907), pp. 545–549.

[16] Kazimierz Bartel (1882–1941), later a professor of descriptive geometry in the Lvov Polytechnic and prime minister of the government of Poland.

16. Zasady geometryi elementarnej do szkół z licznemi ćwiczeniami (podręcznik do gimnazjum) [Principles of elementary geometry for schools with numerous exercises (a textbook for middle schools)], publ. M. Arct, Warszawa, 1908.
17. Rektyfikacja elipsy, Rachunkowe i wykreślne metody przybliżonego wyznaczania obwodu elipsy, [Rectification of an ellipse, Computational and graphical methods of approximate determination of the circumference of an ellipse], *Czasopismo Techniczne*, Lwów, vol. 26 (1908), no. 11, pp. 185–187, no. 12, pp. 200–203.
18. Zasady algebry elementarnej. Podręcznik i zbiór zadań dla szkół, opracowany według najnowszych wymagań pedagogicznych, (704 ss.) [Principles of elementary algebra. A textbook and exercise set, compiled according to the latest pedagogical requirements (704 pp.)], Warszawa 1911.
19. Stoliki wirujące [Table-turning], series Biblioteka Wiedzy Ogólnej [Library of General Knowledge], publ. Kultura i Sztuka, Lwów, 1915, wyd. II przejrzane i uzupełnione (II edition, revised and completed), Lwów - Przemyśl, 1926.[17]
20. Problemat życia pozagrobowego. Nieśmiertelność duszy [The problem of afterlife. Immortality of soul.], publ. Kultura i Sztuka, Lwów, 1915.[18]
21. O zasadzie sprzeczności, [On the principle of contradiction], *Przegląd Filozoficzny*, vol. 30, issue 4 (1927), p. 284.
22. O antynomji Russela [On Russell's antinomy], *Przegląd Filozoficzny*, 30, issue 4 (1927), p. 291–292.

4 Mathematics of Lucjan Emil Böttcher

4.1 Glossary

In order to describe in some detail Böttcher's contributions to the area of iterations of rational functions on the Riemann sphere (which is the classical core of complex dynamics in one variable), let us start with a glossary of basic notions and properties related to his results (following [1, 7, 24, 28, 34]).

Let M be a set and let $f : M \mapsto M$ be a mapping. We define the iterates of f as $f^0 = Id$ and $f^{n+1} = f \circ f^n$, $n = 0, 1, \ldots$. The *orbit* of a point p is the set $\{f^n(p) : n \geq 0\}$. A point $p \in M$ is *periodic* for f if there is an $m \geq 1$ such that $f^m(p) = p$. In particular, fixed points are periodic. An orbit of a periodic point is also called a *periodic cycle*. A point is called preperiodic if there are some $m, n > 0$ such that $f^m(p) = f^{m+n}(p)$. The *backward orbit* of a point p is the set $\bigcup_{n \geq 0} f^{-n}(\{p\})$ (the union of all preimages of p under all iterates of f).

Periodic cycles are examples of sets which are (forward) invariant under f. More precisely, a set $E \subset M$ is *forward invariant* under f if $f(E) = E$, and it is *backward invariant* under f if $f^{-1}(E) = E$.

[17] Estreicher gives the date 1913 for the first edition and 1927 for the second one.
[18] Estreicher gives the publication date as 1916.

When M is a topological space, one can study behavior of iterates of a continuous map f near its fixed points. A fixed point p of f is called *topologically attracting* if it has a neighborhood U such that all iterates f^n are defined in U and the sequence $f^n|_U$ converges uniformly to the constant map with value p in U. The *basin of attraction* of p is the set of all points z such that the sequence $z, f(z), f^2(z), \ldots$ converges to p. A fixed point p is *topologically repelling* if it has a neighborhood U such that for every $p' \neq p$ in U there exists an $n \geq 1$ such that $f^n(p')$ lies outside of U.

A map is *topologically transitive* if for every pair of nonempty open sets U and V in M, there is a nonnegative integer n such that $f^n(U) \cap V \neq \emptyset$.
Two maps f and g on M are *topologically conjugate* if there exists a homeomorphism $h : M \mapsto M$ such that $h^{-1} \circ f \circ h = g$.

Let (M, d) and (N, d') be metric spaces and assume that N is compact. A family $\mathcal{F} \subset C(M, N)$ is *normal* if every infinite sequence in $\mathcal{F}$ contains a subsequence which converges uniformly on every compact subset of M.

Consider now the case when $M = N = \hat{\mathbb{C}}$, the Riemann sphere (with, e.g., the spherical metric), and $f : \hat{\mathbb{C}} \mapsto \hat{\mathbb{C}}$ is a rational map of degree $d \geq 2$, i.e., $f = P/Q$ for coprime complex polynomials P, Q with $\max(\deg P, \deg Q) = d \geq 2$. The properties of a fixed point p of f are determined by its so called multiplier, i.e., the value $f'(p)$ when $p \in \mathbb{C}$; the multiplier at ∞ is understood as the derivative of $1/f(1/z)$ at $z = 0$. We say that p is *superattracting* if $f'(p) = 0$, *attracting* if $0 < |f'(p)| < 1$; *repelling* if $|f'(p)| > 1$; *rationally neutral* if $f'(p)$ is a root of unity (called also parabolic if $f'(p) = 1$); *irrationally neutral* if $|f'(p)| = 1$ but $f'(p)$ is not a root of unity. It can be proved that topologically attracting (resp. repelling) fixed points are exactly those that are attracting or superattracting (resp. repelling) in the sense of the multipliers. By the chain rule, the same classification using multipliers can be applied to periodic orbits.

The notion of a normal family plays a major role in the study of dynamics of rational maps. There are several criteria for the family of holomorphic maps on a subset of the sphere to be normal. For example, a theorem by P. Montel (who introduced the notion) states that a uniformly bounded family of holomorphic functions on a domain in $\mathbb{C}$ is normal. Let us also mention F. Marty's criterion: A family $\mathcal{F}$ of holomorphic maps on a region $U \subset \hat{\mathbb{C}}$ is normal if and only if for every compact $K \subset U$, there exists a constant C_K such that $\frac{|f'(z)|}{1+|f(z)|} \leq C_K$ for all $z \in K$ and all $f \in \mathcal{F}$. The quantity on the left-hand side of this inequality is called the spherical derivative of f.

Using the notion of a normal family, one can divide the sphere into two subsets. The *Fatou set* is the maximal open subset of $\hat{\mathbb{C}}$ on which the family of all iterates of f is normal and the *Julia set* is its complement. The Julia set is nonempty when $d \geq 2$, but the Fatou set may be empty (as demonstrated by Lattès examples). An example of a map with nonempty Fatou set is $f(z) = z^2$: it is straightforward to show that its Julia set is the unit circle. It is not always easy to determine Julia sets, but there are ways to approximate them in the Riemann sphere and to produce computer pictures of them. One such way is to look at the backward orbit of a "typical" point. Namely, if the backward orbit of a point a is infinite (which happens for all but at most two points in the sphere), then its closure contains the Julia set of f. Moreover, if a is in the Julia set of f, then the closure of its backward orbit equals the Julia set.

Attracting and superattracting cycles for f belong to the Fatou set, while the repelling cycles belong to the Julia set. In fact, repelling cycles are dense in the Julia set. In general, periodic cycles in the Fatou set (if any) can be identified by considering limit functions of the sequences of iterates of f. By Sullivan's theorem, every component of the Fatou set is preperiodic. Assume that the component F_0 is forward invariant. First consider the situation in which every limit of a subsequence of f^n is constant. Then all these constant functions are equal to ζ such that $f(\zeta) = \zeta$, and $f^n \to \zeta$ locally uniformly on F_0. Hence, ζ is an attracting or superattracting fixed point. If there are nonconstant limit functions, then the identity is among them, and the map f is analytically conjugate to an irrational rotation of the unit disk or an annulus (depending on whether F_0 is simply or doubly connected). One can also say something when there is a point $\zeta \in \partial F_0$ (hence in the Julia set) such that $f^n \to \zeta$ locally uniformly as $n \to \infty$. Then, necessarily $f'(\zeta) = 1$.

For a rational function f of degree $d \geq 2$ with a neutral fixed point z_0, f is locally conjugate to its linear part if and only if z_0 is in the Fatou set. For f with an irrationally neutral fixed point, $f'(z_0) = e^{2\pi i \xi}$, ξ an irrational real number, there are sufficient and necessary conditions of linearizability formulated in terms of approximation of ξ by rationals (given by H. Cremer, C.L. Siegel, A.D. Bryuno, and J.-C. Yoccoz), but much still remains to be investigated.

By using perturbation methods it can be proved that a rational map of degree d can have at most $2d - 2$ nonrepelling cycles. This upper bound equals the number of critical points of f, which are the zeros and poles of f'. In the 1920s, Fatou conjectured this bound, but was able to prove only a weaker one. It was proved only in the 1980s by M. Shishikura [42] that the estimate is sharp. In 1999, A. Epstein [17] gave a different proof of Fatou–Shishikura inequality.

One can construct maps having only repelling cycles. The *Lattès example* gives such a map satisfying $\mathcal{P}(2z) = f(\mathcal{P}(z))$, where $\mathcal{P}$ is the Weierstrass elliptic function associated with a certain lattice in $\mathbb{C}^2$. The construction proceeds as follows (for the details, see [7], Sect. 4.3). A lattice Λ is defined as $\Lambda = \{m\lambda + n\mu : m, n \in \mathbb{C}\}$, where λ and μ are fixed complex numbers which are not real multiples of each other. One says that a nonconstant function f is elliptic for Λ if it is a meromorphic function on $\mathbb{C}$ and each $\omega \in \Lambda$ is a period of f, e.g., the Weierstrass elliptic function $\mathcal{P}$ is defined as $\mathcal{P}(z) = \frac{1}{z^2} + \sum^* [\frac{1}{(z+\omega^2)} - \frac{1}{\omega^2}]$, where $\sum^*$ is taken over all nonzero elements in Λ. For $\mathcal{P}$, the addition formula $\mathcal{P}(2z) = R(\mathcal{P}(z))$ holds with the rational function $R(z) = \frac{z^4 + g_2 z^2/2 + 2g_3 z + (g_2/4)^2}{4z^3 - g_2 z - g_3}$, where g_2, g_3 are uniquely defined in terms of Λ. This function R has the whole sphere $\hat{\mathbb{C}}$ as its Julia set. One can show that the family R^n is not equicontinuous on any open subset of $\hat{\mathbb{C}}$: let D be any disk in $\mathbb{C}$ and let $U = \mathcal{P}^{-1}(D)$. For n large enough, the dilation $2^n U$ contains a period parallelogram $\Omega = \{z + s\lambda + t\mu : 0 \leq s, t \leq 1\}$ of Λ. By the addition formula and periodicity, for these n we have $R^n(D) = R^n(\mathcal{P}(U)) = \mathcal{P}(2^n U) = \hat{\mathbb{C}}$, which implies that the Fatou set for R is empty. For a comprehensive treatment of Lattès maps, see [35].

Rational maps for which the Julia set is the whole sphere can also be characterized by the following property: $J(R) = \hat{\mathbb{C}}$ if and only if there is some point z whose forward orbit $\{R^n(z) : n \geq 1\}$ is dense in the sphere.

Böttcher was interested in studying iterations of arbitrary (not necessarily integer) exponent and saw the theory of Lie groups as a suitable framework for this. A *Lie group* G is a smooth manifold and a group, with the operations of product $\mu : G \times G \mapsto G$ and inverse $i : G \mapsto G$ such that μ and i are smooth. An example is $\mathbb{R}^n$ considered as an additive group (with an atlas given by the identity map).

A *one-parameter subgroup* of G is a homomorphism of Lie groups $\theta : \mathbb{R} \mapsto G$ (considering $\mathbb{R}$ as a Lie group), i.e, a homomorphism of groups which is a smooth map. There is one-to-one correspondence between one-parameter subgroups of G and vectors in G_e, the tangent space to G at e, where e is the unit in G. It is given by $\theta \mapsto \theta'(0,1)$, where $(0,1)$ is the unit tangent vector to $\mathbb{R}$ at the origin. This correspondence can be used to define the exponential map $\exp : G_e \mapsto G$ as follows: let $v \in G_e$ and let θ_v be the corresponding one-parameter subgroup of G. Then, $\exp(tv) := \theta_v(t)$. The exponential map generates all one-parameter subgroups of G as follows: if θ is a one-parameter subgroup, then $\frac{\theta(t+h)-\theta(t)}{h} = \theta(t)\frac{\theta(t)-e}{h}$. Hence, $\frac{d\theta}{dt}(t) = \frac{d\theta}{dt}(0)\theta(t)$, so $\theta(t) = \exp(t\frac{d\theta}{dt}(0))$.

The problem of embedding (a discrete semigroup of) an iterated function into one-parameter group of transformations, which was of interest to Böttcher, is in general complicated. Some maps admit such embedding while for others there are obstructions. See the survey in [5] for discussion and references. One of the cases for which the question has been answered in full is that of analytic maps with a rational fixed point. We refer to [25, Chap. IV]. Here, we only can say, in very rough terms, that a functional invariant called the Ecalle–Voronin modulus is constructed. The analytic classification theorem for parabolic germs states that every parabolic germ is associated with a unique modulus, which is the same for all analytically conjugate germs, and conversely, two formally conjugate parabolic germs with the same modulus are analytically conjugate. A parabolic germ can be represented as a flow map of a holomorphic vector field if and only if its Ecalle–Voronin modulus is trivial. The modulus is also used to answer other questions about parabolic germs, e.g., the extraction of iterational roots.

A related problem is that of analytic iteration: for a given analytic function $f(z)$, find a function $\Phi(w,z)$ analytic in z and continuous in w such that $\Phi(w+u,z) = \Phi(w,\Phi(u,z))$ and $\Phi(1,z) = f(z)$. See [2] for some historical discussion and references and [13, 16, 29] for the relation between functional equations (including Böttcher equation) and fractional iteration.

4.2 The Works of Lucjan Emil Böttcher

Below we will discuss the content of Böttcher's research in mathematics. We included all publications we could access.

1. "Contributions to the theory of iterational calculus" (Polish)
   In this self-report, Böttcher announces forthcoming publication of his doctoral thesis in German and in Polish. He mentions its main topics such as a general notion of iteration, its relation with groups of transformations and convergence of iterations.
2. Böttcher's doctoral thesis
   In the foreword to his dissertation "Beiträge zu der Theorie der iterationsrechnung ... " written under the supervision of Sophus Lie and published in Leipzig in 1898, Böttcher states that his intention is not to give a complete presentation of the theory of iteration, but only of its most important problems. He sees the significance of theory of iteration in its relation with functional equations. He sets three goals for himself: (1) the study of iterated functions with exponents of iteration that are positive integers; (2) generalizing iteration to admit an arbitrary exponent: negative, rational, real, or complex; (3) the study of borderline between 1 and 2 above. The last point is unclear; perhaps after succeeding in theoretically generalizing iteration to admit an arbitrary exponent, he intended to go on to study the iteration of functions with these arbitrary exponents. This brings him to seek an interpretation of iteration in terms of groups of transformations, the theory of which was developed by Lie.
   Chapter I of the dissertation is devoted to expressing iterations by means of one-parameter continuous groups of transformations. The laws of addition, multiplication, and association are formulated to illustrate that the collection of iterated transformations with an arbitrary basic transformation forms a group. The analogy between arithmetic operations and iterations is highlighted. Iterations with arbitrary exponents are introduced as solutions of certain functional equations or limits of certain expressions. The need to consider complex numbers arises when solving the equation $xa^m = 1$ in order to define iterations with negative exponents or solving $KW(x_1, \ldots, x_n) = WL(x_1, \ldots, x_n)$ to get iterations with fractional exponents. It is observed that the fractional power is not a plain number, but a "manifold." Fundamental theorems are formulated on pp. 20–21. They are identified as special cases of some results by Lie. A relation is given between iterations and solutions of some differential equations.
   Later on, the general considerations give way to the specific study of the rational maps of the Riemann sphere. The question of convergence of iterates appears on p. 33, where the notion of an "iterational chain" is introduced. In modern terminology, it corresponds to the notion of an orbit of a point, where the successive images are additionally joined by arcs. It should be noted that the orbits are studied on their own, and various possibilities for limit functions of orbits are investigated, even though at that time the interest was mainly in the periodic behavior of whole functions rather than single orbits, i.e., in functions satisfying Babbage equation $f^n(z) \equiv z$. Böttcher notices that among the rational functions on the Riemann sphere, the only functions that are "periodic" are the Möbius transforms $z \mapsto (az + b)/(cz + d)$, as the degree of the rational map rises under iteration. The "points of convergence" are identified as fixed points of the iterated transformation, and it is observed that the magnitude of $f'(z)$ plays

a role in the classification of fixed points. The case $|f'(z)| < 1$ gives rise to "neighborhoods of convergence points," or, as they are known nowadays, basins of attraction. In the case $|f'(z)| = 1$, the distinction is made between $\arg f'(z)$ being rational or not, but more details are provided only when $f'(z) = 1$. It is stated that in the case $f'(z) = 1$, the fixed point z lies on the boundary of the "region of convergence." A description of petals (attracting and repelling) is given, following the work of L. Leau.

More interestingly, the boundaries of the regions of convergence (now known to be contained in the Julia set) came into focus. The following theorem is stated on p. 62. The boundary of the subregion associated with the point x, which belongs to the region of convergence, is a closed curve (fully determined in the sphere), which is invariant under the transformation $x_1 + iy_1 = f(x + iy)$. This theorem is not true in general, as the Julia set of a rational map (which equals the boundary of the Fatou set, if the latter is not empty) does not have to be a closed curve (e.g., for $z \mapsto z^2 + i$) or even a connected set (e.g., $z \mapsto z^2 + 4$), but it is true that it is totally invariant under the map f. Additionally, it is stated that

> [t]he whole iteration chain whose starting point is on the boundary of the subregions of convergence is fully registered by this boundary[,]

which may be interpreted as the statement that the map f is topologically transitive on its Julia set. In the process of determining the basin of attraction by iterating backward, a neighborhood of an attracting point (which mentions preimages of attracting periodic points, called by Böttcher "Żorawski points," after Kazimierz Żorawski, who studied them in his work on iteration), the way to approximate the "boundary curve" using backward orbits of points, is outlined. Even though the notion of a normal family was not available at that time, some of Böttcher's insights can be formulated in terms of normal families of holomorphic maps. First of all, he discussed the regions of convergence (taking into account only basins of attraction and parabolic petals) and of nonconvergence for iterates (he gave examples in which such a region is the whole sphere; nowadays we know that the Julia set which is not the whole sphere must be nowhere dense). Also, one of the few theorems for which he sketches a proof (in Sect. 37) reads as follows:

> A fundamental property of the invariant curves of the basic function $f(z)$ which simultaneously form the boundary of arbitrary subregions is that $\lim_{n\to\infty} f_n'(z) \neq 0$.

In view of Marty's criterion, it can also be interpreted as the property that on the Julia set the iterates of f do not form a normal family.

Section 38 starts with the observation that all points of "negative convergence" (i.e., repelling points) correspond to invariant curves of f. On p. 63, there is an example of a rational function without "regions of convergence," i.e., a function whose Fatou set is empty. It is $f(z) = (Az^4 + B(2z^2 - 1))/(B + A(2z^2 - z^4)) = \mathrm{Cn}(2)\mathrm{Cn}_{-1}(z)$, where Cn is the Jacobi elliptic cosine function. Later, Samuel Lattès would independently construct examples of maps whose Julia set is the whole sphere, also starting with elliptic functions.

In Part III, Böttcher studies a problem of defining an iterate with a general base and a general exponent as a solution of a functional equation.

3. "Principles of iterational calculus" (Polish)
There is substantial overlap between this paper and Böttcher's doctoral thesis, published almost simultaneously; perhaps this is what he referred to when announcing publication of his Ph.D. thesis in Polish. However, some new content appeared in this work.
Again, the main idea is to treat iterational calculus as a special chapter of the theory of one-parameter continuous groups of transformations. Böttcher revisits the group laws and their use in defining the iteration with an arbitary exponent. He points out to the role of conjugacy in these definitions and the ways to obtain conjugacy as a solution of a functional equation, attributing the point of view to Babbage. A new detail is the construction of an infinitesimal generator of one-parameter group, as well as of the "iterative logarithm" of a transformation f, following Korkine and Farkas.
As in the thesis, a good deal of attention is devoted to the convergence of iterations of rational functions. Some new light is thrown on the subject. It is made explicit that the investigation of division of the complex plane or sphere into the regions of convergence and divergence of iteration of a rational function is inspired by the work of Arthur Cayley, who succeeded in completely describing such division coming from the study of Newton–Fourier method for a quadratic equation over complex numbers. "Fundamental theorems" concern limits of orbits under iteration and "points of convergence" are classified. A Lattès-type example appears as an illustration of "transcendental convergence," different from the example presented in the thesis. There are more details about the points which belong to the "unbounded limit group." In later terminology, these are properties of the Julia set. First of all, this "group" consists of finite points spread infinitely densely on a certain curve. This, like many Böttcher's statements, is not completely clear, and can be interpreted in two ways: the statement says that repelling periodic points are dense in the Julia set, or perhaps it says that the map is topologically transitive on its Julia set. (Both properties hold for rational maps of the sphere.) The "boundary curves of the convergence regions" for some maps are correctly identified: for $f(z) = z^r, r = 2, 3, \ldots$ the curve is the unit circle, while for $f(z) = 2z^2 - 1$ and for $f(z) = 3z - 4z^3$ (as well as for higher degree Chebyshev polynomials) the curve is the segment $[-1, 1]$ on the real axis. So Böttcher gives first examples of Julia sets. He also presents a new example of an everywhere chaotic rational map

$$f(z) = \frac{4z(1-z)(1-k^2z)}{(1-k^2z^2)^2} = \mathrm{sn}^2 2 \int_0^{\sqrt{z}} \frac{du}{\sqrt{(1-u^2)(1-k^2u^2)}},$$

where sn denotes the elliptic sine function and the path of integration is arbitrary. An argument is sketched that this function displays "transcendental convergence" (i.e., chaotic behavior) in the whole complex plane by noticing that any limit point of the iterates f^n (taken pointwise) would equal $\mathrm{sn}^2(\infty)$, which is not a well defined quantity.

Other novel developments are: (1) a conjecture about behavior of an analytic function around an irrationally neutral fixed point, and (2) the upper bound for a number of "regions of convergence" by the number of critical points of the map f (stated without proof). Böttcher writes in the remark after Sect. 51:

> It seems that a point x giving $f(x) = x,\ f^{(1)}(x) = e^{\mu\sqrt{-1}}$ with μ irrational with respect to π cannot be a limit point of positive, negative, or even singular convergence. [...] Whereas it turns out that the number of all regions of positive and singular convergence, regular as well as rhythmic, cannot exceed the number of the zeros and poles of the derivative of the iterated function, at least in the case when the latter is algebraic rational. Therefore, the number of regions of dependence is definitely finite.

In this paper, Böttcher also introduces the equation which now bears his name:

$$Ff(z) = (F(z))^m,$$

where f is a given function which is analytic in a neighborhood of its fixed point x.

He proposes a method of solving this equation. When

$$f(z) = x + \frac{(z-x)^m}{m!} f^{(m)}(x) + \frac{(z-x)^{m+1}}{(m+1)!} f^{(m+1)}(x) + \ldots,$$

i.e., when the fixed point x is superattracting, he considers what he calls a "basic algorithm"

$$R(z) := \lim_{n\to\infty} \sqrt[m^n]{f_n(z) - z}$$

and proceeds as follows: Let $\Xi(z)$ be a function such that $\Xi f(z) = \Xi(z)$. Then the general solution to the functional equation is given by

$$Q^{\Xi(z)(\log R(z))},$$

where Q is an arbitrary constant. No proof of convergence is given.

4. "Main laws of convergence of iterations and their analytic applications" (Russian)
This is the most known and quoted of Böttcher's papers. Again, it overlaps with the dissertation as well as with "Zasady." The first part starts with discussion of iteratively periodic functions and the observation that the only rational iteratively periodic functions are of the form $f(z) = (Az+B)/(Cz+D)$. Then, the limits of iterates of an analytic functions are considered, the main case being $f^n(z) \to x$ with $f(x) = x$, but it is conjectured (though not investigated) that in some cases the limit of $f^n(z)$ may depend on the point z (i.e., be a nonconstant function). The same Lattès-type function as in "Zasady" (involving elliptic sine) is given as an example of "chaotic convergence." It is stated that for this function any

complex number can be obtained as a limit of iterates of f (i.e., in modern terminology, f has orbits that are dense in the sphere).

Böttcher comes back to the question he had asked in his thesis about the existence of one-parameter continuous group of transformations containing a given function f and all its iterates. He admits a possibility that the solution may exists only in some regions of the plane.

In subsequent parts, he analyzes in detail the dynamic behavior of $f(z) = z^2$: the convergence of f^n to 0 for $|z| < 1$ and to ∞ for $|z| > 1$. He gives a long list of references on "iterative periodicity" and iteration in general, including Newton's Opuscula.

He studies the case of f with an attracting fixed point x and constructs the basin of attraction starting with a disk C such that $f^n(z) \to x$ in C and taking $\bigcup_n f^{-n}(C)$.

Next he considers the auxiliary functions $B(z), D(z)$ and $L(z)$ which are used to solve various functional equations. He studies properties of these functions (domain of existence, zeros, poles, etc.)

Iterative properties of $f(z) = x + \frac{f^{(m)}(x)}{m!}(z-x)^m + \ldots$ are investigated in detail in Chap. II. Böttcher's equation is solved using the solutions to Grévy's equation. First, a function F is defined by the relation $1 + F(z) = \frac{(f(z)-x)\left(\frac{f^{(m)}(x)}{m!}\right)^{-1}}{(z-x)^m}$ and then the infinite product $(1 + F(z))(1 + F(z_1))\ldots$ is used to find solutions of the equations $Yf(z) = \frac{f^{(m)}(x)}{m!}(z-x)^{m-1}Y(z), \quad Y'(x) = 1$ as well as $\Psi f(z) = \frac{f'(x)}{m}\Psi(z)$ and $\beta f(z) = m\beta(z)$. Then, the function $B(z)$ is defined as $\int_z^x \frac{du}{\Psi(u)} + \frac{1}{m-1}\int_x^z \frac{du}{\Psi(u)}$. It is observed that B has logarithmic singularities at x and all its preimages under f. The function $N = e^B$ solves the equation $Nf(z) = [N(z)]^m$. The remaining part of the chapter presents the study of a function with a parabolic fixed point, mainly following Leau.

5. "On properties of some functional determinants" (Polish)
   Böttcher studies certain functional determinants which are analogous to Wrońskians but contain difference operators instead of differential ones.
6. "Rozwijanie na szeregi potegowe funkcji określonej rownaniem algebraicznem nieprzywiedlnem $f(x, y) = 0$"
   This is an expository paper. Böttcher discusses expansion of y in (fractional) powers of $x - \bar{x}$, mainly in the case when $\bar{x}$ satisfies the system $f(x, y) = 0, \quad f_{0,1}^{(1)}(x, y) = 0$. He outlines the procedure based on the construction of a certain polygon in the plane, called Newton–Puiseux polygon. He gives references to the current texts but not to the original works of Newton or Puiseux, and does not mention these names.
7. "Iteration of a linear function" (Polish)
   Böttcher studies iteration of functions of the form $f(z) = (Az+B)/(Cz+D)$. He determines that such a function is iteratively periodic (i.e., it satisfies $f^k(z) \equiv z$ for some k) if and only if it is an elliptic function, i.e., such that the ratio $(Cr_2 + D)/(Cr_1 + D)$ is a number of unit modulus, but not 1. Here, r_1, r_2 are fixed points of f. The period of iteration is rational resp. irrational if $\frac{\arg(Cr_2+D)-\arg(C_1+D)}{2\pi}$ is

rational resp. irrational. An example of such a function is $f(z) = \frac{(a+bi)z+(c+di)}{(-c+di)z+(a-bi)}$. Seeking a general formula for iteration of linear functions, he gives recurrence formulas for A_n, B_n, C_n, D_n in $f_n(z) = (A_n z+B_n)/(C_n z+D_n)$. He also expresses the iterates f_n in terms of symmetric functions of two (distinct) fixed points of f and treats expansion of iterates into continued fractions.

8. "A note on solution of the functional equation $\Psi f(z) - \Psi(z) = F(z)$" (Polish) In this short note, Böttcher finds a particular solution to the equation mentioned in the title under the assumptions that $f(z) = x + \frac{f'(x)}{1!}(z-x)+\ldots$, $0 < |f'(x)| < 1$ and $F(z) = A_0 + \sum_{n=1}^{\infty} [A_{+n}(z-x)^{m/n} + A_{-n}(z-x)^{-m/n}]$ converge in some annulus $r_1 < |z-x| < R_1$. Using "Koenig's algorithm" $B(z) = \lim_{n\to\infty} \frac{f_n(z)-x}{[f'(x)]^n}$, which solves $Bf(z) = f'(x)B(z)$, he changes the coordinate and expands F in fractional powers of B: $F(z) = B_0 + \sum_{n=1}^{\infty} [B_n B^{m/n}(z) + B_{-n} B^{-m/n}(z)]$, convergent in some $r_2 < |B(z)| < R_2$. The function $\psi_0(z) = \frac{B_0 \log B(z)}{\log f'(z)} + \sum_{n=1}^{\infty} \frac{B_n B(z)^{m/n}}{f'(x)^{m/n}-1} + \frac{B_{-n}(B(z))^{-m/n}}{f'(x)^{-m/n}-1}$ is his solution to the title equation.
9. "Nouvelle méthode d'intégration d'un système de n équations fonctionelles linéaires du premier ordre ... " (French)
In this paper, Böttcher gives a general solution to the system of n linear functional equations of first order, of the form

$$U_i(z) = \sum_{j=1}^{n} A_{ij}(z) U_j(f(z)), \quad i = 1, \ldots, n$$

using a solution of the equation $Bf(z) = hB(z)$, $|h| < 1$. He proves the result by constructing a formal Laurent series and investigating its convergence. He considers in detail some special cases of f, e.g., $f(z) = az$ or $f(z) = az^b$. Another result of the paper is a "fundamental law" which concerns the method of finding a complete and general solution of the mentioned system. The final part of the paper is devoted to discussion of some properties of the obtained solution, e.g. its singular points, zeros, and points of ramification.
10. "Iteration of a quadratic function" (Polish) Böttcher considers the iteration of $f(z) = Az^2 + 2Bz + C$, not necessarily with a positive exponent. He discusses the iterations of $f(z) = z^2$ and of $f(z) = 2z^2 - 1$ with an arbitrary exponent by considering them on infinitely sheeted Riemann surfaces. He determines an orbit of an arbitrary point under these iterations. He also introduces the use of "canonical forms" to aid iteration of quadratic functions. For example, he shows that the group of iterations of $f(z) = Az^2 + 2Bz + C$ can be transformed onto the group of iterations of $\varphi(z) = z^2 + T$, $T = AC - B^2 + B$ by $\omega_1 \varphi = f \omega_1$ or $\omega_2 f = \varphi \omega_2$, where $\omega_1(z) = (z-B)/A$, $\omega_2(z) = Az + B$.
11. "A contribution to the calculus of iterations of an algebraic rational entire function" (Polish) Böttcher considers the following problem: given that $n^\nu = p$, $\nu < 1$, $p < n$, find whether the entire function $F(z) = A_0 z^n + \ldots + A_{n-1} z + A_n$ (i.e., a polynomial) of degree n has a νth iterate $F_\nu(z) = a_0 z^p + \ldots + a_{p-1} z + a_p$, which is also an entire function, of degree p. By expanding $[F(z)]^{p/n}$ in powers of z^{-1} in a neighborhood of infinity, he

obtains conditions for solutions (involving certain functions commuting with each other). He also solves an analogous problem for $p > n$.

12. "The notion of iteration ... " (Polish) Böttcher revisits the problem of defining an iterate with an arbitrary "base" f and exponent p. For this, he uses a function F (which he calls an "iterational functial" [sic]) such that $f(z) = \lim_{n\to\infty}(1 + \frac{1}{n}F(z))^{\circ n}$; he takes $f^{\circ p} = \lim_{n\to\infty}(1+\frac{p}{n}F(f))^{\circ n}(z)$ (the superscript $\circ n$ denotes the iteration with exponent n). As in his PhD thesis, he claims that such a function F can be constructed in a basin of a point x in the complex plane such that $f(x) = x$ and $0 < |f'(x)| < 1$ by the method of A. Korkine: $F(z) = q\lim_{n\to\infty}\left[\frac{f^{\circ n}(z)-x}{(f^{\circ n})'(x)}\right]$, $q = \log f'(x)$. He proposes the construction of such a function in the case $f'(x) = 0$ or 1, but gives no proofs.
13. "Iteration $f_x(z)$ of an algebraic function as a metatranscendental function of the exponent x" (Russian) Böttcher studies the conditions under which the iteration $f_x(z)$ is a differentiable function of the exponent x (a problem already outlined in his Ph.D. thesis). This brings him to consider the differential equations of the type

$$\Phi(x, \varphi(x), \varphi'(x), \ldots, \varphi^{(n)}(x)) = 0$$

when Φ is polynomial. (He defines a metatranscendental function as one which cannot satisfy an equation of this type for any n.) An interesting element is the use of the language of set theory and cardinalities: it is stated that for a fixed n, the set of solutions of such an equation is countable if the coefficients of the polynomial Φ are algebraic numbers but of cardinality continuum otherwise.

4.3 *More on Böttcher's Theorem*

Introducing a new type of functional equation, called nowadays Böttcher equation, and solving it, is Böttcher's most important and most known achievement. The equation $Ff(z) = (F(z))^m$ and its generalizations can be considered under various assumptions about the given function f (cf. [32], [38]). The case of a complex analytic f satisfying $f(z) = \frac{z^m}{m!}f^{(m)}(0) + \frac{z^{m+1}}{(m+1)!}f^{(m+1)}(0) + \ldots$ is interesting from the dynamical viewpoint since the solution F provides an analytic conjugacy between f and the monomial z^m (and therefore, it is often called Böttcher's coordinate). Because only local behavior is considered, one can assume that the fixed point is at 0. (By a change of coordinates, this in particular applies to the important example of a polynomial f of degree m in a neighborhood of its superattracting fixed point at infinity.) The following theorem, commonly known as Böttcher's theorem, summarizes the result:

Theorem 1 *Let $f(z) = a_m z^m + a_{m+1}z^{m+1} + \ldots$, $m \geq 2, a_m \neq 0$ be an analytic function in a neighborhood of 0. Then, there exists a conformal map F of a neighborhood of 0 onto the unit disk, $F(z) = z + bz^2 + \ldots$, satisfying Böttcher's equation $Ff(z) = [F(z)]^m$.*

As we already mentioned, Böttcher proposed two (related) ways of constructing such a map (also called Böttcher's function): one using the quantity $\lim_{n\to\infty} \sqrt[m^n]{f_n(z) - z}$ and the other using solutions to Grévy's equation, which were obtained as path integrals. Both approaches were only sketched. Let us quote Böttcher's own summary of the calculations leading to his solution as given in § 34 of his Russian paper from 1904:

> It remains for us to study another important algorithm. We have the formula $B(z) = \frac{\lg(z_n - x)}{m^n}$, $n = \infty$. We write this formula in the form $B(z) = \lg \sqrt[m^n]{(z_n - x)}$ and we get $N(z) = e^{B(z)} = \lim_{n\to\infty} \sqrt[m^n]{(z_n - x)}$. This algorithm satisfies the condition $Nf(z) = [N(z)]^m$, and since
>
> $$B(z) = \lg(z - x) + \frac{\lg \frac{f^{(m)}(x)}{m!}}{m-1} + \frac{A_0}{1}(z - x) + \frac{A_1}{2}(z - x)^2 + \ldots,$$
>
> then
>
> $$N(z) = e^{B(z)} = e^{\lg(z-x)} \cdot e^{\frac{\lg \frac{f^{(m)}(x)}{m!}}{m-1}} \cdot e^{\frac{A_0}{1}(z-x)+\frac{A_1}{2}(z-x)^2+\ldots},$$
>
> i.e.,
>
> $$N(z) = \sqrt[m-1]{f^{(m)}(x)/m!}\,(z - x)\left[1 + \sum_{\nu=1}^{\infty} \frac{\left(\frac{A_0}{1}(z - x) + \frac{A_1}{2}(z - x)^2 + \ldots\right)^{\nu}}{\nu!}\right],$$
>
> from which it follows that
>
> $$N(z) = \sqrt[m-1]{f^{(m)}(x)/m!}\,(z - x) + \frac{N^{(2)}(x)}{2!}(z - x)^2 + \frac{N^{(3)}(x)}{3!}(z - x)^3 + \ldots.$$
>
> The region of convergence is the same as for the function $B(z)$, and if z leaves the region, we compute $B(z)$ on the basis of the remark made in §31, and therefore we compute $N(z)$.

The details of the solution involving taking successive roots of iterates were supplied in 1920, independently by J.F. Ritt [40] and P. Fatou [20]. In his article, Ritt cites the Russian paper by Böttcher, while Fatou informs only that

> (...) the existence of this function (...) seems to have been first proved by Mr. Böttcher (p. 189)

and does not refer to any of the Böttcher's publications. (See [4] for more about early reception of Böttcher's theorem.) Here is Ritt's proof of Böttcher's theorem:

We can assume that $a_m = 1$ and consider a small disk in which $f^p(x)$ approaches zero uniformly as p increases and such that f has no zero other than $x = 0$. (In this disk, every iterate f^p, $p = 1, 2, \ldots$, exists, and vanishes only for $x = 0$, at which point it has a zero of order m^p.) We can thus select a sequence of functions $f^{1/m}, f_2^{1/m^2}, \ldots, f_p^{1/m^p}, \ldots$ which is uniformly bounded. Now, $f_{p+q}(x) = [f^p(x)]^{m^q}(1 + \epsilon_{p,q}(x))$, where $\epsilon_{p,q}(x)$ tends toward zero as p increases, uniformly with respect to x and q. Consequently, $[f_{p+q}(x)]^{1/m^{p+q}} =$

$[f^p(x)^{1/m^p}(1+\eta_{p,q}(x))$ where $\eta_{p,q}(x)$ approaches zero as p increases, uniformly with respect to x and q. The functions in the selected sequence are, for p large enough, less than one in modulus, since they are roots of functions which approach zero, and it follows that they converge uniformly to an analytic function F. Furthermore, since $[f_p(f(x))]^{1/m^p} = [[f_{p+1}(x)]^{1/m^{p+1}}]^m$, we find, as p approaches infinity, that $Ff(x) = [F(x)]^m$.

Fatou's proof is more detailed, but follows the same idea. It was reproduced in several references mentioning Böttcher's function, e.g. in [9, 12, 44]. (The maps considered in [44] are holomorphic self-maps of the unit disk.) See also an account of this proof in [3].

In modern days, new proofs of Böttcher's theorem were given. We will present several such proofs.

J. Milnor's Proof ([34]): This proof is in the classical spirit. The superattracting fixed point is placed at infinity rather that zero, i.e., the expansion of f is

$$f(z) = a_m z^m + a_{m-1}z^{m-1} + \ldots + a_0 + a_{-1}z^{-1} + \ldots, \qquad |z| > R.$$

By considering $z \mapsto \alpha f(z/\alpha)$ with $\alpha^{m-1} = a_m$ one can assume that $a_m = 1$, so that $f(z) = z^m(1 + \mathcal{O}(|1/z|)$ for large z. Substituting $e^z = Z$ lifts f to a continuous map $F(Z) = \log f(e^Z)$, uniquely defined for $\Re(z) > \log R$ up to an addition of some multiple of $2\pi i$. The lift F can be chosen so that $|F(Z) - mZ| < 1$ in some half-plane $\Re(z) > \sigma$ and this half-plane is mapped to itself by F. If $Z_0 \mapsto Z_1 \mapsto \ldots$ is any orbit under F in this half-plane, then $|Z_{k+1} - mZ_k| < 1$ and $|W_{k+1} - mW_k| < 1/m^{k+1}$, where $W_k := Z_k/m^k$. Hence, the sequence of holomorphic functions $W_k = W_k(Z_0)$ converges uniformly as $k \to \infty$ to a holomorphic function Φ. This function satisfies $\Phi(F(z)) = m\Phi(Z)$ and $\Phi(Z + 2\pi i) = \Phi(Z) + 2\pi i$. Therefore, the mapping $\phi(z) = \exp(\Phi(\log z))$ is well defined near infinity and satisfies $\phi(f(z)) = (\phi(z))^m$.

Here is a nonclassical proof, which uses quasiconformal maps to extend a partially defined topological conjugacy and improve its regularity.

M. Lyubich's Proof ([33]): Consider a small disk U_ε and let W denote the connected component of $f^{-1}U_\varepsilon$ containing 0. Let $V = \overline{W \setminus U_\varepsilon}$ and let $r \in (0, 1)$. Consider a diffeomorphism $h_0 : V \mapsto A[r^k, r]$ such that $h_0(f(z)) = h_0(z)^m, \quad z \in \partial W$. This diffeomorphism extends to a homeomorphism $h : W \mapsto U$ for which $h(f(z)) = h(z)^m, \quad z \in W$. The conformal structure $h_*\sigma$ extends naturally to a conformal structure on the unit disk U which is invariant under the map $G : z \mapsto z^m$ in a neighborhood of an arbitrary point different from zero. By the measurable Riemann mapping theorem, there exists a quasiconformal homeomorphism $\psi : \bar{U} \mapsto \bar{U}$ with $\psi(0) = 0, \psi(1) = 1$ and $\psi_*\mu = \sigma$. Then, the mapping $g = \psi \circ G \circ \psi^{-1}$ is locally conformal outside 0 and is an m-sheeted covering of U^* by itself. Then $g : z \mapsto z^m$.

We mention that there is another proof of Böttcher's theorem using quasiconformal maps (combined with holomorphic motions), due to Y. Jiang [26].

Yet another proof uses operator theory after observing that the exponential of the Böttcher function is an eigenvector of a composition operator.

T. Gamelin's Proof ([23]): Let U_δ be a disk centered at 0 with radius δ such that $0 < \delta < 1/2$. Let A_0 be the space of functions g analytic in U_δ satisfying $g(0) = 0$ and let T_0 be the restriction to A_0 of the composition operator $(Tg)(z) = g(f(z))$. We can assume that $|f(z)| \leq 2|z|^m$ in U_δ, so $|f^n(z)| \leq (2\delta)^{m^n}$. If $g \in A_0$ satisfies $\|g\| \leq 1$, then by Schwarz lemma $|(T_0^n g)(z)| \leq |f^n(z)| \leq (2\delta)^{m^n}$, so $\|T_0^n\| \leq (2\delta)^{m^n}$. Hence, the Neumann series $(\lambda I - T_0)^{-1} = \sum_{k=0}^{\infty} \frac{T_0^k}{\lambda^{k+1}}$ converges for all $\lambda \neq 0$. Note that Böttcher equation is equivalent to the resolvent equation $(mI - T_0)G = h$, where $G(z) = \log(F(z)/z)$, $h(z) = \log(f(z)/z^m)$. It is thus enough to solve the resolvent equation for $G \in A_0$ and then set $F(z) = z\exp(G(z))$. Such a solution can be obtained using Banach contraction principle.

The operator-theoretic approach was also applied in [30], where the (somewhat more general) Böttcher function was obtained as the unique fixed point of a certain contraction operator.

A construction of the inverse map to Böttcher's function for the family of quadratic maps $f_\lambda : z \mapsto (z - \lambda)^2$ by an iterational process based on Caratheodory's theorem on convergence of domains can be found in [6].

It should be mentioned that the classical method of proving Böttcher's theorem was used recently in [21] to give the classification (up to holomorphic conjugacy) of attracting rigid germs in $\mathbb{C}^2$, and in [27] in the study of local stable manifolds of a dominant meromorphic self-map $f : X \dashrightarrow X$, where X is a compact Kähler manifold of dimension $n > 1$. (In the latter, the authors also consider a Böttcher-like function in a neighborhood of a "saddle type" invariant circle that is superattracting in $n - 1$ complex dimensions and repelling in one real dimension, getting a very general, sharp theorem. Their methods work in fact for Hermitian manifolds.) As far as other applications and generalizations are concerned, an analog of Böttcher's theorem for a class of transcendental entire functions was proved in [39]. An analog of Böttcher coordinate for mappings of the form $f = g \circ h$, where g is a polynomial of degree $d \geq 2$ and h is an affine mapping of the complex plane to itself, was found in [22]. The existence of a local Böttcher coordinate for a holomorphic map in several complex variables with a superattracting fixed point was proved in [10]; some special cases and examples were introduced and studied in [43]. An analog of Böttcher coordinate for a map with a superattracting hyperplane was considered in [8].

Appendix: Archival Materials

The documents reproduced here (Figs. 1, 2, 3, 4, 5, and 6) come from: L. Böttcher, Teka osobowa (personal file), fond 27, op. 4-40, Lvov District Archive, and

L. Böttcher, Universitätsarchiv, Lepizig, Phil. Fak., Prom., 714, Bl. 7. We thank these institutions. All translations are by the second named author.

4.4 A.1 Böttcher's Mathematical Studies in Leipzig

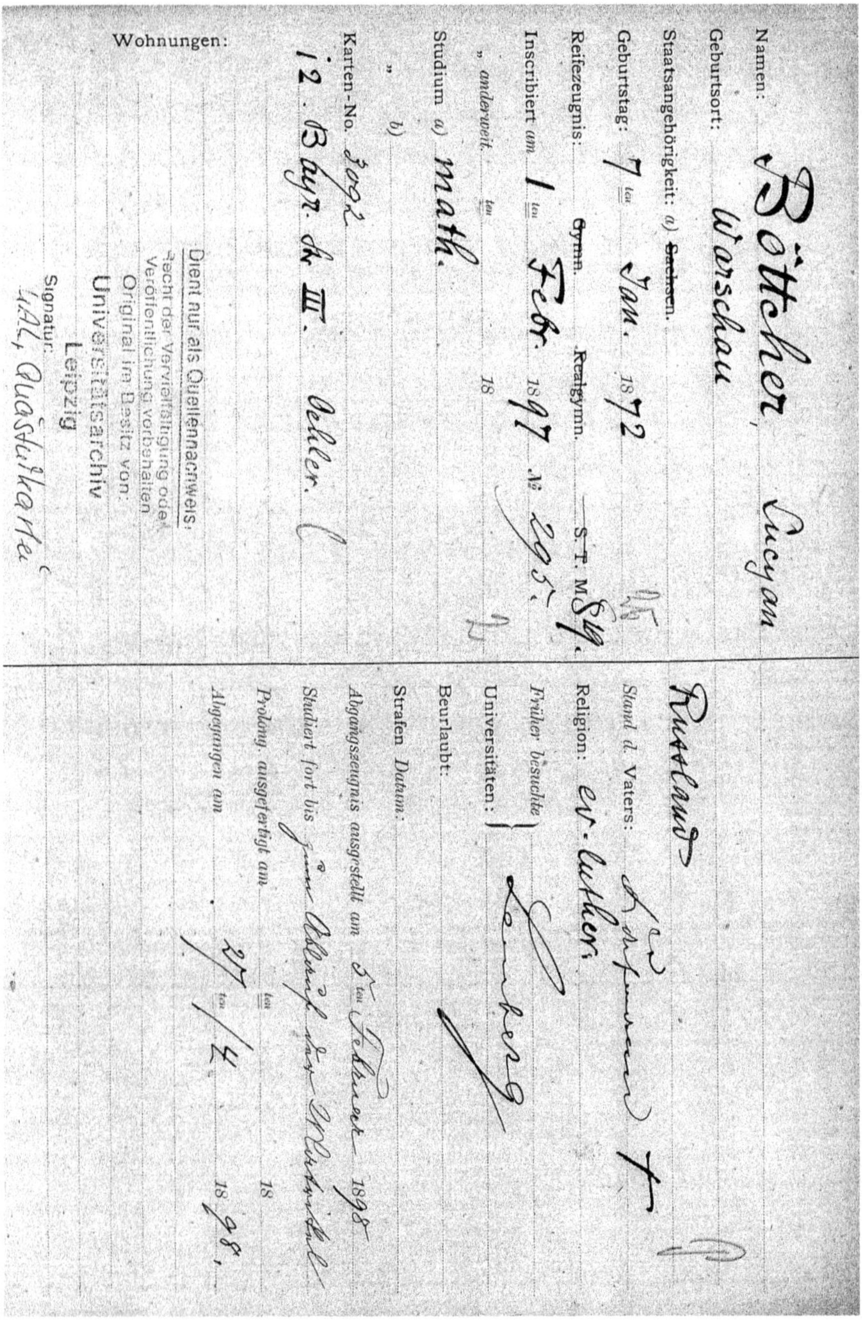

Fig. 1 Matriculation in the university courses at Leipzig

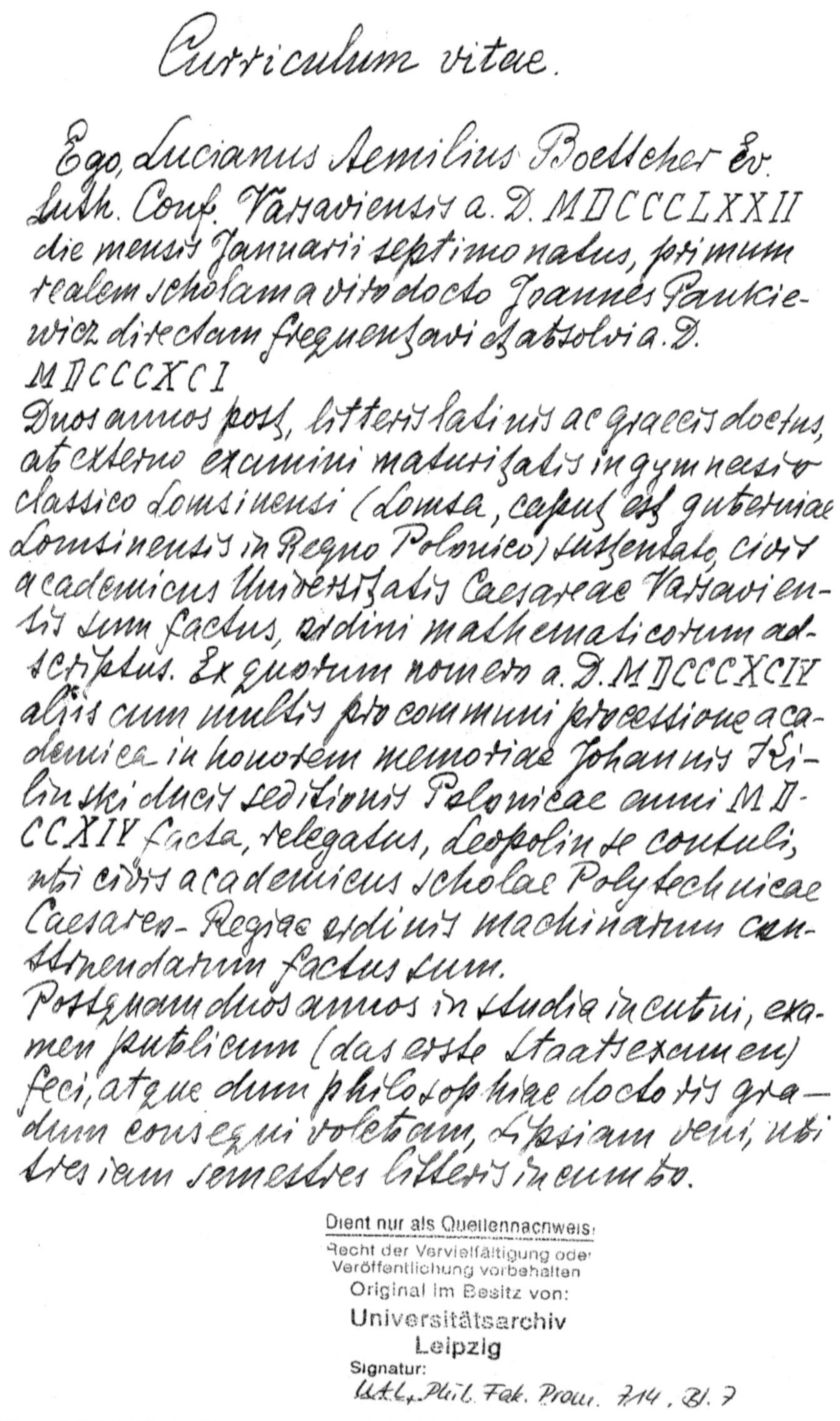
Curriculum vitae.

Ego, Lucianus Aemilius Boettcher Ev.
Luth. Conf. Varsaviensis a. D. MDCCCLXXII
die mensis Januarii septimo natus, primum
realem scholam a viro docto Joannes Pankie-
wicz directam frequentavi et absolvi a. D.
MDCCCXCI
Duos annos post, litteris latinis ac graecis doctus,
ab externo examini maturitatis in gymnasio
classico Lomsinensi (Lomsa, caput est guberniae
Lomsinensis in Regno Polonico) sustentato, civis
academicus Universitatis Caesareae Varsavien-
sis sum factus, ordini mathematicorum ad-
scriptus. Ex quorum numero a. D. MDCCCXCIV
aliis cum multis pro communi processione aca-
demica in honorem memoriae Johannis Ki-
linski ducis seditionis Polonicae anni MD-
CCXIV facta, relegatus, Leopolin se contuli,
ubi civis academicus scholae Polytechnicae
Caesareo-Regiae ordinis machinarum con-
struendarum factus sum.
Postquam duos annos in studia incubui, exa-
men publicum (das erste Staatsexamen)
feci, atque dum philosophiae doctoris gra-
dum consequi volebam, Lipsiam veni, ubi
tres iam semestres litteris incumbo.

Fig. 2 L.E. Böttcher's CV in Latin, printed with his thesis

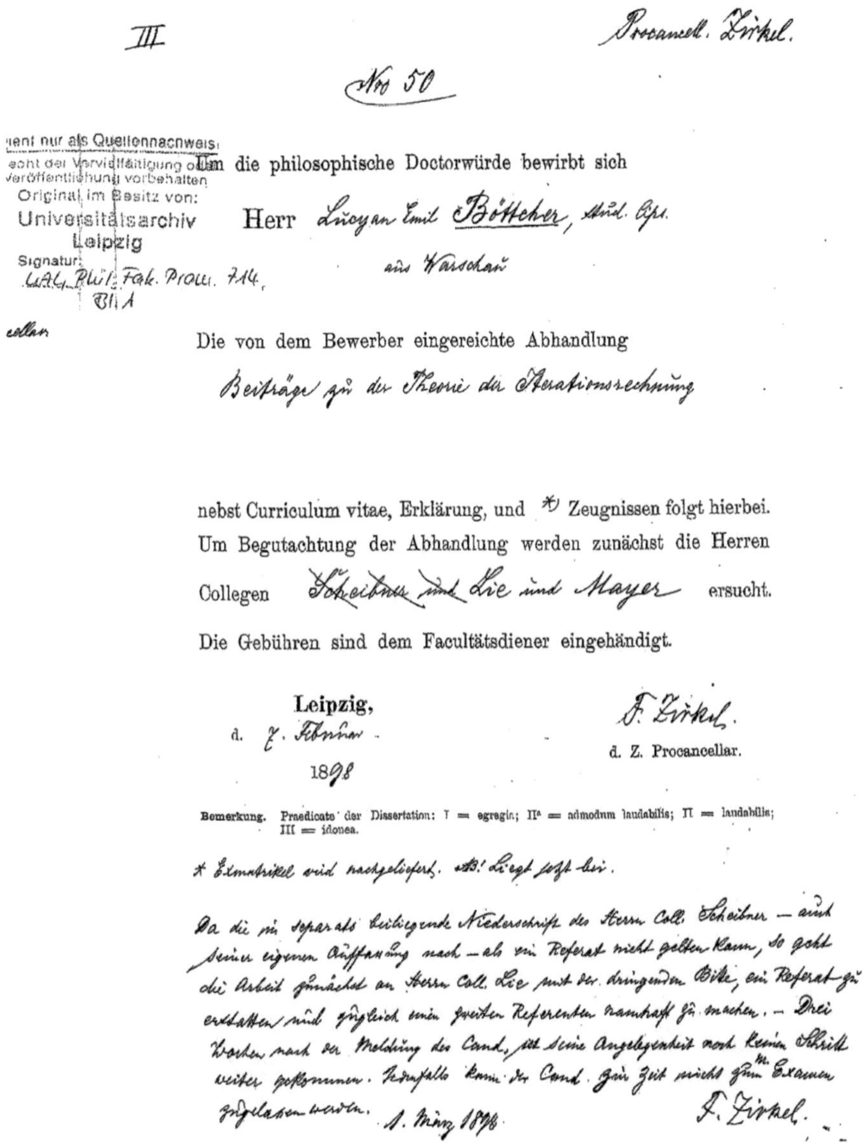

III

Procancell. Zirkel.

No 50

Um die philosophische Doctorwürde bewirbt sich

Herr Lucyan Emil Böttcher, stud. Spr.

aus Warschau

Die von dem Bewerber eingereichte Abhandlung

Beiträge zu der Theorie der Iterationsrechnung

nebst Curriculum vitae, Erklärung, und *) Zeugnissen folgt hierbei.
Um Begutachtung der Abhandlung werden zunächst die Herren Collegen ~~Scheibner und~~ Lie und Mayer ersucht.
Die Gebühren sind dem Facultätsdiener eingehändigt.

Leipzig, d. 7. Februar 1898

F. Zirkel.
d. Z. Procancellar.

Bemerkung. Praedicate der Dissertation: I = egregia; IIa = admodum laudabilis; II = laudabilis; III = idonea.

* Exmatrikel wird nachgeliefert. NB.! Liegt jetzt bei.

Da die in separato beiliegende Niederschrift des Herrn Coll. Scheibner – auch seiner eigenen Auffassung nach – als ein Referat nicht gelten kann, so geht die Arbeit zunächst an Herrn Coll. Lie mit der dringenden Bitte, ein Referat zu erstatten und zugleich einen zweiten Referenten namhaft zu machen. – Drei Wochen nach der Meldung des Cand. ist seine Angelegenheit noch keinen Schritt weiter gekommen. Jedenfalls kann der Cand. zur Zeit nicht zum Examen zugelassen werden. 1. März 1898

F. Zirkel.

Fig. 3 Formal evaluation of Böttcher's thesis and examinations

4.5 L. Böttcher's Habilitation at the Lvov University

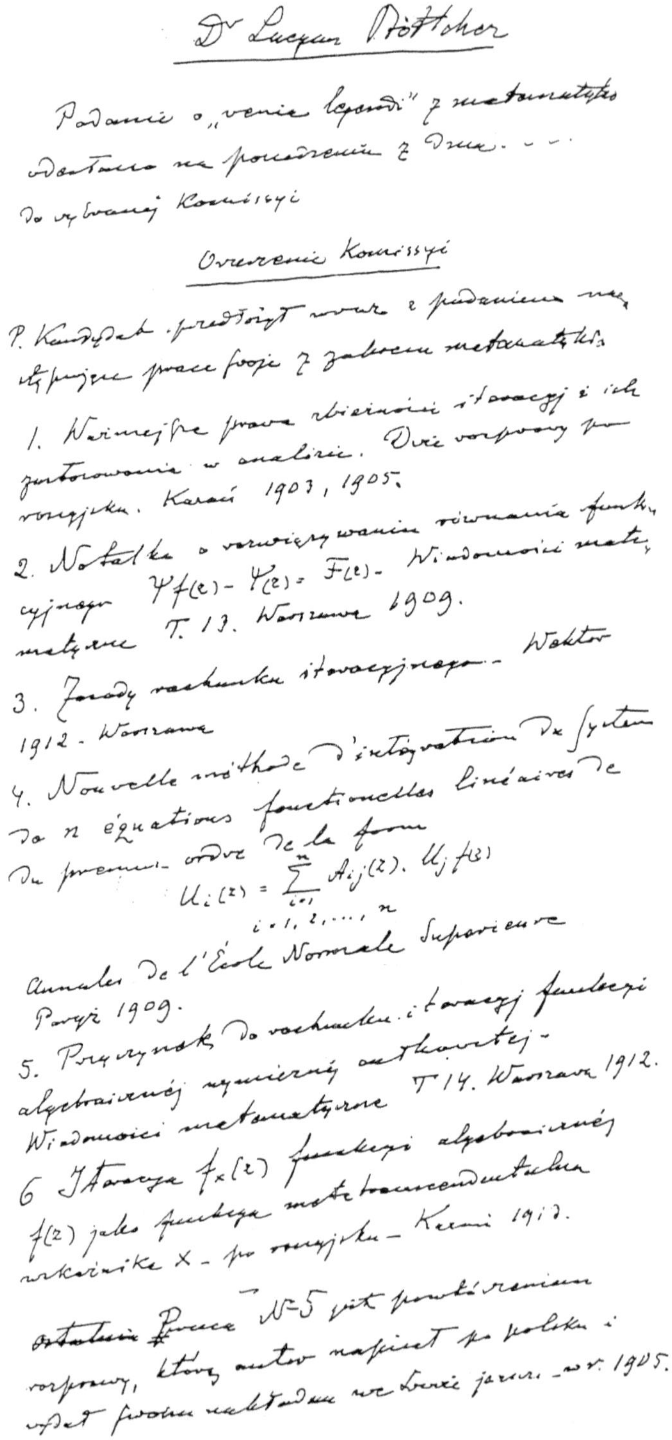

Fig. 4 Böttcher's application for veniam legendi in mathematics: attempt 3, 1918

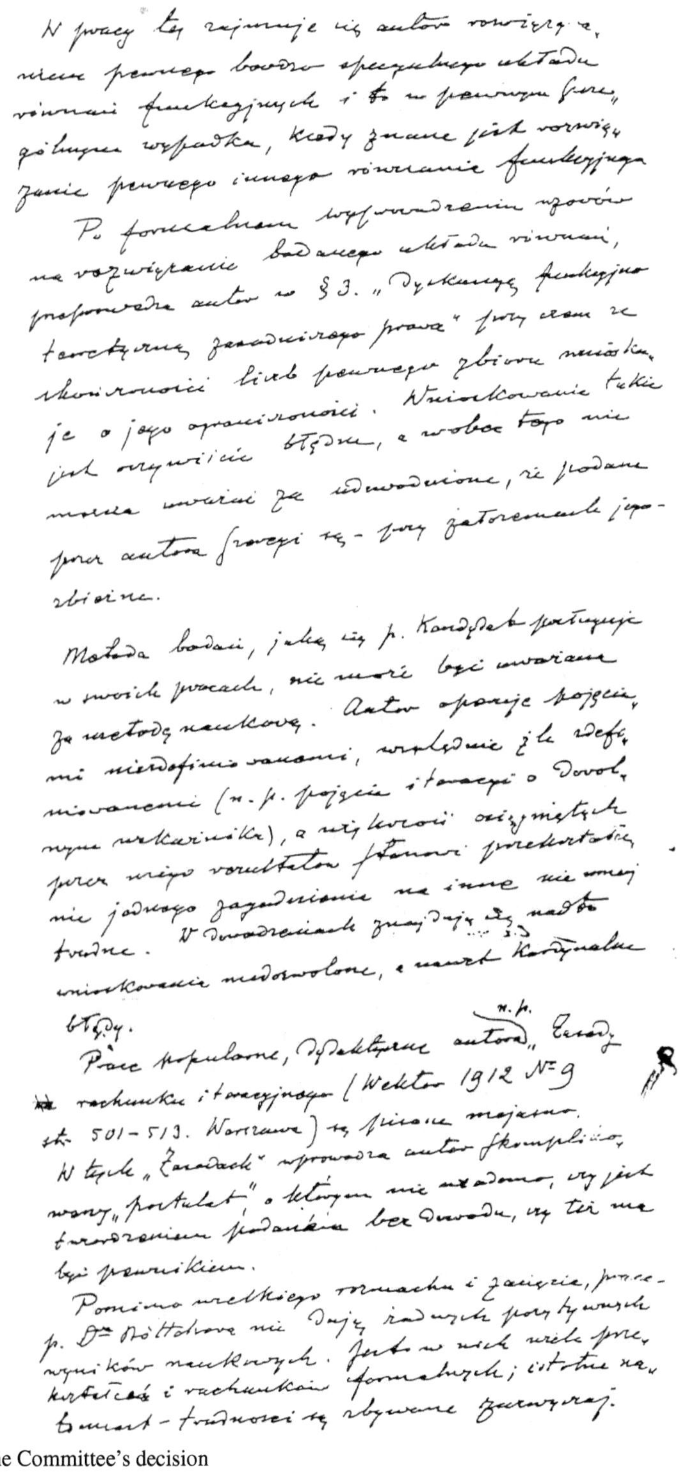
W pracy tej zajmuje się autor rozwiązaniem pewnego bardzo specjalnego układu równań funkcyjnych i to w pewnym szczególnym wypadku, kiedy znane jest rozwiązanie pewnego innego równania funkcyjnego.

Po formalnym wyprowadzeniu wzorów na rozwiązanie badanego układu równań, przeprowadza autor w § 3. „Dyskusję funkcyjno teoretyczną zasadniczego prawa" przy czem ze skończoności liczb pewnego zbioru wnioskuje o jego ograniczoności. Wnioskowanie takie jest oczywiście błędne, a wobec tego nie można uważać za udowodnione, że podane przez autora funkcje są - przy założeniach jego - zbieżne.

Metoda badań, jaką się p. Kandydat posługuje w swoich pracach, nie może być uważana za metodę naukową. Autor operuje pojęciami niedefiniowanymi, względnie źle zdefiniowanymi (n.p. pojęcie iteracji o dowolnym wskaźniku), a większość osiągniętych przez niego rezultatów stanowi przekształcenie jednego zagadnienia na inne nie mniej trudne. W dowodzeniach znajdują się nadto wnioskowania niedozwolone, a nawet kardynalne błędy.

Prace popularne, dydaktyczne autora, n.p. „Zasady rachunku iteracyjnego (Wektor 1912 N°9 str. 501-513. Warszawa) są pisane mgliście. W tych „Zasadach" wprowadza autor skomplikowany „postulat", o którym nie wiadomo, czy jest twierdzeniem podanym bez dowodu, czy też ma być pewnikiem.

Pomimo wielkiego rozmachu i zacięcia, prace p. Dra Böttchera nie dają żadnych pozytywnych wyników naukowych. Jest w nich wiele przekształceń i rachunków formalnych; istotne natomiast trudności są zbywane zazwyczaj.

Fig. 5 The Committee's decision

The candidate submitted along with the application, the following works in mathematics:

1. Major laws of convergence of iterations and their applications in analysis. Two papers in Russian, Kazan, 1903, 1905
2. A note of solving the functional equation $\Psi f(z) - \Psi(z) = F(z)$, *Wiadomości Matematyczne*, vol. 13, Warsaw 1909
3. Principles of iterational calculus, *Wektor*, 1912, Warsaw
4. Nouvelle méthode d'intégration d'un système de n équations fonctionelles linéairés du premier ordre de la forme $U_i(z) = \sum_{j=1}^{j=n} A_{i,j}(z)U_j F(z)$, Annales l'Ecole Normale Supérieure, Paris, 1909
5. A contribution to the calculus of iteration of a rational entire function, *Wiadomości Matematyczne*, vol.14, Warsaw 1912
6. Iteration $f_x(z)$ of an algebraic function $f(z)$ metatranscendental in the index x, in Russian, Kazan 1912

Paper no. 5 duplicates one written by the author in Polish and self-published already in 1905.[19] In it, the author deals with solving a very specific system of functional equations, in a very particular case in which a solution of some other functional equation is known. After formal deduction of formulas for a solution to the system of equations under investigation the author proceeds to give in § 3 "A functional-theoretical discussion of the fundamental law," concluding boundedness of a certain set from finiteness of the numbers in it. Such reasoning is obviously erroneous, and therefore one cannot consider it to be proven that the series given by the author are—under his conditions—convergent.

The method used by the candidate in his works cannot be considered scientific. The author works with undefined, or ill-defined, notions (e.g., the notion of an iteration with an arbitrary exponent), and the majority of the results he achieves are transformations of one problem into another, no less difficult. In the proofs, there are moreover illegitimate conclusions, or even fundamental mistakes. The author's popular, instructional works, e.g., "Principles of iterational calculus" (*Wektor* 1912, no. 9, pp. 501–513, Warsaw), are written in an unclear manner. In these "Principles," the author introduces a complicated new "postulate," about which it is not known whether it is a theorem stated without a proof or it is supposed to be an axiom.

Despite great verve and determination, Dr. Böttcher's works do not yield any positive scientific results. There are many formal manipulations and computations in them; essential difficulties are usually dismissed with a few words without deeper treatment. The content and character diverges significantly from modern research. One should also add:

1. The shortcoming, or rather lack of rigor of the definition of iteration with an arbitrary exponent introduced by the candidate met with justified and clearly written criticism by Dr. Stanisław Ruziewicz in *Wektor*, Warsaw 1912, no. 5 [On a problem concerning commuting functions].

---

[19] According to [19], the duplicate is paper number 4, not 5.

-Politechnika Lwowska.-

Karta ewidencyjna.

1. Imiona i nazwisko : Lucjan Emil Böttcher
2. Tytuły dyplomowe : doktor filozofji,
3. Tytuły funkcyjne : docent matematyki,

4. Tytuły honorowe :

5. Odznaczenia :

6. Rektor w latach :

7. Dziekan Wydziału w latach :

8. Dzień, miesiąc i rok urodzenia : 7 stycznia 1872.

9. Miejsce urodzenia : Warszawa powiat: detto Województwo: detto

10. Obywatelstwo : polskie narodowość: polska wyznanie : ewang-augsb.

11. Miejsce przynależności : Lwów powiat : detto Województwo : detto

12. Dokładny adres : Sodowa 4. № telefonu : ./. (________)

13. Stosunek do służby wojskowej :

14. Stan cywilny : żonaty, data ślubu : 26 grudnia 1900.

15. Imiona i nazwisko; data urodzenia; ew. zatrudnienie żony : Marja Wolle,

16. Imiona i nazwisko; data urodzenia; stan cywilny; zatrudnienie dzieci : Adolf, Władysław 23.X.1901, st. Pol., Marjan 14.V.1911, ucz. Szk. realnej.

17. Zatrudnienie przed służbą państwową :

18. Zatrudnienie poza służbą państwową : ./.

19. Starszeństwo po :

20. Dane, dotyczące uposażenia :

21. Daty i liczby ważniejszych aktów :

22. Wydział : Katedra :

Fig. 6 Böttcher's registry card from the Lvov Polytechnics (including the information about his family)

Fig. 7 A current photograph of the house in Sodowa 4 in Lvov where Böttcher lived. (Source: S. Domoradzki)

2. Dr. Böttcher applies for a second time for veniam legendi in mathematics. The first time the candidate was advised to withdraw his application because of the faults that the Committee at that time found with the candidate's works. These faults and inadequacies were of the same nature which characterize the candidate's work also today.

The Committee's decision was passed unanimously on June 21, 1918: Not to admit Dr. Böttcher to further stages of habilitation. (Signature is illegible).

Acknowledgment We would like to thank the following people: Prof. Ludwig Reich (University of Vienna), who inspired the first author's interest in Lucjan Emil Böttcher; also, Dr. Danuta Ciesielska (Pedagogical University in Kraków), Prof. Alexandre Eremenko (Purdue University), Dr. Paweł Polak (Pontifical Uniwersity of John Paul II in Kraków), Prof. Liliana R. Shakirova (Kazan Federal University), and Prof. Mykhaylo Zarichnyy (Lviv University).

This work was partially supported by the Centre for Innovation and Transfer of Natural Sciences and Engineering Knowledge, University of Rzeszów.

References

1. Adams, J.F..: Lectures on Lie Groups. W.A. Benjamin, New York (1969)
2. Alexander, D.S.: A History of Complex Dynamics. From Schröder to Fatou and Julia. Aspects of Mathematics, vol. E24. Vieweg, Braunschweig (1994)
3. Alexander, D.S., Iavernaro, F., Rosa, A.: Early Days in Complex Dynamics: A History of Complex Dynamics in One Variable During 1906–1942. History of Mathematics, vol. 38. American Mathematical Society/London Mathematical Society, Providence/London (2012)
4. Audin, M.: Fatou, Julia, Montel. The Great Prize of Mathematical Sciences of 1918, and Beyond. Lecture Notes in Mathematics, vol. 2014. History of Mathematics Subseries. Springer, Heidelberg (2011) (Translated from the 2009 French original by the author)
5. Balibrea, F., Reich, L., Smítal, J.: Iteration theory: dynamical systems and functional equations. Int. J. Bifurc. Chaos Appl. Sci. Eng. **13**(7), 1627–1647 (2003). (Dynamical systems and functional equations (Murcia, 2000))
6. Barnsley, M.F., Geronimo, J.S., Harrington, A.N.: On the invariant sets of a family of quadratic maps. Comm. Math. Phys. **88**(4), 479–501 (1983)
7. Beardon, A.F..: Iteration of Rational Functions: Complex Analytic Dynamical Systems. Graduate Texts in Mathematics, vol. 132. Springer, New York (1991)
8. Bedford, E., Jonsson, M.: Dynamics of regular polynomial endomorphisms of $\mathbb{C}^k$. Am. J. Math. **122**(1), 153–212 (2000)
9. Blanchard, P.: Complex analytic dynamics on the Riemann sphere. Am. Math. Soc. Bull. New Ser. **11**(1), 85–141 (1984)
10. Buff, X., Epstein, A., Koch, S.: Böttcher coordinates. Preprint, http://arxiv.org/pdf/1104.2981.pdf (2011). Accessed 10 Sept 2013
11. Byelous, I.O..: Vydatni vcheni-matematyky L'vivs'koi Politekhniky (1844–1939) [Distinguished Scholars—Mathematicians of Lvov Polytechnics (1844–1939)]. L'viv Polytechnic, L'viv (2012)
12. Carleson, L., Gamelin, T.W.: Complex Dynamics. Universitext: Tracts in Mathematics. Springer, New York (1993)
13. Cowen, C.C.: Analytic solutions of Böttcher's functional equation in the unit disk. Aequ. Math. **24**(2–3), 187–194 (1982)
14. Domoradzki, S.: Mathematics in Lwów before the Lwów Mathematical School. In: Bečvarova, M., Binder, C. (eds.) Mathematics in the Austrian-Hungarian Empire, pp. 55–73. Matfyzpress, Praha (2010)
15. Domoradzki, S.: The Growth of Mathematical Culture in the Lvov Area in the Autonomy Period (1870–1920). History of Mathematics, vol. 47. Matfyzpress, Praha (2011)
16. Elin, M., Goryainov, V., Reich, S., Shoikhet, D.: Fractional iteration and functional equations for functions analytic in the unit disk. Comput. Methods Funct. Theory **2**(2), 353–366 (2004)
17. Epstein, A.L.: Infinitesimal Thurston rigidity and the Fatou-Shishikura inequality. Preprint, http://arxiv.org/pdf/math/9902158.pdf. Accessed 10 Sept 2013, (1999)
18. Eremenko, A.E., Lyubich, M.Yu.: Dinamika analiticheskikh transformatsii [The dynamics of analytic transformations]. Algebra i Analiz **1**(3), 1–70 (1989) (Translated from Russian in: Leningrad Math. J. **1**(30), 563–634 (1990))
19. Estreicher, K.: Bibliografia Polska 1901–1939, tom 3, pod redakcją J. Wilgat. "Przedruk techniką typograficzną," ["A typographical reprint"]). [Polish Bibliography 1901–1939, vol. 3, edited by J. Wilgat], Wydawnictwa Artystyczne i Filmowe, Warszawa (1977), http://www.bn.org.pl/download/document/1234357120.pdf. Accessed 10 Sept 2013
20. Fatou, P.: Sur les équations fonctionelles [On functional equations]. Bull. Soc. Math. Fr. **47**, 161–271 (1919)
21. Favre, C.: Classification of 2-dimensional contracting rigid germs and Kato surfaces. I. J. Math. Pures Appl. Neuv. Sér. **79**(9), 475–514 (2000)
22. Fletcher, A., Fryer, R.: On Böttcher coordinates and quasiregular maps. In: Jiang, Y., Mitra, S. (eds.) Quasiconformal Mappings, Riemann Surfaces, and Teichmüller Spaces (in Honor of

Clifford Earle's 75th Birthday). Contemporary Mathematics, vol. 575, pp. 53–76. American Mathematical Society, Providence (2012)
23. Gamelin, T.W.: Conjugation theorems via Neumann series. Unpublished, http://www.math.ucla.edu/~twg/mrs1.pdf (2003). Accessed 10 Sept 2013
24. Hu, P.-C., Yang, C.-C.: Differentiable and Complex Dynamics of Several Variables. Mathematics and its Applications, vol. 483. Kluwer Academic, Dordrecht (1999)
25. Ilyashenko, Yu., Yakovenko, S.: Lectures on Analytic Differential Equations. Graduate Studies in Mathematics, vol. 86. American Mathematical Society, Providence (2008)
26. Jiang, Y.: Holomorphic motions and normal forms in complex analysis. In: Lau, K.-S., Xin, Z.-P., Yau, S.-T. (eds.) Third International Congress of Chinese Mathematicians. Part 1, 2. Proceedings of the Congress held at the Chinese University of Hong Kong, Hong Kong, Dec 2004. AMS/IP Studies in Advanced Mathematics, vol. 42, part 1, 2, pp. 457–466. American Mathematical Society, Providence (2008)
27. Kaschner, S.R., Roeder, R.K.W.: Superstable manifolds of invariant circles and co-dimension 1 Böttcher functions. Preprint, http://arxiv.org/pdf/1208.3013v1.pdf (2012) Accessed 10 Sept 2012 (to appear in Ergodic Theory Dynamical Systems)
28. Katok, A., Hasselblatt, B.: Introduction to the Modern Theory of Dynamical Systems. Encyclopedia of Mathematics and its Applications, vol. 54. Cambridge University Press, Cambridge (1995). (With a supplementary chapter by Katok and Leonardo Mendoza)
29. Khatskevich, V., Reich, S., Shoikhet, D.: Abel-Schröder equations for linear fractional mappings and the Koenigs embedding problem. Acta Sci. Math. (Szeged). **69**(1–2), 67–98 (2003)
30. Kline, B.J.: A global Boettcher's theorem. Ph.D. Thesis, http://hdl.handle.net/2142/21910 University of Illinois at Urbana-Champaign, IL (1995). Accessed 10 Sept 2013
31. Kucharzewski, F.: Polskie piśmiennictwo techniczne [Polish technical literature], Przegląd Techniczny **99**(14) 1914
32. Kuczma, M., Choczewski, B., Ger, R.: Iterative Functional Equations. Encyclopedia of Mathematics and its Applications, vol. 32. Cambridge University Press, Cambridge (1990)
33. Lyubich, M.Yu.: Dinamika ratsyonal'nykh pryeobrazheniy: topologicheskaya kartina [Dynamics of rational transformations: a topological picture]. Akademiya Nauk SSSR i Moskovskoye Matematicheskoye Obshchestvo. Uspekhi Matematicheskikh Nauk **41** (4)(250), 35–95, 239 (1986)
34. Milnor, J.: Dynamics in One Complex Variable, 3rd edn. Annals of Mathematics Studies, vol. 160. Princeton University Press, Princeton (2006)
35. Milnor, J.: On Lattès maps. In: Hjorth, P.G., Petersen, C.L. (eds.) Dynamics on the Riemann sphere. A Bodil Branner Festschrift, June 2003, pp. 9–43. European Mathematical Society, Zürich (2006). (Including papers from the Symposium on Holomorphic Dynamics held in Holbæk)
36. Polak, P.: "Byłem Pana przeciwnikiem (profesorze Einstein)..". Relatywistyczna rewolucja naukowa z perspektywy środowiska naukowo-filozoficznego przedwojennego Lwowa. ["I was your adversary (Professor Einstein).." The relativistic scientific revolution from the perspective of scientific and philosophical community of pre-war Lvov.]. Copernicus Center, Kraków (2012)
37. Popławski, Z.: Dzieje Politechniki Lwowskiej [The history of Lvov Polytechnics 1844–1945]. Ossolineum, Wrocław (1992)
38. Reich, L.: Generalized Böttcher equations in the complex domain. In: Aulaskari, R., Heittokangas, J., Latvala, V., Taskinen, J. (eds.) Symposium on Complex Differential and Functional Equations. Papers from the Symposium in honor of Ilpo Laine's 60th Birthday held in Joensuu, October 14–15, 2002, pp. 135–147. University of Joensuu, Department of Mathematics. Report Series, vol. 6. University of Joensuu, Joensuu (2004)
39. Rempe, L.: Rigidity of escaping dynamics for transcendental entire functions. Acta. Math. **203**(2), 235–267 (2009)
40. Ritt, J.F.: On the iteration of rational functions. Trans. Am. Math. Soc. **21**(30), 348–356 (1920)

41. Schiller, J.: Portret zbiorowy nauczycieli warszawskich publicznych szkół średnich 1795–1862. [A Collective Portrait of Teachers of Warsaw Public High Schools 1795–1862.] Polska Akademia Nauk, Instytut Historii Nauki, Zakład Dziejów Oświaty, Warszawa (1998)
42. Shishikura, M.: Surgery of complex analytic dynamical systems. In: Dynamical systems and nonlinear oscillations (Kyoto, 1985), 93–105, World Sci. Adv. Ser. Dynam. Systems, 1, World Sci. Publishing, Singapore (1986)
43. Ushiki, S.: Boettcher's theorem and super-stable manifolds for multidimensional complex dynamical systems. In: Ushiki, S. (ed.) Proceedings of the RIMS Conference 'Structure and Bifurcations of Dynamical Systems', 18–21 Feb 1991, Kyoto. Advanced Series in Dynamical Systems, vol. 11, pp. 168–184. World Scientific, Singapore (1992)
44. Valiron, G.: Fonctions analytiques. Presses Universitaires de France, Paris (1954)
45. Waniurski, J.: Lucjan Edward [sic] Böttcher. In: W. Więsław (ed.), XVIII Ogólnopolska Szkoła Historii Matematyki "Sławne dzieła naukowe i rocznice" [Statewide Polish School in History of Mathematics "Famous Scholarly Works and Anniversaries"], pp. 183–188. Wyższa Szkoła Matematyki i Informatyki Użytkowej w Białymstoku [College of Applied Mathematics and Computer Science in Białystok], Białystok-Supraśl (2004)
46. Wilk, A.: Recenzja: L. Böttcher, *Zasady geometryi elementarnej dla szkół z licznemi ćwiczeniami* [Review: L. Böttcher, *Principles of Elementary Geometry for Schools with Numerous Exercises*]. Muzeum **25**(2), 75–78 (1909)

Spectral Properties of Toeplitz Operators Acting on Gabor Type Reproducing Kernel Hilbert Spaces

H. G. Feichtinger, K. Nowak and M. Pap

Abstract This is a survey presenting an overview of the results describing the behavior of the eigenvalues of compact Gabor–Toeplitz operators and Gabor multipliers. We introduce Gabor–Toeplitz operators and Gabor multipliers as Toeplitz operators defined in the context of general reproducing kernel Hilbert spaces. In the first case the reproducing kernel Hilbert space is derived from the continuous Gabor reproducing formula, and in the second case, out of the discrete Gabor reproducing formula, based on tight Gabor frames. The extended metaplectic representation provides all affine transformations of the phase-space. Both classes of operators satisfy natural transformation properties with respect to this group, and both have natural interpretations from the point of view of phase space geometry. Toeplitz operators defined on the Fock space of several complex variables are at the background of the topic. The Berezin transform of general reproducing kernel Hilbert spaces applied to both kinds of Toeplitz operators shares in both cases the same natural phase-space interpretation of the Fock space model. In the first part of the survey we discuss the dependence of the eigenvalues on symbols and generating functions. Then we concentrate on Szegö type asymptotic formulae in order to analyze the dependence on the symbol and on Schatten class cutoff phenomena in dependence on the generating function. In the second part we restrict attention to symbols which are characteristic functions of phase space domains, called localization domains in the current context. The corresponding Toeplitz operators are called localization operators. We present results expressing mutual interactions between localization domains and generating functions from the point of view of the eigenvalues of the localization operators. In particular, we discuss asymptotic boundary forms quantifying these interactions

H. G. Feichtinger (✉)
Faculty of Mathematics, University Vienna, Oskar-Morgenstern-Platz 1,
1090 Wien, Austria
e-mail: hans.feichtinger@univie.ac.at

K. Nowak
1177 Heyward Road, Wayne, PA, 19087, USA,

Department of Computer Science, Drexel University, 3141 Chestnut St.,
Philadelphia, 19104, PA, USA,
e-mail: kn33@drexel.edu

M. Pap
Faculty of Sciences, University of Pécs, Ifjúság út 6, Pécs 7634 Hungary
e-mail: papm@gamma.ttk.pte.hu

T. M. Rassias, P. M. Pardalos (eds.), *Mathematics Without Boundaries*,
DOI 10.1007/978-1-4939-1106-6_6,

locally at the boundary points of localization domains. Our approach to localization operators is motivated by the principles of the semiclassical limit. We finish the survey with a list of open problems and possible future research directions.

Keywords Compact Gabor–Toeplitz operators · Eigenvalues · Eigenvalue plunge region · Asymptotic boundary form · Fock space · Metaplectic representation

1 Hilbert Spaces with a Reproducing Kernel

Let M be a measure space equipped with a nonnegative measure ν. Let $L^2(M,\ d\nu) = \{f : M \to \mathbb{C} \mid f \text{is measurable and} \int_M |f(z)|^2\ d\nu(z) < \infty\}$ be a Hilbert space consisting of square integrable functions with respect to ν equipped with the inner product

$$\langle f, g\rangle_{L^2(M,\ d\nu)} = \int_M f(z)\overline{g(z)}\, d\nu(z).$$

Let $\mathcal{H} \subset L^2(M,\ d\nu)$ be a closed linear subspace of $L^2(M,\ d\nu)$ consisting of functions f for which pointwise values $f(z)$ are well defined for all $z \in M$, and all functionals $f \mapsto f(z)$ are bounded with respect to $L^2(M,\ d\nu)$ norm.

A closed linear subspace of a Hilbert space is a Hilbert space, and Hilbert spaces are self-dual, therefore for each $z \in M$ there is $E_z \in \mathcal{H}$ such that

$$f(z) = \langle f, E_z\rangle_{L^2(M,\ d\nu)}. \tag{1}$$

The function $E(w,z) = E_z(w)$ is called the reproducing kernel of $\mathcal{H}$, and the space $\mathcal{H}$ is called a *Hilbert space with reproducing kernel*. The orthogonal projection $P_{\mathcal{H}}$ from $L^2(M,\ d\nu)$ onto $\mathcal{H}$ is given by the formula

$$P_{\mathcal{H}} f(z) = \langle f, E_z\rangle_{L^2(M,\ d\nu)}. \tag{2}$$

We observe that

$$||E_z||^2_{L^2(M,\ d\nu)} = \langle E_z, E_z\rangle_{L^2(M,\ d\nu)} = E(z,z),$$

and we introduce the normalized reproducing kernel

$$e_z(w) = \frac{E_z(w)}{||E_z||} = \frac{E(w,z)}{E(z,z)^{1/2}}, \tag{3}$$

and correspondingly the normalized measure

$$d\mu(z) = E(z,z)\, d\nu(z). \tag{4}$$

We may interpret the reproducing property (1) via the following integral formulae. For every $f \in \mathcal{H}$

$$f = \int_M \langle f, E_z\rangle_{L^2(M,\ d\nu)} E_z\, d\nu(z) = \int_M \langle f, e_z\rangle_{L^2(M,\ d\nu)} e_z\, d\mu(z), \tag{5}$$

where the integral is understood in the weak sense. It is often more convenient to deal with the normalized kernel e_z instead of the kernel E_z and with the normalized measure μ instead of the original measure ν. In the case of the Bergman space of holomorhic functions defined on a bounded symmetric domain the resulting normalized measure μ is Möbius invariant. In the case of the Fock space of holomorphic functions defined on $\mathbb{C}^n$ the normalized measure μ is just the Lebesgue measure.

2 Toeplitz Operators Defined on Hilbert Spaces with a Reproducing Kernel and the Berezin Transform

Let $\mathcal{H} \subset L^2(M, \ d\nu)$ be a Hilbert space with normalized reproducing kernel of the form $\{e_z\}_{z\in M}$ and normalized measure constructed out of ν, of the form $d\mu(z) = E(z,z)\,d\nu(z)$. Let $b \in L^\infty(M, \ d\nu)$, i.e., the essential supremum of $|b(z)|$ with respect to ν is finite. The operator $T_b : \mathcal{H} \to \mathcal{H}$ defined as

$$T_b = P_{\mathcal{H}} M_b, \tag{6}$$

where M_b denotes the operator of pointwise multiplication by function b, is called the Toeplitz operator with symbol b or the Berezin operator with contravariant symbol b. The mapping

$$b \mapsto T_b \tag{7}$$

is called the *Berezin quantization*. It is a quantization procedure expressing a transition from classical to quantum mechanics. The *Berezin covariant symbol* of the operator $T \in B(\mathcal{H})$, the Banach algebra of all bounded linear operators on $\mathcal{H}$, is defined as the diagonal with respect to the normalized reproducing kernel, i.e.,

$$T \mapsto \langle Te_z, e_z\rangle. \tag{8}$$

The Berezin transform maps $L^\infty(M, \ d\nu)$ into $L^\infty(M, \ d\nu)$. It is defined as a composition of the Berezin quantization (7) followed by the covariant symbol map (8), i.e.,

$$b \mapsto \tilde{b}, \text{ where } \tilde{b}(z) = \langle T_b e_z, e_z\rangle_{L^2(M,\ d\nu)}. \tag{9}$$

The orthogonal decomposition $L^2(M, \ d\nu) = \mathcal{H} \oplus \mathcal{H}^\perp$ produces the matrix representation

$$P_{\mathcal{H}} M_b P_{\mathcal{H}} = \begin{bmatrix} T_b & 0 \\ 0 & 0 \end{bmatrix}. \tag{10}$$

The formula (10) implies that as far as spectral properties are concerned we may identify operator T_b acting on $\mathcal{H}$ with the operator $P_{\mathcal{H}} M_b P_{\mathcal{H}}$ acting on $L^2(M, \ d\nu)$.

We use the name Toeplitz with respect to both operators T_b and $P_{\mathcal{H}} M_b P_{\mathcal{H}}$. The operator $P_{\mathcal{H}} M_b P_{\mathcal{H}}$ may be interpreted as the linear mapping

$$f \mapsto \int_M b(z) \langle f, e_z \rangle_{L^2(M, \, d\nu)} e_z \, d\mu(z), \tag{11}$$

with the convergence of the integral understood in a weak sense. Indeed, since the formula (2) holds, and the normalized reproducing kernel and the normalized measure have the form (3), and (4), one has for any $f, g \in L^2(M, \, d\nu)$:

$$\begin{aligned} \langle P_{\mathcal{H}} M_b P_{\mathcal{H}} f, g \rangle_{L^2(M, \, d\nu)} &= \int_M b(z) \langle f, E_z \rangle_{L^2(M, \, d\nu)} \overline{\langle g, E_z \rangle}_{L^2(M, \, d\nu)} \, d\nu(z) \\ &= \int_M b(z) \langle f, e_z \rangle_{L^2(M, \, d\nu)} \overline{\langle g, e_z \rangle}_{L^2(M, \, d\nu)} \, d\mu(z). \end{aligned}$$

Let $L^p(M, \, d\mu)$, $1 \leq p < \infty$, denote the Lebesgue space consisting of functions integrable with pth power with respect to measure μ, and $S^p(\mathcal{H})$ the space of compact operators acting on $\mathcal{H}$ with l^p summable singular values. The symbol $L^\infty(M, \, d\mu)$ stands for the space of functions with finite essential supremum of their absolute value with respect to measure μ, $S^\infty(\mathcal{H})$ denotes the algebra of compact operators acting on $\mathcal{H}$, and $B(\mathcal{H})$ the algebra of bounded operators acting on $\mathcal{H}$. It is a direct consequence of the formulae (11) and (5) that the Berezin quantization (7) is

(i) A linear contraction from $L^\infty(M, \, d\mu)$ into $B(\mathcal{H})$, i.e.$||T_b||_{B(\mathcal{H})} \leq ||b||_{L^\infty(M, \, d\mu)}$
(ii) A linear contraction from $L^1(M, \, d\mu)$ into $S^1(\mathcal{H})$ i.e. $||T_b||_{S^1(\mathcal{H})} \leq ||b||_{L^1(M, \, d\mu)}$
(iii) Positivity preserving, i.e., if b≥ 0, then$T_b \geq 0$.

In (i) and (ii) we have to use Minkowski integral inequality. Another argument making use of (11) and (5) shows that the Berezin covariant symbol map (8) is

(iv) A linear contraction from $B(\mathcal{H})$ into $L^\infty(M, \, d\mu)$, i.e. $\sup_{z \in M} |\langle T e_z, e_z \rangle| \leq ||T||_{B(\mathcal{H})}$
(v) A linear contraction from $S^1(\mathcal{H})$ into $L^1(M, \, d\mu)$, i.e.,

$$\int_M |\langle T e_z, e_z \rangle| \, d\mu(z) \leq ||T||_{S^1(\mathcal{H})}.$$

Complex interpolation allows us to conclude that for all $1 \leq p < \infty$ mappings (7), (8), (9) are linear contractions from $L^p(M, \, d\mu)$ into $S^p(\mathcal{H})$, from $S^p(\mathcal{H})$ into $L^p(M, \, d\mu)$, and from $L^p(M, \, d\mu)$ into $L^p(M, \, d\mu)$, respectively.

The Berezin transform is an important tool, frequently occurring in the context of Toeplitz operators. Papers by Axler and Zheng [1] and Isralowitz and Zhu [19] are good illustrations of how it is used. The results concerning eigenvalues and singular values in the current survey are based on a direct approach, not making use of the Berezin transform.

3 Fock Space and its Toeplitz Operators

A general reference for this section is Folland's book [13]. Let $\mathcal{F}$ be the *Fock space*, also called *Segal–Bargmann space*, consisting of holomorphic functions defined on $\mathbb{C}^n$ that are square integrable with respect to the measure $e^{-\pi|z|^2}$, i.e.,

$$\mathcal{F} = \left\{ f \in L^2\left(\mathbb{C}^n, e^{-\pi|z|^2}\right) \mid f \text{is entire on } \mathbb{C}^n \right\}. \tag{12}$$

The reproducing kernel (1) of $\mathcal{F}$ has the form $E_w(z) = E(z,w) = e^{\pi z\overline{w}}$, and the normalized measure (4) obtained out of $d\nu(z) = e^{-\pi|z|^2}\,dz$ is $d\mu(z) = dz$.

Write $\phi(x) = 2^{n/4}e^{-\pi x^2}$, with $\phi_{q,p}(x) = e^{2\pi ipx - \pi ipq}\phi(x-q)$, where $x, q, p \in \mathbb{R}^n$, and assume $h \in L^2(\mathbb{R}^n)$. A direct computation shows that

$$\langle h, \phi_{q,p}\rangle_{L^2(\mathbb{R}^n)} = e^{-\frac{\pi}{2}|z|^2} \underbrace{2^{n/4} \int_{\mathbb{R}^n} h(x) e^{2\pi zx - \pi x^2 - \frac{\pi}{2}z^2}\,dx}_{Bh(z)}, \tag{13}$$

where $z = q - ip$. The map $B : L^2(\mathbb{R}^n) \to \mathcal{F}$, defined in (13), is a unitary map, called the *Bargmann transform*, establishing the correspondence between $\phi_{q,p}$ and e_z the normalized reproducing kernel (3) of the Fock space $\mathcal{F}$ defined in (12), and between the Fock space Toeplitz operator with symbol b and the Gabor–Toeplitz operator (15) with the generating function ϕ and the symbol b, i.e., between

$$f \mapsto \int_{\mathbb{C}^n} b(z)\langle f, e_z\rangle\, e_z\, d\mu(z), \text{and}$$

$$h \mapsto \int_{\mathbb{R}^{2n}} b(q,p)\langle h, \phi_{q,p}\rangle \phi_{q,p}\, dq\, dp,$$

where again $z = q - ip$.

Fock space Toeplitz operators and the variations on their theme were examined from various angles. Papers by Berger and Coburn [3], Grudsky and Vasilevsky [18], Isralowitz and Zhu [19], Janas [20], Janas and Stochel [21], Perälä et al. [27], and the book by Zhu [30] present the recent progress done in the field. In the current survey we concentrate on the properties of the eigenvalues of compact, nonnegative Toeplitz operators with integrable symbols.

4 Gabor Reproducing Formula and Gabor–Toeplitz Operators

Gabor reproducing formula has the form

$$f = \int_{\mathbb{R}^n}\int_{\mathbb{R}^n} \langle f, \phi_{q,p}\rangle \phi_{q,p}\, dq\, dp, \tag{14}$$

where $f, \phi \in L^2(\mathbb{R}^n)$, $||\phi||_{L^2(\mathbb{R})} = 1$, $\phi_{q,p}(x) = e^{2\pi ipx}\phi(x-q), q, p \in \mathbb{R}^n$, and the convergence of the integral is understood in the weak sense. The function ϕ is called the *generating function* of the reproducing formula (14). By introducing into (14) a weight function $b \in L^\infty(\mathbb{R}^{2n})$ we construct a bounded operator

$$T_{\phi,b} f = \int_{\mathbb{R}^n} \int_{\mathbb{R}^n} b(q,p) \langle f, \phi_{q,p} \rangle \phi_{q,p} \, dq \, dp. \tag{15}$$

The operator $T_{\phi,b}$ is called a *Gabor–Toeplitz operator*, and function b is called its (upper) *symbol*. It follows directly from the reproducing formula (14) that the transform $W_\phi : L^2(\mathbb{R}^n) \to L^2(\mathbb{R}^{2n})$ given by

$$W_\phi f(q,p) = \langle f, \phi_{q,p} \rangle \tag{16}$$

is an isometry and that the integral operator $P_\phi : L^2(\mathbb{R}^{2n}) \to L^2(\mathbb{R}^{2n})$ defined as

$$P_\phi F(q,p) = \int_{\mathbb{R}^n} \int_{\mathbb{R}^n} F(s,r) \langle \phi_{s,r}, \phi_{q,p} \rangle \, ds \, dr \tag{17}$$

is an orthogonal projection onto the closed linear subspace $W_\phi \left(L^2(\mathbb{R}^n) \right)$ of $L^2(\mathbb{R}^{2n})$. We conclude that the space $W_\phi \left(L^2(\mathbb{R}^n) \right) \subset L^2(\mathbb{R}^{2n})$ is a Hilbert space with a reproducing kernel and that the operator $P_\phi M_b P_\phi$ acting on $L^2(\mathbb{R}^{2n})$, where M_b denotes the operator of pointwise multiplication by b, is a reproducing kernel Hilbert space Toeplitz operator defined by the symbol function b and the orthogonal projection P_ϕ. The orthogonal decomposition $L^2(\mathbb{R}^{2n}) = W_\phi \left(L^2(\mathbb{R}^n) \right) \oplus W_\phi \left(L^2(\mathbb{R}^n) \right)^\perp$ generates the matrix representation

$$P_\phi M_b P_\phi = \begin{bmatrix} W_\phi T_{\phi,b} W_\phi^* & 0 \\ 0 & 0 \end{bmatrix}. \tag{18}$$

We infer out of (18) that as far as spectral properties are concerned we may identify Gabor–Toeplitz operator $T_{\phi,b}$ with the Toeplitz operator $P_\phi M_b P_\phi$. In fact the relationship (18) is the reason for introducing the name Gabor–Toeplitz operator.

The classical approach to time–frequency localization is due to Landau, Pollak, Slepian and Widom. Daubechies' book [5] contains a very nice account of the initial results. Papers by Landau and Widom [23] and Landau [22] present the essential stages of progress in the the study of the eigenvalues of classical time–frequency localization operators. The study of localization operators in the context of the Gabor reproducing formula and the Gabor frames was initialized by Ingrid Daubechies. Her book [5] discusses also this part of the story.

5 Tight Gabor Frames and Gabor Multipliers

Let $\mathcal{H}$ be a Hilbert space. A family of functions $\{\phi_j\}_{j \in J} \subset \mathcal{H}$ is called a frame of $\mathcal{H}$ if there are constants $0 < A, B < \infty$ such that for any $f \in \mathcal{H}$

$$A \|f\|_{\mathcal{H}}^2 \leq \sum_{j \in J} \left| \langle f, \phi_j \rangle_{\mathcal{H}} \right|^2 \leq B \|f\|_{\mathcal{H}}^2. \tag{19}$$

For any frame of $\mathcal{H}$ it is possible to choose the largest constant A and the smallest constant B for which (19) holds. These two extreme values of A and B are called the lower and the upper frame bounds of the frame $\{\phi_j\}_{j\in J}$. If the lower and the upper frame bounds are equal then the frame is called tight. Any tight frame gives rise to a discrete reproducing formula. We may renormalize a tight frame $\{\phi_j\}_{j\in J}$ and obtain the norm equality

$$||f||^2_{\mathcal{H}} = \sum_{j\in J} \left|\langle f, \phi_j\rangle_{\mathcal{H}}\right|^2,$$

which in turn via the polarization identity may be interpreted as a reproducing formula

$$f = \sum_{j\in J} \langle f, \phi_j\rangle_{\mathcal{H}} \phi_j, \tag{20}$$

with the convergence of the sum understood in the weak sense. There is a canonical way of constructing a tight frame out of a frame. For any frame $\{\phi_j\}_{j\in J}$ we define the frame operator via

$$Sf = \sum_{j\in J} \langle f, \phi_j\rangle_{\mathcal{H}} \phi_j.$$

Condition (19) guarantees that the above sum is weakly convergent for any $f \in \mathcal{H}$ and that the resulting linear operator S is bounded and invertible on $\mathcal{H}$. It is straightforward to verify that the family $\{S^{-1/2}\phi_j\}_{j\in J}$ is a tight frame on $\mathcal{H}$.

Gabor frames have the form $\{\phi_\gamma\}_{\gamma\in\Gamma}$, where $\phi \in L^2(\mathbb{R}^n)$, $\gamma = (q, p)$, $\phi_{q,p}(x) = e^{2\pi i p x}\phi(x - q)$, and Γ is a lattice in $\mathbb{R}^{2n}$, i.e., it is a discrete, cocompact subgroup of $\mathbb{R}^{2n}$. The function ϕ is called the *generating function of the Gabor frame* $\{\phi_\gamma\}_{\gamma\in\Gamma}$. We assume further on that the system $\{\phi_\gamma\}_{\gamma\in\Gamma}$ is a tight Gabor frame normalized in such a way that the reproducing formula (20) holds. We construct a bounded operator

$$G_{\phi,b} f = \sum_{\gamma\in\Gamma} b(\gamma)\langle f, \phi_\gamma\rangle \phi_\gamma \tag{21}$$

by introducing a weight $b \in l^\infty(\Gamma)$ into (20). The operator $G_{\phi,b}$ is called a *Gabor multiplier* with *symbol b*. It is a discrete analogue of the Gabor–Toeplitz operator $T_{\phi,b}$. The reproducing formula (20) implies that the transform $V_\phi : L^2(\mathbb{R}^n) \to l^2(\Gamma)$ given by

$$V_\phi f(\gamma) = \langle f, \phi_\gamma\rangle \tag{22}$$

is an isometry and that the operator $Q_\phi : l^2(\Gamma) \to l^2(\Gamma)$ defined as

$$Q_\phi H(\gamma) = \sum_{\lambda\in\Gamma} H(\lambda)\langle \phi_\lambda, \phi_\gamma\rangle \tag{23}$$

is an orthogonal projection onto the closed linear subspace $V_\phi\left(L^2(\mathbb{R}^n)\right)$ of $l^2(\Gamma)$). In a way similar to the continuous case, we conclude that the space $V_\phi(L^2(\mathbb{R}^n)) \subset l^2(\Gamma)$ is a Hilbert space with a reproducing kernel and that the operator $Q_\phi M_b Q_\phi$ acting on

$l^2(\Gamma)$, where M_b is the operator of pointwise multiplication by b, and is a reproducing kernel Hilbert space Toeplitz operator defined by the symbol function b and the orthogonal projection Q_ϕ. The orthogonal decomposition $l^2(\Gamma) = V_\phi\left(L^2(\mathbb{R}^n)\right) \oplus V_\phi\left(L^2(\mathbb{R}^n)\right)^\perp$ produces the matrix representation

$$Q_\phi M_b Q_\phi = \begin{bmatrix} V_\phi G_{\phi,b} V_\phi^* & 0 \\ 0 & 0 \end{bmatrix}. \tag{24}$$

As in the continuous case also in the discrete case we may identify Gabor multplier $G_{\phi,b}$ with the Toeplitz operator $Q_\phi M_b Q_\phi$. In conclusion we may summarize. Continuous formulae (15), (16), (17), and (18) have their exact discrete counterparts (21), (22), (23), and (24).

Gabor frames were studied extensively over the last 30 years. Books by Daubechies [5], Gröchenig [16], collected volumes edited by Feichtinger and Strochmer [10, 12], and the survey paper by Casazza [4] present most of the results obtained over the first 20 years. Papers by Balan [2], Gröchenig and Stöckler [17] are two selected important contributions to the theory that took place over the last 10 years period. Several further aspects of Gabor multipliers, not treated in the current survey, are discussed in the papers by Feichtinger and Nowak [8], and Dörfler and Torrésani [7].

6 The Extended Metaplectic Representation

A general reference for this section is Folland's book [13]. Wigner distribution $W_f(x,\xi)$, defined for $f \in L^2(\mathbb{R}^n)$, is the quantum substitute of the classical joint probability distribution of position and momentum, position represented by variable x, and momentum represented by variable ξ,

$$W_f(x,\xi) = \int_{\mathbb{R}^n} e^{-2\pi i\langle \xi, y\rangle} f\left(x + \frac{y}{2}\right) \overline{f\left(x - \frac{y}{2}\right)}\, dy. \tag{25}$$

The Heisenberg group $\mathbb{H}^n$ is the group obtained by defining on $\mathbb{R}^{2n+1}$ the product

$$(z,t)\cdot(w,s) = \left(z + w, t + s - \frac{1}{2}\omega(z,w)\right),$$

where $z, w \in \mathbb{R}^{2n}$, $t, s \in \mathbb{R}$ and ω is the symplectic form defined on $\mathbb{R}^{2n}$, i.e.,

$$\omega(z,w) = z^t J w,\ J = \begin{bmatrix} 0 & I_n \\ -I_n & 0 \end{bmatrix}.$$

The translation and modulation operators acting on $L^2(\mathbb{R}^n)$ are defined by

$$T_x f(y) = f(y-x) \text{ and } M_\xi f(y) = e^{2\pi i \langle \xi, y\rangle} f(y),$$

where $x, \xi \in \mathbb{R}^n$, $f \in L^2(\mathbb{R}^n)$. The Schrödinger representation of the group $\mathbb{H}^n$, acting on $L^2(\mathbb{R}^n)$, is then defined by

$$\rho(x,\xi,t)f(y) = e^{2\pi i t} e^{\pi i \langle x,\xi\rangle} e^{2\pi i \langle \xi, y-x\rangle} f(y-x) = e^{2\pi i t} e^{\pi i \langle x,\xi\rangle} T_x M_\xi f(y),$$

where we write $z = (x, \xi)$ when we separate the position component x from the momentum component ξ of a point z of the phase space $\mathbb{R}^{2n}$. The symplectic group $Sp(n,\mathbb{R})$ consists of those nonsingular $2n \times 2n$ matrices with real entries that preserve the symplectic form ω. The symplectic group acts on $\mathbb{H}^n$ via automorphisms that leave the center $\{(0,t) \mid t \in \mathbb{R}\}$ of $\mathbb{H}^n$ pointwise fixed,

$$A \cdot (z,t) = (Az, t).$$

We conclude that for any fixed $A \in Sp(n,\mathbb{R})$ there is a unitary representation of $\mathbb{H}^n$, acting on $L^2(\mathbb{R}^n)$, defined as the composition

$$\rho_A(z,t) = \rho(A \cdot (z,t)),$$

with its restriction to the center of $\mathbb{H}^n$ being a multiple of the identity. By the Stone–von Neumann theorem ρ_A is unitary equivalent to ρ, i.e., there is an intertwining unitary operator $\mu(A)$ acting on $L^2(\mathbb{R}^n)$ such that for all $(z,t) \in \mathbb{H}^n$

$$\rho_A(z,t) = \mu(A) \circ \rho(z,t) \circ \mu(A)^{-1}.$$

By Schur's lemma, μ is determined up to a phase factor e^{is}, $s \in \mathbb{R}$. It turns out that the phase ambiguity is really a sign, so that μ lifts to a representation of the double cover of the symplectic group $Sp(n,\mathbb{R})$. The constructed representation of the double cover of $Sp(n,\mathbb{R})$ is the well known metaplectic or Shale–Weil representation.

The representations ρ and μ can be combined and give rise to the extended metaplectic representation of the group $\mathbb{H}^n \rtimes Sp(n,\mathbb{R})$, the semidirect product of $\mathbb{H}^n$ and $Sp(n,\mathbb{R})$. The group law is

$$((z,t),A) \cdot (w,s),B) = ((z,t) \cdot (Aw,s)\,, AB) \tag{26}$$

and the extended metaplectic representation μ_e is

$$\mu_e\,((z,t),A) = \rho(z,t) \circ \mu(A). \tag{27}$$

From the point of view of the interpretation as a phase space action, the phase factor of μ_e does not matter, therefore we remove it and treat μ_e as a projective representation of the semidirect product $\mathbb{R}^{2n} \rtimes Sp(n,\mathbb{R})$. Group $\mathbb{R}^{2n} \rtimes Sp(n,\mathbb{R})$ consists of all affine transformations of the phase space $\mathbb{R}^{2n}$. Its action on $\mathbb{R}^{2n}$ is

$$(z,A)\,w = Aw + z, \tag{28}$$

where $z, w \in \mathbb{R}^{2n}$, $A \in Sp(n, \mathbb{R})$. The interpretation of the extended metaplectic representation as an affine action on phase space is justified by the invariance properties of the Wigner distribution (25), namely

$$W_{\mu_e(z,A)f}(x,\xi) = W_f\,(z,A)^{-1}\,(x,\xi), \tag{29}$$

for all $x, \xi \in \mathbb{R}^n$. It occurs that only affine maps (z, A) with $A \in Sp(n, \mathbb{R})$ preserve the class of Wigner distributions. Therefore, the group $\mathbb{R}^{2n} \rtimes Sp(n, \mathbb{R})$ contains all affine transformations of the phase space $\mathbb{R}^{2n}$.

As the Wigner distribution (25) satisfies (29) also Gabor–Toeplitz operators (15) and Gabor multipliers (21) have natural invariance properties with respect to the extended metaplectic representation (27).

Group law (26) and affine action rule (28) allow us to identify the parametrization space of (15) consisting of $q, p \in \mathbb{R}^n$ with the subgroup $\{(w, I_{2n}) \mid w \in \mathbb{R}^{2n}\}$ of $\mathbb{R}^{2n} \rtimes Sp(n, \mathbb{R})$. Let $g = (v, A) \in \mathbb{R}^{2n} \rtimes Sp(n, \mathbb{R})$. Taking into account the definition of the extended metaplectic representation (27) we obtain

$$\begin{aligned}
\mu_e(g)T_{\phi,b}\mu_e(g)^* f &= \int_{\mathbb{R}^n}\int_{\mathbb{R}^n} b(q,p)\langle f, \mu_e(g)\phi_{q,p}\rangle \mu_e(g)\phi_{q,p}\, dq\, dp \\
&= \int_{\mathbb{R}^{2n}} b(w)\langle f, \mu_e(g(w, I_{2n}))\phi\rangle \mu_e(g(w, I_{2n}))\phi\, dw \\
&= \int_{\mathbb{R}^{2n}} b(w)\langle f, \mu_e\left(g(w, I_{2n})g^{-1}\right)\mu_e(g)\phi\rangle \mu_e\left(g(w, I_{2n})g^{-1}\right) \\
&\qquad \mu_e(g)\phi\, dw \\
&= \int_{\mathbb{R}^{2n}} b\left(A^{-1}z\right)\langle f, \mu_e(z, I_{2n})\mu_e(g)\phi\rangle \mu_e(z, I_{2n})\mu_e(g)\phi\, dz \\
&= T_{\mu_e(g)\phi, b\circ A^{-1}} f.
\end{aligned}$$

We conclude that for $g = (v, A) \in \mathbb{R}^{2n} \rtimes Sp(n, \mathbb{R})$

$$\mu_e(g)T_{\phi,b}\mu_e(g)^* = T_{\mu_e(g)\phi, b\circ A^{-1}}. \tag{30}$$

A similar calculation shows that also for a Gabor multiplier (21) an appropriate invariance property is available, namely

$$\mu_e(g)G_{\phi,b}\mu_e(g)^* = G_{\mu_e(g)\phi, b\circ A^{-1}}, \tag{31}$$

where the right hand side operator is defined on the lattice $A\Gamma$, with its symbol $b \circ A^{-1} \in l^\infty(A\Gamma)$, and $\mu_e(g)\phi$ the generating function of a tight Gabor frame on $A\Gamma$.

7 Dependence on the Symbol Function

If the symbol function b is integrable, with values in the closed interval $[0, 1]$, then the operator $T_{\phi,b}$ is called a Gabor–Toeplitz operator of localization type. Asymptotic properties as $R \to \infty$ of the symbolic calculus of Gabor–Toeplitz operators T_{ϕ,b_R}, where b is integrable, $0 \leq b(q,p) \leq 1$, and $b_R(q,p) = b\left(\frac{q}{R}, \frac{p}{R}\right)$ is the L^∞ normalized dilation of b, were studied in [11]. For h a continuous function defined on the closed interval [0, 1] operator $h\left(T_{\phi,b_R}\right)$ is defined via the spectral decomposition of T_{ϕ,b_R}. A Szegö type formula for operators of the form $h\left(T_{\phi,b_R}\right)$ was obtained in [11],

$$\lim_{R\to\infty} \frac{\mathrm{tr}\left(T_{\phi,b_R} h(T_{\phi,b_R})\right)}{R^{2n}} = \int_{\mathbb{R}^n}\int_{\mathbb{R}^n} b(q,p)h(b(q,p))\,dq\,dp. \tag{32}$$

We need to multiply $h\left(T_{\phi,b_R}\right)$ by T_{ϕ,b_R} in (32) in order to make sure that the operator to which trace is applied is trace class. As a consequence of (32) it has been concluded in [11] that in the normalized limit with factor R^{2n} in the denominator both the distribution of the eigenvalues of T_{ϕ,b_R}

$$N(\delta, R) = \left|\left\{i \mid \lambda_i\left(T_{\phi,b_R}\right) > \delta\right\}\right|, \; 0 < \delta < 1, \tag{33}$$

and the size of their plunge region

$$M(\delta_1, \delta_2, R) = \left|\left\{i \mid \delta_1 < \lambda_i\left(T_{\phi,b_R}\right) < \delta_2\right\}\right|, \; 0 < \delta_1 < \delta_2 < 1, \tag{34}$$

are expressed directly via the corresponding quantities of the symbol function, i.e., the distribution of b

$$\left|\{(q,p) \mid b(q,p) > \delta\}\right|, \tag{35}$$

and the Lebesgue measure of its plunge region

$$\left|\{(q,p) \mid \delta_1 < b(q,p) < \delta_2\}\right|. \tag{36}$$

In the passage to the normalized limit it was necessary to assume that the level sets $\{(q,p) | \, b(q,p) = \tau\}$, $\tau = \delta, \delta_1, \delta_2$, have Lebesgue measure 0. It follows from the correspondence of the eigenvalue distribution and the symbol distribution indicated above that asymptotically the best localization properties, i.e., asymptotically there are no eigenvalues in the open interval (0, 1), occur for symbols being characteristic functions of Lebesgue measurable sets $\Omega \subset \mathbb{R}^{2n}$.

If the symbol b is summable, with values in the closed interval [0, 1], then the operator $G_{\phi,b}$ is called a Gabor multiplier of localization type. A discrete analogue of the Szegö type formula (32) applied to G_{ϕ,b_R} requires more assumptions on the the symbol function b. It is enough to assume that b is defined on $\mathbb{R}^{2n}$, that it is Riemann integrable, has compact support, and that $0 \leq b(q,p) \leq 1$. Also in the discrete setup the Szegö type formula captures essential asymptotic properties of the symbolic calculus of Gabor multipliers of localization type as $R \to \infty$, where R is the parameter of the L^∞ normalized dilation $b_R(q,p) = b\left(\frac{q}{R}, \frac{p}{R}\right)$. For any continuous function h defined on [0, 1] the operator $h(G_{\phi,b_R})$ is defined in terms of the spectral decomposition of G_{ϕ,b_R}. The discrete analogue of (32) was obtained in [8]. It states that if $\Gamma \subset \mathbb{R}^{2n}$ is a lattice, $\phi \in L^2(\mathbb{R}^n)$, the system $\{\phi_\gamma\}_{\gamma\in\Gamma}$ is a tight

Gabor frame, and the symbol function b defined on $\mathbb{R}^{2n}$ is Riemann integrable, has compact support, and satisfies $0 \leq b(q,p) \leq 1$, then

$$\lim_{R\to\infty} \frac{\operatorname{tr}\left(G_{\phi,b_R} h(G_{\phi,b_R})\right)}{R^{2n}} = \frac{||\phi||^2_{L^2(\mathbb{R}^n)}}{\mathrm{Vol}_\Gamma} \int_{\mathbb{R}^{2n}} b(q,p)h(b(q,p))\,dq\,dp, \tag{37}$$

where Vol_Γ is the volume of the fundamental domain of the lattice Γ. The same normalized limit results as for the eigenvalue distribution of (33) and the size of their plunge region of (34) can be concluded for Gabor multipliers of localization type. In the discrete setup we need to insert the multiplicative constant $\frac{||\phi||^2_{L^2(\mathbb{R}^n)}}{\mathrm{Vol}_\Gamma}$ in front of (35) and (36), the same constant as in (37), but otherwise the normalized limits are same as for Gabor–Toeplitz localization operators. We can also conclude that asymptotically the best localization properties occur for symbols being characteristic functions of bounded Riemann measurable sets $\Omega \subset \mathbb{R}^{2n}$.

There are many variants of Szegö type results occurring in the literature. Our approach to Gabor–Toeplitz operators and Gabor multipliers was motivated by the paper by Widom [29].

8 Dependence on the Generating Function

To get started let us assume that the symbol function b of the Gabor–Toeplitz operator (15)

$$T_{\phi,b} f = \int_{\mathbb{R}^n}\int_{\mathbb{R}^n} b(q,p)\langle f, \phi_{q,p}\rangle \phi_{q,p}\,dq\,dp$$

is continuous and nonnegative, and that it has compact support. Let $f(q,p) = b(q,p)^{1/2}$, and let us define the operator $R_{\phi,f} : L^2(\mathbb{R}^n) \to L^2(\mathbb{R}^{2n})$ by the integral kernel

$$K_{\phi,f}(q,p,y) = f(q,p)\overline{\phi_{q,p}(y)}, \tag{38}$$

where variables q, p represent the the range of $R_{\phi,f}$ and variable y its domain. It follows directly from the form of the kernel (38) that

$$T_{\phi,b} = R^*_{\phi,f} R_{\phi,f}.$$

We observe that the eigenvalues of $T_{\phi,b}$ are the squares of the singular values of $R_{\phi,f}$, therefore from the spectral theory point of view we may use $T_{\phi,b}$ and $R_{\phi,f}$ interchangeably. For $0 < p \leq \infty$ we define S^p as the class of compact operators acting from $L^2(\mathbb{R}^n)$ into $L^2(\mathbb{R}^{2n})$ with l^p summable singular values. By S^∞ we denote the class of compact operators equipped with the operator norm. The study of the singular values of $R_{\phi,f}$ was done in [26]. Necessary and sufficient criteria for the membership of $R_{\phi,f}$ in S^p for all $0 < p \leq \infty$ were obtained.

Let $R_0 = [-1/2, 1/2]$, $R_1 = [0, 1]$. For $k \in \mathbb{Z}^n$ and $\tau = (i_1, \ldots, i_n)$, $i_j = 0, 1$, we define the cubes

$$Q_k^\tau = k + R_{i_1} \times \ldots \times R_{i_n}.$$

We take two nonnegative C^∞ functions h_0, h_1 satisfying conditions $\operatorname{supp} h_0 \subset (-1/2, 1/2)$, $\operatorname{supp} h_1 \subset (0, 1)$, $\sum_{k\in\mathbb{Z}} (h_0(x-k) + h_1(x-n)) \equiv 1$. We define

$$m_k^\tau(x) = \prod_{j=1}^{n} h_{i_j}\left(x_j - k_j\right) \text{ and } m^\tau(x) = \sum_{k\in\mathbb{Z}^n} m_k^\tau(x).$$

It is clear that $\operatorname{supp} m_k^\tau \subset Q_k^\tau$, $\operatorname{dist}\left(\operatorname{supp} m_k^\tau, \partial Q_k^\tau\right) \geq \delta > 0$, $\sum_\tau m^\tau \equiv 1$. Each space $L^2\left(Q_k^\tau\right)$ is equipped with the orthonormal trigonometric basis

$$e_l(x) = e^{2\pi i\langle l,x\rangle},\ l \in \mathbb{Z}^n.$$

The process of introducing τ and the property of separated supports of functions m_k^τ together with the application of Schur multipliers allowed us to obtain two-sided estimates for the quantity $||R_{\phi,f}||_{S^p}$. It has been shown in [26] that for any nonzero, continuous function f with compact support, and any $0 < p \leq \infty$

$$||R_{\phi,f}||_{S^p}^p \cong \sum_{\tau,k,\rho,l} ||m_l^\rho \left(m_k^\tau \phi\right)^\wedge ||_{L^2}^p, \tag{39}$$

where $\hat{f}(\xi) = \int_{\mathbb{R}^n} f(x) e^{-2\pi i\langle x,\xi\rangle}$ is the Fourier transform. The constants of equivalence in (39) do not depend on ϕ.

9 Two-Sided Control of the Eigenvalue Plunge Region

Szegö type asymptotic formulae (32) and (37) indicated that localization type Gabor–Toeplitz operators and Gabor multipliers are the closest to orthogonal projections if their symbols are characteristic functions of sets of finite measure. In the next step operators $T_{\phi,\Omega}$, $G_{\phi,\Omega}$, with symbols b of the form χ_Ω, where Ω is a set of finite measure, were studied directly without the asymptotic limit. If the symbol function is a characteristic function of a set of finite measure, then we call the corresponding operator $T_{\phi,\Omega}$ or $G_{\phi,\Omega}$ a *localization operator*.

Two-sided estimates of the size of the eigenvalue plunge region

$$M(\delta_1, \delta_2, \phi, \Omega) = \left|\left\{i \mid \delta_1 < \lambda_i\left(T_{\phi,\Omega}\right) < \delta_2\right\}\right|, \tag{40}$$

expressed in terms of the area of a strip of fixed size R around the boundary $\partial\Omega$, and uniform with respect to generating functions ϕ and localization domains Ω, of the form

$$c_1 \left|(\partial\Omega)^R\right| \leq M\left(\delta_1, \delta_2, \phi, \Omega\right) \leq c_2 \left|(\partial\Omega)^R\right|, \tag{41}$$

where $(\partial\Omega)^R = \{(q,p) \mid \mathrm{dist}((q,p),\partial\Omega)| < R\}$, were obtained in [6]. In order to get two-sided uniform positive constants c_1, c_2 it was necessary to assume uniform decay and nondegeneracy of the reproducing kernels derived from properties of the generating functions ϕ and uniform access to localization domains Ω and their complements Ω^c from the points near their boundaries $\partial\Omega$. The size R of the strip $(\partial\Omega)^R$ around $\partial\Omega$ is one of the uniform parameters controlling the mutual interactions between the generating functions and the domains of localization.

Two-sided uniform estimates of the size of the eigenvalue plunge region, uniform with respect to a class of tight Gabor frames and a family of localization domains, a discrete analog of (40), were obtained in [8]. In the discrete context the localization domains Ω are finite subsets of Γ, a fixed lattice of $\mathbb{R}^{2n}$ parameterizing tight Gabor frames under consideration, and the strips of size R around the boundary $\partial\Omega$ occurring in (41) are substituted by

$$(\partial\Omega)^R = \left\{\gamma \in \Omega \mid \mathrm{dist}(\gamma, \Omega^c) < R\right\} \cup \left\{\gamma \in \Omega^c \mid \mathrm{dist}(\gamma, \Omega) < R\right\}. \tag{42}$$

Finally the Lebesgue measure of $|(\partial\Omega)^R|$ occurring in (41) has to be replaced by the number of points of Γ inside $(\partial\Omega)^R$. Uniform decay and nondegeneracy of the reproducing kernels obtained from tight Gabor frames are expressed in terms of the Euclidean metric that is inherited by the lattice $\Gamma \subset \mathbb{R}^{2n}$ from $\mathbb{R}^{2n}$, the same metric as the one that occurs in (42).

10 The Projection Functional and the Asymptotic Boundary Form

The uniform two-sided control of the eigenvalue plunge region (40) of Gabor–Toeplitz operators in terms of the size of the strip around the boundary of the localization domain (41), and its analogue for Gabor multipliers, suggested further steps in the study of the interactions between the generating function of either the Gabor reproducing formula or the tight Gabor frame and the localization domain. It was time to go back to the asymptotic limit. The philosophy of the approach came out of the semiclassical limit. After taking the limit the phenomena representing phase space uncertainty disappear and the interactions between the generating function of either the Gabor reproducing formula or the tight Gabor frame and the boundary of the localization domain can be expressed pointwise and interpreted geometrically. Here is the technical side of the story. An appropriate change of coordinates places the dilation factor on the reproducing kernel. In the limit the reproducing kernel is represented by a directional form defined on the unit sphere, called the surface tension for historical reasons, and the form is applied to the unit normal vector of the current boundary point. The next step in the study of mutual interactions between generating functions and domains of localization from the point of view of estimating the size of the eigenvalue plunge region of $T_{\phi,\Omega,}$ was accomplished in [25]. The eigenvalues of Gabor–Toeplitz localization operators $\lambda_i\left(T_{\phi,\Omega,}\right)$ satisfy estimates

$0 \leq \lambda_i \left(T_{\phi,\Omega,}\right) \leq 1$. The *projection functional*

$$PF(T_{\phi,\Omega}) = \sum_{i=0}^{\infty} \lambda_i \left(T_{\phi,\Omega}\right) \left(1 - \lambda_i(T_{\phi,\Omega})\right) \tag{43}$$

measures the extent by which the operator $T_{\phi,\Omega}$, fails to be an orthogonal projection. The projection functional takes non-negative values, and it vanishes on the space consisting of finite dimensional orthogonal projections. Only eigenvalues $0 < \lambda_i \left(T_{\phi,\Omega,}\right) < 1$, i.e., the plunge region eigenvalues, contribute to its value. We assume that $\Omega \subset \mathbb{R}^{2n}$ is a bounded domain with C^1 boundary. The symbol σ denotes the surface measure defined on $\partial\Omega$. Here the function n denotes the Gauss map, i.e., $n(r,s)$ is the unit normal vector at $(r,s) \in \partial\Omega$ directed outside Ω. We write $P_v = \{w \in \mathbb{R}^{2n} \mid w \cdot v = 0\}$ for the linear subspace of $\mathbb{R}^{2n}$ consisting of vectors orthogonal to v, and by $U_v = \{w \in \mathbb{R}^{2n} \mid w \cdot v \geq 0\}$ the half space inside $\mathbb{R}^{2n}$ with v being the inner normal vector. The boundary form is defined as

$$BF(\phi, \Omega) = \int_{\partial\Omega} \int_{U_{n(r,s)}} |\langle \phi, \phi_{q,p} \rangle|^2 \text{dist}((q,p), P_{n(r,s)}) dq\, dp\, d\sigma(r,s). \tag{44}$$

Under the integrability condition imposed on ϕ that makes formula (44) well defined it has been shown in [25] that

$$\lim_{R\to\infty} \frac{PF\left(T_{\phi,R\Omega,}\right)}{R^{2n-1}} = BF(\phi, \Omega). \tag{45}$$

Formula (45) expresses the limit behavior of the projection functional in terms of the boundary form. It provides a very explicit, quantitative way of describing the interactions between the boundary $\partial\Omega$ and the reproducing kernel obtained out of the generating function ϕ.

Let us fix a volume V and a generating function ϕ for which the boundary form (44) is well defined, and let us ask for what localization domains Ω of volume V the boundary form $BF(\phi, \Omega)$ takes the smallest possible value. Let us introduce surface tension M_ϕ, defined on the unit sphere in phase space $\mathbb{R}^{2n}$, given by the formula

$$M_\phi(v) = \int_{U_v} |\langle \phi, \phi_{q,p} \rangle|^2 \text{dist}((q,p), P_v) dq\, dp. \tag{46}$$

The function M_ϕ determines the Wulff shape

$$K_\phi = \bigcap_{v \in S^{2n-1}} \{w \in \mathbb{R}^{2n} : w \cdot v \leq M_\phi(v)\}. \tag{47}$$

For the given generating function ϕ let us construct K_ϕ and let us scale it with a constant c, so that the volume of cK_ϕ is V. It occurs that, up to translation, the domain cK_ϕ is the only domain among all domains Ω with finite perimeter and volume V for which $BF(\Omega, \phi)$ is minimal. The shape cK_ϕ is the optimal domain of localization. The investigation of the Wulff shape has a long history. The concept itself came out of the problem of finding the shape of crystals in equilibrium. By the Gibbs–Curie principle, the equilibrium shape of the crystal minimizes the surface energy. Wulff proposed his construction in 1901. It took, however, a long time to prove in full generality that the Wulff shape represents the global minimum. We refer the reader to Fonseca and Müller [14], McCann [24], Taylor [28], and the survey by Gardner [15] for more information on the subject. It is a fascinating fact that the classical construction of crystals in equilibrium also produces optimal domains of localization. In the context of Gabor–Toeplitz operators the surface tension M_ϕ has the form (46) and it occurs that it is a uniformly convex and uniformly smooth norm. The invariance formula (30) is an indispensable geometric tool in the context of projection functionals and boundary forms. Not only is it helpful in finding the right form of the results to observe that all ingredients should satisfy proper invariance properties, but it also allows a suitable positioning of the localization domain before switching to local coordinates.

The discrete analogue of the limit formula for the normalized projection functional (45) has been worked out so far only in dimension 1. The work was done in [9]. The limit was directly computed in the case of the separable lattice of the form $a\mathbb{Z} \times b\mathbb{Z}$, where $a, b > 0$, $ab < 1$, and then transferred to a general lattice $\Gamma \subset \mathbb{R}^2$ via the extended metaplectic representation and the invariance formula (31).

A collection of line segments $l_i \subset \mathbb{R}^2$, $i = 1, 2, \ldots, C$, each of them starting and ending at a lattice point of $\Gamma \subset \mathbb{R}^2$ is called a *lattice cycle* if the union $l_1 \cup l_2 \cup \ldots \cup l_C$ forms a closed continuous line without self intersections. A bounded, connected, and closed subset $\Omega \subset \mathbb{R}^2$ is called a *lattice domain* if its boundary consists of a finite family of lattice cycles C_i, $i = 1, 2, \ldots, B$ satisfying condition $\operatorname{dist}\left(C_i, C_j\right) > 0$ for $i \neq j$. For a lattice domain $\Omega \subset \mathbb{R}^2$ and a generating function ϕ of a tight Gabor frame $\{\phi_\gamma\}_{\gamma \in \Gamma}$ the boundary form $\mathrm{BF}(\phi, \Omega)$ is defined by the formula

$$BF(\phi, \Omega) = \sum_{i=1}^{N} \operatorname{length}(l_i) \sum_{\gamma \in U_i} \operatorname{dist}(\gamma, P_i) |\langle \phi, \phi_\gamma \rangle|^2, \tag{48}$$

where l_i, $i = 1, \ldots, N$ are the line segments constituting the boundary of Ω, n_i is the unit vector orthogonal to l_i directed outside Ω, $P_i = \{w \in \mathbb{R}^2 \mid w \cdot n_i = 0\}$, $U_i = \{w \in \Gamma \mid w \cdot n_i \geq 0\}$. Under the summability condition imposed on ϕ that makes the boundary form (48) well defined it has been shown in [9] that

$$\lim_{R \to \infty} \frac{PF\left(G_{\phi, R\Omega}\right)}{R} = \frac{1}{\mathrm{Vol}_\Gamma} BF(\phi, \Omega), \tag{49}$$

where Vol_Γ is the area of the fundamental domain of the lattice Γ.

11 Open Problems and Possible Future Research Directions

Let us recall that Toeplitz operators defined on the Fock space (12) and the Bargmann transform (13), connecting coherent states $\phi_{q,p}$ with the normalized reproducing kernel e_z of the Fock space, are at the origin of both Gabor–Toeplitz operators and

Gabor multipliers. Gabor–Toeplitz operators show up once we dare to change the Gaussian to some other generating function. Gabor multipliers come out as Riemann sum approximations on a lattice, with an upgrade to a reproducing formula done on the level of the generating function. For all new developments on the Fock space it makes sense to ask for their analogues in the context of Gabor–Toeplitz operators and Gabor multipliers.

The functions $N(\delta, R)$, $M(\delta_1, \delta_2, R)$, $\delta_1 < \delta_2$, count the eigenvalues larger than δ, and the eigenvalues in the range (δ_1, δ_2), respectively. Asymptotic expansions of $N(\delta, R)$ and $M(\delta_1, \delta_2, R)$, as $R \to \infty$ and the localization domain Ω is dilated by a factor of R are of big interest for both $T_{\phi,\Omega}$ and $G_{\phi,\Omega}$.

Both Gabor–Toeplitz operators and Gabor multipliers express fascinating interconnections between the geometry of the localization domain Ω, discrete geometry in the case of Gabor multipliers, and the functional analytic properties of the generating function. Explaining these interconnections from the point of view of semiclassical analysis is an important task for the future.

Let us formulate specific problems related to the results presented in Sect. 10 of the current survey. We want to quantify the relationship between the localization domain Ω and the projection functional PF from the point of view of the boundary form $BF(\phi, \Omega)$ representing the semiclassical limit with respect to $R \to \infty$. For a fixed volume V of a localization Ω of a Gabor–Toeplitz operator $T_{\phi,\Omega}$ we would like to know the optimal values of the boundary form $BF(\phi, \Omega)$:

(i) The global minimum over both the localization domains and the generating functions (the answer seems still open)

(ii) The minimum over the localization domains with the generating function fixed (the answer is known, the Wulff shape is the optimal localization domain)

(iii) The minimum over the generating functions with the localization domain fixed (the answer is open)

All analogue questions for Gabor multipliers are open. Let us assume that the number of points of the localization domain Ω of a Gabor multiplier $G_{\phi,\Omega}$ is V. Again one would like to investigate the optimal values of the boundary form $BF(\phi, \Omega)$:

(iv) The global minimum over both the localization domains and the generating functions

(v) The minimum over the localization domains with the generating function fixed (the resulting optimal localization domains should be called phase space crystals)

(vi) The minimum over the generating functions with the localization domain fixed

References

1. Axler, S., Zheng, D.: Compact operators via the Berezin transform. Indiana Univ. Math. J. **47**, 387–400 (1998)
2. Balan, R.: The noncommutative Wiener lemma, linear independence, and spectral properties of the algebra of time-frequency shift operators. Trans. Am. Math. Soc. **360**, 3921–3941 (2008)

3. Berger, C.A., Coburn, L.A.: Toeplitz operators on the Segal-Bargmann space. Trans. Am. Math. Soc. **301**, 813–829 (1987)
4. Casazza, P.G.: The art of frame theory. Taiwanese J. Math. **4**, 129–201 (2000)
5. Daubechies, I.: Ten lectures on wavelets. CBMS-NSF regional conference series, no. 6, SIAM, Philadelphia, 1992
6. DeMari, F., Feichtinger, H.G., Nowak, K.: Uniform eigenvalue estimates for time-frequency localization operators. J. Lond. Math. Soc. **65**, 720–732 (2002)
7. Dörfler, M., Torrésani, B.: Representations of operators in the time-frequency domain and generalized Gabor multipliers. J. Fourier Anal. Appl. **16**, 261–293 (2010)
8. Feichtinger, H.G., Nowak, K.: A first survey of Gabor multipliers. In: Feichtinger, H.G., Strohmer, T. (eds.) Applied and Numerical Harmonic Analysis, pp. 99–128. Birkhauser (2003)
9. Feichtinger, H.G., K. Nowak, M.: Pap *Asymptotic boundary forms for tight Gabor frames and lattice localization domains*, in preparation
10. Feichtinger, H.G., Strohmer, T.: Gabor Analysis and Algorithms, Theory and Applications. Birkhäuser, Boston (1998)
11. Feichtinger, H.G., Nowak, K.: A Szegö-type theorem for Gabor-Toeplitz localization operators. Michigan Math. J. **49**, 13–21 (2001)
12. Feichtinger, H.G., Strohmer, T.: Advances in Gabor Analysis. Birkhäuser, Boston (2002)
13. Folland, G.: Harmonic Analysis in Phase Space. Princeton University Pres, Princeton (1989)
14. Fonseca, I., Müller, S.: The Wulff Theorem Revisited. Proc. Roy. Soc. Edinb. Sect. A **119**, 125–136 (1991)
15. Gardner, R.J.: The Brunn-Minkowski inequality. Bull. Am. Math. Soc. **39**(3), 355–405 (2002)
16. Gröchenig, K.: Foundations of Time-Frequency Analysis. Birkhäuser, Boston (2001)
17. Gröchenig, K., Stöckler, J.: Gabor Frames and Totally Positive Functions. Duke Math. J. **162**(6), 1003–1031 (2013)
18. Grudsky, S., Vasilevski, N.: Toeplitz operators on the Fock space: Radial component effects. Integr. Equ. Oper. Theory **44**, 10–37 (2002)
19. Isralowitz, J., Zhu, K.: Toeplitz operators on the Fock space. Integr. Equ. Oper. Theory **66**, 593–611 (2010)
20. Janas, J.: Unbounded Toeplitz operators in the Segal-Bargmann space. Studia Math. **99**, 87–99 (1991)
21. Janas, J., Stochel, J.: Unbounded Toeplitz operators in the Segal-Bargmann space II. J. Funct. Anal. **126**, 418–447 (1994)
22. Landau, H.J.: On the density of phase-space expansions. IEEE Trans. Inf. Theory **39**, 1152–1156 (1993)
23. Landau, H.J., Widom, H.: Eigenvalue distribution of time and frequency limiting. J. Math. Anal. Appl. **77**, 469–481 (1980)
24. McCann, R.J.: Equilibrium shapes for planar crystals in an external field. Comm. Math. Phys. **195**, 699–723 (1998)
25. Nowak, K.: Optimal localization domains for Gabor-Toeplitz operators. unpublished manuscript.
26. Nowak, K.: Local Toeplitz operators based on wavelets: Phase space patterns for rough wavelets. Studia Math. **119**, 37–64 (1996)
27. Perälä, A., Taskinen, J., Virtanen, J.: Toeplitz operators with distributional symbols on Fock spaces. Funct. Approx. Comment. Math. **44**, 203–213 (2011)
28. Taylor, J.: Crystalline variational problems. Bull. Am. Math. Soc. **84**, 568–588 (1978)
29. Widom, H.: Szegö limit theorem: The higher-dimensional matrix case. J. Funct. Anal. **39**, 182–198 (1980)
30. Zhu, K.: Analysis on Fock Spaces. Graduate Texts in Mathematics 263. Springer, New York (2012)

Which Numbers Simplify Your Problem?

Paolo Giordano

The author has been supported by FWF grants M1247-N13 and P25116-N25.

Abstract An extension of the ring of scalar quantities, from the usual field of real numbers to a non-Archimedean, sometimes permits to simplify some problems which, at a first sight, may seem not correlated with infinitesimal and infinite numbers. We present four simple cases, each one at the level of possibility for the creativity of a motivated student. The ring of Fermat reals and its applications to physics and differential geometry, the ring of Colombeau generalized numbers and its applications to the foundations of generalized functions, the Levi-Civita field and the derivation of complicated computer functions and the Surreals numbers as a universal non-Archimedean ring. The definition of each one of these rings is strongly motivated at elementary level and some open problems and ideas are introduced in the first two cases.

Keywords Non-Archimedean rings · Infinitesimal numbers · Infinite numbers · Fermat reals · Colombeau generalized numbers · Levi-Civita field · Surreal numbers

1 Introduction

Usually one thinks that the field of scalar quantities is fixed once and for all, and there are no reasons to find something better. Probably, this is also due to the teaching of real numbers at first university courses, so that it sounds too unusual adding something to it and at the same time simplify your life. Of course, there remains the frequent problem that some students found inconsistencies passing from the course of physics to that of calculus and vice versa: do they use the same ring of scalars? Its very positive that someone feels the problem, because problems stimulate the searching for solutions, but frequently the adopted solution appeals to the fact that different courses require different mentalities, and hence different answers to different professors.

P. Giordano (✉)
Fakultät für Mathematik, University of Vienna, Oskar-Morgenstern-Platz 1, 1090 Wien, Austria
e-mail: paolo.giordano@univie.ac.at

T. M. Rassias, P. M. Pardalos (eds.), *Mathematics Without Boundaries*,
DOI 10.1007/978-1-4939-1106-6_7,

In the present work, we want to show that there are very simple extensions $R \supseteq \mathbb{R}$ of the real field containing also infinitesimal or infinite numbers (or both) that frequently permit to simplify your problems and to open new surprising perspectives. For example, infinitesimals and infinities permit to define new mathematical models, which are otherwise very hard to solve using the usual finite numbers: this is what is frequently done in several applied sciences.

Nowadays, there are plenty of possible extensions of the real field: nonstandard analysis [1, 35, 55] and its constructive version [53, 54], Synthetic differential geometry [3, 42, 48, 52], the Levi-Civita field [49, 50, 60], surreal numbers [2, 12, 14, 22, 41], formal power series [6], super-real fields [25], Colombeau generalized numbers [11, 36], Fermat reals [26, 27], and others. There are also theories introducing infinitesimals methods without focusing on an extension of $\mathbb{R}$, like Weil functors [43–44] and differential geometry over a base ring [7]. See [20] for an interesting presentation of the early history of some models of non-Archimedean continua.

We therefore need to fix a meaningful criterion to focus our attention only on some of them.

We will present only those number systems whose construction is *simple*: a strongly motivated student can surely arrive by herself/himself to define and study some of these new numbers, of course with the usual help of a few inspiration and a lot of transpiration. Our main thesis is that some of these new rings of numbers can drastically *simplify* some of your problems, so that they do not have to radically change your usual way of doing mathematics. As a partial proof of their simplicity, we will see that three among four of the extensions $R \supseteq \mathbb{R}$ we will present in this work have a corresponding computer implementation. For the remaining one, we think that a useful implementation is possible.

Of course, sometimes a radical change of perspective is necessary, but this is out of scope for the present work. On the contrary, a certain change of mentality is necessary to work in both the most powerful theories of infinitesimals: nonstandard analysis and synthetic differential geometry. In both cases a strong formal control of the mathematics we are doing is necessary: in the first case, for example, to use the transfer principle, and in the second one to check whether our proof is valid in intuitionistic logic or not (see e.g., [3–5, 31] for more details). Moreover, the models of both these theories are of course completely understandable by a graduate student, but are, generally speaking, outside the possibilities of its creativity and hence of our main criterion of simplicity.

In our opinion a good mathematical theory is the one that is able to create a good dialectic between some form of intuition and the corresponding rigorous version. Using this dialogue we can guess intuitively true properties and translate them into formal statements, and, vice versa, we can interpret formulas in an intuitive way helping our building of knowledge. Both worlds, the intuitive and the formal one, are important, and the lacking of one of them conducts only to a partial knowledge. Therefore, a general distinction criterion to classify a theory of infinitesimals is the possibility to establish a dialogue between potential infinitesimals and actual infinitesimals. On one hand of this dialogue there are potential infinitesimals, represented by some kind of functions $i : E \longrightarrow \mathbb{R}$ defined on a directed set $(E, \leq)$, like sequences $i : \mathbb{N} \longrightarrow \mathbb{R}$ or functions defined on a subset E of $\mathbb{R}$, and such that

$$\lim_{(E,\leq)} i = 0. \tag{1}$$

Classical example are, of course, $i(n) = \frac{1}{n}$ for $n \in \mathbb{N}_{>0}$ and $i(t) = t$ for $t \in \mathbb{R}_{\geq 0}$. On the other hand, there are actual infinitesimals as elements $d \in R \supseteq \mathbb{R}$ of a suitable ordered ring $(R, \leq)$ such that

$$\forall r \in \mathbb{R}_{>0} : \ -r \leq i \leq r.$$

The dialogue can be realized, if any, in several ways, using, for example, the standard part and the limit (1), or through some connection between the order relation defined on R and the order of the directed set $(E, \leq)$, or through the ring operations of R and pointwise operations on the set of potential infinitesimals. From our point of view, it is very natural to see this dialogue as an advantage, if the theory permits this possibility. First of all, it is a dialogue between two different, but from several aspects equivalent, instruments to formalize natural phenomena and mathematical problems, and hence it seems natural to expect a close relation between them. Secondly, this dialogue can remarkably increase our intuition on actual infinitesimals and can suggest further generalizations. For example, in the context of Fermat reals (see Sect. 2), it seems very natural to try a generalization taking some stochastic processes $(x_t)_{t\in\mathbb{R}_{\geq 0}}$ instead of little-oh polynomials, creating in this way "stochastic infinitesimals."

Theories with a more or less strong dialogue between potential infinitesimals and actual infinitesimals are: nonstandard analysis, the theory of Colombeau generalized numbers, the theory of surreal numbers and the theory of Fermat reals (see [31] for a study of the relationships between these two notions of infinitesimals).

This dialogue, and hence the consequent generalizations or intuitions, are more difficult in formal algebraic approaches to infinitesimals. Very roughly, these approaches can be summarized following the spirit of J. Conway's opinion reported in [13]: If one needs some kind of infinitesimal d, add this new symbol to $\mathbb{R}$ and impose to it the properties you need, for example, $d^2 = 0$. In this class of theories we can inscribe all the other theories: synthetic differential geometry, Weil functors, differential geometry over a base ring, and Levi-Civita field. They can be thought of as theories generated by two different elementary ideas: the ring of dual numbers $\mathbb{R}[\varepsilon]/\langle\varepsilon^2 = 0\rangle$ (firstly generalized by the strongly stimulating and influential article [72]) and the fields of formal power series, see for example [6]. The distinction between these two classes of theories, those that try a dialogue with potential infinitesimals and those approaching formally the problem, is essentially philosophical and at the end choosing one of them rather than the other one is more of a personal opinion than a rational choice. First of all, the distinction is not always so crisp, and (non constructive) nonstandard analysis represents a case where the above mentioned dialogue cannot always be performed, see [31]. Moreover, it is also surely important to note that formal theories of infinitesimals are able to reach a great formal power and flexibility, and sometimes through them a sort of a posteriori intuition about actual infinitesimals can be gained.

Of course, these number systems can be considered interesting by themselves. However, we will also motivate their introduction starting from a problem that, at first sight, seems not directly related to the idea of extending $\mathbb{R}$. Of course, our presentation of these problems cannot be exhaustive, but our references permit to jump outside these boundaries.

2 Which Numbers to Have Infinitesimals in Physics and Differential Geometry?

As we have already said in the introduction, several students experienced the unpleasant feeling to switch from the classical calculus lectures, where usually everything is $\varepsilon - \delta$, to the physics ones where the lecturer report, for example, Einstein's formulas like

$$\frac{1}{\sqrt{1-\frac{v^2}{c^2}}} = 1 + \frac{v^2}{2c^2} \quad , \quad \sqrt{1-h_{44}(x)} = 1 - \frac{1}{2}h_{44}(x) \tag{2}$$

with explicit use of infinitesimals $v/c \ll 1$ or $h_{44}(x) \ll 1$, such that, for example, $h_{44}(x)^2 = 0$. Einstein in [23] wrote the formula (using the equality sign rather than the approximate equality sign $\simeq$)

$$f(x, t+\tau) = f(x, \tau) + \tau \cdot \frac{\partial f}{\partial t}(x, t), \tag{3}$$

justifying it with the words "since τ is very small." Formulas (2) are a particular case of the general formula (3). Dirac in [18] wrote an analogous equality when studying the Newtonian approximation in general relativity. Using this type of infinitesimals, we can write an equality, in some infinitesimal neighborhood, between a smooth function and its tangent straight line, or, in other words, a Taylor formula without remainder.

Looking at this seeming inconsistency, if student's passion is still alive, she can start a long journey passing through some of the different theories we mentioned above or, the best we can hope for her, she can also try her own solution. Indeed, this is a natural step, since two different languages ($\varepsilon - \delta$ and informal infinitesimals) which are able to describe a sufficiently large part of nature, must have a stringent relationship among each other. Using only elementary analysis, maybe after a couple of tens of years, she can also be successful in creating a new useful theory. The solution ${}^\bullet\mathbb{R}$ is essentially unique, and indeed it can be described axiomatically.

Axiom, Decomposition ${}^\bullet\mathbb{R}$ is a commutative ring with unity, called *ring of Fermat reals*. Every Fermat reals $x \in^\bullet \mathbb{R}$ can be written, in a unique way, as

$$x = {}^\circ x + \sum_{i=1}^{N} \alpha_i \cdot \mathrm{d}t_{a_i}, \tag{4}$$

where $^\circ x, \alpha_i, a_i \in \mathbb{R}$ are standard reals, $a_1 > a_2 > \cdots > a_N \geq 1$, $\alpha_i \neq 0$. The term $^\circ x \in \mathbb{R}$ is called *standard part* of x, and $\alpha_i =: {^\circ x_i}$ its i-th standard part. Vice versa, any writing of the type (4), which is called the *decomposition* of x, gives a Fermat real, so that e.g., $^\bullet\mathbb{R} \supseteq \mathbb{R}$ and $^\circ r = r$ for all $r \in \mathbb{R}$.

The motivations for the name *Fermat reals* can be found in [27, 28], but we can summarize them saying that, in our opinion, P. Fermat would had liked the ring $^\bullet\mathbb{R}$!

Axiom, Base Infinitesimals The terms $\mathrm{d}t_a$ verify the following properties

$$\begin{aligned} \mathrm{d}t_a \cdot \mathrm{d}t_b &= \mathrm{d}t_{\frac{ab}{a+b}} \\ (\mathrm{d}t_a)^p &= \mathrm{d}t_{\frac{a}{p}} \quad \forall p \in \mathbb{R}_{\geq 1} \\ \mathrm{d}t_a &= 0 \quad \forall a \in \mathbb{R}_{<1}. \end{aligned} \tag{5}$$

Therefore, among Fermat reals we also have nilpotent infinitesimals, like $x = 3\,\mathrm{d}t_2$, since $x^3 = 27\,\mathrm{d}t_{\frac{2}{3}} = 0$. These are exactly the same type of infinitesimals used by Einstein in formulas like (2). We will simply use the symbol $\mathrm{d}t$ for $\mathrm{d}t_1$. Intuitively speaking, looking at the first of (5) we can say that $\mathrm{d}t_a$ behaves like $t^{\frac{1}{a}}$. Thinking at the other two properties of (5) we can also intuitively say that the greater is a and the greater is the nilpotent infinitesimal $\mathrm{d}t_a$. Our axiom on the total order relation $<$ will confirm this intuition. This relation has not to be confused with the following notion.

Axiom, Order of Infinitesimals The *order* $\omega(x) =: a_1$ (see (4)) can be interpreted as the leading term in the decomposition and hence it has the following expected properties

$$\begin{aligned} \omega(x+y) &\leq \max\left[\omega(x), \omega(y)\right] \\ \frac{1}{\omega(x \cdot y)} &= \frac{1}{\omega(x)} + \frac{1}{\omega(y)}, \end{aligned}$$

whenever x, y are infinitesimals such that $x + y \neq 0$, respectively $x \cdot y \neq 0$. In the decomposition (4), the term $\alpha_i =: \omega_i(x)$ will be called the i-th order of x. Directly from (4) it is not hard to prove that if $k \in \mathbb{N}_{>1}$, then $x^k = 0$ if and only if $\omega(x) < k$. Nilpotent Fermat reals can be thought as non-zero numbers which are so small that a suitable power of them gives zero.

Axiom, Ideals of Infinitesimals For $a \in \mathbb{R}_{\geq 0} \cup \{\infty\}$, the set

$$D_a := \{x \in {^\bullet\mathbb{R}} \mid {^\circ x} = 0,\ \omega(x) < a + 1\}$$

is an ideal. Moreover, for $k \in \mathbb{N}_{\geq 1}$ we have that $D_k = \left\{x \in {^\bullet\mathbb{R}} \mid x^{k+1} = 0\right\}$.

We will simply use D for D_1. The ideal D_k is therefore a perfect candidate as the k-th order infinitesimal neighborhood of zero, where every k-th order Taylor formula doesn't have remainder, since $x^{k+1} = 0$. This is indeed the subject of the next

Axiom, Taylor Formulas Set ${}^\bullet\mathbb{R}^d := {}^\bullet\mathbb{R} \times \ldots\ldots^d \times {}^\bullet\mathbb{R}$, then every ordinary smooth function $f \in \mathcal{C}^\infty(A, \mathbb{R})$ defined on an open set $A \subseteq \mathbb{R}^d$ can be extended to the set

$$ {}^\bullet A := \left\{ x \in {}^\bullet\mathbb{R}^d \mid {}^\circ x \in A \right\}, \tag{6} $$

$$ {}^\bullet f : {}^\bullet A \longrightarrow {}^\bullet\mathbb{R}, $$

obviously obtaining a true extension, i.e., ${}^\bullet f(x) = f(x)$ if $x \in A$. Moreover, the following Taylor formula

$$ \forall h \in D_k^d : {}^\bullet f(x+h) = \sum_{\substack{j \in \mathbb{N}^d \\ |j| \le k}} \frac{h^j}{j!} \cdot \frac{\partial^{|j|} f}{\partial x^j}(x) \tag{7} $$

holds, where $x \in A$ is a standard point, and $D_k^d = D_k \times \ldots\ldots^d \times D_k$. Therefore, smooth functions becomes exactly equal to polynomials of degree k in the infinitesimal k-th order neighbourhood $x + D_k$. In particular $f(x+k) = f(x) + h \cdot f(x)$ for $h \in D$, that is, every smooth functions is equal to its tangent line in a first order infinitesimal neighborhood. Einstein's formulas (2) are particular cases of this infinitesimal Taylor formula. Let us note that applying this formula to the function $f(x) = x^2$ in D we obtain that $h^2 = 0$ for every $h \in D$: if we want to write Einstein's formulas exactly as he did (i.e., with the equality sign and not with an approximate equality sign), we are necessarily forced to work in a ring with nilpotent infinitesimals and not in a field.

Axiom, Cancellation Laws Let $h_1, \ldots, h_n \in D_\infty$, $i_1, \ldots, i_n \in \mathbb{N}$, $x \in^\bullet \mathbb{R}$, then we have

1. $h_1^{i_1} \cdot \ldots \cdot h_n^{i_n} = 0$ if and only if $\sum_{k=1}^n \frac{i_k}{\omega(h_k)} > 1$.
2. x is invertible if and only if ${}^\circ x \neq 0$.
3. If $x \cdot r = x \cdot s$ in ${}^\bullet\mathbb{R}$, where $r, s \in \mathbb{R}$ and $x \neq 0$, then $r = s$.

If you are scared by working in a ring instead of a field, these laws permit to effectively work with this type of infinitesimals. If x is invertible, and one proceeds as in the case of formal power series, it is not hard to prove that

$$ \frac{1}{x} = \frac{1}{{}^\circ x} \cdot \sum_{j=0}^{+\infty} (-1)^j \cdot \left(\sum_{i=1}^{N} \frac{\alpha_i}{{}^\circ x} \cdot \mathrm{d}t_{a_i} \right), $$

where the series is really a finite sum due to nilpotency.

Axiom, Total Order Relation The ring of Fermat reals is totally ordered by the relation $<$. This relation verifies the following properties: let $x, y \in {}^\bullet\mathbb{R}$, if ${}^\circ x \neq {}^\circ y$, then

$$ x < y \Longleftrightarrow {}^\circ x < {}^\circ y. $$

Vice versa, if $^\circ x = {}^\circ y$, then

1. If $\omega(x) > \omega(y)$, then $x > y$ if and only if $^\circ x_1 > 0$.
2. If $\omega(x) = \omega(y)$, then

$$^\circ x_1 > {}^\circ y_1 \Longrightarrow x > y$$
$$^\circ x_1 < {}^\circ y_1 \Longrightarrow x < y.$$

The axiom gives an effective criterion to decide whether $x < y$ or not. Indeed:

1. First of all $x < y$ is equivalent to $0 < y - x$, so we can describe the algorithm for the case $0 < x$, $x \in {}^\bullet\mathbb{R} \setminus \mathbb{R}$ only (from the first part of the previous axiom it follows that $<$ extends the usual order relation on $\mathbb{R}$). If the standard part $^\circ x \neq 0$, then the order relation can be decided on the basis of this standard part only, for example, $2 + \mathrm{d}t_2 > 0$ and $1 + \mathrm{d}t_2 < 3 + \mathrm{d}t$.
2. Otherwise, if the standard part $^\circ x = 0$, we look at the order $\omega(x)$ and at the first standard part $^\circ x_1$, which is the coefficient of the biggest infinitesimals in the decompositions of x: because $\omega(x) > \omega(0) = 0$, we have that $x > 0$ iff $^\circ x_1 > 0$. E.g. $3\,\mathrm{d}t_2 > 0$; $\mathrm{d}t_2 > a\,\mathrm{d}t$ for every $a \in \mathbb{R}$; $0 < \mathrm{d}t < \mathrm{d}t_2 < \mathrm{d}t_3 < \ldots < \mathrm{d}t_k$ for every k.

More advanced axioms are needed to deal with (quasi-standard) smooth functions that are more general than extension $^\bullet f$ of standard smooth functions f, for example, like the very simple $g(x) = x + \mathrm{d}t$; see [26, 27] for more details. Therefore, the previously listed axioms are surely incomplete, on the one hand, but also redundant, from another point of view. However, they permit to characterize the structure $({}^\bullet\mathbb{R}, +, \cdot, <, {}^\circ(-), \mathrm{d}t_{(-)})$.

Theorem 1 *Let us assume that the structure* $(R, \oplus, \odot, \prec, {}^\diamond(-), \delta t_{(-)})$ *verifies the previous axioms, then there exists one and only one isomorphism of ordered rings*

$$f : {}^\bullet\mathbb{R} \longrightarrow R$$

such that $f(\alpha \cdot \mathrm{d}t_a) = \alpha \odot \delta t_a$ *and* $^\diamond f(x) = {}^\circ x$. *Moreover, this isomorphism preserves also the order function* $\omega(-)$.

Proof L Let $x = r + \sum_{j=1}^{N} \alpha_j \cdot \mathrm{d}t_{a_j}$ be the decomposition of $x \in {}^\bullet\mathbb{R}$. The only possibility to define the searched isomorphism is clearly

$$f(x) := r \oplus \bigoplus_{j=1}^{N} \alpha_j \odot \delta t_{a_j} \in R. \tag{8}$$

We have to prove that f is a morphism of ordered rings. Let $y = s + \sum_{i=1}^{M} \beta_i \cdot \mathrm{d}t_{b_i}$ be the decomposition of $y \in {}^\bullet\mathbb{R}$, then to find the decomposition of $x + y$ and apply our Definition 8, we firstly have to consider the sets of all the orders appearing in these decompositions:

$$O_x := \{a_j \mid j = 1, \ldots, N\}$$

$$O_y := \{b_i \mid i = 1, \dots, M\}$$
$$O := O_x \cup O_y.$$

Secondly, we have to sum all the coefficients of the two decompositions corresponding to the same order: for every $q \in O$ set

$$c_q := \sum \left[\{\alpha_j \mid j = 1, \dots, N,\ a_j = q\} \cup \{\beta_i \mid i = 1, \dots, M,\ b_i = q\} \right].$$

Let us note that, by the definition of decomposition, if $q \in O_x \setminus O_y$, then $c_q = \alpha_j$, where $j = 1, \dots, N$ is the unique index such that $a_j = q$. Analogously if $q \in O_y \setminus O_x$, whereas, if $q \in O_x \cap O_y$, then $c_q = \alpha_j + \beta_i$, where i and j are the unique indexes such that $a_j = q = b_j$. Now, we have

$$x + y = r + s + \sum_{q \in O} c_q \cdot \mathrm{d}t_q.$$

Finally, we have to enumerate all the orders in $O \setminus \{q \in O \mid c_q = 0\}$ in increasing way

$$\{q_1, \dots, q_K\} = O \setminus \{q \in O \mid c_q = 0\}$$

$$q_1 > q_2 > \dots > q_K \geq 1.$$

Therefore

$$x + y = r + s + \sum_{k=1}^{K} c_{q_k} \cdot \mathrm{d}t_{q_k}$$

is the decomposition of $x + y$. Applying the Definition 8 of f we get

$$f(x + y) = r \oplus s \oplus \bigoplus_{k=1}^{K} c_{q_k} \odot \delta t_{q_k}$$

$$\begin{aligned} f(x) + f(y) &= r \oplus \bigoplus_{j=1}^{N} \alpha_j \odot \delta t_{a_j} \oplus s \oplus \bigoplus_{i=1}^{M} \beta_i \odot \delta t_{b_i} = \\ &= r \oplus s \oplus \bigoplus_{q \in O} c_q \odot \delta t_q = \\ &= f(x + y). \end{aligned}$$

In order to prove the preservation of the product, we can proceed in a similar way:

$$x \cdot y = rs + \sum_{i=1}^{M} r\beta_i \cdot \mathrm{d}t_{b_i} + \sum_{j=1}^{N} s\alpha_j \cdot \mathrm{d}t_{a_j} + \sum_{i,j} \alpha_j \beta_i \cdot \mathrm{d}t_{a_j} \cdot \mathrm{d}t_{b_i}.$$

Now, we can use property (5) obtaining $\mathrm{d}t_{a_j} \cdot \mathrm{d}t_{b_i} = \mathrm{d}t_{\frac{a_j b_i}{a_j + b_i}}$. To the resulting sum, we can apply the method used above to obtain the decomposition of $x \cdot y$, that is, the sum of all the coefficients corresponding to the same order, the deletion of the terms for which $c_q = 0$ or $q < 1$ and, finally, the ordering of the remaining summands. The proof proceeds exactly as above for the sum, noting that we also have to use (5) for the structure $(R, \oplus, \odot, \prec, ^\diamond(-), \delta t_{(-)})$, but in the reverse order with respect to the previous application. Finally, the inverse morphism is necessarily defined as

$$g\left(r \oplus \bigoplus_{j=1}^{N} \alpha_j \odot \delta t_{a_j}\right) = r + \sum_{j=1}^{N} \alpha_j \cdot \mathrm{d}t_{a_j}.$$

Exactly as described above, we can prove that g is indeed a morphism of ordered rings. □

2.1 The Model

Surprisingly, the model of the previous list of axioms is quite simple. For proofs and motivations related to this section, see [27]. We firstly need the following class of functions

Definition 1 We say that x *is a little-oh polynomial*, and we write $x \in \mathbb{R}_o[t]$ iff

1. $x : \mathbb{R}_{\geq 0} \longrightarrow \mathbb{R}$
2. We can write

$$x(t) = r + \sum_{i=1}^{k} \alpha_i \cdot t^{a_i} + o(t) \quad \text{as} \quad t \to 0^+$$

for suitable

$$k \in \mathbb{N}$$
$$r, \alpha_1, \ldots, \alpha_k \in \mathbb{R}$$
$$a_1, \ldots, a_k \in \mathbb{R}_{\geq 0}.$$

Hence, a little-oh polynomial $x \in \mathbb{R}_o[t]$ is a polynomial function with real coefficients, in the real variable $t \geq 0$, with generic positive powers of t, and up to a little-oh function as $t \to 0^+$. Simple examples of little-oh polynomials are the following: $x(t) = 1+t+t^{1/2}+t^{1/3}+o(t)$ and $x(t) = r+o(t)$. For little-oh polynomials we will also use the notation $x_t := x(t)$.

Definition 2 Let $x, y \in \mathbb{R}_o[t]$, then we say that $x \sim y$ or that $x = y$ in ${}^\bullet\mathbb{R}$ iff $x_t = y_t + o(t)$ as $t \to 0^+$. Because it is easy to prove that $\sim$ is an equivalence relation, we can define the quotient ring ${}^\bullet\mathbb{R} := \mathbb{R}_o[t]/\sim$, where in $\mathbb{R}_o[t]$ we consider the pointwise ring operations. We will use the notation $x = [x_t] \in {}^\bullet\mathbb{R}$ for the equivalence class generated by the little-oh polynomial $t \in \mathbb{R}_{\geq 0} \mapsto x_t \in \mathbb{R}$. Moreover:

1. We define the standard part map as $^\circ(-) : x \in {}^\bullet\mathbb{R} \mapsto {}^\circ x = x(0) \in \mathbb{R}$
2. $\mathrm{d}t_a := [t^{\frac{1}{a}}] \in {}^\bullet\mathbb{R}$ for all $a \in \mathbb{R}_{\geq 0}$.
3. Let $x = [x_t]$, $y = [y_t] \in {}^\bullet\mathbb{R}$, then we say that $x < y$ iff $x \neq y$ and there exists $z \in \mathbb{R}_o[t]$ such that

$$\exists \delta \in \mathbb{R}_{>0}\, \forall t \in (0, \delta] : \ x_t \leq y_t + z_t$$

$$z_t = o(t) \quad \text{as } t \to 0^+.$$

4. Let $A \subseteq \mathbb{R}^n$ be an open subset, $f \in \mathcal{C}^\infty(A, \mathbb{R})$ a smooth function. Define ${}^\bullet A$ like in (6) (or, equivalently, as ${}^\bullet A = A_o[t]/\sim$, where $A_o[t]$ is the set of little-oh polynomials taking values in A). For $x \in {}^\bullet A$ define

$${}^\bullet f(x) := [f(x_t)] \in {}^\bullet\mathbb{R}.$$

It is not hard to show that this is indeed a model for our axioms (see [27]). Clearly, we can define the absolute value, powers, and logarithms of invertible Fermat reals and generalize their usual properties, [27]. We can also define meaningful metrics on ${}^\bullet\mathbb{R}$ and roots of (nilpotent!) infinitesimals, and prove applications to fractional derivatives, [32].

2.2 *Geometrical Representation*

In the introduction we argued that a good mathematical theory should always try to maintain a clear intuitive meaning, in the sense to keep a good dialectic between provable formal properties and their intuitive meaning. In this direction we can see the possibility to find a geometrical representation of Fermat reals.

The idea is that to any Fermat real $x \in {}^\bullet\mathbb{R}$ we can associate the function

$$t \in \mathbb{R}_{\geq 0} \mapsto {}^\circ x + \sum_{i=1}^{N} {}^\circ x_i \cdot t^{1/\omega_i(x)} \in \mathbb{R} \tag{9}$$

where N is, of course, the number of addends in the decomposition of x. Therefore, a geometric representation of this function is also a geometric representation of the number x, because different Fermat reals have different decompositions, see "*Axiom, decomposition.*" Finally, we can guess that, because the notion of equality in ${}^\bullet\mathbb{R}$ depends only on the germ generated by each little-oh polynomial (see Definition 2), we can represent each $x \in {}^\bullet\mathbb{R}$ using only the first small part of the function (9).

Definition 3 If $x \in {}^\bullet\mathbb{R}$ and $\delta \in \mathbb{R}_{>0}$, then

$$\mathrm{graph}_\delta(x) := \left\{ ({}^\circ x + \sum_{i=1}^{N} {}^\circ x_i \cdot t^{1/\omega_i(x)}, t) \mid 0 \leq t < \delta \right\}$$

where N is the number of addends in the decomposition of x.

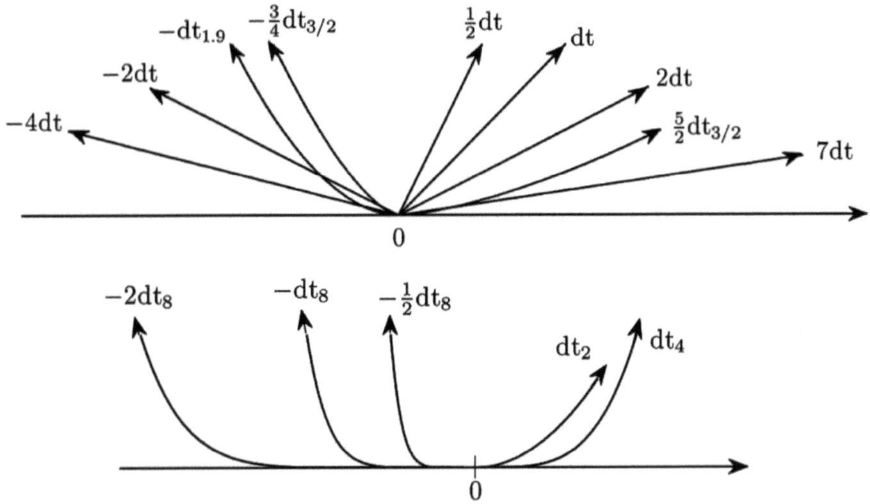

Fig. 1 Some first and higher order infinitesimals

Note that the value of the function is placed in the abscissa position, so that the correct representation of $\text{graph}_\delta(x)$ is given by Fig. 1.
This inversion of abscissa and ordinate in the $\text{graph}_\delta(x)$ permits to represent this graph as a line tangent to the classical straight line $\mathbb{R}$ and hence to have a better graphical picture. Finally, note that if $x \in \mathbb{R}$ is a standard real, then $N = 0$ and the $\text{graph}_\delta(x)$ is a vertical line passing through $^\circ x = x$.
The following theorem permits to represent geometrically the Fermat reals.

Theorem 2 *If* $\delta \in \mathbb{R}_{>0}$, *then the function*

$$x \in {}^\bullet\mathbb{R} \mapsto \text{graph}_\delta(x) \subset \mathbb{R}^2$$

is injective. Moreover if $x, y \in {}^\bullet\mathbb{R}$, *then we can find* $\delta \in \mathbb{R}_{>0}$ *(depending on* x *and* y*) such that*

$$x < y$$

if and only if

$$\forall p, q, t: \quad (p, t) \in \text{graph}_\delta(x),\ (q, t) \in \text{graph}_\delta(y) \Longrightarrow p < q \tag{10}$$

For a proof of this theorem, see [28]. See Fig. 2 for the meaning of condition (10).

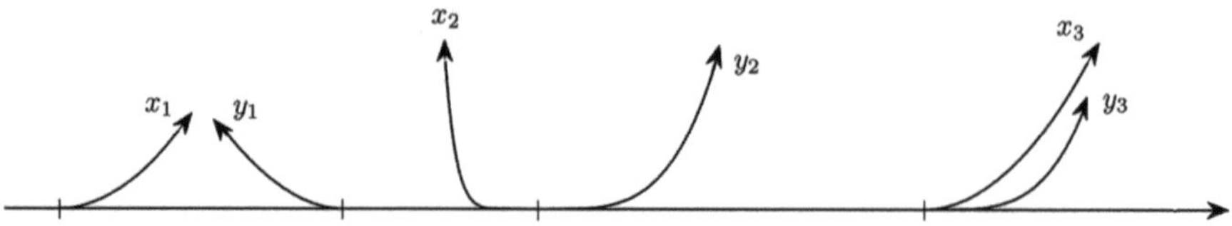

Fig. 2 Different cases in which $x_i < y_i$

2.3 Computer Implementation

The definition of the ring of Fermat reals is highly constructive. Therefore, using object oriented programming, it is not hard to write a computer code corresponding to ${}^\bullet\mathbb{R}$. We realized a first version of this software using Matlab R2010b.

The constructor of a Fermat real is `x=FermatReal(s,w,r)`, where `s` is the $n+1$ double vector of standard parts (`s(1)` is the standard part $^\circ x$) and `w` is the double vector of orders (`w(1)` is the order $\omega(x)$ if $x \in {}^\bullet\mathbb{R} \setminus \mathbb{R}$, otherwise `w=[]` is the empty vector). The last input `r` is a logical variable and assumes the value `true` if we want that the display of the number `x` is realized using the Matlab `rats` function for both its standard parts and orders. In this way, the number will be displayed using continued fraction approximations and therefore, in many cases, the calculations will be exact. These inputs are the basic methods of every Fermat real, and can be accessed using the `subsref`, and `subsasgn`, notations `x.stdParts`, `x.orders`, `x.rats`. The function `w=orders(x)` gives exactly the double vector `x.orders` if $x \in {}^\bullet\mathbb{R} \setminus \mathbb{R}$ and `0` otherwise.

The function `dt(a)`, where `a` is a double, constructs the Fermat real $\mathrm{d}t_a$. Because we have overloaded all the algebraic operations, like `x+y`, `x*y`, `x-y`, `-x`, `x==y`, `x~=y`, `x<y`, `x<=y`, `x^y`, we can define a Fermat real, for example, using an expression of the form `x=2+3*dt(2)-1/3*dt(1)`, which corresponds to `x=FermatReal([2 3 -1/3],[2 1],true)`.

We have also realized the function `y=decomposition(x)`, which gives the decomposition of the Fermat real `x`, `abs(x)`, `log(x)`, `exp(x)`, `isreal(x)`, `isinfinitesimal(x)`, `isinvertible(x)`.

The ratio `x/y` has been implemented for `x` and `y` infinitesimals and `y~=0`, or in case `y` is invertible. Finally, the function `y=ext(f,x)`, corresponds to ${}^\bullet f(x)$ and has been realized using the evaluation of the symbolic Taylor formula of the inline function `f`.

The functions `x^p`, `sqrt(x)` and `nthroot(x,n)` have been realized both for `x` infinitesimal or invertible using the formulas derived in [32].

Using these tools, we can easily find, e.g., that

$$\frac{\sin\left(\sqrt{\mathrm{d}t_3 + 2\,\mathrm{d}t_2}\right)}{\cos\left(\sqrt[3]{-4\,\mathrm{d}t}\right)} = \mathrm{d}t_6 + \mathrm{d}t_3 - \frac{2}{3}\,\mathrm{d}t_2 + \frac{1096}{2787}\,\mathrm{d}t_{\frac{6}{5}} + \frac{1234}{913}\,\mathrm{d}t.$$

This corresponds to the following Matlab code:

```
>> x=sqrt(dt(3)+2*dt(2))
   x =
   dt_6 + dt_3 - 1/2*dt_2 + 1/2*dt_3/2 - 5/8*dt_6/5
>> y=nthroot(-4*dt(1),3)
   y =
   -1008/635*dt_3
>> g=inline('cos(y)')
   g =
   Inline function: g(y) = cos(y)
   >> f=inline('sin(x)')
   f =
   Inline function: f(x) = sin(x)
>> decomposition(ext(f,x)/ext(g,y))
   ans =
   dt_6 + dt_3 - 2/3*dt_2 + 1096/2787*dt_6/5 + 1234/913*dt
```

The Matlab source code is freely available under open-source license, and can be requested from the author of the present article.

2.4 *Applications and Developments*

The simplicity of the ring of Fermat reals permits to easily extend this approach along different directions.

We can see at the plenty of informal uses of infinitesimals in Physics from a new point of view. Frequently, these informal calculations can be faithfully repeated using $^\bullet\mathbb{R}$, without any changes. This can be considered as a partial proof that this theory is a meaningful way to help us in solving our modelling problems without forcing us to change completely our methodological approach. Formalizing these physical models, is a good way to learn properties of Fermat reals. On the other hand, working in a rigorous mathematical theory of infinitesimals permits to gain a great formal power and to discover the physical meaning of several informal approximations. See, for example, [28], where it is proved that the wave equation is *equivalent* to a suitable condition of infinitesimal oscillations of a string. In [28] other elementary examples are developed: the deduction of the heat equation, a study of the electric dipole, the Newtonian limit in relativity, the curvature of a smooth curve, the area of the circle and volumes of revolution, the stretching of a spring and others.

The theory of Fermat reals takes strong inspiration from synthetic differential geometry and indeed, it can be deeply developed so as to have at disposal infinitesimal methods valid both in finite and infinite dimensional spaces. Every smooth manifold M can be extended using this type of infinitesimals analogously to what we shown before for the extension $^\bullet\mathbb{R} \supseteq \mathbb{R}$, and obtaining a smooth embedding $^\bullet M \supseteq M$. More generally, this extension is applicable to every diffeological space [39] obtaining a functor with very good preservation properties. The category of diffeological spaces is cartesian closed and embeds the category of smooth manifolds, so that these Fermat extensions can also be applied to infinite dimensional function spaces. In this framework, we can define tangent vectors of the space M as smooth maps of the form $t : D \longrightarrow {}^\bullet M$, that is, as infinitesimal linear curves traced on ${}^\bullet M$. We can define intrinsically the sum of two tangent vectors as the diagonal of the infinitesimal parallelogram generated by these tangent vectors. We can prove that a vector field can be defined as an infinitesimal smooth transformation of the space ${}^\bullet M$ into itself, that is, as $V : D \longrightarrow {}^\bullet M^{{}^\bullet M}$ such that $V(0) = 1_{{}^\bullet M}$. We can define infinitesimal integral curves and proves their existence even in infinite dimensional spaces of functions like $\mathcal{C}^\infty({}^\bullet M, {}^\bullet N)$. These are only few examples of a very long ongoing project. For more details, see [26, 29, 30].

2.5 Open Problems and Ideas

The following are some open problems and ideas the reader can try to face. Please, feel free to contact me in case you want to have more feedbacks, considering also that some of these ideas are inside ongoing projects P25116-N25 and P25311-N25 of the Austrian Science Fund FWF.

Computer Implementation 2.0 Up to now, the computer code implementing the ring of Fermat reals has been written only to show concretely the possibilities of the ring ${}^\bullet\mathbb{R}$. On the other hand, it is clear that it is possible to write it with a more specific aim. For example, like in case of the Levi-Civita field (see references in Sect. 4) possible applications of a specifically rewritten code can be thought for automatic differentiation theory. Let us note that, even if the theory of Fermat reals applies to smooth functions, a full treatment of right and left derivatives is possible, [26], so that the theory can be applied consistently also to piecewise smooth functions. Finally, the use of nilpotent elements permits to fully justify that every derivative estimation of a computer function (see Sect. 4) reduces to a finite number of algebraic calculations.

Infinities and Nilpotent Infinitesimals Almost every theory of actual infinitesimals is usually well coupled with a corresponding theory of infinite numbers. In case of a field, this coupling is naturally given by the reciprocal function. These infinitely large numbers found fruitful applications in formulating integrals as infinite sums, limits at infinite points, hyperfinite subdivisions of the real line, study of singularities, generalized functions, to cite only a few. Of course, in case of nilpotent infinitesimals, the possibility to trivially have infinities as reciprocals of infinitesimals does not hold because in that case, for example, $h^2 = 0$ would imply $h = 0$. Is there some meaningful way to bypass the impossibility to have infinities as reciprocals of nilpotent infinitesimals? What property of multiplication of numbers should be weakened to allow coexistence between these infinities and nilpotent infinitesimals? Is the corresponding formalism sufficiently easy to use and intuitively meaningful? Can these infinities be used to define a suitable class of generalized functions? Hint: the reciprocal of $h = [t] \in {}^\bullet\mathbb{R}$ is necessarily something generated by $t \in \mathbb{R}_{>0} \mapsto t^{-1} \in \mathbb{R}$.

First Order Infinitesimals Whose Product Is Not Zero In the ring ${}^\bullet\mathbb{R}$ the product of any two first order infinitesimals $h, k \in D$ is zero: $h \cdot k = 0$. As proved in [28, Theorem 24] this is a general consequence of the total order property, but to have infinitesimals h, k such that $h^2 = 0 = k^2$ and $h \cdot k \neq 0$ would be useful to study the double tangent bundle (see [42]). An idea, inspired by rings like $\mathbb{R}[t,s]/\langle t^2 = 0, s^2 = 0\rangle$ we can try to explore, can be roughly stated saying that "two first order infinitesimals $(h_t)_t$ and $(k_s)_s$ have a non-zero product $(h_t \cdot k_s)_{t,s}$ if they depend on two independent variables t and s." A possible formalization of this idea can be sketched in the following way. Instead of little-oh polynomials, let us consider maps of the form $x : \mathbb{R}^v_{\geq 0} \longrightarrow \mathbb{R}$ (v depending on x) such that

$$x(t_1, \dots, t_v) = r + \sum_{j=1}^{k} \alpha_j \cdot t_1^{a_{1j}} \cdot \ldots \cdot t_v^{a_{vj}} + o(t_1) + \ldots + o(t_v), \tag{11}$$

The analogue of the equality in ${}^\bullet\mathbb{R}$ is now that $x \sim y$ if and only if x and y are both defined on the same domain $\mathbb{R}^v_{\geq 0}$ and $x(t_1, \dots, t_v) = y(t_1, \dots, t_v) + o(t_1) + \ldots + o(t_v)$ as $t_k \to 0^+$ for all k. This idea seems positive for two reasons: firstly, if we define a new Fermat reals ring in this way, considering only the subring of all the maps $\mathbb{R}_o[t_i]$ which only depend on one variable t_i, we obtain a ring ${}^\bullet\mathbb{R}[t_i]$ isomorphic to the present ${}^\bullet\mathbb{R}$. Secondly, let us consider $h_{t_1} := t_1$ and $k_{t_2} := t_2$, then we have that $h^2 \sim 0$ and $k^2 \sim 0$, but not $h \cdot k \sim 0$. Of course, from [28, Theorem 24] it follows that every subring ${}^\bullet\mathbb{R}[t_i]$ is totally ordered, but the whole ring cannot be totally ordered.

Perturbation Theory Several classical methods used in perturbation theory seem to become simpler if we take the small perturbation parameter ε as a non zero nilpotent infinitesimal in the ring of Fermat reals, that is, if $\varepsilon \in {}^\bullet\mathbb{R}_{\neq 0}$, $\varepsilon^n = 0$. For example, we can more easily use the nilpotency property of ε instead of using big-oh asymptotic estimates; we can also take advantage of exact Taylor formulas in ${}^\bullet\mathbb{R}$, that is, with nilpotent infinitesimal increments and no remainder. More generally, we can use finite sums in the ring ${}^\bullet\mathbb{R}$ instead of (convergent or formal) power series, and also we can take advantage of the existence of arbitrary roots of nilpotent infinitesimals (see [32]). Do these properties permit to gain an easier calculus of classical perturbation methods?

Other open problems and ideas can be found in [26, Chap. 14].

3 Which Numbers Permit to Derive Locally Integrable Functions?

Exactly as real numbers can be seen as equivalence classes of sequences $(q_n)_{n\in\mathbb{N}}$ of rationals, it is very natural to try to define a non-Archimedean extension of $\mathbb{R}$ using a suitable quotient ring $\mathcal{R}/\sim$, where $\mathcal{R} \subseteq \mathbb{R}^I$ is a subalgebra of nets $(x_\varepsilon)_{\varepsilon\in I} \in \mathbb{R}^I$

defined on a directed set $(I, \leq)$. We can consider $I = \mathbb{N}$ with the usual order, i.e., for $n \to +\infty$, or, like for the ring of Fermat reals $I = \mathbb{R}_{\geq 0}$, corresponding to $t \to 0^+$. Even if this is not an intrinsic choice, in this section we will consider the customary $I = (0, 1]$, corresponding to $\varepsilon \to 0^+$. In this section we will denote ε-dependent nets simply by $(X_\varepsilon) := (X_\varepsilon)_{\varepsilon \in I}$.

If we are trying to construct such a quotient ring $R = \mathcal{R}/\sim$ so that it contains infinitesimals and infinities, it is sufficiently natural to face the following observation. Let us assume that $[z_\varepsilon] = 0 \in R$, where $[z_\varepsilon] := [\varepsilon \in I \mapsto z_\varepsilon \in \mathbb{R}]_\sim$, and $[J_\varepsilon] \in R$ is generated by a positive infinite net, i.e., $\lim_{\varepsilon \to 0^+} J_\varepsilon = +\infty$. Then we would have

$$\begin{aligned}[z_\varepsilon] \cdot [J_\varepsilon] &= 0 \cdot [J_\varepsilon] = 0 \\ &= [z_\varepsilon \cdot J_\varepsilon]. \end{aligned} \tag{12}$$

Finally, let us assume that

$$\forall [w_\varepsilon] \in R : \; [w_\varepsilon] = 0 \Longrightarrow \lim_{\varepsilon \to 0^+} w_\varepsilon = 0 \tag{13}$$

(for the reader having some interest in nonstandard Analysis, we recall that this intuitively natural property is false for the hyperreals, see [31]). Under these assumptions, (12) yields $\lim_{\varepsilon \to 0^+} z_\varepsilon \cdot J_\varepsilon = 0$, and hence

$$\exists \varepsilon_0 \in I \, \forall \varepsilon \in (0, \varepsilon_0] : \; |z_\varepsilon| \leq J_\varepsilon^{-1}. \tag{14}$$

We can state this property saying that *the nets* (z_ε) *generating* 0, *that is, such that* $(z_\varepsilon) \sim 0$, *must be dominated by the reciprocals of every infinite* $[J_\varepsilon] \in R$. It is not hard to prove that if every infinite net (J_ε) is in the subalgebra $\mathcal{R}$, so that it generates a number $[J_\varepsilon] \in R$, then (14) implies that the equivalence relation $\sim$ must be trivial:

$$\exists \varepsilon_0 \in I \, \forall \varepsilon \in (0, \varepsilon_0] : \; z_\varepsilon = 0. \tag{15}$$

This situation corresponds to the Schmieden–Laugwitz model, [56].

If we do not want to have the trivial model (15), the only possibility we have is hence either to negate the natural property (13) or to restrict the class of all the nets (J_ε) generating infinite numbers in R. A first natural idea to explore is surely to consider only the following infinite nets

$$\mathcal{I} := \left\{ (\varepsilon^{-n}) \mid n \in \mathbb{N}_{>0} \right\}. \tag{16}$$

and hence to consider the subalgebra $\mathcal{R} \subseteq \mathbb{R}^I$ so that it does not contain every infinite net $(J_\varepsilon) \in \mathbb{R}^I$, but only nets bounded by some $(J_\varepsilon) \in \mathcal{I}$:

Definition 4

1. We say that (x_ε) *is moderate*, and we write $(x_\varepsilon) \in \mathbb{R}_M$ if

$$\exists (J_\varepsilon) \in \mathcal{I} : \; x_\varepsilon = O(J_\varepsilon) \text{ as } \varepsilon \to 0^+,$$

i.e., if

$$\exists N \in \mathbb{N}\, \exists \varepsilon_0 \in I\, \forall \varepsilon \in (0, \varepsilon_0] : \; |x_\varepsilon| \le \varepsilon^{-N}.$$

2. Let (x_ε), $(y_\varepsilon) \in \mathbb{R}^I$, then we say that $(x_\varepsilon) \sim (y_\varepsilon)$ if

$$\forall (J_\varepsilon) \in \mathcal{I} : \; x_\varepsilon = y_\varepsilon + O(J_\varepsilon^{-1}) \text{ as } t \to 0^+,$$

that is if

$$\forall n \in \mathbb{N}\, \exists \varepsilon_0 \in I\, \forall \varepsilon \in (0, \varepsilon_0] : \; |x_\varepsilon - y_\varepsilon| \le \varepsilon^n.$$

This is a congruence relation on the ring $\mathbb{R}_M$ of moderate nets with respect to the pointwise operations, and we can hence define

$$\widetilde{\mathbb{R}} := \mathbb{R}_M / \sim$$

which is called *Colombeau ring of generalized numbers*.

3. More generally, the space of generalized points in $\Omega \subseteq \mathbb{R}^n$ is $\widetilde{\Omega} := \Omega_M / \sim$, where $\Omega_M = \{(x_\varepsilon) \in \Omega^I \mid \exists N \in \mathbb{N} : |x_\varepsilon| = O(\varepsilon^{-N})\}$ is called the set of moderate nets in Ω.
4. We can also define a order relation on $\widetilde{\mathbb{R}}$ with the usual $[x_\varepsilon] \le [y_\varepsilon]$ iff there exists $(z_\varepsilon) \in \mathbb{R}^I$ such that

$$\exists \varepsilon_0 \in I\, \forall \varepsilon \in (0, \varepsilon_0] : \; x_\varepsilon \le y_\varepsilon + z_\varepsilon$$

$$(z_\varepsilon) \sim 0.$$

Even if this order is not total, we still have the possibility to define the minimum $\min([x_\varepsilon], [y_\varepsilon]) := [\min(x_\varepsilon, y_\varepsilon)]$, and analogously the maximum function and the absolute value $|[x_\varepsilon]| := [|x_\varepsilon|] \in \widetilde{\mathbb{R}}$.

5. Usually on $\widetilde{\mathbb{R}}^n$ the so-called *sharp topology* is considered. This is the topology generated by balls $B_\rho(x) = \{y \in \widetilde{\mathbb{R}}^n \mid |y - x| < \rho\}$, where $| - |$ is the natural extension of the Euclidean norm in $\widetilde{\mathbb{R}}^n$, i.e., $|[x_\varepsilon]| := [|x_\varepsilon|]$, and $\rho \in \widetilde{\mathbb{R}}_{>0}$ is positive invertible.

Theorem 3 *$\widetilde{\mathbb{R}}$ is a partially ordered ring.*

Proof See e.g., [11, 36]. □

Therefore, we restrict our construction only on nets growing polynomially in ε, and we say that two of these nets are equivalent if their difference is eventually under every polynomial in ε. This permit to give a clear meaning to (12) and (14) and hence to obtain a congruence relation.

The usual real numbers $r \in \mathbb{R}$ are trivially embedded in $\widetilde{\mathbb{R}}$ considering constant nets $[r] \in \widetilde{\mathbb{R}}$. Of course $[\varepsilon^n] \in \widetilde{\mathbb{R}}$ is an invertible infinitesimal in this ring, whose reciprocal is $[\varepsilon^{-n}]$ which is necessarily a positive infinite number, that is, $[\varepsilon^{-n}] \ge r$ for all $r \in \mathbb{R}_{>0}$. Once again, it seems not hard to extend the field of real numbers so as to obtain a ring containing infinitesimals and infinities, that is, a non-Archimedean ring.

Remark 1

1. The interested reader is invited to abstract from the particular example (16) as far as it seems possible. See also [15, 16] and references therein.
2. There is a conceptual deep difference between the ring of Fermat reals and the Colombeau ring of generalized numbers. In the former, we have considered as nets $(x_t)_t$ representing new numbers $[x_t] \in {}^\bullet\mathbb{R}$, only the particular class of little-oh polynomials; this choice is essential to have, for example, the decomposition of $[x_t]$, the total order property and the cancellation laws. In the latter we have considered the algebra $\mathbb{R}_M$ of moderate nets (x_ε) without any other restriction, and hence all these good properties are not provable anymore. On the other hand, in the ring $\widetilde{\mathbb{R}}$ we have a greater flexibility to solve equations: Let us assume that $f \in \mathcal{C}^\infty([a,b],\mathbb{R})$ is a smooth function and that $[y_\varepsilon] \in \widetilde{\mathbb{R}}$ is a Colombeau generalized number in the interval $[f(a), f(b)]$, that is, such that $f(a) \le [y_\varepsilon] \le f(b)$. Proceeding for every ε, we can easily say that there surely exists a generalized number $[x_\varepsilon] \in \widetilde{\mathbb{R}}$ such that $a \le [x_\varepsilon] \le b$ and $[f(x_\varepsilon)] = [y_\varepsilon]$. This is not possible, in general, if we consider more restricted subalgebras $\mathcal{R} \subset \mathbb{R}_M$, that is, considering nets with continuous ε-dependency; see [33] for more details.

Why is the ring $\widetilde{\mathbb{R}}$ related to the derivatives of locally integrable functions (i.e., function $f : \mathbb{R} \longrightarrow \mathbb{R}$ such that $\int_K |f|$ exists finite on every compact set $K \Subset \mathbb{R}$)? It is well known that the problem to consider (generalized) derivatives of locally integrable functions arises frequently in physics considering idealized models like in shock mechanics, material points mechanics, charged particles in electrodynamics, gravitational waves in general relativity, etc. The conceptual, even if only intuitive, solution by Cauchy, Poisson, Kirchhoff, Helmholtz, Kelvin, Heaviside, Dirac et al. ([17, 40, 47]) to find, for example, the derivative of the Heaviside function

$$H(x) = \begin{cases} 1 & \text{if} x \ge 0 \\ 0 & \text{otherwise}' \end{cases} \tag{17}$$

was to consider a Gaussian with an infinitesimal standard deviation. Using our ring $\widetilde{\mathbb{R}}$ (or any other ring with invertible infinitesimals) we can formalize this idea considering, e.g., $\sigma = [\varepsilon^n] \in \widetilde{\mathbb{R}}$ and

$$\delta(x) = \frac{1}{\sqrt{2\pi}\sigma} e^{-\frac{x^2}{2\sigma^2}} \in \widetilde{\mathbb{R}}. \tag{18}$$

Actually, this idea conduct us to consider a net of smooth functions $(\delta_\varepsilon) = \left(\frac{1}{\sqrt{2\pi}\varepsilon^n} e^{-\frac{x^2}{2\varepsilon^{2n}}} \right)$ having properties like

$$\int_{-\infty}^{+\infty} \delta_\varepsilon = 1 \tag{19}$$

$$\int_{-\infty}^{+\infty} f(y) \cdot \delta_\varepsilon(x-y)\,\mathrm{d}y \to f(x) \qquad \text{almost everywhere as } \varepsilon \to 0^+ \tag{20}$$

for every locally integrable function $f : \mathbb{R} \longrightarrow \mathbb{R}$. In particular, this gives a well known regularization of our Heaviside function with the net of smooth functions $H_\varepsilon : x \in \mathbb{R} \mapsto \int_{-\infty}^{+\infty} H(y) \cdot \delta_\varepsilon(x-y)\,\mathrm{d}y \in \mathbb{R}$. Since $H_\varepsilon \in \mathcal{C}^\infty(\mathbb{R}, \mathbb{R})$ we can define the derivative of the heaviside function as the Colombeau generalized number $H'(0) := [H'_\varepsilon(0)] \in \widetilde{\mathbb{R}}$. We can say that we solved the problem to find the derivatives of H at the origin, by approximating it with a net of smooth function. The careful reader can object that this definition of derivative depends on the chosen infinitesimal standard deviation $\sigma = [\varepsilon^n]$ and on the net of smooth functions (δ_ε) derived from the Gaussian, in the sense that there are many other nets verifying properties (19) and (20) which can be used following the same thread of ideas. For example, if $\varphi \in \mathcal{C}^\infty(\mathbb{R}, \mathbb{R})$ is a smooth function which is zero outside a suitable interval (i.e., it has compact support, see e.g., [11, 57]), but with total mass $\int_{\mathbb{R}} \varphi = 1$, then

$$\varphi_\varepsilon(x) = \frac{1}{\varepsilon}\varphi\left(\frac{x}{\varepsilon}\right) \tag{21}$$

verifies (19) and (20). This important doubt divides the scientific community in two parts: A first group aims at finding an intrinsic definition of *derivation*, that is, of the operation $H \mapsto H'$. This point of view conducted to Schwartz distributions theory, see [57], a wonderful theory which stimulated strongly the growing of modern mathematics and functional analysis. It is possible to argue, see e.g., [59], that this definition of derivation is only formal and without any geometrical meaning. For example, following Schwartz approach ‚*the value of the derivative* $H'(0)$ has no meaning. A second research thread states that it could be interesting to study different regularizations of locally integrable functions using different nets of smooth functions, for example, to have the possibility to study infinitesimal differences of these regularizations facing the numerical solution of partial differential equations. Once we have chosen the regularization suitable for our aims, which is not always a clear choice, we can say that the derivative $H'(0) := [H'_\varepsilon(0)] \in \widetilde{\mathbb{R}}$ has a clear geometrical meaning of an ε-net of slopes of tangent straight lines moving, as $\varepsilon \to 0^+$, toward an infinite number in $\widetilde{\mathbb{R}}$, see e.g., [11, 33, 36] and references therein. Of course, these two points of view has interesting common threads of research, see [36] for a comprehensive presentation.

3.1 Generalized Smooth Functions

Starting from the previous ideas, we can think to define in general terms an interesting class of functions f generated by nets (u_ε) of smooth functions, $u_\varepsilon \in \mathcal{C}^\infty(\mathbb{R}^n, \mathbb{R}^d)$, exactly like (18). These functions f will be ordinary set-theoretical maps, but they will be defined on, and with values in, subsets of $\widetilde{\mathbb{R}}$. We will call them *generalized smooth functions* (GSF). One of the basic ideas in defining a GSF is to ask the minimal logical conditions so that the prescription $f(x) := [u_\varepsilon(x_\varepsilon)]$ is well-defined. In other words, we want to impose a condition of moderateness $(u_\varepsilon(x_\varepsilon)) \in \mathbb{R}_M$ and the $\sim$-independence (see Definition 4) from the representatives of the generalized point $x = [x_\varepsilon] \in \widetilde{\mathbb{R}}$. We will see that the former condition suffices to prove the latter. For all the proofs of this section, see [33, 71] and references therein.

The first notion we introduce concerns membership for ε-dependent objects; it assures that the nets we will consider are always closed with respect to different representatives of the same generalized point and permits to define a well-behaved set $(\Omega_\varepsilon)_M$ of moderate nets (x_ε) with values in (Ω_ε). This generalizes the step $\mathbb{R} \mapsto \mathbb{R}_M$.

Definition 5 Let (A_ε) be a net of subsets of $\mathbb{R}^n$, and let (x_ε) be a net of points in $\mathbb{R}^n$, then we say that (x_ε) *strongly belongs to* (A_ε) and we write

$$x_\varepsilon \in_\varepsilon A_\varepsilon$$

if

1. $x_\varepsilon \in A_\varepsilon$ for ε sufficiently small, i.e., $\exists \varepsilon_0 \in I \, \forall \varepsilon \in (0, \varepsilon_0] : \, x_\varepsilon \in A_\varepsilon$;
2. If $(x'_\varepsilon) \sim (x_\varepsilon)$, then also $x'_\varepsilon \in A_\varepsilon$ for ε sufficiently small.

The following result provides a certain geometrical intuition about this notion of strong membership and justifies its name.

Proposition 1 *For $\varepsilon \in I$, let $A_\varepsilon \subseteq \mathbb{R}^n$ and let $x_\varepsilon \in \mathbb{R}^n$. Then $x_\varepsilon \in_\varepsilon A_\varepsilon$ if and only if there exists some $q > 0$ such that $d(x_\varepsilon, A^c_\varepsilon) > \varepsilon^q$ for ε sufficiently small, where $A^c_\varepsilon := \mathbb{R}^n \setminus A_\varepsilon$. Therefore, if $(d(x_\varepsilon, A^c_\varepsilon)) \in \mathbb{R}_M$, then $x_\varepsilon \in_\varepsilon A_\varepsilon$ if and only if $[d(x_\varepsilon, A^c_\varepsilon)] \in \widetilde{\mathbb{R}}$ is invertible.*

Definition 6 Let (A_ε) be a net of subsets of $\mathbb{R}^n$, then

$$(A_\varepsilon)_M := \{(x_\varepsilon) \in \mathbb{R}^n_M \mid x_\varepsilon \in_\varepsilon A_\varepsilon\}.$$

Let us note that $(A_\varepsilon)_M \subseteq \mathbb{R}^n_M$ so that we can still talk of the generalized point $[x_\varepsilon] \in \widetilde{\mathbb{R}}^n$. Therefore, we can consider the so-called *strongly internal* set defined by the net (A_ε):

$$\langle A_\varepsilon \rangle := \{[x_\varepsilon] \in \widetilde{\mathbb{R}}^n \mid x_\varepsilon \in_\varepsilon A_\varepsilon\}. \tag{22}$$

In order to define the class of GSF, we will first introduce the notion of a net (u_ε) defining a generalized smooth map of the type $X \longrightarrow Y$, $X \subseteq \widetilde{\mathbb{R}}^n$, $Y \subseteq \widetilde{\mathbb{R}}^d$. This is a net of smooth functions $u_\varepsilon \in \mathcal{C}^\infty(\Omega_\varepsilon, \mathbb{R}^d)$ which induces well defined maps of the form $[\partial^\alpha u_\varepsilon(-)] : \langle \Omega_\varepsilon \rangle \longrightarrow \mathbb{R}^d$, for all multi-index α.

Definition 7 Let $X \subseteq \widetilde{\mathbb{R}}^n$ and $Y \subseteq \widetilde{\mathbb{R}}^d$ be generic subsets of generalized points. Let (Ω_ε) be a net of open sets of $\mathbb{R}^n$, and (u_ε) be a net of smooth functions, with $u_\varepsilon \in \mathcal{C}^\infty(\Omega_\varepsilon, \mathbb{R}^d)$. Then we say that

$$(u_\varepsilon) \textit{ defines a generalized smooth map} : X \longrightarrow Y$$

if:

1. $X \subseteq \langle \Omega_\varepsilon \rangle$ and $[u_\varepsilon(x_\varepsilon)] \in Y$ for all $x \in X$
2. $\forall [x_\varepsilon] \in X \, \forall \alpha \in \mathbb{N}^n : \ (\partial^\alpha u_\varepsilon(x_\varepsilon)) \in \mathbb{R}^d_M$;

For any property $\mathcal{P}(\,-\,)$, the notation $\forall [x_\varepsilon] \in X : \ \mathcal{P}([x_\varepsilon])$ means: for all $(x_\varepsilon) \in (\Omega_\varepsilon)_M$, if $[x_\varepsilon] \in X$, then $\mathcal{P}([x_\varepsilon])$ holds.

A generalized smooth map is simply a function of the form $f = [u_\varepsilon(\,-\,)]|_X$:

Definition 8 Let $X \subseteq \widetilde{\mathbb{R}}^n$ and $Y \subseteq \widetilde{\mathbb{R}}^d$ be generic subsets of generalized points, then we say that

$$f : X \longrightarrow Y \text{is a } \textit{generalized smooth function}$$

if there exists a net $u_\varepsilon \in \mathcal{C}^\infty(\Omega_\varepsilon, \mathbb{R}^d)$ defining a generalized smooth map of type $X \longrightarrow Y$, in the sense of Definition 7, such that f is the map

$$f = [u_\varepsilon(\,-\,)]|_X. \tag{23}$$

We will also say that f *is generated (or defined) by the net of smooth functions* (u_ε). The set of all these GSF will be denoted by $\mathcal{GC}^\infty(X, Y)$.

Let us note explicitly that Definitions 7 and 8 really state minimal logical conditions to have a set-theoretical map defined by a net of smooth functions. In particular, the following Theorem 4 states that the equality (23) is meaningful, i.e., that we have independence from the representatives for all derivatives $[x_\varepsilon] \in X \mapsto [\partial^\alpha u_\varepsilon(x_\varepsilon)] \in \widetilde{\mathbb{R}}^d$, $\alpha \in \mathbb{N}^n$.

Theorem 4 *Let $X \subseteq \widetilde{\mathbb{R}}^n$ and $Y \subseteq \widetilde{\mathbb{R}}^d$ be generic subsets of generalized points. Let (Ω_ε) be a net of open sets of $\mathbb{R}^n$, and (u_ε) be a net of smooth functions, with $u_\varepsilon \in \mathcal{C}^\infty(\Omega_\varepsilon, \mathbb{R}^d)$. Assume that (u_ε) defines a generalized smooth map of the type $X \longrightarrow Y$, then*

$$\forall \alpha \in \mathbb{N}^n \, \forall (x_\varepsilon), (x'_\varepsilon) \in (\Omega_\varepsilon)_M : \ [x_\varepsilon] = [x'_\varepsilon] \in X \ \Rightarrow \ (\partial^\alpha u_\varepsilon(x_\varepsilon)) \sim (\partial^\alpha u_\varepsilon(x'_\varepsilon)).$$

To make clear the relationships between these GSF and Schwartz distributions, we introduce the (special) Colombeau algebra on the open set $\Omega \subseteq \mathbb{R}^n$ as the quotient $\mathcal{G}(\Omega) := \mathcal{E}_M(\Omega)/\mathcal{N}(\Omega)$ of *moderate nets* over *negligible nets*, where the former is

$$\begin{aligned} \mathcal{E}_M(\Omega) := \{(u_\varepsilon) \in \mathcal{C}^\infty(\Omega)^I \mid \forall K \Subset \Omega \, \forall \alpha \in \mathbb{N}^n \\ \exists N \in \mathbb{N} : \sup_{x \in K} |\partial^\alpha u_\varepsilon(x)| = O(\varepsilon^{-N})\} \end{aligned}$$

and the latter is

$$\begin{aligned} \mathcal{N}(\Omega) := \{(u_\varepsilon) \in \mathcal{C}^\infty(\Omega)^I \mid \forall K \Subset \Omega \, \forall \alpha \in \mathbb{N}^n \\ \forall m \in \mathbb{N} : \sup_{x \in K} |\partial^\alpha u_\varepsilon(x)| = O(\varepsilon^{m})\}. \end{aligned}$$

Here, as usual, every asymptotic relation is for $\varepsilon \to 0^+$. For this quotient algebra, the interested reader can find by herself how to reformulate the motivations we saw to introduce $\widetilde{\mathbb{R}}$. The mapping $\Omega \mapsto \mathcal{G}(\Omega)$ is a fine and supple sheaf of differential algebras and the space of Schwartz distributions $\mathcal{D}'(\Omega)$ can be embedded into $\mathcal{G}(\Omega)$ by convolution $T \in \mathcal{D}'(\Omega) \mapsto [T * \varphi_\varepsilon] \in \mathcal{G}(\Omega)$, where φ_ε is defined like in (21) (see [11, 36] for more details). Since $\mathcal{G}(\Omega) \supseteq \mathcal{D}'(\Omega)$ is an algebra, this construction represents a possible solution to Schwartz impossibility theorem for multiplication of distributions, see [58].

The space of compactly supported generalized points $\widetilde{\Omega}_c$ is defined by $\Omega_c / \sim$, where $\Omega_c := \{(x_\varepsilon) \in \Omega^I \mid \exists K \Subset \Omega \, \exists \varepsilon_0 \, \forall \varepsilon < \varepsilon_0 : x_\varepsilon \in K\}$ and $\sim$ is the same equivalence relation as in the case of $\widetilde{\Omega}$. Any Colombeau generalized function $u \in \mathcal{G}(\Omega)$ acts on generalized points from $\widetilde{\Omega}_c$ by $u(x) := [u_\varepsilon(x_\varepsilon)]$ and is uniquely determined by its point values (in $\widetilde{\mathbb{R}}$) on compactly supported points, [36], but not on standard points (elements of Ω).

We first state that Colombeau generalized functions (CGF) are included in our GSF with the following

Theorem 5 *A Colombeau generalized function $u \in \mathcal{G}(\Omega)$ (in particular a Schwartz distribution) defines a generalized smooth map $u : \widetilde{\Omega}_c \longrightarrow \widetilde{\mathbb{R}}$. This assignment provides an injection of $\mathcal{G}(\Omega)$ for every open set Ω.*

On the contrary to what happens for distributions and CGF, there is no problem in considering the composition of two GSF:

Theorem 6 *Subsets $S \subseteq \widetilde{\mathbb{R}}^s$ with the trace of the sharp topology, and generalized smooth maps as arrows form a subcategory of the category of topological spaces. We will call this category $\mathcal{GC}^\infty$, the category of GSF.*

From this theorem and the embedding $\mathcal{C}^\infty(\Omega) \subseteq \mathcal{D}'(\Omega) \subseteq \mathcal{G}(\Omega) \subseteq \mathcal{GC}^\infty(\widetilde{\Omega}_c, \widetilde{\mathbb{R}})$ it follows that a classical way to define a GSF is to follow the original idea of Cauchy, Poisson, Kirchhoff, Helmholtz, Kelvin, Heaviside, Dirac ([17, 40, 47]), that is, to fix an infinitesimal or infinite parameter in a suitable standard smooth function $\varphi \in \mathcal{C}^\infty(P \times X, Y)$ considering $u_\varepsilon := \varphi(p_\varepsilon, -)$, see [33] for more details.

3.2 Differential and Integral Calculus for Generalized Smooth Functions

Looking at the definition of generalized smooth map, even in the one dimensional case, it seems natural to define $f'(x) := [u'_\varepsilon(x_\varepsilon)]$, even though proving the independence of this definition from the net (u_ε) doesn't seem immediate. We can arrive at this goal with the so-called Fermat–Reyes theorem. In order to state it, we need the following notion.

Definition 9 Let $U \subseteq \widetilde{\mathbb{R}}^n$, then we say that $T \subseteq \widetilde{\mathbb{R}}^{2n}$ is *a thickening of U* if

1. $\forall x \in U : (x, 0) \in T$
2. For all $(x, h) \in T$ there exist $a, b \in \widetilde{\mathbb{R}}_{>0}$ invertible, with $b < a$, such that:
   a) $|h| < b$
   b) $B_a(x) \subseteq U$

c) $B_a(x) \times B_b(0) \subseteq T$.

The Fermat–Reyes theorem in the present setting is the following. It represents the first of a long list of results showing the large similarities between ordinary smooth functions and GSF. For example, if $U \subseteq \mathbb{R}^n$ is an ordinary open set, then $\text{th}(\Omega) := \{(x,h) \in \mathbb{R}^{2n} \mid [x, x+h] \subseteq U\}$ is a thickening of U.

Theorem 7 *Let $U \subseteq \widetilde{\mathbb{R}}$ be a sharply open set, and $f \in \mathcal{GC}^\infty(U, \widetilde{\mathbb{R}})$ be a generalized smooth map generated by the net of smooth functions $u_\varepsilon \in \mathcal{C}^\infty(\Omega_\varepsilon, \mathbb{R})$. Then*

1. *If S is a thickening of U such that $S \subseteq \langle \text{th}(\Omega_\varepsilon) \rangle$, then there exists a thickening $T \subseteq S$ of U and a generalized smooth map $r \in \mathcal{GC}^\infty(T, \widetilde{\mathbb{R}})$, called the generalized incremental ratio of f, such that*

$$f(x+h) = f(x) + h \cdot r(x,h) \qquad \forall (x,h) \in T.$$

   Moreover $r(x,0) = [u'_\varepsilon(x_\varepsilon)]$ for every $x \in U$ and we can thus define $f'(x) := r(x,0)$, so that $f' \in \mathcal{GC}^\infty(U, \widetilde{\mathbb{R}})$.
2. *Any two generalized incremental ratios of f coincide on the intersection of their domains.*

By this theorem and by the inclusion of CGF into generalized smooth maps (Theorem 5) it follows that in case f is induced by an ordinary Schwartz distribution our notion of derivative and the usual notion of derivative of a distribution coincide (see e.g., [36]). From the analogous properties for the nets of smooth functions generating our GSF, the usual properties of derivatives follow chain rule without limitations included. Since we think at our generalized smooth functions as generalizing ordinary smooth functions, but including distributions and (co)domains with infinitesimals and infinities numbers, it is natural to inquire existence and uniqueness of primitives.

Theorem 8 *Let $f \in \mathcal{GC}^\infty(U, \widetilde{\mathbb{R}})$ be a generalized smooth function defined in the sharply open set $U \subseteq \widetilde{\mathbb{R}}$. Let $a, b, c \in \widetilde{\mathbb{R}}$, with $a < b$ and $c \in [a,b] := \{x \in \widetilde{\mathbb{R}} \mid a \leq x \leq b\} \subseteq U$. Then there exists one and only one generalized smooth function $F \in \mathcal{GC}^\infty([a,b], \widetilde{\mathbb{R}})$ such that $F(c) = 0$ and $F'(x) = f(x)$ for all $x \in [a,b]$. Moreover, defining $\int_c^{(-)} f := \int_c^{(-)} f(s)\,\mathrm{d}s \in \mathcal{GC}^\infty([a,b], \widetilde{\mathbb{R}})$ as this F all the elementary rules of integration follows, integration by substitution included.*

Finally, it is natural to expect that several classical theorem of differential and integral calculus can be extended from the ordinary smooth case to our generalized smooth framework.

Theorem 9 *Let $f \in \mathcal{GC}^\infty(U, \widetilde{\mathbb{R}})$ be a GSF defined in the sharply open set $U \subseteq \widetilde{\mathbb{R}}$. Let $a, b \in \widetilde{\mathbb{R}}$, with $a < b$, such that $[a,b] = \{x \in \widetilde{\mathbb{R}} \mid a \leq x \leq b\} \subseteq U$. Then*

1. *$\exists m, M \in [a,b]\, \forall x \in [a,b] : f(m) \leq f(x) \leq f(M)$ (extreme value theorem)*
2. *$\exists c \in [a,b] : f(b) - f(a) = (b-a) \cdot f'(c)$ (mean value theorem)*
3. *$\exists c \in [a,b] : \int_a^b f = (b-a) \cdot f(c)$ (integral mean value theorem)*
4. *Assume that $f(a) < f(b)$. Then $\forall y \in \widetilde{\mathbb{R}} : f(a) < y < f(b) \Rightarrow \exists c \in [a,b] : y = f(c)$ (intermediate value theorem).*

3.3 Open Problems and Ideas

The theory of GSF is only at the beginning even if, in our point of view, it seems promising. Its first aim is to define a class of generalized functions more near to the informal use we can find in physics. Therefore we need to include Schwartz distributions, but considering composable set-theoretical functions defined on a non-Archimedean ring and permitting all polynomial operations.

Some open problems in this framework are the following:

Not Only Polynomials Generalize the definition of GSF using a generic family of infinities $\mathcal{J} \subseteq \mathbb{R}^I$ instead of polynomials (16). Let $\mathcal{GC}^\infty_{\mathcal{J}}$ be the corresponding category of GSF. What families $\mathcal{J}$ always permit to embed Schwartz distributions: $\mathcal{D}'(\Omega) \subseteq \mathcal{GC}^\infty_{\mathcal{J}}(\widetilde{\Omega}_c)$? What families $\mathcal{J}$ permits to identify an ordinary smooth function $f \in \mathcal{C}^\infty(\mathbb{R}, \mathbb{R})$ with the corresponding GSF generated by the constant net $u_\varepsilon = f$ and also with the Schwartz distribution corresponding to f (coherence of embeddings)? The family $\mathcal{I}$ defined in (16) permits to extend to GSF (and hence to Schwartz distributions) any polynomially growing operation; for any smooth operation $\omega \in \mathcal{C}^\infty(\mathbb{R}^n, \mathbb{R})$ does there exists a family $\mathcal{J} = \mathcal{J}(\omega)$ such that ω is extendibles to the corresponding GSF in $\mathcal{GC}^\infty_{\mathcal{J}}$?

The Levi-Civita–Colombeau Ring Study the subring of $\widetilde{\mathbb{R}}$ of all the generalized numbers $[x_\varepsilon] \in \widetilde{\mathbb{R}}$ generated by representatives of the form

$$x_\varepsilon = \beta_1 \varepsilon^{-b_1} + \ldots + \beta_N \varepsilon^{-b_N} + \alpha_0 + \alpha_1 \varepsilon^{a_1} + \ldots + \alpha_n \varepsilon^{a_n} + o(\varepsilon^{a_n}) \text{ as } \varepsilon \to 0^+,$$

where β_j, $\alpha_i \in \mathbb{R}$ and $(b_j)_{j=1}^N$, $(a_i)_{i=1}^n$ are suitable increasing sequences of real numbers. Call this subring the Levi-Civita–Colombeau ring, or briefly LC^2. Is LC^2 a subfield? Is it totally ordered? Can we defined GSF using only LC^2? Are Schwartz distributions embedded in this new class of GSF based on LC^2? Can we prove coherence of embeddings (see previous problem)?

Computer Implementation of $\widetilde{\mathbb{R}}$ Can we code a computer implementation of the ring of Fermat reals $\widetilde{\mathbb{R}}$ and define some useful graphical representation of some elements of the Colombeau algebra $\mathcal{G}$? For example, can we graphically represent a Cauchy–Dirac generalized function $f = [u_\varepsilon(-)]$, where $u_\varepsilon := \varphi(p_\varepsilon, -)$ and $\varphi \in \mathcal{C}^\infty(P \times X, Y)$ as $\varepsilon \to 0^+$? Can we write an algorithm to establish that, up to the machine precision, two moderate nets (x_ε), $(x'_\varepsilon) \in \mathbb{R}_M$ (of course defined by two user defined computer functions) are equivalent, that is, they generate the same number $[x_\varepsilon] \in \widetilde{\mathbb{R}}$? Do these graphical visualizations permit to guess what are the meaningful properties characterizing different derivatives defined by different regularizations of locally integrable functions? Are these properties useful in the numerical solution of singular partial differential equations?

4 Which Numbers Help to Compute Derivatives?

The Levi-Civita field (from now on: LCF) originally appeared in [49, 50]. For an account of Levi-Civita's work see also [46]. For detailed works in this topic and the proofs of the theorems we will state in this section, see [60–65] and references therein.

To motivate the introduction of the LCF, we start from the idea to add to the real field $\mathbb{R}$ a single new invertible infinitesimal number d to which we want to apply all the ring operations but also arbitrary roots. Hence, we would like to be able to form numbers like $3 + d + 2d^2$, but also like d^{-1}, $d^{1/2}$, $d^{-1/2}$, $2 - d^3 + 4d^{3/5} - \frac{1}{2}d^{-2/3}$. This can be easily obtained using purely algebraic methods, for example, considering formal power series of the form

$$x = \sum_{q \in \mathbb{Q}} x_q \cdot d^q. \tag{24}$$

There is no problem in defining the sum pointwise

$$x + y := \sum_{q \in \mathbb{Q}} x_q d^q + \sum_{q \in \mathbb{Q}} y_q d^q := \sum_{q \in \mathbb{Q}} \left(x_q + y_q\right) d^q,$$

but we can recognize a first limitation considering the product, which is defined, as usual for formal power series and polynomials, as

$$x \cdot y := \sum_{q \in \mathbb{Q}} \left(\sum_{r+s=q} x_r \cdot y_s \right) \cdot d^q. \tag{25}$$

Indeed, the sum $\sum_{r+s=q} x_r \cdot y_s$ can have an infinite number of addends, depending on how many non-zero coefficients x_r and y_s we have in the factors x and y, and hence the previous definition of product $x \cdot y$ can be meaningless for generic formal power series x, y. Because we want that the Definition (25) works for every pair x, y, we must introduce a limitation on the number of coefficients in our formal power series. In other words, we must limit the number of non-zero coefficients in the formal series. For example, we can have

$$\bar{x} = d^{-3} + d^{-2} + d^{-1} + 1 + d + d^2 + d^3 + d^4 + d^5 + \dots,$$

or

$$\bar{y} = \dots + d^{-5} + d^{-4} + d^{-3} + d^{-2} + d^{-1} + 1 + d + d^2.$$

In the LCF the choice fall on power series with a finite number of exponents "on the left," such that

$$\forall q \in \mathbb{Q} : \text{ card}\,\{r \in \mathbb{Q} \mid x_r \neq 0\,,\, r \leq q\} \text{ is finite.} \tag{26}$$

This choice may recall moderate nets in the Colombeau ring of generalized numbers, but we have to remember that here we are using a formal/algebraic point of view, and that we do not have a notion of convergence for our series.

Leaving the intuitive motivations to arrive to a more formal mathematics, we can introduce our formal power series thinking of the corresponding definition for polynomials: identifying a polynomial $a_0 + a_1x + a_2x^2 + \ldots + a_nx^n$ with the n-tuple of its coefficients $(a_0, a_1, a_2, \ldots, a_n)$ is equivalent to say that a polynomial is a function of the form $a : \{0, 1, 2, \ldots, n\} \longrightarrow \mathbb{R}$. Analogously we can define

Definition 10 The support set $\mathcal{R}$ of the LCF is the set of all the functions $x : \mathbb{Q} \longrightarrow \mathbb{R}$ with *left-finite support*, such that

$$\forall q \in \mathbb{Q} : \ card\,\{r \in \mathbb{Q} \mid x_r \neq 0 \ r \leq q\} \text{ is finite.}$$

So, the value[1] $x(q) =: x[q]$ has to be thought as the coefficient of the addend $x[q] \cdot d^q$. The ring operations are defined for $q \in \mathbb{Q}$ as

$$(x + y)[q] := x[q] + y[q]$$
$$(x \cdot y)[q] := \sum_{\substack{r,s \in \mathbb{Q} \\ r+s=q}} x[r] \cdot y[s],$$

and we can verify that $(\mathcal{R}, +, \cdot)$ becomes a field. In a formal power series like (24) the leading term $x_m \cdot d^m$ with the lower value $m \in \mathbb{Q}$ of the exponent determines the behavior of the number from several points of view, for example, with respect to order. For this reason, we introduce the following notations:

Definition 11 Let $x, y \in \mathcal{R}$, then

1. $\text{Supp}(x) := \{q \in \mathbb{Q} \mid x[q] \neq 0\}$
   a) $\lambda(x) := \min(\text{supp}(x))$ for $x \neq 0$ and $\lambda(0) := +\infty$. The term $\lambda(x)$ is called *order of magnitude*.
   b) $x \sim y \quad :\Longleftrightarrow \quad \lambda(x) = \lambda(y)$. This relation is called *agreement of order of magnitude*.
   c) $x \approx y \quad :\Longleftrightarrow \quad \lambda(x) = \lambda(y)$ and $x\,[\lambda(x)] = y\,[\lambda(y)]$
   d) $x =_r y \quad :\Longleftrightarrow \quad \forall q \in \mathbb{Q}_{\leq r} : \ x[q] = y[q]$

If $r > 0$, the relation $=_r$ can be interpreted as equality up to k-th order infinitesimals: if $x =_r y$, then $x - y$ is given by sum of infinitesimals $a_q d^q$ of order $d > r$. E.g. we have $1+d+d^2-2d^3 =_2 1+d+d^2+4d^5$ and $d^{n+1} =_n 0$. So, it appears sufficiently clear that, even if we do not have a ring with nilpotent elements, the equivalence relation $=_r$ can supply a possible alternative language useful for the use of this type of infinitesimals made in differential geometry.

In the LCF we can prove the existence of roots:

Theorem 10 *Let $x \in \mathcal{R}$ be nonzero and $n \in \mathbb{N}_{>0}$. If n is even and the leading term $x[\lambda(x)] > 0$, then x has two n-th roots in $\mathcal{R}$. If n is even and $x[\lambda(x)] < 0$, then x has no n-th roots in $\mathcal{R}$. If n is odd, then x has a unique n-th root in $\mathcal{R}$.*

---

[1] The notation with square brackets $x[q]$ permits to avoid confusion when one consider functions defined on the LCF $\mathcal{R}$.

Exactly like in the study of polynomials, the particular number $d[q] := 1$ for $q = 1$ and $d[q] := 0$ otherwise, works as the independent variable in our formal power series, and the equality (24) can now be proved for every $x \in \mathcal{R}$ because $(d^r)[q] = 1$ if $q = r$ and $(d^r)[q] = 0$ otherwise. Let us note explicitly that if $r = \frac{p}{q} \in \mathbb{Q}$ with $p, q \in \mathbb{Z}, q > 0$, then $d^r = \sqrt[q]{d^p}$, so we need the previous Theorem 10.

Obviously, the embedding of the reals is given by $r \in \mathbb{R} \mapsto r[-] \in \mathcal{R}$, where $r[0] = r$ and $r[q] = 0$ otherwise, but it is now also clear that formal Laurent series (and hence also D. Tall's superreal numbers, see e.g., [70]), i.e., numbers of the form

$$x = \sum_{k=-N}^{+\infty} x_k \cdot d^k$$

are embedded in the LCF.

Essential for the development of the LCF as an ordered field but also for the different notions of continuity and differentiability of functions $f : \mathcal{R} \longrightarrow \mathcal{R}$ is the order relation. As hinted above, we can define the order relation by comparison of the leading terms

Definition 12 If $x, y \in \mathcal{R}$, we define

$$x > 0 \quad :\Longleftrightarrow \quad x[\lambda(x)] > 0$$

$$x > y \quad :\Longleftrightarrow \quad x - y > 0$$

With this relation the LCF becomes a totally ordered field extending the real field.

In the LCF, due to the formal approach, it is not so clear what functions $f : \mathbb{R} \longrightarrow \mathbb{R}$ can be extended to the whole $\mathcal{R}$ and hence the approach is different from the cases of ${}^{\bullet}\mathbb{R}$ and $\widetilde{\mathbb{R}}$ and mimics the classical approach of calculus. The next step is hence to use the order relation to define a corresponding order topology.

Definition 13 Because the order relation on $\mathcal{R}$ is total, we can define the absolute value in the usual way

$$|x| := \begin{cases} x & \text{if } x \geq 0 \\ -x & \text{if } x < 0 \end{cases}$$

Moreover, we say that a subset $U \subseteq \mathcal{R}$ is *open in the order topology* iff

$$\forall u \in U\ \exists \delta \in \mathcal{R}_{>0} : \quad \{x \in \mathcal{R} : |x - u| < \delta\} \subseteq U.$$

For example, the sequence $(d^n)_{n \in \mathrm{N}}$ converges to the zero sequence in the order topology. Using the same idea, that is the formal analogy with the reals $\mathbb{R}$, we can define continuity, differentiability, and convergence of sequences.

Definition 14 Let $D \subseteq \mathcal{R}$ and $f : D \longrightarrow \mathcal{R}$, then we say that f *is topologically continuous at* $x_0 \in D$ iff

$$\forall \varepsilon \in \mathcal{R}_{>0}\, \exists \delta \in \mathcal{R}_{>0} :\; \forall x \in D :\; |x - x_0| < \delta \Rightarrow |f(x) - f(x_0)| < \varepsilon.$$

Definition 15 Let $D \subseteq \mathcal{R}$ and $f : D \longrightarrow \mathcal{R}$, then we say that f *is topologically differentiable at* $x_0 \in D$ iff there exists a number $l \in \mathcal{R}$ such that

$$\forall \varepsilon \in \mathcal{R}_{>0}\, \exists \delta \in \mathcal{R}_{>0}\, \forall x \in D :\; 0 < |x - x_0| < \delta \Rightarrow \left| \frac{f(x) - f(x_0)}{x - x_0} - l \right| < \varepsilon.$$

Definition 16 Let $s : \mathbb{N} \longrightarrow \mathcal{R}$, then we say that s *converges strongly to* $l \in \mathcal{R}$ *iff*

$$\forall \varepsilon \in \mathcal{R}_{>0}\, \exists N \in \mathbb{N} :\; \forall n \in \mathbb{N} :\; n \geq N \Rightarrow |s_n - l| < \varepsilon.$$

It is interesting that now, using this notion of convergence, we can associate to our formal power series a notion of convergence:

Theorem 11 *Let* $x \in \mathcal{R}$ *and define recursively*

$$q_0 := \lambda(x)$$

$$q_{k+1} := \begin{cases} \min\left(\operatorname{supp}(x) \setminus \{q_0, q_1, \ldots, q_k\}\right) & if\, \operatorname{supp}(x) \supset \{q_0, q_1, \ldots, q_k\} \\ 0 & otherwise \end{cases}$$

then the sequence $n \in \mathbb{N} \mapsto \sum_{k=0}^{n} x[q_k] \cdot d^{q_k} \in \mathcal{R}$ *converges strongly to* x*, so that we can write*

$$x = \sum_{k=0}^{+\infty} x[q_k] \cdot d^{q_k}.$$

Note that this theorem does not realize the dialogue between potential infinitesimals and actual infinitesimals we mentioned in the introduction, because, trivially, in its statement there is no mention of any such potential infinitesimal, that is, of a function of the form $i : E \longrightarrow \mathbb{R}$ convergent to zero, instead in the statement convergence is understood in the LCF sense.

Using the notion of skeleton group (see e.g., [66]), it is possible to give some motivations for the choice of the domain $\mathbb{Q}$ for the elements of the LCF $\mathcal{R}$: $\mathbb{Q}$ is the smallest set of exponents that permits to have a non-Archimedean field and roots of positive elements.

However, the idea to follow formal analogies to define continuity and differentiability (see Definitions 14 and 15) presents several problems (not characteristic of the LCF, but common to every non-Archimedean totally ordered field): e.g., the function $f : [0, 1] \longrightarrow \mathcal{R}$ defined by

$$f(x) := \begin{cases} 0 & \text{if } x \text{ is infinitely small} \\ 1 & \text{if } x \text{ is finite} \end{cases}$$

is topologically continuous and topologically differentiable, but it does not assume the value $d \in [0,1]$ even though $f(0) < d < f(1)$, hence it does not verifies the intermediate value theorem. Moreover, $f'(x) = 0$ for all $x \in [0,1]$ but f is not constant on $[0,1]$. Therefore, also the simplest initial value problem $y' = 0$, $y(0) = 0$ does not have a unique solution. This is due to the fact that infinitesimals are totally disconnected from finite numbers because $n \cdot d < r$ for every $r \in \mathbb{R}_{>0}$ and for every $n \in \mathbb{N}$, and this is a general problem of non-Archimedean fields. The solution adopted in the framework of the LCF is to change the notion of topological continuity introducing a Lipschitz condition:

Definition 17 Let $a < b$ be given in $\mathcal{R}$, let $I \in \{(a,b),[a,b],(a,b],[a,b]\}$ be an interval of $\mathcal{R}$ and $f : I \longrightarrow \mathcal{R}$, then we say that *f is continuous in I* iff

$$\exists M \in \mathcal{R} : \forall x \in I : |f(x) - f(y)| \leq M \cdot |x - y| .$$

The corresponding definition of differentiability recalls the approach with smooth incremental ratios we have seen for GSF.

Definition 18 Under the same hypothesis as in the previous definition we say that *f is differentiable on I* iff there exists a continuous function

$$f'[-,-] : I \times I \longrightarrow \mathcal{R}$$

such that

$$\forall x, y \in I : f(y) = f(x) + f'[x,y] \cdot (y - x).$$

We will set $f'(x) := f'[x,x]$.

This definition is also very interesting, because, even if the approach of the LCF is a formal approach to infinitesimals, some of the chosen solutions are very similar to those adopted in nonformal approaches to infinitesimals. For example, note the requirement that $f'[-,-]$ should be continuous, and hence it should verify a Lipschitz condition, in analogy to the Lipschitz differentiability introduced in [45].

A natural problem dealing with LCF is what class of functions $f : \mathbb{R} \longrightarrow \mathbb{R}$ can be extended to a meaningful subclass of $\mathcal{R}$ with the possibility to generalize to them some properties, like the intermediate value theorem, an inverse function theorem, the extreme value theorem, the mean value theorem, Rolle's theorem or the existence and uniqueness of primitive functions. Because of the left-finiteness of our formal power series $x \in \mathcal{R}$, the most natural class of functions are those locally expandable in power series. The solution adopted is to define a notion of convergence of power series with coefficients in $\mathcal{R}$, to prove for them the above mentioned theorems, and hence to show that standard power series in $\mathbb{R}$ are included as a particular case of this notion of convergence in $\mathcal{R}$. It is also interesting to note that this concept of convergence is not the one derived from the formal analogies with the real case (see e.g., the Definition 14) but it is rather derived from a family of seminorms. For more details on this development, see the above mentioned references.

4.1 *Using the LCF to Derive Computationally Complicated Functions*

The left-finiteness of the Levi-Civita numbers permits to represent them on a computer. Indeed, for every $r \in \mathbb{Q}$ the amount of information we have to store in the power series (24), up to the terms $x_q d^q$ with $q \le r$, is finite and we can represent all these numbers with the usual precision available in a computer. Therefore, using the equivalence relation $=_r$ (see Definition 11) we can implement a calculus of Levi-Civita numbers on a computer. This founds useful applications in automatic differentiation (see e.g., [8]). Here the problem is to find a computer algorithm to calculate the derivatives, up to a given order, of computationally complicated functions like (see [66])

$$f(x) = \frac{\sin(x^3+2x+1) + \dfrac{3+\cos(\sin(\ln|1+x|))}{\exp\left(\tanh\left(\sinh\left(\cosh\left(\frac{\sin(\cos(\tan(\exp(x))))}{\cos(\sin(\exp(\tan(x+2))))}\right)\right)\right)\right)}}{2+\sin(\sinh(\cos(\tan^{-1}(\ln(\exp(x)+x^2+3)))))},$$

obtained by composition of elementary functions like sin, cos, exp, ..., the Heaviside function (17) and of the field operations. These are called *computer functions*, and can be extended to a suitable subset of $\mathcal{R}$ using their expansion in power series. The property that permits to compute these derivatives is expressed in the following theorem:

Theorem 12 *Let f be a computer function continuous at $x_0 \in \mathbb{R}$ and extendible to $x_0 \pm d$. Then f is n-times differentiable at x_0 if and only if there exist real numbers $\alpha_1, \dots, \alpha_n$ such that*

$$f(x_0+d) =_n \sum_{j=0}^{n} \frac{\alpha_j}{j!} \cdot d^j$$

$$f(x_0-d) =_n \sum_{j=0}^{n} (-1)^j \frac{\alpha_j}{j!} \cdot d^j.$$

Moreover, in this case we have $f^{(j)}(x_0) = \alpha_j$ for $j = 0, \dots, n$.

In the cited references, one can find several examples of computation of derivatives using these formulas, and of non smooth functions whose regularity is proved using this theorem. A software, called COSY INFINITY, has also been created, which is suitable for the computation of derivatives of functions using the LCF (see [10] and references therein also for the comparison with other methods of computation of derivatives).

5 What is the Universal Non-Archimedean Field?

Surreal numbers has been introduced by J.H. Conway and presented in [41] and in [12].[2] One of the most surprising features of surreal numbers is that starting from a simple set of rules it is possible to construct a rich algebraic structure containing the real numbers as well as infinite and infinitesimals, but also all the ordinal numbers, the hyperreals of NSA, the Levi-Civita field and the field of rational functions. Indeed, in a precise sense we will see later, the ordered field **No** of surreal numbers is the largest possible ordered field or, in other words, the above mentioned simple rules for the construction of surreal numbers represent the most general way to obtain a notion of number culminating in an ordered field.

There are two basic ideas to introduce surreal numbers: the first is to have the possibility to construct numbers in a transfinite-recursive way using a notion analogous to that of Dedekind cut (called *Conway cut*). If we have a totally ordered set $(N, <)$, a Conway cut is simply a pair (L, R) of subsets $L, R \subseteq N$ such that

$$\forall l \in L \, \forall r \in R: \ \ l < r. \tag{27}$$

In this case we will simply write $L < R$. This is exactly the notion of Dedekind cut without the condition that the subsets L, R have to be contiguous (i.e., without the condition that $\forall \varepsilon > 0 \, \exists l \in L \, \exists r \in R : \ |l - r| < \varepsilon$). Exactly because we do not have this further condition, we need another condition for a pair (L, R) to identify a unique "number." Indeed, the second idea, intuitively stated, is that every Conway cut identifies uniquely *the simplest number* x between L and R:

$$\forall l \in L \, \forall r \in R: \ \ l < x < r. \tag{28}$$

We can intuitively represent a Conway cut and the associated simplest number in the following way

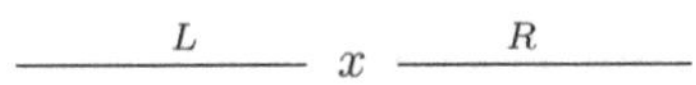

A little more formally, the class **No** of surreal numbers is introduced by Conway using a suitable set of rules. We can think at these rules as axioms defining a suitable structure $(\mathbf{No}, \le, \{- \mid -\})$. In the following, as usual, $x < y$ means $x \le y$ and $x \ne y$.

Construction If $L, R \subseteq \mathbf{No}$ and $L < R$, then $\{L \mid R\} \in \mathbf{No}$, that is starting from a Conway cut (L, R) we can construct a surreal with $\{L \mid R\} \in \mathbf{No}$.

Surjectivity If $x \in \mathbf{No}$, then there exist $L, R \subseteq \mathbf{No}$ such that $L < R$ and $x = \{L \mid R\}$, that is all surreal numbers can be constructed starting from a Conway cut.

Inequality If $x = \{L_x \mid R_x\}$ and $y = \{L_y \mid R_y\}$ are well defined,[3] then $x \le y$ if and only if $L_x < \{y\}$ and $\{x\} < R_y$, i.e., $l_x < y$ and $x < r_y$ for every

[2] Really, the same field of numbers has been predate by [14] (in Spanish) and [37] (in German).
[3] That is $L_x < R_x$ and $L_y < R_y$. Let us note that using a notation like $x = \{L_x \mid R_x\}$ we do not mean that a number $x \in \mathbf{No}$ uniquely determines the subsets L_x and R_x.

$l_x \in L_x$ and every $r_y \in R_y$. This rule can be represented in the following way

$$\overset{L_x}{\overline{\qquad\qquad}} \; x \; \overset{R_x}{\overline{\qquad\qquad}} \; \cdots \; \overset{L_y}{\overline{\qquad\qquad}} \; y \; \overset{R_y}{\overline{\qquad\qquad}}$$

Equality If $x, y \in \mathbf{No}$, then $x = y$ if and only if $x \leq y$ and $y \leq x$, that is equality between surreal numbers can be defined starting from the order relation $\leq$.

Starting from these simple rules/axioms we can already construct several meaningful examples of numbers in **No**. From the Definition 27 of $L < R$ we see that always $L < \emptyset$ and $\emptyset < R$ for every $L, R \subseteq \mathbf{No}$. So we have $\emptyset < \emptyset$ and from the construction rule $\{\emptyset \mid \emptyset\} \in \mathbf{No}$. Therefore, **No** is not empty and we can iterate the process. For simplicity, we will write $\{|\} := \{\emptyset \mid \emptyset\}$, $\{L \,|\} := \{L \mid \emptyset\}$, $\{|\, R\} := \{\emptyset \mid R\}$, $\{x_1, \ldots, x_n \mid R\} := \{\{x_1, \ldots, x_n\} \mid R\}$ and $\{L \mid x_1, \ldots, x_n\} := \{L \mid \{x_1, \ldots, x_n\}\}$. Hence we have, e.g.,

$$\{|\} \in \mathbf{No}$$

$$x \in \mathbf{No} \implies \{|x\},\ \{x|\} \in \mathbf{No}.$$

However, the understanding of the class **No** has a great improvement if we introduce the above mentioned interpretation of simplicity. Conway's idea is that a number $x \in \mathbf{No}$ is simpler than $y \in \mathbf{No}$ if x is defined before y in the previous iterative process (using Conway's terminology: x was born before y). So, $\{|\}$ is the simplest number[4] and this justify the definition $\{|\} =: 0$. On the next step, we have e.g., $\{|0\}$ and $\{0|\}$ which have the same degree of simplicity (because they are both defined in the second step of the iterative process). To interpret these numbers we have hence to use the idea of simplicity expressed in (28): $\{0|\}$ is the simplest number greater than 0 and $\{|0\}$ is the simplest number less than 0. This justifies the definition $1 := \{0|\}$ and $-1 := \{|0\}$. Up to isomorphisms we can hence affirm that $\mathbb{N}, \mathbb{Z} \subseteq \mathbf{No}$. Another meaningful example based on this interpretation is the number $\{0 \mid 1\}$ which has to be thought as the simplest number between 0 and 1, i.e., $\frac{1}{2} := \{0 \mid 1\}$. From the Inequality rule we can prove that $\{|\, x\} \leq 0 \leq \{x \,|\}$ for every $x \in \mathbf{No}$, and that $1 \not\leq 0$, hence $0 < 1$ follows from the equality rule. Analogously, one can prove that $\ldots < -3 < -2 < -1 < 0 < 1 < 2 < 3 < \ldots$ Moreover, we can also easily see that e.g., $\{-1 \mid 1\} = 0$ so we note that different subsets $L, R \subseteq \mathbf{No}$ can define the same number $x = \{L \,|, R\}$.

But now we also have that $\mathbb{N} \subseteq \mathbf{No}$ and hence we can form the number $\mathbb{N} + 1 := \{\mathbb{N} \,|\}$ and this catch a glimpse of the possibility to extend all this using transfinite induction.

Instead of further proceeding with Conway's approach to **No** we want to sketch his point of view to the foundational questions arising from his construction. These ideas are precisely stated in the appendix to part zero of [12]. The mainstream's approach to a topic like **No**, where one must use e.g., transfinite induction, is to

[4] Of course, at this stage of development and using this not-strictly formal point of view, our use of the notion of "simplicity" is only informal and it is natural to ask for a more formal definition, considering, moreover, its uniqueness. This will be done in the next section.

fix a formal theory of sets (like Zermelo–Fraenkel **ZF** theory or von Neumann–Bernay–Gödel **NBG** theory) and to formalize every construction inside that theory. Sometimes, this formalization can conduct to a theory far from the original intuition, because different formalizations are possible of a given informal construction. Let us note explicitly that here J.H. Conway's term of judgment is the searching for a good dialectic between informal constructions and their formal counterpart, which is exactly the criterion to define a good mathematical theory we talked about in the introduction. Conway's foundational point of view can be summarized citing, as in [12]:

> It seems to us, however, that mathematics has now reached the stage where formalization within some particular axiomatic theory is irrelevant, even for foundational studies. It should be possible to specify conditions on a mathematical theory which would suffice for embeddability within **ZF** (supplemented by additional axioms of infinity if necessary), but which do not otherwise restrict the possible constructions in that theory.

5.1 *The Cuesta Dutari Definition of* No

From our point of view, one has the best result if there exists a formalization respecting the above mentioned good formal-informal dialectic, and hence a great effort has to be dedicated to the searching for this best formalization, if any, before assuming a point of view so general and radical like J.H. Conway's one. This type of formalization for surreal numbers is possible and is given by [14]. In this section we only sketch the first steps of this development; for a complete treatment, see e.g., [2, 19, 22] and references therein. The set theory we will consider is **NBG**.

Definition 19 Let $(T, \leq)$ be a totally ordered set, then we say that (L, R) *is a Cuesta Dutari cut in* T iff

1. $L, M \subseteq T$
2. $L < R$
3. $L \cup R = T$.

Moreover, we will denote by $\text{CD}(T)$ the class of all Cuesta Dutari cuts in T. Essentially a Cuesta Dutari cuts is a Conway cut with the additional condition $L \cup R = T$. Since $(\emptyset, T)$ and $(T, \emptyset)$ are always Cuesta Dutari cuts, we have that $\text{CD}(T)$ is never empty.

If we think than each Cuesta Dutari cut identifies a new number, the union $T \cup \text{CD}(T)$ can be thought of as a *completion* of the totally ordered set $(T, \leq)$ if we can extend the order relation $\leq$ to this union. This is done in the following:

Definition 20 Let $(T, \leq)$ be a totally ordered set, then on the *Cuesta Dutari completion* $\chi(T) := T \cup CD(T)$ *of* T we define the order relation:

1. If $x, y \in T$ then we will say that x is less than or equal to y iff $x \leq y$ in T. Because of this first case, the order relation on $\chi(T)$ will be denoted again by the symbol $\leq$.

2. If $x \in T$ and $y = (L, R) \in CD(T)$, then:
   a) $x \in L \implies x < y$
   b) $x \in R \implies y < x$
3. If $x = (L_x, R_x)$, $y = (L_y, R_y) \in \chi(T)$, then $x < y$ iff $L_x \subset L_y$.

It is indeed possible to prove (see the above cited references) that $(\chi(T), \leq)$ is a totally ordered set. For example if $x = (L, R) \in \mathrm{CD}(T)$, then $L < \{x\} < R$ and, as a further example, $(\emptyset, T)$ is the least element of $\chi(T)$, whereas $(T, \emptyset)$ is the greatest element.

We do not have to think at the Cuesta Dutari completion as a final completion starting from a single given ordered set $(T, \leq)$ but, instead, as a tool for a transfinite-recursive construction:

Definition 21 Let **On** be the class of all ordinals, we define by transfinite recursion the family $(T_\alpha)_{\alpha \in \mathbf{On}}$ of ordered sets given by:

1. T_0 is the empty set ordered with the empty relation,
2. For every $\beta \in \mathbf{On}$:
   a) If $\alpha + 1 = \beta$, then $T_\beta := \chi(T_\alpha)$
   b) If β is a nonzero limit ordinal, then $T_\beta := \bigcup_{\alpha < \beta} T_\alpha$.

Finally we set $\mathbf{No} := \bigcup_{\alpha \in \mathbf{On}} T_\alpha$.

The ordinal index α in the previous transfinite recursive definition gives the notion of simplicity of a number $x \in \mathbf{No}$, that is its *birthday* using Conway's terminology.

Definition 22 If $x \in \mathbf{No}$, we define the *birth-order* function by

$$b(x) := \min_{\leq} \{\alpha \in \mathbf{No} \mid x \in T_\alpha\},$$

where $\leq$ is the order relation defined on **No**.

So we have, e.g., $b(0) = 0$, $b(1) = b(-1) = 1$, $b(2) = b\left(\frac{1}{2}\right) = b\left(-\frac{1}{2}\right) = 2$.

At this point, the most important result is Conway's simplicity theorem: it states that every Conway cut (L, R) determines uniquely the simplest surreal number filling the gap between the subsets L and R:

Theorem 13 *Let $L, R \subseteq \mathbf{No}$ with $L < R$, then there exist one and only one $x \in \mathbf{No}$ such that:*

1. *$L < \{x\} < R$*
2. *If $y \in \mathbf{No} \setminus \{x\}$ verifies $L < \{y\} < R$, then $b(x) < b(y)$.*

This unique x will be denoted by $\{L \mid R\}$.

So, e.g., $0 := \{\emptyset | \emptyset\} \in T_1 = \chi(T_0) \subset \mathbf{No}$.

5.1.1 Ring Operations on No

On the class of surreal numbers we can define all the field operations, the simplest one being subtraction:

$$x = \{L \mid R\} \Longrightarrow -x := \{-R \mid -L\}.$$

The definition of sum and product requires some motivations. Let $x = \{L_x \mid R_x\}$ and $y = \{L_y \mid R_y\}$ be surreal numbers, so that

$$l_x < x < r_x \tag{29}$$

$$l_y < y < r_y \tag{30}$$

for every $l_x \in L_x, r_x \in R_x, l_y \in L_y$ and $r_y \in R_y$. Since we want **No** to be an ordered group with respect to addition, we must have:

$$\begin{aligned} l_x + y &< x + y \\ x + l_y &< x + y \\ x + y &< r_x + y \\ x + y &< x + r_y. \end{aligned}$$

Hence, Conway defines $x + y$ as the simplest number verifying these inequalities, i.e., using transfinite recursion we can define

$$x + y := \left\{(L_x + y) \cup (x + L_y) \,|\, (R_x + y) \cup (x + R_y)\right\},$$

where, e.g., $L_x + y := \{l_x + y \,:\, l_x \in L_x\}$. Analogously, we can proceed to justify the definition of product. From (29) and (30), in the hypothesis that **No** be an ordered group under multiplication, we must have that $x - l_x$, $y - l_y$, $r_x - x$ and $r_y - y$ are all greater than zero. Taking all the products of these terms involving an x and a y we have

$$\begin{aligned} 0 &< (x - l_x) \cdot (y - l_y) = xy - l_x y - x l_y + l_x l_y \\ 0 &< (r_x - x) \cdot (r_y - y) = xy - r_x y - x r_y + r_x r_y \\ 0 &< (x - l_x) \cdot (r_y - y) = -xy + l_x y + x r_y - l_x r_y \\ 0 &< (r_x - x) \cdot (y - l_y) = -xy + r_x y + x l_y + r_x l_y. \end{aligned}$$

As a consequence, from these we get inequalities bounding xy:

$$\begin{aligned} l_x y + x l_y - l_x l_y &< xy < l_x y + x r_y - l_x r_y \\ r_x y + x r_y - r_x r_y &< xy < r_x y + x l_y + r_x l_y. \end{aligned}$$

We can hence define (once again by transfinite recursion):

$$\begin{aligned} L_{x \cdot y} &:= (L_x y + x L_y - L_x L_y) \cup (R_x y + x R_y - R_x R_y) \\ R_{x \cdot y} &:= (L_x y + x R_y - L_x R_y) \cup (R_x y + x L_y + R_x L_y) \\ x \cdot y &:= \left\{L_{x \cdot y} \mid R_{x \cdot y}\right\}, \end{aligned}$$

where e.g., $L_x y := \{l_x \cdot y \,:\, l_x \in L_x\}$ and $L_x L_y := \{l_x \cdot l_y \,|\, l_x \in L_x\,,\, l_y \in L_y\}$. Using these definitions we can prove that **No** verifies the axioms of an ordered field.

5.2 Examples of Surreal Numbers and the Universal Property

As we already sketched, up to isomorphism we have $n = \{0, 1, 2, \dots, n-1 \,|\} \in \mathbb{N} \subseteq$ **No**, $-n = \{|\, -n+1, -n+2, \dots, -2, -1, 0\} \in \mathbb{Z} \subseteq$ **No**, but also $\omega := \{\mathbb{N} \,|\}$. It results $n < \omega$ for every $n \in \mathbb{N}$ and hence **No** is a non-Archimedean field. Moreover, because we have an ordered field containing the integers, we also have $\mathbb{Q} \subseteq$ **No**, i.e., all the rationals can be seen as surreal numbers. Finally, using Dedekind cuts we can also identify $\mathbb{R}$ with a subfield of **No**.

But using transfinite induction we can also define

$$\varphi(0) := 0$$

$$\varphi(\beta) := \{\{\varphi(\alpha) : \alpha < \beta\} \,|\} \in \mathbf{No} \quad \forall \beta \in \mathbf{On},$$

and we can prove that φ is an order-preserving map from the class of all ordinals **On** into **No**, with birthday function verifying $b(\varphi(\beta)) = \beta$. This also proves that **No** is a proper class and not a set because φ, being order preserving, is injective. For this reason usually one says that **No** is a Field, with the capital initial letter to underline that it verifies the axioms of a field, but its support set is a proper class.

We have already seen that in **No** we have infinities like $\omega = \{\mathbb{N} \,|\}$, but we can also easily construct infinitesimals like

$$\varepsilon := \left\{ 0 \,\middle|\, \left\{ \frac{1}{n} : n \in \mathbb{N}_{>0} \right\} \right\},$$

which is strictly positive but smaller than any strictly positive real.

All these examples can conduct us toward the conjecture that the class **No** is some kind of "universal" field containing every possible extension of the real field. Indeed we have the following theorem.

Theorem 14 *The field* ***No*** *verifies the following properties*

1. ***No*** *is an ordered Field*
2. *If A is an ordered subfield of* **No**, *and A is also an ordered subfield of B, with the support sets $|A|$ and $|B|$ which are sets and not proper classes, then there exist an ordered subfield B' of* **No** *and an isomorphism $f : B \longrightarrow B'$ of ordered subfields such that $f|_A = 1_A$.*

Moreover, if F verifies these properties 1 and 2 (like ***No*** *does). Then $F \simeq$* ***No*** *as an ordered field.*

From this point of view the field of surreal numbers is remarkably inclusive.[5] For example applying the previous theorem with $A = \mathbb{R}$ and $B = {}^*\mathbb{R}$, the hyperreal numbers of nonstandard analysis, we obtain that **No** contains, up to isomorphism, the hyperreals.

[5] From this point of view the name "surreal numbers" is less meaningful than the original Conway's "numbers" without any adjective.

The field **No** has many remarkable properties, it is a real closed field, there is the possibility to define exponential and logarithm and even a notion of Riemann integral (see [24]). For applications to game theory, see [12, 38]. On the other hand, like any other non Archimedean ordered field, **No** is totally disconnected, therefore we have examples of functions differentiable on an interval with everywhere zero derivative which are not constant, we do not have the uniqueness of the primitive of a continuous function and we do not have uniqueness in the simplest initial value problem: $y'(x) = 0$, $y(0) = 0$. From a methodological point of view, as we have already sketched above, Conway's construction seems to be based on the search of a theory with strong intuitive meaning, essentially due to Conway's simplicity Theorem 13. Formalization like [2, 14] permit to obtain a good dialectic between formal theory and intuitive interpretation.

References

1. Albeverio, S., Fenstad, J.E., Høegh-Krohn, R., Lindstrøm, T.: Nonstandard Methods in Stochastic Analysis and Mathematical Physics. Pure and applied mathematics, 2nd ed. Academic Press, Orlando (1988). Reprint: Dover (2009)
2. Alling, N.L.: Foundation of Analysis Over Surreal Number Fields, vol. 141 of North-Holland Mathematics Studies. North-Holland Publishing Co., Holland (1987)
3. Bell, J.L.: A Primer of Infinitesimal Analysis. Cambridge University Press, Cambridge (1998)
4. Benci, V., Di Nasso, M.: A ring homomorphism is enough to get nonstandard analysis. Bull. Belg. Math. Soc.–S. Stevin. **10**, 481–490 (2003)
5. Benci, V., Di Nasso, M.: A purely algebraic characterization of the hyperreal numbers. Proc. Am. Math. Soc. **133**(9), 2501–2505 (2005)
6. Berstel, J., Reutenauer, C.: Noncommutative Rational Series with Applications, Encyclopedia of Mathematics and its Applications, vol. 137. Cambridge University Press, Cambridge (2011)
7. Bertram, W.: Differential Geometry, Lie Groups and Symmetric Spaces Over General Base Fields and Rings. American Mathematical Society, Providence (2008)
8. Berz, M.: Automatic differentiation as nonarchimedean analysis, Computer arithmetic and enclosure methods. p. 439. Elsevier Science Publisher, Amsterdam (1992)
9. Berz, M.: Analysis on a nonArchimedean extension of the real numbers. Mathematics Summer Graduate School of the German National Merit Foundation, MSUCL-933, Department of Physics, Michigan State University, 1992 and 1995 edition, (1994)
10. Berz, M., Hoffstatter, G., Wan, W., Shamseddine, K., Makino, K.: COSY INFINITY and its applications to nonlinear dynamics. In: Berz, M., Bischof, C., Corliss, G., Griewank, A. (eds.) Computational Differentiation: Techniques, Applications, and Tools, pp. 363–367. SIAM, Philadelphia (1996)
11. Colombeau, J.F.: Multiplication of Distributions. Springer, Berlin (1992)
12. Conway, J.H.: On Numbers and Games, Number 6 in L.M.S. Monographs. Academic Press, London & New York (1976)
13. Conway, J.: Infinitesimals vs. indivisibles replies: 20. The Math Forum Drexel, Feb. 17 1999. URL http://mathforum.org/kb/message.jspa? messageID=1381465&tstart=0. (1999)
14. Cuesta Dutari, N.: Algebra ordinal, Revista de la Real Academia de Ciencias Exactas. Fisicas y Naturales. **XLVIII**(2), 79–160 (1954)
15. Delcroix, A.: Topology and functoriality in $(\mathcal{C}, \mathcal{E}, \mathcal{P})$-algebras: application to singular differential problems. J. Math. Anal. Appl. **359**, 394–403 (2009)
16. Delcroix, A., Scarpalezos, D.: Asymptotic scales-asymptotic algebras. Integr. Trans. Spec. Funct. **6**, 1–4, 181–190 (1998)

17. Dirac, P.A.M.: The physical interpretation of the quantum dynamics. Proc. R. Soc. Lond. A(113), 621–641 (1926–1927)
18. Dirac, P.A.M.: General Theory of Relativity. Wiley, Hoboken (1975)
19. Ehrlich, P.: An alternative construction of Conway's ordered field No. Algebra Uni. **25**, 7–16 (1988)
20. Ehrlich, P.: The rise of non-Archimedean mathematics and the roots of a misconception I: The emergence of non-Archimedean systems of magnitudes. Arch. Hist. Exact Sci. **60**(1), 1–121 (2006)
21. Ehrlich, P.: Conway names , the simplicity hierarchy and the surreal number tree. J. Logic Anal. **3**(1), 1–26 (2011)
22. Ehrlich, P.: The absolute arithmetic continuum and the unification of all numbers great and small. Bull. Symbol. Logic. **18**(1), 1–45 (2012)
23. Einstein, A.: Investigations on the Theory of the Brownian Movement. Dover, New York (1926)
24. Fornasiero, A.: Integration on surreal numbers. PhD thesis, University of Edinburgh. www.dm.unipi.it/~fornasiero/phd_thesis/thesis_fornasiero_linearized.pdf (2004)
25. Garth Dales, H., Woodin, W. H.: Super-real fields , London Mathematical Society Monographs, New Series, vol. 14. The Clarendon Press Oxford University Press, New York, 1996, Totally ordered fields with additional structure; Oxford Science Publications (1996)
26. Giordano, P.: Fermat reals: Nilpotent infinitesimals and infinite dimensional spaces. arXiv:0907.1872, July (2009)
27. Giordano, P.: The ring of Fermat reals. Adv. Math. **225**(4), 2050–2075 (2010)
28. Giordano, P.: Infinitesimals without logic. Russian J. Math. Phys. **17**(2), 159–191 (2010)
29. Giordano, P.: Infinite dimensional spaces and cartesian closedness. J. Math. Phys. Anal. Geomet. **7**(3), 225–284 (2011)
30. Giordano, P.: Fermat–Reyes method in the ring of Fermat reals. Adv. Math. **228**, 862–893 (2011)
31. Giordano, P., Katz, M.: Potential and actual infinitesimals in models of continuum. www.mat.univie.ac.at/~giordap7/#preprints (2012)
32. Giordano, P., Kunzinger, M.: Topological and algebraic structures on the ring of Fermat reals. Israel J. Math. DOI: 10.1007/s11856-012-0079-z. (2011)
33. Giordano, P., Kunzinger, M.: Generalized functions as smooth set-theoretical maps, article in preparation, www.mat.univie.ac.at/~giordap7/#preprints
34. Giordano, P., Kunzinger, M., Vernaeve, H.: Strongly internal sets and generalized smooth functions. Submitted to Journal of Mathematical Analysis and Applications on February 2014. See www.mat.univie.ac.at/~giordap7/#preprints
35. Goldblatt, R.: Lectures on the Hyperreals: An Introduction to Nonstandard Analysis. Springer, New York, (1998)
36. Grosser, M., Kunzinger, M., Oberguggenberger, M., Steinbauer, R.: Geometric Theory of Generalized Functions. Kluwer, Dordrecht (2001)
37. Harzheim, E.: Beiträge zur Theorie der Ordnungstypen, insbesondere der a η_α-Mengen. Math. Annalen. **154**, 116–134 (1964)
38. Henle, M.: Which Numbers are Real? The Mathematical Association of America, Wasington, DC (2012)
39. Iglesias-Zemmour, P.: Diffeology, AMS , Mathematical Surveys and Monographs 185, to appear on April (2013)
40. Katz, M.G., Tall, D.: A Cauchy–Dirac delta function: foundations of science. http://dx.doi.org/10.1007/s10699-012-9289-4 and http://arxiv.org/abs/1206.0119 (2012)
41. Knuth, D.E.: Surreal Numbers : How Two Ex-students Turned On to Pure Mathematics and Found Total Happiness: A Mathematical Novelette. Addison-Wesley, Massacheuts (1974)
42. Kock, A.: Synthetic Differential Geometry. London Math, Society Lecture Note Series, vol. 51. Cambridge University Press, Cambridge (1981)
43. Kolár, I., Michor, P.W. Slovák, J.: Natural Operations in Differential Geometry. Springer, Berlin (1993)

44. Kriegl, A., Michor, P.W.: Product preserving functors of infinite dimensional manifolds. Archivum Mathematicum (Brno). **32**(4), 289–306 (1996)
45. Kriegl, A., Michor, P.W.: The Convenient Settings of Global Analysis. Mathematical Surveys and monographs, vol. 53. American Mathematical Society, Providence (1997)
46. Laugwitz, D.: Tullio Levi-Civita's work on nonarchimedean structures (with an appendix: Properties of Levi-Civita fields). In Atti dei Convegni Lincei 8: Convegno Internazionale Celebrativo del Centenario della Nascita di Tullio Levi-Civita, Roma, 1975, Accademia Nazionale dei Lincei (1975)
47. Laugwitz, D.: Definite values of infinite sums: aspects of the foundations of infinitesimal analysis around 1820. Arch. Hist. Exact Sci. **39**(3), 195–245 (1989)
48. Lavendhomme, R.: Basic Concepts of Synthetic Differential Geometry. Kluwer Academic Publishers, Dordrecht (1996)
49. Levi-Civita, T.: Sugli infiniti ed infinitesimi attuali quali elementi analitici. Atti del Regio Istituto Veneto di Scienze Lettere ed Arti. **VII**(4), 1765–1815 (1893)
50. Levi-Civita, T.: Sui numeri transfiniti. Rendiconti della Reale Accademia dei Lincei. **VI**(1 Sem.), 113–121 (1898)
51. Mamane, L.E.: Surreal numbers in Coq. In: Types for Proofs and Programs. Lecture Notes in Computer Science, vol. 3839, pp. 170–185. Springer, Berlin (2006)
52. Moerdijk, I., Reyes, G.E.: Models for Smooth Infinitesimal Analysis. Springer, Berlin (1991)
53. Palmgren, E.: Developments in constructive nonstandard analysis. Bull. Symbol. Logic. **4**(3), 233–272 (1998)
54. Palmgren, E.: A note on Brouwer's weak continuity principle and the transfer principle in nonstandard analysis. J. Logic Anal. **3**, pp. 1–7 (2012)
55. Robinson, A.: Non-standard Analysis. North-Holland Publishing Co., Amsterdam (1966)
56. Schmieden, C., Laugwitz, D.: Eine Erweiterung der Infinitesimalrechnung. Math. Zeitschr. **69**, 1–39 (1958)
57. Schwartz, L.: Théorie des distributions, vol. 1–2. Hermann, Paris (1950)
58. Schwartz, L.: Sur l'impossibilité de la multiplications des distributions. C.R. Acad. Sci. Paris. **239**, 847–848 (1954)
59. Sebastião e Silva, J.: Sur une construction axiomatique de la theorie des distributions, Rev. da Fac. de Ciências de Lisboa, 2ª série-A, vol. 4, pp. 79–186 (1954–1955)
60. Shamseddine, K.: One-variable and multi-variable calculus on a non-Archimedean field extension of the real numbers, p-Adic numbers, ultrametric analysis, and applications. In print
61. Shamseddine, K.: On the topological structure of the Levi-Civita field. J. Math. Anal. Appl. **368**, 281–292 (2010)
62. Shamseddine, K.: Absolute and relative extrema, the mean value theorem and the inverse function theorem for analytic functions on a Levi-Civita field. In: Jesus A.-G., Diarra, B., Escassut, A. (eds.) Contemporary Mathematics, Advances in Non-Archimedean Analysis, vol. 551, pp. 257–268. American Mathematical Society, Providence (2011)
63. Shamseddine, K.: Nontrivial order preserving automorphisms of non-Archimedean fields. In: Jarosz, K. (ed.) Contemporary Mathematics. Function Spaces in Modern Analysis, vol. 547, pp. 217–225. American Mathematical Society, Providence (2011)
64. Shamseddine, K.: New results on integration on the Levi-Civita field. Indagationes Math. **24**(1), 199–211 (2013)
65. Shamseddine, K., Berz, M.: Intermediate value theorem for analytic functions on a Levi-Civita field. Bull. Belg. Math. Soc. Simon Stevin. **14**, 1001–1015 (2007)
66. Shamseddine, K., Berz, M.: Analysis on the Levi-Civita field: a brief overview. In: Berz, M., Shamseddine, K. (eds.) Contemporary Mathematics. Advances in p-Adic and Non-Archimedean Analysis, vol. 508, pp. 215–237. American Mathematical Society, Providence (2010). ISBN 978-0-8218-4740-4
67. Shamseddine, K., Grafton, W.: Preliminary notes on Fourier series for functions on the Levi-Civita field. Int. J. Math. Anal. **6**(19), 941–950 (2012)

68. Shamseddine, K., Sierens, T.: On locally uniformly differentiable functions on a complete non-Archimedean ordered field extension of the real numbers. ISRN Math. Anal. **2012**: Article ID 387053 (2012), http://www.hindawi.com/journals/isrn.mathematical.analysis/2012/387053/abs/ they say only the additional information: 20 pages http://dx.doi.org/10.5402/2012/387053
69. Shamseddine, K., Rempel, T., Sierens, T.: The implicit function theorem in a non-Archimedean setting. Indagationes Math. **20**(4), 603–617 (2009),
70. Tall, D.: Looking at graphs through infinitesimal microscopes, windows and telescopes. Math. Gaz. **64**, 22–49 (1980)
71. Vernaeve, H.: Generalized analytic functions on generalized domains, arXiv:0811.1521v1, (2008)
72. Weil, A.: Théorie des points proches sur les variétés différentiables, Colloque de Géometrie Différentielle, pp. 111–117, C.N.R.S. (1953)

Approximation Properties of Phillips Operators

N. K. Govil and Vijay Gupta

Abstract This paper deals with presenting a survey-cum-expository account of some developments concerning the approximation properties of well known Phillips operators. These operators are sometimes called as genuine Szász Durrmeyer operators, because of their property of reproducing constants as well as linear functions. We give the alternate form to present these operators in terms of Hypergeometric functions, which are related to the modified Bessel's function of first kind of index 1. Also, we observe that the r-th moment can be represented in terms of confluent hypergeometric functions, and further it can be written in terms of generalized Laguerre polynomials. In addition, we will present some known results on such operators, which include simultaneous approximation, linear and iterative combinations, global direct and inverse results, rate of convergence for functions of bounded variation, and q-analogues of these operators.

Keywords Phillips operators · Laguerre polynomials · Rate of convergence · Simultaneous approximation · Linear combinations · Iterative combinations · Direct results · q-Phillips operators

1 Introduction

As is well known, the theory of approximation for linear positive operators started by the well known theorem due to Weierstrass, which states that "every continuous function defined on a closed and bounded interval can be uniformly approximated by polynomials". There are lot of proofs in literature of this well known theorem but the most common proof is based on Bernstein polynomials. Rate of convergence, saturation theorems and the inverse problems for Bernstein polynomials have been

N. K. Govil (✉)
Department of Mathematics and Statistics, Auburn University,
AL 36849-5108 Auburn, USA,
e-mail: govilnk@auburn.edu

V. Gupta
Department of Mathematics, Netaji Subhas Institute of Technology,
Sector 3 Dwarka New Delhi 110078, India
e-mail: vijaygupta2001@hotmail.com

T. M. Rassias, P. M. Pardalos (eds.), *Mathematics Without Boundaries*,
DOI 10.1007/978-1-4939-1106-6_8,

intensively studied by several researchers. After that many researchers have studied such results for different operators. The study of direct theorems in the theory of approximation was initiated by the classical work due to Jackson [23] on algebraic and trigonometric polynomials of best approximation. The corresponding inverse theorems were established by Bernstein (see [32]) by an application of his famous inequality. In the trigonometric case the results of Jackson–Bernstein type had an essential gap for the case $\alpha = 1$, which was filled later by Zygmund [38] through the introduction of class Z $(Lip * 1)$. Further generalizations of the Zygmund class have been given with developments of the theory of approximation.

In the theory of approximation, it is well known that the order of approximation for linear positive operator L_n is at best $O(n^{-1})$. Thus, to improve the order of approximation several combinations were appropriately introduced and defined in the literature. One such technique was linear combinations, while by using iterative combinations, one can also improve the order of approximation. Also, rate of convergence for functions of bounded variation has been an interesting area of research in the last few decades. Applications of q-calculus in the area of approximation was also started about 15 years ago when Phillips [34] first considered the q analogue of Bernstein polynomials and after that several researchers worked in this direction. There are several operators available in the literature till date.

In the present paper, we present a survey-cum-expository account of some developments concerning the approximation properties of well known Phillips operators. These operators are sometimes also called as genuine Szász Durrmeyer operators, because of their property of reproducing constants as well as linear functions. Here, we will give an alternate form to present these operators in terms of Hypergeometric functions, which are related to the modified Bessel's function of first kind of index 1. Also, we observe that the r-th moment of these operators can be represented in terms of confluent hypergeometric functions, and further it can be written in terms of generalized Laguerre polynomials. Besides, we will also present some known results on such operators dealing with simultaneous approximation, linear and iterative combinations, global direct and inverse results, rate of convergence for functions of bounded variation and q-analogues of these operators.

We begin with the definition of Phillips operators.

For $f \in C[0,\infty)$, the Phillips operators (see [33]) are defined as

$$P_n(f,x) = n\sum_{k=1}^{\infty} s_{n,k}(x)\int_0^{\infty} s_{n,k-1}(t)f(t)dt + e^{-nx}f(0) \tag{1}$$

where $s_{n,k}(x)$ is the Szász basis function is given by

$$s_{n,k}(x) = e^{-nx}\frac{(nx)^k}{k!}.$$

The Phillips operators are closely related to the Szász operators [37] operators defined by

$$S_n(f,x) \equiv (S_n f)(x) = \sum_{k=0}^{\infty} s_{n,k}(x) f\left(\frac{k}{n}\right).$$

Indeed, if we replace the discrete values $f(k/n)$ in [37] by the integral terms

$$n\int_0^\infty s_{n,k-1}(v)f(v)dv,\ k=1,2,\ldots.$$

then we arrive at (1). Alternatively the operators (1) can be written as

$$P_n(f,x)=\int_0^\infty W(n,x,t)f(t)dt, \tag{2}$$

where $W(n,x,t)$ is the kernel of the operators P_n given by

$$W(n,x,t)=e^{-n(x+t)}\sum_{k=1}^{\infty}\left(\frac{\left(n^2x\right)^k t^{k-1}}{k!(k-1)!}+\delta(t)\right),$$

$\delta(t)$ being the Dirac delta function.

In the year 1985 Mazhar and Totik [30] also defined the operators (1) while defining the integral modification of Szász–Mirakyan operators and they considered some approximation properties of these operators. In the recent years, Srivastava and Gupta [36] introduced a general class of operators and one of the special cases of their operators reduce to the Phillips operators P_n.

For $f\in C[0,\infty)$, Jain [22] introduced the modification of well known Szász–Mirakyan operators as

$$B_n^{\beta}(f,x)=\sum_{k=0}^{\infty}l_{n,k}^{(\beta)}(x)f(k/n), \tag{3}$$

where $0\le\beta<1$ and the basis function is defined as

$$l_{n,k}^{(\beta)}(x)=\frac{nx(nx+k\beta)^{k-1}}{k!}e^{-(nx+k\beta)}. \tag{4}$$

It was observed in [22] that $\sum_{k=0}^{\infty}l_{n,k}^{(\beta)}(x)=1$, and we can use this property to introduce a general form of Phillips operators as

$$P_n^{(\beta)}(f,x)=n\sum_{k=1}^{\infty}l_{n,k}^{(\beta)}(x)\int_0^\infty s_{n,k-1}(t)f(t)\,dt+l_{n,0}^{(\beta)}(x)f(0),$$

where $s_{n,k-1}(t)$ and $l_{n,k}^{(\beta)}(x)$ are as given in (1) and (5) respectively. It may be observed here that for special value $\beta=0$, the operators $P_n^{(\beta)}(f,x)$ reduce to the well-known Phillips operators defined by (1).

2 Alternate Forms and Moments

Let $(n)_k$ represent the Pochhammer symbol given by

$$(n)_k = n(n+1)(n+2)(n+3)\ldots.(n+k-1).$$

It can be easily seen that $k! = (1)_k$ and $(k-1)! = (2)_k$. The hypergeometric function is defined as

$$_2F_1(a,b;c;x) = \sum_{k=0}^{\infty} \frac{(a)_k(b)_k}{(c)_k k!} x^k.$$

Also

$$_0F_1(-;a;x) = \sum_{k=0}^{\infty} \frac{(a)_k}{k!} x^k.$$

The confluent hypergeometric function is a degenerate form of the hypergeometric function $_2F_1(a,b;c;x)$ which arises as a solution of the confluent hypergeometric differential equation which is defined as

$$_1F_1(a;b;x) = \sum_{k=0}^{\infty} \frac{(a)_k}{(b)_k k!} x^k.$$

Here we observe that the above operator (1) can be written in terms of hypergeometric series as

$$\begin{aligned} P_n(f,x) &= n\int_0^{\infty} f(t)\sum_{k=1}^{\infty} e^{-nx}\frac{(nx)^k}{k!}e^{-nt}\frac{(nt)^{k-1}}{(k-1)!}dt + e^{-nx}f(0) \\ &= n^2x\int_0^{\infty} e^{-n(x+t)}f(t)\sum_{k=1}^{\infty}\frac{\left(n^2xt\right)^{k-1}}{k!(k-1)!}dt + e^{-nx}f(0) \\ &= n^2x\int_0^{\infty} e^{-n(x+t)}f(t)\sum_{k=0}^{\infty}\frac{\left(n^2xt\right)^{k}}{(2)_k k!}dt + e^{-nx}f(0) \\ &= n^2x\int_0^{\infty} e^{-n(x+t)}f(t)\,{}_0F_1\left(-;2;n^2xt\right)dt + e^{-nx}f(0), \end{aligned}$$

where $_0F_1\left(-;2;n^2xt\right) = \Gamma(2)\left(n^2xt\right)^{-1/2} I_1\left(2n\sqrt{xt}\right)$ and $I_1\left(2n\sqrt{xt}\right)$ is the modified Bessel's function of first kind of index 1, which is given by

$$I_1\left(2n\sqrt{xt}\right) = \frac{1}{i}J_1\left(i2n\sqrt{xt}\right) = \frac{1}{i}\sum_{j=0}^{\infty}\frac{1}{j!\Gamma(j+2)}\left(in\sqrt{xt}\right)^{2j+1}.$$

The moments estimation of the Phillips operators are available in literature in several forms.

Here we present a different form to estimate the moments in terms of confluent hypergeometric functions, which is given in the following lemma:

Lemma 1 *For $n > 0$ and $r \geq 1$, we have*

$$P_n(t^r, x) = \frac{x\Gamma(r+1)}{n^{r-1}}\, {}_1F_1\left(1-r; 2; -nx\right). \tag{5}$$

Alternatively, we have

$$P_n(t^r, x) = \frac{x\Gamma(r)}{n^{r-1}}\, L^1_{r-1}(-nx), \tag{6}$$

where ${}_1F_1\left(a; b; x\right)$ is Confluent Hypergeometric Functions and $L^1_{r-1}(-nx)$ generalized Laguerre polynomials.

Proof By definition of operator (1), we have

$$\begin{aligned}
P_n(t^r, x) &= n \sum_{k=1}^{\infty} e^{-nx} \frac{(nx)^k}{k!} \int_0^{\infty} t^r e^{-nt} \frac{(nt)^{k-1}}{(k-1)!} dt \\
&= n \sum_{k=1}^{\infty} e^{-nx} \frac{(nx)^k}{k!(k-1)!} \int_0^{\infty} e^{-nt} n^{k-1} t^{k+r-1} dt \\
&= ne^{-nx} \sum_{k=1}^{\infty} \frac{(nx)^k}{k!(k-1)!} \frac{\Gamma(k+r)}{n^{r+1}} \\
&= ne^{-nx} \sum_{k=0}^{\infty} \frac{(nx)^{k+1}}{(k+1)!k!} \frac{\Gamma(k+r+1)}{n^{r+1}}
\end{aligned}$$

Using, $\Gamma(r+k+1) = \Gamma(r+1)(r+1)_k$, we have

$$\begin{aligned}
P_n(t^r, x) &= n^2 x e^{-nx} \sum_{k=0}^{\infty} \frac{(nx)^k}{(k+1)!k!} \frac{(r+1)_k \Gamma(r+1)}{n^{r+1}} \\
&= \frac{x\Gamma(r+1)e^{-nx}}{n^{r-1}} \sum_{k=0}^{\infty} \frac{(r+1)_k}{(2)_k k!} (nx)^k \\
&= \frac{x\Gamma(r+1)e^{-nx}}{n^{r-1}}\, {}_1F_1\left(r+1; 2; nx\right).
\end{aligned}$$

Now, if we use Kummer's transformations ${}_1F_1(a; b; x) = e^x\, {}_1F_1(b-a; b; -x)$, we get

$$P_n(t^r, x) = \frac{x\Gamma(r+1)}{n^{r-1}}\, {}_1F_1\left(1-r; 2; -nx\right),$$

which completes the proof of (5).

It is obvious that the confluent hypergeometric functions can be related with generalized Laguerre polynomials $L_n^m(x)$ with the relation

$$L_n^m(x) = \frac{(m+n)!}{m!n!}\, {}_1F_1\left(-n; m+1; x\right).$$

Therefore (6) follows immediately using the above relation.

Remark 1 By Lemma 1, we have the first three moments of the Phillips operators as

$$P_n(1,x) = 1,\, P_n(t,x) = x,\, P_n(t^2,x) = x^2 + \frac{2x}{n}.$$

The higher order moments can be obtained easily by Lemma 1.

It was observed in [29] that the kernel of Phillips operators satisfy varying property, that is, for $\mathrm{P} = x\left(1+\frac{D}{n}\right)^2 = x\left(1+\frac{2}{n}D+\frac{1}{n^2}D^2\right),\ D = \frac{\partial}{\partial x}$

$$\mathrm{P}(W(n,x,t)) = tW(n,x,t). \tag{7}$$

We observe here that by using (7), we have

$$P_n(f(t)t,x) = \mathrm{P}P_n(f(t),x) \equiv x\left(1+\frac{2}{n}D+\frac{1}{n^2}D^2\right)P_n(f(t),x). \tag{8}$$

Remark 2 By definition of the Phillips operators, we have $P_n(1,x) = 1$, and thus if we put $f(t) \equiv 1$ in (8), we get

$$P_n(t,x) = x\left(1+\frac{2}{n}D+\frac{1}{n^2}D^2\right)P_n(f(t),x) = x\left(1+\frac{2}{n}D+\frac{1}{n^2}D^2\right)1 = x$$

Similarly, if we put $f(t) = t$ in (8), we get

$$P_n(t^2,x) = x\left(1+\frac{2}{n}D+\frac{1}{n^2}D^2\right)x = x\left(1+\frac{2}{n}D+\frac{1}{n^2}D^2\right)x = x^2+\frac{2x}{n}.$$

The higher order moments can be obtained easily by using (8).

Also in analysis the recurrence formula for central moments are important. There are several recurrence relations for these moments. May [29] established:

Lemma 2 *Let $A_m(n,x)$ be defined by*

$$\Lambda_m(n,x) = n^m \int_0^\infty W(n,x,t)((t-x)^m dt.$$

Then $A_m(n,x)$ is a polynomial in nx; the degree of $A_m(n,x)$ in nx is $[m/2]$, while the degree of $P_n(t^m,x)$ in x is exactly m.

Recently Gupta [16] established the following recurrence formula for moments

Lemma 3 *For $m \in N \bigcup\{0\}$, if the m-th order moment is defined by*

$$\mu_{n,m}(x) = n \sum_{k=0}^{\infty} s_{n,k}(x) \int_0^{\infty} s_{n,k-1}(t)(t-x)^m dt + e^{-nx}(-x)^m,$$

then

$$n\mu_{n,m+1}(x) = x[\mu_{n,m}^{(1)}(x) + 2\, m\mu_{n,m-1}(x)] + m\mu_{n,m}(x).$$

Govil, Gupta and Noor [13] also estimated recurrence formula for moment in some other form to study simultaneous approximation.

3 Simultaneous Approximation

By simultaneous approximation, we mean the approximation of derivative of functions by the corresponding order moments of the operators. In this direction Agrawal and Gupta [3] estimated the rate of convergence in simultaneous approximation and they obtained the pointwise convergence, asymptotic formula and an estimate of error in terms of higher order modulus of continuity for the Phillips operators (see also [13] for other properties in simultaneous approximation).
Let us consider $C_N[0,\infty) =: \{f \in C[0,\infty) : |f(t)| \leq Ce^{Nt}, C > 0\}$. The norm $||.||_{C_N}$ for functions belonging to the class $C_N[0,\infty)$ is defined as

$$||f||_{C_N} = \sup_{x\in[0,\infty)} |f(t)|e^{-Nt}.$$

Theorem 1 [3] *(Point-wise convergence) If $p = 1, 2, 3, \ldots,$ $f \in C_N[0,\infty)$ for some $N > 0$ and $f^{(p)}$ exist at a point $x \in [0,\infty)$, then*

$$P_\lambda^{(p)}(f(t), x) = f^{(p)}(x) + o(1), as\ \lambda \to \infty. \tag{9}$$

Further, if $f^{(p)}$ exists and is continuous on $(a-\eta, b+\eta) \subset (0,\infty), \eta > 0$, then (9) *holds uniformly in $x \in [a,b]$.*

Theorem 2 [3] *(Asymptotic formula) Let $f \in C_N[0,\infty)$ for some $n > 0$. If $f^{(p+2)}$ exists at a point $x \in (0,\infty)$, then*

$$\lim_{\lambda\to\infty} \lambda \left[P_\lambda^{(p)}(f(t), x) - f^{(p)}(x)\right] = pf^{(p+1)}(x) + xf^{(p+2)}(x). \tag{10}$$

Further, if $f^{(p+2)}$ exists and is continuous on $(a-\eta, b+\eta) \subset (0,\infty), \eta > 0$, then (10) *holds uniformly in $x \in [a,b]$.*

Theorem 3 [3] *(Error estimation) Let $f \in C_N[0,\infty)$ for some $n > 0$ and $p \leq q \leq p+2$. If $f^{(q)}$ exists and is continuous on $(a-\eta, b+\eta) \subset (0,\infty), \eta > 0$, then for λ sufficiently large*

$$||P_\lambda^{(p)}(f(t), .) - f^{(p)}|| \leq \max\{C\lambda^{(-(q-p)/2}\omega(f^{(q)}, \lambda^{-1/2}), C'\lambda^{-1}\}.$$

where $C = C(p), C' = C'(p,f)$, $\omega\left(f^{(q)}, \delta\right)$ is the modulus of continuity of $f^{(q)}$ on $(a-\eta, b+\eta)$, and $||.||$ denotes the sup-norm on $[a,b]$.

4 Linear Combinations

It is observed that the order of approximation for these operators is $O(n^{-1})$. More than 3 decades ago, May [29] applied the combinations due to Butzer [5] to study approximation properties of these operators and he was able to achieve the better approximation. The k-th Butzer combinations $P_n(f,k,x)$ of the operators $P_{2^j n}(f,x)$ are defined as

$$P_n(f,k,x) = \sum_{j=0}^{k} C(j,k) P_{2^j n}(f,x),$$

where

$$C(j,k) = \prod_{\substack{i=0 \\ i \neq j}}^{k} \frac{2^j}{2^j - 2^i}.$$

In the literature most saturation and inverse results are for positive linear operators. Because of the Korovkin's theorem, the optimal rate of convergence for any sequence of linear positive operator L_n is at most $O(n^{-2})$. In order to achieve faster convergence one has to slacken positivity condition of the operators. Ditzian and May [8] established a saturation theorem for the Bernstein polynomial of the linear combinations due to Butzer [5]. Also May [29] established some direct, inverse and saturations theorems for the combinations due to Butzer [5], using Steklov's mean. We mention here the main results obtained by May in [29].

Theorem 4 [29] *(Error estimation) Let $f \in C_N[0,\infty)$ and $0 < a < a_1 < b_1 < b < \infty$. Then there exists a constant M such that*

$$||P_n(f,k,.) - f||_{C[a_1,b_1]} \leq M\{\omega_{2k+2}\left(f, n^{-1/2}, a, b\right) + n^{-k+1}||f||_{C_N}\},$$

where the k-th order modulus of continuity on $[a,b]$ is defined by

$$\omega_k(f,\delta,a,b) = \sup_{|h|<\delta}\{|\Delta_h^k f(x)| : x, x+kh \in [a,b]\}.$$

Although, May [29] considered the linear combinations due to Butzer, he established the inverse result for $P_n(f,0,x) = P_n(f,x)$, with the statement that the inverse theorem for the combinations can be proved analogously.

Let $0 < \alpha < 2$ and let $Lip^*(\alpha, C[a,b])$ be the Zygmund class of functions:

$$Lip^*(\alpha, C[a,b]) = \{f \in C[0,1] : \omega_2(f,h,a,b) \leq Mh^{\alpha}\},$$

where $\omega_2(f,\delta,a,b)$ is the second order modulus of smoothness of f.

Theorem 5 [29] *(Inverse theorem) Let $0 < a_i < a_{i+1} < b_{i+1} < b_i < \infty$, $i = 1,2;$ $0 < \alpha < 2$, and suppose $f \in C_N[0,\infty)$. Then in the following statements, the implications $(i) \Longrightarrow (ii) \Longleftrightarrow (iii) \Longrightarrow (iv)$ hold.*

(i) $||P_n(f,.) - f||_{C[a_1,b_1]} = O\left(n^{-\alpha/2}\right)$;
(ii) $f \in Lip^*(\alpha, C[a_2,b_2])$;
(iii) $f \in Lip(\alpha, C[a_2,b_2])$, *if* $0 < \alpha < 1$, $f \in Lip^*(1, C[a_2,b_2])$, *if* $\alpha = 1$, $f' \in C^1[a_2,b_2])$ *and* $f' \in Lip(\alpha - 1, C[a_2,b_2])$, *if* $1 < \alpha < 2$;
(iv) $||P_n(f,.) - f||_{C[a_2,b_2]} = O(n^{-\alpha/2})$.

Theorem 6 [29] *(Saturation theorem) Let* $f \in C_N[0,\infty)$ *and let* $0 < a < a_1 < b_1 < b < \infty$. *Denote* $I(f,n,k,a,b) = n^{k+1}||P_n(f,k,.) - f||_{C[a,b]}$. *Then in the following statements, the implications* $(i) \Rightarrow (ii) \Rightarrow (iii)$ *and* $(iv) \Rightarrow (v) \Rightarrow (vi)$ *hold true:*

(i) $I(f,n,k,a,b) = O(1)$;
(ii) $f^{(2k+1)} \in A.C.(a,b)$ *and* $f^{(2k+2)} \in L_\infty[a,b]$;
(iii) $I(f,n,k,a,b) = O(1)$;
(iv) $I(f,n,k,a,b) = o(1)$;
(v) $f \in C^{2k+2}(a,b)$ *and* $\sum_{j=k+1}^{2k+2} Q(j,k,x) f^{(j)}(x) = 0, x \in (a,b)$ *where* $Q(j,k,x)$ *are polynomials in* x.
(vi) $I(f,n,k,a_1,b_1) = o(1)$. *Here, all* $O(1)$ *and* $o(1)$ *terms are with respect to* n *when* $n \to \infty$.

Rathore [35] in the year 1973, and May [28] in 1976, introduced independently the general linear combinations to improve the order of approximation. May [28] estimated direct, inverse and saturation results for exponential type operators which include Bernstein polynomials, Baskakov, Szász, Post–Widder and Gauss Weierstrass operators. He considered a more general combinations than that of Butzer [5].

The linear combinations $P_n(f,k,x)$ of the operators $P_{d_jn}(f,x)$, considered by May [28] and Rathore [35] are given by

$$P_n(f,k,x) = \sum_{j=0}^{k} C(j,k) P_{d_jn}(f,x),$$

where

$$C(j,k) = \prod_{i=0, i\neq j}^{k} \frac{d_j}{d_j - d_i}, \ k \neq 0; \ C(0,0) = 1,$$

and $d_0, d_1, \ldots.., d_k$ are arbitrary, but fixed distinct positive integers.

In an alternate form the linear combinations $P_n(f,k,x)$ can be defined as:

$$P_n(f,k,x) = \frac{1}{\Delta} \begin{vmatrix} P_{d_0n}(f,x) & d_0^{-1} & d_0^{-2} & \ldots\ldots. & d_0^{-k} \\ P_{d_1n}(f,x) & d_1^{-1} & d_1^{-2} & \ldots\ldots & d_1^{-k} \\ .. & .. & .. & .. & .. \\ .. & .. & .. & .. & .. \\ P_{d_kn}(f,x) & d_k^{-1} & d_k^{-2} & \ldots\ldots. & d_k^{-k} \end{vmatrix}$$

where Δ is the Vandermonde determinant obtained by replacing the operator column of the above determinant by the entries 1.

Gupta and Agrawal [17] used the combinations due to Rathore in [35] to prove an asymptotic formula and error estimation in ordinary approximation for Phillips operators.

Theorem 7 *(Asymptotic formula)* [17] *Let $f \in C_N[0,\infty)$. If $f^{(2k+2)}$ exists at a point $x \in (0,\infty)$, then*

$$\lim_{n\to\infty} n^{k+1}[P_n(f,k,x) - f(x)] = \sum_{i=k+1}^{2k+2} \frac{f^{(i)}(x)}{i!} Q(i,k,x) + o(1)$$

and

$$\lim_{n\to\infty} n^{k+1}[P_n(f,k+1,x) - f(x)] = 0,$$

where $Q(i,k,x)$ are given as

$$Q(2k+2,k,x) = (-1)^k \frac{(2k+2)!}{(k+1)!} \frac{x^{k+1}}{\prod_{i=0}^{k} d_i},$$

$$Q(2k+1,k,x) = (-1)^k \frac{(2k+1)!}{(k-1)!} \frac{x^{k}}{\prod_{i=0}^{k} d_i}.$$

Moreover, if $f^{(2k+2)} \in C(0,\infty)$ then the above limits in the theorem hold uniformly.

Theorem 8 [17] *(Error estimation) Let $p \in N$, $1 \le p \le 2k+2$, and $f \in C_N[0,\infty)$. If $f^{(p)}$ exists and is continuous on $[0,a]$, having the modulus of continuity $\omega(f^{(p)}, n^{-1/2})$ on $[0,a]$, then for n sufficiently large*

$$||P_n(f,k.) - f||_{C[0,b]} \le \max\left(C_1 n^{-q/2}\omega(f^{(p)}, n^{-1/2}), C_2 n^{-(k+1)}\right),$$

where $C_1 = C_1(k,p)$ and $C_2 = C_2(k,p,f)$.

For these general linear combinations Gupta and Sahai [19] also established direct and inverse theorems of the type Theorems 4 and 5.

5 Iterative Combinations

Micchelli [31] was first to consider the iterative combinations to improve the order of approximation of the well known Bernstein polynomials.

The iterative combinations for Phillips operators were considered by Agrawal and Gupta [2] and are defined as:

$$T_{\lambda,k}(f,x) = (I - (I - P_\lambda)^k)(f(t),x) = \sum_{p=1}^{k} (-1)^{p+1} \binom{k}{p} P_\lambda^p(f,x),$$

where $P_\lambda^p(f,x)$, $p \in N$ denotes the p-th iterate, and $P_\lambda^0(f,x) = I$.

Theorem 9 [2] *(Asymptotic formula) Let* $f \in C_N[0,\infty)$. *If* $f^{(2k)}$ *exists at a point* $t \in [0,\infty)$, *then*

$$\lim_{\lambda\to\infty} \lambda^k \left[T_{\lambda,k}(f,x) - f(x)\right] = \sum_{j=2}^{2k} \frac{f^{(j)}(t)}{j!} Q(j,k,x)$$

and

$$\lim_{\lambda\to\infty} \lambda^k \left[T_{\lambda,k+1}(f,x) - f(x)\right] = 0$$

where $Q(j,k,x)$ *are certain polynomials in* x. *Further, if* $f^{(2k-1)}$ *exists and is absolutely continuous over* $[0,b]$, *and* $f^{(2k)} \in L_\infty[0,b]$, *then for any* $[c,d] \subset (0,b)$, *there holds*

$$||T_{\lambda,k}(f,.) - f||_{C[c,d]} \leq M\lambda^{-k} \left\{||f||_{C_N} + ||f^{(2k)}||_{L_\infty[0,b]}\right\}.$$

Theorem 10 [2] *(Error estimation) Let* $f \in C_N[0,\infty)$. *Then for sufficiently large* λ, *we have*

$$||T_{\lambda,k}(f,.) - f||_{C(I_2)} \leq M\left(\omega_{2k}\left(f, \lambda^{-1/2,I_1}\right) + \lambda^{-k}||f||_{C_N}\right)$$

where M *is constant independent of* f *and* λ, $I_i = [a_i, b_i], i = 1,2$ *and* $0 < a_1 < a_2 < b_2 < b_1 < \infty$.

Later, Agrawal [1] studied the simultaneous approximation for the iterative combinations and established the asymptotic formula and an error estimation in terms of modulus of continuity.

Theorem 11 [1] *(Asymptotic formula) Let* $f \in C_N[0,\infty)$, *for some* $N > 0$. *If* $f^{(2k+p)}$ *exists at a point* $x \in (0,\infty$, *then*

$$\lim_{\lambda\to\infty} \lambda^k \left[T_{\lambda,k}^{(p)}(f,x) - f^{(p)}(x)\right] = \sum_{j=p}^{2k+p} Q(j,k,p,x) f^{(j)}(x),$$

where $Q(j,k,p,x)$ *are certain polynomials in* x. *Further, if* $f^{(2k+p)}$ *is continuous on* (a_1,b_1), *then the above limit holds uniformly on* $[a_2,b_2]$, *with* $0 < a_1 < a_2 < b_2 < b_1 < \infty$.

Theorem 12 [1] *(Error estimate) Let* $f \in C_N[0,\infty)$, *for some* $N > 0$, $p \leq q \leq 2k+p$, $f^{(q)}$ *exists and be continuous on* $(a-\eta, b=\eta) \subset (0,\infty), \eta > 0$. *Then for sufficiently large* λ, *we have*

$$||T_{\lambda,k}^{(p)}(f,.) - f^{(p)}||_{C[a,b]} \leq \max\left(C_1\lambda^{-(q-p)/2}\omega(f^{(q)},\lambda^{-1/2}), C_2\lambda^{-k}\right),$$

where $C_1 = C_1(k,p)$ *and* $C_2 = C_2(k,p,f)$.

In this continuation, Agrawal, Gupta and Goirala [4] extended the studies on such iterative combinations and established an error estimate on the degree of approximation in terms of higher order modulus of continuity, by proving

Theorem 13 [4] *(Error estimate) Let $f \in C_N[0,\infty)$ for some $N > 0$. If $f^{(r)}$ exists and is continuous on I_1, then for sufficiently large λ, we have*

$$||T_{\lambda,k}^{(r)}(f,.) - f^{(r)}||_{C(I_2)} \leq C\left\{\lambda^{-k}||f||_{C_N} + \omega_{2k}\left(f^{(r)}, \lambda^{-1/2}, I_1\right)\right\},$$

where C is independent of f and λ and $0 < a_1 < a_2 < b_2 < b_1 < \infty, I_i = [a_i, b_i], i = 1, 2$.

Gupta and Agrawal [18] also obtained estimate of error in L_p-approximation in terms of higher order integral modulus of smoothness using Steklov means for iterative combination as follows:

Suppose that

$$0 < a_1 < a_3 < a_2 < a_1 < b_2 < b_3 < b_1 < \infty, \quad I_i = [a_i, b_i], \ i = 1, 2, 3, \ldots.$$

Theorem 14 [18] *(Error estimation) If $f \in L_p[0,\infty), \ p \geq 1$, then for all λ sufficiently large, we have*

$$||T_{\lambda,k}(f,.) - f||_{L_p(I_2)} \leq M_k\left(\omega_{2k}\left(f, \lambda^{-1/2}, p, I_1\right) + \lambda^{-k}||f||_{L_p[0,\infty)}\right)$$

where M_k is a constant independent of f and λ.

6 Global Direct and Converse Results

Let $C_B[0,\infty)$ be the space of all real valued continuous bounded functions f on $[0,\infty)$ endowed with the norm $||f|| = \sup_{x\geq 0}|f(x)|$. For $f \in C_B[0,\infty)$ the Phillips operators are well defined. Taking into account the construction given in [11], Finta and Gupta [12] considered the following modification of Phillips type operator:

$$\left(P_n^t f\right)(x) \equiv P_n^t(f,x) = \frac{1}{G(x,t)} \int_0^\infty g(x,t,\theta) P_n(f,\theta) d\theta \qquad (11)$$

and $P_n^t(f,0) = f(0)$, where $g : [0,\infty) \times (t_0,\infty) \to [0,\infty)$ is a function such that $g(x,t,.) \in C[0,\infty) \cap L^1[0,\infty)$ for all $(x,t) \in [0,\infty) \times (t_0,\infty)$ and $t_0 \geq 0, \ G : (0,\infty) \times (t_0,\infty) \to (0,\infty)$,

$$G(x,t) = \int_0^\infty g(x,t,\theta) d\theta$$

with the assumptions

$$\frac{1}{G(x,t)} \int_0^\infty g(x,t,\theta)(\theta - x) d\theta = 0$$

and

$$\frac{1}{G(x,t)} \int_0^\infty g(x,t,\theta)(\theta - x)^2 d\theta \leq \beta(t) x \qquad (12)$$

for every $(x,t) \in [0,\infty) \times (t_0,\infty)$ and $\beta : (t_0,\infty) \to (0,\infty)$ a given function. The parameter t may depend only on the natural number n. For example, an operator of type (11) is the following:

$$S_n^t(f,x) = \left(1+\frac{n}{t}\right)^{-tx} \left\{ n \sum_{k=1}^{\infty} \left(1+\frac{t}{n}\right)^{-k} \frac{tx(tx+1)\ldots(tx+k-1)}{k!} \right.$$

$$\left. \cdot \int_0^\infty p_{n,k-1}(v) f(v) dv + f(0) \right\},$$

where $f \in C_B[0,\infty)$, $x \geq 0$ and $t > 0$ (see [11])

Here, we shall present the direct and inverse global theorems for the operators defined above in Eqs. (11) and (1), respectively, which were established in [12]. The direct results are formulated by the second order Ditzian–Totik modulus of smoothness given by

$$\omega_\phi^2(f,\delta) = \sup_{0<h\leq\delta} \sup_{x\pm h\varphi(x)\in[0,\infty)} |f(x+h\varphi(x)) - 2f(x) + f(x-h\varphi(x))|, \quad (13)$$

where $\varphi(x) = \sqrt{x}$, $x \geq 0$.

The corresponding K-functional is

$$K_{2,\varphi}\left(f,\delta^2\right) = \inf_{h\in W_\infty^2(\varphi)} \{||f-h|| + \delta^2||\varphi^2 h''||\},$$

where

$$W_\infty^2(\varphi) = \{h \in C_B[0,\infty) : h' \in AC_{loc}[0,\infty), \varphi^2 h'' \in C_B[0,\infty)\}.$$

It is known (see [9]) that $K_{2,\varphi}(f,\delta^2)$ and $W_\infty^2(\varphi)$ are equivalent, that is, there exists an absolute constant $C > 0$, such that

$$C^{-1}\omega_\phi^2(f,\delta) \leq K_{2,\varphi}\left(f,\delta^2\right) \leq C\omega_\phi^2(f,\delta).$$

Below C_0 and K denote absolute positive constants and $C > 0$ is an absolute constant, which can be different at each occurrence. The inverse results are strong converse theorems of type B, in the terminology of [9].

Theorem 15 [12] *(Direct result) Let* $(P_n)_{n\geq1}$ *and* $\left(P_n^t\right)_{n\geq1}$ *be as defined in* [18] *and* [3]. *Then there exists* $C > 0$, *such that*

$$||P_n f - f|| \leq C\omega_\phi^2\left(f, \frac{1}{\sqrt{n}}\right)$$

and

$$||P_n^t f - f|| \leq C\omega_\phi^2\left(f, (n^{-1} + \beta(t))^{1/2}\right)$$

for all $f \in C_B[0,\infty)$ *and* $n = 1, 2, \ldots$.

Theorem 16 [12] *(Converse result of type B) Let* $(P_n)_{n\geq 1}$ *be defined as in* (1). *Then there exist constants K and C such that*

$$\omega_\phi^2\left(f,\frac{1}{\sqrt{n}}\right)\leq C(||P_nf-f||+||P_{Kn}f-f||)$$

for all $f\in C_B[0,\infty)$ *and* $n=1,2,\ldots$. *Furthermore, if* $(P_n^t)_{n\geq 1}$ *is given by* (11), $\beta(t)$ *is given by* (12) *and* $t=t(n)$ *is such that*

$$(6\,K+9C_0)n\beta(t)\leq\widetilde{\beta}<1,$$

where $C_0>0$ *is an absolute constant satisfying the inequality*

$$\frac{1}{n}||\phi^2\,(P_nf)''\,||\leq C_0\,(||P_nf-f||+||P_{Kn}f-f||)\,,$$

then

$$\left(1-\widetilde{\beta}\right)(||P_nf-f||+||P_{Kn}f-f||)\leq||P_n^tf-f||+||P_{Kn}^tf-f||$$

$$\leq\left(1+\widetilde{\beta}\right)(||P_nf-f||+||P_{Kn}f-f||),$$

and there exists $C>0$ *such that*

$$\omega_\phi^2\left(f,\frac{1}{\sqrt{n}}\right)\leq C\left(||P_n^tf-f||+||P_{Kn}^tf-f||\right)$$

for all $f\in C_B[0,\infty)$*, and* $n=1,2,\ldots$

Very recently Heilmann and Tachev [21] studied the Phillips operators and established direct and converse results. They also have shown that these operators satisfy commutativity property. For this, they take

$$C_\alpha[0,\infty)=\{f\in C[0,\infty):|f(t)|\leq Me^{\alpha t},t\in[0,\infty),M>0\},$$

and prove Theorem 17.

Theorem 17 [21] *(Commutativity property) For all* $f\in C_\alpha[0,\infty),\ m>\alpha,\ n>\alpha,\ \frac{mn}{m+n}>\alpha$, *we have*

$$P_m(P_n,f)=P_{\frac{mn}{m+n}}.$$

From the above theorem, one easily gets

Corollary 1 [21] *For all* $f\in C_\alpha[0,\infty),\ m>\alpha,\ n>\alpha,\ \frac{mn}{m+n}>\alpha$, *we have*

$$P_m(P_n,f)=P_n(P_m,f).$$

And, for $l\in N,\ \frac{m}{l}>\alpha$,

$$P_m^l=P_n(P_{\frac{m}{l}}).\tag{14}$$

Theorem 18 [21] *(Second commutativity property) Let $f \in C_\alpha[0,\infty)$ be such that $f', f'' \in C_\alpha[0,\infty)$. Then the operators P_n, $n > \alpha$ and $\widetilde{D}^2$ commute, namely*

$$\left(\widetilde{D}^2 o P_n\right) f = \left(P_n o \widetilde{D}^2\right) f,$$

where an appropriate differential operators $\widetilde{D}$ is given by $\widetilde{D}^2\, f := x D^2\, f$.

Corollary 2 [21] *For $f \in W^2_\infty(\varphi)$ it holds*

$$\|\widetilde{D}^2 o P_n) f\| < ||\widetilde{D}^2\, f||.$$

Theorem 19 [21] *(Strong Voronovskaja type result) Let $g \in C_B[0,\infty)$ be such that $\varphi^2\, g''', \varphi^3\, g''' \in C_B[0,\infty)$ and $n > 0$. Then*

$$\left|P_n g - g - \frac{1}{n}\varphi^2 g''\right|$$

$$< \frac{\sqrt{6}}{2} \cdot \frac{1}{n} \max\left\{\frac{4}{3}\sqrt{1+2c} \cdot \frac{1}{\sqrt{n}}||\varphi^3 g''' \cdot \frac{\sqrt{1+2c}}{c}\frac{1}{n}||\varphi^3 g'''||\right\},$$

where c denotes a positive constant.

Also, Heilmann and Tachev [21] considered the following K-functional

$$K^2_\varphi(f,\delta^2) = \inf_{g \in W^2_\infty(\varphi)} \{||f-g|| + \delta^2 ||\widetilde{D}^2\, g||\},$$

where

$$W^2_\infty(\varphi) = \{h \in C_B[0,\infty) : h' \in AC_{loc}[0,\infty), \varphi^2\, h'' \in C_B[0,\infty)\}.$$

It is known (see [9]) that $K_{2,\varphi}(f,\delta^2)$ and $\omega^2_\phi(f,\delta)$ are equivalent, that is, for the modulus of continuity defined by Eq. (13), there exists an absolute constant $C > 0$ such that

$$C^{-1}\omega^2_\phi(f,\delta) \le K^2_\varphi\left(f,\delta^2\right) \le C\omega^2_\phi(f,\delta).$$

Theorem 20 [21] *(Direct result) Let $f \in C_B[0,\infty)$. Then, for $n > 0$ holds the inequality*

$$||P_n f - f|| \le 2K^2_\varphi\left(f,\frac{1}{n}\right).$$

Theorem 21 [21] *(Strong converse result) Let $f \in C_B[0,\infty)$. Then, for $n > 0$ the following inequality is true.*

$$2K^2_\varphi\left(f,\frac{1}{n}\right) \le 92.16||P_n f - f||. \qquad (15)$$

As a consequence of Theorems 20 and 21, Heilmann and Tachev [21] also obtained the following equivalence result.

Corollary 3 [21] *For* $f \in C_B[0,\infty)$, $n > 0$ *we have the following equivalences*

$$\frac{1}{2}||P_n f - f|| \leq 2K_\varphi^2\left(f, \frac{1}{n}\right) \leq 92.16||P_n f - f||,$$

$$C_1||P_n f - f|| \leq \omega_\varphi^2\left(f, \frac{1}{\sqrt{n}}\right) \leq C_2||P_n f - f||,$$

where $C_1,\ C_2 > 0$ *are absolute constants.*

7 Rate of Convergence

Rate of convergence for functions of bounded variation is also an active area of research on linear positive operators. In this direction Gupta and Shrivastava in [20] estimated the rate of convergence of Phillips operators for functions of bounded variation, using probabilistic approach.

Let $B_N[0,\infty)$, $N \geq 0$, be the class of all measurable functions f such that $|f(t)| \leq Me^{Nt}$ for all $t \in [0,\infty)$ and some positive M for which the operators $P_n(f,t)$ are well defined.

Theorem 22 [20] *Let* $f \in B_N[0,\infty)$ *be a function of bounded variation on every finite subinterval of* $[0,\infty)$, *and let*

$$g_x(t) = \begin{cases} f(t) - f(x+) & if\ \ x < t < \infty, \\ 0 & if\ \ t = x, \\ f(t) - f(x-) & if\ \ 0 \leq t < x \end{cases}$$

We denote by $V_a^b(g_x)$ *the total variation of* g_x *on* $[a,b]$. *Then for all* $x > 0$ *and* $n > 4N$, *we have*

$$\left|P_n(f,x) - \frac{1}{2}\{f(x+) + f(x-)\}\right| \tag{16}$$

$$\leq \frac{7}{nx}\sum_{m=1}^{n} V_{x-x/\sqrt{m}}^{x+x/\sqrt{m}}(g_x) + \frac{32x^2+24x+5}{4\sqrt{nx}}|f(x+) - f(x-)|$$

$$+\sqrt{\frac{2}{nx}}\left(1 + \frac{1}{\sqrt{nx}}\right)e^{2Nx}.$$

As Bézier curve play an important role in computer aided geometric designs (CAGD), in this context Gupta in [15] introduced the the Bézier variant of the Phillips operators.

For $\alpha \geq 1$ and for a function f defined on $[0, \infty)$, the Bézier variant of Phillips operator is defined as

$$P_{n,\alpha}(f, x) = n \sum_{k=1}^{\infty} Q_{n,k}^{(\alpha)}(t) \int_0^{\infty} s_{n,k-1}(u)du + Q_{n,0}^{(\alpha)}(t) f(0), \tag{17}$$

where $Q_{n,k}^{(\alpha)}(t) = J_{n,k}^{\alpha}(x) - J_{n,k+1}^{\alpha}(x),\ J_{n,k}(x) = \sum_{j=k}^{\infty} s_{n,j}(x)$.

Obviously, the operators $P_{n,\alpha}(f, x)$ defined by (17) are linear positive operators. Particularly when $\alpha = 1$, the operators (17) reduce to Phillips operators $P_{n,1} = P_n$. Also $P_{n,\alpha}(1, x) = 1$. Some basic properties of $J_{n,k}$ are as follows:

1. $J_{n,k}(x) - J_{n,k+1}(x) = s_{n,k}(x),\ k = 0, 1, 2, \ldots$.
2. $J'_{n,k}(x) = n s_{n,k-1}(x),\ k = 1, 2, 3, \ldots$.
3. $J_{n,k}(x) = n \int_0^x s_{n,k-1}(t)dt,\ k = 1, 2, 3, \ldots$.
4. $\sum_{k=1}^{\infty} J_{n,k}(x) = n \int_0^x \sum_{k=1}^{\infty} s_{n,k-1}(x) = nx$
5. for every $k \in N,\ 0 \leq J_{n,k}(x) < 1$ and $J_{n,k}(x)$ increases strictly on $[0, \infty)$.

Now, we present the rate of convergence of the Bézier variant of Phillips operators, defined by (17) and established in [15], for functions of bounded variation.

Theorem 23 [15] *Let f be a function of bounded variation on every finite subinterval of $[0, \infty)$ and let $V_a^b(g_x)$ the total variation of g_x on $[a, b]$. Also let $\alpha \geq 1$ and $f(t) = O(t^r),\ t \to \infty$ for some $r > 0$ and n sufficiently large, we have*

$$\left| P_n(f, x) - \left[\frac{1}{\alpha + 1} f(x+) + \frac{\alpha}{\alpha + 1} f(x-) \right] \right| \leq |f(x+) - f(x-)| \frac{\alpha H(j)}{\sqrt{nx}} \tag{18}$$

$$+ \frac{4\alpha + x}{nx} \sum_{k=1}^{n} V_{x-x/\sqrt{k}}^{x+x/\sqrt{k}}(g_x) + O(n^{-r})$$

where $H(j) = \frac{(j+1/2)^{j+1/2}}{j!} e^{-(j+1/2)}$ and

$$g_x(t) = \begin{cases} f(t) - f(x+) & if\ \ x < t < \infty, \\ 0 & if\ \ t = x, \\ f(t) - f(x-) & if\ \ 0 \leq t < x \end{cases}$$

Govil, Gupta and Noor in [13] defined the class $B_{r,\alpha}$ by $B_{r,\alpha} = \{f : f^{(r-1)} \in C[0, \infty),\ f_{\pm}^{(r)}(x)$ exists everywhere and are bounded on every finite subinterval of $[0, \infty)$ and $f_{\pm}^{(r)}(t) = O(e^{\alpha t})(t \to \infty)$, for some $\alpha > 0,\ r = 1, 2, 3, \ldots\}$. By $f_{\pm}^{(0)}(x)$ we mean $f(x\pm)$. They established the rate of convergence in simultaneous approximation, for which they proved

Theorem 24 [13] *Let $f \in B_{r,\alpha},\ r = 1, 2, \ldots, \alpha > 0$. Then for every $x \in (0, \infty)$ and $n \geq \max\{2, 4\alpha\}$*

$$\left| P_n^{(r)}(f, x) - \frac{1}{2} \left\{ f_{+}^{(r)}(x) + f_{-}^{(r)}(x) \right\} \right| \tag{19}$$

$$+\frac{|2r-1|}{\sqrt{8enx}}\left|f_+^{(r)}(x)+f_-^{(r)}(x)\right|+\frac{x+2n}{nx}\sum_{x-x/\sqrt{k}}^{x+x/\sqrt{k}}\left(g_{r,x}\right)+(nx)^{-1/2}\left(n2^r\right)^{1/2}e^{2\alpha x},$$

where $g_{r,x}$ is the auxiliary function defined by

$$g_{r,x}(t)=\begin{cases} f^{(r)}(t)-f_-^{(r)}(x), & 0\le t<x,\\ 0 & t=x,\\ f^{(r)}(t)-f_+^{(r)}(x), & x<t<\infty,\end{cases}$$

and $V_a^b\left(g_{r,x}(t)\right)$ is the total variation of $g_{r,x}(t)$ on $[a,b]$. In particular $g_{0,x}(t)\equiv g_x(t)$ as defined in Theorem 23.

8 q-Phillips Operators

Applications of q-calculus in approximation theory was initiated 25 years ago when the q analogue of Bernstein polynomials was introduced by Lupas [25]. Ten years later Phillips [34] proposed another q analogue of the classical Bernstein polynomials. After that many researchers have done work in this direction. Very recently, Mahmudov in [27] introduced the following q-Szász–Mirakjan operator

$$S_{n,q}\left(f;x\right)=\frac{1}{\prod_{j=0}^{\infty}\left(1+(1-q)q^j\,[n]\,x\right)}\sum_{k=0}^{\infty}f\left(\frac{[k]}{q^{k-2}\,[n]}\right)q^{\frac{k(k-1)}{2}}\frac{[n]^k\,x^k}{[k]!},$$

where $x\in[0,\infty)$, $0<q<1$, $f\in C\,[0,\infty)$, and investigated their approximation properties.

Also, Mahmudov, Gupta and kaffaoglu [26] introduced the following, so called q-Phillips operators.

Definition 1 For $f\in R^{[0,\infty)}$, the q-parametric Phillips operators are defined as

$$P_{n,q}(f;x)=[n]\sum_{k=1}^{\infty}q^{k-1}S_{n,k}(q;qx)\int_0^{\infty/(1-q)}S_{n,k-1}(q;t)f(t)d_qt \qquad (20)$$
$$+e_q\left(-[n]\,qx\right)f(0),$$

where $x\in[0,\infty)$ and $S_{n,k}(q;x)=e_q(-[n]x)q^{\frac{k(k-1)}{2}}\frac{[n]^kx^k}{[k]!}$.

Throughout the present section we employ the standard notations of q-calculus. The q-integer $[n]$, and q-factorial $[n]!$ are respectively defined by

$$[n]:=\begin{cases}\dfrac{1-q^n}{1-q} & \text{if}\quad q\in R^+\backslash\{1\},\\ n & \text{if}\quad q=1\end{cases}\quad \text{for } n\in N\quad\text{and}\quad [0]=0,$$

$$[n]! := [1][2]\ldots[n] \quad \text{for } n \in N \quad \text{and} \quad [0]! = 1.$$

For integers $0 \le k \le n$, the q-binomial is defined as

$$\left[\begin{array}{c} n \\ k \end{array} \right] := \frac{[n]!}{[k]![n-k]!}.$$

There are two q-analogues of the exponential function e^z, which are defined by

$$e_q(z) = \sum_{k=0}^{\infty} \frac{z^k}{[k]!} = \frac{1}{(1-(1-q)z)_q^{\infty}}, \quad |z| < \frac{1}{1-q}, \quad |q| < 1,$$

and

$$E_q(z) = \prod_{j=0}^{\infty} \left(1 + (1-q) q^j z\right) = \sum_{k=0}^{\infty} q^{k(k-1)/2} \frac{z^k}{[k]!} = (1 + (1-q)z)_q^{\infty}, \quad |q| < 1,$$

where $(1-x)_q^{\infty} = \prod_{j=0}^{\infty} \left(1 - q^j x\right)$.

We set

$$S_{n,k}(q;x) = \frac{1}{E_q([n]x)} q^{\frac{k(k-1)}{2}} \frac{[n]^k x^k}{[k]!} = e_q(-[n]x)\, q^{\frac{k(k-1)}{2}} \frac{[n]^k x^k}{[k]!}, \quad n = 1, 2, \ldots$$

It is clear that $S_{n,k}(q;x) \ge 0$ for all $q \in (0,1)$ and $x \in [0, \infty)$ and moreover

$$\sum_{k=0}^{\infty} S_{n,k}(q;x) = e_q(-[n]x) \sum_{k=0}^{\infty} q^{\frac{k(k-1)}{2}} \frac{([n]x)^k}{[k]!} = 1.$$

The q-Jackson integrals and the q-improper integrals are defined as

$$\int_0^a f(t)\, d_q t := a(1-q) \sum_{n=0}^{\infty} f(aq^n)\, q^n, \quad a > 0,$$

and

$$\int_0^{\infty/A} f(t)\, d_q t := (1-q) \sum_{n=-\infty}^{\infty} f\left(\frac{q^n}{A}\right) \frac{q^n}{A}, \quad A > 0.$$

The two q-Gamma functions are defined as

$$\Gamma_q(x) = \int_0^{\frac{1}{1-q}} t^{x-1} E_q(-qt)\, d_q t,$$

and

$$\gamma_q^A(x) = \int_0^{\infty/A(1-q)} t^{x-1} e_q(-qt)\, d_q t.$$

For every $A, x > 0$ one has

$$\Gamma_q(x) = K(A;x)\, \gamma_q^A(x),$$

where $K(A;x) = \frac{1}{1+A} A^x \left(1 + \frac{1}{A}\right)_q^x (1+A)_q^{1-x}$.

In particular, for any positive integer n, (see [6])

$$K(A;n) = q^{\frac{n(n-1)}{2}} \text{ and } \Gamma_q(n) = q^{\frac{n(n-1)}{2}} \gamma_q^A(n).$$

Let $C_B[0,\infty)$ be the space of all real-valued continuous bounded functions f on $[0,\infty)$, endowed with the norm $\|f\| = \sup_{x\in[0,\infty)} |f(x)|$. The Peetre's K-functional is defined by

$$K_2(f;\delta) = \inf_{g\in C^2[0,\infty)} \left\{\|f-g\| + \delta \left\|g''\right\|\right\},$$

where $C_B^2[0,\infty) := \left\{g \in C_B[0,\infty) : g', g'' \in C_B[0,\infty)\right\}$. By [7, Theorem 2.4] there exists an absolute constant $M > 0$ such that

$$K_2(f,\delta) \leq M\omega_2(f;\sqrt{\delta}), \tag{21}$$

where $\delta > 0$ and the second order modulus of smoothness is defined as

$$\omega_2(f;\sqrt{\delta}) = \sup_{0<h\leq\delta} \sup_{x\in[0,\infty)} |f(x+2h) - 2f(x+h) + f(x)|,$$

where $f \in C_B[0,\infty)$ and $\delta > 0$.

Also, note that

$$\omega(f;\delta) = \sup_{0<h\leq\delta} \sup_{x\in[0,\infty)} |f(x+h) - f(x)|.$$

Concerning the error estimate of q-Phillips operators, using the methods of DeVore and Lorentz [7] we have the following results of Mahmudov, Gupta and Kaffaoglu [26].

Theorem 25 [26] *Let $f \in C_B[0,\infty)$. Then, for every $x \in [0,\infty)$, there exists a constant $M > 0$ such that*

$$\left|P_{n,q}(f;x) - f(x)\right| \leq M\omega_2\left(f;\sqrt{\delta_n(x)}\right),$$

where

$$\delta_n(x) = \left(\frac{1-q^2}{q^2}\right)x^2 + \frac{(1+q)}{q^2[n]}x.$$

Theorem 26 [26] *Let $0 < \alpha \leq 1$ and E be any subset of the interval $[0,\infty)$. Then, if $f \in C_B[0,\infty)$ is locally $Lip(\alpha)$, that is, the condition*

$$|f(y) - f(x)| \leq L\,|y-x|^\alpha, y \in E \text{ and } x \in [0,\infty),$$

holds, then for each $x \in [0, \infty)$ *, we have*

$$\left|P_{n,q}(f;x) - f(x)\right| \leq L\left\{\delta_n^{\frac{\alpha}{2}}(x) + 2\left(d\left(x, E\right)\right)^{\alpha}\right\},$$

where L *is a constant depending on* α *and* f; *and* $d\left(x, E\right)$ *is the distance between* x *and* E *defined as*

$$d\left(x, E\right) = \inf\left\{|t - x| : t \in E\right\}.$$

We consider the following classes of functions:

$$C_m[0,\infty) := \left\{ f \in C[0,\infty) : \exists M_f > 0 \ |f(x)| < M_f\left(1 + x^m\right) \right.$$

$$\left. \text{and } \|f\|_m := \sup_{x\in[0,\infty)} \frac{|f(x)|}{1+x^m} \right\},$$

$$C_m^*[0,\infty) := \left\{ f \in C_m[0,\infty) : \lim_{x\to\infty} \frac{|f(x)|}{1+x^m} < \infty \right\}, \quad m \in \mathbb{N}.$$

For every $f \in C_m^*[0,\infty)$ the weighted modulus of continuity (see [24]) is defined as follows:

$$\Omega_m(f,\delta) = \sup_{x\geq 0,\ 0<h\leq\delta} \frac{|f(x+h) - f(x)|}{1+(x+h)^m}$$

The following results of Mahmudov, Gupta and Kaffaoglu [26] concerning rate of convergence also appear to be of interest.

Theorem 27 [26] *If* $f \in C_1^*[0,\infty)$ *then the inequality*

$$\left\|P_{n,q}(f) - f\right\|_2 \leq k(q)\,\Omega_1\left(f; \frac{1}{\sqrt{[n]}}\right),$$

where k *is a constant independent of* f *and* n.

Theorem 28 [26] *Let* $q_n \in (0,1)$. *then the sequence* $\left\{P_{n,q_n}(f)\right\}$ *converges to* f *uniformly on* $[0, A]$ *for each* $f \in C_2^*[0,\infty)$ *if and only if* $\lim_{n\to\infty} q_n = 1$.

Theorem 29 [26] *Assume that* $q_n \in (0,1)$, $q_n \to 1$ *and* $q_n^n \to a$ *as* $n \to \infty$. *For any* $f \in C_2^*[0,\infty)$ *such that* $f', f'' \in C_2^*[0,\infty)$ *the following equality holds*

$$\lim_{n\to\infty} [n]_{q_n}\left(P_{n,q_n}(f;x) - f(x)\right) = \left((1-a)x^2 + x\right) f''(x)$$

uniformly on any $[0, A]$, $A > 0$.

Recently in 2011, Yuksel [39] has studied q-Phillips operators in slightly different form and obtained some direct results. We omit the details.

References

1. Agrawal, P.N..: Simultaneous approximation by Micchelli combination of Phillips operators. In: Singh B., Murari K., Gupta U.S., Prasad G., Sukavanam N.(eds.) Proceedings of Conference "Mathematics and its applications in engineering and industry", University of Roorkee, India, pp. 569–577 (1997)
2. Agrawal, P.N., Gupta, V.: On the iterative combination of Phillips operators. Bull. Inst. Math. Acad. Sinica **18**(4), 361–368 (1990)
3. Agrawal, P.N., Gupta, V.: On convergence of derivatives of Phillips operators. Demonstr. Math. **27**(2), 501–510 (1994)
4. Agrawal, P.N., Gupta, V., Gairola, A. R.: On the derivatives of iterative combinations of Phillips operators. Nonlinear Funct. Anal. Appl. **12**(2), 195–203 (2007)
5. Butzer, P.L.: Linear combinations of Bernstein polynomials. Canad. J. Math. **5**, 559–567 (1953)
6. De Sole, A., Kac, V.: On integral representations of q-gamma and q-beta functions. Atti Accad. Naz. Lincei Cl. Sci. Fis. Mat. Natur. Rend. Lincei (9) Mat. Appl. **16(1)**, 11–29 (2005)
7. DeVore, R.A., Lorentz, G.G.: Constructive approximation. Springer, Berlin (1993)
8. Ditzian, Z., May, C.P.: A saturation result for combinations of Bernstein polynomials. Tôhoku Math. J. **28**, 363–373 (1976)
9. Ditzian, Z., Totik, V.: Moduli of smoothness. Springer, New York, (1987)
10. Ditzian, Z., Ivanov, K.G.: Strong converse inequality. J. Anal. Math. **61**, 61–111 (1993)
11. Finta, Z.: Direct and converse theorems for integral-type operators. Demonstr. Math. **36**, 137–147 (2003)
12. Finta, Z., Gupta, V.: Direct and inverse estimates for Phillips type operators. J. Math. Anal. Appl. **303**(2), 627–642 (2005)
13. Govil, N.K., Gupta, V., Noor, M. A.: Simultaneous approximation for the Phillips operators. Int. J. Math. Math. Sci. **2006**(7), (2006) (1–9 Art Id. **49094**)
14. Gupta, V.: Rate of approximation by a new sequence of linear positive operators. Comput. Math. Appl. **45**, 1895–1904 (2003)
15. Gupta, V.: Rate of convergence by Bézier variant of Phillips operators. Taiwanese J. Math. **8**(2), 183–190 (2004)
16. Gupta, V.: A note on modified Phillips operators. Southeast Asian Bull Math. **34**, 847–851 (2010)
17. Gupta, V., Agrawal, P.N.: Linear combinations of Phillips operators. Indian Acad. Math. **11**(2), 106–114 (1989)
18. Gupta, V., Agrawal, P.N.: Lp-approximation by iterative combination of Phillips operators. Publ. Inst. Math. (Beograd) **52**(66), 101–109 (1992)
19. Gupta, V., Sahai, A.: On linear combination of Phillips operators. Soochow J. Math. **19**(3), 313–323 (1993)
20. Gupta, V., Srivastava, G.S.: On the rate of convergence of Phillips operators for functions of bounded variation. Comment. Math. **36**, 123–130 (1996)
21. Heilmann, M., Tachev, G.: Commutativity, direct and strong converse results for Phillips operators. East J. Approx. Theory **17**(3), 299–317 (2011)
22. Jain, G.C.: Approximation of functions by a new class of linear operators. J. Austral. Math. Soc. **13**(3), 271–276 (1972)
23. Jackson, D.: The theory of approximation, Vol. 11. American Mathematical Society Call Publications, New York (1930)
24. López-Moreno, A.J.: Weighted simultaneous approximation with Baskakov type operators, Acta Math. Hungar. **104**(1–2), 143–151 (2004)
25. Lupas, A. A q-analogue of the Bernstein operator, In: Seminar on numerical and statistical calculus (Cluj-Napoca, 1987), 85–92, Preprint, 87-9 University, Babes-Bolyai, Cluj.
26. Mahmoodov, N., Gupta, V., Kaffaoglu, H.: On certain q-Phillips operators, Rocky Mountain J. Math. **42**(4), 1291–1312 (2012)

27. Mahmudov, N.I.: On q-parametric Szász-Mirakjan operators. Mediterranen J. Math. **7(3)**, 297–311 (2010)
28. May, C.P.: Saturation and inverse theorems for combinations of a class of exponential type operators. Canad. J. Math. **28**, 1224–1250 (1976)
29. May, C.P.: On Phillips operators. J. Approx. Theory. **20**, 315–322 (1977)
30. Mazhar, S.M., Totik, V.: Approximation by modified Szász operators. Acta. Sci. Math. **49**, 257–269 (1985)
31. Micchelli, C.A.: The saturation class and iterates of the Bernstien polynomials. J. Appro. Theory. **8**, 1–18 (1973)
32. Natanson, I.P..: Constructive function theory, vol. 1, Frederic Unger, New York (1964)
33. Phillips, R.S.: An inversion formula and semi-groups of linear operators. Ann. Math. **59**, 325–356 (1954)
34. Phillips, G.M.: Bernstein polynomials based on the q-integers. Ann. Numer. Math. **4**, 511–518 (1997)
35. Rathore, R.K.: Linear combinations of linear positive operators and generating relations in special functions, Ph. D. Thesis submitted to IIT Delhi (1973)
36. Srivastava, H.M., Gupta, V.: A certain family of summation-integral type operators, Math. Comput. Model **19**, 1307–1315 (2003)
37. Szász, O.: Generalizations of S. Bernstein's polynomial to the infinite interval. J. Res. Nat. Bur. Standards. **45**, 239–245 (1950)
38. Zygmund, A.: Trigonometric series. Dover Pub, Inc., New York (1995)
39. Yuksel, I.: Approximation by q-Phillips operators. Hacettepe J. Math. Stat. **40**(2), 191–205 (2011)

Geometric Patterns in Uniform Distribution of Zeros of the Riemann Zeta Function

Mehdi Hassani

Abstract We introduce a geometrical investigation of the distribution of several sequences involving the imaginary parts of the nontrivial zeros of the Riemann zeta function.

Keywords Distribution modulo 1 · Nonreal zeros of the Riemann zeta function

1 Uniform Distribution and Weyl's Criterion

An arbitrary real sequence $(a_n)_{n\geq 1}$ is uniformly distributed modulo 1, if for all real numbers a, b with $0 \leq a < b \leq 1$ we have $\#\{n \leq N : \{a_n\} \in [a, b]\} \sim (b - a)N$ as $N \to \infty$. Here, $\{x\} = x - \lfloor x \rfloor$ denotes the fractional part of the real number x. An efficient criterion to determine uniform distribution modulo 1 of a given sequence, due to Weyl [16], asserts that the sequence $(a_n)_{n\geq 1}$ is uniformly distributed modulo 1 if and only if $\sum_{n\leq N} e(ha_n) = o(N)$ as $N \to \infty$, for every positive integer h. Here, and in what follows, we let $e(x) = \mathrm{e}^{2\pi i x}$.

Uniform distribution modulo 1 of various number theoretic functions is known in literature. Among them, Rademacher [13] observed that under the assumption of the Riemann hypothesis the sequence $(\alpha\gamma_n)_{n\geq 1}$ is uniformly distributed modulo 1, where $\alpha \neq 0$ is a fixed real number and γ_n runs over the imaginary parts of the zeros of $\zeta(s)$. Later Hlawka [7] proved this assertion unconditionally.

2 Geometrical Approach

Dekking and Mendès France [2] introduced an idea to make visible the Weyl sums $\sum_{n\leq N} e(ha_n)$ for a given real sequence $(a_n)_{n\geq 1}$ and given positive integers h and N, by drawing a plane curve generated by successively connected lines segment, which joins the point V_n to V_{n+1}, where $V_n = (S_1(n), S_2(n))$ with

M. Hassani (✉)
University of Zanjan, University Blvd., 45371-38791, Zanjan, Iran
e-mail: mehdi.hassani@znu.ac.ir

T. M. Rassias, P. M. Pardalos (eds.), *Mathematics Without Boundaries*,
DOI 10.1007/978-1-4939-1106-6_9,

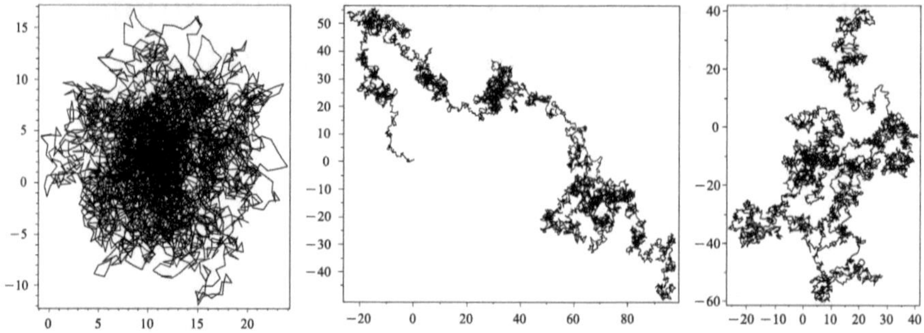

Fig. 1 Graphs of the Weyl sums $\sum_{n\le 5000} e(\alpha\gamma_n)$ for $\alpha = 1, \pi, \mathrm{e}$, respectively from left to right

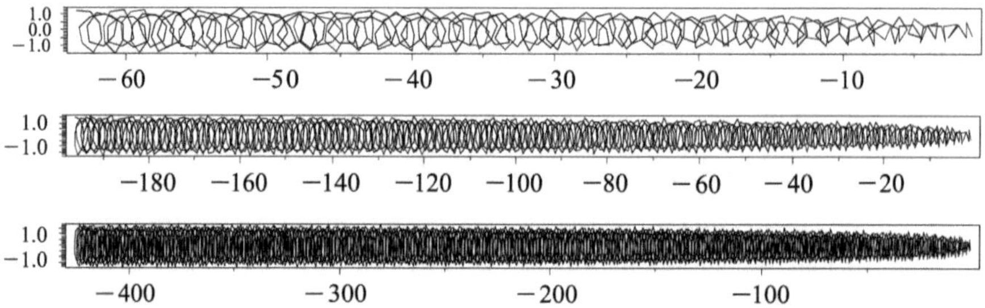

Fig. 2 Graphs of the Weyl sums $\sum_{n\le N} e(\alpha\gamma_n)$ with $\alpha = (\log 2)/2\pi$, for $N = 500, 2000, 5000$ respectively from top to down

$S_1(n) = \sum_{k\le n} \cos(2\pi h a_k)$ and $S_2(n) = \sum_{k\le n} \sin(2\pi h a_k)$, for $1 \le n < N$. Some classical examples are pictured in [2] and [3].

In this paper, we generate several graphs of the Weyl sums involving the sequence $(\alpha\gamma_n)_{n\ge1}$, for certain values of α, as well as, several sequences concerning the values of γ_n. To generate figures appeared in this chapter, we used Maple software to do several computations running over the numbers $(\gamma_n)_{1\le n\le 20000}$, all of which are based on the tables of zeros of the Riemann zeta function due to Odlyzko [11].

We note that in the graphs of the Weyl sums, the length of each line segment is 1. Thus, if $1 \le n < N$, then the frame (rectangular border) that includes the graph of the Weyl sum has the size not exceeding $N \times N$. We denote the height and the width of the frame containing the graph of the Weyl sum $\sum_{n\le N} e(ha_n)$, respectively by $H_h(N)$ and $W_h(N)$. Since $\sum_{n\le N} e(ha_n) = S_1(N) + iS_2(N)$, then the Weyl criterion geometrically asserts that the real sequence $(a_n)_{n\ge1}$ is uniformly distributed modulo 1 if and only if $H_h(N) = o(N)$ and $W_h(N) = o(N)$ as $N \to \infty$, for any positive integer h. Since it is not possible to consider all positive integer values of h, hence we will take $h = 1$ in all graphs, and simply we let $H_1(N) = H(N)$ and $W_1(N) = W(N)$. In this chapter we study two different families of the graphs of the Weyl sums involving the sequence $(\alpha\gamma_n)_{n\ge1}$ according to the ratio $H(N)/W(N)$ of the frame that includes the graph of the Weyl sum, as in Figs. 1 and 2.

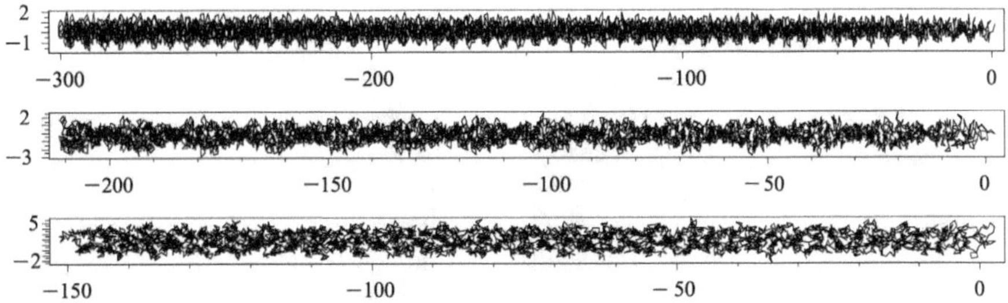

Fig. 3 Graphs of the Weyl sums $\sum_{n\leq 5000} e(\alpha\gamma_n)$ with $\alpha = (\log 2^k)/2\pi$, for $k = 2, 3, 4$ respectively from top to down

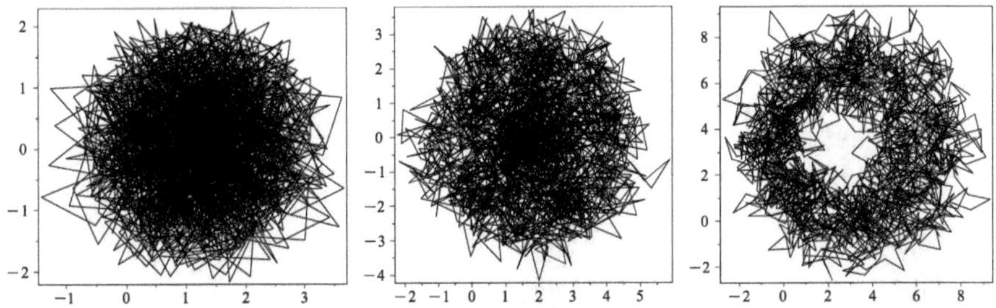

Fig. 4 Graphs of the Weyl sums $\sum_{n\leq 2000} e(\alpha\gamma_n)$ with $\alpha = (\log m)/2\pi$, for $m = 6, 12, 36$ respectively from left to right

We focus on the interesting case $\alpha = (\log m)/2\pi$, where m is a positive integer. In Sect. 3 we will explain why this case is interesting in our investigation. Thanks to the works of Rademacher [13] and Hlawka [7], all of the sequences we pictured their Weyl sums in this section and in Sect. 3, are uniformly distributed modulo 1. More precisely, Hlawka [7] proved that the discrepancy of the set $\{\{\alpha\gamma_n\} : 0 < \gamma_n \leq T\}$ with $\alpha = (\log m)/2\pi$, is $O(1/\log T)$, under the assumption of the Riemann hypothesis, and that it is $O(1/\log\log T)$ unconditionally. Fujii [6] showed unconditionally that this discrepancy is $O(\log\log T/\log T)$. We recall that the discrepancy of the real numbers $a_1, a_2, \ldots, a_N$ is defined by

$$\sup_{0\leq a<b\leq 1} \left| \frac{1}{N}\#\big\{n \leq N : \{a_n\} \in [a, b]\big\} - (b - a) \right| .$$

For more information about discrepancy we refer the reader to the second chapter of the book of Kuipers and Niederreiter [9].

We observe that for the case $\alpha = (\log m)/2\pi$, when m is a prime power (for example as in Figs. 2 and 3), the frame including the graph of the Weyl sum has an unusual long size. Of course, this doesn't happen for other integer values of m, for example in Fig. 4, and even for other noninteger values of m, for example in Fig. 5. In the next section, we introduce a mathematical justification for this phenomenon, and in the last section, we visualize some more noteworthy patterns.

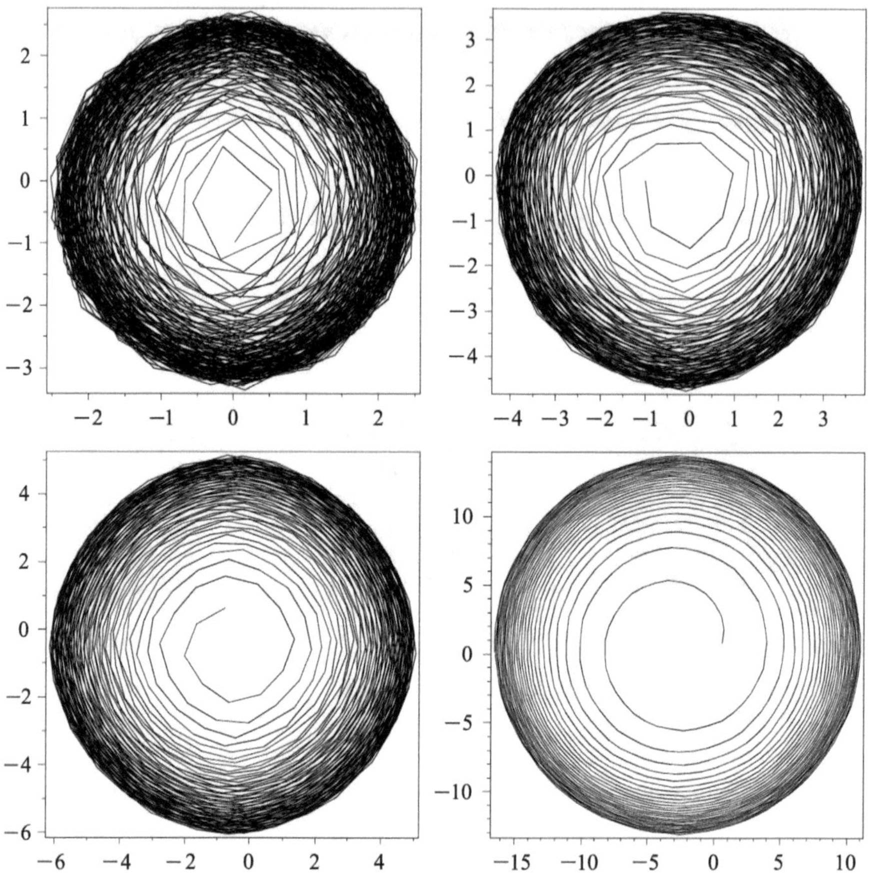

Fig. 5 Graphs of the Weyl sums $\sum_{n\le 2000} e(\alpha\gamma_n)$ with $\alpha = (\log 2^{1/k})/2\pi$, for $k = 2, 3, 4, 10$ respectively from left to right

3 Landau's Formula and Justification of the Frames' Size

The sizes of the frames including the graphs of the Weyl sums in Sect. 2 largely explained by a theorem of Landau [10] which asserts, for each fixed real $x > 1$, that

$$\sum_{0<\gamma\le T} x^{\rho} = -\frac{T}{2\pi}\Lambda(x) + O(\log T),$$

as T tends to infinity, where $\rho = \beta + i\gamma$ runs over the nontrivial zeros of the Riemann zeta function, and $\Lambda(x)$ is the von Mangoldt function for integral $x > 1$ and $\Lambda(x) = 0$ for nonintegral values of x.

Assuming that the Riemann hypothesis holds, as we did to generate present graphs, one has $\rho = 1/2 + i\gamma$, and hence

$$\sum_{0<\gamma\leq T} x^{i\gamma} = -\frac{T}{2\pi}x^{-\frac{1}{2}}\Lambda(x) + O(\log T).$$

We have $m^{i\gamma} = e(\alpha\gamma)$ with $\alpha = (\log m)/2\pi$. As usual, we let $N(T)$ denote the number of the nontrivial zeros $\rho = \beta + i\gamma$ of the Riemann zeta function satisfying $0 < \beta < 1$ and $0 < \gamma \leq T$. By using the von Mangoldt's approximation for $N(T)$ (see [8] and [15]), we obtain the validity of $\gamma_n \sim 2\pi n/\log n$ as $n \to \infty$. Thus, assuming that the Riemann hypothesis holds, if $\alpha = (\log m)/2\pi$ and $N = N(T)$, then we obtain

$$\sum_{n\leq N} e(\alpha\gamma_n) = -\frac{T}{2\pi}m^{-\frac{1}{2}}\Lambda(m) + O(\log T).$$

So, for example as in Fig. 2, we see that the real part of $\sum_{n\leq N} e(\alpha\gamma_n)$ tending to $-\infty$, at a rate $T \cong 2\pi N/\log N$, and the imaginary part of it, is bounded by $O(\log T)$. Although in Fig. 2 the imaginary parts are bounded by 1, we expect that eventually they get larger. One will get similar behaviour for $\alpha = (\log m)/(2\pi)$ whenever m is a power of a prime but in other cases, and in particular when m is not an integer, and consequently $\Lambda(m) = 0$, the right-hand-side of the last relation is just $O(\log T)$, and in this case one would expect a "random walk" of the types in Figs. 1, 4, and 5 with the size bounded by $O(\log T)$.

4 Spiral Patterns and Other Patterns

In this section, we produce the graphs of the Weyl sums of several sequences related to the values of γ_n. We know that some of these sequences are uniformly distributed modulo 1, however, we have no mathematical justification for the details of the presented patterns. Moreover, we observe that some of them have spiral shapes. To clear the situations of these spiral patterns, we will picture the graphs of the Weyl sums as a space curve.

4.1 Sequences Concerning Polynomial Values

Figure 6 shows the graphs of the Weyl sums $\sum_{n\leq 5000} e(\gamma_n^k)$ with $k = 2, 3, 10$. We observe that both of the ratios $H(N)/N$ and $W(N)/N$ for $N = 5000$ are small ($\cong 0.04$). This observation supports the following conjecture.

Conjecture 1 For any nonconstant polynomial $P(x)$ with real coefficients, the sequence $(P(\gamma_n))_{n\geq 1}$ is uniformly distributed modulo 1.

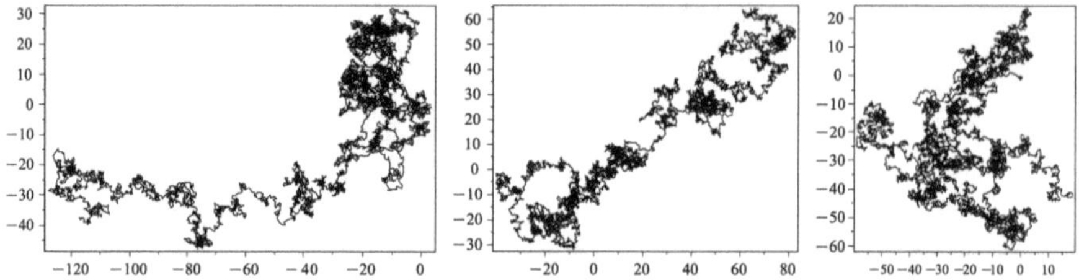

Fig. 6 Graphs of the Weyl sums $\sum_{n \le 5000} e(\gamma_n^k)$ with $k = 2, 3, 10$ respectively from left to right

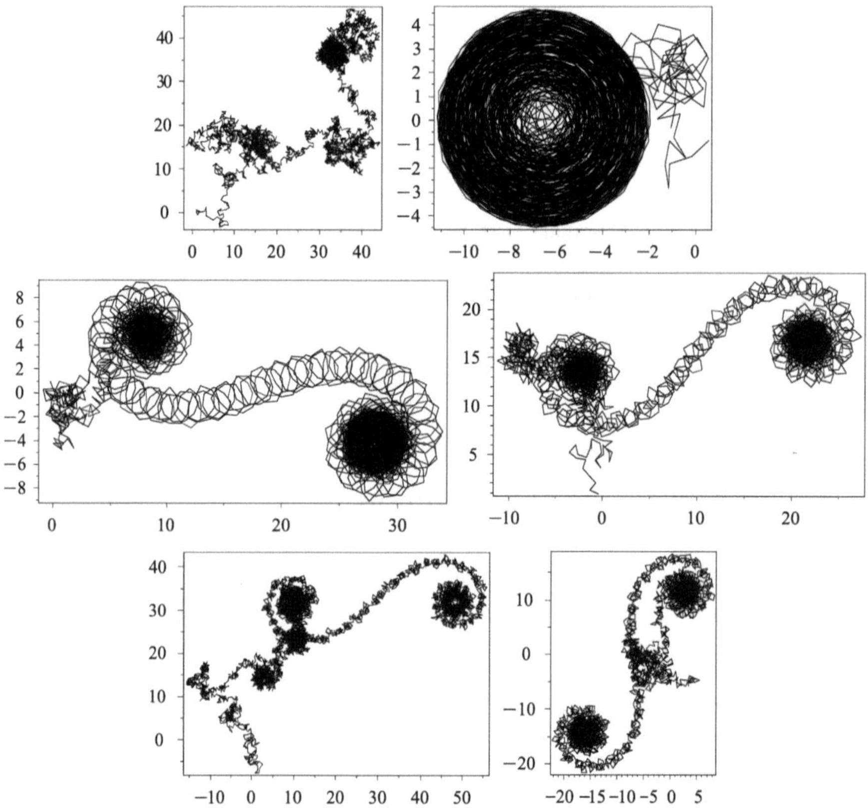

Fig. 7 Graphs of the Weyl sums $\sum_{n \le 5000} e(\gamma_n^r)$ with $r = 10/11, 2/3$ (first row), $r = 3/4, 4/5$ (second row), $r = 7/8, 17/20$ (third row)

4.2 Sequences Concerning Rational Powers

Figures 7 and 8 show the graphs of the Weyl sums $\sum_{n \le 5000} e(\gamma_n^r)$ for several rational values of r.

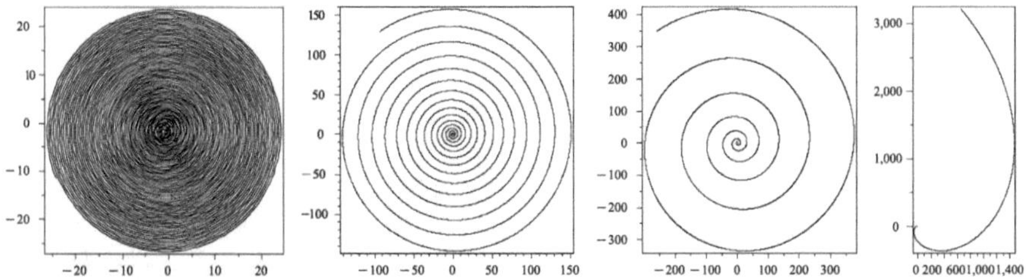

Fig. 8 Graphs of the Weyl sums $\sum_{n\leq 5000} e(\gamma_n^r)$ with $r = 1/2, 1/3, 1/4, 1/10$ respectively from left to right

Fujii [5] developed a method to obtain uniform distribution modulo 1 of a family of sequences $(f(\gamma_n))_{n\geq 1}$ for a wide class of smooth functions f. In particular, his method implies that sequences of the form $(\gamma_n^r)_{n\geq 1}$ for any fixed real $r \in (0, 1)$ are uniformly distributed modulo 1. Although in Fig. 8 the rectangular frame of the rightmost graph seems large, we expect that eventually it gets smaller to become $o(N)$ as $N \to \infty$, when we consider the related Weyl sum over $n \leq N$.

4.3 *Sequences Concerning* $\log \gamma_n$

The above mentioned family of smooth functions f due to Fujii [5], for which the sequences $(f(\gamma_n))_{n\geq 1}$ are uniformly distributed modulo 1 includes a number of logarithmic functions.

More precisely, he asserts that for any fixed arbitrary real number $t > 0$ the sequences with general terms

$$(\log \gamma_n)(\log_k \gamma_n), \qquad (\log \gamma_n)^{1+t}, \qquad \frac{\gamma_n}{(\log \gamma_n)^{t-1}}, \qquad \frac{\gamma_n \log \gamma_n}{\log_k \gamma_n},$$

all are uniformly distributed modulo 1. Here, $k \geq 1$ is an integer, and $\log_k$ denotes the k-fold iterative logarithm function. Also, he remarks that the sequence with general term $\log \gamma_n$ is not uniformly distributed modulo 1. Figure 9 shows the graphs of the Weyl sums $\sum_{n\leq 5000} e(a_n)$ with $a_n = \log \gamma_n$, $a_n = \gamma_n \log \gamma_n$, $a_n = \gamma_n / \log \gamma_n$ and $a_n = n \log \gamma_n$. The sizes of the frames in these graphs led us to the following conjecture.

Conjecture 2 Sequences with general terms $a_n = \gamma_n \log \gamma_n$, and $a_n = n \log \gamma_n$ are uniformly distributed modulo 1.

4.4 *Sequences Concerning the Difference* $\gamma_{n+1} - \gamma_n$

The sequence with general term $a_n = (\gamma_{n+1} - \gamma_n)((\log \gamma_n)/2\pi)/2\pi$ appears in the work of Odlyzko [12], where he studies the distribution of spacing between zeros

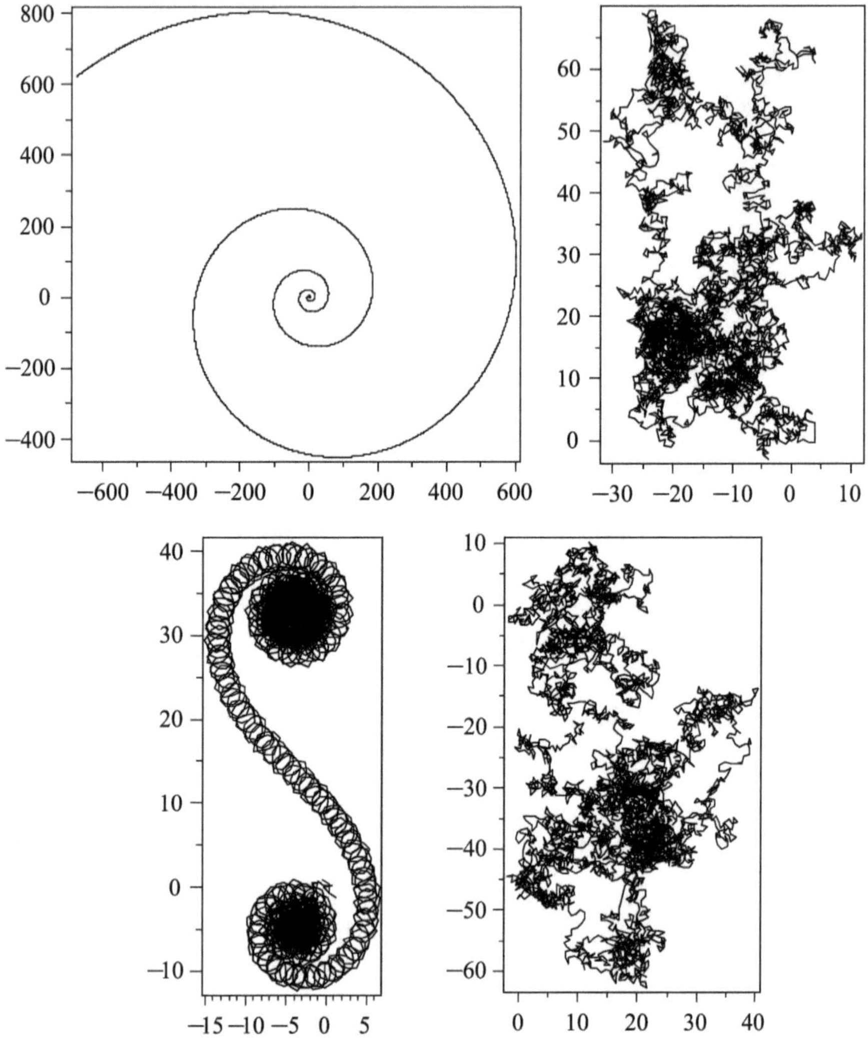

Fig. 9 Graphs of the Weyl sums $\sum_{n \le 5000} e(a_n)$ with $a_n = \log \gamma_n$, $a_n = \gamma_n \log \gamma_n$, $a_n = \gamma_n / \log \gamma_n$ and $a_n = n \log \gamma_n$, respectively from left to right

of the zeta function. He is motivated by the facts that the nontrivial zeros of $\zeta(s)$ become denser and denser, the higher up one goes in the critical strip, and the average vertical spacing between consecutive zeros at height t is $2\pi / \log(t/2\pi)$. Therefore, in order to study quantities that are largely invariant with height, he defines, under the assumption of the Riemann hypothesis, the above mentioned sequence a_n as the normalized spacing between consecutive zeros with imaginary parts γ_n and γ_{n+1}. Moreover, he remarks that the above mentioned sequence a_n has the mean value 1 in the sense that for any positive integers K and L, we have

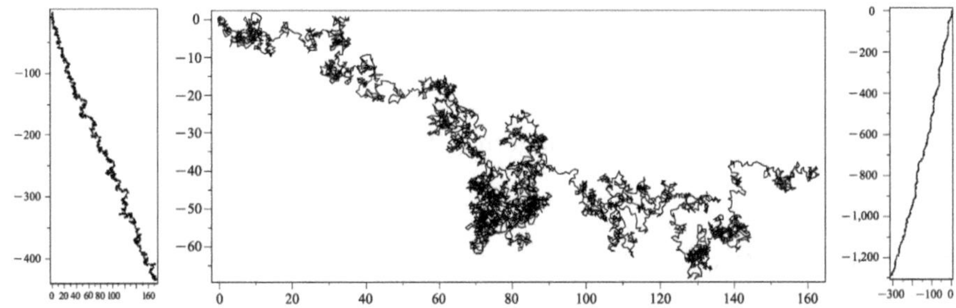

Fig. 10 Graphs of the Weyl sums $\sum_{n\leq 5000} e(a_n)$ with $a_n = (\gamma_{n+1} - \gamma_n)((\log \gamma_n)/2\pi)/2\pi$, $a_n = (\gamma_{n+1} - \gamma_n)(\log n)/2\pi$, and $a_n = 2\pi/((\log n)(\gamma_{n+1} - \gamma_n))$, respectively from left to right

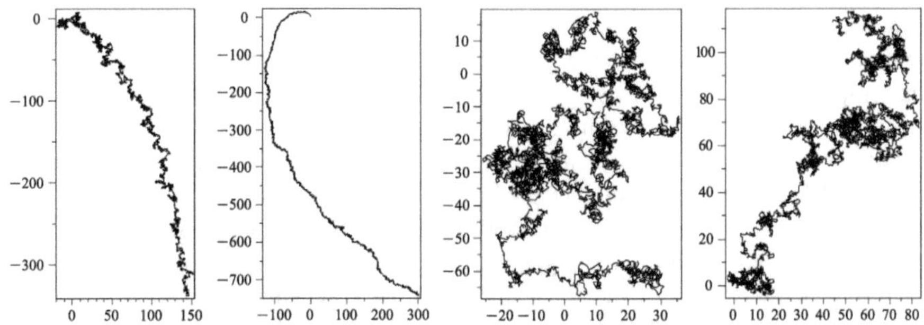

Fig. 11 Graphs of the Weyl sums $\sum_{n\leq 5000} e(a_n)$ with $a_n = \gamma_{n+1} - \gamma_n$, $a_n = 1/(\gamma_{n+1} - \gamma_n)$, $a_n = (\gamma_{n+1} - \gamma_n)\log n$ and $a_n = (\gamma_{n+1} - \gamma_n)n$, respectively from left to right

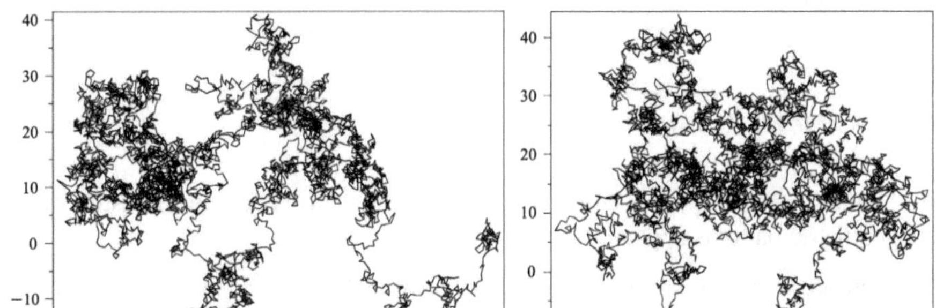

Fig. 12 Graphs of the Weyl sums $\sum_{n\leq 5000} e(a_n)$ with $a_n = (\log n)/(\gamma_{n+1} - \gamma_n)$ and $a_n = n/(\gamma_{n+1} - \gamma_n)$, respectively from left to right

$\sum_{K<n\leq K+L} a_n = L + O(\log (KL))$. Figures 10, 11, and 12 show the graphs of the Weyl sums $\sum_{n\leq 5000} e(a_n)$ with a_n concerning the difference $\gamma_{n+1} - \gamma_n$ in several cases, including the above mentioned sequence studied by Odlyzko in the leftmost graph of Fig. 10.

Considering the sizes of the frames in Figs. 10, 11, and 12, we formulate the following conjecture.

Conjecture 3 All of the sequences in Figs. 10, 11, and 12, except the sequence $a_n = 2\pi/((\log n)(\gamma_{n+1} - \gamma_n))$, are uniformly distributed modulo 1.

4.5 Tornado Patterns in Space Curves

As we mentioned in the above sections, by a theorem of Fujii, sequences with general terms $a_n = \gamma_n^r$ where $r \in (0, 1)$, and $a_n = \gamma_n/\log\gamma_n$ are uniformly distributed modulo 1. Some of the graphs of their corresponding Weyl sums in Fig. 7, and Fig. 9, consist of "S" shape, where many of lines segment snuggle around two holes and generate S-shape graphs. To detect exactly for which values of N the graphs of the Weyl sums $\sum_{n\le N} e(a_n)$ generate such S-shapes, we study the space curve form of the graphs of the Weyl sums $\sum_{n\le N} e(ha_n)$ for a given real sequence $(a_n)_{n\ge 1}$ and given positive integers h and N, by drawing a space curve generated by successively connected lines segment, which joins the point V_n to V_{n+1}, where $V_n = (S_1(n), S_2(n), n/1000)$. Here $S_1(n) = \sum_{k\le n} \cos(2\pi h a_k)$ and $S_2(n) = \sum_{k\le n} \sin(2\pi h a_k)$, for $1 \le n < N$.

In Fig. 13, we generate the space curves of the Weyl sums $\sum_{n\le 20000} e(a_n)$ with $a_n = \gamma_n^{10/11}$, $a_n = \gamma_n^{17/20}$, and $a_n = \gamma_n/\log\gamma_n$. These curves show more details of the related S-shape graphs. They contain some "tornado" patterns, and it seems that as N growths in the related Weyl sums $\sum_{n\le N} e(a_n)$, we expect some more "tornado" patterns around several holes.

4.6 Further Remarks

We end up this chapter with some remarks for further investigations and studies of the geometric patterns appeared here.

1. There are also some spiral and tornado patterns in the classical examples, pictured in Figs. 14 and 15. Studying the mathematical justification for the patterns in these classical examples will help us in understanding of the patterns appeared in the case of the Riemann zeta function.

   We note that the most simple tornado pattern is the space curve of the Weyl sum $\sum_{n\le N} e(a_n)$ with $a_n = \alpha n$ where α is an irrational coefficient. The leftmost graph pictured in Fig. 14 shows the space curve of the Weyl sum $\sum_{n\le 1000} e(n/e)$, as an example. The fundamental fact here is that for the linear sequence $a_n = \alpha n$ with irrational coefficient α, the generating vertices V_n of the related plane curve lie on a circle with radii $1/(2|\sin(\pi\alpha)|)$ and the center located at the point $(-1/2, \cot(\pi\alpha)/2)$. Moreover, the graph of the Weyl sum of the linear sequence $a_n = \alpha n$ is dense in an annulus with radiuses $|\cot(\pi\alpha)|/2$ and $1/(2|\sin(\pi\alpha)|)$.

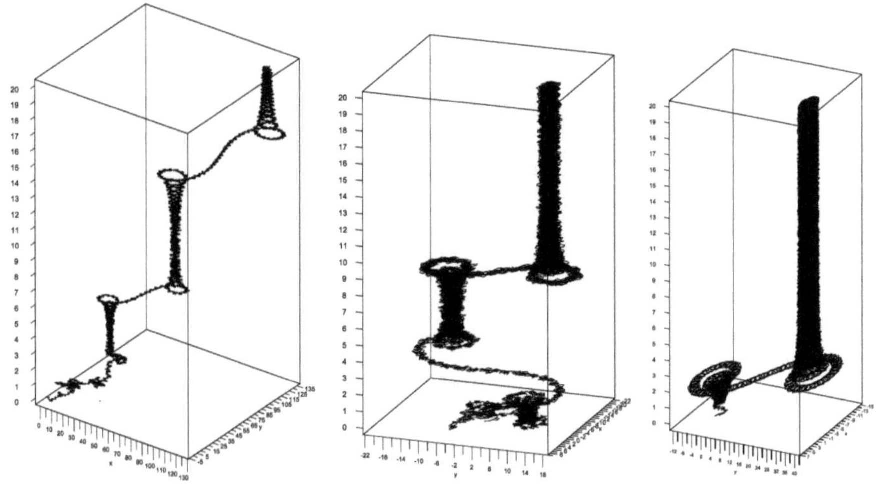

Fig. 13 The space curves of the Weyl sums $\sum_{n\leq 20000} e(a_n)$ with $a_n = \gamma_n^{10/11}$, $a_n = \gamma_n^{17/20}$, and $a_n = \gamma_n / \log \gamma_n$, respectively from left to right

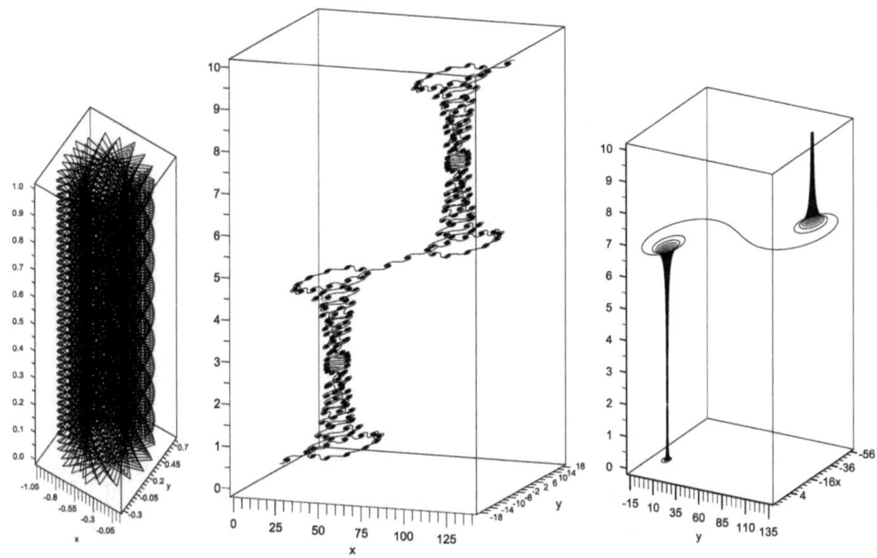

Fig. 14 The space curves of the Weyl sums $\sum_{n\leq 1000} e(n/\mathrm{e})$ (leftmost graph), and $\sum_{n\leq 10000} e(a_n)$ with $a_n = (50/5001)n^2$ (mid graph), and $a_n = (1/125)n^{3/2}$ (rightmost graph)

Based on this fundamental fact, to justify the patterns in Fig. 15, Tenenbaum and Mendès France [14] remark that because of the weak growth of $\log n$, the curve behaves locally like the curve associated with the linear sequence $a_n = c_H n$ where c_H is a local constant with $c_H \approx \{\log H\}$ for $n \approx H$. Thus, the

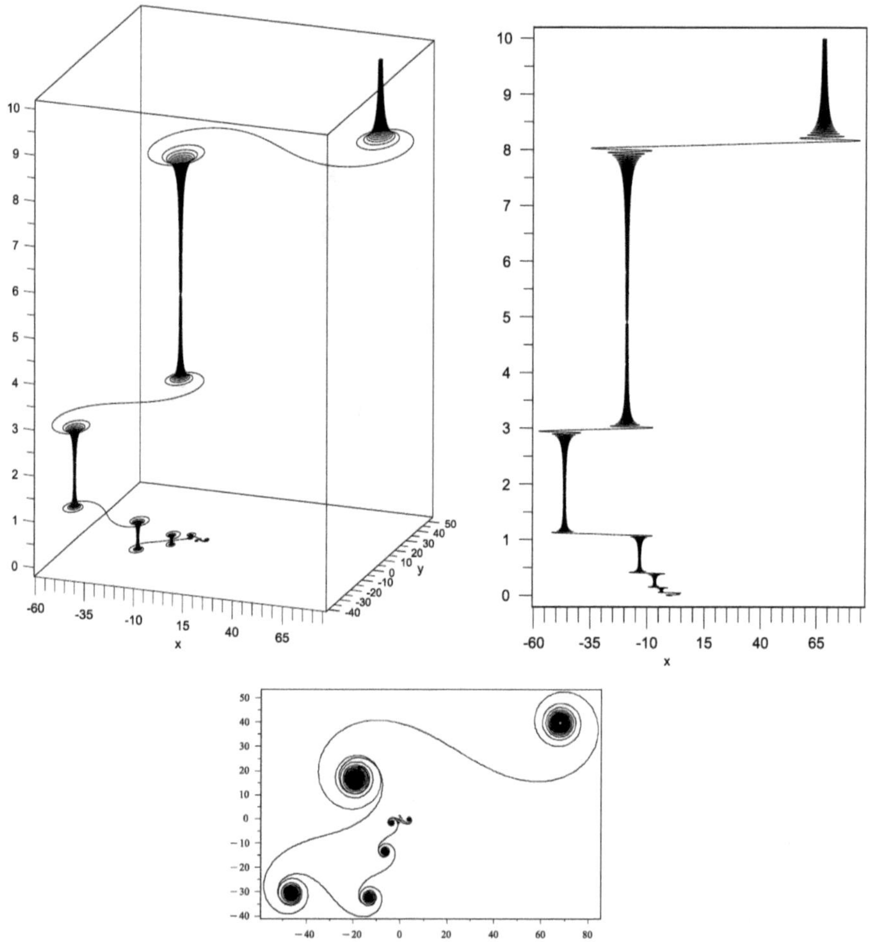

Fig. 15 The plane and the space curves of the Weyl sum $\sum_{n\le 10000} e(a_n)$ with $a_n = n \log n$

graph of the Weyl sum of the sequence $a_n = n \log n$ appears as a succession of annuli, joined by almost straight lines, corresponding to the values of H such that $\{\log H\} \approx 0$. This will happen for the values of $H \approx e^m$ for some positive integers m satisfying $1 \le m \le \log 10000$. In the case of the present example, this will happen at heights $H \approx e^m$ with $m = 1, 2, \ldots, 9$, and more visibly for $H \approx 148, 403, 1096, 2980, 8103$ all with a factor $1/1000$ (because of our definition of V_n), as indicated in the side view in the rightmost graph of Fig. 15.

2. One may do similar experiments with other types of L-functions. For example, Akbary and Murty [1] proved that the imaginary parts of the zeros of Dirichlet L-functions, modular L-functions, and Maass L-functions are also uniformly distributed modulo 1. So it should be possible to do similar experiments if one has access to the zeros of these L-functions.

3. As we see, the analysis of the patterns for the sequence $(\alpha\gamma_n)_{n\geq 1}$ largely depends on the Landau's formula for the zeros of the Riemann zeta function. Ford and Zaharescu [4] obtained the following uniform version of this formula

$$\sum_{0<\gamma\leq T} x^{\rho} = -\frac{\Lambda(n_x)}{2\pi}\frac{e^{iT\log(\frac{x}{n_x})}-1}{i\log(\frac{x}{n_x})} + O\big(\log x \log^2(Tx) + \frac{\log T}{\log x}\big),$$

where $x > 1$, $T \geq 2$, and n_x denotes the nearest prime power to x. This precise forms of Landau's formula maybe helpful in understanding of the geometric patterns appeared in this paper.

Acknowledgements I am greatly indebted to Prof. A. Akbary for pointing out a big number of grammatical mistakes, and for clearing mathematical content of the paper by asking some important questions. Also, I gratefully acknowledge the many helpful suggestions by Prof. J.-M. Deshouillers during the preparation of the chapter, and specially his suggestion for adding Fig. 10, and introducing me the paper [12]. Finally, I deem my duty to thank Prof. R. Heath-Brown for giving very valuable comments on the mathematical justification of the geometric patterns described in this chapter.

References

1. Akbary, A., Murty, M. R.: Uniform distribution of zeros of Dirichlet series. In: de Koninck J. M., Granville A., Luca F. (eds.) Anatomy of integers. CRM proceedings & lecture notes, vol. 46, pp. 143–158. American Mathematical Society, Providence (2008)
2. Dekking, F., Mendès France, M.: Uniform distribution modulo one: a geometrical viewpoint. J. Reine Angew. Math. **329**, 143–153 (1981)
3. Deshouillers, J.-M.: Geometric aspect of Weyl sums. In: Iwaniec H. (ed.) Elementary and analytic theory of numbers, Banach Center Publications. vol. 17, pp. 75–82. PWN-Polish Scientific Publishers, Warsaw (1985)
4. Ford, K., Zaharescu, A.: On the distribution of imaginary parts of zeros of the Riemann zeta function. J. Reine Angew. Math. **579**, 145–158 (2005)
5. Fujii, A.: On the uniformity of the distribution of the zeros of the Riemann zeta function. J. Reine Angew. Math. **302**, 167–205 (1978)
6. Fujii, A.: Some problems of diophantine approximation in the theory of the Riemann zeta function III. Comment. Math. Univ. St. Paul. **42**, 161–187 (1993)
7. Hlawka, E.: Über die Gleichverteilung gewisser Folgen, welche mit den Nullstellen der Zeta-funcktionen zusammenhängen. Österr. Akad. Wiss., Math.-NaturwKl. Abt. II. **184**, 459-471 (1975)
8. Ivić, A.: The Riemann zeta function: theory and applications. John Wiley & Sons, New York (1985)
9. Kuipers, L., Niederreiter, H.: Uniform distribution of sequences. Dover Publications, Mineola (2006)
10. Landau, E.: Über die Nullstellen der ζ-Funktion. Math. Ann. **71**, 548–568 (1911)
11. Odlyzko, A. M.: Home Page. Tables of zeros of the Riemann zeta function. Available at http://www.dtc.umn.edu/~odlyzko/ (2011)
12. Odlyzko, A. M.: On the distribution of spacings between zeros of the zeta function. Math. Comp. **48**, 273–308 (1987)
13. Rademacher, H.: Fourier analysis in number theory, Symposium on harmonic analysis and related integral transforms: Final technical report. Cornell University, Ithica, N.Y., 25 p (1956) (Also in: H. Rademacher, Collected Works, pp. 434–458)

14. Tenenbaum, G., Mendès France, M.: The prime numbers and their distribution (Student Mathematical Library), vol. 6. American Mathematical Society, Providence (2001)
15. Titchmarsh, E. C.: The theory of the Riemann zeta function, 2nd ed. (Revised by D.R. Heath-Brown). Oxford University Press, Oxford (1986)
16. Weyl, H.: Über die Gleichverteilung von Zahlen mod. Eins. Math. Ann. **77**, 313–352 (1916)

Fractional Cauchy Problem in Sense of the Complex Hadamard Operators

Rabha W. Ibrahim

Abstract The theory of fractional calculus has found interesting applications in the theory of analytic functions and geometric functions. We extend the definitions of the Hadamard fractional operators into the open unit disk. A class of nonlinear fractional differential equations (Cauchy-type problem) in the unit disk is studied using these fractional operators. The existence and uniqueness of the solution are established. Some properties of the integral operator are inflict such as boundedness in a space of analytic function and semigroup property. Moreover, we prove that the linear Cauchy problem (homogenous and nonhomogeneous) is solvable in a holomorphic space and its solution approximates to the Mittag–Leffler function. Examples are illustrated.

Keywords Fractional calculus · Fractional differential equation · Fractional operators

1 Introduction

Fractional calculus (real and complex) is a rapidly growing subject of interest for physicists and mathematicians. The reason for this is that problems may be discussed in a much more stringent and elegant way than using traditional methods. Fractional differential equations have emerged as a new branch of applied mathematics which has been used for many mathematical models in science and engineering. In fact, fractional differential equations are considered as an alternative model to nonlinear differential equations. Several different derivatives were introduced: Riemann–Liouville, Hadamard, Grunwald–Letnikov, Riesz, Erdelyi–Kober operators and Caputo [1–8]. There are two fractional derivatives that have been studied widely for their applications: Riemann–Liouville and Hadamard operators.

Fractional Cauchy problems replace the usual first-order time derivative by a fractional derivative. There are various studies imposed in this direction. The most resent study of this equation is the existence and uniqueness of solutions. Diethelm and Ford

R. W. Ibrahim (✉)
Institute of Mathematical Sciences, University Malaya, 50603,
Kuala Lumpur, Malaysia
e-mail: rabhaibrahim@yahoo.com

T. M. Rassias, P. M. Pardalos (eds.), *Mathematics Without Boundaries*,
DOI 10.1007/978-1-4939-1106-6_10,

[9] discussed the existence, uniqueness, and structural stability of solutions of Cauchy fractional order. The differential operators are taken in the Riemann–Liouville sense and the initial conditions are specified according to Caputo's suggestion [9]. Momani and Ibrahim [10, 11] established the existence and uniqueness of fractional Cauchy problem in sense of the Erdelyi–Kober and Weyl–Riesz operators respectively [10, 11]. The author investigated the existence and uniqueness for fractional Cauchy problem in a complex domain [12–14].

Recently, the theory of fractional calculus has found interesting applications in the theory of analytic functions and geometric functions. The classical definition of fractional operators (integral and differential) and their various other generalizations have fruitfully been applied in obtaining, for example, the characterization properties, coefficient estimates, distortion inequalities, and convolution structures for various subclasses of analytic functions and the works in the research monographs. In this note we shall debate the existence and uniqueness of Cauchy problem using the Hadamard operators in the open unit disk.

2 Hadamard Fractional Operators

In this section, we shall impose the Hadamard fractional operators in complex domain. Let $\mathcal{H}(U,\mathbb{C})$ be the space of analytic functions in the open unit disk $U = \{z : |z| < 1\}$. Assume the norm

$$\|f\|_{\alpha,\log} = \sup_{z,\zeta\in U} \left| \left(\log \frac{z}{\zeta}\right)^{\alpha} f(z) \right|$$

then the space $\mathcal{H}_{\alpha,\log}(U,\mathbb{C})$ is given by

$$\mathcal{H}_{\alpha,\log}(U,\mathbb{C}) := \{f \in \mathcal{H}(U,\mathbb{C}) : \|f\|_{\alpha,\log} < \infty\}.$$

We shall consider fractional differential equations containing Hadamard derivatives.

Definition 1 The fractional Hadamard integral of order $\alpha > 0$ is defined, for a function $f(z)$ by

$$\mathfrak{I}_z^{\alpha} f(z) := \frac{1}{\Gamma(\alpha)} \int_0^z \frac{f(\zeta)}{\zeta} \left(\log \frac{z}{\zeta}\right)^{\alpha-1} d\zeta; \quad \alpha > 0,$$

where the function $f(z)$ is analytic in simply-connected region of the complex z-plane ($\mathbb{C}$) containing the origin.

Definition 2 The fractional Hadamard derivative of order $\alpha \in (n-1, n)$ is defined, for a function $f(z)$ by

$$\mathfrak{D}_z^{\alpha} f(z) := \left(z \frac{d}{dz}\right)^n \left(\mathfrak{I}_z^{n-\alpha} f\right)(z)$$

where the function $f(z)$ is analytic in simply-connected region of the complex z-plane $\mathbb{C}$ containing the origin.

2.1 Boundedness in a Space of Analytic Function

For $0 < p < 1$, the Bergman space $\mathcal{A}^p$ is the set of functions f analytic in the unit disk $U := \{z : z \in \mathbb{C}; |z| < 1\}$ with $\|f\|^p_{\mathcal{A}^p} < \infty$, where the norm is defined by

$$\|f\|^p_{\mathcal{A}^p} = \frac{1}{\pi}\int_U |f(z)|^p dA < \infty, \quad z \in U,$$

and dA is denoted Lebesgue area measure.

Theorem 1 *Let $\alpha > 0, 0 < p < \infty$. Then the operator $\Im^\alpha_z$ is bounded in $_{\mathcal{A}^p}$ and*

$$\|\Im^\alpha_z f(z)\|^p_{\mathcal{A}^p} \leq C\|f(z)\|^p_{\mathcal{A}^p},$$

where

$$C := \int_0^1 |\frac{1}{\Gamma(\alpha)}\frac{\left(\log\left(\frac{1}{w}\right)\right)^{\alpha-1}}{w}|^p dw.$$

Proof Assume that $f(z) \in \mathcal{A}^p$. Then we have

$$\begin{aligned}\|\Im^\alpha_z f(z)\|^p_{\mathcal{A}^p} &= \frac{1}{\pi}\int_U |\Im^\alpha_z f(z)|^p dA\\ &= \frac{1}{\pi}\int_0^1 \left|\frac{1}{\Gamma(\alpha)}\int_0^z \frac{f(\zeta)}{\zeta}\left(\log\frac{z}{\zeta}\right)^{\alpha-1} d\zeta\right|^p dA.\end{aligned}$$

Let $w := \frac{\zeta}{z}$, $z \neq 0$, then we obtain

$$\begin{aligned}\|\Im^\alpha_z f(z)\|^p_{\mathcal{A}^p} &= \frac{1}{\pi}\int_0^1 \left|\frac{1}{\Gamma(\alpha)}\int_0^{\frac{\zeta}{w}} \frac{f(wz)}{w}\left(\log\frac{1}{w}\right)^{\alpha-1} dw\right|^p dA\\ &= \frac{1}{\pi}\int_0^1 \left|\frac{1}{\Gamma(\alpha)}\int_U \frac{f(wz)}{w}\left(\log\frac{1}{w}\right)^{\alpha-1} dw\right|^p dA\\ &\leq \int_0^1 \left|\frac{1}{\Gamma(\alpha)}\frac{(\log\frac{1}{w})^{\alpha-1}}{w}\right|^p \left(\frac{1}{\pi}\int_U |f(\zeta)|^p dA\right) dw\\ &:= C\|f\|^p_{\mathcal{A}^p}.\end{aligned}$$

This completes the proof.

2.2 Semigroup Property

Theorem 2 *Let* $f \in \mathcal{H}(U, \mathbb{C})$ *and* $\alpha, \beta > 0$. *Then*

$$\mathfrak{I}_z^{\alpha+\beta} f(z) = \mathfrak{I}_z^{\alpha} \mathfrak{I}_z^{\beta} f(z), \quad \alpha > 0,\ \beta > 0. \tag{1}$$

Proof For function f by using Dirichlet technique yields

$$\begin{aligned}
\mathfrak{I}_z^{\alpha}\mathfrak{I}_z^{\beta} f(z) &= \frac{1}{\Gamma(\alpha)} \int_0^z \frac{\mathfrak{I}_\zeta^{\beta} f(\zeta)}{\zeta} \left(\log \frac{z}{\zeta}\right)^{\alpha-1} d\zeta \\
&= \frac{1}{\Gamma(\alpha)} \int_0^z \left(\log \frac{z}{\zeta}\right)^{\alpha-1} \frac{d\zeta}{\zeta} \left[\frac{1}{\Gamma(\beta)} \int_\xi^\zeta \left(\log \frac{\zeta}{\xi}\right)^{\beta-1} f(\xi) \frac{d\xi}{\xi}\right] \\
&= \frac{1}{\Gamma(\alpha)\Gamma(\beta)} \int_0^z \frac{f(\xi)}{\xi} \Big(\int_\xi^\zeta \left(\log \frac{z}{\zeta}\right)^{\alpha-1} \left(\log \frac{\zeta}{\xi}\right)^{\beta-1} \Big) d\zeta\, d\xi.
\end{aligned} \tag{2}$$

Let $\omega := \frac{\log \frac{\zeta}{\xi}}{\log \frac{z}{\xi}}$, we pose

$$\begin{aligned}
\int_\xi^\zeta \left(\log \frac{z}{\zeta}\right)^{\alpha-1} \left(\log \frac{\zeta}{\xi}\right)^{\beta-1} &= \left(\log \frac{z}{\xi}\right)^{\alpha+\beta-1} \int_0^1 (1-\omega)^{\alpha-1} \omega^{\beta-1} d\omega \\
&= \left(\log \frac{z}{\xi}\right)^{\alpha+\beta-1} \frac{\Gamma(\alpha)\Gamma(\beta)}{\Gamma(\alpha+\beta)}.
\end{aligned} \tag{3}$$

By (2) and (3) we obtain,

$$\begin{aligned}
\mathfrak{I}_z^{\alpha}\mathfrak{I}_z^{\beta} f(z) &= \frac{1}{\Gamma(\alpha)\Gamma(\beta)} \int_0^z \frac{f(\xi)}{\xi} \Big(\int_\xi^\zeta \left(\log \frac{z}{\zeta}\right)^{\alpha-1} \left(\log \frac{\zeta}{\xi}\right)^{\beta-1} \Big) d\zeta\, d\xi \\
&= \frac{1}{\Gamma(\alpha)\Gamma(\beta)} \int_0^z \left(\log \frac{z}{\xi}\right)^{\alpha+\beta-1} \frac{\Gamma(\alpha)\Gamma(\beta)}{\Gamma(\alpha+\beta)} \frac{f(\xi)}{\xi} d\xi \\
&= \frac{1}{\Gamma(\alpha+\beta)} \int_0^z \frac{f(\xi)}{\xi} \left(\log \frac{z}{\xi}\right)^{\alpha+\beta-1} d\xi \\
&= \mathfrak{I}_z^{\alpha+\beta}.
\end{aligned}$$

This completes the proof.

In view of Theorem 2 and Definition 2, we have the following result:

Corollary 1 *Let* $f \in \mathcal{H}(U, \mathbb{C})$ *and* $0 < \alpha < 1$. *Then*

$$\left(\mathfrak{D}_z^{\alpha} \mathfrak{I}_z^{\alpha} f\right)(z) = f(z).$$

Lemma 1 *Let* $\alpha > 0, \mu > -1$. *Then*

$$\mathfrak{I}_z^{\alpha} (\log z)^{\mu} = \frac{\Gamma(\mu+1)}{\Gamma(\mu+\alpha+1)} (\log z)^{\mu+\alpha}$$

$$\mathfrak{D}_z^\alpha (\log z)^\mu = \frac{\Gamma(\mu+1)}{\Gamma(\mu-\alpha+1)} (\log z)^{\mu-\alpha} .$$

Proof By making use of the change of variable $\eta = \frac{\log \zeta}{\log z}$ we pose

$$\begin{aligned}
\mathfrak{J}_z^\alpha (\log z)^\mu &= \frac{1}{\Gamma(\alpha)} \int_0^z (\log \zeta)^\mu \left(\log \frac{z}{\zeta}\right)^{\alpha-1} \frac{d\zeta}{\zeta} \\
&= \frac{(\log z)^{\alpha+\mu}}{\Gamma(\alpha)} \int_0^1 (1-\eta)^{\alpha-1} \eta^\mu d\eta \\
&= \frac{(\log z)^{\alpha+\mu}}{\Gamma(\alpha)} \int_0^1 (1-\eta)^{\alpha-1} \eta^{(\mu+1)-1} d\eta \\
&= \frac{(\log z)^{\alpha+\mu}}{\Gamma(\alpha)} \frac{\Gamma(\alpha)\Gamma(\mu+1)}{\Gamma(\mu+\alpha+1)} \\
&= \frac{\Gamma(\mu+1)}{\Gamma(\mu+\alpha+1)} (\log z)^{\mu+\alpha} .
\end{aligned}$$

For $0 < \alpha < 1$, we have

$$\begin{aligned}
\mathfrak{D}_z^\alpha (\log z)^\mu &= \frac{1}{\Gamma(1-\alpha)} z \frac{d}{dz} \int_0^z (\log \zeta)^\mu \left(\log \frac{z}{\zeta}\right)^{-\alpha} \frac{d\zeta}{\zeta} \\
&= z \frac{d}{dz} \frac{(\log z)^{\mu+1-\alpha}}{\Gamma(1-\alpha)} \int_0^1 (1-\eta)^{-\alpha} \eta^\mu d\eta \\
&= z \frac{d}{dz} \frac{(\log z)^{\mu+1-\alpha}}{\Gamma(1-\alpha)} \int_0^1 (1-\eta)^{(1-\alpha)+1} \eta^{(\mu+1)-1} d\eta \\
&= z \frac{d}{dz} \frac{(\log z)^{\mu-\alpha+1}}{\Gamma(1-\alpha)} \frac{\Gamma(1-\alpha)\Gamma(\mu+1)}{\Gamma(\mu-\alpha+2)} \\
&= \frac{\Gamma(\mu+1)}{\Gamma(\mu-\alpha+1)} (\log z)^{\mu-\alpha} .
\end{aligned}$$

In the same manner of Lemma 1, we have the following result

Lemma 2 *Let $\beta \in \mathbb{C}, \alpha > 0, \mu > -1$. Then*

$$\begin{aligned}
\left[\mathfrak{J}_z^\alpha \zeta^\beta (\log \zeta)^\mu\right](z) &= \frac{\Gamma(\mu+1)}{\Gamma(\mu+\alpha+1)} z^\beta (\log z)^{\mu+\alpha} \\
\left[\mathfrak{D}_z^\alpha \zeta^\beta (\log \zeta)^\mu\right](z) &= \frac{\Gamma(\mu+1)}{\Gamma(\mu-\alpha+1)} z^\beta (\log z)^{\mu-\alpha} .
\end{aligned}$$

Lemma 3 *Let $\alpha > 0$ and $f \in \mathcal{H}(U, \mathbb{C})$. Then*

$$\left(\mathfrak{J}_z^\alpha \mathfrak{D}_z^\alpha f\right)(z) = f(z) - \frac{(\mathfrak{J}_z^{1-\alpha} f)(z))|_{z=0} (\log z)^{\alpha-1}}{\Gamma(\alpha)}$$

where $\left(\mathfrak{J}_z^{1-\alpha} f\right)(z))|_{z=0}$ means that the limit is taken at some points in a neighborhood of the origin.

Proof Since

$$\left(\mathfrak{I}_z^{1-\alpha} f\right)(z) = \left(\mathfrak{I}_z^{1}\mathfrak{D}_z^{\alpha} f\right)(z) + \left(\mathfrak{I}_z^{1-\alpha} f)(z)\right)|_{z=0}.$$

In virtue of the relation that

$$\int_0^z \left(\log\frac{z}{\zeta}\right)^{-\alpha} (\log\zeta)^{\alpha-1}\frac{d\zeta}{\zeta} = \Gamma(\alpha)\Gamma(1-\alpha),$$

yields

$$\begin{aligned}\left(\mathfrak{I}_z^{1-\alpha} f\right)(z) &= \left(\mathfrak{I}_z^{1}\mathfrak{D}_z^{\alpha} f\right)(z) + \frac{\left(\mathfrak{I}_z^{1-\alpha} f)(z)\right)|_{z=0}}{\Gamma(\alpha)\Gamma(1-\alpha)} \times \int_0^z \left(\log\frac{z}{\zeta}\right)^{-\alpha} (\log\zeta)^{\alpha-1}\frac{d\zeta}{\zeta}\\ &= \left(\mathfrak{I}_z^{1-\alpha}\mathfrak{I}_z^{\alpha}\mathfrak{D}_z^{\alpha} f\right)(z) + \frac{\left(\mathfrak{I}_z^{1-\alpha} f)(z)\right)|_{z=0}}{\Gamma(\alpha)} \times \left(\mathfrak{I}_z^{1-\alpha}(\log z)^{\alpha-1}\right)(z).\end{aligned}$$

Thus we receive

$$\mathfrak{I}_z^{1-\alpha}\left(f(z) - (\mathfrak{I}_z^{\alpha}\mathfrak{D}_z^{\alpha} f)(z) - \frac{\left(\mathfrak{I}_z^{1-\alpha} f)(z)\right)|_{z=0}}{\Gamma(\alpha)} \times (\log z)^{\alpha-1}\right) = 0.$$

Hence, this leads to the desired assertion. This completes the proof.

In this paper we concern about the Cauchy problem

$$\mathfrak{D}_z^{\alpha} u(z) = f(z,u), \tag{4}$$

$$u(z)|_{z=0} = b, \tag{5}$$

where $f : U \times \mathbb{C} \to \mathbb{C}$ is analytic in U, $u : U \to \mathbb{C}$ and

$$b := \frac{\left(\mathfrak{I}_z^{1-\alpha} f)(z)\right)|_{z=0}}{\Gamma(\alpha)} \times (\log z)^{\alpha-1}.$$

The existence and uniqueness are established for the problem (4)–(5). The initial condition $u(z)|_{z=0} = b$ signifies that the limit is assumed at some points in a neighborhood of the origin.

3 Existence and Uniqueness

In this csection, we establish the existence and uniqueness of solution of the Cauchy problem (4)–(5) in the class of holomorphic functions $\mathcal{H}_{\alpha,\log}(U,\mathbb{C})$ under the Lipschitzian condition on $f(z,u)$ on the second variable: there exists a constant $K > 0$ such that for all $u, v : U \to \mathbb{C}$ we have

$$\|f(z,u) - f(z,v)\| \le K\|u - v\|, \quad z \in U,\ K > 0. \tag{6}$$

According to Lemma 3, we have the following result:

Lemma 4 *Let* $0 < \alpha < 1$, $f(z,u) \in \mathcal{H}_{\alpha,\log}(U,\mathbb{C})$. *If* $u \in \mathcal{H}_{\alpha,\log}(U,\mathbb{C})$ *then the initial value problem (4)–(5) is equivalent to the Volterra integral equation*

$$u(z) = \mathfrak{I}_z^{\alpha} f(z,u), \quad z \in U. \tag{7}$$

Theorem 3 *Let* $0 < \alpha < 1$, $f(z,u) \in \mathcal{H}_{\alpha,\log}(U,\mathbb{C})$ *and condition (6) are satisfied. If for positive real number* ℓ

$$K < \frac{\Gamma(\alpha)}{\ell}, \quad \int_0^1 |\left(\log\frac{1}{\zeta}\right)^{\alpha-1}\frac{1}{\zeta}|d\zeta \le \ell, \tag{8}$$

then there exists a unique solution $u(z)$ *in the space* $\mathcal{H}_{\alpha,\log}(U,\mathbb{C})$ *solving the Cauchy problem (4)–(5).*

Proof Define an operator $(Tu)(z) = u(z)$ by

$$(Tu)(z) = \frac{1}{\Gamma(\alpha)}\int_0^z \frac{f(\zeta,u(\zeta))}{\zeta}\left(\log\frac{z}{\zeta}\right)^{\alpha-1}d\zeta.$$

Let $u,\ f \in \mathcal{H}_{\alpha,\log}(U,\mathbb{C})$ then it follows that

$$\begin{aligned}
|(Tu)(z)| &= |\frac{1}{\Gamma(\alpha)}\int_0^z \frac{f(\zeta,u(\zeta))}{\zeta}\left(\log\frac{z}{\zeta}\right)^{\alpha-1}d\zeta| \\
&\le \frac{\|f\|_{\mathcal{H}_{\alpha,\log}}}{\Gamma(\alpha)}|\int_0^1\left(\log\frac{1}{\zeta}\right)^{\alpha-1}\frac{d\zeta}{\zeta}| \\
&\le \frac{\ell\|f\|_{\mathcal{H}_{\alpha,\log}}}{\Gamma(\alpha)}, \quad z \to 1.
\end{aligned}$$

Hence $Tu \in \mathcal{H}_{\alpha,\log}(U,\mathbb{C})$. By using the conditions (6) and (7), we obtain

$$\begin{aligned}
|(Tu)(z) - (Tv)(z)| &= |\frac{1}{\Gamma(\alpha)}\int_0^z \frac{f(\zeta,u(\zeta))}{\zeta}\left(\log\frac{z}{\zeta}\right)^{\alpha-1}d\zeta \\
&\quad - \frac{1}{\Gamma(\alpha)}\int_0^z \frac{f(\zeta,v(\zeta))}{\zeta}\left(\log\frac{z}{\zeta}\right)^{\alpha-1}d\zeta| \\
&\le \frac{\|f(z,u) - f(z,v)\|}{\Gamma(\alpha)}|\int_0^1\left(\log\frac{1}{\zeta}\right)^{\alpha-1}\frac{d\zeta}{\zeta}| \\
&\le \frac{\ell K\|u-v\|}{\Gamma(\alpha)}, \quad z \to 1.
\end{aligned}$$

According to the Banach Theorem the problem (4)–(5) has a unique solution $u \in \mathcal{H}_{\alpha,\log}(U,\mathbb{C})$. This completes the proof.

Theorem 4 *Let* $f \in \mathcal{H}_{\alpha,\log}(U,\mathbb{C})$. *Assume that there exists a positive constant* γ *and a positive real-valued function* ϕ *of a complex variable* $z \in U$ *such that* $\phi(z) \in L^{1/\gamma}(U)$ *and* $\|f\|_{\mathcal{H}_{\alpha,\log}(U,\mathbb{C})} \le \phi(z)$. *Then (4)–(5) have a solution.*

Proof Define the operator T as in Theorem 3. We proceed to prove that Tu is closed, bounded, and convex in close domain. By employing the Hölder inequality, we receive

$$\begin{aligned}
|(Tu)(z)| &= |\frac{1}{\Gamma(\alpha)}\int_0^z \frac{f(\zeta,u(\zeta))}{\zeta}\left(\log\frac{z}{\zeta}\right)^{\alpha-1} d\zeta| \\
&\le |\frac{1}{\Gamma(\alpha)}\int_0^z \frac{\phi(\zeta)}{\zeta}\left(\log\frac{z}{\zeta}\right)^{\alpha-1} d\zeta| \\
&\le \frac{1}{\Gamma(\alpha)}\left(|\int_0^1 \left(\log\frac{1}{\zeta}\right)^{\frac{\alpha-1}{1-\gamma}}\frac{d\zeta}{\zeta}|\right)^{1-\gamma}\left(\int_0^z \phi(\zeta)^{\frac{1}{\gamma}} d\zeta\right)^{\gamma} \\
&\le \frac{\ell^{1-\gamma}}{\Gamma(\alpha)}\left(\int_0^1 \phi(\zeta)^{\frac{1}{\gamma}} d\zeta\right)^{\gamma}.
\end{aligned}$$

Next, we claim that T is completely continuous.

$$\begin{aligned}
\|(Tu_k)(z)-(Tu)(z)\| &= \frac{1}{\Gamma(\alpha)}|\int_0^z \frac{f(\zeta,u_k(\zeta))}{\zeta}\left(\log\frac{z}{\zeta}\right)^{\alpha-1} d\zeta \\
&\quad -\int_0^z \frac{f(\zeta,u(\zeta))}{\zeta}\left(\log\frac{z}{\zeta}\right)^{\alpha-1} d\zeta| \\
&\le \frac{1}{\Gamma(\alpha)}|\int_0^1 \left(\log\frac{1}{\zeta}\right)^{\alpha-1}\frac{d\zeta}{\zeta}|\|f(\zeta,u_k(\zeta))-f(\zeta,u(\zeta))\| \\
&= \frac{\ell}{\Gamma(\alpha)}\|f(\zeta,u_k(\zeta))-f(\zeta,u(\zeta))\|.
\end{aligned}$$

Hence $\|(Tu_k)(z)-(Tu)(z)\| \to 0$ when $k\to\infty$. This implies that T is continuous. Moreover, T satisfies $\|Tu\|\le r$, $r>0$; thus T is uniformly bounded. To complete our proof, we must show that T is relatively compact. Let z_1 and $z_2 \in U$ such that $z_1 \neq z_2$ we have

$$\begin{aligned}
\|(Tu)(z_2)-(Tu)(z_1)\| &= \frac{1}{\Gamma(\alpha)}\|\int_0^{z_1}\left[\left(\log\frac{z_2}{\zeta}\right)^{\alpha-1}-\left(\log\frac{z_1}{\zeta}\right)^{\alpha-1}\right] f(\zeta,u(\zeta))d\zeta \\
&\quad +\int_{z_1}^{z_2}\left[\left(\log\frac{z_2}{\zeta}\right)^{\alpha-1}\right] f(\zeta,u(\zeta))d\zeta\| \\
&\le \frac{1}{\Gamma(\alpha)}\int_0^z \|\left(\log\frac{z_2}{\zeta}\right)^{\alpha-1}-\left(\log\frac{z_1}{\zeta}\right)^{\alpha-1}\|\|f(\zeta,u(\zeta))\|d\zeta \\
&\quad +\int_{z_1}^{z_2}\|\left(\log\frac{z_2}{\zeta}\right)^{\alpha-1}\|\|f(\zeta,u(\zeta))\|d\zeta \\
&\le \frac{1}{\Gamma(\alpha)}\int_0^z \|\left(\log\frac{z_2}{\zeta}\right)^{\alpha-1}-\left(\log\frac{z_1}{\zeta}\right)^{\alpha-1}\|\phi(\zeta)d\zeta
\end{aligned}$$

$$
\begin{aligned}
&+\int_{z_1}^{z_2} \| \left(\log \frac{z_2}{\zeta}\right)^{\alpha-1} \|\phi(\zeta)d\zeta \\
&\le \frac{1}{\Gamma(\alpha)}\left(\int_0^{z_1}\left[\left(\log\frac{z_1}{\zeta}\right)^{\alpha-1}-\left(\log\frac{z_2}{\zeta}\right)^{\alpha-1}\right]^{\frac{1}{1-\gamma}} d\zeta\right)^{1-\gamma} \\
&\left(\int_0^{z_1}\phi(\zeta)^{\frac{1}{\gamma}}d\zeta\right)^{\gamma} \\
&+\frac{1}{\Gamma(\alpha)}\left(\int_{z_1}^{z_2}\left(\log\frac{z_2}{\zeta}\right)^{\frac{\alpha-1}{1-\gamma}}d\zeta\right)^{1-\gamma}\left(\int_{z_1}^{z_2}\phi(\zeta)^{\frac{1}{\gamma}}d\zeta\right)^{\gamma} \\
&\le \frac{1}{\Gamma(\alpha)}\left(\int_0^{z_1}[\left(\log\frac{z_1}{\zeta}\right)^{\alpha-1}-\left(\log\frac{z_2}{\zeta}\right)^{\alpha-1}]^{\frac{1}{1-\gamma}}d\zeta\right)^{1-\gamma} \\
&\left(\int_0^{z_1}\phi(\zeta)^{\frac{1}{\gamma}}d\zeta\right)^{\gamma}+\frac{1}{\Gamma(\alpha)}\left(\int_{z_1}^{z_2}\left(\log\frac{z_2}{\zeta}\right)^{\frac{\alpha-1}{1-\gamma}}d\zeta\right)^{1-\gamma} \\
&\left(\int_0^{z_2}\phi(\zeta)^{\frac{1}{\gamma}}d\zeta\right)^{\gamma} \\
&\le \frac{3\Phi}{\Gamma(\alpha)}\ell^{1-\gamma},
\end{aligned}
$$

where

$$
\Phi := \max_{z\in U}\left(\int_0^{z}\phi(\zeta)^{\frac{1}{\gamma}}d\zeta\right)^{\gamma}, \quad z \in U.
$$

As $z_1 \to z_2$ the above inequality tends to zero. Thus $\{Tu : u, \|u\| \le r := \frac{3\Phi}{\Gamma(\alpha)}\ell^{1-\gamma}\}$ is equicontinuous and hence T is relatively compact. By Schauder's fixed point theorem, Eqs. (4)–(5) have a solution.

4 Linear Problem

Here we discuss the existence of a solution for linear cases of Cauchy problem. A function frequently used in the solutions of fractional-order systems is the Mittag-Leffler function, which is similar to the exponential function frequently used in the solutions of integer-order systems; it is defined as

$$
E_\alpha(z) = \sum_{n=0}^{\infty}\frac{z^n}{\Gamma(n\alpha+1)}
$$

where $\alpha > 0$ and Γ is the Gamma function [3]. The Mittag-Leffler function with two parameters has the following form:

$$E_{\alpha,\beta}(z) = \sum_{n=0}^{\infty} \frac{z^n}{\Gamma(n\alpha + \beta)}$$

where $\alpha > 0, \beta > 0$ (see [3]). First, we consider the homogenous Cauchy problem

$$\mathfrak{D}_z^\alpha u(z) = \lambda u(z), \tag{9}$$

$$u(0) = \mathfrak{I}_z^\alpha b. \tag{10}$$

Thus, the Volterra integral type of the Cauchy problem (9)–(10) is given by

$$u(z) = \lambda \mathfrak{I}_z^\alpha u(z), \quad z \in U. \tag{11}$$

Hence, in term of approximation, we pose

$$u_n(z) = \lambda \mathfrak{I}_z^\alpha u_{n-1}(z), \quad z \in U. \tag{12}$$

Without loss of generality, we assume that $u(z)$ converges to z, $z \in U$ and ζ converges to 1, in a neighborhood of the origin. Consider that

$$\begin{aligned}
u_1(z) &= \lambda \mathfrak{I}_z^\alpha u_0(z) = \lambda \mathfrak{I}_z^\alpha \left[\mathfrak{I}_z^\alpha b\right] = \frac{\lambda}{\Gamma(3\alpha)} z\,(\log z)^{3\alpha-1} \\
u_2(z) &= \lambda^2 \mathfrak{I}_z^\alpha u_1(z) = \frac{\lambda^2}{\Gamma(4\alpha)} z\,(\log z)^{4\alpha-1} \\
&\vdots \\
u_n(z) &= z \sum_{k=0}^{n} \frac{\lambda^k}{\Gamma(2\alpha + \alpha k)} (\log z)^{k\alpha+2\alpha-1}.
\end{aligned}$$

By taking a $n \to \infty$ and using the Mittag-Leffler function, we receive

$$u(z) = z E_{(\alpha,2\alpha)} \left(\lambda(\log z)^\alpha\right) \tag{13}$$

We deduce the following result:

Theorem 5 *Let $0 < \alpha < 1$. Then the Cauchy problem (9)–(10) is solvable in $\mathcal{H}_{\alpha,\log}(U, \mathbb{C})$ and its solution approximates to the formula (13).*

Similarly, to the above we deduce the following: result for the nonhomogeneous Cauchy problem

$$\mathfrak{D}_z^\alpha u(z) = \lambda u(z) + f(z), \tag{14}$$

$$u(0) = \mathfrak{I}_z^\alpha b, \tag{15}$$

where f is analytic in the unit disk. Thus the Volterra integral type of the Cauchy problem (14)–(15) is given by

$$u(z) = \lambda \mathfrak{I}_z^{\alpha} u(z) + \mathfrak{I}_z^{\alpha} f(z), \quad z \in U. \tag{16}$$

Consequently, (16) has the approximate solution

$$u_n(z) = z \sum_{k=0}^{n} \frac{\lambda^k}{\Gamma(2\alpha + \alpha k)} (\log z)^{k\alpha+2\alpha-1} + \sum_{k=0}^{n} \left(\lambda^k \mathfrak{I}_z^{\alpha k+\alpha} f\right)(z), \quad (n = 1, 2, \ldots).$$

By taking $n \to \infty$ and making the change of order and summations and integration, we obtain the solution

$$u(z) = zE_{(\alpha,2\alpha)} \left(\lambda(\log z)^{\alpha}\right) + \frac{1}{\Gamma(\alpha)} \int_0^z E_{(\alpha,2\alpha)} \left(\lambda(\log \frac{z}{\zeta})^{\alpha}\right) \frac{f(\zeta)}{\zeta} \left(\log \frac{z}{\zeta}\right)^{\alpha-1} d\zeta, \tag{17}$$

$$(z \in U, 0 < \alpha < 1).$$

Here, we formulate the following result:

Theorem 6 *Let* $0 < \alpha < 1$. *Then the Cauchy problem (14)–(15) is solvable in* $\mathcal{H}_{\alpha,\log}(U, \mathbb{C})$ *and its solution approximates to the formula (17).*

Example 1 The Cauchy type problem for the fractional differential equation of order 1/2:

$$\mathfrak{D}_z^{0.5} u(z) = u(z) + z, \quad z \in U, \tag{18}$$

$$u(0) = \mathfrak{I}_z^{0.5} b \tag{19}$$

has the unique solution of the form

$$u(z) = zE_{(0.5,1)} \left((\log z)^{0.5}\right) + \frac{1}{\sqrt{\pi}} \int_0^z \left(\log \frac{z}{\zeta}\right)^{-0.5} E_{(0.5,1)} \left((\log \frac{z}{\zeta})^{0.5}\right) d\zeta$$

where the solution of the homogenous problem is given by

$$u(z) = zE_{(0.5,1)} \left((\log z)^{0.5}\right).$$

In general we have the following problem:

Example 2 The Cauchy type problem for the fractional differential equation of order 1/2:

$$\mathfrak{D}_z^{0.5} u(z) = \lambda u(z) + f(z), \quad z \in U, \tag{20}$$

$$u(0) = \mathfrak{I}_z^{0.5} b \tag{21}$$

has the unique solution of the form

$$u(z) = zE_{(0.5,1)}\left(\lambda(\log z)^{0.5}\right) + \frac{1}{\sqrt{\pi}} \int_0^z \frac{f(\zeta)}{\zeta} \left(\log \frac{z}{\zeta}\right)^{-0.5} E_{(0.5,1)}\left(\lambda \left(\log \frac{z}{\zeta}\right)^{0.5}\right) d\zeta$$

where the solution of the homogenous problem is given by

$$u(z) = zE_{(0.5,1)}\left(\lambda(\log z)^{0.5}\right).$$

Example 3 The Cauchy type problem for the fractional differential equation of order $0 < \lambda < 1$

$$\mathfrak{D}_z^{\lambda} u(z) = \lambda u(z) + \frac{1}{\lambda}, \quad z \in U, \tag{22}$$

$$u(0) = \mathfrak{I}_z^{\lambda} b \tag{23}$$

has the unique solution of the form

$$u(z) = zE_{(\lambda,2\lambda)}\left(\lambda(\log z)^{\lambda}\right) + \frac{1}{\Gamma(\lambda)} \int_0^z \left(\log \frac{z}{\zeta}\right)^{\lambda-1} E_{(\lambda,2\lambda)}\left(\lambda \left(\log \frac{z}{\zeta}\right)^{\lambda}\right) \frac{d\zeta}{\zeta}$$

where the solution of the homogenous problem is given by

$$u(z) = zE_{(\lambda,2\lambda)}\left(\lambda(\log z)^{\lambda}\right).$$

5 Conclusion

From the above information, we defined the Hadamard operators (differential and integral) of arbitrary order in the unit disk. We established that the fractional integral operator fulfil the semi group property. In addition, their applications manifested in fractional differential equations by finding the sufficient conditions for the existence and uniqueness of Cauchy problem in the unit disk.

References

1. Srivastava, H.M., Owa, S.: Univalent Functions, Fractional Calculus, and Their Applications. Wiley (Halsted Press), New York (1989)
2. Miller, K.S., Ross, B.: An Introduction to the Fractional Integrals and Derivatives—Theory and Application. Wiley, New York (1993)
3. Kiryakova, V.: Generalized Fractional Calculus and Applications. Wiley, New York (1994)
4. Podlubny, I.: Fractional Differential Equations. Academic Press, New York (1999)
5. Hilfer, R.: Applications of Fractional Calculus in Physics. World Scientific, Singapore (2000)
6. West, B.J., Bologna, M., Grigolini, P.: Physics of Fractal Operators. Springer, New York (2003)

7. Kilbas, A.A., Srivastava, H.M., Trujillo, J.J.: Theory and Applications of Fractional Differential Equations. North-Holland Mathematics Studies. Elsevier, New York (2006)
8. Sabatier, J., Agrawal, O.P., Machado, J.A.: Advance in Fractional Calculus: Theoretical Developments and Applications in Physics and Engineering. Springer, Dordrecht (2007)
9. Diethelm, K., Ford, N.J.: Analysis of fractional differential equations. J. Math. Anal. Appl. **265**, 229–248 (2002)
10. Ibrahim, R.W., Momani, S.M.: On the existence and uniqueness of solutions of a class of fractional differential equations. J. Math. Anal. Appl. **334**, 1–10 (2007)
11. Momani, S.M., Ibrahim, R.W.: On a fractional integral equation of periodic functions involving Weyl-Riesz operator in Banach algebras. J. Math. Anal. Appl. **339**, 1210–1219 (2008)
12. Ibrahim, R.W., Darus, M.: Subordination and superordination for univalent solutions for fractional differential equations. J. Math. Anal. Appl., **345**, 871–879 (2008)
13. Ibrahim, R.W., Darus, M.: On analytic functions associated with the Dziok-Srivastava linear operator and Srivastava-Owa fractional integral operator. Arab. J. Sci. Eng. **36**, 441–450 (2011)
14. Ibrahim, R.W.: Existence and uniqueness of holomorphic solutions for fractional Cauchy problem. J. Math. Anal. Appl., **380**, 232–240 (2011)

Studies on Generalized Fractional Operators in Complex Domain

Rabha W. Ibrahim

Abstract In the geometric function theory much attention is paid to various fractional operators (differential and integrals) mapping the class of univalent functions and its subclasses into themselves. Recently these operators have applications in different fields such as mathematical physics and computer sciences. In this note we shall introduce a generalized fractional differential operator for the class of univalent functions by employing the Srivastava-Owa fractional differential operator in the unit disk. Geometric properties such as convexity are discussed. Topological properties such as boundedness and compactness are studied in different spaces. Furthermore, conditions are given for the generalized fractional integral operator to be bounded in Hardy space.

Keywords Fractional calculus · Fractional differential equation · Fractional operators

1 Introduction

Fractional Calculus is the ramification of calculus that generalizes the derivative of a function to non-integer order. It was launched in 1695 when L'Hopital wrote to Leibniz demanding him about a particular notation he had utilized in his publications for the nth-derivative of the linear function $f(x) = \frac{D^n x}{Dx^n}$. L'Hopital's posed the question to Leibniz, what would the result be if $n = 1/2$. Leibniz's response: "An apparent paradox, from which one day useful consequences will be drawn." In these words fractional calculus was born. Later the theory of integrals and derivatives of non-integer order started inflowing by Liouville, Riemann, Grunwald and Letnikov.

Fractional calculus can be viewed as a branch of mathematical analysis which concerns with integro-differential operators and equations where the integrals are of convolution type and demonstrate (weakly singular) kernels of power-law type. It is exactingly linked to the theory of pseudo-differential operators. Related topics include special functions, integral transforms, stochastic processes, control theory, etc.

R. W. Ibrahim (✉)
Institute of Mathematical Sciences, University of Malaya, 50603 Kuala Lumpur, Malaysia
e-mail: rabhaibrahim@yahoo.com

T. M. Rassias, P. M. Pardalos (eds.), *Mathematics Without Boundaries*,
DOI 10.1007/978-1-4939-1106-6_11,

In the past few decades, derivatives and integrals of fractional order have various applications: kinetic theories, statistical mechanics, dynamics in complex media, and many others. Many researchers have pointed out that fractional-order models are more suitable than integer-order models for various real materials.

Fractional derivative provides an excellent tool for the depiction of memory and hereditary properties of various materials and processes. This is the main advantage of the fractional derivatives in comparison with classical integer-order models in which such effects are neglected. The advantages of the fractional derivatives appeared in modeling mechanical and electrical properties of real materials, as well as in the description of rheological properties of rocks, and social science such as food supplements, climate and economics [1, 2, 3, 4, 5, 6, 7, 8, 9, 10]. Furthermore, fractional calculus has been a fruitful field of research in engineering. In fact, many areas are currently paying attention to the fractional calculus concepts such as wave propagation, electromagnetism, chaos and fractals, heat transfer, electronics, signal processing, image processing, robotics, system identification, traffic systems, genetic algorithms, percolation, modeling and identification, telecommunications, irreversibility, physics, control systems, legged robots, redundant robots, heat diffusion, and digital circuit synthesis [11, 12, 13, 14].

In this work we shall provide a generalization of fractional differential operator in the open unit disk. Some geometric and topological properties such as boundedness and compactness are introduced. Moreover, sufficient conditions are given for the fractional integral operator to be bounded in Hardy space.

2 Complex Fractional Operators

Newly, the theory of fractional calculus has organized pleasant applications in the field of analytic function theory. The conventional definitions of fractional operators and their generalizations have fruitfully been employed in obtaining, for example, coefficient estimates, the characteristic properties, distortion inequalities and convolution structures for different subclasses of analytic functions. In [15], Srivastava and Owa, appropriated definitions for fractional operators (derivative and integral) in the complex z-plane $\mathbb{C}$ as follows:

Definition 1 The fractional derivative of order α is defined, for a function $f(z)$ by

$$D_z^\alpha f(z) := \frac{1}{\Gamma(1-\alpha)} \frac{d}{dz} \int_0^z \frac{f(\zeta)}{(z-\zeta)^\alpha} d\zeta; \quad 0 \le \alpha < 1,$$

where the function $f(z)$ is analytic in simply-connected region of the complex z-plane $\mathbb{C}$ containing the origin and the multiplicity of $(z-\zeta)^{-\alpha}$ is removed by requiring $log(z-\zeta)$ to be real when$(z-\zeta) > 0$.

Definition 2 The fractional integral of order α is defined, for a function $f(z)$, by

$$I_z^\alpha f(z) := \frac{1}{\Gamma(\alpha)} \int_0^z f(\zeta)(z-\zeta)^{\alpha-1} d\zeta; \quad \alpha > 0,$$

where the function $f(z)$ is analytic in simply-connected region of the complex z-plane ($\mathbb{C}$) containing the origin and the multiplicity of $(z-\zeta)^{\alpha-1}$ is removed by requiring $log(z-\zeta)$ to be real when $(z-\zeta)>0$.

In [16], the author derived a formula for the generalized fractional integral as follows: for natural $n \in \mathbb{N} = \{1, 2, \dots\}$ and real μ, the n-fold integral of the form

$$I_z^{\alpha,\mu} f(z) = \int_0^z \zeta_1^\mu d\zeta_1 \int_0^{\zeta_1} \zeta_2^\mu d\zeta_2 \dots \int_0^{\zeta_{n-1}} \zeta_n^\mu f(\zeta_n) d\zeta_n.$$

By applying the Dirichlet technique implies

$$\int_0^z \zeta_1^\mu d\zeta_1 \int_0^{\zeta_1} \zeta^\mu f(\zeta) d\zeta = \int_0^z \zeta^\mu f(\zeta) d\zeta \int_\zeta^z \zeta_1^\mu d\zeta_1$$
$$= \frac{1}{\mu+1} \int_0^z (z^{\mu+1} - \zeta^{\mu+1}) \zeta^\mu f(\zeta) d\zeta.$$

Repeating the above step $n-1$ times yields

$$\int_0^z \zeta_1^\mu d\zeta_1 \int_0^{\zeta_1} \zeta_2^\mu d\zeta_2 \dots \int_0^{\zeta_{n-1}} \zeta_n^\mu f(\zeta_n) d\zeta_n = \frac{(\mu+1)^{1-n}}{(n-1)!}$$
$$\int_0^z (z^{\mu+1} - \zeta^{\mu+1})^{n-1} \zeta^\mu f(\zeta) d\zeta,$$

which imposes the fractional operator type

$$I_z^{\alpha,\mu} f(z) = \frac{(\mu+1)^{1-\alpha}}{\Gamma(\alpha)} \int_0^z (z^{\mu+1} - \zeta^{\mu+1})^{\alpha-1} \zeta^\mu f(\zeta) d\zeta, \tag{1}$$

where α and $\mu \neq -1$ are real numbers and the function $f(z)$ is analytic in simply-connected region of the complex z-plane $\mathbb{C}$ containing the origin and the multiplicity of $(z^{\mu+1} - \zeta^{\mu+1})^{-\alpha}$ is removed by requiring $\log(z^{\mu+1} - \zeta^{\mu+1})$ to be real when $(z^{\mu+1} - \zeta^{\mu+1}) > 0$. When $\mu = 0$, we arrive at the standard Srivastava-Owa fractional integral.

It was shown that

$$I_z^{\alpha,\mu} z^\nu = \frac{z^{\alpha(\mu+1)+\nu}}{(\mu+1)^\alpha} \frac{\Gamma(\frac{\nu+\mu+1}{\mu+1})}{\Gamma(\alpha + \frac{\nu+\mu+1}{\mu+1})}.$$

Corresponding to the fractional integral operator, the fractional differential operator is

$$D_z^{\alpha,\mu} f(z) := \frac{(\mu+1)^\alpha}{\Gamma(1-\alpha)} \frac{d}{dz} \int_0^z \frac{\zeta^\mu f(\zeta)}{(z^{\mu+1} - \zeta^{\mu+1})^\alpha} d\zeta; \quad 0 \le \alpha < 1, \tag{2}$$

where the function $f(z)$ is analytic in simply-connected region of the complex z-plane $\mathbb{C}$ containing the origin and the multiplicity of $(z^{\mu+1} - \zeta^{\mu+1})^{-\alpha}$ is removed by requiring $\log(z^{\mu+1} - \zeta^{\mu+1})$ to be real when $(z^{\mu+1} - \zeta^{\mu+1}) > 0$. We obtain

$$D_z^{\alpha,\mu} z^\nu = \frac{(\mu+1)^{\alpha-1} \Gamma(\frac{\nu}{\mu+1} + 1)}{\Gamma(\frac{\nu}{\mu+1} + 1 - \alpha)} z^{(1-\alpha)(\mu+1)+\nu-1}.$$

The author calculated the fractal index of the differential operator (2) (see [17])

$$D_z^{\alpha,\mu} f(z) = \frac{\partial f}{\partial Z} D_z^{\alpha,\mu} Z = \frac{\partial f}{\partial Z} \Theta_{\alpha,\mu},$$

$$\Big(Z := z^{\alpha(\mu+1)}, \quad \Theta_{\alpha,\mu} = \frac{(\mu+1)^{\alpha} \Gamma(\frac{n\alpha}{\mu+1}+1)}{n\Gamma(\frac{n\alpha}{\mu+1}+1-\alpha)}, \quad f(Z) = Z^{\frac{n}{\mu+1}}\Big)$$

as follows:

$$\begin{aligned} D_z^{\alpha,\mu} f(z) &:= \frac{\partial^{\alpha(\mu+1)} f}{\partial z^{\alpha(\mu+1)}} \\ &= \frac{\partial f}{\partial Z} \frac{\partial^{\alpha(\mu+1)} Z^n}{\partial z^{\alpha(\mu+1)}} \\ &= \frac{(\mu+1)^{\alpha-1}\Gamma(\frac{n\alpha}{\mu+1}+1)}{\Gamma(\frac{n\alpha}{\mu+1}+1-\alpha)} z^{(1-\alpha)(\mu+1)+n\alpha-1} \\ &:= \frac{\partial f}{\partial Z}\Theta_{\alpha,\mu} \\ &= \frac{n\Theta_{\alpha,\mu}}{\mu+1} z^{(1-\alpha)(\mu+1)+n\alpha-1}. \end{aligned}$$

Note that the fractal index is a good method for finding exact solutions of fractional differential equations.

Let $\mathcal{A}$ denote the class of functions $f(z)$ normalized by

$$f(z) = z + \sum_{n=2}^{\infty} a_n z^n, \quad z \in U. \tag{3}$$

Also, let $\mathcal{S}$ and $\mathcal{C}$ denote the subclasses of $\mathcal{A}$ consisting of functions which are, respectively, univalent and convex in U. It is well known that; if the function $f(z)$ given by (3) is in the class $\mathcal{S}$, then $|a_n| \le n, \quad n \in \mathbb{N}\backslash\{1\}$. Moreover, if the function $f(z)$ given by (3) is in the class $\mathcal{C}$, then $|a_n| \le 1, \quad n \in \mathbb{N}$.
In our present investigation, we shall also make use of the Fox-Wright generalization ${}_q\Psi_p[z]$ of the hypergeometric ${}_qF_p$ function defined by [18]

$$\begin{aligned} {}_q\Psi_p \begin{bmatrix} (\alpha_1, A_1), \ldots, (\alpha_q, A_q); & \\ & z \\ (\beta_1, B_1), \ldots, (\beta_p, B_p); & \end{bmatrix} &= {}_q\Psi_p[(\alpha_j, A_j)_{1,q}; (\beta_j, B_j)_{1,p}; z] \\ &:= \sum_{n=0}^{\infty} \frac{\Gamma(\alpha_1 + nA_1)\ldots\Gamma(\alpha_q + nA_q)}{\Gamma(\beta_1 + nB_1)\ldots\Gamma(\beta_p + nB_p)} \frac{z^n}{n!} \\ &= \sum_{n=0}^{\infty} \frac{\prod_{j=1}^{q}\Gamma(\alpha_j + nA_j)}{\prod_{j=1}^{p}\Gamma(\beta_j + nB_j)} \frac{z^n}{n!} \end{aligned}$$

where $A_j > 0$ for all $j = 1, \ldots, q$, $B_j > 0$ for all $j = 1, \ldots, p$ and $1 + \sum_{j=1}^{p} B_j - \sum_{j=1}^{q} A_j \ge 0$ for suitable values $|z| < 1$ and α_i, β_j are complex parameters.

It is well known that

$$ {}_q\Psi_p \left[\begin{matrix} (\alpha_1, 1), \ldots, (\alpha_q, 1); \\ z \\ (\beta_1, 1), \ldots, (\beta_p, 1); \end{matrix} \right] = \Lambda^{-1} \, {}_qF_p(\alpha_1, \ldots, \alpha_q, \beta_1, \ldots, \beta_p; z) $$

where

$$ \Lambda := \frac{\prod_{j=1}^{p} \Gamma(\beta_j)}{\prod_{i=1}^{q} \Gamma(\alpha_i)} $$

and ${}_qF_p$ is the generalized hypergeometric function.

3 Geometric Properties

In this section, we proceed to impose a criteria of univalence using the differential fractional operator (2). Now by making use the operator (2), we introduce the following extension operator $\Phi^{\alpha,\mu} : \mathcal{A} \to \mathcal{A}$:

$$ \begin{aligned} \Phi^{\alpha,\mu} f(z) &:= \frac{\Gamma(\frac{1}{\mu+1}+1-\alpha)}{(\mu+1)^{\alpha-1}\Gamma(\frac{1}{\mu+1}+1)} z^{\alpha-\mu+\mu\alpha} D_z^{\alpha,\mu} f(z) \\ &= \frac{\Gamma(\frac{1}{\mu+1}+1-\alpha)}{(\mu+1)^{\alpha-1}\Gamma(\frac{1}{\mu+1}+1)} z^{\alpha-\mu+\mu\alpha} D_z^{\alpha,\mu}(z + \textstyle\sum_{n=2}^{\infty} a_n z^n) \\ &= \frac{\Gamma(\frac{1}{\mu+1}+1-\alpha)}{(\mu+1)^{\alpha-1}\Gamma(\frac{1}{\mu+1}+1)} z^{\alpha-\mu+\mu\alpha} [\frac{(\mu+1)^{\alpha-1}\Gamma(\frac{1}{\mu+1}+1)}{\Gamma(\frac{1}{\mu+1}+1-\alpha)} z^{-\alpha+\mu-\mu\alpha+1} \\ &\quad + \textstyle\sum_{n=2}^{\infty} \frac{\Gamma(\frac{n}{\mu+1}+1)}{\Gamma(\frac{n}{\mu+1}+1-\alpha)} a_n z^{n-\alpha+\mu-\mu\alpha}] \\ &= z + \textstyle\sum_{n=2}^{\infty} \frac{\Gamma(\frac{1}{\mu+1}+1-\alpha)}{\Gamma(\frac{1}{\mu+1}+1)} \frac{\Gamma(\frac{n}{\mu+1}+1)}{\Gamma(\frac{n}{\mu+1}+1-\alpha)} a_n z^n \\ &:= z + \textstyle\sum_{n=2}^{\infty} \phi_n^{\alpha,\mu} a_n z^n. \end{aligned} \tag{4} $$

In term of the Fox-Wright generalized function,

$$ \begin{aligned} \Phi^{\alpha,\mu} f(z) &= z + \sum_{n=2}^{\infty} \frac{\Gamma(n+1)\Gamma(\frac{1}{\mu+1}+1-\alpha)}{\Gamma(\frac{1}{\mu+1}+1)} \frac{\Gamma(\frac{n}{\mu+1}+1)}{\Gamma(\frac{n}{\mu+1}+1-\alpha)} \frac{a_n}{n!} z^n \\ &= \sum_{n=0}^{\infty} \frac{\Gamma(n+1)\Gamma(\frac{1}{\mu+1}+1-\alpha)}{\Gamma(\frac{1}{\mu+1}+1)} \frac{\Gamma(\frac{n}{\mu+1}+1)}{\Gamma(\frac{n}{\mu+1}+1-\alpha)} \frac{a_n}{n!} z^n \\ &= \frac{\Gamma(\frac{1}{\mu+1}+1-\alpha)}{\Gamma(\frac{1}{\mu+1}+1)} {}_2\Psi_1 \left[\begin{matrix} (1,1), (1, \frac{1}{\mu+1}); \\ z \\ (1-\alpha, \frac{1}{\mu+1}); \end{matrix} \right] * f(z) \end{aligned} \tag{5} $$

$$= \frac{\Gamma(\frac{1}{\mu+1}+1-\alpha)}{\Gamma(\frac{1}{\mu+1}+1)} {}_2\Psi_1[z] * f(z),$$

where $a_0 = 0, a_1 = 1$ and $*$ is the Hadamard product. Note that

$$\Phi^{0,0} f(z) = f(z), \quad \Phi^{1,0} f(z) = z f'(z)$$

and

$$(\Phi^{\alpha,0} f)(z) = (L(2, 2-\alpha) f)(z),$$

where $L(a, c)f$ is the Carlson-Shaffer operator.

Recently, various works, such as convolution and inclusion properties, distortion theorem, extreme points, coefficient estimates etc., are proposed by many authors for the operators due to Srivastava involving the Wright function, generalized hypergeometric function and Meijer's G-functions. These operators are Dziok-Srivastava, Srivastava-Wright, Cho-Kwon-Srivastava operator, Cho-Saigo-Srivastava operator, Jung-Kim-Srivastava and Srivastava-Owa operators (see 19–29).

Theorem 1 *Suppose that the following conditions are satisfied:*

1. $\alpha_i > 0,\ \beta_j > 0;\ i = 1, .., q, j = 1, \ldots, p;\ q \le p + 1,$
2. *for* $0 \le \alpha < 1$,

$$\frac{\Gamma(\frac{1}{\mu+1}+1-\alpha)}{\Gamma(\frac{1}{\mu+1}+1)}\Big\{ {}_2\Psi_1\begin{bmatrix}(3,1),(1+\frac{2}{\mu+1},\frac{1}{\mu+1}); \\ 1 \\ (1-\alpha+\frac{2}{\mu+1},\frac{1}{\mu+1});\end{bmatrix} + {}_2\Psi_1\begin{bmatrix}(2,1),(1+\frac{1}{\mu+1},\frac{1}{\mu+1}); \\ 1 \\ (1-\alpha+\frac{1}{\mu+1},\frac{1}{\mu+1});\end{bmatrix}\Big\} < 2.$$

Then for each univalent function $f \in \mathcal{A}$, the operator (5) is also univalent that $\Phi^{\alpha,\mu} : \mathcal{S} \to \mathcal{S}$.

Proof Assume that $f \in \mathcal{S}$ and

$$\Phi^{\alpha,\mu} f(z) = z + \sum_{n=2}^{\infty} b_n z^n,$$

where

$$b_n = \phi^{\alpha,\mu} a_n$$

and $\phi^{\alpha,\mu}$ is as in (4), that the following condition is satisfied (see [22]):

$$\rho_1 := \sum_{n=2}^{\infty} n|b_n| = \sum_{n=2}^{\infty} n\phi^{\alpha,\mu}|a_n| < 1.$$

By using the Brange's Theorem, namely: $|a_n| < n$ we estimate ρ as follows

$$\rho_1 = \sum_{n=2}^{\infty} n\phi^{\alpha,\mu}|a_n| \le \sum_{n=2}^{\infty} n^2\phi^{\alpha,\mu} = \sum_{n=2}^{\infty} \frac{n^2}{(1)_n}[\phi^{\alpha,\mu}(1)_n] = \sum_{n=2}^{\infty} \frac{n^2}{(1)_n}\rho(n) < 1, \quad (6)$$

where

$$\rho(n) := \frac{(1)_n \Gamma(\frac{1}{\mu+1}+1-\alpha)}{\Gamma(\frac{1}{\mu+1}+1)} \frac{\Gamma(\frac{n}{\mu+1}+1)}{\Gamma(\frac{n}{\mu+1}+1-\alpha)}$$

and the symbol $(a)_n$ is the Pochhammer symbol defined by

$$(a)_n = \frac{\Gamma(a+n)}{\Gamma(a)} = \begin{cases} 1, & n=0 \\ a(a+1)\ldots(a+n-1), & n=\{1,2,\ldots\}, \end{cases}$$

with $(1)_n = n!$. We proceed to transform the series in (6) into a sum of two terms using the relation

$$\frac{n^2}{(1)_n} = \frac{n}{(1)_{n-1}} = \frac{1}{(1)_{n-2}} + \frac{1}{(1)_{n-1}}.$$

Therefore, the estimate (6) takes the form

$$\begin{aligned}
\rho_1 \leq & \sum_{n=2}^{\infty} \frac{n^2}{(1)_n}\rho(n) = \sum_{n=2}^{\infty}[\frac{1}{(1)_{n-2}} + \frac{1}{(1)_{n-1}}]\rho(n) = \sum_{n=2}^{\infty}\frac{\rho(n)}{(1)_{n-2}} + \sum_{n=2}^{\infty}\frac{\rho(n)}{(1)_{n-1}} \\
= & \sum_{n=2}^{\infty} \frac{(1)_n \Gamma(\frac{1}{\mu+1}+1-\alpha)}{\Gamma(\frac{1}{\mu+1}+1)} \frac{\Gamma(\frac{n}{\mu+1}+1)}{\Gamma(\frac{n}{\mu+1}+1-\alpha)} \frac{1}{(1)_{n-2}} \\
+ & \sum_{n=2}^{\infty} \frac{(1)_n \Gamma(\frac{1}{\mu+1}+1-\alpha)}{\Gamma(\frac{1}{\mu+1}+1)} \frac{\Gamma(\frac{n}{\mu+1}+1)}{\Gamma(\frac{n}{\mu+1}+1-\alpha)} \frac{1}{(1)_{n-1}} \\
= & \frac{\Gamma(\frac{1}{\mu+1}+1-\alpha)}{\Gamma(\frac{1}{\mu+1}+1)}[\sum_{n=0}^{\infty} \frac{(1)_{n+2}\Gamma(\frac{n+2}{\mu+1}+1)}{\Gamma(\frac{n+2}{\mu+1}+1-\alpha)} \frac{1}{(1)_n} + \sum_{n=1}^{\infty} \frac{(1)_{n+1}\Gamma(\frac{n+1}{\mu+1}+1)}{\Gamma(\frac{n+1}{\mu+1}+1-\alpha)} \frac{1}{(1)_n}] \\
= & \frac{\Gamma(\frac{1}{\mu+1}+1-\alpha)}{\Gamma(\frac{1}{\mu+1}+1)}[\sum_{n=0}^{\infty} \frac{\Gamma(n+3)\Gamma(\frac{n+2}{\mu+1}+1)}{\Gamma(\frac{n+2}{\mu+1}+1-\alpha)} \frac{1}{(1)_n} + \sum_{n=1}^{\infty} \frac{\Gamma(n+2)\Gamma(\frac{n+1}{\mu+1}+1)}{\Gamma(\frac{n+1}{\mu+1}+1-\alpha)} \frac{1}{(1)_n}] \\
= & \frac{\Gamma(\frac{1}{\mu+1}+1-\alpha)}{\Gamma(\frac{1}{\mu+1}+1)}\{ {}_2\Psi_1 \left[\begin{matrix} (3,1),(1+\frac{2}{\mu+1},\frac{1}{\mu+1}); \\ 1 \\ (1-\alpha+\frac{2}{\mu+1},\frac{1}{\mu+1}); \end{matrix} \right] + {}_2\Psi_1 \left[\begin{matrix} (2,1),(1+\frac{1}{\mu+1},\frac{1}{\mu+1}); \\ 1 \\ (1-\alpha+\frac{1}{\mu+1},\frac{1}{\mu+1}); \end{matrix} \right] \\
- & \frac{\Gamma(\frac{1}{\mu+1}+1)}{\Gamma(\frac{1}{\mu+1}+1-\alpha)}\} \\
< & \ 1.
\end{aligned}$$

Thus, we transform the estimate ρ_1 in terms of the Fox-Wright function ${}_2\Psi_1$ at $z = 1$. Under the conditions (1), the series is convergent with finite values of ${}_2\Psi_1[1]$. Hence we have the desired assertion and this completes the proof.

In the similar manner of Theorem 1, we have the following result:

Theorem 2 *Assume the condition (1) holds. If for* $0 \leq \alpha < 1$,

$$\frac{\Gamma(\frac{1}{\mu+1}+1-\alpha)}{\Gamma(\frac{1}{\mu+1}+1)} {}_2\Psi_1 \left[\begin{matrix} (2,1),(1+\frac{1}{\mu+1},\frac{1}{\mu+1}); \\ 1 \\ (1-\alpha+\frac{1}{\mu+1},\frac{1}{\mu+1}); \end{matrix} \right] < 2.$$

Then the operator (5) maps a convex function $f(z)$ into a univalent function that is $\Phi^{\alpha,\mu} : \mathcal{C} \to \mathcal{S}$.

Proof Assume that $f \in \mathcal{C}$ and the operator (4) such that

$$\Phi^{\alpha,\mu} f(z) = z + \sum_{n=2}^{\infty} b_n z^n,$$

where

$$b_n = \phi^{\alpha,\mu} a_n.$$

The following condition is satisfied (see [22]):

$$\rho_2 := \sum_{n=2}^{\infty} n|b_n| = \sum_{n=2}^{\infty} n\phi^{\alpha,\mu}|a_n| < 1.$$

By using the fact that $|a_n| \le 1$ we estimate ρ_2 as follows

$$\begin{aligned}\rho_2 &= \sum_{n=2}^{\infty} n\phi^{\alpha,\mu}|a_n| \le \sum_{n=2}^{\infty} n\phi^{\alpha,\mu}\\ &= \sum_{n=2}^{\infty} \frac{n}{(1)_n}[\phi^{\alpha,\mu}(1)_n] = \sum_{n=2}^{\infty} \frac{n}{(1)_n}\rho(n)\\ &< 1, \end{aligned} \tag{7}$$

where

$$\rho(n) = \frac{(1)_n \Gamma(\frac{1}{\mu+1} + 1 - \alpha)}{\Gamma(\frac{1}{\mu+1} + 1)} \frac{\Gamma(\frac{n}{\mu+1} + 1)}{\Gamma(\frac{n}{\mu+1} + 1 - \alpha)}.$$

Since

$$\frac{n}{(1)_n} = \frac{1}{(1)_{n-1}}$$

therefore, the estimate (7) takes the formula

$$\begin{aligned}\rho_2 &\le \sum_{n=2}^{\infty} \frac{n}{(1)_n}\rho(n) = \sum_{n=2}^{\infty} \frac{1}{(1)_{n-1}}\rho(n) = \sum_{n=2}^{\infty} \frac{\rho(n)}{(1)_{n-1}}\\ &= \sum_{n=2}^{\infty} \frac{(1)_n \Gamma(\frac{1}{\mu+1} + 1 - \alpha)}{\Gamma(\frac{1}{\mu+1} + 1)} \frac{\Gamma(\frac{n}{\mu+1} + 1)}{\Gamma(\frac{n}{\mu+1} + 1 - \alpha)} \frac{1}{(1)_{n-1}}\\ &= \frac{\Gamma(\frac{1}{\mu+1} + 1 - \alpha)}{\Gamma(\frac{1}{\mu+1} + 1)} \sum_{n=1}^{\infty} \frac{(1)_{n+1} \Gamma(\frac{n+1}{\mu+1} + 1)}{\Gamma(\frac{n+1}{\mu+1} + 1 - \alpha)} \frac{1}{(1)_n}\end{aligned}$$

$$= \frac{\Gamma(\frac{1}{\mu+1}+1-\alpha)}{\Gamma(\frac{1}{\mu+1}+1)} \sum_{n=1}^{\infty} \frac{\Gamma(n+2)\Gamma(\frac{n+1}{\mu+1}+1)}{\Gamma(\frac{n+1}{\mu+1}+1-\alpha)} \frac{1}{(1)_n}$$

$$= \frac{\Gamma(\frac{1}{\mu+1}+1-\alpha)}{\Gamma(\frac{1}{\mu+1}+1)} \Big\{ {}_2\Psi_1 \left[\begin{array}{r} (2,1),(1+\frac{1}{\mu+1},\frac{1}{\mu+1}); \\ 1 \\ (1-\alpha+\frac{1}{\mu+1},\frac{1}{\mu+1}); \end{array} \right] - \frac{\Gamma(\frac{1}{\mu+1}+1)}{\Gamma(\frac{1}{\mu+1}+1-\alpha)} \Big\}$$

$$< 1.$$

Hence, we estimate ρ_2 in terms of the Fox-Wright function ${}_2\Psi_1$ at $z = 1$. This completes the proof.

4 Topological properties

In this section, we study the boundedness and compactness of the operator (5) on two spaces. A function $f \in \mathcal{H}$ (the space of analytic functions) is called in the class Σ if and only if it has the norm (see [30])

$$\|f\| = sup_{z\in U}(1-|z|^2)|\frac{f''(z)}{f'(z)}| < \infty, \quad (z \in U). \tag{8}$$

Note that the fraction $T_f := \frac{f''(z)}{f'(z)}$ is called per-Schwarzian derivative which is usually used to discuss the univalency of analytic functions (see [31, 32]). Moreover, the norm in (8) is a modification for one defined in [33].

Consider the space $\mathcal{B}_{\log}$ of all functions $f \in \mathcal{H}$ which are satisfying

$$\|f\|_{\mathcal{B}_{\log}} = \sup_{z\in U}(1-|z|^2)|\frac{f'(z)}{f(z)}| \ln\frac{1}{(1-|z|^2)} < \infty, \quad (z \in U).$$

We have the following results:

Theorem 3 *Assume that f is an analytic function on U. Then $\Phi^{\alpha,\mu} : \mathcal{B}_{\log} \to \mathcal{B}_{\log}$ is compact.*

Proof Let $f(z) \in \mathcal{H}$. Then by using some properties of the convolution product we have,

$$(1-|z|^2)|\frac{(\Phi^{\alpha,\mu}f)'(z)}{(\Phi^{\alpha,\mu}f)(z)}| \ln\frac{1}{(1-|z|^2)} = (1-|z|^2)|\frac{\left(\frac{\Gamma(\frac{1}{\mu+1}+1-\alpha)}{\Gamma(\frac{1}{\mu+1}+1)}\,{}_2\Psi_1[z]*f(z)\right)'}{\left(\frac{\Gamma(\frac{1}{\mu+1}+1-\alpha)}{\Gamma(\frac{1}{\mu+1}+1)}\,{}_2\Psi_1[z]*f(z)\right)}| \ln\frac{1}{(1-|z|^2)}$$

$$= (1-|z|^2)|\frac{{}_2\Psi_1[z]*f'(z)}{{}_2\Psi_1[z]*f(z)}| \ln\frac{1}{(1-|z|^2)}$$

$$\leq (1-|z|^2)|\frac{f'(z)}{f(z)}| \ln\frac{1}{(1-|z|^2)}$$

By taking the supremum for the last assertion over U yields $\Phi^{\alpha,\mu} : \mathcal{B}_{\log} \to \mathcal{B}_{\log}$. Now let $(f_n)_{n\in\mathbb{N}}$ be a sequence in $\mathcal{B}_{\log}$ and that $f_n \to 0$ uniformly on $\overline{U}$ as $n \to \infty$.

$$\begin{aligned}
\|\Phi^{\alpha,\mu} f_n\|_{\mathcal{B}_{\log}} &= \sup_{z\in U}(1-|z|^2)\Big|\frac{(\Phi^{\alpha,\mu} f_n)'(z)}{\Phi^{\alpha,\mu} f_n(z)}\Big|\ln\frac{1}{(1-|z|^2)}\\
&= \sup_{z\in U}(1-|z|^2)\Big|\frac{\big(\frac{\Gamma(\frac{1}{\mu+1}+1-\alpha)}{\Gamma(\frac{1}{\mu+1}+1)}\,{}_2\Psi_1[z]*f_n(z)\big)'}{\big(\frac{\Gamma(\frac{1}{\mu+1}+1-\alpha)}{\Gamma(\frac{1}{\mu+1}+1)}\,{}_2\Psi_1[z]*f_n(z)\big)}\Big|\ln\frac{1}{(1-|z|^2)}\\
&\leq \sup_{z\in U}(1-|z|^2)\Big|\frac{f_n'(z)}{f_n(z)}\Big|\ln\frac{1}{(1-|z|^2)}\\
&\leq \|f_n\|_{\mathcal{B}_{\log}}.
\end{aligned}$$

Since for $f_n \to 0$ on $\overline{U}$ we have $\|f_n\|_{\mathcal{B}_{\log}} \to 0$, and that ε is an arbitrary positive number, by letting $n \to \infty$ in the last inequality, we obtain that $lim_{n\to\infty}\|\Phi^{\alpha,\mu} f_n\| = 0$. Therefore, $\Phi^{\alpha,\mu}$ is compact.

Similarly, we have

Theorem 4 *Assume that f is an analytic function on U. Then $\Phi^{\alpha,\mu} : \Sigma \to \Sigma$ is compact.*

Theorem 5 *Let $\alpha > 0, 0 < p < \infty$ and $\mu \in \mathbb{R}$. Then the operator $I_z^{\alpha,\mu}$ is bounded in $\mathcal{H}^p$ and*

$$\|I_z^{\alpha,\mu} f(z)\|^p_{\mathcal{H}^p} \leq C\|f(z)\|^p_{\mathcal{H}^p},$$

where

$$C := \int_0^1 \Big|\frac{(\mu+1)^{1-\alpha}}{\Gamma(\alpha)}(1-w^{\mu+1})^{\alpha-1}w^{\mu}dw\Big|^p.$$

Proof Assume that $f(z) \in \mathcal{H}^p$. Then we have

$$\begin{aligned}
\|I_z^{\alpha,\mu} f(z)\|^p_{\mathcal{H}^p} &= \int_{\partial U}|I_z^{\alpha,\mu} f(z)|^p d\sigma(z)\\
&= \int_{\partial U}\Big|\frac{(\mu+1)^{1-\alpha}}{\Gamma(\alpha)}\int_0^z (z^{\mu+1}-\zeta^{\mu+1})^{\alpha-1}\zeta^{\mu} f(\zeta)d\zeta\Big|^p d\sigma(z)\\
&= \int_{\partial U}\Big|\frac{(\mu+1)^{1-\alpha}}{\Gamma(\alpha)}\int_0^z (1-\frac{\zeta^{\mu+1}}{z^{\mu+1}})^{\alpha-1}z^{(\mu+1)(\alpha-1)}\zeta^{\mu} f(\zeta)d\zeta\Big|^p d\sigma(z).
\end{aligned}$$

Let $w := \frac{\zeta}{z}$, $|w| < 1$, then we receive

$$\|I_z^{\alpha,\mu} f(z)\|^p_{\mathcal{H}^p} = \int_{\partial U}\Big|\frac{(\mu+1)^{1-\alpha}}{\Gamma(\alpha)}\int_0^{\frac{\zeta}{w}}(1-w^{\mu+1})^{\alpha-1}z^{(\mu+1)(\alpha)}w^{\mu} f(wz)dw\Big|^p d\sigma(\zeta)$$

$$\leq \int_0^1 \Big|\frac{(\mu+1)^{1-\alpha}}{\Gamma(\alpha)} \int_{\partial U} (1-w^{\mu+1})^{\alpha-1} w^{\mu} f(wz) dw\Big|^p d\sigma(\zeta)$$

$$\leq \int_0^1 \Big|\frac{(\mu+1)^{1-\alpha}}{\Gamma(\alpha)} (1-w^{\mu+1})^{\alpha-1} w^{\mu} dw\Big|^p \Big(\int_{\partial U} |f(\zeta)|^p d\sigma(\zeta)\Big)$$

$$:= C\|f\|^p_{\mathcal{H}^p}.$$

This completes the proof.

5 Conclusion

In a normalized subclass of analytic functions in the open unit disk, we defined a fractional deferential operator in the sense of the generalized Srivastava-Owa. We imposed sufficient conditions of this operator to be univalent and convex. Moreover, boundedness and compactness in some well known spaces are discussed. Finally, it was shown that the integral operator is bounded in Hardy space.

References

1. Miller, K.S., Ross, B.: An Introduction to the Fractional Integrals and Derivatives-Theory and Application. Wiley, New York (1993)
2. Kiryakova, V.: Generalized Fractional Calculus and Applications. Wiley, New York (1994)
3. Podlubny, I.: Fractional Differential Equations. Academic, New York (1999)
4. Hilfer, R.: Applications of Fractional Calculus in Physics. World Scientific, Singapore (2000)
5. West, B.J., Bologna, M., Grigolini, P.: Physics of Fractal Operators. Springer, New York (2003)
6. Kilbas, A.A., Srivastava, H.M., Trujillo, J.J.: Theory and Applications of Fractional Differential Equations. Elsevier, Amsterdam (2006)
7. Sabatier, J., Agrawal, O.P., Machado, J.A.: Advance in Fractional Calculus: Theoretical Developments and Applications in Physics and Engineering. Springer, Dordrecht (2007)
8. Lakshmikantham, V., Leela, S., Vasundhara, J.: Theory of Fractional Dynamic Systems. Cambridge Academic Publishers, Cambridge (2009)
9. Baleanu, D., Guvenc, B., Tenreiro, J.A.: New Trends in Nanotechnology and Fractional Calculus Applications. Springer, New York (2010)
10. Diethelm, K.: The Analysis of Fractional Differential Equations. Lecture Notes in Mathematics. Springer, Heidelberg (2010)
11. Machado, J.A., Silva, M.F., Barbosa, R.S., Jesus, I.S., Reis, C.M., Marcos, M.G., Galhano, A.F.: Some applications of fractional calculus in engineering. Math. Probl. Eng. **2010**, 1–34 (2010)
12. Jalab, H.A., Ibrahim, R.W.: Denoising algorithm based on generalized fractional integral operator with two parameters. Discrete Dyn. Nat. Soc. **2012**, 1–14 (2012)
13. Jalab, H.A., Ibrahim, R.W.: Texture feature extraction based on fractional mask convolution with Cesaro means for content-based image retrieval. Anthony, P., Ishizuka, M., Lukose, D. (eds.) PRICAI 2012: Trends in Artificial Intelligence, pp. 170–179. Springer, Heidelberg (2012)
14. Ibrahim, R.W.: On holomorphic solution for space- and time-fractional telegraph equations in complex domain. J. Funct. Space. Appl. **2012**, 1–10 (2012)

15. Srivastava, H.M., Owa, S.: Univalent Functions, Fractional Calculus, and Their Applications. Halsted Press, New York (1989)
16. Ibrahim, R.W.: On generalized Srivastava–Owa fractional operators in the unit disk. Adv. Differ. Equ-NY. **55**, 1–10 (2011)
17. Ibrahim, R.W.: Fractional complex transforms for fractional differential equations. Adv. Differ. Equ-NY. **192**, 1–11 (2012)
18. Srivastava, H.M., Karlsson, P.W.: Multiple Gaussian Hypergeometric Series. Halsted Press, Chichester (1985)
19. Owa, S., Srivastava, H.M.: Univalent and starlike generalized hypergeometric functions. Canad. J. Math. **39**, 1057–1077 (1987)
20. Ibrahim, R.W., Darus, M.: Differential operator generalized by fractional derivatives. Miskolc Math. Notes. **12**, 167-184 (2011)
21. Yang, Y., Tao, Y.Q., Liu, J.L.: Differential subordinations for certain meromorphically multivalent functions defined by Dziok–Srivastava operator. Abstr. Appl. Anal. **2011**,1–9 (2011)
22. Kiryakova, V.: Criteria for univalence of the Dziok-Srivastava and the Srivastava–Wright operators in the class A. Appl. Math. Copmu. **218**, 883–892 (2011)
23. Darus, M., Ibrahim, R.W.: On the existence of univalent solutions for fractional integral equation of Voltera type in complex plane. ROMAI J. **1**, 77–86 (2011)
24. Srivastava, H.M., Darus, M., Ibrahim, R.W.: Classes of analytic functions with fractional powers defined by means of a certain linear operator. Integ. Tranc. Special Funct. **22**, 17–28 (2011)
25. Piejko, K., Sokól, J.: Subclasses of meromorphic functions associated with the Cho–Kwon–Srivastava operator. J. Math. Anal. Appl. **337**, 1261–1266 (2008)
26. Sokól, J.: On some applications of the Dziok–Srivastava operator. Appl. Math. Comp. **201**, 774–780 (2008)
27. Ibrahim, R.W., Darus, M.: On analytic functions associated with the Dziok–Srivastava linear operator and Srivastava–Owa fractional integral operator. Arab. J. Sci. Eng. **36**, 441–450 (2011)
28. Frasin, B.: New properties of the Jung–Kim–Srivastava integral operators. Tamkang J. Math. **42**, 205-215 (2011)
29. Kuroki, K., Owa, S.: Double integral operators concerning starlike of order β. Int. J. Differ. Equat. **2009**, 1–13 (2009)
30. Kim, Y.C., Ponnusamy, S., Sugawa, T.: Mapping properties of nonlinear integral operators and pre-Schwarzian derivatives. J. Math. Anal. Appl. **299**, 433–447 (2004)
31. Kim, Y.C., Sugawa, T.: Norm estimates of the pre-Schwarzian derivative for certain classes of univalent functions. P. Edinburgh Math. Soc. **49**, 131–143 (2006)
32. Ibrahim, R.W., Darus, M.: General properties for Volterra-type operators in the unit disk. ISRN Math Anal. **2011**, 1–11 (2011)
33. Li, S., Stevič, S.: Volterra-type operators on Zygmund spaces. J. Inequal. Appl. **2007**, 1–10 (2007)

On Completely Monotonic and Related Functions

Stamatis Koumandos

Mathematics Subject Classification: Primary 33B15, 33B10, 33E20, 41A60, 11B83; Secondary: 41A80, 26D05, 26D15

Abstract We deal with several classes of functions, such as, completely monotonic functions, absolutely monotonic functions, logarithmically completely monotonic functions, Stieltjes functions, and Bernstein functions. We give several interesting relations among theses classes of functions as well as various examples and applications. We show that several special functions belong to the aforementioned classes. We give a survey of recent results in this area and provide new proofs as well as additional remarks and comments.

Keywords Hausdorff moment sequences · Completely monotonic functions · Bernstein functions · Logarithmically completely monotonic functions · Stieltjes functions · Gamma function · Polygamma functions · Barnes double gamma function · Bernoulli and Euler numbers

1 Introduction

Completely monotonic functions have a long history, going back to the seminal work of F. Hausdorff [31] who called such functions "total monotone". He also discovered their close relation with moment sequences of finite positive measures on $[0, 1]$. In 1928, S. N. Bernstein [19] characterized completely monotonic functions as Laplace transforms of positive measures on $[0, \infty)$ (see Theorem 2 of Sect. 2). Over the years, these functions received attention by many researchers who proved several interesting properties of them and discovered their connections to other classes of functions. For instance, I. J. Schoenberg [69] studied these functions in the context of metric spaces and showed that they can be used to construct rotation invariant positive definite functions on Euclidean spaces.

S. Koumandos (✉)
Department of Mathematics and Statistics, University of Cyprus,
P.O. Box 20537, 1678 Nicosia, Cyprus
e-mail: skoumand@ucy.ac.cy

T. M. Rassias, P. M. Pardalos (eds.), *Mathematics Without Boundaries*,
DOI 10.1007/978-1-4939-1106-6_12, © Springer Science+Business Media, LLC 2014

Completely monotonic functions emerge in many different branches of mathematics and have some remarkable applications. They are of importance in probability theory [17, 21, 26, 36], potential theory [15], mathematical physics [24], numerical analysis [78], asymptotic analysis [28, 38, 40, 48, 49, 61], and combinatorics [10]. For a detailed collection of the most significant properties of completely monotonic functions, we refer the reader to the classical books of D. Widder [77] and W. Feller [26]. The article [14] contains various results demonstrating their relationship with several other classes of functions as well as some historical comments.

The aim of the present paper is to give a systematic account of some important results concerning completely monotonic and related functions, survey some recent developments, provide new proofs and give additional remarks and comments. Several examples of special functions that are completely monotonic as well as examples of special functions that belong to other closely related classes are given.

The paper is organized in the following way. Section 2 contains definitions and basic facts about completely monotonic sequences and functions. Some related classes of functions are also defined, such as, absolutely monotonic functions, completely monotonic functions of positive order and Bernstein functions. We give their characterizations, their interrelation, and their most significant properties. In Sect. 3, we deal with two important subclasses of completely monotonic functions, namely, the logarithmically completely monotonic functions and the Stieltjes functions. We demonstrate the importance of these classes, through characterizations, various properties and applications. The relation of Stieltjes functions with some classes of holomorphic functions is also discussed. We end this section with a result concerning properties of the remainders of the asymptotic expansion of a particular class of Stieltjes functions. In Sect. 4, we give a general result on the complete monotonicity of the remainders of the asymptotic expansion of a class of Laplace transforms. We apply this result in the case of several special functions such as the logarithm of the classical Gamma function, the logarithm of Barnes' double Gamma function, the function $\psi(x)$ and the function $\beta(x)$. The generating functions of Bernoulli and Euler numbers enter naturally in this section and various properties of the remainders of their Maclaurin series expansion are given. Using the Euler–Maclaurin and Boole's summation formulae, we can derive explicit expressions for such remainders. In Sect. 5, we highlight some results on completely monotonic functions involving the Gamma and psi function. In particular, we give examples of bounded Bernstein functions, which can be used to obtain sharp estimates for trigonometric sums of hypergeometric type arising in problems of geometric function theory. A surprising connection of a problem concerning complete monotonicity of some functions involving the function $\psi(x)$ to the asymptotic behavior of a function of Hardy and Littlewood is also described. Finally, in Sect. 6, we present some examples of completely monotonic and Bernstein functions associated with a function introduced by Ramanujan in his study of approximations of the exponential function.

2 Completely Monotonic Sequences and Functions

A sequence $(a_n)_{n=0}^{\infty}$ of nonnegative real numbers is said to be completely monotonic if

$$\Delta^m a_n := \Delta^{m-1} a_n - \Delta^{m-1} a_{n+1} \geq 0, \quad \text{for all} \quad n = 0, 1, 2 \dots, \; m = 1, 2, \dots, \tag{1}$$

where Δ^0 is the identity operator: $\Delta^0 a = a$.

If a sequence $(a_n)_{n=0}^{\infty}$ is completely monotonic and nonconstant then strict inequality holds in (1), see [57].

F. Hausdorff gave the following characterization of completely monotonic sequences (cf. [77, p. 108]).

Theorem 1 *A sequence $(a_n)_{n=0}^{\infty}$ is completely monotonic if and only if*

$$a_n = \int_0^1 t^n \, d\mu(t), \quad n = 0, 1, 2, \dots, \tag{2}$$

where μ is a finite nonnegative Borel measure on $[0, 1]$.

An example of a completely monotonic sequence is

$$a_n := \frac{(\alpha)_n}{n!}, \quad n = 0, 1, 2, \dots, \quad \text{with} \quad 0 < \alpha < 1,$$

where the Pochhammer symbol $(a)_n$ is defined by

$$(a)_0 = 1, \quad (a)_n = a(a+1)\dots(a+n-1) = \frac{\Gamma(n+a)}{\Gamma(a)}, \quad n \geq 1$$

and $\Gamma(x)$ is Euler's gamma function defined by $\Gamma(x) = \int_0^{\infty} t^{x-1} e^{-t} dt$ for $x > 0$. Indeed, we have

$$\frac{(\alpha)_n}{n!} = \frac{\sin \alpha \pi}{\pi} \int_0^1 t^n \, \frac{1}{(1-t)^{\alpha} \, t^{1-\alpha}} \, dt, \quad n = 0, 1, 2, \dots .$$

More examples of completely monotonic sequences will be given in the next sections.

A function $f : (a, b) \to \mathbb{R}$ is said to be completely monotonic on (a, b), where $-\infty \leq a < b \leq \infty$, if it has derivatives of all orders and satisfies

$$(-1)^n f^{(n)}(x) \geq 0 \quad \text{for all} \quad a < x < b \quad \text{and} \quad n \geq 0. \tag{3}$$

A function $f : [a, b) \to \mathbb{R}$ is called completely monotonic on $[a, b)$, if it is continuous at $x = a$ and satisfies 3. Completely monotonic functions on $(a, b]$ and $[a, b]$ are defined in a similar way.

J. Dubourdieu [25] proved that if a nonconstant function f is completely monotonic on (a, ∞), then strict inequality holds in (3). See [73] for a simpler proof of this result.

Completely monotonic functions are of particular importance in the case where $a = 0$ and $b = \infty$ because of the following characterizations given by S. N. Bernstein, see [77, pp. 160–161].

Theorem 2 *(i) The function f is completely monotonic on $[0, \infty)$ if and only if*

$$f(x) = \int_0^\infty e^{-xt} dm(t), \tag{4}$$

for a nonnegative finite Borel measure m on $[0, \infty)$ such that the integral converges for all $x \geq 0$.

(ii) The function f is completely monotonic on $(0, \infty)$ if and only if it is representable in the form (4), where m is a nonnegative measure on $[0, \infty)$ such that the integral converges for all $x > 0$.

An application of Theorem 2 entails the completely monotonicity of the function $f(x) = 1/x^\alpha$, $\alpha > 0$, on $(0, \infty)$. Indeed, we have

$$\frac{1}{x^\alpha} = \frac{1}{\Gamma(\alpha)} \int_0^\infty e^{-xu} u^{\alpha-1} du, \quad \alpha > 0.$$

There is a close relationship between completely monotonic functions and completely monotonic sequences. This is given in the theorem below (cf. [77, p. 158]).

Theorem 3 *If a function $f(x)$ is completely monotonic on $[a, \infty)$ and if δ is any fixed positive number, then the sequence $(f(a + n\,\delta))_{n=0}^\infty$ is completely monotonic.*

Some important subclasses of completely monotonic functions have been introduced in [48]. More specifically, a function $f : (0, \infty) \to \mathbb{R}$ *is called completely monotonic of order* $\alpha \geq 0$ if $x^\alpha f(x)$ is completely monotonic on $(0, \infty)$.

We also recall that the Riemann–Liouville fractional integral $I_\alpha(m)(t)$ of order $\alpha > 0$ of a Borel measure m on $[0, \infty)$ is defined by

$$I_\alpha(m)(t) = \frac{1}{\Gamma(\alpha)} \int_0^t (t-s)^{\alpha-1} dm(s).$$

A characterization of completely monotonic functions of positive order was established in [48, Theorem 1.3]:

Theorem 4 *The function $f : (0, \infty) \to \mathbb{R}$ is completely monotonic of order $\alpha > 0$ if and only if f is the Laplace transform of a fractional integral of order α of a positive Radon measure m on $[0, \infty)$, that is,*

$$f(x) = \int_0^\infty e^{-xt} I_\alpha(m)(t)\, dt$$

and the integral converges for all $x > 0$.

The following special case is of interest in applications.

Corollary 1 *Let r be an integer* ≥ 2. *The function* $f(x)$ *is completely monotonic of order r on* $(0, \infty)$ *if and only if*

$$f(x) = \int_0^\infty e^{-xt}\, p(t)\, dt\,,$$

where the integral converges for all $x > 0$ *and* $p(t)$ *is* $r-2$ *times continuously differentiable on* $[0, \infty)$ *with*

$$p^{(r-2)}(t) = \int_0^t m([0, s])\, ds$$

for some Radon measure m and $p^{(k)}(0) = 0$ *for* $0 \leq k \leq r-2$.

For completely monotonic functions of order 1, we have

Corollary 2 *The function* $x\, f(x)$ *is completely monotonic on* $(0, \infty)$ *if and only if*

$$f(x) = \int_0^\infty e^{-xt}\, p(t)\, dt,$$

where $p(t)$ *is nonnegative, increasing on* $[0, \infty)$ *and the integral converges for all* $x > 0$.

Another characterization of completely monotonic functions of order 1 is the following.

Proposition 1 *The function* $f(x)$ *is completely monotonic of order 1 on* $(0, \infty)$ *if and only if*

$$(-1)^n\, x^{n+1}\, f^{(n)}(x)$$

is nonnegative and decreasing on $(0, \infty)$ *for all* $n = 0, 1, 2, \ldots$

The class of completely monotonic functions of order 1 coincides with the class of strongly completely monotonic functions introduced in [72] (see also [48, Proposition 1.1]). These functions are connected to the important property of superadditivity (cf. [72]).

Several examples of completely monotonic functions of order n, $n \in \mathbb{N}$, will be given is Sect. 4.

A function $f : (a, b) \to \mathbb{R}$ is called absolutely monotonic on (a, b), where $-\infty \leq a < b \leq \infty$, if it has derivatives of all orders and satisfies

$$f^{(n)}(x) \geq 0 \text{ for all } a < x < b \text{ and } n \geq 0. \tag{5}$$

A function $f : [a, b) \to \mathbb{R}$ is called absolutely monotonic on $[a, b)$, if it is continuous at $x = a$ and if it is absolutely monotonic on (a, b). Absolutely monotonic functions on $(a, b]$ and $[a, b]$ are defined in a similar way. It is clear that a power series having nonnegative coefficients defines an absolutely monotonic function on the right half of the interval of convergence. An absolutely monotonic function on $[a, b)$ extends to an analytic function defined in the open disc centered at a and of radius $b - a$. In particular, any function absolutely monotonic on $[0, \infty)$ extends to an entire function. There is an obvious connection between absolutely monotonic and completely monotonic functions: f is completely monotonic on (a, b) if and only if $x \mapsto f(-x)$ is absolutely monotonic on $(-b, -a)$.

A nontrivial example of an elementary function that is absolutely monotonic is the following. Let k be a positive integer. The function

$$\frac{\sinh(k\pi t)}{t^2+1} \tag{6}$$

is absolutely monotonic on $[0, \infty)$. See [50, Lemma 2.1]. This result can be used to prove absolute monotonicity of several special functions related to Euler's gamma function and Barnes' double and triple gamma function. See [50] for details and compare with Proposition 18 of Sect. 4. It is worth noting that the function $\sinh(at)/(t^2+1)$ *is not* absolutely monotonic on $[0, \infty)$ in the case where a is not a positive integer multiple of π.

An important counterpart of completely monotonic functions are the *Bernstein* functions. A function $f : (0, \infty) \to (0, \infty)$ is called *a Bernstein function* if f has derivatives of all orders and f' is completely monotonic on $(0, \infty)$. These functions play an important role in the theory of convolution semigroups of measures supported on the positive half line and related functional calculus, see [67]. It is easy to see, for example, that $x \mapsto x\,t/(x+t)$ is a Bernstein function on $(0, \infty)$ for all $t > 0$.

There is a characterization of Bernstein functions corresponding to Theorem 2 (cf. [20, p. 84]):

Theorem 5 *$f : (0, \infty) \to (0, \infty)$ is a Bernstein function if and only if*

$$f(x) = a\,x + b + \int_0^\infty (1 - e^{-xt})\,d\nu(t), \tag{7}$$

where a, b are nonnegative constants and ν, called the Lévy measure, is a positive measure on $(0, \infty)$ satisfying

$$\int_0^\infty \frac{t}{1+t}\,d\nu(t) < \infty. \tag{8}$$

The expression (7) is called the *Lévy–Khinchine* representation of f. It is easy to see that the condition (8) is equivalent to

$$\int_0^1 t\,d\nu(t) < \infty \quad \text{and} \quad \int_1^\infty d\nu(t) < \infty. \tag{9}$$

Since a Bernstein function is positive and increasing, it has a nonnegative limit $\lim_{x\to 0^+} f(x) := f(0+)$. It follows from the expression (7) that $b = f(0+)$. Suppose that the condition (9) is fulfilled. Since the function $(1-e^{-u})/u$ is bounded on $0 < u < \infty$, the Lebesgue dominated convergence theorem implies $\lim_{x\to\infty} \frac{1}{x} \int_0^\infty (1 - e^{-xt})\,d\nu(t) = \lim_{x\to\infty} \int_0^1 \frac{1-e^{-xt}}{xt}\,t\,d\nu(t) + \lim_{x\to\infty} \frac{1}{x} \int_1^\infty (1 - e^{-xt})\,d\nu(t) = 0.$

Hence, in the representation (7), we have $a = \lim_{x\to\infty} \dfrac{f(x)}{x}$ and in particular $f(x) = O(x)$ as $x \to \infty$.

Bounded Bernstein functions appear often in applications. They can be characterized as follows.

Proposition 2 *A Bernstein function f is bounded on $(0, \infty)$ if and only if in the representation (7) we have $a = 0$ and $\nu((0, \infty)) < \infty$.*

Proof Since $a = \lim_{x\to\infty} \frac{f(x)}{x}$ it suffices to observe that for $a = 0$, we have

$$\sup_{x\in(0,\infty)} f(x) = \lim_{x\to\infty} f(x) = b + \nu((0, \infty)).$$

□

As a consequence of the above we obtain the following.

Proposition 3 *If a Bernstein function f is bounded on $(0, \infty)$ and $c := \sup_{x\in(0,\infty)} f(x)$ then the function $c - f(x)$ is completely monotonic on $(0, \infty)$.*

Proof Using Proposition 2, we have in this case

$$c - f(x) = \int_0^\infty e^{-xt}\, d\nu(t), \quad \text{with} \quad \nu((0, \infty)) < \infty$$

and by virtue of Theorem 2 the desired result follows at once. Note that the assertion of the proposition can also be proved by a straightforward application of the definitions. □

A converse of Proposition 3 holds true:

Proposition 4 *If g is a completely monotonic function on $(0, \infty)$ with $g(0+) < \infty$, then the function $f(x) := g(0+) - g(x)$ is a bounded Bernstein function on $(0, \infty)$.*

Proof The result is obtained by a direct application of the definitions. □

The following relationship between completely monotonic and Bernstein functions is of interest.

Proposition 5 *If $f(x)$ is a Bernstein function on $(0, \infty)$, then the function $\frac{f(x)}{x}$ is completely monotonic on $(0, \infty)$.*

Proof Dividing both sides of (7) by x, we see that it is sufficient to prove that the function

$$g(x) := \int_0^\infty \frac{(1 - e^{-xt})}{x}\, d\nu(t),$$

is completely monotonic on $(0, \infty)$. We have

$$\begin{aligned}(-1)^n g^{(n)}(x) &= \int_0^\infty (-1)^n \left[\frac{(1 - e^{-xt})}{x}\right]^{(n)} d\nu(t)\\ &= \frac{n!}{x^{n+1}} \int_0^\infty \left[1 - e^{-xt} \sum_{k=0}^{n} \frac{(xt)^k}{k!}\right] d\nu(t) > 0.\end{aligned}$$

This completes the proof of the proposition. □

A different proof of Proposition 5 is given in [62]. The converse of this proposition need not be true. Consider, for example, the function $\beta(x)$ defined by

$$\beta(x) := \int_0^1 \frac{t^{x-1}}{t+1}\,dt = \sum_{k=0}^{\infty} \frac{(-1)^k}{k+x}, \quad x > 0. \tag{10}$$

The function $\beta(x)$ is completely monotonic on $(0, \infty)$. Indeed,

$$\beta(x) = \int_0^{\infty} e^{-xt}\left(1 - \frac{1}{e^t+1}\right) dt. \tag{11}$$

Since the function $p(t) := 1 - \frac{1}{e^t+1}$ is positive and strictly increasing on $(0, \infty)$, it follows from Corollary 2 that the function $x\,\beta(x)$ is completely monotonic on $(0, \infty)$ and therefore, it cannot be a Bernstein function. The same argument applies, of course, to *any* function f that is completely monotonic of order 1.

We refer to [14] and [68] for additional results and background information about Bernstein functions.

3 Logarithmically Completely Monotonic and Stieltjes Functions

In this section, we consider two important subclasses of completely monotonic functions:

A function $f : (0, \infty) \to (0, \infty)$ is called *logarithmically completely monotonic* if f has derivatives of all orders and $-(\log f)'$ is completely monotonic on $(0, \infty)$. Applying Leibniz's rule and induction (cf. [28, Lemma 2] or [8, Lemma 2.4 (ii)]) it can be shown that every logarithmically completely monotonic function is completely monotonic. The converse need not be true. Consider, for example, the function $f(x) = e^{-x} + e^{-2x}$. This is clearly a completely monotonic function on $(0, \infty)$ but not logarithmically completely monotonic since $g(x) := -(\log f(x))' = 1 + \frac{e^{-x}}{1+e^{-x}}$ is not completely monotonic because $g^{(3)}(x) = \frac{e^{-x}(-1+4e^{-x}-e^{-2x})}{(1+e^{-x})^4}$ changes sign in $(0, \infty)$. This example also shows that the sum of two logarithmically completely monotonic functions is not always a logarithmically completely monotonic function.

The function $f(x) = 1/\cosh\sqrt{x}$ is logarithmically completely monotonic and therefore, completely monotonic on $[0, \infty)$. Indeed, from the well known infinite product representation

$$\cosh z = \prod_{k=0}^{\infty}\left(1 + \frac{4z^2}{(2k+1)^2\pi^2}\right)$$

it follows that

$$(-\log f(x))' = \sum_{k=0}^{\infty} \frac{4}{4x + (2k+1)^2\pi^2}.$$

Taking into consideration the fact that the pointwise limit of a series of completely monotonic functions is also completely monotonic (cf. [77, p. 151]), the last expression entails the desired result.

In a similar way, it can be shown that the function $\sqrt{x}/\sinh\sqrt{x}$ is logarithmically completely monotonic on $[0,\infty)$. The above examples are particular cases of a more general result obtained in [34, Theorem 1.10], which states that for $\nu > -1$ the function

$$x^{\nu/2}2^{-\nu}\{I_\nu(\sqrt{x})\Gamma(\nu+1)\}^{-1}$$

is logarithmically completely monotonic on $[0, \infty)$. Here, $I_\nu(x)$ is the modified Bessel function of the first kind. All these results are, in turn, special cases of [63, Proposition 2.1].

Let us recall the classical notion of a *logarithmically convex function* as a positive function on an interval such that its logarithm is a convex function. It is immediately obvious that every logarithmically completely monotonic function is logarithmically convex. The stronger statement that *every* completely monotonic function on $(0, \infty)$ is logarithmically convex is also true, see [77, p. 167].

Logarithmically completely monotonic functions can be characterized in the following way (cf. [33, Theorem 4.4]).

Theorem 6 *For a function* $f : (0, \infty) \to (0, \infty)$ *the statements below are equivalent:*

(*i*) f *is logarithmically completely monotonic,*
(*ii*) f^α *is completely monotonic for all* $\alpha > 0$,
(*iii*) $f^{1/n}$ *is completely monotonic for all* $n = 1, 2, \ldots$.

This theorem expresses the relationship between logarithmically completely monotonic functions and infinitely divisible distributions in the convolution sense: A measure μ on $[0, \infty)$ is called *infinitely divisible* if for each $n \in \mathbb{N}$ there exists a measure μ_n on $[0, \infty)$ such that $\mu = \mu_n * \mu_n * \ldots * \mu_n$ (n times), where $*$ denotes the convolution of measures. Applying Theorem 2 and the convolution theorem for Laplace transform (cf. [77, p. 89]), we obtain an equivalent formulation of Theorem 6.

Theorem 7 *A function* $f : (0, \infty) \to (0, \infty)$ *is logarithmically completely monotonic if and only if*

$$f(x) = \int_0^\infty e^{-xt}\, d\mu(t),$$

where μ *is an infinitely divisible measure on* $[0, \infty)$ *and the integral converges for all* $x > 0$.

Using Theorem 6, we easily obtain another characterization of Bernstein functions.

Theorem 8 *A function* $f : (0, \infty) \to (0, \infty)$ *is a Bernstein function if and only if for every* $\alpha > 0$ *the function* $e^{-\alpha f(x)}$ *is completely monotonic.*

We next consider a subclass of logarithmically completely monotonic functions, which is of importance in many applications, see for example [32]:

A function $f : (0, \infty) \to \mathbb{R}$ is called *a Stieltjes function*, if it is of the form

$$f(x) = c + \int_0^\infty \frac{d\mu(t)}{x+t},$$

where c is a nonnegative constant and μ is a positive measure on $[0, \infty)$ making the integral convergent for any $x > 0$. An example of a Stieltjes function is the function

$$\frac{\log(x+1)}{x} = \int_0^\infty \frac{1}{x+t}\,\frac{h(t)}{t}\,dt, \tag{12}$$

where $h(t) = 0$, for $0 \le t < 1$ and $h(t) = 1$ for $t \ge 1$.

It is clear that every Stieltjes function is completely monotonic on $(0, \infty)$. The converse need not be true. For example, the function

$$\frac{1}{x(1+x^2)} = \int_0^\infty e^{-xt}\,(1 - \cos t)\,dt \tag{13}$$

while it is obviously completely monotonic, it *is not* a Stieltjes function. Observe that, every Stieltjes function has an analytic extension in the cut plane $\mathbb{A} := \mathbb{C} \setminus (-\infty, 0]$, therefore the function $\frac{1}{x(1+x^2)}$ cannot be a Stieltjes function, since it has poles at $\pm i$.

Some interesting examples of Stieltjes functions can be found in [7, 13], and [16].

There is a fundamental relationship between Stieltjes functions and Laplace transforms (cf. [77, p. 334]).

Theorem 9

$$F(x) = \int_0^\infty \frac{d\mu(t)}{x+t}, \quad \textit{for all} \ \ x > 0,$$

where μ is a positive measure on $[0, \infty)$, if and only if

$$F(x) = \int_0^\infty e^{-xt}\,f(t)\,dt \ \ \textit{with} \ \ f(t) = \int_0^\infty e^{-ts}\,d\mu(s).$$

It has been shown in [29] that every completely monotonic density is infinitely divisible. Accordingly, combining Theorems 7 and 9, we deduce that every Stieltjes function is logarithmically completely monotonic.

Stieltjes functions can be characterized by means of their analytic extension mentioned above. We have, in fact, the following theorem (cf. [3, p. 127]).

Theorem 10 *A function $f : (0, \infty) \to \mathbb{R}$ is a Stieltjes function if and only if $f(x) \ge 0$ for all $x > 0$ and it has an analytic extension to the cut plane $\mathbb{A} = \mathbb{C} \setminus (-\infty, 0]$ satisfying Im $f(x + iy) \le 0$ for all $y > 0$.*

Applying this theorem we easily derive the following property.

Proposition 6 *If f is a Stieltjes function, then for every $\alpha \in (0, 1)$ the function f^α is also a Stieltjes function.*

Combining this Proposition with Theorem 6 (iii) we have another proof that every Stieltjes function is logarithmically completely monotonic. Of course, these two classes do not coincide. Consider, for instance, the function

$$f(x) = \frac{1}{x^2(1+x^2)} = \int_0^\infty e^{-xt}\,(t - \sin t)\,dt\,.$$

This is logarithmically completely monotonic because

$$(-\log f(x))' = \frac{2}{x} + \frac{2x}{1+x^2} = 2\int_0^\infty e^{-xt}\,(1+\cos t)\,dt.$$

The function $f(x)$ is not a Stieltjes function for the same reason as in the example (13).

There is an important relation between the class of Stieltjes functions and the class of Bernstein functions. This is given in the next proposition.

Proposition 7 *Suppose that f is a Stieltjes function that is not identically zero. Then $1/f$ is a Bernstein function.*

A proof of this result can be found in [16]. The converse of the above proposition need not be true. Although, the reciprocal of a not identically zero Bernstein function is always completely monotonic, see [14, Theorem 5.4], it is not necessarily a Stieltjes function. A typical example is the function $f(x) = 1 - e^{-x}, \ \ x > 0$.

The class of Stieltjes functions is closely related to another class, which is of importance in geometric function theory.

Let $(a_k)_{k=0}^{\infty}$ be a completely monotonic sequence with $a_0 = 1$, equivalently (cf. Theorem 1)

$$a_k = \int_0^1 t^k \, \mathrm{d}\omega(t), \quad k = 0, 1, \dots,$$

where ω is a probability measure on $[0,\ 1]$. We denote by $\mathcal{T}$ the class of functions F of the form

$$F(x) = \sum_{k=0}^{\infty} a_k\, x^k = \int_0^1 \frac{\mathrm{d}\omega(t)}{1 - t\,x}, \quad |x| < 1.$$

Every function F in $\mathcal{T}$ has an analytic extension to the cut plane $\mathbb{L} := \mathbb{C} \setminus [1, \infty)$.

There is an one-to-one correspondence between functions in the class $\mathcal{T}$ and functions f in the class $\mathcal{S}$ of *normalized Stieltjes functions*, that is, functions f of the form

$$f(x) = \int_0^\infty \frac{\mathrm{d}\nu(t)}{t + x}, \quad \text{with } \ f(1) = 1, \ \ x > 0.$$

Indeed, we can easily prove the following.

Proposition 8 *If $F(x) \in \mathcal{T}$ then $f(x) = \dfrac{1}{x} F\left(\dfrac{x-1}{x}\right) \in \mathcal{S}$. Conversely, if $f(x) \in \mathcal{S}$ then $F(x) = \dfrac{1}{1-x} f\left(\dfrac{1}{1-x}\right) \in \mathcal{T}$.*

A function f analytic in $\mathbb{L} = \mathbb{C} \setminus [1, \infty)$, normalized by $f(0) = f'(0) - 1 = 0$ is called *universally convex* if it maps every half-plane containing the point 1 in its boundary and the origin in its interior univalently onto a convex domain.

It can be shown that (cf. [66])

Proposition 9 *If f is universally convex then f is univalent in the whole of $\mathbb{L}$ and maps $\mathbb{L}$ onto a domain starlike with respect to the origin. Such f maps every disc and every half-plane contained in $\mathbb{L}$ onto a convex domain.*

A characterization of universally convex functions is the following (cf. [66]).

Theorem 11 *A function f analytic in $\mathbb{L} = \mathbb{C} \setminus [1, \infty)$, normalized by $f(0) = f'(0) - 1 = 0$, is universally convex if and only if*

$$1 + \frac{z}{2} \frac{f''(z)}{f'(z)} \in \mathcal{T}.$$

It is worth mentioning here some examples of universally convex functions: Consider the polylogarithms defined by

$$\mathrm{Li}_\alpha(z) := \sum_{k=1}^{\infty} \frac{z^k}{k^\alpha}, \alpha > 0.$$

The function $\mathrm{Li}_\alpha(z)$ is universally convex if and only if $\alpha \geq 1$ (cf. [66]).

Let $a, b, c \neq 0$ and $-1 \leq a \leq \min\{1, c\}$, $-1 \leq b \leq c$. Then, the function $f(z) = \frac{c}{a\,b}\left({}_2F_1(a, b\,;\, c\,;\, z) - 1 \right)$ is universally convex (cf. [66]). As usual, ${}_2F_1(a, b\,;\, c\,;\, z)$ denotes the Gaussian hypergeometric function, see [4].

Universally starlike functions can also be defined and characterized in an analogous way by means of the class $\mathcal{T}$, see [66] for details.

For the relation of Stieltjes functions with the class of Nevanlinna–Pick holomorphic functions, see [14].

Next, we consider a subclass of Bernstein functions, which is closely related to Stieltjes functions.

A function $f : (0, \infty) \to \mathbb{R}$ is called *complete Bernstein function* if it is of the form

$$f(x) = x^2 \int_0^{\infty} e^{-xs} \rho(s)\, ds, \qquad x > 0, \tag{14}$$

where ρ is a Bernstein function. An example of a complete Bernstein function is the function

$$x\, e^x\, E_1(x) = 1 - e^x\, E_2(x) = x^2 \int_0^{\infty} e^{-xs} \log(s + 1)\, ds, \quad x > 0,$$

where $E_n(x) := \int_1^{\infty} e^{-xt}/t^n\, dt, \quad n = 0, 1, 2, \ldots, \quad x > 0$ is the exponential integral of order n, see [1, p. 228]. Note that the function $\log(x + 1)$ is a Bernstein function with Lévy–Khinchine representation

$$\log(x + 1) = \int_0^{\infty} (1 - e^{-xt}) \frac{e^{-t}}{t}\, dt. \tag{15}$$

Using Theorems 2, 5, and 9, we derive characterizations of complete Bernstein functions (cf. [67, Theorem 1.5]).

Theorem 12 *The following properties are equivalent:*

(*i*) *The function* $f : (0, \infty) \to \mathbb{R}$ *is complete Bernstein.*
(*ii*) *The function* $f(x)/x$ *is a Stieltjes function.*
(*iii*) f *is a Bernstein function with Lévy–Khinchine representation*

$$f(x) = a\,x + b + \int_0^\infty (1 - e^{-xt})\, g(t)\, dt,$$

where $a,\, b \geq 0$ *and the function* $g(t)$ *is completely monotonic on* $(0, \infty)$.

Applying this Theorem and using either (15) or (12), we conclude that the function $\log(x+1)$ is a complete Bernstein function.

D. V. Widder, see [76, Theorem 10.1], gave the following characterization of Stieltjes functions:

Theorem 13 f *is a Stieltjes function if and only if*

$$(x^n\, f(x))^{(n)}$$

is completely monotonic on $(0, \infty)$ *for all* $n = 0, 1, 2 \ldots$.

In order to demonstrate the usefulness of Theorem 13, let us consider an application obtained in [51].

Proposition 10 *Let* a_0 *be a nonnegative constant. We suppose that for all* $n = 0, 1, 2 \ldots$

$$F(x) = a_0 + \frac{a_1}{x} + \ldots + \frac{a_n}{x^n} + (-1)^n\, R_n(x)$$

and the function $R_n(x)$ *is completely monotonic of order* n *on* $(0, \infty)$ *for all* $n = 0, 1, 2 \ldots$. *Then,* $F(x)$ *is a Stieltjes function on* $(0, \infty)$.

Proof We observe that

$$x^n\, F(x) = a_0\, x^n + a_1\, x^{n-1} + \ldots + a_{n-1}\, x + a_n + (-1)^n\, x^n\, R_n(x)\,, \qquad (16)$$

for all $n = 0, 1, 2 \ldots$. Since the function $R_n(x)$ is completely monotonic of order n on $(0, \infty)$, the function $x^n\, R_n(x)$ is completely monotonic on $(0, \infty)$ and so is the function

$$(-1)^n\, (x^n\, R_n(x))^{(n)},$$

for all $n = 0, 1, 2 \ldots$. It follows from (16) that

$$(x^n\, F(x))^{(n)} = a_0\, n! + (-1)^n\, (x^n\, R_n(x))^{(n)}$$

and this implies that the function $(x^n\, F(x))^{(n)}$ is completely monotonic on $(0, \infty)$ for all $n = 0, 1, 2 \ldots$. Using Theorem 13, we deduce that $F(x)$ is a Stieltjes function on $(0, \infty)$.

It turns out that a converse of Proposition 10 holds true. This has also been established in [51].

Theorem 14 *Suppose that* $F(x) = \int_0^\infty e^{-xt} f(t)dt, \quad x > 0$, *where* $f(t)$ *is completely monotonic on* $[0,\infty)$ *and analytic at the origin. Then*

$$F(x) = \sum_{k=0}^{n-1} \frac{f^{(k)}(0)}{x^{k+1}} + (-1)^n R_n(x)$$

and $R_n(x)$ *is a completely monotonic function of order n on* $(0, \infty)$, *for all* $n \geq 0$ $(R_0(x) = F(x))$. *The remainder* $R_n(x)$ *satisfies the estimate*

$$0 < R_n(x) < (-1)^n \frac{f^{(n)}(0)}{x^{n+1}}, \quad \textit{for all} \;\; n = 0, 1, 2, \dots \;\; x > 0.$$

As an application, consider the asymptotic expansion

$$e^x x^{-a} \int_x^\infty e^{-t} t^{a-1} dt \sim \frac{1}{x} + \frac{a-1}{x^2} + \frac{(a-1)(a-2)}{x^3} + \dots, \text{ as } x \to \infty.$$

For $a < 1$, we define

$$F_a(x) := e^x x^{-a} \int_x^\infty e^{-t} t^{a-1} dt, \quad x > 0$$

and observe that

$$F_a(x) = \int_0^\infty e^{-xu} \frac{1}{(1+u)^{1-a}} du.$$

The function

$$f(u) = \frac{1}{(1+u)^{1-a}} = \sum_{k=0}^\infty \frac{(1-a)_k (-1)^k}{k!} u^k, \quad |u| < 1,$$

is analytic at $u = 0$. It is also completely monotonic on $[0, \infty)$, because

$$f(u) = \frac{1}{\Gamma(1-a)} \int_0^\infty e^{-us} e^{-s} s^{-a} ds.$$

Applying Theorem 14 gives

$$F_a(x) = \sum_{k=0}^{n-1} \frac{(-1)^k (1-a)_k}{x^{k+1}} + (-1)^n \kappa_n(x)$$

and $\kappa_n(x)$ is a completely monotonic function of order n on $(0, \infty)$ satisfying

$$0 < \kappa_n(x) < \frac{(1-a)_n}{x^{n+1}}, \quad \text{for } x > 0 \text{ and } n \geq 0.$$

Of course, $F_a(x)$ is a Stieltjes function because

$$F_a(x) = \frac{1}{\Gamma(1-a)} \int_0^\infty \frac{1}{x+s} e^{-s} s^{-a} ds.$$

Remark 1 A function f is completely monotonic on $[0, \infty)$, although necessarily analytic on $(0, \infty)$, need not be analytic at the end-point $x = 0$, even if all derivatives

$f^{(n)}(0^+)$ are finite for $n = 0, 1, 2 \ldots$. An example of such function is

$$f(x) = \int_0^\infty e^{-xt}\, d\alpha(t), \quad \alpha(t) = 1 - \exp(-t^{1/2}), \tag{17}$$

see [56].

Nevertheless, it is possible to obtain the following characterization (cf. [52]).

Theorem 15 *The function* $f(t) = \int_0^\infty e^{-ts}\, d\mu(s)$*, where* μ *is a finite positive measure on* $[0, \infty)$ *making the integral convergent for all* $t \geq 0$*, is analytic at* $t = 0$ *if and only if*

$$s_n(\mu) := \int_0^\infty t^n\, d\mu(t) < \infty \quad \textit{for all} \quad n = 0, 1, 2, \ldots \quad \textit{and} \quad \limsup_n \frac{[s_n(\mu)]^{1/n}}{n} < \infty.$$

Observe that for the measure α in (17) we have $s_n(\alpha) = (2n)!$ for all $n \geq 0$ and $\lim_{n\to\infty} \frac{[s_n(\alpha)]^{1/n}}{n} = \infty$.

In view of Theorems 9 and 15, Theorem 14 is equivalent to the following.

Theorem 16 *Suppose that* $F(x) = \int_0^\infty \frac{d\mu(t)}{x+t}, x > 0$ *with* $s_n(\mu) < \infty$ *for all* $n = 0, 1, 2, \ldots$ *and* $\limsup_n \frac{[s_n(\mu)]^{1/n}}{n} < \infty$*. Then*

$$F(x) = \sum_{k=0}^{n-1} \frac{(-1)^k\, s_k(\mu)}{x^{k+1}} + (-1)^n R_n(x)$$

and $R_n(x)$ *is a completely monotonic function of order* n *on* $(0, \infty)$*, for all* $n \geq 0$*. The remainder* $R_n(x)$ *satisfies the estimate*

$$0 < R_n(x) < \frac{s_n(\mu)}{x^{n+1}}, \quad \textit{for all} \quad n = 0, 1, 2, \ldots, \quad x > 0.$$

4 Complete Monotonicity of the Remainders of Some Asymptotic Expansions

Motivated by Theorem 14, it is of interest to examine the complete monotonicity of the remainders in other asymptotic expansions and give applications to various known special functions. We begin with a variant of Theorem 14 established in [51].

Theorem 17 *Suppose that*

$$F(x)=\int_0^\infty e^{-xt}g(t)\,dt,\quad x>0,$$

where $g(t)=f(t^2)$ *and* $f(t)$ *is completely monotonic on* $[0,\infty)$ *and analytic at the origin. Then*

$$F(x)=\sum_{k=0}^{n-1}\frac{g^{(2k)}(0)}{x^{2k+1}}+(-1)^n R_n(x),$$

where the remainder $R_n(x)$ *is a completely monotonic function of order* n *on* $(0,\ \infty)$, *for all* $n\geq 0$ $(R_0(x)=F(x))$. *The remainder* $R_n(x)$ *satisfies the estimate*

$$0<R_n(x)<(-1)^n\frac{g^{(2n)}(0)}{x^{2n+1}},\quad \textit{for all}\ \ n=0,1,2,\ldots,\quad x>0.$$

Furthermore, $R_n(x)$ *has the representation*

$$R_n(x)=(-1)^n\int_0^\infty e^{-xt}\left(g(t)-\sum_{k=0}^{n-1}\frac{g^{(2k)}(0)}{(2k)!}t^{2k}\right)dt.$$

Our first application of the above Theorem concerns the remainder $r_n(x)$ of the asymptotic expansion of the logarithm of Euler's gamma function, viz.

$$\log\,\Gamma(x)=\left(x-\frac{1}{2}\right)\log x-x+\frac{1}{2}\log(2\pi)++\sum_{k=1}^{n}\frac{B_{2k}}{(2k-1)\,2k}\frac{1}{x^{2k-1}}$$
$$+(-1)^n r_n(x), \tag{18}$$

where B_k are the Bernoulli numbers defined by

$$\frac{t}{e^t-1}=\sum_{k=0}^\infty B_k\frac{t^k}{k!}=1-\frac{t}{2}+\sum_{j=1}^\infty B_{2j}\frac{t^{2j}}{(2j)!},\quad |t|<2\pi. \tag{19}$$

It has been shown in [6] that for all $n=0,1,2,\ldots,$ the remainder $r_n(x)$ is a completely monotonic function on $(0,\ \infty)$. A simpler proof of this result was given in [38]. Then, the stronger result that the function $r_n(x)$ is completely monotonic of order $k\leq n$ has been established in [48, Theorem 2.1]. Next, we give a new proof of this as an application of Theorem 17. An estimate for the function $r_n(x)$ is given in [40]. We are able to obtain the same estimate by Theorem 17, too. We have, in fact, the following.

Proposition 11 *(i) The remainder* $r_n(x)$ *in the asymptotic expansion 18 is a completely monotonic function of order* n *on* $(0,\ \infty)$, *for all* $n\geq 0$.

(*ii*) *The following inequality holds true*

$$0 < r_n(x) < (-1)^n \frac{B_{2n+2}}{(2n+1)(2n+2)} \frac{1}{x^{2n+1}},$$

for all $x > 0$ *and* $n \geq 0$.

Proof We observe that

$$g(t) := \frac{1}{t^2}\left(\frac{t}{e^t - 1} - 1 + \frac{t}{2}\right) = \sum_{k=1}^{\infty} \frac{2}{t^2 + 4\pi^2 k^2},$$

see [37, p. 419]. The series representation on the right hand side reveals that the function $f(t) := g(\sqrt{t}),\ t \geq 0$ is completely monotonic on $[0, \infty)$. On the other hand, the expansion (19) implies that $f(t)$ is also analytic at $t = 0$ and

$$g^{(2k)}(0) = \frac{B_{2k+2}}{(2k+1)(2k+2)}.$$

Applying Theorem 17 in Binet's formula

$$\log \Gamma(x) = \left(x - \frac{1}{2}\right) \log x - x + \frac{1}{2} \log (2\pi)$$
$$+ \int_0^{\infty} \left(\frac{t}{e^t - 1} - 1 + \frac{t}{2}\right) \frac{e^{-xt}}{t^2}\, dt, \quad x > 0,$$

see [4, p. 28], the assertions of the proposition follow. □

For all $t > 0$, we have

$$\frac{t}{e^t - 1} = 1 - \frac{t}{2} + \sum_{k=1}^{n} \frac{B_{2k}}{(2k)!} t^{2k} + (-1)^n\, t^{2n+2}\, V_n(t). \tag{20}$$

From Binet's formula and the above, we infer that the remainder $r_n(x)$ in the asymptotic expansion (18) has the form

$$r_n(x) = \int_0^{\infty} e^{-xt}\, t^{2n}\, V_n(t)\, dt. \tag{21}$$

It is shown in [71, p. 64], employing methods of complex analysis, that the remainder term $V_n(t)$ is

$$V_n(t) = \sum_{k=1}^{\infty} \frac{2}{(t^2 + 4\pi^2 k^2)(2\pi k)^{2n}}. \tag{22}$$

In [38], it is proven that $V_n(t)$ takes also the form

$$V_n(t) = \frac{1}{(2n+1)!} \frac{1}{e^t - 1} \int_0^1 e^{tu}\, (-1)^n\, B_{2n+1}(u)\, du, \tag{23}$$

where $B_{2n+1}(u)$ are the Bernoulli polynomials defined by

$$\frac{t\,e^{tu}}{e^t-1}=\sum_{k=0}^{\infty}B_k(u)\,\frac{t^k}{k!},\qquad |t|<2\pi.$$

Both (22) and (23) imply that $V_n(t)>0$ for all $t\geq 0$ and $n=0,1,2,\ldots$.

The proof of (23) given in [38] is particularly short: Apply the Euler–Maclaurin summation formula

$$\frac{f(0)+f(1)}{2}-\int_0^1 f(u)\,du=\sum_{j=1}^{n}\frac{B_{2j}}{(2j)!}\Big(f^{(2j-1)}(1)-f^{(2j-1)}(0)\Big)$$
$$+\frac{1}{(2n+1)!}\int_0^1 f^{(2n+1)}(u)\,B_{2n+1}(u)\,du\,, \tag{24}$$

see [4, p. 617], for $f(t)=e^{tu}$ to deduce (23).

In order to prove that (23) implies (22), we recall that

$$B_{2n+1}(u)=2\,(-1)^{n-1}\,(2n+1)!\,\sum_{k=1}^{\infty}\frac{\sin(2k\pi u)}{(2k\pi)^{2n+1}} \tag{25}$$

see [1, p. 805]. Substituting (25) in (23) and using the elementary integral

$$\int_0^1 e^{tu}\,\sin(2k\pi\,u)\,du=-2k\pi\,\frac{e^t-1}{t^2+4\pi^2k^2}$$

we obtain (22).

Suppose that the function $f(u)$ is defined on $[m,\infty)$ and has derivatives of all orders satisfying $\lim_{u\to\infty}f^{(k)}(u)=0$ for all $k=0,1,2,\ldots$. Then the following variant of the Euler-Maclaurin summation formula holds (cf. [71, p. 12]).

$$\sum_{j=m}^{\infty}f(j)=\int_m^{\infty}f(u)\,du+\frac{1}{2}\,f(m)-\sum_{j=1}^{n}\frac{B_{2j}}{(2j)!}\,f^{(2j-1)}(m)$$
$$+\frac{1}{(2n+1)!}\int_m^{\infty}f^{(2n+1)}(u)\,\tilde{B}_{2n+1}(u)\,du\,, \tag{26}$$

where $m=0,1,2\ldots$ and $\tilde{B}_n(x)$, $n=0,1,2\ldots$ are the periodic Bernoulli functions defined by

$$\tilde{B}_n(x)=B_n(x),\ \text{ for }\ 0\leq x<1,\ \text{ and }\ \tilde{B}_n(x+1)=\tilde{B}_n(x),\ \text{ for }\ x\in\mathbb{R}.$$

Applying (26) for $m=0$ and $f(u)=e^{-tu}$ and using (20), we find the representation

$$V_n(t)=\frac{(-1)^{n-1}}{(2n+1)!}\int_0^{\infty}e^{-tu}\,\tilde{B}_{2n+1}(u)\,du. \tag{27}$$

It follows from this that the function $V_n(t)$ is positive and strictly decreasing on $[0, \infty)$ but *not* completely monotonic because the function $(-1)^{n-1}\tilde{B}_{2n+1}(u)$ changes sign in $[0, \infty)$ for all $n = 0, 1, 2, \ldots$.

We note that the Fourier series expansion (25) holds for every $u \in \mathbb{R}$ if we replace $B_{2n+1}(u)$ by $\tilde{B}_{2n+1}(u)$. Substituting this expansion in (27) and taking into consideration the well known Laplace transform

$$\int_0^\infty e^{-tu} \sin(au)\, du = \frac{a}{t^2 + a^2}, \tag{28}$$

we see that (27) implies (22).

We can also prove the following remarkable result.

Proposition 12 *The functions* $v_n(t) := V_n(\sqrt{t})$ *are Stieltjes functions and therefore completely monotonic on* $[0, \infty)$.

Proof Using the expression (22) and bearing in mind that the class of Stieltjes functions is closed under pointwise convergence, see [32], the desired result follows. Notice that

$$v_n(t) = \int_0^\infty e^{-ts}\, p_n(s)\, ds, \quad \text{where } p_n(s) := \sum_{k=1}^\infty \frac{2}{(4k^2\pi^2)^n}\, e^{-s\,4k^2\pi^2}$$

and that the functions $p_n(s)$ are also completely monotonic on $[0, \infty)$. Then apply Theorem 9 to prove the proposition in a different way. □

The functions $V_n(t)$ have many other interesting properties. Some of them are given in the next proposition. We refer to [48–50] for additional properties of these functions.

Proposition 13 *(i)* $(t^{2j} V_n(t))^{(l)} > 0$ *for all* $n \geq 0$ *and all* $1 \leq l \leq j$, $t \geq 0$.
(ii) $(t^{2j-1} V_n(t))^{(l)} > 0$ *for all* $n \geq 0$ *and all* $1 \leq l \leq j-1$, $t \geq 0$.

Proof (i) By Leibniz's rule, we get

$$(t^j\, v_n(t))^{(l)}(0) = 0, \quad 1 \leq l \leq j-1.$$

Combining Proposition 12 with Theorem 13 we infer that

$$(t^j\, v_n(t))^{(j)} > 0, \quad \text{for all } t \geq 0,$$

hence,

$$(t^j\, v_n(t))^{(l)} > 0, \quad \text{for all } t \geq 0 \quad 1 \leq l \leq j.$$

Using the above together with the observation that $V_n(t) = v_n(t^2)$, we complete the proof of assertion (i). The result of (ii) is obtained in a similar way. □

The next result will be useful in the sequel.

Proposition 14 *The functions*

$$F_n(x) := \int_0^\infty e^{-xt}\, [t^{2n} V_n(t)]'\, dt \quad \text{and} \quad G_n(x) := \int_0^\infty e^{-xt}\, t^{2n-1} V_n(t)\, dt$$

are completely monotonic of order $n-1$ *on* $(0, \infty)$ *for all* $n \geq 1$.

Proof It follows from (27) that $\lim_{t\to\infty} V_n(t) = 0$. Then, integrating by parts we get

$$F_n(x) = x \int_0^\infty e^{-xt}\, t^{2n}\, V_n(t)\, dt = x\, r_n(x).$$

On account of Proposition 11 (i) and the above we deduce that the function $F_n(x)$ is completely monotonic of order $n-1$ on $(0, \infty)$. The corresponding assertion for the function $G_n(x)$ follows from Proposition 13 (ii) in combination with the characterization given in Corollary 1. □

Another interesting application of Theorem 17 is obtained when considering the function $g(t) = f(t^2) = t/\sinh t$. As shown in Sect. 3, the function $f(t) = \sqrt{t}/\sinh\sqrt{t}$ is completely monotonic on $[0, \infty)$. It is also analytic at $t = 0$ because of the expansion

$$\frac{t}{\sinh t} = \sum_{k=0}^{\infty} \frac{(2-2^{2k})B_{2k}}{(2k)!}\, t^{2k}, \quad |t| < \pi. \tag{29}$$

Moreover, it is well known that

$$\int_0^\infty e^{-xt} \frac{t}{\sinh t}\, dt = \frac{1}{2}\psi'\left(\frac{x+1}{2}\right),$$

where $\psi(x) = \Gamma'(x)/\Gamma(x)$ is the psi or digamma function. In view of the above, Theorem 17 yields the following result (cf. [51, Theorem 9.9]).

Proposition 15 *The expansion*

$$\frac{1}{2}\psi'\left(\frac{x+1}{2}\right) = \sum_{k=0}^{n-1} \frac{(2-2^{2k})B_{2k}}{x^{2k+1}} + (-1)^n h_n(x)$$

holds, where the remainder $h_n(x)$ is a completely monotonic function of order n on $(0, \infty)$ and satisfies the estimate

$$0 < h_n(x) < (-1)^n \frac{(2-2^{2n})\, B_{2n}}{x^{2n+1}}, \quad \textit{for all} \ \ n = 0, 1, 2, \ldots, \quad x > 0.$$

Of course, we have

$$h_n(x) = \int_0^\infty e^{-xt}\, t^{2n} \omega_n(t)\, dt,$$

where the functions $\omega_n(t)$ are defined by

$$\frac{t}{\sinh t} = \sum_{k=0}^{n-1} \frac{(2-2^{2k})B_{2k}}{(2k)!}\, t^{2k} + (-1)^n\, t^{2n}\, \omega_n(t). \tag{30}$$

An explicit expression for the functions $\omega_n(t)$ has been found in [51] using the Cauchy residue theorem. Here, we can derive the same formula in a different way as follows: We observe that

$$\frac{t}{\sinh t} = \frac{2t}{e^t - 1} - \frac{2t}{e^{2t} - 1}.$$

In view of (20), we have

$$\omega_n(t) = 2^{2n} V_{n-1}(2t) - 2\, V_{n-1}(t).$$

From (27) and the above, we obtain

$$\omega_n(t) = \frac{(-1)^n}{(2n-1)!} \int_0^{\infty} e^{-tu} \left[2^{2n-1} \tilde{B}_{2n-1}\left(\frac{u}{2}\right) - 2\, \tilde{B}_{2n-1}(u) \right] du. \tag{31}$$

Using the Fourier series expansion (25) for the periodic function $\tilde{B}_{2n-1}(x)$, we infer that

$$2^{2n-1} \tilde{B}_{2n-1}\left(\frac{u}{2}\right) - 2\, \tilde{B}_{2n-1}(u) = 2(-1)^n (2n-1)! \sum_{k=1}^{\infty} (-1)^{k-1} \frac{\sin(k\pi u)}{(k\pi)^{2n-1}}.$$

Substituting this in (31) and using (28), we conclude that

$$\omega_n(t) = \sum_{k=1}^{\infty} \frac{2(-1)^{k-1}}{(k\pi)^{2n-2}(t^2 + k^2\pi^2)}. \tag{32}$$

By this representation and (31), we can easily prove the following.

Proposition 16 *For every $n \geq 1$, the function $\sigma_n(t) := \omega_n(\sqrt{t})$ is completely monotonic on $[0, \infty)$. The function $\omega_n(t)$ is positive and strictly decreasing on $[0, \infty)$ but not completely monotonic on this interval.*

It should be noted that we can obtain analogous results to Proposition 15, *for every* polygamma function $\psi^{(n)}(x)$, $n = 0, 1, 2, \ldots$: Differentiate (18), use (21), and apply Proposition 13 in combination with the characterization of Corollary 1.

Our next application concerns the remainder of the asymptotic expansion of the logarithm of the double gamma function.

The multiple zeta and gamma functions have been introduced by E. W. Barnes in the beginning of the twentieth century, see [11] and [12]. For instance, the double zeta function $\zeta_2(x, s \mid a_1, a_2)$ is defined by the double infinite series

$$\zeta_2(x, s \mid a_1, a_2) = \sum_{m,n=0}^{\infty} \frac{1}{(x + m\, a_1 + n\, a_2)^s},$$

$\mathrm{Re}\, x > 0, \mathrm{Re}\, s > 2,\ a_1, a_2 > 0$. Barnes showed that $\zeta_2(x, s \mid a_1, a_2)$ has a meromorphic continuation in s to the whole complex plane, with simple poles at $s = 1, 2$.

The double gamma function is defined by the relation

$$\log \Gamma_2(x \mid a_1, a_2) = \frac{\partial \zeta_2(x, s \mid a_1, a_2)}{\partial s}|_{s=0} - C,$$

where

$$C := \lim_{x \to 0} \left[\frac{\partial \zeta_2(x, s \mid a_1, a_2)}{\partial s}|_{s=0} + \log x \right].$$

Barnes discovered many properties of these functions and applied them to the theory of elliptic and theta functions. The multiple zeta and gamma functions have been studied in recent years by many researchers, due to their important role in analytic number theory and mathematical physics. We refer the readers to the papers [40, 48, 59, 61, 65] and the references given therein for background information on the subject and various applications of these functions.

In 2000, S. N. M. Ruijsenaars [65] gave complete asymptotic expansions for multiple zeta and gamma functions in terms involving the powers of $1/x$.

Let $\Gamma_2(x) := \Gamma_2(x \mid 1, 1)$. In this case, Ruijsenaars's asymptotic formula for the logarithm of the function $\Gamma_2(x)$ takes the form

$$\log \Gamma_2(x) = -C - \left(\frac{x^2}{2} - x + \frac{5}{12} \right) \log x + \frac{3}{4} x^2 - x$$

$$+ \sum_{k=3}^{N} (-1)^k \frac{B_{2,k}}{(k-2)(k-1)k} \frac{1}{x^{k-2}} + R_{2,N}(x),$$

with

$$R_{2,N}(x) := \int_0^\infty \frac{e^{-xt}}{t^3} \left(\frac{t^2}{(1-e^{-t})^2} - \sum_{k=0}^{N} \frac{(-1)^k}{k!} B_{2,k} t^k \right) dt, \tag{33}$$

$N \geq 2, \mathrm{Re} x > 0$, where $B_{2,k} = B_{2,k}(0)$ and $B_{2,k}(x)$ are *the double* Bernoulli polynomials defined by

$$\left(\frac{t}{e^t - 1} \right)^2 e^{xt} = \sum_{k=0}^{\infty} B_{2,k}(x) \frac{t^k}{k!}, \qquad |t| < 2\pi.$$

The relation between $B_{2,k}(x)$ and $B_k(x)$ is given by a formula of N. E. Nörlund [60], *viz.*

$$B_{2,k}(x) = k(k-1) \left\{ (x-1) \frac{B_{k-1}(x)}{k-1} - \frac{B_k(x)}{k} \right\}.$$

From this, it follows that

$$B_{2,2k} = -(2k-1) B_{2k}, \quad B_{2,2k-1} = -(2k-1) B_{2k-2}, \tag{34}$$

for $k = 2, 3, \ldots$ and $B_{2,0} = 1, \;\; B_{2,1} = -1, \;$ and $\; B_{2,2} = 5/6$.

The next proposition was obtained in [48], see also [40] and [61]. Here, we show that it can be derived from the corresponding results for the ordinary Gamma function.

Proposition 17 *The remainder* $(-1)^{n-1}R_{2,2n}(x)$ *is a completely monotonic function of order* $n-1$ *on* $(0,\infty)$, *for all* $n \geq 1$. *Moreover, for any* $n \geq 1$ *and* $x > 0$, *we have*

$$0 < (-1)^{n-1}R_{2,2n}(x) < \frac{(-1)^n B_{2,2n+1}}{(2n-1)2n(2n+1)}\frac{1}{x^{2n-1}} + \frac{(-1)^{n-1}B_{2,2n+2}}{2n(2n+1)(2n+2)}\frac{1}{x^{2n}}. \tag{35}$$

Proof We observe that

$$\begin{aligned}(-1)^{n-1}\frac{1}{t^3}&\left(\frac{t^2}{(1-e^{-t})^2} - \sum_{k=0}^{2n}\frac{(-1)^k}{k!}B_{2,k}\,t^k\right)\\ &= t^{2n-2}V_{n-1}(t) + t^{2n-1}V_n(t) + [t^{2n}V_n(t)]',\end{aligned} \tag{36}$$

(cf. [40, (2.8)] or [48, (6)]. Taking into consideration (33) and using Proposition 14 together with Proposition 11 (i) and the relation (21), we conclude that the function $(-1)^{n-1}R_{2,2n}(x)$ is completely monotonic of order $n-1$ on $(0,\infty)$. Taking the Laplace transform in (36) and using Proposition 11(ii), (21), and (34) we obtain (35). □

An analogue of this Proposition for the case of triple gamma function has been established in [49].

Let us now define the functions

$$f_n(t) := t^{2n+2}V_n(t)\sinh\frac{t}{2} = (-1)^n\left(\frac{t}{e^t-1} - 1 + \frac{t}{2} - \sum_{k=1}^{n}\frac{B_{2k}}{(2k)!}t^{2k}\right)\sinh\frac{t}{2}.$$

The next result is also of interest.

Proposition 18 *The function* $f_n(t)$ *is absolutely monotonic on* $[0,\infty)$ *for all* $n \geq 1$.

Proof Use the expression (22) and the absolute monotonicity of the functions (6). □

The corresponding proposition for the generating function of double Bernoulli numbers is the following.

Proposition 19 *Let*

$$g_n(t) = (-1)^{n-1}\left(\frac{t^2}{(1-e^{-t})^2} - \sum_{k=0}^{2n}\frac{(-1)^k}{k!}B_{2,k}\,t^k\right)\sinh^2(t/2).$$

The function $g_n(t)$ *is absolutely monotonic on* $[0,\infty)$ *for all* $n \geq 1$.

This was conjectured in [40], then proved in [50, Theorem 1.1].

There is an analogue of Proposition 15, which can be obtained by applying Theorem 17 for $g(t) = f(t^2) = 1/\cosh t$. The function $f(t) = 1/\cosh\sqrt{t}$ is completely monotonic on $[0, \infty)$ (see Sect. 3) and analytic at $t = 0$; we have, in fact, the expansion

$$\frac{1}{\cosh t} = \sum_{k=0}^{\infty} \frac{E_k}{k!} t^k = \sum_{k=0}^{\infty} \frac{E_{2k}}{(2k)!} t^{2k}, \quad \text{for} \quad |t| < \pi/2,$$

where E_n are the Euler numbers of the first kind. The numbers E_n are integers, in contrast with the Bernoulli numbers. For instance, we have $E_0 = 1, E_2 = -1, E_4 = 5, E_6 = -61$, while those with odd index are zero: $E_{2n+1} = 0, n = 0, 1, 2, \dots$. We refer to [71] for various formulas involving the numbers E_n.

A straightforward computation shows that

$$\beta\left(\frac{x+1}{2}\right) = \int_0^{\infty} e^{-xt} \frac{1}{\cosh t} \, dt,$$

where $\beta(x)$ is defined in (10). Applying Theorem 17, we immediately obtain the following.

Proposition 20 *The expansion*

$$\beta\left(\frac{x+1}{2}\right) = \sum_{k=0}^{n-1} \frac{E_{2k}}{x^{2k+1}} + (-1)^n w_n(x), \tag{37}$$

holds for all $n \geq 0$ and all $x > 0$. The function $w_n(x)$ is completely monotonic of order n on $(0, \infty)$ and satisfies the estimate

$$0 < w_n(x) < \frac{(-1)^n E_{2n}}{x^{2n+1}}, \quad \textit{for all} \ \ x > 0, \ \ n \geq 0.$$

Clearly, we have

$$w_n(x) = \int_0^{\infty} e^{-xt} \, t^{2n} \, \theta_n(t) \, dt,$$

where the functions $\theta_n(t)$ are defined by

$$\frac{1}{\cosh t} = \sum_{k=0}^{n-1} \frac{E_{2k}}{(2k)!} t^{2k} + (-1)^n t^{2n} \theta_n(t), \quad n \geq 1.$$

In order to find an expression for the functions $\theta_n(t)$ similar to (22) and (32), we employ an analogue of formula (26), which is known as Boole's summation formula (cf. [71, p. 17]). Suppose that the function $f : [0, \infty) \to \mathbb{R}$ has k continuous derivatives ($k = 0, 1, 2, \dots$) and that $\lim_{u\to\infty} f^{(j)}(u) = 0$ for each $j = 0, 1, \dots k$. Let $\xi \in [0, 1]$. Then

$$\sum_{j=0}^{\infty} (-1)^j f(\xi + j) = \frac{1}{2} \sum_{j=0}^{k-1} \frac{E_j(\xi)}{j!} f^{(j)}(0) + \frac{1}{2} \int_0^{\infty} \frac{\tilde{E}_{k-1}(\xi - u)}{(k-1)!} f^{(k)}(u) \, du, \tag{38}$$

where $E_n(x)$ are the Euler polynomials defined by

$$\frac{2\,e^{xt}}{e^t+1}=\sum_{n=0}^{\infty}\frac{E_n(x)}{n!}\,t^n,\quad |t|<\pi.$$

The relation between Euler polynomials and numbers is

$$E_n=2^n\,E_n\left(\frac{1}{2}\right). \tag{39}$$

We denote by $\tilde{E}_n(x)$ the periodic Euler functions defined by

$$\tilde{E}_n(x)=E_n(x),\ \text{ for }\ 0\le x<1,\ \text{ and }\ \tilde{E}_n(x+1)=-\tilde{E}_n(x),\ \text{ for }\ x\in\mathbb{R}.$$

Applying formula (38) for $f(u)=e^{-2tu}$ and $\xi=1/2$ and taking into account (39), we obtain

$$\theta_n(t)=(-1)^n\,\frac{2^{2n-1}}{(2n-1)!}\int_0^{\infty}e^{-tu}\,\tilde{E}_{2n-1}\left(\frac{u-1}{2}\right)du\,. \tag{40}$$

For the functions $\tilde{E}_{2n-1}(x)$, we have the Fourier series expansion

$$\tilde{E}_{2n-1}(x)=4\,(-1)^n\,(2n-1)!\sum_{k=0}^{\infty}\frac{\cos\left[(2k+1)\pi\,x\right]}{[(2k+1)\,\pi]^{2n}}, \tag{41}$$

see [1, p. 805]. Substituting (41) in (40) and using (28), we conclude that

$$\theta_n(t)=\sum_{k=0}^{\infty}\frac{2\,(-1)^k}{\left[(2k+1)\,\frac{\pi}{2}\right]^{2n-1}\left[t^2+((2k+1)\,\frac{\pi}{2})^2\right]}. \tag{42}$$

This has been proven [51] by complex analysis methods. From (42) and (40), we easily derive an analogue of Proposition 16:

Proposition 21 *For every $n\ge 1$, the function $\tau_n(t):=\theta_n(\sqrt{t})$ is completely monotonic on $[0,\infty)$. The function $\theta_n(t)$ is positive and strictly decreasing on $[0,\infty)$ but not completely monotonic on this interval.*

An asymptotic expansion similar to (37) holds for the function $\beta(x)$ itself. We recall that the Euler numbers of the second kind E_n^* are defined by

$$\frac{2}{e^t+1}=\sum_{k=0}^{\infty}\frac{E_k^*}{k!}\,t^k=1+\sum_{k=1}^{\infty}\frac{E_{2k-1}^*}{(2k-1)!}\,t^{2k-1},\quad \text{for}\quad |t|<\pi$$

and they are related to Bernoulli numbers by the formula

$$E_{2k-1}^*=\frac{(1-2^{2k})\,B_{2k}}{k},\quad k\ge 1.$$

We have, in fact, the following result (cf. [51, Theorem 9.7]).

Proposition 22 *The expansion*

$$2\beta(x) = \frac{1}{x} - \sum_{k=1}^{n} \frac{E^*_{2k-1}}{x^{2k}} + (-1)^n q_n(x),$$

holds for all $n \geq 1$ *and* $x > 0$. *The function* $q_n(x)$ *is completely monotonic of order* $n+1$ *on* $(0, \infty)$ *and satisfies*

$$0 < q_n(x) < \frac{(-1)^{n+1} E^*_{2n+1}}{x^{2n+2}}, \quad \textit{for all} \;\; x > 0, \;\; n \geq 0.$$

Proof Integrating by parts in (11), we obtain

$$\beta(x) = \frac{1}{2x} + \frac{1}{x} \int_0^\infty e^{-xt} \frac{1}{4\cosh^2 \frac{t}{2}} \, dt.$$

We observe that the function

$$g(t) = \frac{1}{4\cosh^2 \frac{t}{2}} = \left(1 - \frac{1}{e^t + 1}\right)'$$

satisfies the conditions of Theorem 17. Then, by straightforward calculations we complete the proof of the proposition. □

Let the functions $\varphi_n(t)$ be defined by

$$\frac{2}{e^t + 1} = 1 + \sum_{k=1}^{n} \frac{E^*_{2k-1}}{(2k-1)!} t^{2k-1} + (-1)^{n-1} t^{2n+1} \varphi_n(t).$$

The function $q_n(x)$ has the representation

$$q_n(x) = \int_0^\infty e^{-xt} t^{2n+1} \varphi_n(t) \, dt.$$

In order to find an expression for the functions $\varphi_n(t)$, we use Boole's summation formula (cf. [71, p. 17]):

$$f(1) = \frac{1}{2} \sum_{j=0}^{k-1} \frac{E_j(1)}{j!} \left[f^{(j)}(1) + f^{(j)}(0) \right] + \frac{(-1)^{k-1}}{2(k-1)!} \int_0^1 f^{(k)}(u) \, E_{k-1}(u) \, du. \tag{43}$$

We also recall the following relations between Euler polynomials and numbers (cf. [1, p. 805]):

$$E_{2k-1}(0) = -E_{2k-1}(1) = E^*_{2k-1}, \quad k = 1, 2, \ldots$$

$$E_0(0) = E_0(1) = 1, \quad E_{2k}(0) = E_{2k}(1) = 0, \quad k = 1, 2, \ldots$$

Applying (43) for $f(u) = e^{tu}$, we obtain

$$\varphi_n(t) = \frac{(-1)^{n-1}}{(2n-1)! \, t \, (e^t + 1)} \int_0^1 e^{tu} \, E_{2n-1}(u) \, du$$

$$= \frac{(-1)^n}{(2n)!\,(e^t+1)} \int_0^1 e^{tu}\, E_{2n}(u)\, du\,. \tag{44}$$

By the Fourier series expansion (41) and the elementary integral

$$\int_0^1 e^{tu} \cos[(2k+1)\pi\, u]\, du = -\frac{t\,(e^t+1)}{t^2+[(2k+1)\pi]^2}$$

we find that

$$\varphi_n(t) = \sum_{k=0}^{\infty} \frac{4}{[(2k+1)\pi]^{2n}} \frac{1}{t^2+[(2k+1)\pi]^2}. \tag{45}$$

Alternatively, using (38) for $f(u) = e^{-tu}$ and $\xi = 0$, we deduce that

$$\varphi_n(t) = \frac{(-1)^n}{(2n)!} \int_0^{\infty} e^{-tu}\, \tilde{E}_{2n}(u)\, du\,. \tag{46}$$

The Fourier series expansion for the functions $\tilde{E}_{2n}(x)$ is (see [1, p. 805])

$$\tilde{E}_{2n}(x) = 4\,(-1)^n\,(2n)! \sum_{k=0}^{\infty} \frac{\sin[(2k+1)\pi\, x]}{[(2k+1)\,\pi]^{2n+1}}\,.$$

Substituting this in (46) and using (28), we obtain again (45). Finally, by a similar argument, we can establish a corresponding result to Proposition 12.

Proposition 23 *For every $n \geq 1$, the function $\rho_n(t) := \varphi_n(\sqrt{t})$ is a Stieltjes function. The function $\varphi_n(t)$ is positive and strictly decreasing on $[0, \infty)$ but not completely monotonic on this interval.*

5 Complete Monotonicity of Functions Involving the Gamma and Psi Functions

There is a number of interesting examples of functions that belong to the classes defined in Sects. 2 and 3 involving the Gamma function $\Gamma(x)$, the psi function $\psi(x)$, and their derivatives. The bibliography on the topic is very extensive. It should be noted that a large number of relevant papers has been published during the last few years. We refer the reader to [7–9, 16, 22, 28, 30, 34, 35, 39, 41, 44, 45, 47, 64] and the references given therein for related results, applications, and helpful information on the subject. Here, we highlight some results of particular interest.

Starting from the well known formula, see for example [4, p. 615],

$$\frac{\Gamma(x+t)}{\Gamma(x+s)} = \frac{1}{\Gamma(s-t)} \int_0^{\infty} e^{-xu}\, e^{-tu}\, (1-e^{-u})^{s-t-1}\, du, \quad 0 < s-t,$$

we see that the ratio $\frac{\Gamma(x+t)}{\Gamma(x+s)}$ is a completely monotonic function on $(0, \infty)$ for $s > t$.

Moreover, using the formula [4, p. 26]

$$\psi(x) = -\gamma + \int_0^\infty \frac{e^{-u} - e^{-xu}}{1 - e^{-u}}\, du, \tag{47}$$

where $x > 0$ and γ is Euler's constant, we get

$$\psi(x+s) - \psi(x+t) = \int_0^\infty e^{-xu}\ \varphi_{s,t}(u)\, du \tag{48}$$

with

$$\varphi_{s,t}(u) := \frac{e^{(1-t)u} - e^{(1-s)u}}{e^u - 1}, \quad \varphi_{s,t}(0) := s - t.$$

It follows from (48) that the function $\psi(x+s) - \psi(x+t)$ is completely monotonic on $(0, \infty)$ for $s > t$ and therefore, the ratio $\frac{\Gamma(x+t)}{\Gamma(x+s)}$ is a logarithmically completely monotonic function on $(0, \infty)$ for $s > t$. A refinement of these results has been obtained in [35, Theorem 2.4]:

Theorem 18 *For $s, t > 0$ and $x > 0$, let*

$$M_{s,t}(x) := \frac{\Gamma(x+t)}{\Gamma(x+s)}\, x^{s-t}.$$

If $t > s$ and $s+t \geq 1$ the function $M_{s,t}(x)$ is logarithmically completely monotonic on $(0, \infty)$. The result fails to hold when $s+t < 1$ replaces $s+t \geq 1$ in the hypothesis.

In [30], the authors proved that certain products of quotients of shifted Γ-functions are logarithmically completely monotonic.

Using the asymptotic formula (see [1, 6.1.47])

$$\frac{\Gamma(x+t)}{\Gamma(x+s)} x^{s-t} = 1 - \frac{(s-t)(s+t-1)}{2x} + \frac{(s-t)(s-t+1)p(s,t)}{24x^2} + O\left(\frac{1}{x^3}\right), \tag{49}$$

as $x \to \infty$, where

$$p(s,t) := 3\,s^2 + 6st + 3t^2 - 5\,s - 7t + 2, \tag{50}$$

we get $\lim_{x\to\infty} M_{s,t}(x) = 1$.

In view of Theorem 18, we deduce

$$M_{s,t}(x) > 1, \quad \text{for all } \ x > 0, \ \text{ when } \ t > s, \ \ s + t \geq 1. \tag{51}$$

The following subsets of the first quadrant $E := \{(s,\, t) \in \mathbb{R}^2 \,:\, s, t > 0\}$ will be important in our further considerations:

$$P := \{(s,t) \in E : s + t \geq 1\}, \quad H := \{(s,t) \in P : s > t\},$$

$$P_1 := \{(s,t) \in E : s - t > 1\}, \quad P_2 := \{(s,t) \in P : 0 < s - t < 1\}.$$

Furthermore, if D is any subset of $\mathbb{R}^2$, then we will denote by D^* its reflection with respect to the line $\{(s, t) \in \mathbb{R}^2 : s = t\}$.

For $s, t > 0$ and $x > 0$, consider now the function

$$\Delta_{s,t}(x) := \frac{\Gamma(s)}{\Gamma(t)} \left[\frac{1}{x^{s-t}} - \frac{\Gamma(x+t)}{\Gamma(x+s)} \right].$$

It is clear that $\Delta_{s,s}(x) = 0$ for all $s, x > 0$. It is also easy to check that $\Delta_{s,s+1} = -1$. In what follows, we will tacitly assume that for $(s, t) \in E$ we always have $t \neq s$, $s+1$. The set of all $(s, t) \in E$ for which the function $\Delta_{s,t}(x)$ is completely monotonic has been determined in [47]. The result is

Theorem 19 *Let $(s, t) \in E$ with $t \neq s$.*

(i) The function $\Delta_{s,t}(x)$ is completely monotonic on $(0, \infty)$ if and only if $(s, t) \in H$.
(ii) The function $\Delta_{s,t}(x)$ changes sign in $(0, \infty)$ precisely when $(s, t) \in E \setminus P$.
(iii) We have $\Delta_{s,t}(x) < 0$ for all $x > 0$ if and only if $(s, t) \in H^*$.

Observe that the above statement (iii) is equivalent to (51). Sharp bounds for the function $\Delta_{s,t}(x)$ are indispensable in establishing sharp estimates for certain trigonometric sums arising in problems of geometric function theory. See for example [42, 44, 46, 53, 54]. Writing

$$\Delta_{s,t}(x) = \frac{\Gamma(s)}{\Gamma(t)} \frac{1}{x^{s-t+1}} L_{s,t}(x),$$

where

$$L_{s,t}(x) := x - \frac{\Gamma(x+t)}{\Gamma(x+s)} x^{s-t+1},$$

we see that the problem amounts to finding sharp bounds for the function $L_{s,t}(x)$. This was done in various steps of increasing generality in [41, 45, 47]. Interestingly, the set of all (s, t) with $s, t > 0$ for which $L_{s,t}(x)$ is a Bernstein function on $(0, \infty)$ can be determined. Let

$$A := \{(s, t) \in P : p(s, t) \geq 0\}$$

and set

$$A_1 := H \cap A, \quad A_2 := P_2^* \cap A, \quad A_3 := P_1^* \cap A,$$

where $p(s, t)$ is as in (50). We have, in fact, the following (cf. [47, Theorem 3.2]):

Theorem 20

(i) For $(s, t) \in P_1$ the function $L_{s,t}(x)$ is strictly increasing and concave on $(0, \infty)$ but not a Bernstein function. In this case, we have

$$0 < L_{s,t}(x) < \frac{(s-t)(s+t-1)}{2} \quad \text{for all } x > 0. \tag{52}$$

(ii) For $(s,t) \in A_1 \setminus P_1$*, the function* $L_{s,t}(x)$ *is a Bernstein function on* $(0,\infty)$ *and*

$$0 < L_{s,t}(x) < \frac{(s-t)(s+t-1)}{2} \quad \textit{for all } \; x > 0. \tag{53}$$

(iii) For $(s,t) \in A_2$*, the function* $-L_{s,t}(x)$ *is a Bernstein function on* $(0,\infty)$ *and*

$$\frac{(s-t)(s+t-1)}{2} < L_{s,t}(x) < 0 \quad \textit{for all } \; x > 0. \tag{54}$$

(iv) For $(s,t) \in A_3$*, the function* $-L_{s,t}(x)$ *is completely monotonic on* $(0,\infty)$ *and*

$$-\infty < L_{s,t}(x) < \frac{(s-t)(s+t-1)}{2} \quad \textit{for all } \; x > 0. \tag{55}$$

All bounds in the inequalities (52)–(55) are sharp because of Theorem 19 and the fact that (49) yields

$$\lim_{x\to\infty} L_{s,t}(x) = \frac{(s-t)(s+t-1)}{2}.$$

For $(s,\, t) \in E \setminus A$ with $t \neq s,\, s+1$, the derivatives $L'_{s,t}(x)$ and $L''_{s,t}(x)$ change sign in $(0,\infty)$ and inequalities (52)–(55) fail to hold for appropriate $x > 0$.

The combination of Theorems 19 and 20 and a summation by parts entail the following estimate.

Corollary 3 *Suppose that* $(s,t) \in A_1$*. Then*

$$\left|\sum_{k=n}^{\infty} \Delta_{s,t}(k) e^{ikx}\right| \leq \frac{1}{\sin\frac{x}{2}} \frac{1}{n^{s-t+1}} \frac{\Gamma(s)}{\Gamma(t)} \frac{(s-t)(s+t-1)}{2}$$

for all $x \in (0, 2\pi)$ *and* $n \in \mathbb{N}$ *with* $n \geq 2$.

An interesting complement of Theorem 18 was also established in [47].

Theorem 21 *Let* $(s,t) \in E$ *with* $s \neq t$*. The function* $1 - M_{s,t}(x)$ *is completely monotonic on* $(0,\infty)$*, if and only if* $(s,t) \in \overline{P}_2$.

There is another question on completely monotonic functions which has an unexpected and surprising connection to trigonometric estimation. Motivated by the work on infinitely divisible probability distributions (measures) on $\mathbb{R}$ and $[0,\infty)$ and completely monotonic functions related to their Laplace and Fourier transforms, W. E. Clark and M. E. H. Ismail [23] introduced the functions $G_m(x) = x^m \psi(x)$ and $\Phi_m(x) = -x^m \psi^{(m)}(x)$, and they proved that $G_m^{(m+1)}$ is completely monotonic on $(0,\infty)$ for all $m = 1, 2, \ldots$ and that $\Phi_m^{(m)}$ is completely monotonic on $(0,\infty)$ for $m = 1, 2, \ldots, 16$. Clark and Ismail conjectured that $\Phi_m^{(m)}$ is completely monotonic on $(0,\infty)$ for all natural numbers m.

In [9, Theorem 1.1], we disproved this conjecture. Indeed, we showed the following.

Theorem 22 *There exists an integer* m_0 *such that for all* $m \geq m_0$*, the function* $\Phi_m^{(m)}$ *is not completely monotonic on* $(0,\infty)$.

It might be surprising that in the proof of this result an important role is played by the function

$$H(x) := \sum_{k=1}^{\infty} \frac{1}{k} \sin\left(\frac{x}{k}\right),$$

which was studied by Hardy and Littlewood in 1936, in connection with some problems on Lampert summability.

It is worth mentioning here the key result which led us to the disproof of the above stated conjecture of Clark and Ismail. We showed that the function $H(x)$ can attain arbitrarily large negative values. Indeed, there exists a sequence $x_k \to \infty$ such that

$$H(x_k) < -C(\log\log x_k)^{1/2}, \tag{56}$$

where $C > 0$ is a constant independent of k. This is obtained by methods and techniques of analytic number theory.

The integral representation (47) and Leibniz's rule give

$$\Phi_m^{(m)}(x) = \int_0^\infty e^{-xt}\, t^m\, f_m(t)\, dt,$$

where

$$f_m(x) := \frac{d^m}{dx^m} \frac{x^m}{1-e^{-x}}.$$

Hence, the complete monotonicity of the functions $\Phi_m^{(m)}$ reduces to the inequalities $f_m(x) > 0$, for all $x > 0$ and $m = 1, 2, \ldots$.

Next, we showed that for all $x > 0$

$$s(x) := \frac{1}{2} + \frac{1}{\pi} H\left(\frac{x}{2\pi}\right) = \lim_{m\to\infty} \frac{1}{m!} f_m\left(\frac{x}{m}\right), \tag{57}$$

$$f_m(x) = \int_0^\infty e^{-t}\, t^m\, s(xt)\, dt, \quad |x| < 2\pi \tag{58}$$

and

$$f_m(x) \ge 0, \quad x \ge 2\log 2. \tag{59}$$

Combining (57), (58), and (59) we infer that the positivity of the functions $f_m(x)$ on $(0, \infty)$, for all $m = 1, 2, \ldots$, is equivalent to

$$H(x) \ge -\pi/2, \quad \text{for all } x > 0,$$

which is impossible according to (56). Therefore, there exists an integer m_0 such that for all $m \ge m_0$ the function $f_m(x)$ attains negative values on $(0, \infty)$. The precise value of the number m_0 remains unknown and this turns out to be a quite hard problem because it depends on the distribution of prime numbers and the unboundedness of the function $H(x)$. Numerical determination or estimation of m_0 is also very difficult. These are interesting problems and form basis for future research in the area.

In a recent paper, A. Kuznetsov [55] obtained several approximations to $H(x)$ for large values of x, and using these approximations, he determined an explicit value of x for which $H(x) < -\pi/2$.

We complete this section with another closely related result, which was obtained in [63, Corollary 1.4].

Proposition 24 *For each $m \geq 1$, the function*

$$(-1)^{m+1}\, x^{-m} \left(x^{2\,m}\, \psi^{(m)}(x)\right)^{(m)}$$

is completely monotonic on $(0, \infty)$.

The reason that this result is interesting is that it is a special case of a more general theorem about entire functions of finite order. See [63] for details.

6 Completely Monotonic Functions Associated with Ramanujan's θ Sequence

Let the sequence $\theta(n)$, $n = 0, 1, 2, \ldots$ be defined by

$$\frac{e^n}{2} = \sum_{j=0}^{n-1} \frac{n^j}{j!} + \frac{n^n}{n!}\, \theta(n).$$

A famous problem due to Ramanujan is to prove that $\theta(n)$ satisfies the estimates $1/3 < \theta(n) \leq 1/2$, for $n = 0, 1, 2 \ldots$. This has been proved on several occasions, see [18] for details. Among the mathematicians who provided, independently, a complete solution to Ramanujan's problem were Szegő [70] and Watson [75]. They showed in addition that the sequence $(\theta(n))_{n=0}^{\infty}$ is strictly decreasing. An important step in the approach of Szegő and Watson to the Ramanujan's problem for the sequence $(\theta(n))_{n=0}^{\infty}$, is that for all positive integers n, we have

$$\theta(n) = 1 + \frac{n}{2} \left[\int_0^1 (u\, e^{1-u})^n\, du - \int_1^{\infty} (v\, e^{1-v})^n\, dv \right]. \tag{60}$$

This expression is used as the definition of the function $\theta(n)$ for all positive real numbers n.

Adell-Jordá [2] and Volkmer [74] showed independently that the sequence $(\theta(n))_{n=0}^{\infty}$ is completely monotonic. This, in turn, has been strengthened in [43] where it is shown the following.

Theorem 23 *The function $\theta(x)$ is completely monotonic on* $[0, \infty)$.

In view of Theorem 3, Theorem 23 implies the result of Adell–Jordá [2] and Volkmer [74] mentioned above. Its proof uses some of the ideas of Watson [75] and it can be summarized as follows.

Using (60), we show that for all positive real numbers x

$$\theta(x) = \frac{1}{3} + \frac{1}{2}\int_0^{\infty} e^{-xt}\,\varphi(t)\,dt, \tag{61}$$

where

$$\varphi(t) := \frac{v(t)}{(v(t)-1)^3} - \frac{u(t)}{(1-u(t))^3}, \qquad t > 0$$

and the functions $v(t),\ u(t)$ are uniquely determined by the conditions

$$u\,e^{1-u} = e^{-t}, \quad v\,e^{1-v} = e^{-t}, \quad 0 \le u \le 1 \le v.$$

It can be verified that $\varphi(t) > 0$ for all $t \ge 0$. Hence by (61) and Theorem 2 (ii) we deduce that the function $\theta(x)$ is completely monotonic on $(0, \infty)$. Then, it is shown that

$$\int_0^{\infty} \varphi(t)\,dt = \frac{1}{3}. \tag{62}$$

From this and the fact that $\theta(0) = \frac{1}{2}$, we conclude that (61) remains valid for $x = 0$, in other words, that the measure in the Bernstein's representation (61) of the function $\theta(x)$ is finite. Applying Theorem 2 (i) we prove that the function $\theta(x)$ is completely monotonic on $[0, \infty)$.

The function φ defined above has the following remarkable properties:

$$\varphi(0) = \frac{8}{135}, \qquad \lim_{t\to+\infty} \varphi(t) = 0, \qquad \lim_{t\to+\infty} t\,\varphi(t) = 0. \tag{63}$$

Also

$$\varphi'(t) < 0, \quad \text{for all} \quad t \ge 0. \tag{64}$$

Therefore,

$$0 < \varphi(t) \le \frac{8}{135}, \qquad t \ge 0.$$

Let us now define the function

$$\lambda(x) := x\left(\theta(x) - \frac{1}{3}\right), \qquad x \ge 0.$$

It follows from (61) that $\lambda(x) > 0$ for all $x > 0$. We also have $\lambda(0) = 0$. Integrating by parts in (61) and taking into consideration (63), we obtain

$$\lambda(x) = \frac{4}{135} + \frac{1}{2}\int_0^{\infty} e^{-xt}\,\varphi'(t)\,dt\,. \tag{65}$$

Differentiating the above yields

$$\lambda'(x) = \frac{1}{2}\int_0^{\infty} e^{-xt}\,(-\,t\,\varphi'(t))\,dt\,. \tag{66}$$

Using (63) and (62), we obtain using integration by parts

$$\int_0^\infty (-t\,\varphi'(t))\,dt = \int_0^\infty \varphi(t)\,dt = \frac{1}{3}.$$

Thus, (64) and (66) reveal that $\lambda'(x)$ is completely monotonic on $[0, \infty)$. Therefore, we arrive at the following conclusion.

Theorem 24 *The function $\lambda(x)$ is a Bernstein function on $[0, \infty)$ and its Lévy–Khinchine representation is given by*

$$\lambda(x) = -\frac{1}{2}\int_0^\infty (1 - e^{-xt})\,\varphi'(t)\,dt, \quad \textit{for all} \;\; x \geq 0.$$

(cf. [43, Theorem 1]). On account of Proposition 5, we see that Theorem 24 implies Theorem 23.

Integrating by parts once more in (65), we find that the asymptotic behavior of $\theta(x)$ is

$$\theta(x) \sim \frac{1}{3} + \frac{4}{135\,x} - \frac{8}{2835\,x^2} + \dots \quad \text{as } x \to \infty.$$

We refer to [5] and [27] for some interesting refinements of the above formula and sharp estimates for the sequence $\theta(n)$. It should be noted that J. C. W. Marsaglia [58] developed an efficient method to obtain an arbitrary number of terms in the aforementioned asymptotic expansion of $\theta(n)$.

Finally, consider the function

$$\sigma(z) := \frac{1}{\theta(1) - 1/3}\sum_{n=1}^{\infty}(\theta(n) - 1/3)z^n\,.$$

It has been conjectured in [66, p. 293] that the function $\sigma(z)$ is universally convex (see Sect. 3). The conjecture appears to be supported by numerical and graphical experiments. However, this remains an open problem.

References

1. Abramowitz, M., Stegun, I.A.: Handbook of Mathematical Functions with Formulas, Graphs and Mathematical Tables. Dover, New York (1965)
2. Adell, J.A., Jordá, P.: On the complete monotonicity of a Ramanujan sequence connected with e^n. Ramanujan J. **16**, 1–5 (2008)
3. Akhiezer, N.I.: The Classical Moment Problem. Oliver and Boyd, Edinburgh (1965)
4. Andrews, G.E., Askey, R., Roy, R.: Special Functions. Cambridge University Press, Cambridge (1999)
5. Alm, S.E.: Monotonicity of the difference between median and mean of gamma distributions and of a related Ramanujan sequence. Bernoulli **9**, 351–371 (2003)
6. Alzer, H.: On some inequalities for the gamma and psi functions. Math. Comp. **66**, 373–389 (1997)

7. Alzer, H., Berg, C.: Some classes of completely monotonic functions. Ann. Acad. Sci. Fenn. Math. **27**, 445–460 (2002)
8. Alzer, H., Berg, C.: Some classes of completely monotonic functions II. Ramanujan J. **11**, 225–248 (2006)
9. Alzer, H., Berg, C., Koumandos, S.: On a conjecture of Clark and Ismail. J. Approx. Theory **134**, 102–113 (2005)
10. Ball, K.: Completely monotonic rational functions and Hall's marriage theorem. J. Comb. Theory Ser. B **61**, 118–124 (1994)
11. Barnes, E.W.: The theory of the double gamma function. Philos. Trans. R. Soc. Ser. A **196**, 265–388 (1901)
12. Barnes, E.W.: On the theory of the multiple gamma function. Trans. Camb. Philos. Soc. **19**, 374–425 (1904)
13. Berg, C.: Integral representation of some functions related to the Gamma function. Mediterr. J. Math. **1**, 433–439 (2004)
14. Berg, C.: Stieltjes-Pick-Bernstein-Schoenberg and their connection to complete monotonicity. In: Mateu, J., Porcu, E. (eds.) Positive Definite Functions: From Schoenberg to Space-Time challenges. Department of Mathematics, University Jaume I, Castellon, Spain (2008)
15. Berg, C., Forst, G.: Potential Theory on Locally Compact Abelian Groups. Springer, New York (1975)
16. Berg, C., Pedersen, H.L.: A completely monotone function related to the Gamma function. J. Comput. Appl. Math. **133**, 219–230 (2001)
17. Berg, C., Mateu, J., Porcu, E.: The Dagum family of isotropic correlation functions. Bernoulli **14**, 1134–1149 (2008)
18. Berndt, B.C.: Ramanujan's Notebooks: Part II. Springer, New York (1989)
19. Bernstein, S.N.: Sur les fonctions absolument monotones. Acta Math. **52**, 1–66 (1928)
20. Bochner, S.: Harmonic Analysis and the Theory of Probability. Dover, Mineola (2005)
21. Bondesson, L.: Generalized Gamma Convolutions and Related Classes of Distributions and Densities. Lecture Notes in Statistics, vol. 76. Springer, New York (1992)
22. Bustoz, J., Ismail, M.E.H.: On Gamma Function Inequalities. Math. Comp. **47**(176), 659–667 (1986)
23. Clark, W.E., Ismail, M.E.H.: Inequalities involving gamma and psi functions. Anal. Appl. **1**, 129–140 (2003)
24. Day, W.A.: On monotonicity of the relaxation functions of viscoelastic material. Proc. Camb. Phil. Soc. **67**, 503–508 (1970)
25. Dubourdieu, J.: Sur un théorème de M. S. Bernstein relatif á la transformation de Laplace-Stieltjes. Compositio Math. **7**, 96–111 (1939)
26. Feller, W.: An Introduction to Probability Theory and its Applications, vol. 2. Wiley, New York (1966)
27. Flajolet, P., Grabner, P.J., Kirschenhofer, P., Prodinger, H.: On Ramanujan's Q-function. J. Comput. Appl. Math. **58**, 103–116 (1995)
28. Frenzen, C.L.: Error bounds for asymptotic expansions of the ratio of two gamma functions. SIAM J. Math. Anal. **18**, 890–896 (1987)
29. Goldie, C.M.: A class of infinitely divisible random variables. Math. Proc. Camb. Phil. Soc. **63**, 1141–1143 (1967)
30. Grinshpan, A.Z., Ismail, M.E.H.: Completely monotonic functions involving the gamma and q-gamma functions. Proc. Am. Math. Soc. **134**, 1153–1160 (2006)
31. Hausdorff, F.: Summationsmethoden und Momentfolgen I. Math. Z. **9**, 74–109 (1921)
32. Hirsch, F.: Intégrales de résolventes et calcul symbolique. Ann. Inst. Fourier **22**, 239–264 (1972)
33. Horn, R.A.: On infinitely divisible matrices, kernels and functions. Z. Wahrscheinlichkeitstheorie und Verw. Geb. **8**, 219–230 (1967)
34. Ismail, M.E.H., Kelker, D.H.: Special functions, Stieltjes transforms and infinite divisibility. SIAM J. Math. Anal. **10**, 884–901 (1979)

35. Ismail, M.E.H., Lorch, L., Muldoon, M.E.: Completely monotonic functions associated with the Gamma function and its q-analogues. J. Math. Anal. Appl. **116**, 1–9 (1986)
36. Kimberling, C.H.: A probabilistic interpretation of complete monotonicity. Aequat. Math. **10**, 152–164 (1974)
37. Knopp, K.: Theory and Application of Infinite Series. Dover, New York (1947)
38. Koumandos, S.: Remarks on some completely monotonic functions. J. Math. Anal. Appl. **324**, 1458–1461 (2006)
39. Koumandos, S.: Remarks on a paper by Chao-Ping Chen and Feng Qi. Proc. Am. Math. Soc. **134**, 1365–1367 (2006)
40. Koumandos, S.: On Ruijsenaars' asymptotic expansion of the logarithm of the double gamma function. J. Math. Anal. Appl. **341**, 1125–1132 (2008)
41. Koumandos, S.: Monotonicity of some functions involving the gamma and psi functions. Math. Comp. **77**, 2261–2275 (2008)
42. Koumandos, S.: Positive trigonometric sums and starlike functions. In: Gautschi, W., Mastroianni, G., Rassias, Th. (eds.) Approximation and Computation. Springer Optimization and its Applications, vol. 42, pp. 157–183. Springer, Berlin (2010)
43. Koumandos, S.: A Bernstein function related to Ramanujan's approximations of $\exp(n)$. Ramanujan J. **30**, 447–459 (2013)
44. Koumandos, S., Lamprecht, M.: On a conjecture for trigonometric sums and starlike functions II. J. Approx. Theory **162**, 1068–1084 (2010)
45. Koumandos, S., Lamprecht, M.: Some completely monotonic functions of positive order. Math. Comp. **79**, 1697–1707 (2010)
46. Koumandos, S., Lamprecht, M.: The zeros of certain Lommel functions. Proc. Am. Math. Soc. **140**, 3091–3100 (2012)
47. Koumandos, S., Lamprecht, M.: Complete monotonicity and related properties of some special functions. Math. Comp. **82**, 1097–1120 (2013)
48. Koumandos, S., Pedersen, H.L.: Completely monotonic functions of positive order and asymptotic expansions of the logarithm of Barnes double gamma function and Euler's gamma function. J. Math. Anal. Appl. **355**, 33–40 (2009)
49. Koumandos, S., Pedersen, H.L.: On the asymptotic expansion of the logarithm of Barnes triple gamma function. Math. Scand. **105**, 287–306 (2009)
50. Koumandos, S., Pedersen, H.L.: Absolutely monotonic functions related to Euler's gamma function and Barnes' double and triple gamma function. Monatsh. Math. **163**, 51–69 (2011)
51. Koumandos, S., Pedersen, H.L.: Turán type inequalities for the partial sums of the generating functions of Bernoulli and Euler numbers. Math. Nach. 285, 2129–2156 (2012). doi:10.1002/mana.201100299 (Published online 7 Sept 2012)
52. Koumandos, S., Pedersen, H.L.: Stieltjes functions, asymptotic expansions and complete monotonicity of positive order (2013)
53. Koumandos, S., Ruscheweyh, S.: Positive Gegenbauer polynomial sums and applications to starlike functions. Constr. Approx. **23**, 197–210 (2006)
54. Koumandos, S., Ruscheweyh, S.: On a conjecture for trigonometric sums and starlike functions. J. Approx. Theory **149**, 42–58 (2007)
55. Kuznetsov, A.: Asymptotic approximations to the Hardy-Littlewood function. J. Comput. Appl. Math. **237**, 603–613 (2013)
56. Lorch, L., Newman, D.J.: On the composition of completely monotonic functions and completely monotonic sequences and related questions. J. Lond. Math. Soc. **s2-28**, 31–45 (1983)
57. Lorch, L., Moser, L.: A remark on completely monotonic sequences, with an application to summability. Can. Math. Bull. **6**, 171–173 (1963)
58. Marsaglia, J.C.W.: The incomplete Gamma function and Ramanujan's rational approximation to e^x. J. Stat. Comput. Simul. **24**, 163–169 (1986)
59. Matsumoto, K.: Asymptotic series for double zeta, double gamma, and Hecke L-functions. Math. Proc. Camb. Phil. Soc. **123**, 385–405 (1998). Corrigendum and addendum. ibid. **132**, 377–384 (2002)

60. Nörlund, N.E.: Mémoire sur les polynomes de Bernoulli. Acta Math. **43**, 121–196 (1922)
61. Pedersen, H.L.: The remainder in Ruijsenaars' asymptotic expansion of Barnes double gamma function. Mediterr. J. Math. **4**, 419–433 (2007)
62. Pedersen, H.L.: Pre Stieltjes functions. Mediterr. J. Math. **8**, 113–122 (2011)
63. Pedersen, H.L.: Completely monotonic functions related to logarithmic derivatives of entire functions. Anal. Appl. Singap. **9**, 409–419 (2011)
64. Qi, F.: Bounds for the ratio of two gamma functions. J. Inequal. Appl. **2010**, Article ID 493058, 84 p. (2010)
65. Ruijsenaars, S.N.M.: On Barnes' multiple zeta and gamma functions. Adv. Math. **156**, 107–132 (2000)
66. Ruscheweyh, S., Salinas, L., Sugawa, T.: Completely monotone sequences and universally prestarlike functions. Israel J. Math. **171**, 285–304 (2009)
67. Schilling, R.L.: Subordination in the sense of Bochner and related functional calculus. J. Austral. Math. Soc. Ser. A **64**, 368–396 (1998)
68. Schilling, R.L., Song, R., Vondraček, Z.: Bernstein Functions. De Gruyter Studies in Mathematics, vol. 37. De Gruyter, Berlin (2010)
69. Schoenberg, I.J.: Metric spaces and completely monotone functions. Ann. Math. **39**, 811–841 (1938)
70. Szegő, G.: Über einige von S. Ramanujan gestellte Aufgaben. J. Lond. Math. Soc. **3**, 225–232 (1928). Also in: Askey, R. (ed.) Collected Papers of Gabor Szegő, vol. 2, pp. 143–150. Birkhäuser, Basel (1982)
71. Temme, N.M.: Special Functions: An Introduction to Classical Functions of Mathematical Physics. Wiley, New York (1996)
72. Trimble, S.Y., Wells, J., Wright, F.T.: Superadditive functions and a statistical application. SIAM J. Math. Anal. **20**, 1255–1259 (1989)
73. van Haeringen, H.: Completely monotonic and related functions. J. Math. Anal. Appl. **204**, 389–408 (1996)
74. Volkmer, H.: Factorial series connected with the Lambert function, and a problem posed by Ramanujan. Ramanujan J. **16**, 235–245 (2008)
75. Watson, G.N.: Theorems stated by Ramanujan (V): Approximations connected with e^x. Proc. Lond. Math. Soc. (2) **29**, 293–308 (1929)
76. Widder, D.V.: A classification of generating functions. Trans. Am. Math. Soc. **39**, 244–298 (1936)
77. Widder, D.V.: The Laplace Transform. Princeton University Press, Princeton (1946)
78. Wimp, J.: Sequences Transformations and their Applications. Academic, New York (1981)

Hyers–Ulam Stability of Set-Valued Mappings

Jung Rye Lee, Choonkil Park and Themistocles M. Rassias

Abstract In this paper, we define set-valued functional equations and prove the Hyers–Ulam stability of set-valued functional equations.

Keywords Hyers–Ulam stability · Additive set-valued functional equation · Quadratic set-valued functional equation · Cubic set-valued functional equation · Quartic set-valued functional equation

1 Introduction and Preliminaries

Set-valued functions in Banach spaces have been developed in the last decades. The pioneering papers by Aumann [5] and Debreu [12] were inspired by problems arising in control theory and mathematical economics. We can refer to the papers by Arrow and Debreu [3] and McKenzie [24]; the monographs by Hindenbrand [16], Aubin and Frankowska [4], Castaing and Valadier [9] and Klein and Thompson [21] and the survey by Hess [15].

The stability problem of functional equations originated from a question of Ulam [38] concerning the stability of group homomorphisms. Hyers [17] gave a first affirmative partial answer to the question of Ulam for Banach spaces. Hyers' Theorem was generalized by Aoki [2] for additive mappings and by Th.M. Rassias [31] for linear mappings by considering an unbounded Cauchy difference. A generalization of the Th.M. Rassias theorem was obtained by Găvruta [14] by replacing the unbounded Cauchy difference with a general control function in the spirit of Th.M.

J. R. Lee (✉)
Department of Mathematics, Daejin University, Pocheon, South Korea
e-mail: jrlee@daejin.ac.kr

C. Park
Research Institute for Natural Sciences, Hanyang University, Seoul, South Korea
e-mail: baak@hanyang.ac.kr

T. M. Rassias
Department of Mathematics, National Technical University of Athens, Athens, Greece
e-mail: trassias@math.ntua.gr

T. M. Rassias, P. M. Pardalos (eds.), *Mathematics Without Boundaries*,
DOI 10.1007/978-1-4939-1106-6_13,

Rassias' approach. In [32], the Hyers–Ulam stability in the context of set-valued functions was studied by Th.M. Rassias.

The functional equation

$$f(x+y)+f(x-y)=2f(x)+2f(y)$$

is called a *quadratic functional equation*. In particular, every solution of the quadratic functional equation is said to be a *quadratic mapping*. A Hyers–Ulam stability problem for the quadratic functional equation was proved by Skof [37] for mappings $f : X \to Y$, where X is a normed space and Y is a Banach space. Cholewa [10] noticed that the theorem of Skof is still true if the relevant domain X is replaced by an Abelian group. Czerwik [11] proved the Hyers–Ulam stability of the quadratic functional equation. The stability problems of several functional equations have been extensively investigated by a number of authors and there are many interesting results concerning this problem (see [1, 6, 13, 14, 18, 19, 33, 34–36]).

In [20], Jun and Kim considered the following cubic functional equation

$$f(2x+y)+f(2x-y)=2f(x+y)+2f(x-y)+12f(x). \tag{1}$$

It is easy to show that the function $f(x)=x^3$ satisfies the functional Eq. (1), which is called a *cubic functional equation*, and every solution of the cubic functional equation is said to be a *cubic mapping*.

In [23], Lee et al. considered the following quartic functional equation:

$$f(2x+y)+f(2x-y)=4f(x+y)+4f(x-y)+24f(x)-6f(y). \tag{2}$$

It is easy to show that the function $f(x)=x^4$ satisfies the functional Eq. (2), which is called a *quartic functional equation,* and every solution of the quartic functional equation is said to be a *quartic mapping*.

Let Y be a Banach space. We define the following:

2^Y: The set of all subsets of Y;
$C_b(Y)$: The set of all closed bounded subsets of Y;
$C_c(Y)$: The set of all closed convex subsets of Y;
$C_{cb}(Y)$: The set of all closed convex bounded subsets of Y.

On 2^Y we consider the addition and the scalar multiplication as follows:

$$C+C'=\{x+x' : x\in C, x'\in C'\}, \qquad \lambda C=\{\lambda x : x\in C\},$$

where $C, C' \in 2^Y$ and $\lambda \in \mathbf{R}$. Further, if $C, C' \in C_c(Y)$, then we denote by $C \oplus C' = \overline{C+C'}$.

It is easy to check that

$$\lambda C+\lambda C'=\lambda(C+C'), \qquad (\lambda+\mu)C\subseteq \lambda C+\mu C.$$

Furthermore, when C is convex, we obtain $(\lambda + \mu)C = \lambda C + \mu C$ for all $\lambda, \mu \in \mathbf{R}^+$.

For a given set $C \in 2^Y$ the distance function $d(\cdot, C)$ and the support function $s(\cdot, C)$ are respectively defined by

$$d(x, C) = \inf\{\|x - y\| : y \in C\}, \qquad x \in Y,$$

$$s(x^*, C) = \sup\{\langle x^*, x\rangle : x \in C\}, \qquad x^* \in Y^*.$$

For every pair $C, C' \in C_b(Y)$, we define the Hausdorff distance between C and C' by

$$h(C, C') = \inf\{\lambda > 0 : C \subseteq C' + \lambda B_Y, \qquad C' \subseteq C + \lambda B_Y\},$$

where B_Y is the closed unit ball in Y.

The following proposition reveals some properties of the Hausdorff distance.

Proposition 1 *For every $C, C', K, K' \in C_{cb}(Y)$ and $\lambda > 0$, the following properties hold*

(a) $h(C \oplus C', K \oplus K') \leq h(C, K) + h(C', K')$;
(b) $h(\lambda C, \lambda K) = \lambda h(C, K)$.

Let $(C_{cb}(Y), \oplus, h)$ be endowed with the Hausdorff distance h. Since Y is a Banach space, $(C_{cb}(Y), \oplus, h)$ is a complete metric semigroup (see [9]). Debreu [12] proved that $(C_{cb}(Y), \oplus, h)$ is isometrically embedded in a Banach space as follows.

Lemma 1 ([12]) *Let $C(B_{Y^*})$ be the Banach space of continuous real-valued functions on B_{Y^*} endowed with the uniform norm $\|\cdot\|_u$. Then the mapping $j : (C_{cb}(Y), \oplus, h) \to C(B_{Y^*})$, given by $j(A) = s(\cdot, A)$, satisfies the following properties:*

(a) $j(A \oplus B) = j(A) + j(B)$;
(b) $j(\lambda A) = \lambda j(A)$;
(c) $h(A, B) = \|j(A) - j(B)\|_u$;
(d) $j(C_{cb}(Y))$ *is closed in* $C(B_{Y^*})$

for all $A, B \in C_{cb}(Y)$ and all $\lambda \geq 0$.

Let $f : \Omega \to (C_{cb}(Y), h)$ be a set-valued function from a complete finite measure space (Ω, Σ, ν) into $C_{cb}(Y)$. Then f is *Debreu integrable* if the composition $j \circ f$ is Bochner integrable (see [8]). In this case, the Debreu integral of f in Ω is the unique element $(D)\int_\Omega f d\nu \in C_{cb}(Y)$ such that $j((D)\int_\Omega f d\nu)$ is the Bochner integral of $j \circ f$. The set of Debreu integrable functions from Ω to $C_{cb}(Y)$ will be denoted by $D(\Omega, C_{cb}(Y))$. Furthermore, on $D(\Omega, C_{cb}(Y))$, we define $(f+g)(\omega) = f(\omega) \oplus g(\omega)$ for all $f, g \in D(\Omega, C_{cb}(Y))$. Then we obtain that $((\Omega, C_{cb}(Y)), +)$ is an abelian semigroup.

Set-valued functional equations have been extensively investigated by a number of authors and there are many interesting results concerning this problem (see [7, 25–30]).

In this paper, we define the additive set-valued functional equation, the quadratic set-valued functional equation, the cubic set-valued functional equation and the quartic set-valued functional equation, and prove the Hyers–Ulam stability of the set-valued functional equations.

Throughout this paper, let X be a real vector space and Y a Banach space.

2 Stability of the Additive Set-Valued Functional Equation

In this section, we prove the Hyers–Ulam stability of the additive set-valued functional equation.

Definition 1 ([22]) Let $f : X \to C_{cb}(Y)$. The additive set-valued functional equation is defined by

$$f(x+y) = f(x) \oplus f(y) \tag{3}$$

for all $x, y \in X$. Every solution of the additive set-valued functional equation is called an *additive set-valued mapping*.

Note that there are some examples in [22].

Theorem 1 *Let $\varphi : X \times X \to [0, \infty)$ be a function such that*

$$\widetilde{\varphi}(x, y) := \sum_{j=0}^{\infty} \frac{1}{2^j} \varphi(2^j x, 2^j y) < \infty \tag{4}$$

for all $x, y \in X$. Suppose that $f : X \to (C_{cb}(Y), h)$ is a mapping satisfying

$$h(f(x+y), f(x) \oplus f(y)) \leq \varphi(x, y) \tag{5}$$

for all $x, y \in X$. Then there exists a unique additive set-valued mapping $A : X \to (C_{cb}(Y), h)$ such that

$$h(f(x), A(x)) \leq \frac{1}{2} \widetilde{\varphi}(x, x) \tag{6}$$

for all $x \in X$.

Proof Let $y = x$ in (5). Since $f(x)$ is convex, we get

$$h(f(2x), 2f(x)) \leq \varphi(x, x) \tag{7}$$

and if we replace x by $2^n x, n \in \mathbf{N}$, in (7), then we obtain

$$h\left(f(2^{n+1}x), 2f(2^n x)\right) \leq \varphi(2^n x, 2^n x)$$

and

$$h\left(\frac{f(2^{n+1}x)}{2^{n+1}}, \frac{f(2^n x)}{2^n}\right) \leq \frac{1}{2^{n+1}} \varphi(2^n x, 2^n x).$$

So

$$h\left(\frac{f(2^n x)}{2^n}, \frac{f(2^m x)}{2^m}\right) \le \frac{1}{2}\sum_{j=m}^{n-1} \frac{1}{2^j}\varphi(2^j x, 2^j x) \tag{8}$$

for all integers n, m with $n \ge m$. It follows from (4) and (8) that $\left\{\frac{f(2^n x)}{2^n}\right\}$ is a Cauchy sequence in $(C_{cb}(Y), h)$.

Let $A(x) = \lim_{n\to\infty} \frac{f(2^n x)}{2^n}$ for each $x \in X$. Then we claim that A is an additive set-valued mapping. Note that $h\left(\frac{f(2^n x+2^n y)}{2^n}, \frac{f(2^n x)}{2^n} \oplus \frac{f(2^n y)}{2^n}\right) \le \frac{1}{2^n}\varphi(2^n x, 2^n y)$. Since $h(A \oplus B, C \oplus D) \le h(A, C) + h(B, D)$, we have

$$\begin{aligned} h(A(x+y), A(x) \oplus A(y)) &\le h\left(T(x+y), \frac{f(2^n x + 2^n y)}{2^n}\right) \\ &\quad + h\left(\frac{f(2^n x + 2^n y)}{2^n}, \frac{f(2^n x)}{2^n} \oplus \frac{f(2^n y)}{2^n}\right) \\ &\quad + h\left(\frac{f(2^n x)}{2^n} \oplus \frac{f(2^n y)}{2^n}, A(x) \oplus A(y)\right), \end{aligned}$$

which tends to zero as $n \to \infty$. So A is an additive set-valued mapping. Letting $m = 0$ and passing the limit $m \to \infty$ in (8), we get the inequality (6).

Now, let $T : X \to (C_{cb}(Y), h)$ be another additive set-valued mapping satisfying (3) and (6). So

$$\begin{aligned} h(A(x), T(x)) &= \frac{1}{2^n} h(A(2^n x), T(2^n x)) \\ &\le \frac{1}{2^n}\left(h(A(2^n x), f(2^n x)) + h(T(2^n x), f(2^n x))\right) \\ &\le \frac{2}{2^{n+1}}\widetilde{\varphi}(2^n x, 2^n x), \end{aligned}$$

which tends to zero as $n \to \infty$ for all $x \in X$. So we can conclude that $A(x) = T(x)$ for all $x \in X$, which proves the uniqueness of A, as desired.

Corollary 1 *Let $1 > p > 0$ and $\theta \ge 0$ be real numbers, and let X be a real normed space. Suppose that $f : X \to (C_{cb}(Y), h)$ is a mapping satisfying*

$$h(f(x+y), f(x) \oplus f(y)) \le \theta(||x||^p + ||y||^p) \tag{9}$$

for all $x, y \in X$. Then there exists a unique additive set-valued mapping $A : X \to Y$ satisfying

$$h(f(x), A(x)) \le \frac{2\theta}{2 - 2^p}||x||^p$$

for all $x \in X$.

Proof The proof follows from Theorem 1 by taking

$$\varphi(x,y) := \theta(||x||^p + ||y||^p)$$

for all $x, y \in X$.

Theorem 2 *Let $\varphi : X \times X \to [0, \infty)$ be a function such that*

$$\widetilde{\varphi}(x,y) := \sum_{j=1}^{\infty} 2^j \varphi(\frac{x}{2^j}, \frac{y}{2^j}) < \infty$$

for all $x, y \in X$. Suppose that $f : X \to (C_{cb}(Y), h)$ is a mapping satisfying Eq. (5). Then there exists a unique additive set-valued mapping $A : X \to (C_{cb}(Y), h)$ such that

$$h(f(x), A(x)) \leq \frac{1}{2}\widetilde{\varphi}(x,x)$$

for all $x \in X$.

Proof It follows from Eq. (7) that

$$h\left(f(x), 2f\left(\frac{x}{2}\right)\right) \leq \varphi\left(\frac{x}{2}, \frac{x}{2}\right)$$

for all $x \in X$.

The rest of the proof is similar to the proof of Theorem 1.

Corollary 2 *Let $p > 1$ and $\theta \geq 0$ be real numbers, and let X be a real normed space. Suppose that $f : X \to (C_{cb}(Y), h)$ is a mapping satisfying Eq. (9). Then there exists a unique additive set-valued mapping $A : X \to Y$ satisfying*

$$h(f(x), A(x)) \leq \frac{2\theta}{2^p - 2}||x||^p$$

for all $x \in X$.

Proof The proof follows from Theorem 2 by taking

$$\varphi(x,y) := \theta(||x||^p + ||y||^p)$$

for all $x, y \in X$.

3 Stability of the Quadratic Set-Valued Functional Equation

Definition 2 ([22]) Let $f : X \to C_{cb}(Y)$. The quadratic set-valued functional equation is defined by

$$2f(x+y) \oplus 2f(x-y) = f(2x) \oplus f(2y)$$

for all $x, y \in X$. Every solution of the quadratic set-valued functional equation is called a *quadratic set-valued mapping*.

Note that there are some examples in [22].

Theorem 3 *Let $\varphi : X \times X \to [0, \infty)$ be a function such that*

$$\widetilde{\varphi}(x, y) := \sum_{j=0}^{\infty} \frac{1}{4^j}\varphi(2^j x, 2^j y) < \infty \tag{10}$$

for all $x, y \in X$. Suppose that $f : X \to (C_{cb}(Y), h)$ is a mapping satisfying $f(0) = \{0\}$ and

$$h(2f(x + y) \oplus 2f(x - y), f(2x) \oplus f(2y)) \le \varphi(x, y) \tag{11}$$

for all $x, y \in X$. Then there exists a unique quadratic set-valued mapping $Q : X \to (C_{cb}(Y), h)$ such that

$$h(f(x), Q(x)) \le \frac{1}{4}\widetilde{\varphi}(x, 0)$$

for all $x \in X$.

Proof Let $y = 0$ in (11). Since $f(x)$ is convex, we get

$$h(f(2x), 4f(x)) \le \varphi(x, 0) \tag{12}$$

and if we replace x by $2^n x, n \in \mathbf{N}$, in (12), then we obtain

$$h\left(f(2^{n+1}x), 4f(2^n x)\right) \le \varphi(2^n x, 0)$$

and

$$h\left(\frac{f(2^{n+1}x)}{4^{n+1}}, \frac{f(2^n x)}{4^n}\right) \le \frac{1}{4^{n+1}}\varphi(2^n x, 0).$$

So

$$h\left(\frac{f(2^n x)}{4^n}, \frac{f(2^m x)}{4^m}\right) \le \frac{1}{4}\sum_{j=m}^{n-1} \frac{1}{4^j}\varphi(2^j x, 0) \tag{13}$$

for all integers n, m with $n \ge m$. It follows from Eqs. (10) and (13) that $\left\{\frac{f(2^n x)}{4^n}\right\}$ is a Cauchy sequence in $(C_{cb}(Y), h)$.

Let $Q(x) = \lim_{n\to\infty} \frac{f(2^n x)}{4^n}$ for each $x \in X$.

The rest of the proof is similar to the proof of Theorem 1.

Corollary 3 *Let $2 > p > 0$ and $\theta \ge 0$ be real numbers, and let X be a real normed space. Suppose that $f : X \to (C_{cb}(Y), h)$ is a mapping satisfying $f(0) = \{0\}$ and*

$$h(2f(x + y) \oplus 2f(x - y), f(2x) \oplus f(2y)) \le \theta(||x||^p + ||y||^p) \tag{14}$$

for all $x, y \in X$. Then there exists a unique quadratic set-valued mapping $Q : X \to Y$ satisfying

$$h(f(x), Q(x)) \leq \frac{\theta}{4 - 2^p}||x||^p$$

for all $x \in X$.

Proof The proof follows from Theorem 3 by taking

$$\varphi(x, y) := \theta(||x||^p + ||y||^p)$$

for all $x, y \in X$.

Theorem 4 *Let $\varphi : X \times X \to [0, \infty)$ be a function such that*

$$\widetilde{\varphi}(x, y) := \sum_{j=1}^{\infty} 4^j \varphi(\frac{x}{2^j}, \frac{y}{2^j}) < \infty$$

for all $x, y \in X$. Suppose that $f : X \to (C_{cb}(Y), h)$ is a mapping satisfying $f(0) = \{0\}$ and (11). Then there exists a unique quadratic set-valued mapping $Q : X \to (C_{cb}(Y), h)$ such that

$$h(f(x), Q(x)) \leq \frac{1}{4}\widetilde{\varphi}(x, 0)$$

for all $x \in X$.

Proof It follows from Eq. (12) that

$$h\left(f(x), 4f\left(\frac{x}{2}\right)\right) \leq \varphi\left(\frac{x}{2}, 0\right)$$

for all $x \in X$.

The rest of the proof is similar to the proofs of Theorems 1 and 3.

Corollary 4 *Let $p > 2$ and $\theta \geq 0$ be real numbers, and let X be a real normed space. Suppose that $f : X \to (C_{cb}(Y), h)$ is a mapping satisfying $f(0) = \{0\}$ and (14). Then there exists a unique quadratic set-valued mapping $Q : X \to Y$ satisfying*

$$h(f(x), Q(x)) \leq \frac{\theta}{2^p - 4}||x||^p$$

for all $x \in X$.

Proof The proof follows from Theorem 4 by taking

$$\varphi(x, y) := \theta(||x||^p + ||y||^p)$$

for all $x, y \in X$.

4 Stability of the Cubic Set-Valued Functional Equation

In this section, we define a cubic set-valued functional equation and prove the Hyers–Ulam stability of the cubic set-valued functional equation.

Definition 3 Let $f : X \to C_{cb}(Y)$. The cubic set-valued functional equation is defined by

$$f(2x+y) \oplus f(2x-y) = 2f(x+y) \oplus 2f(x-y) \oplus 12f(x)$$

for all $x, y \in X$. Every solution of the cubic set-valued functional equation is called a *cubic set-valued mapping*.

Theorem 5 *Let $\varphi : X \times X \to [0, \infty)$ be a function such that*

$$\widetilde{\varphi}(x, y) := \sum_{j=0}^{\infty} \frac{1}{8^j} \varphi(2^j x, 2^j y) < \infty \tag{15}$$

for all $x, y \in X$. Suppose that $f : X \to (C_{cb}(Y), h)$ is a mapping satisfying $f(0) = \{0\}$ and

$$h(f(2x+y) \oplus f(2x-y), 2f(x+y) \oplus 2f(x-y) \oplus 12f(x)) \le \varphi(x, y) \tag{16}$$

for all $x, y \in X$. Then there exists a unique cubic set-valued mapping $C : X \to (C_{cb}(Y), h)$ such that

$$h(f(x), C(x)) \le \frac{1}{16}\widetilde{\varphi}(x, 0)$$

for all $x \in X$.

Proof Let $y = 0$ in Eq. (16). Since $f(x)$ is convex, we get

$$h(2f(2x), 16f(x)) \le \varphi(x, 0) \tag{17}$$

and if we replace x by $2^n x, n \in \mathbf{N}$, in Eq. (17), then we obtain

$$h\left(2f(2^{n+1}x), 16f(2^n x)\right) \le \varphi(2^n x, 0)$$

and

$$h\left(\frac{f(2^{n+1}x)}{8^{n+1}}, \frac{f(2^n x)}{8^n}\right) \le \frac{1}{2 \cdot 8^{n+1}} \varphi(2^n x, 0).$$

So

$$h\left(\frac{f(2^n x)}{8^n}, \frac{f(2^m x)}{8^m}\right) \le \frac{1}{16} \sum_{j=m}^{n-1} \frac{1}{8^j} \varphi(2^j x, 0) \tag{18}$$

for all integers n, m with $n \geq m$. It follows from Eqs. (15) and (18) that $\left\{\frac{f(2^n x)}{8^n}\right\}$ is a Cauchy sequence in $(C_{cb}(Y), h)$.

Let $C(x) = \lim_{n\to\infty} \frac{f(2^n x)}{8^n}$ for each $x \in X$.

The rest of the proof is similar to the proof of Theorem 1.

Corollary 5 *Let $3 > p > 0$ and $\theta \geq 0$ be real numbers, and let X be a real normed space. Suppose that $f : X \to (C_{cb}(Y), h)$ is a mapping satisfying*

$$h(f(2x+y) \oplus f(2x-y), 2f(x+y) \oplus 2f(x-y) \oplus 12f(x)) \leq \theta(||x||^p + ||y||^p) \tag{19}$$

for all $x, y \in X$. Then there exists a unique cubic set-valued mapping $C : X \to Y$ satisfying

$$h(f(x), C(x)) \leq \frac{\theta}{2(8-2^p)}||x||^p$$

for all $x \in X$.

Proof The proof follows from Theorem 5 by taking

$$\varphi(x, y) := \theta(||x||^p + ||y||^p)$$

for all $x, y \in X$.

Theorem 6 *Let $\varphi : X \times X \to [0, \infty)$ be a function such that*

$$\widetilde{\varphi}(x, y) := \sum_{j=1}^{\infty} 8^j \varphi(\frac{x}{2^j}, \frac{y}{2^j}) < \infty$$

for all $x, y \in X$. Suppose that $f : X \to (C_{cb}(Y), h)$ is a mapping satisfying Eq. (16). Then there exists a unique cubic set-valued mapping $C : X \to (C_{cb}(Y), h)$ such that

$$h(f(x), C(x)) \leq \frac{1}{16}\widetilde{\varphi}(x, 0)$$

for all $x \in X$.

Proof It follows from Eq. (17) that

$$h\left(f(x), 8f\left(\frac{x}{2}\right)\right) \leq \frac{1}{2}\varphi\left(\frac{x}{2}, 0\right)$$

for all $x \in X$.

The rest of the proof is similar to the proofs of Theorems 1 and 5.

Corollary 6 *Let $p > 3$ and $\theta \geq 0$ be real numbers, and let X be a real normed space. Suppose that $f : X \to (C_{cb}(Y), h)$ is a mapping satisfying Eq. (19). Then there exists a unique cubic set-valued mapping $C : X \to Y$ satisfying*

$$h(f(x), C(x)) \leq \frac{\theta}{2(2^p-8)}||x||^p$$

for all $x \in X$.

Proof The proof follows from Theorem 6 by taking

$$\varphi(x, y) := \theta(||x||^p + ||y||^p)$$

for all $x, y \in X$.

5 Stability of the Quartic Set-Valued Functional Equation

In this section, we define a quartic set-valued functional equation and prove the Hyers–Ulam stability of the quartic set-valued functional equation.

Definition 4 Let $f : X \to C_{cb}(Y)$. The quartic set-valued functional equation is defined by

$$f(2x + y) \oplus f(2x - y) \oplus 6f(y) = 4f(x + y) \oplus 4f(x - y) \oplus 24f(x)$$

for all $x, y \in X$. Every solution of the quartic set-valued functional equation is called a *quartic set-valued mapping*.

Theorem 7 *Let $\varphi : X \times X \to [0, \infty)$ be a function such that*

$$\widetilde{\varphi}(x, y) := \sum_{j=0}^{\infty} \frac{1}{16^j} \varphi(2^j x, 2^j y) < \infty \tag{20}$$

for all $x, y \in X$. Suppose that $f : X \to (C_{cb}(Y), h)$ is a mapping satisfying $f(0) = \{0\}$ and

$$h(f(2x + y) \oplus f(2x - y) \oplus 6f(y), 4f(x + y) \oplus 4f(x - y) \oplus 24f(x)) \le \varphi(x, y) \tag{21}$$

for all $x, y \in X$. Then there exists a unique quartic set-valued mapping $T : X \to (C_{cb}(Y), h)$ such that

$$h(f(x), T(x)) \le \frac{1}{32} \widetilde{\varphi}(x, 0)$$

for all $x \in X$.

Proof Let $y = 0$ in Eq. (21). Since $f(x)$ is convex, we get

$$h(2f(2x), 32f(x)) \le \varphi(x, 0) \tag{22}$$

and if we replace x by $2^n x, n \in \mathbf{N}$, in Eq. (22), then we obtain

$$h\left(2f(2^{n+1}x), 32f(2^n x)\right) \le \varphi(2^n x, 0)$$

and

$$h\left(\frac{f(2^{n+1}x)}{16^{n+1}}, \frac{f(2^n x)}{16^n}\right) \le \frac{1}{2 \cdot 16^{n+1}}\varphi(2^n x, 0).$$

So

$$h\left(\frac{f(2^n x)}{16^n}, \frac{f(2^m x)}{16^m}\right) \le \frac{1}{32}\sum_{j=m}^{n-1}\frac{1}{16^j}\varphi(2^j x, 0) \tag{23}$$

for all integers n, m with $n \ge m$. It follows from Eqs. (20) and (23) that $\left\{\frac{f(2^n x)}{16^n}\right\}$ is a Cauchy sequence in $(C_{cb}(Y), h)$.

Let $T(x) = \lim_{n\to\infty}\frac{f(2^n x)}{16^n}$ for each $x \in X$.

The rest of the proof is similar to the proof of Theorem 1.

Corollary 7 *Let* $4 > p > 0$ *and* $\theta \ge 0$ *be real numbers, and let* X *be a real normed space. Suppose that* $f : X \to (C_{cb}(Y), h)$ *is a mapping satisfying*

$$\begin{aligned} &h(f(2x+y) \oplus f(2x-y) \oplus 6f(y), 4f(x+y) \oplus 4f(x-y) \oplus 24f(x)) \\ &\quad \le \theta(||x||^p + ||y||^p) \end{aligned} \tag{24}$$

for all $x, y \in X$. *Then there exists a unique quartic set-valued mapping* $T : X \to Y$ *satisfying*

$$h(f(x), T(x)) \le \frac{\theta}{2(16 - 2^p)}||x||^p$$

for all $x \in X$.

Proof The proof follows from Theorem 7 by taking

$$\varphi(x, y) := \theta(||x||^p + ||y||^p)$$

for all $x, y \in X$.

Theorem 8 *Let* $\varphi : X \times X \to [0, \infty)$ *be a function such that*

$$\widetilde{\varphi}(x, y) := \sum_{j=1}^{\infty} 16^j \varphi\left(\frac{x}{2^j}, \frac{y}{2^j}\right) < \infty$$

for all $x, y \in X$. *Suppose that* $f : X \to (C_{cb}(Y), h)$ *is a mapping satisfying Eq. (21). Then there exists a unique quartic set-valued mapping* $T : X \to (C_{cb}(Y), h)$ *such that*

$$h(f(x), T(x)) \le \frac{1}{32}\widetilde{\varphi}(x, 0)$$

for all $x \in X$.

Proof It follows from Eq. (22) that

$$h\left(f(x), 16f\left(\frac{x}{2}\right)\right) \leq \frac{1}{2}\varphi\left(\frac{x}{2}, 0\right)$$

for all $x \in X$.

The rest of the proof is similar to the proofs of Theorems 1 and 7.

Corollary 8 *Let $p > 4$ and $\theta \geq 0$ be real numbers, and let X be a real normed space. Suppose that $f : X \to (C_{cb}(Y), h)$ is a mapping satisfying Eq. (24). Then there exists a unique quartic set-valued mapping $T : X \to Y$ satisfying*

$$h(f(x), T(x)) \leq \frac{\theta}{2(2^p - 16)}||x||^p$$

for all $x \in X$.

Proof The proof follows from Theorem 8 by taking

$$\varphi(x, y) := \theta(||x||^p + ||y||^p)$$

for all $x, y \in X$.

Acknowledgement C. Park was supported by Basic Science Research Program through the National Research Foundation of Korea funded by the Ministry of Education, Science and Technology (NRF-2012R1A1A2004299). In this section, we prove the Hyers–Ulam stability of the quadratic set-valued functional equation.

References

1. Aczel, J., Dhombres, J.: Functional Equations in Several Variables. Cambridge University Press, Cambridge (1989)
2. Aoki, T.:On the stability of the linear transformation in Banach spaces. J. Math. Soc. Japan. **2**, 64–66 (1950)
3. Arrow, K.J., Debreu, G.: Existence of an equilibrium for a competitive economy. Econometrica. **22**, 265–290 (1954)
4. Aubin, J.P., Frankow, H.: Set-Valued Analysis. Birkhäuser, Boston (1990)
5. Aumann, R.J.: Integrals of set-valued functions. J. Math. Anal. Appl. **12**, 1–12 (1965)
6. Bae, J., Park, W.: A functional equation having monomials as solutions. Appl. Math. Comput. **216**, 87–94 (2010)
7. Cardinali, T., Nikodem, K., Papalini, F.: Some results on stability and characterization of K-convexity of set-valued functions. Ann. Polon. Math. **58**, 185–192 (1993)
8. Cascales, T., Rodrigeuz, J.: Birkhoff integral for multi-valued functions. J. Math. Anal. Appl. **297**, 540–560 (2004)
9. Castaing, C., Valadier, M.: Convex Analysis and Measurable Multifunctions. Lecture Notes in Mathematics, vol. **580**. Springer, Berlin (1977)
10. Cholewa, P.W.: Remarks on the stability of functional equations. Aequ. Math. **27**, 76–86 (1984)
11. Czerwik, S.: On the stability of the quadratic mapping in normed spaces. Abh. Math. Sem. Univ. Hamburg. **62**, 59–64 (1992)

12. Debreu, G.: Integration of correspondences. Proceedings of Fifth Berkeley Symposium on Mathematical Statistics and Probability, vol. II, part I, 351–372 (1966)
13. Eshaghi Gordji, M., Savadkouhi, M.B.: Stability of a mixed type cubic-quartic functional equation in non-Archimedean spaces. Appl. Math. Lett. **23**, 1198–1202 (2010)
14. Găvruta, P.: A generalization of the Hyers–Ulam–Rassias stability of approximately additive mappings. J. Math. Anal. Appl. **184**, 431–436 (1994)
15. Hess, C.: Set-valued integration and set-valued probability theory: an overview. In Handbook of Measure Theory, vols. I and II. North-Holland, Amsterdam (2002)
16. Hindenbrand, W.: Core and Equilibria of a Large Economy. Princeton University Press, Princeton (1974)
17. Hyers, D.H.: On the stability of the linear functional equation. Proc. Natl. Acad. Sci. U S A. **27**, 222–224 (1941)
18. Hyers, D.H., Isac, G., Rassias, T.M.: Stability of Functional Equations in Several Variables. Birkhäuser, Basel (1998)
19. Isac, G., Rassias, T.M.: On the Hyers-Ulam stability of ψ-additive mappings. J. Approx. Theory. **72**, 131–137 (1993)
20. Jun, K., Kim, H.: The generalized Hyers–Ulam–Rassias stability of a cubic functional equation. J. Math. Anal. Appl. **274**, 867–878 (2002)
21. Klein, E., Thompson, A.: Theory of Correspondence. Wiley, New York (1984)
22. Lee, K.: Stability of functional equations related to set-valued functions. (In press)
23. Lee, S., Im, S., Hwang, I.: Quartic functional equations. J. Math. Anal. Appl. **307**, 387–394 (2005)
24. McKenzie, L.W.: On the existence of general equilibrium for a competitive market. Econometrica. **27**, 54–71 (1959)
25. Nikodem, K.: On quadratic set-valued functions. Publ. Math. Debrecen. **30**, 297–301 (1984)
26. Nikodem, K.: On Jensen's functional equation for set-valued functions. Radovi Mat. **3**, 23–33 (1987)
27. Nikodem, K.: Set-valued solutions of the Pexider functional equation. Funkc. Ekvacioj. **31**, 227–231 (1988)
28. Nikodem, K.: K-Convex and K-Concave Set-Valued Functions. Zesz. Nauk. Politech. Łódz. **559** (1989)
29. Piao, Y.J.: The existence and uniqueness of additive selection for (α, β)-(β, α) type subadditive set-valued maps. J. Northeast Norm. Univ. **41**, 38–40 (2009)
30. Popa, D.: Additive selections of (α, β)-subadditive set-valued maps. Glas. Mat. **36**(56), 11–16 (2001)
31. Rassias, T.M.: On the stability of the linear mapping in Banach spaces. Proc. Amer. Math. Soc. **72**, 297–300 (1978)
32. Rassias, T.M.: Stability and set-valued functions. In: Cazacu, C.A., Lehto, O.E., Rassias, T.M. (eds.) Analysis and Topology, pp. 585–614. World Scientific, Singapore (1988)
33. Rassias, J.M.: Solution of a problem of Ulam. J. Approx. Theory. **57**, 268–273 (1989)
34. Rassias, T.M.: Functional Equations and Inequalities. Kluwer Academic, Dordrecht (2000)
35. Rassias, T.M.: On the stability of functional equations in Banach spaces. J. Math. Anal. Appl. **251**, 264–284 (2000)
36. Rassias, T.M.: On the stability of functional equations and a problem of Ulam. Acta Math. Appl. **62**, 23–130 (2000)
37. Skof, F.: Proprietà locali e approssimazione di operatori. Rend. Sem. Mat. Fis. Milano. **53**, 113–129 (1983)
38. Ulam, S.M.: Problems in Modern Mathematics. Wiley, New York (1940) (Chapter VI, Science Editions)

On the Generalized Hyers–Ulam Stability in Multi-Banach Spaces Associated to a Jensen-type Additive Mapping

Fridoun Moradlou and Themistocles M. Rassias

Abstract Using the fixed point method, we prove the generalized Hyers–Ulam–Rassias stability of the following functional equation in multi-Banach spaces:

$$f\left(\frac{\sum_{i=1}^{n} r_i x_i}{k}\right) + \sum_{1\le i<j\le n} f\left(\frac{r_i x_i + r_j x_j}{k}\right) = \frac{n}{k}\sum_{i=1}^{n} r_i f(x_i). \tag{1}$$

The concept of generalized Hyers–Ulam–Rassias stability originated from Th. M. Rassias' stability theorem that appeared in his paper: On the stability of the linear mapping in Banach spaces, Proc. Amer. Math. Soc. **72**, 297–300, 1978.

Keywords 2010 Mathematics subject classification · Primary 39B82 · Secondary 39B52, 46B99

1 Introduction and Preliminaries

A classical question in the theory of functional equations is the following: "When is it true that a function, which approximately satisfies a functional equation $\mathcal{E}$ must be close to an exact solution of $\mathcal{E}$?" If the problem accepts a solution, we say that the equation $\mathcal{E}$ is stable. Such a problem was formulated by Ulam [49] in 1940 and solved the next year for the Cauchy functional equation by Hyers [16]. It gave rise to the stability theory for functional equations. Hyers' result was extended by Aoki [1] in 1950 by considering the unbounded Cauchy differences. In 1978, Th. M. Rassias [41] proved that the additive mapping T, obtained by Hyers or Aoki, is linear if, in addition, for each $x \in E$ the mapping $f(tx)$ is continuous in $t \in \mathbb{R}$. Găvruta [14] generalized the Rassias' result. Beginning around the year 1980, the stability problems of several functional equations and approximate homomorphisms have

F. Moradlou (✉)
Department of Mathematics, Sahand University of Technology, Tabriz, Iran
e-mail: moradlou@sut.ac.ir; fridoun.moradlou@gmail.com

T. M. Rassias
Department of Mathematics, National Technical University of Athens,
Zografou Campus, 15780 Athens, Greece
e-mail: trassias@math.ntua.gr

T. M. Rassias, P. M. Pardalos (eds.), *Mathematics Without Boundaries*,
DOI 10.1007/978-1-4939-1106-6_14,

been extensively investigated by a number of authors and there are many interesting results concerning this problem (see [2, 4, 7, 8, 20, 22–29, 32–40, 43–47]).

In 2003 Cădariu and Radu applied the fixed point method to the investigation of the Jensen functional equation [5] (see also [6, 7, 19, 38]). They could present a short and a simple proof (different from the "*direct method*," initiated by Hyers in 1941) for the generalized Hyers–Ulam stability of Jensen functional equation [5], for Cauchy functional equation [7] and for quadratic functional equation [6].

The functional equation

$$Q(x+y)+Q(x-y)=2Q(x)+2Q(y), \tag{2}$$

is called a *quadratic functional equation,* and every solution of Eq. (2) is said to be a *quadratic mapping*. F. Skof [48] proved the Hyers–Ulam stability of the quadratic functional Eq. (2) for mappings $f: E_1 \to E_2$, where E_1 is a normed space and E_2 is a Banach space. Also S. Czerwik [10] proved the Hyers-Ulam-Rassias stabilty of the quadratic functional Eq. (2). C. Borelli and G.L. Forti [3] generalized the stability result of the quadratic functional Eq. (2).

Jun and Kim [17] introduced the following functional equation

$$f(2x+y)+f(2x-y)=2f(x+y)+2f(x-y)+12f(x) \tag{3}$$

and they established the general solution and the generalized Hyers–Ulam–Rassias stability problem for the functional Eq. (3).

It is easy to see that the function $f(x)=cx^3$ is a solution of the above functional Eq. (3). Thus, it is natural that (3) is called a *cubic functional equation* and every solution of the cubic functional Eq. (3) is said to be a *cubic mapping*.

The functional equation

$$f(2x+y)+f(2x-y)=4f(x+y)+4f(x-y)+24f(x)-6f(y) \tag{4}$$

is called a *quartic functional equation* since the function $f(x)=cx^4 \quad (c \in \mathbb{R})$ is a solution of the above functional Eq. (4). Every solution of the quartic functional Eq. (4) is said to be a *quartic mapping*. The stability problem for a quartic functional equation first was considered by Th. M. Rassias [42] for mappings $f: X \to Y$, where X is a real normed space and Y is a Banach space. In 2003, Chung and Sahoo [9] determined the general solution of the quartic functional equation. Lee et al. [21] proved Hyers–Ulam stability of the quartic functional equation. In 2008, Najati [29] proved generalized Hyers-Ulam-Rassias stability of the quartic functional equation.

Recently, Dales and Polyakov [11] introduced the notion of multinormed spaces. This concept is somewhat similar to operator sequence spaces and has some connections with operator spaces and Banach lattices. Dales and Moslehian [12] investigated stability of Cauchy functional equation in multi-Banach spaces (see also [28, 50]).

In this paper, for a fixed positive integer $n \geq 2$, we introduce the following additive functional equation of the Jensen type:

$$f\left(\frac{\sum_{i=1}^{n} r_i x_i}{k}\right)+\sum_{1\le i<j\le n} f\left(\frac{r_i x_i + r_j x_j}{k}\right)=\frac{n}{k}\sum_{i=1}^{n} r_i f(x_i), \tag{5}$$

where $r_1, \cdots, r_n \in \mathbb{R}$ and k is a fixed nonzero integer. Every solution of the functional Eq. (5) is said to be a *generalized additive mapping of the Jensen type.*

This paper is organized as follows: In Sect. 2, we present the some known methods which are used for investigation of Hyers–Ulam–Rassias stability of functional equations. In Sect. 3, we will adopt the idea of Cădariu and Radu [5, 7, 38], to prove the generalized Hyers–Ulam–Rassias stability of additive functional equation of the Jensen type on multi-Banach spaces.

2 Methods for Investigation of Stability of Functional Equation

2.1 Direct Method

The most important and widely used method for investigation of stability of functional equation is *direct method* which is based on iteration process. At first, this method is used by Hyers [16] to answer the problem of Ulam.

Theorem 1 *(Hyers–Rassias) Let X be a normed space and Y be a Banach space. Assume that $f : X \to Y$ is a mapping satisfying*

$$\|f(x+y) - f(x) - f(y)\| \leq \epsilon,$$

for some $\epsilon > 0$ and all $x, y \in X$. Then there exists a unique additive mapping $T : X \to Y$ such that

$$\|f(x) - T(x)\| \leq \epsilon$$

for all $x \in X$. If, in addition, $f(tx)$ is continuous in t for each fixed $x \in X$, then the function T is linear.

As you see, there exist two essential conditions for proving the stability of functional equations in this method:

- Completeness of the range and
- Commutative domain.

2.2 Sandwich Method

In 1999, K. Nikodem et al. [30] proved the following theorem, which is the base of *Sandwich method.* In this theorem, they give a necessary and sufficient condition for existence of separating additive mapping.

Theorem 2 *Let $(S, +)$ be a commutative semigroup and $p, q : S \to [-\infty, \infty)$ be two mappings. Then there exists a mapping $T : S \to [-\infty, \infty)$ such that $p \leq T \leq q$*

if and only if for all positive integers m and n and all $x_1, \cdots, x_m, y_1, \cdots, y_n \in S$ *satisfying* $x_1 + \cdots + x_m = y_1 + \cdots + y_n$*, we have*

$$p(x_1) + \cdots + p(x_m) = p(y_1) + \cdots + p(y_n).$$

Two years later, Páles [31] proved Hyers–Ulam stability of Cauchy functional equation by using the above theorem.

Theorem 3 *Let* $(S, +)$ *be a commutative semigroup and* $p, q : S \to [-\infty, \infty)$ *be two mappings. Assume that* $f : S \to \mathbb{R}$ *is a mapping such that*

$$p(y) \leq f(x + y) - f(x) \leq q(y),$$

for all $x, y \in S$*. Then there exists an additive mapping* $T : S \to [-\infty, \infty)$ *such that*

$$p(x) \leq f(x) \leq q(x),$$

for all $x \in S$*.*

Corollary 1 *Let E be a Banach space and* $f : E \to \mathbb{R}$ *be a mapping such that*

$$|f(x + y) - f(x) - f(y)| \leq \epsilon,$$

for some $\epsilon > 0$ *and all* $x, y \in E$*. Then there exists an additive mapping* $T : E \to \mathbb{R}$ *such that*

$$|f(x) - T(x)| \leq \epsilon,$$

for all $x \in E$*.*

2.3 Fixed Point Method

In 2003 Cădariu and Radu presented a short and a simple proof for the generalized Hyers–Ulam stability of Jensen functional equation [5], for Cauchy functional equation [7] and for the quadratic functional equation [6], which is called *fixed point method.*

Now, we recall two fundamental results in fixed point theory.

Theorem 4 ([5]) *Let* (X, d) *be a complete metric space and let* $J : X \to X$ *be strictly contractive, i.e.,*

$$d(Jx, Jy) \leq Lf(x, y), \qquad \forall x, y \in X$$

for some Lipschitz constant $L < 1$*. Then*

(1) The mapping J has a unique fixed point $x^* = Jx^*$*;*
(2) The fixed point x^* *is globally attractive, i.e.,*

$$\lim_{n\to\infty} J^n x = x^*,$$

for any starting point $x \in X$*;*

(3) One has the following estimation inequalities:

$$d(J^n x, x^*) \leq L^n d(x, x^*),$$

$$d(J^n x, x^*) \leq \frac{1}{1-L} d(J^n x, J^{n+1} x),$$

$$d(x, x^*) \leq \frac{1}{1-L} d(x, Jx),$$

for all nonnegative integers n and all $x \in X$.

Definition 1 Let X be a set. A function $d : X \times X \to [0, \infty]$ is called a *generalized metric* on X if d satisfies

(1) $d(x, y) = 0$ if and only if $x = y$;
(2) $d(x, y) = d(y, x)$ for all $x, y \in X$;
(3) $d(x, z) \leq d(x, y) + d(y, z)$ for all $x, y, z \in X$.

Theorem 5 ([13]) *Let (X, d) be a complete generalized metric space and let $J : X \to X$ be a strictly contractive mapping with Lipschitz constant $L < 1$. Then for each given element $x \in X$, either*

$$d(J^n x, J^{n+1} x) = \infty,$$

for all nonnegative integers n or there exists a positive integer n_0 such that

(1) $d(J^n x, J^{n+1} x) < \infty, \qquad \forall n \geq n_0$;
(2) *The sequence $\{J^n x\}$ converges to a fixed point y^* of J;*
(3) *y^* is the unique fixed point of J in the set $Y = \{y \in X \mid d(J^{n_0} x, y) < \infty\}$;*
(4) *$d(y, y^*) \leq \frac{1}{1-L} d(y, Jy)$ for all $y \in Y$.*

Theorem 6 *Let X be a normed space and Y be a Banach space. Assume that $f : X \to Y$ is a mapping such that*

$$\| f(x + y) - f(x) - f(y) \| \leq \epsilon,$$

for some $\epsilon > 0$ and for all $x, y \in X$. Then there exists a unique additive mapping $T : X \to Y$, such that

$$\| f(x) - T(x) \| \leq \epsilon,$$

for all $x \in X$.

2.4 Weighted Space Method

Recently P. Găvruta and L. Găvruta [15] used a new method for investigation of Hyers–Ulam–Rassias stability of a nonlinear functional equation, Volterra integral operator and Fredholm operator. This method generalized the *fixed point method.*

Theorem 7 *Let S be a nonempty set and (X, d) be a complete metric space. Assume that $\eta : S \to S$, $\varphi : S \to (0, \infty)$ and $F : S \times X \to X$ are mappings. We suppose that there exists $\alpha \in [0, 1)$ so that*

$$\varphi(\eta(x))d(F(x, u(\eta(x))), F(x, v(\eta(x)))) \leq \alpha\varphi(x)d(u(\eta(x)), v(\eta(x))), \quad x \in S. \quad (6)$$

If $y : S \to X$ is so that:

$$d(y(x), F(x, y(\eta(x)))) \leq \varphi(x), \quad x \in S. \quad (7)$$

Then there exists a unique $y_0 : S \to X$ such that

$$y_0(x) = F(x, y_0(\eta(x))), \quad (8)$$

and

$$d(y(x), y_0(x)) \leq \frac{1}{1-\alpha}\varphi(x), \quad x \in S. \quad (9)$$

Let $I = [a, b]$ and $c \in I$. Denote by $C(I)$ the space of all complex-valued continuous functions on I.

Theorem 8 *Assume that the functions $L : I \to [0, \infty)$ is integrable, $g \in C(I)$, $f : I \times \mathbb{C} \to \mathbb{C}$ and $\varphi : I \to (0, \infty)$ are continuous. Suppose that there exists a unique $\alpha \in [0, 1)$ so that*

$$|\int_c^x L(t)\varphi(t)dt| \leq \alpha\varphi(x), \; x \in I; \quad (10)$$

$$|f(t, u(t)) - f(t, v(t))| \leq L(t)|u(t) - v(t)|, \; t \in I, \forall u, v \in C(I). \quad (11)$$

If $y \in C(I)$ is so that

$$|y(x) - g(x) - \int_c^x f(t, y(t))dt| \leq \varphi(x), \; x \in I,$$

then there exists a unique $y_0 \in C(I)$:

$$y_0(x) = g(x) + \int_c^x f(t, y_0(t))dt,$$

and

$$|y(x) - y_0(x)| \leq \frac{\varphi(x)}{1-\alpha}, \quad x \in I.$$

Theorem 9 *Let $g \in C(I)$ and assume that $\varphi : I \to (0, \infty)$ is continuous, $L : I \times I \to [0, \infty)$ is integrable and $K : I \times I \times \mathbb{C} \to \mathbb{C}$ is continuous. Suppose that there exists $\beta > 0$, such that*

$$\int_I L(x,t)\varphi(t)dt \leq \beta\varphi(x), \quad x \in I;$$

$$|K(x,t,u(t)) - K(x,t,v(t))| \leq L(x,t)|u(t) - v(t)|, \ u,v \in C(I).$$

Let $y \in C(I)$ be so that

$$|y(x) - g(x) - \lambda \int_I K(x,t,y(t))dt| \leq \varphi(x), \quad x \in I.$$

If $|\lambda| < \frac{1}{\beta}$, then there exists a unique $y_0 \in C(I)$:

$$y_0(x) = g(x) + \lambda \int_I K(x,t,y_0(t))dt,$$

and

$$|y(x) - y_0(x)| \leq \frac{\varphi(x)}{1 - |\lambda|\beta}, \quad x \in I.$$

3 Stability of Jensen-Type Functional Equation: Fixed Point Method

Assume that $(E, \|.\|)$ is a complex linear space, and let $m \in \mathbb{N}$. We denote by E^m the linear space $E \oplus E \oplus \cdots \oplus E$ consisting of m-tuples $(x_1, \cdots, x_m)$, where $x_1, \cdots, x_m \in E$. The linear operations on E^m are defined coordinate-wise. When we write $(0, \ldots, 0, x_i, 0, \ldots, 0)$ for an element in E^m, we understand that x_i appears in the ith coordinate. The zero element of either E or E^m is denoted by 0. We denote by $\mathbb{N}_m$ the set $\{1, 2, \cdots, m\}$ and by σ_m the group of permutations on m symbols.

In this section, we recall the notion of a multinormed space and some preliminaries concerning multinormed spaces. For more details, we refer readers to the book of Dales and Polyakov [11].

Definition 2 Let $(E, \|.\|)$ be a complex normed space and let $m \in \mathbb{N}$. A multinorm of level m on $\{E^s \ : \ s \in \mathbb{N}_m\}$ is a sequence

$$(\|.\|_s) = (\|.\|_s : s \in \mathbb{N}_m),$$

such that $\|.\|_s$ is a norm on E^s for each $s \in \mathbb{N}_m$ such that $\|x\|_1 = \|x\|$ for each $x \in E$, and such that the following Axioms $(A1)$–$(A4)$ are satisfied for each $s \in \mathbb{N}_m$ with $s \geq 2$:

$(A1)$ For each $\sigma \in \sigma_s$ and $x_1, \cdots, x_s \in E$, we have

$$\|(x_{\sigma(1)}, \cdots, x_{\sigma(s)})\|_s = \|(x_1, \cdots, x_s)\|_s;$$

$(A2)$ For each $\alpha_1, \cdots, \alpha_s \in \mathbb{C}$ and $x_1, \cdots, x_s \in E^s$, we have

$$\|(\alpha_1 x_1, \cdots, \alpha_s x_s)\|_s \le (\max_{i \in \mathbb{N}_s} |\alpha_i|) \|(x_1, \cdots, x_s)\|_s;$$

$(A3)$ For each $x_1, \cdots, x_s \in E$, we have

$$\|(x_1, \cdots, x_{s-1}, 0)\|_s = \|(x_1, \cdots, x_{s-1})\|_{s-1};$$

$(A4)$ For each $x_1, \cdots, x_s \in E$, we have

$$\|(x_1, \cdots, x_{s-1}, x_{s-1})\|_s = \|(x_1, \cdots, x_{s-1})\|_{s-1};$$

In this case, we say that $((E^s, \|.\|_s) : s \in \mathbb{N}_m)$ is a *multinormed space of level m*.

Definition 3 A multinorm on $\{E^s : s \in \mathbb{N}\}$ is a sequence

$$(\|.\|_s) = (\|.\|_s : s \in \mathbb{N}),$$

such that $(\|.\|_s : s \in \mathbb{N}_m)$ is a multinorm of level m for each $m \in \mathbb{N}$. In this case, we say that $((E^s, \|.\|_s) : s \in \mathbb{N})$ is a multinormed space.

It is natural to ask whether the four Axioms $(A1)$–$(A4)$ are independent. Dales and Polyakov give following examples to show that this is indeed the case.

Example 1 Let $(E, \|.\|)$ be a nonzero normed space. Set $\|x\|_1 = \|x\|$ for all $x \in X$ and for each $n \in \mathbb{N}$ with $n \ge 2$, define

$$\|(x_1, \cdots, x_n)\|_n = \max\left\{\|x_1\|, \frac{\|x_2\|}{2}, \cdots, \frac{\|x_n\|}{2}\right\} \quad ((x_1, \cdots, x_n) \in E^n).$$

It is easy to see that $\|.\|_n$ is a norm on E^n for each $n \in \mathbb{N}$ and that $\|.\|_n$ is a sequence that satisfies Axioms $(A2)$, $(A3)$ and $(A4)$ for each $n \in \mathbb{N}$. However, take $x \in E$ with $\|x\| = 1$. Then $\|(2x, 3x)\|_2 = 2$, but $\|(3x, 2x)\|_2 = 3$ and so $\|.\|_2$ does not satisfy Axiom $(A1)$.

Example 2 Let $E = \mathbb{C}$. For $z \in \mathbb{C}$, set $\|z\|_1 = |z|$. Next, for $(z, w) \in \mathbb{C}^2$, set

$$r((z, w)) = \frac{1}{2}(|z - w| + |z + w|).$$

Then r is a norm on $\mathbb{C}^2$. Further,

$$r((z, z)) = r((z, 0)) = |z| \quad (z \in \mathbb{C}),$$

and also

$$r((z, w)) = r((w, z)) \ge \max\{|z|, |w|\} \quad ((z, w) \in \mathbb{C}^2).$$

For $n \in \mathbb{N}$ with $n \ge 2$, put

$$\|(z_1, \cdots, z_n)\|_n = \max\{r((z_i, z_j)) \ : i, j \in \mathbb{N}_n\} \ ((z_1, \cdots, z_n) \in \mathbb{C}^n),$$

such that for all $(z, w) \in \mathbb{C}^2$, $\|(z, w)\|_2 = r((z, w))$ and

$$\|(z_1, \cdots, z_n)\|_n \geq \max_{i \in \mathbb{N}_n} |z_i| \qquad ((z_1, \cdots, z_n) \in \mathbb{C}^n).$$

It is easy to see that $\|.\|_n$ is a norm on $\mathbb{C}^n$ and that the sequence $(\|.\|_n)$ satisfies $(A1)$, $(A3)$ and $(A4)$ for each $n \in \mathbb{N}$.

However, $\|.\|_2$ does not satisfy Axiom $(A2)$. Indeed,

$$\|(1, i)\|_2 = \frac{1}{2}(|1 - i| + |1 + i|) = \sqrt{2} > 1 = \|(1, 1)\|_2.$$

Example 3 Assume that $E = \mathbb{C}$. Put $\|z\|_1 = |z| \quad (z \in \mathbb{C})$ and

$$\|(z, w)\|_2 = \frac{1}{2}(|z| + |w|) \qquad (z, w \in \mathbb{C}).$$

Then $\|.\|_2$ is a norm on $\mathbb{C}^2$, and $\|.\|_2$ satisfies Axioms $(A1)$, $(A2)$ and $(A4)$ for n = 2. However $\|(1, 0)\|_2 = \frac{1}{2} < 1 = \|1\|_1$, and so Axiom $(A3)$ does not hold.

Example 4 Let $(E, \|.\|)$ be a nonzero normed space. For each $n \in \mathbb{N}$, put

$$\|(x_1, \cdots, x_n)\|_n = \Big(\sum_{j=1}^{n} \|x_j\|^p\Big)^{\frac{1}{p}} \quad ((x_1, \cdots, x_n) \in E^n),$$

where $p \geq 1$. Then it is immediately checked that, for each p, the function $\|.\|_n$ is a norm on E^n and that $(\|.\|_n)$ is a sequence that satisfies Axioms $(A1)$, $(A2)$ and $(A3)$ for each $n \in \mathbb{N}$, but $\|.\|_2$ does not satisfy Axiom $(A4)$.

Lemma 1 *Let $((E^s, \|.\|_s) : s \in \mathbb{N})$ be a multinormed space. The following properties are immediate consequences of the axioms for multinormed spaces.*

(i) Let $k \in \mathbb{N}_{n-1}$ and $x_1, \cdots, x_{k+1} \in E$. Then

$$\|(x_1, \cdots, x_k)\|_k \leq \|(x_1, \cdots, x_k, x_{k+1})\|_{k+1}.$$

(ii) Let $j, k \in \mathbb{N}$ with $j + k \leq n$ and let $x_1, \cdots, x_j, y_1, \cdots, y_k \in E$. Then

$$\|(x_1, \cdots, x_j, y_1, \cdots, y_k)\|_{j+k} \leq \|(x_1, \cdots, x_j)\|_j + \|(y_1, \cdots, y_k)\|_k.$$

(iii) Let $k \in \mathbb{N}_n$ with $k \geq 2$ and let $x_1, \cdots, x_k \in E$. Put $x = \dfrac{x_{k-1} + x_k}{2}$, Then

$$\|(x_1, \cdots, x_{k-2}, x, x)\|_k \leq \|(x_1, \cdots, x_{k-2}, x_{k-1}, x_k)\|_k.$$

(iv) For all $x \in E$ and $s \in \mathbb{N}$, we have

$$\|(x, \cdots, x)\|_s = \|x\|.$$

(v) For all $s \in \mathbb{N}$ and all $x_1, \cdots, x_s \in E$, we have

$$\max_{i \in \mathbb{N}_s} \|x_i\| \leq \|(x_1, \cdots, x_s)\|_s \leq \sum_{i=1}^{s} \|x_i\| \leq s \max_{i \in \mathbb{N}_s} \|x_i\|.$$

The following Lemma is a consequence of (v):

Lemma 2 *Suppose that $(E, \|.\|)$ is a Banach space. Then $(E^s, \|.\|_s)$ is a Banach space for each $s \in \mathbb{N}$.*

Definition 4 Let $((E^s, \|.\|_s) : s \in \mathbb{N})$ be a multinormed space for which $(E, \|.\|)$ is a Banach space. Then $((E^s, \|.\|_s) : s \in \mathbb{N})$ is called a multi-Banach space.

Now, we recall two important examples of multinorms for arbitrary space $(E, \|.\|)$.

Example 5 Let $(E, \|.\|)$ be a normed space. For $m \in \mathbb{N}$, define $\|.\|_m$ on E^m by

$$\|(x_1, \cdots, x_m)\|_m = \max_{i \in \mathbb{N}_m} \|x_i\| \qquad (x_1, \cdots, x_m \in E).$$

It is immediate that $((E^s, \|.\|_s) : s \in \mathbb{N})$ is a multinormed space. The sequence $(\|.\|_m \ : \ m \in \mathbb{N})$ is called *minimum multinorm*. The terminology "minimum" is justified by Lemma 1.

Example 6 Let $(E, \|.\|)$ be a normed space and let $\{(\|.\|_m^\alpha \ : \ m \in \mathbb{N}) \ : \ \alpha \in A\}$ be the (nonempty) family of all multinorms on $\{E^s \ : s \in \mathbb{N}\}$. For $s \in \mathbb{N}$, define

$$|||(x_1, \cdots, x_s)|||_s = \sup_{\alpha \in A} \|(x_1, \cdots, x_s)\|_s^\alpha \quad (x_1, \cdots, x_s \in E).$$

Then $(|||.|||_m \ : \ m \in \mathbb{N})$ is a multinorm on $\{E^s \ : s \in \mathbb{N}\}$, which is called *maximum multinorm*.

Throughout this paper, n will be a positive integer such that $n \geq 2$, k will be a fixed nonzero integer and $r_1, \cdots, r_n$ will be real numbers such that $r_i, r_j \neq 0$ for fixed $1 \leq i < j \leq n$.

Lemma 3 *Let $\mathcal{X}$ and $\mathcal{Y}$ be linear spaces and suppose that $r_1, \cdots, r_n$ are real numbers with $\sum_{i=1}^n r_i \neq \dfrac{(n^2 - n + 2)k}{2n}$. Assume that a mapping $L : \mathcal{X} \to \mathcal{Y}$ satisfies the functional Eq. (5) for all $x_1, \cdots, x_n \in \mathcal{X}$. Then the mapping L is Cauchy additive. Moreover, $L\left(\frac{r_j x}{k}\right) = \frac{r_j}{k} L(x)$ for all $x \in \mathcal{X}$ and all $1 \leq j \leq n$.*

Proof Since $\displaystyle\sum_{i=1}^{n} r_i \neq \frac{(n^2 - n + 2)k}{2n}$, putting $x_1 = \cdots = x_n = 0$ in (5), we get $L(0) = 0$. Letting $x_m = 0$ in (5) for all $1 \leq m \leq n$ with $m \neq i, j$, we get

$$2L\left(\frac{r_i x_i + r_j x_j}{k}\right) + (n-2)L\left(\frac{r_i x_i}{k}\right) + (n-2)L\left(\frac{r_j x_j}{k}\right) = \frac{n}{k}(r_i L(x_i) + r_j L(x_j)), \qquad (12)$$

for all $x_i, x_j \in \mathcal{X}$. Letting $x_j = 0$ in (12), we have

$$L\left(\frac{r_i x_i}{k}\right) = \frac{r_i}{k} L(x_i), \tag{13}$$

for all $x_i \in \mathcal{X}$. Similarly, by putting $x_i = 0$ in (12), we get

$$L\left(\frac{r_j x_j}{k}\right) = \frac{r_j}{k} L(x_j), \tag{14}$$

for all $x_j \in \mathcal{X}$. It follows from (12), (13) and (14) that

$$2L\left(\frac{r_i x_i + r_j x_j}{k}\right) + (n-2)L\left(\frac{r_i x_i}{k}\right) + (n-2)L\left(\frac{r_j x_j}{k}\right) = n\left(L(\frac{r_i x_i}{k}) + L(\frac{r_j x_j}{k})\right), \tag{15}$$

for all $x_i, x_j \in \mathcal{X}$. Replacing x_i and x_j by kx/r_i and ky/r_j in (15), respectively, we get

$$2L(x+y) + (n-2)L(x) + (n-2)L(y) = n(L(x) + L(y)), \tag{16}$$

for all $x, y \in \mathcal{X}$. Therefore, L is additive. Moreover, let $x \in \mathcal{X}$ and $1 \le i \le n$. Setting $x_i = x$ and $x_j = 0$ for all $1 \le j \le n$ with $j \ne i$ in (5), we get that $L\left(\frac{r_j x}{k}\right) = \frac{r_j}{k} L(x)$.

Using the same proof as Lemma 3, we have an alternative result of Lemma 3 when $\sum_{i=1}^n r_i = \dfrac{(n^2 - n + 2)k}{2n}$.

Lemma 4 *Let $\mathcal{X}$ and $\mathcal{Y}$ be linear spaces. Assume that a mapping $L : \mathcal{X} \to \mathcal{Y}$ with $L(0) = 0$ satisfies the functional Eq. (5) for all $x_1, \cdots, x_n \in \mathcal{X}$. Then the mapping L is Cauchy additive. Moreover, $L(\frac{r_j x}{k}) = \frac{r_j}{k} L(x)$ for all $x \in \mathcal{X}$ and all $1 \le j \le n$.*

We will use the following Lemma in the proof of the next theorems.

Lemma 5 ([25]) *Let $\mathcal{X}$ and $\mathcal{Y}$ be vector spaces and let $f : \mathcal{X} \to \mathcal{Y}$ be an additive mapping such that $f(\mu x) = \mu f(x)$ for all $x \in \mathcal{X}$ and all $\mu \in \mathbb{T}^1 := \{\lambda \in \mathbb{C} : |\lambda| = 1\}$. Then the mapping $f : \mathcal{X} \to \mathcal{Y}$ is $\mathbb{C}$-linear.*

Let X and Y be vector spaces. For a given mapping $f : X \to Y$, we define

$$D_{\mu, r_1, \cdots, r_n} f(x_1, \cdots, x_n) = f\left(\frac{\sum_{i=1}^n \mu r_i x_i}{k}\right) + \sum_{1 \le i < j \le n} f\left(\frac{\mu r_i x_i + \mu r_j x_j}{k}\right) - \frac{n}{k} \sum_{i=1}^n \mu r_i f(x_i)$$

for all $\mu \in \mathbb{T}^1$ and all $x_1, \cdots, x_n \in X$.

Now, we prove the generalized Hyers–Ulam–Rassias stability of Jensen-type additive mapping on multi-Banach spaces for the functional equation $D_{\mu, r_1, \cdots, r_n} f(x_1, \cdots, x_n) = 0$.

Theorem 10 *Let E be a linear space and let $\{(F^l, \|.\|_l) : l \in \mathbb{N}\}$ be a multi - Banach space. Suppose that $s \in \mathbb{N}$ and $f : E \to F$ is a mapping satisfying $f(0) = 0$ for which there exists a control function $\varphi : E^{ns} \to [0, \infty)$ such that*

$$\|(D_{\mu,r_1,\cdots,r_n} f(X^{(1)}), \cdots, D_{\mu,r_1,\cdots,r_n} f(X^{(s)}))\|_s \leq \varphi(X^{(1)}, \cdots, X^{(s)}) \tag{17}$$

for all $\mu \in \mathbb{T}^1$ and all $X^{(1)} = (x_1^{(1)}, \cdots, x_n^{(1)}), \cdots, X^{(s)} = (x_1^{(s)}, \cdots, x_n^{(s)}) \in E^n$. If there exists a Lipschitz constant $L < 1$ such that

$$\varphi\big(X^{(1)}, \cdots, X^{(s)}\big) \leq 2L\varphi\Big(\frac{X^{(1)}}{2}, \cdots, \frac{X^{(s)}}{2}\Big) \tag{18}$$

for all $X^{(1)}, \cdots, X^{(s)} \in E^n$, then there exists a unique linear mapping $T : E \to F$ such that

$$\begin{aligned}
&\|(f(x_1) - T(x_1), \cdots, f(x_s) - T(x_s))\|_s \\
&\leq \frac{1}{4-4L}\Big[\varphi\left(X_{i,j}(\frac{kx_1}{r_i}, \frac{kx_1}{r_j}), X_{i,j}(\frac{kx_2}{r_i}, \frac{kx_2}{r_j}), \cdots, X_{i,j}(\frac{kx_s}{r_i}, \frac{kx_s}{r_j})\right) \qquad (19)\\
&+ \varphi\left(X_{i,j}(\frac{kx_1}{r_i}, 0), X_{i,j}(\frac{kx_2}{r_i}, 0), \cdots, X_{i,j}(\frac{kx_s}{r_i}, 0)\right) \\
&+ \varphi\left(X_{i,j}(0, \frac{kx_1}{r_j}), X_{i,j}(0, \frac{kx_2}{r_j}), \cdots, X_{i,j}(0, \frac{kx_s}{r_j})\right)\Big],
\end{aligned}$$

for all $x_1, \cdots, x_s \in E$, where

$$X_{i,j}(x, y) = (0, \cdots, 0, \underbrace{x}_{i\,th}, 0, \cdots, 0, \underbrace{y}_{j\,th}, 0, \cdots, 0),$$

for all $x, y \in E$.

Proof For convenience, set

$$\begin{aligned}
\varphi_{i,j}(x_1, x_2, \cdots, x_s) = &\ \varphi\left(X_{i,j}(\frac{kx_1}{r_i}, \frac{kx_1}{r_j}), X_{i,j}(\frac{kx_2}{r_i}, \frac{kx_2}{r_j}), \cdots, X_{i,j}(\frac{kx_s}{r_i}, \frac{kx_s}{r_j})\right) \\
&+ \varphi\left(X_{i,j}(\frac{kx_1}{r_i}, 0), X_{i,j}(\frac{kx_2}{r_i}, 0), \cdots, X_{i,j}(\frac{kx_s}{r_i}, 0)\right) \\
&+ \varphi\left(X_{i,j}(0, \frac{kx_1}{r_j}), X_{i,j}(0, \frac{kx_2}{r_j}), \cdots, X_{i,j}(0, \frac{kx_s}{r_j})\right),
\end{aligned}$$

where $x_1, \cdots, x_s \in E$ and $1 \leq i < j \leq n$. Consider the set $\mathcal{X} := \{g : E \to F, \quad g(0) = 0\}$ and introduce the *generalized metric* on $\mathcal{X}$:

$$\begin{aligned}
d(g, h) = \inf\{C \in \mathbb{R}_+ : &\|(g(x_1) - h(x_1), \cdots, g(x_s) - h(x_s))\|_s \\
&\leq C\varphi_{i,j}(x_1, x_2, \cdots, x_s), \quad \forall x_1, x_2, \cdots, x_s \in E\}.
\end{aligned}$$

It is easy to show that $(\mathcal{X}, d)$ is complete. Now we consider the linear mapping $J : \mathcal{X} \to \mathcal{X}$ such that $Jg(x) := \frac{1}{2}g(2x)$ for all $x \in E$. For any $g, h \in \mathcal{X}$, we have

$$\begin{aligned}
&d(g,h) < C \\
&\qquad \Longrightarrow \|\big(g(x_1) - h(x_1), \cdots, g(x_s) - h(x_s)\big)\|_s \le C\varphi_{i,j}(x_1, x_2, \cdots, x_s), \\
&\qquad\qquad (x_1, x_2, \cdots, x_s \in E) \\
&\qquad \Longrightarrow \|\Big(\frac{1}{2}g(2x_1) - \frac{1}{2}h(2x_1), \cdots, \frac{1}{2}g(2x_s) - \frac{1}{2}h(2x_s)\Big)\|_s \\
&\qquad\quad \le \frac{1}{2}C\varphi_{i,j}(2x_1, 2x_2, \cdots, 2x_s), \\
&\qquad \Longrightarrow \|\Big(\frac{1}{2}g(2x_1) - \frac{1}{2}h(2x_1), \cdots, \frac{1}{2}g(2x_s) - \frac{1}{2}h(2x_s)\Big)\|_s \\
&\qquad\quad \le LC\varphi_{i,j}(x_1, x_2, \cdots, x_s), \\
&\qquad \Longrightarrow d(Jg, Jh) \le LC.
\end{aligned}$$

Therefore, we see that

$$d(Jg, Jh) \le L\, d(g,h), \qquad \forall g, h \in \mathcal{X}.$$

This means J is a strictly contractive self-mapping of $\mathcal{X}$, with the Lipschitz constant L.

Letting $\mu = 1$ and for each $1 \le r \le n$ with $r \ne i, j$ and each $1 \le m \le s$, let $x_i^{(m)} = x_m$, $x_j^{(m)} = y_m$ and $x_r^{(m)} = 0$ in (17), we get

$$\begin{aligned}
&\Big\|\Big(2f\left(\frac{r_i x_1 + r_j y_1}{k}\right) + (n-2)f\left(\frac{r_i x_1}{k}\right) + (n-2)f\left(\frac{r_j y_1}{k}\right) \\
&\qquad - \frac{n}{k}(r_i f(x_1) + r_j f(y_1)\Big), \\
&\quad 2f\left(\frac{r_i x_2 + r_j y_2}{k}\right) + (n-2)f\left(\frac{r_i x_2}{k}\right) + (n-2)f\left(\frac{r_j y_2}{k}\right) \\
&\qquad - \frac{n}{k}(r_i f(x_2) + r_j f(y_2)), \cdots, \\
&\quad 2f\left(\frac{r_i x_s + r_j y_s}{k}\right) + (n-2)f\left(\frac{r_i x_s}{k}\right) + (n-2)f\left(\frac{r_j y_s}{k}\right) \\
&\qquad - \frac{n}{k}\Big(r_i f(x_s) + r_j f(y_s)\Big)\Big\|_s \\
&\le \varphi\big(X_{i,j}(x_1, y_1), X_{i,j}(x_2, y_2), \cdots, X_{i,j}(x_s, y_s)\big),
\end{aligned} \tag{20}$$

for all $x_1, y_1, x_2, y_2 \cdots, x_s, y_s \in E$. Letting $y_1 = y_2 = \cdots = y_s = 0$ in (20), we have

$$\left\|\left(n\left(f\left(\frac{r_i x_1}{k}\right)-\frac{r_i}{k}f(x_1)\right), n\left(\frac{r_i x_2}{k}\right)-\frac{r_i}{k}f(x_2)\right),\cdots, nf\left(\frac{r_i x_s}{k}\right)-\frac{r_i}{k}f(x_s)\right)\right)\right\|_s$$
$$\leq \varphi\big(X_{i,j}(x_1,0), X_{i,j}(x_2,0),\cdots, X_{i,j}(x_s,0)\big), \tag{21}$$

for all $x_1, x_2, \cdots, x_s \in E$. Similarly, letting $x_1 = x_2 = \cdots = x_s = 0$ in (20), we get

$$\left\|\left(n\left(f\left(\frac{r_i y_1}{k}\right)-\frac{r_i}{k}f(y_1)\right), n\left(f\left(\frac{r_i y_2}{k}\right)-\frac{r_i}{k}f(y_2)\right),\cdots, nf\left(\frac{r_i x_s}{k}\right)-\frac{r_i}{k}f(y_s)\right)\right)\right\|_s$$
$$\leq \varphi\big(X_{i,j}(0,y_1), X_{i,j}(0,y_2),\cdots, X_{i,j}(0,y_s)\big), \tag{22}$$

for all $y_1, y_2, \cdots, y_s \in E$. It follows from (20), (21), and (22) that

$$\begin{aligned}
&\Bigg\|\Bigg(2f\left(\frac{r_i x_1 + r_j y_1}{k}\right) - 2f\left(\frac{r_i x_1}{k}\right) - 2f\left(\frac{r_j y_1}{k}\right),\\
&2f\left(\frac{r_i x_2 + r_j y_2}{k}\right) - 2f\left(\frac{r_i x_2}{k}\right) - 2f\left(\frac{r_j y_2}{k}\right),\cdots,\\
&2f\left(\frac{r_i x_s + r_j y_s}{k}\right) - 2f\left(\frac{r_i x_s}{k}\right) - 2f\left(\frac{r_j y_s}{k}\right)\Bigg)\Bigg\|_s\\
&\leq \varphi\left(X_{i,j}(x_1,y_1), X_{i,j}(x_2,y_2),\cdots, X_{i,j}(x_s,y_s)\right) \qquad (23)\\
&+\varphi\left(X_{i,j}(x_1,0), X_{i,j}(x_2,0),\cdots, X_{i,j}(x_s,0)\right)\\
&+\varphi\left(X_{i,j}(0,y_1), X_{i,j}(0,y_2),\cdots, X_{i,j}(0,y_s)\right),
\end{aligned}$$

for all $x_1, y_1, x_2, y_2, \cdots, x_s, y_s \in E$. Replacing x_m and y_m by kx_m/r_i and kx_m/r_j, for all $1 \leq m \leq s$ in (23), respectively, we get

$$\Big\|\Big(2f(2x_1) - 4f(x_1), 2f(2x_2) - 4f(x_2),\cdots, 2f(2x_s) - 4f(x_s)\Big)\Big\|_s$$
$$\leq \varphi_{i,j}(x_1, x_2,\cdots, x_s), \tag{24}$$

for all $x_1, x_2, \cdots, x_s \in E$. So

$$\left\|\left(\frac{1}{2}f(2x_1) - f(x_1), \frac{1}{2}f(2x_2) - f(x_2),\cdots, \frac{1}{2}f(2x_s) - f(x_s)\right)\right\|_s$$
$$\leq \frac{1}{4}\varphi_{i,j}(x_1, x_2,\cdots, x_s),$$

for all $x_1, x_2, \cdots, x_s \in E$. Hence $d(f, Jf) \leq \dfrac{1}{4}$.

By Theorem 5, there exists a mapping $T : E \to F$ such that

(1) T is a fixed point of J, i.e.,

$$T(x) = \frac{1}{2}T(2x), \tag{25}$$

for all $x \in E$. The mapping T is a unique fixed point of J in the set

$$Y = \{g \in \mathcal{X} : d(f, g) < \infty\}.$$

This implies that T is a unique mapping satisfying (25) such that there exists $C \in (0, \infty)$ satisfying

$$\|T(x_1) - f(x_1), \cdots, T(x_s) - f(x_s)\|_s \leq C\varphi_{i,j}(x_1, x_2, \cdots, x_s),$$

for all $x_1, x_2, \cdots, x_s \in E$.

(2) $d(J^m f, T) \to 0$ as $m \to \infty$. This implies the equality

$$\lim_{m\to\infty} \frac{1}{2^m} f\left(2^m x\right) = T(x), \tag{26}$$

for all $x \in E$.

(3) $d(f, T) \leq \frac{1}{1-L} d(f, Jf)$, which implies the inequality

$$d(f, T) \leq \frac{1}{4-4L}.$$

This implies that the inequality (19) holds.

Letting $\mu = 1$ and replacing $X^{(1)} = \cdots = X^{(s)} = (x_1, x_2, \cdots, x_n) := X$ in (17), by using the properties of norm in multinormed spaces and (26), we have

$$\begin{aligned}
&\|(D_{1,r_1,\cdots,r_n} T(x_1, \cdots, x_n), \cdots, D_{1,r_1,\cdots,r_n} T(x_1, \cdots, x_n))\|_s \\
&= \lim_{m\to\infty} \frac{1}{2^m} \|(D_{1,r_1,\cdots,r_n} f(2^m x_1, \cdots, 2^m x_n), \cdots, D_{1,r_1,\cdots,r_n} f(2^m x_1, \cdots, 2^m x_n))\|_s \\
&= \lim_{m\to\infty} \frac{1}{2^m} \|D_{1,r_1,\cdots,r_n} f(2^m x_1, \cdots, 2^m x_n)\| \\
&\leq \lim_{m\to\infty} \frac{1}{2^m} \varphi\left(2^m X, \cdots, 2^m X\right) \\
&\leq \lim_{m\to\infty} L^m \varphi\left(X, \cdots, X\right) = 0
\end{aligned}$$

for all $x_1, x_2, \cdots, x_n \in E$. So

$$T\left(\frac{\sum_{i=1}^n r_i x_i}{k}\right) + \sum_{1\leq i<j\leq n} T\left(\frac{r_i x_i + r_j x_j}{k}\right) = \frac{n}{k} \sum_{i=1}^n r_i T(x_i).$$

By Lemma 3, the mapping $T : E \to F$ is Cauchy additive, i.e., $T(x + y) = T(x) + T(y)$ for all $x, y \in E$.

Now, we show that the mapping $T : E \to F$ is $\mathbb{C}-$linear. Letting $X^{(1)} = \cdots = X^{(s)} = X = (0, \cdots, 0, \underbrace{x}_{ith}, 0, \cdots, 0)$ in (17) and using the properties of norm in multinormed spaces, we have

$$\|D_{\mu,r_1,\cdots,r_n} T(0, \cdots, 0, \underbrace{x}_{ith}, 0, \cdots, 0)\|$$

$$
\begin{aligned}
&= \left\| \left(D_{\mu,r_1,\cdots,r_n} T(X^{(1)}), \cdots, D_{\mu,r_1,\cdots,r_n} T(X^{(s)}) \right) \right\|_s \\
&= \left\| \left(nT\left(\frac{\mu r_i x}{k}\right) - \frac{n\mu r_i}{k} T(x), \cdots, nT\left(\frac{\mu r_i x}{k}\right) - \frac{n\mu r_i}{k} T(x) \right) \right\|_s \\
&= \lim_{m\to\infty} \frac{1}{2^m} \left\| \left(nf\left(\frac{\mu r_i 2^m x}{k}\right) - \frac{n\mu r_i}{k} f(2^m x), \cdots, nf\left(\frac{\mu r_i 2^m x}{k}\right) - \frac{n\mu r_i}{k} f(2^m x) \right) \right\|_s \\
&= \lim_{m\to\infty} \frac{1}{2^m} \left\| nf\left(\frac{\mu r_i 2^m x}{k}\right) - \frac{n\mu r_i}{k} T(2^m x) \right\| \\
&\le \lim_{m\to\infty} \frac{1}{2^m} \varphi(2^m X, \cdots, 2^m X) \\
&\le \lim_{m\to\infty} L^m \varphi(X, \cdots, X) = 0,
\end{aligned}
$$

for all $x \in E$ and all $\mu \in \mathbb{T}^1$. So

$$T\left(\frac{\mu r_i x}{k}\right) = \frac{\mu r_i}{k} T(x),$$

for all $x \in E$ and all $\mu \in \mathbb{T}^1$. Since

$$T(\frac{r_i x}{k}) = \frac{r_i}{k} T(x),$$

for all $x \in E$ and all $r_i \neq 0$. Therefore by Lemma 5, the mapping T is $\mathbb{C}-$linear.

Corollary 2 *Let E be a linear space and let $\{(F^l, \|.\|_l) : l \in \mathbb{N}\}$ be a multi-Banach space. Suppose that $s \in \mathbb{N}$ and $0 < p < 1$ and $f : E \to F$ is a mapping with $f(0) = 0$ satisfying*

$$\|(D_{\mu,r_1,\cdots,r_n} f(X^{(1)}), \cdots, D_{\mu,r_1,\cdots,r_n} f(X^{(s)}))\|_s \le \epsilon \sum_{m=1}^{s} \sum_{t=1}^{n} \|x_t^{(m)}\|^p, \tag{27}$$

for all $\mu \in \mathbb{T}^1$ and all $X^{(1)}, \cdots, X^{(s)} \in E^n$. Then there exists a unique linear mapping $T : E \to F$ such that

$$\|(f(x_1) - T(x_1), \cdots, f(x_s) - T(x_s))\|_s \le \frac{\epsilon k^p}{2 - 2^p} \left(\frac{1}{r_i^p} + \frac{1}{r_j^p} \right) \sum_{m=1}^{s} \|x_m\|^p,$$

for all $x_1, \cdots, x_s \in E$.

Proof The proof follows from Theorem 10 by taking

$$\varphi(X^{(1)}, \cdots, X^{(s)}) = \epsilon \sum_{m=1}^{s} \sum_{t=1}^{n} \|x_t^{(m)}\|^p,$$

for all $X^{(1)}, \cdots, X^{(s)} \in E^n$. We can choose $L = \dfrac{1}{2^{1-p}}$ to get the desired result.

Theorem 11 *Let E be a linear space and let $\{(F^l, \|.\|_l) : l \in \mathbb{N}\}$ be a multi - Banach space. Suppose that $s \in \mathbb{N}$ and $f : E \to F$ is a mapping satisfying $f(0) = 0$ for which there exists a control function $\varphi : E^{ns} \to [0, \infty)$ satisfying (17) for all $\mu \in \mathbb{T}^1$ and all $X^{(1)}, \cdots, X^{(s)} \in E^n$. If there exists a Lipschitz constant $L < 1$ such that*

$$\varphi\big(X^{(1)}, \cdots, X^{(s)}\big) \leq \frac{1}{2} L\varphi\big(2X^{(1)}, \cdots, 2X^{(s)}\big), \tag{28}$$

for all $X^{(1)}, \cdots, X^{(s)} \in E^n$, then there exists a unique linear mapping $T : E \to F$ such that

$$\begin{aligned} &\|(f(x_1) - T(x_1), \cdots, f(x_s) - T(x_s))\|_s \\ &\leq \frac{L}{4-4L}\Big[\varphi\Big(X_{i,j}(\frac{kx_1}{r_i}, \frac{kx_1}{r_j}), X_{i,j}(\frac{kx_2}{r_i}, \frac{kx_2}{r_j}), \cdots, X_{i,j}(\frac{kx_s}{r_i}, \frac{kx_s}{r_j})\Big) \\ &+ \varphi\Big(X_{i,j}(\frac{kx_1}{r_i}, 0), X_{i,j}(\frac{kx_2}{r_i}, 0), \cdots, X_{i,j}(\frac{kx_s}{r_i}, 0)\Big) \\ &+ \varphi\Big(X_{i,j}(0, \frac{kx_1}{r_j}), X_{i,j}(0, \frac{kx_2}{r_j}), \cdots, X_{i,j}(0, \frac{kx_s}{r_j})\Big)\Big], \end{aligned} \tag{29}$$

for all $x_1, \cdots, x_s \in E$, where

$$X_{i,j}(x, y) = (0, \cdots, 0, \underbrace{x}_{ith}, 0, \cdots, 0, \underbrace{y}_{jth}, 0, \cdots, 0),$$

for all $x, y \in E$.

Proof Similar to the proof of Theorem 10, we consider the linear mapping $J : \mathcal{X} \to \mathcal{X}$ such that $Jg(x) := 2\, g(\frac{1}{2}x)$ for all $x \in E$. We can conclude that J is a strictly contractive self-mapping of $\mathcal{X}$, with the Lipschitz constant L.

It follows from (24) that

$$\Big\|\Big(f(x_1) - 2f(\frac{1}{2}x_1), f(x_2) - 2f(\frac{1}{2}x_2), \cdots, f(x_s) - 2f(\frac{1}{2}x_s)\Big)\Big\|_s \leq \frac{L}{4}\varphi_{i,j}(x_1, x_2, \cdots, x_s),$$

for all $x_1, x_2, \cdots, x_s \in E$. Hence $d(f, Jf) \leq \frac{L}{4}$.

By Theorem 5, there exists a mapping $T : E \to F$ such that

(1) T is a fixed point of J, i.e.,

$$T(x) = 2\, T(\frac{x}{2}), \tag{30}$$

for all $x \in E$. The mapping T is a unique fixed point of J in the set

$$Y = \{g \in \mathcal{X} : d(f, g) < \infty\}.$$

This implies that T is a unique mapping satisfying (30) such that there exists $C \in (0, \infty)$ satisfying

$$\|(f(x_1) - T(x_1), \cdots, f(x_s) - T(x_s))\|_s \leq C\varphi_{i,j}(x_1, x_2, \cdots, x_s)$$

for all $x_1, x_2, \cdots, x_s \in E$.

(2) $d(J^m f, T) \to 0$ as $m \to \infty$. This implies the equality

$$\lim_{m\to\infty} 2^m f\left(\frac{1}{2^m}x\right) = T(x) \tag{31}$$

for all $x \in E$.

(3) $d(f, T) \leq \dfrac{1}{1-L}d(f, Jf)$, which implies the inequality

$$d(f, T) \leq \frac{L}{4-4L}.$$

This implies that the inequality (29) holds.

The rest of the proof is similar to the proof of Theorem 10.

Corollary 3 *Let E be a linear space and let $\{(F^l, \|.\|_l) : l \in \mathbb{N}\}$ be a multi-Banach space. Suppose that $s \in \mathbb{N}$ and $p > 1$ and $f : E \to F$ is a mapping with $f(0) = 0$ satisfying*

$$\|(D_{\mu,r_1,\cdots,r_n} f(X^{(1)}), \cdots, D_{\mu,r_1,\cdots,r_n} f(X^{(s)}))\|_s \leq \epsilon \sum_{m=1}^{s}\sum_{t=1}^{n} \|x_t^{(m)}\|^p, \tag{32}$$

for all $\mu \in \mathbb{T}^1$ and all $X^{(1)}, \cdots, X^{(s)} \in E^n$. Then there exists a unique linear mapping $T : E \to F$ such that

$$\|(f(x_1) - T(x_1), \cdots, f(x_s) - T(x_s))\|_s \leq \frac{\epsilon k^p}{2^p - 2}\left(\frac{1}{r_i^p} + \frac{1}{r_j^p}\right)\sum_{m=1}^{s} \|x_m\|^p,$$

for all $x_1, \cdots, x_s \in E$.

Proof The proof follows from Theorem 11 by taking

$$\varphi(X^{(1)}, \cdots, X^{(s)}) = \epsilon \sum_{m=1}^{s}\sum_{t=1}^{n} \|x_t^{(m)}\|^p,$$

for all $X^{(1)}, \cdots, X^{(s)} \in E^n$. We can choose $L = \dfrac{1}{2^{p-1}}$ to get the desired result.

References

1. Aoki, T.: On the stability of the linear transformation in Banach spaces. J. Math. Soc. Japan **2**, 64–66 (1950)
2. Baak, C., Boo, D.-H., Rassias, T.M.: Generalized additive mapping in Banach modules and isomorphisms between C* - algebras. J. Math. Anal. Appl. **314**(1), 150–151 (2006)
3. Borelli, C., Forti, G.L.: On a general Hyers–Ulam stability result. Internat. J. Math. Math. Sci. **18**, 229–236 (1995)
4. Bourgin, D.G.: Classes of transformations and bordering transformations. Bull. Amer. Math. Soc. **57**, 223–237 (1951)
5. Cădariu, L., Radu, V.: Fixed points and the stability of Jensen's functional equation. J. Inequal. Pure Appl. Math. **4**(1), Art. 4 (2003)
6. Cădariu, L., Radu, V.: Fixed points and the stability of quadratic functional equations. An. Univ. Timişoara Ser. Mat. Inform. **41**, 25–48 (2003)
7. Cădariu, L., Radu, V.: On the stability of the Cauchy functional equation: a fixed point approach. Grazer Math. Ber. **346**, 43–52 (2004)
8. Cho, Y.J., Rassias, T.M., Saadati, R.: Stability of Functional Equations in Random Normed Spaces. Springer, New York (2013)
9. Chung, J.K., Sahoo P.K.: *On the general solution of a quartic functional equation*. Bull. Korean Math. Soc. **40**, No. 4, 565–576 (2003)
10. Czerwik, S.: The stability of the quadratic functional equation. In: Rassias, T.M., Tabor, J. (eds.) Stability of Mappings of Hyers–Ulam Type, pp. 81–91. Hadronic Press, Palm Harbor (1994)
11. Dales, H.G., Polyakov, M.E.: Multi-normed spaces and multi-Banach algebras, arxiv:1112.5148v2 (2012)
12. Dales, H.G., Moslehian, M.S.: Stability of mappings on multi-normed spaces. Glasg. Math. J. **49**(2), 321—332 (2007)
13. Diaz, J., Margolis, B.: A fixed point theorem of the alternative for contractions on a generalized complete metric space. Bull. Amer. Math. Soc. **74**, 305–309 (1968)
14. Găvruta, P.: A generalization of the Hyers–Ulam–Rassias stability of approximately additive mappings. J. Math. Anal. Appl. **184**, 431–436 (1994)
15. Găvruta, P., Găvruta, L.: A new method for the generalized Hyers–Ulam–Rassias stability. Int. J. Nonlinear Anal. Appl. **1**(2), 11–18 (2010)
16. Hyers, D.H.: On the stability of the linear functional equation. Proc. Nat. Acad. Sci. U S A. **27**, 222–224 (1941)
17. Jun, K., Kim, H.: The generalized Hyers–Ulam–Rassias stability of a cubic functional equation. J. Math. Anal. Appl. **274**, 867–878 (2002)
18. Jung, S.-M.: Hyers–Ulam–Rassias Stability of Functional Equations in Nonlinear Analysis. Springer, New York (2011)
19. Jung, S.-M., Rassias, J.M.: A fixed point approach to the stability of a functional equation of the spiral of Theodorus. Fixed Point Theory A. **2008**, Art. ID: 945010 (2008)
20. Kannappan, P.: Functional Equations and Inequalities with Applications. Springer, New York (2009)
21. Lee, S.H., Im, S.M., Hwang, I.S.: Quartic functional equations, J. Math. Anal. Appl. **307**, 387–394 (2005)
22. Moradlou, F., Vaezi, H., Park, C.: Fixed points and stability of an additive functional equation of n-Apollonius type in C^*-algebras, Abstract and Applied Analysis, vol. 2008, Article ID 672618, 13 pp. (2008). doi:10.1155/2008/672618
23. Moradlou,F., Najati, A., Vaezi, H.: Stability of homomorphisms and derivations on C^*-ternary rings associated to an Euler–Lagrange type additive mapping. Results Math. **55**, 469–486 (2009)
24. Moradlou, F., Vaezi, H., Eskandani, G.Z.: Hyers–Ulam–Rassias stability of a quadratic and additive functional equation in quasi-Banach spaces. Mediterr. J. Math. **6**(2), 233–248 (2009)

25. Moradlou, F.: Additive functional inequalities and derivations on Hilbert C^*-modules. Glasg. Math. J. **55**, 341–348 (2013)
26. Moradlou, F.: Approximate Euler–Lagrange–Jensen type additive mapping in multi-Banach spaces: a fixed point approach. Commun. Korean Math. Soc. **28**(2), 319–333 (2013)
27. Moradlou, F., Eshaghi Gordji, M.: Approximate Jordan derivations on Hilbert C^*- modules. Fixed Point Theory. 14(2), 413–426 (2013)
28. Moslehian, M.S., Nikodem, K., Popa, D.: Asymptotic aspect of the quadratic functional equation in multi-normed spaces. J. Math. Anal. Appl. **355**, 717–724 (2009)
29. Najati, A.: On the stability of quartic functional equation. J. Math. Anal. Appl. **340**, 569–574 (2008)
30. Nikodem, K., Páles, Z., Wąsowicz, S.: Abstract separation theorems of rod type and their applications. Ann. Polon. Math. **72**(3), 207–217 (1999)
31. Páles, Z.: Hyers–Ulam stability of the Cauchy functional equation on square-symmetric groupoids. Publ. Math. Debrecen **58**(4), 651–666 (2001)
32. Pardalos, P.M., Georgiev, P.G., Srivastava, H.M. (eds.): Nonlinear Analysis : Stability, Approximation and Inequalities. In honor of Themistocles M. Rassias on the accasion of his 60th birthday. Springer, New York (2012)
33. Park, C.-G., Rassias, T.M.: Hyers-–Ulam stability of a generalized Apollonius type quadratic mapping. J. Math. Anal. Appl. **322**(1), 371–381 (2006)
34. Park, C.: On the stability of the linear mapping in Banach modules. J. Math. Anal. Appl. **275**, 711–720 (2002)
35. Park, C.: Fixed points and Hyers–Ulam–Rassias stability of Cauchy–Jensen functional equations in Banach algebras. Fixed Point Theory A. **2007**, Art. ID 50175 (2007)
36. Park, C.: Hyers–Ulam–Rassias stability of a generalized Apollonius–Jensen type additive mapping and isomorphisms between C^*-algebras. Math. Nachr. **281**, 402–411 (2008)
37. Park,C., Rassias, T.M.: Hyers–Ulam stability of a generalized Apollonius type quadratic mapping. J. Math. Anal. Appl. **322**, 371–381 (2006)
38. Radu, V.: The fixed point alternative and stability of functional equations. Fixed Point Theory. **4**(1), 91–96 (2003)
39. Rassias, J.M.: On approximation of approximately linear mappings by linear mappings. J. Funct. Anal. **46**, 126–130 (1982)
40. Rassias, J.M.: On approximation of approximately linear mappings by linear mappings. Bull. Sci. Math. **108**, 445–446 (1984)
41. Rassias, T.M.: On the stability of the linear mapping in Banach spaces. Proc. Amer. Math. Soc. **72**, 297–300 (1978)
42. Rassias, T.M.: Solution of the Ulam stability problem for quartic mappings. Glas. Mat. Ser. III. **34**, 243–252 (1999)
43. Rassias, T.M.: The problem of S.M. Ulam for approximately multiplicative mappings, J. Math. Anal. Appl. **246**, 352–378 (2000)
44. Rassias, T.M.: On the stability of functional equations in Banach spaces. J. Math. Anal. Appl. **251**, 264–284 (2000)
45. Rassias, T.M. (ed.): Survey on Classical Inequalities. Kluwer Academic, Dordrecht (2000)
46. Rassias, T.M., Tabor, J. (eds.): Stability of Mappings of Hyers-Ulam Type. Hadronic Press, Palm Harbor (1994)
47. Rassias, T.M., Brzdek, J. (eds.): Functional Equations in Mathematical Analysis. Springer, New York (2012)
48. Skof, F.: Local properties and approximations of operators. Rend. Sem. Mat. Fis. Milano. **53**, 113–129 (1983)
49. Ulam, S.M.: A Collection of the Mathematical Problems, Interscience, New York (1960)
50. Zhou Xu, T., Rassias, J.M., Xin Xu, W.: Generalized Ulam-Hyers stability of a general mixed AQCQ-functional equation in multi-Banach spaces: a fixed point approach. Eur. J. Pure Appl. Math. **3**(6), 1032–1047 (2010)

Elliptic Problems with Nonhomogeneous Differential Operators and Multiple Solutions

Dumitru Motreanu and Patrick Winkert

Abstract The present survey aims to report on recent advances in the study of nonlinear elliptic problems whose differential part is expressed by a general operator in divergence form. The pattern of such differential operator is the p-Laplacian Δ_p with $1 < p < +\infty$. More general operators can be considered, possibly having completely different properties, for instance not satisfying any homogeneity requirement. A major objective of our work is to provide existence theorems of multiple solutions for boundary value problems governed by such general operators. In this direction, a three nontrivial solutions theorem is presented. In the case of problems determined by the p-Laplacian, we give a theorem ensuring the existence of at least four nontrivial solutions. Moreover, a complete sign information is available: two positive solutions, a negative solution and a nodal (sign-changing) solution. Finally, we provide a theorem guaranteeing the existence of a positive solution for a problem involving the (p,q)-Laplacian operator $\Delta_p+\Delta_q$, with $1 < q < p$, and a nonlinearity depending on the solution and its gradient.

Keywords Nonlinear elliptic boundary value problem · Nonhomogeneous differential operator · p-Laplacian · Eigenvalue problem · Multiple solutions · Critical point · Minimizer · Regularity · Maximum principle

1 Introduction

In this survey we focus on nonlinear elliptic boundary value problems with Dirichlet and Neumann boundary conditions presenting recent existence results of multiple solutions with precise sign information such as positive, negative and nodal (i.e., sign-changing) solutions. Specifically, we address some advances related to the nonlinear

D. Motreanu (✉)
Département de Mathématiques, Université de Perpignan, Avenue Paul Alduy 52, 66860 Perpignan, France
e-mail: motreanu@univ-perp.fr

P. Winkert
Technische Universität Berlin, Institut für Mathematik, Straße des 17. Juni 136, 10623 Berlin, Germany
e-mail: winkert@math.tu-berlin.de

T. M. Rassias, P. M. Pardalos (eds.), *Mathematics Without Boundaries*,
DOI 10.1007/978-1-4939-1106-6_15, © Springer Science+Business Media, LLC 2014

Dirichlet boundary value problem

$$\begin{cases} -\text{div}\, A(x, \nabla u) = f(x, u) & \text{in } \Omega, \\ u = 0 & \text{on } \partial\Omega, \end{cases} \tag{1}$$

and the nonlinear Neumann boundary value problem

$$\begin{cases} -\text{div}\, A(x, \nabla u) = f(x, u) & \text{in } \Omega, \\ \frac{\partial u}{\partial n} = 0 & \text{on } \partial\Omega, \end{cases} \tag{2}$$

where n denotes the outward unit normal vector on $\partial\Omega$, $1 < p < +\infty$ is a given number, and $\Omega \subset \mathbb{R}^N$ is a bounded domain with C^2 boundary $\partial\Omega$. For a later use we denote by $\overline{\Omega}$ the closure of Ω in $\mathbb{R}^N$.

The nonlinearity in the right-hand side of the elliptic equation in (1) and (2) is required to satisfy the following assumption:

(F) $f : \Omega \times \mathbb{R} \to \mathbb{R}$ is a Carathéodory function, which means that $f(x,t)$ is measurable in x and continuous in t, with $f(x, 0) = 0$ for a.e. $x \in \Omega$, such that the subcritical growth condition for $f(x, \cdot)$ holds

$$|f(x,t)| \le C(1 + |t|^{r-1}) \quad \text{for every } t \in \mathbb{R}, \text{ a.e. } x \in \Omega,$$

where $C > 0$ and $1 \le r < p^*$ are constants. Here p^* denotes the critical Sobolev exponent, that is

$$p^* = \begin{cases} \frac{Np}{N-p} & \text{if} p < N, \\ +\infty & \text{if} p \ge N. \end{cases}$$

The left-hand side of (1) and (2) is expressed in the form of divergence $\text{div} A(x, \nabla (\cdot))$ and is also nonlinear. Therein we have a continuous map $A : \overline{\Omega} \times \mathbb{R}^N \to \mathbb{R}^N$ satisfying the growth condition

$$|A(x, y)| \le C(1 + |y|^{p-1}) \quad \text{for all } (x, y) \in \overline{\Omega} \times \mathbb{R}^N,$$

with a constant $C > 0$. In the above statement the number $1 < p < +\infty$ is the same as the one in assumption (F).

We say that $u \in W_0^{1,p}(\Omega)$ is a *(weak) solution* of problem (1) if

$$\int_\Omega A(x, \nabla u) \cdot \nabla v \, dx = \int_\Omega f(x, u) v \, dx \quad \text{for all } v \in W_0^{1,p}(\Omega). \tag{3}$$

Similarly, we say that $u \in W^{1,p}(\Omega)$ is a *(weak) solution* of problem (2) if

$$\int_\Omega A(x, \nabla u) \cdot \nabla v \, dx = \int_\Omega f(x, u) v \, dx \quad \text{for all } v \in W^{1,p}(\Omega). \tag{4}$$

We note that owing to the above growth conditions, the integrals involved in (3) and (4) exist.

In the following we describe the topics studied in this chapter. Section 2 discusses the hypotheses on the differential operator in (1) and (2) and important properties of it. We stress that we are dealing with a general notion, for instance no homogeneity assumptions are supposed. We point out that under suitable assumptions, the problems (1) and (2) have a variational structure. Looking for the critical points of the Euler functionals associated to problems (1) and (2), which coincide with the weak solutions of these problems, the obvious candidates are the local minimizers. In this respect, Sect. 3 sets forth the celebrated relation of C^1-minimizers versus $W^{1,p}$-minimizers for problem (2) in the case of our general differential operator determined by the possibly nonhomogeneous map $A(x, y)$.

As noted above, a major difficulty in handling problems (1) and (2) is the lack of homogeneity for operator $A(x, y)$. This is seen for example in the study of the spectrum of the corresponding differential operator $-\text{div}A(x, \nabla u)$ driving the principal part in the problems (1) and (2). In Sect. 4, in order to overcome the lack of homogeneity, asymptotic $(p-1)$-homogeneity at zero and infinity conditions for the nonhomogeneous operator $A(x, \cdot)$ are introduced, under which a basic result on the spectrum of the operator $-\text{div}A(x, \nabla u)$ is stated. Next, in Sect. 5 it is presented a multiplicity theorem ensuring the existence of at least three nontrivial solutions for problem (2). Two of the solutions are of opposite constant signs and are obtained by minimization, whereas the existence of the third nontrivial solution is deduced through the application of the mountain pass theorem. The proof strongly relies on the relation involving C^1-minimizers versus $W^{1,p}$-minimizers.

More insight on multiple solutions with complete sign information for problem (1) is obtained in the case where $\text{div}A(x, \nabla u)$ is the p-Laplacian operator. Here a striking particularity of the studied problem is that the right-hand side nonlinearity contains a $(p-1)$-sublinear term (usually, called "concave term") $\beta(x)|u(x)|^{q-2}u(x)$ with $\beta \in L^\infty(\Omega) \setminus \{0\}$, $\beta \geq 0$, and $1 < q < p$. In Sect. 6 we state a result guaranteeing the existence of at least two positive solutions provided $\|\beta\|_\infty$ is sufficiently small. Section 7 is devoted to the existence of opposite constant sign extremal solutions, which means the smallest positive solution and the biggest negative solution. This is done by applying the method of sub-supersolutions. Section 8 contains our main result on multiple solutions with precise sign information, which ensures the existence of four nontrivial solutions: two positive, one negative and one nodal (sign changing). The proof is based on the existence of extremal solutions of opposite constant sign exploiting the observation that every nontrivial solution situated between opposite constant sign extremal solutions is necessarily nodal. This technique originates in [4] (see also [5]).

Finally, a new problem is formulated in Sect. 9 which is conducted by the (p, q)-Laplacian operator $\Delta_p + \Delta_q$, with $1 < q < p$, and whose right-hand side depends on the solution and its gradient. The dependence of the nonlinearity of the gradient of the solution prevents the problem from having a variational structure. Our result ensures the existence of a positive solution. The approach combines approximation, fixed point argument and an adequate comparison principle.

2 Properties of the Differential Operator in (1) and (2)

The map A entering the left-hand side of problems (1) and (2) is assumed to satisfy the following conditions:

(H) $A(x,y) = a(x,|y|)y$, where $a(x,t) > 0$ for all $(x,t) \in \overline{\Omega} \times (0,+\infty)$, and
(i) $A \in C^0(\overline{\Omega} \times \mathbb{R}^N, \mathbb{R}^N) \cap C^1(\overline{\Omega} \times (\mathbb{R}^N \setminus \{0\}), \mathbb{R}^N)$;
(ii) There exists a constant $C_1 > 0$, such that

$$|D_y A(x,y)| \le C_1 |y|^{p-2} \quad \text{for every } x \in \overline{\Omega}, \text{and } y \in \mathbb{R}^N \setminus \{0\};$$

(iii) There exists a constant $C_0 > 0$ such that

$$D_y A(x,y)\xi \cdot \xi \ge C_0 |y|^{p-2} |\xi|^2 \quad \text{for every } x \in \overline{\Omega}, y \in \mathbb{R}^N \setminus \{0\}, \text{and } \xi \in \mathbb{R}^N;$$

(iv) There exists a constant $C_2 > 0$ such that

$$|D_x A(x,y)| \le C_2(1 + |y|^{p-1}) \quad \text{for every } x \in \overline{\Omega}, \text{and } y \in \mathbb{R}^N \setminus \{0\};$$

(v) There exist constants $C_3 > 0$ and $1 \ge t_0 > 0$ such that

$$|D_x y A(x,y)| \le C_3 |y|^{p-1}(-\log|y|) \quad \text{for every } x \in \overline{\Omega},$$
$$y \in \mathbb{R}^N \setminus \{0\} \text{ with } 0 < |y| < t_0.$$

The notation $D_y A$ means the differential of the mapping $A(x,y)$ with respect to the variable $y \in \mathbb{R}^N$, and here $1 < p < +\infty$ is given in assumption (F). Notice that if A does not depend on $x \in \Omega$, then conditions (H) (iv), (v) are automatically satisfied.

We provide some examples of maps A complying with hypothesis (H).

Example 2.1 Let $\theta \in C^1(\overline{\Omega})$ with $\theta > 0$. Then the following mappings fulfil the conditions stated in (H).

(i) For $1 < p < +\infty$: $A(x,y) = \theta(x)|y|^{p-2}y$;
(ii) For $p \ge 2$: $A(x,y) = \theta(x)(|y|^{p-2}y + \ln(1 + |y|^{p-2})y)$;
(iii) For $1 < \tau \le p \le q$ and $\tau \ne 2$:

$$A(x,y) = \begin{cases} \theta(x)(|y|^{p-2}y + |y|^{q-2}y) & \text{if } |y| \le 1 \\ \theta(x)\left(|y|^{p-2}y + \frac{q-2}{\tau-2}|y|^{\tau-2}y - \frac{q-\tau}{\tau-2}y\right) & \text{if } |y| > 1; \end{cases}$$

(iv) The weighted sum of p-Laplacian and a generalized mean curvature operator:

$$A(x,y) = \theta(x)\left(|y|^{p-2}y + c\,\frac{|y|^{p-2}y}{1+|y|^p}\right),$$

with $0 < c < 4\,p(p-1)$ if $1 < p < 2$ and $0 < c < \frac{4\,p}{(p-1)^2}$ if $p \ge 2$.

The form of the operator A allows us to build a variational setting for problems (1) and (2). To this end, we note that the continuity of A guaranteed by hypothesis (H) (i) implies that the function $(x,t) \mapsto a(x,t)t$ is continuous for all $x \in \overline{\Omega}$ and $t \geq 0$. This can be seen from the equality

$$A(x,(t,0,\dots,0)) = (a(x,t)t,0,\dots,0) \quad \text{for all } x \in \overline{\Omega},\ t > 0.$$

Therefore for every $x \in \bar{\Omega}$ and $y \in \mathbb{R}^N$ it is well defined:

$$G(x,y) := \int_0^{|y|} a(x,t)t\,dt.$$

A straightforward computation shows that the gradient $\nabla_y G(x,y)$ with respect to the variable $y \in \mathbb{R}^N$ is given by

$$\nabla_y G(x,y) = A(x,y) \ \text{ for all } x \in \bar{\Omega} \text{ and } y \in \mathbb{R}^N.$$

Moreover, due to hypothesis (H) (iii), $G(x,y)$ is convex in y for all x and satisfies:

$$A(x,y)\cdot y \geq G(x,y) \quad \text{for all } (x,y) \in \overline{\Omega} \times \mathbb{R}^N. \tag{5}$$

As a consequence of (H) and by means of (5) we have the following proposition that summarizes some significant facts regarding the operator A and the corresponding potential G.

Proposition 2.1 *Let hypothesis (H) be satisfied. Then there hold:*

(i) The map $y \mapsto A(x,y)$ is maximal monotone and strictly monotone for all $x \in \overline{\Omega}$.

(ii)

$$|A(x,y)| \leq \tfrac{C_1}{p-1}|y|^{p-1} \quad \textit{for all } (x,y) \in \overline{\Omega} \times \mathbb{R}^N.$$

(iii)

$$A(x,y)\cdot y \geq \tfrac{C_0}{p-1}|y|^p \quad \textit{for all } (x,y) \in \overline{\Omega} \times \mathbb{R}^N.$$

(iv)

$$\tfrac{C_0}{p(p-1)}|y|^p \leq G(x,y) \leq \tfrac{C_1}{p(p-1)}|y|^p \quad \textit{for all } (x,y) \in \overline{\Omega} \times \mathbb{R}^N.$$

The following result mentioning an essential property of the differential operator $\operatorname{div} A(x,\nabla(\cdot))$ corresponding to the map A was proven in [20, Proposition 10]. This result is a key tool in checking the Palais–Smale condition for the Euler functional associated to problems (1) and (2).

Proposition 2.2 *Assume hypothesis (H). Then the map $V : W^{1,p}(\Omega) \to \left(W^{1,p}(\Omega)\right)^*$ defined by*

$$\langle V(u), v\rangle = -\langle \operatorname{div} A(x,\nabla u), v\rangle = \int_\Omega A(x,\nabla u)\cdot \nabla v dx$$

for all $u, v \in W^{1,p}(\Omega)$, *is maximal monotone and has the* $(S)_+$ *property, that is, any sequence* $(u_n) \subset W^{1,p}(\Omega)$ *such that* $u_n \rightharpoonup u$ *in* $W^{1,p}(\Omega)$ *and*

$$\limsup_{n\to\infty} \langle V(u_n), u_n - u\rangle \leq 0,$$

for some $u \in W^{1,p}(\Omega)$, *fulfills* $u_n \to u$ *in* $W^{1,p}(\Omega)$.

Another useful related result is the following one whose proof is given in [23].

Proposition 2.3 *Assume hypothesis* (H). *For any number* $\lambda > 0$, *let the mapping* $T_\lambda : W^{1,p}(\Omega) \to (W^{1,p}(\Omega))^*$ *be defined by*

$$\langle T_\lambda(u), v\rangle = \int_\Omega A(x, \nabla u) \cdot \nabla v \, dx + \lambda \int_\Omega |u|^{p-2} uv \, dx.$$

Then the inverse $T_\lambda^{-1} : (W^{1,p}(\Omega))^* \to W^{1,p}(\Omega)$ *of* T_λ *exists and is continuous.*

3 C^1-Minimizers Versus $W^{1,p}$-Minimizers for Problem (2)

This section is devoted to the famous result dealing with the comparison of local C^1-minimizers and local $W^{1,p}$-minimizers for Euler functionals associated to some elliptic boundary value problems. Precisely, we set forth such a result in the case of our problem (2). A similar result holds for the Dirichlet problem (1) with $W_0^{1,p}(\Omega)$ in place of $W^{1,p}(\Omega)$.

Recall that the function $f : \Omega \times \mathbb{R} \to \mathbb{R}$ entering problem (2) is subject to hypothesis (F) asserting the subcritical growth condition for f. Setting $F(x,t) = \int_0^t f(x,s)ds$, we define the functional $\varphi : W^{1,p}(\Omega) \to \mathbb{R}$ by

$$\varphi(u) = \int_\Omega G(x, \nabla u)dx - \int_\Omega F(x, u)dx, \tag{6}$$

with the function G introduced in Sect. 2. Under assumptions (H) and (F), the functional φ is well defined on $W^{1,p}(\Omega)$ and is of class C^1. Moreover, its critical points coincide with the (weak) solutions of problem (2), that is, we have $\varphi'(u) = 0$ if and only if (4) holds true.

The main result of this section is the following.

Theorem 3.1 *Assume that the conditions* (H) *and* (F) *are fulfilled. If* $u_0 \in W^{1,p}(\Omega)$ *is a local* $C^1(\overline{\Omega})$*-minimizer of* φ *in (6), that is, there exists* $r_0 > 0$ *such that*

$$\varphi(u_0) \leq \varphi(u_0 + h) \quad \text{for all } h \in C^1(\overline{\Omega}) \text{ with } \|h\|_{C^1(\overline{\Omega})} \leq r_0,$$

then $u_0 \in C^1(\overline{\Omega})$ *with* $\frac{\partial u_0}{\partial n} = 0$ *on* $\partial\Omega$, *and it is a local* $W^{1,p}(\Omega)$*-minimizer of* φ, *that is, there exists* $r_1 > 0$ *such that*

$$\varphi(u_0) \leq \varphi(u_0 + h) \quad \text{for all } h \in W^{1,p}(\Omega) \text{ with } \|h\|_{W^{1,p}(\Omega)} \leq r_1.$$

Proof (sketch)

Since $C^1(\overline{\Omega})$ is dense in $W^{1,p}(\Omega)$, it follows that $\varphi'(u_0) = 0$, or equivalently

$$-\mathrm{div} A(x, \nabla u_0(x)) = f(x, u_0(x)) \text{ in } \Omega, \quad \frac{\partial u_0}{\partial n} = 0 \text{ on } \partial\Omega.$$

Suppose that u_0 is not a local $W^{1,p}(\Omega)$-minimizer of φ. Because the functional φ is sequentially weakly lower semicontinuous, for $\varepsilon > 0$ small we can find $h_\varepsilon \in \overline{B}^r_\varepsilon := \{u \in W^{1,p}(\Omega) : \|u\|_{L^r(\Omega)} \leq \varepsilon\}$ such that

$$\varphi(u_0 + h_\varepsilon) = \inf\{\varphi(u_0 + u) : u \in \overline{B}^r_\varepsilon\} < \varphi_0(u_0). \tag{7}$$

The Lagrange multiplier rule yields $\lambda_\varepsilon \leq 0$ with

$$\varphi'(u_0 + h_\varepsilon) = \lambda_\varepsilon |h_\varepsilon|^{r-2} h_\varepsilon,$$

which results in

$$-\mathrm{div} A(x, \nabla(u_0 + h_\varepsilon)(x)) = f(x, (u_0 + h_\varepsilon)(x)) + \lambda_\varepsilon |h_\varepsilon(x)|^{r-2} h_\varepsilon(x) \quad \text{in } \Omega.$$

Setting $\tilde{A}(x, y) := A(x, \nabla u_0(x) + y) - A(x, \nabla u_0(x))$, it is seen that $\tilde{A}$ satisfies

$$-\mathrm{div} \tilde{A}(x, \nabla h_\varepsilon) = f(x, u_0 + h_\varepsilon) - f(x, u_0) + \lambda_\varepsilon |h_\varepsilon|^{r-2} h_\varepsilon \quad \text{in } \Omega.$$

Now we can show that there exists a constant $d_0 > 0$ such that $|\lambda_\varepsilon||h_\varepsilon(x)|^{r-1} \leq d_0$ for all $x \in \Omega$ and $\varepsilon > 0$ small.

At this point, in view of Lieberman [17, Theorem 2] there exist $\theta \in (0, 1)$ and $M > 0$ such that

$$h_\varepsilon \in C^{1,\theta}(\overline{\Omega}) \text{ and } \|h_\varepsilon\|_{C^{1,\theta}(\overline{\Omega})} \leq M \text{ for all } \varepsilon > 0 \text{ small.} \tag{8}$$

Exploiting that $C^{1,\theta}(\overline{\Omega})$ is embedded compactly in $C^1(\overline{\Omega})$, we infer that $u_0 + h_\varepsilon \to u_0$ in $C^1(\overline{\Omega})$ as $\varepsilon \to 0$. Consequently, assertion (7) yields $\varphi(u_0) \leq \varphi_0(u_0 + h_\varepsilon)$ for ε sufficiently small, which contradicts the choice of h_ε. This completes the proof.

Theorem 3.1 was proven in [20, Proposition 24] (see also [22, Theorem 3.1]). We emphasize that if $C_0 = C_1 = p - 1$ in assumption (H), the operator $V = \mathrm{div} A(x, \nabla(\cdot)) : W^{1,p}(\Omega) \to \left(W^{1,p}(\Omega)\right)^*$, with A in hypothesis (H), becomes the p-Laplacian operator on $W^{1,p}(\Omega)$. For p-Laplacian operator and Neumann problems, the result can be found in [26]. The counterpart result for Dirichlet problems in the case $p = 2$ is due to Brezis and Nirenberg [3], which was extended to $1 < p < +\infty$ by García Azorero, Peral Alonso, and Manfredi [10] (see also [14] for $p \geq 2$). For different extensions and related results we refer to [2, 8, 12, 15, 16, 21, 31, 32]. Theorem 3.1 is very useful in the study of problem (2), especially for obtaining the existence of multiple solutions through variational methods, for example by using the mountain pass theorem, combined with the method of sub-supersolutions.

4 Asymptotically Homogeneous Case

The eigenvalue problem with Dirichlet boundary condition for the nonlinear operator $\operatorname{div} A(x, \nabla(\,\cdot\,))$ means the existence of $\lambda \in \mathbb{R}$ for which

$$\begin{cases} -\operatorname{div} A(x, \nabla u) = \lambda |u|^{p-2}u & \text{in } \Omega, \\ u = 0 & \text{on } \partial\Omega, \end{cases} \qquad (EV;\lambda)$$

has a nontrivial solution. The eigenvalue problem $(EV;\lambda)$ is hard because the operator $A(x,\cdot)$, so $\operatorname{div} A(x, \nabla(\cdot))$, is not homogeneous. Even in the case of the p-Laplacian Δ_p, which is $(p-1)$-homogeneous, the spectrum is not completely known (it is known what is called the beginning of the spectrum of $-\Delta_p$ as can be seen in [7], see also [25, 24]).

In order to overcome the lack of homogeneity in $(EV;\lambda)$ it was introduced in [23] an asymptotic $(p-1)$-homogeneity at zero condition for the nonhomogeneous operator $A(x,\cdot)$:

$(AH0)$ There exist a positive function $a_0 \in C^1(\overline{\Omega}, \mathbb{R})$ and a continuous function $\tilde{a}_0(x,t)$ on $\overline{\Omega} \times [0, +\infty)$ such that

$$A(x,y) = a_0(x)|y|^{p-2}y + \tilde{a}_0(x,|y|)y \quad \text{for all } x \in \Omega,\ y \in \mathbb{R}^N,$$

$$\lim_{t \to 0^+} \frac{\tilde{a}_0(x,t)}{t^{p-2}} = 0 \quad \text{uniformly in } x \in \overline{\Omega}.$$

Under assumption $(AH0)$, with the weight function a_0 we define

$$\lambda_1(a_0) := \inf \left\{ \int_\Omega a_0(x)|\nabla u|^p\,dx\,;\ \|u\|_p = 1 \right\}.$$

Also in order to overcome the lack of homogeneity in $(EV;\lambda)$ it was introduced in [23] an asymptotic $(p-1)$-homogeneity at infinity condition for the nonhomogeneous operator $A(x,\cdot)$:

(AH) There exist a positive function $a_\infty \in C^1(\overline{\Omega}, \mathbb{R})$ and a continuous function $\tilde{a}(x,t)$ on $\overline{\Omega} \times \mathbb{R}$ such that

$$A(x,y) = a_\infty(x)|y|^{p-2}y + \tilde{a}(x,|y|)y \quad \text{for all } x \in \Omega,\ y \in \mathbb{R}^N,$$

$$\lim_{t \to +\infty} \frac{\tilde{a}(x,t)}{t^{p-2}} = 0 \quad \text{uniformly in } x \in \overline{\Omega}.$$

Under assumption (AH), with the weight function a_∞, we define

$$\lambda_1(a_\infty) := \inf \left\{ \int_\Omega a_\infty(x)|\nabla u|^p\,dx\,;\ \|u\|_p = 1 \right\}.$$

We cite from [23] the following result on the spectrum of the nonlinear operator $\operatorname{div} A(x, \nabla(\,\cdot\,))$.

Theorem 4.1 *Assume* (H), $(AH0)$, (AH) *and that* $\lambda_1(a_0) \neq \lambda_1(a_\infty)$. *Then, for every* λ *between* $\lambda_1(a_0)$ *and* $\lambda_1(a_\infty)$, *problem* $(EV;\lambda)$ *has a nontrivial positive solution, therefore* λ *is an eigenvalue of the operator* $-\mathrm{div} A(x, \nabla(\cdot))$.

Theorem 4.1 is important because it shows that generally the spectrum of the operator $-\mathrm{div}(A(x, \nabla u))$ on $W_0^{1,p}(\Omega)$ is not discreet. However, Theorem 4.1 cannot be applied to the negative p-Laplacian $-\Delta_p$ on $W_0^{1,p}(\Omega)$ because in this case $\lambda_1(a_0) = \lambda_1(a_\infty)$.

5 A Three Solutions Theorem

In this section we present an existence and multiplicity result for the Neumann problem (2), on a bounded domain $\Omega \subset \mathbb{R}^N$ with a C^2-boundary $\partial\Omega$, involving the nonhomogeneous operator A introduced in Sect. 2. We suppose that the nonlinearity $f(x, u)$ in the right-hand side of the equation in (2) satisfies the following conditions:

$(H_f)_1$ $f : \Omega \times \mathbb{R} \to \mathbb{R}$ is a Carathéodory function such that $f(x, 0) = 0$ a.e. in Ω and

(i) $|f(x,s)| \leq a(x) + C|s|^{r-1}$ for a.a. $x \in \Omega$, all $s \in \mathbb{R}$, with $a \in L^\infty(\Omega)_+, C > 0$, and $1 \leq r < p^*$;

(ii) If $F(x,t) = \int_0^t f(x,s)ds$, there exists $\theta \in L^\infty(\Omega), \theta \leq 0, \theta \neq 0$, such that

$$\limsup_{|s|\to\infty} \frac{pF(x,s)}{|s|^p} \leq \theta(x) \quad \text{uniformly for a.a. } x \in \Omega;$$

(iii) There exist $\delta_0 > 0$ and $\eta > \lambda_1$ such that (with C_1 as in (H))

$$\frac{C_1 \eta}{p(p-1)}|s|^p \leq F(x,s) \quad \text{for a.a. } x \in \Omega, \text{all } |s| \leq \delta_0;$$

(iv) There exists $\lambda > 0$ such that

$$(f(x,s) + \lambda|s|^{p-2}s)s \geq 0 \quad \text{for a.a. } x \in \Omega, \text{all } s \in \mathbb{R}.$$

Example 5.1 A simple example of function satisfying assumption $(H_f)_1$ is the following one:

$$f(s) = \begin{cases} \hat{\eta}|s|^{p-2}s & \text{if } |s| \leq 1 \\ C|s|^{q-2}s - (C - \hat{\eta})|s|^{p-2}s & \text{if } |s| > 1, \end{cases}$$

where $\hat{\eta} > \frac{C_1\lambda_1}{p-1}, C > \hat{\eta}$, and $1 < q < p$.

The subsequent result on multiple solutions to problem (2) can be found in [22, Theorem 4.2]. In the statement of it we use the space

$$C_n^1(\overline{\Omega}) = \{u \in C^1(\overline{\Omega}) : \frac{\partial u}{\partial n} = 0 \text{ on } \partial\Omega\}.$$

The Banach space $C^1(\overline{\Omega})$ is an ordered Banach space with the positive cone

$$C_+ = \{u \in C^1(\overline{\Omega}) : \ u(x) \geq 0 \text{ for all } x \in \overline{\Omega}\}.$$

This cone has a nonempty interior which is given by

$$\text{int}C_+ = \{u \in C_+ : \ u(x) > 0 \text{ for all } x \in \overline{\Omega}\}.$$

Theorem 5.1 *Let the hypotheses* (H) *and* $(H_f)_1$ *be satisfied. Then problem (2) possesses at least three nontrivial smooth solutions* $u_0 \in \text{int}C_+, v_0 \in -\text{int}C_+$ *and* $y_0 \in C^1_n(\overline{\Omega})$.

Proof (sketch)

For $\lambda > 0$ as in $(H_f)_1$(iv), we introduce the function

$$f_+^\lambda(x,s) = \begin{cases} 0 & \text{if } s \leq 0 \\ f(x,s) + \lambda s^{p-1} & \text{if } s > 0. \end{cases}$$

Let $F_+^\lambda(x,t) = \int_0^t f_+^\lambda(x,s)\,ds$. Consider the C^1-functional $\varphi_+^\lambda : W^{1,p}(\Omega) \to \mathbb{R}$ defined by

$$\varphi_+^\lambda(u) = \int_\Omega G(x,\nabla u)\,dx + \frac{\lambda}{p}\|u\|_p^p - \int_\Omega F_+^\lambda(x,u)\,dx \text{ for all } u \in W^{1,p}(\Omega).$$

By hypotheses $(H_f)_1$(i), (ii) and Proposition 2.1, φ_+^λ is coercive and so we can find $u_0 \in W^{1,p}(\Omega)$ such that

$$\varphi_+^\lambda(u_0) = \inf_{W^{1,p}(\Omega)} \varphi_+^\lambda.$$

From $(H_f)_1$(iii) we have that $\varphi_+^\lambda(u_0) < 0$, hence $u_0 \neq 0$. Also we note that

$$V(u_0) + \lambda |u_0|^{p-2}u_0 = f_+^\lambda(x,u_0).$$

Therefore u_0 is a nontrivial solution of problem (2) with $u \in L^\infty(\Omega)$ (see [33]).

On account of the nonlinear regularity theory (see [17]), we infer that $u_0 \in C_+\setminus\{0\}$, while hypothesis $(H_f)_1$ (iv) implies

$$-\text{div}A(x,\nabla u_0(x)) + \lambda u_0(x)^{p-1} = f(x,u_0(x)) + \lambda u_0(x)^{p-1} \geq 0 \text{ a.e. in } \Omega.$$

Invoking the nonlinear maximum principle in [20, Theorem A], we conclude that $u_0 \in \text{int}C_+$. Similarly, we get $v_0 \in -\text{int}C_+$.

Recall the C^1-energy functional $\varphi : W^{1,p}(\Omega) \to \mathbb{R}$ associated with problem (2) defined in (6) and note that $\varphi|_{C_+} = \varphi_+^\lambda|_{C_+}$. So $u_0 \in \text{int}C_+$ is a local $C^1(\overline{\Omega})$-minimizer of φ. On the basis of Theorem 3.1, we infer that u_0 is a local $W^{1,p}(\Omega)$-minimizer of φ. Similarly, we have that $v_0 \in -\text{int}C_+$ is a local $W^{1,p}(\Omega)$-minimizer of φ.

Now we will provide a third nontrivial solution of problem (2). Without any loss of generality, we may assume that φ has a finite critical set. Then we can find $\rho > 0$ such that

$$\varphi(v_0) \leq \varphi(u_0) < \inf\,[\varphi(u) : \|u - u_0\| = \rho] =: m_\rho. \tag{9}$$

Moreover, we have that φ is coercive; hence it satisfies the Palais–Smale condition. Consequently, the mountain pass theorem can be applied giving $y_0 \in W^{1,p}(\Omega)$ such that $\varphi(y_0) \geq m_\rho$ and $\varphi'(y_0) = 0$. As before, through the nonlinear regularity theory, it follows that $y_0 \in C_n^1(\overline{\Omega})$ is a solution of problem (2). By (9), it turns out that $y_0 \notin \{u_0, v_0\}$.

It remains to show that $y_0 \neq 0$. From the mountain pass theorem we have the minimax characterization

$$\varphi(y_0) = \inf_{\gamma\in\Gamma} \max\nolimits_{t\in[0,1]}\varphi(\gamma(t)),$$

where

$$\Gamma = \{\gamma \in C([0,1], W^{1,p}(\Omega)) : \gamma(0) = v_0,\ \gamma(1) = u_0\}.$$

If we can produce a path $\gamma_0 \in \Gamma$ such that $\varphi|_{\gamma_0} < 0$, then $\varphi(y_0) < 0 = \varphi(0)$ and so $y_0 \neq 0$ and we are done.

In what follows we generate such a path $\gamma_0 \in \Gamma$. Consider the sets $M = W^{1,p}(\Omega)\cap\partial B_1^{L^p}$ endowed with the relative $W^{1,p}(\Omega)$-topology and $M_c = M\cap C^1(\overline{\Omega})$ furnished with the relative $C^1(\overline{\Omega})$-topology. We notice that M_c is dense in M. So, if we set

$$\hat{\Gamma} = \{\hat{\gamma} \in C([-1,1], M) : \hat{\gamma}(-1) = -\hat{u}_0,\ \hat{\gamma}(1) = \hat{u}_0\},$$

and

$$\hat{\Gamma}_c = \{\hat{\gamma} \in C([-1,1], M_c) : \hat{\gamma}(-1) = -\hat{u}_0,\ \hat{\gamma}(1) = \hat{u}_0\},$$

we have that $\hat{\Gamma}_c$ is dense in $\hat{\Gamma}$. Then, from [1], given $\varepsilon > 0$ we can find $\hat{\gamma}_0 \in \hat{\Gamma}_c$ with

$$\max[\|\nabla u\|_p^p : u \in \hat{\gamma}_0([-1,1])] \leq \lambda_1 + \varepsilon.$$

Here λ_1 stands for the second eigenvalue (the first nontrivial eigenvalue) of $-\Delta_p$ on $W^{1,p}(\Omega)$. The corresponding result for the Dirichlet problem can be found in [7]. For $\eta > \lambda_1$ in $(H_f)_1$(iii), let $\varepsilon > 0$ with $\lambda_1 + \varepsilon < \eta$. Also let $\xi > 0$ be such that $\xi|u(x)| \leq \delta_0$ for all $x \in \overline{\Omega}$, all $u \in \hat{\gamma}_0([-1,1])$, with $\delta_0 > 0$ in $(H_f)_1$(iii). Then by $(H_f)_1$(iii) we deduce that $\varphi(\xi u) < 0$ for all $u \in \hat{\gamma}_0([-1,1])$. It follows that $\overline{\gamma}_0 = \xi\hat{\gamma}_0$ is a path from $-\xi\hat{u}_0$ to $\xi\hat{u}_0$ with $\varphi|_{\overline{\gamma}_0} < 0$.

Without loss of generality we may assume that $\{0, u_0\}$ are the only critical points of φ_+^λ (if $y \in W^{1,p}(\Omega)$ is another critical point of φ_+^λ, then $y \in \mathrm{int}C_+$, thus it is a third nontrivial smooth solution). Set $a := \varphi_+^\lambda(u_0) = \varphi(u_0) < 0 = \varphi_+^\lambda(0)$. Hence we can apply the second deformation theorem (see, e.g., [11, p. 628]) and obtain a homotopy $h : [0,1] \times ((\varphi_+^\lambda)^0 \setminus K^0_{\varphi_+^\lambda}) \to (\varphi_+^\lambda)^0$ satisfying the properties stated in that theorem, where $(\varphi_+^\lambda)^0 = (\varphi_+^\lambda)^{-1}(-\infty, 0]$ and $K^0_{\varphi_+^\lambda}$ denotes the set of critical points at the level 0. Let $\gamma_+(t) = (h(t, \xi\hat{u}_0))^+$ for all $t \in [0,1]$. Then $\gamma_+(0) = \xi\hat{u}_0$ ($\hat{u}_0$ is the L^p-normalized eigenfunction of the negative Neumann p-Laplacian corresponding to the eigenvalue 0, that is, $\hat{u}_0(x) = 1/|\Omega|_N^{1/p}$) and $\gamma_+(1) = u_0$ because $(\varphi_+^\lambda)^a = (\varphi_+^\lambda)^{-1}(-\infty, a] = \{u_0\}$. Consequently, we get

$$\varphi|_{\gamma_+} < 0.$$

In a similar fashion we produce a path γ_- joining v_0 and $-\xi\hat{u}_0$ such that

$$\varphi|_{\gamma_-} < 0.$$

Finally, we concatenate γ_-, $\overline{\gamma}_0$ and γ_+ to construct a path $\gamma_0 \in \Gamma$ satisfying $\varphi|_{\gamma_0} < 0$, which completes the proof.

6 Positive Solutions

In this section we seek multiple positive solutions for the special case of problem (1) where in place of the general operator $\operatorname{div} A(x, \nabla u)$ we take the p-Laplacian Δ_p with $1 < p < +\infty$.

Let $\Omega \subset \mathbb{R}^N$ be a bounded domain with a C^2-boundary $\partial\Omega$. Given $\beta \in L^\infty(\Omega) \setminus \{0\}$, $\beta \geq 0$, we consider for $1 < q < p$ the nonlinear Dirichlet problem:

$$\begin{cases} -\Delta_p u(x) = \beta(x)|u(x)|^{q-2}u(x) + f(x, u(x)) & \text{in } \Omega, \\ u = 0 & \text{on } \partial\Omega, \end{cases} \tag{10}$$

where Δ_p denotes the p-Laplace differential operator defined on $W_0^{1,p}(\Omega)$ by

$$\Delta_p u = \operatorname{div}(|\nabla u|^{p-2}\nabla u).$$

Note that in (10), the right-hand side nonlinearity contains a $(p-1)$-sublinear term (usually, called "concave term") $\beta(x)|u(x)|^{q-2}u(x)$, with $1 < q < p$, and an additional Carathéodory term $f(x, u)$.

Denote by λ_1 the first eigenvalue of the negative Dirichlet p-Laplacian operator $(-\Delta_p, W_0^{1,p}(\Omega))$ and by $\hat{u}_1$ the L^p-normalized positive eigenfunction corresponding to λ_1. Through the nonlinear regularity theory (see [17]) and the nonlinear strong maximum principle of Vazquez [30], we have that $\hat{u}_1 \in \operatorname{int} C_+$, where this time we denote

$$C_+ = \{u \in C^1(\overline{\Omega}) : u(x) \geq 0 \text{ for all } x \in \overline{\Omega} \text{ and } u(x) = 0 \text{ on } \partial\Omega\}. \tag{11}$$

This cone has a nonempty interior equal to

$$\operatorname{int} C_+ = \{u \in C_+ : u(x) > 0 \text{ for all } x \in \Omega \text{ and } \tfrac{\partial u}{\partial n}(x) < 0 \text{ for all } x \in \partial\Omega\}.$$

As before, n stands for the outward unit normal on $\partial\Omega$.

The hypotheses for the nonlinearity $f(x,u)$ in (10) are as follows:

$(H_f)_2$ $f : \Omega \times \mathbb{R} \to \mathbb{R}$ is a function such that

(i) For all $s \in \mathbb{R}$, $x \mapsto f(x,s)$ is measurable;

(ii) For a.a. $x \in \Omega$, $x \mapsto f(x,s)$ is continuous and $f(x,0) = 0$;

(iii) For a.a. $x \in \Omega$ and all $s \in \mathbb{R}$, we have $|f(x,s)| \leq a(x) + c|s|^{r-1}$, with $a \in L^\infty(\Omega)_+$, $c > 0$ and $p < r < p^*$;

(iv) If $F(x,s) = \int_0^s f(x,t)\,dt$, then

(iv.a) $\lim_{s\to+\infty} \frac{F(x,s)}{s^p} = +\infty$ uniformly for a.a. $x \in \Omega$,

(iv.b) there exist $\tau \in (\max\{\frac{r-p}{p}N, r-p\}, p^*)$ and $\gamma_0 > 0$, such that $\tau > q$ and

$$\liminf_{s\to+\infty} \frac{f(x,s)s - pF(x,s)}{s^\tau} > \gamma_0 \text{ uniformly for a.a. } x \in \Omega;$$

(v) There exist $\xi, \theta \in L^\infty(\Omega)_+$, such that $\theta(x) \leq \lambda_1$ a.e. on Ω, with strict inequality on a set of positive measure, and

$$-\xi(x) \leq \liminf_{s\to 0^+} \frac{f(x,s)}{s^{p-1}} \leq \limsup_{s\to 0^+} \frac{f(x,s)}{s^{p-1}} \leq \theta(x) \text{ uniformly for a.a. } x \in \Omega.$$

The following theorem can be found in [28].

Theorem 6.1 *If hypotheses $(H_f)_2$ hold, then there exists $\xi^* > 0$ such that if $\|\beta\|_\infty < \xi^*$, then problem (10) has at least two distinct (nontrivial) positive solutions $u_0, \hat{u} \in \mathrm{int}C_+$.*

Proof (sketch)

We introduce the following truncation of the right-hand side of the equation in (10):

$$\hat{f}_+(x,s) = \begin{cases} 0 & \text{if } s \leq 0 \\ \beta(x)s^{q-1} + f(x,s) & \text{if } s > 0, \end{cases}$$

which is a Carathéodory function.

Next we set $\hat{F}_+(x,s) = \int_0^s \hat{f}_+(x,t)\,dt$ and then consider the functional $\hat{\varphi}_+ \in C^1(W_0^{1,p}(\Omega))$ defined by

$$\hat{\varphi}_+(u) = \frac{1}{p}\|\nabla u\|_p^p - \int_\Omega \hat{F}_+(x,u(x))\,dx \text{ for all } u \in W_0^{1,p}(\Omega).$$

From hypotheses $(H_f)_2$ (i)–(iv) we derive that $\hat{\varphi}_+$ satisfies the Cerami condition. Then the first positive solution u_0 is obtained through the mountain pass theorem, which leads to

$$\hat{\varphi}_+(u_0) \geq \inf_{\partial B_\rho} \hat{\varphi}_+ =: \hat{\eta}_+ > 0, \tag{12}$$

where ∂B_ρ denotes the sphere in $W_0^{1,p}(\Omega)$ centered at 0 and of a small radius $\rho > 0$.

Now we generate a second positive solution for problem (10). By $(H_f)_2$(v), we find constants $\hat{c}_0 > 0$ and $\delta > 0$ such that

$$F(x,s) \geq -\frac{\hat{c}_0}{p} s^p \text{ for a.a. } x \in \Omega, \text{ all } s \in [0,\delta].$$

Let $v \in C_+$ with $v > 0$ in Ω and $\|v\| = \|\nabla v\|_p = 1$. For $t > 0$ small such that $tv(x) \in [0,\delta]$ for all $x \in \overline{\Omega}$, we have the estimate

$$\hat{\varphi}_+(tv) \leq t^q \left[\frac{t^{p-q}}{p} \left(1 + \frac{\hat{c}_0}{\lambda_1}\right) - \frac{1}{q} \int_\Omega \beta(x)\, v(x)^q\, dx \right]$$

since $1 = \|\nabla v\|_p^p \geq \lambda_1 \|v\|_p^p$. Choosing $t > 0$ possibly smaller and recalling that $p > q$, we have

$$\hat{\varphi}_+(tv) < 0 \text{ and } t\|v\| = t \leq \rho. \tag{13}$$

By (13) we have that

$$\eta := \hat{\eta}_+ - \inf_{\overline{B}_\rho} \hat{\varphi}_+ > 0,$$

where $\overline{B}_\rho = \{u \in W_0^{1,p}(\Omega) : \|u\| \leq \rho\}$.

Let $\varepsilon \in (0,\eta)$. By the Ekeland's variational principle, there exists $u_\varepsilon \in \overline{B}_\rho$ such that

$$\hat{\varphi}_+(u_\varepsilon) \leq \inf_{\overline{B}_\rho} \hat{\varphi}_+ + \varepsilon \tag{14}$$

and

$$\hat{\varphi}_+(u_\varepsilon) \leq \hat{\varphi}_+(y) + \varepsilon\|y - u_\varepsilon\| \text{ for all } y \in \overline{B}_\rho. \tag{15}$$

From inequality (14) and since $\varepsilon < \eta$, we obtain that $\hat{\varphi}_+(u_\varepsilon) < \inf_{\partial B_\rho} \hat{\varphi}_+$. It follows that $u_\varepsilon \in B_\rho = \{u \in W_0^{1,p}(\Omega) : \|u\| < \rho\}$, so for every $h \in W_0^{1,p}(\Omega)$ one has $u_\varepsilon + th \in B_\rho$ whenever $t > 0$ is sufficiently small. Inserting $y = u_\varepsilon + th$ in (15), dividing by t and then letting $t \to 0$ lead to $\|\hat{\varphi}'_+(u_\varepsilon)\| \leq \varepsilon$.

Fix now a sequence $\varepsilon_n \downarrow 0$ and denote for simplicity $u_n = u_{\varepsilon_n}$. Then one has $\hat{\varphi}'_+(u_n) \to 0$ and also $(1 + \|u_n\|)\hat{\varphi}'_+(u_n) \to 0$ as $n \to \infty$. Through the Cerami condition, this guarantees the existence of $\hat{u} \in W_0^{1,p}(\Omega)$ such that, along a relabeled subsequence, $u_n \to \hat{u}$ in $W_0^{1,p}(\Omega)$, which enables us to pass to limit in (14) to obtain

$$\hat{\varphi}_+(\hat{u}) = \inf_{\overline{B}_\rho} \hat{\varphi}_+ < 0 = \hat{\varphi}_+(0),$$

thereby $\hat{u} \neq 0$. Moreover, from (12) we see that $\hat{u} \neq u_0$. Since $\hat{\varphi}'_+(u_n) \to 0$ and $u_n \to \hat{u}$, we infer that $\hat{\varphi}'_+(\hat{u}) = 0$. Then, as for u_0, we obtain that $\hat{u}$ solves problem (10) and $\hat{u} \in \text{int} C_+$.

Remark 1 If the right-hand side of the equation in problem (10) has the special form

$$\lambda|u|^{q-2}u + (u^+)^{\tau-1} \text{ with } \lambda > 0 \text{ and } q < p < \tau < p^*,$$

Theorem 6.1 recovers the existence result in the case of p-Laplacian with $1 < p < +\infty$ due to García Azorero et al. [10] and the analogous result of Guo and Zhang [14] for $p \geq 2$.

7 Extremal Solutions

Consider the particular form of problem (10) restricting β to constants $\lambda > 0$, that is

$$\begin{cases} -\Delta_p u(x) = \lambda|u(x)|^{q-2}u(x) + f(x,u(x)) & \text{in } \Omega, \\ u = 0 & \text{on } \partial\Omega. \end{cases} \tag{16}$$

We recall that here we have, as in Sect. 6, $1 < q < p$. We slightly modify our assumptions:

$(H_f)_3$ $f : \Omega \times \mathbb{R} \to \mathbb{R}$ satisfies hypotheses $(H_f)_2$ (i)–(iv), and in addition

(vi) $\lim_{s\to 0^+} \frac{f(x,s)}{s^{p-1}} = 0$ uniformly for a.a. $x \in \Omega$;

(vii) (local sign condition) there is a number $b > 0$ such that

$$f(x,s) \geq 0 \text{ for a.a. } x \in \Omega \text{ and all } s \in (0,b).$$

The aim of this section is to study the extremal solutions for problem (16), which means the biggest negative solution and the smallest positive solution of (16). To this end we recall that λ_1 denotes the first eigenvalue of the negative Dirichlet p-Laplacian $(-\Delta_p, W_0^{1,p}(\Omega))$ and $\hat{u}_1$ stands for the L^p-normalized positive eigenfunction corresponding to λ_1. There holds $\hat{u}_1 \in \text{int}C_+$, with C_+ given in (11). The result below can be found in [28].

Proposition 7.1 *If hypotheses $(H_f)_3$ hold, then there is $\lambda^* > 0$ such that for all $\lambda \in (0,\lambda^*)$, problem (16) has a smallest positive solution $u_0 \in \text{int}C_+$ satisfying $\|u_0\|_\infty < b$.*

Proof

First, we look for suitable positive sub-supersolutions of problem (16). Hypothesis $(H_f)_3$ (vi) implies that for any $\varepsilon > 0$, we can find $\hat{c}_\varepsilon > 0$ such that

$$f(x,s) \leq \varepsilon s^{p-1} + \hat{c}_\varepsilon s^{r-1} \text{ for a.a. } x \in \Omega \text{ and all } s > 0.$$

Let $e \in \text{int}C_+$ solve

$$\begin{cases} -\Delta_p e(x) = 1 & \text{in } \Omega, \\ e = 0 & \text{on } \partial\Omega. \end{cases}$$

Notice that there exists $\lambda^* > 0$ such that for all $\lambda \in (0, \lambda^*)$ there is $\eta_1 = \eta_1(\lambda) \in (0, b/\|e\|_\infty)$ with

$$\lambda(\eta_1\|e\|_\infty)^{q-1} + \varepsilon(\eta_1\|e\|_\infty)^{p-1} + \hat{c}_\varepsilon(\eta_1\|e\|_\infty)^{r-1} < \eta_1^{p-1}.$$

Fix $\lambda \in (0, \lambda^*)$ and set $\overline{u} = \eta_1 e$. Then we see that

$$\begin{aligned} -\Delta_p \overline{u} = \eta_1^{p-1} &> \lambda(\eta_1\|e\|_\infty)^{q-1} + \varepsilon(\eta_1\|e\|_\infty)^{p-1} + \hat{c}_\varepsilon(\eta_1\|e\|_\infty)^{r-1} \\ &\geq \lambda\overline{u}^{q-1} + f(z, \overline{u}). \end{aligned}$$

This expresses that $\overline{u} \in \operatorname{int} C_+$ is a (strict) supersolution for problem (16). Moreover, we have $\|\overline{u}\|_\infty < b$.

Choose $\tilde{\varepsilon} > 0$ so small that $\tilde{\varepsilon}\hat{u}_1(x) < \overline{u}(x)$ for all $x \in \Omega$ and $\lambda_1\tilde{\varepsilon}^{p-q}\hat{u}_1^{p-q} < \lambda$. Set $\underline{u} = \tilde{\varepsilon}\hat{u}_1$. By hypothesis $(H_f)_3$ (vii) we infer that

$$-\Delta_p \underline{u} = \lambda_1\tilde{\varepsilon}^{p-1}\hat{u}_1^{p-1} < \lambda\tilde{\varepsilon}^{q-1}\hat{u}_1^{q-1} \leq \lambda\underline{u}^{q-1} + f(x, \underline{u}).$$

It follows that $\underline{u} \in \operatorname{int} C_+$ is a (strict) subsolution for problem (16), and in addition we have $\underline{u} \leq \overline{u}$.

Consider now the sequence $\underline{u}_n = \tilde{\varepsilon}_n\hat{u}_1$ with $\tilde{\varepsilon}_n \downarrow 0$. Since the problem (16) has a smallest solution u_*^n in the ordered interval $[\underline{u}_n, \overline{u}]$ and it satisfies $u_*^n \in \operatorname{int} C_+$, it follows that the sequence $\{u_*^n\}_{n\geq 1} \subset W_0^{1,p}(\Omega)$ is bounded. Acting with $u_*^n - u_+ \in W_0^{1,p}(\Omega)$ on

$$-\Delta_p u_*^n = \lambda(u_*^n)^{q-1} + f(\cdot, u_*^n(\,\cdot\,))$$

results in

$$\lim_{n\to\infty} \langle -\Delta_p u_*^n, u_*^n - u_+ \rangle = 0.$$

The $(S)_+$ property for the p-Laplacian (see also Proposition 2.2) enables us to get $u_*^n \to u_+$ in $W_0^{1,p}(\Omega)$ as $n \to \infty$. It follows that u_+ is a solution of the problem (16). The nonlinear regularity theory and strong maximum principle imply that $u_+ \in \operatorname{int} C_+$ provided $u_+ \neq 0$.

Let us prove that $u_+ \neq 0$. To this end we introduce the function $\psi : W_0^{1,p}(\Omega) \to \mathbb{R}$ by

$$\psi(u) = \frac{1}{p}\|\nabla u\|_p^p - \frac{\lambda}{q}\|u^+\|_q^q \quad \text{for all } u \in W_0^{1,p}(\Omega).$$

Here u^+ denotes the positive part of the function u. Since ψ coercive and sequentially weakly lower semicontinuous, and taking into account that $q < p$, there is $u \in W_0^{1,p}(\Omega)$ such that

$$\psi(u) = \inf_{W_0^{1,p}(\Omega)} \psi < 0,$$

so $u \neq 0$ and

$$-\Delta_p u - \lambda(u^+)^{q-1} = 0.$$

Acting with $-u^- \in W_0^{1,p}(\Omega)$, where u^- denotes the negative part of the function u, ensures that $u \geq 0$, thus u solves the problem

$$\begin{cases} -\Delta_p u(x) = \lambda u(x)^{q-1} & \text{in } \Omega, \\ u = 0 & \text{on } \partial\Omega. \end{cases}$$

This renders $\Delta_p u(x) \leq 0$ a.e. in Ω, which entails $u \in \text{int}C_+$ due to the strong maximum principle [30].

Since $u_*^n \in \text{int}C_+$, there is the biggest $\xi_n > 0$ such that $\xi_n u \leq u_*^n$ for all $n \geq 1$. If $\xi_n < 1$, then we have

$$-\Delta_p u_*^n(x) \geq \lambda u_*^n(x)^{q-1} \geq \lambda \xi_n^{q-1} u(x)^{q-1} > \lambda \xi_n^{p-1} u(x)^{q-1} = -\Delta_p(\xi_n u)(x).$$

By means of the comparison principle in [13], we arrive at $u_*^n - \xi_n u \in \text{int}C_+$, which contradicts the maximality of ξ_n.

Therefore $\xi_n \geq 1$, which implies that $u \leq u_*^n$. We conclude that $u \leq u_+$, so $u_+ \neq 0$. Now it is straightforward to check that u_+ is indeed the smallest positive solution of (16). The proof is thus complete.

Next we deal with the biggest negative solution. We work under asymmetrical situations on $\mathbb{R}_+$ and $\mathbb{R}_-$. We impose the hypotheses:

$(H_f)_4$ $f : \Omega \times \mathbb{R} \to \mathbb{R}$ satisfies hypotheses $(H_f)_2$ (i)–(iii) and in addition

(viii) There exists $\theta \in L^\infty(\Omega)_+$ such that $\theta(x) \leq \lambda_1$ a.e. on Ω, with strict inequality on a set of positive measure, and

$$\limsup_{s \to -\infty} \frac{f(x,s)}{|s|^{p-2}s} \leq \theta(x) \text{ uniformly for a.a. } x \in \Omega;$$

(ix)

$$\lim_{s \to 0^-} \frac{f(x,s)}{|s|^{p-2}s} = 0 \text{ uniformly for a.a. } x \in \Omega;$$

(x) (local sign condition) There is a number $b > 0$ such that $f(x,s) \leq 0$ for a.a. $x \in \Omega$ and all $s \in (-b, 0)$.

Remark 2 Assumption $(H_f)_4$ (viii) guarantees that for a.a. $x \in \Omega$, $f(x,\cdot)$ has a $(p-1)$-(sub)linear growth near $-\infty$.

We have the following result from [28].

Proposition 7.2 *If hypotheses* $(H_f)_4$ *hold, then for every* $\lambda > 0$, *problem (16) has a biggest negative solution* $v_- \in -\text{int}C_+$.

Proof (sketch)

First we search for negative sub-supersolutions of problem (16). By $(H_f)_2$(iii), $(H_f)_4$(viii) we have that given $\varepsilon > 0$ we can find $\xi_\varepsilon \in L^\infty(\Omega)_+ \setminus \{0\}$ such that

$$f(x,s) \geq (\theta(x) + \varepsilon)|s|^{p-2}s - \xi_\varepsilon(x) \text{ for a.a. } x \in \Omega \text{ and all } s \leq 0.$$

Define the Carathéodory function

$$g_\lambda(x,s) = \begin{cases} \lambda|s|^{q-2}s + (\theta(x)+\varepsilon)|s|^{p-2}s - \xi_\varepsilon(x) & \text{if } s \leq 0 \\ -\xi_\varepsilon(x) & \text{if } s > 0, \end{cases}$$

and consider the C^1-functional $\sigma : W_0^{1,p}(\Omega) \to \mathbb{R}$ defined by

$$\sigma(v) = \frac{1}{p} \|\nabla v\|_p^p - \int_\Omega G_\lambda(x, v(x))\, dx \text{ for all } v \in W_0^{1,p}(\Omega),$$

where $G_\lambda(x,s) = \int_0^s g_\lambda(x,t)\, dt$. The function σ is coercive and sequentially weakly lower semicontinuous, so there is $\underline{v} \in W_0^{1,p}(\Omega)$ such that

$$\sigma(\underline{v}) = \inf_{W_0^{1,p}(\Omega)} \sigma.$$

We act with $\underline{v}^+ \in W_0^{1,p}(\Omega)$ on the problem

$$\begin{cases} -\Delta_p \underline{v} + \lambda(\underline{v}^-)^{q-1} + (\theta + \varepsilon)(\underline{v}^-)^{p-1} + \xi_\varepsilon = 0 & \text{in } \Omega \\ \underline{v} = 0 & \text{on } \partial\Omega, \end{cases}$$

which implies that $\underline{v} \leq 0$, $\underline{v} \neq 0$, $\Delta_p(-\underline{v}) \leq 0$, so $\underline{v} \in -\text{int}C_+$ according to the strong maximum principle in [30]. Furthermore, $\underline{v}$ is a lower solution of problem (16). Given $\tilde{\varepsilon} > 0$ small, set $\overline{v} = -\tilde{\varepsilon}\hat{u}_1 \in -\text{int}C_+$. Hypothesis $(H_f)_4$ (ix) ensures that $\overline{v}$ is an upper solution of (16) and $\underline{v} \leq \overline{v}$.

Consider the sequence of functions $\overline{v}_n = -\tilde{\varepsilon}_n\hat{u}_1$ corresponding to a sequence $\tilde{\varepsilon}_n \downarrow 0$. Then, for every integer $n \geq 1$, we get the biggest solution $v_*^n \in -\text{int}C_+$ of problem (16) in the ordered interval $[\underline{v}, \overline{v}_n]$. Taking into account this maximality property and using once again the strong maximum principle (which is possible thanks to hypothesis $(H_f)_4$ (x)), we establish that $v_*^n \to v_-$ in $W_0^{1,p}(\Omega)$ for some $v_- \in -\text{int}C_+$. Finally, we can show that v_- is the biggest negative solution of (16), which completes the proof.

8 Nodal Solution

Our main result on problem (16) asserts the existence of an additional solution which is nodal (sign changing). The complete proof of the theorem below as well as other related results can be found in [28].

Theorem 8.1 *If hypotheses $(H_f)_2$ (i)–(iv), $(H_f)_3$ (vi)–(vii) and $(H_f)_4$ (viii)–(x) hold, then there exists $\lambda^* > 0$ such that for all $\lambda \in (0, \lambda^*)$, problem (16) has at least four distinct nontrivial smooth solutions: $u_0, \hat{u} \in \text{int}C_+$ with $u_0 \neq \hat{u}$, $v_0 \in -\text{int}C_+$ and $y_0 \in C_0^1(\overline{\Omega}) \setminus \{0\}$ nodal.*

Proof (sketch)

Theorem 6.1 and Proposition 7.2 provide a number $\lambda^* > 0$ such that for every $\lambda \in (0, \lambda^*)$ there exist three constant sign solutions $u_0, \hat{u} \in \text{int}C_+$ and $v_0 \in -\text{int}C_+$ of problem (16). Fix any $\lambda \in (0, \lambda^*)$. In view of Propositions 7.1 and 7.2 there exist the smallest positive solution $u_+ \in \text{int}C_+$ and the biggest negative solution $v_- \in -\text{int}C_+$ of problem (16). Consider the truncation:

$$\hat{f}_\lambda(x,s) = \begin{cases} \lambda|v_-(x)|^{q-2}v_-(x) + f(x, v_-(x)) & \text{if } s < v_-(x) \\ \lambda|s|^{q-2}s + f(x,s) & \text{if } v_-(x) \leq s \leq u_+(x) \\ \lambda u_+(x)^{q-1} + f(x, u_+(x)) & \text{if } u_+(x) < s. \end{cases}$$

Associated with this truncation we introduce the C^1 functional $\hat{\varphi}_\lambda : W_0^{1,p}(\Omega) \to \mathbb{R}$ defined by

$$\hat{\varphi}_\lambda(u) = \frac{1}{p} \|\nabla u\|_p^p - \int_\Omega \hat{F}_\lambda(x, u(x))\, dx \quad \text{for all } u \in W_0^{1,p}(\Omega),$$

where

$$\hat{F}_\lambda(x,t) = \int_0^t \hat{f}_\lambda(x,s) ds \quad \text{for all } (x,t) \in \Omega \times \mathbb{R}.$$

Using the definition of the truncation function given above, it is straightforward to prove that every critical point of $\hat{\varphi}_\lambda$ is located in the ordered interval $[v_-, u_+]$. This can be checked by testing the equation $\hat{\varphi}_\lambda'(u) = 0$ with $(u - u_+)^+$ and $(u - v_-)^-$. Then we can show that v_- and u_+ are local minimizers of the functional $\hat{\varphi}_\lambda$. Moreover, we may assume that they are isolated critical points because otherwise we obtain infinitely many nodal solutions of problem (16).

Notice that the imposed assumptions permit to apply the mountain pass theorem to the functional $\hat{\varphi}_\lambda$. It follows that there exists a critical point $y_0 \in [v_-, u_+]$ of $\hat{\varphi}_\lambda$ with $y_0 \neq v_-$ and $y_0 \neq u_+$. A major step in the proof is to establish that $y_0 \neq 0$. In this respect we make use of the critical groups associated to the function $\hat{\varphi}_\lambda$ at y_0 and 0. For the theory of critical groups we refer to [6] and [19].

The fact that y_0 is a critical point of $\hat{\varphi}_\lambda$ of mountain pass type implies that the critical group $C_1(\hat{\varphi}_\lambda, y_0)$ is nontrivial, that is $C_1(\hat{\varphi}_\lambda, y_0) \neq 0$. On the other hand, we can prove that for all $u \in W_0^{1,p}(\Omega)$ with $\|u\| \leq \rho$ and $\hat{\varphi}_\lambda(u) \leq 0$, we have $\hat{\varphi}_\lambda(tu) \leq 0$ whenever $t \in [0, 1]$. In turn, this allows to develop a homotopy argument revealing that the critical groups $C_k(\hat{\varphi}_\lambda, 0)$ are all of them trivial, that is $C_k(\hat{\varphi}_\lambda, 0) = 0$ for all k. Consequently, comparing the critical groups $C_1(\hat{\varphi}_\lambda, y_0)$ and $C_1(\hat{\varphi}_\lambda, 0)$, we derive that $y_0 \neq 0$. Now we can conclude that the nontrivial solution y_0 of problem (16) is nodal. Indeed, relying on the location property $y_0 \in [v_-, u_+]$, the solution y_0 must be sign changing because otherwise it is contradicted the maximality of v_- among the negative solutions or the minimality of u_+ among the positive solutions. The proof is thus complete.

9 New Problem

In this section we consider the nonlinear Dirichlet problem

$$\begin{cases} -\Delta_p u - \mu \Delta_q u = f(x, u, \nabla u) & \text{in } \Omega, \\ u > 0 & \text{in } \Omega, \\ u = 0 & \text{on } \partial\Omega, \end{cases} \qquad (NP)$$

on a bounded domain Ω in $\mathbb{R}^N$ with a $C^{1,\alpha}$-boundary $\partial\Omega$, for some $0 < \alpha \leq 1$. In the left-hand side of the equation in (NP) we have the p-Laplacian Δ_p and the

q-Laplacian Δ_q with $1 < q < p < +\infty$, and a constant $\mu \geq 0$. So, basically, the principal part of the elliptic equation is determined by the (p,q)-Laplacian operator $-\Delta_p - \Delta_q$. The right-hand side of the elliptic equation in problem (NP) depends on the gradient ∇u of the solution u. This is a novelty with respect to problem (1). Problems where the right-hand side depends on the gradient of the solution are sometimes called with convection term.

We emphasize that the (p,q)-Laplacian operator is not homogeneous. However, it does not belong to the class of possibly nonhomogeneous operators described in statement (H) (in the sense to be in the divergence form $\operatorname{div} A(x, \nabla(\cdot))$ with some $A(x,\cdot)$ satisfying (H)). Actually, all the conditions in (H) are satisfied excepting (H) (ii) that to fulfill our purpose should be replaced with

(ii') There exists a constant $C_1 > 0$ such that

$$|D_y A(x,y)| \leq C_1(|y|^{p-2} + 1) \quad \text{for every } x \in \overline{\Omega} \text{ and } y \in \mathbb{R}^N \setminus \{0\}.$$

If this is the case, then we can set $A(x,y) = |y|^p + \mu |y|^q$ for all $(x,y) \in \overline{\Omega} \times \mathbb{R}^N$ to comply with (H).

We suppose that the nonlinearity $f(x,u,\nabla u)$ satisfies the hypothesis:

(FG) $f : \Omega \times \mathbb{R} \times \mathbb{R}^N \to \mathbb{R}$ is a Carathéodory function such that $f(x,0,\xi) = 0$ a.e. in Ω, for all $\xi \in \mathbb{R}^N$, and

$$b_0|t|^{r_0} \leq f(x,t,\xi) \leq b_1(1 + |t|^{r_1} + |\xi|^{r_2})$$

for all $(x,t,\xi) \in \Omega \times \mathbb{R} \times \mathbb{R}^N$, with constants $b_0, b_1 > 0$, $r_1, r_2 \in [0, p-1)$, $r_0 \in [0, p-1)$ if $\mu = 0$, and $r_0 \in [0, q-1)$ if $\mu > 0$.

The following result can be found in [9].

Theorem 9.1 *Under assumption (FG), problem (NP) admits a (positive) solution $u \in C_0^1(\overline{\Omega})$.*

Proof (sketch)
First, it is shown that for every $\varepsilon \in (0,1)$, the approximate problem

$$\begin{cases} -\Delta_p u - \mu \Delta_q u = f(x, u+\varepsilon, \nabla u) & \text{in } \Omega, \\ u > 0 & \text{in } \Omega, \\ u = 0 & \text{on } \partial\Omega, \end{cases} \qquad (P_\varepsilon)$$

has at least a (positive) solution $u_\varepsilon \in C_0^1(\overline{\Omega})$. This conclusion is achieved through a Schauder basis of $W_0^{1,p}(\Omega)$ by using Brouwer's fixed point theorem on the corresponding finite dimensional spaces and then passing to the limit. For the solution $u_\varepsilon \in C_0^1(\overline{\Omega})$ of problem (P_ε), from hypothesis (FG) and

$$\int_\Omega (|\nabla u_\varepsilon|^p + \mu |\nabla u_\varepsilon|^q) dx = \int_\Omega f(x, u_\varepsilon + \varepsilon, \nabla u_\varepsilon) u_\varepsilon dx,$$

we get the estimate

$$\|u_\varepsilon\|^p \le C(\|u_\varepsilon\| + \|u_\varepsilon\|^{r_1+1} + \|u_\varepsilon\|^{r_2+1}),$$

with a constant $C > 0$ which is independent of ε. Since $1, r_1+1, r_2+1 < p$, we infer that

$$\|u_\varepsilon\| \le C_0, \tag{17}$$

for a constant $C_0 > 0$ independent of ε. In view of (17), we can find $\varepsilon_n \to 0^+$ such that the corresponding sequence $\{u_n = u_{\varepsilon_n}\}$ is strongly convergent

$$u_n \to u \text{ in } W_0^{1,p}(\Omega) \text{ as } n \to \infty, \tag{18}$$

with some $u \in W_0^{1,p}(\Omega)$. On the basis of (18) and because u_n solves (P_{ε_n}), it turns out that u is a solution of the equation

$$\begin{cases} -\Delta_p u - \mu\Delta_q u = f(x, u, \nabla u) & \text{in } \Omega \\ u = 0 & \text{on } \partial\Omega. \end{cases}$$

The regularity up to the boundary in [17, Theorem 1] and [18, p. 320] ensures that $u \in C^{1,\beta}(\overline{\Omega})$ with some $\beta \in (0,1)$. At this point we cannot guarantee that u is nontrivial. In order to prove that $u > 0$, we develop a comparison argument.

We note that there exists a (positive) solution $\underline{u} \in C_0^1(\overline{\Omega})$ to the problem

$$\begin{cases} -\Delta_p u - \mu\Delta_q u = b_0 u^{r_0} & \text{in } \Omega \\ u > 0 & \text{in } \Omega \\ u = 0 & \text{on } \partial\Omega, \end{cases} \tag{19}$$

where b_0 and r_0 are the positive constants in hypothesis (FG). In the sequel, $\underline{u}$ will be regarded as a subsolution of (19). We also observe that hypothesis (FG) implies that u_ε is a supersolution of problem (19) for each $\varepsilon \in (0,1)$.

Now we apply the comparison principle in [9] to problem (19), by taking the subsolution $\underline{u}$ and the supersolution u_ε. We emphasize that for applying this comparison principle it is essential to know that $u_\varepsilon > 0$ in Ω as well as that $r_0 < q-1$ if $\mu > 0$, and $r_0 < p-1$ if $\mu = 0$. In order to apply the comparison principle we also need to check that

$$\frac{u_\varepsilon}{\underline{u}}, \frac{\underline{u}}{u_\varepsilon} \in L^\infty(\Omega).$$

To this end it suffices to show that whenever $x \to x_0 \in \partial\Omega$ with $x \in \Omega$, one has

$$\max\left\{\limsup_{x\to x_0} \frac{\underline{u}(x)}{u_\varepsilon(x)}, \limsup_{x\to x_0} \frac{u_\varepsilon(x)}{\underline{u}(x)}\right\} < +\infty. \tag{20}$$

The property stated in (20) is established on the basis of the Hopf boundary point lemma in the strong maximum principle applied to both Dirichlet problems (19) and (P_ε) with corresponding solutions $\underline{u}$ and u_ε, which amounts to saying that

$$\frac{\partial \underline{u}}{\partial n}(x_0) < 0, \quad \frac{\partial u_\varepsilon}{\partial n}(x_0) < 0 \text{ for all } x_0 \in \partial\Omega, \tag{21}$$

where n denotes the exterior normal unit vector on $\partial\Omega$. The Hopf boundary point lemma holds true for problems (19) and (P_ε) by virtue of [29, Theorem 5.5.1]. Recalling that $u_\varepsilon, \underline{u} \in C^1(\overline{\Omega})$, it is clear from the L'Hôpital theorem and (21) that the property required in (20) is fulfilled. Therefore we can do the comparison of the solution $\underline{u}$ (regarded as a subsolution) of (19) with the supersolution u_ε of (19) implying that

$$u_\varepsilon(x) \geq \underline{u}(x) > 0 \text{ for all } x \in \Omega \text{ and } \varepsilon \in (0,1). \tag{22}$$

Using (18), we can pass to the limit in (22) along a sequence $\varepsilon_n \to 0$. This leads to $u(x) \geq \underline{u}(x) > 0$ for all $x \in \Omega$, so u is a (positive) solution of problem (NP), which completes the proof.

References

1. Aizicovici, S., Papageorgiou, N.S., Staicu, V.: The spectrum and an index formula for the Neumann p-Laplacian and multiple solutions for problems with a crossing nonlinearity. Discret. Contin. Dyn. Syst. **25**(2), 431–456 (2009)
2. Barletta, G., Papageorgiou, N.S.: A multiplicity theorem for the Neumann p-Laplacian with an asymmetric nonsmooth potential. J. Glob. Optim. **39**(3), 365–392 (2007)
3. Brezis, H., Nirenberg, L.: H^1 versus C^1 local minimizers. C. R. Acad. Sci. Paris Sér. I Math. **317**(5), 465–472 (1993)
4. Carl, S., Perera, K.: Sign-changing and multiple solutions for the p-Laplacian. Abstr. Appl. Anal. **7**(12), 613–625 (2002)
5. Carl, S., Le, V.K., Motreanu, D.: Nonsmooth Variational Problems and Their Inequalities. Comparison Principles and Applications. Springer Monographs in Mathematics. Springer, New York (2007)
6. Chang, K.-C.: Infinite-Dimensional Morse Theory and Multiple Solution Problems. Progress in Nonlinear Differential Equations and Their Applications, vol. 6. Birkhäuser, Boston, (1993)
7. Cuesta, M., de Figueiredo, D.G., Gossez, J.-P.: The beginning of the Fučik spectrum for the p-Laplacian. J. Differ. Equ. **159**(1), 212–238 (1999)
8. Fan, X.: On the sub-supersolution method for $p(x)$-Laplacian equations. J. Math. Anal. Appl. **230**(1), 665–682 (2007)
9. Faria, L.F.O., Miyagaki, O.H., Motreanu, D.: Comparison and positive solutions for problems with (p,q)-Laplacian and convection term. Proc. Edinburgh Math. Soc. (2014). doi:10.1017/S0013091513000576
10. García Azorero, J.P., Peral Alonso, I., Manfredi, J.J.: Sobolev versus Hölder local minimizers and global multiplicity for some quasilinear elliptic equations. Commun. Contemp. Math. **2**(3), 385–404 (2000)
11. Gasinski, L., Papageorgiou, N.S.: Nonlinear Analysis. Chapman Hall/CRC Press, Boca Raton (2006)
12. Gasiński, L., Papageorgiou, N.S.: Anisotropic nonlinear Neumann problems. Calc. Var. Partial Differ. Equ. **42**(3–4), 323–354 (2011)
13. Guedda, M., Véron, L.: Quasilinear elliptic equations involving critical Sobolev exponents. Nonlinear Anal. **13**(8), 879–902 (1989)
14. Guo, Z., Zhang, Z.: $W^{1,p}$ versus C^1 local minimizers and multiplicity results for quasilinear elliptic equations. J. Math. Anal. Appl. **286**(1), 32–50 (2003)
15. Iannizzotto, A., Papageorgiou, N.S.: Existence of three nontrivial solutions for nonlinear Neumann hemivariational inequalities. Nonlinear Anal. **70**(9), 3285–3297 (2009)

16. Khan, A.A., Motreanu, D.: Local minimizers versus X-local minimizers. Optim. Lett. (2012). doi:10.1007/s11590-012-0474-8
17. Lieberman, G.M.: Boundary regularity for solutions of degenerate elliptic equations. Nonlinear Anal. **12**(11), 1203–1219 (1988)
18. Lieberman, G.M.: The natural generalization of the natural conditions of Ladyzhenskaya and Ural'tseva for elliptic equations. Commun. Partial Differ. Equat. 16(2–3), 311–361 (1991)
19. Mawhin, J., Willem, M.: Critical Point Theory and Hamiltonian Systems. Applied Mathematical Sciences, vol. 74. Springer-Verlag, New York (1989)
20. Miyajima, S., Motreanu, D., Tanaka, M.: Multiple existence results of solutions for the Neumann problems via super- and sub-solutions, J. Funct. Anal. **262**(4), 1921–1953 (2012)
21. Motreanu, D., Papageorgiou, N.S.: Multiple solutions for nonlinear elliptic equations at resonance with a nonsmooth potential. Nonlinear Anal. **56**(8), 1211–1234 (2004)
22. Motreanu, D., Papageorgiou, N.S.: Multiple solutions for nonlinear Neumann problems driven by a nonhomogeneous differential operator. Proc. Amer. Math. Soc. **139**(10), 3527–3535 (2011)
23. Motreanu, D., Tanaka, M.: Generalized eigenvalue problems of nonhomogeneous elliptic operators and their application. Pacific J. Math. **265**(1), 151–184 (2013)
24. Motreanu, D., Winkert, P.: On the Fučik spectrum for the p-Laplacian with Robin boundary condition. Nonlinear Anal. **74**(14), 4671–4681 (2011)
25. Motreanu, D., Winkert, P.: The Fučík spectrum for the negative p-Laplacian with different boundary conditions. Nonlinear Analysis. Springer Optimization and Its Applications, vol.68, pp. 471–485. Springer, New York (2012)
26. Motreanu, D., Motreanu, V.V., Papageorgiou, N.S.: Nonlinear Neumann problems near resonance. Indiana Univ. Math. J. **58**(3), 1257–1279 (2009)
27. Motreanu, D., Motreanu, V.V., Papageorgiou, N.S.: Multiple constant sign and nodal solutions for nonlinear Neumann eigenvalue problems. Ann. Scuola Norm. Sup. Pisa Cl. Sci. (5). **10**(3), 729–755 (2011)
28. Motreanu, D., Motreanu, V.V., Papageorgiou, N.S.: On p-Laplace equations with concave terms and asymmetric perturbations. Proc. Roy. Soc. Edinb. A. **141A** (1), 171–192 (2011)
29. Pucci, P., Serrin, J.: The Maximum Principle. Birkhäuser, Basel (2007)
30. Vazquez, J.L.: A strong maximum principle for some quasilinear elliptic equations. Appl. Math. Optim. **12**(3), 191–202 (1984)
31. Winkert, P.: Constant-sign and sign-changing solutions for nonlinear elliptic equations with Neumann boundary values. Adv. Differ. Equat. **15**(5–6), 561–599 (2010)
32. Winkert, P.: Local $C^1(\overline{\Omega})$-minimizers versus local $W^{1,p}(\Omega)$-minimizers of nonsmooth functionals. Nonlinear Anal. **72**(11), 4298–4303 (2010)
33. Winkert, P., Zacher, R.: A priori bounds of solutions to elliptic equations with nonstandard growth. Discret. Contin. Dyn. Syst. Ser. S **5**(4), 865–878 (2012)

Isotone Retractions onto the Positive Cone of the Ordered Euclidean Space

A. B. Németh and S. Z. Németh

Abstract The isotone retraction onto the positive cone of an ordered Euclidean space is a continuous retraction onto this cone preserving the order relation this cone engenders. The existence of such a retraction imposes rather strong conditions on the cone. This chapter contains a comprehensive treatment of the problem, reviewing and completing the results of the authors on this theme appeared in the literature.

Keywords Convex cones · Retractions · Isotone mappings

1 Introduction

We begin with some terminology. Let $\mathbb{R}^m$ be the m-dimensonal Euclidean space endowed with the scalar product $\langle \cdot, \cdot \rangle$.

The closed subset $K \subset \mathbb{R}^m$ is called a *cone*, if it satisfies the conditions:

(i) $K + K \subset K$
(ii) $tK \subset K, \forall\, t \in \mathbb{R}_+ = [0, +\infty)$
(iii) $K \cap (-K) = \{0\}$
(iv) $K - K = \mathbb{R}^m$.

The *dual* of K is $K^* = \{y \in \mathbb{R}^m : \langle x, y \rangle \geq 0\}$ and the *polar* of K is $K^\perp = -K^*$. It is known that if $L = K^\perp$, then $L^\perp = K$. In this case, K and L are called *mutually polar*. The cone K is called *subdual* if $K^* \subset K$ and *self-dual* if $K^* = K$.

By putting $x,\ y \in \mathbb{R}^m$, $x \leq_K y \Leftrightarrow y - x \in K$, the cone K induces an *order relation* $\leq_K$ in $\mathbb{R}^m$, that is, a binary relation, which is reflexive, transitive, and antisymmetric. This order relation is *translation invariant*, in the sense that $x \leq_K y$ implies $x + z \leq_K y + z$ for all $z \in \mathbb{R}^m$, and *scale invariant*, in the sense that $x \leq_K y$

S. Z. Németh (✉)
School of Mathematics, University of Birmingham,
Watson Building, Edgbaston, Birmingham, B15 2TT UK
e-mail: s.nemeth@bham.ac.uk

A. B. Németh
Faculty of Mathematics and Computer Science, Babeş Bolyai University,
Str. Kogălniceanu nr. 1–3, 400084 Cluj-Napoca, Romania
e-mail: nemab@math.ubbcluj.ro

T. M. Rassias, P. M. Pardalos (eds.), *Mathematics Without Boundaries*,
DOI 10.1007/978-1-4939-1106-6_16, © Springer Science+Business Media, LLC 2014

implies $tx \leq_K ty$ for any $t \in \mathbb{R}_+$. If $\leq$ is a translation invariant and scale invariant order relation on $\mathbb{R}^m$, then $\leq = \leq_K$ with $K = \{ x \in \mathbb{R}^m : 0 \leq x\}$.

The cone $K \subset \mathbb{R}^m$ is called *polyhedral* if there exists a finite set of vectors $v_1, \ldots, v_k \in \mathbb{R}^m$ such that

$$K = \left\{x \in \mathbb{R}^m : \langle x, v_j\rangle \geq 0,\ j = 1, \ldots, k\right\}. \tag{1}$$

Since K must be a cone, then k must be $\geq m$ and among v_j must exist m linearly independent ones.

The order relation induced by the cone K in $\mathbb{R}^m$ is called *latticial* if for each pair of points $x,\ y \in \mathbb{R}^m$ there exists their supremum, that is the smallest upper bound (with respect to $\leq_K$) of them, denoted by, $\sup\{x, y\} = x \vee y$. In this case, K is called a *simplicial cone.*

A classical result of Youdine [18] says that for a cone $K \subset \mathbb{R}^m$ the induced order relation $\leq_K$ is latticial if and only if K is a polyhedral cone in whose representation (1) $k = m$ and $v_1, \ldots, v_m$ are linearly independent vectors. (Whence the denomination "simplicial" for a such cone. A simplicial cone is also called a Youdine cone.)

The mapping $\rho : \mathbb{R}^m \to \mathbb{R}^m$ *is called* isotone retraction onto K, *if it is a retraction onto K (i. e.,* $\rho(\mathbb{R}^m) = K,\ \rho(x) = x$ if $x \in K$), *it is continuous, and whenever* $x \leq_K y$ *it follows that* $\rho(x) \leq_K \rho(y)$.

If K is a simplicial cone, then the *lattice operation* $\rho : x \to 0 \vee x := x^+$ is an isotone retraction.

The notion of the isotone retraction is due to S. Z. Németh. Historically, the subject arose from some application-oriented investigations.

G. Isac and A. B. Németh have characterized a cone in the Euclidean space which admits an isotone projection onto it [3], where isotonicity is considered with respect to the order induced by the cone. They called such a cone isotone projection cone. G. Isac and A. B. Németh [4] and S. J. Bernau [1] considered the similar problem for the Hilbert space. It turns out that the isotonicity of the projection is a very strong requirement, which implies the latticiality of the order induced by the cone. Thus, the investigation of the isotone projection cones becomes part of the theory of latticially ordered Hilbert spaces. In this context, it has been related to other lattice theoretic notions by G. Isac and L.-E. Persson [8, 9].

An isotone projection is another particular form of the isotone retraction. It is related to the projection onto the polar of the cone by a duality property [16].

Both the resolvability and the approximation of solutions of nonlinear complementarity problems can be handled by using the metric projection onto the cone defining the problem. The isotonicity of the projection provides new existence results and iterative methods in this theory [5, 6, 7, 11].

Although rather particular, the isotone projection cones play an important role in other applications. The positive monotone cone used in the Euclidean distance geometry (see [2]) is an isotone projection one. A simple method of projection on isotone projection cones proposed by us (see [14]) has become important in the effective handling of the problem of map-making from relative distance information, e.g., stellar cartography (see www.convexoptimization.com/wikimization/index.php/Projection_on_Polyhedral_Cone and Sect. 5.13.2.4 in [2]).

Exploring the theoretical background of the nonlinear complementarity, S. Z. Németh have introduced in the notes [12] and [13] the notion of the isotone retraction onto a cone. He observed the surprising fact that the seemingly mild requirement of the existence of an isotone retraction onto a cone in a Hilbert space (i.e., the existence of a continuous retraction onto that cone, preserving the order relation this cone defines), implies very strong conditions on the structure of the cone. It was shown there that together with the so called sharpness (another rather mild condition) of the additive complementary of the retraction, the existence of such a retraction is equivalent with the simpliciality of the cone. (If X is a vector space, the additive complementary of the mapping $F : X \to X$ is $G = I - F$ with I the identity mapping; the mapping $G : X \to X$ was called sharp if $G(X) \cap (-G(X)) = \{0\}$.) The isotone projection and the positive part operation for a simplicial cone both satisfy the sharpness condition of their additive complementary.

The projection is strongly related to the linear structure of $\mathbb{R}^n$. Together with the positive part mapping for a simplicial cone, which is also strongly related to the linear structure, the isotone projection has particularity to be idempotent in the sense that applied twice is the same thing as applied once. In the above cited papers of G. Isac and A. B. Németh, it was shown that there is a strong connection between some simplicial cones and the isotonicity of mappings defined by the aid of the projection. Thus, we arrive to the apparent conclusion that the geometric properties of the projection, together with its isotonicity, respectively, the isotonicity of some mappings defined with it, are at the origin of the phenomena inducing the latticiality type strong conditions on the positive cone of the space.

The above cited results of S. Z. Németh have put into evidence the surprising fact, that not the geometric properties, but their topological properties are prevalent to the fact that together with isotonicity these mappings impose strong conditions on the positive cone of the space. This way the obtained results exhibit a strong relation between the ordering and the topology of the Hilbert and Euclidean spaces. This relationship has been further explored in [15].

In this chapter, we shall consider the problem of the isotone retraction in the Euclidean space only.

2 The Polyhedrality of the Isotone Retraction Cone

We begin with some well-known notions and some well-known easily verifiable facts from the convex geometry.

The subset K_0 of the cone K is called its *face*, if it is a cone and if from $x \in K_0,\ y \in K$, and $x - y \in K$, it follows that $y \in K_0$. Using the order relation $\leq_K$, the above definition can be formulated as follows. The cone $K_0 \subset K$ is a face of K if from $x \in K_0$ and $0 \leq_K y \leq_K x$ it follows that $y \in K_0$. The cone K itself is its face. A face $K_0 \subset K$ is called *proper* face of K if $K_0 \neq K$.

If $x \in K$, then

$$\mathrm{fce}(x) := \{y \in K : \exists t \in (0, +\infty) \text{ with } ty \leq_K x\}$$

is a face of K, called the *face engendered* by x.

If H is a supporting hyperplane of K, that is a set of form

$$H = \{x \in \mathbb{R}^m : \langle u, x \rangle = 0 \text{ with } u \neq 0\}$$

with the property that

$$K \subset H_- = \{y \in \mathbb{R}^m : \langle u, y \rangle \leq 0\},$$

then $H \cap K$ is a face of K, called *exposed face*. All the proper faces of a polyhedral cone are exposed faces.

Let $K \subset \mathbb{R}^m$ be a cone in $\mathbb{R}^m$.

Since $K - K = \mathbb{R}^m$, int $K \neq \emptyset$, and $K+$int K =int K. Each proper face of a cone in $\mathbb{R}^m$ is on its boundary bdr K. If $x \in$ bdr K, then there always exists an exposed face of K containing it.

Let the cone $K \subset \mathbb{R}^m$ be polyhedral, represented as the intersection of a minimal finite number of closed halfspaces. The intersection of the cone with the hyperplanes defining these halfspaces are exactly its maximal proper faces.

The next lemma exhibits some simple, but fundamental properties of an isotone retraction. These will be used to show that every isotone retraction cone in $\mathbb{R}^m$ is polyhedral.

Lemma 1 *Let $\rho : \mathbb{R}^m \to K$ be an isotone retraction. Then,*

1. *For $x \in \mathbb{R}^m \backslash K$ we have $\rho(x) \in$ bdr K and $\rho(x - K) \subset$ fce $\rho(x)$.*
2. *If K is polyhedral, K_0 is its maximal face, L is the subspace engendered by K_0, $K \subset L^-$, where L^- is the closed subspace determined by L and containing K, then $\rho(L^+) \subset K_0$.*

Proof

1. Since $K - K = \mathbb{R}^m$, there exist the elements $u, v \in K$ such that, $x = u - v$. Hence, $x \leq u$. Then, there exists some $t \in [0, 1[$ such that $w = tx + (1 - t)u \in$ bdr K. Obviously, $x \leq w$ (since $(1 - t)x \leq (1 - t)u$). Hence,

$$0 \leq \rho(x) \leq \rho(w) = w.$$

Consequently, $\rho(x) \in$ fce$(w) \subset$ bdr$K$. For every $y \in x - K$ we have $0 \leq \rho(y) \leq \rho(x) \in$ bdr K. Hence,

$$\rho(x - K) \subset \text{fce}\rho(x).$$

2. Now, suppose that K is polyhedral and let $x \in L$. Since K_0 engenders L, there exist $u \in K_0$ such that $u - x \in K_0$. Hence, $x \leq u$ and then $\rho(x) \leq u$ and since $u \in K_0$, it follows that $\rho(x) \in K_0$.
Let $x \in L^+$. Take $u \in K$ with $x \leq u$. Since L separates L^+ and K, there exists $t \in [0, 1[$ such that $w = tx + (1 - t)u \in L$. Then, according to our above proof $\rho(w) \in K_0$. Since $x \leq w$, it follows that $0 \leq \rho(x) \leq \rho(w)$. In conclusion, $\rho(x) \in K_0$.

Corollary 1 *If K_1 and K_2 with the halfspaces L_1^+, respectively, L_2^+ defined as in the item 2 of Lemma 1 are two maximal faces of the polyhedral cone K and $\rho : \mathbb{R}^m \to K$ is an isotone retraction, then $L_1^+ \cap L_2^+$ is mapped by ρ onto the face $K_1 \cap K_2$.*

The next theorem shows that every isotone retraction cone in $\mathbb{R}^m$ is polyhedral. We shall show later that there are isotone retraction cones, which are polyhedral, but not simplicial.

Theorem 1 *If K is an isotone retraction cone in $\mathbb{R}^m$, then it is polyhedral.*

Proof (a) Denote by B_ε the open ball with the center in 0 and the radius ε, that is, the set

$$B_\varepsilon = \{x \in \mathbb{R}^m : \|x\| < \varepsilon\}.$$

We shall show that $\forall\, \varepsilon > 0 \;\exists\, \delta > 0$ such that

$$\forall\, v \in B_\delta \;\; \exists x \in K \cap (v + K) \text{ with } x \in B_\varepsilon.$$

To this end, take the nonempty open set $U \subset \operatorname{int} K \cap B_\varepsilon$, and let $x \in U$. Then, $x - U$ is a neighborhood of 0. Hence, there exists $\delta > 0$ with $B_\delta \subset x - U$. If $v \in B_\delta$, then $v \in x - U$ and hence, $x \in v + U \subset v + K$, and we are done.

(b) From the assertion in (a), it follows that for a sequence (v_n) with $v_n \to 0$, there exists a sequence (x_n) with $x_n \in K \cap (v_n + K)$ and $x_n \to 0$.

(c) Let L be a hyperplane meeting K in 0. Then, it can be easily seen that for $k \in K\backslash\{0\}$ the set $B = (k+L) \cap K$ will be a bounded closed convex subset of $k+L$, which is a so called *base of* K, which means that each nonzero element of K is the positive multiple of an element in $k + L$.

Consider the set B as a subset of $\mathbb{R}^{n-1}$. It is well known that B is the convex hull of their extremal points, and the set of exposed points of B is dense in the set of its extremal points by the theorem of Straszewicz [17]. (The point $z \in B$ is its exposed point if there exists a hyperplane in $\mathbb{R}^{n-1}$ which meets B in the single point z.)

The exposed points of B have the particularity that they generate exposed rays of K; that is, rays on the boundary of K along which K is "supported" by some hyperplane, in the sense that this hyperplane meets K in that ray. Such a ray is a face of K which can meet any other proper face of its at 0 only.

(d) The assertion that K is polyhedral is equivalent to the fact that B possesses a finite set of extremal points in $k + L$.

Assume that K is not polyhedral. Then, B has an infinity of extremal points. Let x_0 be an accumulation point of these extremal points. Then, according to the observation in (c), there is a sequence (x_n) of exposed points of B such that $x_n \to x_0$.

From (b), we have a sequence (y_n) with $y_n \in (x_0 - K) \cap (x_n - K)$ and $y_n \to x_0$.

(e) Suppose that $\rho : R^m \to K$ is an isotone retraction. Then, by Lemma 1, $\rho(y_n)$ is on the face of x_0 and on the face of x_n. Since the face of x_n is an exposed ray, by (c), we must have that $\rho(y_n) = 0\ \forall n$. On the other hand, from $y_n \to x_0$ and the continuity of ρ, it follows that $\rho(y_n) \to \rho(x_0) = x_0 \neq 0$. The obtained contradiction shows that K must be polyhedral.

3 Isotone Retractions and Simpliciality

The next lemma exhibits some more simple, but useful properties of isotone retractions onto cones. They will be used to prove items 1, 2, and 3 of Theorem 2.

Lemma 2 *Let $K \subset \mathbb{R}^m$ be a cone and $\rho : \mathbb{R}^m \to K$ isotone. We have as follows:*

1. *If $\rho(0) = 0$, then $\rho(-K) = \{0\}$ and $-K \subset \operatorname{im}(I - \rho)$. In particular, this is the case if ρ is a retraction.*
2. *Suppose that ρ is a retraction.*
   a) *If there is an $x \in K$ such that $x \leq \rho(x)$, then there exists $x^+ = 0 \vee x$ and $\rho(x) = x^+$.*
   b) *If $K \subset \mathbb{R}^m$ is a simplicial cone, then $\rho(x) \leq x^+$, for all $x \in \mathbb{R}^m$.*

Proof

1. Let $v \in -K$ arbitrary, or equivalently $v \leq 0$. By the isotonicity of ρ, we have $0 \leq \rho(v) \leq \rho(0) = 0$. Hence, $\rho(v) = 0$, or equivalently $\rho(-K) = \{0\}$. We also have

$$v = \rho(v) + (I - \rho)(v) = (I - \rho)(v) \in \operatorname{im}(I - \rho).$$

   Hence, $-K \subset \operatorname{im}(I - \rho)$.
2. a) Suppose that there is an $x \in \mathbb{R}^m$ such that $x \leq \rho(x)$. It is easy to see that $\rho(x)$ is an upper bound of the set $\{0, x\}$. Let u be an arbitrary upper bound of the set $\{0, x\}$. Then, $0 \leq u$ and $x \leq u$. Since ρ is a retraction, by the isotonicity of ρ, we have $\rho(x) \leq \rho(u) = u$. Hence, $\rho(x) = 0 \vee x = x^+$.
   b) Suppose that K is simplicial. Since $x \leq x^+$, by the isotonicity of ρ, we have $\rho(x) \leq \rho(x^+) = x^+$, for all $x \in \mathbb{R}^m$.

Although an isotone retraction cone is not necessarily simplicial, item 4 of the next theorem shows that in a still very general case the answer is yes. For this, we have to require an extra condition of sharpness for the complement of the isotone retraction ρ onto the cone, that is, the condition $(I - \rho)(\mathbb{R}^m) \cap -(I - \rho)(\mathbb{R}^m) = \{0\}$.

Theorem 2 *Let $K \subset \mathbb{R}^m$ be a cone and $\leq$ the order defined by K. We have as follows:*

1. *If K is simplicial and $\rho : \mathbb{R}^m \to K$ is defined by $\rho(x) = x^+$, then ρ is a continuous isotone retraction such that $x \leq \rho(x)$, for all $x \in \mathbb{R}^m$, and $I - \rho$ is isotone and sharp.*
2. *If there is an isotone retraction $\rho : \mathbb{R}^m \to K$ such that $x \leq \rho(x)$, for all $x \in \mathbb{R}^m$, then K is simplicial and $\rho(x) = x^+$, for all $x \in K$.*
3. *If there is a retraction $\rho : \mathbb{R}^m \to K$ such that ρ and $I - \rho$ are isotone, then K is simplicial and $\rho(x) = x^+$, for all $x \in K$.*
4. *If K is a cone and there is a continuous isotone retraction $\rho : \mathbb{R}^m \to K$ such that the complement $I - \rho$ of ρ is sharp, then K is simplicial.*

Proof

1. Suppose that K is simplicial. Let $\rho : \mathbb{R}^m \to K$ defined by $\rho(x) = x^+$. Obviously, ρ is a continuous isotone retraction, $x \leq \rho(x)$. Moreover, since
$$(I - \rho)(x) = x - x^+ = -x^- = -(-x)^+,$$
$I - \rho$ is isotone and sharp.
2. Suppose that there is an isotone retraction $\rho : \mathbb{R}^m \to K$ such that $x \leq \rho(x)$, for all $x \in \mathbb{R}^m$. By item 2a of Lemma 2, it follows that there exists $x^+ = 0 \vee x$ and $\rho(x) = x^+$, for all $x \in \mathbb{R}^m$. By using standard arguments, it can be easily shown that $x \vee y = (x - y)^+ + y$ for all $x, y \in \mathbb{R}^m$. Thus, K is simplicial.
3. Suppose that $\rho : \mathbb{R}^m \to K$ is a retraction such that ρ and $I - \rho$ are isotone. By item 2, it is sufficient to show that $x \leq \rho(x)$, for all $x \in \mathbb{R}^m$. Let $x \in \mathbb{R}^m$ be arbitrary. Since $K - K = \mathbb{R}^m$, it follows that there are $u, v \in K$ such that $x = u - v$. Hence, $x \leq u$. Since $I - \rho$ is isotone and ρ is a retraction, we have $x - \rho(x) \leq u - \rho(u) = 0$. Hence, $x \leq \rho(x)$.
4. Suppose that $\rho : \mathbb{R}^m \to K$ is a continuous isotone retraction. For all $y \in \mathbb{R}^m$ define $\rho_y : \mathbb{R}^m \to \mathbb{R}^m$ by $\rho_y(x) = y + \rho(x - y)$. Since ρ is continuous and isotone, ρ_y is also continuous and isotone. Moreover, $y \leq \rho_y(x)$. Let u and v be arbitrary elements in $\mathbb{R}^m$. We have to show that there exists $u \vee v$. If u and v are comparable the statement is trivial. Suppose that they are not comparable. First we remark that the set $\{u, v\}$ has an upper bound. Indeed, since $K - K = \mathbb{R}^m$, there exist $u_1, u_2, v_1, v_2 \in K$ such that $u = u_1 - u_2$ and $v = v_1 - v_2$. Hence, $u_1 + v_1$ is an upper bound of the set $\{u, v\}$. Let w be an arbitrary upper bound of the set $\{u, v\}$; that is, an arbitrary element of $(u + K) \cap (v + K)$. The mappings ρ_u and ρ_v are isotone. Moreover, $\rho_u(w) = u + \rho(w - u) = u + (w - u) = w$ and similarly $\rho_v(w) = w$. Consider the operators $\sigma = \rho_u \circ \rho_v$ and $\tau = \rho_v \circ \rho_u$. They are isotone because ρ_u and ρ_v are. Moreover, $\sigma(w) = \tau(w) = w$. Put $v_n = \tau^n(v)$, $u_1 = \rho_u(v)$ and $u_n = \sigma^{n-1}(u_1)$. We have $u \leq \rho_u(v) = u_1$. Also, $u \leq \rho_u(v)$ implies $v \leq \rho_v(u) \leq \rho_v \circ \rho_u(v) = v_1$ and therefore $u_1 = \rho_u(v) \leq \rho_u(v_1)$, or equivalently $u_1 \leq \rho_u \circ \rho_v \circ \rho_u(v) = \sigma(u_1) = u_2$. Bearing in mind that σ, τ are isotone $\sigma(w) = \tau(w) = w$, the relations
$$v \leq v_1 \leq \cdots \leq v_n \leq \cdots \leq w$$
and
$$u \leq u_1 \leq u_2 \leq \cdots \leq u_n \leq \cdots \leq w$$
can be verified by using mathematical induction. We further have
$$v_n = \tau^n(v) = (\rho_v \circ \rho_u)^n (v) = \rho_v \circ (\rho_u \circ \rho_v)^{n-1} \circ \rho_u(v)$$
$$= \rho_v \circ \sigma^{n-1}(u_1) = \rho_v(u_n) \tag{2}$$
and
$$u_{n+1} = \sigma(u_n) = \rho_u \circ \rho_v(u_n) = \rho_u(v_n). \tag{3}$$

It is well known that the cones in Euclidean spaces are also regular, that is, each monotone increasing and bounded sequence with respect to the order induced by the cones is also convergent. The cone K is also regular, hence as the sequences $\{u_n\}$ and $\{v_n\}$ are increasing and bounded above by w, and $u \leq u_n \leq w$ and $v \leq v_n \leq w$, there exists the limit

$$u^* = \lim_{n\to+\infty} u_n \text{ and } v^* = \lim_{n\to+\infty} v_n \tag{4}$$

such that

$$u \leq u^* \leq w \text{ and } v \leq v^* \leq w. \tag{5}$$

From the continuity of the mappings ρ_u and ρ_v and the relations (2), (3), and (4), it follows that $v^* = \rho_v(u^*)$ and $u^* = \rho_u(v^*)$. Hence, we have $v^* - u^* + u^* - v = \rho(u^* - v)$ and $u^* - v^* + v^* - u = \rho(v^* - u)$, or equivalently $(I - \rho)(a) = c$ and $(I - \rho)(b) = -c$, where $a = u^* - v$, $b = v^* - u$ and $c = u^* - v^*$. Thus, we have $c \in im(I - \rho) \cap im(\rho - I)$. Since $I - \rho$ is sharp, we have $c = 0$, or equivalently $u^* = v^*$. Hence, from relation (5), it follows that $u^* = v^* = u \vee v$.

Theorem 2 also shows that in case of a continuous isotone retraction ρ onto the cone, the requirement for $I - \rho$ to be isotone completely determines ρ. It is still an open question what would imply the isotonicity of $I - \rho$ only.

Lemma 3 *Let $K, L \subset \mathbb{R}^m$ be cones. If $\rho : \mathbb{R}^m \to K$ is a continuous K-isotone retraction such that $im(I - \rho) \subset L$, then K is simplicial and $-K \subset L$.*

Proof $\operatorname{im}(I - \rho) \subset L$ implies that $I - \rho$ is sharp. Hence, from item 1 of Theorem 2, we get that K is simplicial. By item 1 of Lemma 2, we also get $-K \subset L$.

The projection mapping onto a cone is characterized by the following theorem of Moreau [10].

Theorem [Moreau] *Let $K, L \subset \mathbb{R}^m$ be two mutually polar cones. Then, the following statements are equivalent:*

(i) $z = x + y,\ x \in K,\ y \in L$ *and* $\langle x, y\rangle = 0$,
(ii) $x = P_K(z)$ *and* $y = P_L(z)$.

Since the projection is a particular retraction, it is an interesting question whether it is possible to give extensions of Moreau's theorem for retractions or not. The next two theorems are a partial answer to this question in case of particular simplicial cones only.

Theorem 3 *Let $K, L \subset \mathbb{R}^m$ be mutually polar, cones. Then, K is a subdual simplicial cone if and only if there is a $\rho_K : \mathbb{R}^m \to K$ continuous K-isotone retraction and a $\rho_L : \mathbb{R}^m \to L$ mapping such that $x = \rho_K(x) + \rho_L(x)$.*

Proof

1. Suppose that K is a subdual simplicial cone. Choose $\rho_K : \mathbb{R}^m \to K$ defined by $\rho_K(x) = x^+$ and $\rho_L : \mathbb{R}^m \to L$ defined by $\rho_L(x) = -x^-$ (with respect to the simplicial structure of K). It is easy to see that ρ_K is K-isotone. Since K is subdual, it follows that ρ_L is well defined.

2. Conversely, suppose that there is a $\rho_K : \mathbb{R}^m \to K$ continuous K-isotone retraction and a $\rho_L : \mathbb{R}^m \to L$ mapping such that $x = \rho_K(x) + \rho_L(x)$. Then, $im(I - \rho_K) \subset L$. Hence, Lemma 3 implies that K is simplicial and subdual.

Theorem 4 *Let $K, L \subset \mathbb{R}^m$ be mutually polar cones. Then, K is a self-dual simplicial cone if and only if there is a $\rho_K : \mathbb{R}^m \to K$ continuous K-isotone retraction and a $\rho_L : \mathbb{R}^m \to L$ continuous L-isotone retraction such that $x = \rho_K(x) + \rho_L(x)$. In this case, $\rho_K(x) = x^+$ (with respect to the simplicial structure of K).*

Proof

1. Suppose that K is a self-dual simplicial cone. Choose $\rho_K : \mathbb{R}^m \to K$ defined by $\rho_K(x) = x^+$ and $\rho_L : \mathbb{R}^m \to L$ defined by $\rho_L(x) = -x^-$ (with respect to the simplicial structure of K). It is easy to see that ρ_K is K-isotone. Since K is self-dual, it follows that ρ_L is well defined. Moreover, since $\rho_L(x) = -(-x)^+$, it follows that ρ_L is K-isotone. But, since K is self-dual, it follows that $-K = L$. Hence, ρ_L is L-isotone too.
2. Conversely, suppose that there is a $\rho_K : \mathbb{R}^m \to K$ continuous K-isotone retraction and a $\rho_L : \mathbb{R}^m \to L$ continuous L-isotone retraction such that $x = \rho_K(x) + \rho_L(x)$. By Theorem 3, we have that both K and L are simplicial subdual cones. Hence, K is a self-dual simplicial cone. Since $I - \rho_K = \rho_L$ is L-isotone and $K = -L$, it follows that $I - \rho_K$ is K-isotone too. Hence, from item 3 of Theorem 2 it follows that $\rho_K(x) = x^+$ (with respect to the simplicial structure of K).

Example 1 Let $K \subset \mathbb{R}^n$ be a subdual simplicial cone. Then, the mapping $\rho : \mathbb{R}^n \to K$ defined by

$$\rho(x) = \begin{cases} \left(1 - \dfrac{\|x^-\|}{\|x^+\|}\right) x^+ & \text{if} \quad \|x^+\| > \|x^-\|, \\ 0 & \text{if} \quad \|x^+\| \le \|x^-\|. \end{cases}$$

is a continuous isotone retraction, but in general $I - \rho$ is not sharp.

Proof Obviously $\rho(x) \in K$, for all $x \in \mathbb{R}^n$. Hence, ρ is well defined. If $x \in K \setminus \{0\}$, then $x^+ = x$ and $x^- = 0$. Hence, $\|x^+\| > \|x^-\|$. By the definition of ρ we have $\rho(x) = x$. If $x = 0$, then $x^+ = x^- = 0$. Hence, $\|x^+\| \le \|x^-\|$. By the definition of ρ we have $\rho(x) = 0 = x$. Thus, $\rho(x) = x$, for all $x \in K$. It follows that ρ is a retraction. Obviously, ρ is continuous. Next, we prove that ρ is isotone. Suppose that $x, y \in \mathbb{R}^n$ such that $x \le y$. We have to show that $\rho(x) \le \rho(y)$. Since

$$0 \le x^+ \le y^+ \tag{6}$$

and $0 \le y^- \le x^-$, we have

$$\|x^+\| \le \|y^+\| \tag{7}$$

and

$$\|y^-\| \leq \|x^-\|. \tag{8}$$

We consider three different cases:

1. $\|y^+\| > \|y^-\|$ and $\|x^+\| > \|x^-\|$. Then, by the definition of ρ, (6), (7), and (8) we have $\rho(x) \leq \rho(y)$.
2. $\|y^+\| > \|y^-\|$ and $\|x^+\| \leq \|x^-\|$. Then, by the definition of ρ we have $\rho(x) = 0 \leq \rho(y)$.
3. $\|y^+\| \leq \|y^-\|$. Then, by (7) and (8) we have $\|x^+\| \leq \|y^+\| \leq \|y^-\| \leq \|x^-\|$. Hence, by the definition of ρ we have $\rho(x) = 0 \leq 0 = \rho(y)$.

In general, $I - \rho$ is not sharp. Indeed, if $K = \mathbb{R}^2_+$, $x = (1, -1) \notin K$, $y = (-1, 1) \notin K$, then by the definition of ρ we have $x + y = 0 = 0 + 0 = \rho(x) + \rho(y)$. Hence, $0 \neq (I - \rho)(x) = -(I - \rho)(y)$, which shows that $\operatorname{im}(I - \rho) \cap \operatorname{im}(-(I - \rho)) \neq \{0\}$. Therefore, $I - \rho$ is not sharp.

The next example shows the surprising fact that there are isotone retractions ρ onto cones whose complement $I - \rho$ is not sharp, but for an arbitrarily small "homotopic deformation" of ρ, we get an isotone retraction whose complement is sharp.

Example 2 Consider the previous example with $\mathbb{R}^n = \mathbb{R}^2$ and $K = \mathbb{R}^2_+$. Let $t \in [0, 1[$ and $\rho_t : \mathbb{R}^2 \to \mathbb{R}^2_+$ be defined by $\rho_t(x) = t\rho(x) + (1 - t)x^+$. Then, ρ_t is a continuous isotone retraction such that $I - \rho_t$ is sharp.

Proof If $t = 0$ the assertion is obvious. Therefore, we can suppose that $t \in]0, 1[$. By the previous example ρ is a continuous isotone retraction. The mapping $x \mapsto x^+$ bears the same properties. Hence, ρ_t bears the same properties too. It remains to prove that $I - \rho$ is sharp. First, note that

$$x^+ - \rho(x) = \frac{\|x^-\|}{\|x^+\|} x^+$$

implies that

$$\langle x^+ - \rho(x), x^- \rangle = 0 \tag{9}$$

and

$$\|x^+ - \rho(x)\| = \|x^-\|. \tag{10}$$

By using (9) and (10), we get

$$\begin{aligned}
&\langle x - \rho_t(x), x^+ - \rho_t(x) \rangle \\
&= \langle tx + (1 - t)x - t\rho(x) - (1 - t)x^+, tx^+ + (1 - t)x^+ - t\rho(x) - (1 - t)x^+ \rangle \\
&= \langle t(x - \rho(x)) - (1 - t)x^-, t(x^+ - \rho(x)) \rangle = t^2 \langle x - \rho(x), x^+ - \rho(x) \rangle \\
&= t^2 \langle x^+ - \rho(x) - x^-, x^+ - \rho(x) \rangle = t^2 \|x^+ - \rho(x)\|^2 = t^2 \|x^-\|^2.
\end{aligned}$$

By tidying up, we have

$$\langle x - \rho_t(x), x^+ - \rho_t(x)\rangle = t^2\|x^-\|^2. \tag{11}$$

By *reductio ad absurdum*, suppose that $I - \rho_t$ is not sharp. Then, there exists $x, y \notin K$ such that

$$x + y = \rho_t(x) + \rho_t(y). \tag{12}$$

Denote the one-dimensional face containing the element $z \neq 0$ of the boundary of K by $fce(z)$. We can suppose that $x, y \notin -K$. Indeed, suppose that $x \in -K$ (the argument is similar if we suppose that $y \in -K$). Then, by item 1 of Lemma 2 we have $\rho_t(x) = 0$. Thus, (12) implies that $y \in K$, which is a contradiction. Hence, $x^+, y^+ \neq 0$. Since ρ_t is an isotone retraction, item 2b of Lemma 2 implies $0 \leq \rho_t(x) \leq x^+$. Hence, $0 \leq x^+ - \rho_t(x) \leq x^+$. Then, $\rho_t(x), x^+ - \rho_t(x) \in \mathrm{fce}(x^+)$. Similarly, $\rho_t(y), y^+ - \rho_t(y) \in \mathrm{fce}(y^+)$. We consider two cases.

1. $\mathrm{fce}(x^+) = \mathrm{fce}(y^+)$
   Then, $x^+ - \rho_t(x) = r(y^+ - \rho_t(y))$ for some $r > 0$. By using the latter relation and (11), it follows that

$$\langle x - \rho_t(x), y^+ - \rho_t(y)\rangle > 0 \tag{13}$$

   and

$$\langle y - \rho_t(y), y^+ - \rho_t(y)\rangle > 0 \tag{14}$$

   By summing up (13) and (14) we get

$$0 = \langle x + y - (\rho_t(x) + \rho_t(y)), y^+ - \rho_t(y) > 0$$

   which is a contradiction.
2. $\mathrm{fce}(x^+) \perp \mathrm{fce}(y^+)$
   Then, $y^+ - y = r(x^+ - \rho_t(x))$ and $x^+ - x = s(y^+ - \rho_t(y))$, for some $r, s > 0$. From (12) and the last two relations we get

$$\rho_t(x) + \rho_t(y) = x + y = (1 - r)x^+ + (1 - s)y^+ + r\rho_t(x) + s\rho_t(y),$$

   or equivalently $(1 - r)\left(x^+ - \rho_t(x)\right) = (1 - s)(\rho_t(y) - y^+)$. The left hand side of the latter equality is in $K \setminus \{0\}$ and the right hand side is in $-K \setminus \{0\}$. But, this contradicts the pointedness of K.

The contradictions obtained in both of the cases show that $I - \rho_t$ is sharp.

Next, we present an even simpler example than Example 1 for an isotone retraction ρ onto a cone whose complement $I - \rho$ in general is not sharp.

Example 3 Let $K \subset \mathbb{R}^n$ be a subdual simplicial cone. Then, the mapping $\rho : \mathbb{R}^n \to K$ defined by

$$\rho(x) = \frac{x^+}{1 + \|x^-\|}$$

is a continuous isotone retraction, but in general $I - \rho$ is not sharp.

Proof Obviously, ρ is a continuous retraction. From (6) to (8), it follows that ρ is isotone. In general, $I - \rho$ is not sharp. Indeed, if $K = \mathbb{R}^2_+$, $x = (3, -2) \notin K$ and $y = (-2, 3) \notin K$, then by the definition of ρ we have

$$\rho(x) + \rho(y) = \frac{(3,0)}{1+2} + \frac{(0,3)}{1+2} = (1,1) = x + y.$$

Similarly to the proof of the previous example, we obtain that $I - \rho$ is not sharp.

4 Construction of Nonsimplicial Isotone Retraction Cones

4.1 A Preparatory Result in the Plane

Let be $\mathbb{R}^2$ the Euclidean plane endowed with a Cartesian reference system sOt.

Take $c, d \in \mathbb{R}^2$ with $c = (-h, j)^T$, $d = (k, l)^T$ and h, j, k, l positive reals.

Let $\kappa \subset \mathbb{R}^2$ be the cone defined by

$$\kappa = \left\{x = (s,t)^T \in \mathbb{R}^2 : c^T x \geq 0;\ d^T x \geq 0\right\}.$$

κ is placed in the closed superior half-plane and is generated by a ray placed, except 0, in the first, and a ray placed in the second open quadrant of the reference system. We can take $c' = (j, h)^T$ and $d' = (-l, k)^T$ as vectors generating κ.

A cone κ of this form will be said to be a *standard position cone* in the reference system sOt.

We define the mapping $\varphi : \mathbb{R}^2 \to \kappa$ by projecting parallel with the horizontal axis onto κ, considering that the closed lower half-plane is projected into 0.

$$\varphi((s,t)^T) = \begin{cases} (s,t)^T, & \text{if } -hs + jt \geq 0;\quad ks + lt \geq 0; \\ \left(-\frac{tl}{k}, t\right)^T, & \text{if } ks + lt < 0;\quad t > 0; \\ \left(\frac{tj}{h}, t\right)^T, & \text{if } -hs + jt < 0;\quad t > 0; \\ 0 = (0,0)^T, & \text{if } t < 0. \end{cases}$$

Proposition 1 *The mapping φ is isotone with respect to the order induced by κ, that is, $x \leq y$ implies $\varphi(x) \leq \varphi(y)$ with $\leq$ the order relation induced by κ.*

Proof From the meaning of φ, as well as from the above formulas, it follows that it is continuous.

By the geometric picture, the isotonicity of φ is a rather intuitive. An analytic verification implies the consideration of some cases.

Suppose that $(s_1, t_1)^T = x \leq y = (s_2, t_2)^T$.

(a) Since $(s,t)^T \in \kappa$ implies $t \geq 0$, from $x \leq y$ it follows that $t_1 \leq t_2$.

(b) The condition $x \in \kappa$ implies $y \in \kappa$, whereby

$$\varphi(x) = x \leq y = \varphi(y).$$

If $t_1 \leq 0$, then $\varphi(x) = 0 \leq \varphi(y)$ for any y.

Hence, the only cases to be considered are those with $x \notin \kappa$ and $t_1 > 0$.

The open upper half-plane $t > 0$ of the reference system sOt is partitioned in the domains: $\kappa \setminus \{0\}$, $M_1 = \{(s,t)^T : ks + lt < 0,\ t > 0\} = \{d^T x < 0,\ t > 0\}$ and $M_2 = \{(s,t)^T : -hs + jt < 0,\ t > 0\} = \{c^T x < 0,\ t > 0\}$.

(c) Let us see that x and y cannot be in different domains M_i, $i = 1, 2$. Assuming for instance that $x \in M_1$ and $y \in M_2$, we will have $d^T x < 0$ and $c^T y < 0$. From the hypothesis on x and y, we also have $c^T(y - x) \geq 0$, and hence, it follows $c^T x \leq c^T y < 0$, whereby $x \in -\kappa$ contradicting the assumption $t_1 > 0$. The case $x \in M_2$ and $y \in M_1$ similarly yields a contradiction.

(d) According to (b), we can suppose that $t_1 > 0$.

Consider the case $x \in M_1$. Due to (c), we have to consider two cases.

(d1) $y \in M_1$. Then, we have

$$\varphi(x) = \left(-\frac{lt_1}{k}, t_1\right)^T,\ \varphi(y) = \left(-\frac{lt_2}{k}, t_2\right)^T.$$

Thus,

$$\varphi(y) = \frac{t_2}{t_1}\varphi(x).$$

By assertion (a) and the fact that $\varphi(x) \in \kappa$, it follows that

$$\varphi(x) \leq \frac{t_2}{t_1}\varphi(x) = \varphi(y).$$

(d2) $y \in \kappa$. Observe that, by the continuity of φ, the relation $\varphi(x) \leq \varphi(y)$ holds also in the case when y is on the bounding ray $F = \{\lambda d' : \lambda \geq 0\} \subset \kappa$ of M_1.

If $y \in int\kappa$, then let z be the intersection point of the line segment xy with F. Then, $x \leq z \leq y$, whereby using the above observation it follows

$$\varphi(x) \leq \varphi(z) = z \leq y = \varphi(y).$$

The above considered cases cover by symmetry all the situations which can occur.

Remark 1 Observe that the isotone retraction φ constructed in the above proposition can also be written as

$$\varphi = (\varphi_1, \varphi_2)^T,$$

where

$$\varphi_1\left((s,t)^T\right) = \begin{cases} s, & \text{if } -hs + jt \geq 0;\ \ ks + lt \geq 0; \\ -\frac{tl}{k}, & \text{if } ks + lt < 0;\ \ t > 0; \\ \frac{tj}{h}, & \text{if } -hs + jt < 0;\ \ t > 0; \\ 0, & \text{if } t < 0, \end{cases}$$

and

$$\varphi_2\left((s,t)^T\right) = \begin{cases} t & \text{if } t > 0, \\ 0 & \text{if } t \leq 0. \end{cases}$$

4.2 A Nonsimplicial Isotone Retraction Cone in $\mathbb{R}^3$

Let $\mathbb{R}^3$ be the three dimensional Euclidean space endowed with the Cartesian reference system $Ox_1x_2x_3$. Consider the sets W_1 and W_2 in $\mathbb{R}^3$ defined by

$$W_1 = \left\{(x_1,x_2,x_3)^T \in \mathbb{R}^3 : \ (x_1,x_3) \in \kappa_1\right\},$$
$$W_2 = \left\{(x_1,x_2,x_3)^T \in \mathbb{R}^3 : \ (x_2,x_3) \in \kappa_2\right\},$$

where κ_i is a standard position cone in the reference system $x_i O x_3, \ i = 1, 2$. Denote with $\varphi^i = \left(\varphi_1^i, \varphi_2^i\right)^T : x_i O x_3 \to \kappa_i$ the isotone mapping defined similarly with φ in the preceding section, $i = 1, 2$.

Remark 2 According to Remark 1, the second coordinate functions of φ^1 and φ^2 depend only on x_3 and $\varphi_2^1 = \varphi_2^2$. This will be of crucial importance in the constructions that follows. To simplify the exposition, we shall use for these coordinate functions the notation $\varphi_2^1\left((x_1,x_3)^T\right) = \varphi_2^2\left((x_2,x_3)^T\right) = \varphi_3(x_3)$.

Let

$$K = W_1 \cap W_2.$$

Then, K is polyhedral cone in $\mathbb{R}^3$ with four proper maximal faces. The relation $(x_1,x_2,x_3)^T = x \le y = (y_1,y_2,y_3)^T$ induced by K is equivalent with the pair of relations $(y_1,y_3)^T - (x_1,x_3)^T \in \kappa_1$ and $(y_2,y_3)^T - (x_2,x_3)^T \in \kappa_2$.

Consider the mapping $\rho : \mathbb{R}^3 \to K$ defined by

$$\rho\left((x_1,x_2,x_3)^T\right) = \left(\varphi_1^1((x_1,x_3)^T), \varphi_1^2\left((x_2,x_3)^T\right), \varphi_3(x_3)\right)^T. \tag{15}$$

Proposition 2 *The mapping $\rho : \mathbb{R}^3 \to K$ is a continuous retraction, isotone with respect to the order induced by K.*

Proof The continuity of ρ is the consequence of the continuity of φ^1 and φ^2.

The relation

$$u = (u_1,u_2,u_3)^T \le (v_1,v_2,v_3)^T = v$$

is equivalent with the following two relations:

$$(v_1,v_3)^T - (u_1,u_3)^T \in \kappa_1, \tag{16}$$

and

$$(v_2,v_3)^T - (u_2,u_3)^T \in \kappa_2. \tag{17}$$

Let us apply Proposition 1 in $x_1 O x_3$ to κ_1, and φ^1 to conclude from (16) that

$$\left(\varphi_1^1((v_1,v_3)^T), \varphi_3(v_3)\right)^T - \left(\varphi_1^1((u_1,u_3)^T), \varphi_3(u_3)\right)^T \in \kappa_1. \tag{18}$$

Similarly, by (17) in x_2O, x_3 the same proposition concludes for κ_2 and φ^2 that

$$\left(\varphi_1^2((v_2, v_3)^T), \varphi_3(v_3)\right)^T - \left(\varphi_1^2((u_2, u_3)^T), \varphi_3(u_3)\right)^T \in \kappa_2. \tag{19}$$

The relations (18) and (19) show that

$$\left(\varphi_1^1((v_1, v_3)^T), \varphi_1^2\left((v_2, v_3)^T\right), \varphi_3(v_3)\right)^T \leq \left(\varphi_1^1((v_1, v_3)^T), \varphi_1^2\left((v_2, v_3)^T\right), \varphi_3(v_3)\right)^T.$$

The obtained relation proves, via the expression (15), the isotonicity of ρ and completes the proof.

Remark 3 The isotone retraction $\rho : \mathbb{R}^3 \to K$ constructed in the above proposition has its null set contained in the lower half space $\left\{x = (x_1, x_2, x_3)^T \in \mathbb{R}^3 : \ x_3 \leq 0\right\}$.

4.3 The Case of the Polyhedral Cone in $\mathbb{R}^3$ with Four Maximal Faces

Using the construction described above we consecrate this section to prove the following.

Theorem 5 *In the three dimensional Euclidean space $\mathbb{R}^3$, every polyhedral cone K having at most four proper maximal faces admits an isotone retraction ρ, whose null set is contained in a closed half-space.*

Proof We can consider only the case when K possesses exactly four maximal faces, in the case of three maximal faces the cone is latticial and the positive part mapping is the required retraction.

We begin by pointing out some simple facts about polyhedral cones.

If W is a polyhedral cone in $\mathbb{R}^3$ and A is a nonsingular linear map, then $A(W)$ is a polyhedral cone with the same configuration of faces as W.

If ρ is an isotone retraction onto W, then $A \circ \rho \circ A^{-1}$ is an isotone retraction with respect to the order relation defined by the cone $A(W)$ onto $A(W)$.

Indeed, if $w - v \in A(W)$, then $A^{-1}w - A^{-1}v \in W$ and hence $\rho \circ A^{-1}w - \rho \circ A^{-1}v \in W$ and thus $A \circ \rho \circ A^{-1}w - A \circ \rho \circ A^{-1}v \in A(W)$, that is $A \circ \rho \circ A^{-1}$ is $A(W)$-isotonic.

If $w \in A(W)$, then $A^{-1}w \in W$, hence $\rho \circ A^{-1}w = A^{-1}w$ and thus $A \circ \rho \circ A^{-1}w = w$.

Finally, for each $x \in \mathbb{R}^m$, $\rho \circ A^{-1}x \in W$ accordingly $A \circ \rho \circ A^{-1}x \in A(W)$.

Let K be the cone in the statement of the theorem. Consider a reference system in $\mathbb{R}^3$ determined as follows. Let two axes of the system be the lines obtained by intersection of the hyperplanes determined by opposite maximal proper faces of K, and the third axis be a line through the origin which contains in its positive semiaxis a point in the interior of K. Let us transform this reference system by the linear map B into a Cartesian one. With $B(K)$ represented in this Cartesian system we get a configuration as in Proposition 2. Hence, $B(K)$ admits a $B(K)$-isotone retraction $r : \mathbb{R}^3 \to B(K)$ with the set of zeros contained in a closed half-space (see Remark 3).

Now, if $A = B^{-1}$, then $K = A \circ B(K)$ and $\rho = A \circ r \circ A^{-1} : \mathbb{R}^3 \to K$ is an isotone retraction with respect to the order relation defined by K onto K. The null set of r is transformed by $A \circ r \circ A^{-1}$ in the null set of ρ. Thus, the retraction ρ has the set of its zeros in a closed half-space of $\mathbb{R}^3$.

References

1. Bernau, S.J.: Isotone projection cones. In: Martinez, J. (ed.) Ordered Algebraic Structures, pp. 3–11. Springer, Netherlands (1991)
2. Dattorro, J.: Convex Optimization and Euclidean Distance Geometry. Meboo, USA (2010). (COandEDG version 2010.06.04)
3. Isac, G., Németh, A.B.: Monotonicity of metric projections onto positive cones of ordered Euclidean spaces. Arch. Math. **46**(6), 568–576 (1986)
4. Isac, G., Németh, A.B.: Every generating isotone projection cone is latticial and correct. J. Math. Anal. Appl. **147**(1), 53–62 (1990)
5. Isac, G., Németh, A.B.: Isotone projection cones in Hilbert spaces and the complementarity problem. Boll. Un. Mat. Ital. **B7**(4), 773–802 (1990)
6. Isac, G., Németh, A.B.: Projection methods, isotone projection cones, and the complementarity problem. J. Math. Anal. Appl. **153**(1), 258–275 (1990)
7. Isac, G., Németh, S.Z.: Regular exceptional family of elements with respect to isotone projection cones in Hilbert spaces and complementarity problems. Optim. Lett. **2**(4), 567–576 (2008)
8. Isac, G., Persson, L.-E.: On an inequality related to the isotonicity of the projection operator. J. Approx. Theory **86**, 129–143 (1996)
9. Isac, G., Persson, L.-E.: Inequalities related to isotonicity of projection and antiprojection operators. Math. Inequal. Appl. **1**, 85–97 (1998)
10. Moreau, J.J.: Décomposition orthogonale d'un espace hilbertien selon deux cônes mutuellement polaires. C. R. Acad. Sci. **255**, 238–240 (1962)
11. Németh, S.Z.: Itèrative methods for nonlinear complementarity problems on isotone projection cones. J. Math. Anal. Appl. **350**(1), 340–347 (2009)
12. Németh, S.Z. Isotone retraction cones in Hilbert spaces. Nonlinear Anal. Theory Meth. Appl. **73**(2), 495–499 (2010)
13. Németh, S.Z.: Characterization of latticial cones in Hilbert spaces by isotonicity and generalized infimum. Acta Math. Hung. **127**(4), 376–390 (2010)
14. Németh, S.Z., Németh, A.B.: How to project onto an isotone projection cone. Linear Alg. Appl. **433**(1), 41–51 (2010)
15. Németh, S.Z., Németh, A.B.: About the existence of an isotone retraction onto a convex cone. J. Convex Anal. **18**, 707–720 (2011)
16. Németh, S.Z., Németh, A.B.: A duality between the metric projection onto a convex cone and the metric projection onto its dual. J. Math. Anal. Appl. **392**(2) 171–178 (2012)
17. Straszewicz, S.: Über exponierte Punkte abgeschlossener Punktmengen. Fund. Math. **24**, 139–143 (1935)
18. Youdine, A.: Solution des deux problèmes de la theéorie des espaces semi-ordonnés. Comptes Rendus (Doklady) Acad. Sci. URSS **23**, 418–422 (1939)

On the Circle Preserving Property of Möbius Transformations

Nihal Yılmaz Özgür

Abstract This paper is mainly concerned with the study of circle-preserving property of Möbius transformations acting on $\widehat{\mathbf{R}}^n = \mathbf{R}^n \cup \{\infty\}$. The circle-preserving property is the most known invariant characteristic property of Möbius transformations. Obviously, a Möbius transformation acting on $\widehat{\mathbf{R}}^n$ is circle-preserving. Recently, for the converse statement, some interesting and nice results have been obtained. Here, we investigate these studies. We consider the relationships between Möbius transformations and sphere-preserving maps in $\widehat{\mathbf{R}}^n$ since the studies about the circle-preserving property of maps in $\widehat{\mathbf{R}}^n$ are related to the study of sphere-preserving maps. For the case $n = 2$, we also consider the problem whether or not the circle-preserving property is an invariant characteristic property of Möbius transformations for the circles corresponding to any norm function $\|.\|$ on $\mathbf{C}$.

Keywords Möbius transformation · Circle-preserving property · Circle-preserving map · Sphere-preserving map

1 Introduction

Let $\mathbf{C}$ and $\widehat{\mathbf{C}} = \mathbf{C} \cup \{\infty\}$ be the complex plane and the extended complex plane, respectively. Möbius transformations, also known as linear fractional transformations, are rational functions of the form

$$T(z) = \frac{az+b}{cz+d}, \tag{1}$$

where the coefficients a, b, c, d are complex numbers satisfying $ad - bc \neq 0$. They are the automorphisms of $\widehat{\mathbf{C}}$, that is, the meromorphic bijections $T : \widehat{\mathbf{C}} \to \widehat{\mathbf{C}}$. Each Möbius transformation T is also a directly conformal homeomorphism of $\widehat{\mathbf{C}}$ onto itself.

Möbius transformations have lots of interesting geometric properties. Their most basic geometric property is that Möbius transformations map circles to circles where

N. Yılmaz Özgür (✉)
Faculty of Science and Literature, Department of Mathematics,
Balıkesir University, 10145 Balıkesir, Turkey
e-mail: nihal@balikesir.edu.tr

T. M. Rassias, P. M. Pardalos (eds.), *Mathematics Without Boundaries*,
DOI 10.1007/978-1-4939-1106-6_17,

straight lines are considered to be circles through ∞, (see [5, 9, 10, 20, 21, 25]). It is well known that the circle-preserving property is an invariant characteristic property of Möbius transformations.

For example the following results are known.

Theorem 1 [8] *Every arbitrary one to one correspondence between the points of a circular disc C and a bounded point set C′ by which circles lying completely in C are transformed into circles lying in C′ must always be either a Möbius transformation* $T(z)$ *or* $T(\overline{z})$.

Theorem 2 [1] *If* $f : \widehat{\mathbf{C}} \to \widehat{\mathbf{C}}$ *is a circle-preserving map, then* f *is a Möbius transformation if and only if* f *is a bijection.*

Theorem 3 [13] *If* $f : \widehat{\mathbf{C}} \to \widehat{\mathbf{C}}$ *is a circle-preserving map, then* f *is a Möbius transformation if and only if* f *is a nonconstant meromorphic function.*

Recently, several new invariant characteristic properties of Möbius transformations have been given using some new geometric concepts (see [6, 7, 14–17, 24, 26, 27, 31, 32]). In [32], it was given a survey of such characterizations by means of the notions of "Apollonius points of $(2n-1)$-gons" and "Apollonius $2n$-gons."

Let $\mathbf{R}^n$ denote real n dimensional space and $\widehat{\mathbf{R}}^n$ denote $\mathbf{R}^n \cup \{\infty\}$. In this paper, we focus on the circle-preserving property of Möbius transformations acting on $\widehat{\mathbf{R}}^n = \mathbf{R}^n \cup \{\infty\}$. A map $f : \widehat{\mathbf{R}}^n \to \widehat{\mathbf{R}}^n$ is said to be r-sphere preserving if f maps every r-dimensional sphere onto an r-dimensional sphere. As a particular case, when $r = 1$ we call the corresponding map f as a circle-preserving map in $\widehat{\mathbf{R}}^n$. From now on we assume that $n \geq 2$ and $1 \leq r < n$.

Obviously a Möbius transformation acting on $\widehat{\mathbf{R}}^n$ is circle preserving. Recently, for the converse statement, some interesting and nice results have been obtained (see [4, 19, 22, 28]). For example, in [19], Jason Jeffers extended Theorem 2 to $\widehat{\mathbf{R}}^n$ and obtained the following theorem.

Theorem 4 [19] *Suppose that* $f : \widehat{\mathbf{R}}^n \to \widehat{\mathbf{R}}^n$ $(n \geq 2)$ *is a bijection that preserves circles. Then,* f *is a Möbius transformation.*

In [22], Baokui Li and Yuefei Wang introduced the notion of a degenerate map and proved the equivalences of a nondegenerate circle-preserving map and a Möbius transformation in $\widehat{\mathbf{R}}^n$. Also they conjectured that degenerate circle-preserving maps do not exist or equivalently, f is a Möbius transformation if and only if f is a circle-preserving map. In [28], Guowu Yao showed that the conjecture is not true. Here, we consider the nondegenerate maps in $\widehat{\mathbf{R}}^n$ and then we construct degenerate circle-preserving maps.

On the other hand, we give a brief account of the relationships between Möbius transformations and sphere-preserving maps in $\widehat{\mathbf{R}}^n$ since the studies about the circle-preserving property of maps in $\widehat{\mathbf{R}}^n$ are related to the study of sphere-preserving maps.

In Sect. 3, we consider the case $n = 2$ separately. The circle-preserving property of Möbius transformations have been investigated extensively in this case, (see [1, 8, 11, 12, 13, 25]). As we know all norms on $\mathbf{C}$ are equivalent, our second problem is to find the answer to the question whether or not the circle preserving property is an invariant characteristic property of Möbius transformations for the circles corresponding to any norm function $\|.\|$ on $\mathbf{C}$. We consider the real linear space structure of the complex plane $\mathbf{C}$ and investigate the answer of the above question. We see that the circle-preserving property is not an invariant characteristic property of Möbius transformations except in the Euclidean case. Then we give the complete classification of the images of circles under the Möbius transformations. We introduce a new class of plane curves which are called as MC-curves.

2 Möbius Transformations Acting on $\widehat{\mathbf{R}}^n$

In [12], Joel Gibbons and Cary Webb obtained an n-dimensional theorem which has no injectivity or continuity hypothesis and no surjectivity hypothesis other than the existence of $2n+2$ image points. This theorem states that if a circle-preserving function from S^n to S^{n+m}, $m \geq 0$, has at least $2n+2$ image points distributed regularly (see Theorem 3 for the precise meaning of this term) throughout the sphere S^{n+m}, then that function must be a Möbius transformation $T(z)$ acting on S^n or $T(\overline{z})$. As it has been noted that an improvement of this theorem is possible by reducing the number of required image points from $2n+2$ to $2n+1$ (see [12] for more details).

In a very recent article [22], Baokui Li and Yuefei Wang gave a condition for a circle-preserving map to be a Möbius transformation. To do this, they introduced two new classes of maps called as degenerate and nondegenerate. At first we begin to investigate the relationships between these new classes of maps and the circle-preserving property of Möbius transformations.

2.1 Nondegenerate Maps and the Circle-Preserving Property

Definition 1 [22] A circle-preserving map $f : \widehat{\mathbf{R}}^n \to \widehat{\mathbf{R}}^n$ $(n \geq 2)$ is called degenerate if its image $f(\widehat{\mathbf{R}}^n)$ is a circle. Otherwise f is called nondegenerate.

In [22], Baokui Li and Yuefei Wang proved that a circle-preserving map is a Möbius transformation as long as its image set of $\widehat{\mathbf{R}}^n$ is not a circle. More precisely they proved the following theorem.

Theorem 5 [22] *For any circle-preserving map $f : \widehat{\mathbf{R}}^n \to \widehat{\mathbf{R}}^n$ $(n \geq 2)$, the followings are equivalent:*

(1) *f is a nondegenerate map.*

(2) *f is a Möbius transformation.*

At first we consider the proof of Theorem 5 in the case $n = 2$. We can restate the theorem for this case as

Theorem 6 [22] *A circle-preserving map $f : \widehat{\mathbf{C}} \to \widehat{\mathbf{C}}$ is a Möbius transformation if and only if f is nondegenerate.*

In this case, the proof of the theorem is depend on the following lemmas.

Lemma 1 [22] *If a circle-preserving map* $f : \widehat{\mathbf{C}} \to \widehat{\mathbf{C}}$ *is nondegenerate, then* f *is surjective.*

Lemma 2 [22] *If a circle-preserving map* $f : \widehat{\mathbf{C}} \to \widehat{\mathbf{C}}$ *is nondegenerate, then* f *is injective.*

Let f be a circle-preserving map. By Lemmas 1 and 2, if f is nondegenerate then f is bijective and by Theorem 2 f is a Möbius transformation. The converse of the theorem is obvious. Hence, we obtain the proof of Theorem 6.

For the case $n > 2$, the proof of Theorem 5 is related to the study of sphere-preserving maps in $\widehat{\mathbf{R}}^n$.

In [22], for the proof of Theorem 5 it was used the following theorem.

Theorem 7 [4] *A map* $f : \widehat{\mathbf{R}}^n \to \widehat{\mathbf{R}}^n$ *is a Möbius transformation if and only if it is locally sphere preserving.*

In this case, the proof depends on the following lemma.

Lemma 3 [22] *If a circle-preserving map* $f : \widehat{\mathbf{R}}^n \to \widehat{\mathbf{R}}^n$ *is nondegenerate, then* f *is* r*-sphere preserving* $(2 \le r \le n)$.

Now the proof of Theorem 5 follows by Lemma 3 and Theorem 7.

So we have seen the equivalences of a nondegenerate circle-preserving map and a Möbius transformation in $\widehat{\mathbf{R}}^n$.

In [22], Baokui Li and Yuefei Wang conjectured that degenerate circle-preserving maps do not exist, or equivalently, f is a Möbius transformation if and only if f is a circle-preserving map.

In [28], Guowu Yao showed that the above conjecture is not true and gave the construction of degenerate circle-preserving maps.

Example 1 [28] Let us define a map $g_1 : \widehat{\mathbf{R}}^n \to \mathbf{R}^1$ as follows:

$$g_1(x) = \begin{cases} \sum_{i=1}^{n} |x_i|; & x \in \mathbf{R}^n, \\ 0; & x = \infty, \end{cases} \tag{2}$$

where $x = (x_1, \ldots, x_n) \in \widehat{\mathbf{R}}^n$. Now we define an equivalence relation on $\mathbf{R}$: x and y in $\mathbf{R}$ are equivalent if and only if $x - y \in \mathbf{Q}$. Let $\widetilde{x}$ denote the set of all elements $y \in \mathbf{R}$ equivalent to x and $\mathbf{R_Q}$ denote the set of all equivalence classes $\widetilde{x}$. We show that the cardinality $card(\mathbf{R_Q})$ of $\mathbf{R_Q}$ is $\aleph$ instead of $\aleph_0$. On the contrary suppose that $card(\mathbf{R_Q})$ is $\aleph_0$, that is, $\mathbf{R_Q}$ is a countable set. Let us choose one representative element from every equivalence class $\widetilde{x}$ in $\mathbf{R_Q}$ to form a subset W in $\mathbf{R}$. Clearly W is countable. We can assume that the set of rational numbers $\mathbf{Q} = \{r_1, r_2, \ldots, r_k, \ldots\}$. Notice that we have $\mathbf{R}^1 = \bigcup_{i=1}^{\infty} (W + \{r_k\})$. This means that $\mathbf{R}^1$ is a countable set, which is a contradiction. Hence, our claim is true.

There exists a canonical quotient map $g_2 : \mathbf{R} \to \mathbf{R_Q}$ defined as

$$g_2(x) = \widetilde{x}, x \in \mathbf{R}. \tag{3}$$

For any nontrivial interval L in $\mathbf{R}$, we claim that $g_2(L) = \mathbf{R_Q}$. Indeed, given a point $\widetilde{y}$ in $\mathbf{R_Q}$, let $g_2^{-1}(\widetilde{y})$ be the set of all the inverse images of $\widetilde{y}$ in $\mathbf{R}$. By the definition of g_2, it is easy to see that $g_2^{-1}(\widetilde{y})$ is dense in $\mathbf{R}$. Thus, there exists at least a point x in L such that $g_2(x) = \widetilde{y}$. So, we have $g_2(L) = \mathbf{R_Q}$ since $\widetilde{y}$ is arbitrary in $\mathbf{R_Q}$.

Now suppose S is a circle in $\widehat{\mathbf{R}}^n$. As a set of points in $\widehat{\mathbf{R}}^n$, $card(S) = \aleph$. Since the cardinality of $\mathbf{R_Q}$ is also $\aleph$, there exists a bijective map $g_3 : \mathbf{R_Q} \to S$.

If we define a map $f : \widehat{\mathbf{R}}^n \to \widehat{\mathbf{R}}^n$ by $f = g_3 \circ g_2 \circ g_1$, then we see that f is a degenerate circle-preserving map which maps every circle in $\widehat{\mathbf{R}}^n$ onto S. Actually, for any given circle T in $\widehat{\mathbf{R}}^n$, $g_1(T)$ must contain an interval L in $\mathbf{R}$. Then, we have $g_2(L) = \mathbf{R_Q}$ and by the definition of g_3 we obtain $g_3 \circ g_2(L) = S$, that is, $f(T) = S$.

For a degenerate circle-preserving map f, since the inverse images of every point in $f\left(\widehat{\mathbf{R}}^n\right)$ are dense in $\widehat{\mathbf{R}}^n$, f is necessarily nowhere continuous. However, in [28], it was shown that certain real analytic degenerate line-preserving maps exist in $\mathbf{R}^n$. Now we give two examples of degenerate line-preserving maps.

Example 2 [22] Let us define

$$f(x) = \begin{cases} \tan(\pi\,|x|)\,; & |x| + \frac{1}{2} \notin \mathbf{N}, \\ 0; & |x| + \frac{1}{2} \in \mathbf{N}, \end{cases} \tag{4}$$

where $x \in \mathbf{R}^n$, $x = (x_1, \ldots, x_n)$, $|x| = \left(\sum_{i=1}^n |x_i|^2\right)^{1/2}$. Clearly $f(\mathbf{R}^n) = \mathbf{R}$ and any line is mapped under f onto the line $\mathbf{R}$.

To give the second example we need the following lemma.

Lemma 4 [28] *Let ϕ be the function*

$$\phi(x) = x \sin x, x \in [0, \infty)\,. \tag{5}$$

Then, for any given $c \geq 0$, $\phi(\,[c, \infty)\,) = \mathbf{R}^1$.

Example 3 [28] Let us define

$$f(x) = |x|^2 \sin|x|^2\,, x \in \mathbf{R}^n. \tag{6}$$

It is easy to see that the map $\varphi : x \to |x|^2$ maps every line in $\mathbf{R}^n$ onto $[c, \infty)$ for some $c \geq 0$. Then, by Lemma 4, we can conclude that f maps every line in $\mathbf{R}^n$ onto $\mathbf{R}^1$ viewed as a line in $\mathbf{R}^n$. Hence, f is a degenerate line-preserving map. In addition, f is real-analytic on $\mathbf{R}^n$.

2.2 *Möbius Transformations and Sphere-Preserving Maps in* $\widehat{\mathbf{R}}^n$

As we have seen in the previous section, the studies about the circle-preserving property of maps in $\widehat{\mathbf{R}}^n$ are related to the studies of sphere-preserving maps. It is known that a Möbius transformation acting on $\widehat{\mathbf{R}}^n$ is r-sphere preserving. The converse question has been investigated by many authors. In this section, we only sketch the studies about sphere-preserving maps; for a complete treatment, see e.g., [4, 18, 19, 23, 29] and the references therein. In [19], Jeffers proved the following theorem.

Theorem 8 [19] *Suppose that* $f : \widehat{\mathbf{R}}^n \to \widehat{\mathbf{R}}^n$ $(n > 1)$ *is a bijection that preserves* r*-dimensional spheres. Then,* f *is a Möbius transformation.*

In [4], Alan F. Beardon and David Minda proved stronger results. We have the following theorem.

Theorem 9 [4] *For any* $f : \widehat{\mathbf{R}}^n \to \widehat{\mathbf{R}}^n$ $(n \geq 2)$, *the following are equivalent.*

(1) f *is Möbius;*
(2) *The restriction of* f *to every sphere is Möbius;*
(3) f *is sphere-preserving and bijective;*
(4) f *is sphere-preserving and injective;*
(5) f *is locally sphere-preserving.*

In a recent paper, Baokui Li and Guowu Yao obtained a new characterization of Möbius transformations by means of the sphere-preserving property. Following the studies obtained in [22], they called an r-sphere preserving map f as degenerate if its image $f(\widehat{\mathbf{R}}^n)$ is an r-dimensional sphere. Otherwise, f is called nondegenerate. The following theorem generalizes Theorem 5.

Theorem 10 [23] *Suppose that there is some* r *such that* $f : \widehat{\mathbf{R}}^n \to \widehat{\mathbf{R}}^n$ *is an* r*-sphere preserving map. Then,* f *is a Möbius transformation if and only if* f *is nondegenerate.*

Using the construction given in Example 1, in [23], it was given as an example of a degenerate r-sphere preserving map. Following the terminology used in Example 1, let S be an r-dimensional sphere in $\widehat{\mathbf{R}}^n$. There exists a bijective map $g_3 : \mathbf{R}_\mathbf{Q} \to S$. If we define a map $f : \widehat{\mathbf{R}}^n \to \widehat{\mathbf{R}}^n$ by $f = g_3 \circ g_2 \circ g_1$, then we see that f is a degenerate r-sphere preserving map.

Note that for a degenerate r-sphere preserving map f, the inverse images of every point in $f(\widehat{\mathbf{R}}^n)$ are dense in $\widehat{\mathbf{R}}^n$ and hence, f is necessarily nowhere continuous. As an immediate consequence we have the following corollary from Theorem 10.

Corollary 1 [23] *Suppose* $f : \widehat{\mathbf{R}}^n \to \widehat{\mathbf{R}}^n$ *is an* r*-sphere preserving map. If* f *continuous at some point* $p \in \widehat{\mathbf{R}}^n$, *then* f *is a Möbius transformation.*

In [23], it was given the following conjecture.

Conjecture 1 [23] Suppose that $f : \widehat{\mathbf{R}}^n \to \widehat{\mathbf{R}}^n$ is a surjective map and maps every r-dimensional sphere into an r-dimensional sphere. Then, f is a Möbius transformation.

In the following theorem, Guowu Yao proved the above conjecture for the case $r = n - 1$. The other cases remain open.

Theorem 11 [29] *Suppose that* $f : \widehat{\mathbf{R}}^n \to \widehat{\mathbf{R}}^n$ *is a surjective map and maps every* $(n-1)$*-sphere into an* $(n-1)$*-sphere. Then,* f *is a Möbius transformation.*

3 Möbius Transformations Acting on $\widehat{\mathbf{C}}$

For convenience, we denote $\widehat{\mathbf{R}}^2$ by $\widehat{\mathbf{C}}$. Recall that the transformations of the form

$$T(z) = \frac{a\bar{z} + b}{c\bar{z} + d}, \tag{7}$$

with complex coefficients that satisfy $ad - bc \neq 0$, are called conjugate Möbius transformations. It is well known that transformations of type (1) and (7) together form a group. Transformations of type (1) form the subgroup of "linear fractional transformations."

There are several results about the circle-preserving property of Möbius transformations acting on $\widehat{\mathbf{C}}$. After the studies given in [1] and [8], Joel Gibbons and Cary Webb proved a global generalization of these local results by removing the bijectivity assumption. They proved that if a function of $\widehat{\mathbf{C}}$ sends every circle into a circle (not necessarily unique) and the image set of this function either contains a set of positive Lebesgue measure or contains at least two circles, then this function is either a Möbius transformation or a conjugate Möbius transformation [11]. In [12], Joel Gibbons and Cary Webb also obtained a local theorem that generalizes all of the earlier results. They proved that if a circle-preserving function on an open subset of $\widehat{\mathbf{C}}$ to the standard sphere S^{2+m}, $m \geq 0$, possesses six image points, no five of which lie on a circle, then that function is either a Möbius transformation or a conjugate Möbius transformation. For more details, one can see [12]. It was noted that a (2D) improvement of this theorem is possible by reducing the number of required image points from six to five.

In [2], Alan F. Beardon obtained two strong local versions of the "only if" part of Theorem 3 using the effect of a conformal map on the curvature of a plane curve and making extensive use of properties of the Schwarzian derivative. Let a function f be meromorphic in a domain D in the extended complex plane $\widehat{\mathbf{C}}$ and for each $n = 1, 2, \ldots, \infty$, let A_n denote the set of z in D with the property that there are at least n circular arcs through z (in different directions) that are mapped to circular arcs by f. With this notation, the following theorems were proved.

Theorem 12 [2] *If either*

- *(i)* A_3 *has a point of accumulation in* D, *or*
- *(ii)* A_∞ *has at least three points, then* f *is Möbius and* $A_3 = A_\infty = \widehat{\mathbf{C}}$.
  This is sharp in the sense that
- *(iii)* *there exists a non-Möbius* f *with* $A_2 = D$, *and*
- *(iv)* *there exists a non-Möbius* f *with* A_∞ *containing exactly two points.*

Theorem 13 [2] *If* $A_\infty \neq \emptyset$, *then either* f *is Möbius (and* $A_\infty = \widehat{\mathbf{C}}$*) or* f *is Möbius equivalent to some map* $z \to z^p (p \geq 2)$ *and* A_∞ *has precisely two points.*

Note that two functions f_1 and f_2 are said to be Möbius equivalent if $f_2 = h \circ f_1 \circ g$ for some Möbius transformations g and h [2].

On the other hand, as circles and lines are the only plane curves whose curvature functions are constant and identically zero, respectively, the following theorem was noted in [33].

Theorem 14 [33] *Let $w = f(z)$ be a nonconstant meromorphic function. Then $w = f(z)$ maps the curves with constant curvature onto the curves with constant curvature in the plane if and only if $w = f(z)$ is a Möbius transformation.*

In the rest of this section, our concern will be the problem whether or not the circle-preserving property is an invariant characteristic property of Möbius transformations for the circles corresponding to any norm function $\|.\|$ on $\mathbf{C}$.

3.1 *Circles Corresponding to Any Norm Function on* C

Our investigations are based on the well-known facts that all Möbius transformations form a group under composition and that every Möbius transformation T of the form (1) is a composition of finitely many similarities and inversions. Recall that the Möbius transformations with $c = 0$ form the subgroup of *similarities* and such transformations have the form $z \to \alpha z + \beta; \alpha, \beta \in \mathbf{C}, \alpha \neq 0$. Also the transformation $z \to \frac{1}{z}$ is called an *inversion* (see [3, 10, 20, 21, 25] for the basic properties of Möbius transformations).

Let $\|.\|$ be any norm function on $\mathbf{C}$. A circle (or a sphere) whose center is at z_0 and of radius r is denoted by $S_r(z_0)$ and defined by $S_r(z_0) = \{z \in \mathbf{C} : \|z - z_0\| = r\}$. We consider the straight lines among the circles corresponding to the Euclidean norm function to be circles through ∞. By a similar way, if $-\frac{d}{c} \in S_r(z_0)$, we call the image $T(S_r(z_0))$ as a degenerate circle corresponding to any norm function on $\mathbf{C}$, where $T(z) = \frac{az+b}{cz+d}$. Throughout the paper $S_r(z_0)$ represents a nondegenerate circle (circle).

The following lemma plays a very important role for our investigations.

Lemma 5 [30] *Let $\|.\|$ be any norm function on the complex plane. Then, for every $\phi \in \mathbf{R}$, the following function defines a norm on the complex plane:*

$$\|z\|_\phi = \left\|e^{-i\phi} z\right\| . \tag{8}$$

Let $r > 0$ be any fixed real number. It can be easily seen that the function

$$\|z\| = r\,|z| \tag{9}$$

defines a norm on $\mathbf{C}$ for every $r > 0$. Clearly, circles of this new norm are the Euclidean circles. We will call these cases as the trivial cases. From now on we assume that $\|.\|$ be any norm function except the trivial cases.

We note that all Möbius transformations do not map circles to circles corresponding to any norm function on $\mathbf{C}$. We can give the following theorem from [30].

Theorem 15 [30] *Let $w = T(z) = \alpha z + \beta$; $\alpha \neq 0$, $\alpha, \beta \in \mathbf{C}$. Then for every circle $S_r(z_0)$ corresponding to any norm function $\|.\|$ on $\mathbf{C}$, $T(S_r(z_0))$ is a circle corresponding to the same norm function or corresponding to the norm function*

$$\|z\|_\phi = \left\|e^{-i\phi}.z\right\|, \tag{10}$$

where $\phi = \arg(\alpha)$.

Hence, we conclude that the circle-preserving property is not an invariant characteristic property of Möbius transformations except in the Euclidean case.

In Theorem 15, we can write $T(z) = |\alpha| e^{i\phi}z + \beta$; $\alpha \neq 0$, $\phi = \arg(\alpha)$ and so we have $T(z) = (T_1 \circ T_2)(z)$, where $T_1(z) = e^{i\phi}z$ and $T_2(z) = |\alpha| z + e^{-i\phi}\beta$. In [30], it was shown that the transformation $T_2(z)$ maps circles to circles corresponding to the same norm function and that the rotation transformation $T_1(z) = e^{i\phi}z$ ($\phi \neq k\pi$, $k \in \mathbf{Z}$) maps the circle $S_r(z_0)$ to the circle $S_r^{\phi}(z_0)$ corresponding to the norm function $\|z\|_\phi$ defined in (10).

Now we ask the question whether or not the inversion transformation $z \to \frac{1}{z}$ maps circles to circles corresponding to the same norm function.

We give the following theorem.

Theorem 16 *Let $\|.\|$ be any norm on $\mathbf{C}$, $S_r(z_0)$ any circle corresponding to this norm with $0 \notin S_r(z_0)$ and $w = T(z) = \frac{1}{z}$. Then, the image $T(S_r(z_0))$ cannot be a circle corresponding to the same norm function.*

Proof Without loss of generality, we consider the unit circle $S_1(0)$ corresponding to any norm $\|.\|$. For any real number θ, the point $z = \frac{e^{i\theta}}{\|e^{i\theta}\|}$ is on the circle $S_1(0)$. The image of the point $z = \frac{e^{i\theta}}{\|e^{i\theta}\|}$ under the transformation $T(z) = \frac{1}{z}$ is $w = \left\|e^{i\theta}\right\| e^{-i\theta}$. If $T(S_1(0))$ is the circle centered at 0 and of radius ρ, then it should be

$$\|w\| = \left\|e^{i\theta}\right\| \left\|e^{-i\theta}\right\| = \rho, \tag{11}$$

for all real numbers θ.

Case 1 Let $\|i\| \neq \|1\|$. By (11), for the points $\frac{1}{\|1\|}, \frac{i}{\|i\|} \in S_1(0)$, we get

$$\rho = \|1\|^2 = \|i\|^2, \tag{12}$$

and hence $\|i\| = \|1\|$, which is a contradiction. Therefore, $T(S_1(0))$ cannot be a circle corresponding to the same norm in this case.

Case 2 Let $\|i\| = \|1\|$. By similar arguments used in the Case 1, the proof follows easily.

In the following theorem, we see that the images of circles under the transformation $T(z) = \frac{1}{z}$ cannot be circles corresponding to any other norm function.

Theorem 17 *Let $\|.\|$ be any norm on $\mathbf{C}$, $S_r(z_0)$ any circle corresponding to this norm with $0 \notin S_r(z_0)$, and $w = T(z) = \frac{1}{z}$. Then, the image $T(S_r(z_0))$ cannot be a circle corresponding to any other norm function.*

Proof Conversely, assume that $T(S_1(0))$ is the circle centered at 0 and of radius ρ corresponding to a different norm function $\|.\|_*$. By similar arguments used in the proof of Theorem 16, then it should be

$$\left\|e^{i\theta}\right\| \left\|e^{-i\theta}\right\|_* = \rho, \tag{13}$$

for all real numbers θ. Then the proof follows easily.

We will call two curves C_1 and C_2 in $\widehat{\mathbf{C}}$ "Möbius equivalent" if there exists a Möbius transformation T such that $C_2 = T(C_1)$. We can give the following corollary.

Corollary 2 *Let U and V be arbitrary circles corresponding to different norms and T_1, T_2 arbitrary Möbius transformations. If $T_1(U)$ is Möbius equivalent to $T_2(V)$, then U is a rotation of V.*

Proof If $T_1(U)$ is Möbius equivalent to $T_2(V)$, then there exists a Möbius transformation T such that $T(T_1(U)) = T_2(V)$. This implies $V = T_2^{-1}(T(T_1(U)))$. By Theorems 15–17, the composition $T_2^{-1} \circ T \circ T_1$ is a rotation transformation, so U is a rotation of V.

Now we give a definition to classify the images of circles under the Möbius transformations.

Definition 2 Let T be any Möbius transformation of the form $T(z) = \frac{az+b}{cz+d}$; $a, b, c, d \in \mathbf{C}$ with $c \neq 0$ and $\|.\|$ any norm on $\mathbf{C}$ except the trivial cases. Then, we call the image $T(S_r(z_0))$ as an MC-curve for any circle $S_r(z_0)$.

It can be seen examples of MC-curves in Figs. 1, 2, 3, and 4.

Before giving the classification theorems, we note that the image of an MC-curve under any similarity transformation is again an MC-curve. Indeed, if U is an MC-curve then by definition we have $U = T(S_r(z_0))$ with some Möbius transformation of the form (1) with $c \neq 0$ and some circle $S_r(z_0)$. For any similarity transformation S of the form $z \to \alpha z + \beta; \alpha, \beta \in \mathbf{C}, \alpha \neq 0$, we have $S(U) = (S \circ T)(S_r(z_0))$. An easy calculation shows that

$$(S \circ T)(z) = \frac{(a\alpha + c\beta)z + (b\alpha + d\beta)}{cz + d} \tag{14}$$

and hence, we see that $S(U)$ is an MC-curve by the definition.

We can give the following theorem.

Theorem 18 *Let $\|.\|$ be any norm on $\mathbf{C}$, $S_r(z_0)$ any circle corresponding to this norm and T any Möbius transformation of the form (1). Then, $T(S_r(z_0))$ can be a degenerate circle or a circle or an MC-curve.*

Proof Let T be any Möbius transformation of the form (1) and $S_r(z_0)$ any circle. If $-\frac{d}{c} \in S_r(z_0)$, then we know that $T(S_r(z_0))$ is a degenerate circle by definition. Assume that $-\frac{d}{c} \notin S_r(z_0)$. We have two cases:

Case 1 Let $c = 0$. From Theorem 15, we know that $T(S_r(z_0))$ is a circle corresponding to the same norm function or corresponding to the norm function $\|z\|_\phi = \left\|e^{-i\phi}z\right\|$, where $\phi = \arg\left(\frac{a}{d}\right)$.

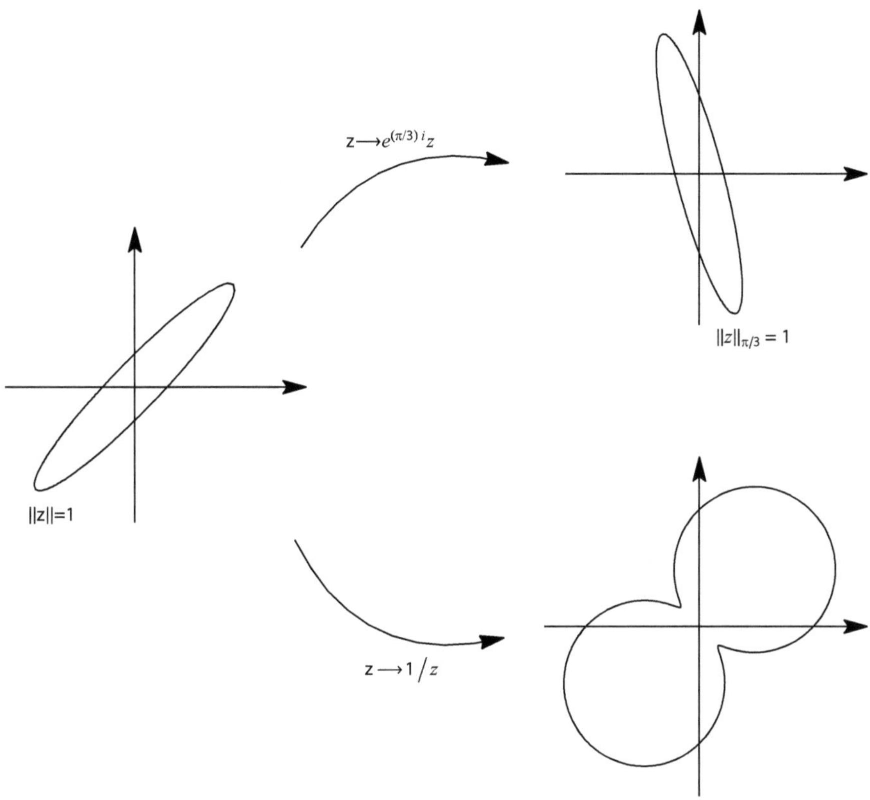

Fig. 1 Images of the unit circle corresponding to the norm defined in Example 4

Case 2 Let $c \neq 0$. It is well known that every Möbius transformation T of the form (1) can be written as the composition of the transformations $T_1(z) = c^2z$, $T_2(z) = z + cd$, $T_3(z) = \frac{1}{z}$, $T_4(z) = (bc - ad)z$ and $T_5(z) = z + \frac{a}{c}$ since we have

$$T(z) = \frac{az + b}{cz + d} = \frac{a}{c} + \frac{bc - ad}{c^2z + cd}. \tag{15}$$

That is, we can write $T(z) = T_5 \circ T_4 \circ T_3 \circ T_2 \circ T_1(z)$. $(T_2 \circ T_1)(S_r(z_0))$ is again a circle corresponding to the same norm function or corresponding to the norm function $\|z\|_\phi = \left\|e^{-i\phi}z\right\|$, where $\phi = \arg\left(c^2\right)$. $T_3((T_2 \circ T_1)(S_r(z_0)))$ is an MC-curve by definition. As we have seen above, the image of this MC-curve under the transformation $T_5 \circ T_4$ is again an MC-curve. Thus, $T(S_r(z_0))$ is an MC-curve.

Example 4 Let us consider the norm function $\|z\| = \sqrt{\frac{(x+y)2}{9} + 4(x - y)^2}$ and the unit circle $S_1(0)$ corresponding to this norm function. The image of $S_1(0)$ under the transformation $T_1(z) = e^{i(\pi/3)}z$ is not a circle corresponding to the same norm function. $T_1(S_1(0))$ is the circle $S_1^{\pi/3}(0)$ corresponding to the norm function

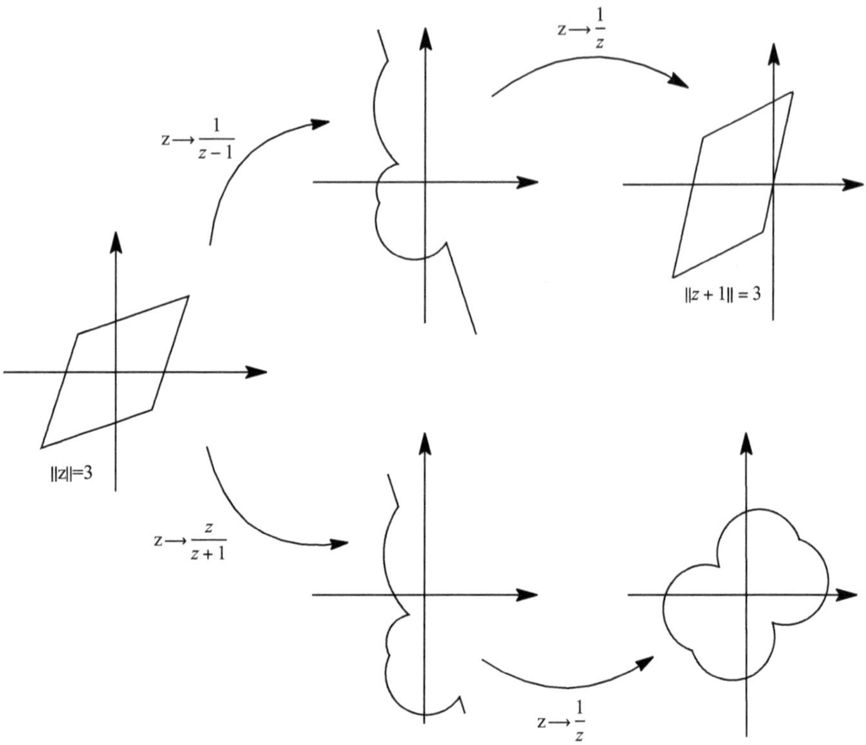

Fig. 2 Images of the circle $S_3(0)$ corresponding to the norm defined in Example 5

$\|z\|_{\pi/3} = \sqrt{\frac{\left[(1-\sqrt{3})x+(1+\sqrt{3})y\right]^2}{36} + \left[(1+\sqrt{3})x + (\sqrt{3}-1)y\right]^2}$. On the other hand, the image of $S_1(0)$ under the transformation $T_2(z) = \frac{1}{z}$ is not a circle corresponding to any norm function on $\mathbf{C}$. $T_2(S_1(0))$ is an MC-curve, (see Fig. 1).

Now we consider the degenerate circles. We have the following theorem.

Theorem 19 *Let U be any degenerate circle corresponding to any norm and T any Möbius transformation. Then, $T(U)$ can be a circle or a degenerate circle or an MC-curve.*

Proof Let U be any degenerate circle corresponding to any norm. Then $\infty \in U$ and by definition, U is the image of a circle $S_r(z_0)$ under some Möbius transformation $S(z) = \frac{az+b}{cz+d}$ with $-\frac{d}{c} \in S_r(z_0)$. If $T(z)$ is a similarity transformation of the form $T(z) = \alpha z + \beta$; $\alpha, \beta \in \mathbf{C}$, $\alpha \neq 0$, then the image of U under T is again a degenerate circle corresponding to the same norm function, since we have $T(U) = (T \circ S)(S_r(z_0))$ and $(T \circ S)\left(-\frac{d}{c}\right) = \infty$.

Let $T(z) = \frac{\gamma z+\delta}{\varepsilon z+\theta}$ where $\varepsilon \neq 0$, then we have $(T \circ S)(z) = \frac{(\gamma a+\delta c)z+(\gamma b+\delta d)}{(\varepsilon a+\theta c)z+(\varepsilon b+\theta d)}$ where $T(\infty) = \frac{\gamma a+\delta c}{\varepsilon a+\theta c}$. If $-\frac{\theta}{\varepsilon} \in U$, then it must be $\varepsilon a + \theta c \neq 0$ since T is a univalent

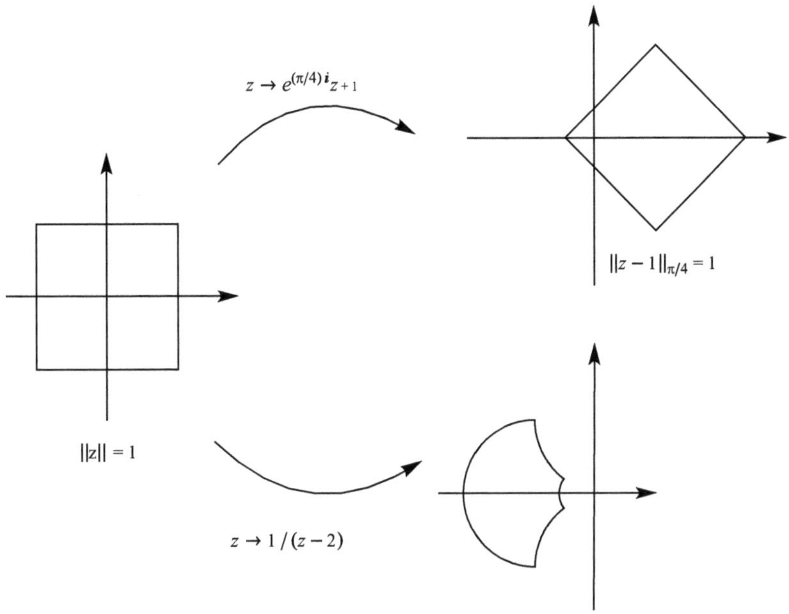

Fig. 3 Images of the unit circle corresponding to the norm defined in Example 6

function. In this case, $T(U)$ is again a degenerate circle. Let $-\frac{\theta}{\varepsilon} \notin U$. If $T \circ S$ is a similarity transformation (in which case we have $\varepsilon a + \theta c = 0$), then $T(U)$ is a circle corresponding to the same norm or to the norm function $\|.\|_\phi$ defined in (8) for some $\phi \in \mathbf{R}$. If $T \circ S$ is not a similarity transformation (in which case we have $\varepsilon a + \theta c \neq 0$), then $T(U)$ cannot be a circle corresponding to any norm function, it is an MC-curve.

Example 5 Let us consider the norm function $\|z\| = |x + y| + 2\,|x - y|$. The image of $S_3(0)$ under the transformation $S(z) = \frac{z}{z+1}$ is a degenerate circle. If we apply the transformation $T(z) = \frac{1}{z}$ (notice that $\varepsilon a + \theta c = 1 \neq 0$) to this image, we see that $(T \circ S)(S_3(0))$ cannot be a circle corresponding to any norm function. $(T \circ S)(S_3(0))$ is an MC-curve. On the other hand, the image of $S_3(0)$ under the transformation $S(z) = \frac{1}{z-1}$ is again a degenerate circle. If we apply the transformation $T(z) = \frac{1}{z}$ (notice that $\varepsilon a + \theta c = 0$) to this image, we see that $(T \circ S)(S_3(0))$ is the circle $S_3(-1)$ corresponding to the same norm, (see Fig. 2).

3.2 *Norms with Property C*

If the norm function has the property $\|z\| = \|\overline{z}\|$ for all $z \in \mathbf{C}$, we can obtain some more nice results. Now we consider the norm functions having this property.

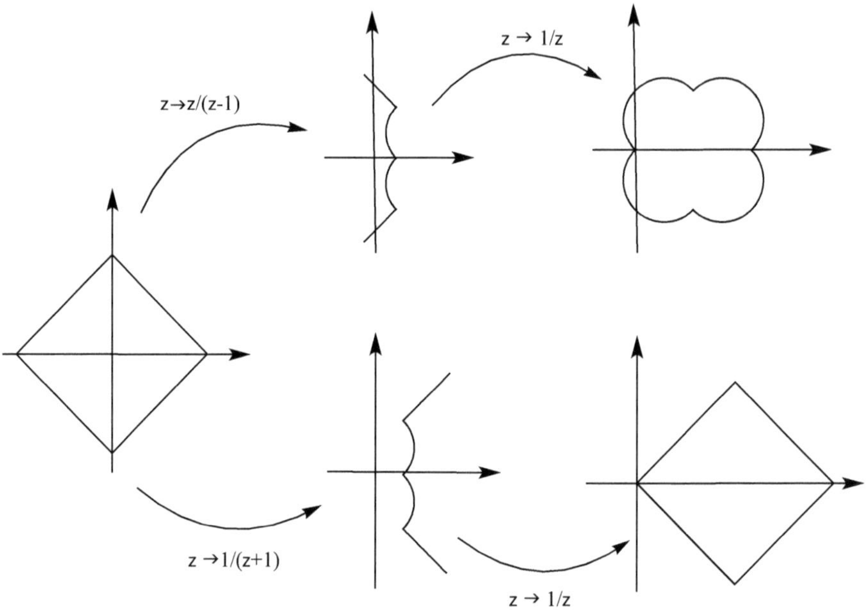

Fig. 4 Images of the unit circle corresponding to the norm defined in Example 7

Definition 3 We say that a norm function $\|.\|$ defined on $\mathbf{C}$ has the Property $\mathcal{C}$ if it satisfies the property $\|z\| = \|\overline{z}\|$ for all $z \in \mathbf{C}$.

Let $\|.\|$ be any norm function on $\mathbf{C}$ with the Property $\mathcal{C}$. In [30], it was shown that the circles corresponding to this norm function and their images under the Möbius transformations have nice properties by means of Apollonius circles. We have the following theorem.

Theorem 20 [30] *Let $\|.\|$ be any norm on $\mathbf{C}$ with the Property $\mathcal{C}$ and let $w = T(z)$ be a Möbius transformation. Then, the points of any circle $S_r(z_0)$ are inverse with respect to the two circle of Apollonius U and V as well as the points of the image $T(S_r(z_0))$ are also inverse with respect to the image circles $U' = T(U)$ and $V' = T(V)$.*

In the following theorem, we see that the points of MC-curves corresponding to any norm with the Property $\mathcal{C}$ are inverse with respect to an inversion in an Euclidean circle. Recall that for an Euclidean circle in $\mathbf{C}$ with center $q \in \mathbf{C}$ and radius $r > 0$, the formula for the inversion in that circle is $R(z) = \frac{r^2}{\overline{z}-\overline{q}} + q$, (see [5] and [20]).

Theorem 21 *Let $\|.\|$ be any norm on $\mathbf{C}$, $S_r(z_0)$ any circle corresponding to this norm and T any Möbius transformation. Assume that the norm function has the Property $\mathcal{C}$. If $T(S_r(z_0))$ is an MC-curve, then it is the image of a circle (corresponding to the same norm function or corresponding to the norm function $\|z\|_\phi = \left\|e^{-i\phi}.z\right\|$ for some $\phi \in \mathbf{R}$) under the inversion in the Euclidean circle with center $\frac{a}{c} \in \mathbf{C}$ and radius* 1.

Proof Let T be any Möbius transformation of the form $T(z) = \frac{az+b}{cz+d}; a, b, c, d \in \mathbf{C}$ with $c \neq 0$ and $T(S_r(z_0))$ any MC-curve. We can compose T with an inversion that sends $\frac{a}{c}$ back to ∞ since we have $T(\infty) = \frac{a}{c}$. Let $R(z) = \frac{1}{\bar{z}-\overline{(\frac{a}{c})}} + \frac{a}{c}$ be the inversion in the Euclidean circle with center $\frac{a}{c} \in \mathbf{C}$ and radius 1. Then an easy calculation shows that

$$R \circ T(z) = -\frac{\overline{c}^2}{\overline{(ad-bc)}}\bar{z} - \frac{\overline{dc}}{\overline{(ad-bc)}} + \frac{a}{c} = \alpha\bar{z} + \beta \tag{16}$$

where $\alpha = -\frac{\overline{c}^2}{\overline{(ad-bc)}} \neq 0$ and $\beta = \frac{a}{c} - \frac{\overline{dc}}{\overline{(ad-bc)}}$. The transformation $z \to \alpha\bar{z} + \beta$ can be written as the composition of the transformations $T_3(z) = \alpha z + \beta$ and $T_2(z) = \bar{z}$. If we apply the transformation $w = T_2(z) = \bar{z}$ to the circle $S_r(z_0)$, we get

$$\|w - w_0\| = \|\bar{z} - \overline{z_0}\| = \left\|\overline{z - z_0}\right\| = \|z - z_0\| = r, \tag{17}$$

where $w_0 = \overline{z_0}$. This shows that $T_2(S_r(z_0)) = S_r(\overline{z_0})$. Now by Theorem 16, the image of $S_r(\overline{z_0})$ under the transformation $T_3(z) = \alpha z + \beta$ is a circle $S_r(z_0)'$ corresponding to the same norm or to the norm function $\|z\|_\phi = \left\|e^{-i\phi}.z\right\|$, where $\phi = \arg(\alpha)$. Since $R^{-1} = R$, we have $T\left(S_r(z_0)\right) = R\left(S_r(z_0)'\right)$. That is, $T(S_r(z_0))$ is the image of a circle (corresponding to the same norm function or corresponding to the norm function $\|z\|_\phi = \left\|e^{-i\phi}.z\right\|$ for some $\phi \in \mathbf{R}$) under the inversion in the Euclidean circle with center $\frac{a}{c} \in \mathbf{C}$ and radius 1.

Example 6 Let us consider the norm function $\|z\| = \max\{|x|, |y|\}$ and the unit circle $S_1(0)$ corresponding to this norm function. Notice that this norm function has the Property $\mathcal{C}$. The image of $S_1(0)$ under the transformation $T_1(z) = e^{\frac{\pi}{4}i}z + 1$ is not a circle corresponding to the same norm function. $T_1(S_1(0))$ is the circle $S_1(1)$ corresponding to the norm function $\|z\|_{\frac{\pi}{4}} = \frac{1}{\sqrt{2}}\max\{|x+y|, |y-x|\}$.

On the other hand, the image of $S_1(0)$ under the transformation $T_2(z) = \frac{1}{z-2}$ is not a circle corresponding to any norm function on $\mathbf{C}$. $T_2(S_1(0))$ is an MC-curve. Furthermore $T_2(S_1(0))$ is the image of $S_1(-2)$, corresponding to the norm function $\|z\| = \max\{|x|, |y|\}$, under the inversion $R(z) = \frac{1}{\bar{z}}$ in the unit Euclidean circle $|z| = 1$, (see Fig. 3).

Theorem 22 *Let U be any degenerate circle corresponding to any norm and T any Möbius transformation. If the norm function has the Property $\mathcal{C}$ and $T(U)$ is an MC-curve, then $T(U)$ is the image of a circle under the inversion in an Euclidean circle.*

Proof The proof follows easily by similar arguments used in the proof of Theorem 21.

Example 7 Let us consider the norm function $\|z\| = |x| + |y|$. Notice that this norm function has the Property $\mathcal{C}$. The image of $S_1(0)$ under the transformation $S(z) = \frac{z}{z-1}$ is a degenerate circle. If we apply the transformation $T(z) = \frac{1}{z}$ (notice that $\varepsilon a + \theta c = 1 \neq 0$) to this image, we see that $(T \circ S)(S_1(0))$ cannot be a circle corresponding to any norm function. $(T \circ S)(S_1(0))$ is an MC-curve and also is the image of the circle $S_1(1)$, corresponding to the same norm, under the inversion in the Euclidean circle $|z-1|=1$.

On the other hand, the image of $S_1(0)$ under the transformation $S(z)=\frac{1}{z+1}$ is again a degenerate circle. If we apply the transformation $T(z)=\frac{1}{z}$ (notice that $\varepsilon a+\theta c=0$) to this image, we see that $(T\circ S)(S_1(0))$ is the circle $S_1(1)$ corresponding to the same norm, (see Fig. 4).

4 Future Works

Besides the open problems mentioned in the earlier chapters, it is natural to consider the generalizations of Theorem 14 for n-dimensional cases where $n\geq 3$. In this direction, in [33], problem for the case $n=3$ using the Poincare extensions of Möbius transformations in $\mathbf{R}^3$ was considered. In 3D Euclidean space, it is well known that the helix is the only curve whose curvature and torsion are constant. In [33], the authors considered the images of the helix (it was used the term "*helix*" to mean "*right circular helix*") under the Möbius transformations by means of its curvature and torsion functions. The following theorem is obtained.

Theorem 23 [33] *Given any arbitrary helices* $\alpha,\alpha'\subseteq\mathbf{R}^3$, *if the Möbius transformations* T *satisfies* $T(\alpha)=\alpha'$ *then* T *is a similarity transformation.*

Finally, we note that it is possible to extend the results obtained in the Sects. 3.1 and 3.2 to the n-dimensional case where $n\geq 3$.

Acknowledgments The author would like to thank Prof. Th. M. Rassias for his encouragement to prepare this survey.

References

1. Aczel, J., Mckiernan, M.A.: On the characterization of plane projective and complex Moebius-transformations. Math. Nachr. **33**, 315–337 (1967)
2. Beardon, A.F.: Curvature, circles, and conformal maps. Am. Math. Mon. **94**(1), 48–53 (1987)
3. Beardon, A.F..: The Geometry of Discrete Groups. Graduate Texts in Mathematics, 91. Springer-Verlag, New York (1995)
4. Beardon, A.F., Minda, D.: Sphere-preserving maps in inversive geometry. Proc. Am. Math. Soc. **130**(4), 987–998 (2002)
5. Blair, D.: Inversion Theory and Conformal Mapping. Student Mathematical Library, issue 9. American Mathematical Society, Providence (2000)
6. Bulut, S., Özgür, N.Y.: A new characteristic of Möbius transformations by use of Apollonius points of pentagons. Turk. J. Math. **28**(4), 299–305 (2004)
7. Bulut, S., Özgür, N.Y.: A new characterization of Möbius transformations by use of apollonius points of $(2n-1)$-gons. Acta Math. Sinica Engl. Ser. **21**(3), 667–672 (2005)
8. Caratheodory, C.: The most general transformations of plane regions which transform circles into circles. Bull. Am. Math. Soc. **43**, 573–579 (1937)
9. Coxeter, H.S.M.: Introduction to geometry. Wiley, New York (1969)

10. Ford, L.R.: Automorphic functions. Chelsea, New York (1951)
11. Gibbons, J., Webb, C.: Circle-preserving transformations. Notices Am. Math. Soc. **22**, 682 (1975). (Abstract)
12. Gibbons, J., Webb, C.: Circle-preserving functions of spheres. Trans. Am. Math. Soc. **248**(1), 67–83 (1979)
13. Haruki, H.: A proof of the principle of circle-transformations by use of a theorem on univalent functions. L'enseignement Math. **18**, 145–146 (1972)
14. Haruki, H., Rassias, T.M.: A new invariant characteristic property of Möbius transformations from the standpoint of conformal mapping. J. Math. Anal. Appl. **181**(2), 320–327 (1994)
15. Haruki, H., Rassias, T.M.: A new characteristic of Möbius transformations by use of Apollonius points of triangles. J. Math. Anal. Appl. **197**(1), 14–22 (1996)
16. Haruki, H., Rassias, T.M.: A new characteristic of Möbius transformations by use of Apollonius quadrilaterals. Proc. Am. Math. Soc. **126**, 2857–2861 (1998)
17. Haruki, H., Rassias, T.M.: A new characterization of Möbius transformations by use of Apollonius hexagons. Proc. Am. Math. Soc. **128**(7), 2105–2109 (2000)
18. Höfer, R.: A characterization of Möbius transformations. Proc. Am. Math. Soc. **128**(4), 1197–1201 (1999)
19. Jeffers, J.: Lost theorems of geometry. Am. Math. Mon. **107**, 800–812 (2000)
20. Jones, G.A., Singerman, D.: Complex Functions. An Algebraic and Geometric Viewpoint. Cambridge University Press, Cambridge (1987)
21. Lehner, J.: Discontinuous Groups and Automorphic Functions. Mathematical Surveys. 8. American Mathematical Society, Providence (1964)
22. Li, B., Wang, Y.: Transformations and non-degenerate maps. Sci. China Ser. A. **48**, 195–205 (2005)
23. Li, B., Yao, G.: On characterizations of sphere-preserving maps. Math. Proc. Camb. Philos. Soc. **147**(2), 439–446 (2009)
24. Li, L., Yao, G.: A new characterization for isometries by triangles. N. Y. J. Math. **15**, 423–429 (2009)
25. Nehari, Z.: Conformal Mapping. McGraw-Hill, New York (1952)
26. Niamsup, P.: A note on the characteristics of Möbius transformations. J. Math. Anal. Appl. **248**(1), 203–215 (2000)
27. Samaris, N.: A new characterization of Möbius transformation by use of 2n points. J. Nat. Geom. **22**(1–2), 35–38 (2002)
28. Yao, G.: On existence of degenerate circle-preserving maps. J. Math. Anal. Appl. **334**(2), 950–953 (2007)
29. Yao, G.: Transformations of spheres without the injectivity assumption. Quasiconformal mappings, Riemann surfaces, and Teichmüller spaces. Contemp. Math. Am. Math. Soc. Providence **575**, 371–375 (2012)
30. Yılmaz Özgür, N.: On some mapping properties of Möbius transformations. Aust. J. Math. Anal. Appl. **6**(13), 8 (2010)
31. Yılmaz Özgür, N., Bulut, S.: A note on the characteristic properties of Möbius transformations. Rad. Mat. **12**, 129–133 (2004)
32. Yılmaz Özgür, N., Bulut, S.: On the invariant characteristic properties of Möbius transformations. Nonlinear Funct. Anal. Appl. **10**(5), 885–894 (2005)
33. Yılmaz Özgür, N., Bulut, S., Özgür, C.: On the images of the helix under the Möbius transformations. Nonlinear Funct. Anal. Appl. **10**(5), 743–749 (2005)

Extended Crystal PDEs

Agostino Prástaro

Abstract In this paper we show that between partial differential equations (PDEs) and crystallographic groups there is an unforeseen relation. In fact we prove that integral bordism groups of PDEs can be considered extensions of crystallographic subgroups. In this respect we can consider PDEs as *extended crystals*. Then an algebraic topological obstruction (*crystal obstruction*), characterizing existence of global smooth solutions for smooth boundary value problems, is obtained. Applications of this new theory to the Ricci flow equation and Navier–Stokes equation are given, which solve some well-known fundamental problems. These results are also extended to singular PDEs (introducing *extended crystal singular PDEs*). An application to singular magnetohydrodynamics partial differential equations (MHD-PDEs) is given following some our previous results on such equations, and showing existence of (finite times stable smooth) global solutions crossing the critical nuclear energy production zone.

Keywords Integral (co)bordism groups in PDEs · Existence of local and global solutions in PDEs · Conservation laws · Crystallographic groups · Singular PDEs

1 Introduction

New points of view were recently introduced by us in the geometric theory of partial differential equations (PDEs) by adopting some algebraic topological approaches. In particular, integral (co)bordism groups are seen as very useful to characterize global solutions. The methods developed by us, in the category of (non)commutative PDEs, in order to find integral bordism groups, allowed us to obtain, as a by-product, existence theorems for global solutions, in a pure geometric way. Another result that is directly related to the knowledge of PDE's integral bordism groups, is the possibility to characterize PDEs by means of some important algebras related to the conservation laws of these equations (PDE's Hopf algebras). Moreover, thanks to an algebraic characterization of PDEs, one has also a natural way to recognize quantized

A. Prástaro (✉)
Department SBAI-Mathematics, University of Roma La Sapienza, Via A. Scarpa, 16-10-00166 Roma, Italy
e-mail: agostino.prastaro@uniroma1.it

T. M. Rassias, P. M. Pardalos (eds.), *Mathematics Without Boundaries*,
DOI 10.1007/978-1-4939-1106-6_18,

PDEs as quantum PDEs, i.e., PDEs in the category of quantum (super)manifolds, in the sense introduced by us in some previous works. These results have opened a new sector in algebraic topology, that we can formally define the *PDE's algebraic topology*. (See Refs. [46–54, 56, 57–86] and related works [2, 3, 38, 87–89].)

The aim of the present chapter is, now, to show that PDEs can be considered as extended crystals, in the sense that their integral bordism groups, characterizing the geometrical structure of PDEs, can be considered as extended groups of crystallographic subgroups. This fundamental relation gives new insights in the PDE's geometrical structure understanding, and also opens new possible mathematical and physical interpretations of the same PDE's structure. In particular, we get a new general workable criterion for smooth global solutions existence satisfying smooth boundary value problems. In fact, we identify the obstruction for existence of such global solutions, with an algebraic topologic object (*crystal obstruction*). Since it is easy to handle this method in all the concrete PDEs of interest, it sheds a lot of light on all the PDE's theory.

Finally we extend above results also to singular PDEs, and we recognize *extended crystal singular PDEs*. For such equations we identify algebraic topological obstructions to the existence of global (smooth) solutions solving boundary value problems and crossing singular points, too. Applications to magnetohydrodynamics partial differential equations (MHD-PDEs), as introduced in some our previous papers [70, 74], and encoding anisotropic incompressible nuclear plasma dynamics are given.

The chapter, after this Introduction, has three more sections and four appendices. In Sect. 2 we consider some fundamental mathematical properties of crystallographic groups that will be used in Sect. 3. There we recall some of our results about PDE characterization by means of integral bordism groups. Furthermore we relate these groups to crystallographic groups. The main results are Theorems 7, 8, and 9. The first two relate formal integrability and complete integrability of PDEs to crystallographic groups. (It is just this theorem that allows us to consider PDEs as *extended crystallographic structures*.) The third main theorem identifies an obstruction (*crystal obstruction*) characterizing existence of global smooth solutions in PDEs. Applications to some important PDEs of Riemannian geometry (e.g., Ricci flow equation) and mathematical physics (e.g., Navier–Stokes equation) are given that solve some well-known fundamental mathematical problems. (Further applications are given in Refs. [69, 70, 73, 74].) Section 4 is devoted to extend above results also to singular PDEs. The main result in this section is Theorem 18 that identifies conditions in order to recognize global (smooth) solutions of singular PDEs crossing singular points. There we characterize 0-*crystal singular PDEs*, i.e., singular PDEs having smooth global solutions crossing singular points, stable at finite times. Applications of these results to singular MHD-PDEs encoding anisotropic incompressible nuclear plasma dynamics are given in Example 17. Here, by using some our previous recent results on MHD-PDEs, we characterize global (smooth) solutions crossing the critical nuclear zone, i.e., where solutions pass from states without nuclear energy production to states where there is nuclear energy production. The stability of such solutions is also considered on the grounds of our recent geometric theory on the PDE's stability and stability of global solutions of PDEs [68, 69, 70, 73, 74].

In appendices we have collected and organized some standard information about crystallographic groups and their subgroups in order to give to the reader a general map for an easier understanding of their usage in the examples considered in the chapter.

2 Crystallographic Groups

Definition 1 Let $(E, (\mathbf{E}, \bar{g}), \alpha)$ be a d-dimensional Euclidean affine space, where $\alpha : \mathbf{E} \times E \to E$ is the action mapping of the n-dimensional vector space $\mathbf{E}$ on the set E of points, and $\bar{g}$ is an Euclidean metric on $\mathbf{E}$. Let us denote by $A(E) = \mathbf{E} \rtimes GL(\mathbf{E})$ the *affine group* of E, i.e., the symmetry group of the above affine structure and by $M(E) = \mathbf{E} \rtimes O(\mathbf{E})$ the *group of Euclidean motions* of E, i.e., the symmetry group of the above Euclidean affine structure. (The symbol $\rtimes$ denotes *semidirect product*, i.e., the set is the Cartesian product and the multiplication is defined as $(a, u)(b, v) = (ab, u + av)$.)[1] Let us denote by $R(E) = \mathbf{E} \rtimes SO(\mathbf{E})$ the group of all rigid motions of E, i.e., the symmetry group of the above oriented Euclidean affine structure, where the orientation is the canonical one induced by the metric. One has the following monomorphisms of inclusions: $R(E) < M(E) < A(E)$. One has the natural split exact commutative diagram

$$\begin{array}{ccccccccc}
0 & \longrightarrow & \mathbf{E} & \longrightarrow & A(E) & \rightleftarrows & GL(\mathbf{E}) & \longrightarrow & 1 \\
 & & \| & & \uparrow & & \uparrow & & \\
0 & \longrightarrow & \mathbf{E} & \longrightarrow & M(E) & \rightleftarrows & O(\mathbf{E}) & \longrightarrow & 1 \\
 & & \| & & \uparrow & & \uparrow & & \\
0 & \longrightarrow & \mathbf{E} & \longrightarrow & R(E) & \rightleftarrows & SO(\mathbf{E}) & \longrightarrow & 1
\end{array} \quad (1)$$

The quotient groups $A(E)/\mathbf{E} \cong GL(\mathbf{E})$, $M(E)/\mathbf{E} \cong O(\mathbf{E})$ and $R(E)/\mathbf{E} \cong SO(\mathbf{E})$, are called *point groups* of the corresponding groups $A(E)$, $M(E)$ and $R(E)$ respectively. $\mathbf{E}$ is called the *translations group*.

Definition 2 A *crystallographic group* is a cocompact[2], discrete subgroup of the isometries of some Euclidean space.

Definition 3 A d-dimensional *affine crystallographic group* $G(d)$ is a subgroup of $M(E)$, such that its subgroup $\mathbf{T} \equiv G(d) \cap \mathbf{E}$ of all pure translations is a discrete normal subgroup of finite index.

---

[1] Recall that given two groups A and B and an homomorphism $\alpha : B \to Aut(A)$, the *semidirect product* is a group denoted by $A \rtimes_\alpha B$, that is the Cartesian product $A \times B$, with product given by $(a, b).(\bar{a}, \bar{b}) = (a.\alpha(b)(\bar{a}), b.\bar{b})$. The semidirect product reduces to the direct product, i.e., $A \rtimes_\alpha B = A \times B \equiv A \oplus B$, when $\alpha(b) = 1$, for all $b \in B$. In the following we will omit the symbol α.

[2] A discrete subgroup $H \subset G$ of a topological group G, is *cocompact* if there is a compact subset $K \subset G$ such that $HK = G$.

If the rank of $\mathbf{T}$ is d, i.e., $\mathbf{T} \cong \mathbb{Z}^d$, $G(d)$ is called a *space group*. The point group $G \equiv G(d)/\mathbf{T}$ of a space group is finite, and isomorphic to a subgroup of $O(\mathbf{E})$: $G < O(\mathbf{E})$.

Remark 1 Note that the point group G of a crystallographic group $G(d)$ cannot necessarily be identified with a subgroup of $G(d)$. In other words, $G(d)$ is in general an *upward extension* of $\mathbf{T}$, as well as an *downward extension* of G, i.e., for a crystallographic group $G(d)$ one has the following short exact sequence:[3]

$$0 \longrightarrow \mathbf{T} \xrightarrow{i} G(d) \xrightarrow{\pi} G \longrightarrow 1 \tag{2}$$

Thus we can give the following definition.

Definition 4 We call *symmorphic crystallographic groups* ones $G(d)$ such that the exact sequence (2) splits.

Theorem 1 (Characterization of Symmorphic Crystallographic Groups) *The following propositions are equivalent.*

(i) *$G(d)$ is a d-dimensional symmorphic crystallographic group.*
(ii) *$G(d)$ has a subgroup $\widetilde{G} < G(d)$ mapped by π isomorphically onto G, i.e., $G(d) = i(\mathbf{T}).G$ and $i(\mathbf{T}) \cap \widetilde{G} = \{1\}$.*
(iii) *$G(d)$ has a subgroup $\widetilde{G} < G(d)$ such that every element $a \in G(d)$ is uniquely expressible in the form $a = i(h)\tilde{g}$, $h \in \mathbf{T}$, $\tilde{g} \in \widetilde{G}$.*
(iv) *The short exact sequence (2) is equivalent to the extension*

$$0 \longrightarrow \mathbf{T} \xrightarrow{i'} \mathbf{T} \rtimes G \xrightarrow{\pi'} G \longrightarrow 1 \tag{3}$$

Proof The equivalence of the propositions (i)–(iv) follows from the Definition 4 and standard results of algebra. (See, e.g., [10].)

Theorem 2 (Cohomology Symmorphic Crystallographic Groups Classes) *The symmorphic crystallographic groups $G(d)$, having point group G and translations group $\mathbf{T}$, are classified by $\mathbf{T}$-conjugacy classes that are in 1-1 correspondence with the elements of $H^1(G;\mathbf{T})$.*

Proof Even if this result refers to standard subjects in homological algebra, let us enter into some details of the proof in order to better specify how the theorem works. In fact these details will be useful in the following. In the case G acts trivially on $\mathbf{T}$, so that the group $G(d) = \mathbf{T} \times G$, the splitting of (2) are in one-to-one correspondence with homomorphisms $G \to \mathbf{T}$. In general case the splittings correspond to derivations (*crossed homomorphisms*) $d : G \to \mathbf{T}$ satisfying $d(ab) = da + a.db$, for all $a, b \in G$. In fact, let us consider the extension (3). A section $s : G \to \mathbf{T} \rtimes G$ has the form $s(g) = (dg, g)$, where d is a function $G \to \mathbf{T}$. One

[3] Usually one denotes such an extension simply with $G(d)/\mathbf{T}$, whether we are not interested to emphasize the notation for $G \cong G(d)/\mathbf{T}$.

has $s(g)s(g') = (dg + g.dg', gg')$. So s will be a homomorphism iff d is a derivation. Two splitting s_1, s_2, are called **T**-*conjugate* if there is an element $h \in \mathbf{T}$ such that $s_1(g) = i(h)s_2 i(h)^{-1}$, for all $g \in G$. Since $(h, 1)(k, g)(h, 1)^{-1} = (h + k - gh, g)$ in $\mathbf{T} \rtimes G$, the conjugacy relation becomes $d_1 g = h + d_2 g - gh$, in terms of the derivations d_1 and d_2 corresponding to s_1 and s_2, respectively. Thus d_1 and d_2 correspond to **T**-conjugate splittings iff their difference $d_2 - d_1$ is a function (*principal derivation*) $G \to \mathbf{T}$ of the form $g \mapsto gh - a$ for some fixed $h \in \mathbf{T}$. Therefore, **T**-conjugacy classes of splittings of a split extension of G by **T** correspond to the elements of the quotient group $Der(G, \mathbf{T})/P(G, \mathbf{T})$, where $Der(G, \mathbf{T})$ is the Abelian group of derivations $G \to \mathbf{T}$, and $P(G, \mathbf{T})$ is the group of principal derivations. On the other hand, considering the cochain complex $C^\bullet(G, \mathbf{T})$, we see that $Der(G, \mathbf{T})$ is the group of 1-cocycles and $P(G, \mathbf{T})$ is the group of 1-coboundaries. Thus we get the theorem.

Theorem 3 (Cohomology Crystallographic Group Classes) *The cohomological classification of d-dimensional crystallographic groups $G(d)$, with point group G, is made by means of the first cohomology group $H^1(G; \mathbb{R}^d/\mathbb{Z}^d)$. One has the natural isomorphism: $H^1(G; \mathbb{R}^d/\mathbb{Z}^d) \cong H^2(G; \mathbb{Z}^d)$. Two cohomology classes define equivalent crystallographic groups iff they are transformed one another by the normalizer of G in $GL_d(\mathbb{Z})$. Two crystallographic groups $G(d)$, $G(d)'$, belong to the same class (arithmetical class) if their point groups, respectively G, G', are conjugate in $GL_d(\mathbb{R})$, (in $GL_d(\mathbb{Z})$).*

Proof The proof follows directly by the following standard (co)homological algebra lemmas.

Lemma 1 **([11])**. *Let G be a group and A a G-module. Let $\mathcal{E}(G, A)$ be the set of equivalence classes of extensions of G by A corresponding to the fixed action of G on A. Then, there is a bijection $\mathcal{E}(G, A) \cong H^2(G, A)$.*

Lemma 2 **([11])**. *For any exact sequence*

$$0 \longrightarrow M' \longrightarrow M \longrightarrow M'' \longrightarrow 0 \tag{4}$$

of G-modules and any integer n there is a natural map $\delta : H^n(G; M'') \to H^{n+1}(G; M')$ such that the sequence

$$0 \longrightarrow H^0(G; M') \longrightarrow H^0(G; M) \longrightarrow H^0(G; M'') \xrightarrow{\delta} H^1(G; M') \longrightarrow \dots \tag{5}$$

is exact. Furthermore if P is a projective (resp. if Q is an injective) $\mathbb{Z}G$-module then $H_n(G; P) = 0$ (resp. $H^n(G; Q) = 0$) for $n > 0$.[4]

In fact, it is enough to take $A = \mathbf{T} = \mathbb{Z}^d$, $M' = \mathbb{Z}^d$, $M = \mathbb{R}^d$ and $M'' = \mathbb{R}^d/\mathbb{Z}^d$.

[4] $\mathbb{Z}G$ is the free $\mathbb{Z}$-module generated by the elements of G. The multiplication in G extends uniquely to a $\mathbb{Z}$-bilinear product $\mathbb{Z}G \times \mathbb{Z}G \to \mathbb{Z}G$ so that $\mathbb{Z}G$ becomes a ring called the *integral group ring* of G. A G-*module* A, is just a (left)$\mathbb{Z}G$-module on the Abelian group A. This action can be identified by a group homomorphism $G \to Hom_{group}(A)$.

Definition 5 We call *lattice subgroups* of a crystallographic group $G(d)$ the set of its subgroups.[5]

Definition 6 We call *d-dimensional crystal* any topological space ${}^{\maltese}E \subset E$, i.e., contained in the d-dimensional Euclidean affine space E, having the crystallographic group $G(d)$ as (extrinsic) symmetry group. We call *unit cell* of ${}^{\maltese}E$, the compact quotient space ${}^{\maltese}E/\mathbf{T}$, having the point group as (extrinsic) symmetry group.[6]

Theorem 4 (Hilbert's 18th Problem) *For any dimension d, there are, up to equivalence, only finitely many-dimensional crystallographic space groups. These are called space-group types. We denote by* $[G(d)]$ *the space-group type identified by the space group* $G(d)$.[7]

Proof Hilbert's 18th problem, introduced at the beginning of 1900, has been first proved in 1911 by L. Bieberbach [8]. For modern proofs see also L. S. Charlap. [15].[8]

Example 1 (3-dimensional crystal-group types). Up to isomorphisms, there are 17 crystallographic groups in dimension 2 and 219 in dimension 3. However, if the spatial groups are considered up to conjugacy with respect to orientation-preserving affine transformations, their number is 230. These last can be called *affine space-group types*. These are just the F. S. Fedorov and A. Schönflies groups [21, 96]. The corresponding translation-subgroup types, or *Bravais lattices*, are 14. In Appendix A, Table 5, are reported the 32 affine crystallographic point-group types, and in Table 7 the 230 affine crystallographic space-group types. There are 66 symmorphic space-group types in $[G(3)]$. For the other 164 $G(3)$ cannot be identified with the semidirect product $\mathbf{T} \rtimes G(3)/\mathbf{T}$. It is useful, also, to know the subgroups corresponding to the 32 crystallographic point groups, i.e., their subgroup lattices, in relation to the results of the following section where integral bordism groups of PDEs will be related to crystallographic subgroups. For this we report in Appendix B a full list for such subgroups, and in Appendix C a list of amalgamated free products in $[G(3)]$.[9] See also Appendix A, Table 8 for the 2-dimensional 17 crystal-group types, and Appendix D for the corresponding list of subgroups.[10]

---

[5] The lattice subgroups of a group G is a lattice under inclusion. The identity $1 \in G$ identifies the *minimum* $\{1\} < G$ and the *maximum* is just G. In the following we will denote also by $e = 1$ the identity of a group G. The subgroup $\mathbf{T} \subset G(d)$ is called also the *Bieberbach lattice* of $G(d)$.

[6] For example if ${}^{\maltese}E$ is identified by means of an infinite graph, (see, e.g., [100]), then the unit cell is the corresponding fundamental finite graph. Of course d-dimensional chains can be associated to such graphs, so that also the corresponding unit cells can be identified with compact d-chains.

[7] Note that a space group is characterized other than by translational and point symmetry, also by the metric-parameters characterizing the unit cell. Thus the number of space groups is necessarily infinite.

[8] See also Refs. [1, 4, 6, 7, 32, 45, 91, 95, 97], and works by M. Gromov [25] and E. Ruh [95] on the almost flat manifolds, that are related to such crystallographic groups.

[9] Let A, B and C groups and $B \lhd A$, $B \lhd C$, the *amalgamated free product* $A\bigstar_B C$ is generated by the elements of A and C with the common elements from B identified.

[10] Let us recall that the index of a subgroup $H \subset G$, denoted $[G : H]$, is the number of *left cosets* $aH = \{ah : h \in H\}$, (resp. *right cosets* $Ha = \{ha : h \in H\}$), of H. For a finite group G one

Definition 7 (Generalized Crystallographic Group) Let K be a principal ideal domain, H a finite group, and M a KH-module which as K-module is free of finite rank, and on which H acts faithfully. A *generalized crystallographic group* is a group $G(d)$ which has a normal subgroup isomorphic to M such that the following conditions are satisfied:

(i) $G(d)/M \cong H$.
(ii) Conjugation in $G(d)$ gives the same action of H on M.
(iii) The extension

$$0 \longrightarrow M \longrightarrow G(d) \longrightarrow H \longrightarrow 0 \tag{6}$$

does not split. We define *dimension* of $G(d)$ the K-free rank d of M. We define *holonomy group* of $G(d)$ the group H.

Example 2 With $K = \mathbb{Z}$ one has the crystallographic groups.

Example 3 A *complex crystallographic group* is a discrete group of affine transformations of a complex affine space $(V, \mathbf{V})$, such that the quotient $X \equiv V/G$ is compact. This is a generalized crystallographic group with $K = \mathbb{C}$.[11]

We now give some examples of crystallographic subgroups in dimension $d = 3$. In fact, it will be useful to know such subgroups in relation to results of the next section.

Example 4 (The Group $\mathbb{Z} \bigoplus \mathbb{Z}$) This is the crystallographic subgroup, of the crystallographic group $G(3) = \mathbb{Z}^3 \times \mathbb{Z}_1$, with point group $C_1 = \mathbb{Z}_1$ (triclinic syngony). Let us emphasize that $\mathbb{Z} \bigoplus \mathbb{Z}$ is crystallographic since it can be identified with the 2-dimensional crystallographic group $G(2) = \mathbb{Z}^2 \times \mathbb{Z}_1 = p1$, generated by translations parallel to the x and y-axes, with point group $\mathbb{Z}_1$.

Example 5 (The Groups $\mathbb{Z}^2 \bigoplus \mathbb{Z}_n$, $n = 2, 3, 4, 6$) These are subgroups of the crystallographic groups $G(3) = \mathbb{Z}^3 \rtimes \mathbb{Z}_n$, $n = 2, 3, 4, 6$, (point groups $\mathbb{Z}_n$, monoclinic, hexagonal, tetragonal, trigonal syngony respectively). Furthermore, $\mathbb{Z}^2 \times \mathbb{Z}_n$, $n = 2, 3, 4, 6$, can be considered also subgroups of the 2-dimensional crystallographic groups $G(2) = \mathbb{Z}^2 \rtimes \mathbb{Z}_n = pn$, $n = 2, 3, 4, 6$, (point groups $\mathbb{Z}_n$, oblique, trigonal, square, hexagonal syngony respectively), generated by translations parallel to the x and y-axes, and a rotation by π/n, $n = 2, 3, 4, 6$, about the origin.

Example 6 (The Groups $\mathbb{Z}_2 \bigoplus \mathbb{Z}_2$) This group coincides with the amalgamated free product $\mathbb{Z}_2 \bigstar_e \mathbb{Z}_2$ that is generated by reflection over x-axis and reflection over y-axis. It is a subgroup of the crystallographic group $G(3) = \mathbb{Z}^3 \rtimes D_4$, (point group D_4, tetragonal syngony). (See Appendixes A and C.) The group $\mathbb{Z}_2 \bigoplus \mathbb{Z}_2$ is also a

---

has the following formula: (*Lagrange's formula*) $[G : H] = \frac{o(G)}{o(H)}$, where $o(G)$, resp. $o(H)$, is the order of G, resp. H. If $aH = Ha$, for any $a \in G$, then H is said to be a *normal subgroup*. Every subgroup of index 2 is normal, and its cosets are the subgroup and its complement.

[11] See also the recent work by Bernstein and Schwarzman on the complex crystallographic groups [7].

subgroup of the 2-dimensional crystallographic groups $G(2) = \mathbb{Z}^2 \rtimes D_4 = p4m$ and $G(2) = \mathbb{Z}^2 \rtimes D_4 = p4g$ that have both point group $D_4 = \mathbb{Z}_2 \times \mathbb{Z}_2$, and square syngony. (See in Appendix A, Tables 6 and 8.)

Example 7 (The Groups $\mathbb{Z}_4 \bigstar_{\mathbb{Z}_2} D_2$) (Note that $D_2 \cong \mathbb{Z}_2$, see in Appendix A, Table 6). This group is generated by rotation of π, rotation by $\pi/2$, and reflection over the x-axis. It can be considered a subgroup of some crystallographic group $G(3)$, (see Appendix C).

Definition 8 The subgroups of $GL_d(\mathbb{Z})$ that are lattice symmetry groups are called *Bravais subgroups*. Every maximal finite subgroup of $GL_d(\mathbb{Z})$ is a *Bravais subgroup*.

Definition 9 The *geometrical (arithmetical) holohedry* of a crystallographic group $G(d)$ is the smallest Bravais subgroup $\widehat{G}(d)$ containing the point group $G \equiv G(d)/\mathbf{T}$ of $G(d)$.

Definition 10 A crystallographic group $G(d)$ is said in *general position* if there is no affine transformation $\phi \in A(E)$ such that $\phi G(d)\phi^{-1} \equiv {}^{\phi}G(d)$ is a crystallographic group whose lattice of parallel translations has lower symmetry.

Proposition 1 *If the crystallographic group $G(d)$ is in general position, then its holohedry $\widehat{G}(d)$ is the lattice symmetry group of the parallel translations* $\mathbf{T}$ *of $G(d)$.*

Definition 11 Two crystallographic groups belong to the same *syngony (Bravais type)* if their geometrical (arithmetical) holohedries coincide.

Example 8 For the 3-dimensional case there are 73 arithmetic crystal classes. The space group types with the same point group symmetry and the same type of centering belong to the same arithmetic crystal class. An arithmetic crystal class is indicated by the crystal symbol of the corresponding point group followed by the symbol of the lattice. Furthermore $G(3)$ has seven syngonies and 14 Bravais types of crystallographic groups. (See Appendix A, Table 7. The number between brackets () after the symbol of the point group is the number of space-group types with that point group.) The geometric crystal classes with the point symmetries of the lattices are called *holohedries* and are 7. The other 25 geometric crystal classes are called *merhoedries*. The *Bravais classes* (or *Bravais arithmetic crystallographic classes*) are the arithmetic crystal classes with the point symmetry of the lattice. The Bravais types of lattices and the Bravais classes have the same point symmetry.

3 Integral Bordism Groups vs. Crystallographic Groups

In the following we shall relate above crystallographic groups to the geometric structure of PDEs. More precisely in some previous works we have characterized the structure of global solutions of PDEs by means of *integral bordism groups*. Let us start with the bordism groups.[12]

[12] For general information on bordism groups and related problems in differential topology, see, e.g., Refs. [31, 39, 41, 52, 90, 94, 98, 99, 102–105, 107, 108].

Theorem 5 (Bordism Groups vs. Crystallographic Groups) *Bordism groups of closed compact smooth manifolds can be considered as subgroups of crystallographic ones. More precisely one has the following:*

(i) *To each nonoriented bordism group Ω_n, can be canonically associated a crystallographic group $G(q)$, (crystal group of Ω_n), for a suitable integer q, (crystal dimension of Ω_n), such that one has the following split short exact sequence:*

$$0 \longrightarrow \mathbb{Z}^q \rightleftarrows G(q) \rightleftarrows \Omega_n \longrightarrow 0 \tag{7}$$

So Ω_n is at the same time a subgroup of its crystal group, as well as an extension of this last. If there are many crystal groups satisfying the condition in (7), then we call respectively crystal dimension and crystal group of Ω_n the littlest one.

(ii) *To each oriented bordism group ${}^+\Omega_n$, $n \not\equiv 0$ mod 4, can be canonically associated a crystallographic group $G(d)$, (crystal-group) of ${}^+\Omega_n$, for a suitable integer d, crystal dimension) of ${}^+\Omega_n$, such that one has the following short exact sequence:*

$$0 \longrightarrow {}^+\Omega_n \longrightarrow G(q) \longrightarrow G(d)/{}^+\Omega_n \longrightarrow 0 \tag{8}$$

So ${}^+\Omega_n$ is a subgroup of its crystal group.

Proof Let us recall the structure of bordism groups.

Lemma 3 (Pontrjagin–Thom–Wall [99, 102, 107]) *A closed n-dimensional smooth manifold V, belonging to the category of smooth differentiable manifolds, is bordant in this category, i.e., $V = \partial M$, for some smooth $(n+1)$-dimensional manifold M, iff the Stiefel–Whitney numbers $< w_{i_1} \cdots w_{i_p}, \mu_V >$ are all zero, where $i_1 + \cdots + i_p = n$ is any partition of n and μ_V is the fundamental class of V. Furthermore, the bordism group Ω_n of n-dimensional smooth manifolds is a finite Abelian torsion group of the form:*

$$\Omega_n \cong \mathbb{Z}_2 \bigoplus \cdots_q \cdots \bigoplus \mathbb{Z}_2, \tag{9}$$

where q is the number of nondyadic partitions of n.[13] *Two smooth closed n-dimensional manifolds belong to the same bordism class iff all their corresponding Stiefel-Whitney numbers are equal. Furthermore, the bordism group ${}^+\Omega_n$ of closed n-dimensional oriented smooth manifolds is a finitely generated Abelian group of the form:*

$${}^+\Omega_n \cong \mathbb{Z} \bigoplus \cdots \bigoplus \mathbb{Z} \bigoplus \mathbb{Z}_2 \bigoplus \cdots \bigoplus \mathbb{Z}_2, \tag{10}$$

[13] A partition $(i_1, \cdots, i_r)$ of n is nondyadic if none of the i_β are of the form $2^s - 1$.

where infinite cyclic summands can occur only if $n \equiv 0$ mod 4. Two smooth closed oriented n-dimensional manifolds belong to the same bordism class iff all their corresponding Stiefel–Whitney and Pontrjagin numbers are equal.[14]

Let us recall that a group G is *cyclic* iff it is generated by a single element. All cyclic groups G are isomorphic either to $\mathbb{Z}_p$, $p \in \mathbb{Z}^+$ or $\mathbb{Z}$. $G \cong \mathbb{Z}_p$ iff there is some finite integer q such that $g^q = e$, for each $g \in G$. Here e is the unit of G. A group G is *virtually cyclic* if it has a cyclic subgroup H of finite index. All finite groups are virtually cyclic, since the trivial subgroup $H = \{e\}$ is cyclic. Therefore, to compute the finite virtually cyclic subgroups of the d-dimensional crystallographic groups is equivalent to compute the finite subgroups. Of course, there can also be infinite virtually cyclic subgroups. For these we can use the following lemma.

Lemma 4 (Scott and Wall [98]) *Given a group G, the only infinite virtually cyclic subgroups of G will be semidirect products $H \rtimes_\alpha \mathbb{Z}$ and amalgamated free products $A\bigstar_B C$ for $H, A, B, C < G$.*

The finite subgroups of the crystallographic groups may be derived exclusively from their point groups. The following lemmas are useful to gain the proof.

Lemma 5 *(a) Let $f = (0, b)$ be a finite-ordered element of a crystallographic group $G(d)$. Furthermore, let $x = (v, 1)$ be a translation. Then, f commutes with x iff $b(v) = v$, i.e., b fixes v.*

(b) $F \times \mathbb{Z}$ is a subgroup of some d-dimensional crystallographic group iff F is a subgroup of some $(d-1)$-dimensional crystallographic group.

Proof (a) It follows directly by computation.

(b) Let us assume that F is a subgroup of some $(d-1)$-dimensional crystallographic group $G(d-1)$, then $F \times \mathbb{Z}$ is a subgroup of the d-dimensional crystallographic group $G(d-1) \times \mathbb{Z}$. Vice versa, let $F \times \mathbb{Z}$ be a subgroup of some d-dimensional crystallographic group $G'(d)$. Since in this direct product the generator $x \in \mathbb{Z}$ commutes with all $f \in F$ and x belongs to the Bieberbach lattice $\mathbf{T}'$ of $G'(d)$, there are at most $(d-1)$ independent elements $t \in \mathbf{T}'$ which do not commute with a given $f \in F$. Therefore, F must be a subgroup of some $(d-1)$-dimensional crystallographic group.

Lemma 6 *If a crystallographic group $G(d)$ admits a subgroup $F \rtimes_\alpha \mathbb{Z}$ for some finite group F and some homomorphism $\alpha : \mathbb{Z} \to Aut(F)$, then $G(d)$ also admits a subgroup $F \times \mathbb{Z}$.*

Proof Since $\mathbb{Z}$ is cyclic, α is completely determined by the $\alpha(1)$. Since F is finite, $Aut(F)$ is also finite, hence one has $\alpha(1)^q = 1$ for some finite q. Therefore, the elements $y = x^{pq} \in \mathbb{Z}$, $\forall p \in \mathbb{Z}$, commute with any $f \in F$, hence $F \rtimes_\alpha \mathbb{Z}$ contains as a subgroup $F \times \mathbb{Z}$, identified with the couples (f, y), were y are above defined elements.

---

[14] *Pontrjagin numbers* are determined by means of homonymous characteristic classes belonging to $H^\bullet(BG, \mathbb{Z})$, where BG is the classifying space for G-bundles, with $G = S_p(n)$. See, e.g., Refs. [39, 41, 94, 99, 102–104, 107, 108].

Lemma 7 *Let $H = F \rtimes_\alpha \mathbb{Z}$, $\mathbb{Z} = < x >$, be a subgroup of a crystallographic group $G(d)$. The elements $f \in F$ fix the shift vectors $x^q \in \mathbf{T}$.*

Proof According to Lemma 6 we can consider the subgroup $K = F \times \mathbb{Z} < H$. Then the elements $z \in \mathbb{Z}$ of K correspond to translation vectors $x^q \in \mathbb{Z} < H$. On the other hand, since K is a direct product $F \times \mathbb{Z}$, all $f \in F$ commute with all $z \in K$. By Lemma 5(a), all $f \in F$ fix all $x^q \in H$.

Lemma 8 (Alperin and Bell [4]) *Let F and H be groups, let $\alpha : H \to Aut(F)$ b a homomorphism, and $\phi \in Aut(F)$. If $\hat{\phi}$ is the inner automorphism of $Aut(F)$ induced by ϕ, then $F \rtimes_{\hat{\phi}\circ\alpha} H \cong F \rtimes_\alpha H$.*

(This means that if we identify the conjugacy classes of authomorphisms of a given group, we need to consider only one element of each class to evaluate the candidacy of all automorphisms in that class.)

Lemma 9 *If the presentation of the amalgamated free product contains two or more elements of order two that do not commute, then the amalgamated free product is not a subgroup of any 3-dimensional crystallographic group.*

Proof In three dimensions there are only three possible elements of order two, inversions, π rotation, and reflection. All these simmetries commute. Therefore, an amalgamated free product with two order two elements that do not commute cannot exist in three dimensions.

Let us first note that for bordism groups identified with some finite or infinite cyclic groups, theorem is surely true by considering the following two standard lemmas.

Lemma 10 *If H is a finite subgroup of a group G, every element $a \in H$ generates a finite cyclic subgroup $< a > \equiv \mathbb{Z}_n \subset H$, where n is the order of a, and $a^{-1} = a^{n-1}$, or equivalently $a^n = e$, where e is the unit of H (and also that of G.)*

Lemma 11 *Every element a of a group G generates a cyclic subgroup $< a > \subset G$. If a has infinite order, then $< a > \cong \mathbb{Z}$.*

Thus if $\Omega_p \cong \mathbb{Z}_2$, or ${}^+\Omega_p \cong \mathbb{Z}$, it follows that theorem is proved.

Now, let us consider the more general situation. We shall consider the following lemma.

Lemma 12 *The group $\mathbb{Z}^s \rtimes \mathbb{Z}_2^s$ can be considered a crystallographic group in the Euclidean space $\mathbb{R}^s$.*

Proof The $\mathbb{Z}^s$-conjugacy classes of splittings of the split extension

$$0 \longrightarrow \mathbb{Z}^s \longrightarrow \mathbb{Z}^s \rtimes \mathbb{Z}_2^s \longrightarrow \mathbb{Z}_2^s \longrightarrow 0 \tag{11}$$

are in one-to-one correspondence with the elements of

$$\left\{\begin{array}{ll} H^1(\mathbb{Z}_2^s;\mathbb{Z}^s) & = Hom_{\mathbb{Z}}(H_1(\mathbb{Z}_2^s;\mathbb{Z});\mathbb{Z}^s) \bigoplus Ext_{\mathbb{Z}}(H_0(\mathbb{Z}_2^s;\mathbb{Z});\mathbb{Z}^s) \\ & \cong \left(Hom_{\mathbb{Z}}(H_1(\mathbb{Z}_2^s;\mathbb{Z});\mathbb{Z})\right)^s \bigoplus Ext_{\mathbb{Z}}(\mathbb{Z};\mathbb{Z})^s \\ & \cong Hom_{\mathbb{Z}}(\mathbb{Z}_2;\mathbb{Z})^{s^2}. \end{array}\right. \tag{12}$$

We have used the fact that for any finite cyclic group K of order i one has

$$H_r(K;\mathbb{Z}) = \left\{\begin{array}{ll}\mathbb{Z} & \text{if } r=0\\ \mathbb{Z}_i & \text{if } r=\text{odd}\\ 0 & \text{if } r>0 \text{ even.}\end{array}\right\} \Rightarrow H_r(\mathbb{Z}_2;\mathbb{Z}) = \left\{\begin{array}{ll}\mathbb{Z} & \text{if } r=0\\ \mathbb{Z}_2 & \text{if } r=\text{odd}\\ 0 & \text{if } r>0 \text{ even.}\end{array}\right\}. \tag{13}$$

Furthermore, the Künneth theorem for groups allows us to write the unnatural isomorphism:

$$H_r(G_1\times G_2;\mathbb{Z}) \cong \bigoplus_{p+q=r} H_p(G_1;\mathbb{Z})\otimes_{\mathbb{Z}} H_q(G_2;\mathbb{Z}) \bigoplus_{p+q=r-1} Tor^{\mathbb{Z}}(H_p(G_1;\mathbb{Z}), H_q(G_2;\mathbb{Z})) \tag{14}$$

for any two groups G_1 and G_2. Taking into account that $Tor^{\mathbb{Z}}(A,B)=0$ for projective $\mathbb{Z}$-module A (or B), and $Ext_{\mathbb{Z}}(\mathbb{Z};\mathbb{Z})=0$ since $\mathbb{Z}$ is a projective $\mathbb{Z}$-module, we get that $H^1(\mathbb{Z}_2^s;\mathbb{Z}^s) = Hom_{\mathbb{Z}}(\mathbb{Z}_2;\mathbb{Z})^{s^2}$. Thus, we conclude that there is not an unique $\mathbb{Z}^s$-conjugacy class in the splitting in (11). Furthermore, the set of equivalence classes of s-dimensional crystallographic groups with such a point group are in 1-1 correspondence with

$$\left\{\begin{array}{ll} H^2(\mathbb{Z}_2^s;\mathbb{Z}^s) & \cong Hom_{\mathbb{Z}}(H_2(\mathbb{Z}_2^s;\mathbb{Z});\mathbb{Z}^s)\bigoplus Ext_{\mathbb{Z}}(H_1(\mathbb{Z}_2^s;\mathbb{Z});\mathbb{Z}^s)\\ & \cong Hom_{\mathbb{Z}}(H_2(\mathbb{Z}_2^s;\mathbb{Z});\mathbb{Z})^s \bigoplus Ext_{\mathbb{Z}}(\mathbb{Z}_2^s;\mathbb{Z}^s)\\ & \cong \bigoplus_{1\le r\le (s-1)} Hom_{\mathbb{Z}}(\mathbb{Z}_2\otimes_{\mathbb{Z}}\mathbb{Z}_2^{s-r};\mathbb{Z})\bigoplus \mathbb{Z}_2^{s^2}. \end{array}\right. \tag{15}$$

The above calculation holds for $s>1$. Instead, for $s=1$, one has $H^2(\mathbb{Z}_2;\mathbb{Z}) = \mathbb{Z}_2$.

Therefore, if $\Omega_n = \mathbb{Z}_2^s = \mathbb{Z}_2\times\cdots_s\cdots\times\mathbb{Z}_2$ it can be identified with a point group G of a crystallographic group $G(q)$, belonging to one of these equivalence classes, such that $G(q) < M(\mathbb{R}^q)$. So one has the exact sequence:

$$0\longrightarrow \mathbb{Z}^q \longrightarrow G(q)\longrightarrow \Omega_n \longrightarrow 1 \tag{16}$$

that proves that Ω_n admits a crystallographic group as its extension. Now, since $G(q)$ contains also as subgroup $\mathbb{Z}^q\times\mathbb{Z}_2^q$, that contains as subgroup $\mathbb{Z}_2^q$, it follows that $G(q)$ contains also as subgroup Ω_n. So one has also the following exact sequence:

$$0\longrightarrow \Omega_n \longrightarrow G(q)\longrightarrow G(q)/\Omega_n \longrightarrow 0 \tag{17}$$

Since $G(q)/\Omega_n\cong\mathbb{Z}^q$, it follows that sequence (17) is the split sequence of (16), and vice versa.

Let us consider, now, the oriented case. Let us exclude the case $n\equiv 0 \bmod 4$. Then we can write ${}^+\Omega_n = \mathbb{Z}^r\times\mathbb{Z}_2^s$. Let us assume $r\ge s$. Then we can consider in $\mathbb{R}^r$ the crystallographic group $G(r)=\mathbb{Z}^r\rtimes\mathbb{Z}_2^s$. This contains as subgroup $\mathbb{Z}^r\times\mathbb{Z}_2^s = {}^+\Omega_n$.

Therefore one has the following exact sequence:

$$0 \longrightarrow {}^{+}\Omega_n \longrightarrow G(r) \longrightarrow G(r)/{}^{+}\Omega_n \longrightarrow 0 \tag{18}$$

Let us assume, now, that $r < s$, then one has the following sequence of subgroups:

$${}^{+}\Omega_n = \mathbb{Z}^r \times \mathbb{Z}_2^s < \mathbb{Z}^s \times \mathbb{Z}_2^s < \mathbb{Z}^s \rtimes \mathbb{Z}_2^s = G(s). \tag{19}$$

So we have the following short exact sequence:

$$0 \longrightarrow {}^{+}\Omega_n \longrightarrow G(s) \longrightarrow G(s)/{}^{+}\Omega_n \longrightarrow 0 \tag{20}$$

Therefore, we can conclude that ${}^{+}\Omega_n$ is a subgroup of the crystallographic group $G(d) = \mathbb{Z}^d \rtimes \mathbb{Z}_2^d$, with $d = \max\{r, s\}$.

Remark 2 It is important to emphasize that crystallographic groups and bordism groups, even if related by above theorem, are in general different groups. In other words, it is impossible to identify any crystallographic group with some bordism group, since the first one is in general non-Abelian; instead the bordism groups are Abelian groups.

We can extend the above proof also by including bordism groups relatively to some manifold.

Theorem 6 *Bordism groups relative to smooth manifolds can be considered as extensions of crystallographic subgroups.*

Proof Let us recall that a *k-cycle* of M be a couple (N, f), where N is a k-dimensional closed (oriented) manifold and $f : N \to M$ is a differentiable mapping. A *group of cycles* (N, f) of an n-dimensional manifold M is the set of formal sums $\sum_i (N_i, f_i)$, where (N_i, f_i) are cycles of M. The quotient of this group by the cycles equivalent to zero, i.e., the boundaries, gives the *bordism groups* $\underline{\Omega}_s(M)$. We define *relative bordisms* $\underline{\Omega}_s(X, Y)$, for any pair of manifolds (X, Y), $Y \subset X$, where the boundaries are constrained to belong to Y. Similarly we define the *oriented bordism groups* ${}^{+}\underline{\Omega}_s(M)$ and ${}^{+}\underline{\Omega}_s(X, Y)$. One has $\underline{\Omega}_s(*) \cong \Omega_s$ and ${}^{+}\underline{\Omega}_s(*) \cong {}^{+}\Omega_s$. For bordisms, the theorem of invariance of homotopy is valid. Furthermore, for any CW-pair (X, Y), $Y \subset X$, one has the isomorphisms: $\underline{\Omega}_s(X, Y) \cong \Omega_s(X/Y)$, $s \geq 0$. One has a natural group homomorphism $\underline{\Omega}_s(X) \to H_s(X; \mathbb{Z}_2)$. This is an isomorphism for $s = 1$. In general, $\underline{\Omega}_s(X) \neq H_s(X; \mathbb{Z}_2)$. In fact one has the following lemma.

Lemma 13 (Quillen [90]) *One has the canonical isomorphism:*

$$\underline{\Omega}_p(X) \cong \bigoplus_{r+s=p} H_r(X; \mathbb{Z}_2) \otimes_{\mathbb{Z}_2} \Omega_s. \tag{21}$$

In particular, as $\Omega_0 = \mathbb{Z}_2$ and $\Omega_1 = 0$, we get $\underline{\Omega}_1(X) \cong H_1(X; \mathbb{Z}_2)$. Note that for contractible manifolds, $H_s(X) = 0$, for $s > 0$, but $\underline{\Omega}_s(X)$ cannot be trivial for any $s > 0$. So, in general, $\underline{\Omega}_s(X) \neq H_s(X; \mathbb{Z}_2)$.

So we get the following short exact sequence:

$$0 \longrightarrow \underline{K}_p(X) \longrightarrow \underline{\Omega}_p(X) \longrightarrow \Omega_p \longrightarrow 0 \tag{22}$$

where $\underline{K}_p(X) \cong \bigoplus_{r+s=p,r>0} H_r(X;\mathbb{Z}_2) \otimes_{\mathbb{Z}_2} \Omega_s$. Therefore by using Theorem 5 we get the proof soon.

Let us, now, consider a relation between PDEs and crystallographic groups. This will give us also a new classification of PDEs on the ground of their integral bordism groups.

Definition 12 We say that a PDE $E_k \subset J^k_n(W)$ is an *extended 0-crystal PDE*, if its integral bordism group is zero.

The first main theorem is the following one relating the integrability properties of a PDE to crystallographic groups.

Theorem 7 (Crystal Structure of PDEs) *Let $E_k \subset J^k_n(W)$ be a formally integrable and completely integrable PDE, such that* $\dim E_k \geq 2n+1$. *Then its integral bordism group $\Omega^{E_k}_{n-1}$ is an extension of some crystallographic subgroup. Furthermore if W is contractible, then E_k is an extended* 0*-crystal PDE, when* $\Omega_{n-1} = 0$.

Proof Let us first recall some definitions and results about integral bordism groups of PDE's. (For details see Refs. [52, 54, 56, 57, 60, 62, 63], and the following related works [2, 3].) Let W be an $(n+m)$-dimensional smooth manifold with fiber structure $\pi : W \to M$ over a n-dimensional smooth manifold M. Let $E_k \subset J^k_n(W)$ be a PDE of order k, for n-dimensional submanifolds of W. For an "admissible" p-dimensional, $p \in \{0, \cdots, n-1\}$, integral manifold $N \subset E_k$, we mean a p-dimensional smooth submanifold of E_k, contained in an admissible integral manifold $V \subset E_k$, of dimension n, i.e., a solution of E_k, that can be deformed into V, in such a way that the deformed manifold $\widetilde{N}$ is diffeomorphic to its projection $\widetilde{X} \equiv \pi_{k,0}(\widetilde{N}) \subset W$. In such a case the k-prolongation $\widetilde{X}^{(k)} = \widetilde{N}$. The existence of p-dimensional admissible integral manifolds $N \subset E_k$ is obtained solving Cauchy problems of dimension $p \in \{0, \cdots, n-1\}$, i.e., finding n-dimensional admissible integral manifolds (solutions) of a PDE $E_k \subset J^k_n(W)$, that contain N. Existence theorems for such solutions can be studied in the framework of the geometric theory of PDEs. For a modern approach founded on webs structures see also [2, 3]. A geometric way to study the structure of global solutions of PDEs, is to consider their integral bordism groups. Let $N_i \subset E_k$, $i = 1, 2$ be two $(n-1)$-dimensional compact closed admissible integral manifolds. Then, we say that they are *E_k-bordant* if there exists a solution $V \subset E_k$, such that $\partial V = N_1 \sqcup N_2$ (where $\sqcup$ denotes disjoint union). We write $N \sim_{E_k} N_2$. This is an equivalence relation and we will denote by $\Omega^{E_k}_{n-1}$ the set of all E_k-bordism classes $[N]_{E_k}$ of $(n-1)$-dimensional compact closed admissible integral submanifolds of E_k. The operation of taking disjoint union defines a structure of Abelian group on $\Omega^{E_k}_{n-1}$. In order to distinguish between integral bordism groups where the bording manifolds are smooth, (resp. singular, resp. weak), we shall use also the following symbols $\Omega^{E_k}_{n-1}$, (resp. $\Omega^{E_k}_{n-1,s}$, resp. $\Omega^{E_k}_{n-1,w}$).[15] Let us first consider the integral bordism group $\Omega^{E_k}_{n-1,w}$ (or $\Omega^{E_k}_{n-1,s}$) for weak solutions (or for singular solutions). We shall use Theorem 2.15 in [62], that we report below to be more direct. (See also [2, 3].)

Theorem 2.15 *in [62]. Let $E_k \subset J^k_n(W)$ be a formally and also completely integrable PDE, such that* $\dim E_k \geq 2n + 1$. *Then one has the following canonical isomorphism:*

$$\Omega^{E_k}_{p,w} \cong \bigoplus_{r,s,r+s=p} H_r(W;\mathbb{Z}_2) \oplus_{\mathbb{Z}_2} \Omega_s. \tag{23}$$

Furthermore, if W is an affine fiber bundle $\pi : W \to M$ over a n-dimensional manifold M, one has the isomorphisms:[16]

$$\Omega^{E_k}_{p,w} \cong \underline{\Omega}_p(M) \cong \bigoplus_{r,s,r+s=p} H_r(M;\mathbb{Z}_2) \oplus_{\mathbb{Z}_2} \Omega_s. \tag{24}$$

So we can write for the weak integral bordism group

$$\Omega^{E_k}_{n-1,w} \cong \bigoplus_{r+s=n-1} H_r(W;\mathbb{Z}_2) \otimes_{\mathbb{Z}_2} \Omega_s. \tag{25}$$

Then one has that $\Omega^{E_k}_{n-1,w}$ is an extension of the bordism group Ω_{n-1}. Hence by using Theorem 5 we get the proof.

Let us now consider the integral bordism group $\Omega^{E_k}_{n-1}$ for smooth solutions. This bordism group is related to previous one, and to singular bordism group $\Omega^{E_k}_{n-1,s}$, by means of the exact commutative diagram (26). Furthermore, the relation between $\Omega^{E_k}_{n-1}$, $\Omega^{E_k}_{n-1,w}$ and Ω_{n-1} is given by means of the exact commutative diagram (27), where $K^{E_k}_{n-1,w} = \ker(a)$, $K^{E_k}_{n-1,w;n-1} = \ker(b)$, $\overline{K}^{E_k}_{n-1} = \ker(c)$, with $c = b \circ a$. From this we get that also $\Omega^{E_k}_{n-1}$ can be considered an extension of Ω_{n-1} if $\Omega^{E_k}_{n-1,w}$ is so. Therefore we can apply Theorem 5 also to the integral bordism group for smooth solutions whether it can be applied to the integral bordism group for weak (or singular) solutions. Therefore, the theorem is proved.

---

[15] Let us recall that *weak solutions*, are solutions V, where the set $\Sigma(V)$ of singular points of V, contains also discontinuity points, $q, q' \in V$, with $\pi_{k,0}(q) = \pi_{k,0}(q') = a \in W$, or $\pi_k(q) = \pi_k(q') = p \in M$. We denote such a set by $\Sigma(V)_S \subset \Sigma(V)$, and, in such cases we shall talk more precisely of *singular boundary* of V, like $(\partial V)_S = \partial V \setminus \Sigma(V)_S$. However, for abuse of notation we shall denote $(\partial V)_S$, (resp. $\Sigma(V)_S$), simply by (∂V), (resp. $\Sigma(V)$), also if no confusion can arise. Solutions with such singular points are of great importance and must be included in a geometric theory of PDEs, too [62]. Let us also emphasize that singular solutions can be identified with integral n-chains in E_k, and in this category can be considered also *fractal solutions*, i.e., solutions with sectional fractal or multifractal geometry. (For fractal geometry see, e.g., [20, 35, 40].)

[16] The bording solutions considered for the bordism groups are singular solutions if the symbols g_k and g_{k+1} are different from zero, and for singular weak solutions in the general case. Here we have denoted $\underline{\Omega}_p(X)$ the p-bordism group of a manifold X. For information on such structure of the algebraic topology see, e.g., [31, 52, 90, 94, 99, 101–104, 107, 108].

$$
\begin{array}{ccccccccc}
 & & 0 & & 0 & & 0 & & \\
 & & \downarrow & & \downarrow & & \downarrow & & \\
0 & \rightarrow & K^{E_k}_{n-1,w/(s,w)} & \rightarrow & K^{E_k}_{n-1,w} & \rightarrow & K^{E_k}_{n-1,s,w} & \rightarrow & 0\\
 & & \downarrow & & \downarrow & & \downarrow & & \\
0 & \rightarrow & K^{E_k}_{n-1,s} & \rightarrow & \Omega^{E_k}_{n-1} & \rightarrow & \Omega^{E_k}_{n-1,s} & \rightarrow & 0\\
 & & \downarrow & & \downarrow & & \downarrow & & \\
 & & 0 & \rightarrow & \Omega^{E_k}_{n-1,w} & \rightarrow & \Omega^{E_k}_{n-1,w} & \rightarrow & 0\\
 & & & & \downarrow & & \downarrow & & \\
 & & & & 0 & & 0 & &
\end{array}
\tag{26}
$$

$$
\begin{array}{ccccccccc}
 & 0 & & & & 0 & & & \\
 & & \searrow & & & \downarrow & & & \\
 & & & \overline{K}^{E_k}_{n-1} & \longrightarrow & K^{E_k}_{n-1,w;n-1} & \rightarrow & 0 & \\
 & & & \uparrow & \searrow & \downarrow & & & \\
0 & \rightarrow & & K^{E_k}_{n-1,w} \rightarrow \Omega^{E_k}_{n-1} & \xrightarrow{a} & \Omega^{E_k}_{n-1,w} & \rightarrow & 0 & \\
 & & & \uparrow & \overset{c}{\searrow} & \downarrow b & & & \\
 & & & 0 & \longrightarrow & \Omega_{n-1} & & & \\
 & & & & & \downarrow & & & \\
 & & & & & 0 & & &
\end{array}
\tag{27}
$$

Definition 13 We say that a PDE $E_k \subset J^k_n(W)$ is an *extended crystal PDE*, if conditions of above theorem are verified. We define *crystal group of* E_k the littlest crystal group such Theorem 7 is satisfied. The corresponding dimension will be called *crystal dimension of* E_k.

In the following we relate crystal structure of PDEs to the existence of global smooth solutions, identifying an algebraic-topological obstruction.

Theorem 8 *Let $E_k \subset J^k_n(W)$ be a formally integrable and completely integrable PDE. Then, in the algebra $\mathbf{H}_{n-1}(E_k) \equiv Map(\Omega^{E_k}_{n-1}; \mathbb{R})$, (Hopf algebra of E_k), there is a subalgebra, (crystal Hopf algebra) of E_k. On such an algebra we can represent* *the algebra* $\mathbb{R}^{G(d)}$ *associated to the crystal group* $G(d)$ *of* E_k. *(This justifies the name.) We call crystal conservation laws of* E_k *the elements of its crystal Hopf algebra.*

Proof In fact the short exact sequence

$$0 \longrightarrow \overline{K}_{n-1}^{E_k} \xrightarrow{d} \Omega_{n-1}^{E_k} \xrightarrow{e} \Omega_{n-1} \longrightarrow 0 \tag{28}$$

obtained from the commutative diagram in (27), identifies for duality the following sequence

$$0 \longleftarrow \mathbf{K}_{n-1}^{E_k} \xleftarrow{d_*} \mathbf{H}_{n-1}(E_k) \xleftarrow{e_*} \mathbb{R}^{\Omega_{n-1}} \longleftarrow 0 \tag{29}$$

On the other hand from the short exact sequence

$$0 \longrightarrow \Omega_{n-1} \xrightarrow{f} G(d) \xrightarrow{g} G(d)/\Omega_{n-1} \longrightarrow 1 \tag{30}$$

identifying Ω_{n-1} with a crystallographic subgroup, we get for duality the following sequence

$$0 \longleftarrow \mathbb{R}^{\Omega_{n-1}} \xleftarrow{f_*} \mathbb{R}^{G(d)} \xleftarrow{g_*} \mathbb{R}^{G(d)/\Omega_{n-1}} \longleftarrow 0. \tag{31}$$

So we can identify the crystal Hopf algebra of E_k with $\mathbb{R}^{\Omega_{n-1}}$. On such an algebra we can represent all the Hopf algebra $\mathbb{R}^{G(d)}$ associated to the crystal group $G(d)$ of E_k.[17]

Theorem 9 *Let* $E_k \subset J_n^k(W)$ *be a formally integrable and completely integrable PDE. Then, the obstruction to find global smooth solutions of* E_k *can be identified with the quotient* $\mathbf{H}_{n-1}(E_\infty)/\mathbb{R}^{\Omega_{n-1}}$.

Proof Let us first consider the following lemma that gives some criteria to recognize $(n-1)$-dimensional admissible integral manifolds in E_k.

Lemma 14 (Cauchy Problem Solutions Criteria) (1) *Let* $E_k \subset J_n^k(W)$ *be a formally integrable and completely integrable PDE on the fiber bundle* $\pi : W \to M$, $\dim W = m+n$, $\dim M = n$. *Let* $N \subset E_k$ *be a smooth* $(n-1)$*-dimensional integral manifold, diffeomorphic to* $X \equiv \pi_{k,0}(N) \subset W$, *such that* $X \subset Y$, *where* Y *is a smooth* n*-dimensional submanifold of* W, *satisfying the condition that its* k*-order prolongation* $Y^{(k)} \subset J_{n-1}^k(W)$, *contains* N, *(0-order admissibility condition). Then, there exists a (weak, or singular, or smooth) solution* $V \subset E_k$, *such that* $N \subset V$.

(2) *Furthermore, if the symbols* g_k *and* g_{k+1}, *of* E_k *and* E_{k+1} *respectively, are different from zero, then* V *can be a singular or smooth solution. Moreover, if there*

---

[17] Let us remark that sequences (29) and (31) do not necessitate to be exact but are always partially exact. In fact, even if it does not necessitate that $d_\circ e_* = 0$ and $f_* \circ g_* = 0$, (hence neither $\mathrm{im}(e_*) = \ker(d_*)$ and $\mathrm{im}(g_*) = \ker(f_*)$), one has that e_* and g_* are monomorphisms and d_* and f_* are epimorphisms. This is enough for our proof. Let us recall also that $\mathbf{H}_{n-1}(E_k) \equiv Map(\Omega_{n-1}^{E_k}; \mathbb{R})$ is an Hopf algebra in extended sense, i.e. it contains the Hopf algebra $\mathbb{R}^{\Omega_{n-1}}$ as a subalgebra. (See also [56].)

exists a nonzero smooth vector field $\zeta : E_k \to TE_k$, *transversal to* N, *and characteristic, at least for a subequation of* E_k, *then a smooth solution* V *passing through* N *can be built by means of the flow* ϕ *associated to* ζ. *For suitable initial Cauchy integral manifolds, solutions can be built by using infinitesimal symmetries of the closed ideal encoding* E_k. *(For details see Theorem 2.10 in [56].)*

(3) *Finally, let* $E_k \to E_{k-1} \equiv \pi_{k,k-1}(E_k)$, *be an affine subbundle of* $\pi_{k,k-1} : J^k_n(W) \to J^{k-1}_n(W)$, *with associated vector bundle* $\bar{\pi}_{k-1} : g_k \to E_{k-1}$, *where* $\bar{\pi}_{k,k-1} = \pi_{k,k-1} \circ \bar{\pi}_k$, *with* $\bar{\pi}_k : g_k \to E_k$ *is the canonical projection. Let* $N \subset E_k$ *be a smooth* $(n-1)$*-dimensional integral manifold, diffeomorphic to* $X \equiv \pi_{k,0}(N) \subset W$ *and satisfying the* 0*-order admissibility condition. Then, there exists a (singular, or smooth) solution* $V \subset E_k$, *such that* $N \subset V$.

An integral manifold $N \subset E_k$, *as above defined and contained into a solution* $V \subset E_k$, *is called admissible.*

Proof Since N is an $(n-1)$-dimensional smooth integral manifold, diffeomorphic to $X \equiv \pi_{k,0}(N) \subset W$, satisfying the 0-order admissibility condition, we can consider $N \subset Y^{(k)} \subset J^k_n(W)$, where Y is just a n-dimensional smooth manifold of W, containing X. In general $Y^{(k)} \not\subset E_k$, but taking into account that E_k is formally integrable and completely integrable, we get that $(E_k)_{+r}$ is a strong retract of $J^{k+r}_n(W)$, $\forall r > 0$. Then, we can deform $Y^{(k+r)} \subset J^{k+r}_n(W)$ into E_{k+r}, obtaining a (weak, or singular, or smooth) solution $\tilde{Y} \subset E_{k+r}$, passing for $N^{(r)} \cong X^{(k+r)}$. Then $\pi_{k+r,k}(\tilde{Y}) \equiv V \subset E_k$ is a solution of E_k, passing through N. In particular, if $g_k \neq 0$, and $g_{k+r} \neq 0$, $\tilde{Y}$ is a singular (or smooth) solution, and so can be also V. Moreover, in the case that N is transversal to a characteristic smooth vector field $\zeta : E_k \to TE_k$, then $V = \bigcup_{\lambda \in]-\epsilon,\epsilon[} \phi_\lambda(N)$ is a smooth solution of E_k passing through N. Under suitable conditions on the Cauchy integral manifold, the vector field used to build the solution can be an infinitesimal symmetry. (For a detailed proof see the one of Theorem 2.10 in [56].)

Finally, whether $E_k \to E_{k-1} \equiv \pi_{k,k-1}(E_k)$, is an affine subbundle of $\pi_{k,k-1} : J^k_n(W) \to J^{k-1}_n(W)$, with associated vector bundle $\bar{\pi}_{k-1} : g_k \to E_{k-1}$, then also E_k is a strong retract of $J^k_n(W)$, so we can reproduce above strategy used to build a solution passing for N, without the necessity to prolong E_k, (if it is not differently required by the structure of this equation).

Example 9 (Fourier's Heat Equation) Let us consider the second-order PDE

$$(F) \subset JD^2(W) \subset J^2_2(W) : F \equiv u_t - u_{xx} = 0 \tag{32}$$

on the fiber bundle $\pi : W \equiv \mathbb{R}^3 \to \mathbb{R}^2$, $(t, x, u) \mapsto (t, x)$. In this case $(F) \to \pi_{2,1}(F) \equiv (F)_{-1} = JD(W)$ is an affine fiber subbundle of the affine fiber bundle $\pi_{2,1} : JD^2(W) \to JD(W)$ with associated vector bundle identified with the symbol g_2: $\zeta = \zeta_{tt}\partial u_{tt} + \zeta_{tx}\partial u_{tx} \in g_2$. So we can apply Lemma 14(3). Therefore, if $N \subset (F)$ is a (compact) 1-dimensional integral manifold, diffeomorphic to its projection $\pi_{2,0}(N) \equiv X \subset W$, we can find solutions $V \subset (F)$, passing from N. In particular, whether N is diffeomorphic to a smooth spacelike curve $u = h(x)$, (at $t = 0$), we get that N is the image of a mapping, say $\gamma : I \equiv [0, 1] \subset \mathbb{R} \to (F)$, iff $\gamma^*\mathcal{C}_2 = 0$, where $\mathcal{C}_2 =< dF, \omega \equiv du - u_t dt - u_x dx, \omega_t \equiv du_t - u_{tt} dt - u_{tx} dx, \omega_x \equiv du_x - u_{xt} dt - u_{xx} dx >$ is the Pfaffian contact ideal encoding solutions of (F). Then one can see that the integral curve γ is represented in coordinates $(t, x, u, u_t, u_x, u_{tt}, u_{tx}, u_{xx})$ on $JD^2(W)$, by the following equations:

$$\begin{cases} t\circ\gamma = 0; \quad x\circ\gamma = x; \quad u\circ\gamma = h(x); \quad u_t\circ\gamma = \frac{d^2h}{dx^2}(x); \quad u_x\circ\gamma = \frac{dh}{dx}(x); \\ u_{xx}\circ\gamma = \frac{d^2h}{dx^2}(x); \quad u_{tx}\circ\gamma = \frac{d^3h}{dx^3}(x); \quad u_{tt}\circ\gamma = \kappa(x). \end{cases} \tag{33}$$

where $\kappa(x)$ is an arbitrary smooth function. Then, by considering that Fourier's heat equation is a formally integrable and completely integrable PDE, we can see, by taking the first and second prolongations of (F), that must necessarily be $\kappa(x) = \frac{d^4h}{dx^4}(x)$. In fact, from the first prolongation we get that must be $u_{tt} = u_{xxt}$. From the second prolongation of (F) one has $u_{txx} = u_{xxxx}$. Since must be $u_{xxt} = u_{txx}$ we get $u_{tt} = u_{xxxx}$. By restriction on $t = 0$, one has $u_{tt}|_{t=0} = u_{xxxx}|_{t=0} = h_{xxxx}$. So we can build solution also by using Lemma 14(1). For example, if we are interested to a solution $V_S \subset (F)$, obtained by means of a rigid propagation of the initial Cauchy spacelike integral curve (33), one can see that must necessarily be $h(x) = \alpha x + \beta$, with $\alpha, \beta \in \mathbb{R}$, $x \in [0, 1] \subset \mathbb{R}$. In other words such a type of steady-state solution, $u(t, x) = \alpha x + \beta$, determines also the admissible structure of the integral initial Cauchy line $\Gamma \subset (F)_{t=0}$. In such a case this must be given by the following parametric equations in $JD^2(W)$:

$$(\Gamma) : \begin{cases} t\circ\gamma = 0; \quad x\circ\gamma = x; \quad u\circ\gamma = \alpha x + \beta; \quad u_t\circ\gamma = 0; \quad u_x\circ\gamma = \alpha; \\ u_{xx}\circ\gamma = 0; \quad u_{tx}\circ\gamma = 0; \quad u_{tt}\circ\gamma = 0. \end{cases} \tag{34}$$

Remark that $\zeta = \partial t$ is the smooth vector field propagating the initial Cauchy curve Γ in V_S, i.e., ζ is the *characteristic vector field* of V_S. This is not a characteristic vector field for Fourier's heat equation or for the steady-state subequation $\{u_{xx} = 0\}$. Instead, $\zeta = \partial t$ is an infinitesimal symmetry for (F). (See Lemma 14(2).) Let us also emphasize that the above steady-state solution is not the unique solution passing for Γ. In order to see this it is enough to generalize the concept of solution and to consider also weak-singular solutions. In fact, let us find regular perturbations of the steady-state solution V_S that at $t = 0$, in correspondence of the boundary points $\{A, B\} = \partial\Gamma$ of Γ, have values respectively a and $a + b$. These can be obtained by considering Fourier's heat equation, which is a linear equation, as its linearized at the steady-state solution V_S, $(F)[V_S] = (F)$, i.e., the equation for perturbations of V_S. Then we get:

$$\nu(t, x|\mu, a, b) \equiv e^{-\mu^2 t}\left[\frac{a + b - b\cos(\mu)}{\sin(\mu)}\sin(\mu x) + b\cos(\mu x)\right]. \tag{35}$$

$\nu(t, x|\mu, a, b)$ depends on an arbitrary positive parameter $\mu > 0$, and two other parameters $a, b \in \mathbb{R}$, and has the following limits:

$$\lim_{t\to+\infty} \nu(t, x|\mu, a, b) = 0; \lim_{\mu\to 0} \nu(t, x|\mu, a, b) = ax + b; \lim_{a,b\to 0} \nu(t, x|\mu, a, b) = 0.$$

This means that with respect to these perturbations the steady-state solution is asymptotically stable, and all such perturbations produce deformations of the steady-state solution, with deformation parameters just (μ, a, b). Let us denote by $V[\mu, a, b]$ the 2-dimensional integral manifold, contained in (F), representing the deformed steady-state solution by $v(t, x|\mu, a, b)$, $0 \leq x \leq 1$. More precisely $V[\mu, a, b]$ is the integral manifold of (F) corresponding to the solution $u(t, x|\mu, a, b) \equiv \alpha x + \beta + v(t, x|\mu, a, b)$. Set $V = V_S \bigcup V[\mu, a, b]$. V is a weak-singular solution of (F), that for $t = 0$ passes for $\Xi^{(2)} = \Gamma^{(2)} \bigcup \Gamma[\mu, a, b]^{(2)}$, a 1-dimensional admissible integral manifold of (F), such that $\pi_{2,0}(\Xi^{(2)}) \equiv \Xi = \Gamma \bigcup \Gamma[\mu, a, b]$, where $\Gamma[\mu, a, b]$ is the curve identified by $V[\mu, a, b]$ for $t = 0$. Ξ is a space-like curve contained in W, that is projected, by means of π, on the interval $[0, 1]$ of the x-axis. The 2-dimensional manifold $Y \equiv \pi_{2,0}(V) = \pi_{2,0}(V_S) \bigcup \pi_{2,0}(V[\mu, a, b]) \equiv Y_S \bigcup Y[\mu, a, b]$, passes for Ξ. Furthermore V converges, for $t \to +\infty$, to the steady-state solution V_S.[18]

Let us, now, show also an explicit construction of solution for (F) by means of the method of the retraction given in Lemma 14. For example, let $f(P) \equiv V \subset J^2_2(W)$, $P \subset \mathbb{R}^2$, be a 2-dimensional integral manifold representing the smooth function $u(t, x) \equiv \sum_{n \geq 0} t^n a_n(x)$, passing for the integral curve given in (33). The coefficients $a_n(x)$ are suitable functions on $x \in [0, 1]$, such that the series converges in $t \in \mathbb{R}^+$, and such that $a_0(x) = h(x)$, $a_1(x) = h_{xx}$, $a_2(x) = \frac{1}{2} h_{xxxx}$. These last conditions assure that V passes for the integral curve (33). The integral manifold V is not a solution of (F) since there are no restrictions on the other coefficients a_n, $n \geq 3$. The parametric equations of V are given in (36(a)).

$$
\text{(a)} \left\{
\begin{array}{l}
t \circ f = t \\
x \circ f = x \\
u \circ f = \sum_{0 \leq n \leq 2} t^n \frac{1}{n!} \frac{d^{2n} h}{dx^{2n}}(x) + \sum_{n \geq 3} t^n a_n(x) \\
u_t \circ f = \sum_{1 \leq n \leq 2} t^{n-1} \frac{d^{2n} h}{dx^{2n}}(x) + \sum_{n \geq 3} n t^{n-1} a_n(x) \\
u_x \circ f = \sum_{0 \leq n \leq 2} \frac{t^n}{n!} \frac{d^{2n+1} h}{dx^{2n+1}}(x) + \sum_{n \geq 3} t^n \frac{da_n}{dx}(x) \\
u_{xx} \circ f = \sum_{0 \leq n \leq 2} \frac{t^n}{n!} \frac{d^{2n+2} h}{dx^{2n+2}}(x) + \sum_{n \geq 3} t^n \frac{d^2 a_n}{dx^2}(x) \\
u_{tx} \circ f = \sum_{1 \leq n \leq 2} t^{n-1} \frac{d^{2n+1} h}{dx^{2n+1}}(x) + \sum_{n \geq 3n} n t^{n-1} \frac{da_n}{dx}(x) \\
u_{tt} \circ f = \frac{d^4 h}{dx^4}(x) + \sum_{n \geq 3} n(n-1) t^{n-2} a_n(x).
\end{array}
\right\}
\Rightarrow \text{(b)} \left\{
\begin{array}{l}
t \circ r \circ f = t \\
x \circ r \circ f = x \\
u \circ r \circ f = \widetilde{u}(t, x) \\
u_t \circ r \circ f = \widetilde{u}_t(t, x) \\
u_x \circ r \circ f = \widetilde{u}_x(t, x) \\
u_{tt} \circ r \circ f = \widetilde{u}_{tt}(t, x) \\
u_{tx} \circ r \circ f = \widetilde{u}_{tx}(t, x)
\end{array}
\right\}. \tag{36}
$$

Taking into account that on (F) we can consider the following coordinate functions $\{t, x, u, u_t, u_x, u_{tt}, u_{tx}\}$, it follows that the retraction mapping $r(f(P)) = r(V) \equiv \widetilde{V} \subset (F)$ has the parametric equation given in (36(b)), where the function $\widetilde{u}(t, x)$ is

[18] Singular solutions, like those described in this example, are very important in many physical applications too, since they represent complex phenomena related to perturbations of some fixed dynamic background. For example, for suitable values of the parameters a and b in (35), manifolds $Y[\mu, a, b]$ intersect Y_S along common characteristic lines, and the singular solutions V are piecewise $\mathbb{Z}_2$-manifolds. (For complementary information on such singular manifolds and singular solutions of PDEs, see also [13, 42, 56, 72].)

determined starting from the function $u(t,x)$, by imposing the condition to belong to (F). So we get $\widetilde{u}(t,x)=\sum_{n\geq 0}t^n\widetilde{a}_n(x)$, with $\widetilde{a}_n\equiv\frac{1}{n!}\frac{d^{2n}h}{dx^{2n}}(x)$.[19] $\widetilde{V}\subset(F)$, represents the solution $\widetilde{u}(t,x)$. This can be considered as a deformation of $u(t,x)$. In fact, the functions defined in (37), is just the explicit deformation connecting $\widetilde{u}(t,x)$ with $u(t,x)$.

$$\begin{array}{l}\widetilde{u}(t,x|\lambda)=\sum_{n\geq 0}t^n\widetilde{a}_n(\lambda|x),\widetilde{a}_n(\lambda|x)\equiv a_n(x)-\lambda[a_n(x)+\widetilde{a}_n(x)]\\ \Downarrow\\ \begin{cases}\widetilde{u}(t,x|0)=\sum_{n\geq 0}t^n a_n(x)\\ \widetilde{u}(t,x|1)=\sum_{n\geq 0}t^n\widetilde{a}_n(x).\end{cases}\end{array}\tag{37}$$

The integral manifolds $\widetilde{V}_\lambda\subset J^2_2(W)$, $\lambda\in[0,1]$, corresponding to $\widetilde{u}(t,x|\lambda)$, are not contained in (F) for all $\lambda\in[0,1]$, hence are not solutions of (F). Therefore, $\widetilde{V}\subset(F)$ is a solution for the corresponding Cauchy problem, obtained by means of the retraction method given in Lemma 14. The parametric equation of $\widetilde{V}$ is given in (38).

$$\left\{\begin{array}{l}t\circ r\circ f=t\\ x\circ r\circ f=x\\ u\circ r\circ f=\sum_{n\geq 0}t^n\frac{1}{n!}\frac{d^{2n}h}{dx^{2n}}(x)\\ u_t\circ r\circ f=\sum_{n\geq 1}t^{n-1}\frac{1}{(n-1)!}\frac{d^{2n}h}{dx^{2n}}(x)\\ u_x\circ r\circ f=\sum_{n\geq 0}t^n\frac{1}{n!}\frac{d^{2n+1}h}{dx^{2n+1}}(x)\\ u_{tt}\circ r\circ f=\sum_{n\geq 2}t^{n-2}\frac{1}{(n-2)!}\frac{d^{2n}h}{dx^{2n}}(x)\\ u_{tx}\circ r\circ f=\sum_{n\geq 1}t^{n-1}\frac{1}{(n-1)!}\frac{d^{2n+1}h}{dx^{2n+1}}(x)\end{array}\right\}\ x\in[0,1].\tag{38}$$

Let us emphasize, that also in this case the solution so obtained is not unique. In fact, similarly to the previous case, where we have considered weak-singular solutions by means of perturbations of Cauchy data for the steady state solution V_S, now we have the following weak-singular solution $\widehat{V}\equiv\widetilde{V}\bigcup\widetilde{V}[\mu,a,b]\subset(F)$, where $\widetilde{V}[\mu,a,b]$ is the 2-dimensional integral manifold identified by the solution of (F) given in (39).

$$\widetilde{u}(t,x|\mu,a,b)=\sum_{n\geq 0}t^n\frac{1}{n!}\frac{d^{2n}h}{dx^{2n}}(x)+e^{-\mu^2 t}\left[\frac{a+b-b\cos(\mu)}{\sin(\mu)}\sin(\mu x)+b\cos(\mu x)\right].\tag{39}$$

---

[19] This function $\widetilde{u}(t,x)$, is well defined and limited in all $(t,x)\in\mathbb{R}^+\times[0,1]$, since the function $h(x)$ is smooth in $[0,1]$ and with $|\frac{d^{2n}h}{dx^{2n}}(x)|\leq C\in\mathbb{R}$, $\forall x\in[0,1]$. In fact one has $\sum_{n\geq 0}t^n\widetilde{a}_n\leq C\sum_{n\geq 0}\frac{t^n}{n!}\equiv C\sum_{n\geq 0}b_n$. The last series converges, with convergence radius $r=\infty$ since $\lim_{n\to\infty}\left|\frac{b_{n+1}}{a_n}\right|=\lim_{n\to\infty}\frac{t}{n+1}=0=\frac{1}{r}$. Therefore the series $\sum_{n\geq 0}t^n\widetilde{a}_n(x)$ is convergent as it is absolutely convergent.

Let us denote by $\widetilde{\Gamma}[\mu, a, b] \subset W_{t=0}$ the spacelike curve identified by $\widetilde{u}(0, x|\mu, a, b)$. Then $\widetilde{V}[\mu, a, b]$ passes for

$$\widetilde{\Gamma}^{(2)}[\mu, a, b] \subset (F)_{t=0},$$

and the singular solution $\widehat{V}$ passes for the spacelike curve

$$\widetilde{\Xi}^{(2)} = \widetilde{\Gamma}^{(2)} \bigcup \widetilde{\Gamma}^{(2)}[\mu, a, b] \subset \widehat{V} \subset (F),$$

where

$$\widetilde{\Xi} = \widetilde{\Gamma} \bigcup \widetilde{\Gamma}[\mu, a, b] \subset W_{t=0}$$

is the space-like curve containing the fixed Cauchy data, i.e., the curve $\widetilde{\Gamma}$. Furthermore the weak-singular solution $\widehat{V}$ asymptotically converges $(t \to +\infty)$ to the regular solution $\widetilde{V}$. Since the singular solution $\widehat{V}$ depends on the parameters $(\mu, a, b) \in \mathbb{R}^+ \times \mathbb{R}^2$, the Cauchy problem identified by the curve $\widetilde{\Gamma}$ has more than one solution.

Let, now, $N_0, N_1 \subset E_k$ be two closed compact $(n-1)$-dimensional admissible integral manifolds of E_k. Then there exists a weak, (resp. singular, resp. smooth) solution $V \subset E_k$, such that $\partial V = N_0 \sqcup N_1$, iff $X \equiv N_0 \sqcup N_1 \in [0] \in \Omega^{E_k}_{n-1,w}$, (resp. $X \in [0] \in \Omega^{E_k}_{n-1,s}$, resp. $X \in [0] \in \Omega^{E_k}_{n-1}$). On the other hand there exists such a smooth solution iff $X \in [0] \in \Omega_{n-1}$ and X has zero all its integral characteristic numbers, i.e., are zero on X all the conservation laws of E_k. Since these last can be identified with the Hopf algebra $\mathbf{H}_{n-1}(E_k) \cong \mathbf{H}_{n-1}(E_\infty)$, where E_∞ is the infinity prolongation of E_k, it follows that the quotient $\mathbf{H}_{n-1}(E_\infty)/\mathbb{R}^{\Omega_{n-1}}$ measures the amount of how the conservation laws of E_k differ from the crystal conservation laws, identified with the elements of the Hopf algebra $\mathbb{R}^{\Omega_{n-1}}$.

Definition 14 We define *crystal obstruction* of E_k the above quotient of algebras, and put: $cry(E_k) \equiv \mathbf{H}_{n-1}(E_\infty)/\mathbb{R}^{\Omega_{n-1}}$. We call 0-*crystal PDE* one $E_k \subset J^k_n(W)$ such that $cry(E_k) = 0$.[20]

Corollary 10 *Let $E_k \subset J^k_n(W)$ be a 0-crystal PDE. Let $N_0, N_1 \subset E_k$ be two closed compact $(n-1)$-dimensional admissible integral manifolds of E_k such that $X \equiv N_0 \sqcup N_1 \in [0] \in \Omega_{n-1}$. Then there exists a smooth solution $V \subset E_k$ such that $\partial V = X$.*

Example 10 (The Ricci flow equation). The Ricci flow equation

$$F_{ij} \equiv (\partial t.g_{ij}) - \kappa R_{ij} = 0 \tag{40}$$

on a Riemannian n-dimensional manifold (M, g), can be encoded by means of a second-order differential equation $(RF) \subset JD^2(E) \subset J^2_{n+1}(E)$ over the following

[20] An extended 0-crystal PDE $E_k \subset J^k_n(W)$ does not necessitate to be a 0-crystal PDE. In fact E_k is an extended 0-crystal PDE if $\Omega^{E_k}_{n-1,w} = 0$. This does not necessarily imply that $\Omega^{E_k}_{n-1} = 0$.

fiber bundle $\pi : E \equiv \mathbb{R} \times \widetilde{S_2^0 M} \to \mathbb{R} \times M$, where $\widetilde{S_2^0 M} \subset S_2^0 M$ is the open subbundle of nondegenerate Riemannian metrics on M. In [62] we have calculated the integral bordism group of the equation $(RF) \subset J^2_{n+1}(E)$. In particular if M is a 3-dimensional closed, compact, smooth, simply connected manifold, we get $\Omega^{(RF)}_{3,w} \cong \Omega^{(RF)}_{3,s} \cong \mathbb{Z}_2$. Thus, (RF) is not an extended 0-crystal PDE. Taking into account exact commutative diagram (27), we get also the short exact sequence $0 \to K^{(RF)}_{3,w} \to \Omega^{(RF)}_3 \to \mathbb{Z}_2 \to 0$. Taking into account Example 5 we can consider $\Omega^{(RF)}_3$ as an extension of a subgroup of the crystallographic group $G(3) = \mathbb{Z}^3 \rtimes \mathbb{Z}_2$ or $G(2) = \mathbb{Z}^2 \rtimes \mathbb{Z}_2$. Therefore, the integral bordism group of the Ricci flow equation on S^3 is an extended crystal PDE, with crystal group $G(2) = \mathbb{Z}^2 \rtimes \mathbb{Z}_2 = p2$ and crystal dimension 2. Furthermore, from the exact commutative diagram (27) we get also the following short exact sequence $0 \to \overline{K}^{(RF)}_3 \to \Omega^{(RF)}_3 \to \Omega_3 \to 0$. Taking into account that $\Omega_3 = 0$, we can have $cry(RF) \neq 0$. On the other hand let us consider admissible only spacelike integral Cauchy manifolds satisfying the following conditions: (i) They are diffeomorphic to S^3 or to M, assumed any smooth 3-dimensional Riemannian, compact, closed, orientable, simply connected manifold; (ii) M is homotopy equivalent to S^3. Such integral manifolds surely exist, since we can embed in E both manifolds M and S^3 and identify these, for example, with spacelike smooth admissible integral manifolds of the subequation $(RF)_t \subset (RF)$, where $(RF)_t = \hat{\pi}_2^{-1}(t)$, $t \in \mathbb{R}$. Here $\hat{\pi}_2$ is the canonical projection $(RF) \to \mathbb{R}$, induced by $\pi_2 : JD^2(E) \to \mathbb{R} \times M$, i.e., $\hat{\pi}_2 \equiv pr_1 \circ \pi_2 : JD^2(E) \to \mathbb{R}$, where pr_1 is the canonical projection $pr_1 : \mathbb{R} \times M \to \mathbb{R}$. In fact, in (RF), $(\partial t.g_{ij})$ is solved with respect to R_{ij}. More precisely, starting from a 3-dimensional compact closed, orientable, simply connected Riemannian manifold (M, γ), we can identify a spacelike integral Cauchy 3-dimensional manifold $N_0 \subset (RF)_{t=t_0=0}$, diffeomorphic to its projection $\pi_{2,0}(N_0) \equiv Y_0 \subset W_{t=t_0=0} \cong \widetilde{S_2^0 M}$, by means of the mapping $f : M \to JD^2(E)$ defined by the parameter equation in (41).

$$\left\{\begin{array}{l} t \circ f = t_0 = 0 \\ x^k \circ f = x^k \\ g_{ij} \circ f = \gamma_{ij} \\ g_{ij,t} \circ f = \kappa R_{ij}(\gamma) \\ g_{ij,h} \circ f = \gamma_{ij,h} \\ g_{ij,th} \circ f = \kappa R_{ij}(\gamma)_{,h} \\ g_{ij,hk} \circ f = \gamma_{ij,hk} \\ g_{ij,tt} \circ f = \Phi_{ij}(\gamma), \end{array}\right\}, 1 \leq i, j, h, k \leq 3. \tag{41}$$

where $\{t, x^k, g_{ij}, g_{ij,t}, g_{ij,h}, g_{ij,th}, g_{ij,hk}, g_{ij,tt}\}_{1 \leq i,j,h,k \leq 3}$ are local coordinates on $JD^2(E)$, $\Phi_{ij}(\gamma)$, $i, j \in \{1, 2, 3\}$, are known analytic functions of γ_{ij} and its derivatives up to fourth order, symmetric in the indexes. In fact, since (RF) is a formally integrable and completely integrable PDE [62], from its first prolongation we get

$$\{g_{ij,tt} = \kappa R_{ij}(g)_{,t}, g_{ij,th} = \kappa R_{ij}(g)_{,h}\}_{1 \leq i,j,h \leq 3},$$

where $R_{ij}(g)_{,t}$ and $R_{ij}(g)_{,h}$ denote the first prolongation of $R_{ij}(g)$, with respect the t-variable and x^h-variable respectively. Since $R_{ij}(g) = R_{ij}(g_{rs}, g_{rs,h}, g_{rs,hk})$,

$i, j \in \{1, 2, 3\}$, are analytic functions, we get also the following analytic functions

$$R_{ij}(g)_{,t} = K_{ij}(g_{rs}, g_{rs,h}, g_{rs,hk}, g_{rs,t}, g_{rs,ht}, g_{rs,hkt}), 1 \leq i, j \leq 3$$

where K_{ij} are analytic functions of their arguments. Taking into account the expression of (RF) and second order prolongation, we get the initial values (3-dimensional Cauchy integral manifold), given in (42).

$$\left\{\begin{array}{l} g_{rs,h}|_{t=0} = \gamma_{rs,h} \\ g_{rs,t}|_{t=0} = \kappa R_{rs}(\gamma) \\ g_{rs,th}|_{t=0} = \kappa R_{rs,h}|_{t=0} \\ g_{rs,hk}|_{t=0} = \gamma_{rs,hk} \\ g_{rs,tt}|_{t=0} = \Phi_{ij}(\gamma) \end{array}\right\}_{1\leq r,s,h,k\leq 3} \tag{42}$$

where $R_{rs,h}$ is the first prolongation of R_{rs} with respect to the coordinates x^h. Taking into account that one has the following functional dependence $R_{rs} = R_{rs}(g_{ij}, g_{ij,p}, g_{ij,pq})$, we get

$$R_{rs,h} = (\partial g^{ij}.R_{rs})g_{ij,h} + (\partial g^{ij,p}.R_{rs})g_{ij,ph} + (\partial g^{ij,pq}.R_{rs})g_{ij,pqh}.$$

As a by-product we get that $R_{rs,h}|_{t=0}$ are functions of γ_{ij} and their derivative up to third order, that we shortly denote by $R_{rs,h}|_{t=0}$. Furthermore, from the first prolongation of (RF) we get also

$$\left\{\begin{array}{l} g_{rs,tt} = \kappa R_{rs,t} = \kappa[(\partial g^{ij}.R_{rs})g_{ij,t} + (\partial g^{ij,p}.R_{rs})g_{ij,pt} + (\partial g^{ij,pq}.R_{rs})g_{ij,pqt}] \\ \qquad = \kappa[(\partial g^{ij}.R_{rs})\kappa R_{ij} + (\partial g^{ij,p}.R_{rs})\kappa R_{ij,p} + (\partial g^{ij,pq}.R_{rs})\kappa R_{ij,pq}] \end{array}\right\}_{1\leq r,s,h,k\leq 3} \tag{43}$$

Then taking $t = 0$, we get that $g_{rs,tt}|_{t=0}$ are functions that depend on γ_{ij} and their derivatives up to fourth order, that we shortly denote by $\Phi_{ij}(\gamma)$.

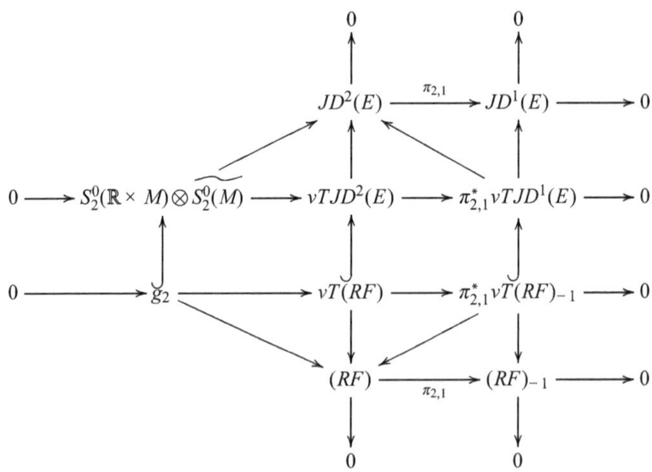

(44)

Similarly one can identify the Riemannian manifold $(S^3, \bar{\gamma})$, with another space-like integral Cauchy 3-dimensional manifold $N_1 \subset (RF)_{t=t_1\neq 0}$, diffeomorphic to its projection $\pi_{2,0}(N_1) \equiv Y_1 \subset W_{t=t_1} \cong \widetilde{S_2^0 M}$. Let us emphasize that, fixed the space-like fiber $(RF)_t$, above integral Cauchy manifolds are uniquely identified by the Riemannian manifolds (M, γ) and $(S^3, \bar{\gamma})$, respectively. For such integral manifolds N_0 and N_1, necessarily pass solutions of (RF), (hence they are admissible).[21] We shall prove this by considering Lemma 14(2). First let us note that, similar to Fourier's heat equation, the Ricci flow equation is an affine subbundle of the affine bundle $\pi_{2,1} : JD^2(E) \to JD^1(E)$, since the fiber $\pi_{2,1}^{-1}(\bar{q}) \equiv (RF)_{\bar{q}}$, for $\bar{q} \in (RF)_{-1}$, is an affine space. In fact, $T_q(RF)_{\bar{q}} \cong (g_2)_q$, and the vector space $(g_2)_q$ is constant on the fiber $\bar{q} \in (RF)_{-1}$. The situation is shown by the exact commutative diagram in (44) and by the fact that a vector field $\zeta = \zeta_{ij}^{\alpha\beta} \partial g_{\alpha\beta}^{ij}$ belonging to the symbol must satisfy Eq. (46). (One has used the usual space–time indexes numbering for the coordinates $(t, x^k)_{1 \leq k \leq 3} = (x^\alpha)_{0 \leq \alpha \leq 3}$ on $\mathbb{R} \times M$.) That equation is constant on the points of the fiber $(RF)_{\bar{q}}$, since it depends only on the derivatives of first-order $g_{ij,h}$ and zero-order g_{ij}.

$$\zeta_{rs}^{\alpha\beta}(\partial g_{\alpha\beta}^{rs}.F_{ij})(q) = 0, q \in (RF). \tag{46}$$

This follows soon from the expression of the Ricci tensor as a differential polynomial. See equation (47).

---

[21] In general a steady state solution of (RF) is not admitted since this should imply that (M, γ) is Ricci flat. Furthermore, regular solutions $g_{ij}(t, x^k) = h(t)\bar{g}_{ij}(x^k)$, with separated time variable from space ones, imply that (M, γ) is an Einstein manifold. In fact $R_{ij}(g) = R_{ij}(h\bar{g}) = R_{ij}(\bar{g})$. The Ricci flow equation becomes $h_t(t)\bar{g}_{ij}(x^k) = \kappa R_{ij}(\bar{g})$. Therefore, must be $h_t(t) = \omega = \kappa R_{ij}(\bar{g})/\bar{g}_{ij}(x^k)$, with $\omega \in \mathbb{R}$. By imposing the initial condition $g_{ij}(0, x^k) = \gamma_{ij}(x^k)$, we get that must be $h(t) = \omega t + 1$ and $\bar{g}_{ij}(x^k) = \gamma_{ij}(x^k)$, hence $R_{ij}(\gamma) = \gamma_{ij}\omega/\kappa$. Vice versa, if the Ricci flow equation is considered only for Einstein manifolds, then solutions with separated variables like above, are admitted. This means that in general, i.e., starting with any (M, γ), we cannot assume solutions $g_{ij}(t, x^k)$ with the above separated variables structure, even if these solutions "arrive" to S^3, that is just an Einstein manifold. The same results can be obtained by considering metrics $g(t, x^k)$, obtained deforming $\gamma(x^k)$, under a space–time flow ϕ_λ of $M \times \mathbb{R}$, i.e., $\{t \circ \phi_\lambda = t + \lambda, x^k \circ \phi_\lambda = \phi_\lambda^k(t, x^i)\}_{1 \leq i,k \leq 3}$, and with initial condition $\{\phi_0^k = x^k\}_{k=1,2,3}$. In fact if we assume that $\phi_\lambda^r = h(\lambda)\bar{\phi}^r(x^k)$, then the Ricci flow equation becomes as given in (45).

$$\frac{\dot{h}^2(\lambda)}{h^2(\lambda)} = \kappa \frac{\bar{\phi}_i^r \bar{\phi}_i^r (R_{rs}(\gamma) \circ \phi_\lambda)}{\bar{\phi}_i^r \bar{\phi}_i^r (\gamma_{rs}(\gamma) \circ \phi_\lambda)} = \omega \in \mathbb{R}^+ \Rightarrow \left\{ \begin{array}{ll} \text{(a)} & \dot{h}(\lambda) - \pm\sqrt{\omega} h(\lambda) = 0 \\ \text{(b)} & \bar{\phi}_i^r \bar{\phi}_i^r [R_{rs}(\gamma) - \frac{\omega}{\kappa}\gamma_{rs}] \circ \phi_\lambda = 0 \end{array} \right\} \tag{45}$$

The integration of the Eq. (45(a)) gives $h(\lambda) = Ce^{\pm\lambda\sqrt{\omega}}$. Furthermore, if (M, γ) is an Einstein manifold, i.e., there exists $\mu \in \mathbb{R}$, such that $R_{rs}(\gamma) = \mu\gamma_{rs}$, then taking $\omega = \mu\kappa$, one has the following solution, uniquely identified by the initial condition, (up to rigid flows): $\phi_\lambda = e^{\pm\lambda\sqrt{\mu\kappa}}x^r$. If (M, γ) is not Einstein, or equivalently, assuming $\omega \neq \kappa\mu$, we see that the solutions of Eq. (45(b)) are $\bar{\phi}^r = a^r \in \mathbb{R}$, since the metric $\bar{\gamma}_{rs} \equiv R_{rs}(\gamma) - \frac{\omega}{\kappa}\gamma_{rs}$ is not degenerate, hence must necessarily be $\bar{\phi}_i^r = 0$. But such a flow does not satisfy initial condition $\{\phi_0^r = x^r\}_{r=1,2,3}$. In conclusion a metric $g_{ij}(t, x^k)$, obtained deforming γ with a space-time flow ϕ_λ, where $\phi_\lambda^r = h(\lambda)\bar{\phi}^r(x^k)$, is a solution of the Ricci flow equation iff (M, γ) is Einstein, or Ricci flat. This last case corresponds to take $\omega = 0$ in Eq. (45) and has as solution the unique flow, up to rigid ones, $\phi_\lambda^r = x^r$.

$$R_{jn} = g^{rp} R_{rjnp} \left\{ \begin{array}{l} R_{rjnp} = \frac{1}{2}(g_{rp,jn} + g_{jn,rp} - g_{rn,jp} - g_{jp,rn}) \\ +g^{ts}([jn,s][rp,t] - [jp,s][rn,t]) \end{array} \right\} \left\{ \begin{array}{l} [ij,k] = \frac{1}{2}(g_{ik,j} + g_{jk,i} - g_{ij,k}) \\ g^{rp} \equiv [g_{rp}]/|g| \\ |g| \equiv \det(g_{rp}) \\ [g_{rp}]: \text{algebraic complement of } g_{rp}. \end{array} \right. \tag{47}$$

Since the derivatives of second order $g_{ij,hk}$ appear in (47) multiplied by zero-order terms only, it follows that Eq. (46) is constant on each fiber $(RF)_{\bar{q}}$. As a by-product it follows that we can apply the retraction method given in Lemma 14 directly on (RF).

In this respect, let us see more explicitly, and less formally, how we can build, a smooth solution of the Cauchy problem (RF) identified by the 3-dimensional integral manifold given in (42). Since, fixed the Riemannian manifold (M, γ), we uniquely identify a 3-dimensional integral manifold $N_0 \subset (RF)$, diffeomorphic to M, (resp. S^3), then in order to build a solution passing for N_0, (resp. N_1), it is enough to prove that we are able to identify a timelike vector field, tangent to (RF), transversal to N_0 that besides the tangent space TN_0 generate an integral planes of (RF). In fact, integral planes $L_{\widehat{q}} \subset (\mathbf{E}_2)_q \cong L_{\widehat{q}} \bigoplus (g_2)_q \subset T_q(RF)$, $\widehat{q} \in (RF)_{+4}$, $q = \pi_{6,2}(\widehat{q})$, are generated by the following *horizontal vectors* $\zeta_\alpha(q) = [\partial x_\alpha + \sum_{0 \leq |\beta| \leq 2} g_{ij,\alpha\beta} \partial g^{ij,\beta}]_{\widehat{q} \in (RF)_{+4}} \in T_q(RF)$. *($|\beta|$ denotes the multiindex length and $0 \leq \alpha \leq 3$.)* In (48) is given a more explicit expression of the horizontal vectors ζ_0.

$$\left\{ \begin{array}{l} \zeta_0 = \partial t + g_{ij,t} \partial g^{ij} + g_{ij,th} \partial g^{ij,h} + g_{ij,tt} \partial g^{ij,t} + g_{ij,thk} \partial g^{ij,hk} + g_{ij,tth} \partial g^{ij,th} + g_{ij,ttt} \partial g^{ij,tt} \\ g_{ij,t} = \kappa R_{ij}(\gamma), g_{ij,th} = \kappa R_{ij,h}(\gamma) \\ g_{ij,tt} = \kappa R_{ij,t}(\gamma) = \kappa \left[(\partial g^{rs}.R_{ij}) \kappa R_{rs}(\gamma) + (\partial g^{rs,p}.R_{ij}) \kappa R_{rs,p}(\gamma) + (\partial g^{rs,pq}.R_{ij}) \kappa R_{rs,pq}(\gamma) \right] \\ g_{ij,thk} = \kappa R_{ij,hk}(\gamma), g_{ij,tth} = \kappa R_{ij,th}(\gamma), g_{ij,ttt} = \kappa R_{ij,tt}(\gamma) \end{array} \right\}. \tag{48}$$

For $\widehat{q} \in (N_0)^{(4)} = D^6\gamma(M) \subset [(RF)_{+4}]_{t=0} \subset (RF)_{+4}$, at the corresponding point $q \in N_0$, one has $\zeta = \zeta_{(S)} + \zeta_{(T)} \in (\mathbf{E}_2)_q$, with $\zeta_{(S)} = X^k \zeta_k(q) \in T_q N_0$, and $\zeta_{(T)} = X^0 \zeta_0(q)$ transverse to N_0. Thus the 4-dimensional integral manifold V_q, tangent to such an integral plane, is tangent to N_0 too, and has the vector ζ_0 as characteristic timelike vector at $q \in N_0$. By varying q in N_0 we generate a 4-dimensional manifold V that is the *envelopment manifold* of the family $\{V_q\}_{q \in N_0}$ of solutions of (RF), formally defined as $V \equiv \bigcup_{q \in N_0} X_{0,q}$, where $X_{0,q}$ is the integral line transversal to N_0, starting from q, tangent to $\zeta_0(q)$ and future directed. Thus V is a line bundle over N_0, containing N_0, with $T_q N_0 \subset T_q V$, $\forall q \in N_0$. This is enough to claim that V is contained in (RF) and it is a timelike integral manifold of (RF), whose tangent space is an horizontal one. In other words, V is not only a viscosity solution, but just a solution of the Cauchy problem identified by N_0 (resp. N_1).[22] (See Fig. 1 where the 3-dimensional manifold N_0 is reduced to dimension 2 (figure on the left side) or dimension 1 (figure on the right side) for graphic necessities.)

[22] Generalized solutions of PDEs, called *viscosity solutions* were introduced by Pierre-Louis Lions and Michael Crandall in the paper [36]. Such mathematical objects do not necessitate to be solutions, but are envelope manifolds of solutions.

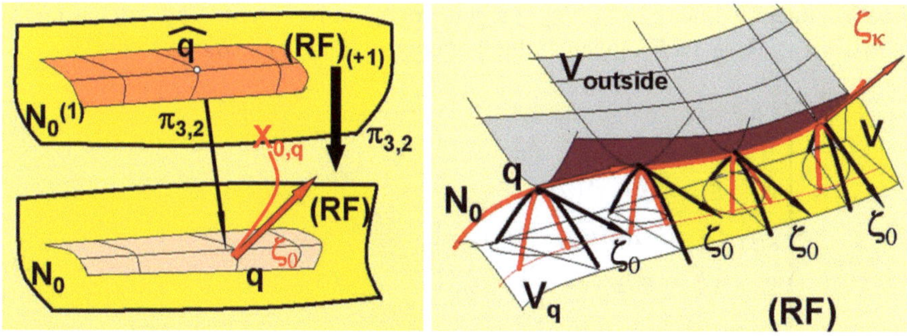

Fig. 1 Solution Cauchy problem for (RF), represented by the integral manifold V, *envelopment manifold*, generated by local solutions represented by manifolds V_q tangent to N_0, $\forall q \in N_0$, and identified by means of 4-dimensional integral planes $L_{\widehat{q}}$, for any $\widehat{q} \in N_0^{(1)}$. $V_{outside} \subset JD^2(E)$ represents a 4-dimensional integral manifold, passing for N_0, but $V_{outside} \not\subset (RF)$. $V_{outside}$ can be deformed in V, taking fixed N_0. (RF) and $(RF)_{+1}$ are represented with frames (in the left-hand picture), containing respectively N_0 and $N_0^{(1)}$

Let us emphasize that, on N_0, (resp. N_1), the components of the characteristic vector field ζ_0 in (48) are uniquely characterized by the metric γ_{ij} and its derivatives up to sixth order. This proves that in order to characterize a solution of the Cauchy problem, identified by $N_0 \subset (RF)$, it is necessary to consider the prolongation $(RF)_{+4}$, of (RF), up to fourth order. Then the 4-dimensional manifold $V = \bigcup_{t\in[0,\epsilon[} \phi_t(N_0) \subset (RF)$, obtained by means of the local flow ϕ_t generated by $\zeta_0 = \partial\phi$, is a local integral manifold, solution of the Cauchy problem identified by N_0. (So ζ_0 is the characteristic vector field for such a solution.)[23] Let us remark that such a solution does not necessitate to be smooth. In fact in this process we have used only the prolongation of (RF) up to fourth order, i.e., we have considered only the projections given in (49).

$$(RF)_{+4} \longrightarrow (RF)_{+3} \longrightarrow (RF)_{+2} \longrightarrow (RF)_{+1} \longrightarrow (RF). \qquad (49)$$

However, we can obtain V as a continuous manifold. In fact, let $g_{ij}(x^\alpha)$ denote a local solution identified by $L_{\widehat{q}} = T_qV_q \subset T_q(RF)$, and $T_qN_0 \subset L_{\widehat{q}}$. Then the

[23] Let us emphasize also that an integral 4-plane, where the components $g_{ij,\beta}$, $0 \le |\beta| \le 3$, satisfy conditions in (42), but not all conditions in (48), has the corresponding integral manifold, say $\widetilde{V}$, that passes for N_0, but it is not contained in (RF). (This is represented by $V_{outside}$ in Fig. 1.) Let us denote the corresponding metric with $\widetilde{g}$. Therefore, the retraction method imposes also to $g_{ij,\beta}$, with $0 \le |\beta| \le 3$, to satisfy conditions reported in (48). The relation between $\widetilde{V} \equiv V_{outside} \subset JD^2(E)$, and $V \subset (RF)$, can be realized with a deformation, g_λ, connecting the corresponding metrics $\widetilde{g}$ and g. More precisely, $g_\lambda = \widetilde{g} + \lambda[g - \widetilde{g}]$, $\lambda \in [0,1]$. The integral manifolds $V_\lambda \subset JD^2(E)$, generated by g_λ, are not contained in (RF) for any $\lambda \in [0,1]$, but all pass for N_0. In fact, since $(g_\lambda)_{ij,\beta} = \widetilde{g}_{ij,\beta} + \lambda[g_{ij,\beta} - \widetilde{g}_{ij,\beta}]$, $|\beta| \ge 0$, we get that $(g_\lambda)_{ij,\beta}|_{N_0} = \widetilde{g}_{ij,\beta}|_{N_0} = g_{ij,\beta}|_{N_0}$, for $0 \le |\beta| \le 2$, but $(g_\lambda)_{ij,\beta}|_{N_0}$ do not satisfy conditions in (48) for $|\beta| = 3$. This is just the meaning of the retraction method considered in the proof of Lemma 14.

timelike integral line $X_{0,q} \subset V_q$, tangent to ζ_0, is represented by a local curve $\xi_q : [0, \epsilon_q[\subset \mathbb{R} \to (RF)$, given by the parametric Eq. (50) in $JD^2(E)$.

$$\begin{cases} t \circ \xi = t \\ x^k \circ \xi = x^k(q) \\ g_{ij} \circ \xi = g_{ij}(t, x^k(q)) \\ g_{ij,t} \circ \xi = \kappa R_{ij}(g(t, x^k))|_{x^k = x^k(q)} \\ g_{ij,h} \circ \xi = g_{ij,h}(t, x^k)|_{x^k = x^k(q)} \\ g_{ij,hk} \circ \xi = g_{ij,hk}(t, x^k)|_{x^k = x^k(q)} \\ g_{ij,th} \circ \xi = \kappa R_{ij,h}(g(t, x^k))|_{x^k = x^k(q)} \\ g_{ij,tt} \circ \xi = \kappa^2 \sum_{0 \leq |p| \leq 2} [(\partial g^{rs,p}.R_{ij}) R_{rs,p}(g(t, x^k))]|_{x^k = x^k(q)} \end{cases} \tag{50}$$

Since ζ_0 continuously changes on N_0, it follows that curves $X_{0,q}$, $q \in N_0$, continuously change with $q \in N_0$, if the solutions $g_{ij}(t, x^k)$ continuously change with $q \in N_0$ too. This is surely realized by the fact that points $q' \in N_0$, near to q, come from points $\widehat{q}' \in N_0^{(1)}$, near to $\widehat{q}$, if $\pi_{3,2}(\widehat{q}) = q$. In fact, $N_0 \cong N_0^{(1)}$ and $\pi_{3,2}(N_0^{(1)}) = N_0$, thus $\pi_{3,2}|_{N_0^{(1)}} : N_0^{(1)} \to N_0$ is necessarily a continuous mapping. Moreover, we can continuously transform any (local) solution $g_{ij}(t, x^k)$ of (RF) into other ones by means of space-like (local) 1-parameter group of diffeomorphisms ϕ_λ, $\lambda \in [0, \epsilon[\subset \mathbb{R}$ of M. In fact, such transformations are locally represented by the following functions $\{t \circ \phi_\lambda = t, x^k \circ \phi_\lambda = \phi_\lambda^k(x^i)\}$. As a by-product it follows that (RF) is invariant under such transformations. In fact $(\phi_\lambda^* g)_{ij,t} - \kappa R_{ij}(\phi_\lambda^* g) = \phi_\lambda^*[g_{ij,t} - \kappa R_{ij}(g)] = 0$. Thus we can continuously transform an integral manifold V_q, into other solutions along the spacelike coordinate lines of N_0, passing for q and identify, in the points $q' \in N_0$, the timelike curves $X_{0,q'}$ that result continuously transformed of $X_{0,q}$. In this way $V = \bigcup_{q \in N_0} X_{0,q}$ is a smooth manifold in a suitable tubular neighborhood of $N_0 \times [0, \epsilon[$, that is the integral manifold of a metric $g_{ij}(t, x^k)$, of class C^3, solution of the Ricci flow equation (RF). This proves that the envelopment manifold V is more regular than a viscosity solution.

In order to obtain a smooth solution of the Cauchy problem given by the integral manifold N_0 it is necessary to repeat the above process on the infinity prolongation $(RF)_{+\infty} \subset JD^\infty(E)$. In fact, also on $(RF)_{+\infty}$, the 3-dimensional integral manifold $N_0 \cong D^\infty\gamma(M) \equiv N_0^{(\infty)} \subset (RF)_{+\infty}$ is uniquely identified by (M, γ), and identifies also a unique transversal timelike characteristic vector field $\zeta_0^{(\infty)}$, tangent to $(RF)_{+\infty}$. More precisely, $\zeta_0^{(\infty)} = \zeta_0 + \sum_{|\alpha|>2} g_{ij,t\alpha} \partial g^{ij,\alpha}$, with $g_{ij,t\alpha} = \kappa R_{ij,\alpha}(\gamma)$, where $\alpha = 0, 1, 2, 3$.

The situation is resumed in the commutative diagram (51).

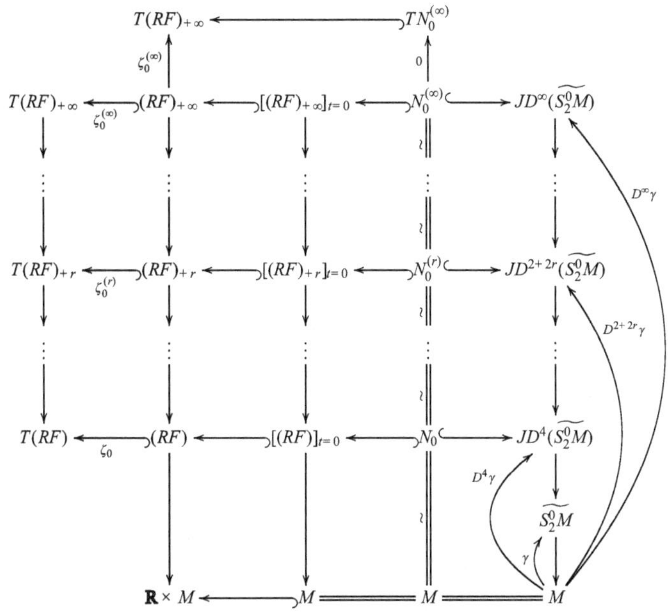

$$(51)$$

Similarly to what made in the Fourier's heat equation, (see Example 9), we can prove that to the above solution one can associate weak-singular ones by using perturbations of the initial Cauchy data. In fact, if $V \subset (RF)$ is a regular solution passing for the integral manifold N_0, we can consider perturbations like solutions $\nu : \mathbb{R}\times M \to E$, of the linearized equation $(RF)[V] \subset JD^2(E)$. Since also $(RF)[V]$ is formally integrable and completely integrable, in neighborhoods of N_0 there exist perturbations. These deform V *(background solution)* giving some new solutions $\widetilde{V} \subset (RF)$. Then $\widehat{V} \equiv V \bigcup \widetilde{V} \subset (RF)$ is a weak-singular solutions of the type just considered in Example 9 for Fourier's heat equation.[24]

Now, for any of two of such integral manifolds, N_0 and N_1, we can find a smooth solution V bording them, $V = N_0 \sqcup N_1$ iff their integral characteristic numbers are equal, i.e., all the conservation laws of (RF) valued on them give equal numbers. By the way, under our assumptions we can consider N_0 and N_1 homotopy equivalent. Let $f : N_1 \to N_0$ be such an homotopy equivalence. Let ω be any conservation law for (RF). Then one has Eq. (52).

$$<[N_0],[\omega]>=\int_{N_0}\omega=\int_{N_1}f^*\omega=<[N_1],[f^*\omega]>=<[N_1],[\omega]>=\int_{N_1}\omega. \tag{52}$$

[24] This generalizes a previous result by Hamilton [27], and after separately by De Turk [19] and Chow and Knopp [16], that proved existence and uniqueness of nonsingular solution for Cauchy problem in some Ricci flow equation. Let us also emphasize that our approach to find solutions for Cauchy problems, works also when $N_0 \subset (RF)$ is diffeomorphic to a 3-dimensional spacelike submanifold of W, that is not necessarily representable by a section of $E_{t=0} \cong \widetilde{S_2^0M} \to M$.

So any possible integral characteristic number of N_0 must coincide with ones with N_1 and vice versa. Thus we can say, that with this meaning of admissibility (*full admissibility hypothesis*) on the Cauchy integral manifolds, one has $cry(RF) = 0$, i.e., (RF) becomes a 0-crystal. Therefore there are not obstructions on the existence of smooth solutions V of (RF) bording N_0 and N_1, $\partial V = N_0 \sqcup N_1$, i.e., solutions without singular points.

This has as a by-product that M and S^3 are homeomorphic manifolds. Hence the Poincaré conjecture is proved. With this respect we can say that this proof of the Poincaré conjecture is related to the fact that under suitable conditions of admissibility for the Cauchy integral manifolds, the Ricci flow equation becomes a 0-crystal PDE.[25]

Example 11 (The d'Alembert Equation $\frac{\partial^2 \log f}{\partial x \partial y} = 0$ on the 2-Dimensional Torus) The d'Alembert equation on a 2-dimensional manifold M can be encoded by a second-order differential equation

$$(d'A)_2 \subset JD^2(W) : \{uu_{xy} - u_x u_y = 0\} \tag{53}$$

with $W \equiv M \times \mathbb{R}$. In [62] we have calculated the integral bordism groups of such an equation. In particular for $M = T^2$, the 2-dimensional torus, one has $\Omega_1^{(d'A)_2} \cong \mathbb{Z}_2 \bigoplus \mathbb{Z}_2$. Taking into account Example 6 we see that $\Omega_1^{(d'A)_2}$ is isomorphic to the crystal group $G(2) = \mathbb{Z}^2 \rtimes D_4 = p4m$. Therefore, $(d'A)_2$ on T^2 is an extended crystal PDE, with crystal dimension 2.

Example 12 (The Tricomi Equation on 2-Dimensional Manifolds) In [62] we have considered the integral bordism groups of the *Tricomi equation* (T):

$$(T) \subset JD^2(W) : \{u_{yy} - yu_{xx} = 0\} \tag{54}$$

defined on a 2-dimensional manifold M, i.e. with $W \equiv M \times \mathbb{R}$. For example, on the 2-dimensional torus T^2 one has $\Omega_1^{(T)} \cong \mathbb{Z}_2 \oplus \mathbb{Z}_2$. Furthermore on $\mathbb{R}P^2$ we obtains $\Omega_1^{(T)} \cong \mathbb{Z}_2$. Thus the Tricomi equation on T^2, (resp. S^2), is an extended crystal PDE with crystal group $p4m$, (resp. $p2$), and crystal dimension 2, (resp. 2).

Example 13 (The Navier–Stokes PDE) The nonisothermal Navier–Stokes equation can be encoded by a second order PDE $\widehat{(NS)} \subset JD^2(W)$ on a 9-dimensional affine fiber bundle $\pi : W \to M$ on the Galilean space-time M. In Table 1 is reported its polynomial differential structure.

[25] The proof of the Poincaré conjecture given here refers to the Ricci flow equation, according to some ideas pioneered by Hamilton [26–30], and followed also by Perelman [43, 44]. However the arguments used here are completely different from ones used by Hamilton and Perelman. (For general information on the relations between Poincaré conjecture and Ricci flow equation, see, e.g., Refs.[5, 17, 18] and papers quoted there.) Here we used our general PDE's algebraic topological theory, previously developed in some works. Compare also with our previous proof given in [2, 3], where, instead was not yet introduced the relation between PDEs and crystallographic groups. Let us note also that whether M is homeomorphic to S^3, then M is necessarily diffeomorphic to S^3. (See [78, 80] for details. For related subjects see also [77, 81–86].)

Table 1 Completely integrable Navier–Stokes equation: $\widehat{(NS)} \subset JD^2(W)$ defined by differential polynomials

(A): $F^0 \equiv \dot{x}^k G^j_{jk} + \dot{x}^i_s \delta^s_i = 0$	$(F^0 \in A[\dot{x}^k, \dot{x}^i_s])$ (continuity equation)
(B): $F^0_\alpha \equiv \dot{x}^k (\partial x_\alpha . G^j_{jk}) + \dot{x}^k_\alpha G^j_{jk} + \dot{x}^i_{s\alpha} \delta^s_i = 0$	$(F^0_\alpha \in A[\dot{x}^k, \dot{x}^i_\alpha, \dot{x}^i_{s\alpha}])$ (first prolonged continuity equation)
(C): $F^j \equiv \dot{x}^s R^j_s + \dot{x}^s \dot{x}^i \rho G^j_{is} + \dot{x}^s \dot{x}^j_s \rho + \rho \dot{x}^j_0 + \dot{x}^k_s S^{js}_k + \dot{x}^j_{is} T^{is} + p_i g^{ij} + \rho(\partial x_i . f) g^{ij} = 0$	$(F^j \in A[\dot{x}^s, \dot{x}^i_s, \dot{x}^j_{is}, p_i])$ (motion equation)
(D): $F^4 \equiv \theta_0 \rho C_p + \rho C_p \dot{x}^k \theta_k + \theta_{is} \overline{E}^{is} + \dot{x}^k \dot{x}^p W_{kp} + \dot{x}^k \dot{x}^s_p \overline{W}^p_{ks} + \dot{x}^k_i \dot{x}^s_p Y^{ip}_{ks} = 0$	$(F^4 \in A[\dot{x}^k, \dot{x}^s_p, \theta_0, \theta_k, \theta_{is}])$ (energy equation)

Functions belonging to $A \subset \mathbb{R}[[x^1, x^2, x^3]]$

$R^j_s \equiv \chi[(\partial x_p . G^j_{is}) + G^j_{pq} G^q_{is} - G^q_{pi} G^j_{qs}] g^{pi}$
$S^{js}_k \equiv \chi[2G^j_{ik} g^{si} - G^s_{qi} \delta^j_k g^{qi}]$
$T^{is} \equiv \chi g^{is} = T^{si}$
$\overline{E}^{is} \equiv -\nu g^{is} = \overline{E}^{si}$
$W_{kp} \equiv -\chi G^a_{jk} G^b_{sp} (g_{ba} g^{sj} + \delta^j_b \delta^s_a) = -2\chi G^s_{bk} G^b_{sp} = W_{pk}$
$\overline{W}^p_{ks} \equiv -2\chi[G^j_{ik} g_{js} g^{ip} + G^p_{sk}] = -4\chi G^p_{sk} = \overline{W}^p_{sk}$
$Y^{ip}_{ks} \equiv -\chi[g_{ks} g^{ip} + \delta^p_k \delta^i_s] = Y^{pi}_{sk}$

$0 \leq \alpha \leq 3, 1 \leq i,j,k,p,s \leq 3. (\partial x_0 . G^j_{jk}) = 0.$

There $\{x^\alpha, \dot{x}^k, p, \theta\}$ are fibered coordinates on W, adapted to the inertial frame, and G^k_{ij} are the canonical connection symbols on M. Furthermore $A \subset \mathbb{R}[[x^1, x^2, x^3]]$ is the algebra of real valued analytic functions of (x^k).[26] We have proved in Refs. [56, 72] that the singular integral bordism groups of such an equation are trivial, i.e., $\Omega^{\widehat{(NS)}}_{3,s} \cong \Omega^{\widehat{(NS)}}_{3,w} \cong 0$. Furthermore, with respect to the notation used in diagram (26), one has that for the integral bordism group for smooth solutions: $\Omega^{\widehat{(NS)}}_3 \cong K^{\widehat{(NS)}}_{3,w}$. Thus we can conclude that the Navier–Stokes equation is an extended 0-crystal PDE, but not a 0-crystal, i.e., $cry\widehat{(NS)} \neq 0$. Note that if we consider admissible only all the Cauchy integral manifolds $X \subset \widehat{(NS)}$ such that all their integral characteristic numbers are zero, *(full admissibility hypothesis)*, it follows that $cry\widehat{(NS)} = 0$. So, under this condition, $\widehat{(NS)}$ becomes a 0-crystal PDE.

Let us emphasize that, similar to the Ricci flow equation, we can identify 3-dimensional spacelike smooth integral manifold $N_0 \subset \widehat{(NS)}_t$, such that $N_0 \cong M_t$, via the canonical projection $\pi_2 : JD^2(W) \to M$, with some smooth spacelike section $s_t : M_t \subset M \to W_t \subset W$, of the configuration bundle $\pi : W \to M$. In

[26] $\mathbb{R}[[x^1, \ldots, x^n]]$ denotes the algebra of formal series $\sum_{i_1,\ldots,i_n} a_{i_1 \cdots i_n} (x^1)^{i_1} \cdots (x^n)^{i_n}$, with $a_{i_1 \cdots i_n} \in \mathbb{R}$. Real analytic functions in the indeterminates $(x^1, \ldots, x^n)$, are identified with above formal series having nonzero converging radius. Thus real analytic functions belong to a subalgebra of $\mathbb{R}[[x^1, \ldots, x^n]]$. This last can be also called the *algebra of real formal analytic functions* in the indeterminates $(x^1, \ldots, x^n)$.

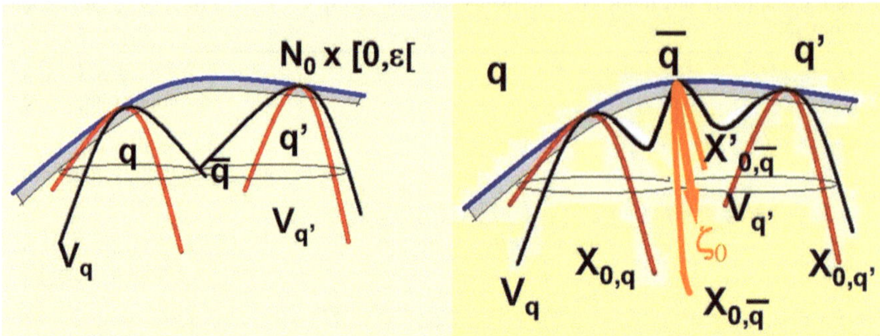

Fig. 2 Construction of envelopment solution V for Cauchy problem $N_0^{(\infty)} \subset \widehat{(NS)}_{+\infty}$, on the infinity prolongation of $\widehat{(NS)}_{+\infty}$ of $\widehat{(NS)}$. One has $V = \bigcup_{q \in N_0^{(\infty)}} X_{0,q}$, with $X_{0,q}$ the unique timelike curve passing for q, tangent to the vector $\zeta_0(q)$ and contained in the unique analytic solution passing for q

fact, all the coordinates in $\widehat{(NS)}$, containing time derivatives, can be expressed by means of the other derivatives containing only space coordinates. (Warn for equation $\widehat{(NS)}$! Even if there are not restrictions on the time derivatives of pressure, between space derivatives of pressure arise some constraints, as well as there are constraints between space derivatives of velocity components, hence sections s_t that identify above considered integral manifolds, are not arbitrary ones. For details see below.) Then we can solve the corresponding Cauchy problem applying Lemma 14, similarly to what made in the Ricci flow equation. It is useful to remark that in order to build envelopment solutions $V = \bigcup_{q \in N_0} X_{0,q}$, it does not necessitate to handle with PDEs that admit any spacelike symmetry. This of course does not happen for any smooth boundary value problem in $\widehat{(NS)}$. With this respect, it is useful also to underline that the popular request to maximize entropy cannot be a sufficient criterion to realize a smooth envelopment solution. (For complementary results on variational problems constrained by the Navier–Stokes equation see [72].) The existence of such a smooth envelopment manifold, can be proved by working on $\widehat{(NS)}_{+\infty}$. In fact, let $q, q' \in N_0^{(\infty)} \subset \widehat{(NS)}_{+\infty}$, and let V_q and $V_{q'}$ be two smooth solutions passing for the initial conditions q and q' respectively. We claim that their timelike integral curves $X_{0,q}$ and $X_{0,q'}$ cannot intersect for suitable short times, i.e., $t \in [0, \epsilon[$, if $q \neq q'$. Really if $\bar{q} \in V_q \bigcap V_{q'} \neq \varnothing$, then $T_{\bar{q}} V_q = T_{\bar{q}} V_{q'} = (\mathbf{E}_\infty)_{\bar{q}}$. This means that in such a point $\bar{q}$, V_q and $V_{q'}$ must have a contact of infinity order with the Navier–Stokes equation and between them. We can assume that $\bar{q}$ is outside a suitable tubular neighborhood $N_0^{(\infty)} \times [0, \epsilon[$ of $N_0^{(\infty)}$, otherwise we should admit that $\bar{q} \in N_0^{(\infty)}$. In this last case V_q and $V_{q'}$ should have in $\bar{q}$ time-like curves $X_{0,\bar{q}}$ and $X'_{0,\bar{q}}$, transversal to $N_0^{(\infty)}$, and with a common tangent vector $\zeta_0(\bar{q})$. Furthermore, V_q and $V_{q'}$ should be tangent to $N_0^{(\infty)}$ at $\bar{q}$: $T_{\bar{q}} V_q = T_{\bar{q}} V_{q'} = (\mathbf{E}_\infty)_{\bar{q}} \supset T_{\bar{q}} N_0^{(\infty)}$. Let us assume that in a suitable neighborhood $N_0^{(\infty)} \times [0, \epsilon[$, the manifold V_q identifies two separated pieces, say and $V_{q|1}$ and $V_{q|2}$, tangent to $N_0^{(\infty)}$ at q and $\bar{q}$, respectively. In other words $V_q \bigcap N_0^{(\infty)} \times [0, \epsilon[= V_{q|1} \bigcup V_{q|2}$. (If this condition is not satisfied, then V_q is necessarily a smooth solution of the Cauchy problem at least in a subset $Y \subset N_0^{(\infty)}$. Then we can write $V = V_1 \bigcup V_2 \equiv \{\bigcup_{q\in Y} X_{0,q}\} \bigcup \{\bigcup_{q\in Z} X_{0,q}\}$, $Z \equiv N_0^{(\infty)} \setminus Y$, and we can continue to similarly discuss about the V_2-part of V.) Then, for the timelike coordinate lines $X_{0,q|1}$ and $X_{0,q|2}$, one has $X_{0,q|1} \bigcap X_{0,q|2} = \varnothing$ in $N_0^{(\infty)} \times [0,\epsilon[$. The same circumstance can be verified between $V_{q'}$ and $V_{\bar{q}}$, eventually by reducing ϵ. This proves that taking ϵ little enough, we can consider $V \equiv \bigcup_{q\in N_0^{(\infty)}} X_{0,q} \subset \widehat{(NS)}_{+\infty}$, a fiber bundle over $N_0^{(\infty)}$ that is a 4-dimensional integral manifold representing a local solution of the Cauchy problem identified by the smooth 3-dimensional integral manifold $N_0 \subset \widehat{(NS)}$. (See Fig. 2.) On V we can recognize a natural smooth fiber bundle structure. In fact, since $\widehat{(NS)}$ is an analytic equation and any point $q \in N_0^{(\infty)}$ identifies an analytic solution in a suitable neighborhood U of $p \equiv \pi_\infty(q) \in M$, it follows that if two such solutions s and s' are defined at $q \in N_0^{(\infty)}$ they should coincide in a suitable neighborhood of p, denote again it by U. Then we can assume that s and s' coincide also in $\Omega \equiv \pi_\infty^{-1}(U) \bigcap N_0^{(\infty)}$. Therefore, we can cover $N_0^{(\infty)}$ by means of a covering set $\{\Omega_\alpha\}_{\alpha\in J}$, where each Ω_α is such that if $q, q' \in \Omega_\alpha$, the corresponding analytic solutions s and s' are well defined in $U_\alpha \subset M$. In this way each timelike curve $X_{0,q}$ is uniquely identified for $q \in \Omega_\alpha$. Then we can define smooth functions (*transition functions*) $\psi_{\alpha\beta} : \Omega_\alpha \bigcap \Omega_\beta \to \mathbb{R}^\bullet \equiv \mathbb{R}\setminus\{0\}$, by $\tau_\alpha = \psi_{\alpha\beta}\tau_\beta$, where τ_α are nowhere vanishing sections of $V \to N_0^{(\infty)}$ on $\Omega_\alpha \subset N_0^{(\infty)}$. These functions satisfy the following three conditions: (a) $\psi_{\alpha\alpha} = 1$; (b) $\psi_{\alpha\beta} = \psi_{\beta\alpha}$; (c) $\psi_{\alpha\beta}\psi_{\beta\gamma}\psi_{\gamma\alpha} = 1$ on $\Omega_\alpha \bigcap \Omega_\beta \bigcap \Omega_\gamma$ (*cocycle condition*). This is enough to claim that the line bundle $V \to N_0^{(\infty)}$ has a smooth structure. Let us emphasize that Lemma 14 can be applied here also to 3-dimensional spacelike smooth Cauchy integral manifolds $N \subset \widehat{(NS)}_t$, that are diffeomorphic to their projections $Y \equiv \pi_{2,0}(N) \subset W_t$, but are not necessarily holonomic images of smooth sections of $\pi : W \to M$. In fact, by using the embeddings $N \subset \widehat{(NS)} \subset JD^2(W) \subset J_4^2(W)$, we can repeat above construction to build smooth envelopment solutions of the Cauchy problem $N^{(\infty)} \subset \widehat{(NS)}_{+\infty} \subset J_4^\infty(W)$. Really, any smooth 3-dimensional spacelike submanifold $Y \subset W_t$, identified by a spacelike smooth integral manifold $Z \subset \widehat{(NS)}_{+\infty} \subset J_4^\infty(W)$, such that $Z \cong Y$, via the canonical projection $\pi_{\infty,0} : J_4^\infty(W) \to W$, can be locally identified by some smooth implicit functions $\{f_I(x^k, y^j) = 0\}_{1\leq I\leq 5}$, with a Jacobian matrix of rank five, where $\{x^k, y^j\}_{1\leq k\leq 3; 1\leq j\leq 5}$ are coordinates in the 8-dimensional affine space W_t, and we write $Z = Y^{(\infty)}$. Then, as a by-product we get also a cohomology criterion to classify envelopment solutions according to the first cohomology space $H^1(N_0, \mathbb{Z}_2)$. In fact, a line bundle $V \to N_0^{(\infty)}$ is classified by the first *Stiefel–Whitney class* of V, that belongs to $H^1(N_0^{(\infty)}, \mathbb{Z}_2) \cong H^1(N_0, \mathbb{Z}_2)$. The classifying space is $\mathbb{R}P^\infty$ and the universal principal bundle is $S^\infty \to \mathbb{R}P^\infty \equiv S^\infty/\mathbb{Z}_2 \cong \mathbb{R}^\infty/\mathbb{R}^\bullet \cong Gr_1(\mathbb{R}^\infty)$, where $Gr_1(\mathbb{R}^\infty)$ denotes the Grassmannian of 1-dimensional vector subspaces in $\mathbb{R}^\infty$, and the nonzero element of $\mathbb{Z}_2$ acts by $v \mapsto -v$. Since S^∞ is contractible one has $\pi_i(S^\infty) \cong \pi_i(\mathbb{R}P^\infty) = 0$, $i > 1$ and $\pi_1(\mathbb{R}P^\infty) \cong \mathbb{Z}_2$. $\mathbb{R}P^\infty$ is the Eilenberg–Maclane space $K(\mathbb{Z}_2, 1)$ [60]. Hence $[N_0, Gr_1] \cong H^1(N_0; \mathbb{Z}_2)$, $f \mapsto f^*\mu$, where μ is the generator of $H^1(\mathbb{R}P^\infty; \mathbb{Z}_2) \cong \mathbb{Z}_2$. Since one has the bijection $[N_0; Gr_1] \to \mathbf{V}_1(N_0)$, where $\mathbf{V}_1(N_0)$ denotes the set of 1-dimensional vector bundles over N_0, we get the bijection $w_1 : \mathbf{V}_1(N_0) \to H^1(N_0; \mathbb{Z}_2)$ that is just the Stiefel–Whitney class for line bundles over $N_0^{(\infty)}$. In conclusion an envelopment solution

$V \supset N$, of an admissible Cauchy integral manifolds $N \subset \widehat{(NS)}$, can be identified with some cohomology class of $H^1(N;\mathbb{Z}_2)$. In particular V is orientable iff its first Stiefel–Whitney cohomology class $w_1(V) = 0$. Of course it is not necessary that all cohomology classes of $H^1(N;\mathbb{Z}_2)$ should be represented by some envelopment solution V passing for N. However, when this happens we say that N is a *wholly cohomologic Cauchy manifold* of $\widehat{(NS)}$. This is surely the case when $N \simeq D^3$, i.e., when the spacelike, smooth, 3-dimensional, integral manifold N is homotopy equivalent to the 3-dimensional disk D^3. In fact in such a case $H^1(N;\mathbb{Z}_2) = 0$, hence there is an unique cohomology type of envelopment solution passing for $N^{(\infty)}$, (or N), the orientable one. Therefore, in such a case N is a wholly cohomologic Cauchy manifold. For example, when N is identified by a smooth spacelike section $s_t : M_t \subset M \to W_t \subset W$, N is a wholly cohomologic Cauchy manifold. It is important to remark also that, fixed some smooth Cauchy problem $N \subset \widehat{(NS)}$, it does not necessitate that a local smooth solution V_ϵ should be unique. In fact, even if the Cartan distribution $\mathbf{E}_\infty \subset \widehat{(NS)}_{+\infty}$ is a 4-dimensional involutive distribution, the manifold $\widehat{(NS)}$ is not finite dimensional. Therefore, for any point $q \in \widehat{(NS)}$ pass infinity 4-dimensional integral manifolds tangent to $(\mathbf{E}_\infty)_q \subset T_q(\widehat{(NS)}_{+\infty})$. However, we can see that if there are two smooth solutions $V_\epsilon, V'_\epsilon \subset \widehat{(NS)}_{+\infty}$ for a fixed smooth Cauchy problem $N \subset \widehat{(NS)}$, their Stiefel–Whitney classes must coincide with a same cohomology class of $H^1(N,\mathbb{Z}_2)$. In fact, let us denote by $\mathbf{N}_\infty(N)_\epsilon$ the *integral* ϵ*-normal bundle* of $N^{(\infty)}$, i.e., $\mathbf{N}_\infty(N)_\epsilon \equiv \bigcup_{q\in N^{(\infty)}} \mathbf{N}_\infty(N)_{\epsilon,q}$, with $\mathbf{N}_\infty(N)_{\epsilon,q} \equiv ((\mathbf{E}_\infty)_q/T_qN)_\epsilon \equiv \{[0,\epsilon[\zeta_0(q)\}$. Then one has the canonical isomorphism of line bundles over $N^{(\infty)}$: $\mathbf{N}_\infty(N)_\epsilon \cong V_\epsilon$, $\lambda\zeta_0(q) \mapsto X_{0,q}(\lambda)$, $\lambda \in [0,\epsilon[$. A similar isomorphism can be recognized between $\mathbf{N}_\infty(N)_\epsilon$ and V'_ϵ, taking into account that also V'_ϵ can be considered a line bundle: $V'_\epsilon \equiv \bigcup_{q\in N^{(\infty)}} X'_{0,q}$. This proves that must necessarily be $\omega_1(V_\epsilon) = \omega_1(\mathbf{N}_\infty(N)_\epsilon) = \omega_1(V'_\epsilon)$.

So we have shown that any spacelike, smooth, 3-dimensional integral manifold $N \subset \widehat{(NS)}$ that is diffeomorphic to its projection on $M_t \subset M$, (for some t), via the canonical projection $\pi_2 : JD^2(W)_t \to M_t$, or is diffeomorphic to its projection on W_t, via the canonical projection $\pi_{2,0} : JD^2(W)_t \to W_t$, admits smooth solutions, $V \supset N$. Since such integral manifolds (Cauchy manifolds) are not arbitrary ones, but satisfy some constraints. Here we shall explicitly prove that such Cauchy manifolds exist. From Table 1 we can write $\widehat{(NS)}$ in the form reported in (55)

$$\left\{\begin{array}{ll}
(\mathrm{A}) & \dot{x}^k G^j_{jk} + \dot{x}^s_s = 0 \\
(\mathrm{B})_0 & \dot{x}^k_0 G^j_{jk} + \dot{x}^s_{s0} = 0 \\
(\mathrm{B})_h & \dot{x}^k(\partial x_h.G^j_{jk}) + \dot{x}^k_h G^j_{jk} + \dot{x}^s_{sh} = 0 \\
(\mathrm{C}) & \dot{x}^j_0 = \frac{1}{\rho}\left[-\dot{x}^s R^j_s - \dot{x}^s\dot{x}^i\rho G^j_{is} - \dot{x}^s\dot{x}^j_s\rho - \dot{x}^k_s S^{js}_k - \dot{x}^j_{is}T^{is} - p_i g^{ij} - \rho(\partial x_i.f)g^{ij}\right] \\
 & \equiv H^j \in A[[\dot{x}^k, \dot{x}^k_h, \dot{x}^k_{rs}, p_i]] \\
(\mathrm{D}) & \theta_0 = \frac{1}{\rho C_p}\left[-\rho C_p\dot{x}^k\theta_k - \theta_{is}\overline{E}^{is} - \dot{x}^k\dot{x}^p W_{kp} - \dot{x}^k\dot{x}^s_p\overline{W}^p_{ks} - \dot{x}^k_i\dot{x}^s_p Y^{ip}_{ks}\right] \\
 & \equiv K \in A[[\dot{x}^k, \dot{x}^k_h, \theta_k, \theta_{is}]]
\end{array}\right. \tag{55}$$

From the prolongation of Eq. (55)(C) with respect to space coordinates x^k, we get

$$\left\{\begin{array}{ll}(\mathrm{C})_q \dot{x}^j_{0q} & = \frac{1}{\rho}\Big[-\dot{x}^s_q R^j_s - \dot{x}^s_q R^j_{s,q} - 2\dot{x}^s_q \dot{x}^i \rho G^j_{is} - \dot{x}^s \dot{x}^i \rho G^j_{is,q} - \dot{x}^s_q \dot{x}^j_s \rho - \dot{x}^s \dot{x}^j_{sq} \rho \\ & -\dot{x}^k_{sq} S^{js}_k - \dot{x}^k_s S^{js}_{k,q} - \dot{x}^j_{isq} T^{is} - \dot{x}^j_{is} T^{is}_{,q} - p_{iq} g^{ij} - p_i g^{ij}_{,q} \\ & -\rho(\partial x_q \partial x_i . f) g^{ij} - \rho(\partial x_i . f) g^{ij}_{,q}\Big] \\ & \equiv H^j_{,q} \in A[[\dot{x}^k, \dot{x}^k_h, \dot{x}^k_{rs}, \dot{x}^k_{rsq}, p_i, p_{iq}]]\end{array}\right. \tag{56}$$

By using Eqs. (55)(C) and (56)$(C)_q$, we can rewrite Eq. (55)$(B)_0$ in the form reported in (57).

$$(\mathrm{E}) G^j_{jk} H^k(x^k, \dot{x}^s, \dot{x}^s_k, \dot{x}^s_{hk}, p_i) + H^s_{,s}(x^k, \dot{x}^s, \dot{x}^s_k, \dot{x}^s_{hk}, \dot{x}^s_{hkq}, p_i, p_{iq}) = 0. \tag{57}$$

Therefore, the parametric equations of a spacelike integral analytic 3-dimensional manifold $N \subset \widehat{(NS)}$, diffeomorphic to M_t, are given in (58).

$$\left\{\begin{array}{ll} x^0 & = t \\ x^k & = x^k \\ \dot{x}^j & = v^j(x^k) \\ p & = p(x^k) \\ \theta & = \theta(x^k) \\ \dot{x}^j_h & = v^j_{,h}(x^k) \\ \dot{x}^j_0 & = H^j(x^k) \\ p_0 & = p_0(x^k) \\ p_k & = p_{,k}(x^k) \\ \theta_0 & = K(x^k) \\ \theta_k & = \theta_{,k}(x^k) \\ \dot{x}^j_{0\,h} & = H^j_{,h}(x^k) \\ \dot{x}^j_{hk} & = v^j_{,hk}(x^k) \\ \dot{x}^j_{00} & = H^j_{,0} = (\partial \dot{x}_s . H^j)\dot{x}^s_0 + (\partial \dot{x}^h_s . H^j)\dot{x}^s_{h0} + (\partial \dot{x}^{hk}_s . H^j)\dot{x}^s_{hk0} + (\partial p^i . H^j) p_{i0} \\ & = (\partial \dot{x}_s . H^j) H^s + (\partial \dot{x}^h_s . H^j)(\partial x_h . H^s) + (\partial \dot{x}^{hk}_s . H^j)(\partial x_k \partial x_h . H^s) \\ & +(\partial p^i . H^j)(\partial x_i . p_0) \\ p_{0k} & = (\partial x_k . p_0) \\ p_{hk} & = (\partial x_h \partial x_k . p) \\ p_{00} & = p_{00}(x^k) \\ \theta_{hk} & = (\partial x_h \partial x_k . \theta) \\ \theta_{0k} & = (\partial x_k . K) \\ \theta_{00} & = (\partial \dot{x}_k . K)\dot{x}^k_0 + (\partial \dot{x}^p_k . K)\dot{x}^k_{p0} + (\partial \theta^k . K)\theta_{k0} + (\partial \theta^{is} . K)\theta_{is0} \\ & = (\partial \dot{x}_k . K) H^k + (\partial \dot{x}^p_k . K)\dot{x}_p . H^k) + (\partial \theta^k . K)(\partial x_k . K) + (\partial \theta^{is} . K)(\partial x_i \partial x_s . K) \end{array}\right. \tag{58}$$

where $v^j = \dot{x}^j(x^k)$ are solutions of the continuity Eq. (55)(A). Therefore, Eq. (55)(A) can be considered a first order equation on the fiber bundle $E_t \equiv M_t \times \mathbf{I} \to M_t$, reported in (59).

$$E_1 \subset JD(E_t) : \{\dot{x}^k G^j_{jk} + \dot{x}^s_s = 0\}. \tag{59}$$

One can see that E_1 is an involutive, formally integrable, and completely integrable PDE. In fact, one has $\boxed{\dim (E_1)_{+1} = 29} = \boxed{\dim E_1 = 14} + \boxed{\dim (g_1)_{+1} = 15}$,

hence the mapping $(E_1)_{+1} \to E_1$ is surjective. Furthermore, we get

$$\boxed{\dim (g_1)_{+1} = 15} = \boxed{\dim g_1 = 8} + \boxed{\dim (g_1)^{(1)} = 5} + \boxed{\dim (g_1)^{(2)} = 2} + \boxed{\dim (g_1)^{(3)} = 0}.$$

This is enough to state that g_1 is an involutive symbol. Therefore, E_1 is formally integrable and, since it is analytic it is also completely integrable. As a by-product, we get that for any analytic solution $\dot{x}^k = \dot{x}^k(x^i)$ of E_1, we can write:

$$\begin{cases} \dot{x}_0^j = A^j(x^s) - g^{ij} p_i \\ \dot{x}_{q0}^j = A^j(x^s)_{,q} - \frac{1}{\rho} g^{ij} p_{iq} - \frac{1}{\rho} g^{ij}_{,q} p_i \end{cases} \tag{60}$$

where $A^j = A^j(x^s)$ are suitable analytic functions. So Eq. (57)(E) can be rewritten as a second order equation E_2 for a function $p = p(x^k)$ as section of the trivial fiber bundle $F_t \equiv M_t \times \mathbb{R} \to M_t$:

$$E_2 \subset JD^2(F_t) : \{g^{is} p_{is} + p_i C^i(x^k) - B(x^k) = 0\} \tag{61}$$

where C^i and B are given analytic functions of x^k. This is an involutive, formally integrable PDE, hence it is completely integrable, since it is analytic. In fact,

$$\boxed{\dim(E_2)_{+1} = 19} = \boxed{\dim(E_2) = 12} + \boxed{\dim(g_2)_{+1} = 7}.$$

Therefore, the map $(E_2)_{+1} \to E_2$ is surjective. Furthermore,

$$\boxed{\dim(g_2)_{+1} = 7} = \boxed{\dim(g_2) = 5} + \boxed{\dim(g_2)^{(1)} = 2} + \boxed{\dim(g_2)^{(2)} = 0},$$

hence g_2 is an involutive symbol. This concludes the proof. Thus, for any point $q \in E_1$, passing solutions of the continuity equation and fixing a point q on the infinity prolongation $(E_1)_{+\infty}$, one identifies an analytic solution defined in a suitable neighborhood of $p \equiv \pi_\infty(q) \in M_t$. (Of course also for equation E_1 we can identify smooth solutions by solving lower dimension Cauchy problems, by means of Lemma 14, i.e., by using envelopment solutions.) Similar considerations can be applied to the equation E_2 to identify analytic and smooth solutions of the pressure functions $p = p(x^k)$. This assures that one can identify spacelike smooth Cauchy manifolds in $\widehat{(NS)}$, that are diffeomorphic with their canonical projections on W_t, for any time t.

In [72] it is proved that the set of full admissible Cauchy integral manifolds is not empty. This result gives us a general criterion to characterize global smooth solutions of the Navier–Stokes equation and completely solves the well-known problem on the existence of global smooth solutions of the Navier–Stokes equation. (For complementary characterizations of the Navier–Stokes equation, see also [55, 58, 68–70 72–74]. There a geometric method to study stability of PDEs and their solutions related to integral and quantum bordism groups of PDEs. It has been introduced and applied to the Navier–Stokes equation too. See also [89] for related subjects.)

4 Extended Crystal Singular PDEs

Singular PDEs can be considered singular submanifolds of jet-derivative spaces. The usual formal theory of PDEs works, instead, on smooth or analytic submanifolds. However, in many mathematical problems and physical applications, it is necessary to work with singular PDEs. (See, e.g., the book by Gromov [25] where he talks of "partial differential relations," i.e., subsets of jet-derivative spaces.) So it is useful to formulate a general geometric theory for such more general mathematical structures. On the other hand, in order to build a formal theory of PDEs it is necessary to assume some regularity conditions. So a geometric theory of singular PDEs must in some sense weaken the regularity conditions usually adopted in formal theory and admit existence of subsets where these regularity conditions are not satisfied. With this respect, and by using our formulation of geometric theory of PDEs and singular PDEs, we study criteria to obtain global solutions of singular PDEs crossing singular points. In particular, some applications concerning singular MHD-PDEs encoding anisotropic incompressible nuclear plasma dynamics are given following some our recent works on this subject. The origin of singularities comes from the fact that there are two regions corresponding to different component PDEs having different Cartan distributions with different dimensions. However, by considering their natural embedding into the same PDE, we can build physically acceptable solutions, i.e., satisfying the second principle of the thermodynamics and crossing the nuclear critical zone of nuclear energy production. A characterization of such solutions by means of algebraic topological methods is given too.

The main result of this section is Theorem 18 that relates singular integral bordism groups of singular PDEs to global solutions passing through singular points, and Example 17 that for some MHD-PDEs characterizes global solutions crossing the nuclear critical zone and satisfying the entropy production energy thermodynamics condition.

Let us, now, resume some fundamental definitions and results on the geometry of PDE's in the category of commutative manifolds, emphasizing some our recent results on the algebraic geometry of PDE's, that allowed us to characterize singular PDE's.[27]

Definition 15 (Algebraic Formulation of PDEs) Let $\pi : W \to M$ be a smooth fiber bundle, $\dim W = m + n$, $\dim M = n$. We denote by $J^k_n(W)$ the *space of all k-jets of submanifolds* of dimension n of W and by $J^k(W)$ the *k-jet-derivative space of sections* of π. Furthermore we denote by $JD^k(W)$ the *k-jet-derivative space* for sections of π. One has $JD^k(W) \cong J^k(W) \subset J^k_n(W)$. $J^k(W)$ is an open subset of $J^k_n(W)$. Let $\mathfrak{A}_k$ be the sheaf of germs of differentiable functions $JD^k(W) \to \mathbb{R}$.

[27] For general information on the geometric theory of PDEs see, e.g., [9, 12, 14, 22–25, 33, 34, 38, 103–105]. In particular, for singular PDE geometry, see the book [60] and the recent papers [3, 79] where many boundary value problems are explicitly considered. For basic information on differential topology and algebraic topology, see, e.g., [9, 24, 31, 39, 41, 47, 94, 99, 98, 102–105, 107, 108].

It is a sheaf of rings, but also a sheaf of $\mathbb{R}$-modules. A subsheaf of ideals $\mathfrak{B}_k$ of $\mathfrak{A}_k$ that is also a subsheaf of $\mathbb{R}$-modules is a *PDE of order k* on the fiber bundle $\pi : W \to M$. A *regular solution* of $\mathfrak{B}_k$ is a section $s : M \to W$ such that $f \circ D^k s = 0, \forall f \in \mathfrak{B}_k$. The set of *integral points* of $\mathfrak{B}_k$ (i.e., the zeros of $\mathfrak{B}_k$ on $JD^k(W)$ is denoted by $J(\mathfrak{B}_k)$. The first *prolongation* $(\mathfrak{B}_k)_{+1}$ of $\mathfrak{B}_k$ is defined as the system of order $k+1$ on $W \to M$, defined by the $f \circ \pi_{k,k-1}$ and $f^{(1)}$, where $f^{(1)}$ on $D^{k+1}s(p)$ is defined by $f^{(1)}(D^{k+1}s(p)) = (\partial x_\alpha.(f \circ D^k s(p)))$. In local coordinates $(x^\alpha, y^j, y^j_\alpha)$ the *formal derivative* $f^{(1)}$ is given by $f^{(1)}(x^\alpha, y^j, y^j_\alpha) = (\partial x_\alpha.f) + \sum_{[\beta]\leq k} y^j_{\beta\alpha}(\partial y^\beta_j.f)$. The system $\mathfrak{B}_k$ is said to be *involutive* at an integral point $q \in JD^k(W)$ if the following two conditions are satisfied: *(i)* $\mathfrak{B}_k$ is a *regular local equation* for the zeros of $\mathfrak{B}_k$ at q (i.e., there are local sections $F_1, \dots, F_t \in \Gamma(U, \mathfrak{B}_k)$ of $\mathfrak{B}_k$ on an open neighborhood U of q, such that the integral points of $\mathfrak{B}_k$ in U are precisely the points q' for which $F_j(q') = 0$ and $dF_1 \wedge \cdots \wedge dF_t(q) \neq 0$, that is $F_1, \cdots, F_t$ are linearly independent at q; and *(ii)* there is a neighborhood U of q such that $\pi^{-1}_{k+1,k}(U) \bigcap J((\mathfrak{B}_k)_{+1})$ is a fibered manifold over $U \bigcap J(\mathfrak{B}_k)$ (with projection $\pi_{k+1,k}$). For a system $\mathfrak{B}_k$ generated by linearly independent Pfaffian forms $\theta^1, \cdots, \theta^k$ (i.e., a Pfaffian system) this is equivalent to the involutiveness defined for distributions.

Theorem 11 [(34)] *Let $\mathfrak{B}_k$ be a system defined on $JD^k(W)$, and suppose that $\mathfrak{B}_k$ is involutive at $q \in J(\mathfrak{B}_k)$. Then, there is a neighborhood U of q satisfying the following. If $\widetilde{q} \in J((\mathfrak{B}_k)_{+s})$ and $\pi_{k+s,k}(\widetilde{q})$ is in U, then there is a regular solution s of $\mathfrak{B}_k$ defined on a neighborhood $p = \pi_{k+s,-1}(\widetilde{q})$ of M such that $D^{k+s}s(p) = \widetilde{q}$.*

Theorem 12 (Cartan-Kuraniski Prolongation Theorem [34, 60]) *Suppose that there exists a sequence of integral points $q^{(s)}$ of $(\mathfrak{B}_k)_{+s}, s = 0, 1, \cdots$, projecting onto each other, $\pi_{k+s,k+s-1}(q^{(s)}) = q^{(s-1)}$, such that:* (a) *$(\mathfrak{B}_k)_{+s}$ is a regular local equation for $J((\mathfrak{B}_k)_{+s})$ at $q^{(s)}$; and* (b) *there is a neighborhood $U^{(s)}$ of $q^{(s)}$ in $J((\mathfrak{B}_k)_{+s})$ such that its projection under $\pi_{k+s,k+s-1}$ contains a neighborhood of $q^{(s-1)}$ in $J((\mathfrak{B}_k)_{+(s-1)})$ and such that $\pi_{k+s,k+s-1} : U^{(s)} \to \pi_{k+s,k+s-1}(U^{(s)})$ is a fibered manifold. Then, $(\mathfrak{B}_k)_{+s}$ is involutive at $q^{(s)}$ for s large enough.*

The algebraic characterization of singular PDE's can be given by adopting the methods of the algebraic geometry, combined with the differential algebra. (See, e.g., [60].) Let us go here in some details about.

Definition 16 A *differential ring* is a ring A with a finite number n of commutating derivations $d_1, \cdots, d_n$, $d_i d_j - d_j d_i = 0, \forall i, j = 1, \cdots, n$. A *differential ideal* is an ideal $\mathfrak{a} \subset A$ which is stable by each d_i, $i = 1, \cdots, n$.

A differential ring $(A, \{d_j\}_{1 \leq j \leq n})$ identifies a subring (*subring of constants*): $C \equiv cst(A) \equiv \{a \in A | d_j a = 0, \forall j = 1, \cdots, n\} \subset A$. We may extend each d_i to a derivation of the full ring of fractions, $Q(A)$, still denoted by d_i and such that $d_i(a/r) = (r d_i a - a d_i r)/r^2$, for any $0 \neq r, a \in A$.

Example 14 If K is a differential field with derivations $\partial_1, \cdots, \partial_\mu$ and y^k, $k = 1, \cdots, m$, are indeterminates over K, we set $y^k_0 = y^k$. Then the polynomial ring $K[y]_d = K[y^k_\mu, k = 1, \cdots, m, \mu = \mu_1 \cdots \mu_s, |\mu| \geq 0]$, can be endowed with a structure of differential ring by defining the *formal derivations* $d_i \equiv \partial_i + y^k_{\mu+1i}\partial y^\mu_k$. Of course $K[y]_d$ is not a Noetherian ring. We write $K[y_q]_d = K[y^k_\mu | k = 1, \cdots, m; 0 \leq |\mu| \leq q]$ and one has $K(y_q)_d = Q(K[y_q]_d)$. We set also $K(y)_d = Q(K[y]_d)$.

Definition 17 A *differential subring* A of a differential ring B is a subring which is stable under the derivations of B. Similarly we can define a *differential extension* L/K of differential fields, and such an extension is said to be *finitely generated* if one can find elements $\eta^1, \cdots, \eta^m \in L$ such that $L = K(\eta^1, \cdots, \eta^m)$. Then the *evaluation epimorphism* is defined by $K[y]_d \to K[\eta]_d \subset L$, $y^k \mapsto \eta^k$. Its kernel is a prime differential ideal.

Proposition 2 [60] *Let* $< S >_d$ *denote the differential ideal generated by the subset* $S \subset A$, *where* A *is a differential ring. If* A *is a differential ring and* $a, b \in A$, *then one has the following:*

(i) $a^{|\mu|+1} d_\mu b \in < d_\nu(ab) | |\nu| \leq |\mu| >$.
(ii) $(d_i a)^{2r-1} \in < a^r >_d$.
(iii) *If* $\mathfrak{a}$ *is a radical differential ideal of the differential ring* A *and* S *is any subset of* A, *then* $\mathfrak{a} : S \equiv \{a \in A | aS \subset \mathfrak{a}\}$ *is again a radical differential ideal of* A.[28]
(iv) *If* $\mathfrak{a}$ *is a differential ideal of a differential ring* A, *then* $rad(\mathfrak{a})$ *is a differential ideal too.*
(v) *One has the following inclusion:* $a.rad < S >_d \subset rad < aS >_d$, $\forall a \in A$, *and for all subset* $S \subset A$.
(vi) *If* S *and* T *are two subsets of a differential ring* A, *then*

$$rad < S >_d .rad < T >_d \subset rad < ST >_d = rad < S >_d \cap rad < T >_d .$$

(vii) *If* S *is any subset of a differential ring* A, *then we have:*

$$rad < S, a_1, \cdots, a_r >_d = rad < S, a_1 >_d \cap \cdots \cap rad < S, a_r >_d .$$

Definition 18 A *differential vector space* is a vector space V over a differential field $(K, \partial_i)_{1 \leq i \leq n}$ such that are defined n homomorphisms d_i, $i = 1, \cdots, n$, of the additive group V such that: $d_i(av) = (\partial_i a)v + a(d_i v), \forall a \in K, \forall v \in V$. Then we say that K is a *differential field of definition*.

Proposition 3 [60] *Let* V *be a differential vector space over a differential field* K, *with derivations* d_i, $i = 1, \cdots, n$, *and let* $\{e_j\}_{j \in I}$ *be a basis of* V. *Then the field*

[28] If $\mathfrak{a}$ is any ideal of A, the *radical* of $\mathfrak{a}$ is the following ideal $r(\mathfrak{a}) \equiv \sqrt{\mathfrak{a}} \equiv \{x \in A | x^n \in \mathfrak{a}$ for some $n > 0\} \equiv rad(\mathfrak{a})$. If $\sqrt{\mathfrak{a}} = \mathfrak{a}$, then $\mathfrak{a}$ is called *radical ideal* or *perfect*. One has also that $r(\mathfrak{a})$ is the intersection of all prime ideals $\mathfrak{p} \subset A$, containing $\mathfrak{a}$. In particular, the radical of the zero ideal $< 0 >$ is the *nilradical*, $nil(A)$, of A, i.e., the set of all nilpotent elements of A. Therefore $nil(A)$ is the intersection of all prime ideals, (since all ideals must contain 0). One has also $nil(A) \subset rad(A)$, where $rad(A)$ is the ideal of A defined by intersection of all maximal ideals $\mathfrak{m} \subset A$. If $\mathfrak{a}$ is a radical ideal, then $A/\mathfrak{a}$ is *reduced*, i.e., the set of its nilpotent elements is reduced to $\{0\}$. In particular $A/nil(A)$ is reduced. If $\pi : A \to A/\mathfrak{a}$ is the canonical projection, then $\pi^{-1}(nil(A/\mathfrak{a})) = r(\mathfrak{a})$.

of definition κ *of a differential subspace* $W \subset V$ *is a differential subfield of* K *if it contains the field of definition of each* $d_1 e_i, \cdots, d_n e_i$ *with respect to* $\{e_i\}$.

Definition 19 A family $\eta = (\eta^1, \cdots, \eta^m)$ of elements in a differential extension of the differential field K is said to be *differentially algebraically independent* (or a *family of differential indeterminates*) over K, if the kernel of the evaluation epimorphism $K[y]_d \to K[\eta]_d$ is zero. Otherwise the family is said to be *differentially algebraically dependent* (or *differentially algebraic*) over K.

Proposition 4 *If* K/κ *and* L/κ *are two given differential extensions with respective derivations* d_K *and* d_L*, there always exists a differential free composite field of* K *and* L *over* κ*.*

Proof The ring $K \bigotimes_\kappa L$ has a natural differential structure given by $d(a \otimes b) = (d_K a) \otimes b + a \otimes (d_L b)$, as $d_K|_\kappa = d_L|_\kappa = \partial$. On the other hand there is a finite number of prime ideals $\mathfrak{p}_i \subset K \bigotimes_\kappa L$ such that $\bigcap_i \mathfrak{p}_i = 0$ and $\mathfrak{p}_i + \mathfrak{p}_j =< 1 >$, $\forall i \neq j$. Now we have the following lemma.

Lemma 15 *If* $\mathfrak{a}_1, \cdots, \mathfrak{a}_r$ *are ideals of a differential ring* A *such that* $\mathfrak{a}_i + \mathfrak{a}_j = A$, $\forall i \neq j$*, and* $\mathfrak{a}_1 \cap \cdots \cap \mathfrak{a}_r$ *is a differential ideal of* A*, then each* $\mathfrak{a}_i$ *is a differential ideal too.*

Therefore, we can conclude that each $\mathfrak{p}_i$ is a differential ideal, hence the proposition is proved.

Lemma 16 *A family* η *is differentially algebraic over* K *iff a differential polynomial* $P \in \mathfrak{p}$ *exists such that* $(\partial y_P.P) \notin \mathfrak{p}$*, where* y_P *is the highest power of* y_p *appearing in* P*.* $S_P \equiv (\partial y_P.P)$ *is called the separout of* P*. (The initial of* P *is the coefficient of the highest power of* y_P *appearing in* P *and it is denoted by* I_P*. More precisely one has* $P = I_P(y_P)^r$ + *terms of lower degree.)*

Proposition 5 [60] *If* S *is any subset of a differential ring* A *and* $r \geq 0$ *is any integer, we call* r*-prolongation of* S*, the ideal*

$$(S)_{+r} =< d_\nu a | a \in S, 0 \leq |\nu| \leq r >\subset A.$$

One has the following properties: (i) $(S)_{+(r+s)} = ((S)_{+s})_{+r}$*. (ii)* $(S)_{+\infty} =< S >_d$*.*

(iii) Let $\mathfrak{a}$ *be a differential ideal of the differential ring* $K[y]_d$*. We set* $\mathfrak{a}_q = \mathfrak{a} \cap K[y_q]_d$, $\mathfrak{a}_0 = \mathfrak{a} \cap K[y]_d$, $\mathfrak{a}_\infty = \mathfrak{a}$*. We call the* r*-prolongation of* $\mathfrak{a}_q$*, the following ideal:*

$$(\mathfrak{a}_q)_{+r} =< d_\nu P | P \in \mathfrak{a}_q, 0 \leq |\nu| \leq r >\subset K[y_{q+r}]_d.$$

One has:

$$(\mathfrak{a}_q)_{+r} \subseteq \mathfrak{a}_{q+r}, \quad (\mathfrak{a}_q)_{+\infty} \subseteq \mathfrak{a}, \quad (\mathfrak{a}_q)_{+r} \cap K[y_q]_d = \mathfrak{a}_q, \forall q, r \geq 0.$$

With algebraic sets it is better to consider radical ideals. Hence if $\mathfrak{r} \subset K[y]_d$ is a radical differential ideal, then $\mathfrak{r}_q$ is a radical ideal of $K[y_q]_d$, for all $q \geq 0$. Then if $E_q = Z(\mathfrak{r}_q)$ is the algebraic set defined over K by $\mathfrak{r}_q = I(E_q)$, we call *r-prolongation* of E_q the following algebraic set: $(E_q)_{+r} = Z((\mathfrak{r}_q)_{+r})$. In general one has $(\mathfrak{r}_q)_{+r} \subseteq \mathfrak{r}_{q+r}$, hence $rad((\mathfrak{r}_q)_{+r}) \subseteq \mathfrak{r}_{q+r}$. Therefore, in general one has: $E_{q+r} \subseteq (E_q)_{+r}$.

Proposition 6 [60] *Let $\mathfrak{p} \subset K[y]_d$ be a prime differential ideal. Then we can identify each field $L_q = Q(K[y_q]_d/\mathfrak{p}_q)$ with a non-differential subfield of $L = Q(K[y]_d/\mathfrak{p})$ and we have: $K \subseteq L_0 \subseteq \cdots \subseteq L_\infty = L$. Then there are vector spaces R_q over L_q or L defined by the following linear system:*

$$(\partial y_k^\mu . P_\tau)(\eta) v_\mu^k = 0, \{1 \le \tau \le t, \quad 1 \le k \le m, \quad |\mu| = q\},$$

where η is a generic solution of $\mathfrak{p}$ and $P_1, \cdots, P_t$ are generating $\mathfrak{p}_q$. Such result does not depend on the generating polynomials. We can also define the vector space g_q (symbol) over L_q or L, by means of the linear system:

$$(\partial y_k^\mu . P_\tau)(\eta) v_\mu^k = 0, \{1 \le \tau \le t, \quad 1 \le k \le m, \quad 0 \le |\mu| \le q\}.$$

For the prolongations $(g_q)_{+r}$ one has, in general, $g_{q+r} \subseteq (g_q)_{+r}$, $\forall q, r \ge 0$.

Definition 20 We say that R_q or g_q is *generic* over E_q, if one can find a certain number of maximum rank determinants D_α that cannot be all zero at a generic solution of $\mathfrak{p}$.

Proposition 7 *R_q or g_q is generic if we may find polynomials $A_\alpha, B_\tau \in K[y_q]_d$ such that:*

$$\sum_\alpha A_\alpha D_\alpha + \sum_\tau B_\tau P_\tau = 1.$$

Furthermore, R_q or g_q are projective modules over the ring $K[y_q]_d/\mathfrak{p}_q \subset K[y]_d/\mathfrak{p}$.

Proof It follows directly from the Hilbert theorem of zeros. (See [60].)

Theorem 13 (**Primality Criterion** [60]) *Let $\mathfrak{p}_q \subset K[y_q]_d$ and $\mathfrak{p}_{q+1} \subset K[y_{q+1}]_d$ be prime ideals such that $\mathfrak{p}_{q+1} = (\mathfrak{p}_q)_{+1}$ and $\mathfrak{p}_{q+1} \cap K[y_q]_d = \mathfrak{p}_q$. If the symbol g_q of the variety R_q defined by $\mathfrak{p}_q$ is 2-acyclic and its first prolongation g_{q+1} is generic over E_q, then $\mathfrak{p} = (\mathfrak{p}_q)_{+\infty}$ is a prime differential ideal with $\mathfrak{p} \cap K[y_{q+r}]_d = (\mathfrak{p}_q)_{+r}$, for all $r \ge 0$.*

Let $\mathfrak{r}_q \subset K[y_q]_d$ and $\mathfrak{r}_{q+1} \subset K[y_{q+1}]_d$ be radical ideals such that $\mathfrak{r}_{q+1} = (\mathfrak{r})_{+1}$ and $\mathfrak{r}_{q+1} \cap K[y_q]_d = \mathfrak{r}_q$. If the symbol g_q of the algebraic set E_q defined by $\mathfrak{r}_q$ is 2-acyclic and its first prolongation g_{q+1} is generic over E_q, then $\mathfrak{r} = (\mathfrak{r}_q)_{+\infty}$ is a radical differential ideal with $\mathfrak{r} \cap K[y_{q+r}]_d = (\mathfrak{r}_q)_{+r}$, for all $r \ge 0$.

Theorem 14 (**Differential Basis**) *If $\mathfrak{r}$ is a differential ideal of $K[y]_d$, then $\mathfrak{r} = rad((\mathfrak{r}_q)_{+\infty})$ for q sufficiently large.*

Proof In fact one has the following lemma.

Lemma 17 *If $\mathfrak{p}$ is a prime ideal of $K[y]_d$, then for q sufficiently large, there is a polynomial $P \in K[y_q]_d$ such that $P \notin \mathfrak{p}_q$ and $P\mathfrak{p}_{q+r} \subset rad((\mathfrak{p}_q)_{+r}) \subset \mathfrak{p}_{q+r}$, for all $r \ge 0$.*

After above lemma the proof follows directly.

Every radical differential ideal of $K[y]$ can be expressed in a unique way as the nonredundant intersection of a finite number of prime differential ideals. The smallest field of definition κ of a prime differential ideal $\mathfrak{p} \subset K[y]$ is a finitely generated differential extension of $\mathbb{Q}$.

Example 15 With $n = 2, m = 2, q = 1$. Let us consider the differential polynomial $P = y_1^1 y_2^2 - y_1^2 y_2^1 - 1$. We obtain for the symbol g_1: $y_1^1 v_2^2 + y_2^2 v_1^1 - y_2^1 v_1^2 - y_1^2 v_2^1 = 0$. Setting $v_i^k = y_l^k w_i^l$ we obtain $(y_1^1 y_2^2 - y_1^2 y_2^1)(w_2^2 + w_1^1) = 0$ and thus $w_2^2 + w_1^1 = 0$ on E_1. Hence g_1 is generic. One can also set $P_1 = y_2^1$, $P_2 = y_1^2$ and we get the relation: $y_2^2 P_1 - y_2^1 P_2 - P \equiv 1$. A similar result should hold for E_1. g_1 is involutive and the differential ideal generated by P in $\mathbb{Q} < y^1, y^2 >$ is therefore a prime ideal.

Definition 21 A *differentially algebraic extension* L over of a differential field K is a differential extension over K where every element of L is differentially algebraic over K.

The *differential transcendence degree* of a differential extension L/K is the number of elements of a maximal subset S of elements of L that are differentially transcendental over K and such that L becomes differentially algebraic over $K(S)$. We shall denote such number by $trd_d(L/K)$.

Theorem 15 [60] *One has the following formula:*

$$\dim(\mathfrak{p}_{q+r}) = \dim(\mathfrak{p}_{q-1}) + \sum_{1 \le i \le n} \frac{(r+i)!}{r!i!} \alpha_q^i, \forall r \ge 0,$$

where α_q^i is the character of the corresponding system of PDEs. The character α_q^i of a q-order PDE $E_q \subset JD^q(W)$, $\pi : W \to M$, $\dim M = n$, with symbol g_q, is the integer $\alpha_q^i \equiv \dim(g_q^{(i-1)})_p - \dim(g_q^{(i)})_p$, $p \in E_q$, where $(g_q^{(i)})_p \equiv \{\zeta \in (g_q)_p | \zeta(v_1) = \cdots = \zeta(v_i) = 0\}$, where $(v_1, \cdots, v_n)$ is the natural basis in $T_{\pi_k(p)} M$.

The character α_q^n and the smallest nonzero character only depend on the differential extension L/K and not on the generators. In particular, one has: $trd_d(L/K) = \alpha_q^n$.

If ζ is differentially algebraic over $K(\eta)_d$ and η is differentially algebraic over K, then ζ is differentially algebraic over K.

If L/K is a differential extension and $\xi, \eta \in L$ are both differentially algebraic over K, then $\xi + \eta$, $\xi\eta$, ξ/η, $(\eta \neq 0)$, and $d_i\xi$ are differentially algebraic over K.

Theorem 16 *Let $(A, \{\partial_j\}_{1 \le j \le n})$ be a differential ring. The set $D(A)$ of differential operators over $(A, \{\partial_j\}_{1 \le j \le n})$ is a non-commutative filtered ring and a filtered bimodule over A.*

Proof If y is a differential indeterminate over A, we may introduce the *formal derivatives* $d_1, \cdots, d_n$ which are such that $d_i d_j - d_j d_i = 0$, $\forall i, j = 1, \cdots, n$, and are defined by: $d_i(ay) = (\partial_i a)y + a(d_i y)$. We shall write $d_i y = y_i$, $d_i y_\mu = y_{\mu + 1i}$, where μ is the multiindex $\mu = (\mu_1, \cdots, \mu_n)$ with length $|\mu| = \mu_1 + \cdots + \mu_n$. If $y = (y^1, \cdots, y^m)$, we set $d_\mu = (d_1)^{\mu_1} \cdots (d_n)^{\mu_n}$ and $d_\mu y^k = y_\mu^k$. Any differential operator of order q over A can be written in the form $P = \sum_{0 \le \mu \le q} a^\mu d_\mu$, $a^\mu \in A$. Set $ord(P) = q$. Then, we can write $D(A) \cong A[d_1, \cdots, d_n] \equiv A[d]$ the *ring of*

Table 2 Examples of singular PDEs defined by differential polynomials.

Name	Singular PDE
PDE with node and triple point $E_1 \subset JD(E)$	$p_1 \equiv (u^1_x)^4 + (u^2_y)^4 - (u^1_x)^2 = 0$ $p_2 \equiv (u^2_x)^6 + (u^1_y)^6 - u^2_x u^1_y = 0$
PDE with cusp and tacnode $\bar{E}_1 \subset JD(E)$	$q_1 \equiv (u^1_x)^4 + (u^2_y)^4 - (u^1_x)^3 + (u^2_y)^2 = 0$ $q_2 \equiv (u^2_x)^4 + (u^1_y)^4 - (u^2_x)^2(u^1_y) - (u^2_x)(u^1_y)^2 = 0$
PDE with conical double point, double line, and pinch point $\tilde{E}_1 \subset JD(F)$	$r_1 \equiv (u^1)^2 - (u^1_x)(u^2_y)^2 = 0$ $r_2 \equiv (u^2)^2 - (u^2_x)^2 - (u^1_y)^2 = 0$ $r_3 \equiv (u^3)^3 + (u^3_x)^3 + (u^2_x)(u^3_y) = 0$

$\pi : E \equiv \mathbb{R}^4 \to \mathbb{R}^2, \quad (x, y, u^1, u^2) \mapsto (x, y)$. $\bar{\pi} : F \equiv \mathbb{R}^5 \to \mathbb{R}^2, \quad (x, y, u^1, u^2, u^3) \mapsto (x, y)$.
$\mathfrak{a} \equiv < p_1, p_2 > \subset A$, $\mathfrak{b} \equiv < q_1, q_2 > \subset A$, $\mathfrak{c} \equiv < r_1, r_2, r_3 > \subset B$.

partial differential operators over A with derivatives $d_1, \cdots, d_n$. The addition rule is clear. The multiplication rule comes from the Leibniz formula:

$$\left\{\begin{array}{l} \partial_\nu(ab) = \sum_{\lambda+\mu=\nu} \frac{\nu!}{\lambda!\mu!}(\partial_\lambda a)(\partial_\mu b) \\ d_\nu(ay) = \sum_{\lambda+\mu=\nu} \frac{\nu!}{\lambda!\mu!}(\partial_\lambda a) d_\mu y \end{array}\right\} \Rightarrow d_\nu a = \sum_{\lambda+\mu=\nu} \frac{\nu!}{\lambda!\mu!}(\partial_\lambda a) d_\mu.$$

Here we have put $\mu! = \mu_1! \cdots \mu_n!$. With these rules $D(A)$ becomes a noncommutative ring and a bimodule over A. In fact, the previous formula defines the right action of A on $D(A)$. The left action of A on $D(A)$ is simply the multiplication on the left by A, that is $aP = a(\sum_{0\le\mu\le q} a^\mu d_\mu) = \sum_{0\le\mu\le q} aa^\mu d_\mu$. Now, the filtration of $D(A)$ is naturally induced by filtration of spaces of differential operators. More precisely $D_q(A) = \{P \in D(A) | ord(P) \le q\}$, where $ord(P) = sup\{|\mu| | a^\mu \neq 0\}$. We set $D_{-1}(A) = 0$ and $D_0(A) = A$. Then, $D_q(A) \subset D_{q+1}(A)$, $D(A) = \bigcup_{q\ge0} D_q(A)$ and $D_q(A)D_p(A) \subseteq D_{p+q}(A)$.

Theorem 17 (Algebraic Criterion for Formal Integrability [60]) *Let $Z_q = Z(\mathfrak{p}_q)$ be the variety defined by means of ideal $\mathfrak{p}_q \subset K[y_q]_d$ such that the following conditions are verified:*

(i) $(\mathfrak{p}_q)_{+1} = \mathfrak{p}_{q+1} \subset K[y_{q+1}]_d$ *is also a prime ideal.*
(ii) $\mathfrak{p}_{q+1} \cap K[y_q]_d = \mathfrak{p}_q$.
(iii) g_{q+1} *is generic over* E_q.
(iv) g_q *is 2-acyclic.*

Then $(\mathfrak{p}_q)_{+\infty} = \mathfrak{p} \subset K[y]_d$ is a prime differential ideal, where $\mathfrak{p}$ is the differential ideal generated by a finite number of differential polynomials $P_1, \cdots, P_t$, defining E_q, and E_q is formally integrable. If one of these conditions is not satisfied we get that $\mathfrak{p}$ is not a prime ideal, hence we have a factorization of $\mathfrak{p}$. In other words the PDE is not formally integrable.

Proof For a detailed proof see [60].

Example 16 (Some singular PDEs) In Table 2 we report some singular PDEs having some algebraic singularities. These are singular PDEs of first order defined on $JD(E) \cong \mathbb{R}^8$ for the first two and on $JD(F) \cong \mathbb{R}^{11}$ for the third. To the ideals $\mathfrak{a} \equiv < p_1, p_2 > \subset A$, $\mathfrak{b} \equiv < q_1, q_2 > \subset A$ and $\mathfrak{c} \equiv < r_1, r_2, r_3 > \subset B$, where $A \equiv \mathbb{R}[u^1, u^2, u^1_x, u^1_y, u^2_x, u^2_y]$ and $B \equiv \mathbb{R}[u^1, u^2, u^3, u^1_x, u^1_y, u^2_x, u^2_y, u^3_x, u^3_y]$, one associates the corresponding algebraic sets

$$\left\{\begin{array}{l} E_1 = \{q \in \mathbb{R}^8 | f(q) = 0, \forall f \in \mathfrak{a}\} \subset \mathbb{R}^8 \\ \bar{E}_1 = \{q \in \mathbb{R}^8 | f(q) = 0, \forall f \in \mathfrak{b}\} \subset \mathbb{R}^8 \\ \tilde{E}_1 = \{q \in \mathbb{R}^{11} | f(q) = 0, \forall f \in \mathfrak{c}\} \subset \mathbb{R}^{11}. \end{array}\right. \tag{62}$$

These algebraic sets are in bijective correspondence with the corresponding radicals: $r(\mathfrak{a}) = \{g \in A | g(p) = 0, \forall p \in E_1\} \supset \mathfrak{a}$, $r(\mathfrak{b}) = \{g \in A | g(p) = 0, \forall p \in \bar{E}_1\} \supset \mathfrak{b}$, $r(\mathfrak{c}) = \{g \in B | g(p) = 0, \forall p \in \tilde{E}_1\} \supset \mathfrak{c}$. (This follows from the Hilbert theorem of zeros [60].)

Let us consider in some details the singular PDE $\tilde{E}_1 \subset JD(F)$, in order to see existence of global algebraic singular solutions and characterize their stability. We have the following representation: $\tilde{E}_1 = \bigcup A_1 \bigcup A_2 \bigcup A_3$, where

$$A_j \equiv \left\{\begin{array}{ll|l} 1) & u^1_x = \frac{(u^1)^2}{(u^2_y)^2} & (u^2_x)^2 \leq (u^2)^2 \\ 2) & u^1_y = s(j)\sqrt{(u^2)^2 - (u^2_x)^2} & u^2_y \neq 0 \\ 3) & u^2_x = -\frac{(u^3)^3}{u^3_y} - (u^3_y)^2 & u^3_y \neq 0 \end{array}\right\} \subset \tilde{E}_1 \subset JD(F) \tag{63}$$

$$A_3 \equiv \left\{q \in JD(F) \cong \mathbb{R}^{11} | u^1 = u^2 = u^3 = u^2_x = u^1_y = u^2_y = u^3_y = 0\right\}. \tag{64}$$

In (63) we put $s(1) = 1$ and $s(2) = -1$. $A_3 \cong \mathbb{R}^4$ is the set of singular points of $\tilde{E}_1$. Instead A_1 and A_2 are formally integrable and completely integrable PDEs. In fact, one has the exact commutative diagrams (65), with $j = 1, 2$:

$$\begin{array}{ccccccccc} & 0 & & & & 0 & & & \\ & \downarrow & & & & \downarrow & & & \\ & A_j & \longrightarrow & F & \longrightarrow 0 & A_3 & \longrightarrow & M & \longrightarrow 0 \\ & \downarrow & & \| & & \downarrow & & \| & \\ & JD(F) & \xrightarrow[\pi_{1,0}]{} & F & \longrightarrow 0 & JD(F) & \xrightarrow[\pi_1]{} & M & \longrightarrow 0 \end{array} \tag{65}$$

This can be seen rewriting equations in (63) in the following equivalent way:

$$A_j \equiv \left\{\begin{array}{ll|l} 1) & u^1_x = \frac{(u^1)^2}{(u^2_y)^2} & (u^2_x)^2 \leq (u^2)^2 \\ 2) & u^1_y = s(j)\sqrt{(u^2)^2 - [\frac{(u^3)^3}{u^3_y} + (u^3_y)^2]^2} & u^2_y \neq 0 \\ 3) & u^2_x = -[\frac{(u^3)^3}{u^3_y} + (u^3_y)^2] & u^3_y \neq 0 \end{array}\right\} \subset \tilde{E}_1 \subset JD(F). \tag{66}$$

The first prolongation $(A_j)_{+1}$, $j=1,2$, is given by the following equations:

$$\left.\begin{cases} 1) & r_1=0\\ 2) & r_2=0\\ 3) & r_3=0\\ 4) & u^1_{xx}=[2u^1u^1_x(u^2_y)^2-(u^1)^2 2u^2_y u^2_{yx}]/(u^2_y)^4\\ 5) & u^1_{xy}=[2u^1u^1_y(u^2_y)^2-(u^1)^2 2u^2_y u^2_{yy}]/(u^2_y)^4\\ 6) & u^1_{xy}=s(j)\frac{a}{2}/\sqrt{(u^2)^2-[\frac{(u^3)^3}{u^3_y}+(u^3_y)^2]^2}\\ 7) & u^1_{yy}=s(j)\frac{b}{2}/\sqrt{(u^2)^2-[\frac{(u^3)^3}{u^3_y}+(u^3_y)^2]^2}\\ 8) & u^2_{xx}=-[\frac{3(u^3)^2u^3_xu^3_y-(u^3)^3u^3_{yx}}{(u^3_y)^2}+2u^3_yu^3_{yx}]\\ 9) & u^2_{xy}=-[\frac{3(u^3)^2(u^3_y)^2-(u^3)^3u^3_{yy}}{(u^3_y)^2}+2u^3_yu^3_{yy}] \end{cases}\quad\middle|\;\begin{matrix}(u^2_x)^2\le(u^2)^2\\ u^2_y\ne0\\ u^3_y\ne0\end{matrix}\right\}\equiv(A_j)_{+1}\subset JD^2(F) \tag{67}$$

where

$$\begin{cases} a=2u^2u^2_x-2[\frac{(u^3)^3}{u^3_y}+(u^3_y)^2][\frac{3(u^3)^2u^3_xu^3_y-(u^3)^3u^3_{yx}}{(u^3_y)^2}+2u^3_yu^3_{yx}]\\ b=2u^2u^2_y-2[\frac{(u^3)^3}{u^3_y}+(u^3_y)^2][\frac{3(u^3)^2(u^3_y)^2-(u^3)^3u^3_{yy}}{(u^3_y)^2}+2u^3_yu^3_{yy}] \end{cases} \tag{68}$$

Then by using (67)(9) to substitute u^2_{yx} in (67)(1), and by using the two different expressions of u^1_{xy} in (67)(5) and (67)(6) to obtain an explicit expression of u^2_{yy}, we get the following equations for the first prolongation of $(A_j)_{+1}$:

$$\left.\begin{cases} 1) & r_1=0\\ 2) & r_2=0\\ 3) & r_3=0\\ 4) & u^1_{xx}=[2u^1u^1_x(u^2_y)^2-(u^1)^2 2u^2_y c]/(u^2_y)^4\\ 5) & u^1_{xy}=s(j)\frac{a}{2}/\sqrt{(u^2)^2-[\frac{(u^3)^3}{u^3_y}+(u^3_y)^2]^2}\\ 6) & u^1_{yy}=s(j)\frac{b}{2}/\sqrt{(u^2)^2-[\frac{(u^3)^3}{u^3_y}+(u^3_y)^2]^2}\\ 7) & u^2_{xx}=-[\frac{3(u^3)^2u^3_xu^3_y-(u^3)^3u^3_{yx}}{(u^3_y)^2}+2u^3_yu^3_{yx}]\\ 8) & u^2_{xy}=-[\frac{3(u^3)^2(u^3_y)^2-(u^3)^3u^3_{yy}}{(u^3_y)^2}+2u^3_yu^3_{yy}]\\ 9) & u^2_{yy}=[\frac{2u^1u^1_y}{(u^2_y)^2}-s(j)\frac{1}{2}\frac{a}{\sqrt{(u^2)^2-[\frac{(u^3)^3}{u^3_y}+(u^3_y)^2]^2}}]\frac{1}{2}\frac{(u^2_y)^3}{(u^1)^2} \end{cases}\quad\middle|\;\begin{matrix}(u^2_x)^2\le(u^2)^2\\ u^2_y\ne0\\ u^3_y\ne0\end{matrix}\right\}\equiv(A_j)_{+1}\subset JD^2(F) \tag{69}$$

with

$$c = -[\frac{3(u^3)^2(u^3_y)^2 - (u^3)^3 u^3_{yy}}{(u^3_y)^2} + 2u^3_y u^3_{yy}]. \tag{70}$$

Therefore, also $(A_j)_{+1}$ are analytic submanifolds of $JD^2(F)$, for $j = 1, 2$. Furthermore, since $(\dim(A_j)_{+1} = 11) = (\dim(A_j = 8) + (\dim(g_1)_{+1} = 3)$, we see that the canonical projections $\pi_{2,1} : (A_j)_{+1} \to A_j$, $j = 1, 2$, are affine subbundles of $JD^2(F) \to JD(F)$, with associated vector bundles $(g_1)_{+1} \to A_j$, $j = 1, 2$. Finally the symbol $\dim(g_1)_{q\in A_j} = 3$, $j = 1, 2$, and $\dim(\partial x \rfloor (g_1)_{q\in A_j}) = 0$. Therefore, $(g_1)_{q\in A_j}$ are involutive. This is enough to conclude that A_j are formally integrable, and since they are also analytic they are completely integrable, too. Furthermore, taking into account that $\dim A_j = 8 > 2\times 2+1 = 5$, we can apply Theorem 2.15 in [62], (here reported in Sect. 3), to calculate the weak and singular integral bordism groups of A_j. One has $\Omega^{A_j}_{1,w} = \Omega^{A_j}_{1,s} = 0$, $j = 1, 2$. Therefore, A_j are extended 0-crystal PDEs. So for any two admissible closed 1-dimensional smooth integral manifolds $N_0, N_1 \subset A_j$, there exists a (singular) 2-dimensional integral manifold, solution $V \subset A_j$, such that $\partial V = N_0 \sqcup N_1$.[29] Such a solution is smooth iff all the integral characteristic numbers of N_0 are equal to ones of N_1.

The Cartan distribution on $\tilde{E}_1$ is given by the following vector fields

$$\zeta = X^x(\partial x + u^k_x \partial u_k) + X^y(\partial y + u^k_y \partial u_k) + Y^x_k \partial u^k_x + Y^y_k \partial u^k_y \tag{71}$$

such that the following equations are satisfied:

$$\left\{\begin{array}{l} Y^x_1 (u^2_y)^2 + Y^y_2 2u^1_x u^2_y - 2u^1(u^1_x X^x + u^1_y X^y) = 0 \\ Y^y_1 2u^1_y + Y^x_2 2u^2_x - 2u^2(u^2_x X^x + u^2_y X^y) = 0 \\ Y^x_2 u^3_y + Y^y_3 (u^2_x + 3(u^3_y)^2) + 3(u^3)^2(u^3_x X^x + u^3_y X^y) = 0. \end{array}\right. \tag{72}$$

Therefore, $\dim(\mathbf{E}_1)_{q\in A_j} = 5$, $j = 1, 2$. For example on A_1, one has the following expression of the Cartan vector field:

$$\left\{\begin{array}{l} \zeta = X^x[\partial x + u^k_x \partial u_k + \frac{2u^1 u^1_x}{(u^2_y)^2} \partial u^1_x + \frac{u^2 u^2_x}{u^1_y} \partial u^1_y - \frac{3(u^3)^2 u^3_x}{[u^2_x + 3(u^3_y)^2]} \partial u^3_y] \\ \quad + X^y[\partial y + u^k_y \partial u_k + \frac{2u^1 u^1_y}{(u^2_y)^2} \partial u^1_x + \frac{u^2 u^2_y}{u^1_y} \partial u^1_y - \frac{3(u^3)^2 u^3_y}{[u^2_x + 3(u^3_y)^2]} \partial u^3_y] \\ \quad + Y^x_2[\partial u^2_x - \frac{u^2_x}{u^1_y} \partial u^1_y - \frac{u^3_y}{[u^2_x + 3(u^3_y)^2]} \partial u^3_y] + Y^x_3 \partial u^3_x + Y^y_2[-\frac{2u^1_x}{u^2_y} \partial u^1_x + \partial u^2_y]. \end{array}\right. \tag{73}$$

Instead, since Eqs. (72) are identities for $q \in A_3$, we get that $(\mathbf{E}_1)_{q\in A_3} = \mathbf{E}_1(F)_{q\in A_3}$, i.e., in the singular points the Cartan distribution of $\tilde{E}_1$ coincides with

[29] Let us emphasize that by using Lemma 14 we can identify admissible smooth 1-dimensional integral manifolds in A_j, $j = 1, 2$. In fact, A_j are formally integrable and completely integrable PDEs. So we can use Lemma 14(1), but also Lemma 14(3), since $\pi_{1,0}(A_j) = F$, and $A_j \to F$ are affine subbundles of $JD(F) \to F$, with associated vector bundle the symbol g_1.

Fig. 3 Algebraic singular solution $V' = V \bigcup_{N_1} Z \subset \tilde{E}_1 \subset JD(F)$, passing through a singular point $q_0 \in A_3$

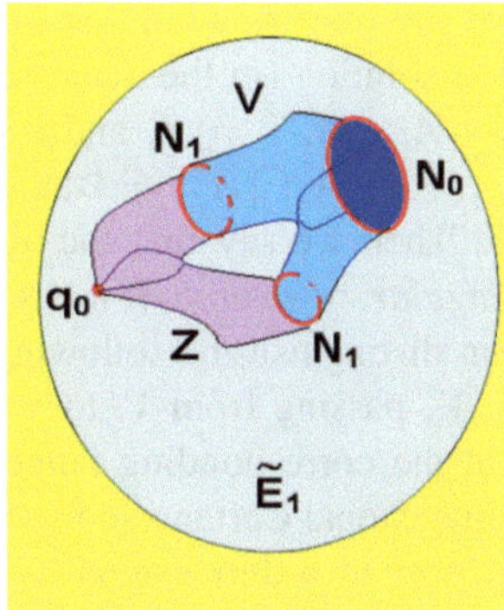

the Cartan distribution of $JD(F)$ that is just given by vector fields given in (71) for arbitrary functions $X^x, X^y, Y_k^x, Y_k^y : JD(F) \to \mathbb{R}$, $k = 1,2,3$. Thus, we can prolong any solution $V \subset A_j$, $j = 1,2$, $\partial V = N_0 \sqcup N_1$ to a solution Z, such that $\partial Z = N_1 \sqcup \{q_0\} \cong N_1$, where $q_0 \in A_3$ also. In other words, $V' \equiv V \bigcup_{N_1} Z$ is an algebraic singular solution of $\tilde{E}_1$. (See Fig. 3.) In fact, we can always find a solution $\tilde{V} \subset J_2^1(F)$ of the trivial equation $J_2^1(F) \subseteq J_2^1(F)$, such that $\partial \tilde{V} = N_1 \sqcup \{q_0\} \cong N_1$, and such that there exists a disk $D_\epsilon^2 \subset \tilde{V}$, centered on q_0, with radius ϵ and boundary $\partial D_\epsilon^2 \equiv N_\epsilon \subset A_j$. (Let us emphasize that $\dim F = 5$, hence we can embed in F any 2-dimensional smooth compact manifold. See, e.g., [31].) Let $\hat{V}$ be the submanifold of $\tilde{V}$ such that $\partial \hat{V} = N_\epsilon \sqcup N_1$. Then, since A_j is a strong retract of $J_2^1(F)$, we can deform $\hat{V}$, obtaining a solution $V''_\epsilon \subset A_j$, such that $\partial V''_\epsilon = N_\epsilon \sqcup N_1$. By taking the limit $\epsilon \to 0$, we can see that the solution V''_ϵ identifies a solution $V'' \subset A_j$, such that $V'' \sqcup \{q_0\} = V' \subset \tilde{E}_1$ is just an algebraic singular solution of the singular equation $\tilde{E}_1$.

Note that the symbol $(g_1)_{q \in A_3} = G_1 \equiv TM \otimes F$, i.e., $(g_1)_{q \in A_3}$ coincides with the symbol of the trivial PDE $JD(F) \subseteq JD(F)$. In fact the components of the symbol on $\tilde{E}_1$, must satisfy the following equations:

$$\left\{\begin{array}{l} Y_1^x (u_y^2)^2 + Y_2^y 2u_x^1 u_y^2 = 0; \\ Y_1^y 2u_y^1 + Y_2^x 2u_x^2 = 0; \\ Y_2^x u_y^3 + Y_3^y [u_x^2 + 3(u_y^3)^2] = 0. \end{array}\right\} \tag{74}$$

These equations are identities on $q \in A_3$. As a by-product, we get that above considered algebraic singular solutions are, in general, unstable in finite times.

Definition 22 We define *extended crystal singular PDE*, a singular PDE $E_k \subset J_n^k(W)$ that splits in irreducible components A_i, i.e., $E_k = \bigcup_i A_i$, where each A_i is an extended crystal PDE. Similarly we define *extended* 0-*crystal singular PDE*, (resp. 0-*crystal singular PDE*), an extended crystal singular PDE where each component A_i is an extended 0-crystal PDE, (resp. 0-crystal PDE).

Definition 23 (Algebraic Singular Solutions of Singular PDEs) Let $E_k \subset J_n^k(W)$ be a singular PDE that splits in irreducible components A_i, i.e., $E_k = \bigcup_i A_i$. Then, we say that E_k admits an *algebraic singular solution* $V \subset E_k$, if $V \bigcap A_r \equiv V_r$ is a solution (in the usual sense) in A_r for at least two different components A_r, say A_i, A_j, $i \neq j$, and such that one of following conditions are satisfied: (a) ${}_{(ij)}E_k \equiv A_i \bigcap A_j \neq \varnothing$; (b) ${}^{(ij)}E_k \equiv A_i \bigcup A_j$ is a connected set, and ${}_{(ij)}E_k = \varnothing$. Then we say that the algebraic singular solution V is in the case (a), *weak*, *singular*, or *smooth*, if it is so with respect to the equation ${}_{(ij)}E_k$. In the case (b), we can distinguish the following situations: *(weak solution):* There is a discontinuity in V, passing from V_i to V_j; *(singular solution):* there is not discontinuity in V, but the corresponding tangent spaces TV_i and TV_j do not belong to a same n-dimensional Cartan subdistribution of $J^k_n(W)$, or alternatively TV_i and TV_j belong to a same n-dimensional Cartan subdistribution of $J^k_n(W)$, but the kernel of the canonical projection $(\pi_{k,0})_* : TJ^k_n(W) \to TW$, restricted to V is larger than zero; *(smooth solution):* there is no discontinuity in V and the tangent spaces TV_i and TV_j belong to a same n-dimensional Cartan subdistribution of $J^k_n(W)$ that projects diffeomorphically on W via the canonical projection $(\pi_{k,0})_* : TJ^k_n(W) \to TW$. Then we say that a solution passing through a critical zone *bifurcate*.[30]

Definition 24 (Integral Bordism for Singular PDEs) Let $N_1, N_2 \subset E_k \subset J^k_n(W)$ be two $(n-1)$-dimensional admissible closed integral manifolds. We say that N_1 *algebraic integral bords* with N_2, if N_1 and N_2 belong to two different irreducible components, say $N_1 \subset A_i$, $N_2 \subset A_j$, $i \neq j$, such that there exists an algebraic singular solution $V \subset E_k$ with $\partial V = N_1 \sqcup N_2$.

In the integral bordism group $\Omega^{E_k}_{n-1}$ (resp. $\Omega^{E_k}_{n-1,s}$, resp. $\Omega^{E_k}_{n-1,w}$) of a singular PDE $E_k \subset J^k_n(W)$, we call *algebraic class* a class $[N] \in \Omega^{E_k}_{n-1}$, (resp. $[N] \in \Omega^{E_k}_{n-1,s}$, resp. $[N] \in \Omega^{E_k}_{n-1},$), with $N \subset A_j$, such that there exists a closed $(n-1)$-dimensional admissible integral manifolds $X \subset A_i \subset E_k$, algebraic integral bording with N, i.e., there exists a smooth (resp. singular, resp. weak) algebraic singular solution $V \subset E_k$, with $\partial V = N \sqcup X$.

Theorem 18 (Singular Integral Bordism Group of Singular PDE) *Let $E_k \equiv \bigcup_i A_i \subset J^k_n(W)$ be a singular PDE. Then under suitable conditions, algebraic singular solutions integrability conditions, we can find (smooth) algebraic singular solutions bording assigned admissible closed smooth $(n-1)$-dimensional integral manifolds N_0 and N_1 contained in some component A_i and A_j, $i \neq j$.*

Proof In fact, we have the following lemmas.

Lemma 18 *Let $E_k \equiv \bigcup_i A_i \subset J^k_n(W)$ be a singular PDE with ${}_{(ij)}E_k \equiv A_i \bigcap A_j \neq \varnothing$. Let us assume that $A_i \subset J^k_n(W)$, $A_j \subset J^k_n(W)$ and ${}_{(ij)}E_k \subset J^k_n(W)$ be formally integrable and completely integrable PDE's such that* $\dim A_i > 2n+1$, $\dim A_j > 2n+1$, $\dim_{(ij)} E_k > 2n+1$. *Then, one has the following isomorphisms:*

$$\Omega^{A_i}_{n-1,w} \cong \Omega^{A_j}_{n-1,w} \cong \Omega^{(ij)E_k}_{n-1,w}. \tag{75}$$

[30] Note that the bifurcation does not necessarily imply that the tangent planes in the points of $V_{ij} \subset V$ to the components V_i and V_j, should be different.

So we can find a weak algebraic singular solution $V \subset E_k$ *such that* $\partial V = N_0 \sqcup N_1$, *with* $N_0 \subset A_i$, $N_1 \subset A_j$, *iff* $N_1 \in [N_0]$.

Furthermore, if $g_k(A_i) \neq 0$, $g_{k+1}(A_i) \neq 0$, $g_k(A_j) \neq 0$, $g_{k+1}(A_j) \neq 0$, $g_k({}_{(ij)}E_k) \neq 0$, $g_{k+1}({}_{(ij)}E_k) \neq 0$, *then one has also the following isomorphisms:*

$$\Omega_{n-1,s}^{A_i} \cong \Omega_{n-1,s}^{A_j} \cong \Omega_{n-1,s}^{(ij)E_k}. \tag{76}$$

So we can find a singular algebraic singular solution $V \subset E_k$ *such that* $\partial V = N_0 \sqcup N_1$, $N_0 \subset A_i$, $N_1 \subset A_j$, *iff* $N_1 \in [N_0]$.

Proof In fact, under the previous hypotheses one has that we can apply Theorem 7 to each component A_i, A_j, and ${}_{(ij)}E_k$ to state that all their weak integral bordism groups of dimension $(n-1)$ are isomorphic to $\bigoplus_{r+s=n-1} H_r(W;\mathbb{Z}_2) \otimes_{\mathbb{Z}_2} \Omega_s$.

Furthermore, under the above hypotheses on nontriviality of symbols, we can apply Theorem 2.1 in [56]. So we can state that weak integral bordism groups are isomorphic to the corresponding singular ones.

Lemma 19 *Let* $E_k = \bigcup_i A_i$ *be a* 0*-crystal singular PDE. Let* ${}^{(ij)}E_k \equiv A_i \bigcup A_j$ *be connected, and* ${}_{(ij)}E_k \equiv A_i \bigcap A_j \neq \varnothing$. *Then* $\Omega_{n-1,s}^{(ij)E_k} = 0$.[31]

Proof In fact, let $Y \subset {}_{(ij)}E_k$ be an admissible closed $(n-1)$-dimensional closed integral manifold, then there exists a smooth solution $V_i \subset A_i$ such that $\partial V_i = N_0 \sqcup Y$ and a solution $V_j \subset A_j$ such that $\partial V_j = Y \sqcup N_1$. Then, $V = V_i \bigcup_Y V_j$ is an algebraic singular solution of E_k. This solution is singular in general.

After above lemmas the proof of the theorem can be considered done besides the algebraic singular solutions integrability conditions.

Example 17 (Extended Crystal Singular MHD-PDEs) In the recent paper [73] we introduced a new PDE of the magnetohydrodynamics encoding the dynamic of anisotropic incompressible plasmas able to describe nuclear energy production. This equation is denoted $\widetilde{(MHD)} \subset JD^2(\widetilde{W})$ and it is reported in Table 3. The fiber bundle considered there is $\pi : \widetilde{W} \to M$, over the Galilean space–time M, with $\widetilde{W} \equiv M \times \mathbf{I} \times \mathbf{S}^3 \times \mathbb{R}^4$, where $\mathbf{I}$ is an affine 3-dimensional space (timelike flow velocities space) and $\mathbf{S}$ is a 3-dimensional Euclidean vector space. A section $s = (v, p, \theta, E, H, I, \bar{\rho}, \bar{h})$ represents flow velocity, isobaric pressure, temperature, electric vector field, magnetic vector field, electric current density, electric charge density, and nuclear energy production density. In that paper it is also proved that equation $\widetilde{(MHD)}$ is formally integrable, completely integrable and an extended 0-crystal. Now, from Theorem 7 we can see that for any two spacelike admissible Cauchy hypersurfaces $N_0 \subset \widetilde{(MHD)}_{t_0}$, $N_1 \subset \widetilde{(MHD)}_{t_1}$, $t_1 \neq t_2$, there exists a (singular) solution $V \subset \widetilde{(MHD)}$, passing through N_0 and N_1. The admissibility requires that N_0 and N_1 are smooth 3-dimensional regular manifolds with respect to the embedding $\widetilde{(MHD)} \subset J_4^2(\widetilde{W})$, locally satisfying the Cauchy problem and with orientable

[31] But, in general, it is $\Omega_{n-1}^{(ij)E_k} \neq 0$.

Table 3 Nuclear energy producing magnetohydrodynamics equation $\{\widetilde{F}^{(s)} = 0; 1 \leq s \leq 7\}$: $\widetilde{(MHD)} \subset JD^2(\widetilde{W})$

Maxwell	$\widetilde{F}^{(1)} \equiv F^{(1)} \equiv B^k{}_{/k} = 0$ (no magnetic monopoles)
	$\widetilde{F}^{(2)} \equiv F^{(2)} \equiv D^k{}_{/k} - \bar{\rho}4\pi = 0$ (Gauss's law of electrostatic)
	$\widetilde{F}^{(3)} \equiv F^{(3)i} \equiv \epsilon^{ijs} E_{j/s} + \frac{1}{c}(\partial t \cdot B^i) = 0$ (Faraday's law)
	$\widetilde{F}^{(4)} \equiv F^{(4)i} \equiv \epsilon^{ijs} H_{j/s} - \frac{1}{c}(\partial t \cdot D^i) - \frac{4\pi}{c} I^i = 0$ (Ampere's law)
	$D^k = \epsilon^{ki} E_i$, ($\epsilon =$dielectric permeability tensor)
	$\epsilon^{ik} = -\epsilon_0 g^{ik} + \bar{\epsilon}(v_{r/s} + v_{s/r}) g^{ri} g^{sk}$
	$B^k = \mu^{ki} H_i$, ($\mu =$magnetic permeability tensor)
	$\mu^{ik} = -\mu_0 g^{ik} + \bar{\mu}(v_{r/s} + v_{s/r}) g^{ri} g^{sk}$
Navier–Stokes	$\widetilde{F}^{(5)} \equiv F^{(5)} \equiv v^k{}_{/k} = 0$ (continuity equation)
	$\widetilde{F}^{(5)} \equiv F^{(5)}_\alpha \equiv \dot{x}^k(\partial x_\alpha . G^j_{jk}) + \dot{x}^k_\alpha G^j_{jk} + \dot{x}^s_{s\alpha} = 0$,
	(first-prolonged continuity equation)
	$\widetilde{F}^{(6)} \equiv F^{(6)i} \equiv \rho \frac{\delta v^i}{\delta t} - P^{ik}{}_{/k} - F^i_{(body)} = 0$, (motion equation)
	$\widetilde{F}^{(7)} \equiv \rho C_v \frac{\delta \theta}{\delta t} + \frac{\delta w}{\delta t} - \nu(\theta_{/i})_{/k} g^{ik} + S^k_{/k} - [2\chi \dot{e}^{ik} + \frac{1}{4\pi}(B^i B^k + E^i E^k)] v_{i/k}$
	$+ I^i E^j g_{ij} - \rho \bar{h} = 0$, (energy equation)
	$q^k = -\nu \theta_{/i} g^{ik}$, (heat flow)
	$P^{ik} = {}^{(rh)}P^{ik} + M^{ik} = -p g^{ik} + \chi(v_{r/s} + v_{s/r}) g^{ri} g^{sk} + M^{ik}$
	$= -g^{ik}[p + \frac{1}{8\pi}(B^s B_s + E^s E_s)] + \wp^{ik}$ (full stress tensor)
	$\wp^{ki} = \chi(v_{r/s} + v_{s/r}) g^{ri} g^{sk} + \frac{1}{4\pi}(B^i B^j + E^i E^j)$ (deviatory stress)
	(magnetic stress tensor) ${}^{(B)}M^{ij} \equiv \frac{1}{4\pi}(-\frac{1}{2} B^k B_k g^{ij} + B^i B^j)$
	(electric stress tensor) ${}^{(E)}M^{ij} \equiv \frac{1}{4\pi}(-\frac{1}{2} E^k E_k g^{ij} + E^i E^j)$
	(electromagnetic stress tensor) $M^{ij} \equiv {}^{(B)}M^{ij} + {}^{(E)}M^{ij}$.
	(body force): $F^i_{(body)} = -\rho(\partial x_k . f) g^{ki} + \bar{\rho} E^i + \epsilon^{ijk} I_j B_k$

$\bar{h}$ body energy source density

boundaries $X_0 \equiv \partial N_0$, $X_1 \equiv \partial N_1$, bording by means of a suitable 3-dimensional time-like integral manifold P.

Then the solution V has boundary $\partial V = N_0 \bigcup_{X_0} P \bigcup_{X_1} N_1$. (For details see Refs. [74, 70] There can be also found the explicit expressions of the differential polynomials defining $\widetilde{(MHD)}$.) In [70] we proved that we can identify a subequation $\widehat{(MHD)} \subset \widetilde{(MHD)}$ such that in some neighborhood of its points, there exist entropy-regular solutions passing from such initial conditions. More precisely, we can represent $\theta^2 \mathcal{R}$ as a polynomial differential of first order on $JD(\widetilde{W})$, belonging to $A[\dot{x}^a, \dot{x}^a_b, H^p, E^p, \theta, \theta_j, \bar{h}]$, where $A \equiv [[x^1, x^2, x^3]]$. This can be seen taking into account that

$$\theta^2 \mathcal{R}(s) = \theta \left[\frac{1}{\rho 4\pi}(B^i B^j + E^i E^j) \dot{e}_{ij} + \frac{2\chi}{\rho} \dot{e}^{ij} \dot{e}_{ij} + \bar{h} \right] + \frac{\nu}{\rho}(\text{grad}\theta)^2. \tag{77}$$

Table 4 Polynomial differential expression for $\theta^2\mathcal{R}$

$$\theta^2\mathcal{R}=\theta\left[\dot{x}^a_b\dot{x}^c_d\mathcal{R}(1)^{bd}_{ac}+\dot{x}^a_b\dot{x}^c_d H^pH^q\mathcal{R}(2)^{bd}_{acpq}+\dot{x}^a_b\dot{x}^c\mathcal{R}(3)^b_{ac}+\dot{x}^a_b\dot{x}^c H^pH^q\mathcal{R}(4)^b_{acpq}\right.$$
$$+\dot{x}^a\dot{x}^b\mathcal{R}(5)_{ab}+\dot{x}^a\dot{x}^bH^pH^q\mathcal{R}(6)_{abpq}+\dot{x}^m_sH^sH^r\mathcal{R}(7)_{mr}+\dot{x}^m_sE^iE^j\mathcal{R}(8)^s_{mij}$$
$$+\dot{x}^tH^qH^r\mathcal{R}(9)_{tqr}+\dot{x}^tE^iE^j\mathcal{R}(10)_{tij}+\dot{x}^r_u\dot{x}^a_b\dot{x}^c_dH^pH^q\mathcal{R}(11)^{ubd}_{racpq}+\dot{x}^r_u\dot{x}^a_b\dot{x}^cH^pH^q\mathcal{R}(12)^{ub}_{racpq}$$
$$\left.+\dot{x}^r_u\dot{x}^a\dot{x}^bH^pH^q\mathcal{R}(13)^u_{rabpq}+\dot{x}^w\dot{x}^a\dot{x}^bH^pH^q\mathcal{R}(14)_{wabpq}+\bar{h}\right]+\theta_j\theta_k\mathcal{R}(15)^{jk}$$
$$\mathcal{R}(1)^{bd}_{ac}=\tfrac{\chi}{\rho}(g^{bd}g_{ac}+\delta^d_a\delta^b_c)$$
$$\mathcal{R}(2)^{bd}_{acpq}=-\tfrac{\mu_0\bar{\mu}}{4\pi\rho}(2\delta^d_a\delta^b_qg_{pc}+\delta^b_q\delta^d_pg_{ca}+g^{db}g_{qa}g_{pc})$$
$$\mathcal{R}(3)^b_{ac}=\tfrac{4\chi}{\rho}G^b_{ac}$$
$$\mathcal{R}(4)^b_{acpq}=-\tfrac{\mu_0\bar{\mu}}{\pi\rho}(\delta^b_qg_{pk}G^k_{ac}+\delta^b_qg_{ka}G^k_{pc}+g_{ap}G^b_{qc}+g^{bj}g_{aq}g_{kp}G^k_{cj})$$
$$\mathcal{R}(5)_{ab}=\tfrac{2\chi}{\rho}G^i_{sa}G^s_{ib}$$
$$\mathcal{R}(6)_{abpq}=-\tfrac{\mu_0\bar{\mu}}{4\pi\rho}(g_{kj}G^j_{qa}G^k_{pb}+g_{qt}G^t_{ka}G^k_{pb}+2g_{pk}G^j_{qa}G^k_{jb})$$
$$\mathcal{R}(7)_{mr}=\tfrac{\mu_0^2}{4\pi\rho}g_{rm}$$
$$\mathcal{R}(8)^s_{mij}=\tfrac{1}{8\pi\rho}(\delta^s_jg_{im}+\delta^s_ig_{jm})$$
$$\mathcal{R}(9)_{tqr}=\tfrac{\mu_0^2}{4\pi\rho}g_{mr}G^m_{qt}$$
$$\mathcal{R}(10)_{tij}=\tfrac{1}{8\pi\rho}(g_{im}G^m_{jt}+g_{jm}G^m_{it})$$
$$\mathcal{R}(11)^{ubd}_{racpq}=\tfrac{\bar{\mu}^2}{8\pi\rho}[(\delta^d_a\delta^b_qg_{ic}+\delta^b_q\delta^d_ig_{ca}+g^{db}g_{qa}g_{ic})(\delta^i_r\delta^u_p+g^{iu}g_{rp})+\delta^b_c(g_{aq}\delta^d_r\delta^u_p+g^{ud}g_{rp}g_{aq})]$$
$$\mathcal{R}(12)^{ub}_{racpq}=\tfrac{\bar{\mu}^2}{4\pi\rho}[\text{polynomial function of } G^i_{jk}\text{ and } g_{ij}]$$
$$\mathcal{R}(13)^u_{rabpq}=\tfrac{\bar{\mu}^2}{4\pi\rho}[\text{polynomial function of } G^i_{jk}\text{ and } g_{ij}]$$
$$\mathcal{R}(14)_{wabpq}=\tfrac{\bar{\mu}^2}{4\pi\rho}[\text{polynomial function of } G^i_{jk}\text{ and } g_{ij}]$$
$$\mathcal{R}(15)^{jk}=\tfrac{\nu}{\rho}g^{kj}$$

and by calculating the corresponding explicit differential polynomial expression for $\theta^2\mathcal{R}:JD(\widetilde{W})\to\mathbb{R}$. (See Table 4. Let us emphasize that for what follows it is not necessary to explicitly calculate the coefficients $mathcalR(\alpha)^{\cdots}_{\cdots}$, $1\le\alpha\le 15$, appearing therein. These are analytical functions of the metric, and in general have a very complex aspect. However, in Table 4 are reported only some of these in order to give an idea of their structure.) $\underline{\widehat{(MHD)}}\subset\widetilde{(MHD)}$ is an *extended* 0-*crystal singular PDE*, in the sense of Definition 22.

Let us resume the proof given in [70], since this is necessary to understand the further developments reported below. Let us define

$$\underline{(MHD)}\equiv\left\{q\in\widetilde{(MHD)}|\theta(q)>0,\bar{h}(q)\ge 0,\mathcal{R}(q)\ge 0\right\}.\tag{78}$$

Then, $\underline{(MHD)}$ is a connected, simply connected bounded domain in $\widetilde{(MHD)}$. In fact, we can split $\underline{(MHD)}$ in the following way

$$\underline{(MHD)}={}_{(+,+)}\widetilde{Y}_1\bigcup{}_{(+,+)}\widetilde{Y}_2\bigcup\widetilde{Y}_3\bigcup{}_{(+,+,+)}\underline{(MHD)},\tag{79}$$

where

$${}_{(+,+,+)}\underline{(MHD)}\equiv\left\{q\in\widetilde{(MHD)}|\bar{h}(q)>0,\theta(q)>0,\mathcal{R}(q)>0\right\}\tag{80}$$

is an open submanifold of $\widetilde{(MHD)}$, and

$$\begin{cases} {}_{(+,+)}\widetilde{Y}_1 & \equiv \left\{q \in \widetilde{(MHD)} | \bar{h}(q) = 0, \theta(q) > 0, \mathcal{R}(q) > 0\right\} \\ {}_{(+,+)}\widetilde{Y}_2 & \equiv \left\{q \in \widetilde{(MHD)} | \mathcal{R}(q) = 0, \theta(q) > 0, \bar{h}(q) > 0\right\} \end{cases} \tag{81}$$

are codimension 1 submanifolds of $\widetilde{(MHD)}$. Furthermore,

$$\widetilde{Y}_3 \equiv \left\{q \in {}_{(+)}\widetilde{(MHD)} | \bar{h}(q) = 0, \mathcal{R}(q) = 0\right\} \tag{82}$$

is a codimension 2 submanifold of

$${}_{(+)}\widetilde{(MHD)} \equiv \left\{q \in \widetilde{(MHD)} | \theta(q) > 0\right\}.$$

Let us study the integrability properties of such submanifolds of $\widetilde{(MHD)}$. First note that since ${}_{(+,+,+)}\underline{(MHD)}$ is an open submanifold in $\widetilde{(MHD)}$ and in $JD^2({}_{(+,+)}\widetilde{W})$, where

$${}_{(+,+)}\widetilde{W} \equiv M \times \mathbf{I} \times \mathbb{R} \times \mathbb{R}^+ \times \mathbf{S}^3 \times \mathbb{R} \times \mathbb{R}^+,$$

it follows that if $q \in {}_{(+,+,+)}\underline{(MHD)}$ there exist solutions belonging to ${}_{(+,+,+)}\underline{(MHD)}$, i.e., entropy-regular solutions passing through q, just follows from the fact that ${}_{(+,+,+)}\underline{(MHD)}$ is an open submanifold of $\widetilde{(MHD)}$. So ${}_{(+,+,+)}\underline{(MHD)}$ is completely integrable. Furthermore, since

$$\begin{cases} {}_{(+,+,+)}\underline{(MHD)}_{+r} & = JD^r({}_{(+,+,+)}\underline{(MHD)}) \bigcap JD^{2+r}({}_{(+,+)}\widetilde{W}) \\ & \subset JD^r(\widetilde{(MHD)}) \bigcap JD^{2+r}(\widetilde{W}) = \widetilde{(MHD)}_{+r} \end{cases} \tag{83}$$

it follows that ${}_{(+,+,+)}\underline{(MHD)}$ is also formally integrable. Let us consider the integrability properties of ${}_{(+,+)}\widetilde{Y}_1 \subset \partial\underline{(MHD)}$. If $q \in {}_{(+,+)}\widetilde{Y}_1$ it follows that $q \in {}_{(+)}\widehat{(MHD)}$. Since this last equation is formally integrable and completely integrable, and ${}_{(+,+)}\widetilde{Y}_1$ is an open submanifold of ${}_{(+)}\widehat{(MHD)}$, it follows that also ${}_{(+,+)}\widetilde{Y}_1$ is formally integrable and completely integrable.

Let us, now, study the integrability properties of

$${}_{(+,+)}\widetilde{Y}_2 \subset JD^2({}_{(+,+)}\widetilde{W}) : \left\{\mathcal{R} = 0, \widetilde{F}^{(s)} = 0\right\}. \tag{84}$$

One can see that this equation is not formally integrable, but becomes so if we add the first prolongation of $\mathcal{R} = 0$. So we can prove that the following equation

$$\widehat{{}_{(+,+)}\widetilde{Y}_2} \subset JD^2({}_{(+,+)}\widetilde{W}) : \left\{\mathcal{R} = 0, \mathcal{R}_\alpha = 0, \widetilde{F}^{(s)} = 0\right\} \tag{85}$$

is formally integrable and completely integrable. Similar considerations can be made on the part $\widetilde{Y}_3$, and we can identify, the corresponding formally integrable PDE $\widehat{\widetilde{Y}}_3$. We skip on the details. By conclusion we get that

$$\underline{\widehat{(MHD)}} \equiv {}_{(+,+)}\widetilde{Y}_1 \bigcup \widehat{{}_{(+,+)}\widetilde{Y}_2} \bigcup \widehat{\widetilde{Y}}_3 \bigcup {}_{(+,+,+)}\underline{(MHD)} \subset \widetilde{(MHD)} \tag{86}$$

is the formally integrable and completely integrable constraint in $\widetilde{(MHD)}$, where for any initial condition, passes an entropy-regular solution for $\widetilde{(MHD)}$. Then we can apply Theorem 7 to conclude that $\underline{\widehat{(MHD)}}$ is an extended crystal singular PDE. Furthermore, taking into account that it results

$$\Omega_{3,w}^{\underline{\widehat{(MHD)}}} \cong \Omega_{3,s}^{\underline{\widehat{(MHD)}}} \cong \Omega_3 = 0,$$

we get that $\underline{\widehat{(MHD)}}$ is also an extended 0-crystal singular PDE. This assures that for any two spacelike admissible Cauchy integral manifolds $N_1 \subset \underline{\widehat{(MHD)}}_{t_1}$, $N_2 \subset \underline{\widehat{(MHD)}}_{t_2}$, $t_1 \neq t_2$, such that if both N_i, $i = 1, 2$, belong to the same component, in the split given in (86), there exist (singular) entropy-regular solutions V such that $\partial V = N_1 \bigcup_{X_1} P \bigcup_{X_2} N_2$, and such that their boundaries $X_1 \equiv \partial N_1$, $X_2 \equiv \partial N_2$, should be orientable and propagating with an admissible 3-dimensional integral manifold P. (For details see [74].)[32] The stability properties of $\underline{\widehat{(MHD)}}$ and its solutions, can be studied by utilizing our recent geometric theory on the stability of PDEs [68–70, 73, 74]. More precisely, $\underline{\widehat{(MHD)}}$ is a *functionally stable singular PDE*, in the sense that it splits in components that are functionally stable PDEs. Furthermore, smooth entropy-regular solutions, i.e., smooth solutions of $\underline{\widehat{(MHD)}}$, do not necessitate to be stable. However, all they can be stabilized and the *stable extended crystal singular PDE* of $\underline{\widehat{(MHD)}}$, i.e., a singular PDE splits in components that are stable extended crystal PDEs. This last is just ${}^{(S)}\underline{\widehat{(MHD)}} = \underline{\widehat{(MHD)}}_{(+\infty)}$. There, all smooth entropy-regular solutions, belonging to only one of the components, in the split representation (86) are stable at finite times.

Let us, now, find global solutions of $\underline{\widehat{(MHD)}}$, crossing the singular sets corresponding to states without nuclear energy production, to ones where $\bar{h} > 0$, i.e., passing from ${}_{(+,+)}\tilde{Y}_1$ to ${}_{(+,+,+)}\underline{(MHD)}$. Let $V \subset \widetilde{(MHD)}$ be a timelike solution such that the following conditions are satisfied:

(i) V is a regular entropy solution;
(ii) $V \bigcap {}_{(+,+)}\tilde{Y}_1 \equiv {}_{(+)}V \neq \varnothing$;
(iii) $V \bigcap {}_{(+,+,+)}\underline{(MHD)} \equiv {}_{(+,+)}V \neq \varnothing$.
(iv) $\partial V = N_0 \bigcup P \bigcup N_1$, $N_0 \subset {}_{(+)}V$, $N_1 \subset {}_{(+,+)}V$, $\partial_{(+)}V = N_0 \bigcup {}_{(+)}P \bigcup Y$.

Then we can give the following split surgery representation of V:

$$V = {}_{(+)}V \bigcup_{Y} Z \tag{87}$$

[32] Let us emphasize that admissible Cauchy integral manifolds can be found in each component of $\underline{\widehat{(MHD)}}$, thanks to Lemma 14. More precisely, the proceeding followed for the Navier–Stokes equation in Example 13, to solve Cauchy problems there, can be applied also to $\underline{\widehat{(MHD)}}$. In fact, envelopment solutions can be built also for all the components of this last equation, since they are formally integrable and completely integrable PDEs.

with $Z = Y \bigcup_{(+,+)} V$. We call a solution V of $\widetilde{(MHD)}$, such that holds the surgery property given in (87) just a *crossing nuclear critical zone solution*. Then, for any two admissible spacelike Cauchy hypersurfaces $N_1 \subset \underline{\widehat{(MHD)}}_{t_1}$, $N_2 \subset \underline{\widehat{(MHD)}}_{t_2}$, $t_1 \neq t_2$, such that the following properties are satisfied:

(a) $N_1 \subset {}_{(+,+)}\tilde{Y}_1$;
(b) $N_2 \subset {}_{(+,+,+)}\underline{(MHD)}$;

there exists a crossing nuclear critical zone solution $V \subset \underline{\widetilde{(MHD)}}$, such that $\partial V = N_1 \bigcup P \bigcup N_2$, where P is a timelike admissible 3-dimensional integral manifold of $\underline{\widetilde{(MHD)}}$. In general such a solution is a singular solution. Therefore, under the above condition of admissibility, one has algebraic classes in $\Omega_{2,s}^{\widehat{(MHD)}}$. Furthermore, V is represented by a smooth integral manifold at Y, with respect to the split given in (87), if the tangent space $T_{(+)}V|_Y = TZ|_Y \subset \mathbf{E}_2(\widetilde{MHD})|_Y$, where $\mathbf{E}_2(\widetilde{MHD})$ is the Cartan distribution of $\widetilde{(MHD)}$. In fact the dimension of the Cartan distribution $\mathbf{E}_2$ of $\underline{\widehat{(MHD)}}$ in the points $q \in {}_{(+,+,+)}\underline{(MHD)}$ is higher than in the points $q \in {}_{(+,+)}\tilde{Y}_1$). This can be seen by direct computation. Let us denote by

$$\zeta = X^{\alpha}[\partial x_{\alpha} + \sum_{0\leq|\beta|\leq 1} y^j_{\alpha\beta}\partial y^{\beta}_j] + Z^j_{\beta_1\beta_2}\partial y^{\beta_1\beta_2}_j \tag{88}$$

the generic vector field of the Cartan distribution $\mathbf{E}_2(\widetilde{W})$, where

$$\{y^j\}_{1\leq j\leq 16} = \{v^i, p, \theta, E^i, H^i, I^i, \bar{\rho}, \bar{h}\}$$

are the vertical coordinates of the fiber bundle $\pi : \widetilde{W} \to M$. The Cartan distribution on the boundary ${}_{(+,+)}\tilde{Y}_1 \subset \underline{\widehat{(MHD)}}$ is given by the vector fields (88) such that the following equations are satisfied: $\zeta.\bar{F}^I = 0$, where $\bar{F}^I$ are the functions defining equation ${}_{(+,+)}\tilde{Y}_1$. These are just the functions defining ${}_{(+,+,+)}\underline{(MHD)}$, i.e., $\widetilde{F}^{(s)}$, $1 \leq s \leq 7$, but with the condition $\bar{h} = 0$. So we get

$$\left\{\dim\mathbf{E}_2(\widetilde{MHD}) = \dim\mathbf{E}_2({}_{(+,+,+)}\underline{(MHD)}) = 148 > \dim\mathbf{E}_2({}_{(+,+)}\tilde{Y}_1) = 138.\right\}. \tag{89}$$

Note that even if the points of ${}_{(+,+)}\tilde{Y}_1$ can be considered singular one, with respect to the Cartan distribution of $\underline{\widehat{(MHD)}}$, the embedding $\underline{\widehat{(MHD)}} \subset \widetilde{(MHD)}$ allows us to prolong a solution from ${}_{(+,+)}\tilde{Y}_1$ to ${}_{(+,+,+)}\underline{(MHD)}$, according to Theorem 18 and surgery representation (87). Then such solution V in general bifurcates along Y. By summarizing, we can say that the crossing nuclear critical zone solution $V \subset \underline{\widehat{(MHD)}}$ is represented by an integral manifold $V \subset \underline{\widehat{(MHD)}}$ that is smooth in a neighborhood of $X \subset Y$, iff for its tangent space TV, the following condition is satisfied: $T_{(+)}V|_X = TZ|_X \subset \mathbf{E}_2(\widetilde{MHD})|_X$.

Now, in order to surgery a smooth solution $V' \subset {}_{(+++)}\underline{(MHD)}$, passing through a compact, smooth, spacelike 3-dimensional manifold N_1, with $X_1 \equiv \partial N_1$ orientable, with suitable smooth solutions of ${}_{(+,+)}\tilde{Y}_1$ it is enough that the following conditions should be satisfied:

$$\begin{cases} \lim_{\bar{h}\to 0} V' = {}_{(+,+)}V \subset {}_{(+++)}\underline{(MHD)} \\ T_{(+,+)}V \subset \mathbf{E}_2({}_{(+++)}\underline{(MHD)})_{\bar{h}=0}. \end{cases} \quad (90)$$

Since ${}_{(+,+)}V$ is a smooth solution we can prolong it to ∞, obtaining a solution ${}_{(+,+)}V^{(\infty)} \subset {}_{(+,+,+)}\underline{(MHD)}_{(+\infty)}$. There it identifies a horizontal 4-plane $\mathbf{H} \subset \mathbf{E}_2({}_{(+,+,+)}\underline{(MHD)}_{(+\infty)})$, contained in the Cartan distribution of ${}_{(+,+,+)}\underline{(MHD)}_{(+\infty)}$. Then $\mathbf{H}$ identifies also a horizontal 4-plane, which we continue to denote with $\mathbf{H}$, in the Cartan distribution of $\widetilde{(MHD)}_{+\infty}$, that on $({}_{(+,+)}\tilde{Y}_1)_{+\infty}$, coincides with the Cartan distribution of this last equation. Then we can smoothly prolong ${}_{(+,+)}V^{(\infty)}$ into $({}_{(+,+)}\tilde{Y}_1)_{+\infty}$, identifying there a smooth solution ${}_{(+)}V^{(\infty)}$. The projection of this algebraic singular solution on ${}_{(+,+)}\tilde{Y}_1 \bigcup {}_{(+,+,+)}\underline{(MHD)}$ identifies a smooth solution V that has the split surgery property (87). Therefore, it is enough to prove that solutions with the property (90) exist in ${}_{(+,+,+)}\underline{(MHD)}$. Now, we can see that a Cartan vector field ζ of ${}_{(+,+,+)}\underline{(MHD)}$ is given in *(88)* and subject to the condition $\zeta.\tilde{F}^{(s)} = 0$. Then conditions (90) are satisfied iff $X^\alpha \bar{h}_\alpha = 0$. Since this condition can be satisfied for suitable functions X^α on ${}_{(+,+,+)}\underline{(MHD)}$, we recognize that in the set of smooth solutions of ${}_{(+,+,+)}\underline{(MHD)}$ exist solutions that smoothly surgery with smooth solutions of ${}_{(+,+)}\tilde{Y}_1$. Such solutions are not finite times stable in $\underline{\widehat{(MHD}}$ since there the symbol is not trivial in each components ${}_{(+,+)}\tilde{Y}_1$ and ${}_{(+,+,+)}\underline{(MHD)}$. However, by using their formal integrability and complete integrability properties, we can state that in $\underline{\widehat{(MHD)}}_{+\infty}$ such solutions are finite times stable. So $\underline{\widehat{(MHD)}}_{+\infty}$ is the *stable extended crystal singular PDE* associated to $\underline{\widehat{(MHD}}$.

5 Appendix A: The Affine Crystallographic Group Types $[G(3)]$ and $[G(2)]$

Table 5 The 32 crystallographic point groups of the space-group types $[G(3)]$

Type	Schoenflies symbol	International symbol
Nonaxial (2)	$C_i = S_2, C_s = C_{1h}$	$\bar{1}, m$
Cyclic (5)	$C_n, n = 1,2,3,4,6$	$n, n = 1,2,3,4,6$
Cyclic with horizontal planes (4)	$C_{nh}, n = 2,3,4,6$	$2/m, \bar{6}, 4/m, 6/m$
Cyclic with vertical planes (4)	$C_{nv}, n = 2,3,4,6$	$mm2, 3m, 4mm, 6mm$
Dihedral (4)	$D_n, n = 2,3,4,6$	$222, 32, 422, 622$
Dihedral with horizontal planes (4)	$D_{nh}, n = 2,3,4,6$	$mmm, \bar{6}m2, 4/mmm, 6/mmm$
Dihedral with planes between axes (2)	$D_{nd}, n = 2,3$	$\bar{4}2m, \bar{3}m$
Improper rotation (2)	$S_n, n = 4,6$	$\bar{4}, \bar{3}$
Cubic groups (5)	T, T_h, T_d, O, O_h	$23, m\bar{3}, \bar{4}3m, 432, m\bar{3}m$

n $(n = 1,2,3,4,6)$: rotations of $2\pi/n$ about a symmetry axis
$\bar{n}$ $(n = 1,2,3,4,6)$: rotation n composed with inversion about symmetry centre; $m = \bar{2}$
$C_n = \mathbb{Z}_n \cong \{0,1,2,3,\cdots,n-1\}$ (cyclic Abelian groups); $C_i = S_2$; $C_s = m$; $C_{3i} = S_6$
D_n: group with an n-fold axis plus a twofold axis perpendicular to that axis
D_n is a non-Abelian group for $n > 2$. The group order is $2n$
D_{nh}: D_n with a mirror plane symmetry perpendicular to the n-fold axis
D_{nv}: D_n with mirror plane symmetries parallel to the n-fold axis
O: symmetry group of the octahedron. The group order is 24. One has the isomorphism $O \cong T_d$
O_h: O with improper operations (those that change orientation). The group order is 48
T: symmetry group of the tetrahedron, isomorphic to the alternating group A_4.The group order is 12
(See Fig. 4 for representations of octahedron and tetrahedron 3-chains.)
T_d: T with improper operations. Non-Abelian group of order 24
T_h: T with the addition of an inversion
A_n: group of even permutations on a set of length n. The group order is $\frac{n!}{2}$

Table 6 Diehdral groups $D_{2m}, m \geq 1$

m	D_{2m}
1	$D_2 \cong \mathbb{Z}_2$
$m \geq 2$	$D_{2m} \cong \mathbb{Z}_m \rtimes \mathbb{Z}_2$ The generator of $\mathbb{Z}_2$ acts on $\mathbb{Z}_m$ as multiplication by -1
2	$D_4 \cong \mathbb{Z}_2 \times \mathbb{Z}_2$ (Klein four-group)

Warn that in crystallography diehdral groups are usually denoted by $D_m = D_{2m}, m \geq 2$
This is just the notation used in Table 5

Table 7 The 230 affine crystallographic space-group types $[G(3)]$

Syngonies (7)	**Geometric classes (32 = point groups)**	**Bravais types (14)**
Triclinic	$C_i = S_2(1), C_1(1)$	$2P$
Monoclinic	$C_2(3), C_{1h}(4), C_{2h}(6)$	$8P, 5C$
Orthorhombic	$D_2(9), C_{2v}(22), D_{2h}(28)$	$30P, 15C, 9I, 5F$
Tetragonal	$C_4(6), S_4(2), C_{4h}(6), D_4(10), C_{4v}(12), D_{2d}(12), D_{4h}(20)$	$49P, 19I$
Trigonal	$C_6(4), S_6(2), D_3(7), C_{3v}(6), D_{3d}(6)$	$18P, 7R$
Hexagonal	$C_3(6), C_{3h}(1), C_{6h}(2), D_6(6), C_{6v}(4), D_{3h}(4), D_{6h}(4)$	$27P$
Cubic	$T(5), T_h(7), O(8), T_d(6), O_h(10)$	$15P, 11F, 9I$

(Crystal class systems = Syngonies): 7. (Triclinic = Anorthic), (Trigonal = Rhombohedral)
Bravais lattice centering types: *P* primitive, *I* body, *F* face, *A/B/C* side, *R* rhombohedral.
(For the body-centered case $(B = I)$, the infinite translation group is $\{\mathbb{Z}^3, \mathbb{Z}^3 + (\frac{1}{2},\frac{1}{2},\frac{1}{2})\}$)

Table 8 The 17 affine crystallographic space-group types $[G(2)]$

Syngony (5)	**International symbols**	**Geometric classes (10 = point groups)**
Oblique	$p1, p2$	$\mathbb{Z}_1, \mathbb{Z}_2$
Rectangular	pm, pg, cm, pmm	D_1, D_1, D_1, D_2
Rhombic	pmg, pgg, cmm	D_2, D_2, D_2
Square	$p4, p4m, p4g$	$\mathbb{Z}_4, D_4, D_4$
Trigonal	$p3, p3m1, p31m$	$\mathbb{Z}_3, D_3, D_3$
Hexagonal	$p6, p6m$	$\mathbb{Z}_6, D_6$

All planar crystallographic groups $G(2)$ are subgroups of $p4m$ or $p6m$, or both (See Appendix D)
p primitive, c centered, m mirror plane, g glide reflection
A glide reflection is an isometry of the Euclidean plane that combines reflection in a line with a translation along that line

Table 9 The four holohedries (lattice symmetries) in $[G(2)]$

Holohedry	**Lattice system**
2	Monoclinic = oblique
$2mm$	Orthorohmbic = {rectangular,rhombic}
$4mm$	Tetragonal = square
$6mm$	Hexagonal = {trigonal,hexagonal}

Table 10 The seven holohedries (lattice symmetries) in $[G(3)]$

Holohedry	**Lattice system**
$\bar{1}$	Triclinic
$2/m$	Monoclinic
mmm	Orthorohmbic
$4/mmm$	Tetragonal (square)
$\bar{3}m$	Trigonal (rombohedric)
$6/mmm$	Hexagonal
$m\bar{3}m$	Cubic

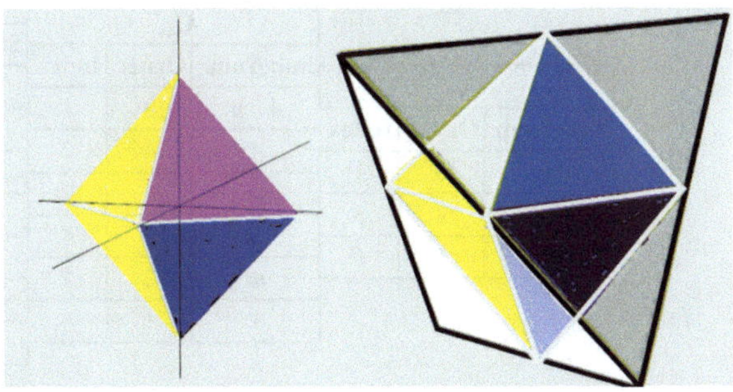

Fig. 4 Octahedron and tetrahedron 3-chains in $\mathbb{R}^3$ (The tetrahedron is a 3-chain in $\mathbb{R}^3$, identified with a regular 4-faced polyhedron, where each face is an equilateral triangle. It can be derived from the octahedron by extending alternate faces until they meet)

6 Appendix B: The Subgroups of the Affine Crystallographic Point-Group Types [G(3)]

$C_i = S_2$		
Subgroup	Order	Index
−1	2	1
1	1	2

$C_s = C_{1h}$		
Subgroup	Order	Index
m	2	1
1	1	2

C_1		
Subgroup	Order	Index
1	1	1

C_2		
Subgroup	Order	Index
2	2	1
1	1	2

C_3		
Subgroup	Order	Index
3	3	1
2	2	3

C_4		
Subgroup	Order	Index
4	4	1
2	2	2
1	1	4

C_6		
Subgroup	Order	Index
6	6	1
3	3	2
2	2	3
1	1	6

C_{2h}		
Subgroup	Order	Index
2/*m*	4	1
2	2	2
m	2	2
−1	2	2
1	1	4

C_{3h}		
Subgroup	Order	Index
−6	6	1
3	3	2
m	2	3
1	1	6

C_{4h}		
Subgroup	Order	Index
4/*m*	8	1
4	4	2
−4	4	2
2/*m*	4	2
2	2	4
m	2	4
−1	2	4
1	1	8

C_{6h}		
Subgroup	Order	Index
6/*m*	12	1
−6	6	2
6	6	2
−3	6	2
3	3	4
2/*m*	4	3
m	2	6
−1	2	6
1	1	12

C_{2v}		
Subgroup	Order	Index
*mm*2	4	1
2	2	2
m	2	2
1	1	4

C_{3v}		
Subgroup	Order	Index
3*m*	6	1
m	2	3
1	1	6

C_{4v}		
Subgroup	Order	Index
4*mm*	8	1
4	4	2
*mm*2	4	2
2	2	4
m	2	4
1	1	8

C_{6v}		
Subgroup	Order	Index
6*mm*	12	1
6	6	2
3*m*	6	2
*mm*2	4	3
2	2	6
m	2	6
1	1	12

D_2		
Subgroup	**Order**	**Index**
222	4	1
2	2	2
1	1	4

D_3		
Subgroup	**Order**	**Index**
32	6	1
3	3	2
2	2	3
1	1	6

D_4		
Subgroup	**Order**	**Index**
422	8	1
4	4	2
222	4	2
2	2	4
1	1	8

D_6		
Subgroup	**Order**	**Index**
622	12	1
6	6	2
32	6	1
3	3	4
2	2	6
m	2	6
1	1	12

D_{2h}		
Subgroup	**Order**	**Index**
mmm	8	1
*mm*2	4	2
222	4	2
2/*m*	4	2
2	2	4
m	2	4
−1	2	4
1	1	8

D_{3h}		
Subgroup	**Order**	**Index**
−6	6	1
3	3	2
m	2	3
1	1	6

D_{4h}		
Subgroup	**Order**	**Index**
4/*mmm*	16	1
−42*m*	8	2
422	8	2
4/*m*	8	2
4	4	4
−4	4	4
mmm	8	2
*mm*2	4	4
222	4	4
2/*m*	4	4
em	2	8
2	2	8
−1	2	8
1	1	16

D_{6h}		
Subgroup	**Order**	**Index**
6/*mm*	24	1
−62*m*	12	2
6*mm*	12	2
622	12	2
6/*m*	12	2
−3*m*	12	2
−6	6	4
6	6	4
3*m*	6	4
32	6	4
−3	6	4
6/*m*	12	2
3	3	8
mmm	8	3
*mm*2	4	6
222	4	6
2/*m*	4	6
2	2	12
m	2	12
−1	2	12
1	1	24

D_{2d}		
Subgroup	**Order**	**Index**
−42*m*	8	1
−4	4	2
*mm*2	4	2
222	4	2
2	2	4
m	2	4
1	1	8

D_{3d}		
Subgroup	**Order**	**Index**
−3*m*	12	1
3*m*	6	2
32	6	2
−3	6	2
3	3	4
2/*m*	4	3
m	2	6
−1	2	6
1	1	12

S_4		
Subgroup	**Order**	**Index**
−4	4	1
2	2	2
1	1	4

$S_6 = C_{3i}$		
Subgroup	Order	Index
−3	6	1
3	3	2
−1	2	3
1	1	6

T		
Subgroup	**Order**	**Index**
23	12	1
3	3	4
222	4	3
2	2	6
1	1	12

T_h		
Subgroup	**Order**	**Index**
m − 3	24	1
23	12	2
−3	6	4
3	3	8
mmm	8	3
*mm*2	4	6
222	4	6
2/*m*	4	6
2	2	12
m	2	12
−1	2	12
1	*e*1	24

T_d		
Subgroup	**Order**	**Index**
−43*m*	24	1
23	12	2
3*m*	6	4
3	3	8
−42*m*	8	3
−4	4	6
*mm*2	4	6
222	4	6
2	2	12
m	2	12
1	1	24

O		
Subgroup	Order	Index
432	24	1
23	12	2
32	6	4
422	8	3
4	4	6
3	3	8
222	4	6
2	2	12
1	1	24

O_h		
Subgroup	Order	Index
$m-3m$	48	1
$-43m$	24	2
432	24	2
$m-3$	24	2
23	12	4
$-3m$	12	4
$3m$	6	8
32	6	8
-3	6	8
3	3	16
$4/mmm$	16	3
$-42m$	8	6
$4mm$	8	6
422	8	6
$4/m$	8	6
-4	4	12
4	4	12
mmm	8	6
222	4	12
$2/m$	4	12
2	2	24
m	2	24
-1	2	24
1	1	48

7 Appendix C: Amalgamated Free Products in the Affine Crystallographic Space-Group Types [G(3)]

Amalgamated free products in $[G(3)]$
$\mathbb{Z}_2 \bigstar_e \mathbb{Z}_2 = < \mathbf{x}, \mathbf{y} >$
$\mathbb{Z}_4 \bigstar_{\mathbb{Z}_2} \mathbb{Z}_4 = < 4, \bar{4} >$
$\mathbb{Z}_4 \bigstar_{\mathbb{Z}_2} D_2 = < 2, 4, \mathbf{x} >$
$\mathbb{Z}_6 \bigstar_{\mathbb{Z}_3} D_3 = < 3, 6, \mathbf{x} >$
$D_4 \bigstar_{D_2} D_4 = < 4, \bar{4}, \mathbf{x} >$
$D_2 \times \mathbb{Z}_2 \bigstar_{D_2} D_4 = < 2, 4, \bar{1}, \mathbf{x} >$
$D_6 \bigstar_{D_3} D_6 = < 6, \bar{6}, \mathbf{x} >$
$D_3 \times \mathbb{Z}_2 \bigstar_{D_3} D_6 = < 2, 6, \bar{1}, \mathbf{x} >$

x reflection over x-axis, y reflection over y-axis

8 Appendix D: The Subgroups of the Affine Crystallographic Space-Group Types [G(2)]

p2

Subgroup	Index
*p*1	2

pm

Subgroup	Index
pg	2
cm	2
*p*1	

pg

Subgroup	Index
*p*1	2

cm

Subgroup	Index
pg	2
pm	2
*p*2	
*p*1	

pmm

Subgroup	Index
cmm	2
pmg	2
pgg	
pm	
cm	
*p*2	
pg	
*p*1	

pgg

Subgroup	Index
*p*2	2
pg	2
*p*1	

p4

Subgroup	Index
*p*2	2
*p*1	

cmm

Subgroup	Index
pmm	2
pmg	2
cm	4
pgg	
pm	2
*p*2	
pg	
*p*1	

p4m

Subgroup	Index
*p*4*g*	2
pmm	2
cmm	2
pmg	4
*p*4	2
pgg	4
pm	4
cm	4
*p*2	4
pg	8
*p*1	8

p4g	
Subgroup	**Index**
pmm	4
cmm	2
pmg	
p4	2
cm	
pgg	
pm	
p2	
pg	
p1	

p6m	
Subgroup	**Index**
pmm	6
cmm	3
pmg	6
pgg	6
p3m1	2
p31m	2
pm	12
cm	6
pg	12
p6	2
p3	4
p2	6
p1	12

p3	
Subgroup	**Index**
p1	3

p3m1	
Subgroup	**Index**
pm	6
cm	3
p3	2
p31m	
pg	
p1	

p31m	
Subgroup	**Index**
p3m1	3
p3	2
p3	2
pm	6
cm	3
pg	3
p1	

p6	
Subgroup	**Index**
p3	2
p2	3
p1	

Acknowledgements I would like to thank the editors for inviting me to write my contribution to this Springer book. A similar version of this work was earlier put on `arXiv:0811.3693v18`.

References

1. Abels, H.: Discrete groups of affine isometries. J. Lie Theory **9**, 321–349 (1999)
2. Agarwal, R.P., Prástaro, A.: Geometry of PDE's.III(I): Webs on PDE's and integral bordism groups. The general theory. Adv. Math. Sci. Appl. **17**(1), 239–266 (2007); Geometry of PDE's.III(II): Webs on PDE's and integral bordism groups. Applications to Riemannian geometry PDE's. Adv. Math. Sci. Appl. **17**(1), 267–281 (2007)
3. Agarwal, R.P., Prástaro, A.: Singular PDE's geometry and boundary value problems. J. Nonlinear Conv. Anal. **9**(3), 417–460 (2008); On singular PDE's geometry and boundary value problems. Appl. Anal. **88**(8), 1115–1131 (2009). doi:10.1080/00036810902943612
4. Alperin, J.L., Bell, R.B.: Groups and Representations. Graduate Text in Mathematics, vol 162. Springer, New York (1995)
5. Anderson, M.T.: Geometrization of 3-manifolds via the Ricci flow. Not. Amer. Math. Soc. **51**, 184–193 (2004)
6. Arfken, G.: Crystallographic Point and Space Groups. Mathematical Methods for Physicists, 3rd edn. Academic Press, Orlando (1985)

7. Bernstein, J., Schwarzman, O.: Complex crystallographic Coxeter groups and affine root systems. J. Nonlinear Math. Phys. **13**(2), 163–182 (2006); Chevally's theorem for the complex crystallographic groups. J. Nonlinear Math. Phys. **13**(3), 323–351 (2006)
8. Bieberbach, L.: Über die bewegungsgruppen der Euklidschen räume.I. Math. Ann. **70**, 297–336 (1911). Über die bewegungsgruppen der Euklidschen räume.II. Math. Ann. 72, 400–412 (1912)
9. Boardman, J.M.: Singularities of differentiable maps. Publ. Math. Inst. Hautes Études Sci. **33**, 21–57 (1967)
10. Bourbaki, N.: Lie Groups and Lie Algebras (Chapters 4–6). Springer-Verlag, Berlin (2002); Lie Groups and Lie Algebras (Chapters 7–9). Springer-Verlag, Berlin (2005)
11. Brown, K.S.: Cohomology of Groups. Springer-Verlag, Paris (1969)
12. Bryant, R.L., Chern, S.S., Gardner, R.B., Goldschmidt, H.L., Griffiths, P.A.: Exterior Differential Systems. Springer-Verlag, New York (1991)
13. Buoncristiano, S., Rourke, C.P., Sanderson, B.J.: A Geometric Approach to Homology Theory (London Math. Soc. Lecture Note Ser. 18), Cambridge University Press, Cambridge; Springer-Verlag, New York (1976)
14. Cappel, S., Ranicki, A., Rosenberg, J. (eds.): Surveys on Surgery Theory: Volume 1, Papers Dedicated to C. T. C. Wall. Princeton University Press, Princeton (1999)
15. Charlap, L.S.: Bieberbach Groups and Flat Manifolds. Springer-Verlag, New York (1986)
16. Chow, B., Knopp, D.: The Ricci Flow: An Introduction, Mathematical Surveys and Monographs, vol. 110. American Mathematical Society (2004)
17. Chow, B., Chu, S.-C., Gliekenstein, D., Guenther, C., Isenberg, J., Ivey, T., Knopf, D., Lu, P., Luo, F., Ni, L.: The Ricci flow: Techniques and Applications: Part I: Geometric Aspects, Mathematical Surveys Monographs, vol. 135. American Mathematical Society (2008)
18. Chow, B., Chu, S.-C., Gliekenstein, D., Guenther, C., Isenberg, J., Ivey, T., Knopf, D., Lu, P., Luo, F., Ni, L.: The Ricci Flow: Techniques and Applications: Part II: Analytic Aspects, Mathematical Surveys Monographs, vol. 144. American Mathematical Society (2008)
19. De Turk, D.M.: Deforming metrics in the direction of their Ricci tensors. J. Differ. Geom. **18**(1), 157–162 (1983)
20. Falconer, K.J.: Fractal Geometry. Mathematical Foundations and Applications, 2nd revised edn. Wiley (2003)
21. Fedorov, E.S.: The Symmetry and Structure of Crystals. Fundamenta Works, Moscow (1949), 111–255. (In Russian).
22. Goldshmidt, H.: Integrability criteria for systems of non-linear partial differential equations. J. Differ. Geom. **1**, 269–307 (1967)
23. Goldshmidt, H., Spencer, D.: Submanifolds and Over-determined Differential Operators. Complex Analysis & Algebraic Geometry, pp. 319–356. Cambridge University Press, Cambridge (1977)
24. Golubitsky, M., Guillemin, V.: Stable Mappings and Their Singularities. Springer-Verlag, New York (1973)
25. Gromov, M.: Almost flat manifolds. J. Differ. Geom. **13**, 231–241 (1978); Almost Flat Manifolds, Astérisque **81**, Soc. Math. France, Paris, 1981 (1986).
26. Hamilton, R.S.: Three-manifolds with positive Ricci curvature. J. Differ. Geom. **17**, 255–306 (1982)
27. Hamilton, R.S.: Four-manifolds with positive Ricci curvature operator. J. Differ. Geom. **24**, 153–179 (1986)
28. Hamilton, R.S.: Eternal solutions to the Ricci flow. J. Differ. Geom. **38**, 1–11 (1993)
29. Hamilton, R.S.: The Formation of Singularities in the Ricci Flow, Surveys in Differential Geometry vol. 2, pp. 7–136. International Press(1995)
30. Hamilton, R.S.: A compactness property for solutions of the Ricci flow on three-manifolds. Comm. Anal. Geom. **7**, 695–729 (1999)
31. Hirsh, M.W.: Differential Topology. Springer-Verlag, Berlin (1976)
32. Kobayashi, S., Nomizu, K.: Foundations of Differential Geometry, vol. I, II, (Interscience Publishers). Wiley, New York (1996) (Reprint (1966))

33. Krasilšhchik, I.S., Lychagin, V., Vinogradov, A.M.: Geometry of Jet Spaces and Nonlinear Partial Differential Equations. Gordon and Breach Science Publishers S.A. , Amsterdam (1986)
34. Kuranishi, M.: On the Cartan prolongation theorem of exsterior differential systems. Amer. J. Math. **79**(1), 1–47 (1957)
35. Lapidus, M.L., van Frankenhuijsen M.: Fractal Geoemetry and Applications. A Jubilee of Benoit Mandelbrot. Proceedings of Simposia in Pure Mathematics **72**, Amer. Math. Soc. (2005)
36. Lions, P.L., Crandall, M.: Viscosity solutions of Hamilton-Jacobi equations. Trans. Amer. Math. Soc. **277**, 1–42 (1983)
37. Ljapunov, A.M.: Stability of Motion with an contribution by V. A. Pliss and an introduction by V. P. Basov. Mathematics in Science and Engineering, vol. 30. Academic Press, New York-London (1966)
38. Lychagin, V., Prástaro, A.: Singularities of Cauchy data, characteristics, cocharacteristics and integral cobordism. Differ. Geom. Appl. **4**, 287–300. doi: 10.1016/0926-2245(94)00017-4 (1994)
39. Madsen, I.B., Milgram, R.J.: The Classifying Spaces for Surgery and Bordism of Manifolds (Ann. Math. Stud.), vol. 92. Princeton University Press, New Jersey 1979
40. Mandelbrot, B.B.: The Fractal Geometry of Nature. W. H. Freeman Co., New York (1982)
41. Milnor, J., Stasheff, J.: Characteristic Classes (Ann. Math. Studies), vol. 76. Princeton University Press (1974)
42. Morgan, J., Sullivan, D.: The transversality characteristic class and linking sycles in surgery theory. Ann. Math. **99**, 463–544 (1974)
43. Perelman, G.: The entropy formula for the Ricci flow and its geometry applications. `arXiv:math/0211159` (2002)
44. Perelman, G.: Ricci flow with surgery on three-manifolds. `arXiv:math/0303109` (2003)
45. Plesken, W., Pest, M.: On maximal finite irreducible subgroup of $GL(n, \mathbb{Z})$, I, II. Math. Comput. **31**, 536–551; 552–573 (1977)
46. Prástaro, A.: Dynamic conservation laws. In: Prástaro, A. (ed.) Geometrodynamics Proceedings. World Scientific Publishing Co., Singapore (1985)
47. Prástaro, A.: Cobordism of PDE's. Boll. Unione Mat. Ital. **30**(5-B), 977–1001 (1991)
48. Prástaro, A.: Quantum geometry of PDE's. Rep. Math. Phys. **30**(3), 273–354. doi:10.1016/0034-4877(91)90063-S (1991)
49. Prástaro, A.: Geometry of super PDE's. In: Prástaro, A., Rassias, T.M. (eds.) Geometry in Partial Differential Equations, pp. 259–315. World Scientific Publishing Co., Singapore (1994)
50. Prástaro, A.: Geometry of quantized super PDE's. Amer. Math. Soc. Transl. **167**(2), 165–192 (1995)
51. Prástaro, A.: Quantum geometry of super PDE's. Rep. Math. Phys. **37**(1) , 23–140. doi:10.1016/0034-4877(96)88921-X (1996)
52. Prástaro, A.: Geometry of PDEs and Mechanics. World Scientific Publishing Co., Singapore. pp. 760 (1996)
53. Prástaro, A.: (Co)bordisms in PDE's and quantum PDE's. Rep. Math. Phys. **38**(3), 443–455. doi:S003448779784894X (1996)
54. Prástaro, A.: Quantum and integral (co)bordism groups in partial differential equations. Acta Appl. Math. **51**(3), 243–302. doi: 10.1023/A:1005986024130 (1998)
55. Prš taro A.: Quantum and integral bordism groups in the Navier-Stokes equation. In: Szenthe, J. (ed.) New Developments in Differential Geometry, Budapest 1996, pp. 343–360. Kluwer Academic Publishers, Dordrecht (1998)
56. Prástaro, A.: (Co)bordism groups in PDE's. Acta Appl. Math. **59**(2),111–202. doi: 10.1023/A:1006346916360 (1999)
57. Prástaro, A.: (Co)bordism groups in quantum PDE's. Acta Appl. Math. **64**(2/3), 111–217. doi: 10.1023/A:1010685903329 (2000)

58. Prástaro, A.: Local and global solutions of the Navier-Stokes equation, Steps in Differential Geometry, Proceedings of the Colloquium on Differential Geometry, 25–30 July, 2000. Debrecen, Hungary, L. Kozma, P. T. Nagy & L. Tomassy (eds.), Univ. Debrecen 263–271 (2001)
59. Prástaro, A.: Quantum manifolds and integral (co)bordism groups in quantum partial differential equations. Nonlinear Anal. **47**(4) 2609–2620. doi: 10.1016/S0362-546X(01)00382-0 (2001)
60. Prástaro, A.: Quantized Partial Differential Equations, p. 500. World Scientific Publishing Co. (2004)
61. Prástaro, A.: Quantum super Yang-Mills equations: Global existence and mass-gap (eds. G. S. Ladde, N. G. Madhin & M. Sambandham). Proc.Dynamic Syst Appl. **4**, 227–234 (2004)
62. Prástaro, A.: Geometry of PDE's. I: Integral bordism groups in PDE's. J. Math. Anal. Appl. **319**, 547–566. doi: 10.1016/j.jmaa.2005.06.044 (2006)
63. Prástaro, A.: Geometry of PDE's. II: Variational PDE's and integral bordism groups. J. Math. Anal. Appl. **321**, 930–948. doi: 10.1016/j.jmaa.2005.08.037 (2006)
64. Prástaro, A.: Conservation laws in quantum super PDE's, Proceedings of the Conference on Differential & Difference Equations and Applications, (eds. R. P. Agarwal & K. Perera), 943–952. Hindawi Publishing Corporation, New York (2006)
65. Prástaro, A.: (Co)bordism groups in quantum super PDE's. I: Quantum supermanifolds.k Nonlinear Anal. Real World Appl. **8**(2), 505–533. doi: 10.1016/j.nonrwa.2005.12.008 (2007)
66. Prástaro, A.: (Co)bordism groups in quantum super PDE's. II: Quantum super PDE's. Nonlinear Anal. Real World Appl. **8**(2), 480–504. doi: 10.1016/j.nonrwa.2005.12.007 (2007)
67. Prástaro, A.: (Co)bordism groups in quantum super PDE's. III: Quantum super Yang-Mills equations. Nonlinear Anal. Real World Appl. **8**(2), 447–479. doi: 10.1016/j.nonrwa.2005.12.006 (2007)
68. Prástaro, A.: (Un)stability and bordism groups in PDE's. Banach J. Math. Anal. **1**(1), 139–147 (2007)
69. Prástaro, A.: On the extended crystal PDE's stability.I: The n-d'Alembert extended crystal PDE's, Appl. Math. Comput. **204**(1), 63–69. doi: 10.1016/j.amc.2008.05.141 (2008)
70. Prástaro, A.: On the extended crystal PDE's stability.II: Entropy-regular-solutions in MHD-PDE's. Appl. Math. Comput. **204**(1), 82–89. doi: 10.1016/j.amc.2008.05.142 (2008)
71. Prástaro, A.: On quantum black-hole solutions of quantum super Yang-Mills equations. Proc. Dynamic Syst. Appl. **5**, 407–414 (2008)
72. Prástaro, A.: Geometry of PDE's. IV: Navier-Stokes equation and integral bordism groups. J. Math. Anal. Appl. **338**(2), 1140–1151. doi:10.1016/j.jmaa.2007.06.009 (2008)
73. Prástaro, A.: Extended crystal PDE's stability. I: The general theory. Math. Comput. Modelling, **49**(9–10), 1759–1780. doi: 10.1016/j.mcm.2008.07.020 (2009)
74. Prástaro, A.: Extended crystal PDE's stability.II: The extended crystal MHD-PDE's. Math. Comput. Modelling, **49**(9–10), 1781–1801. doi: 10.1016/j.mcm.2008.07.021 (2009)
75. Prástaro, A.: Surgery and bordism groups in quantum partial differential equations. I: The quantum Poincaré conjecture. Nonlinear Anal. Theory Methods Appl. **71**(12), 502–525. doi: 10.1016/j.na.2008.11.077 (2009)
76. Prástaro, A.: Surgery and bordism groups in quantum partial differential equations. II: Variational quantum PDE's. Nonlinear Anal. Theory Methods Appl. **71**(12), 526–549. doi: 10.1016/j.na.2008.10.063 (2009)
77. Prástaro, A.: Quantum extended crystal PDE's. Nonlinear Stud. **18**(3), 1–39. arXiv:1105.0166 [math.AT] (2011)
78. Prástaro, A.: Exotic heat PDE's. Commun. Math. Anal. **10**(1), 64–81. arXiv: 1006.4483[math.GT] (2011)
79. Prástaro, A.: Quantum extended crystal super PDE's, arXiv: 0906.1363[math.AT] (2012)
80. Prástaro, A.: Exotic heat PDE's.II. Essays in Mathematics and its Applications. (Dedicated to Stephen Smale.) (Eds. P. M. Pardalos and Th. M. Rassias.) Springer, New York, 369–419 arXiv: 1009.1176[math.AT] (2012)

81. Prástaro, A.: Exotic n-d'Alembert PDE's and stability. Approximation and Inequalities. (Dedicated to Themistocles M. Rassias for his 60th birthday.) (Eds. G. Georgiev (USA), P. Pardalos (USA) and H. M. Srivastava (Canada)), Springer, New York, 571–586. arXiv: 1011.0081[math.AT] (2012)
82. Prástaro, A.: Exotic PDE's. Mathematics Without Boundaries. Surveys in Interdisciplinary Research, (P. Pardalos (USA) & Th. M. Rassias (Greece) eds.), Springer, New York, (to appear). arXiv:1101.0283[math.AT].
83. Prástaro, A.: Quantum exotic PDE's. Nonlinear Analysis. Real World Appl. 14(2), 893–928 doi: 10.1016/j.nonrwa.2012.04.001. arXiv: 1106.0862[math.AT] (2013)
84. Prástaro, A.: Strong reactions in quantum super PDE's. I: Quantum hypercomplex exotic super PDE's. arXiv: 1205.2894[math.AT]. (Part I and Part II are unified in arXiv.) (2013)
85. Prástaro, A.: Strong reactions in quantum super PDE's. II: Nonlinear quantum propagators. arXiv: 1205.2894[math.AT]. (Part I and Part II are unified in arXiv.) (2013)
86. Prástaro, A.: Strong reactions in quantum super PDE's. III: Exotic quantum supergravity. arXiv: 1206.4856[math.AT] (2014)
87. Prástaro, A., Rassias, T.M.: A geometric approach of the generalized d'Alembert equation. J. Comput. Appl. Math. **113**(1–2), 93–122 (2000)
88. Prástaro, A., Rassias, T.M.: A geometric approach to a noncommutative generalized d'Alembert equation. C. R. Acad. Sc. Paris **330**(I–7), 545–550 (2000)
89. Prástaro, A., Rassias, T.M.: Ulam stability in geometry of PDE's. Nonlinear Funct. Anal. Appl. **8**(2), 259–278 (2003)
90. Quillen, D.: Elementary proofs of some results of cobordism theory using Steenrod operators. Adv. Math. **7**, 29–56 (1971)
91. Raghunathan, M.S.: Discrete Subgroups of Lie Groups. Springer-Verlag, New York (1972)
92. Ranicki, A.: Noncommutative Localization in Topolgy. Proceedings of the Workshop, Edinburgh, April 29–30 (2002). Cambridge, Cambridge University Press. London Math. Soc. Lecture Note Series **330**, 81–102 (2006)
93. Rotman, J.J.: An Introduction to Homological Algebra. Academic Press, San Diego (1979)
94. Rudyak, Y.B.: On Thom Spectra, Orientability and Cobordism. Springer-Verlag, Berlin (1998)
95. Ruh, E.A.: Almost flat manifolds. J. Differ. Geom. **17**, 1–14 (1982)
96. Schönflies, A.: Krystallsysteme und Kristallstruktur. Teubner, Leipzig (1891)
97. Schwartzenberger, R.L.E.: N-dimensional Crystallography. Pitman (1980)
98. Scott, P., Wall, T.: Topological Methods in Group Theory, Homological Group Theory, Proceedings from Durham (1977), Lecture Notes, London Matt. Soc. **36**, 137–203 (1979)
99. Stong, R.E.: Notes on Bordism Theories. Princeton University Press (Ann. Math. Stud.) (1968)
100. Sunada, T.: Crystals that nature might miss creating. Notices Amer. Math. Soc. **55**(2), 209–215 (2008)
101. Sweedler, M.E.: Hopf Algebras. Benjamin, New York (1969)
102. Switzer, R.: Algebraic Topology-homotopy and Homology. Springer-Verlag, Berlin (1975)
103. Thom, R.: Quelques propriété globales des variétés différentieles. Comm. Math. Helv. **28**, 17–86 (1954)
104. Thom, R.: Remarques sur les problemes comportant des inéqualities différentielles globales. Bull. Soc. Math. France. **87**, 455–468 (1954)
105. Thom, R.: Les singularités des applications differentiables. Ann. Inst. Fourier(Grenoble) **6**, 43–87 (1955–1956)
106. Ulam, S.M.: A Collection of Mathematical Problems. Interscience Publication, New York (1960)
107. Wall, C.T.C.: Determination of the cobordism ring. Ann. of Math. **72**, 292–311 (1960)
108. Wall, C.T.C.: Surgery on compact manifolds. London Math. Soc. Monographs **1**. Academic Press, New York (1970); 2nd edn. (ed. A. A. Ranicki), Amer. Math. Soc. Surveys and monographs **69**, Amer. Math. Soc. (1999)

Multiplicative Arithmetic Functions of Several Variables: A Survey

László Tóth

Abstract We survey general properties of multiplicative arithmetic functions of several variables and related convolutions, including the Dirichlet convolution and the unitary convolution. We introduce and investigate a new convolution, called gcd convolution. We define and study the convolutes of arithmetic functions of several variables, according to the different types of convolutions. We discuss the multiple Dirichlet series and Bell series and present certain arithmetic and asymptotic results of some special multiplicative functions arising from problems in number theory, group theory, and combinatorics. We give a new proof to obtain the asymptotic density of the set of ordered r-tuples of positive integers with pairwise relatively prime components and consider a similar question related to unitary divisors.

Keywords Arithmetic function of several variables · Multiplicative function · Greatest common divisor · Least common multiple · Relatively prime integers · Unitary divisor · Arithmetic convolution · Dirichlet series · Mean value · Asymptotic density · Asymptotic formula

1 Introduction

Multiplicative arithmetic functions of a single variable are very well known in the literature. Their various properties were investigated by several authors and they represent an important research topic up to now. Less known are multiplicative arithmetic functions of several variables of which detailed study was carried out by R. VAIDYANATHASWAMY [73] more than 80 years ago. Since then many, sometimes scattered results for the several variables case were published in papers and monographs, and some authors of them were not aware of the paper [73]. In fact, there are two different notions of multiplicative functions of several variables, used in the last decades, both reducing to the usual multiplicativity in the one-variable case. For the other concept, we use the term firmly multiplicative function.

L. Tóth (✉)
Institute of Mathematics, Universität für Bodenkultur, Gregor Mendel-Straße 33, A-1180, Vienna, Austria

Department of Mathematics, University of Pécs, Ifjúság u. 6, H-7624 Pécs, Hungary
e-mail: ltoth@gamma.ttk.pte.hu

T. M. Rassias, P. M. Pardalos (eds.), *Mathematics Without Boundaries*,
DOI 10.1007/978-1-4939-1106-6_19, © Springer Science+Business Media, LLC 2014

In this chapter we survey general properties of multiplicative arithmetic functions of several variables and related convolutions, including the Dirichlet convolution and the unitary convolution. We introduce and investigate a new convolution, called gcd convolution. We define and study the convolutes of arithmetic functions of several variables, according to the different types of convolutions. The concept of the convolute of a function with respect to the Dirichlet convolution was introduced by R. VAIDYANATHASWAMY [73]. We also discuss the multiple Dirichlet series and Bell series. We present certain arithmetic and asymptotic results of some special multiplicative functions arising from problems in number theory, group theory, and combinatorics. We give a new proof to obtain the asymptotic density of the set of ordered r-tuples of positive integers with pairwise relatively prime components. Furthermore, we consider a similar question, namely the asymptotic density of the set of ordered r-tuples with pairwise unitary relatively prime components, that is, the greatest common unitary divisor of each two distinct components is 1.

For general properties of (multiplicative) arithmetic functions of a single variable see, e.g., the books of T. M. APOSTOL [4], G. H. HARDY, E. M. WRIGHT [27], P. J. MC.CARTHY [42], W. SCHWARZ, J. SPILKER [53], and R. SIVARAMAKRISHNAN [55]. For algebraic properties of the ring of arithmetic functions of a single variable with the Dirichlet convolution, we refer to H. N. SHAPIRO [54]. Incidence algebras and semilattice algebras concerning arithmetic functions of a single variable were investigated by D. A. SMITH [57]. For properties of certain subgroups of the group of multiplicative arithmetic functions of a single variable under the Dirichlet convolution, we refer to the papers by T. B. CARROLL, A. A. GIOIA [7], P.-O. DEHAYE [14], J. E. DELANY [17], T. MACHENRY [41], and R. W. RYDEN [48]. Algebraical and topological properties of the ring of arithmetic functions of a single variable with the unitary convolution were given by J. SNELLMAN [58, 59]. See also J. SÁNDOR, B. CRSTICI [50, Sect. 2.2] and H. SCHEID [51].

2 Notations

Throughout the paper we use the following notations.

General notations:

- $\mathbb{N} = \{1, 2, \dots\}$, $\mathbb{N}_0 = \{0, 1, 2, \dots\}$,
- the prime power factorization of $n \in \mathbb{N}$ is $n = \prod_p p^{\nu_p(n)}$, the product being over the primes p, where all but a finite number of the exponents $\nu_p(n)$ are zero,
- $d \,||\, n$ means that d is a unitary divisor of n, i.e., $d \mid n$ and $\gcd(d, n/d) = 1$,
- $\operatorname{gcud}(n_1, \dots, n_k)$ denotes the greatest common unitary divisor of $n_1, \dots, n_k \in \mathbb{N}$,
- $\mathbb{Z}_n = \mathbb{Z}/n\mathbb{Z}$ is the additive group of residue classes modulo n,
- ζ is the Riemann zeta function,
- γ is Euler's constant.

Arithmetic functions of a single variable:

- δ is the arithmetic function given by $\delta(1) = 0$ and $\delta(n) = 0$ for $n > 1$,
- id is the function $\operatorname{id}(n) = n$ ($n \in \mathbb{N}$),

- ϕ_k is the Jordan function of order k given by $\phi_k(n) = n^k \prod_{p|n} (1 - 1/p^k)$ $(k \in \mathbb{C})$,
- $\phi = \phi_1$ is Euler's totient function,
- ψ is the Dedekind function given by $\psi(n) = n \prod_{p|n} (1 + 1/p)$,
- μ is the Möbius function,
- τ_k is the Piltz divisor function of order k, $\tau_k(n)$ representing the number of ways of expressing n as a product of k factors,
- $\tau(n) = \tau_2(n)$ is the number of divisors of n,
- $\sigma_k(n) = \sum_{d|n} d^k$ $(k \in \mathbb{C})$,
- $\sigma(n) = \sigma_1(n)$ is the sum of divisors of n,
- $\omega(n) = \# \{p : \nu_p(n) \neq 0\}$ stands for the number of distinct prime divisors of n,
- $\mu^{\times}(n) = (-1)^{\omega(n)}$
- $\Omega(n) = \sum_p \nu_p(n)$ is the number of prime power divisors of n,
- $\lambda(n) = (-1)^{\Omega(n)}$ is the Liouville function,
- $\xi(n) = \prod_p \nu_p(n)!$,
- $c_n(k) = \sum_{1 \le q \le n, \gcd(q,n)=1} \exp(2\pi i q k/n)$ $(n, k \in \mathbb{N})$ is the Ramanujan sum, which can be viewed as a function of two variables.

Arithmetic functions of several variables:

- $\mathcal{A}_r$ is the set of arithmetic functions of r variables $(r \in \mathbb{N})$, i.e., of functions $f : \mathbb{N}^r \to \mathbb{C}$,
- $\mathcal{A}_r^{(1)} = \{f \in \mathcal{A}_r : f(1, \dots, 1) \neq 0\}$,
- $\mathbf{1}_r$ is the constant 1 function in $\mathcal{A}_r$, i.e., $\mathbf{1}_r(n_1, \dots, n_r) = 1$ for every $n_1, \dots, n_r \in \mathbb{N}$,
- $\delta_r(n_1, \dots, n_r) = \delta(n_1) \cdots \delta(n_r)$, that is $\delta_r(1, \dots, 1) = 1$ and $\delta_r(n_1, \dots, n_r) = 0$ for $n_1 \cdots n_r > 1$,
- for $f \in \mathcal{A}_r$ the function $\overline{f} \in \mathcal{A}_1$ is given by $\overline{f}(n) = f(n, \dots, n)$ for every $n \in \mathbb{N}$.

Other notations will be fixed inside the paper.

3 Multiplicative Functions of Several Variables

In what follows, we discuss the notions of multiplicative, firmly multiplicative, and completely multiplicative functions. We point out that properties of firmly and completely multiplicative functions of several variables reduce to those of multiplicative, respectively completely multiplicative functions of a single variable. Multiplicative functions can not be reduced to functions of a single variable. We also present examples of such functions.

3.1 Multiplicative Functions

A function $f \in \mathcal{A}_r$ is said to be multiplicative if it is not identically zero and

$$f(m_1 n_1, \dots, m_r n_r) = f(m_1, \dots, m_r) f(n_1, \dots, n_r)$$

holds for any $m_1,\dots,m_r,n_1,\dots,n_r \in \mathbb{N}$ such that $\gcd(m_1\cdots m_r, n_1\cdots n_r)=1$.

If f is multiplicative, then it is determined by the values $f(p^{\nu_1},\dots,p^{\nu_r})$, where p is prime and $\nu_1,\dots,\nu_r \in \mathbb{N}_0$. More exactly, $f(1,\dots,1)=1$ and for any $n_1,\dots,n_r \in \mathbb{N}$,

$$f(n_1,\dots,n_r)=\prod_p f\left(p^{\nu_p(n_1)},\dots,p^{\nu_p(n_r)}\right).$$

If $r=1$, i.e., in the case of functions of a single variable we reobtain the familiar notion of multiplicativity: $f \in \mathcal{A}_1$ is multiplicative if it is not identically zero and $f(mn)=f(m)f(n)$ for every $m,n \in \mathbb{N}$ such that $\gcd(m,n)=1$.

Let $\mathcal{M}_r$ denote the set of multiplicative functions in r variables.

3.2 *Firmly Multiplicative Functions*

We call a function $f \in \mathcal{A}_r$ firmly multiplicative (following P. Haukkanen [28]) if it is not identically zero and

$$f(m_1n_1,\dots,m_rn_r)=f(m_1,\dots,m_r)f(n_1,\dots,n_r)$$

holds for any $m_1,\dots,m_r,n_1,\dots,n_r \in \mathbb{N}$ such that $\gcd(m_1,n_1)=\dots=\gcd(m_r,n_r)=1$. Let $\mathcal{F}_r$ denote the set of firmly multiplicative functions in r variables.

A firmly multiplicative function is completely determined by its values at $(1,\dots,1,p^\nu,1,\dots,1)$, where p runs through the primes and $\nu \in \mathbb{N}_0$. More exactly, $f(1,\dots,1)=1$ and for any $n_1,\dots,n_r \in \mathbb{N}$,

$$f(n_1,\dots,n_r)=\prod_p \left(f(p^{\nu_p(n_1)},1,\dots,1)\cdots f(1,\dots,1,p^{\nu_p(n_r)})\right).$$

If a function $f \in \mathcal{A}_r$ is firmly multiplicative, then it is multiplicative. Also, if $f \in \mathcal{F}_r$, then $f(n_1,\dots,n_r)=f_1(n_1,1,\dots,1)\cdots f_r(1,\dots,1,n_r)$ for every $n_1,\dots,n_r \in \mathbb{N}$. This immediately gives the following property:

Proposition 1 *A function $f \in \mathcal{A}_r$ is firmly multiplicative if and only if there exist multiplicative functions $f_1,\dots,f_r \in \mathcal{M}_1$ (each of a single variable) such that $f(n_1,\dots,n_r)=f_1(n_1)\cdots f_r(n_r)$ for every $n_1,\dots,n_r \in \mathbb{N}$. In this case $f_1(n)=f(n,1,\dots,1)$, ..., $f_r(n)=f(1,\dots,1,n)$ for every $n \in \mathbb{N}$.*

In the case of functions of a single variable, the notion of firmly multiplicative function reduces to that of multiplicative function. For $r>1$ the concepts of multiplicative and firmly multiplicative functions are different.

3.3 Completely Multiplicative Functions

A function $f \in \mathcal{A}_r$ is called completely multiplicative if it is not identically zero and

$$f(m_1n_1, \dots, m_rn_r) = f(m_1, \dots, m_r)f(n_1, \dots, n_r)$$

holds for any $m_1, \dots, m_r, n_1, \dots, n_r \in \mathbb{N}$. Note that R. VAIDYANATHASWAMY [73] used for such a function the term "linear function."

Let $\mathcal{C}_r$ denote the set of completely multiplicative functions in r variables. If $f \in \mathcal{C}_r$, then it is determined by its values at $(1, \dots, 1, p, 1, \dots, 1)$, where p runs through the primes. More exactly, $f(1, \dots, 1) = 1$ and for any $n_1, \dots, n_r \in \mathbb{N}$,

$$f(n_1, \dots, n_r) = \prod_p \left(f(p, 1, \dots, 1)^{\nu_p(n_1)} \cdots f(1, \dots, 1, p)^{\nu_p(n_r)} \right).$$

In the case of functions of a single variable, we reobtain the familiar notion of completely multiplicative function: $f \in \mathcal{A}_1$ is completely multiplicative if it is not identically zero and $f(mn) = f(m)f(n)$ for every $m, n \in \mathbb{N}$.

It is clear that if a function $f \in \mathcal{A}_r$ is completely multiplicative, then it is firmly multiplicative. Also, similar to Proposition 1:

Proposition 2 *A function $f \in \mathcal{A}_r$ is completely multiplicative if and only if there exist completely multiplicative functions $f_1, \dots, f_r \in \mathcal{C}_1$ (each of a single variable) such that $f(n_1, \dots, n_r) = f_1(n_1) \cdots f_r(n_r)$ for every $n_1, \dots, n_r \in \mathbb{N}$. In this case $f_1(n) = f(n, 1, \dots, 1), \dots, f_r(n) = f(1, \dots, 1, n)$ for every $n \in \mathbb{N}$.*

3.4 Examples

The functions $(n_1, \dots, n_r) \mapsto \gcd(n_1, \dots, n_r)$ and $(n_1, \dots, n_r) \mapsto \operatorname{lcm}(n_1, \dots, n_r)$ are multiplicative for every $r \in \mathbb{N}$, but not firmly multiplicative for $r \geq 2$.

The functions $(n_1, \dots, n_r) \mapsto \tau(n_1) \cdots \tau(n_r)$, $(n_1, n_2) \mapsto \tau(n_1)\sigma(n_2)$ are firmly multiplicative, but not completely multiplicative.

The functions $(n_1, \dots, n_r) \mapsto n_1 \cdots n_r$, $(n_1, n_2) \mapsto n_1\lambda(n_2)$ are completely multiplicative.

According to Propositions 1 and 2, firmly multiplicative and completely multiplicative functions reduce to multiplicative, respectively completely multiplicative functions of a single variable. There is no similar characterization for multiplicative functions of several variables.

Let $h \in \mathcal{M}_1$. Then the functions $(n_1, \dots, n_r) \mapsto h(\gcd(n_1, \dots, n_r))$, $(n_1, \dots, n_r) \mapsto h(\operatorname{lcm}(n_1, \dots, n_r))$ are multiplicative. The product and the quotient of (nonvanishing) multiplicative functions are multiplicative.

If $f \in \mathcal{M}_r$ is multiplicative and we fix one (or more, say s) variables, then the resulting function of $r - 1$ (or $r - s$) variables is not necessary multiplicative. For example, $(k, n) \mapsto c_n(k)$ is multiplicative as a function of two variables, see Sect. 4.1, but for a fixed n, the function $k \mapsto c_n(k)$ is in general, not multiplicative (it is multiplicative if and only if $\mu(n) = 1$).

If $f \in \mathcal{M}_r$ is multiplicative, then the function $\overline{f}$ of a single variable is multiplicative.

Other examples from number theory, group theory, and combinatorics:

Example 1 Let $N_{g_1,\ldots,g_r}(n_1,\ldots,n_r)$ denote the number of solutions x (mod n), with $n = \operatorname{lcm}(n_1,\ldots,n_r)$, of the simultaneous congruences $g_1(x) \equiv 0 \pmod{n_1}, \ldots, g_r(x) \equiv 0 \pmod{n_r}$, where $g_1,\ldots,g_r$ are polynomials with integer coefficients. Then the function $(n_1,\ldots,n_r) \mapsto N_{g_1,\ldots,g_r}(n_1,\ldots,n_r)$ is multiplicative. See L. Tóth [64, Sect. 2] for a proof.

Example 2 For a fixed integer $r \geq 2$, let

$$\rho(n_1,\ldots,n_r) = \begin{cases} 1, \text{if } n_1,\ldots,n_r \text{ are pairwise relatively prime,} \\ 0, \text{otherwise.} \end{cases} \tag{1}$$

This function is multiplicative, which follows from the definition, and for every $n_1,\ldots,n_r \in \mathbb{N}$,

$$\rho(n_1,\ldots,n_r) = \sum_{d_1|n_1,\ldots,d_r|n_r} \tau(d_1\cdots d_r)\mu(n_1/d_1)\cdots\mu(n_r/d_r), \tag{2}$$

cf. Sect. 4.1.

Example 3 For $r \geq 2$ let

$$\rho^{\times}(n_1,\ldots,n_r) = \begin{cases} 1, \text{if } \operatorname{gcud}(n_i,n_j) = 1 \text{ for every } i \neq j, \\ 0, \text{otherwise.} \end{cases} \tag{3}$$

This is the characteristic function of the set of ordered r-tuples $(n_1,\ldots,n_r) \in \mathbb{N}^r$ such that $n_1,\ldots,n_r$ are pairwise unitary relatively prime, i.e., for every prime p, there are no $i \neq j$ with $\nu_p(n_i) = \nu_p(n_j) \geq 1$. This function is also multiplicative (by the definition).

Example 4 Consider the group $G = \mathbb{Z}_{n_1} \times \cdots \times \mathbb{Z}_{n_r}$. Let $s(n_1,\ldots,n_r)$ and $c(n_1,\ldots,n_r)$ denote the total number of subgroups of the group G and the number of its cyclic subgroups, respectively. Then the functions $(n_1,\ldots,n_r) \mapsto s(n_1,\ldots,n_r)$ and $(n_1,\ldots,n_r) \mapsto c(n_1,\ldots,n_r)$ are multiplicative. For every $n_1,\ldots,n_r \in \mathbb{N}$,

$$c(n_1,\ldots,n_r) = \sum_{d_1|n_1,\ldots,d_r|n_r} \frac{\phi(d_1)\cdots\phi(d_r)}{\phi(\operatorname{lcm}(d_1,\ldots,d_r))}, \tag{4}$$

see L. Tóth [64, Theorem 3], [65, Theorem 1]. In the case $r = 2$ this gives

$$c(n_1,n_2) = \sum_{d_1|n_1,d_2|n_2} \phi(\gcd(d_1,d_2)).$$

Also,

$$s(n_1, n_2) = \sum_{d_1 \mid n_1, d_2 \mid n_2} \gcd(d_1, d_2), \tag{5}$$

for every $n_1, n_2 \in \mathbb{N}$. See M. HAMPEJS, N. HOLIGHAUS, L. TÓTH, C. WIESMEYR [26] and M. HAMPEJS, L. TÓTH [25].

Example 5 We define the sigma function of r variables by

$$\sigma(n_1, \ldots, n_r) = \sum_{d_1 \mid n_1, \ldots, d_r \mid n_r} \gcd(d_1, \ldots, d_r) \tag{6}$$

having the representation

$$\sigma(n_1, \ldots, n_r) = \sum_{d \mid \gcd(n_1, \ldots, n_r)} \phi(d) \tau(n_1/d) \cdots \tau(n_r/d), \tag{7}$$

valid for every $n_1, \ldots, n_r \in \mathbb{N}$. Note that for $r = 1$, this function reduces to the sum-of-divisors function and in the case $r = 2$, we have $\sigma(m, n) = s(m, n)$ given in Example 4.

We call $(n_1, \ldots, n_r) \in \mathbb{N}^r$ a perfect r-tuple if $\sigma(n_1, \ldots, n_r) = 2\gcd(n_1, \ldots, n_r)$. If $2^r - 1 = p$ is a Mersenne prime, then $(p, p, \ldots, p)$ is a perfect r-tuple. For example, $(3, 3)$ is a perfect pair and $(7, 7, 7)$ is perfect triple. We formulate as an open problem: Which are all the perfect r-tuples?

Example 6 The Ramanujan sum $(k, n) \mapsto c_n(k)$ having the representation

$$c_n(k) = \sum_{d \mid \gcd(k,n)} d\mu(n/d) \tag{8}$$

is multiplicative as a function of two variables. This property was pointed out by K. R. JOHNSON [34], see also Sect. 4.1.

Example 7 For $n_1, \ldots, n_r \in \mathbb{N}$ let $n := \operatorname{lcm}(n_1, \ldots, n_r)$. The function of r variables

$$E(n_1, \ldots, n_r) = \frac{1}{n} \sum_{j=1}^{n} c_{n_1}(j) \cdots c_{n_r}(j)$$

has combinatorial and topological applications, and was investigated in the papers of V. A. LISKOVETS [40] and L. TÓTH [66]. All values of $E(n_1, \ldots, n_r)$ are nonnegative integers and the function E is multiplicative. Furthermore, it has the following representation ([66, Proposition 3]):

$$E(n_1, \ldots, n_r) = \sum_{d_1 \mid n_1, \ldots, d_r \mid n_r} \frac{d_1 \mu(n_1/d_1) \cdots d_r \mu(n_r/d_r)}{\operatorname{lcm}(d_1, \ldots, d_r)}, \tag{9}$$

valid for every $n_1, \ldots, n_r \in \mathbb{N}$. See also L. TÓTH [67, 70] for generalizations of the function E.

Example 8 Another multiplicative function, similar to E is

$$A(n_1,\dots,n_r)=\frac{1}{n}\sum_{k=1}^{n}\gcd(k,n_1)\cdots\gcd(k,n_r),$$

where $n_1,\dots,n_r\in\mathbb{N}$ and $n:=\operatorname{lcm}(n_1,\dots,n_r)$, as above. One has for every $n_1,\dots,n_r\in\mathbb{N}$,

$$A(n_1,\dots,n_r)=\sum_{d_1|n_1,\dots,d_r|n_r}\frac{\phi(d_1)\cdots\phi(d_r)}{\operatorname{lcm}(d_1,\dots,d_r)}, \tag{10}$$

see L. Tóth [63, Eq. (45)], [66, Proposition 12].

See the paper of M. Peter [47] for properties of recurrent multiplicative arithmetical functions of several variables.

4 Convolutions of Arithmetic Functions of Several Variables

In this section, we survey the basic properties of the Dirichlet and unitary convolutions of arithmetic functions of several variables. We also define and discuss the gcd, lcm, and binomial convolutions, not given in the literature in the several variables case. We point out that the gcd convolution reduces to the unitary convolution in the one variable case, but they are different for r variables with $r>1$. For other convolutions, we refer to the papers of J. Sándor, A. Bege [49], E. D. Schwab [52], and M. V. Subbarao [60].

4.1 Dirichlet Convolution

For every $r\in\mathbb{N}$, the set $\mathcal{A}_r$ of arithmetic functions of r variables is a $\mathbb{C}$-linear space with the usual linear operations. With the Dirichlet convolution defined by

$$(f*g)(n_1,\dots,n_r)=\sum_{d_1|n_1,\dots,d_r|n_r} f(d_1,\dots,d_r)g(n_1/d_r,\dots,n_r/d_r)$$

the space $\mathcal{A}_r$ forms a unital commutative $\mathbb{C}$-algebra, the unity being the function δ_r, and $(\mathcal{A}_r,+,*)$ is an integral domain. Moreover, $(\mathcal{A}_r,+,*)$ is a unique factorization domain, as pointed out by T. Onozuka [45]. In the case $r=1$, this was proved by E. D. Cashwell, C. J. Everett [9]. The group of invertible functions is $\mathcal{A}_r^{(1)}$. The inverse of f will be denoted by f^{-1*}. The inverse of the constant 1 function $\mathbf{1}_r$ is μ_r, given by $\mu_r(n_1,\dots,n_r)=\mu(n_1)\cdots\mu(n_r)$ (which is firmly multiplicative, where μ is the classical Möbius function).

The Dirichlet convolution preserves the multiplicativity of functions. This property, well known in the one variable case, follows easily from the definitions. Using this fact, the multiplicativity of the functions $c(n_1, \ldots, n_r)$, $s(n_1, n_2)$, $\sigma(n_1, \ldots, n_r)$, $E(n_1, \ldots, n_r)$, and $A(n_1, \ldots, n_r)$ is a direct consequence of the convolutional representations (4), (5), (6), (9) and (10), respectively. The multiplicativity of the Ramanujan sum $(k,n) \mapsto c_n(k)$ follows in a similar manner from (8), showing that $c_n(k)$ is the convolution of the multiplicative functions f and g defined by $f(m,n) = m$ for $m = n$, $f(m,n) = 0$ for $m \neq n$ and $g(m,n) = \mu(m)$ for every $m, n \in \mathbb{N}$.

If $f, g \in \mathcal{F}_r$, then

$$(f * g)(n_1, \ldots, n_r) = (f_1 * g_1)(n_1) \cdots (f_r * g_r)(n_r)$$

and

$$f^{-1*}(n_1, \ldots, n_r) = f_1^{-1*}(n_1) \cdots f_r^{-1*}(n_r),$$

with the notations of Proposition 1, hence $f * g \in \mathcal{F}_r$ and $f^{-1*} \in \mathcal{F}_r$. We deduce

Proposition 3 *One has the following subgroup relations:*

$$(\mathcal{F}_r, *) \leq (\mathcal{M}_r, *) \leq \left(\mathcal{A}_r^{(1)}, *\right).$$

The set $\mathcal{C}_r$ does not form a group under the Dirchlet convolution. If $f \in \mathcal{C}_r$, then $f^{-1*} = \mu_r f$ (well known in the case $r = 1$). Note that for every $f \in \mathcal{C}_r$, one has $(f * f)(n_1, \ldots, n_r) = f(n_1, \ldots, n_r)\tau(n_1) \cdots \tau(n_r)$, in particular $(\mathbf{1}_r * \mathbf{1}_r)(n_1, \ldots, n_r) = \tau(n_1) \cdots \tau(n_r)$.

R. Vaidyanathaswamy [73] called the Dirichlet convolution "composition of functions."

Other convolutional properties known in the one-variable case, for example Möbius inversion, can easily be generalized. As mentioned in Sect. 3.4, Example 2, the characteristic function ρ of the set of ordered r-tuples $(n_1, \ldots, n_r) \in \mathbb{N}^r$ such that $n_1, \ldots, n_r \in \mathbb{N}$ are pairwise relatively prime is multiplicative. One has for every $n_1, \ldots, n_r \in \mathbb{N}$,

$$\sum_{d_1 | n_1, \ldots, d_r | n_r} \rho(d_1, \ldots, d_r) = \tau(n_1 \cdots n_r), \tag{11}$$

since both sides are multiplicative and in the case of prime powers $n_1 = p^{\nu_1}, \ldots, n_r = p^{\nu_r}$ both sides of (11) are equal to $1 + \nu_1 + \ldots + \nu_r$. Now, Möbius inversion gives the formula (2).

For further algebraic properties of the $\mathbb{C}$-algebra $\mathcal{A}_r$ and more generally, of the R-algebra $A_r(R) = \{f : \mathbb{N}^r \mapsto R\}$, where R is an integral domain and using the concept of firmly multiplicative functions, see E. Alkan, A. Zaharescu, M. Zaki [1]. That paper includes, among others, constructions of a class of derivations and of a family of valuations on $A_r(R)$. See also P. Haukkanen [28] and A. Zaharescu, M. Zaki [75].

4.2 Unitary Convolution

The linear space $\mathcal{A}_r$ forms another unital commutative $\mathbb{C}$-algebra with the unitary convolution defined by

$$(f \times g)(n_1,\ldots,n_r) = \sum_{d_1||n_1,\ldots,d_r||n_r} f(d_1,\ldots,d_r)g(n_1/d_1,\ldots,n_r/d_r).$$

Here the unity is the function δ_r again. Note that $(\mathcal{A}_r,+,\times)$ is not an integral domain, there exist divisors of zero. The group of invertible functions is again $\mathcal{A}_r^{(1)}$. The unitary r variables Möbius function $\mu_r^{\times}$ is defined as the inverse of the function $\mathbf{1}_r$. One has $\mu_r^{\times}(n_1,\ldots,n_r) = \mu^{\times}(n_1)\cdots\mu^{\times}(n_r) = (-1)^{\omega(n_1)+\ldots+\omega(n_r)}$ (and it is firmly multiplicative). Similar to Proposition 3,

Proposition 4 *One has the following subgroup relations:*

$$(\mathcal{F}_r,\times) \le (\mathcal{M}_r,\times) \le (\mathcal{A}_r^{(1)},\times).$$

If $f \in \mathcal{F}_r$, then its inverse is $f^{-1\times} = \mu_r^{\times} f$ and

$$(f \times f)(n_1,\ldots,n_r) = f(n_1,\ldots,n_r)2^{\omega(n_1)+\ldots+\omega(n_r)},$$

in particular $(\mathbf{1}_r \times \mathbf{1}_r)(n_1,\ldots,n_r) = 2^{\omega(n_1)+\ldots+\omega(n_r)}$.

R. Vaidyanathaswamy [73] used the term "compounding of functions" for the unitary convolution.

For further algebraic properties of the R-algebra $A_r(R) = \{f : \mathbb{N}^r \mapsto R\}$, where R is an integral domain with respect to the unitary convolution and using the concept of firmly multiplicative functions, see E. Alkan, A. Zaharescu, M. Zaki [2].

4.3 Gcd Convolution

We define a new convolution for functions $f, g \in \mathcal{A}_r$, we call it gcd convolution, given by

$$(f \odot g)(n_1,\ldots,n_r) = \sum_{\substack{d_1e_1=n_1,\ldots,d_re_r=n_r \\ \gcd(d_1\cdots d_r, e_1\cdots e_r)=1}} f(d_1,\ldots,d_r)g(e_1,\ldots,e_r), \qquad (12)$$

which is in concordance with the definition of multiplicative functions.

In the case $r=1$, the unitary and gcd convolutions are identic, i.e., $f \times g = f \odot g$ for every $f, g \in \mathcal{A}_1$, but they differ for $r > 1$.

Main properties: $\mathcal{A}_r$ forms a unital commutative $\mathbb{C}$-algebra with the gcd convolution defined by (12). The unity is the function δ_r and there exist divisors of zero. The group of invertible functions is again $\mathcal{A}_r^{(1)}$. Here the inverse of the constant 1 function is $\mu_r^{\odot}(n_1,\ldots,n_r) = (-1)^{\omega(n_1\cdots n_r)}$. More generally, the inverse

$f^{-1\odot}$ of an arbitrary multiplicative function f is given by $f^{-1\odot}(n_1,\ldots,n_r) = (-1)^{\omega(n_1\cdots n_r)} f(n_1,\ldots,n_r)$. Also, if $f \in \mathcal{M}_r$, then

$$(f \odot f)(n_1,\ldots,n_r) = f(n_1,\ldots,n_r) 2^{\omega(n_1\cdots n_r)}.$$

Proposition 5 *One has*

$$(\mathcal{M}_r,\odot) \leq (\mathcal{A}_r^{(1)},\odot).$$

The set $\mathcal{F}_r$ does not form a group under the gcd convolution. To see this, note that $(\mathbf{1}_r \odot \mathbf{1}_r)(n_1,\ldots,n_r) = 2^{\omega(n_1\cdots n_r)}$, but this function is not firmly multiplicative, since $\omega(n_1\cdots n_r) = \omega(n_1) + \ldots + \omega(n_r)$ does not hold for every $n_1,\ldots,n_r \in \mathbb{N}$ (cf. Proposition 1).

4.4 Lcm Convolution

We define the lcm convolution of functions of r variables by

$$(f \oplus g)(n_1,\ldots,n_r) = \sum_{\operatorname{lcm}(d_1,e_1)=n_1,\ldots,\operatorname{lcm}(d_r,e_r)=n_r} f(d_1,\ldots,d_r) g(e_1,\ldots,e_r).$$

In the case $r = 1$, this convolution originates by R. D. von Sterneck [74], was investigated and generalized by D. H. Lehmer [37–39], and is also called von Sterneck-Lehmer convolution. See also R. G. Buschman [5].

Note that the lcm convolution can be expressed by the Dirichlet convolution. More exactly,

Proposition 6 *For every $f,g \in \mathcal{A}_r$,*

$$f \oplus g = (f * \mathbf{1}_r)(g * \mathbf{1}_r) * \mu_r. \tag{13}$$

Proof Write

$$\begin{aligned}
&\sum_{a_1|n_1,\ldots,a_r|n_r} (f \oplus g)(a_1,\ldots,a_r) \\
&= \sum_{\operatorname{lcm}(d_1,e_1)|n_1,\ldots,\operatorname{lcm}(d_r,e_r)|n_r} f(d_1,\ldots,d_r) g(e_1,\ldots,e_r) \\
&= \sum_{d_1|n_1,\ldots,d_r|n_r} f(d_1,\ldots,d_r) \sum_{e_1|n_1,\ldots,e_r|n_r} g(e_1,\ldots,e_r) \\
&= (f * \mathbf{1}_r)(n_1,\ldots,n_r)(g * \mathbf{1}_r)(n_1,\ldots,n_r),
\end{aligned}$$

and by Möbius inversion we obtain (13). □

In the case $r = 1$, Proposition 6 is due to R. D. VON STERNECK [74] and D. H. LEHMER [38].

Note that $(\mathbf{1}_r \oplus \mathbf{1}_r)(n_1, \ldots, n_r) = \tau(n_1^2) \cdots \tau(n_r^2)$. It turns out that the lcm convolution preserves the multiplicativity of functions, but $\mathcal{M}_r$ does not form a group under the lcm convolution. The unity for the lcm convolution is the function δ_r again. Here $(\mathcal{A}_r, +, \oplus)$ is a unital commutative ring having divisors of zero. The group of invertible functions is $\widetilde{\mathcal{A}}_r = \{f \in \mathcal{A}_r : (f * \mathbf{1}_r)(n_1, \ldots, n_r) \neq 0 \text{for every} (n_1, \ldots, n_r) \in \mathbb{N}^r\}$ and we deduce

Proposition 7

i) Let $\widetilde{\mathcal{M}}_r = \{f \in \mathcal{M}_r : (f * \mathbf{1}_r)(n_1, \ldots, n_r) \neq 0$ for every $(n_1, \ldots n_r) \in \mathbb{N}^r\}$. Then $\widetilde{\mathcal{M}}_r$ is a subgroup of $\widetilde{\mathcal{A}}_r$ with respect to the lcm convolution.

ii) The inverse of the function $\mathbf{1}_r \in \widetilde{\mathcal{A}}_r$ is $\mu_r^{\oplus} = \mu_r * 1/(\mathbf{1}_r * \mathbf{1}_r)$. The function $\mu_r^{\oplus}$ is multiplicative and for every prime powers $p^{\nu_1}, \ldots, p^{\nu_r}$,

$$\mu_r^{\oplus}(p^{\nu_1}, \ldots, p^{\nu_r}) = \frac{(-1)^r}{\nu_1(\nu_1 + 1) \cdots \nu_r(\nu_r + 1)}.$$

Proof Here (ii) follows from $(\mathbf{1}_r * \mathbf{1}_r)(n_1, \ldots, n_r) = \tau(n_1) \cdots \tau(n_r)$, already mentioned in Sect. 4.1. □

4.5 Binomial Convolution

We define the binomial convolution of the functions $f, g \in \mathcal{A}_r$ by

$$(f \circ g)(n_1, \ldots, n_r)$$

$$= \sum_{d_1 | n_1, \ldots d_r | n_r} \left(\prod_p \binom{\nu_p(n_1)}{\nu_p(d_1)} \cdots \binom{\nu_p(n_r)}{\nu_p(d_r)} \right) f(d_1, \ldots, d_r) g(n_1/d_1, \ldots, n_r/d_r),$$

where $\binom{a}{b}$ is the binomial coefficient. It is remarkable that the binomial convolution preserves the complete multiplicativity of arithmetical functions, which is not the case for the Dirichlet convolution and other convolutions. Let ξ_r be the firmly multiplicative function given by $\xi_r(n_1, \ldots, n_r) = \xi(n_1) \cdots \xi(n_r)$, that is, $\xi_r(n_1, \ldots, n_r) = \prod_p (\nu_p(n_1)! \cdots \nu_p(n_r)!)$. Then for every $f, g \in \mathcal{A}_r$,

$$f \circ g = \xi_r \left(\frac{f}{\xi_r} * \frac{g}{\xi_r} \right), \tag{14}$$

leading to the next result.

Proposition 8 *The algebras $(\mathcal{A}_r, +, \circ, \mathbb{C})$ and $(\mathcal{A}_r, +, *, \mathbb{C})$ are isomorphic under the mapping $f \mapsto \frac{f}{\xi_r}$.*

Formula (14) also shows that the binomial convolution preserves the multiplicativity of functions. Furthermore, for any fixed $r \in \mathbb{N}$, the structure $(\mathcal{A}_r, +, \circ)$ is an integral domain with unity δ_r. The group of invertible functions is again $\mathcal{A}_r^{(1)}$. If $f, g \in \mathcal{F}_r$, then with the notations of Proposition 1,

$$(f \circ g)(n_1, \dots, n_r) = (f_1 \circ g_1)(n_1) \cdots (f_r \circ g_r)(n_r)$$

and the inverse of f is

$$f^{-1\circ}(n_1, \dots, n_r) = f_1^{-1\circ}(n_1) \cdots f_r^{-1\circ}(n_r),$$

hence $f \circ g \in \mathcal{F}_r$ and $f^{-1\circ} \in \mathcal{F}_r$. The inverse of the function $\mathbf{1}_r$ under the binomial convolution is the function λ_r given by $\lambda_r(n) = \lambda(n_1) \cdots \lambda(n_r)$, i.e., $\lambda_r(n) = (-1)^{\Omega(n_1)+\dots+\Omega(n_r)}$. We deduce

Proposition 9 *One has the following subgroup relations:*

$$(\mathcal{C}_r, \circ) \leq (\mathcal{F}_r, \circ) \leq (\mathcal{M}_r, \circ) \leq \left(\mathcal{A}_r^{(1)}, \circ\right).$$

In the case $r = 1$, properties of this convolution are discussed in the paper by L. Tóth, P. Haukkanen [71]. The proofs are similar in the multivariable case.

5 Generating Series

As generating series for multiplicative arithmetic functions of r variables, we present certain properties of the multiple Dirichlet series, used earlier by several authors and the Bell series, which constituted an important tool of R. Vaidyanathaswamy [73].

5.1 Dirichlet Series

The multiple Dirichlet series of a function $f \in \mathcal{A}_r$ is given by

$$D(f; z_1, \dots, z_r) = \sum_{n_1, \dots, n_r = 1}^{\infty} \frac{f(n_1, \dots, n_r)}{n_1^{z_1} \cdots n_r^{z_r}}.$$

Similar to the one variable case, if $D(f; z_1, \dots, z_r)$ is absolutely convergent in $(s_1, \dots, s_r) \in \mathbb{C}^r$, then it is absolutely convergent in every $(z_1, \dots, z_r) \in \mathbb{C}^r$ with $\Re z_j \geq \Re s_j$ $(1 \leq j \leq r)$.

Proposition 10 *Let $f, g \in \mathcal{A}_r$. If $D(f; z_1, \dots, z_r)$ and $D(g; z_1, \dots, z_r)$ are absolutely convergent, then $D(f * g; z_1, \dots, z_r)$ is also absolutely convergent and*

$$D(f * g; z_1, \dots, z_r) = D(f; z_1, \dots, z_r) D(g; z_1, \dots, z_r).$$

Also, if $f \in \mathcal{A}_r^{(1)}$, then

$$D\left(f^{-1_*}; z_1, \ldots, z_r\right) = D\left(f; z_1, \ldots, z_r\right)^{-1},$$

formally or in the case of absolute convergence.

If $f \in \mathcal{M}_r$ is multiplicative, then its Dirichlet series can be expanded into a (formal) Euler product, that is,

$$D\left(f; z_1, \ldots, z_r\right) = \prod_p \sum_{\nu_1,\ldots,\nu_r=0}^{\infty} \frac{f(p^{\nu_1}, \ldots, p^{\nu_r})}{p^{\nu_1 z_1 + \ldots + \nu_r z_r}}, \tag{15}$$

the product being over the primes p. More exactly,

Proposition 11 *Let $f \in \mathcal{M}_r$. For every $(z_1, \ldots, z_r) \in \mathbb{C}^r$, the series $D(f; z_1, \ldots, z_r)$ is absolutely convergent if and only if*

$$\prod_p \sum_{\substack{\nu_1,\ldots,\nu_r=0 \\ \nu_1+\ldots+\nu_r \geq 1}}^{\infty} \frac{|f(p^{\nu_1}, \ldots, p^{\nu_r})|}{p^{\nu_1 \Re z_1 + \ldots + \nu_r \Re z_r}} < \infty$$

and in this case the equality (15) *holds.*

Next we give the Dirichlet series representations of certain special functions discussed above. These follow from the convolutional identities given in Sect. 3.4. For every $g \in \mathcal{A}_1$ we have formally,

$$\sum_{n_1,\ldots,n_r=1}^{\infty} \frac{g(\gcd(n_1, \ldots, n_r))}{n_1^{z_1} \cdots n_r^{z_r}} = \frac{\zeta(z_1) \cdots \zeta(z_r)}{\zeta(z_1 + \ldots + z_r)} \sum_{n=1}^{\infty} \frac{g(n)}{n^{z_1+\ldots+z_r}}. \tag{16}$$

In particular, taking $g(n) = n$ (16) gives for $r \geq 2$ and $\Re z_1 > 1, \ldots, \Re z_r > 1$,

$$\sum_{n_1,\ldots,n_r=1}^{\infty} \frac{\gcd(n_1, \ldots, n_r)}{n_1^{z_1} \cdots n_r^{z_r}} = \frac{\zeta(z_1) \cdots \zeta(z_r) \zeta(z_1 + \ldots + z_r - 1)}{\zeta(z_1 + \ldots + z_r)}$$

and taking $g = \delta$, one obtains for $r \geq 2$ and $\Re z_1 > 1, \ldots, \Re z_r > 1$,

$$\sum_{\substack{n_1,\ldots,n_r=1 \\ \gcd(n_1,\ldots,n_r)=1}}^{\infty} \frac{1}{n_1^{z_1} \cdots n_r^{z_r}} = \frac{\zeta(z_1) \cdots \zeta(z_r)}{\zeta(z_1 + \ldots + z_r)}, \tag{17}$$

where the identity (17) is the Dirichlet series of the characteristic function of the set of points in $\mathbb{N}^r$, which are visible from the origin (cf. T. M. Apostol [4, p. 248, Exercise 15]).

The Dirichlet series of the characteristic function ρ concerning r pairwise relatively prime integers is

$$\sum_{n_1,\ldots,n_r=1}^{\infty} \frac{\rho(n_1, \ldots, n_r)}{n_1^{z_1} \cdots n_r^{z_r}}$$

$$= \zeta(z_1)\cdots\zeta(z_r)\prod_p \left(\prod_{j=1}^r \left(1-\frac{1}{p^{z_j}}\right) + \sum_{j=1}^r \frac{1}{p^{z_j}} \prod_{\substack{k=1\\k\neq j}}^r \left(1-\frac{1}{p^{z_k}}\right) \right) \tag{18}$$

$$= \zeta(z_1)\cdots\zeta(z_r)\prod_p \left(1+\sum_{j=2}^r (-1)^{j-1}(j-1) \sum_{1\le i_1<\ldots<i_j\le r} \frac{1}{p^{z_{i_1}+\ldots+z_{i_j}}}\right), \tag{19}$$

valid for $\Re z_1 > 1, \ldots, \Re z_r > 1$. Here (18) follows at once from the definition (1) of the function ρ. For (19), see L. TÓTH [68, Eq. (4.2)]. Note that in the paper [68] certain other Dirichlet series representations are also given, leading to generalizations of the Busche–Ramanujan identities.

Concerning the Ramanujan sum $c_n(k)$ and the functions $s(m,n)$ and $c(m,n)$, cf. Section 3.4, Example 4 we have for $\Re z > 1, \Re w > 1$,

$$\sum_{k,n=1}^{\infty} \frac{c_n(k)}{k^z n^w} = \frac{\zeta(w)\zeta(z+w-1)}{\zeta(z)},$$

$$\sum_{m,n=1}^{\infty} \frac{s(m,n)}{m^z n^w} = \frac{\zeta^2(z)\zeta^2(w)\zeta(z+w-1)}{\zeta(z+w)}, \tag{20}$$

$$\sum_{m,n=1}^{\infty} \frac{c(m,n)}{m^z n^w} = \frac{\zeta^2(z)\zeta^2(w)\zeta(z+w-1)}{\zeta^2(z+w)}. \tag{21}$$

The formulae (20) and (21) were derived by W. G. NOWAK, L. TÓTH [44]. Also, for $\Re z > 2, \Re w > 2$,

$$\sum_{m,n=1}^{\infty} \frac{\operatorname{lcm}(m,n)}{m^z n^w} = \frac{\zeta(z-1)\zeta(w-1)\zeta(z+w-1)}{\zeta(z+w-2)}.$$

As a generalization of (20), for $\Re z_1 > 1, \ldots, \Re z_r > 1$,

$$\sum_{n_1,\ldots,n_r=1}^{\infty} \frac{\sigma(n_1,\ldots,n_r)}{n_1^{z_1}\cdots n_r^{z_r}} = \frac{\zeta^2(z_1)\cdots\zeta^2(z_r)\zeta(z_1+\ldots+z_r-1)}{\zeta(z_1+\ldots+z_r)}.$$

5.2 Bell Series

If f is a multiplicative function of r variables, then its (formal) Bell series to the base p (p prime) is defined by

$$f_{(p)}(x_1,\ldots,x_r) = \sum_{e_1,\ldots,e_r=0}^{\infty} f\left(p^{e_1},\ldots,p^{e_r}\right) x_1^{e_1}\cdots x_r^{e_r},$$

where the constant term is 1. The main property is the following: for every $f, g \in \mathcal{M}_r$,

$$(f * g)_{(p)}(x_1, \ldots, x_r) = f_{(p)}(x_1, \ldots, x_r) g_{(p)}(x_1, \ldots, x_r).$$

The connection of Bell series to Dirichlet series and Euler products is given by

$$D(f; z_1, \ldots, z_r) = \prod_p f_{(p)}\left(p^{-z_1}, \ldots, p^{-z_r}\right), \tag{22}$$

valid for every $f \in \mathcal{M}_r$. For example, the Bell series of the gcd function $f(n_1, \ldots, n_r) = \gcd(n_1, \ldots, n_r)$ is

$$f_{(p)}(x_1, \ldots, x_r) = \frac{1 - x_1 \cdots x_r}{(1 - x_1) \cdots (1 - x_r)(1 - p x_1 \cdots x_r)}.$$

The Bell series of other multiplicative functions, in particular of $c(m, n)$, $s(m, n)$, $\sigma(n_1, \ldots, n_r)$, and $c_n(k)$ can be given from their Dirichlet series representations and using the relation (22).

Note that in the one variable case, the Bell series to a fixed prime of the unitary convolution of two multiplicative functions is the sum of the Bell series of the functions, that is

$$(f \times g)_{(p)}(x_1) = f_{(p)}(x_1) + g_{(p)}(x_1),$$

see R. Vaidyanathaswamy [73, Theorem XII]. This is not valid in the case of r variables with $r > 1$.

6 Convolutes of Arithmetic Functions of Several Variables

Let $f \in \mathcal{A}_r$. By choosing $n_1 = \ldots = n_r = n$ we obtain the function of a single variable $n \mapsto \overline{f}(n) = f(n, \ldots, n)$. If $f \in \mathcal{M}_r$, then $\overline{f} \in \mathcal{M}_1$, as already mentioned. Less trivial ways to retrieve from f functions of a single variable is to consider for $r > 1$,

$$\Psi_{\text{dir}}(f)(n) = \sum_{d_1 \cdots d_r = n} f(d_1, \ldots, d_r), \tag{23}$$

$$\Psi_{\text{unit}}(f)(n) = \sum_{\substack{d_1 \cdots d_r = n \\ \gcd(d_i, d_j) = 1, i \neq j}} f(d_1, \ldots, d_r), \tag{24}$$

$$\Psi_{\text{gcd}}(f)(n) = \sum_{\substack{d_1 \cdots d_r = n \\ \gcd(d_1, \ldots, d_r) = 1}} f(d_1, \ldots, d_r), \tag{25}$$

$$\Psi_{\text{lcm}}(f)(n) = \sum_{\text{lcm}(d_1, \ldots, d_r) = n} f(d_1, \ldots, d_r), \tag{26}$$

$$\Psi_{\text{binom}}(f)(n) = \sum_{d_1\cdots d_r=n} \left(\prod_p \binom{\nu_p(n)}{\nu_p(d_1),\ldots,\nu_p(d_r)} \right) f(d_1,\ldots,d_r), \tag{27}$$

where the sums are over all ordered r-tuples $(d_1,\ldots,d_r) \in \mathbb{N}^r$ with the given additional conditions, the last one involving multinomial coefficients. For (24) the condition is that $d_1\cdots d_r = n$ and $d_1,\ldots,d_r$ are pairwise relatively prime. Note that (24) and (25) are the same for $r = 2$, but they differ in the case $r > 2$.

Assume that there exist functions $g_1,\ldots,g_r \in \mathcal{A}_1$ (each of a single variable) such that $f(n_1,\ldots,n_r) = g_1(n_1)\cdots g_r(n_r)$ (in particular this holds if $f \in \mathcal{F}_r$ by Proposition 1). Then (23), (24), (26), and (27) reduce to the Dirichlet convolution, unitary convolution, lcm convolution, and binomial convolution, respectively, of the functions $g_1,\ldots,g_r$. For $r = 2$, we have the corresponding convolutions of two given functions of a single variable.

Note the following special case of (26):

$$\sum_{\text{lcm}(d_1,\ldots,d_r)=n} \phi(d_1)\cdots\phi(d_r) = \phi_r(n) \qquad (n \in \mathbb{N}),$$

due to R. D. von Sterneck [74].

We remark that (23), with other notation, appears in [73, p. 591–592], where $\Psi_{\text{dir}}(f)$ is called the "convolute" of f (obtained by the convolution of the arguments). We will call $\Psi_{\text{dir}}(f)$, $\Psi_{\text{unit}}(f)$, $\Psi_{\text{gcd}}(f)$, $\Psi_{\text{lcm}}(f)$, and $\Psi_{\text{binom}}(f)$ the Dirichlet convolute, unitary convolute, gcd convolute, lcm convolute, and binomial convolute, respectively of the function f.

Some special cases of convolutes of functions which are not the product of functions of a single variable are the following. Special Dirichlet convolutes are

$$g_r(n) = \sum_{d_1\cdots d_r=n} \gcd(d_1,\ldots,d_r), \tag{28}$$

$$\ell_r(n) = \sum_{d_1\cdots d_r=n} \text{lcm}(d_1,\ldots,d_r). \tag{29}$$

$$N(n) = \sum_{d\mid n} \phi(\gcd(d, n/d)). \tag{30}$$

For $r = 2$, (28) and (29) are sequences A055155 and A057670, respectively in [56]. The function $N(n)$ given by (30) represents the number of parabolic vertices of $\gamma_0(n)$ (sequence A001616 in [56]), cf. S. Finch [23].

In the case $r = 2$, the lcm convolute of the gcd function

$$c(n) = \sum_{\text{lcm}(d,e)=n} \gcd(d,e), \tag{31}$$

represents the number of cyclic subgroups of the group $\mathbb{Z}_n \times \mathbb{Z}_n$, as shown by A. Pakapongpun, T. Ward [46, Example 2]. That is, $c(n) = c(n,n)$ for every $n \in \mathbb{N}$, with the notation of Sect. 3.4, Example 4 (it is sequence A060648 in [56]).

The Dirichlet, unitary, and lcm convolutes of the Ramanujan sums are

$$a(n) = \sum_{d \mid n} c_d(n/d), \tag{32}$$

$$b(n) = \sum_{d \parallel n} c_d(n/d), \tag{33}$$

$$h(n) = \sum_{\mathrm{lcm}(d,e)=n} c_d(e). \tag{34}$$

All the functions $g_r, \ell_r, N, c, a, b, h$ defined above are multiplicative, as functions of a single variable. See Corollary 1.

6.1 General Results

The Dirichlet, unitary, gcd, lcm, and binomial convolutions preserve the multiplicativity of functions of a single variable, cf. Section 4. As a generalization of this property, we prove the next result.

Proposition 12 *Let $f \in \mathcal{M}_r$ be an arbitrary multiplicative function. Then all the functions $\Psi_{\mathrm{dir}}(f)$, $\Psi_{\mathrm{unit}}(f)$, $\Psi_{\mathrm{gcd}}(f)$, $\Psi_{\mathrm{lcm}}(f)$, and $\Psi_{\mathrm{binom}}(f)$ are multiplicative.*

Note that for the Dirichlet covolute, this property was pointed out by R. Vaidyanathaswamy in [73, pp. 591–592].

Proof By the definitions. Let $n, m \in \mathbb{N}$ such that $\gcd(n,m) = 1$. If $d_1 \cdots d_r = nm$, then there exist unique integers $a_1, b_1, \ldots, a_r, b_r \in \mathbb{N}$ such that $a_1, \ldots, a_r \mid n$, $b_1, \ldots, b_r \mid m$ and $d_1 = a_1 b_1, \ldots, d_r = a_r b_r$. Here $\gcd(a_1 \cdots a_r, b_1 \cdots b_r) = 1$. Using the multiplicativity of f, we obtain

$$\begin{aligned}
\Psi_{\mathrm{dir}}(f)(nm) &= \sum_{\substack{a_1 \cdots a_r = n \\ b_1 \cdots b_r = m}} f(a_1 b_1, \ldots, a_r b_r) \\
&= \sum_{a_1 \cdots a_r = n} f(a_1, \ldots, a_r) \sum_{b_1 \cdots b_r = m} f(b_1, \ldots, b_r) \\
&= \Psi_{\mathrm{dir}}(f)(n) \Psi_{\mathrm{dir}}(f)(m),
\end{aligned}$$

showing the multiplicativity of $\Psi_{\mathrm{dir}}(f)$. The proof in the case of the other functions is similar. For the function $\Psi_{\mathrm{gcd}}(f)$,

$$\Psi_{\mathrm{gcd}}(f)(nm) = \sum_{\substack{a_1 \cdots a_r = n \\ b_1 \cdots b_r = m \\ \gcd(a_1 b_1, \ldots, a_r b_r) = 1}} f(a_1 b_1, \ldots, a_r b_r),$$

where $1 = \gcd(a_1b_1, \ldots, a_rb_r) = \gcd(a_1, \ldots, a_r)\gcd(b_1, \ldots, b_r)$, since the gcd function in r variables is multiplicative. Hence

$$\Psi_{\gcd}(f)(nm) = \sum_{\substack{a_1\cdots a_r=n\\ \gcd(a_1,\ldots,a_r)=1}} f(a_1, \ldots, a_r) \sum_{\substack{b_1\cdots b_r=m\\ \gcd(b_1,\ldots,b_r)=1}} f(b_1, \ldots, b_r)$$

$$= \Psi_{\gcd}(f)(n)\Psi_{\gcd}(f)(m).$$

In the case of the function $\Psi_{\text{lcm}}(f)$, use that the lcm function in r variables is multiplicative, whence $nm = \text{lcm}(a_1b_1, \ldots, a_rb_r) = \text{lcm}(a_1, \ldots, a_r)\text{lcm}(b_1, \ldots, b_r)$ and it follows that $\text{lcm}(a_1, \ldots, a_r) = n$, $\text{lcm}(b_1, \ldots, b_r) = m$. □

Remark 1 Alternative proofs for the multiplicativity of $\Psi_{\text{unit}}(f)$, $\Psi_{\gcd}(f)$, and $\Psi_{\text{lcm}}(f)$ can be given as follows. In the first two cases, the property can be reduced to that of $\Psi_{\text{dir}}(f)$. Let

$$f^{\flat}(n_1, \ldots, n_r) = \begin{cases} f(n_1, \ldots, n_r), & \text{if } \gcd(n_1, \ldots, n_r) = 1, \\ 0, & \text{otherwise.} \end{cases}$$

Then $\Psi_{\gcd}(f) = \Psi_{\text{dir}}(f^{\flat})$. If f is multiplicative, then $f^{\flat}$ is also multiplicative and the multiplicativity of $\Psi_{\gcd}(f)$ follows by the same property of the Dirichlet convolute. Similarly for $\Psi_{\text{unit}}(f)$. Furthermore,

$$\Psi_{\text{lcm}}(f) = (\overline{f * \mathbf{1}_r}) * \mu. \tag{35}$$

Indeed, we have similar to the proof of (13) given above,

$$\sum_{d\mid n} \Psi_{\text{lcm}}(f)(d) = \sum_{\text{lcm}(d_1,\ldots,d_r)\mid d} f(d_1, \ldots, d_r) = \sum_{d_1\mid n,\ldots,d_r\mid n} f(d_1, \ldots, d_r)$$

$$= (f * \mathbf{1}_r)(n, \ldots, n) = (\overline{f * \mathbf{1}_r})(n).$$

If f is multiplicative, so is $f * \mathbf{1}_r$ (as a function of r variables). Therefore, $\overline{f * \mathbf{1}_r}$ is multiplicative (as a function of a single variable) and deduce by (35) that $\Psi_{\text{lcm}}(f)$ is multiplicative.

Remark 2 Note that if f is completely multiplicative, then $\Psi_{\text{binom}}(f)$ is also completely multiplicative.

Corollary 1 *The functions $g_r, \ell_r, N, c, a, b, h$ defined by* (28)–(34) *are multiplicative.*

Proposition 13 *The convolutes of the function $f = \mathbf{1}_r$, that is the number of terms of the sums defining the convolutes are the following multiplicative functions:*

(i) $\Psi_{\text{dir}}(\mathbf{1}_r) = \tau_r$, *the Piltz divisor function of order r given by* $\tau_r(n) = \prod_p \binom{\nu_p(n)+r-1}{r-1}$,
(ii) $\Psi_{\text{unit}}(\mathbf{1}_r) = H_r$ *given by* $H_r(n) = r^{\omega(n)}$,

(iii) $\Psi_{\mathrm{gcd}}(\mathbf{1}_r) = N_r$, *where* $N_r(n) = \sum_{a^r b=n} \mu(a)\tau_r(b) = \prod_p \left(\binom{\nu_p(n)+r-1}{r-1} - \binom{\nu_p(n)-1}{r-1}\right)$ *with* $\binom{\nu-1}{r-1} = 0$ *for* $\nu < r$,

(iv) $\Psi_{\mathrm{lcm}}(\mathbf{1}_r) = M_r$, *where* $M_r(n) = \sum_{ab=n} \mu(a)\tau(b)^r = \prod_p \left((\nu_p(n)+1)^r - \nu_p(n)^r\right)$,

(v) $\Psi_{\mathrm{binom}}(\mathbf{1}_r) = Q_r$, *where* $Q_r(n) = r^{\Omega(n)}$.

Proof (i) and (ii) are immediate from the definitions

(iii) by the property of the Möbius function,

$$\Psi_{\mathrm{gcd}}(\mathbf{1}_r)(n) = \sum_{d_1\cdots d_r=n} \sum_{a \mid \gcd(d_1,\ldots,d_r)} \mu(a) = \sum_{a^k b_1\cdots b_r=n} \mu(a) = \sum_{a^r b=n} \mu(a)\tau_r(b).$$

(iv) follows from (35) in the case $f = \mathbf{1}_r$

(v) by Remark 2. See also L. Tóth, P. Haukkanen [71, Cor. 3.2]. □

Here i) and iv) of Proposition 13 can be generalized as follows.

Proposition 14 *Assume that there is a function* $g \in \mathcal{A}_1$ *(of a single variable) such that* $f(n_1,\ldots,n_r) = g(\gcd(n_1,\ldots,n_r))$ *for every* $n_1,\ldots,n_r \in \mathbb{N}$. *Then for every* $n \in \mathbb{N}$,

$$\Psi_{\mathrm{dir}}(f)(n) = \sum_{d_1\cdots d_r=n} g\left(\gcd(d_1,\ldots,d_r)\right) = \sum_{a^r b=n} (\mu * g)(a)\tau_r(b), \tag{36}$$

$$\Psi_{\mathrm{lcm}}(f)(n) = \sum_{\mathrm{lcm}(d_1,\ldots,d_r)=n} g(\gcd(d_1,\ldots,d_r)) = (g * \mu * \mu * \tau^r)(n). \tag{37}$$

Proof The identity (36) is given by E. Krätzel, W. G. Nowak, L. Tóth [35, Proposition 5.1]. We recall its proof, which is simple and similar to that of (iii) of Proposition 13: for $f(n_1,\ldots,n_r) = g(\gcd(n_1,\ldots,n_r))$,

$$\begin{aligned}\Psi_{\mathrm{dir}}(f)(n) &= \sum_{d_1\cdots d_r=n} \sum_{a \mid \gcd(d_1,\ldots,d_r)} (\mu * g)(a) \\ &= \sum_{a^r b_1\cdots b_r=n} (\mu * g)(a) = \sum_{a^r b=n} (\mu * g)(a)\tau_r(b).\end{aligned}$$

Now for (37),

$$\begin{aligned}\Psi_{\mathrm{lcm}}(f)(n) &= \sum_{\mathrm{lcm}(d_1,\ldots,d_r)=n} \sum_{a \mid \gcd(d_1,\ldots,d_r)} (\mu * g)(a) \\ &= \sum_{a \mid n} (\mu * g)(a) \sum_{\mathrm{lcm}(b_1,\ldots,b_r)=n/a} 1 = \sum_{a \mid n} (\mu * g)(a)(\mu * \tau^r)(n/a) \\ &= (g * \mu * \mu * \tau^r)(n),\end{aligned}$$

using (iv) of Proposition 13. □

Proposition 15 *For every $f, g \in \mathcal{A}_r$,*

$$\Psi_{\mathrm{dir}}(f * g) = \Psi_{\mathrm{dir}}(f) * \Psi_{\mathrm{dir}}(g),$$
$$\Psi_{\mathrm{unit}}(f \times g) = \Psi_{\mathrm{unit}}(f) \times \Psi_{\mathrm{unit}}(g),$$
$$\Psi_{\mathrm{gcd}}(f \odot g) = \Psi_{\mathrm{gcd}}(f) \times \Psi_{\mathrm{gcd}}(g),$$
$$\Psi_{\mathrm{lcm}}(f \oplus g) = \Psi_{\mathrm{lcm}}(f) \oplus \Psi_{\mathrm{lcm}}(g),$$
$$\Psi_{\mathrm{binom}}(f \circ g) = \Psi_{\mathrm{binom}}(f) \circ \Psi_{\mathrm{binom}}(g).$$

Proof For the Dirichlet convolute Ψ_{dir},

$$\begin{aligned}
\Psi_{\mathrm{dir}}(f * g)(n) &= \sum_{d_1 \cdots d_r = n} (f * g)(d_1, \ldots, d_r) \\
&= \sum_{d_1 \cdots d_r = n} \sum_{a_1 b_1 = d_1, \ldots, a_r b_r = d_r} f(a_1, \ldots, a_r) g(b_1, \ldots, b_r) \\
&= \sum_{a_1 b_1 \cdots a_r b_r = n} f(a_1, \ldots, a_r) g(b_1, \ldots, b_r) \\
&= \sum_{xy = n} \sum_{a_1 \cdots a_r = x} f(a_1, \ldots, a_r) \sum_{b_1 \cdots b_r = y} g(b_1, \ldots, b_r) \\
&= \sum_{xy = n} \Psi_{\mathrm{dir}}(f)(x) \Psi_{\mathrm{dir}}(g)(y) = (\Psi_{\mathrm{dir}}(f) * \Psi_{\mathrm{dir}}(g))(n),
\end{aligned}$$

and similar for the other ones. □

Proposition 16 *Let $r \geq 2$. The following maps are surjective algebra homomorphisms: $\Psi_{\mathrm{dir}} : (\mathcal{A}_r, +, \cdot, *) \to (\mathcal{A}_1, +, \cdot, *)$, $\Psi_{\mathrm{unit}} : (\mathcal{A}_r, +, \cdot, \times) \to (\mathcal{A}_1, +, \cdot, \times)$, $\Psi_{\mathrm{gcd}} : (\mathcal{A}_r, +, \cdot, \odot) \to (\mathcal{A}_1, +, \cdot, \times)$, $\Psi_{\mathrm{lcm}} : (\mathcal{A}_r, +, \cdot, \oplus) \to (\mathcal{A}_1, +, \cdot, \oplus)$, $\Psi_{\mathrm{binom}} : (\mathcal{A}_r, +, \cdot, \circ) \to (\mathcal{A}_1, +, \cdot, \circ)$.*

Proof Use Proposition 15. For the surjectivity: for a given $f \in \mathcal{A}_1$ consider $F \in \mathcal{A}_r$ defined by $F(n, 1, \ldots, 1) = f(n)$ for every $n \in \mathbb{N}$ and $F(n_1, \ldots, n_r) = 0$ otherwise, i.e., for every $n_1, \ldots, n_r \in \mathbb{N}$ with $n_2 \cdots n_r > 1$. Then $\Psi_{\mathrm{dir}}(F) = \Psi_{\mathrm{unit}}(F) = \Psi_{\mathrm{gcd}}(F) = \Psi_{\mathrm{lcm}}(F) = \Psi_{\mathrm{binom}}(F) = f$. □

Corollary 2 *Let $r \geq 2$.*

(i) The maps $\Psi_{\mathrm{dir}} : (\mathcal{M}_r, *) \to (\mathcal{M}_1, *)$, $\Psi_{\mathrm{unit}} : (\mathcal{M}_r, \times) \to (\mathcal{M}_1, \times)$, $\Psi_{\mathrm{gcd}} : (\mathcal{M}_r, \odot) \to (\mathcal{M}_1, \times)$ and $\Psi_{\mathrm{binom}} : (\mathcal{M}_r, \circ) \to (\mathcal{M}_1, \circ)$ are surjective group homomorphisms.
(ii) The maps $\Psi_{\mathrm{dir}} : (\mathcal{F}_r, *) \to (\mathcal{M}_1, *)$, $\Psi_{\mathrm{unit}} : (\mathcal{F}_r, \times) \to (\mathcal{M}_1, \times)$ and $\Psi_{\mathrm{binom}} : (\mathcal{F}_r, \circ) \to (\mathcal{M}_1, \circ)$ are surjective group homomorphisms.
(iii) The map $\Psi_{\mathrm{binom}} : (\mathcal{C}_r, \circ) \to (\mathcal{C}_1, \circ)$ is a surjective group homomorphism.

Proof Follows from Propositions 12, 15 and from the fact that for every (completely) multiplicative f, the function F constructed in the proof of Proposition 16 is (completely) multiplicative. For (iii) use also Remark 2. □

6.2 Special Cases

We present identities for the convolutes of some special functions. For Dirichlet convolutes we have the next result.

Corollary 3 *For every $k \in \mathbb{C}$,*

$$\sum_{d_1\cdots d_r=n} (\gcd(d_1,\ldots,d_r))^k = \sum_{a^r b=n} \phi_k(a)\tau_r(b).$$

$$\sum_{d_1\cdots d_r=n} \sigma_k(\gcd(d_1,\ldots,d_r)) = \sum_{a^r b=n} a^k\tau_r(b). \tag{38}$$

Proof Follows from the first identity of Proposition 14. □

The function (38) is for $r = 2$ and $k = 0$ the sequence A124315 in [56], and for $r = 2$, $k = 1$, it is sequence A124316 in [56]. See also [35, Sect. 5].

Special cases of lcm convolutes which do not seem to be known are the following. Let $\beta = \mathrm{id} * \lambda$ be the alternating sum-of-divisors function, see L. Tóth [69].

Corollary 4 *For every $n \in \mathbb{N}$,*

$$\sum_{\mathrm{lcm}(d_1,\ldots,d_r)=n} \gcd(d_1,\ldots,d_r) = (\phi * M_r)(n),$$

where the function M_r is given in Proposition 13,

$$\sum_{\mathrm{lcm}(d_1,\ldots,d_r)=n} \tau(\gcd(d_1,\ldots,d_r)) = \tau(n)^r,$$

$$\sum_{\mathrm{lcm}(d,e)=n} \phi(\gcd(d,e)) = \psi(n),$$

$$\sum_{\mathrm{lcm}(d,e)=n} \sigma(\gcd(d,e)) = (\tau * \psi)(n) = (\phi * \tau^2)(n), \tag{39}$$

$$\sum_{\mathrm{lcm}(d,e)=n} \beta(\gcd(d,e)) = \sigma(n),$$

$$\sum_{\mathrm{lcm}(d,e)=n} \mu(\gcd(d,e)) = \mu^2(n),$$

$$\sum_{\mathrm{lcm}(d,e)=n} \lambda(\gcd(d,e)) = 1.$$

Proof Follow from the second identity of Proposition 14. □

Here (39) is $s(n,n)$, representing the number of all subgroups of the group $\mathbb{Z}_n \times \mathbb{Z}_n$ (sequence A060724 in [56]).

For the convolutes of the Ramanujan sum we have

Proposition 17

$$\sum_{de=n} c_d(e) = \begin{cases} \sqrt{n}, & n \text{ perfect square}, \\ 0, & \text{otherwise}, \end{cases} \tag{40}$$

$$\sum_{\substack{de=n \\ \gcd(d,e)=1}} c_d(e) = \begin{cases} 1, & n \text{ squarefull}, \\ 0, & \text{otherwise}, \end{cases} \tag{41}$$

$$\sum_{\operatorname{lcm}(d,e)=n} c_d(e) = \phi(n) \quad (n \in \mathbb{N}). \tag{42}$$

Proof According to Corollary 1, all these functions are multiplicative and it is enough to compute their values for prime powers. Alternatively, the formula (8) can be used. □

Formulae (40) and (41) are well known, see, e.g., P. J. McCarthy [42, pp. 191–192], while (42) seems to be new.

7 Asymptotic Properties

We discuss certain results concerning the mean values of arithmetic functions of r variables and the asymptotic densities of some sets in $\mathbb{N}^r$. We also present asymptotic formulae for special multiplicative functions given in the previous sections.

7.1 Mean Values

Let $f \in \mathcal{A}_r$. The mean value of f is

$$M(f) = \lim_{x_1,\dots,x_r\to\infty} \frac{1}{x_1\cdots x_r} \sum_{n_1\le x_1,\dots,n_r\le x_r} f(n_1,\dots,n_r),$$

where $x_1,\dots,x_r$ tend to infinity independently, provided that this limit exists. As a generalization of Wintner's theorem (valid for the case $r=1$), N. Ushiroya [72, Theorem 1] proved the next result.

Proposition 18 *If $f \in \mathcal{A}_r$ $(r \ge 1)$ and*

$$\sum_{n_1,\dots,n_r=1}^{\infty} \frac{|(\mu_r * f)(n_1,\dots,n_r)|}{n_1\cdots n_r} < \infty,$$

then the mean value $M(f)$ exists and

$$M(f) = \sum_{n_1,\dots,n_r=1}^{\infty} \frac{(\mu_r * f)(n_1,\dots,n_r)}{n_1\cdots n_r}.$$

For multiplicative functions we have the following result due to N. Ushiroya [72, Theorem 4] in a slightly different form.

Proposition 19 *Let* $f \in \mathcal{M}_r$ $(r \geq 1)$*. Assume that*

$$\sum_p \sum_{\substack{\nu_1,\ldots,\nu_r=0 \\ \nu_1+\ldots+\nu_r \geq 1}}^{\infty} \frac{|(\mu_r * f)(p^{\nu_1},\ldots,p^{\nu_r})|}{p^{\nu_1+\ldots+\nu_r}} < \infty.$$

Then the mean value $M(f)$ *exists and*

$$M(f) = \prod_p \left(1-\frac{1}{p}\right)^r \sum_{\nu_1,\ldots,\nu_r=0}^{\infty} \frac{f(p^{\nu_1},\ldots,p^{\nu_r})}{p^{\nu_1+\ldots+\nu_r}}.$$

Corollary 5 (N. Ushiroya [72, Theorem 7]) *Let* $g \in \mathcal{M}_1$ *be a multiplicative function and denoted by* a_g*, the absolute convergence abscissa of the Dirichlet series* $D(g;z)$*. Then for every* $r > 1$, $r > a_g$ *the main value of the function* $(n_1,\ldots,n_r) \mapsto g(\gcd(n_1,\ldots,n_r))$ *exists and*

$$M(f) = \frac{1}{\zeta(r)} \sum_{n=1}^{\infty} \frac{g(n)}{n^r}.$$

Proof Follows from Proposition (18) and the identity (16). □

For example, the mean value of the function $(n_1,\ldots,n_r) \mapsto \gcd(n_1,\ldots,n_r)$ is $\zeta(r-1)/\zeta(r)$ $(r \geq 3)$, the mean value of the function $(n_1,\ldots,n_r) \mapsto \phi(\gcd(n_1,\ldots,n_r))$ is $\zeta(r-1)/\zeta^2(r)$ $(r \geq 3)$.

The analog of Proposition 18 for the unitary convolution is the next result (see W. Narkiewicz [43] in the case $r = 1$).

Proposition 20 *If* $f \in \mathcal{A}_r$ $(r \geq 1)$ *and*

$$\sum_{n_1,\ldots,n_r=1}^{\infty} \frac{|(\mu_r^{\times} \times f)(n_1,\ldots,n_r)|}{n_1\cdots n_r} < \infty,$$

then the mean value $M(f)$ *exists and*

$$M(f) = \sum_{n_1,\ldots,n_r=1}^{\infty} \frac{(\mu_r^{\times} \times f)(n_1,\ldots,n_r)\phi(n_1)\cdots\phi(n_r)}{n_1^2\cdots n_r^2}.$$

For further results on the mean values of multiplicative arithmetic functions of several variables and generalizations to the several variables case of results of G. Halász [24], we refer to the papers of O. Casas [8], H. Delange [15, 16], E. Heppner [29, 30] and K-H. Indlekofer [33]. See also E. Alkan, A. Zaharescu, M. Zaki [3].

7.2 Asymptotic Densities

Let $S \subset \mathbb{N}^r$. The set S possesses the asymptotic density d_S if the characteristic function χ_S of S has the mean value $M(\chi_S) = d_S$. In what follows we consider the densities of certain special sets.

Let $g \in \mathcal{M}_1$ be a multiplicative function such that $g(n) \in \{0, 1\}$ for every $n \in \mathbb{N}$. Let $S_g = \{(n_1, \dots, n_r) \in \mathbb{N}^r : g(\gcd(n_1, \dots, n_r)) = 1\}$. It follows from Corollary 5 that for $r \geq 2$ the set S_g has the asymptotic density given by

$$d_{S_g} = \frac{1}{\zeta(r)} \sum_{n=1}^{\infty} \frac{g(n)}{n^r}.$$

In particular for $r \geq 2$, the set of points $(n_1, \dots, n_r) \in \mathbb{N}^r$ which are visible from the origin, i.e., such that $\gcd(n_1, \dots, n_r) = 1$ holds has the density $1/\zeta(r)$ (the case $g = \delta$). Another example: the set of points $(n_1, \dots, n_r) \in \mathbb{N}^r$ such that $\gcd(n_1, \dots, n_r)$ is squarefree has the density $1/\zeta(2r)$ (the case $g = \mu^2$). These results are well known.

Now consider the set of points $(n_1, \dots, n_r) \in \mathbb{N}^r$ such that $n_1, \dots, n_r$ are pairwise relatively prime. The next result was first proved by L. Tóth [62], giving also an asymptotic formula for $\sum_{n_1,\dots,n_r \leq x} \rho(n_1, \dots, n_r)$ and by J.-Y. Cai, E. Bach [6, Theorem 3.3]. Here we give a simple different proof.

Proposition 21 *Let $r \geq 2$. The asymptotic density of the set of points in $\mathbb{N}^r$ with pairwise relatively prime coordinates is*

$$A_r = \prod_p \left(1 - \frac{1}{p}\right)^{r-1} \left(1 + \frac{r-1}{p}\right). \tag{43}$$

Proof Apply Proposition (18) for the function $f = \rho$ defined by (1). Then according to the Dirichlet series representation (18), the density is

$$\sum_{n_1,\dots,n_r=1}^{\infty} \frac{(\mu_r * \rho)(n_1, \dots, n_r)}{n_1 \cdots n_r} = \prod_p \left(\left(1 - \frac{1}{p}\right)^r + \frac{r}{p} \left(1 - \frac{1}{p}\right)^{r-1} \right),$$

which equals A_r, given by (43). □

See the quite recent paper by J. Hu [31] for a generalization of Proposition 21. See J. L. Fernández, P. Fernández [19–22] for various statistical regularity properties concerning mutually relatively prime and pairwise relatively prime integers.

Unitary analogs of the problems of above are the following.

Proposition 22 *Let $r \geq 2$. The asymptotic density of the set of points $(n_1, \dots, n_r) \in \mathbb{N}^r$ such that $\operatorname{gcud}(n_1, \dots, n_r) = 1$ is*

$$\prod_p \left(1 - \frac{(p-1)^r}{p^r(p^r - 1)}\right).$$

Proof The characteristic function of this set is given by

$$\delta(\text{gcud}(n_1,\ldots,n_r)) = \sum_{d||\text{gcud}(n_1,\ldots,n_r)} \mu^\times(d) = \sum_{d_1||n_r,\ldots,d_r||n_r} G(d_1,\ldots,d_r),$$

where

$$G(n_1,\ldots,n_r) = \begin{cases} \mu^\times(n), & \text{if } n_1 = \ldots = n_r = n, \\ 0, & \text{otherwise.} \end{cases}$$

We deduce from Proposition 20 that the density in question is

$$\sum_{n_1,\ldots,n_r=1}^{\infty} \frac{G(n_1,\ldots,n_r)\phi(n_1)\cdots\phi(n_r)}{n_1^2\cdots n_r^2} = \sum_{n=1}^{\infty} \frac{\mu^\times(n)\phi(n)^r}{n^{2r}}$$

$$= \prod_p \left(1 - \frac{(p-1)^r}{p^r(p^r-1)}\right).$$

□

Proposition 22 was proved by L. Tóth [61] using different arguments. See [61] for other related densities and asymptotic formulae.

Corollary 6 (r=2) *The set of points* $(m,n) \in \mathbb{N}^2$ *such that* $\text{gcud}(m,n) = 1$ *has the density*

$$\prod_p \left(1 - \frac{p-1}{p^2(p+1)}\right).$$

Proposition 23 *Let* $r \geq 2$. *The asymptotic density of the set of points in* $\mathbb{N}^r$ *with pairwise unitary relatively prime coordinates is*

$$A_r^\times = \prod_p \sum_{\nu_1,\ldots,\nu_r=0}^{\infty} \frac{Q(p^{\nu_1},\ldots,p^{\nu_r})\phi(p^{\nu_1})\cdots\phi(p^{\nu_r})}{p^{2\nu_1+\ldots+2\nu_r}},$$

where Q *is the multiplicative function of* r *variables given as follows: Let* $p^{\nu_1},\ldots,p^{\nu_r}$ *be arbitrary powers of the prime* p *with* $\nu_1,\ldots,\nu_r \in \mathbb{N}_0$, $\nu_1+\ldots+\nu_r \geq 1$. *Assume that the exponents* $\nu_1,\ldots,\nu_r$ *have* q $(1 \leq q \leq r)$ *distinct positive values, taken* $t_1,\ldots,t_q$ *times* $(1 \leq t_1+\ldots+t_q \leq r)$. *Then*

$$Q(p^{\nu_1},\ldots,p^{\nu_r}) = (-1)^r(1-t_1)\cdots(1-t_q).$$

Proof We use Proposition 20 for the function $\rho^\times$ defined by (3). The density is

$$A_r^\times = \sum_{n_1,\ldots,n_r=1}^{\infty} \frac{Q(n_1,\ldots,n_r)\phi(n_1)\cdots\phi(n_r)}{n_1^2\cdots n_r^2},$$

where $Q_r = \mu_r^{\times} \times \rho^{\times}$ and the Euler product formula can be used by the multiplicativity of the involved functions. □

Note that for $r = 2$,

$$Q(p^{\nu_1}, p^{\nu_2}) = \begin{cases} 1, & \nu_1 = \nu_2 = 0, \\ -1, & \nu_1 = \nu_2 \geq 1, \\ 0, & \text{otherwise} \end{cases}$$

and reobtain Corollary 6.

Corollary 7 $(r = 3, 4)$

$$A_3^{\times} = \zeta(2)\zeta(3) \prod_p \left(1 - \frac{4}{p^2} + \frac{7}{p^3} - \frac{9}{p^4} + \frac{8}{p^5} - \frac{2}{p^6} - \frac{3}{p^7} + \frac{2}{p^8}\right).$$

$$A_4^{\times} = \zeta^2(2)\zeta(3)\zeta(4) \prod_p \left(1 - \frac{8}{p^2} + \frac{3}{p^3} + \frac{27}{p^4} - \frac{24}{p^5} - \frac{14}{p^6} - \frac{3}{p^7} \right.$$
$$\left. + \frac{37}{p^8} - \frac{30}{p^9} + \frac{42}{p^{10}} - \frac{33}{p^{11}} - \frac{41}{p^{12}} + \frac{78}{p^{13}} - \frac{44}{p^{14}} + \frac{9}{p^{15}}\right).$$

Proof According to Proposition 23,

$$Q(p^{\nu_1}, p^{\nu_2}, p^{\nu_3}) = \begin{cases} 1, & \nu_1 = \nu_2 = \nu_3 = 0, \\ -1, & \nu_1 = \nu_2 \geq 1, \nu_3 = 0 \text{ and symmetric cases}, \\ 2, & \nu_1 = \nu_2 = \nu_3 \geq 1, \\ 0, & \text{otherwise} \end{cases}$$

and

$$Q(p^{\nu_1}, p^{\nu_2}, p^{\nu_3}, p^{\nu_4}) = \begin{cases} 1, & \nu_1 = \nu_2 = \nu_3 = 0, \\ -1, & \nu_1 = \nu_2 \geq 1, \nu_3 = \nu_4 = 0 \text{ and symmetric cases}, \\ -2, & \nu_1 = \nu_2 = \nu_3 \geq 1, \nu_4 = 0 \text{ and symmetric cases}, \\ -3, & \nu_1 = \nu_2 = \nu_3 = \nu_4 \geq 1, \\ 1, & \nu_1 = \nu_2 > \nu_3 = \nu_4 \geq 1 \text{ and symmetric cases}, \\ 0, & \text{otherwise}, \end{cases}$$

and direct computations lead to the given infinite products. □

We refer to the papers by J. CHRISTOPHER [12], H. DELANGE [15], and N. USHIROYA [72] for related density results.

7.3 Asymptotic Formulae

One has

$$\sum_{m,n \leq x} \gcd(m,n) = \frac{x^2}{\zeta(2)} \left(\log x + 2\gamma - \frac{1}{2} - \frac{\zeta(2)}{2} - \frac{\zeta'(2)}{\zeta(2)}\right) + O\left(x^{1+\theta+\varepsilon}\right), \quad (44)$$

for every $\varepsilon > 0$, where θ is the exponent appearing in Dirichlet's divisor problem, that is

$$\sum_{n\le x} \tau(n) = x\log x + (2\gamma - 1)x + O\left(x^{\theta+\varepsilon}\right). \tag{45}$$

It is known that $1/4 \le \theta \le 131/416 \approx 0.3149$, where the upper bound, the best up to date, is the result of M. N. HUXLEY [32].

If $r \ge 3$, then

$$\sum_{n_1,\dots,n_r\le x} \gcd(n_1,\dots,n_r) = \frac{\zeta(r-1)}{\zeta(r)} x^r + O(R_r(x)), \tag{46}$$

where $R_3(x) = x^2\log x$ and $R_r(x) = x^{r-1}$ for $r \ge 4$, which follows from the representation

$$\sum_{n_1,\dots,n_r\le x} \gcd(n_1,\dots,n_r) = \sum_{d\le x} \phi(d)\,[x/d]^r .$$

Furthermore,

$$\sum_{m,n\le x} \operatorname{lcm}(m,n) = \frac{\zeta(3)}{4\zeta(2)} x^4 + O\left(x^3\log x\right). \tag{47}$$

The formulae (44), (46), and (47) can be deduced by elementary arguments and go back to the work of E. CESÀRO [10], E. COHEN [13] and P. DIACONIS, P. ERDŐS [18]. See also L. TÓTH [63, Eq. (25)]. A formula similar to (44), with the same error term holds for $\sum_{m,n\le x} g(\gcd(m,n))$, where $g = h * \mathrm{id}$, $h \in \mathcal{A}_1$ is bounded, including the cases $g = \phi, \sigma, \psi$. See L. TÓTH [63, p. 7]. See J. L. FERNÁNDEZ, P. FERNÁNDEZ [19–21] for statistical regularity properties of the gcd's and lcm's of positive integers.

Consider next the function $g_2(n) = \sum_{d\mid n} \gcd(d, n/d)$, which is the Dirichlet convolute of the gcd function for $r = 2$.

Proposition 24

$$\sum_{n\le x} g_2(n) = \frac{3}{2\pi^2} x \left(\log^2 x + c_1\log x + c_2\right) + R(x),$$

where c_1, c_2 are constants and $R(x) = O\left(x^{\theta}(\log x)^{\theta'}\right)$ with $\theta = \frac{547}{832} = 0.65745\dots$, $\theta' = \frac{26947}{8320}$.

This was proved using analytic tools (Huxley's method) by E. KRÄTZEL, W. G. NOWAK, L. TÓTH [35, Theorem 3.5]. See M. KÜHLEITNER, W. G. NOWAK [36] for omega estimates on the function $g_2(n)$. The papers [35] and [36] contain also results for the function $g_r(n)$ ($r \ge 3$), defined by (28) and related functions.

For the function $\ell_2(n) = \sum_{d\mid n} \operatorname{lcm}(d, n/d)$, representing the Dirichlet convolute of the lcm function for $r = 2$, one can deduce the next asymptotics.

Proposition 25

$$\sum_{n\le x} \ell_2(n)/n = \frac{\zeta(3)}{\zeta(2)} x \left(\log x + 2\gamma - 1 - \frac{2\zeta'(2)}{\zeta(2)} + \frac{2\zeta'(3)}{\zeta(3)} \right) + O\left(x^{\theta+\varepsilon}\right),$$

where θ *is given by* (45).

A similar formula can be given for the function $\ell_r(n)$ $(r \ge 3)$ defined by (29).

For the functions $s(m,n)$ and $c(m,n)$ defined in Sect. 3.4, Example 4, W. G. NOWAK, L. TÓTH [44] proved the following asymptotic formulae.

Proposition 26 *For every fixed* $\varepsilon > 0$,

$$\sum_{m,n\le x} s(m,n) = \frac{2}{\pi^2} x^2 \left(\log^3 x + a_1 \log^2 x + a_2 \log x + a_3\right) + O\left(x^{\frac{1117}{701}+\varepsilon}\right),$$

$$\sum_{m,n\le x} c(m,n) = \frac{12}{\pi^4} x^2 \left(\log^3 x + b_1 \log^2 x + b_2 \log x + b_3\right) + O\left(x^{\frac{1117}{701}+\varepsilon}\right),$$

where $1117/701 \approx 1.5934$ *and* $a_1, a_2, a_3, b_1, b_2, b_3$ *are explicit constants.*

See the recent paper by T. H. CHAN, A. V. KUMCHEV [11] concerning asymptotic formulae for $\sum_{n\le x, k\le y} c_n(k)$.

Acknowledgement The author gratefully acknowledges support from the Austrian Science Fund (FWF) under the project Nr. M1376-N18.

References

1. Alkan, E., Zaharescu, A., Zaki, M.: Arithmetical functions in several variables. Int. J. Number Theory **1**, 383–399 (2005)
2. Alkan, E., Zaharescu, A., Zaki, M.: Unitary convolution for arithmetical functions in several variables. Hiroshima Math. J. **36** 113–124 (2006)
3. Alkan, E., Zaharescu, A., Zaki, M.: Multidimensional averages and Dirichlet convolution. Manuscr. Math. **123**, 251–267 (2007)
4. Apostol, T.M.: Introduction to Analytic Number Theory. Springer, New York (1976)
5. Buschman, R.G.: lcm-products of number-theoretic functions revisited. Kyungpook Math. J. **39**, 159–164 (1999)
6. Cai, J.-Y. Bach, E.: On testing for zero polynomials by a set of points with bounded precision. Theor. Comp. Sci. **296**, 15–25 (2003)
7. Carroll, T.B., Gioia, A.A.: On a subgroup of the group of multiplicative arithmetic functions. J. Aust. Math. Soc. **20**(Series A), 348–358 (1975)
8. Casas, O.: Arithmetical functions in two variables. An analogue of a result of Delange. Lect. Mat. **27**, 5–12 (2006). (Número especial)
9. Cashwell, E.D., Everett, C.J.: The ring of number-theoretic functions. Pac. J. Math. **9**, 975–985 (1959)
10. Cesàro, E.: Étude moyenne du plus grand commun diviseur de deux nombres. Annali di Matematica Pura ed Applicata **13**, 235–250 (1885)

11. Chan, T.H., Kumchev, A.V.: On sums of Ramanujan sums. Acta Arith. **152**, 1–10 (2012)
12. Christopher, J.: The asymptotic density of some k dimensional sets. Amer. Math. Mon. **63**, 399–401 (1956)
13. Cohen, E.: Arithmetical functions of a greatest common divisor, III. Cesàro's divisor problem. Proc. Glasg. Math. Assoc. **5**, 67–75 (1961–1962)
14. Dehaye, P.-O.: On the structure of the group of multiplicative arithmetical functions. Bull. Belg. Math. Soc. **9**, 15–21 (2002)
15. Delange, H.: On some sets of pairs of positive integers. J. Number Theory **1**, 261–279 (1969)
16. Delange, H.: Sur les fonctions multiplicatives de plusieurs entiers. Enseignement Math. **(2)**16 (1970). (219–246 (1971), Errata: **(2)** 17 (1971), 186)
17. Delany, J.E.: Groups of arithmetical functions. Math. Mag. **78**, 83–97 (2005)
18. Diaconis, P., Erdős, P.: On the Distribution of the greatest common divisor, in A festschrift for Herman Rubin, IMS lecture notes monograph series institute of mathematical statistics, vol **45**, (2004), 56–61 (original version: Technical Report No. 12, Department of Statistics, Stanford University, Stanford, 1977)
19. Fernández, J.L., Fernández, P.: Asymptotic normality and greatest common divisors, Preprint, arXiv:1302.2357 (2013) [math.PR]
20. Fernández, J.L., Fernández, P.: On the probability distribution of the gcd and lcm of r-tuples of integers, Preprint, arXiv:1305.0536 (2013) [math.NT]
21. Fernández, J.L., Fernández, P.: Equidistribution and coprimality, Preprint, arXiv:1310.3802 (2013) [math.NT]
22. Fernández, J.L., Fernández, P.: Random index of codivisibility, Preprint, arXiv:1310.4681v1 (2013) [math.NT]
23. Finch, S.: Modular forms on $SL_2(\mathbb{Z})$, manuscript. http://www.people.fas.harvard.edu/sfinch/csolve/frs.pdf (2005)
24. Halász, G.: Über die Mittelwerte multiplikativer zahlentheoretischer Funktionen. Acta Math. Acad. Sci. Hung. **19**, 365–403 (1968)
25. Hampejs, M., Tóth, L.: On the subgroups of finite Abelian groups of rank three. Ann. Univ. Sci. Bp. Sect. Comp. **39**, 111–124 (2013)
26. Hampejs, M., Holighaus, N., Tóth, L., Wiesmeyr, C.: On the subgroups of the group $\mathbb{Z}_m \times \mathbb{Z}_n$, Preprint, arXiv:1211.1797 (2012) [math.GR]
27. Hardy, G.H., Wright, E.M.: An Introduction to the Theory of Numbers, 6th edn. In: Heath-Brown D. R., Silverman, J. H. (eds.) Oxford University Press, Oxford (2008)
28. Haukkanen, P.: Derivation of arithmetical functions under the Dirichlet convolution, manuscript.
29. Heppner, E.: Über benachbarte multiplikative zahlentheoretische Funktionen mehrerer Variablen. Arch. Math. (Basel) **35**, 454–460 (1980)
30. Heppner, E.: Über Mittelwerte multiplikativer zahlentheoretischer Funktionen mehrerer Variablen. Monatshefte Math. **91**, 1–9 (1981)
31. Hu, J.: The probability that random positive integers are k-wise relatively prime. Int. J. Number Theory **9**, 1263–1271 (2013)
32. Huxley, M.N.: Exponential sums and lattice points III. Proc. Lond. Math. Soc. **87**, 591–609 (2003)
33. Indlekofer K.-H.: Multiplikative Funktionen mehrerer Variablen. J. Reine Angew. Math. **256**, 180–184 (1972)
34. Johnson, K.R.: Reciprocity in Ramanujan's sum. Math. Mag. **59**, 216–222 (1986)
35. Krätzel, E., Nowak, W.G., Tóth, L.: On certain arithmetic functions involving the greatest common divisor. Cent. Eur. J. Math. **10**, 761–774 (2012)
36. Kühleitner, M., Nowak, W.G.: On a question of A. Schinzel: Omega estimates for a special type of arithmetic functions. Cent. Eur. J. Math. **11**, 477–486 (2013)
37. Lehmer, D.H.: A new calculus of numerical functions. Amer. J. Math. **53**(4), 843–854 (1931)
38. Lehmer, D.H.: On a theorem of von Sterneck. Bull. Am. Math. Soc. **37**(10), 723–726 (1931)
39. Lehmer, D.H.: Arithmetic of double series. Trans. Am. Math. Soc. **33**, 945–957 (1931)

40. Liskovets, V.A.: A multivariate arithmetic function of combinatorial and topological significance. Integers **10**, 155–177 (2010)
41. MacHenry, T.: A subgroup of the group of units in the ring of arithmetic functions. Rocky Mt. J. Math. **29**, 1055–1065 (1999)
42. McCarthy, P.J.: Introduction to Arithmetical Functions. Springer, New York (1986)
43. Narkiewicz, W.: On a summation formula of E. Cohen. Colloq. Math. **11**, 85–86 (1963)
44. Nowak, W.G., Tóth, L.: On the average number of subgroups of the group $\mathbb{Z}_m \times \mathbb{Z}_n$, Int. J. Number Theory, **10**, 363–374 (2014) (Accepted, arXiv:1307.1414 [math.NT])
45. Onozuka, T.: The multiple Dirichlet product and the multiple Dirichlet series, manuscript, 2013
46. Pakapongpun, A., Ward, T.: Functorial orbit counting. J. Integer Seq. **12**, 20 (2009) (Article 09.2. 4)
47. Peter, M.: Rekurrente zahlentheoretische Funktionen in mehreren Variablen. Arch. Math. (Basel) **68**, 202–213 (1997)
48. Ryden, R.W.: Groups of arithmetic functions under Dirichlet convolution. Pac. J. Math. **44**, 355–360 (1973)
49. Sándor, J., Bege, A.: The Möbius function: Generalizations and extensions. Adv. Stud. Contemp. Math. (Kyungshang), **6**(2), 77–128 (2003)
50. Sándor, J., Crstici, B.: Handbook of Number Theory II. Kluwer Academic Publishers, Dordrecht (2004)
51. Scheid, H.: Einige Ringe zahlentheoretischer Funktionen. J. Reine Angew. Math. **237**, 1–11 (1969)
52. Schwab, E.D.: Generalized arithmetical functions of three variables. Int. J. Number Theory **6**, 1689–1699 (2010)
53. Schwarz, W., Spilker, J.: Arithmetical functions, An introduction to elementary and analytic properties of arithmetic functions and to some of their almost-periodic properties, London Mathematical Society Lecture Note Series, 184. Cambridge University Press, Cambridge (1994)
54. Shapiro, H.N.: On the convolution ring of arithmetic functions. Comm. Pure Appl. Math. **25**, 287–336 (1972)
55. Sivaramakrishnan, R.: Classical Theory of Arithmetic Functions, Monographs and Textbooks in Pure and Applied Mathematics, vol. 126. Marcel Dekker (1989)
56. Sloane, N.J.A.: The On-Line Encyclopedia of Integer Sequences. http://oeis.org
57. Smith, D.A.: Generalized arithmetic function algebras. In: Gioia, A.A., Goldsmith, D.L. (eds.) The Theory of Arithmetic Functions, vol. 251, pp. 205–245. Springer, Berlin (1972) (Proceedings of the Conference at Western Michigan University, Kalamazoo, Mich., 1971, Lecture Notes in Math)
58. Snellman, J.: Truncations of the ring of arithmetical functions with unitary convolution. Int. J. Math. Game Theory Algebra **13**, 485–519 (2003)
59. Snellman, J.: The ring of arithmetical functions with unitary convolution: Divisorial and topological properties. Arch. Math. (Brno), **40**, 161–179 (2004)
60. Subbarao, M.V.: On some arithmetic convolutions, In: The Theory of Arithmetic Functions, Lecture Notes in Mathematics No. 251, 247–271. Springer (1972)
61. Tóth, L.: On the asymptotic densities of certain subsets of $\mathbb{N}^k$. Riv. Mat. Univ. Parma (6)4 121–131 (2001)
62. Tóth, L.: The probability that k positive integers are pairwise relatively prime. Fibonacci Quart. **40**, 13–18 (2002)
63. Tóth, L.: A survey of gcd-sum functions. J. Integer Seq. **13**, 23 (2010). (Article 10.8. 1)
64. Tóth, L.: Menon's identity and arithmetical sums representing functions of several variables. Rend. Sem. Mat. Univ. Politec. Torino **69**, 97–110 (2011)
65. Tóth, L.: On the number of cyclic subgroups of a finite Abelian group. Bull. Math. Soc. Sci. Math. Roum. (N.S.) **55**(103), 423–428 (2012)
66. Tóth, L.: Some remarks on a paper of V. A. Liskovets. Integers **12**, 97–111 (2012)
67. Tóth, L.: Sums of products of Ramanujan sums. Ann. Univ. Ferrara **58**, 183–197 (2012)

68. Tóth, L.: Two generalizations of the Busche-Ramanujan identities. Int. J. Number Theory **9**, 1301–1311 (2013)
69. Tóth, L.: A survey of the alternating sum-of-divisors function. Acta Univ. Sapientiae Math. 5, 93–107 (2013) (accepted, arXiv:1111.4842 [math.NT])
70. Tóth, L.: Averages of Ramanujan sums: Note on two papers by E. Alkan, accepted, arXiv:1305.6018 [math.NT].
71. Tóth, L., Haukkanen, P.: On the binomial convolution of arithmetical functions. J. Comb. Number Theory **1**, 31–47 (2009)
72. Ushiroya, N.: Mean-value theorems for multiplicative arithmetic functions of several variables. Integers **12**, 989–1002 (2012)
73. Vaidyanathaswamy, R.: The theory of multiplicative arithmetic functions. Trans. Am. Math. Soc. **33**, 579–662 (1931)
74. von Sterneck, R.D.: Ableitung zahlentheoretischer Relationen mit Hilfe eines mehrdimensionalen Systemes von Gitterpunkten. Monatsh. Math. Phys. **5**, 255–266 (1894)
75. Zaharescu, A., Zaki, M.: Derivations and generating degrees in the ring of arithmetical functions. Proc. Indian Acad. Sci. (Math. Sci.) **117**(2), 167–175 (2007)

Sequential Maximality Principles

Mihai Turinici

Abstract In Sect. 1, we show that all variational results deductible from the Bernays–Tarski Dependent Choices Principle and including Ekeland's variational statement (J Math Anal Appl 47:324–353, 1974) are equivalent to each other. Moreover, in Sect. 2, one establishes that such a conclusion comprises as well the "asymptotic" versions of the Brezis-Browder ordering principle (Adv Math 21:355–364, 1976). The objective of Sect. 3 is to emphasize that the vector variational principles in Goepfert et al. (Nonlinear Anal 39:909–922, 2000) are obtainable from their scalar versions. Finally, in Sect. 4, we stress that the quasiorder version of the "smooth" variational principle in Borwein and Preiss (Trans Am Math Soc 303:517–527, 1987) is also entering in the logical segment we just described.

Keywords Quasiordered structure · Brezis-Browder principle · Dependent Choice · Ekeland variational principle · logical equivalence · asymptotic maximality · vector variational statement · gauge function · smooth variational principle

1 Brezis–Browder Principles

1.1 Introduction

Let M be a nonempty set. Take a *quasiorder* $(\leq)$ (i.e., reflexive and transitive relation) over it; and a function $x \mapsto \psi(x)$ from M to $R_+ := [0, \infty[$. Call $z \in M$, $(\leq, \psi)$-*maximal* when: $z \leq w \in M$ implies $\psi(z) = \psi(w)$. A basic result about such points is the 1976 Brezis–Browder ordering principle [9] (in short: BB).

Theorem 1 *Suppose that*

(a01) $(M, \leq)$ is sequentially inductive:
each ascending sequence has an upper bound (modulo $(\leq)$)
(a02) ψ is $(\leq)$-decreasing ($x \leq y \Longrightarrow \psi(x) \geq \psi(y)$).

M. Turinici (✉)
"A. Myller" Mathematical Seminar, "A. I. Cuza" University,
700506 Iaşi, Romania
e-mail: mturi@uaic.ro

T. M. Rassias, P. M. Pardalos (eds.), *Mathematics Without Boundaries*,
DOI 10.1007/978-1-4939-1106-6_20, © Springer Science+Business Media, LLC 2014

Then, for each $u \in M$ there exists a $(\leq, \psi)$-maximal $v \in M$ with $u \leq v$.

This statement, including (cf. Sect. 1.3) Ekeland's variational principle [21] (in short: EVP), found some useful applications to convex and nonconvex analysis. So, it was the subject of many extensions; see, for instance, Hyers et al. [30, Chap. 5]. These are interesting from a technical perspective. However, in all concrete situations when a variational principle of this type (VP, say) is to be applied, a substitution by the Brezis–Browder's is always possible. This raises the question of to what extent are the logical inclusions VP $\Longrightarrow$ BB $\Longrightarrow$ EVP effective. As results from the developments below, the former inclusion is sometimes reversible; i.e., many statements VP including BB are but logical equivalents of BB. Concerning the latter inclusion, we show (in Sect 1.2) that BB is deductible from the dependent choices principle (in short: DC) due independently to Bernays [5] and Tarski [46]. So, to close the circle between these, it will suffice proving that EVP includes DC. An early result of this type was provided in 1987 by Brunner [12]. It is our aim to show (in Sect. 1.4) that a further extension of it is possible, in the sense: DC is deductible from a certain discrete Lipschitz countable version of EVP. Summing up, (BB) and (EVP) are both equivalent with (DC); and, as such, mutually equivalent. This tells us that all variational statements VP with DC $\Longrightarrow$ VP $\Longrightarrow$ BB are equivalent to each other.

1.2 (DC) Implies (BBe)

Let M be a nonempty set; and $\mathcal{R} \subseteq M \times M$ stand for a (nonempty) relation over it. For each $x \in M$, denote $M(x, \mathcal{R}) = \{y \in M; x\mathcal{R}y\}$. The following DC is to be noted:

Proposition 1 *Suppose that*

(b01) $M(c, \mathcal{R})$ is nonempty, for each $c \in M$.

Then, for each $a \in M$ there exists $(x_n) \subseteq M$ with $[x_0 = a,\ x_n \mathcal{R} x_{n+1},\ \forall n]$.

This principle—proposed independently by Bernays [5] and Tarski [46]—is deductible from AC (= the Axiom of Choice), but not conversely; cf. Wolk [54]. Moreover, by the developments in Blair [6], Goldblatt [26], Moskhovakis [38, Chap. 8], and Schechter [43, Chap. 6], it suffices to cover a large part of the "usual" mathematics; see also Moore [37, Appendix 2, Table 4].

Let M be a nonempty set. Take a *quasiorder* $(\leq)$ over it, as well as a function $x \mapsto \varphi(x)$ from M to $R \cup \{-\infty\} \cup \{\infty\}$. Define the $(\leq, \varphi)$-maximal property of some $z \in M$ as before. The following "extended" variant of (BB) (denoted as: (BBe)) is to be considered.

Theorem 2 *Assume that (a01) and (a02) are valid (in this extended setting). Then, conclusion of Theorem 1 holds (modulo φ).*

Proof **(i)** Suppose that φ is bounded on M. Define the function $\beta : M \to R$ as: $\beta(v) := \inf [\varphi(M(v, \leq))]$, $v \in M$. Clearly, β is increasing; and

$$\varphi(v) \geq \beta(v), \text{ for all } v \in M. \tag{1}$$

Further, (a02) gives at once a characterization like

$$v \text{ is } (\leq, \varphi)\text{-maximal iff } \varphi(v) = \beta(v). \tag{2}$$

Now, assume by contradiction that the conclusion in this statement is false; i.e., (in combination with (2)) there must be some $u \in M$ such that:

(b02) for each $v \in M_u := M(u, \leq)$, one has $\varphi(v) > \beta(v)$.

Consequently (for all such v), $\varphi(v) > (1/2)(\varphi(v) + \beta(v)) > \beta(v)$; hence

$$v \leq w \text{ and } (1/2)(\varphi(v) + \beta(v)) > \varphi(w), \tag{3}$$

for at least one w (belonging to M_u). The relation $\mathcal{R}$ over M_u introduced via (3) fulfills $M_u(v, \mathcal{R}) \neq \emptyset$, for all $v \in M_u$. So, by (DC), there must be a sequence (u_n) in M_u with $u_0 = u$ and

$$u_n \leq u_{n+1}, (1/2)(\varphi(u_n) + \beta(u_n)) > \varphi(u_{n+1}), \text{ for all } n. \tag{4}$$

We have thus constructed an ascending sequence (u_n) in M_u for which the real sequence $(\varphi(u_n))$ is (via (b02)) strictly descending and bounded below; hence $\lambda := \lim_n \varphi(u_n)$ exists in R. By (a01), (u_n) is bounded from above in M: there exists $v \in M$ such that $u_n \leq v$, $\forall n$ [hence, $v \in M_u$]. From (a02), $\varphi(u_n) \geq \varphi(v)$, $\forall n$; and (as β=increasing) $\varphi(v) > \beta(v) \geq \beta(u_n)$, $\forall n$. The former of these relations gives $\lambda \geq \varphi(v)$ (passing to limit as $n \to \infty$). On the other hand, the latter of these relations yields (via (4)) $(1/2)(\varphi(u_n)+\beta(v)) > \varphi(u_{n+1})$, $\forall n$. Passing to limit as $n \to \infty$ gives $(\varphi(v) \geq)\beta(v) \geq \lambda$; so, combining with the preceding one, $\varphi(v) = \beta(v)(= \lambda)$, contradiction. Hence, (b02) cannot be accepted; and the conclusion follows. **(ii)** Assume now that φ is unbounded. Define the auxiliary function $\chi : M \to [0, \pi]$ as $\chi(x) = A(\varphi(x)), x \in M$ (i.e., $\chi = A \circ \varphi$); where

$$A(t) = \pi/2 + \text{arctg}(t) \text{ if } t \in R; A(-\infty) = 0; A(\infty) = \pi.$$

The function χ is decreasing and bounded from below on M. Therefore, by the preceding step, for each $u \in M$ there exists a $(\leq, \chi)$-maximal $v \in M$ with $u \leq v$. This, along with $\chi(v) = \chi(w) \iff \varphi(v) = \varphi(w)$ tells us that v is $(\leq, \varphi)$-maximal too; and we are done.

Note that, by the argument above, (DC) $\Longrightarrow$ (BBe) $\iff$ (BB). For a slightly different proof, we refer to Cârjă et al. [13, Chap. 2, Sect 2.1]. Further metrical extensions of (BBe) may be found in Turinici [51].

1.3 (BB) Implies (EVP)

In the following, the relationships between BB and some other maximal results in the area is discussed. And then, we show that EVP is deductible from all these.

(A) Let $(M, \leq)$ be a quasiordered structure; and $x \mapsto \varphi(x)$ stand for a function between M and $R_+ \cup \{\infty\}$.

Theorem 3 *Assume that (a01) and (a02) are true, as well as*

(c01) $(M, \leq)$ *is almost regular (modulo* φ*):*
$\forall x \in M, \forall \varepsilon > 0, \exists y = y(x, \varepsilon) \geq x$ *with* $\varphi(y) \leq \varepsilon$.

Then, for each $u \in M$ *there exists* $v \in M$ *with* $u \leq v$ *and* $\varphi(v) = 0$ *(hence* v *is* $(\leq, \varphi)$*-maximal).*

Proof By (c01), there exists $z \geq u$ with $\varphi(z) < \infty$. Clearly, (a01)+(a02) apply to $M(z, \leq)$ and $(\leq, \varphi)$. So (via BB), for $z \in M(z, \leq)$ there must be some $v \in M(z, \leq)$ with **(i)** $z \leq v$ (hence $u \leq v$) and **(ii)** v is $(\leq, \varphi)$-maximal in $M(z, \leq)$. Suppose by contradiction that $\gamma := \varphi(v) > 0$; and fix some β in $]0, \gamma[$. By (c01) again, there exists $y = y(v, \beta) \geq v$ (hence $y \in M(z, \leq)$) with $\varphi(y) \leq \beta < \gamma (= \varphi(v))$; impossible, by the second conclusion above. Hence, $\varphi(v) = 0$.

So far, Theorem 3 is deductible from (BB). But, the converse inclusion is also true; to verify it, we need some conventions. By a (generalized) *pseudometric* over M we shall mean any map $d : M \times M \to R_+ \cup \{\infty\}$. Fix such an object; supposed to be *reflexive* $[d(x, x) = 0, \forall x \in M]$. Call $z \in M$, $(\leq, d)$*-maximal*, if: $u, v \in M$ and $z \leq u \leq v$ imply $d(u, v) = 0$. Note that, if d is (in addition) *sufficient* $[d(x, y) = 0 \Longrightarrow x = y]$, the $(\leq, d)$-maximal property becomes: $z \leq w \in M \Longrightarrow z = w$ (and reads: z is *strongly* $(\leq)$ - *maximal*). So, existence results involving such points are "metrical" versions of the Zorn–Bourbaki maximal principle (cf. Moore [37, Chap. 4, Sect. 4]). A natural way of deriving them is to start from the fact that, in terms of the associated function $[\varphi_d(x) = \sup\{d(u, v); x \leq u \leq v\}, x \in M]$, this property may be characterized as: $\varphi_d(z) = 0$. So, a basic source for determining such elements is Theorem 3 (applied to the introduced function). To do this, note that (a02) is fulfilled by φ_d. On the other hand, the almost regularity (modulo φ_d) condition (c01) may be written as

(c02) $(M, \leq)$ is weakly regular (modulo d): $\forall x \in M, \forall \varepsilon > 0$,
$\exists y = y(x, \varepsilon) \geq x$ such that $y \leq u \leq v \Longrightarrow d(u, v) \leq \varepsilon$.

Putting these together, it results (via Theorem 3) the following maximality statement involving these data (cf. Kang and Park [33]):

Theorem 4 *Assume that* $(M, \leq; d)$ *is such that (a01) and (c02) hold. Then, for each* $u \in M$ *there exists a* $(\leq, d)$*-maximal* $v \in M$ *with* $u \leq v$.

Clearly, Theorem 4 is a logical consequence of (BB). The reciprocal of this is also true. In fact, put $d(x, y) = |\psi(x) - \psi(y)|, x, y \in M$; and let $\beta : M \to R$ stand for the function of Theorem 2 (with ψ in place of φ). Note that, if $\psi(x) > \beta(x)$, any point $y \in M(x, \leq)$ with $\beta(x) \leq \psi(y) < \beta(x) + \varepsilon$ is like in (c02); as results from

$$x \leq y \leq u \leq v \Longrightarrow \psi(y) \geq \psi(u) \geq \psi(v) \geq \beta(v) \geq \beta(u) \geq \beta(y) \geq \beta(x).$$

So, Theorem 4 applies to these data; and, from its conclusion, all is clear. We therefore established the inclusion chain: BB $\Longrightarrow$ Th 3 $\Longrightarrow$ Th 4 $\Longrightarrow$ BB. Hence, all these ordering principles are nothing but logical equivalents of (BB). Note finally that, by the developments in Turinici [48], the obtained conclusions comprise as well the 1974 "uniform" maximal principle in Brøndsted [10]; as well as the 1983 "metrical" one in Dancs et al. [17]. Further aspects may be found in Bae et al. [2]; see also Szaz [44].

(B) A useful application of these facts is to "monotone" VPs. Let $(M, \leq)$ be a quasiordered structure; and $d : M \times M \to R_+$ be a *metric* (i.e., sufficient semimetric) over M. Call the subset Z of M, $(\leq)$-*closed* when the limit of each $(\leq)$-ascending sequence in Z belongs to Z. In particular, we say that $(\leq)$ is *self-closed* provided $M(x, \leq)$ is $(\leq)$-closed, for each $x \in M$; or, equivalently: the limit of each ascending sequence is an upper bound of it (modulo $(\leq)$). For example, this holds if $(\leq)$ is *semiclosed* (according to Nachbin [39, Appendix]): $M(x, \leq)$ is closed, for each $x \in M$. Finally, call d, $(\leq)$-*complete* provided each $(\leq)$-ascending d-Cauchy sequence is d-convergent.

We are now in position to state the announced result. Assume that

(c03) $(\leq)$ is self-closed and d is $(\leq)$-complete;

and take a function $\varphi : M \to R \cup \{\infty\}$ in accordance with

(c04) φ is inf-proper ($\mathrm{Dom}(\varphi) \neq \emptyset$ and $\varphi_* := \inf[\varphi(M)] > -\infty$)
(c05) φ is $(\leq, d)$-lsc: $\{x \in M; \varphi(x) \leq t\}$ is $(\leq)$-closed, for each $t \in R$.

Theorem 5 *Let these conditions hold. Then, for each $u \in \mathrm{Dom}(\varphi)$ there exists $v \in \mathrm{Dom}(\varphi)$ with*

$$u \leq v, d(u, v) \leq \varphi(u) - \varphi(v) \text{ (hence } \varphi(u) \geq \varphi(v)) \tag{5}$$

$$d(v, x) > \varphi(v) - \varphi(x), \text{ for each } x \in M(v, \leq)\setminus\{v\}. \tag{6}$$

Proof Denote for simplicity $M[u] = \{x \in M; u \leq x, \varphi(u) \geq \varphi(x)\}$. Clearly, $\emptyset \neq M[u] \subseteq \mathrm{Dom}(\varphi)$; moreover, by (c03) + (c05),

$$M[u] \text{ is } (\leq)\text{-closed; hence } d \text{ is } (\leq)\text{-complete on } M[u]. \tag{7}$$

Let $(\preceq)$ stand for the relation: $x \preceq y$ iff $[x \leq y,\ d(x, y) + \varphi(y) \leq \varphi(x)]$. Clearly, $(\preceq)$ acts as an *order* (antisymmetric quasiorder) on $\mathrm{Dom}(\varphi)$; so, it remains as such on $M[u]$. We claim that conditions of BB are fulfilled on $M[u]$ and $(\preceq, \varphi)$. In fact, by this very definition, φ is $(\preceq)$-decreasing on $M[u]$. On the other hand, let (x_n) be a $(\preceq)$-ascending sequence in $M[u]$:

(c06) $x_n \leq x_m$ and $d(x_n, x_m) \leq \varphi(x_n) - \varphi(x_m)$, if $n \leq m$.

The sequence $(\varphi(x_n))$ is descending and (by (c04)) bounded from below; hence a Cauchy one. This, along with (c06), tells us that (x_n) is a $(\leq)$-ascending d-Cauchy sequence in $M[u]$; wherefrom (by (7)), there must be some $y \in M[u]$ with $x_n \to y$. Passing to limit as $m \to \infty$ in (c06) one derives (via (c03) + (c05))

$$x_n \leq y, d(x_n, y) \leq \varphi(x_n) - \varphi(y), \text{ (i.e.: } x_n \preceq y\text{), for all } n.$$

In other words, $y \in M[u]$ is an upper bound (modulo $(\preceq)$) of (x_n); and this shows that $(M[u], \preceq)$ is sequentially inductive. From BB it then follows that, for the starting $u \in M[u]$ there exists $v \in M[u]$ with **j)** $u \preceq v$ and **jj)** $v \preceq x \in M[u]$ implies $\varphi(v) = \varphi(x)$. The former of these is just (5). And the latter one gives at once (6). In fact, let $y \in M$ be such that $v \leq y$, $d(v, y) \leq \varphi(v) - \varphi(y)$. As a consequence, $y \in M[u]$ and $v \preceq y$; so that (by **jj)** above) $\varphi(v) = \varphi(y)$. Combining with the previous relation gives $d(v, y) = 0$ (hence $v = y$); and we are done.

A basic particular case of our developments corresponds to $(\leq) = M \times M$ (= the *trivial* quasiorder on M); when (c05) may be written as

(c07) φ is d-lsc ($\liminf_n \varphi(x_n) \geq \varphi(x)$, provided $x_n \to x$).

The corresponding version of Theorem 5 under this choice of our data is just the 1974 EVP [21].

Theorem 6 *Let the metric space (M, d) be complete and the function $\varphi : M \to R \cup \{\infty\}$ be as in (c04) + (c07). Then, for each $u \in \mathrm{Dom}(\varphi)$ there exists $v = v(u) \in \mathrm{Dom}(\varphi)$ with*

$$d(u, v) \leq \varphi(u) - \varphi(v) \text{ (hence } \varphi(u) \geq \varphi(v)\text{)} \tag{8}$$

$$d(v, x) > \varphi(v) - \varphi(x), \text{ for all } x \in M \setminus \{v\}. \tag{9}$$

Returning to the initial framework, note that (c05) also holds under (a02) (modulo φ) and the self-closeness of $(\leq)$. For this reason, Theorem 5 will be called the *monotone* version of EVP (in short: (EVPm)). For concrete applications of these, we refer to Hyers et al. [30, Chap. 5].

1.4 (EVPdLc) Implies (DC)

By the developments above, we have the implications: (DC) $\Longrightarrow$ (BBe) $\Longrightarrow$ (BB) $\Longrightarrow$ (EVPm) $\Longrightarrow$ (EVP). So, it is natural asking whether these may be reversed. Clearly, the natural setting for solving this problem is (ZF)(= the *standard* Zermelo–Fraenkel system) without (AC) (= the axiom of choice); referred to in the following as the *reduced* Zermelo–Fraenkel system.

Let $(X, \leq)$ be a partially ordered structure. We say that $(\leq)$ has the *inf-lattice* property, provided: $x \wedge y := \inf(x, y)$ exists, for all $x, y \in X$. Further, call $z \in$

X, $(\leq)$-*maximal* if $X(z, \leq) = \{z\}$; the class of all these points will be denoted as $\max(X, \leq)$. In this case, $(\leq)$ is termed a *Zorn order* when $\max(X, \leq)$ is nonempty and *cofinal* in X [for each $u \in X$ there exists a $(\leq)$-maximal $v \in X$ with $u \leq v$]. Further aspects are to be described in a metrical setting. Let $d : X \times X \to R_+$ be a metric over X; and $\varphi : X \to R_+$ be some function. Then, the natural choice for $(\leq)$ above is

$$x \leq_{(d,\varphi)} y \text{ iff } d(x, y) \leq \varphi(x) - \varphi(y);$$

referred to as the Brøndsted order [11] attached to (d, φ). Denote $X(x, \rho) = \{u \in X; d(x, u) < \rho\}$, $x \in X$, $\rho > 0$ [the *open sphere* with center x and radius ρ]. Call the ambient metric space (X, d), *discrete* when for each $x \in X$ there exists $\rho = \rho(x) > 0$ such that $X(x, \rho) = \{x\}$. Note that, under such a hypothesis, any function $\psi : X \to R$ is continuous over X. However, the Lipschitz property ($|\psi(x) - \psi(y)| \leq Ld(x, y)$, $x, y \in X$, for some $L > 0$) cannot be assured, in general.

Now, the statement below is a particular case of EVP:

Theorem 7 *Let the metric space (X, d) and $\varphi : X \to R_+$ satisfy*

(d01) (X, d) is discrete bounded and complete
(d02) $(\leq_{(d,\varphi)})$ has the inf-lattice property
(d03) φ is d-nonexpansive and $\varphi(X)$ is countable.

Then, $(\leq_{(d,\varphi)})$ is a Zorn order.

We shall refer to it as: the discrete Lipschitz-countable version of EVP (in short: (EVPdLc)). Clearly, (EVP) $\Longrightarrow$ (EVPdLc). The remarkable fact to be added is that this last principle yields (DC); so, it completes the circle between all these.

Proposition 2 *We have (EVPdLc) $\Longrightarrow$ (DC) (in the reduced Zermelo–Fraenkel system). So (by the above), the maximal/variational principles (BB), (BBe), (EVPm) and (EVP) are all equivalent with (DC); hence, mutually equivalent.*

Proof Let M be a nonempty set; and $\mathcal{R}$ stand for some relation over it, with the property (b01). Fix in the following $a \in M$, $b \in M(a, \mathcal{R})$. For each $p \geq 2$ in N (= the set of natural numbers), let $N(p, >) := \{0, \ldots, p-1\}$ stand for the *initial segment* determined by p. Denote by X_p the class of all finite sequences $x : N(p, >) \to M$ with: $x(0) = a$, $x(1) = b$, and $x(n)Rx(n + 1)$ for $0 \leq n \leq p - 2$. In this case, $N(p, >)$ is just $\mathrm{Dom}(x)$ (the *domain* of x); and $p = \mathrm{card}(N(p, >))$ will be referred to as the *order* of x [denoted as $\omega(x)$]. The effectiveness of this construction is to be judged from

Lemma 1 *Under the premise (b01) of (DC), we have*

$$\textit{(for each } p \geq 2\textit{):} \ \ X_p \textit{ is nonempty.} \tag{10}$$

Proof The case $p = 2$ is clear. Assume that (10) is true for all $p \in \{2, \ldots, q\}$, where $q \geq 2$; we claim that it holds as well for $p = q + 1$. As X_q is nonempty, there exists a function $F : N(q; >) \to M$ with the properties $[F(0) = a, F(1) = b, F(n)\mathcal{R}F(n+1)$, for $0 \leq n \leq q-2]$. Taking (b01) into account, it follows that the subset $M(F(q-1);\mathcal{R})$ is nonempty. Letting z be an element of it, define a function $G : N(q+1,>) \to M$ as: $[G(n) = F(n), n < q; G(q) = z]$. This object is an element of X_{q-1}; and the conclusion follows.

Having these precise, denote $X = \cup\{X_n; n \geq 2\}$. Let $\preceq$ stand for the partial order (on X)

(d04) $x \preceq y$ iff $\mathrm{Dom}(x) \subseteq \mathrm{Dom}(y)$ and $x = y|_{\mathrm{Dom}(x)}$;

and $\prec$ denote its associated strict order. All we have to prove is that $(X, \preceq)$ has strictly ascending infinite sequences. To this end, we need some conventions and auxiliary facts.

(A) Let $x, y \in X$ be arbitrary fixed. Denote

$K(x,y) := \{n \in \mathrm{Dom}(x) \cap \mathrm{Dom}(y); x(n) \neq y(n)\}$.

If x and y are *comparable* (i.e., either $x \preceq y$ or $y \preceq x$; written as: $x <> y$) then $K(x,y) = \emptyset$. Conversely, if $K(x,y) = \emptyset$, then $x \preceq y$ if $\mathrm{Dom}(x) \subseteq \mathrm{Dom}(y)$ and $y \preceq x$ if $\mathrm{Dom}(y) \subseteq \mathrm{Dom}(x)$; hence $x <> y$. Summing up,

$$(x, y \in X):\ x <> y \text{ if and only if } K(x,y) = \emptyset. \tag{11}$$

The negation of this property means: x and y are *not comparable* (denoted as: $x||y$). By the characterization above, it is equivalent with $K(x,y) \neq \emptyset$. Note that, in such a case, $k(x,y) := \min(K(x,y))$ is well defined as an element of $N(2,\leq)$; and $N(k(x,y),>)$ is the largest initial interval of $\mathrm{Dom}(x) \cap \mathrm{Dom}(y)$ where x and y are identical.

Lemma 2 *The partial order $(\preceq)$ has the inf-lattice property. Moreover, $x \mapsto \omega(x)$ is strictly increasing ($x \prec y$ implies $\omega(x) < \omega(y)$) and*

$$2 \leq \omega(x \wedge y) \leq \min\{\omega(x) - 1, \omega(y) - 1\}, \text{ whenever } x||y. \tag{12}$$

Proof **(of Lemma 2) (i)** Let $x, y \in X$ be arbitrary fixed. The case $x <> y$ is clear; so, without loss, one may assume that $x||y$. Note that, by the remark above, $K(x,y) \neq \emptyset$ and $k := k(x,y)$ exists as an element of $N(2,\leq)$. Let the finite sequence $z \in X_k$ be introduced as $z = x|_{N(k,>)} = y|_{N(k,>)}$. For the moment $z \preceq x$ and $z \preceq y$. Suppose that $w \in X_h$ (where $h \geq 2$) fulfills the same properties. Then, the restrictions of x and y to $N(h,>)$ are identical; wherefrom (see above) $h \leq k$ and $w \preceq z$. **(ii)** Evident. **(iii)** As $x||y$, we must have $\omega(x), \omega(y) \geq 3$ and $\omega(x \wedge y) = k(x,y) \geq 2$. This, and $k(x,y) \in \mathrm{Dom}(x) \cap \mathrm{Dom}(y)$, give the desired relation.

(B) Our next objective is to introduce a metrical structure as well as an associated objective function over X, which should have all required properties. Put $\varphi(x) = 3^{-\omega(x)}$, $x \in X$; and note that $\varphi(X) = \{3^{-n}; n \geq 2\}$ (hence, φ has countable many strictly positive values). Then, define

(d05) $d(x,y) = |\varphi(x) - \varphi(y)|$, if $x <> y$; $d(x,y) = \varphi(x \wedge y)$, when $x||y$.

Lemma 3 *The mapping $(x,y) \mapsto d(x,y)$ is a metric on X.*

Proof **(of Lemma 3)** Clearly, d is reflexive and symmetric $[d(x,y) = d(y,x), x, y \in X]$. On the other hand, d is sufficient. In fact, assume $d(x,y) = 0$. By a previous evaluation of $\varphi(X)$, it results that x and y are comparable and $\omega(x) = \omega(y)$; wherefrom, $x = y$. Finally, let us verify the triangular property: $d(x,z) \leq d(x,y)+d(y,z)$, for all $x, y, z \in X$. Without loss, one may assume that $\omega(x) \leq \omega(z)$; for, otherwise, we simply interchange x and z. Two alternatives are open before us.

(a) The points x and z are comparable; that is, $x \preceq z$ (by the hypothesis above). We start from the obvious relation

$$|\varphi(s) - \varphi(t)| \leq \max\{\varphi(s), \varphi(t)\} \leq \varphi(s \wedge t),\ s,t \in X.$$

Combining with $d(x,z) = |\varphi(x) - \varphi(z)| \leq |\varphi(x) - \varphi(y)| + |\varphi(y) - \varphi(z)|$ yields the desired fact, for all possible cases concerning (x, y) and (y, z).

(b) The points x and z are not comparable $(x||z)$. Four subcases appear:

Subcase (b1): Suppose that $x <> y$, $y <> z$. The alternatives $[x \preceq y, y \preceq z]$ and $[y \preceq x, z \preceq y]$ give $x <> z$; contradiction. So, it remains to discuss the alternatives:

(b11) $x \preceq y, z \preceq y$. Then, x and z are the restrictions of y to $\mathrm{Dom}(x)$ and $\mathrm{Dom}(z)$ respectively; wherefrom $x \preceq z$, contradiction.

(b12) $y \preceq x$, $y \preceq z$. We start from the direct consequence of (12)

$$3 \max\{\varphi(s), \varphi(t)\} \leq \varphi(s \wedge t) \leq 3^{-2},\ \ s,t \in X, s||t.$$

The relation to be checked becomes $\varphi(x \wedge z) \leq 2\varphi(y) - \varphi(x) - \varphi(z)$. By the imposed conditions, $y \preceq x \wedge z$; wherefrom $\varphi(y) \geq \varphi(x \wedge z)$. A sufficient condition for the desired relation to be true is $\varphi(x \wedge z) \leq 2\varphi(x \wedge z) - \varphi(x) - \varphi(z)$; i.e., $\varphi(x) + \varphi(z) \leq \varphi(x \wedge z)$; evident, by the precise consequence.

Subcase (b2): Suppose that $x||y$, $y <> z$. Two logical possibilities occur:

(b21) $x||y$, $y \preceq z$. We have to establish that: $\varphi(x \wedge z) \leq \varphi(x \wedge y) + \varphi(y) - \varphi(z)$. But, evidently, $x \wedge z \succeq x \wedge y$; wherefrom $\varphi(x \wedge z) \leq \varphi(x \wedge y)$; and then, all is clear.

(b22) $x||y$, $z \preceq y$ (or, equivalently: $z \preceq y$, $y||x$). The desired relation becomes: $\varphi(x \wedge z) \leq \varphi(x \wedge y) + \varphi(z) - \varphi(y)$. For the moment, $x \wedge z \preceq x \wedge y$. If $x \wedge z \prec x \wedge y$, we must get $q := \omega(x \wedge z) < \omega(x \wedge y)$; so (by definition) $x(q) = y(q)$. As $z = y|_{\mathrm{Dom}(z)}$ and $q \in \mathrm{Dom}(x) \cap \mathrm{Dom}(z)$, this yields $y(q) = z(q)$; hence $x(q) = z(q)$, contradiction. Consequently, $x \wedge z = x \wedge y$; and conclusion follows.

Subcase (b3): $x <> y$, $y||z$. As before, two logical possibilities occur:

(b31) $x \preceq y$, $y||z$. This is just the alternative **(b22)**, with (x, y, z) in place of (z, y, x).

(b32) $y \preceq x$, $y||z$. The triangular relation becomes $\varphi(x \wedge z) \leq \varphi(y) - \varphi(x) + \varphi(y \wedge z)$. But, from hypothesis, $x \wedge z \succeq y \wedge z$; wherefrom, all is clear.

Subcase (b4): $x||y$, $y||z$. We have to establish that: $\varphi(x \wedge z) \leq \varphi(x \wedge y) + \varphi(y \wedge z)$. As before, the alternative $[\omega(x \wedge z) \geq \omega(x \wedge y)$ or $\omega(x \wedge z) \geq \omega(y \wedge z)]$ gives the desired fact. On the other hand, the alternative $q := \omega(x \wedge z) < \min\{\omega(x \wedge y), \omega(y \wedge z)\}$ yields $x(q) = y(q)$, $y(q) = z(q)$; hence $x(q) = z(q)$, contradiction.

Having discussed all possible cases, the conclusion follows.

(C) Note that, by a previous remark involving $\varphi(X)$, one has $\mathrm{diam}(X) \leq 3^{-2}$. Further properties of the triplet $(X, d; \varphi)$ are contained in

Lemma 4 *Under the notations above, one has (for each $m \geq 2$)*

$$x, y \in X,\ \omega(x) \leq m,\ d(x, y) < 2 \cdot 3^{-m-1} \Longrightarrow x = y; \tag{13}$$

so that, the metric space (X, d) is discrete.

Proof **(of Lemma 4)** Assume that $x \neq y$. We show that this cannot be in agreement with our hypothesis. Two cases are open before us:

(i) Let x and y be comparable: either $x \prec y$ or $y \prec x$. If $x \prec y$, we have $\omega(x) + 1 \leq \omega(y)$; and then $(1/3)\varphi(x) \geq \varphi(y)$; hence (by definition) $d(x, y) = \varphi(x) - \varphi(y) \geq (2/3)\varphi(x) \geq 2 \cdot 3^{-m-1}$, contradiction. If $y \prec x$ then (by the same way as before) $d(x, y) \geq (2/3)\varphi(y) \geq (2/3)\varphi(x) \geq 2 \cdot 3^{-m-1}$, again a contradiction.

(ii) Suppose that x and y are not comparable. Then (by definition) $d(x, y) = \varphi(x \wedge y) \geq \varphi(x) \geq 3 \cdot 3^{-m-1}$, contrary to the hypothesis.

Lemma 5 *Under the same notations above,*

$$|\varphi(x) - \varphi(y)| \leq d(x, y),\ \forall x, y \in X; \tag{14}$$

so, φ is d-nonexpansive (hence, all the more d-Lipschitz).

Proof **(of Lemma 5)** If x and y are comparable, then $d(x, y) = |\varphi(x) - \varphi(y)|$; and we are done. If x and y are not comparable then, without loss, one may assume $\omega(x) \leq \omega(y)$; hence $\varphi(x) \geq \varphi(y)$. As $x \wedge y \preceq x$, we have $d(x, y) = \varphi(x \wedge y) \geq \varphi(x) \geq \varphi(x) - \varphi(y) = |\varphi(x) - \varphi(y)|$; and the conclusion follows.

(D) Given the couple (d, φ) as before, we introduce the Brøndsted order $(\leq_{(d,\varphi)})$ on X; also denoted as $(\leq)$, for simplicity. It is natural to ask which is the relationship between it and the initial order $(\preceq)$ on X.

Lemma 6 *We necessarily have (under these conventions)*

$$x \preceq y \text{ if and only if } x \leq y. \tag{15}$$

That is: these partial orders coincide over X.

Proof **(of Lemma 6)** Clearly, $x \preceq y$ gives $\omega(x) \leq \omega(y)$; wherefrom $d(x, y) = \varphi(x) - \varphi(y)$; i.e., $x \leq y$. Conversely, assume that $x \leq y$. For the moment, x and y are comparable; since otherwise, the imposed condition gives $\varphi(x \wedge y) \leq \varphi(x) - \varphi(y) \leq \varphi(x)$ [hence $\omega(x \wedge y) \geq \omega(x)$]; in contradiction with (12). The alternative $y \preceq x$ yields (by the first part) $y \leq x$; wherefrom (as $(\leq)$ is order) $x = y$. Hence, anyway $x \preceq y$.

(E) We are now in position to complete the argument. As (b01) holds, we necessarily have $\max(X, \preceq) = \emptyset$; i.e.: for each $x \in X$ there exists $y \in X$ with $x \prec y$. This, along with (EVPdLc) tells us that (X, d) is not complete; i.e., there exists at least one d-Cauchy sequence (x_n) in X which is not d-convergent. Note that both these properties are transferable to all subsequences of (x_n). So (passing to a subsequence if necessary), one may take this sequence in such a way that, for all m,

(d06) $d(x_p, x_q) < 3^{-m-1}$, whenever $p, q \geq m$.

In fact, the d-Cauchy property assures us (with $\varepsilon = 3^{-m-1}$) that

$$C(m) := \{n \in N; d(x_p, x_q) < 3^{-m-1}, \text{for all } p, q \geq n\} \neq \emptyset,\ \forall m \in N.$$

In addition, $n \mapsto C(n)$ is $(\subseteq)$-decreasing; hence $n \mapsto g(n) := \min[C(n)]$ is $(\leq)$-increasing. This finally tells us that $n \mapsto h(n) := n + g(n)$ is strictly $(\leq)$-increasing;

wherefrom $(y_n = x_{h(n)}; n \in N)$ is a subsequence of (x_n) fulfilling (d06), in view of: $p, q \geq m \Longrightarrow h(p), h(q) \geq h(m) \geq g(m)$.

Now, by the divergence of (x_n), we must have

$$B(h) := \{n \in N(h, <); \omega(x_n) > \omega(x_h)\} \neq \emptyset, \text{ for all } h \in N. \tag{16}$$

For, otherwise, there exists $h \in N$ such that $\omega(x_n) \leq m := \omega(x_h)$, $\forall n \geq h$; wherefrom (denoting $r := \max(m, h)$ and taking (d06) into account) $d(x_r, x_n) < 3^{-r-1} \leq 3^{-m-1}$, $\forall n \geq r$; and this, by Lemma 4, yields $x_n = x_r$, $\forall n \geq r$; hence, (x_n) is convergent, contradiction. As a consequence of this, the selfmap (of N) $F(n) = \min B(n)$, $n \in N$, is well defined (without any use of DC); in addition, by the very definition above,

$$F(h) > h, \ \omega(x_{F(h)}) > \omega(x_h), \ \text{ for all } h \in N. \tag{17}$$

By the first half of this relation, $(p(n) := F^n(1); n \geq 0)$ is strictly ascending, with $p(0) = 1$; hence, $p(n) > n$, for all n. It therefore generates a subsequence $(y_n := x_{p(n)}; n \in N)$ of (x_n) with the supplementary property (deductible from (d06))

$$d(y_k, y_h) < 3^{-k-1}, \text{ whenever } k, h \in N \text{ satisfy } k \leq h. \tag{18}$$

Further, by (17) (the second half), $(q(n) := \omega(y_n); n \in N)$ is strictly ascending too, with $q(0) \geq 2$; hence, $q(n) \geq n + 2$, for all n. The sequence $(z_n := y_n|_{N(n+2,>)}; n \geq 0)$ is therefore well defined in X. We claim that

$$(z_n; n \geq 0) \text{ is strictly ascending: } z_k \prec z_h, \text{ for } k < h.$$

In fact, $y_k <> y_h$ gives $y_k \prec y_h$ (as $\omega(y_k) < \omega(y_h)$); and, from this, $z_k \prec z_h$. And, if $y_k || y_h$ we have (by (18)), $\omega(y_k \wedge y_h) \geq k + 2$; so that

$$z_k = y_k|_{N(k+2,>)} = y_h|_{N(k+2,>)} = z_h|_{N(k+2,>)};$$

wherefrom $z_k \prec z_h$. But then, the sequence $(c_n = z_n(n); n \geq 0)$ is well defined in M; moreover, by this ascending property, $c_0 = a, c_1 = b$ and $c_n \mathcal{R} c_{n+1}$, for all n. This gives us the desired conclusion.

In particular, when the specific assumptions (d02) and (d03) are ignored in Theorem 7, Proposition 2 reduces to the result in Brunner [12]. For a different proof of the same (with respect to the standard version of EVP) see Dodu and Morillon [19].

2 Asymptotic Maximal Statements

2.1 Introduction

Let M be a nonempty set. Take a *quasiorder* $(\leq)$ over it; as well as a function $\psi : M \to R_+$. Call $z \in M$, *$(\leq, \psi)$-maximal* when: ψ is constant on $M(z, \leq)$

$:= \{w \in M; z \leq w\}$. The following result about such elements (referred to as: the Asymptotic BB (in short: (BBa)) is our starting point. Let us introduce the associated function (from M to R_+) $\beta(x) := \inf \psi[M(x, \leq)]$, $x \in M$; note that $[\beta(x) \leq \psi(x), \forall x \in M]$ and β is increasing $[x \leq y \Rightarrow \beta(x) \leq \beta(y)]$.

Theorem 8 *Suppose that ψ is $(\leq)$-decreasing and*

> *(a01) $(M, \leq)$ is sequentially inductive (modulo ψ): each ascending sequence (x_n) in M with $\psi(x_n) - \beta(x_n) \to 0$ has an upper bound (modulo $(\leq)$).*

Then, for each $u \in M$ there exists a $(\leq, \psi)$-maximal $v \in M$ with $u \leq v$.

In particular, (a01) holds under

> (a02) $(M, \leq)$ is sequentially inductive:
> each ascending sequence in M has an upper bound (modulo $(\leq)$);

when (BBa) reduces to the BB [9]; which, in turn, includes (as already established) EVP [22]. Some other statements in the area (which also extend BB) may be found in Turinici [49]; see also Altman [1]. As remarked in a previous place, in all concrete situations when a maximality principle of this type (MP, say) is to be applied, a substitution of it by (BB) is always possible. In other words, from a practical perspective, such enlargements of (BB) are not effective. Concerning their theoretical confirmation, some partial results were made precise in Turinici [53]. On the other hand, there do exist maximality principles (including EVP) which, at a first glance, are either noncomparable with (BB) (cf. Kang and Park [33]) or include it in a strict sense (cf. Du [20]). It is our first aim to stress (in Sects. 2.2 and 2.4) that all these are obtainable from a "uniform" version of (BBa). As a second aim, we show that

(i) all these maximal principles (MP) (hence, in particular, (BBa)) are deductible (cf. Sect. 2.2) from the "diagonal dependent choices" principle (in short: DDC); which, in turn, includes the DC due to Bernays [5] and Tarski [46];

(ii) (DDC) (hence, a fortiori, DC) is deductible (cf. Sect. 2.3) from a certain discrete Lipschitz countable version of (EVP).

So, combining these, all maximal principles in question are equivalent with both (DDC) and (EVP).

2.2 (DDC) Implies (BBa)

Let M be a nonempty set. By a *relation* over it we mean any (nonempty) part $\mathcal{R}$ of $M \times M$. For each $x \in M$, denote $M(x, \mathcal{R}) = \{y \in M; x\mathcal{R}y\}$ (the *x-section* of $\mathcal{R}$ in M).

(A) The following "diagonal" type statement is our starting point.

Proposition 3 *Let the sequence of relations $(\mathcal{R}_n; n \in N)$ fulfill*

> *(b01) $(\forall n \in N)$: $M(c, \mathcal{R}_n) \neq \emptyset$, $\forall c \in M$ [i.e.: $\mathrm{Dom}(\mathcal{R}_n) = M$].*

Then, for each $a \in M$ there exists $(x_n) \subseteq M$ with $[x_0 = a,\ x_n \mathcal{R}_n x_{n+1}, \forall n]$.

We shall term it, the *Diagonal Dependent Choices* Principle (in short: DDC). In particular, when $(\mathcal{R}_n)$ is constant, (DDC) is just the DC due to Bernays [5] and Tarski [46]. Note that (DDC) is deductible from the axiom of choice; but not conversely.

(B) Let again M be some nonempty set. Take a quasiorder $(\leq)$ over it, as well as a function $\rho : M \to R_+ \cup \{\infty\}$. Define the $(\leq, \rho)$-maximal property of some $z \in M$ as before. The following "uniform" variant of (BBa) (referred to as: the uniform asymptotic BB, in short, (BBau)) is deductible from (DDC) above.

Theorem 9 *Assume that ρ is $(\leq)$-decreasing, and*

(b02) $((M, \leq)$ is almost regular (modulo ρ))
$\forall x \in M, \forall \varepsilon > 0, \exists y = y(x, \varepsilon) \geq x : \rho(y) \leq \varepsilon$
(b03) $(M, \leq)$ is sequentially inductive (modulo ρ): each ascending sequence (x_n) with $\rho(x_n) \to 0$ has an upper bound (modulo $(\leq)$).

Then, for each $u \in M$ there exists $v \in M$ with $u \leq v$ and $\rho(v) = 0$ (hence v is $(\leq, \rho)$-maximal).

Proof For the arbitrary fixed ε in $R_+^0 :=]0, \infty[$, let $\mathcal{R}(\varepsilon)$ denote the relation (over M): $x\mathcal{R}(\varepsilon)y$ iff $x \leq y$ and $\rho(y) \leq \varepsilon$; note that, from (b02),

$$M(c, \mathcal{R}(\varepsilon)) \neq \emptyset, \text{ for all } c \in M \text{ and all } \varepsilon > 0. \tag{19}$$

Now, let (ε_n) be a strictly descending sequence in R_+^0 with $\varepsilon_n \to 0$. From (19), the sequence of relations $(\mathcal{R}_n := \mathcal{R}(\varepsilon_n); n \in N)$ fulfills (b01). So, by (DDC), there must be a sequence (u_n) in M with $u_0 = u$ and

$$u_n \leq u_{n+1}, \ \rho(u_{n+1}) \leq \varepsilon_n, \text{ for all } n. \tag{20}$$

The obtained sequence is therefore $(\leq)$-ascending and $\rho(u_n) \to 0$ as $n \to \infty$. So, combining with (b03), (u_n) has at least one upper bound (modulo $(\leq)$), say v; i.e., $u_n \leq v$, for all n. This, along with (20) and [ρ = decreasing] gives $\rho(v) \leq \varepsilon_n$, for all n; wherefrom $\rho(v) = 0$.

(B) A basic particular case of these facts corresponds to the construction below. By a (generalized) *pseudometric* over M we shall mean any map $d : M \times M \to R_+ \cup \{\infty\}$. Fix such an object; supposed to be *reflexive* $[d(x, x) = 0, \forall x \in M]$. Call $z \in M$, *$(\leq, d)$-maximal*, if: $u, v \in M$ and $z \leq u \leq v$ imply $d(u, v) = 0$. Note that, in terms of the associated function (from M to $R_+ \cup \{\infty\}$)

(b04) $\rho_d(x) = \sup\{d(u, v); x \leq u \leq v\}, x \in M,$

this property may be characterized as: $\rho_d(z) = 0$. So, a basic source for determining such elements is Theorem 9 (applied to this function). To do this, we start from the obvious fact that ρ_d is decreasing. Concerning the remaining properties, let us say that $(M, \leq)$ is *almost regular* (modulo d), when it fulfills (b02) relative to ρ_d; that is,

(b05) $\forall x \in M, \forall \varepsilon > 0, \exists y = y(x, \varepsilon) \geq x$: $(y \leq u \leq v \Longrightarrow d(u, v) \leq \varepsilon)$.

Further, call the sequence (x_n), *strongly $(\leq, d)$-Cauchy* provided $\rho_d(x_n) \to 0$ as $n \to \infty$. Then, condition (b03) relative to ρ_d means:

(b06) each ascending strongly $(\le, d)$-Cauchy sequence has an upper bound;

and will be termed as: $(M, \le)$ is *weakly sequentially inductive* (modulo d). Putting these together, it results (via (BBau)):

Theorem 10 *Assume that $(M, \le)$ is almost regular and weakly sequentially inductive (modulo d). Then, for each $u \in M$ there exists a $(\le, d)$-maximal $v \in M$ with $u \le v$.*

Now, evidently, each (ascending) strongly $(\le, d)$-Cauchy sequence (x_n) is d-*Cauchy* $[\forall \varepsilon > 0, \exists n(\varepsilon)$ such that $n(\varepsilon) \le p \le q \Longrightarrow d(x_p, x_q) \le \varepsilon]$. Hence, (b06) holds under the stronger counterpart

(b07) each ascending d-Cauchy sequence has an upper bound.

As a direct consequence of this, we have the following statement in Turinici [49] (see also Kang and Park [33]):

Theorem 11 *Let the quasiordered structure $(M, \le)$ be such that (b07) holds, as well as*

(b08) $(M, \le)$ is regular (modulo d):
each ascending sequence is d-Cauchy.

Then, conclusions of Theorem 10 are retainable.

Proof It will suffice verifying that (b08) implies (b05). Assume by contradiction that this last condition is not fulfilled; that is (for some $x \in M$, $\varepsilon > 0$): $\forall y \in M(x, \le)$, $\exists u, v \in M$: $[y \le u \le v, d(u, v) > \varepsilon]$. This, by definition, yields (for the same (x, ε)):

$$\forall y \in M(x, \le), \exists\, (u, v) \in \mathrm{gr}(\le):\ y \le u,\, d(u, v) > \varepsilon.$$

[Here, for simplicity, we denoted $\mathrm{gr}(\le) = \{(a, b) \in X \times X; a \le b\}$]. Fix $y_0, y_1 \in X$ with $x \le y_0 \le y_1$; hence, $w_0 := (y_0, y_1) \in \mathrm{gr}(\le)$. Define a relation $\mathcal{R} = \mathcal{R}(\varepsilon)$ on $M := \{(a, b) \in \mathrm{gr}(\le); y_1 \le a\}$ as:

$$(a_1, b_1)\mathcal{R}(a_2, b_2) \text{ if and only if } b_1 \le a_2,\, d(a_2, b_2) > \varepsilon.$$

From the imposed condition, $M((a, b), \mathcal{R}) \neq \emptyset$, $\forall (a, b) \in M$. So, by (DC), there exists a sequence $(w_n := (y_{2n}, y_{2n+1}); n \ge 0)$ in M with: $[w_0 = (y_0, y_1)$, $w_n \mathcal{R} w_{n+1}$, $\forall n]$; hence, by definition,

$$y_{2n+1} \le y_{2n+2},\quad d(y_{2n+2}, y_{2n+3}) > \varepsilon, \quad \text{for all } n \ge 0.$$

As a consequence, $(y_n; n \ge 0)$ is ascending and not d-Cauchy; in contradiction with (b08); hence the claim. Summing up, Theorem 10 applies to our data; wherefrom, we are done.

In particular, when $d(.,.)$ is a (generalized) metric on X, (b07) holds under

(b09) d is $(\le)$-complete: each ascending d-Cauchy sequence is d-convergent
(b10) $(\le)$ is d-self-closed:
the d-limit of each ascending sequence is an upper bound of it.

Note that these conditions hold whenever

(b11) d is complete and $(\leq)$ is semi-closed ($M(x, \leq)$ is closed, $\forall x \in M$).

The obtained variants of Theorem 11 include the results in Conserva and Rizzo [15]; see also Granas and Horvath [27].

(C) Returning to the introductory setting, let $(M, \leq)$ be a quasiordered structure and $\psi : M \to R_+$ be decreasing; remember that the function $\beta(x) := \inf \psi[M(x, \leq)]$, $x \in M$ is increasing and $\beta(x) \leq \psi(x)$, $x \in M$.

Proposition 4 *We have Theorem 10* $\Longrightarrow$ *(BBa).*

Proof Let the premises of (BBa) hold; and put $g(x, y) = |\psi(x) - \psi(y)|$, $x, y \in M$. Clearly, g is reflexive symmetric and triangular; hence, it is a semimetric on M. We show that conditions of Theorem 10 are fulfilled for $(M, \leq)$ and g. In fact,

$$\rho_g(x) = \sup\{\psi(u) - \psi(v); x \leq u \leq v\} = \psi(x) - \beta(x);$$

and this (via (a01)) shows that (b06) (modulo g) holds. Moreover, let $x \in M$ and $\varepsilon > 0$ be arbitrary fixed. By definition, there exists $y = y(x, \varepsilon)$ in $M(x, \leq)$ with $\beta(x) + \varepsilon > \psi(y)$; and this, combined with $\psi(y) \geq \beta(y) \geq \beta(x)$, yields $\rho_g(y) \leq \varepsilon$; wherefrom, (b02) (modulo ρ_g)—or, equivalently, (b05) (modulo g)—holds too. The proof is complete.

2.3 (EVPdL) Implies (DDC)

By the developments above, we have the chain of implications: (DDC) $\Rightarrow$ (BBau) $\Rightarrow$ Th 10 $\Longrightarrow$ Th 11 $\Longrightarrow$ (EVP), Th 10 $\Longrightarrow$ (BBa) $\Longrightarrow$ (BB) $\Rightarrow$ (EVP). So, it is natural asking whether these implications may be reversed. Clearly, the natural setting for solving this problem is (ZF) without (AC) (the *reduced* Zermelo–Fraenkel system).

Let X be a nonempty set; and $(\leq)$ be an order on it. We say that $z \in X$ is a $(\leq)$-*maximal* element if $X(z, \leq) = \{z\}$; the class of all these points will be denoted as $\max(X, \leq)$. In this case, $(\leq)$ is called a *Zorn order* when $\max(X, \leq)$ is nonempty and *cofinal* in X [for each $u \in X$ there exists a $(\leq)$-maximal $v \in X$ with $u \leq v$]. Further aspects are to be described in a metric setting. Let $d : X \times X \to R_+$ be a metric over X; and $\varphi : X \to R_+$ be some function. Then, the natural choice for $(\leq)$ above is

$$x \leq_{(d,\varphi)} y \text{ iff } d(x, y) \leq \varphi(x) - \varphi(y);$$

referred to as the *Brøndsted* order [11] attached to (d, φ). Denote $X(x, \rho) = \{u \in X; d(x, u) < \rho\}$, $x \in X$, $\rho > 0$ [the *open* sphere with center x and radius ρ]. Call the ambient metric space (X, d), *discrete* when for each $x \in X$ there exists $\rho = \rho(x) > 0$ such that $X(x, \rho) = \{x\}$. Note that, under such an assumption, any function $\psi : X \to R$ is continuous over X. However, the d-*Lipschitz* property ($|\psi(x) - \psi(y)| \leq Ld(x, y)$, $x, y \in X$, for some $L > 0$) cannot be assured, in general.

Now, the statement below is a particular case of EVP:

Theorem 12 *Let the metric d and the function $\varphi : X \to R_+$ satisfy*

(c01) (X, d) is discrete bounded and complete
(c02) φ is d-nonexpansive (hence, d-Lipschitzian).

Then, $(\leq_{(d,\varphi)})$ is a Zorn order.

We shall refer to it as: the discrete Lipschitz version of EVP (in short: (EVPdL)). Clearly, (EVP) $\Longrightarrow$ (EVPdL). The remarkable fact to be added is that this last principle yields (DDC); so, it completes the circle between all these. To do this, the following version of (DDC) will be considered. Let J and M be a couple of nonempty sets.

Proposition 5 *Suppose that the relation $\mathcal{Q}$ in J and the J-indexed family of relations $(\mathcal{R}(j); j \in J)$ in M fulfill*

(c03) $(\forall k \in J)$, $(\forall c \in M)$: $J(k, \mathcal{Q}) \neq \emptyset$ and $M(c, \mathcal{R}(k)) \neq \emptyset$.

Then, for each $(i, a) \in J \times M$ there exists a sequence $((k_n, x_n); n \geq 0)$ in $J \times M$ such that $(k_0, x_0) = (i, a)$ and $[k_n \mathcal{Q} k_{n+1},\ x_n \mathcal{R}(k_n) x_{n+1}]$, for all n.

We shall term it, the *iterated dependent choices* principle (in short: IDC). In particular, when $J = N$ and the relation $\mathcal{Q}$ is taken as $[s\mathcal{Q}t$, iff $s + 1 = t]$, then (IDC) is just (DDC).

Proposition 6 *We have (in the reduced Zermelo–Fraenkel system) (EVPdL) $\Rightarrow$ (DC) $\Longrightarrow$ (IDC) $\Longrightarrow$ (DDC). So (by the above), the maximal/variational principles (BBau), Th 10, (BBa), (BB) and (EVP) are all equivalent with (DDC); hence, mutually equivalent.*

Proof The first inclusion is already known (see a previous place); and the third one was already verified; so, it remains to establish the second inclusion. Let the relation $\mathcal{Q}$ in J and the J-indexed family of relations $(\mathcal{R}(j); j \in J)$ in M be such that (c03) holds. Define a relation $\mathcal{S}$ over $J \times M$ as: $(i, x)\mathcal{S}(j, y)$ iff $i\mathcal{Q}j$ and $x\mathcal{R}(i)y$. Clearly, $[J \times M]((k, c), \mathcal{S}) \neq \emptyset$, for each $(k, c) \in J \times M$. So, by (DC), there exists. for the starting $(i, a) \in J \times M$ a sequence $((k_n, x_n))$ in $J \times M$ with $(k_0, x_0) = (i, a)$ and $(k_n, x_n)\mathcal{S}(k_{n+1}, x_{n+1})$, for all n; that is (by definition): $k_n \mathcal{Q} k_{n+1}$ and $x_n \mathcal{R}(k_n) x_{n+1}$, for all n. This gives all conclusions we need.

2.4 Further Aspects

In what follows, a lot of additional asymptotic maximal principles equivalent with (BBa) will be introduced.

(A) Let $(M, \leq)$ be a quasiordered structure; and $\mathcal{P}_0(M)$ stand for the class of all nonempty subsets in M. According to Du [20], any function $\mu : \mathcal{P}_0(M) \to R_+ \cup \{\infty\}$ with

(d01) μ is monotone: $\mu(A) \leq \mu(B)$ if $A \subseteq B$

will be called *sizing-up*. In this case, for each $A \in \mathcal{P}_0(M)$, $\mu(A)$ will be referred to as the *μ-diameter* of A.

A natural construction of this type is the following. Remember that, by a (generalized) *pseudometric* over M we shall mean any map $d : M \times M \to R_+ \cup \{\infty\}$. Fix such an object; supposed to be *reflexive* [$d(x, x) = 0, \forall x \in M$]. Then, the function

(d02) $\mu(A) = \sup\{d(x, y); x, y \in A, x \leq y\}$, $A \in \mathcal{P}_0(M)$

is sizing-up, as it can be readily seen; it will be referred to as the $(\leq, d)$*-diameter* of A.

Returning to the general setting (with an arbitrary μ), the function ρ_μ from M to $R_+ \cup \{\infty\}$ given as

(d03) $\rho_\mu(x) = \mu(M(x, \leq))$, $x \in M$,

is decreasing. Concerning the remaining properties to be checked (for an application of (BBau)), let us say that $(M, \leq)$ is *almost regular* (modulo μ), when it fulfills (b02) relative to ρ_μ: i.e.,

(d04) $\forall x \in M, \forall \varepsilon > 0$, $\exists y = y(x, \varepsilon) \geq x$: $\mu(M(y, \leq)) \leq \varepsilon$.

Further, call the ascending sequence (x_n) of M, *strongly* $(\leq, \mu)$-Cauchy provided $\mu(M(x_n, \leq)) \to 0$ as $n \to \infty$. Then, (b03) (modulo ρ_μ) means:

(d05) each ascending strongly $(\leq, \mu)$-Cauchy sequence has an upper bound;

and will be termed as: $(M, \leq)$ is sequentially inductive (modulo μ). Finally, call the point $z \in M$, μ*-maximal* provided $\mu(M(z, \leq)) = 0$. An application of (BBau) to these data yields the following maximal principle:

Theorem 13 *Suppose that $(M, \leq; \mu)$ is such that $(M, \leq)$ is almost regular and sequentially inductive (modulo μ). Then, for each $u \in M$ there exists a μ-maximal $v \in M$ with $u \leq v$.*

Now, by definition, each strongly μ-Cauchy sequence (x_n) is μ*-Cauchy* [in the sense: $\mu(\{x_n, x_{n+1}, \ldots\}) \to 0$ as $n \to \infty$]. Hence, (d05) holds under

(d06) each ascending μ-Cauchy sequence has an upper bound.

In this case, Theorem 13 is just the maximal principle in Du [20]; which, among others, includes (BB) (hence, (EVP) as well). Further "structural" aspects may be found in Zhu and Li [55].

A natural question is that of the reverse implication above being valid: (BBau) is reducible to Theorem 13. As we shall see, a positive answer to this is available. In fact, let the function $\varphi : M \to R_+ \cup \{\infty\}$ be decreasing. The associated function $\mu := \mu_\varphi$ introduced as

(d07) $\mu(A) = \sup \varphi(A)$, $A \in \mathcal{P}_0(M)$,

is sizing-up, as it can be readily seen. Moreover, its associated by (d03) function ρ_μ is identical with φ. This tells us that an application of Theorem 13 to these data yields (BBau); and proves our claim. Summing up, Theorem 13 is a logical equivalent of (BBau); whence (by Proposition 6) it is equivalent with both (DDC) and (EVP).

(B) Let $(M, \leq)$ be a quasiordered structure. Taking the decreasing function $\psi : M \to R_+$ and the reflexive pseudometric $d : M \times M \to R_+ \cup \{\infty\}$, define the $(\leq, \psi, d)$-maximal property of some $z \in M$ as: $z \leq w$ implies $\psi(z) = \psi(w)$, $d(z, w) = 0$. The following pseudometric variant of (BBa) (referred to as: the pseudometric asymptotic BB; and denoted as: (BBap)) is to be considered.

Theorem 14 *Assume that (a01) and (b05)+(b06) hold. Then, for each $u \in M$ there exists a $(\leq, \psi, d)$-maximal $v \in M$ with $u \leq v$.*

Proof From (BBa) it follows that, for the starting $u \in M$ there exist a $(\leq, \psi)$-maximal $v \in M$ with $u \leq v$. On the other hand, by Theorem 10, for the obtained $v \in M$ there exists a $(\leq, d)$-maximal $w \in M$ such that $v \leq w$. Clearly, w has all desired properties; hence the conclusion.

Note that, as (BBau)$\Longrightarrow$Theorem 10$\Longrightarrow$(BBa), we have (BBau)$\Rightarrow$Theorem 14. Moreover, Theorem 14 $\Rightarrow$ (BBa); just take $d = \rho_g$, where g is the associated pseudometric (see above). Hence (by Proposition 6), Theorem 14 is equivalent with both (BBau) and (BBa). Further extensions may be found in Turinici [51]; see also Cârjă et al. [13, Chap. 2, Sect. 2.1].

3 Vector Variational Principles

3.1 Introduction

Let Y be a (real) separated *locally convex space*; and K, some (convex) cone of it ($\alpha K + \beta K \subseteq K$ for each $\alpha, \beta \geq 0$). The relation $(\leq_K)$ on Y defined as

(a01) $(y_1, y_2 \in Y)$: $y_1 \leq_K y_2$ if and only if $y_2 - y_1 \in K$

is reflexive and transitive; hence a *quasiorder* [also denoted as $(\leq)$, when K is understood]; moreover, it is *compatible* with the linear structure of Y. Let H be another (convex) cone of Y with $K \subseteq H$; and pick some $k^0 \in K \setminus (-H)$. Further, take some complete metric space (X, d). Define the quasiorder $(\succeq)$ on $X \times Y$ according to

(a02) $(x_1, y_1) \succeq (x_2, y_2)$ iff $k^0 d(x_1, x_2) \leq y_1 - y_2$;

and take some nonempty part A of $X \times Y$. For both practical and theoretical reasons, it would be useful to determine sufficient conditions under which $(A, \succeq)$ has points with certain *maximal* properties. A basic result in this direction, obtained by Goepfert, Tammer, and Zălinescu [24, Theorem 1] (in short: GTZ) deals with the case $H = \text{cl}(K)$ (= the *closure* of K), under

(a03) $P_Y(A)$ is bounded below (modulo K) ($\exists \widetilde{y} \in Y$: $P_Y(A) \subseteq \widetilde{y} + K$)
(a04) if $((x_n, y_n)) \subseteq A$ is $(\succeq)$-ascending and $x_n \to x$ then $x \in P_X(A)$ and there exists $y \in A(x)$ such that $(x_n, y_n) \succeq (x, y)$, for all n.

(Here, by P_X and P_Y we denoted the *projection* operators from $X \times Y$ to X and Y respectively).

Theorem 15 *Let these assumptions hold. Then, for each* $(x_0, y_0) \in A$ *there exists* $(\bar{x}, \bar{y}) \in A$ *with*
(I) $(x_0, y_0) \succeq (\bar{x}, \bar{y})$, **(II)** $(\bar{x}, \bar{y}) \succeq (x', y') \in A \Longrightarrow \bar{x} = x'$.

The authors' (constructive) argument is based on the Cantor intersection principle and a separation theorem involving the pair $(k^0, -\mathrm{cl}(K))$. Note that, the former of these requires a denumerable version of AC; and the latter one is based on the Zorn maximality principle (equivalent with (AC)). On the other hand, Hamel [29, Chap. 4, Sect. 4.7] established that GTZ is reducible to the ordering principle in Brezis and Browder [9] (in short: BB). But, his argument (again based on separation techniques) is restricted to the normed case (modulo Y); so, we may ask whether this is removable. The answer is positive (cf. Sect. 3.2): a metrical version of (BB) obtainable from the generic dependent choices principle (in short: GDC) will suffice for deducing an extended version of GTZ for arbitrary (nontopological) vector spaces Y.

A relevant application of GTZ is the couple of vector type variational principles obtained by the quoted authors [24, Corollaries 2 and 3]. These include the maximality results in Isac [31] and Nemeth [40]; which, in turn, extend Ekeland's variational principle [21] (in short: EVP); so, it is natural to denote their union as (EVPv). (An extended variant of all these (also denoted in this way) is given in Sect. 3.4; and the main tool for establishing it is the concept of *gauge function*, discussed in Sect. 3.3). Concerning the converse inclusion, note that (see below) (EVP) $\Longrightarrow$ (GDC); and this makes all these principles be equivalent to each other.

3.2 *(GDC)* $\Longrightarrow$ *(BBm)* $\Longrightarrow$ *(EVP)* $\Longrightarrow$ *(GDC)*

Let M be a nonempty set; and $\mathcal{R} \subseteq M \times M$ stand for a (nonempty) relation over it. For each $x \in M$, denote $M(x, \mathcal{R}) = \{y \in M; x\mathcal{R}y\}$. The following GDC is our starting point:

Proposition 7 *Suppose that*

(b01) $M(c, \mathcal{R})$ *is nonempty, for each* $c \in M$.

Then, there exists a sequence $(x_n) \subseteq M$ *with* $x_n \mathcal{R} x_{n+1}$, *for all* n.

Technically speaking, (GDC) is deductible from (DC); which says that, under (b01), one has a conclusion like: for each $a \in M$, there exists a sequence $(x_n) \subseteq M$ with $x_0 = a$ and $x_n \mathcal{R} x_{n+1}$, for all n. [The reciprocal is also valid; this will be verified a bit further].

(A) Let M be a nonempty set. Take a *quasiorder* $(\leq)$ over it, and a function $\varphi : M \to R \cup \{\infty\}$. Call $z \in M$, *$(\leq, \varphi)$-maximal* when φ is constant on $M(z, \leq)$.

Proposition 8 *Assume that*

(b02) φ *is inf-proper* ($\mathrm{Dom}(\varphi) \neq \emptyset$ *and* $\inf [\varphi(M)] > -\infty$)

(b03) φ *is* $(\leq)$*-decreasing* $(x \leq y \Longrightarrow \varphi(x) \geq \varphi(y))$
(b04) each ascending sequence in $\mathrm{Dom}(\varphi)$ has an upper bound (in M).

Then, for each $u \in \mathrm{Dom}(\varphi)$ *there exists a* $(\leq, \varphi)$*-maximal* $v \in M$ *with* $u \leq v$.

Proof Define a function $\beta : M \to R \cup \{\infty\}$ as: $\beta(v) := \inf [\varphi(M(v, \leq))]$, $v \in M$. Clearly, β is increasing; and $\varphi(v) \geq \beta(v)$, for all $v \in M$. Moreover, (b03) gives at once a characterization like: $v \in \mathrm{Dom}(\varphi)$ is $(\leq, \varphi)$-maximal iff $\varphi(v) = \beta(v)$. Now, assume by contradiction that the conclusion in this statement is false; i.e., (by the remark above) there must be some $u \in \mathrm{Dom}(\varphi)$ such that

(b05) for each $v \in M_u := M(u, \leq)$, one has $\varphi(v) > \beta(v)$.

Consequently (for all such v), $\varphi(v) > (1/2)(\varphi(v) + \beta(v)) > \beta(v)$; hence

$$v \leq w \text{ and } (1/2)(\varphi(v) + \beta(v)) > \varphi(w), \tag{21}$$

for at least one w (belonging to M_u). The relation $\mathcal{R}$ over M_u introduced via (21) fulfills $M_u(v, \mathcal{R}) \neq \emptyset$, for all $v \in M_u$. So, by (GDC), there must be a sequence (u_n) in M_u with

$$u_n \leq u_{n+1}, (1/2)(\varphi(u_n) + \beta(u_n)) > \varphi(u_{n+1}), \text{ for all } n. \tag{22}$$

We have thus constructed an ascending sequence (u_n) in $M_u \subseteq \mathrm{Dom}(\varphi)$ for which $(\varphi(u_n))$ is (strictly) descending and bounded below; hence $\lambda := \lim_n \varphi(u_n)$ exists in R. According to (b04), there exists $v \in \mathrm{Dom}(\varphi)$ such that $u_n \leq v$, for all n. From (b03) (and β=increasing) $\varphi(u_n) \geq \varphi(v)$ and $\varphi(v) \geq \beta(v) \geq \beta(u_n)$, $\forall n$. The former of these gives $\lambda \geq \varphi(v)$; and the latter one yields (via (22)) $(1/2)(\varphi(u_n) + \beta(v)) > \varphi(u_{n+1})$, for all $n \in N$. Passing to limit as $n \to \infty$, we have $(\varphi(v) \geq)\beta(v) \geq \lambda$; so, combining with the preceding one, $\varphi(v) = \beta(v)(= \lambda)$, contradiction.

In particular, when $\mathrm{Dom}(\varphi) = M$, Proposition 8 is identical with (BB); for simplicity reasons, it will be also denoted in this way.

(B) A useful application of these developments is to be given under the lines below. Let M be a nonempty set; and $(\leq)$, some *quasiorder* over it. By a *pseudometric* over M we shall mean any map $d : M \times M \to R_+$. If, in addition, d is *reflexive* $[d(x,x) = 0, \forall x \in M]$, *triangular* $[d(x,z) \leq d(x,y) + d(y,z), \forall x,y,z \in M]$ and *symmetric* $[d(x,y) = d(y,x), \forall x,y \in M]$, we say that it is a *semimetric* (on M). Suppose that we fixed such an object. Call the sequence (x_n) (in M), *d-Cauchy* when $[\forall \varepsilon > 0, \exists n(\varepsilon)$: $n(\varepsilon) \leq p \leq q \Longrightarrow d(x_p, x_q) \leq \varepsilon]$. Further, let us say that (x_n), *d-converges* to $x \in M$ when $d(x_n, x) \to 0$ as $n \to \infty$. In this case, x is called the *d-limit* of (x_n); and if such elements exist, we say that (x_n) is *d-convergent* (in M). Note that (for any sequence) d-convergent $\Longrightarrow$ d-Cauchy; the reciprocal of this is not in general true.

Now, let the function $\varphi : M \to R \cup \{\infty\}$ be as in (b02) + (b03). Denote by $(\leq_{(d,\varphi)})$ the Brøndsted quasiorder [11] attached to (d, φ):

$x \leq_{(d,\varphi)} y$ iff $d(x,y) + \varphi(y) \leq \varphi(x)$.

The natural condition (with a practical finality) to be imposed is

(b06) $(\leq)$ is coarser than $(\leq_{(d,\varphi)})$ over $\mathrm{Dom}(\varphi)$:
$x, y \in \mathrm{Dom}(\varphi)$, $x \leq y$ imply $d(x, y) \leq \varphi(x) - \varphi(y)$.

Proposition 9 *Assume that (in addition to (b02) + (b03) + (b06))*

(b07) each ascending d-Cauchy sequence in $\mathrm{Dom}(\varphi)$
is bounded above (modulo $(\leq)$*).*

Then, for each $u \in \mathrm{Dom}(\varphi)$ *there exists* $v \in \mathrm{Dom}(\varphi)$ *with*
(k1) $u \leq v$, **(k2)** $v \leq x \in M$ *imply* $d(v, x) = 0$, $\varphi(v) = \varphi(x)$.

Proof We claim that (b04) holds for $(M_u, \leq)$ and φ; where $M_u := M(u, \leq)$. Let (x_n) be an ascending (modulo $(\leq)$) sequence in M_u:

(b08) $x_n \leq x_m$ (hence $d(x_n, x_m) \leq \varphi(x_n) - \varphi(x_m)$), if $n \leq m$.

The (real) sequence $(\varphi(x_n))$ is descending and bounded; hence a Cauchy one. This tells us that (x_n) is an ascending (modulo $(\leq)$) d-Cauchy sequence in $M_u \subseteq \mathrm{Dom}(\varphi)$. So, from (b07), $x_n \leq x$, for all n and some $x \in M$ (fulfilling $x \in M_u$); hence the claim. From (BB) it then follows that, for the starting $u \in M_u$ there is another $v \in M_u$ with
(h1) $u \leq v$, **(h2)** $v \leq x \in M_u$ implies $\varphi(v) = \varphi(x)$.
The obtained point has, via (b06), all properties we need.

Call Proposition 9, the *(semi) metrical* version of (BB) (in short: (BBm)). It is natural to ask what happens with the above conclusions when (b06) is no longer accepted. A positive answer to this is available, with respect to the (coarser) "product" quasiorder $(\preceq) := (\leq) \cap (\leq_{(d,\varphi)})$.

Proposition 10 *Assume that (in addition to (b02) + (b03))*

(b09) each $(\leq)$*-ascending d-Cauchy sequence in* $\mathrm{Dom}(\varphi)$ *is d-convergent*
(b10) the d-limit of each ascending d-convergent sequence in $\mathrm{Dom}(\varphi)$
is an upper bound of it (modulo $(\leq)$*).*

Then, for each $u \in \mathrm{Dom}(\varphi)$ *there exists* $v \in \mathrm{Dom}(\varphi)$ *with*
(k3) $u \preceq v$, **(k4)** $v \preceq x \in M$ *imply* $d(v, x) = 0$, $\varphi(v) = \varphi(x)$.

Proof As before, we claim that (b04) holds for $(M_u, \preceq)$ and φ; where $M_u := M(u, \preceq)$. Let (x_n) be an ascending (modulo $(\preceq)$) sequence in M_u

(b11) $x_n \leq x_m$, $d(x_n, x_m) \leq \varphi(x_n) - \varphi(x_m)$, if $n \leq m$.

The (real) sequence $(\varphi(x_n))$ is descending and bounded; hence a Cauchy one. This tells us that (x_n) is an ascending (modulo $(\leq)$) d-Cauchy sequence in M_u; so, from (b09), $x_n \to x$ as $n \to \infty$, for some $x \in M$. Taking (b10) into account, $x_n \leq x$ (hence $\varphi(x_n) \geq \varphi(x)$), for all n; note that, as a consequence, $x \in \mathrm{Dom}(\varphi)$. On the other hand, fix some rank n. From (b11), $d(x_n, x) \leq d(x_n, x_m) + d(x_m, x) \leq \varphi(x_n) - \varphi(x) + d(x_m, x)$, $\forall m \geq n$. Passing to limit as $m \to \infty$ yields (combining with the above) $x_n \preceq x$, $\forall n$; and the claim follows. From (BB) it then follows that, for the starting $u \in M_u$, there is another $v \in M_u$ with

(h3) $u \preceq v$, **(h4)** $v \preceq x \in M_u$ implies $\varphi(v) = \varphi(x)$.

The obtained point has all properties we need.

This result is nothing else than the monotone version of EVP (in short: (EVPm)); cf. Turinici [50]. Note that, if (b06) holds, ($\leq$) is identical with ($\preceq$) on Dom(φ); and conclusions of Proposition 10 are identical with the ones of Proposition 9.

In particular, when ($\leq$) is just ($\leq_{(d,\varphi)}$), (b10) holds under

(b12) φ is d-lsc over M: $\liminf_n \varphi(x_n) \geq \varphi(x)$, whenever $x_n \to x$;

so, by simply combining with (b09) + (b10) $\Longrightarrow$ (b07), one derives, from either of the above statements

Proposition 11 *Assume that (M, d) is complete and (b02) + (b12) hold. Then, for each $u \in$ Dom(φ) there exists $v \in$ Dom(φ) with*

(k5) $d(u, v) \leq \varphi(u) - \varphi(v)$,

(k6) $x \in M$, $d(v, x) \leq \varphi(v) - \varphi(x)$ *imply* $d(v, x) = 0$, $\varphi(v) = \varphi(x)$.

This is just EVP (the semimetric variant). For pseudometric extensions, we refer to Hyers et al. [30, Chap. 5]; see also Turinici [51].

Summing up, we have the inclusional chains (GDC) $\Longrightarrow$ (BB) $\Longrightarrow$ (BBm) $\Longrightarrow$ (EVP), (BB) $\Longrightarrow$ (EVPm) $\Longrightarrow$ (EVP). On the other hand we have (according to some previous developments) (EVP $\Longrightarrow$ (DC). This, along with (DC) $\Longrightarrow$ (GDC), tells us that all these maximal/variational principles are mutually equivalent.

3.3 Gauge Functions

Let Y be a (real) *vector space*. Take a convex cone H of Y; and let ($\leq$) stand for its induced quasiorder. Pick some $k^0 \in H \setminus (-H)$; and put (for $y \in Y$)

(c01) $\Gamma(H; k^0; y) = \{s \in R_+; k^0 s \leq y\}$, $\gamma(H; k^0; y) = \sup \Gamma(H; k^0; y)$.

(By convention, $\sup(\emptyset) = -\infty$). We therefore defined a couple of functions $\Gamma(.) := \Gamma(H; k^0; .)$ and $\gamma(.) := \gamma(H; k^0; .)$ from Y to $\mathcal{P}(R_+)$ and $R_+ \cup \{-\infty, \infty\}$ respectively; the latter of these will be referred to as the *gauge* function attached to $(H; k^0)$. Such objects were introduced (in the normed context) by Gerth (Tammer) and Weidner [23]. Some refinements of these (to the locally convex setting) were provided by Goepfert et al. [24]. The present developments may be viewed as a non-topological extension of the preceding ones.

(i) We start by noting that, for each $y \in Y$, $\Gamma(y)$ is *hereditary* ($s \in \Gamma(y) \Longrightarrow [0, s] \subseteq \Gamma(y)$). In addition, we have (by definition)

$$y \notin H \Leftrightarrow \Gamma(y) = \emptyset \Leftrightarrow \gamma(y) = -\infty; \ y \in H \Leftrightarrow \Gamma(y) \neq \emptyset \Leftrightarrow \gamma(y) \in [0, \infty].$$

(ii) Further, Γ and γ are *positively homogeneous*: $\Gamma(ty) = t\Gamma(y)$, $\gamma(ty) = t\gamma(y)$, $\forall t > 0$, $\forall y \in Y$. (Here, $t\emptyset = \emptyset$, $\forall t > 0$). Moreover (as $\gamma(0) = 0$, $\gamma(k^0) = 1$), γ is *H-proper*: $\mathrm{Dom}_H(\gamma) := \mathrm{Dom}(\gamma) \cap H$ is nonempty.

(iii) A natural question is that of γ being *totally H-proper*: $\mathrm{Dom}_H(\gamma) = H$. According to Cristescu [16, Chap. 5, Sect. 1], we say that H is *Archimedean*, if

(c02) $[h \in Y, v \in H$ and $\Gamma(H; h; v) = R_+]$ imply $h \in -H$.

Lemma 7 *Assume that H is Archimedean. Then*
(a) $\infty \notin \gamma(H)$ *(hence γ is totally H-proper)*
(b) $\Gamma(y)$ *is (bounded) closed (hence $\gamma(y) \in \Gamma(y)$), for each $y \in H$.*

Proof The first part is clear, by the choice of k^0 (and the accepted hypotheses). For the second one, let $y \in H$ be arbitrary fixed. By the hereditary property, $k^0\gamma(y) - y \le k^0 t$, for all $t > 0$; wherefrom $\Gamma(H; k^0\gamma(y) - y; k^0) = R_+$. This, and the Archimedean property of H, give the desired fact.

(iv) Returning to the general case, note that γ is increasing: $y_1, y_2 \in Y$, $y_1 \le y_2$ implies $\gamma(y_1) \le \gamma(y_2)$. As a consequence, $\gamma(y) \le t$, whenever $y \le k^0 t$; so, $\mathrm{Dom}_H(\gamma)$ is "large" enough. Some other properties are stated in

Lemma 8 *The gauge function γ is super-additive: $\gamma(y_1 + y_2) \ge \gamma(y_1) + \gamma(y_2)$, whenever the right member exists. As a consequence, γ is subtractive: $\gamma(y_1 - y_2) \le \gamma(y_1) - \gamma(y_2)$, whenever $y_1 \ge y_2$ and $\gamma(y_2) < \infty$.*

Proof Without loss, one may assume that $\gamma(y_1) > 0$, $\gamma(y_2) > 0$. By the hereditary property, $y_1 \ge k^0 t_1, y_2 \ge k^0 t_2$, whenever $0 \le t_1 < \gamma(y_1), 0 \le t_2 < \gamma(y_2)$; and this yields (for all such (t_1, t_2)) $y_1 + y_2 \ge k^0[t_1 + t_2]$ (that is, $\gamma(y_1 + y_2) \ge t_1 + t_2$). Combining with the arbitrariness of the precise couple gives the first part. The second part is directly obtainable from this one, in a standard way.

3.4 (BBm) $\Longrightarrow$ (GTZ) $\Longrightarrow$ (EVPv)

With this information, we may return to the question of the introductory part. Let Y be a (real) vector space; and K, some (convex) cone of it. Denote by $(\le_K)$ the induced quasiorder; also expressed as $(\le)$, when K is understood. Let H be another (convex) cone of Y with $K \subseteq H$ and H is Archimedean. Further, take some $k^0 \in K \setminus (-H)$; and let $\gamma(.) = \gamma(H; k^0; .)$ stand for the gauge function attached to (H, k^0). Note that (by the developments above and imposed conditions)

$$0 \le \gamma(y) < \infty, \text{ for all } y \in H \text{ (hence } \mathrm{Dom}_H(\gamma) = H) \tag{23}$$

$$y_1, y_2 \in Y, y_1 \le_H y_2 \Longrightarrow \gamma(y_1) \le \gamma(y_2). \tag{24}$$

Finally, take a semimetric space (X, d).

(A) Let $(\succeq)$ stand for the quasiorder (a02) on $X \times Y$; and A be some part of $X \times Y$. As before, we are interested to determine sufficient conditions under which $(A, \succeq)$ has points with certain maximal properties. For the working context we just described, these are as follows:

(d01) $P_Y(A)$ is bounded below (modulo H) $[\exists\widetilde{y} \in Y\colon P_Y(A)\subseteq\widetilde{y} + H]$
(d02) each ascending e-Cauchy sequence $((x_n, y_n))\subseteq A$
is bounded above in A (modulo $(\succeq)$).

Here, e stands for the semimetric on $X \times Y$ introduced as

(d03) $e((x_1, y_1), (x_2, y_2)) = d(x_1, x_2),\ (x_1, y_1), (x_2, y_2) \in X \times Y.$

Theorem 16 *Let the precise assumptions be in force. Then, for each starting* $(x_0, y_0) \in A$ *there exists* $(\bar{x}, \bar{y}) \in A$ *with the properties*
(III) $(x_0, y_0) \succeq (\bar{x}, \bar{y})$,
(IV) $(\bar{x}, \bar{y}) \succeq (x', y') \in A \Longrightarrow d(\bar{x}, x') = 0,\ \gamma(\bar{y} - \widetilde{y}) = \gamma(y' - \widetilde{y})$ *(hence* $\gamma(\bar{y} - y') = 0$*).*

Proof Without loss, one may assume that (d01) may be written as

(d04) $P_Y(A)\subseteq H$ (or, equivalently: $\widetilde{y} = 0$ in that condition).

For otherwise, passing to $\widetilde{A} = \{(x, y) \in X \times Y; (x, y + \widetilde{y}) \in A\}$, this is fulfilled, in view of $P_Y(\widetilde{A}) = P_Y(A) - \widetilde{y}$; as well as (d02). And from the conclusion involving $\widetilde{A}$ it clearly follows the one relative to A. Having these precise, let us introduce the function (from $X \times Y$ to $R_+ \cup \{-\infty, \infty\}$)

$$\Phi(x, y) = \gamma(y), (x, y) \in X \times Y \text{ (that is: } \Phi = \gamma \circ P_Y);$$

its restriction to A will be also denoted in this way. We claim that (BBm) is applicable to $(A, \succeq; e)$ and Φ. In fact, Lemma 7 gives us [via (d04)] $0 \le \Phi(x, y) = \gamma(y) < \infty$, for each $(x, y) \in A$;. wherefrom, Φ (restricted to A) is bounded from below on A and proper ($\mathrm{Dom}(\Phi) = A$). In addition, by the monotonicity of the gauge function, $(x_1, y_1) \succeq (x_2, y_2)\Longrightarrow y_1 \ge y_2\Longrightarrow\gamma(y_1) \ge \gamma(y_2)\Longrightarrow\Phi(x_1, y_1) \ge \Phi(x_2, y_2)$; which tells us that Φ (restricted to A) is $(\succeq)$-decreasing. Further, let (x_1, y_1), (x_2, y_2) in A be such that $(x_1, y_1) \succeq (x_2, y_2)$ [that is: $k^0 d(x_1, x_2) \le y_1 - y_2$]. By Lemma 8 and (23), this yields $d(x_1, x_2) \le \gamma(y_1 - y_2) \le \gamma(y_1) - \gamma(y_2)$; whence

$$(x_1, y_1) \succeq (x_2, y_2)\Longrightarrow e((x_1, y_1), (x_2.y_2)) \le \Phi(x_1, y_1) - \Phi(x_2, y_2);$$

and (b06) holds. Finally, (b07) is identical with (d02); and our claim is proved. By (BBm) it then follows that, for $a_0 := (x_0, y_0) \in A$ there exists $\bar{a} := (\bar{x}, \bar{y}) \in A$ with the properties
(k7) $a_0 \succeq \bar{a}$
(k8) if $a' := (x', y') \in A$ fulfills $\bar{a} \succeq a'$ then $e(\bar{a}, a') = 0,\ \Phi(\bar{a}) = \Phi(a')$.
This gives all conclusions in the statement.

In particular, when Y is a (real) separated locally convex space, the convex cone $H = \mathrm{cl}(K)$ is Archimedean; and Theorem 16 reduces to (GTZ). This inclusion is technically strict; because the last relation in **(IV)** cannot be deduced from the one in **(II)**. However, for simplicity reasons, this statement will be still denoted in this way. Further aspects may be found in Turinici [51]; see also Goepfert et al. [25, Chap. 3, Sect. 10].

(B) Under the same initial setting involving the (convex) cones K and H, let us adjoin to Y an element $\infty \notin Y$ taken as

(d05) $(\forall b \in Y, \forall \lambda > 0)$: $\infty = b + \infty = \infty + b$; $\infty = \lambda\infty$; $b \leq \infty$ and $\infty \not\leq b$.

Further, take a function $F : X \to Y \cup \{\infty\}$ according to

(d06) F is inf-proper (modulo H): $\mathrm{Dom}(F) \neq \emptyset$ and there exists $\widetilde{y} \in Y$ such that $F(\mathrm{Dom}(F)) \subseteq \widetilde{y} + H$.

It is our intention to formulate a vector EVP involving these data. To this end, let $(\geq)$ stand for the quasiorder on X

(d07) $(x_1, x_2 \in X)$: $x_1 \geq x_2$ iff $k^0 d(x_1, x_2) + F(x_2) \leq F(x_1)$.

Note that, as a direct consequence, one has (via (a02))

$$(x_1, x_2 \in \mathrm{Dom}(F)): x_1 \geq x_2 \text{ iff } (x_1, F(x_1)) \succeq (x_2, F(x_2)). \tag{25}$$

Theorem 17 *Assume that (in addition), (X, d) is complete and*

(d08) the d-limit of each ascending sequence in $\mathrm{Dom}(F)$ *is an upper bound of it (modulo $(\geq)$).*

Then, for each $x_0 \in \mathrm{Dom}(F)$ there exists $\bar{x} \in \mathrm{Dom}(F)$ with
(V) $x_0 \geq \bar{x}$,
(VI) $\bar{x} \geq x' \in X \Longrightarrow d(\bar{x}, x') = 0$, $\gamma(F(\bar{x}) - \widetilde{y}) = \gamma(F(x') - \widetilde{y})$ *(hence* $\gamma(F(\bar{x}) - F(x')) = 0$*).*

Proof Put $A = \{(x, F(x)); x \in \mathrm{Dom}(F)\}$. We claim that Theorem 16 applies to $(A, \succeq; e)$ [where $(\succeq)$ and e are the above ones]; wherefrom, all is clear. In fact, (d01) is retainable for these data, via (d06). Further, let $((x_n, F(x_n)))$ be an ascending (modulo $(\succeq)$) e-Cauchy sequence in A. By (25), (x_n) is an ascending (modulo $(\geq)$) d-Cauchy sequence in $\mathrm{Dom}(F)$; so (by completeness) $x_n \to x$ as $n \to \infty$, for some $x \in X$. Combining with (d08) yields $x_n \geq x$, $\forall n$ (hence, in particular, $x \in \mathrm{Dom}(F)$). This, again via (25), yields $(x_n, F(x_n)) \succeq (x, F(x)) \in A$, for all n; and (d02) is retainable as well. The proof is thereby complete.

Note that, with the convention $\gamma(\infty) = \infty$, we may extend the gauge function γ attached to (H, k^0) up to some function $\gamma : Y \cup \{\infty\} \to R_+ \cup \{-\infty, \infty\}$, which still fulfills (23) + (24). Further, define the function (from X to $R \cup \{\infty\}$) $\varphi(x) = \gamma(F(x) - \widetilde{y})$, $x \in X$. It is not without importance to emphasize that an alternate way of proving Theorem 17 is to establish that conditions of (BBm) are fulfilled over $(X, \geq; d)$ and φ. This is essentially based on the following technical facts

$$\varphi(x) = \infty \text{ iff } F(x) = \infty \quad (\text{i.e., } \mathrm{Dom}(\varphi) = \mathrm{Dom}(F)) \tag{26}$$

$$\varphi(x) \geq 0, \text{ for all } x \in X \quad (\text{hence } \inf[\varphi(X)] \geq 0)); \tag{27}$$

$$x_1, x_2 \in \mathrm{Dom}(F), x_1 \geq x_2 \Longrightarrow d(x_1, x_2) \leq \varphi(x_1) - \varphi(x_2); \tag{28}$$

we do not give details.

Now, a sufficient condition for (d08) above is evidently

(d09) $X(x, \geq)$ is closed, for all $x \in \mathrm{Dom}(F)$.

Another circumstance for the same is the couple of requirements

(d10) for each $(\geq)$-ascending sequence (x_n) in $\mathrm{Dom}(F)$
with $x_n \to x \in X$, we have $F(x_n) \geq F(x)$, for all n
(d11) $\Gamma(K; k^0; y)$ is closed, for each $y \in K$.

In fact, let (x_n) be a $(\geq)$-ascending sequence in $\mathrm{Dom}(F)$ with $x_n \to x$ (for some $x \in X$). By definition, $k^0 d(x_n, x_m) \leq F(x_n) - F(x_m)$, whenever $n \leq m$; and, from (d10), $F(x_n) \geq F(x)$, for all n (hence, in particular, $x \in \mathrm{Dom}(F)$). As a consequence, $k^0 d(x_n, x_m) \leq F(x_n) - F(x)$, whenever $n \leq m$; and this, along with (d11), gives (passing to limit as $m \to \infty$) $k^0 d(x_n, x) \leq F(x_n) - F(x)$ (i.e.: $x_n \geq x$), for all n; hence the claim. In particular, when Y is a (real) separated locally convex space, Theorem 17 includes (EVPv). However, as before, the obtained result will be still denoted in this way. Further structural aspects may be found in Chen and Huang [14], Ha [28], Li and Zhang [36], Qiu [41], Rozoveanu [42], and Tammer [45]; see also Khanh and Quy [34].

4 Smooth Variational Principles

4.1 Introduction

Let $(M; \leq; d)$ be a quasiordered metrical structure fulfilling

(a01) $(\leq)$ is self-closed and d is $(\leq)$-complete;

and take a function $\varphi : M \to R \cup \{\infty\}$ in accordance with

(a02) φ is inf-proper ($\mathrm{Dom}(\varphi) \neq \emptyset$ and $\inf[\varphi(M)] > -\infty$)
(a03) φ is $(\leq, d)$-lsc: $\liminf_n \varphi(x_n) \geq \varphi(x)$,
whenever (x_n) is $(\leq)$-ascending and $x_n \to x$.

Take a couple of numbers $\varepsilon > 0$, $\lambda > 0$; and put $M(\varphi, \varepsilon) = \{x \in M; \varphi(x) < \inf[\varphi(M)] + \varepsilon\}$ (hence $M(\varphi, \varepsilon) \subseteq \mathrm{Dom}(\varphi)$).

Theorem 18 *Let these conditions hold. Then*
(i) *for each $u \in M(\varphi, \varepsilon)$, there exists $v = v(u) \in \mathrm{Dom}(\varphi)$ with*

$$u \leq v, \ (\varepsilon/\lambda) d(u, v) \leq \varphi(u) - \varphi(v) \tag{29}$$

$$(\varepsilon/\lambda) d(v, x) > \varphi(v) - \varphi(x), \ \forall x \in M(v, \leq) \setminus \{v\} \tag{30}$$

(ii) *for each $u \in M(\varphi, \varepsilon)$, there exists $v = v(u) \in \mathrm{Dom}(\varphi)$, fulfilling the property (30), as well as*

$$u \leq v,\ \varphi(u) \geq \varphi(v) \text{ and } d(u, v) < \lambda. \tag{31}$$

In fact, we have a couple of variational principles under this statement. Precisely, the first part, Th 18 (i), is nothing else than the monotone version of EVP [21], denoted as (EVPm); while the second part, Th 18 (ii), is a "local" version of the same, denoted EVPm(loc). Clearly, EVPm $\Longrightarrow$ EVPm(loc); so, it remains to discuss the converse inclusion. An appropriate answer is contained in

Proposition 12 *We have EVPm(loc) $\Longrightarrow$ EVPm; so that, these variational principles are mutually equivalent.*

Proof Let the premises of (EVPm) be admitted. Define a quasiorder (reflexive and transitive relation) over M according to

(a04) $x \preceq y$ iff $x \leq y$ and $(\varepsilon/\lambda)d(x, y) + \varphi(y) \leq \varphi(x)$.

Given $u \in M(\varphi, \varepsilon)$, put $M_u := M(u, \preceq)$; and let φ_u stand for the restriction of φ to M_u. We claim that the conditions of EVPm(loc) are fulfilled on $(M_u; \preceq; d)$ and φ_u. In fact, as $M_u \subseteq \mathrm{Dom}(\varphi)$, we have

$$\mathrm{Dom}(\varphi_u) = M_u,\ \ \inf \varphi_u(M_u) \geq \inf \varphi(M) > -\infty; \tag{32}$$

and this shows that (a02) holds for such elements. Moreover, as $(\preceq)$ is coarser than $(\leq)$ on $\mathrm{Dom}(\varphi)$, (a03) holds as well for these data. Finally, let (x_n) be a $(\preceq)$-ascending sequence in M_u:

(a05) $x_n \leq x_m$ and $(\varepsilon/\lambda)d(x_n, x_m) \leq \varphi(x_n) - \varphi(x_m)$, if $n \leq m$.

The sequence $(\varphi(x_n))$ is descending and bounded from below; hence a Cauchy one. This, along with (a05), tells us that (x_n) is a $(\leq)$-ascending d-Cauchy sequence in $M_u \subseteq \mathrm{Dom}(\varphi)$; wherefrom (by (a01) (the second half), $x_n \to y$ for some $y \in M$. Passing to limit as $m \to \infty$ in (a05) one derives (by means of (a01) (the first half) and (a03))

$$x_n \leq y,\ d(x_n, y) \leq \varphi(x_n) - \varphi(y),\ \text{(i.e.: } x_n \preceq y\text{), for all } n;$$

note that, as a consequence of this, $y \in M_u$. Putting these facts together, it follows that (a01) holds for our elements, as claimed. By EVPm(loc) it then results that, for the starting $u \in M_u = \mathrm{Dom}(\varphi_u)$ there exists $v = v(u) \in M_u$ with the properties (30) + (31), relative to $(M_u; \preceq; d)$ and φ_u. This element has all the properties needed in (EVPm).

A basic particular case of these developments corresponds to the choice $(\leq) = M \times M$ (= the *trivial* quasiorder on M). Let (M, d) be a complete metric space; and $\varphi : M \to R \cup \{\infty\}$, taken as in (a02) and

(a06) φ is d-lsc ($\liminf_n \varphi(x_n) \geq \varphi(x)$, whenever $x_n \to x$).

Theorem 19 *Let* $\varepsilon, \lambda > 0$ *be arbitrary fixed. Then*

(i) *for each* $u \in M(\varphi, \varepsilon)$, *there exists* $v = v(u) \in \mathrm{Dom}(\varphi)$ *with*

$$(\varepsilon/\lambda)d(u, v) \leq \varphi(u) - \varphi(v) \tag{33}$$

$$(\varepsilon/\lambda)d(v, x) > \varphi(v) - \varphi(x), \forall x \in M\backslash\{v\}. \tag{34}$$

(ii) *for each and* $u \in M(\varphi, \varepsilon)$, *there exists* $v = v(u) \in \mathrm{Dom}(\varphi)$. *with the property (34), as well as*

$$\varphi(u) \geq \varphi(v) \text{ and } d(u, v) < \lambda. \tag{35}$$

As before, we have a couple of variational principles under this statement. Precisely, Th 19 (i) is just EVP. Further, denote EVP(loc):=Th 19 (ii); this will be referred to as the *local* variant of EVP. Clearly, EVP $\Longrightarrow$ EVP(loc). On the other hand, by Proposition 12, EVP(loc) $\Longrightarrow$ EVP; so that, these principles are mutually equivalent; cf. Bao and Khanh [3].

This principle found some basic applications to control and optimization, generalized differential calculus, critical point theory and global analysis; we refer to the 1979 paper by Ekeland [22] for a survey of these. So, it cannot be surprising that, soon after its formulation, many extensions of EVP were proposed. For example, the abstract (order) one starts from the fact that, with respect to the (quasi)order

(a07) $(x, y \in M)\ x \leq y$ iff $(\varepsilon/\lambda)d(x, y) + \varphi(y) \leq \varphi(x)$

the point $v \in M$ appearing in (34) is *maximal*; so that, EVP is nothing but a denumerable version of the Zorn–Bourbaki maximal statement [8]; note that, an appropriate extension of this type is just the 1976 Brezis–Browder ordering principle [9] (in short: BB) and its variants we already stated. Further, the dimensional way of extension refers to the ambient space (R) of $\varphi(M)$ being substituted by a (topological or not) vector space. However, as shown in a previous place, it is obtainable from a semi-metric version of BB; equivalent (see above) with EVP. On the other hand, the (pseudo) metrical one consists in the conditions imposed to the metric d over M being relaxed. The basic result in this direction was obtained in 1992 by Tataru [47]; subsequent extensions of it may be found in the 1996 paper by Kada et al. [32]. But, as established in Turinici [52], all such results are again deductible from BB; or, in other words, from EVP. Finally, we must add to this list the 1987 "smooth" variational principle due to Borwein and Preiss [7] (in short: BP). Note that, in contrast with the preceding developments, the constructive reasoning proposed there is not (formally) reducible to "maximal" techniques.

Now, it is legitimate to ask whether BP is a genuine extension of EVP. As we shall see (in Sect. 4.2), the answer to this is negative; and it will remain as such for its (extended) "monotone" version, constructed under the lines in Turinici [50].

4.2 (DDC) $\Longrightarrow$ (BPm) $\Longrightarrow$ (EVPm)

Let us return to the setting of (EVP). A practical disadvantage of the variational relation (34) is that, even if the ambient metric d is generated by a norm, $||.||$, and the original function $x \mapsto \varphi(x)$ is differentiable, its (related) perturbed function $x \mapsto \varphi(x) + \delta||x||$ is not differentiable. However, when the norm in question is generated by a scalar product, this property is retainable for the perturbed function $x \mapsto \varphi(x)+\delta||x||^2$; so, we may ask of to what extent is this useful in order to remove that drawback.

An appropriate answer is obtainable from the developments of (BP). In fact, the involved reasoning may be carried out over a quasiorder setting. Some preliminary facts are needed. Let $(M; \leq; d)$ be a quasiordered metrical structure taken as in (a01); and $p \geq 1$ be fixed in the sequel. Denote by $\Gamma_p(\leq)$ the class of all functions

(b01) $\psi(x) = \sum_n \mu_n(d(x, v_n))^p, \ x \in M,$

where (μ_n) is a sequence of positive numbers with $\sum_n \mu_n = 1$ and (v_n) is an ascending (modulo $(\leq)$) convergent sequence in X. Note that, by the Weierstrass criterion, the convergence of this functional series is uniform over bounded subsets of X. Hence, the (partial sum) functional sequence

(b02) $\psi_n(x) = \sum_{k<n} \mu_k(d(x, v_k))^p, \ x \in M, n \geq 1$

converges toward the function ψ, uniformly on bounded subsets of M. This, in particular, says that ψ is continuous; and, moreover,

$$\psi_n(x_n) \to \psi(x) \text{ as } n \to \infty, \text{ whenever } x_n \to x. \tag{36}$$

Further, let $\mathcal{F}(M)$ stand for the class of all functions $\chi : M \to R \cup \{\infty\}$; and $\mathcal{F}_0(M)$, the subclass of all $\chi \in \mathcal{F}(M)$ fulfilling (a02)+(a03). Take the function $\varphi \in \mathcal{F}_0(M)$; as well as the couple of numbers $\varepsilon > 0$, $\lambda > 0$. The following "monotone" Borwein–Preiss principle (in short: BPm) is available.

Theorem 20 *Let these conditions hold. Then, for each $u \in M(\varphi, \varepsilon)$, there exist $\psi \in \Gamma_p(\leq)$ and $v \in M(\varphi, \varepsilon)$ so that*

$$u \leq v, \ d(u, v) < \lambda, \ \varphi(u) \geq \varphi(v) \tag{37}$$

$$\varphi(w) - \varphi(v) + (\varepsilon/\lambda^p)(\psi(w) - \psi(v)) \geq 0, \ \forall w \in M(v, \leq). \tag{38}$$

Proof Define the constants $(\varepsilon_1, \varepsilon_2)$ and (θ, μ, δ), as

(b03) $\varphi(u) - \inf [\varphi(X)] < \varepsilon_2 < \varepsilon_1 < \varepsilon, 0 < \mu < 1 - (\varepsilon_1/\varepsilon),$
$0 < \theta < \mu[1 - (\varepsilon_2/\varepsilon_1)^{1/p}]^p, \delta = (1 - \mu)(\varepsilon/\lambda^p).$

For each $n \in N$, we introduce a relation $\mathcal{S}_n$ over $\mathcal{F}_0(M) \times M$ according to

(b04) $(\chi, w)\mathcal{S}_n(\chi^*, w^*)$ iff $[\chi^*(x) = \chi(x) + \delta\mu^n(d(x, w))^p, x \in M]$,
$w \leq w^*$ and $\chi^*(w^*) \leq \theta\chi(w) + (1 - \theta) \inf [\chi^*(M(w, \leq))]$.

To see its effectiveness, note that $\chi(w) = \chi^*(w) \geq \inf[\chi^*(M(w, \leq))]$. So, if equality holds (i.e., the infimum is attained at w), one simply takes $w^* = w$; otherwise (when the strict inequality appears) use the definition of infimum to get the desired point, $w^* \geq w$. As a consequence, $\text{Dom}(\mathcal{S}_n) = \mathcal{F}_0(M) \times M$, for each n. This, along with the diagonal dependent choices principle (DDC), tells us that, for the starting $(\varphi_0 = \varphi, u_0 = u)$ in $\mathcal{F}_0(M) \times M$, there must be a sequence (φ_n, u_n) in $\mathcal{F}_0(M) \times M$ in such a way that $[(\varphi_n, u_n)\mathcal{S}_n(\varphi_{n+1}, u_{n+1}), \forall n]$; that is (for all such n)

$$\varphi_{n+1}(x) = \varphi_n(x) + \delta\mu^n(d(x, u_n))^p, \quad x \in M$$

$$u_n \leq u_{n+1}, \varphi_{n+1}(u_{n+1}) \leq \theta\varphi_n(u_n) + (1-\theta)\inf[\varphi_{n+1}(M(u_n, \leq))].$$

Clearly, (u_n) is ascending (modulo $(\leq)$). To establish its convergence, denote

$$\alpha_n = \varphi_n(u_n)(= \varphi_{n+1}(u_n)), \quad \sigma_n = \inf[\varphi_n(M(u_n, \leq))], \quad n \in N.$$

By the above definitions, one has

$$\sigma_n \leq \sigma_{n+1} \leq \alpha_{n+1} \leq \theta\alpha_n + (1-\theta)\sigma_{n+1} \leq \alpha_n, \quad n \in N.$$

(The last inequality is a direct consequence of

$$(\forall n): \ \alpha_{n+1} \leq \theta\alpha_n + (1-\theta)\alpha_{n+1} \Longrightarrow \alpha_{n+1} \leq \alpha_n;$$

we do not give details). In other words, (σ_n) is ascending, (α_n) is descending and $\sigma_n \leq \alpha_n$, for all n. Moreover, $\alpha_{n+1} - \sigma_{n+1} \leq \theta(\alpha_n - \sigma_n)$, $n \in N$; wherefrom (via (b03))

$$\alpha_n - \sigma_n \leq \theta^n(\alpha_0 - \sigma_0) \leq \theta^n \varepsilon_2, \quad n \in N; \tag{39}$$

which, in particular, says that (α_n) and (σ_n) are convergent (in R) sequences and have the same limit. On the other hand, for all n,

$$\alpha_n \geq \alpha_{n+1} = \varphi_n(u_{n+1}) + \delta\mu^n(d(u_n, u_{n+1}))^p \geq \sigma_n + \delta\mu^n(d(u_n, u_{n+1}))^p;$$

so that, combining with (39),

$$d(u_n, u_{n+1}) \leq ((\alpha_n - \sigma_n)/\delta\mu^n)^{1/p} \leq (\theta/\mu)^{n/p}(\varepsilon_2/\delta)^{1/p}, \quad n \in N; \tag{40}$$

which shows us that (u_n) is an ascending d-Cauchy sequence. By the order-completeness hypothesis, (u_n) is thus convergent; and then, the formula

(b05) $\psi(x) = (1-\mu)\sum_n \mu^n(d(x, u_n))^p, \ x \in M$

defines an element of $\Gamma_p(\leq)$. Let v stand for the (unique) limit of (u_n). We claim that the couple (ψ, v) satisfies the written conclusions. In fact, (a01) (the first half) gives $u \leq v$. Further, by (40) and (b03)

$$d(u,v) \leq \sum_n d(u_n,u_{n+1}) \leq \\ (\varepsilon_2/\delta)^{1/p}/[1-(\theta/\mu)^{1/p}] < (\varepsilon_1/\delta)^{1/p} < [(1-\mu)\varepsilon/\delta]^{1/p} = \lambda.$$

We also have [the (partial sum) functional sequence (ψ_n) being that of (b02), where (μ_n) is the sequence appearing in (b05)]

$$\varphi_n(x) = \varphi(x) + (\varepsilon/\lambda^p)\psi_n(x),\ x \in M, \text{ for all } n \geq 1.$$

This clearly gives

$$\varphi(u_n) + (\varepsilon/\lambda^p)\psi_n(u_n) = \alpha_n \leq \alpha_0 = \varphi(u),\ \ n \geq 1.$$

Combining this with (36) and the $(\leq, d)$-lsc property of φ, one has

$$\varphi(v) \leq \varphi(v) + (\varepsilon/\lambda^p)\psi(v) \leq \lim_n (\alpha_n) \text{ (hence } \varphi(v) \leq \varphi(u)).$$

Finally, let $w \in M(v, \leq)$ be arbitrary fixed. We may write

$$\varphi(w) + (\varepsilon/\lambda^p)\psi(w) = \lim_n \varphi_n(w) \geq \lim_n (\sigma_n) = \lim_n (\alpha_n).$$

This, along with the relations above, yields the conclusions we need.

In particular, when $(\leq)$ is the trivial quasiorder on M, (BPm) reduces to the 1987 smooth variational principle due to Borwein and Preiss [7] (in short: BP). Another particular case is $p = 1$. Note that each element ψ of $\Gamma_1(\leq)$ is nonexpansive:

$$|\psi(x) - \psi(y)| \leq d(x,y),\ x, y \in M. \tag{41}$$

As a consequence of this, the second conclusion of (BPm) gives

$$\varphi(w) - \varphi(v) + (\varepsilon/\lambda)d(w,v) \geq 0, \text{ for all } w \in M(v, \leq). \tag{42}$$

which partially covers (30). However, by exploiting a device in Bao and Khanh [3], it may be shown that this covering is complete.

Proposition 13 *Under the above conventions, we have*

(i) *(BPm) $\Longrightarrow$ EVPm(loc) $\Longleftrightarrow$ (EVPm)*

(ii) *(BPm), (EVPm), and (EVP) are equivalent to each other.*

Proof Let $\varepsilon > 0$, $\lambda > 0$ and $u \in M(\varphi, \varepsilon)$, be taken as in the premise of Theorem 18 (ii). Denote by $(\preceq)$ the quasiorder on M introduced as

(b06) $x \preceq y$ iff $x \leq y$, $(\varepsilon/\lambda)d(x,y) + \varphi(y) \leq \varphi(x)$;

clearly, it is an order on Dom(φ). Put $M_u := M(u, \preceq)$; and let again φ stand for the restriction of φ to M_u. By the imposed conditions, M_u is $(\leq)$-closed; hence all the more $(\leq)$-complete. In addition, the restriction of $(\leq)$ to M_u is self-closed; hence, (a01) holds over M_u. Moreover, as $\varphi(u) < \inf[\varphi(M)] + \varepsilon \leq \inf[\varphi(M_u)] + \varepsilon$, u is an element of $M_u(\varphi, \varepsilon)$; wherefrom, φ is proper on M_u. Finally, by the $(\leq)$-closeness of M_u, φ is $(\leq, d)$-lsc (on M_u). Summing up, (BPm) is applicable to $(M_u, \leq; d)$, the function φ and $p = 1$. Hence, for the starting $\varepsilon > 0$, $\nu > \lambda$ and $u \in M_u(\varphi, \varepsilon)$, there must be some $v \in M_u(\varphi, \varepsilon)$ with the properties (37) and (38), with (M_u, ν) in place of (M, λ); whence (taking (42) into account)

$$u \leq v,\ d(u, v) < \nu,\ \varphi(u) \geq \varphi(v) \tag{43}$$

$$(\varepsilon/\nu)d(v, w) \geq \varphi(v) - \varphi(w),\ \text{if } v \leq w \in M_u. \tag{44}$$

Now, the localizing relation $v \in M_u$ gives (via $u \in M_u(\varphi, \varepsilon)$)

$$u \leq v,\ (\varepsilon/\lambda)d(u, v) \leq \varphi(u) - \varphi(v) < \varepsilon.$$

This yields $d(u, v) < \lambda$; so, combining with (43), one derives conclusion (31) in Theorem 18. It remains to show that the conclusion (30) of this statement holds too. Assume by contradiction that it would be false:

(b07) $(\varepsilon/\lambda)d(v, w) \leq \varphi(v) - \varphi(w)$, for some $w \in M(v, \leq)$, $w \neq v$.

This firstly gives $\varphi(v) \geq \varphi(w)$; hence $w \in \text{Dom}(\varphi)$. Secondly, by these relations, $v \preceq w$; and this, along with $u \preceq v$ gives $w \in M_u$. Taking (44) into account, gives

$$(\varepsilon/\lambda)d(v, w) \leq (\varepsilon/\nu)d(v, w),\quad v \neq w;$$

so that, $\lambda \geq \nu$, contradiction. This proves the first part of the statement. The second one follows by Theorem 20 and the relation (DDC) $\Longleftrightarrow$ (EVP), established in a previous place.

In particular, when $(\leq)$ is the trivial quasiorder of M, this means that (BP) includes (EVP) in a complete manner; so—by the above—the discussed variational principles are equivalent to each other. These conclusions also comprise the smooth variational statement in Li and Shi [35]; we do not give details. Further aspects may be found in Bejancu [4]; see also Deville et al. [18].

References

1. Altman, M.: A generalization of the Brezis–Browder principle on ordered sets. Nonlinear Anal. **6**, 157–165 (1982)
2. Bae, J.S., Cho, E.W., Yeom, S.H.: A generalization of the Caristi–Kirk fixed point theorem and its applications to mapping theorems. J. Korean Math. Soc. **31**, 29–48 (1994)
3. Bao, T.Q., Khanh, P.Q.: Are several recent generalizations of Ekeland's variational principle more general than the original principle?. Acta Math. Vietnam. **28**, 345–350 (2003)
4. Bejancu, A.: On the Ekeland and Borwein–Preiss principles in finite dimensions. An. Şt. Univ. "A. I. Cuza" Iaşi (S. I-a, Mat.) **40**, 63–67 (1994)
5. Bernays, P.: A system of axiomatic set theory: Part III. Infinity and enumerability analysis. J. Symb. Logic **7**, 65–89 (1942)

6. Blair, C.E.: The Baire Category Theorem implies the principle of dependent choice. Bull. Acad. Pol. Sci. (Sér. Math. Astronom. Phys.) **10**, 933–934 (1977)
7. Borwein, J.M., Preiss, D.: A smooth variational principle with applications to subdifferentiability and to differentiability of convex functions. Trans. Am. Math. Soc. **303**, 517–527 (1987)
8. Bourbaki, N.: Sur le théorème de Zorn. Archiv Math. **2**, 434–437 (1949/1950)
9. Brezis, H., Browder, F.E.: A general principle on ordered sets in nonlinear functional analysis. Adv. Math. **21**, 355–364 (1976)
10. Brøndsted, A.: On a lemma of Bishop and Phelps. Pac. J. Math. **55**, 335–341 (1974)
11. Brøndsted, A.: Fixed points and partial orders. Proc. Am. Math. Soc. **60**, 365–366 (1976)
12. Brunner, N.: Topologische Maximalprinzipien. Zeitschr. Math. Logik Grundl. Math. **33**, 135–139 (1987)
13. Cârjă, O., Necula, M., Vrabie, I.I.: Viability, Invariance and Applications. North-Holland Mathematics Studies, vol. 207. Elsevier, Amsterdam (2007)
14. Chen, G.Y., Huang, X.X.: A unified approach to the existing three types of variational principles for vector valued functions. Math. Meth. Oper. Res. **48**, 349–357 (1998)
15. Conserva, V., Rizzo, S.: Maximal elements in a class of order complete metric subspaces. Math. Jpn. **37**, 515–518 (1992)
16. Cristescu, R.: Topological Vector Spaces. Noordhoff, Leyden (1977)
17. Dancs, S., Hegedus, M., Medvegyev, P.: A general ordering and fixed-point principle in complete metric space. Acta Sci. Math. (Szeged) **46**, 381–388 (1983)
18. Deville, R., Godefroy, G., Zizler, V.: A smooth variational principle with applications to Hamilton–Jacobi equations in infinite dimensions. J. Funct. Anal. **111**, 197–212 (1993)
19. Dodu, J., Morillon, M.: The Hahn–Banach property and the axiom of choice. Math. Log. Q. **45**, 299–314 (1999)
20. Du, W.S.: On some nonlinear problems induced by an abstract maximal element principle. J. Math. Anal. Appl. **347**, 391–399 (2008)
21. Ekeland, I.: On the variational principle. J. Math. Anal. Appl. **47**, 324–353 (1974)
22. Ekeland, I.: Nonconvex minimization problems. Bull. Amer. Math. Soc. (New Ser.) **1**, 443–474 (1979)
23. Gerth (Tammer), C., Weidner, P.: Nonconvex separation theorems and some applications in vector optimization. J. Optim. Theory Appl. **67**, 297–320 (1990)
24. Goepfert, A., Tammer, C., Zălinescu, C.: On the vectorial Ekeland's variational principle and minimal points in product spaces. Nonlinear Anal. **39**, 909–922 (2000)
25. Goepfert, A., Riahi, H., Tammer, C., Zălinescu, C.: Variational Methods in Partially Ordered Spaces. Canadian Mathematical Society Books in Mathematics, vol. 17. Springer, New York (2003)
26. Goldblatt, R.: On the role of the Baire category theorem and dependent choice in the foundation of logic. J. Symb. Logic **50**, 412–422 (1985)
27. Granas, A., Horvath, C.D.: On the order-theoretic Cantor theorem. Taiwan. J. Math. **4**, 203–213 (2000)
28. Ha, T.X.D.: Some variants of the Ekeland variational principle for a set-valued map. J. Optim. Theory Appl. **124**, 187–206 (2005)
29. Hamel, A.: Variational Principles on Metric and Uniform Spaces. Habilitation Thesis, Martin-Luther University, Halle-Wittenberg (Germany) (2005)
30. Hyers, D.H., Isac, G., Rassias, T.M.: Topics in Nonlinear Analysis and Applications. World Scientific, Singapore (1997)
31. Isac, G.: The Ekeland's principle and Pareto ε-efficiency. In: Tamiz, M. (ed.) Multi-Objective Programming and Goal Programming. Lecture Notes in Economic and Mathematical Systems, vol. 432, pp. 148–163. Springer, Berlin (1996)
32. Kada, O., Suzuki, T., Takahashi, W.: Nonconvex minimization theorems and fixed point theorems in complete metric spaces. Math. Jap. **44**, 381–391 (1996)
33. Kang, B.G., Park, S.: On generalized ordering principles in nonlinear analysis. Nonlinear Anal. **14**, 159–165 (1990)

34. Khanh, P.Q., Quy, D.N.: On generalized Ekeland's variational principle and equivalent formulations for set-valued mappings. J. Glob. Optim. **49**, 381–396 (2011)
35. Li, Y., Shi, S.: A generalization of Ekeland's ε-variational principle and its Borwein–Preiss smooth variant. J. Math. Anal. Appl. **246**, 308–319 (2000)
36. Li, S.J., Zhang, W.Y.: A minimization theorem for a set-valued mapping. Appl. Math. Lett. **21**, 769–773 (2008)
37. Moore, G.H.: Zermelo's Axiom of Choice: Its Origin, Development and Influence. Springer, New York (1982)
38. Moskhovakis, Y.: Notes on Set Theory. Springer, New York (2006)
39. Nachbin, L.: Topology and Order. Van Nostrand, Princeton (1965)
40. Nemeth, A.B.: A nonconvex vector minimization problem. Nonlinear Anal. **10**, 669–678 (1986)
41. Qiu, J.H.: Local completeness, drop theorem and Ekeland variational principle. J. Math. Anal. Appl. **311**, 23–39 (2005)
42. Rozoveanu, P.: Ekeland's variational principle for vector valued functions. Math. Reports [St. Cerc. Mat.] **2**(52), 351–366 (2000)
43. Schechter, E.: Handbook of Analysis and its Foundation. Academic, New York (1997)
44. Szaz, A.: An improved Altman type generalization of the Brezis Browder ordering principle.. Math. Commun. **12**, 155–161 (2007)
45. Tammer, C.: A generalization of Ekeland's variational principle. Optimization **25**, 129–141 (1992)
46. Tarski, A.: Axiomatic and algebraic aspects of two theorems on sums of cardinals. Fund. Math. **35**, 79–104 (1948)
47. Tataru, D.: Viscosity solutions of Hamilton–Jacobi equations with unbounded nonlinear terms. J. Math. Anal. Appl. **163**, 345–392 (1992)
48. Turinici, M.: A generalization of Brezis–Browder's ordering principle. An. Şt. Univ. "A. I. Cuza" Iaşi (S I-a: Mat) **28**, 11–16 (1982)
49. Turinici, M.: Metric variants of the Brezis–Browder ordering principle. Demonstr. Math. **22**, 213–228 (1989)
50. Turinici, M.: A monotone version of the variational Ekeland's principle. An. Şt. Univ. "A. I. Cuza" Iaşi (S. I-a: Mat) **36**, 329–352 (1990)
51. Turinici, M.: Minimal points in product spaces. An. Şt. Univ. "Ovidius" Constanţa (Ser. Math) **10**, 109–122 (2002)
52. Turinici, M.: Variational statements on KST-metric structures. An. Şt. Univ. "Ovidius" Constanţa (Mat) **17**, 231–246 (2009)
53. Turinici, M.: Functional variational principles and coercivity over normed spaces. Optimization **59**, 199–222 (2010)
54. Wolk, E.S.: On the principle of dependent choices and some forms of Zorn's lemma. Can. Math. Bull. **26**, 365–367 (1983)
55. Zhu, J., Li, S.J.: Generalization of ordering principles and applications. J. Optim. Theory Appl. **132**, 493–507 (2007)

Univalence Conditions and Properties for Some New Integral Operators

Nicoleta Ularu and Daniel Breaz

Abstract In this paper we introduce three new integral operators. For these operators we prove the univalence conditions, the convexity order and some other properties for some special classes of functions. In view of starlike and convex functions classes, we prove for these operators convexity properties, starlike properties and some other properties that are obtained in analytic form. *Subject Classifications*: 30C45.

Keywords Univalence · Integral operators · Starlike functions · Convex functions · Analytical functions

1 Preliminaries and Definitions

Let $\mathcal{U} = \{z : |z| < 1\}$ be the unit disk and $\mathcal{A}$ be the class of all functions of the form

$$f(z) = z + \sum_{n=2}^{\infty} a_n z^n \tag{1}$$

which are analytic in $\mathcal{U}$ and satisfy the condition

$$f(0) = f'(0) - 1 = 0.$$

We note by $\mathcal{S}$ the class of univalent and regular functions. A function $f(z)$ is starlike of order α if

$$\text{Re}\left\{\frac{zf'(z)}{f(z)}\right\} > \alpha, z \in \mathcal{U}$$

for $0 \leq \alpha < 1$, and we denote this class by $\mathcal{S}^*(\alpha)$. We denote by $\mathcal{K}(\alpha)$ the subclass of $\mathcal{A}$ that consists of all the functions that are convex of order α and satisfy the

N. Ularu (✉)
University of Piteşti, Târgul din Vale Str., No. 1, Piteşti, Romania
e-mail: nicoletaularu@yahoo.com

D. Breaz
"1 Decembrie 1918" University of Alba Iulia, N. Iorga Str., No. 11-13, 510009 Alba Iulia, Romania
e-mail: dbreaz@uab.ro

T. M. Rassias, P. M. Pardalos (eds.), *Mathematics Without Boundaries*,
DOI 10.1007/978-1-4939-1106-6_21, © Springer Science+Business Media, LLC 2014

inequality

$$\mathrm{Re}\left\{1+\frac{zf''(z)}{f'(z)}\right\}>\alpha,\ \ z\in\mathcal{U}$$

for $0 \le \alpha < 1$.
A function $f \in \mathcal{A}$ belongs to the class $\mathcal{R}(\alpha)$ if

$$\mathrm{Re}(f'(z)) > \alpha,\ z \in \mathcal{U}.$$

$\mathcal{B}(\mu,\alpha), \mu \ge 0, 0 \le \alpha < 1$ is a family that consists all the functions $f(z) \in \mathcal{A}$ that satisfy the inequality

$$\left|f'(z)\left(\frac{z}{f(z)}\right)^{\mu}-1\right|<1-\alpha,\ \ z\in\mathcal{U}.$$

This family is a comprehensive class of analytic functions that includes various classes of analytic univalent functions. Some interesting subclasses are $\mathcal{B}(1,\alpha) \subset \mathcal{S}^*(\alpha), \mathcal{B}(0,\alpha) \equiv \mathcal{R}(\alpha)$ and $\mathcal{B}(2,\alpha) \equiv \mathcal{B}(\alpha)$, which have been studied by Frasin and Darus in [4]. Uralegaddi et al. [9] and Owa and Srivastava [6] studied the class $\mathcal{N}(\beta)$. So, $\mathcal{N}(\beta)$ is a subclass of $\mathcal{A}$ which consists of all the functions $f(z)$ that satisfy the inequality

$$\mathrm{Re}\left\{\frac{zf''(z)}{f'(z)}+1\right\}<\beta,\ z\in\mathcal{U}.$$

F. Rønning [8] has introduced the class of univalent functions $\mathcal{SP}(\alpha,\beta), \alpha > 0, \beta \in [0,1)$, the class of all functions f that satisfy the inequality

$$\left|\frac{zf'(z)}{f(z)}-(\alpha+\beta)\right|\le \mathrm{Re}\frac{zf'(z)}{f(z)}+\alpha-\beta$$

for all $z \in \mathcal{U}$ and the class $\mathcal{SP}$ if and only if

$$\mathrm{Re}\frac{zf'(z)}{f(z)}>\left|\frac{zf'(z)}{f(z)}-1\right|$$

for all $z \in \mathcal{U}$. The class $\beta - Sp(\alpha)$ is the class of uniformly $\beta-$starlike functions of order α that satisfy the inequality

$$\mathrm{Re}\left\{\frac{zf'(z)}{f(z)}-\alpha\right\}\ge\beta\left|\frac{zf'(z)}{f(z)}-1\right|$$

for $-1 \le \alpha \le 1, \beta > 0$ and $z \in \mathcal{U}$.

Theorem 1 [7] *Let $f \in A$ and $m > 0$ such that*

$$\left|\left(\frac{z^2 f'(z)}{f^2(z)}-1\right)-\frac{m-1}{2}|z|^{m+1}\right|\le\frac{m+1}{2}|z|^{m+1} \tag{2}$$

for all $z \in \mathcal{U}$. Then the function f is analytic and univalent in $\mathcal{U}$.

Lemma 1 [5] *(General Schwarz Lemma) Let f be the function regular in the disk $\mathcal{U}_R = \{z \in \mathbb{C} : |z| < R\}$, with $|f(z)| < M$, M fixed. If f has in $z = 0$ one zero with multiply $\geq m$, then*

$$|f(z)| \leq \frac{M}{R^m}|z|^m,\ z \in \mathcal{U}_R \tag{3}$$

the equality (in the inequality (3) *for $z \neq 0$) can hold only if $f(z) = e^{i\theta}\frac{M}{R^m}z^m$, where θ is constant.*

Theorem 2 [2] *If the function f is regular in unit disk $\mathcal{U}$, $f(z) = z + a_2z^2 + \ldots$ and*

$$\left(1 - |z|^2\right)\left|\frac{zf''(z)}{f'(z)}\right| \leq 1$$

for all $z \in \mathcal{U}$, then the function f is univalent in $\mathcal{U}$.

L.V. Ahlfors [1] and J. Becker [3] have obtained the following univalence criterion:

Theorem 3 *Let $|c| \leq 1, c \neq -1$. If $f(z) = z + a_2z^2 + \ldots$ is a regular function in $\mathcal{U}$ and*

$$\left|c|z|^2 + (1 - |z|^2)\frac{zf''(z)}{f'(z)}\right| \leq 1,$$

for all $z \in \mathcal{U}$, then the function f is regular and univalent in $\mathcal{U}$.

J.W. Alexander has introduced for first time an integral operator on a subclass of univalent functions in 1915. In the paper *Functions which map the interior of the unit circle upon simple regions*, he introduced the operator

$$J_1(f)(z) = \int_0^z \frac{f(t)}{t}dt. \tag{4}$$

In this paper we introduce three new integral operators:

$$I(f,g)(z) = \int_0^z \prod_{i=1}^n \frac{f_i(t)}{g_i(t)}dt \tag{5}$$

$$I_1(f,g)(z) = \int_0^z \prod_{i=1}^n \left(\frac{f_i(t)}{g_i(t)}\right)^{\gamma_i} dt \tag{6}$$

$$I_2(f,g)(z) = \int_0^z \prod_{i=1}^n \frac{(f_i(t))^{\gamma_i}}{(g_i(t))^{\sigma_i}}dt \tag{7}$$

Remark 1 If we consider $\gamma_1 = \cdots = \gamma_n = 1$ in the operator given by (6), we obtain the operator given by (5).

Remark 2 For $n = 1$ and the function $g(t) = t$ in (5), we obtain the integral operator that was introduced by Alexander, given by (4).

2 Main Results

2.1 Univalence Conditions

Theorem 4 *Let $f_i \in \mathcal{A}, g_i \in \mathcal{P}$, and M_i, N_i be two positive real numbers $(M_i, N_i \geq 1)$ for $i = \overline{1,n}$. If*

$$\left|\frac{f_i'(z)}{f_i(z)}\right| \leq M_i \text{ and } \left|\frac{g_i'(z)}{g_i(z)}\right| \leq N_i$$

for $i = \overline{1,n}, z \in \mathcal{U}$ and $\sum_{i=1}^{n}(M_i + N_i) \leq \frac{3\sqrt{3}}{2}$, then the integral operator $I(f,g)$ defined by (5) is in the univalent function class $\mathcal{S}$.

Proof From (5), it follows that

$$\frac{I''(f,g)(z)}{I'(f,g)(z)} = \sum_{i=1}^{n}\left(\frac{f_i'(z)}{f_i(z)} - \frac{g_i'(z)}{g_i(z)}\right) \tag{8}$$

and

$$\begin{aligned}(1-|z|^2)\left|\frac{zI''(f,g)(z)}{I'(f,g)(z)}\right| &= (1-|z|^2)\left|z\left[\sum_{i=1}^{n}\left(\frac{f_i'(z)}{f_i(z)} - \frac{g_i'(z)}{g_i(z)}\right)\right]\right| \\ &\leq (1-|z|^2)|z|\left[\sum_{i=1}^{n}\left(\left|\frac{f_i'(z)}{f_i(z)}\right| + \left|\frac{g_i'(z)}{g_i(z)}\right|\right)\right] \\ &\leq (1-|z|^2)|z|\sum_{i=1}^{n}(M_i + N_i).\end{aligned} \tag{9}$$

Since $\max_{|z|\leq 1}\left[(1-|z|^2)|z|\right] = \frac{2}{3\sqrt{3}}$, from the hypothesis and (9) we get

$$(1-|z|^2)\left|\frac{zI''(f,g)(z)}{I'(f,g)(z)}\right| \leq 1, \text{ for } z \in \mathcal{U}.$$

Using Theorem 2 we obtain that the operator $I(f,g)$ is in the class $\mathcal{S}$.

Corollary 1 *Let $f \in \mathcal{A}, g \in \mathcal{P}$, and $M \geq 1, N \geq 1$ be two positive real numbers. If*

$$\left|\frac{f'(z)}{f(z)}\right| \leq M, \text{ and } \left|\frac{g'(z)}{g(z)}\right| \leq N$$

for $z \in \mathcal{U}$ and $M + N \leq \frac{3\sqrt{3}}{2}$, then the integral operator $I(f,g)(z) = \int_0^z \frac{f(t)}{g(t)}dt$ is in the univalent function class $\mathcal{S}$.

Proof We consider $n = 1$ in Theorem 4.

Remark 3 If we consider $n = 1$ and the function $g(t) = t$ in Theorem 4, we obtain the univalence criterion for the Alexander operator $J_1(f)(z)$ defined by (4).

Theorem 5 *Let $f_i \in \mathcal{A}, g_i \in \mathcal{P}$, and M_i, N_i be two positive real numbers (M_i, $N_i \geq 1$) for $i = \overline{1,n}$. If*

$$\left|\frac{f_i'(z)}{f_i(z)}\right| \leq M_i \text{ and } \left|\frac{g_i'(z)}{g_i(z)}\right| \leq N_i$$

for $i = \overline{1,n}, z \in \mathcal{U}$ and $\sum\limits_{i=1}^{n} (\gamma_i M_i + \sigma_i N_i) \leq \frac{3\sqrt{3}}{2}$, then the integral operator $I_2(f,g)$ defined by (7) is in the univalent function class $\mathcal{S}$.

Proof From (7) we obtain that

$$\frac{I_2''(f,g)(z)}{I_2'(f,g)(z)} = \sum_{i=1}^{n} \gamma_i \frac{f_i'(z)}{f_i(z)} - \sum_{i=1}^{n} \sigma_i \frac{g_i'(z)}{g_i(z)} \tag{10}$$

and

$$\begin{aligned}(1-|z|^2)\left|\frac{zI_2''(f,g)(z)}{I_2'(f,g)(z)}\right| &= (1-|z|^2)\left|z\left[\sum_{i=1}^{n}\left(\gamma_i\frac{f_i'(z)}{f_i(z)} - \sigma_i\frac{g_i'(z)}{g_i(z)}\right)\right]\right| \\ &\leq (1-|z|^2)\,|z|\left[\sum_{i=1}^{n}\left(\gamma_i\left|\frac{f_i'(z)}{f_i(z)}\right| + \sigma_i\left|\frac{g_i'(z)}{g_i(z)}\right|\right)\right] \qquad (11)\\ &\leq (1-|z|^2)\,|z|\sum_{i=1}^{n}(\gamma_i M_i + \sigma_i N_i).\end{aligned}$$

Since $\max\limits_{|z|\leq 1}\left[(1-|z|^2)|z|\right] = \frac{2}{3\sqrt{3}}$, the hypothesis and (11) result in

$$(1-|z|^2)\left|\frac{zI''(f,g)(z)}{I'(f,g)(z)}\right| \leq 1, \text{ for } z \in \mathcal{U}.$$

Using Theorem 2 results that the operator $I_2(f,g)$ is in the class $\mathcal{S}$.

For $n = 1$ in Theorem 5 we obtain

Corollary 2 *Let $f \in \mathcal{A}, g \in \mathcal{P}$, and M, N be two positive real numbers ($M, N \geq 1$). If*

$$\left|\frac{f'(z)}{f(z)}\right| \leq M \text{ and } \left|\frac{g'(z)}{g(z)}\right| \leq N$$

and $\gamma M + \sigma N \leq \frac{3\sqrt{3}}{2}$, then the integral operator $\int\limits_0^z \frac{f(t)^\gamma}{g(t)^\sigma}dt$ is in the univalent function class $\mathcal{S}$.

Theorem 6 *Let $f_i \in \mathcal{A}, g_i \in \mathcal{P}$ where f_i satisfy the inequality (2) and $M_i \geq 1, N_i \geq 1$ two positive real numbers for $i = \overline{1,n}$. If*

$$|f_i(z)| \leq M_i \text{ and } \left|\frac{zg_i'(z)}{g_i(z)}\right| \leq N_i \tag{12}$$

and

$$|c| \leq 1 - \sum_{i=1}^{n} |\gamma_i|[(m+1)M_i + N_i], \tag{13}$$

then the integral operator $I_1(f,g)$ is in the univalent function class $\mathcal{S}$.

Proof From (6) we obtain that

$$\frac{I_1''(f,g)(z)}{I_1'(f,g)(z)} = \sum_{i=1}^{n} \gamma_i \left(\frac{f_i'(z)}{f_i(z)} - \frac{g_i'(z)}{g_i(z)}\right)$$

and

$$\begin{aligned}\left|c|z|^2 + (1-|z|^2)\frac{zI_1''(f,g)}{I_1'(f,g)}\right| &= \left|c|z|^2 + \left(1-|z|^2\right) z \sum_{i=1}^{n} \gamma_i \left(\frac{f_i'(z)}{f_i(z)} - \frac{g_i'(z)}{g_i(z)}\right)\right| \\ &\leq |c| + \sum_{i=1}^{n} |\gamma_i| \left(\left|\frac{zf_i'(z)}{f_i(z)}\right| + \frac{zg_i'(z)}{g_i(z)}\right) \\ &\leq |c| + \sum_{i=1}^{n} |\gamma_i| \left(\left|\frac{z^2 f_i'(z)}{f_i^2(z)}\right|\left|\frac{f_i(z)}{z}\right| + \left|\frac{zg_i'(z)}{g_i(z)}\right|\right)\end{aligned} \tag{14}$$

From (12) using the general Schwarz lemma we obtain that $\left|\frac{f(z)}{z}\right| \leq M_i$ for $i = \overline{1,n}$ and

$$\begin{aligned}\left|c|z|^2 + \left(1-|z|^2\right)\frac{zI_1''(f,g)}{I_1'(f,g)}\right| &\leq |c| + \sum_{i=1}^{n} |\gamma_i| \cdot \left(\left|\left(\frac{z^2 f_i'(z)}{f_i(z)} - 1\right) - \frac{m-1}{2}|z|^{m+1}\right| M_i\right) \\ &+ \sum_{i=1}^{n} |\gamma_i| \left(\left(1 + \frac{m-1}{2}|z|^{m+1}\right) M_i\right) + \sum_{i=1}^{n} |\gamma_i| N_i\end{aligned} \tag{15}$$

Using (2) in (15) we obtain

$$\left|c|z|^2 + \left(1-|z|^2\right)\frac{zI_1''(f,g)}{I_1'(f,g)}\right| \leq |c| + \sum_{i=1}^{n} |\gamma_i|[(m+1)M_i + N_i].$$

From (13) we obtain that

$$\left|c|z|^2 + (1-|z|^2)\frac{zI_1''(f,g)}{I_1'(f,g)}\right| \leq 1,$$

so, from Theorem 3 we get that the operator $I_1(f,g) \in \mathcal{S}$.

For $n = 1$ in Theorem 6 we obtain

Corollary 3 *Let $f \in \mathcal{A}, g \in \mathcal{P}$ where f satisfy the inequality (2) and $M \geq 1, N \geq 1$ are two positive real numbers. If*

$$|f(z)| \leq M \text{ and } \left|\frac{zg'(z)}{g(z)}\right| \leq N$$

and

$$|c| \leq 1 - |\gamma|[(m+1)M + N],$$

then the integral operator $I_1(f,g) = \int_0^z \left(\frac{f(t)}{g(t)}\right)^\gamma dt$ is in the univalent function class $\mathcal{S}$.

2.2 Convexity and Other Properties

Theorem 7 *Let $f_i \in \mathcal{A}, g_i \in \mathcal{P}$, with $f_i \in \mathcal{B}(\mu_i, \alpha_i)$ and $g_i \in \mathcal{S}^*(\beta_i)$, for $i = \overline{1,n}, \mu_i \geq 0, 0 \leq \alpha_i, \beta_i < 1$. If $|f_i(z)| \leq M_i, (M_i \geq 1)$, for $z \in \mathcal{U}$ and $0 \leq \sum_{i=1}^{n}\left[\beta_i - (2-\alpha_i)M_i^{\mu_i - 1}\right] < 1$ then the integral operator $I(f,g)$ defined by (5) has the convexity order $\delta = 1 - \sum_{i=1}^{n}\left[\beta_i - (2-\alpha_i)M_i^{\mu_i-1}\right]$, with $\beta_i \geq 0$ for all $i = \overline{1,n}$*

Proof From (5) we obtain

$$\frac{I''(f,g)(z)}{I'(f,g)(z)} = \sum_{i=1}^{n}\left(\frac{f_i'(z)}{f_i(z)} - \frac{g_i'(z)}{g_i(z)}\right)$$

and

$$\begin{aligned}
\operatorname{Re}\left\{\frac{zI''(f,g)(z)}{I'(f,g)(z)} + 1\right\} &= \operatorname{Re}\left(\sum_{i=1}^{n}\left(\frac{zf_i'(z)}{f_i(z)} - \frac{zg_i'(z)}{g_i(z)}\right) + 1\right) \\
&= 1 + \operatorname{Re}\left(\sum_{i=1}^{n}\frac{zf_i'(z)}{f_i(z)}\right) - \operatorname{Re}\left(\sum_{i=1}^{n}\frac{zg_i'(z)}{g_i(z)}\right) \qquad (16)
\end{aligned}$$

Using the hypothesis that $f_i \in \mathcal{B}(\mu_i, \alpha)$ and $g_i \in \mathcal{S}^*(\beta_i)$ from (16) results:

$$\operatorname{Re}\left\{\frac{zI''(f,g)(z)}{I'(f,g)(z)} + 1\right\} \geq 1 - \sum_{i=1}^{n}\left|f_i'(z)\left(\frac{z}{f_i(z)}\right)^{\mu_i}\right|\left|\frac{f_i(z)}{z}\right|^{\mu_i - 1} - \sum_{i=1}^{n}\beta_i. \qquad (17)$$

From the general Schwarz lemma, we obtain that $\left|\frac{f_i(z)}{z}\right| \leq M_i$, for $i = \overline{1,n}$ and the inequality (17) is equivalent with

$$\operatorname{Re}\left\{\frac{zI''(f,g)(z)}{I'(f,g)(z)}+1\right\} \geq 1-\sum_{i=1}^{n}\left(\left|f_i(z)\left(\frac{z}{f_i(z)}\right)^{\mu_i}-1\right|+1\right)M_i^{\mu_i-1}-\sum_{i=1}^{n}\beta_i$$

$$\geq 1-\sum_{i=1}^{n}\left[\beta_i-(2-\alpha_i)M_i^{\mu_i-1}\right]=\delta.$$

This completes the proof.

For $n=1$ in Theorem 7 we obtain:

Corollary 4 *Let $f\in\mathcal{A}, g\in\mathcal{P}$, with $f\in\mathcal{B}(\mu,\alpha)$ for $\mu\geq 0, 0\leq\alpha,\beta<1$ and $g\in\mathcal{S}^*(\beta)$. If $|f(z)|\leq M,(M\geq 1)$, for $z\in\mathcal{U}$ and $0\leq[\beta-(2-\alpha)M^{\mu-1}]<1$, then the integral operator $I(f,g)(z)=\int\limits_0^z \frac{f(t)}{g(t)}dt$ has the convexity order $\delta=1-[\beta-(2-\alpha)M^{\mu-1}]$, with $\beta\geq 0$.*

For $\mu=1$ in Theorem 7 we obtain:

Corollary 5 *Let $f_i\in\mathcal{A}, g_i\in\mathcal{P}$, with $f_i\in\mathcal{B}(1,\alpha_i)\subset\mathcal{S}^*(\alpha_i)$ and $g_i\in\mathcal{S}^*(\beta_i)$, for $i=\overline{1,n}, 0\leq\alpha_i,\beta_i<1$. If $|f_i(z)|\leq M_i,(M_i\geq 1)$, for $z\in\mathcal{U}$ and $0\leq\sum\limits_{i=1}^{n}[\beta_i-(2-\alpha_i)]<1$, then the integral operator $I(f,g)$ defined by (5) has the convexity order $\delta=1-\sum\limits_{i=1}^{n}[\beta_i-(2-\alpha_i)]$ with $\beta_i\geq 0$ for all $i=\overline{1,n}$.*

Theorem 8 *Let $f_i\in\mathcal{A}, g_i\in\mathcal{P}$, with $f_i\in\beta-S_p(\alpha_i)$ for $\beta>0,-1\leq\alpha_i<1$ and $g_i\in\mathcal{S}^*\left(\frac{1}{1+\beta_i}\right)$, for $i=\overline{1,n}$ and $\beta_i\neq -1$. Then the integral operator $I(f,g)\in\mathcal{K}(\rho)$, where $\rho=1-\sum\limits_{i=1}^{n}\left(\frac{1}{1+\beta_i}-\alpha_i\right)$, with $0\leq\sum\limits_{i=1}^{n}\left(\frac{1}{1+\beta_i}-\alpha_i\right)<1$ for $z\in\mathcal{U}$ and $i=\overline{1,n}$.*

Proof From (5) we obtain

$$\frac{I''(f,g)(z)}{I'(f,g)(z)}=\sum_{i=1}^{n}\left(\frac{f_i'(z)}{f_i(z)}-\frac{g_i'(z)}{g_i(z)}\right)$$

and

$$\operatorname{Re}\left\{\frac{zI''(f,g)(z)}{I'(f,g)(z)}\right\}=\operatorname{Re}\left\{\sum_{i=1}^{n}\left(\frac{zf_i'(z)}{f_i(z)}-\frac{zg_i'(z)}{g_i(z)}\right)\right\}$$

$$=\operatorname{Re}\frac{zf_1'(z)}{f_1(z)}+\cdots+\operatorname{Re}\frac{zf_n'(z)}{f_n(z)}-\operatorname{Re}\left(\sum_{i=1}^{n}\frac{zg_i'(z)}{g_i(z)}\right) \tag{18}$$

$$=\operatorname{Re}\left(\frac{zf_1'(z)}{f_1(z)}-\alpha_1\right)+\alpha_1+\cdots+\operatorname{Re}\left(\frac{zf_n'(z)}{f_n(z)}-\alpha_n\right)$$

$$+\alpha_n - \mathrm{Re}\left(\sum_{i=1}^{n} \frac{zg_i'(z)}{g_i(z)}\right).$$

Since $f_i \in \beta - S_p(\alpha_i)$ and $g_i \in \mathcal{S}^*(\frac{1}{1+\beta_i})$ using (18) we obtain

$$\mathrm{Re}\left\{\frac{zI''(f,g)(z)}{I'(f,g)(z)}\right\} \geq \beta_1 \left|\frac{zf_1'(z)}{f_1(z)} - 1\right| + \alpha_1 + \cdots + \beta_n \left|\frac{zf_n'(z)}{f_n(z)} - 1\right| + \alpha_n$$
$$-\sum_{i=1}^{n} \frac{1}{1+\beta_i} \geq \sum_{i=1}^{n}\left(\alpha_i - \frac{1}{1+\beta_i}\right)$$

Thus we obtain that

$$\mathrm{Re}\left\{\frac{zI''(f,g)(z)}{I'(f,g)(z)} + 1\right\} \geq 1 - \sum_{i=1}^{n}\left(\frac{1}{1+\beta_i} - \alpha_i\right) = \rho$$

which implies that $I(f,g)$ is in $\mathcal{K}(\rho)$.

For $n = 1$ in Theorem 8 we get:

Corollary 6 *Let $f \in \mathcal{A}, g \in \mathcal{P}$, with $f \in \beta - S_p(\alpha)$ for $\beta > 0, -1 \leq \alpha < 1$ and $g \in \mathcal{S}^*\left(\frac{1}{1+\beta}\right)$ for $\beta \neq -1$. Then the integral operator $I(f,g)(z) = \int_0^z \frac{f(t)}{g(t)}dt \in \mathcal{K}(\rho)$, where $\rho = 1 - \left(\frac{1}{1+\beta} - \alpha\right)$, with $0 \leq \frac{1}{1+\beta} - \alpha < 1$ for $z \in \mathcal{U}$.*

Theorem 9 *Let $f_i \in \mathcal{A}, g_i \in \mathcal{P}, \alpha_i, \beta_i > 0, \alpha_i, \beta_i \neq 1$, with $f_i \in \mathcal{S}^*\left(\frac{1+\alpha_i}{1-\alpha_i}\right)$ and $g_i \in \mathcal{S}^*\left(\frac{1-\alpha_i}{1-\beta_i}\right)$, for $i = \overline{1,n}$. Then the integral operator $I(f,g) \in \mathcal{K}(\eta)$, where $\eta = 1 - \sum_{i=1}^{n}\left(\frac{1-\alpha_i}{1-\beta_i} - \frac{1+\alpha_i}{1-\alpha_i}\right)$, with $0 \leq \sum_{i=1}^{n}\left(\frac{1-\alpha_i}{1-\beta_i} - \frac{1+\alpha_i}{1-\alpha_i}\right) < 1$ for $z \in \mathcal{U}$ and $i = \overline{1,n}$.*

Proof From (5) we obtain

$$\frac{I''(f,g)(z)}{I'(f,g)(z)} = \sum_{i=1}^{n}\left(\frac{f_i'(z)}{f_i(z)} - \frac{g_i'(z)}{g_i(z)}\right)$$

and

$$\mathrm{Re}\left\{\frac{zI''(f,g)(z)}{I'(f,g)(z)}\right\} = \mathrm{Re}\left\{\sum_{i=1}^{n}\left(\frac{zf_i'(z)}{f_i(z)} - \frac{zg_i'(z)}{g_i(z)}\right)\right\}$$
$$= \mathrm{Re}\left(\sum_{i=1}^{n}\frac{zf_i'(z)}{f_i(z)}\right) - \mathrm{Re}\left(\sum_{i=1}^{n}\frac{zg_i'(z)}{g_i(z)}\right) \quad (19)$$

Since $f_i \in \mathcal{S}^*\left(\frac{1+\alpha_i}{1-\alpha_i}\right)$ and $g_i \in \mathcal{S}^*\left(\frac{1-\alpha_i}{1-\beta_i}\right)$ using (19) we obtain

$$\mathrm{Re}\left\{\frac{zI''(f,g)(z)}{I'(f,g)(z)} + 1\right\} \geq 1 + \sum_{i=1}^{n}\left(\frac{1+\alpha_i}{1-\alpha_i}\right) - \sum_{i=1}^{n}\left(\frac{1-\alpha_i}{1-\beta_i}\right)$$

$$\geq 1-\sum_{i=1}^{n}\left(\frac{1-\alpha_i}{1-\beta_i}-\frac{1+\alpha_i}{1-\alpha_i}\right)=\eta$$

which implies that $I(f,g)$ is in $\mathcal{K}(\eta)$.

For $\alpha_1=\alpha_2=\cdots=\alpha_n=\alpha$ we obtain:

Corollary 7 *Let $f_i\in\mathcal{A}, g_i\in\mathcal{P}, \alpha,\beta_i>0, \alpha,\beta_i\neq 1$, with $f_i\in\mathcal{S}^*\left(\frac{1+\alpha}{1-\alpha}\right)$ and $g_i\in\mathcal{S}^*\left(\frac{1-\alpha}{1-\beta_i}\right)$, for $i=\overline{1,n}$. Then the integral operator $I(f,g)\in\mathcal{K}(\eta)$, where $\eta=1-\sum_{i=1}^{n}\left(\frac{1-\alpha}{1-\beta_i}-\frac{1+\alpha}{1-\alpha_i}\right)$, with $0\leq\sum_{i=1}^{n}\left(\frac{1-\alpha}{1-\beta_i}-\frac{1+\alpha}{1-\alpha}\right)<1$ for $z\in\mathcal{U}$ and $i=\overline{1,n}$.*

Theorem 10 *Let $f_i\in\mathcal{A}, g_i\in\mathcal{P}$ and $0\leq a\leq 1, -1\leq b\leq 0$, with $f_i\in\mathcal{B}(\mu_i,\alpha_i)$ for $\mu_i\geq 0, 0\leq\alpha_i<1$ and $g_i\in\mathcal{S}^*\left(\frac{1-a}{1-b}\right)$, for $i=\overline{1,n}$. If $|f_i(z)|\leq M_i$, $(M_i\geq 1)$, for $z\in\mathcal{U}$ and $i=\overline{1,n}$ then the integral operator $I_1(f,g)$ defined by (6) has the convexity order $\phi=1-\sum_{i=1}^{n}\left[\gamma_i\left(\frac{1-a}{1-b}\right)-|\gamma_i|(2-\alpha_i)M_i^{\mu_i-1}\right]$, with $0\leq\sum_{i=1}^{n}\left[\gamma_i\left(\frac{1-a}{1-b}\right)-|\gamma_i|(2-\alpha_i)M_i^{\mu_i-1}\right]<1$ for all $i=\overline{1,n}$.*

Proof From (6) we obtain:

$$\frac{I_1''(f,g)(z)}{I_1'(f,g)(z)}=\sum_{i=1}^{n}\gamma_i\left(\frac{f_i'(z)}{f_i(z)}-\frac{g_i'(z)}{g_i(z)}\right)$$

and

$$\operatorname{Re}\left\{\frac{zI''(f,g)(z)}{I'(f,g)(z)}+1\right\}=\operatorname{Re}\left(\sum_{i=1}^{n}\gamma_i\left(\frac{zf_i'(z)}{f_i(z)}-\frac{zg_i'(z)}{g_i(z)}\right)+1\right)$$

$$=1+\operatorname{Re}\left(\sum_{i=1}^{n}\gamma_i\frac{zf_i'(z)}{f_i(z)}\right)-\operatorname{Re}\left(\sum_{i=1}^{n}\gamma_i\frac{zg_i'(z)}{g_i(z)}\right) \tag{20}$$

Using the hypothesis that $f_i\in\mathcal{B}(\mu_i,\alpha_i)$ and $g_i\in\mathcal{S}^*(\frac{1-a}{1-b})$ from (16) results:

$$\operatorname{Re}\left\{\frac{zI''(f,g)(z)}{I'(f,g)(z)}+1\right\}\geq 1-\sum_{i=1}^{n}|\gamma_i|\left|f_i'(z)\left(\frac{z}{f_i(z)}\right)^{\mu_i}\right|\left|\frac{f_i(z)}{z}\right|^{\mu_i-1}-\sum_{i=1}^{n}\gamma_i\left(\frac{1-a}{1-b}\right) \tag{21}$$

From the general Schwarz lemma we obtain that $\left|\frac{f_i(z)}{z}\right|\leq M_i$, for $i=\overline{1,n}$ and the inequality (17) is equivalent with:

$$\operatorname{Re}\left\{\frac{zI''(f,g)(z)}{I'(f,g)(z)}+1\right\}\geq 1+\sum_{i=1}^{n}|\gamma_i|\left(\left|f_i(z)\left(\frac{z}{f_i(z)}\right)^{\mu_i}-1\right|+1\right)M_i(\mu_i-1)$$

$$-\sum_{i=1}^{n}\gamma_i\left(\frac{1-a}{1-b}\right)\geq 1-\sum_{i=1}^{n}\left[\gamma_i\left(\frac{1-a}{1-b}\right)-|\gamma_i|(2-\alpha_i)M_i^{\mu_i-1}\right]=\phi.$$

This completes the proof.

For $\mu = 1$ and $\alpha_1 = \cdots = \alpha_n = \alpha$ in Theorem 10 we obtain:

Corollary 8 *Let $f_i \in \mathcal{A}, g_i \in \mathcal{P}$ and $0 \le a \le 1, -1 \le b \le 0$, with $f_i \in \mathcal{B}(1,\alpha) \subset \mathcal{S}^*(\alpha)$ and $g_i \in \mathcal{S}^*\left(\frac{1-a}{1-b}\right)$, for $i = \overline{1,n}, 0 \le \alpha < 1$. If $|f_i(z)| \le M_i, (M_i \ge 1)$, for $z \in \mathcal{U}$ and $i = \overline{1,n}$ then the integral operator $I_1(f,g)$ defined by (6) has the convexity order $\phi = 1 - \sum_{i=1}^{n}\left[\gamma_i\left(\frac{1-a}{1-b}\right) - |\gamma_i|(2-\alpha)\right]$, with $0 \le \sum_{i=1}^{n}\left[\gamma_i\left(\frac{1-a}{1-b}\right) - |\gamma_i|(2-\alpha)\right] < 1$ for all $i = \overline{1,\ n}$.*

For $\mu = 0$ in Theorem 10 we obtain:

Corollary 9 *Let $f_i \in \mathcal{A}, g_i \in \mathcal{P}$ and $0 \le a \le 1, -1 \le b \le 0$, with $f_i \in \mathcal{R}(\alpha_i)$ and $g_i \in \mathcal{S}^*\left(\frac{1-a}{1-b}\right)$, for $i = \overline{1,n}$. If $|f_i(z)| \le M_i, (M_i \ge 1)$, for $z \in \mathcal{U}$ and $i = \overline{1,n}$ then the integral operator $I_1(f,g)(z)$ defined by (6) has the convexity order $\phi = 1 - \sum_{i=1}^{n}\left[\gamma_i\left(\frac{1-a}{1-b}\right) - |\gamma_i|(2-\alpha_i)\frac{1}{M_i}\right]$, with $0 \le \sum_{i=1}^{n}\left[\gamma_i\left(\frac{1-a}{1-b}\right) - |\gamma_i|(2-\alpha_i)\frac{1}{M_i}\right] < 1$ for all $i = \overline{1,n}$.*

Theorem 11 *Let the functions $f_i \in \mathcal{A}, g_i \in \mathcal{P}$ and $\gamma_i, \beta_i > 0$ real numbers for $i = \overline{1,n}$ and*

$$\sum_{i=1}^{n}\frac{\gamma_i}{\beta_i} \le 1.$$

We suppose that the functions f_i for $i = \overline{1,n}$ are starlike functions and g_i are starlike of order $\frac{1}{\beta_i}$, for $i = \overline{1,n}$. Then the integral operator $I_1(f,g)$ defined by (6) is convex by order $1 - \sum_{i=1}^{n}\frac{\gamma_i}{\beta_i}$.

Proof From (6) we obtain that

$$\frac{I_1''(f,g)(z)}{I_1'(f,g)(z)} = \sum_{i=1}^{n}\gamma_i\left(\frac{f_i'(z)}{f_i(z)} - \frac{g_i'(z)}{g_i(z)}\right)$$

and $\frac{zI_1''(f,g)(z)}{I_1'(f,g)(z)} = \sum_{i=1}^{n}\gamma_i\left(\frac{zf_i'(z)}{f_i(z)} - \frac{zg_i'(z)}{g_i(z)}\right)$. From here we get

$$\begin{aligned}\frac{zI_1''(f,g)(z)}{I_1'(f,g)(z)} + 1 &= \gamma_1\frac{zf_1'(z)}{f_1(z)} + \gamma_2\frac{zf_2'(z)}{f_2(z)} + \cdots + \gamma_n\frac{zf_n'(z)}{f_n(z)}.\\ &\quad - \left(\gamma_1\frac{zg_1'(z)}{g_1(z)} + \gamma_2\frac{zg_2'(z)}{g_2(z)} + \cdots + \gamma_n\frac{zg_n'(z)}{g_n(z)}\right) + 1 \qquad (22)\end{aligned}$$

But $f_i \in \mathcal{S}^*$, so $\mathrm{Re}\frac{zf_i'(z)}{f_i(z)} > 0$, for all $i = \overline{1,n}$ and $g_i \in \mathcal{S}^*\left(\frac{1}{\beta_i}\right)$, so $\mathrm{Re}\frac{g_i'(z)}{g_i(z)} > \frac{1}{\beta_i}$, for all $i = \overline{1,n}$.

Using this affirmations in (22) we obtain:

$$\mathrm{Re}\left(\frac{zI_1''(f,g)(z)}{I_1'(f,g)(z)}+1\right) > \gamma_1\cdot 0+\cdots+\gamma_n\cdot 0-\left(\gamma_1\cdot\frac{1}{\beta_1}+\cdots+\gamma_n\cdot\frac{1}{\beta_n}\right)+1$$

$$> 1-\sum_{i=1}^{n}\frac{\gamma_i}{\beta_i}.$$

So the function $I_1(f,g)$ is convex by order $1-\sum_{i=1}^{n}\frac{\gamma_i}{\beta_i}$, for all $i=\overline{1,n}$.

For $n=1$ in Theorem 11 we get:

Corollary 10 *Let the functions* $f\in\mathcal{A}, g\in\mathcal{P}$ *and* $\gamma,\beta>0$ *real numbers and* $\frac{\gamma}{\beta}\le 1$. *We suppose that the functions* f *is starlike function and* g *is starlike of order* $\frac{1}{\beta}$. *Then the integral operator* $\int_0^z\left(\frac{f(t)}{g(t)}\right)^{\gamma}dt$ *is convex by order* $1-\frac{\gamma}{\beta}$.

Theorem 12 *Let the functions* $f_i\in\mathcal{A}, g_i\in\mathcal{P}$ *and the real numbers* $\gamma_i,\beta_i,\alpha>0$, $\beta\in[0,1), \alpha\neq\beta$ *for* $i=\overline{1,n}$ *and*

$$\sum_{i=1}^{n}\gamma_i<\frac{1}{\beta-\alpha}, \text{ and } \sum_{i=1}^{n}\frac{\gamma_i}{\beta_i}\le 1$$

and $(\beta-\alpha)\sum_{i=1}^{n}\gamma_i-\sum_{i=1}^{n}\frac{\gamma_i}{\beta_i}+1\in(0,1)$. *We suppose that* $f_i\in\mathcal{SP}(\alpha,\beta)$ *and* $g_i\in\mathcal{S}^*\left(\frac{1}{\beta_i}\right)$, *for all* $i=\overline{1,n}$. *Then the operator* $I_1(f,g)$ *defined by (6) is convex by order* $(\beta-\alpha)\sum_{i=1}^{n}\gamma_i-\sum_{i=1}^{n}\frac{\gamma_i}{\beta_i}+1$.

Proof From (6) we get

$$\begin{aligned}\frac{zI_1''(f,g)(z)}{I_1'(f,g)(z)}&=\sum_{i=1}^{n}\gamma_i\left(\frac{zf_i'(z)}{f_i(z)}-\frac{zg_i'(z)}{g_i(z)}\right)\\&=\sum_{i=1}^{n}\gamma_i\left(\frac{zf_i'(z)}{f_i(z)}+\alpha-\beta\right)+(\beta-\alpha)\sum_{i+1}^{n}\gamma_i- \\&-\left(\gamma_1\frac{zg_1'(z)}{g_1(z)}+\cdots+\gamma_n\frac{zg_n'(z)}{g_n(z)}\right)\end{aligned}\tag{23}$$

From (23), it follows that

$$\begin{aligned}\mathrm{Re}\left(\frac{zI_1''(f,g)(z)}{I_1'(f,g)(z)}+1\right)&=\mathrm{Re}\left(\sum_{i=1}^{n}\gamma_i\left(\frac{zf_i'(z)}{f_i(z)}\right)\right)+(\beta-\alpha)\sum_{i+1}^{n}\gamma_i\\&-\mathrm{Re}\left(\gamma_1\frac{zg_1'(z)}{g_1(z)}+\cdots+\gamma_n\frac{zg_n'(z)}{g_n(z)}\right)+1\end{aligned}\tag{24}$$

Since $f_i \in \mathcal{SP}(\alpha, \beta)$ and $g_i \in \mathcal{S}^*\left(\frac{1}{\beta_i}\right)$, from (24) we obtain

$$\operatorname{Re}\left(\frac{zI_1''(f,g)(z)}{I_1'(f,g)(z)}+1\right) \geq \sum_{i=1}^{n} \gamma_i \left|\frac{zf_i'(z)}{f_i(z)}+\alpha-\beta\right| + (\beta-\alpha)\sum_{i=1}^{n}\gamma_i - \sum_{i=1}^{n}\frac{\gamma_i}{\beta_i}+1 \tag{25}$$

But $\gamma_i \left|\frac{zf_i'(z)}{f_i(z)}+\alpha-\beta\right| > 0$ for all $i=\overline{1,n}$, so from (25), we have

$$\operatorname{Re}\left(\frac{zI_1''(f,g)(z)}{I_1'(f,g)(z)}+1\right) \geq (\beta-\alpha)\sum_{i=1}^{n}\gamma_i - \sum_{i=1}^{n}\frac{\gamma_i}{\beta_i}+1 > 0.$$

Thus, and from the hypothesis we obtain that the operator $I_1(f,g)$ is convex by order $(\beta-\alpha)\sum_{i=1}^{n}\gamma_i - \sum_{i=1}^{n}\frac{\gamma_i}{\beta_i}+1$.

For $\beta_1 = \beta_2 = \cdots = \beta_n = \beta$ in Theorem 12 we obtain:

Corollary 11 *Let the functions $f_i \in \mathcal{A}$, $g_i \in \mathcal{P}$ and the real numbers $\gamma_i, \beta > 0$ for $i=\overline{1,n}$ and*

$$\sum_{i=1}^{n}\gamma_i < \frac{1}{\beta-\alpha}, \text{ and } \sum_{i=1}^{n}\frac{\gamma_i}{\beta} \leq 1$$

and $(\beta-\alpha)\sum_{i=1}^{n}\gamma_i - \sum_{i=1}^{n}\frac{\gamma_i}{\beta}+1 \in (0,1)$. We suppose that $f_i \in \mathcal{SP}(\alpha,\beta)$ and $g_i \in \mathcal{S}^*\left(\frac{1}{\beta}\right)$, for all $i=\overline{1,n}$. Then the operator $I_1(f,g)$ defined by (6) is convex by order $(\beta-\alpha)\sum_{i=1}^{n}\gamma_i - \sum_{i=1}^{n}\frac{\gamma_i}{\beta}+1$.

Theorem 13 *Let $f_i \in \mathcal{A}$, $g_i \in \mathcal{P}$ and $0 \leq \alpha_i < 1, \beta_i > 0$, with $f_i \in \mathcal{B}(\mu_i,\alpha_i)$ for $\mu_i \geq 0$ and $g_i \in \mathcal{S}^*(\frac{1-\alpha_i}{1+\beta_i})$, for all $i=\overline{1,n}$. If*

$$|f_i(z)| \leq M_i \tag{26}$$

for all $i=\overline{1,n}$, then the integral operator $I(f,g) \in \mathcal{N}(\delta)$, where

$$\delta = 1 + \sum_{i=1}^{n}(2-\alpha_i)M_i^{\mu_i-1} + \sum_{i=1}^{n}\left(\frac{1-\alpha_i}{1+\beta_i}\right).$$

Proof From (5) we obtain

$$\frac{I''(f,g)(z)}{I'(f,g)(z)} = \sum_{i=1}^{n}\left(\frac{f_i'(z)}{f_i(z)} - \frac{g_i'(z)}{g_i(z)}\right)$$

and

$$\operatorname{Re}\left\{\frac{zI''(f,g)(z)}{I'(f,g)(z)}\right\} = \operatorname{Re}\left[\sum_{i=1}^{n}\left(\frac{zf_i'(z)}{f_i(z)} - \frac{zg_i'(z)}{g_i(z)}\right)\right]$$

$$\leq \left|\sum_{i=1}^{n} \frac{zf_i'(z)}{f_i(z)} - \sum_{i=1}^{n} \frac{zg_i'(z)}{g_i(z)}\right|$$
$$\leq \sum_{i=1}^{n} \left|\frac{zf_i'(z)}{f_i(z)}\right| + \sum_{i=1}^{n} \left|\frac{zg_i'(z)}{g_i(z)}\right| \tag{27}$$
$$\leq \sum_{i=1}^{n} \left|f_i'(z)\left(\frac{z}{f_i(z)}\right)^{\mu_i}\right| \left|\frac{f_i(z)}{z}\right|^{\mu_i-1} + \sum_{i=1}^{n} \left|\frac{zg_i'(z)}{g_i(z)}\right|.$$

From (26), using Schwarz lemma we obtain that $\left|\frac{f_i(z)}{z}\right| \leq M_i$.

Thus, from (27) and using the hypothesis that $f_i \in \mathcal{B}(\mu_i, \alpha_i)$, $g_i \in \mathcal{S}^*(\frac{1-\alpha_i}{1+\beta_i})$, we have

$$\operatorname{Re}\left\{\frac{zI''(f,g)(z)}{I'(f,g)(z)}\right\} \leq \sum_{i=1}^{n} \left(\left|f_i'(z)\left(\frac{z}{f_i(z)}\right)^{\mu_i} - 1\right| + 1\right) M_i^{\mu_i-1}$$
$$+ \sum_{i=1}^{n} \left(\frac{1-\alpha_i}{1-\beta_i}\right) \tag{28}$$
$$\leq \sum_{i=1}^{n} (2-\alpha_i) M_i^{\mu_i-1} + \sum_{i=1}^{n} \left(\frac{1-\alpha_i}{1-\beta_i}\right).$$

From (28) we have

$$\operatorname{Re}\left\{\frac{zI''(f,g)(z)}{I'(f,g)(z)} + 1\right\} \leq 1 + \sum_{i=1}^{n} (2-\alpha_i) M_i^{\mu_i-1} + \sum_{i=1}^{n} \left(\frac{1-\alpha_i}{1-\beta_i}\right) = \delta.$$

So, the operator $I(f,g) \in \mathcal{N}(\delta)$.

For $\mu = 0$ and $\alpha_1 = \cdots = \alpha_n = \alpha$ in Theorem 13 we obtain:

Corollary 12 *Let* $f_i \in \mathcal{A}, g_i \in \mathcal{P}$ *and* $\alpha \geq 0, \beta_i > 0$, *with* $f_i \in \mathcal{R}(\alpha)$ *and* $g_i \in \mathcal{S}^*(\frac{1-\alpha}{1+\beta_i})$, *for all* $i = \overline{1,n}$. *If*

$$|f_i(z)| \leq M_i \tag{29}$$

for all $i = \overline{1,n}$, *then the integral operator* $I(f,g) \in \mathcal{N}(\delta)$, *where*

$$\delta = 1 + \sum_{i=1}^{n} (2-\alpha)\frac{1}{M_i} + \sum_{i=1}^{n} \left(\frac{1-\alpha}{1+\beta_i}\right).$$

Theorem 14 *Let* $f_i \in \mathcal{A}, g_i \in \mathcal{P}$ *and* $\alpha_i \in [0,1), \beta_i \geq 0$, *with* $f_i \in \mathcal{B}(\mu_i, \alpha_i)$, $g_i \in \mathcal{S}^*\left(\frac{1+\alpha_i}{1+\alpha_i\beta_i}\right)$ *for* $i = \overline{1,n}$, $\mu_i \geq 0$. *If*

$$|f_i(z)| \leq M_i \tag{30}$$

for all $i = \overline{1,n}$, then the integral operator $I_2(f,g) \in \mathcal{N}(\rho)$, where

$$\rho = 1 + \sum_{i=1}^{n} |\gamma_i|(2-\alpha_i)M_i^{\mu_i-1} + \sum_{i=1}^{n} \sigma_i \left(\frac{1+\alpha_i}{1+\alpha_i\beta_i}\right).$$

Proof From (7) we get that

$$\frac{I_2''(f,g)(z)}{I_2'(f,g)(z)} = \sum_{i=1}^{n} \gamma_i \frac{f_i'(z)}{f_i(z)} - \sum_{i=1}^{n} \sigma_i \frac{g_i'(z)}{g_i(z)}$$

and

$$\begin{aligned}
\operatorname{Re}\left\{\frac{zI_2''(f,g)(z)}{I_2'(f,g)(z)}\right\} &= \operatorname{Re}\left\{\sum_{i=1}^{n} \gamma_i \frac{zf_i'(z)}{f_i(z)} - \sum_{i=1}^{n} \sigma_i \frac{zg_i'(z)}{g_i(z)}\right\} \\
&\le \left|\left(\sum_{i=1}^{n} \gamma_i \frac{zf_i'(z)}{f_i(z)}\right) - \left(\sum_{i=1}^{n} \sigma_i \frac{zg_i'(z)}{g_i(z)}\right)\right| \qquad (31) \\
&\le \sum_{i=1}^{n} |\gamma_i| \left|\frac{zf_i'(z)}{f_i(z)}\right| + \sum_{i=1}^{n} |\sigma| \left|\frac{zg_i'(z)}{g_i(z)}\right| \\
&\le \sum_{i=1}^{n} |\gamma_i| \left|f_i'(z)\left(\frac{z}{f_i(z)}\right)^{\mu_i}\right| \left|\frac{f_i(z)}{z}\right|^{\mu_i-1} + \sum_{i=1}^{n} |\sigma_i| \left|\frac{zg_i'(z)}{g_i(z)}\right|.
\end{aligned}$$

Using Schwarz lemma, from (26) we obtain that $\left|\frac{f_i(z)}{z}\right| \le M_i$.
From the above *inequality* and from the hypothesis, we obtain that (31) is equivalent with

$$\begin{aligned}
\operatorname{Re}\left\{\frac{zI_2''(f,g)(z)}{I_2'(f,g)(z)}\right\} &\le \sum_{i=1}^{n} |\gamma_i| \left(\left|f_i'(z)\left(\frac{z}{f_i(z)}\right)^{\mu_i} - 1\right| + 1\right) M_i^{\mu_i-1} \\
&\quad + \sum_{i=1}^{n} |\sigma_i| \left(\frac{1+\alpha_i}{1+\alpha_i\beta_i}\right) \qquad (32) \\
&\le \sum_{i=1}^{n} |\gamma_i|(2+\alpha_i)M_i^{\mu_i-1} + \sum_{i=1}^{n} |\sigma| \left(\frac{1+\alpha_i}{1+\alpha_i\beta_i}\right).
\end{aligned}$$

From (32) we obtain that

$$\operatorname{Re}\left\{\frac{zI_2''(f,g)(z)}{I_2'(f,g)(z)} + 1\right\} \le \sum_{i=1}^{n} |\gamma_i|(2+\alpha_i)M_i^{\mu_i-1} + \sum_{i=1}^{n} |\sigma_i| \left(\frac{1+\alpha_i}{1+\alpha_i\beta_i}\right) + 1 = \rho.$$

So the operator $I_2(f,g) \in \mathcal{N}(\rho)$.

For $\mu = 0$ and $\gamma_1 = \cdots = \gamma_n = \gamma, \sigma_1 = \cdots = \sigma_n = \sigma$ in Theorem 14 we obtain:

Corollary 13 *Let $f_i \in \mathcal{A}, g_i \in \mathcal{P}$ and $\alpha_i, \beta_i \geq 0$, with $f_i \in \mathcal{R}(\alpha_i), g_i \in \mathcal{S}^*\left(\frac{1+\alpha_i}{1+\alpha_i\beta_i}\right)$ for $i = \overline{1,n}$. If*

$$|f_i(z)| \leq M_i \tag{33}$$

for all $i = \overline{1,n}$, then the integral operator $I_2(f,g) \in \mathcal{N}(\rho)$, where

$$\rho = 1 + \sum_{i=1}^{n} |\gamma|(2-\alpha_i)\frac{1}{M_i} + \sum_{i=1}^{n} \sigma\left(\frac{1+\alpha_i}{1+\alpha_i\beta_i}\right).$$

For $\mu = 1$ and $\alpha_1 = \cdots = \alpha_n = \alpha, \beta_1 = \cdots = \beta_n = \beta$ in Theorem 14 we obtain:

Corollary 14 *Let $f_i \in \mathcal{A}, g_i \in \mathcal{P}$ and $\alpha \in [0,1), \beta \geq 0$, with $f_i \in \mathcal{B}(1,\alpha) \subset \mathcal{S}^*(\alpha), g_i \in \mathcal{S}^*\left(\frac{1+\alpha}{1+\alpha\beta}\right)$ for $i = \overline{1,n}$. If*

$$|f_i(z)| \leq M_i \tag{34}$$

for all $i = \overline{1,n}, z \in \mathcal{U}$, then the integral operator $I_2(f,g) \in \mathcal{N}(\rho)$, where

$$\rho = 1 + \sum_{i=1}^{n} |\gamma_i|(2-\alpha) + \sum_{i=1}^{n} \sigma_i\left(\frac{1+\alpha}{1+\alpha\beta}\right).$$

Theorem 15 *Let $f_i \in \mathcal{A}, g_i \in \mathcal{P}$, with $f_i(z) \in \mathcal{B}(\mu_i,\alpha_i)$ for $\mu_i \geq 0, \alpha_i \in [0,1)$ and $g_i(z) \in \mathcal{S}^*\left(\frac{\alpha_i}{1+\alpha_i}\right)$, for $\alpha_i > 0$ and M_i positive real numbers, for all $i = \overline{1,n}$ and $z \in \mathcal{U}$. If*

$$|f_i(z)| \leq M_i \tag{35}$$

then the integral operator $I_2(f,g) \in \mathcal{K}(\lambda)$, where

$$\lambda = 1 - \sum_{i=1}^{n}\left[|\sigma|\left(\frac{\alpha_i}{1+\alpha_i}\right) - |\gamma_i|(2-\alpha_i)M_i^{\mu_i-1}\right]$$

and $0 \leq \sum_{i=1}^{n}\left[|\sigma|\left(\frac{\alpha_i}{1+\alpha_i}\right) - |\gamma_i|(2-\alpha_i)M_i^{\mu_i-1}\right] < 1$.

Proof From (7) results

$$\frac{I_2''(f,g)(z)}{I_2'(f,g)(z)} = \sum_{i=1}^{n} \gamma_i \frac{f_i'(z)}{f_i(z)} - \sum_{i=1}^{n} \sigma_i \frac{g_i'(z)}{g_i(z)}$$

and

$$\mathrm{Re}\left(\frac{zI_2''(f,g)(z)}{I_2'(f,g)(z)} + 1\right) \geq 1 - \sum_{i=1}^{n} |\gamma_i| \left|f_i'(z)\left(\frac{z}{f_i(z)}\right)^{\mu_i}\right| \left|\frac{f_i(z)}{z}\right|^{\mu_i-1} - \sum_{i=1}^{n} |\sigma|\left(\frac{\alpha_i}{1+\alpha_i}\right). \tag{36}$$

From (35), applying Schwarz lemma we obtain that $\left|\frac{f_i(z)}{z}\right| \leq M_i$.
Using the hypothesis we have that (36) is equivalent with

$$\operatorname{Re}\left(\frac{zI_2''(f,g)(z)}{I_2'(f,g)(z)}+1\right) \geq 1+\sum_{i=1}^{n}|\gamma_i|\left(\left|f_i'(z)\left(\frac{z}{f_i(z)}\right)^{\mu_i}-1\right|+1\right)M_i^{\mu_i-1}$$

$$-\sum_{i=1}^{n}|\sigma|\left(\frac{\alpha_i}{1+\alpha_i}\right)$$

$$\geq 1-\sum_{i=1}^{n}\left[|\sigma|\left(\frac{\alpha_i}{1+\alpha_i}\right)-|\gamma_i|(2-\alpha_i)M_i^{\mu_i-1}\right]=\lambda.$$

Thus, we see that the operator $I_2(f,g)$ has the convexity order λ.
For $\alpha_1 = \cdots = \alpha_n = \alpha$ in Theorem 15 we obtain:

Corollary 15 *Let $f_i \in \mathcal{A}, g_i \in \mathcal{P}$, with $f_i(z) \in \mathcal{B}(\mu_i, \alpha)$ for $\mu_i \geq 0, \alpha \in [0,1)$ and $g_i(z) \in \mathcal{S}^*\left(\frac{\alpha}{1+\alpha}\right)$, for $\alpha > 0$ and M_i positive real numbers, for all $i = \overline{1,n}$ and $z \in \mathcal{U}$. If*

$$|f_i(z)| \leq M_i \tag{37}$$

then the integral operator $I_2(f,g) \in \mathcal{K}(\lambda)$, where

$$\lambda = 1-\sum_{i=1}^{n}\left[|\sigma|\left(\frac{\alpha}{1+\alpha}\right)-|\gamma_i|(2-\alpha)M_i^{\mu_i-1}\right]$$

and $0 \leq \sum_{i=1}^{n}\left[|\sigma|\left(\frac{\alpha}{1+\alpha}\right)-|\gamma_i|(2-\alpha)M_i^{\mu_i-1}\right] < 1.$

For $\mu = 0$ in Theorem 15 we obtain:

Corollary 16 *Let $f_i \in \mathcal{A}, g_i \in \mathcal{P}$, with $f_i(z) \in \mathcal{R}(\alpha_i)$ and $g_i(z) \in \mathcal{S}^*(\frac{\alpha_i}{1+\alpha_i})$, for $\alpha_i > 0$ and M_i positive real numbers, for all $i = \overline{1,n}$ and $z \in \mathcal{U}$. If*

$$|f_i(z)| \leq M_i \tag{38}$$

then the integral operator $I_2(f,g) \in \mathcal{K}(\lambda)$, where

$$\lambda = 1-\sum_{i=1}^{n}\left[|\sigma|\left(\frac{\alpha_i}{1+\alpha_i}\right)-|\gamma_i|(2-\alpha_i)\frac{1}{M_i}\right]$$

and $0 \leq \sum_{i=1}^{n}\left[|\sigma|\left(\frac{\alpha_i}{1+\alpha_i}\right)-|\gamma_i|(2-\alpha_i)\frac{1}{M_i}\right] < 1.$

Theorem 16 *Let $f_i \in \mathcal{A}, g_i \in \mathcal{P}$ and $\alpha_i, \beta_i, \gamma_i, \sigma_i$ be real numbers with $\alpha_i, \beta_i, \gamma_i, \sigma_i > 0$ for $i = \overline{1,n}$ and*

$$\sum_{i=1}^{n}\left(\sigma_i\beta_i - \frac{\gamma_i}{\alpha_i}\right) \le 1$$

We suppose that the functions $f_i \in \mathcal{S}^*\left(\frac{1}{\alpha_i}\right)$ and $g_i \in \mathcal{S}^*(\beta_i)$, for all $i = \overline{1,n}$. In this conditions the integral operator $I_2(f,g)$ defined by (7) is convex by order $1 - \sum\limits_{i=1}^{n}\left(\sigma_i\beta_i - \frac{\gamma_i}{\alpha_i}\right)$.

Proof From (7), it follows that

$$\frac{I_2''(f,g)(z)}{I_2'(f,g)(z)} = \sum_{i=1}^{n}\gamma_i\frac{f_i'(z)}{f_i(z)} - \sum_{i=1}^{n}\sigma_i\frac{g_i'(z)}{g_i(z)}$$

and

$$\frac{zI_2''(f,g)(z)}{I_2'(f,g)(z)} = \gamma_1\frac{zf_1'(z)}{f_1(z)} + \cdots + \gamma_n\frac{zf_n'(z)}{f_n(z)} - \left(\sigma_1\frac{zg_1'(z)}{g_1(z)} + \cdots + \sigma_n\frac{zg_n'(z)}{g_n(z)}\right) \tag{39}$$

Then (39) is equivalent with

$$\mathrm{Re}\left(\frac{zI_2''(f,g)(z)}{I_2'(f,g)(z)} + 1\right) = \gamma_1 Re\frac{zf_1'(z)}{f_1(z)} + \cdots + \gamma_n Re\frac{zf_n'(z)}{f_n(z)}$$
$$- \left(\sigma_1 Re\frac{zg_1'(z)}{g_1(z)} + \cdots + \sigma_n Re\frac{zg_n'(z)}{g_n(z)}\right) + 1 \tag{40}$$

Since $f_i \in \mathcal{S}^*\left(\frac{1}{\alpha_i}\right)$ and $g_i \in \mathcal{S}^*(\beta_i)$ result that $\mathrm{Re}\frac{zf_i(z)}{f_i(z)} > \frac{1}{\alpha_i}$, respectively $\mathrm{Re}\frac{zg_i(z)}{g_i(z)} > \beta_i$, for all $i = \overline{1,n}$ and $z \in \mathcal{U}$.
Using this in the inequality (40) we obtain that

$$\mathrm{Re}\left(\frac{zI_2''(f,g)(z)}{I_2'(f,g)(z)} + 1\right) > \gamma_1\frac{1}{\alpha_1} + \cdots + \gamma_n\frac{1}{\alpha_n} - (\sigma_1\beta_1 + \cdots + \sigma_n\beta_n) + 1$$
$$= 1 - \sum_{i=1}^{n}\left(\sigma_i\beta_i - \frac{\gamma_i}{\alpha_i}\right). \tag{41}$$

So, $\mathrm{Re}\left(\frac{zI_2''(f,g)(z)}{I_2'(f,g)(z)} + 1\right) > 0$, implies that $I_2(f,g)$ is convex.

If we consider in Theorem 16, $\alpha_1 = \cdots = \alpha_n = \alpha$ and $\beta_1 = \cdots = \beta_n = \beta$ we get:

Corollary 17 *Let $f_i \in \mathcal{A}, g_i \in \mathcal{P}$ and $\alpha, \beta, \gamma_i, \sigma_i$ real numbers numbers with $\alpha, \beta, \gamma_i, \sigma_i > 0$ for $i = \overline{1,n}$ and*

$$\sum_{i=1}^{n}\left(\sigma_i\beta - \frac{\gamma_i}{\alpha}\right) \le 1$$

We suppose that the functions $f_i \in \mathcal{S}^*\left(\frac{1}{\alpha}\right)$ *and* $g_i \in \mathcal{S}^*(\beta)$, *for all* $i = \overline{1,n}$. *In this conditions the integral operator* $I_2(f,g)$ *defined by (7) is convex by order* $1 - \sum_{i=1}^{n}\left(\sigma_i\beta - \frac{\gamma_i}{\alpha}\right)$.

Theorem 17 *Let* $f_i \in \mathcal{A}, g_i \in \mathcal{P}$, *with* $f_i \in \beta - Sp(\alpha_i)$ *and* $g_i \in \mathcal{S}^*(\frac{1-\alpha_i}{1-\beta_i})$, *for* $\alpha_i, \beta_i \geq 0, \beta_i \neq 1$ *and* $i = \overline{1,n}$. *Then the integral operator* $I_2(f,g) \in \mathcal{K}(\theta)$, *where* $\theta = 1 - \sum_{i=1}^{n}\left(\sigma_i\left(\frac{1-\alpha_i}{1-\beta_i}\right) - \gamma_i\alpha_i\right)$ *and* $0 \leq \sum_{i=1}^{n}\left(\sigma_i\left(\frac{1-\alpha_i}{1-\beta_i}\right) - \gamma_i\alpha_i\right) < 1$, *for* $z \in \mathcal{U}$ *and* $i = \overline{1,n}$.

Proof From (7) we obtain

$$\frac{I_2''(f,g)(z)}{I_2'(f,g)(z)} = \sum_{i=1}^{n}\gamma_i\frac{f_i'(z)}{f_i(z)} - \sum_{i=1}^{n}\sigma_i\frac{g_i'(z)}{g_i(z)}$$

and

$$\begin{aligned}
\mathrm{Re}\left\{\frac{zI_2''(f,g)(z)}{I_2'(f,g)(z)}\right\} &= \mathrm{Re}\left\{\sum_{i=1}^{n}\gamma_i\frac{zf_i'(z)}{f_i(z)} - \sum_{i=1}^{n}\sigma_i\frac{zg_i'(z)}{g_i(z)}\right\}\\
&= \mathrm{Re}\gamma_1\frac{zf_1'(z)}{f_1(z)} + \cdots + \mathrm{Re}\gamma_n\frac{zf_n'(z)}{f_n(z)}\\
&\quad - \mathrm{Re}\sigma_1\frac{zg_1'(z)}{g_1(z)} + \cdots + \mathrm{Re}\sigma_n\frac{g_n'(z)}{g_n(z)}\\
&= \mathrm{Re}\gamma_1\left(\frac{zf_1'(z)}{f_1(z)} - \alpha_1\right) + \gamma_1\alpha_1 + \cdots + \mathrm{Re}\gamma_2\\
&\quad \left(\frac{zf_n'(z)}{f_n(z)} - \alpha_n\right) + \gamma_n\alpha_n\\
&\quad - \mathrm{Re}\left(\sum_{i=1}^{n}\frac{zg_i'(z)}{g_i(z)}\right).
\end{aligned} \tag{42}$$

Since $f_i \in \beta - Sp(\alpha_i)$ and $g_i \in \mathcal{S}\left(\frac{1-\alpha_i}{1-\beta_i}\right)$ using (42) we obtain

$$\begin{aligned}
\mathrm{Re}\left\{\frac{zI_2''(f,g)(z)}{I_2'(f,g)(z)}\right\} &\geq \gamma_1\beta_1\left|\frac{zf_1'(z)}{f_1(z)} - 1\right| + \gamma_1\alpha_1 + \cdots + \gamma_n\beta_n\left|\frac{zf_n'(z)}{f_n(z)} - 1\right| + \gamma_n\beta_n\\
&\quad - \sum_{i=1}^{n}\sigma_i\left(\frac{1-\alpha_i}{1-\beta_i}\right) \geq \sum_{i=1}^{n}\left(\gamma_i\alpha_i - \sigma_i\left(\frac{1-\alpha_i}{1-\beta_i}\right)\right)
\end{aligned}$$

Thus we obtain that

$$\mathrm{Re}\left\{\frac{zI''(f,g)(z)}{I'(f,g)(z)} + 1\right\} \geq 1 - \sum_{i=1}^{n}\left(\sigma_i\left(\frac{1-\alpha_i}{1-\beta_i}\right) - \gamma_i\alpha_i\right) = \theta$$

which implies that $I_2(f,g)$ is in $\mathcal{K}(\theta)$.

If we consider $\alpha_1 = \cdots = \alpha_n$ in Theorem 17 we obtain:

Corollary 18 *Let* $f_i \in \mathcal{A}, g_i \in \mathcal{P}$, *with* $f_i \in \beta - Sp(\alpha)$ *and* $g_i \in \mathcal{S}^*\left(\frac{1-\alpha}{1-\beta_i}\right)$, *for* $\alpha, \beta_i \geq 0, \beta_i \neq 1$ *and* $i = \overline{1,n}$. *Then the integral operator* $I_2(f,g) \in \mathcal{K}(\theta)$, *where* $\theta = 1 - \sum_{i=1}^{n}\left(\sigma_i\left(\frac{1-\alpha}{1-\beta_i}\right) - \gamma_i\alpha\right)$ *and* $0 \leq \sum_{i=1}^{n}\left(\sigma_i\left(\frac{1-\alpha}{1-\beta_i}\right) - \gamma_i\alpha\right) < 1$, *for* $z \in \mathcal{U}$ *and* $i = \overline{1,n}$.

For $n = 1$ in Theorem 17 we get:

Corollary 19 *Let* $f \in \mathcal{A}, g \in \mathcal{P}$, *with* $f \in \beta - Sp(\alpha)$ *and* $g \in \mathcal{S}^*\left(\frac{1-\alpha}{1-\beta}\right)$, *for* $\alpha, \beta \geq 0, \beta \neq 1$. *Then the integral operator* $I_2(f,g)(z) = \int_0^z \frac{f(t)^\gamma}{g(t)^\sigma}dt \in \mathcal{K}(\theta)$, *where* $\theta = 1 - \left(\sigma\left(\frac{1-\alpha}{1-\beta}\right) - \gamma\alpha\right)$ *and* $0 \leq \left(\sigma\left(\frac{1-\alpha}{1-\beta}\right) - \gamma\alpha\right) < 1$, *for* $z \in \mathcal{U}$.

Acknowledgement This work was partially supported by the strategic project POSDRU 107/1.5/S/77265, inside POSDRU Romania 2007–2013 co-financed by the European Social Fund: Investing in People.

References

1. Ahlfors, L.V.: Sufficient conditions for quasiconformal extension. Ann. Math. Stud. **79**, 23–29 (1973) (Proceedings of the 1973 Conference at the University of Maryland)
2. Becker, J.: Löwnersche Differentialgleichung und quasikonform fortsezbare schichte Functionen. J. Reine Angew. Math. **255**, 23–43 (1972)
3. Becker, J.: Löwnersche Differentialgleichung und Schlichtheits-Kriterion. Math. Ann. **202**(4), 321–335 (1973)
4. Frasin, B.A., Darus, M.: On certain analytic univalent functions. Int. J. Math. Math. Sci. **25**(5), 305–310 (2001)
5. Mayer, O.: The functions theory of one variable complex, *Acad. Ed.*, Bucureşti (1981)
6. Owa, S., Srivastava, H.M.: Some generalized convolution properties associated with certain subclasses of analytic functions. J. Inequal. Pure Appl. Math. **3**(3), Article ID 42, 13 p. (2002)
7. Răducanu, D., Radomir, I., Gageonea, M.E., Pascu, N.R. A generalization of Ozaki–Nunokawa's univalence criterion. J. Inequal. Pure Appl. Math. **5**(4), Article ID 95 (2004)
8. Rønning, F.: Integral representations of bounded starlike functions. Ann. Pol. Math. **LX**(3), 289–297 (1995)
9. Uralegaddi, A., Ganigi, M.D., Sarangi, S.M.: Univalent functions with positive coefficients. Tamkang J. Math. **25**(3), 225–230 (1994)

On the Study of Hyperbolic Triangles and Circles by Hyperbolic Barycentric Coordinates in Relativistic Hyperbolic Geometry

Abraham Albert Ungar

Abstract Barycentric coordinates are commonly used in Euclidean geometry. Following the adaptation of barycentric coordinates for use in hyperbolic geometry in recently published books on analytic hyperbolic geometry, known and novel results concerning triangles and circles in the hyperbolic geometry of Lobachevsky and Bolyai are discovered. Among the novel results are the hyperbolic counterparts of important theorems in Euclidean geometry. These are: (i) the Inscribed Gyroangle Theorem, (ii) the Gyrotangent–Gyrosecant Theorem, (iii) the Intersecting Gyrosecants Theorem, and (iv) the Intersecting Gyrochord Theorem. Here in gyrolanguage, the language of analytic hyperbolic geometry, we prefix a gyro to any term that describes a concept in Euclidean geometry and in associative algebra to mean the analogous concept in hyperbolic geometry and nonassociative algebra. Outstanding examples are *gyrogroups* and *gyrovector spaces*, and Einstein addition being both *gyrocommutative* and *gyroassociative*. The prefix "gyro" stems from "gyration," which is the mathematical abstraction of the special relativistic effect known as "Thomas precession."

Keywords Einstein addition · Einstein gyrovector space · Relativistic hyperbolic geometry · Gyrotrigonometry · Gyrobarycentric coordinates · Gyrotriangle · Gyrocircle

1 Introduction

A barycenter in astronomy is the point between two objects where they balance each other. It is the center of gravity where two or more celestial bodies orbit each other. In 1827, Möbius published a book whose title, *Der Barycentrische Calcul*, translates as *The Barycentric Calculus*. The word *barycentric* means center of gravity, but the book is entirely geometrical and, hence, called by Jeremy Gray [5], *Möbius's Geometrical Mechanics*. The 1827 Möbius book is best remembered for introducing a new system of coordinates, the *barycentric coordinates*. The historical contribution of Möbius' barycentric coordinates to vector analysis is described by Crowe [1, pp. 48–50].

A. A. Ungar (✉)
Department of Mathematics, North Dakota State University, Fargo, ND 58108, USA
e-mail: Ungar.Abraham@ndsu.edu

T. M. Rassias, P. M. Pardalos (eds.), *Mathematics Without Boundaries*,
DOI 10.1007/978-1-4939-1106-6_22, © Springer Science+Business Media, LLC 2014

Commonly used as a tool in the study of Euclidean geometry, barycentric coordinates have been adapted for use as a tool in the study of the hyperbolic geometry of Lobachevsky and Bolyai as well, given in several recently published books [20, 22, 25, 26].

Relativistic hyperbolic geometry is a model of analytic hyperbolic geometry in which Einstein addition plays the role of vector addition. Einstein addition is a binary operation in the ball of vector spaces, which is neither commutative nor associative. However, Einstein addition is both gyrocommutative and gyroassociative, giving rise to gyrogroups and gyrovector spaces. The latter, in turn, form the algebraic setting for relativistic hyperbolic geometry, just as vector spaces form the algebraic setting for the standard model of Euclidean geometry.

Relativistic hyperbolic geometry admits the notion of relativistic hyperbolic barycentric coordinates, just as Euclidean geometry admits the notion of Euclidean barycentric coordinates. Relativistic hyperbolic barycentric coordinates and classical Euclidean barycentric coordinates share remarkable analogies. In particular, they are both covariant. Indeed, Relativistic barycentric coordinate representations are covariant with respect to the Lorentz coordinate transformation group, just as classical, Euclidean barycentric coordinate representations are covariant with respect to the Galilean coordinate transformation group. The remarkable analogies suggest that hyperbolic barycentric coordinates can prove useful in the study of hyperbolic geometry, just as Euclidean barycentric coordinates prove useful in the study of Euclidean geometry.

Indeed, following the adaptation of Euclidean barycentric coordinates for use in hyperbolic geometry, where they are called *gyrobarycentric coordinates*, we employ here the technique of gyrobarycentric coordinates to rediscover and discover known and new results in hyperbolic geometry. An introduction to hyperbolic barycentric coordinates and their application in hyperbolic geometry is found in [29]. Some familiarity with relativistic hyperbolic geometry as studied in [29] is assumed. Relativistic hyperbolic geometry is studied extensively in [20, 22, 25, 26]; see also [12, 13, 19, 23, 30] and [16–18, 21, 24, 28].

Among the novel results in hyperbolic geometry that are discovered, here are the following outstanding results:

1. The Inscribed Gyroangle Theorem, which is the hyperbolic counterpart of the well-known Inscribed Angle Theorem in Euclidean geometry (Sects. 9–10).
2. The Gyrotangent–Gyrosecant Theorem, which is the hyperbolic counterpart of the well-known Tangent–Secant Theorem in Euclidean geometry (Sect. 13).
3. The Intersecting Gyrosecants Theorem, which is the hyperbolic counterpart of the well-known Intersecting Secants Theorem in Euclidean geometry (Sect. 14).
4. The Intersecting Gyrochords Theorem, which is the hyperbolic counterpart of the well-known Intersecting Chords Theorem in Euclidean geometry (Sect. 26).

The prefix "gyro" that we extensively use stems from the term "gyration" [27], which is the mathematical abstraction of the special relativistic effect known as "Thomas precession."

The use of nonassociative algebra and hyperbolic trigonometry (gyrotrigonometry) involves straightforward, but complicated calculations. Hence, computer algebra, like Mathematica, for algebraic manipulations is an indispensable tool in this work.

2 Einstein Addition

Our journey into the fascinating world of relativistic hyperbolic geometry begins in Einstein addition and passes through important novel theorems that capture remarkable analogies between Euclidean and hyperbolic geometry. Einstein addition, in turn, is the binary operation that stems from Einstein's composition law of relativistically admissible velocities that he introduced in his 1905 paper [2, 3, p. 141] that founded the special theory of relativity.

Let c be an arbitrarily fixed positive constant and let $\mathbb{R}^n = (\mathbb{R}^n, +, \cdot)$ be the Euclidean n-space, $n = 1, 2, 3, \ldots$, equipped with the common vector addition, +, and inner product, ·. The home of all n dimensional Einsteinian velocities is the c-ball

$$\mathbb{R}^n_c = \{\mathbf{v} \in \mathbb{R}^n : \|\mathbf{v}\| < c\} \tag{1}$$

It is the open ball of radius c, centered at the origin of $\mathbb{R}^n$, consisting of all vectors $\mathbf{v}$ in $\mathbb{R}^n$ with norm smaller than c.

Einstein addition and scalar multiplication play in the ball $\mathbb{R}^n_c$, the role that vector addition and scalar multiplication play in the Euclidean n-space $\mathbb{R}^n$.

Definition 1 Einstein addition is a binary operation, $\oplus$, in the c-ball $\mathbb{R}^n_c$ given by the equation, [19], ([14], Eq. 2.9.2), ([11], p. 55], [4],

$$\mathbf{u}\oplus\mathbf{v} = \frac{1}{1+\dfrac{\mathbf{u}\cdot\mathbf{v}}{c^2}}\left\{\mathbf{u} + \frac{1}{\gamma_{\mathbf{u}}}\mathbf{v} + \frac{1}{c^2}\frac{\gamma_{\mathbf{u}}}{1+\gamma_{\mathbf{u}}}(\mathbf{u}\cdot\mathbf{v})\mathbf{u}\right\} \tag{2}$$

for all $\mathbf{u}, \mathbf{v} \in \mathbb{R}^n_c$, where $\gamma_{\mathbf{u}}$ is the Lorentz gamma factor given by the equation

$$\gamma_{\mathbf{v}} = \frac{1}{\sqrt{1 - \dfrac{\|\mathbf{v}\|^2}{c^2}}} \tag{3}$$

where $\mathbf{u} \cdot \mathbf{v}$ and $\|\mathbf{v}\|$ are the inner product and the norm in the ball, which the ball $\mathbb{R}^n_c$ inherits from its space $\mathbb{R}^n$.

A frequently used identity that follows immediately from (3) is

$$\frac{\mathbf{v}^2}{c^2} = \frac{\|\mathbf{v}\|^2}{c^2} = \frac{\gamma_{\mathbf{v}}^2 - 1}{\gamma_{\mathbf{v}}^2} \tag{4}$$

A nonempty set with a binary operation is called a *groupoid* so that, accordingly, the pair $(\mathbb{R}^n_c, \oplus)$ is an *Einstein groupoid.*

In the Newtonian limit of large c, $c \to \infty$, the ball $\mathbb{R}^n_c$ expands to the whole of its space $\mathbb{R}^n$, as we see from (1), and Einstein addition $\oplus$ in $\mathbb{R}^n_c$ reduces to the ordinary vector addition $+$ in $\mathbb{R}^n$, as we see from (2) and (3).

In applications to velocity spaces, $\mathbb{R}^n = \mathbb{R}^3$ is the Euclidean three-space, which is the space of all classical, Newtonian velocities, and $\mathbb{R}^n_c = \mathbb{R}^3_c \subset \mathbb{R}^3$ is the c-ball of $\mathbb{R}^3$ of all relativistically admissible, Einsteinian velocities. The constant c represents in special relativity the vacuum speed of light. Since we are interested in geometry, we allow n to be any positive integer and, sometimes, replace c by s.

We naturally use the abbreviation $\mathbf{u}\ominus\mathbf{v} = \mathbf{u}\oplus(-\mathbf{v})$ for Einstein subtraction, so that, for instance, $\mathbf{v}\ominus\mathbf{v} = \mathbf{0}$, $\ominus\mathbf{v} = \mathbf{0}\ominus\mathbf{v} = -\mathbf{v}$. Einstein addition and subtraction satisfy the equations

$$\ominus(\mathbf{u}\oplus\mathbf{v}) = \ominus\mathbf{u}\ominus\mathbf{v} \tag{5}$$

and

$$\ominus\mathbf{u}\oplus(\mathbf{u}\oplus\mathbf{v}) = \mathbf{v} \tag{6}$$

for all $\mathbf{u}, \mathbf{v}$ in the ball $\mathbb{R}^n_c$, in full analogy with vector addition and subtraction in $\mathbb{R}^n$. Identity (5) is called the *gyroautomorphic inverse property* of Einstein addition, and Identity (6) is called the *left cancellation law* of Einstein addition. We may note that Einstein addition does not obey the naive right counterpart of the left cancellation law (6) since, in general,

$$(\mathbf{u}\oplus\mathbf{v})\ominus\mathbf{v} \neq \mathbf{u}. \tag{7}$$

However, this seemingly lack of a *right cancellation law* of Einstein addition is repaired, for instance, in [26, Sect. 1.9].

Finally, as demonstrated in [29] and in [19, 20, 22, 23, 25, 26], Einstein addition admits scalar multiplication, giving rise to Einstein gyrovector spaces. These, in turn, form the algebraic setting for relativistic model of hyperbolic geometry, just as vector spaces form the algebraic setting for the standard model of Euclidean geometry. A brief description of the road from Einstein addition to gyrogroups and gyrovector spaces, necessary for a fruitful reading of this chapter, is found in [29].

3 Gyrocircles

Assuming familiarity with Einstein gyrovector spaces, as studied in [29], and particularly with the concepts of *gyrobarycentric independence*, *gyroflats*, and *hyperbolic span* in [29, Definition 13], the gyrocircle definition follows.

Definition 2 (Gyrocircle) *Let* $S = \{A_1, A_2, A_3\}$ *be a gyrobarycentrically independent set in an Einstein gyrovector space* $(\mathbb{R}^n_s, \oplus, \otimes)$, $n \geq 2$, *and let*

$$\mathbb{A}_3 = A_1\oplus\text{Span}\{\ominus A_1\oplus A_2, \ominus A_1\oplus A_3\} \subset \mathbb{R}^n\,. \tag{8}$$

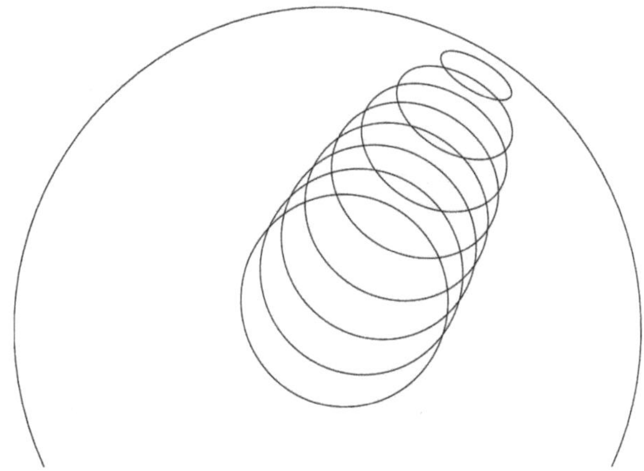

Fig. 1 A sequence of gyrocircles with gyroradius $\frac{1}{3}$ in an Einstein gyrovector plane $\mathbb{R}^2_{s=1}$ with gyrocenters approaching the boundary of the open unit disc $\mathbb{R}^2_{s=1}$ is shown. The *center* of the disc is conformal. Hence, the gyrocircle with gyrocenter at the center of the disc coincides with Euclidean circle. The Euclidean circle is increasingly flattened as its gyrocenter approaches the boundary of the disc

The locus of a point in $\mathbb{A}_3 \cap \mathbb{R}^n_s$ which is at a constant gyrodistance r from a fixed point $O \in \mathbb{A}_3 \cap \mathbb{R}^n_s$ is a gyrocircle $C(r, O)$ with gyrocenter O and gyroradius r.

The gyrocircle $C(r, O)$ with gyroradius r, $0 < r < s$, and gyrocenter $O \in \mathbb{R}^2_s$ in an Einstein gyrovector plane $(\mathbb{R}^2_s, \oplus, \otimes)$ is the set of all points $P \in \mathbb{R}^2_s$ such that $\|\ominus P \oplus O\| = r$. It is given by the equation

$$C(r, O, \theta) = O \oplus \begin{pmatrix} r\cos\theta \\ r\sin\theta \end{pmatrix}, \tag{9}$$

$0 \le \theta < 2\pi$. Indeed, by the left cancellation law we have

$$\|\ominus O \oplus C(r, O, \theta)\| = \left\| \begin{pmatrix} r\cos\theta \\ r\sin\theta \end{pmatrix} \right\| = r. \tag{10}$$

A sequence of gyrocircles of gyroradius $\frac{1}{3}$ in an Einstein gyrovector plane $\mathbb{R}^2_{s=1}$ with gyrocenters approaching the boundary of the open unit disc $\mathbb{R}^2_{s=1}$ is shown in Fig. 1. The center of the disc in Fig. 1 is conformal, as explained in [26, Sect. 6.2]. Accordingly, a gyrocircle with gyrocenter at the center of the disc is identical to Euclidean circle. This Euclidean circle is increasingly flattened in the Euclidean sense when the gyrocircle gyrocenter approaches the boundary of the disc.

4 Gyrotriangle Circumgyrocenter

Definition 3 (Gyrotriangle Circumgyrocircle, Circumgyrocenter, Circumgyroradius) *Let $A_1A_2A_3$ be a gyrotriangle in an Einstein gyrovector space $(\mathbb{R}^n_s, \oplus, \otimes)$, $n \ge 2$, and let $\mathbb{A}_3$ be the set*

$$\mathbb{A}_3 = A_1 \oplus \mathrm{Span}\{\ominus A_1 \oplus A_2, \ominus A_1 \oplus A_3\} \subset \mathbb{R}^n. \tag{11}$$

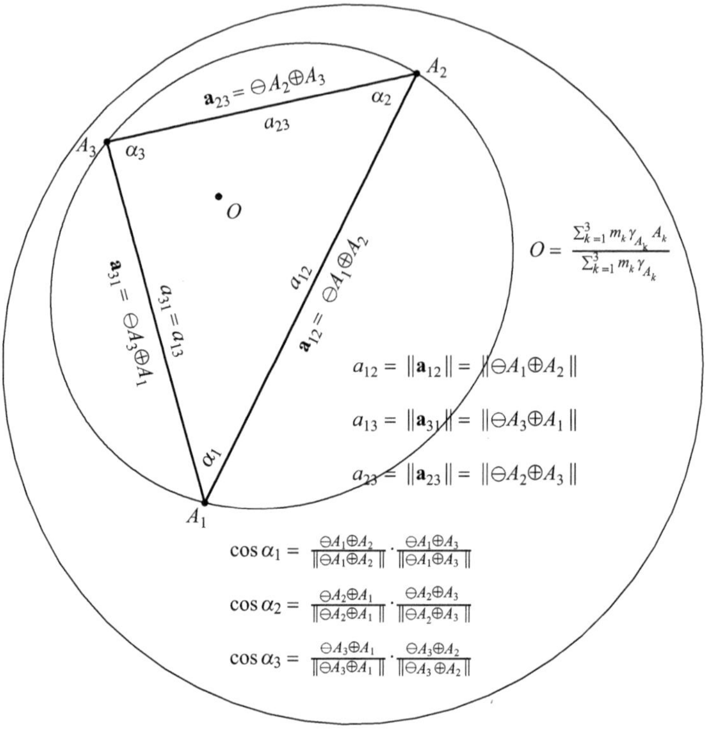

Fig. 2 The circumgyrocircle, and the circumgyrocenter O, of gyrotriangle $A_1A_2A_3$ in an Einstein gyrovector space $(\mathbb{R}^n_s, \oplus, \otimes)$, $n = 2$, is shown along with its associated index notation. Here $\|\ominus A_1 \oplus O\| = \|\ominus A_2 \oplus O\| = \|\ominus A_3 \oplus O\|$, where O is the gyrotriangle circumgyrocenter, given by its gyrobarycentric representation (39), with respect to the gyrobarycentrically independent set $S = \{A_1, A_2, A_3\}$. The Euclidean counterpart of this figure is shown in Fig. 3, p. 14

The circumgyrocenter of the gyrotriangle is the point O, $O \in \mathbb{A}_3 \cap \mathbb{R}^n_s$, equigyrodistant from the three gyrotriangle vertices. The gyrodistance from O to each vertex A_k, $k = 1, 2, 3$, of the gyrotriangle is the gyrotriangle circumgyroradius, and the gyrocircle with gyrocenter O and gyroradius r in $\mathbb{A}_3 \cap \mathbb{R}^n_s$ is the gyrotriangle circumgyrocircle.

It should be noted that not every gyrotriangle in $\mathbb{R}^n_s$ possesses a circumgyrocenter.

Let $A_1A_2A_3$ be a gyrotriangle in an Einstein gyrovector space $(\mathbb{R}^n_s, \oplus, \otimes)$ that possesses a circumgyrocircle, and let $O \in \mathbb{A}_3 \cap \mathbb{R}^n_s$ be the circumgyrocenter of the gyrotriangle, as shown in Fig. 2. Then, O possesses a gyrobarycentric representation,

$$O = \frac{m_1\gamma_{A_1} A_1 + m_2\gamma_{A_2} A_2 + m_3\gamma_{A_3} A_3}{m_1\gamma_{A_1} + m_2\gamma_{A_2} + m_3\gamma_{A_3}}, \tag{12}$$

with respect to the gyrobarycentrically independent set $S = \{A_1, A_2, A_3\}$. The gyrobarycentric coordinates m_1, m_2, and m_3 are to be determined in (20) below, in terms of gamma factors of the gyrotriangle sides and, alternatively in (34), in terms of the gyrotriangle gyroangles.

Following the *Gyrobarycentric representation Gyrocovariance Theorem*, [26, Theorem 4.6, pp. 90–91], with a left gyrotranslation by $X = \ominus A_1$, and using the index notation

$$\mathbf{a}_{ij} = \ominus A_i \oplus A_j, \qquad a_{ij} = \|\mathbf{a}_{ij}\|, \qquad \gamma_{ij} = \gamma_{\mathbf{a}_{ij}} = \gamma_{a_{ij}}, \tag{13}$$

noting that $a_{ij} = a_{ji}$, $\gamma_{ij} = \gamma_{ji}$, $\mathbf{a}_{ii} = \mathbf{0}$, $a_{ii} = 0$, and $\gamma_{ii} = 1$. We have

$$\begin{aligned} \gamma_{\ominus A_1 \oplus O} &= \frac{m_1 \gamma_{\ominus A_1 \oplus A_1} + m_2 \gamma_{\ominus A_1 \oplus A_2} + m_3 \gamma_{\ominus A_1 \oplus A_3}}{m_O} \\ &= \frac{m_1 + m_2 \gamma_{12} + m_3 \gamma_{13}}{m_O}, \end{aligned} \tag{14}$$

where by [26, Eq. (4.27), p. 90] and the *Gyrobarycentric representation Gyrocovariance Theorem*, [26, Theorem 4.6, pp. 90–91], the circumgyrocenter gyrobarycentric representation constant $m_O > 0$ with respect to the set of the gyrotriangle vertices is given by the equation

$$m_O^2 = m_1^2 + m_2^2 + m_3^2 + 2(m_1 m_2 \gamma_{12} + m_1 m_3 \gamma_{13} + m_2 m_3 \gamma_{23}). \tag{15}$$

Similarly, by the *Gyrobarycentric representation Gyrocovariance Theorem*, [26, Theorem 4.6, pp. 90–91], with left gyrotranslations by $X = \ominus A_1$, by $X = \ominus A_2$, and by $X = \ominus A_3$, we have, respectively,

$$\begin{aligned} \gamma_{\ominus A_1 \oplus O} &= \frac{m_1 + m_2 \gamma_{12} + m_3 \gamma_{13}}{m_O} \\ \gamma_{\ominus A_2 \oplus O} &= \frac{m_1 \gamma_{12} + m_2 + m_3 \gamma_{23}}{m_O} \\ \gamma_{\ominus A_3 \oplus O} &= \frac{m_1 \gamma_{13} + m_2 \gamma_{23} + m_3}{m_O}. \end{aligned} \tag{16}$$

The condition that the circumgyrocenter O is equigyrodistant from its gyrotriangle vertices A_1, A_2, and A_3 implies

$$\gamma_{\ominus A_1 \oplus O} = \gamma_{\ominus A_2 \oplus O} = \gamma_{\ominus A_3 \oplus O}. \tag{17}$$

Equations (16) and (17), along with the normalization condition $m_1 + m_2 + m_3 = 1$, yield the following system of three equations for the three unknowns m_1, m_2, and m_3,

$$\begin{aligned} m_1 + m_2 + m_3 &= 1 \\ m_1 + m_2 \gamma_{12} + m_3 \gamma_{13} &= m_1 \gamma_{13} + m_2 \gamma_{23} + m_3 \\ m_1 \gamma_{12} + m_2 + m_3 \gamma_{23} &= m_1 \gamma_{13} + m_2 \gamma_{23} + m_3, \end{aligned} \tag{18}$$

which can be written as the matrix equation,

$$\begin{pmatrix} 1 & 1 & 1 \\ 1-\gamma_{13} & \gamma_{12}-\gamma_{23} & \gamma_{13}-1 \\ \gamma_{12}-\gamma_{13} & 1-\gamma_{23} & \gamma_{23}-1 \end{pmatrix} \begin{pmatrix} m_1 \\ m_2 \\ m_3 \end{pmatrix} = \begin{pmatrix} 1 \\ 0 \\ 0 \end{pmatrix} . \tag{19}$$

Solving (19) for the unknowns m_1, m_2, and m_3, we have:

$$\begin{aligned} m_1 &= \frac{1}{D}(\ \gamma_{12}+\gamma_{13}-\gamma_{23}-1)(\gamma_{23}-1) \\ m_2 &= \frac{1}{D}(\ \gamma_{12}-\gamma_{13}+\gamma_{23}-1)(\gamma_{13}-1) \\ m_3 &= \frac{1}{D}(-\gamma_{12}+\gamma_{13}+\gamma_{23}-1)(\gamma_{12}-1)\,, \end{aligned} \tag{20}$$

where D is the determinant of the 3×3 matrix in (19),

$$\begin{aligned} D &= 2(\gamma_{12}\gamma_{13}+\gamma_{12}\gamma_{23}+\gamma_{13}\gamma_{23}) \\ &\quad -(\gamma_{12}^2-1)-(\gamma_{13}^2-1)-(\gamma_{23}^2-1)-2(\gamma_{12}+\gamma_{13}+\gamma_{23}) \\ &= 1+2\gamma_{12}\gamma_{13}\gamma_{23}-\gamma_{12}^2-\gamma_{13}^2-\gamma_{23}^2-2(\gamma_{12}-1)(\gamma_{13}-1)(\gamma_{23}-1)\,. \end{aligned} \tag{21}$$

Gyrotrigonometric substitutions into the extreme right-hand side of (21) from [26, Sect. 7.12] give the gyrotrigonometric representation of D,

$$\begin{aligned} D &= \frac{16F^2}{\sin^2\alpha_1 \sin^2\alpha_2 \sin^2\alpha_3} - \frac{16F}{\sin^2\alpha_1 \sin^2\alpha_2 \sin^2\alpha_3}\sin^2\tfrac{\delta}{2} \\ &= \frac{16F}{\sin^2\alpha_1 \sin^2\alpha_2 \sin^2\alpha_3}(F-\sin^2\tfrac{\delta}{2}) \\ &= \frac{16F}{\sin^2\alpha_1 \sin^2\alpha_2 \sin^2\alpha_3}\sin\tfrac{\delta}{2} \\ &\quad \times \{\sin(\alpha_1+\tfrac{\delta}{2})\sin(\alpha_2+\tfrac{\delta}{2})\sin(\alpha_3+\tfrac{\delta}{2})-\sin\tfrac{\delta}{2}\}, \end{aligned} \tag{22}$$

where $\delta = \pi - \alpha_1 - \alpha_2 - \alpha_3$ is the defect of gyrotriangle $A_1A_2A_3$, and where

$$F = F(\alpha_1, \alpha_2, \alpha_3) = \sin\tfrac{\delta}{2}\sin(\alpha_1+\tfrac{\delta}{2})\sin(\alpha_2+\tfrac{\delta}{2})\sin(\alpha_3+\tfrac{\delta}{2})\,. \tag{23}$$

The extreme right-hand side of (22) is the product of two factors, the first of which is positive. The second factor is positive if and only if the gyrotriangle circumgyrocenter $O \in \mathbb{R}^n$ lies inside the ball $\mathbb{R}^n_s$, as we will see in Theorem 1, p. 12. This factor vanishes if and only if O lies on the boundary of the ball $\mathbb{R}^n_s$, and it is negative if and only if O lies outside the closure of the ball $\mathbb{R}^n_s$, as we will see in Theorem 1.

The circumgyrocenter O of gyrotriangle $A_1A_2A_3$ is given by (12) where the gyrobarycentric coordinates m_1, m_2, and m_3 are given by (20). Since in gyrobarycentric coordinates, only ratios of coordinates are relevant, the gyrobarycentric coordinates,

m_1, m_2, and m_3 in (20) can be simplified by removing their common nonzero factor $1/D$.

Gyrobarycentric coordinates, m_1, m_2, and m_3, of the circumgyrocenter O of gyrotriangle $A_1A_2A_3$ are thus given by the equations

$$\begin{aligned} m_1' &= (\gamma_{12} + \gamma_{13} - \gamma_{23} - 1)(\gamma_{23} - 1) \\ m_2' &= (\gamma_{12} - \gamma_{13} + \gamma_{23} - 1)(\gamma_{13} - 1) \\ m_3' &= (-\gamma_{12} + \gamma_{13} + \gamma_{23} - 1)(\gamma_{12} - 1)\,. \end{aligned} \tag{24}$$

Hence, by (15) along with the gyrobarycentric coordinates in (24), we have

$$\begin{aligned} m_O^2 &= (1 + 2\gamma_{12}\gamma_{13}\gamma_{23} - \gamma_{12}^2 - \gamma_{13}^2 - \gamma_{23}^2) \\ &\quad \times \{1 + 2\gamma_{12}\gamma_{13}\gamma_{23} - \gamma_{12}^2 - \gamma_{13}^2 - \gamma_{23}^2 - 2(\gamma_{12} - 1)(\gamma_{13} - 1)(\gamma_{23} - 1)\} \\ &= \{(\gamma_{12} + \gamma_{13} + \gamma_{23} - 1)^2 - 2(\gamma_{12}^2 + \gamma_{13}^2 + \gamma_{23}^2 - 1)\} \\ &\quad \times (1 + 2\gamma_{12}\gamma_{13}\gamma_{23} - \gamma_{12}^2 - \gamma_{13}^2 - \gamma_{23}^2). \end{aligned} \tag{25}$$

In order to emphasize comparative patterns that the gyrotriangle circumgyrocircle and the gyrotetrahedron circumgyrosphere possess, we present the first equation in (25) in the determinantal form

$$m_O^2 = D_3(D_3 - H_3)\,, \tag{26}$$

where D_3 is the determinant

$$D_3 = \begin{vmatrix} 1 & \gamma_{12} & \gamma_{13} \\ \gamma_{12} & 1 & \gamma_{23} \\ \gamma_{13} & \gamma_{23} & 1 \end{vmatrix} \tag{27}$$

and where

$$H_3 = 2(\gamma_{12} - 1)(\gamma_{13} - 1)(\gamma_{23} - 1)\,. \tag{28}$$

The gyrotriangle $A_1A_2A_3$ in Fig. 2 possesses a circumgyrocenter if and only if $m_O^2 > 0$.

The factor D_3 of m_O^2 in (25) and in (26),

$$D_3 = 1 + 2\gamma_{12}\gamma_{13}\gamma_{23} - \gamma_{12}^2 - \gamma_{13}^2 - \gamma_{23}^2 = \begin{vmatrix} 1 & \gamma_{12} & \gamma_{13} \\ \gamma_{12} & 1 & \gamma_{23} \\ \gamma_{13} & \gamma_{23} & 1 \end{vmatrix} > 0\,, \tag{29}$$

is positive for any gyrotriangle $A_1A_2A_3$ in an Einstein gyrovector space. Hence, as we see from (25), $m_O^2 > 0$ if and only if the points A_1, A_2, and A_3 obey the *circumgyrocircle existence condition*

$$(\gamma_{12} + \gamma_{13} + \gamma_{23} - 1)^2 > 2(\gamma_{12}^2 + \gamma_{13}^2 + \gamma_{23}^2 - 1) \tag{30a}$$

or, equivalently, the circumgyrocircle existence condition

$$4(\gamma_{12}-1)(\gamma_{13}-1) > (\gamma_{12}+\gamma_{13}-\gamma_{23}-1)^2\,, \tag{30b}$$

or, equivalently, the circumgyrocircle existence condition

$$D_3 > H_3\,. \tag{30c}$$

Gamma factors of gyrotriangle side gyrolengths are related to its gyroangles by the *AAA* to *SSS Conversion Law* [26, Theorem 6.5, p. 137]

$$\begin{aligned}\gamma_{23} &= \frac{\cos\alpha_1+\cos\alpha_2\cos\alpha_3}{\sin\alpha_2\sin\alpha_3}\\ \gamma_{13} &= \frac{\cos\alpha_2+\cos\alpha_1\cos\alpha_3}{\sin\alpha_1\sin\alpha_3}\\ \gamma_{12} &= \frac{\cos\alpha_3+\cos\alpha_1\cos\alpha_2}{\sin\alpha_1\sin\alpha_2}\,.\end{aligned} \tag{31}$$

Substituting these from (31) into (24) we obtain

$$\begin{aligned}m'_1 &= F'\sin\left(\frac{-\alpha_1+\alpha_2+\alpha_3}{2}\right)\sin\alpha_1\\ m'_2 &= F'\sin\left(\frac{\alpha_1-\alpha_2+\alpha_3}{2}\right)\sin\alpha_2\\ m'_3 &= F'\sin\left(\frac{\alpha_1+\alpha_2-\alpha_3}{2}\right)\sin\alpha_3\,,\end{aligned} \tag{32}$$

where the common factor $F' = F'(\alpha_1,\alpha_2,\alpha_3)$ in (32) is given by the equation

$$F' = 2^3\frac{\cos^2(\frac{\alpha_1+\alpha_2+\alpha_3}{2})\cos(\frac{-\alpha_1+\alpha_2+\alpha_3}{2})\cos(\frac{\alpha_1-\alpha_2+\alpha_3}{2})\cos(\frac{\alpha_1+\alpha_2-\alpha_3}{2})}{\sin\alpha_1\sin\alpha_2\sin\alpha_3}\,. \tag{33}$$

Since in gyrobarycentric coordinates only ratios of coordinates are relevant, the gyrobarycentric coordinates, m'_1, m'_2, and m'_3 in (32) can be simplified by removing a common nonzero factor. Hence, convenient gyrobarycentric coordinates, m''_1, m''_2, and m''_3, of the circumgyrocenter O of gyrotriangle $A_1A_2A_3$, expressed in terms of the gyrotriangle gyroangles are given by the equations

$$\begin{aligned}m''_1 &= -\sin\left(\frac{-\alpha_1+\alpha_2+\alpha_3}{2}\right)\sin\alpha_1 = \cos(\alpha_1+\tfrac{\delta}{2})\sin\alpha_1\\ m''_2 &= -\sin\left(\frac{\alpha_1-\alpha_2+\alpha_3}{2}\right)\sin\alpha_2 = \cos(\alpha_2+\tfrac{\delta}{2})\sin\alpha_2\\ m''_3 &= -\sin\left(\frac{\alpha_1+\alpha_2-\alpha_3}{2}\right)\sin\alpha_3 = \cos(\alpha_3+\tfrac{\delta}{2})\sin\alpha_3\,,\end{aligned} \tag{34}$$

where $\delta = \pi-\alpha_1-\alpha_2-\alpha_3$ is the defect of gyrotriangle $A_1A_2A_3$.

The circumgyrocenter O, (12), lies in the interior of its gyrotriangle $A_1A_2A_3$ if and only if its gyrobarycentric coordinates are all positive or all negative. Hence, we see from the gyrobarycentric coordinates (34) of O that:

1. the circumgyrocenter O lies in the interior of its gyrotriangle $A_1A_2A_3$ if and only if the largest gyroangle of the gyrotriangle has measure less than the sum of the measures of the other two gyroangles. This result is known in hyperbolic geometry; see, for instance, [6, p. 132], where the result is proved synthetically. Similarly, we also see from the gyrobarycentric coordinates (34) of O in (12) that
2. the circumgyrocenter O lies in the interior of its gyrotriangle $A_1A_2A_3$ if and only if all the three gyroangles $\alpha_1+\delta/2$, $\alpha_2+\delta/2$ and $\alpha_3+\delta/2$ are acute.

Expressing Inequality (30) gyrotrigonometrically, by means of (31), it can be shown from (25) that $m_O^2>0$ if and only if

$$\cos\frac{3\alpha_1-\alpha_2-\alpha_3}{2}+\cos\frac{-\alpha_1+3\alpha_2-\alpha_3}{2}+\cos\frac{-\alpha_1-\alpha_2+3\alpha_3}{2} > 3\cos\frac{\alpha_1+\alpha_2+\alpha_3}{2} \tag{35}$$

or, equivalently, if and only if

$$\sin(2\alpha_1+\tfrac{\delta}{2})+\sin(2\alpha_2+\tfrac{\delta}{2})+\sin(2\alpha_3+\tfrac{\delta}{2})>3\sin\tfrac{\delta}{2} \tag{36}$$

Inequality (36) is an elegant condition for the existence of a circumgyrocenter. In a different approach, we will discover below in (38) the circumgyrocenter existence condition in a different elegant form.

Gyrotrigonometric substitutions into the second equation in (25) from [26, Sect. 7.12] yield the following gyrotrigonometric expression for the constant $m_O>0$ of the circumgyrocenter gyrobarycentric representation (12):

$$\begin{aligned} m_O^2 &= \frac{4^2F^2}{\sin^2\alpha_1\sin^2\alpha_2\sin^2\alpha_3}\left\{\frac{4^2F^2}{\sin^2\alpha_1\sin^2\alpha_2\sin^2\alpha_3}-\frac{4^2F\sin^2\frac{\delta}{2}}{\sin^2\alpha_1\sin^2\alpha_2\sin^2\alpha_3}\right\} \\ &= \frac{4^4F^3}{\sin^4\alpha_1\sin^4\alpha_2\sin^4\alpha_3}(F-\sin^2\tfrac{\delta}{2}) \\ &= \frac{4^4F^3\sin\frac{\delta}{2}}{\sin^4\alpha_1\sin^4\alpha_2\sin^4\alpha_3}\{\sin(\alpha_1+\tfrac{\delta}{2})\sin(\alpha_2+\tfrac{\delta}{2})\sin(\alpha_3+\tfrac{\delta}{2})-\sin\tfrac{\delta}{2}\}\,. \end{aligned} \tag{37}$$

Hence, $m_O^2>0$ if and only if $F>\sin^2(\delta/2)$ or, equivalently,

$$\sin(\alpha_1+\tfrac{\delta}{2})\sin(\alpha_2+\tfrac{\delta}{2})\sin(\alpha_3+\tfrac{\delta}{2})>\sin\tfrac{\delta}{2}\,. \tag{38}$$

One may demonstrate directly that the two circumgyrocenter existence conditions (36) and (38) are equivalent.

Formalizing the main results of this section, we have the following theorem:

Theorem 1 (Circumgyrocenter Theorem) *Let* $S = \{A_1, A_2, A_3\}$ *be a gyrobarycentrically independent set of three points in an Einstein gyrovector space* $(\mathbb{R}_s^n, \oplus, \otimes)$. *The circumgyrocenter* $O \in \mathbb{R}^n$ *of gyrotriangle* $A_1A_2A_3$, *shown in* Fig. 2, *possesses the gyrobarycentric representation*

$$O = \frac{m_1\gamma_{A_1}A_1 + m_2\gamma_{A_2}A_2 + m_3\gamma_{A_3}A_3}{m_1\gamma_{A_1} + m_2\gamma_{A_2} + m_3\gamma_{A_3}} \tag{39}$$

with respect to the set $S = \{A_1, A_2, A_3\}$, *with gyrobarycentric coordinates* $(m_1 : m_2 : m_3)$ *given by*

$$\begin{aligned} m_1 &= (\gamma_{12} + \gamma_{13} - \gamma_{23} - 1)(\gamma_{23} - 1) \\ m_2 &= (\gamma_{12} - \gamma_{13} + \gamma_{23} - 1)(\gamma_{13} - 1) \\ m_3 &= (-\gamma_{12} + \gamma_{13} + \gamma_{23} - 1)(\gamma_{12} - 1) \end{aligned} \tag{40}$$

or, equivalently, by the gyrotrigonometric gyrobarycentric coordinates

$$\begin{aligned} m_1 &= \cos(\alpha_1 + \tfrac{\delta}{2}) \sin\alpha_1 \\ m_2 &= \cos(\alpha_2 + \tfrac{\delta}{2}) \sin\alpha_2 \\ m_3 &= \cos(\alpha_3 + \tfrac{\delta}{2}) \sin\alpha_3\,. \end{aligned} \tag{41}$$

The circumgyrocenter gyrobarycentric representation constant m_O *with respect to the set* $S = \{A_1, A_2, A_3\}$ *is an elegant product of two factors, given by the equation*

$$\begin{aligned} m_O^2 =& \{(\gamma_{12} + \gamma_{13} + \gamma_{23} - 1)^2 - 2(\gamma_{12}^2 + \gamma_{13}^2 + \gamma_{23}^2 - 1)\} \\ &\times (1 + 2\gamma_{12}\gamma_{13}\gamma_{23} - \gamma_{12}^2 - \gamma_{13}^2 - \gamma_{23}^2) = D_3(D_3 - H_3)\,, \end{aligned} \tag{42}$$

where D_3 *and* H_3 *are given by* (27) and (28).

The circumgyrocenter lies in the ball, $O \in \mathbb{R}_s^n$, *if and only if* $m_O^2 > 0$ *or, equivalently, if and only if one of the following mutually equivalent inequalities, each of which is a circumgyrocircle existence condition, is satisfied*:

$$\sin(2\alpha_1 + \tfrac{\delta}{2}) + \sin(2\alpha_2 + \tfrac{\delta}{2}) + \sin(2\alpha_3 + \tfrac{\delta}{2}) > 3\sin\tfrac{\delta}{2} \tag{43}$$

$$\sin(\alpha_1 + \tfrac{\delta}{2})\sin(\alpha_2 + \tfrac{\delta}{2})\sin(\alpha_3 + \tfrac{\delta}{2}) > \sin\tfrac{\delta}{2} \tag{44}$$

$$(\gamma_{12} + \gamma_{13} + \gamma_{23} - 1)^2 > 2(\gamma_{12}^2 + \gamma_{13}^2 + \gamma_{23}^2 - 1) \tag{45}$$

$$4(\gamma_{12} - 1)(\gamma_{13} - 1) > (\gamma_{12} + \gamma_{13} - \gamma_{23} - 1)^2. \tag{46}$$

5 Triangle Circumcenter

In this section the gyrotriangle circumgyrocenter in Fig. 2, p. 6, will be translated into its Euclidean counterpart in Fig. 3.

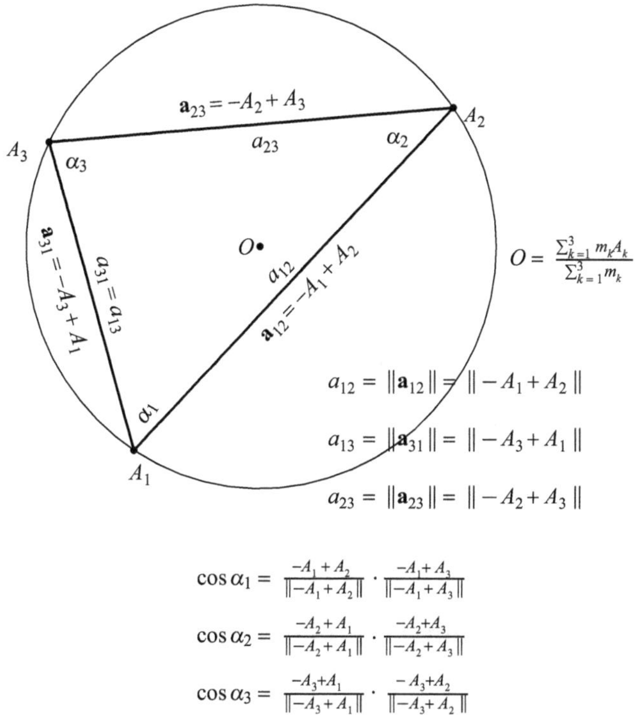

Fig. 3 The circumcircle, and the circumcenter O, of triangle $A_1A_2A_3$ in the Euclidean vector space $\mathbb{R}^n$, $n = 2$, is shown along with its standard notation. Here $R = \| -A_1 + O\| = \| -A_2 + O\| = \| -A_3 + O\|$, where R is the triangle circumradius and O is the triangle circumcenter, given by its barycentric coordinate representation (48) with respect to the barycentrically independent set $S = \{A_1, A_2, A_3\}$. The hyperbolic counterpart of this figure is shown in Fig. 2, p. 6

The gyrobarycentric representation (39) with gyrotrigonometric gyrobarycentric coordinates $(m_1 : m_2 : m_3)$ given by (41) remains invariant in form under the Euclidean limit $s \to \infty$, so that it is valid in Euclidean geometry as well, where $\delta = 0$. Hence, in the transition from hyperbolic geometry, where $\delta > 0$, to Euclidean geometry, where $\delta = 0$, the gyrobarycentric coordinates (41) reduce to the barycentric coordinates

$$\begin{aligned} m_1 &= \cos\alpha_1 \sin\alpha_1 = \tfrac{1}{2}\sin 2\alpha_1 \\ m_2 &= \cos\alpha_2 \sin\alpha_2 = \tfrac{1}{2}\sin 2\alpha_2 \qquad \text{(Euclidean Geometry)} \qquad (47) \\ m_3 &= \cos\alpha_3 \sin\alpha_3 = \tfrac{1}{2}\sin 2\alpha_3\,. \end{aligned}$$

Hence, finally, a trigonometric barycentric representation of the circumcenter O of triangle $A_1A_2A_3$ in $\mathbb{R}^n$, Fig. 3, with respect to the barycentrically independent set $S = \{A_1, A_2, A_3\} \subset \mathbb{R}^n$ is given by Result (48) of the following corollary of Theorem 1, which recovers a well-known result in Euclidean geometry [7].

Corollary 1 *Let* α_k, $k = 1, 2, 3$ *and* O *be the angles and circumcenter of a triangle* $A_1A_2A_3$ *in the Euclidean space* $\mathbb{R}^n$. *Then,*

$$O = \frac{\sin 2\alpha_1 A_1 + \sin 2\alpha_2 A_2 + \sin 2\alpha_3 A_3}{\sin 2\alpha_1 + \sin 2\alpha_2 + \sin 2\alpha_3} . \tag{48}$$

Theorem 1 and its Corollary 1 form an elegant example that illustrates the result that:

1. gyrotrigonometric gyrobarycentric coordinates of a point in an Einstein gyrovector space $\mathbb{R}^n_s$ survive unimpaired in Euclidean geometry, where they form
2. trigonometric barycentric coordinates of a point in a corresponding Euclidean vector space $\mathbb{R}^n$.
   The converse is, however, not valid since
3. trigonometric barycentric coordinates of a point in a Euclidean vector space $\mathbb{R}^n$ may embody the Euclidean condition that the triangle angle sum in π, so that they need not survive in hyperbolic geometry.

6 Gyrotriangle Circumgyroradius

The circumgyroradius R of gyrotriangle $A_1A_2A_3$ with circumgyrocenter O in an Einstein gyrovector space $(\mathbb{R}^n_s, \oplus, \otimes)$, shown in Figs. 2 and 4, is given by

$$R = \|{\ominus}A_1{\oplus}O\| = \|{\ominus}A_2{\oplus}O\| = \|{\ominus}A_3{\oplus}O\| . \tag{49}$$

By (14), the circumgyroradius R satisfies the equation

$$\gamma_{_R} = \gamma_{_{\|\ominus A_1 \oplus O\|}} = \gamma_{_{\ominus A_1 \oplus O}} = \frac{m_1 + m_2\gamma_{12} + m_3\gamma_{13}}{m_O} , \tag{50}$$

where m_1, m_2, and m_3 are given by (40), and where m_O is given by (42).

Hence, following (40), (42), and (50), with the notation for D_3 and H_3 in (27) and (28), we have

$$\begin{aligned}\gamma_R^2 &= \frac{1 + 2\gamma_{12}\gamma_{13}\gamma_{23} - \gamma_{12}^2 - \gamma_{13}^2 - \gamma_{23}^2}{1 + 2\gamma_{12}\gamma_{13}\gamma_{23} - \gamma_{12}^2 - \gamma_{13}^2 - \gamma_{23}^2 - 2(\gamma_{12} - 1)(\gamma_{13} - 1)(\gamma_{23} - 1)} \\ &= \frac{2\gamma_{12}\gamma_{13}\gamma_{23} - (\gamma_{12}^2 + \gamma_{13}^2 + \gamma_{23}^2 - 1)}{(\gamma_{12} + \gamma_{13} + \gamma_{23} - 1)^2 - 2(\gamma_{12}^2 + \gamma_{13}^2 + \gamma_{23}^2 - 1)} \\ &= \frac{D_3}{D_3 - H_3} .\end{aligned} \tag{51}$$

Interestingly, from (51) and (26), we obtain an elegant relationship between (i) the constant m_O of the gyrobarycentric representation of the circumgyrocenter O of a gyrotriangle $A_1A_2A_3$ with respect to the gyrotriangle vertices and (ii) the gamma factor of the circumgyroradius R,

$$m_O\gamma_R = D_3 . \tag{52}$$

Following (51) we have, by (4),

$$R^2 = s^2\frac{\gamma_R^2-1}{\gamma_R^2} = 2s^2\frac{(\gamma_{12}-1)(\gamma_{13}-1)(\gamma_{23}-1)}{1+2\gamma_{12}\gamma_{13}\gamma_{23}-\gamma_{12}^2-\gamma_{13}^2-\gamma_{23}^2} = s^2\frac{H_3}{D_3} . \tag{53}$$

Hence, finally, the circumgyroradius R of gyrotriangle $A_1A_2A_3$ in Figs. 2 and 4 is given by

$$R = \sqrt{2}s\sqrt{\frac{(\gamma_{12}-1)(\gamma_{13}-1)(\gamma_{23}-1)}{1+2\gamma_{12}\gamma_{13}\gamma_{23}-\gamma_{12}^2-\gamma_{13}^2-\gamma_{23}^2}} , \tag{54}$$

implying

$$\sqrt{\frac{(\gamma_{12}+1)(\gamma_{13}+1)(\gamma_{23}+1)}{2}}\,R = s\sqrt{\frac{(\gamma_{12}^2-1)(\gamma_{13}^2-1)(\gamma_{23}^2-1)}{1+2\gamma_{12}\gamma_{13}\gamma_{23}-\gamma_{12}^2-\gamma_{13}^2-\gamma_{23}^2}} . \tag{55}$$

Identity (55) captures a remarkable analogy between the law of gyrosines and the law of sines. Indeed, following (55), the law of gyrosines [26, Theorem 6.9, p. 140] for gyrotriangle $A_1A_2A_3$ in Fig. 4 is linked to the circumgyroradius R of the gyrotriangle by the equation

$$\frac{\gamma_{23}a_{23}}{\sin\alpha_1} = \frac{\gamma_{13}a_{13}}{\sin\alpha_2} = \frac{\gamma_{12}a_{12}}{\sin\alpha_3} = \sqrt{\frac{(\gamma_{12}+1)(\gamma_{13}+1)(\gamma_{23}+1)}{2}}\,R , \tag{56}$$

called the *extended law of gyrosines*.

Following the gamma-gyrotrigonometric identity [26, Eq. (7.154), p. 189], the extended law of gyrosines (56) can be written as

$$\frac{\gamma_{23}a_{23}}{\sin\alpha_1} = \frac{\gamma_{13}a_{13}}{\sin\alpha_2} = \frac{\gamma_{12}a_{12}}{\sin\alpha_3} = 2\frac{\sin(\alpha_1+\frac{\delta}{2})\sin(\alpha_2+\frac{\delta}{2})\sin(\alpha_3+\frac{\delta}{2})}{\sin\alpha_1\sin\alpha_2\sin\alpha_3}\,R , \tag{57}$$

where $\delta = \pi - (\alpha_1+\alpha_2+\alpha_3)$ is the defect of gyrotriangle $A_1A_2A_3$.

In the Euclidean limit of large s, $s\to\infty$, gamma factors tend to 1 and gyrotriangle defects tend to 0. Hence, in that limit, the extended law of gyrosines (56) tends to the well-known extended law of sines [9, p. 87],

$$\frac{a_{23}}{\sin\alpha_1} = \frac{a_{13}}{\sin\alpha_2} = \frac{a_{12}}{\sin\alpha_3} = 2R \qquad \text{(Euclidean Geometry)} . \tag{58}$$

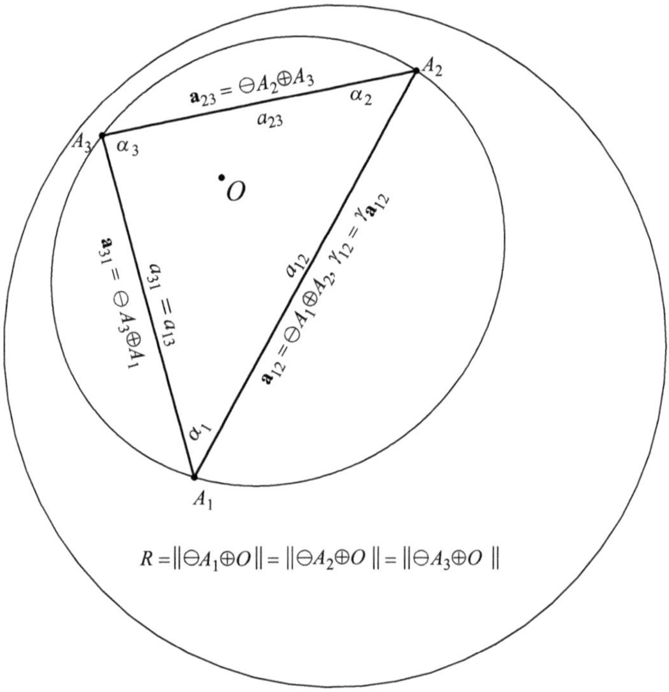

Fig. 4 The circumgyrocircle of gyrotriangle $A_1A_2A_3$ in the Einstein gyrovector space $(\mathbb{R}^n_s, \oplus, \otimes)$ is shown for $n = 2$. Its gyrocenter, O, is the gyrotriangle circumgyrocenter, given by its gyrobarycentric representation (39), p. 12, and its gyroradius R is the gyrotriangle circumgyroradius, given by each of the equations $R = \|\ominus A_k \oplus O\|$, $k = 1, 2, 3$. The gyrocircle is a flattened Euclidean circle, as shown in Fig. 1

Formalizing the results in (56) and (57) we have the following theorem:

Theorem 2 (Extended Law of Gyrosines) *Let $A_1A_2A_3$ be a gyrotriangle in an Einstein gyrovector space $(\mathbb{R}^n_s \oplus, \otimes)$ with gyroangles $\alpha_1, \alpha_2, \alpha_3$, side-gyrolengths a_{23}, a_{13}, a_{12}, and circumgyroradius R,* Fig. 4.

Then,

$$\frac{\gamma_{23}a_{23}}{\sin\alpha_1} = \frac{\gamma_{13}a_{13}}{\sin\alpha_2} = \frac{\gamma_{12}a_{12}}{\sin\alpha_3} = \sqrt{\frac{(\gamma_{12}+1)(\gamma_{13}+1)(\gamma_{23}+1)}{2}}\,R \tag{59}$$

and

$$\frac{\gamma_{23}a_{23}}{\sin\alpha_1} = \frac{\gamma_{13}a_{13}}{\sin\alpha_2} = \frac{\gamma_{12}a_{12}}{\sin\alpha_3} = 2\frac{\sin(\alpha_1+\frac{\delta}{2})\sin(\alpha_2+\frac{\delta}{2})\sin(\alpha_3+\frac{\delta}{2})}{\sin\alpha_1\sin\alpha_2\sin\alpha_3}\,R\,. \tag{60}$$

Interestingly, the gyrotriangle circumgyroradius R has an elegant representation in terms of its gyrotriangle gyroangles. Indeed, expressing the gamma factors in (53) in terms of the gyrotriangle gyroangles α_k, $k = 1, 2, 3$, takes the gyrotrigonometric form

$$\frac{R^2}{s^2} = \frac{\cos\frac{\alpha_1+\alpha_2+\alpha_3}{2}}{\cos\frac{\alpha_1-\alpha_2-\alpha_3}{2}\cos\frac{-\alpha_1+\alpha_2-\alpha_3}{2}\cos\frac{-\alpha_1-\alpha_2+\alpha_3}{2}}$$
$$= \frac{\sin\frac{\delta}{2}}{\sin(\alpha_1+\frac{\delta}{2})\sin(\alpha_2+\frac{\delta}{2})\sin(\alpha_3+\frac{\delta}{2})} \tag{61}$$
$$= \frac{F(\alpha_1,\alpha_2,\alpha_3)}{\sin^2(\alpha_1+\frac{\delta}{2})\sin^2(\alpha_2+\frac{\delta}{2})\sin^2(\alpha_3+\frac{\delta}{2})}$$

in any Einstein gyrovector space $(\mathbb{R}^n_s, \oplus, \otimes)$, where $F = F(\alpha_1, \alpha_2, \alpha_3)$ is given by (23) [26, Eq. (7.144), p. 187]. Equation (61) can be written conveniently as

$$\frac{R}{s} = \frac{\sqrt{F}}{\sin(\alpha_1+\frac{\delta}{2})\sin(\alpha_2+\frac{\delta}{2})\sin(\alpha_3+\frac{\delta}{2})}. \tag{62}$$

Following (62), the first equation in [26, Eq. (7.143), p. 187] and [26, Eq. (7.150), p. 188] we have the equation

$$\sin(\alpha_3+\tfrac{\delta}{2}) = \frac{\gamma_{12}a_{12}}{R(\gamma_{12}+1)}, \tag{63}$$

which will prove useful in (102), p. 28.

In the Euclidean limit, $s \to \infty$, the gyrotriangle defect tends to 0, $\delta \to 0$, so that each side of (61) tends to 0.

An important relation that results from (61) is formalized in the following theorem:

Theorem 3 *Let α_k, $k = 1, 2, 3$, and R be the gyroangles and circumgyroradius of a gyrotriangle $A_1A_2A_3$ in an Einstein gyrovector space $(\mathbb{R}^n_s \oplus, \otimes)$. Then*

$$s^2 \sin\tfrac{\delta}{2} = R^2 \sin(\alpha_1+\tfrac{\delta}{2})\sin(\alpha_2+\tfrac{\delta}{2})\sin(\alpha_3+\tfrac{\delta}{2}) \tag{64}$$

and

$$s\sqrt{F(\alpha_1,\alpha_2,\alpha_3)} = R\sin(\alpha_1+\tfrac{\delta}{2})\sin(\alpha_2+\tfrac{\delta}{2})\sin(\alpha_3+\tfrac{\delta}{2}) \tag{65}$$

where $F(\alpha_1, \alpha_2, \alpha_3)$ is given by (23) [26, *Eq.* (7.144), *p.* 187].

Proof Identity (64) follows immediately from (61), and Identity (65) follows from (64) and the definition of F.

The Euclidean limit, $s \to \infty$, of the left-hand side of each of the Eqs. (64) and (65) of Theorem 3 is indeterminate, being a limit of type $\infty \cdot 0$. In that limit, the gyrotriangle defect δ tends to zero, so that the right-hand side of (64) tends to $R^2 \sin\alpha_1 \sin\alpha_2 \sin\alpha_3$.

Theorem 3 gives rise to useful results, as indicated by the following theorem.

Theorem 4 *Let $A_1A_2A_3$ be a gyrotriangle that possesses a circumgyroradius R in an Einstein gyrovector space $(\mathbb{R}^n_s \oplus, \otimes)$,* Fig. 2, *p. 6. Then, in the gyrotriangle index notation* (13),

$$
\begin{aligned}
a_{12} &= \frac{2R\sin(\alpha_1+\frac{\delta}{2})\sin(\alpha_2+\frac{\delta}{2})\sin(\alpha_3+\frac{\delta}{2})}{\cos\alpha_3+\cos\alpha_1\cos\alpha_2}\\
a_{13} &= \frac{2R\sin(\alpha_1+\frac{\delta}{2})\sin(\alpha_2+\frac{\delta}{2})\sin(\alpha_3+\frac{\delta}{2})}{\cos\alpha_2+\cos\alpha_1\cos\alpha_3}\\
a_{23} &= \frac{2R\sin(\alpha_1+\frac{\delta}{2})\sin(\alpha_2+\frac{\delta}{2})\sin(\alpha_3+\frac{\delta}{2})}{\cos\alpha_1+\cos\alpha_2\cos\alpha_3}
\end{aligned}
\tag{66}
$$

and

$$
\begin{aligned}
\gamma_{12}a_{12} &= \frac{2R\sin(\alpha_1+\frac{\delta}{2})\sin(\alpha_2+\frac{\delta}{2})\sin(\alpha_3+\frac{\delta}{2})}{\sin\alpha_1\sin\alpha_2}\\
\gamma_{13}a_{13} &= \frac{2R\sin(\alpha_1+\frac{\delta}{2})\sin(\alpha_2+\frac{\delta}{2})\sin(\alpha_3+\frac{\delta}{2})}{\sin\alpha_1\sin\alpha_3}\\
\gamma_{23}a_{23} &= \frac{2R\sin(\alpha_1+\frac{\delta}{2})\sin(\alpha_2+\frac{\delta}{2})\sin(\alpha_3+\frac{\delta}{2})}{\sin\alpha_2\sin\alpha_3},
\end{aligned}
\tag{67}
$$

where $a_{ij} = \|\ominus A_i \oplus A_j\|$, $1 \leq i, j \leq 3$.

Proof The identities in (66) and (67) follow from Identity (65) of Theorem 3 and from the AAA to SSS conversion law [26, Theorem 6.8, p. 140].

Comparing the first equation in each of (66) and (67) yields the gyrotriangle identity

$$
\cos\alpha_3+\cos\alpha_1\cos\alpha_2 = \gamma_{12}\sin\alpha_1\sin\alpha_2 \qquad \text{(Hyperbolic Geometry)} \tag{68}
$$

that any gyrotriangle $A_1A_2A_3$ in an Einstein gyrovector space $(\mathbb{R}_s^n, \oplus, \otimes)$ obeys.

With $\delta = 0$, Theorem 4 specializes to the following corresponding theorem in Euclidean geometry:

Theorem 5 *Let* $A_1A_2A_3$ *be a triangle with a circumradius* R *in the Euclidean vector space* $\mathbb{R}^n$, Fig. 3, *p. 14. Then, in the triangle index notation,*

$$
\begin{aligned}
a_{12} &= \frac{2R\sin\alpha_1\sin\alpha_2\sin\alpha_3}{\cos\alpha_3+\cos\alpha_1\cos\alpha_2}\\
a_{13} &= \frac{2R\sin\alpha_1\sin\alpha_2\sin\alpha_3}{\cos\alpha_2+\cos\alpha_1\cos\alpha_3}\\
a_{23} &= \frac{2R\sin\alpha_1\sin\alpha_2\sin\alpha_3}{\cos\alpha_1+\cos\alpha_2\cos\alpha_3}
\end{aligned}
\tag{69}
$$

and

$$
\begin{aligned}
a_{12} &= \frac{2R\sin\alpha_1\sin\alpha_2\sin\alpha_3}{\sin\alpha_1\sin\alpha_2}\\
a_{13} &= \frac{2R\sin\alpha_1\sin\alpha_2\sin\alpha_3}{\sin\alpha_1\sin\alpha_3}\\
a_{23} &= \frac{2R\sin\alpha_1\sin\alpha_2\sin\alpha_3}{\sin\alpha_2\sin\alpha_3},
\end{aligned}
\tag{70}
$$

where $a_{ij} = \|-A_i + A_j\|$, $1 \le i, j \le 3$.

As expected from Identities (69) and (70) of Theorem 5, the triangle angles obey the triangle identity $\alpha_1 + \alpha_2 + \alpha_3 = \pi$, so that

$$\cos\alpha_3 + \cos\alpha_1 \cos\alpha_2 = \sin\alpha_1 \sin\alpha_2 \qquad \text{(Euclidean Geometry)} \tag{71}$$

for any triangle $A_1A_2A_3$ in the Euclidean space $\mathbb{R}^n$. Indeed, the triangle identity (71) is the Euclidean counterpart of the gyrotriangle identity (68).

7 Triangle Circumradius

The circumradius R_{euc} of triangle $A_1A_2A_3$ with circumcenter O in the Euclidean space $\mathbb{R}^n$, shown in Fig. 17, p. 52, is given by

$$R_{euc} = \|-A_1 + O\| = \|-A_2 + O\| = \|-A_3 + O\| \,. \tag{72}$$

We will determine the triangle circumradius in the Euclidean space $\mathbb{R}^n$ as the Euclidean limit of the circumgyroradius of gyrotriangle $A_1A_2A_3$ in an Einstein gyrovector space $(\mathbb{R}^n_s, \oplus, \otimes)$, that is, $R_{euc} = \lim_{s\to\infty} R$. Accordingly, let $A_1A_2A_3$ be a gyrotriangle that possesses a circumgyrocircle in the Einstein gyrovector space, and let R be the gyrotriangle circumgyroradius.

Then, R is given by (53), which can be written equivalently as

$$R^2 = \frac{2\,s^2(\gamma_{23} - 1)}{D} \tag{73}$$

where

$$\begin{aligned} D = 2&\left\{1 + \frac{\gamma_{23} - 1}{\gamma_{13} - 1} + \frac{\gamma_{23} - 1}{\gamma_{12} - 1} + (\gamma_{23} - 1)\right\} \\ &- \left\{\frac{\gamma_{12} - 1}{\gamma_{23} - 1} + \frac{\gamma_{23} - 1}{\gamma_{12} - 1} + \frac{(\gamma_{23} - 1)^2}{(\gamma_{12} - 1)(\gamma_{13} - 1)}\right\}, \end{aligned} \tag{74}$$

as one can check by straightforward algebra.

Expressing R by (73) and (74), rather than (53), enables its Euclidean limit to be determined manifestly.

Indeed, we have the following Lemma about Euclidean limits:

Lemma 1 *Let* $A_i, A_j \in \mathbb{R}^n_s \subset \mathbb{R}^n$ *be two distinct points in an Einstein gyrovector space* $(\mathbb{R}^n_s, \oplus, \otimes)$, *and let*

$$\begin{aligned} a^{ein}_{ij} &= \|\ominus A_i \oplus A_j\| \\ \gamma_{ij} &= \gamma_{\ominus A_i \oplus A_j} \\ a_{ij} &= \|-A_i + A_j\| \,. \end{aligned} \tag{75}$$

Then,

$$\lim_{s\to\infty} s^2(\gamma_{ij} - 1) = \tfrac{1}{2}a_{ij}^2 \tag{76}$$

and

$$\lim_{s\to\infty} s^2(\gamma_{ij}^2 - 1) = a_{ij}^2 \,. \tag{77}$$

Proof In the Euclidean limit, $s \to \infty$, gamma factors tend to 1. Hence, by (4), p. 3,

$$\begin{aligned}\lim_{s\to\infty} s^2(\gamma_{ij} - 1) &= \lim_{s\to\infty} s^2(\gamma_{ij} - 1) \times \tfrac{1}{2} \lim_{s\to\infty} (\gamma_{ij} + 1) \\ &= \tfrac{1}{2} \lim_{s\to\infty} s^2(\gamma_{ij}^2 - 1) \\ &= \tfrac{1}{2} \lim_{s\to\infty} \gamma_{ij}^2 (a_{ij}^{ein})^2 \\ &= \tfrac{1}{2} a_{ij}^2 \,,\end{aligned} \tag{78}$$

as desired. The proof of (77) is similar.

Following Lemma 1, we have the Euclidean limit

$$\lim_{s\to\infty} D = 2\left\{1 + \frac{a_{23}^2}{a_{13}^2} + \frac{a_{23}^2}{a_{12}^2} + 0\right\} - \left\{\frac{a_{12}^2}{a_{13}^2} + \frac{a_{13}^2}{a_{12}^2} + \frac{a_{23}^4}{a_{12}^2 a_{13}^2}\right\} . \tag{79}$$

Hence, by (73), (78), and (79), and straightforward algebra,

$$R_{euc}^2 := \lim_{s\to\infty} R^2 = \frac{a_{12}^2 a_{13}^2 a_{23}^2}{2(a_{12}^2 a_{13}^2 + a_{12}^2 a_{23}^2 + a_{13}^2 a_{23}^2) - (a_{12}^4 + a_{13}^4 + a_{23}^4)} \,, \tag{80}$$

where $a_{ij} = \| -A_i + A_j \|$.

Formalizing the result of this section we obtain the following theorem.

Theorem 6 *The circumradius R of a triangle $A_1A_2A_3$ in the Euclidean space $\mathbb{R}^n$, $n \geq 2$, is given by the equation*

$$R^2 = \frac{a_{12}^2 a_{13}^2 a_{23}^2}{2(a_{12}^2 a_{13}^2 + a_{12}^2 a_{23}^2 + a_{13}^2 a_{23}^2) - (a_{12}^4 + a_{13}^4 + a_{23}^4)} \,, \tag{81}$$

where

$$a_{ij} = \| -A_i + A_j \| \,, \tag{82}$$

$1 \leq i, j \leq 3$, *are the sidelengths of the triangle.*

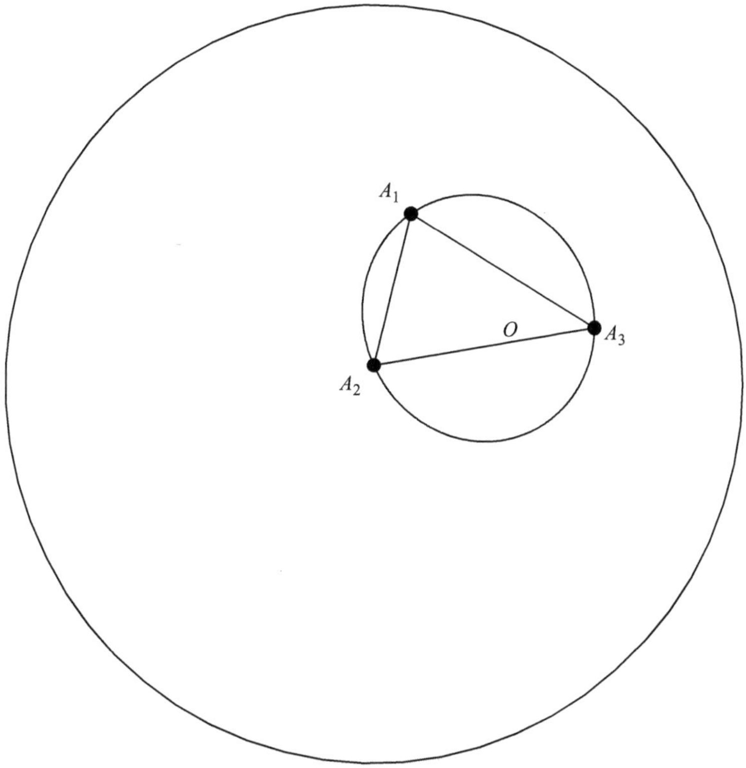

Fig. 5 Here A_1, A_2, and A_3 are arbitrarily selected three points of an Einstein gyrovector space $(\mathbb{R}^n_s, \oplus, \otimes)$ that satisfy the circumgyrocircle existence condition (83). Accordingly, there exists a unique gyrocircle that passes through these points

8 The Gyrocircle Through Three Points

Theorem 7 (The Gyrocircle Through Three Points) *Let A_1, A_2, and A_3 be any three points that form a gyrobarycentrically independent set in an Einstein gyrovector space $(\mathbb{R}^n_s, \oplus, \otimes)$, shown in* Figs. 5 and 6. *There exists a unique gyrocircle that passes through these points if and only if gyrotriangle $A_1A_2A_3$ obeys the circumgyrocircle existence condition,* (30),

$$(\gamma_{12} + \gamma_{13} + \gamma_{23} - 1)^2 > 2(\gamma_{12}^2 + \gamma_{13}^2 + \gamma_{23}^2 - 1) \tag{83}$$

or, equivalently, if and only if gyrotriangle $A_1A_2A_3$ obeys the gyrotrigonometric circumgyrocircle existence condition, (36),

$$\sin(2\alpha_1 + \tfrac{\delta}{2}) + \sin(2\alpha_2 + \tfrac{\delta}{2}) + \sin(2\alpha_3 + \tfrac{\delta}{2}) > 3\sin\tfrac{\delta}{2}\,, \tag{84}$$

where we use the index notation (13) *for gyrotriangle $A_1A_2A_3$.*

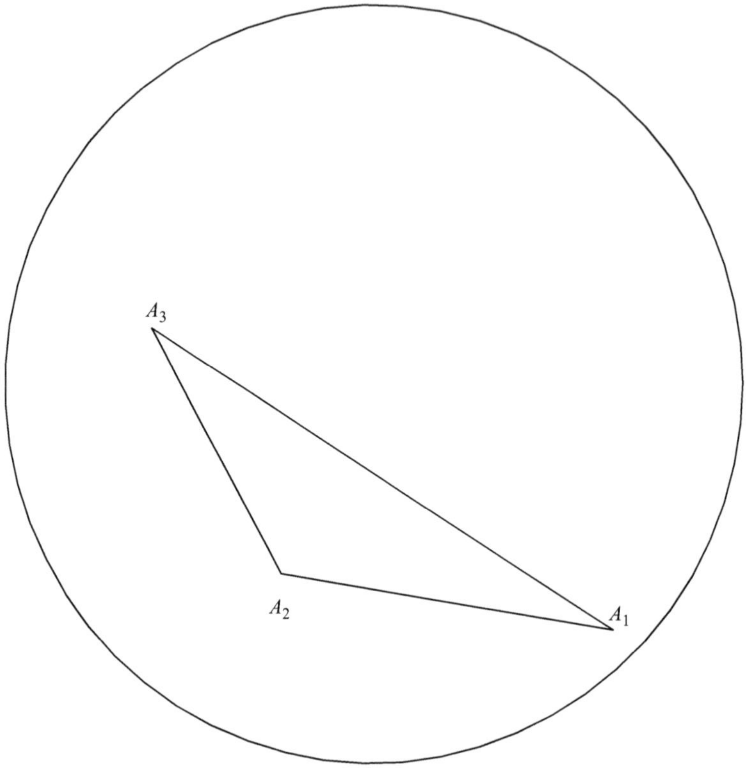

Fig. 6 Here A_1, A_2, and A_3 are arbitrarily selected three points of an Einstein gyrovector space $(\mathbb{R}_s^n, \oplus, \otimes)$ that do not satisfy the circumgyrocircle existence condition (83). Accordingly, there exists no gyrocircle that passes through these points

When a gyrocircle through the three points exists, it is the unique gyrocircle with gyrocenter O given by, (39),

$$O = \frac{m_1 \gamma_{A_1} A_1 + m_2 \gamma_{A_2} A_2 + m_3 \gamma_{A_3} A_3}{m_1 \gamma_{A_1} + m_2 \gamma_{A_2} + m_3 \gamma_{A_3}}, \tag{85}$$

where

$$\begin{aligned} m_1 &= (\gamma_{12} + \gamma_{13} - \gamma_{23} - 1)(\gamma_{23} - 1) \\ m_2 &= (\gamma_{12} - \gamma_{13} + \gamma_{23} - 1)(\gamma_{13} - 1) \\ m_3 &= (-\gamma_{12} + \gamma_{13} + \gamma_{23} - 1)(\gamma_{12} - 1) \end{aligned} \tag{86}$$

and with gyroradius R given by, (54),

$$R = \sqrt{2}s \sqrt{\frac{(\gamma_{12} - 1)(\gamma_{13} - 1)(\gamma_{23} - 1)}{1 + 2\gamma_{12}\gamma_{13}\gamma_{23} - \gamma_{12}^2 - \gamma_{13}^2 - \gamma_{23}^2}} = s\sqrt{\frac{H_3}{D_3}}\ . \tag{87}$$

Proof The gyrocircle in the theorem, if exists, is the circumgyrocircle of gyrotriangle $A_1A_2A_3$. The gyrocenter O of the gyrocircle is, therefore, given by (85) and (86), as we see from Theorem 1, p. 12; and the gyroradius, R, of the gyrocircle is given by (54), p. 16.

Finally, the circumgyrocircle of gyrotriangle $A_1A_2A_3$ exists if and only if the gyrotriangle satisfies the circumgyrocircle existence condition (83), or, equivalently, (85), as explained in the paragraph of Inequality (30), p. 10, and in (35) and (36).

Example 1 If the three points A_1, A_2, and A_3 in Theorem 7 are not distinct, a gyrocircle through these points is not unique. Indeed, in this case we have

$$(\gamma_{12} + \gamma_{13} + \gamma_{23} - 1)^2 = 2(\gamma_{12}^2 + \gamma_{13}^2 + \gamma_{23}^2 - 1), \tag{88}$$

as one can readily check, thus violating the circumgyrocircle existence condition (83).

Example 2 If the three points A_1, A_2, and A_3 in Theorem 7 are distinct and gyrocollinear, there is no gyrocircle through these points. Hence, in this case the circumgyrocircle existence condition (83) must be violated. Hence, these points must satisfy the inequality

$$(\gamma_{12} + \gamma_{13} + \gamma_{23} - 1)^2 \le 2(\gamma_{12}^2 + \gamma_{13}^2 + \gamma_{23}^2 - 1). \tag{89}$$

Example 3 Let the three points A_1, A_2, and A_3 in Theorem 7 be the vertices of an equilateral gyrotriangle with side gyrolengths a. Then, $\gamma_{12} = \gamma_{13} = \gamma_{23} = \gamma_a$, so that the circumgyrocircle existence condition (83) reduces to

$$\gamma_a > 1, \tag{90}$$

which is satisfied by any side gyrolength a, $0 < a < s$. Hence, by Theorem 7, any equilateral gyrotriangle in an Einstein gyrovector space possesses a circumgyrocircle. Moreover, the circumgyrocenter of an equilateral gyrotriangle lies on the interior of the gyrotriangle since the circumgyrocenter gyrobarycentric coordinates m_k, $k = 1, 2, 3$, in (86) are all positive.

More about the unique gyrocircle that passes through three given points of an Einstein gyrovector space is studied in Sect. 15.

A generalization of results in this section from the gyrotriangle circumgyrocircle to the gyrotetrahedron circumgyrosphere is presented in Chap. 10. Remarkably, the pattern that D_3 and H_3 exhibit in this section remains the same pattern in Chap. 10, exhibited by D_4 and H_4.

9 The Inscribed Gyroangle Theorem I

The Inscribed Gyroangle Theorem appears in two distinct, interesting versions, each of which reduces in the Euclidean limit of large s to the well-known Inscribed Angle Theorem in Euclidean geometry. Version I is presented in this section, and Version II is presented in Sect. 10.

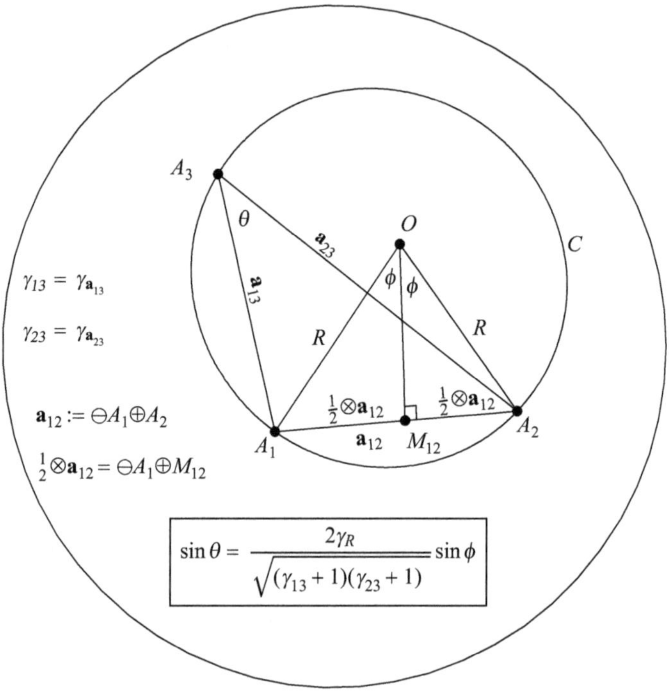

Fig. 7 The Inscribed Gyroangle Theorem I. Gyroangle $\theta = \angle A_1 A_3 A_2$ is inscribed in a gyrocircle C of gyroradius R (the circumgyroradius of gyrotriangle $A_1 A_2 A_3$) centered at O in an Einstein gyrovector plane $(\mathbb{R}_s^2, \oplus, \otimes)$, and $\phi = \angle A_1 O M_{12} = \angle A_2 O M_{12}$, where M_{12} is the gyromidpoint of the gyrosegment $A_1 A_2$. Accordingly, $2\phi = \angle A_1 O A_2$ is a gyrocentral gyroangle, and both θ and 2ϕ subtend on the same gyroarc on the gyrocircle C. The relationship between θ and ϕ, (91), is shown. In the Euclidean limit of large s, $s \to \infty$, gamma factors tend to 1 and, hence, the relationship between θ and ϕ in Euclidean geometry reduces to $\sin\theta = \sin\phi$ or, equivalently, $\theta = \phi$ (if A_3 and O lie on the same side of gyrosegment $A_1 A_2$) and $\theta = \pi - \phi$ (if A_3 and O lie on opposite sides of gyrosegment A_1 A_2; see Fig. 9)

Figure 7 presents a gyrotriangle $A_1 A_2 A_3$ and its circumgyrocircle with gyrocenter O at the gyrotriangle circumgyrocenter, given by (39), p. 12, and with gyroradius R, given by the gyrotriangle circumgyroradius (54), p. 16. The gamma factor γ_R of R is given by (51), p. 15.

A gyrocentral gyroangle of a gyrocircle is a gyroangle whose vertex is located at the gyrocenter of the gyrocircle. For instance, gyroangle $\angle A_1 O A_2$ in Figs. 7, 8, and 9 is gyrocentral.

An inscribed gyroangle of a gyrocircle is a gyroangle whose vertex is on the gyrocircle and whose sides each intersects the gyrocircle at another point. For instance, gyroangle $\angle A_1 A_3 A_2$ in Figs. 7, 8, and 9 is inscribed. The Inscribed Gyroangle Theorem gives a relation between inscribed gyroangles and a gyrocentral gyroangle of a gyrocircle that subtend on the same gyroarc of the gyrocircle in an Einstein gyrovector space.

Theorem 8 (The Inscribed Gyroangle Theorem I) *Let θ be a gyroangle inscribed in a gyrocircle of gyroradius R, and let 2ϕ be the gyrocentral gyroangle of the gyrocircle such that both θ and 2ϕ subtend on the same gyroarc $\widehat{A_1A_2}$ on the gyrocircle, as shown in* Fig. 7. *Then, in the notation of* Fig. 7 *and in* (13), *p. 7,*

$$\sin\theta = \frac{2\gamma_R}{\sqrt{(\gamma_{13}+1)(\gamma_{23}+1)}}\sin\phi\,. \tag{91}$$

Proof Under the conditions of the theorem, as described in Fig. 7, let M_{12} be the gyromidpoint of gyrosegment A_1A_2, implying

$$\phi := \angle A_1OM_{12} = \angle A_2OM_{12} = \tfrac{1}{2}\angle A_1OA_2 \tag{92}$$

so that 2ϕ is the gyrocentral gyroangle $\angle A_1OA_2$ shown in Fig. 7.

Furthermore, let

$$\mathbf{a}_{12} = \ominus A_1 \oplus A_2 \tag{93}$$

so that [26, p. 100],

$$\ominus A_1 \oplus M_{12} = \tfrac{1}{2}\otimes\mathbf{a}_{12} \tag{94}$$

and

$$\gamma_{\frac{1}{2}\otimes\mathbf{a}_{12}}(\tfrac{1}{2}\otimes\mathbf{a}_{12}) = \frac{\gamma_{12}\mathbf{a}_{12}}{\sqrt{2}\sqrt{1+\gamma_{12}}}\,. \tag{95}$$

Taking magnitudes of both sides of (95) and noting the homogeneity property ($V9$) of Einstein gyrovector spaces [25, Sect. 2.7], we have

$$\gamma_{\frac{1}{2}\otimes a_{12}}(\tfrac{1}{2}\otimes a_{12}) = \frac{\gamma_{12}a_{12}}{\sqrt{2}\sqrt{1+\gamma_{12}}}\,. \tag{96}$$

Applying the extended law of gyrosines (56), p. 16, to gyrotriangle $A_1A_2A_3$ and its circumgyroradius R in Fig. 7, we have

$$\frac{\gamma_{12}a_{12}}{\sin\theta} = \sqrt{\frac{(\gamma_{12}+1)(\gamma_{13}+1)(\gamma_{23}+1)}{2}}\,R\,, \tag{97}$$

implying

$$\sin\theta = \frac{\sqrt{2}\gamma_{12}a_{12}}{\sqrt{(1+\gamma_{12})(1+\gamma_{13})(1+\gamma_{23})}\,R}\,. \tag{98}$$

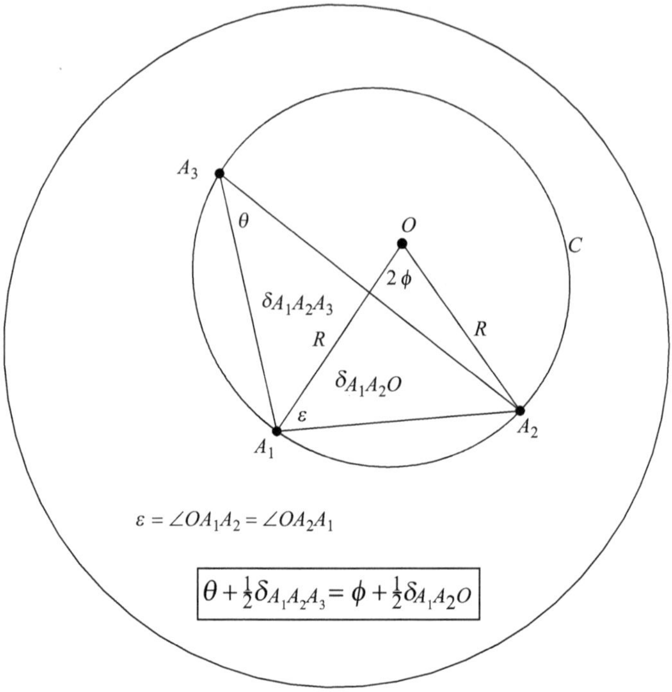

Fig. 8 The Inscribed Gyroangle Theorem II. Unlike the Inscribed Gyroangle Theorem described in Fig. 7, here the relation between the inscribed gyroangle θ and the gyrocentral gyroangle 2ϕ is expressed in terms of gyrotriangle defects. The latter vanish in Euclidean geometry, reducing the relation between θ and ϕ to the equation $\theta = \phi$

Applying the elementary gyrosine definition in gyrotrigonometry, illustrated in [26, Fig. 6.5, p. 147], to the right gyroangled gyrotriangle $A_1M_{12}O$ in Fig. 7, we obtain the first equation in (99),

$$\sin\phi = \frac{\gamma_{\frac{1}{2}\otimes a_{12}}(\frac{1}{2}\otimes a_{12})}{\gamma_R R} = \frac{\gamma_{12}a_{12}}{\sqrt{2}\sqrt{1+\gamma_{12}}\;\gamma_R R}\,. \tag{99}$$

The second equation in (99) follows from (96).

Finally, the desired Identity (91) follows by eliminating the factor $\gamma_{12}a_{12}$ between (98) and (99).

10 The Inscribed Gyroangle Theorem II

Theorem 9 (The Inscribed Gyroangle Theorem II) *Let θ be a gyroangle inscribed in a gyrocircle with gyrocenter O, and let 2ϕ be the gyrocentral gyroangle of the gyrocircle such that both θ and 2ϕ subtend on the same gyroarc $\widehat{A_1A_2}$ on the gyrocircle, as shown in* Fig. 8. *Furthermore, in the notation in* Fig. 8, *let $\delta_{A_1A_2A_3}$*

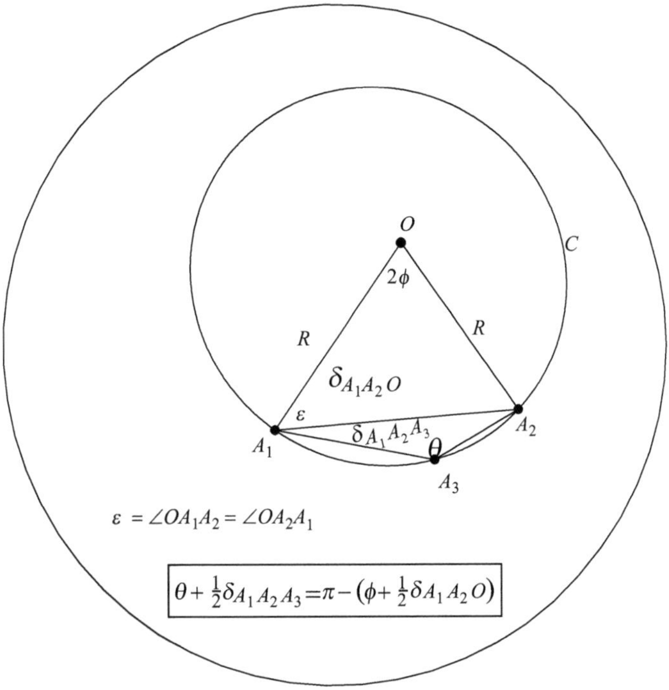

Fig. 9 The Inscribed Gyroangle Theorem II. This figure is similar to Fig. 8 except that here the points O and A_3 lie on opposite sides of the chord A_1A_2 of gyrocircle C in the gyroplane through the points A_1, A_2, and A_3. As in Fig. 8, the relation between the inscribed gyroangle θ and the gyrocentral gyroangle 2ϕ is expressed in terms of gyrotriangle defects. The latter vanish in Euclidean geometry, reducing the relation between θ and ϕ to the equation $\theta = \pi - \phi$

be the defect of gyrotriangle $A_1A_2A_3$ *and, similarly, let* $\delta_{A_1A_2O}$ *be the defect of gyrotriangle* A_1A_2O*. Then,*

$$\sin(\theta + \tfrac{1}{2}\delta_{A_1A_2A_3}) = \sin(\phi + \tfrac{1}{2}\delta_{A_1A_2O}), \tag{100}$$

that is, either

$$\theta + \tfrac{1}{2}\delta_{A_1A_2A_3} = \phi + \tfrac{1}{2}\delta_{A_1A_2O}, \tag{101a}$$

as in Fig. 8, *or*

$$\theta + \tfrac{1}{2}\delta_{A_1A_2A_3} = \pi - (\phi + \tfrac{1}{2}\delta_{A_1A_2O}), \tag{101b}$$

as in Fig. 9.

Proof In the gyrotriangle index notation (13), p. 7, with $\alpha_3 = \theta$ and $\delta = \delta_{A_1A_2A_3}$ to conform with the notation for gyrotriangle $A_1A_2A_3$ in Figs. 8 and 9, we have by (63), p. 18,

$$\sin(\theta + \tfrac{1}{2}\delta_{A_1A_2A_3}) = \frac{\gamma_{12}a_{12}}{R(\gamma_{12}+1)}. \tag{102}$$

Here $\delta_{A_1A_2A_3}$ is the defect of gyrotriangle $A_1A_2A_3$ and R is the circumgyroradius of gyrotriangle $A_1A_2A_3$ in Figs. 7 and 8. Accordingly,

$$R = \|\ominus O \oplus A_1\| = \|\ominus O \oplus A_2\| . \tag{103}$$

The expression $\gamma_{12}a_{12}/(R(\gamma_{12}+1))$ is expressed in (102) in terms of gyroangles of gyrotriangle $A_1A_2A_3$, shown in Figs. 8 and 9. We now wish to express it in terms of gyroangles of gyrotriangle A_1A_2O, also shown in Figs. 8 and 9. Accordingly, let

$$\begin{aligned} \epsilon &= \angle A_1A_2O = \angle A_2A_1O \\ a_{12} &= \|\ominus A_1 \oplus A_2\| \\ \gamma_{12} &= \gamma_{a_{12}} \end{aligned} \tag{104}$$

be parameters of gyrotriangle A_1A_2O and let $\delta = \delta_{A_1A_2O}$ be the defect of the gyrotriangle, as shown in Figs. 8 and 9.

Then,

$$\delta = \delta_{A_1A_2O} = \pi - 2\phi - 2\epsilon , \tag{105}$$

so that

$$\epsilon = \frac{\pi}{2} - (\phi + \frac{\delta}{2}), \tag{106}$$

implying

$$\epsilon + \frac{\delta}{2} = \frac{\pi}{2} - \phi , \tag{107}$$

so that

$$\sin(\epsilon + \frac{\delta}{2}) = \cos\phi . \tag{108}$$

Applying the first equation in [26, Eq. (7.143), p. 187] to gyrotriangle A_1A_2O in Fig. 8, we have

$$\frac{1}{s}\gamma_{12}a_{12} = \frac{2\sqrt{F}}{\sin^2\epsilon} \tag{109}$$

where, by (23), p. 9, and by (108),

$$\begin{aligned} F &= \sin\tfrac{\delta}{2}\sin^2(\epsilon + \tfrac{\delta}{2})\sin(2\phi + \tfrac{\delta}{2}) \\ &= \sin\tfrac{\delta}{2}\cos^2\phi\sin(2\phi + \tfrac{\delta}{2}) . \end{aligned} \tag{110}$$

Applying [26, Eq. (7.150), p. 188] to gyrotriangle A_1A_2O in Fig. 8, we have

$$\gamma_{12} + 1 = \frac{2\sin^2(\epsilon + \frac{\delta}{2})}{\sin^2\epsilon} . \tag{111}$$

Following (108), (109), and (111) we have

$$\frac{1}{s}\frac{\gamma_{12}a_{12}}{\gamma_{12}+1} = \frac{\sqrt{F}}{\sin^2(\epsilon+\frac{\delta}{2})} = \frac{\sqrt{F}}{\cos^2\phi}\,. \tag{112}$$

We now turn to calculate R/s. Applying the *AAA* to *SSS Conversion Law* [26, Theorem 6.5, p. 137] to gyrotriangle A_1A_2O in Fig. 8, we have

$$\begin{aligned}\gamma_{_R} &= \frac{\cos\epsilon+\cos\epsilon\cos 2\phi}{\sin\epsilon\sin 2\phi} = \frac{\cos\epsilon}{\sin\epsilon}\frac{1+\cos 2\phi}{\sin 2\phi}\\ &= \cot\epsilon\frac{2}{\sin 2\phi}\frac{1+\cos 2\phi}{2} = \cot\epsilon\frac{\cos^2\phi}{\sin\phi\cos\phi}\\ &= \cot\epsilon\cot\phi\,.\end{aligned} \tag{113}$$

By (4), p. 3, (113), and by (110) we have

$$\begin{aligned}\frac{R^2}{s^2} &= \frac{\gamma_{_R}^2-1}{\gamma_{_R}^2} = \frac{\cot^2\epsilon\cot^2\phi-1}{\cot^2\epsilon\cot^2\phi}\\ &= \frac{\sin\frac{\delta}{2}\sin(2\phi+\frac{\delta}{2})}{\cos^2\phi\sin^2(\phi+\frac{\delta}{2})} = \frac{F}{\cos^4\phi\sin^2(\phi+\frac{\delta}{2})}\,,\end{aligned} \tag{114}$$

where $\delta = \delta_{A_1A_2O}$ is the defect of gyrotriangle A_1A_2O given by (105), thus obtaining the equation

$$\frac{R}{s} = \frac{\sqrt{F}}{\cos^2\phi\sin(\phi+\frac{\delta}{2})}\,. \tag{115}$$

Equations (112) and (115), along with the notation for δ in (105), imply

$$\frac{\gamma_{12}a_{12}}{R(\gamma_{12}+1)} = \frac{\frac{1}{s}\frac{\gamma_{12}a_{12}}{\gamma_{12}+1}}{\frac{R}{s}} = \frac{\frac{\sqrt{F}}{\cos^2\phi}}{\frac{\sqrt{F}}{\cos^2\phi\sin(\phi+\frac{\delta}{2})}} = \sin(\phi+\tfrac{\delta}{2}) = \sin(\phi+\tfrac{1}{2}\delta_{A_1A_2O}). \tag{116}$$

Finally, (102) and (116) imply

$$\sin(\theta+\tfrac{1}{2}\delta_{A_1A_2A_3}) = \sin(\phi+\tfrac{1}{2}\delta_{A_1A_2O})\,, \tag{117}$$

which, in turn, implies the results of the Theorem in (100) and in (101a) and (101b).

Remark 1 In the proof of Theorem 9, we take for granted that the gyroline through the points A_1 and A_2 has two sides in the gyroplane through the points A_1, A_2, and A_3 so that in Fig. 8 the points A_3 and O lie on the same side of the gyroline while in Fig. 9 these points lie on opposite sides of the gyroline. The need for a careful study of the very intuitive idea that every line in a Euclidean plane has "two sides" was pointed out by Millman and Parker in [10, p. 63], resulting in what they call the *Plane Separation Axiom* (PSA). An analogous *Gyroplane Separation Axiom* (GPSA) for a gyroplane in an Einstein gyrovector space is studied by Sönmez and Ungar in [15].

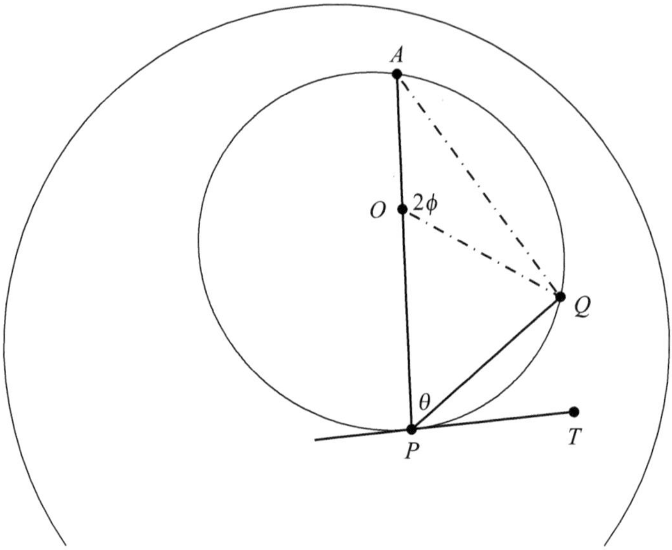

Fig. 10 The circumgyrocircle of gyrotriangle APQ with circumgyrocenter O in an Einstein gyrovector plane. The circumgyrocircle is shown along with (i) its gyrodiameter AP that extends its gyroradius OP and (ii) its gyrotangent gyroline at the gyrotangency point P. When Q approaches P, the gyroline through P and Q approaches the gyrotangent gyroline through P and T. Shown are also an inscribed gyroangle θ and a gyrocentral gyroangle 2ϕ, which are related by the Inscribed Gyroangle Theorem II, Theorem 9, p. 26

11 Gyrocircle Gyrotangent Gyrolines

Employing the Inscribed Gyroangle Theorem II, Theorem 9, p. 26, we prove the following theorem.

Theorem 10 *Let C be a gyrocircle with gyrocenter O and let L_{PT} be a gyrotangent gyroline of C at a tangency point P in an Einstein gyrovector space $(\mathbb{R}_s^n, \oplus, \otimes)$, shown in* Fig. 10. *Then, the gyrotangent gyroline L_{PT} is perpendicular to the gyroradius OP terminating at P.*

Proof The proof is given by the following chain of equations, which are numbered for subsequent derivation. Let Q be a point close to P on gyrocircle C, shown in Fig. 10. Then,

$$
\begin{aligned}
\angle APT &\overset{(1)}{\overbrace{=\!=\!=}} \lim_{Q\to P} \angle APQ \\
&\overset{(2)}{\overbrace{=\!=\!=}} \lim_{Q\to P} \left\{\tfrac{1}{2}\angle AOQ + \tfrac{1}{2}\delta_{AOQ} - \tfrac{1}{2}\delta_{APQ}\right\} \\
&\overset{(3)}{\overbrace{=\!=\!=}} \tfrac{1}{2}\angle AOP + \tfrac{1}{2}\delta_{AOP} - \tfrac{1}{2}\delta_{APP} \\
&\overset{(4)}{\overbrace{=\!=\!=}} \frac{\pi}{2}\,.
\end{aligned}
\qquad (118)
$$

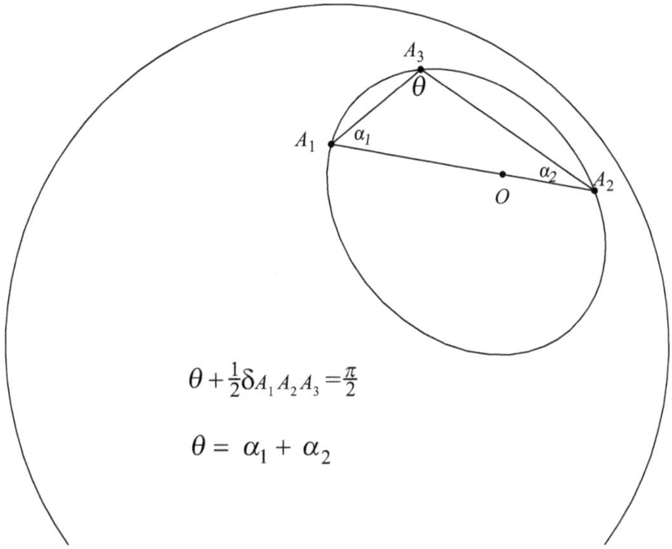

Fig. 11 The semigyrocircle gyrotriangle

Derivation of the numbered equalities in (118) follows:

1. This limit is clear from Fig. 10. Indeed, when Q approaches P, the gyroline through P and Q approaches the gyrotangent gyroline through P and T.
2. Follows from (1) by the Inscribed Gyroangle Theorem II, Theorem 9, p. 26, with $\theta = \angle APQ$.
3. Follows from (2) by obvious limits as $Q \to P$, noting that the defect of a degenerate gyrotriangle (that is, a gyrotriangle whose vertices are gyrocollinear) vanishes.
4. Follows from (3) since $\angle AOP = \pi$.

12 Semigyrocircle Gyrotriangles

In the special case when the gyrocircle gyrochord A_1A_2 in Figs. 8 and 9 is the gyrocircle gyrodiameter, as shown in Fig. 11, we have $2\phi = \pi$ and $\delta_{A_1A_2O} = 0$ so that, by Theorem 9, (101a),

$$\theta + \tfrac{1}{2}\delta_{A_1A_2A_3} = \frac{\pi}{2}\,. \tag{119}$$

By definition, the gyrotriangular defect of gyrotriangle $A_1A_2A_3$ in Fig. 11 is given by the equation

$$\delta_{A_1A_2A_3} = \pi - (\alpha_1 + \alpha_2 + \theta)\,. \tag{120}$$

Hence, by (119) and (120),

$$\theta = \alpha_1 + \alpha_2 \,. \tag{121}$$

In Euclidean geometry triangle, defects vanish, so that (119) reduces to the well known result according to which $\theta = \pi/2$ in Euclidean geometry.

13 The Gyrotangent–Gyrosecant Theorem

A gyrotangent gyroline of a gyrocircle is a gyroline that intersects the gyrocircle in exactly one point. The point of contact is called the point of tangency.

A gyrosecant gyroline of a gyrocircle is a gyroline that intersects the gyrocircle in two different points. The gyrosegment that links these points is a gyrochord of the gyrocircle.

By Theorem 10, p. 30, the gyroradius of a gyrocircle drawn to the point of tangency of a gyrotangent gyroline of the gyrocircle is perpendicular to the gyrotangent gyroline. Accordingly, the gyrotangent gyrosegment A_1P of the gyrocircle in Fig. 12, with the tangency point A_1 is perpendicular to the gyroradius OA_1 drawn from the gyrocircle gyrocenter O to the tangency point A_1. In full analogy with the well-known tangent–secant theorem of Euclidean geometry, we present the gyrotangent–gyrosecant theorem of hyperbolic geometry.

Theorem 11 (The Gyrotangent–Gyrosecant Theorem, I) *Let $A_1, A_2, A_3 \in \mathbb{R}_s^n$ be three distinct points that lie on a gyrocircle in an Einstein gyrovector space $\mathbb{R}_s^n = (\mathbb{R}_s^n, \oplus, \otimes)$ such that the gyrosecant gyroline of the gyrocircle through the points A_2, A_3 and the gyrotangent gyroline of the gyrocircle with the tangency point A_1 share a point P not on the gyrocircle, as shown in* Fig. 12.

Furthermore, let

$$\begin{aligned} d_1 &= \|\ominus A_1 \oplus P\| \\ d_2 &= \|\ominus A_2 \oplus P\| \\ d_3 &= \|\ominus A_3 \oplus P\| \,. \end{aligned} \tag{122}$$

and

$$d_{23} = \|\ominus A_2 \oplus A_3\| = d_3 \ominus d_2 \,. \tag{123}$$

Then

$$\gamma_{d_1}^2 d_1^2 = \frac{2}{\gamma_{d_3 \ominus d_2} + 1} \gamma_{d_2} d_2 \gamma_{d_3} d_3 \,. \tag{124}$$

Proof Using the gyrotriangle index notation (13), p. 7, we have, in particular,

$$\gamma_{23} = \gamma_{\ominus A_2 \oplus A_3} = \gamma_{\|\ominus A_2 \oplus A_3\|} \,. \tag{125}$$

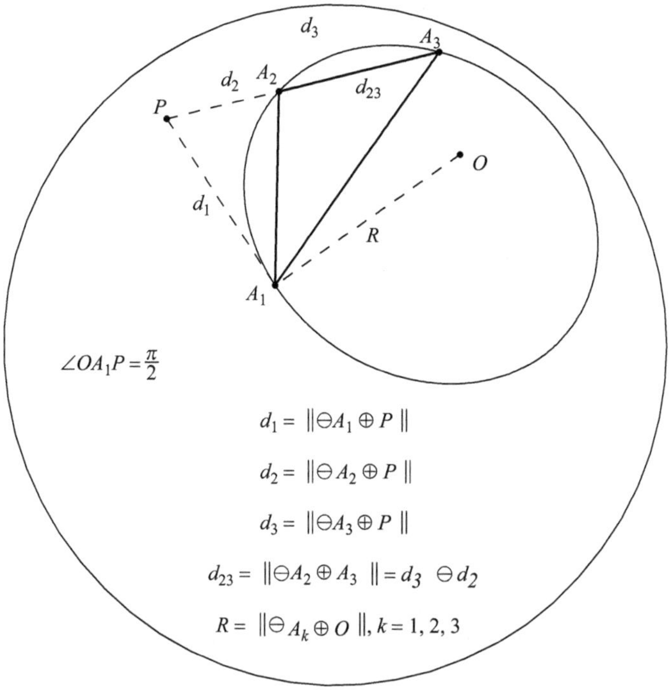

Fig. 12 Illustrating the Gyrotangent–Gyrosecant Theorem 11, a gyrotriangle $A_1A_2A_3$ in an Einstein gyrovector space $(\mathbb{R}^n_s, \oplus, \otimes)$ is shown for $n = 2$, along with its circumgyrocircle and its circumgyrocenter O and circumgyroradius $R = |OA_1|$. The gyrotangent gyrosegment PA_1 is tangent to the circumgyrocircle at the tangency point A_1. Gyrosecant gyrosegment PA_3 intersects the circumgyrocircle at the points A_2 and A_3. Gyrodistances d_k, $k = 1, 2, 3$ between various points that illustrate the gyrotangent–gyrosecant theorem are also shown. Thus, in particular, d_{23} is the gyrodistance between A_2 and A_3. Owing to the gyrotriangle equality $d_3 = d_2 \oplus d_{23}$, so that d_{23} equals the gyrodifference between d_3 and d_2

The points P, A_2, and A_3 are gyrocollinear, as shown in Fig. 12. Hence, by the gyrotriangle equality [26, Sect. 2.4],

$$d_2 \oplus \|\ominus A_2 \oplus A_3\| = d_3 \,, \tag{126}$$

implying

$$\|\ominus A_2 \oplus A_3\| = d_3 \ominus d_2 \,, \tag{127}$$

as stated in the theorem, (123). Hence, by (125) and (127),

$$\gamma_{23} = \gamma_{d_3 \ominus d_2} \,. \tag{128}$$

Let O be the circumgyrocenter of gyrotriangle $A_1A_2A_3$, as shown in Fig. 12. Gyrosegment PA_1 is tangent to the circumgyrocircle of the gyrotriangle at the tangency point A_1. Hence, gyrotriangle PA_1O is right gyroangled, with $\angle PA_1O = \pi/2$, so that, by the Einstein–Pythagoras Identity [26, Eq. (6.57), p. 144],

$$\gamma_{\ominus A_1 \oplus P} \gamma_{\ominus A_1 \oplus O} = \gamma_{\ominus P \oplus O} \cdot \tag{129}$$

The circumgyrocenter O of gyrotriangle $A_1A_2A_3$ in Fig. 12 possesses the gyrobarycentric representation

$$O = \frac{m_{1,O}\gamma_{A_1} A_1 + m_{2,O}\gamma_{A_2} A_2 + m_{3,O}\gamma_{A_3} A_3}{m_{1,O}\gamma_{A_1} + m_{2,O}\gamma_{A_2} + m_{3,O}\gamma_{A_3}} \tag{130}$$

with respect to the gyrobarycentrically independent set $\{A_1, A_2, A_3\}$ where, by (40), p. 12,

$$\begin{aligned} m_{1,O} &= (\gamma_{12} + \gamma_{13} - \gamma_{23} - 1)(\gamma_{23} - 1) \\ m_{2,O} &= (\gamma_{12} - \gamma_{13} + \gamma_{23} - 1)(\gamma_{13} - 1) \\ m_{3,O} &= (-\gamma_{12} + \gamma_{13} + \gamma_{23} - 1)(\gamma_{12} - 1) . \end{aligned} \tag{131}$$

Hence, by the Gyrobarycentric Representation Gyrocovariance Theorem [26, Theorem 4.6, pp. 90–91] with $X = \ominus A_1$, we have the following two Eqs. (132) and (133). The first equation is

$$\begin{aligned} &\ominus A_1 \oplus O \\ &= \ominus A_1 \oplus \frac{m_{1,O}\gamma_{A_1} A_1 + m_{2,O}\gamma_{A_2} A_2 + m_{3,O}\gamma_{A_3} A_3}{m_{1,O}\gamma_{A_1} + m_{2,O}\gamma_{A_2} + m_{3,O}\gamma_{A_3}} \\ &= \frac{m_{1,O}\gamma_{\ominus A_1 \oplus A_1}(\ominus A_1 \oplus A_1) + m_{2,O}\gamma_{\ominus A_1 \oplus A_2}(\ominus A_1 \oplus A_2) + m_{3,O}\gamma_{\ominus A_1 \oplus A_3}(\ominus A_1 \oplus A_3)}{m_{1,O}\gamma_{\ominus A_1 \oplus A_1} + m_{2,O}\gamma_{\ominus A_1 \oplus A_2} + m_{3,O}\gamma_{\ominus A_1 \oplus A_3}} \\ &= \frac{m_{2,O}\gamma_{12}\mathbf{a}_{12} + m_{3,O}\gamma_{13}\mathbf{a}_{13}}{m_{1,O} + m_{2,O}\gamma_{12} + m_{3,O}\gamma_{13}} . \end{aligned} \tag{132}$$

The second equation is

$$\gamma_R = \gamma_{\ominus A_1 \oplus O} = \frac{m_{1,O} + m_{2,O}\gamma_{12} + m_{3,O}\gamma_{13}}{m_O} \tag{133}$$

where $R = \|\ominus A_1 \oplus O\|$ is the circumgyroradius of gyrotriangle $A_1A_2A_3$. Here, the constant $m_O > 0$ of the gyrobarycentric representation (130) of O is given by the equation

$$\begin{aligned} m_O^2 &= m_{1,O}^2 + m_{2,O}^2 + m_{3,O}^2 \\ &\quad + 2(m_{1,O}m_{2,O}\gamma_{12} + m_{1,O}m_{3,O}\gamma_{13} + m_{2,O}m_{3,O}\gamma_{23}) , \end{aligned} \tag{134}$$

as we see from the *Gyrobarycentric representation constant* associated with the Gyrobarycentric Representation Gyrocovariance Theorem [26, Theorem 4.6, pp. 90–91]. There will be no need to use the right-hand side of (134) in the proof.

Let the point P, shown in Fig. 12, be given by its gyrobarycentric representation

$$P = \frac{m_2\gamma_{_{A_2}}A_2 + m_3\gamma_{_{A_3}}A_3}{m_2\gamma_{_{A_2}} + m_3\gamma_{_{A_3}}} \tag{135}$$

with respect to the gyrobarycentrically independent set $\{A_1, A_2\}$, where the gyrobarycentric coordinates m_2 and m_3 are to be determined in (146) below. The gyrobarycentric representation (135) with respect to the set $\{A_1, A_2\}$ exists since the point P lies on the gyroline that passes through the points A_2 and A_3.

By the Gyrobarycentric Representation Gyrocovariance Theorem [26, Theorem 4.6, pp. 90–91] with $X = \ominus A_1$ we have

$$\ominus A_1 \oplus P = \frac{m_2\gamma_{12}\mathbf{a}_{12} + m_3\gamma_{13}\mathbf{a}_{13}}{m_2\gamma_{12} + m_3\gamma_{13}} \tag{136}$$

and

$$\gamma_{\ominus A_1 \oplus P} = \frac{m_2\gamma_{12} + m_3\gamma_{13}}{m_P} . \tag{137}$$

Here, $m_P > 0$ is the constant of the gyrobarycentric representation (135) of P, given by the equation

$$m_P^2 = m_2^2 + m_3^2 + 2m_2m_3\gamma_{23} , \tag{138}$$

as we see from the representation constant associated with the Gyrobarycentric Representation Gyrocovariance Theorem [26, Theorem 4.6, pp. 90–91].

Similarly to (136) and (137), we have

$$\begin{aligned} \ominus A_2 \oplus P &= \frac{m_3\gamma_{23}\mathbf{a}_{23}}{m_2 + m_3\gamma_{23}} \\ \ominus A_3 \oplus P &= \frac{m_2\gamma_{23}\mathbf{a}_{32}}{m_2\gamma_{23} + m_3} \end{aligned} \tag{139}$$

and

$$\begin{aligned} \gamma_{\ominus A_2 \oplus P} &= \frac{m_2 + m_3\gamma_{23}}{m_P} \\ \gamma_{\ominus A_3 \oplus P} &= \frac{m_2\gamma_{23} + m_3}{m_P} . \end{aligned} \tag{140}$$

Following (130), we have, by the Gyrobarycentric Representation Gyrocovariance Theorem [26, Theorem 4.6, pp. 90–91],

$$\ominus P \oplus O = \frac{m_{1,O}\gamma_{\ominus A_1 \oplus P}(\ominus P \oplus A_1) + m_{2,O}\gamma_{\ominus A_2 \oplus P}(\ominus P \oplus A_2) + m_{3,O}\gamma_{\ominus A_3 \oplus P}(\ominus P \oplus A_3)}{m_{1,O}\gamma_{\ominus A_1 \oplus P} + m_{2,O}\gamma_{\ominus A_2 \oplus P} + m_{3,O}\gamma_{\ominus A_3 \oplus P}}, \tag{141}$$

noting that $\gamma_{\ominus P\oplus A_k} = \gamma_{\ominus A_k\oplus P}$, $k = 1, 2, 3$.

Hence, by a gamma factor identity of the Gyrobarycentric Representation Gyrocovariance Theorem [26, Theorem 4.6, pp. 90–91], applied to (141), and by (137) and (139), we have

$$\begin{aligned}\gamma_{\ominus P\oplus O} &= \frac{m_{1,O}\gamma_{\ominus A_1\oplus P} + m_{2,O}\gamma_{\ominus A_2\oplus P} + m_{3,O}\gamma_{\ominus A_3\oplus P}}{m_O} \\ &= \frac{m_{1,O}(m_2\gamma_{12} + m_3\gamma_{13}) + m_{2,O}(m_2 + m_3\gamma_{23}) + m_{3,O}(m_2\gamma_{23} + m_3)}{m_O m_P}\,.\end{aligned} \tag{142}$$

Substituting (133), (137), and (142) into the Einstein–Pythagoras Identity (129), we obtain the equation

$$\begin{aligned}&m_{1,O}(m_2\gamma_{12} + m_3\gamma_{13}) + m_{2,O}(m_2 + m_3\gamma_{23}) + m_{3,O}(m_2\gamma_{23} + m_3) \\ &= (m_2\gamma_{12} + m_3\gamma_{13})(m_{1,O} + m_{2,O}\gamma_{12} + m_{3,O}\gamma_{13})\,.\end{aligned} \tag{143}$$

The latter, in turn, can be written as

$$\begin{aligned}0 &= m_2\{m_{2,O}(\gamma_{12}^2 - 1) + m_{3,O}(\gamma_{12}\gamma_{13} - \gamma_{23})\} \\ &\quad + m_3\{m_{2,O}(\gamma_{12}\gamma_{13} - \gamma_{23}) + m_{3,O}(\gamma_{13}^2 - 1)\}\,.\end{aligned} \tag{144}$$

Solving (144) for m_2 and m_3, we obtain the equations

$$\begin{aligned}m_2 &= K\{m_{2,O}(\gamma_{12}\gamma_{13} - \gamma_{23}) + m_{3,O}(\gamma_{13}^2 - 1)\} \\ m_3 &= -K\{m_{2,O}(\gamma_{12}^2 - 1) + m_{3,O}(\gamma_{12}\gamma_{13} - \gamma_{23})\}\end{aligned} \tag{145}$$

for any nonzero factor K.

Owing to the homogeneity of gyrobarycentric coordinates, the factor K is irrelevant. Hence, we select $K = 1$, obtaining the equations

$$\begin{aligned}m_2 &= m_{2,O}(\gamma_{12}\gamma_{13} - \gamma_{23}) + m_{3,O}(\gamma_{13}^2 - 1) \\ m_3 &= -m_{2,O}(\gamma_{12}^2 - 1) - m_{3,O}(\gamma_{12}\gamma_{13} - \gamma_{23})\,,\end{aligned} \tag{146}$$

where $m_{2,O}$ and $m_{3,O}$ are given by (131).

Equation (146) for the gyrobarycentric coordinates m_2 and m_3 of the point P in (135) results from the Einstein–Pythagoras Identity (129). Hence, they insure that gyrosegments A_1P and A_1O are perpendicular to each other, as desired.

By the first equation in (122), and by (4), p. 3, we have

$$\frac{1}{s^2}\gamma_{d_1}^2 d_1^2 = \gamma_{d_1}^2 - 1 = \gamma_{\ominus A_1\oplus P}^2 - 1\,, \tag{147}$$

where the gamma factor $\gamma_{\ominus A_1\oplus P}$ is given by (137), (138), and (146).

Substituting (137) into (147), we have

$$\frac{1}{s^2}\gamma_{d_1}^2 d_1^2 = \frac{(m_2\gamma_{12} + m_3\gamma_{13})^2 - m_P^2}{m_P^2}, \tag{148}$$

where m_2, m_3, and m_P^2 are given by (146) and (138).

Hence, by (146), (148) and by straightforward algebra,

$$\frac{1}{s^2}\gamma_{d_1}^2 d_1^2 = \frac{2}{s^2 D^2 m_P^2}(\gamma_{12} - 1)(\gamma_{13} - 1)(\gamma_{23} - 1), \tag{149}$$

where

$$D = \gamma_{12}^2 + \gamma_{13}^2 + \gamma_{23}^2 - 2\gamma_{12}\gamma_{13}\gamma_{23}. \tag{150}$$

Similarly to (147), by (122) and (4), p. 3, we have

$$\frac{1}{s^2}\gamma_{d_2}^2 d_2^2 = \gamma_{d_2}^2 - 1 = \gamma_{\ominus A_2 \oplus P}^2 - 1, \tag{151}$$

where the gamma factor $\gamma_{\ominus A_2 \oplus P}$ is given by the first equation in (140) and by (138) and (146).

Substituting the first equation in (140) into (151), we have

$$\frac{1}{s^2}\gamma_{d_2}^2 d_2^2 = \frac{(m_2 + m_3\gamma_{23})^2 - m_P^2}{m_P^2}, \tag{152}$$

where m_2, m_3, and m_P^2 are given by (146) and (138).

Hence, by (148), (152), and straightforward algebra,

$$\frac{1}{s^2}\gamma_{d_2}^2 d_2^2 = \frac{1}{s^2 D^2 m_P^2}(\gamma_{12} - 1)^2(\gamma_{23}^2 - 1). \tag{153}$$

Similarly to (147) and (151), by (122) and (4), p. 3, we have

$$\frac{1}{s^2}\gamma_{d_3}^2 d_3^2 = \gamma_{d_3}^2 - 1 = \gamma_{\ominus A_3 \oplus P}^2 - 1, \tag{154}$$

where the gamma factor $\gamma_{\ominus A_3 \oplus P}$ is given by the second equation in (140), and by (138) and (146).

Substituting the second equation in (140) into (154), we have

$$\frac{1}{s^2}\gamma_{d_3}^2 d_3^2 = \frac{(m_2\gamma_{23} + m_3)^2 - m_P^2}{m_P^2}, \tag{155}$$

where m_2, m_3, and m_P^2 are given by (146) and (138).

Hence, by (146), (155), and straightforward algebra,

$$\frac{1}{s^2}\gamma_{d_3}^2 d_3^2 = \frac{1}{s^2 D^2 m_P^2}(\gamma_{13} - 1)^2(\gamma_{23}^2 - 1). \tag{156}$$

Finally, following (149), (153), (156), and straightforward algebra, noting (128), we have

$$\frac{(\gamma_{d_1}^2 d_1^2)^2}{\gamma_{d_2}^2 d_2^2 \gamma_{d_3}^2 d_3^2} = \left(\frac{2}{\gamma_{23}+1}\right)^2 = \left(\frac{2}{\gamma_{d_3 \ominus d_2}+1}\right)^2, \tag{157}$$

thus verifying the result (124) of the theorem.

In order to restate Theorem 11 in a way that emphasizes analogies with its Euclidean counterpart, we introduce the notation

$$|AB| := \|\ominus A \oplus B\| = \|\ominus B \oplus A\| \qquad \text{(Hyperbolic Geometry)} \tag{158}$$

for the gyrodistance between points A and B of an Einstein gyrovector space $(\mathbb{R}_s^n, \oplus, \otimes)$. Accordingly, $|AB|$ is the gyrolength of gyrosegment AB.

In order to emphasize analogies we use, ambiguously, the same notation in the context of Euclidean geometry as well, that is,

$$|AB| := \|-A+B\| = \|-B+A\| \qquad \text{(Euclidean Geometry)} \tag{159}$$

is the distance between points A and B of a Euclidean vector space $\mathbb{R}^n$. Accordingly, $|AB|$ is the length of segment AB. It should always be clear from the context whether $|AB|$ represents the gyrolength of a gyrosegment in hyperbolic geometry, or the length of a segment in Euclidean geometry.

Using the notation in (158) and (159), we restate Theorem 11 as follows, noting that by (127) and (122),

$$d_3 \ominus d_2 = |PA_3| \ominus |PA_2| = |A_2 A_3|\,. \tag{160}$$

Theorem 12 (The Gyrotangent–Gyrosecant Theorem, II) *If a gyrotangent of a gyrocircle from an external point P meets the gyrocircle at A_1, and a gyrosecant from P meets the gyrocircle at A_2 and A_3, as shown in* Fig. 12, *then*

$$\gamma_{|PA_1|}^2 |PA_1|^2 = \frac{2}{\gamma_{|A_2 A_3|}+1} \gamma_{|PA_2|} |PA_2| \gamma_{|PA_3|} |PA_3|. \tag{161}$$

In the Euclidean limit, $s \to \infty$, gyrolengths of gyrosegments tend to lengths of corresponding segments and gamma factors tend to 1. Hence, in that limit, the Gyrotangent–Gyrosecant Theorem 12 reduces to the following well-known Tangent–Secant Theorem of Euclidean geometry:

Theorem 13 (The Tangent–Secant Theorem) *If a tangent of a circle from an external point P meets the circle at A_1, and a secant from P meets the circle at A_2 and A_3, then*

$$|PA_1|^2 = |PA_2||PA_3|. \tag{162}$$

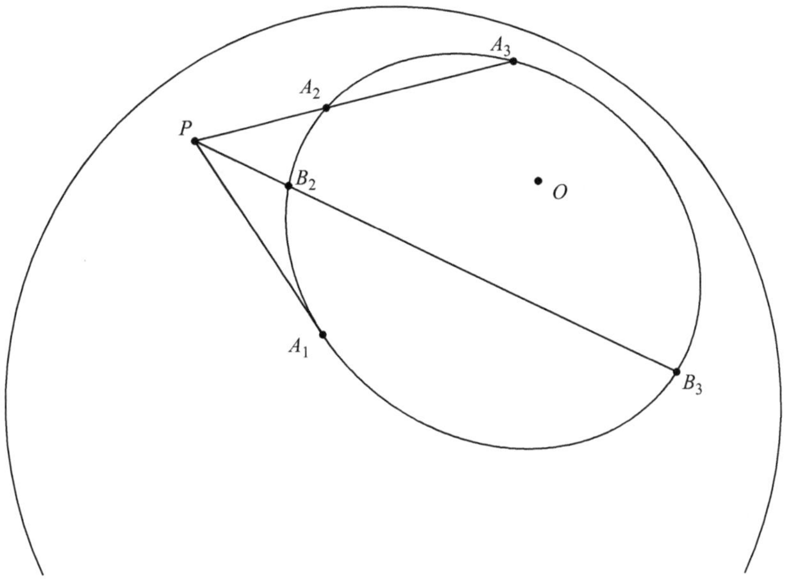

Fig. 13 Illustrating the Intersecting Gyrosecants Theorem 14, two intersecting gyrosecants PA_3 and PB_3 of a gyrocircle are shown. They, respectively, intersect the gyrocircle at the points A_2, A_3 and at the points B_2, B_3

14 The Intersecting Gyrosecants Theorem

As an obvious corollary of the Gyrotangent–Gyrosecant Theorem 12, we have the following theorem for intersecting gyrosecants of a gyrocircle:

Theorem 14 (The Intersecting Gyrosecants Theorem) *If two gyrosecants of a gyrocircle in an Einstein gyrovector space* $(\mathbb{R}_s^n, \oplus, \otimes)$, *drawn to the gyrocircle from an external point* P, *meet the gyrocircle at points* A_2, A_3 *and at points* B_2, B_3, *respectively, as shown in* Fig. 13, *then*

$$\frac{\gamma_{|PA_2|}|PA_2|\gamma_{|PA_3|}|PA_3|}{\gamma_{|A_2A_3|}+1} = \frac{\gamma_{|PB_2|}|PB_2|\gamma_{|PB_3|}|PB_3|}{\gamma_{|B_2B_3|}+1}. \tag{163}$$

Proof Let PA_1 be a gyrotangent gyrosegment of the gyrocircle drawn from P and meeting the gyrocircle at A_1, as shown in Fig. 13. Then, by Theorem 12, each of the two sides of (163) equals half the left-hand side of (161), thus verifying (163).

In the Euclidean limit, $s \to \infty$, gyrolengths of gyrosegments tend to lengths of corresponding segments and gamma factors tend to 1. Hence, in that limit, the Intersecting Gyrosecants Theorem 14 of hyperbolic geometry reduces to the following well-known Intersecting Secants Theorem of Euclidean geometry:

Theorem 15 (The Intersecting Secants Theorem) *If two secants of a circle in a Euclidean vector space* $\mathbb{R}^n$, *drawn to the circle from an external point* P, *meet the* *circle at points A_2, A_3 and at points B_2, B_3, respectively, then*

$$|PA_2||PA_3| = |PB_2||PB_3|. \tag{164}$$

15 Gyrocircle Gyrobarycentric Representation

Let $A_1A_2A_3$ be a gyrotriangle that possesses a circumgyrocircle in an Einstein gyrovector space $(\mathbb{R}^n_s, \oplus, \otimes)$. The circumgyrocenter O of gyrotriangle $A_1A_2A_3$, shown in Figs. 13 and 14, possesses the gyrobarycentric representation

$$O = \frac{m_{1,O}\gamma_{A_1}A_1 + m_{2,O}\gamma_{A_2}A_2 + m_{3,O}\gamma_{A_3}A_3}{m_{1,O}\gamma_{A_1} + m_{2,O}\gamma_{A_2} + m_{3,O}\gamma_{A_3}} \tag{165}$$

with respect to the gyrobarycentrically independent set $\{A_1, A_2, A_3\}$ of the reference gyrotriangle vertices where, by (40), p. 12,

$$\begin{aligned} m_{1,O} &= (\gamma_{12} + \gamma_{13} - \gamma_{23} - 1)(\gamma_{23} - 1) \\ m_{2,O} &= (\gamma_{12} - \gamma_{13} + \gamma_{23} - 1)(\gamma_{13} - 1) \\ m_{3,O} &= (-\gamma_{12} + \gamma_{13} + \gamma_{23} - 1)(\gamma_{12} - 1), \end{aligned} \tag{166}$$

as in (130) and (131).

The constant $m_O > 0$ of the gyrobarycentric representation (165) of O is given by (25) and (26), p. 9,

$$\begin{aligned} m_O^2 &= m_{1,O}^2 + m_{2,O}^2 + m_{3,O}^2 \\ &\quad + 2(m_{1,O}m_{2,O}\gamma_{12} + m_{1,O}m_{3,O}\gamma_{13} + m_{2,O}m_{3,O}\gamma_{23}) \\ &= D_3(D_3 - H_3). \end{aligned} \tag{167}$$

With $\mathbb{A}_3 = A_1 \oplus \mathrm{Span}\{\ominus A_1 \oplus A_2, \ominus A_1 \oplus A_3\} \subset \mathbb{R}^n$, let A be a generic point in $\mathbb{A}_3 \cap \mathbb{R}^n_s$, given by its gyrobarycentric representation

$$A = \frac{m_1\gamma_{A_1}A_1 + m_2\gamma_{A_2}A_2 + m_3\gamma_{A_3}A_3}{m_1\gamma_{A_1} + m_2\gamma_{A_2} + m_3\gamma_{A_3}} \tag{168}$$

with respect to the gyrobarycentrically independent set $S = \{A_1, A_2, A_3\}$. A relationship between the gyrobarycentric coordinates m_1, m_2, and m_3 of A is to be determined in (183) below by the condition that the point A lies on the circumgyrocircle of gyrotriangle $A_1A_2A_3$, as shown in Fig. 14.

By the Gyrobarycentric Representation Gyrocovariance Theorem [26, Theorem 4.6, pp. 90–91] with $X = \ominus A$ we have,

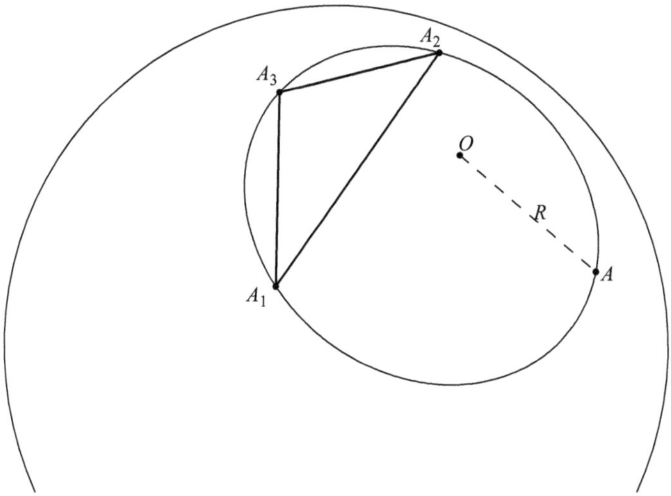

Fig. 14 A generic point A on the circumgyrocircle $C(A_1A_2A_3)$ of a gyrotriangle $A_1A_2A_3$ in an Einstein gyrovector space $(\mathbb{R}_s^n, \oplus, \otimes)$. The point A lies on the circumgyrocircle $C(A_1A_2A_3)$ if and only if it possesses the gyrobarycentric representation (168) with gyrobarycentric coordinates that satisfy (183)

$$\begin{aligned}
&\ominus A \oplus O \\
&= \ominus A \oplus \frac{m_{1,O}\gamma_{A_1} A_1 + m_{2,O}\gamma_{A_2} A_2 + m_{3,O}\gamma_{A_3} A_3}{m_{1,O}\gamma_{A_1} + m_{2,O}\gamma_{A_2} + m_{3,O}\gamma_{A_3}} \\
&= \frac{m_{1,O}\gamma_{\ominus A_1 \oplus A}(\ominus A \oplus A_1) + m_{2,O}\gamma_{\ominus A_2 \oplus A}(\ominus A \oplus A_2) + m_{3,O}\gamma_{\ominus A_3 \oplus A}(\ominus A \oplus A_3)}{m_{1,O}\gamma_{\ominus A_1 \oplus A} + m_{2,O}\gamma_{\ominus A_2 \oplus A} + m_{3,O}\gamma_{\ominus A_3 \oplus A}}
\end{aligned} \tag{169}$$

and

$$\gamma_d = \gamma_{\ominus A \oplus O} = \frac{m_{1,O}\gamma_{\ominus A_1 \oplus A} + m_{2,O}\gamma_{\ominus A_2 \oplus A} + m_{3,O}\gamma_{\ominus A_3 \oplus A}}{m_O}, \tag{170}$$

where $d = \|\ominus A \oplus O\|$ is the gyrodistance from A to O, and where the constant $m_O > 0$ of the gyrobarycentric representation (165) of O is given by (167).

We will now calculate the gamma factors $\gamma_{\ominus A_k \oplus A}$, $k = 1, 2, 3$, that appear in (170).

Applying the Gyrobarycentric Representation Gyrocovariance Theorem [26, Theorem 4.6, pp. 90–91] with $X = \ominus A_1$ to the gyrobarycentric representation (168) of A, we have

$$\ominus A_1 \oplus A = \ominus A_1 \oplus \frac{m_1\gamma_{A_1} A_1 + m_2\gamma_{A_2} A_2 + m_3\gamma_{A_3} A_3}{m_1\gamma_{A_1} + m_2\gamma_{A_2} + m_3\gamma_{A_3}}$$

$$= \frac{m_1\gamma_{\ominus A_1\oplus A_1}(\ominus A_1\oplus A_1) + m_2\gamma_{\ominus A_1\oplus A_2}(\ominus A_1\oplus A_2) + m_3\gamma_{\ominus A_1\oplus A_3}(\ominus A_1\oplus A_3)}{m_1\gamma_{\ominus A_1\oplus A_1} + m_2\gamma_{\ominus A_1\oplus A_2} + m_3\gamma_{\ominus A_1\oplus A_3}}$$

$$= \frac{m_2\gamma_{12}\mathbf{a}_{12} + m_3\gamma_{13}\mathbf{a}_{13}}{m_1 + m_2\gamma_{12} + m_3\gamma_{13}} \tag{171}$$

and

$$\gamma_{\ominus A_1\oplus A} = \frac{m_1 + m_2\gamma_{12} + m_3\gamma_{13}}{m_A}, \tag{172}$$

where $m_A > 0$ is the constant of the gyrobarycentric representation (168) of A, given by the equation

$$\begin{aligned} m_A^2 &= m_1^2 + m_2^2 + m_3^2 + 2(m_1m_2\gamma_{12} + m_1m_3\gamma_{13} + m_2m_3\gamma_{23}) \\ &= (m_1 + m_2 + m_3)^2 \\ &\quad + 2\{m_1m_2(\gamma_{12} - 1) + m_1m_3(\gamma_{13} - 1) + m_2m_3(\gamma_{23} - 1)\}. \end{aligned} \tag{173a}$$

It proves useful to write (173a) as

$$m_A^2 = (m_1 + m_2 + m_3)^2 + 2\,K(A; A_1, A_2, A_3), \tag{173b}$$

where

$$K = K(A; A_1, A_2, A_3) = \sum_{\substack{i,j=1 \\ i<j}}^{3} m_i m_j(\gamma_{ij} - 1), \tag{173c}$$

where m_k, $k = 1, 2, 3$, are the gyrobarycentric coordinates of A in the representation (168) of A with respect to the vertices of gyrotriangle $A_1A_2A_3$ and $\gamma_{ij} = \|\ominus A_i\oplus A_j\|$.

Similarly to (171) and (172), we have

$$\ominus A_2\oplus A = \frac{m_1\gamma_{12}\mathbf{a}_{21} + m_3\gamma_{23}\mathbf{a}_{23}}{m_1\gamma_{12} + m_2 + m_3\gamma_{23}} \tag{174}$$

and

$$\gamma_{\ominus A_2\oplus A} = \frac{m_1\gamma_{12} + m_2 + m_3\gamma_{23}}{m_A}. \tag{175}$$

Similarly to (171) and (172), and (174) and (175), we have

$$\ominus A_3\oplus A = \frac{m_1\gamma_{13}\mathbf{a}_{31} + m_2\gamma_{23}\mathbf{a}_{32}}{m_1\gamma_{13} + m_2\gamma_{23} + m_3} \tag{176}$$

and

$$\gamma_{\ominus A_3 \oplus A} = \frac{m_1\gamma_{13} + m_2\gamma_{23} + m_3}{m_A}. \tag{177}$$

Substituting the gamma factors (172), (175), and (177), as well as (166) and (167), into (170), we obtain the elegant equation

$$\gamma_d^2 = \frac{D_3}{D_3 - H_3}\left(\frac{m_1 + m_2 + m_3}{m_A}\right)^2, \tag{178}$$

where, as in (28) and (29), p. 10,

$$\begin{aligned} D_3 &= 1 + 2\gamma_{12}\gamma_{13}\gamma_{23} - \gamma_{12}^2 - \gamma_{13}^2 - \gamma_{23}^2 \\ &= 2\left\{(\gamma_{12} - 1)(\gamma_{13} - 1) + (\gamma_{12} - 1)(\gamma_{23} - 1) + (\gamma_{13} - 1)(\gamma_{23} - 1)\right. \\ &\quad \left. +(\gamma_{12} - 1)(\gamma_{13} - 1)(\gamma_{23} - 1)\right\} \\ &\quad - \left\{(\gamma_{12} - 1)^2 + (\gamma_{13} - 1)^2 + (\gamma_{23} - 1)^2\right\} \qquad (179a) \\ &= \begin{vmatrix} 1 & \gamma_{12} & \gamma_{13} \\ \gamma_{12} & 1 & \gamma_{23} \\ \gamma_{13} & \gamma_{23} & 1 \end{vmatrix} \end{aligned}$$

and

$$H_3 = 2(\gamma_{12} - 1)(\gamma_{13} - 1)(\gamma_{23} - 1). \tag{179b}$$

As expected, the denominator $D_3 - H_3$ in (178) is positive, by the circumgyrocircle existence condition (30c), p. 10.

The point A in (168) lies on the circumgyrocircle of gyrotriangle $A_1A_2A_3$ (Fig. 14) if and only if $d = R$, or, equivalently, if and only if

$$\gamma_d^2 = \gamma_R^2. \tag{180}$$

Inserting into (180), γ_d^2 from (178) and γ_R^2 from (51), p. 15, we obtain the equation

$$\frac{D_3}{D_3 - H_3}\left(\frac{m_1 + m_2 + m_3}{m_A}\right)^2 = \frac{D_3}{D_3 - H_3}, \tag{181}$$

implying

$$m_A^2 = (m_1 + m_2 + m_3)^2. \tag{182}$$

The latter, in turn, is valid if and only

$$K(A; A_1, A_2, A_3) := \sum_{\substack{i,j=1 \\ i<j}}^{3} m_i m_j(\gamma_{ij} - 1) = 0, \tag{183}$$

as we see from (173). Equation (183) expresses the *circumgyrocircle condition* in the sense that it provides a necessary and sufficient condition that the point A in (168) lies on the circumgyrocircle $C(A_1A_2A_3)$ of gyrotriangle $A_1A_2A_3$.

A gyrotrigonometric version of the circumgyrocircle condition (183) follows from [26, Eq. (7.149), p. 188],

$$\begin{aligned} m_1m_2 \sin\alpha_3 \sin(\alpha_3+\frac{\delta}{2}) + m_1m_3 \sin\alpha_2 \sin(\alpha_2+\frac{\delta}{2}) \\ + m_2m_3 \sin\alpha_1 \sin(\alpha_1+\frac{\delta}{2}) = 0\,. \end{aligned} \tag{184}$$

In order to parameterize the points of the circumgyrocircle $C(A_1A_2A_3)$ of a gyrotriangle $A_1A_2A_3$ in an Einstein gyrovector space $\mathbb{R}^n_s$ by a real parameter, we assume $m_3 \neq 0$, so that the circumgyrocircle condition (183) can be written as

$$\frac{m_1}{m_3}\frac{m_2}{m_3}(\gamma_{12}-1) + \frac{m_1}{m_3}(\gamma_{13}-1) + \frac{m_2}{m_3}(\gamma_{23}-1) = 0\,. \tag{185}$$

Selecting $m_2/m_3 = t$ as a parameter on the *extended real line*, $t \in \mathbb{R} \cup \{-\infty, \infty\}$, a system of parametric gyrobarycentric coordinates of the point A in (168) with respect to the gyrobarycentrically independent set $\{A_1, A_2, A_3\}$ that obeys the circumgyrocircle condition (183) is obtained, given by

$$\begin{aligned} m_1 &= -(\gamma_{23}-1)t \\ m_2 &= (\gamma_{12}-1)t^2 + (\gamma_{13}-1)t = m_3t \\ m_3 &= (\gamma_{12}-1)t + (\gamma_{13}-1)\,. \end{aligned} \tag{186}$$

Several points A of the circumgyrocircle $C(A_1A_2A_3)$ of a gyrotriangle $A_1A_2A_3$, which are given by (168) and (186) and which correspond to the circumgyrocircle parameter values $t = -8, -7, \ldots, -1, 0, 1, \ldots, 7, 8$, are shown in Fig. 15. The parameter values $t = -\infty$ and $t = \infty$ correspond jointly to the point A_2.

The two special values, $t = \pm\infty$, of the parameter t are identified in the sense that they correspond jointly to the same point, A_2, on circumgyrocircle $C(A_1A_2A_3)$, as shown in Fig. 15. This identification suggests the replacement of the parameter $t \in \mathbb{R} \cup \{-\infty, \infty\}$ by the parameter θ, $-\pi \le \theta \le \pi$, according to the bijective (one-to-one) relation

$$t = \tan\frac{\theta}{2}\,, \tag{187}$$

for which a corresponding identification is built in naturally. Indeed, the two values $\theta = \pm\pi$ correspond to the two values $t = \pm\infty$ and, accordingly, $\theta = -\pi$ and $\theta = \pi$ are identified.

Inserting t from (187) in (186), and noting that $\tan\frac{\theta}{2} = \sin\frac{\theta}{2}/\cos\frac{\theta}{2}$, and that gyrobarycentric coordinates are homogeneous, we obtain the following system of parametric gyrobarycentric coordinates of the point A in (168), p. 41, with respect to the set S,

$$m_1 = -\,(\gamma_{23}-1)\sin\frac{\theta}{2}\cos\frac{\theta}{2}$$

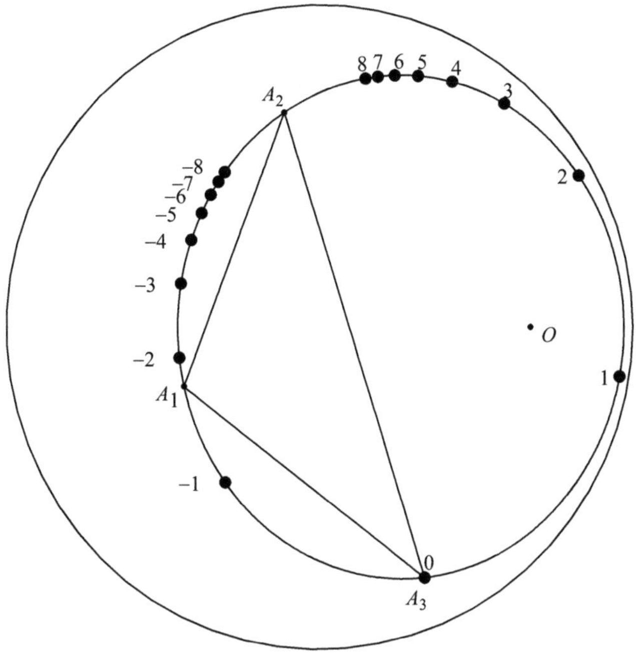

Fig. 15 Several points on the circumgyrocircle $C(A_1A_2A_3)$ of a gyrotriangle $A_1A_2A_3$ in an Einstein gyrovector space $(\mathbb{R}^n_s, \oplus, \otimes)$, $n \geq 2$, determined by several values of the gyrocircle parameter t in (206), $t = -8, -7, \ldots, -1, 0, 1, \ldots, 7, 8$, are presented. Clearly, the parameter value $t = 0$ corresponds to gyrobarycentric coordinates $m_1 = m_2 = 0$, $m_3 \neq 0$, as evidenced from (186). The latter, in turn, corresponds to the point $A = A_3$ on the circumgyrocircle, as shown here and as evidenced from (168). When $t \to \pm\infty$, corresponding points on the circumgyrocircle tend to A_2, as indicated here and as evidenced from (186), where $|m_1|, |m_3| \ll |m_2|$ for large $|t|$. The point A_1 on the circumgyrocircle corresponds to the parameter value $t = -(\gamma_{13} - 1)/(\gamma_{12} - 1)$

$$
\begin{aligned}
m_2 =& (\gamma_{12} - 1)\sin^2\frac{\theta}{2} + (\gamma_{13} - 1)\sin\frac{\theta}{2}\cos\frac{\theta}{2} \\
m_3 =& (\gamma_{12} - 1)\sin\frac{\theta}{2}\cos\frac{\theta}{2} + (\gamma_{13} - 1)\cos^2\frac{\theta}{2}\,.
\end{aligned}
\tag{188}
$$

We now apply to (188) well-known trigonometric half-angle identities, noting that gyrobarycentric coordinates are homogeneous, and translate the parameter interval by π, from the interval $[-\pi, \pi]$ to the interval $[0, 2\pi]$, obtaining the following elegant parametric gyrobarycentric coordinates of A in (168), p. 41.

$$
\begin{aligned}
m_1 =& -(\gamma_{23} - 1)\sin\theta \\
m_2 =& \ \ (\gamma_{13} - 1)\sin\theta + (\gamma_{12} - 1)(1 - \cos\theta) \\
m_3 =& \ \ (\gamma_{12} - 1)\sin\theta + (\gamma_{13} - 1)(1 + \cos\theta)\,,
\end{aligned}
\tag{189}
$$

$0 \leq \theta \leq 2\pi$. The two values of the parameter θ, $\theta = 0$ and $\theta = 2\pi$, are identified, as indicated in Fig. 16, where several points A on the circumference of circumgyrocircle

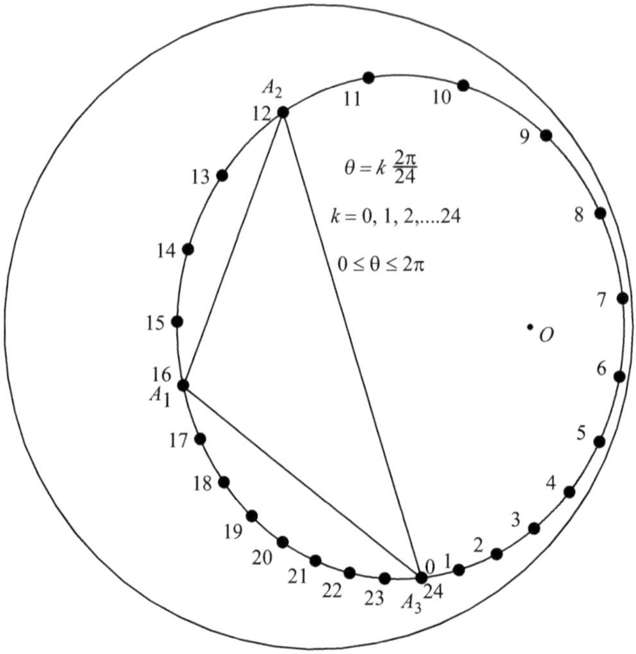

Fig. 16 Several points on the circumgyrocircle $C(A_1A_2A_3)$ of a gyrotriangle $A_1A_2A_3$ in an Einstein gyrovector space $(\mathbb{R}^n_s, \oplus, \otimes)$, $n \geq 2$, determined by several values of the circumgyrocircle parameter θ in (189), $\theta = k(2\pi/24)$, $k = 0, 1, 2, \ldots, 24$, are presented. The parameter values that correspond to $k = 0$ ($\theta = 0$) and $k = 24$ ($\theta = 2\pi$) represent jointly the point A_3 and are, accordingly, identified. The parameter value that corresponds to $k = 12$ ($\theta = \pi$) represents the point A_2

$C(A_1A_2A_3)$, which correspond to several values of the circumgyrocircle parameter θ, $0 \leq \theta \leq 2\pi$, are presented.

16 Gyrocircle Interior and Exterior Points

We are now in the position to present a definition followed by two theorems that characterize the points of the circumgyrocircle of a given gyrotriangle, as well as interior and exterior points of the circumgyrocircle in Einstein gyrovector spaces.

Definition 4 (Gyrocircle Interior and Exterior Points) *Let $A_1A_2A_3$ be a gyrotriangle that possesses a circumgyrocircle $C(A_1A_2A_3)$ with circumgyrocenter O and circumgyroradius R in an Einstein gyrovector space $(\mathbb{R}^n_s, \oplus, \otimes)$. Let P be a generic point in $\mathbb{A}_3 \cap \mathbb{R}^n_s$,*

$$P = \frac{m_1\gamma_{A_1}A_1 + m_2\gamma_{A_2}A_2 + m_3\gamma_{A_3}A_3}{m_1\gamma_{A_1} + m_2\gamma_{A_2} + m_3\gamma_{A_3}} \in \mathbb{A}_3 \cap \mathbb{R}^n_s\,, \tag{190}$$

where, as in (8), *p.* 4,

$$\mathbb{A}_3 = A_1 \oplus \mathrm{Span}\{\ominus A_1 \oplus A_2, \ominus A_1 \oplus A_3\} \subset \mathbb{R}^n\,, \tag{191}$$

and let $d = \|\ominus P \oplus O\|$ *be the gyrodistance of* P *from* O. *Then,* P *lies*

1. *in the interior of* $C(A_1A_2A_3)$ *if* $d < R$;
2. *in the exterior of* $C(A_1A_2A_3)$ *if* $d > R$; *and*
3. *on* $C(A_1A_2A_3)$ *if* $d = R$.

Theorem 16 (Point to Circumgyrocenter Gyrodistance) *Let* $A_1A_2A_3$ *be a gyrotriangle that possesses a circumgyrocircle* $C(A_1A_2A_3)$ *with circumgyrocenter* O *and circumgyroradius* R *in an Einstein gyrovector space* $(\mathbb{R}^n_s, \oplus, \otimes)$, *and let* $A \in \mathbb{A}_3 \cap \mathbb{R}^n_s$ *be a point given by its gyrobarycentric representation*

$$A = \frac{m_1\gamma_{A_1} A_1 + m_2\gamma_{A_2} A_2 + m_3\gamma_{A_3} A_3}{m_1\gamma_{A_1} + m_2\gamma_{A_2} + m_3\gamma_{A_3}} \tag{192}$$

with respect to the gyrobarycentrically independent set $S = \{A_1, A_2, A_3\}$. *Furthermore, let*

$$d = \|\ominus A \oplus O\| \tag{193}$$

be the gyrodistance from A *to* O.

Then,

$$d = \sqrt{1 - \frac{2\,s^2\,K}{M^2\gamma_R^2 R^2}}\,R\,, \tag{194}$$

where

$$K = \sum_{\substack{i,j=1\\ i<j}}^{3} m_i m_j(\gamma_{ij} - 1) \tag{195}$$

and

$$M = \sum_{k=1}^{3} m_k\,. \tag{196}$$

Proof Following (i) (178), p. 43, (ii) (51), p. 15, and (iii) (173), p. 42, we have

$$\begin{aligned}\gamma_d^2 &= \frac{D_3}{D_3 - H_3}\frac{M^2}{m_A^2}\\ \gamma_R^2 &= \frac{D_3}{D_3 - H_3}\end{aligned} \tag{197}$$

$$m_A^2 = M^2 + 2\,K$$

where D_3 and H_3 are given by (179).

Hence, by (197) and (4), p. 3, or (53), p. 15,

$$R^2 = s^2 \frac{\gamma_R^2 - 1}{\gamma_R^2} = s^2 \frac{H_3}{D_3} \tag{198}$$

and

$$d^2 = s^2 \frac{\gamma_d^2 - 1}{\gamma_d^2} = s^2 \frac{M^2 H_3 - 2\,K(D_3 - H_3)}{M^2 D_3}\,, \tag{199}$$

so that, by straightforward algebra,

$$\frac{d^2}{R^2} = 1 - \frac{2\,s^2\,K}{M^2 \gamma_R^2 R^2}\,, \tag{200}$$

as desired.

Theorem 17 (Gyrocircle Gyrobarycentric Representation) *Let $A_1A_2A_3$ be a gyrotriangle that possesses a circumgyrocircle $C(A_1A_2A_3)$ with circumgyrocenter O and circumgyroradius R in an Einstein gyrovector space $(\mathbb{R}_s^n, \oplus, \otimes)$, and let $A \in \mathbb{A}_3 \cap \mathbb{R}_s^n$ be a point given by its gyrobarycentric representation*

$$A = \frac{m_1 \gamma_{A_1} A_1 + m_2 \gamma_{A_2} A_2 + m_3 \gamma_{A_3} A_3}{m_1 \gamma_{A_1} + m_2 \gamma_{A_2} + m_3 \gamma_{A_3}} \tag{201}$$

with respect to the gyrobarycentrically independent set $S = \{A_1, A_2, A_3\}$. Furthermore, let

$$\begin{aligned} K(A; A_1, A_2, A_3) &= m_1 m_2 (\gamma_{12} - 1) + m_1 m_3 (\gamma_{13} - 1) + m_2 m_3 (\gamma_{23} - 1)\,, \\ T(A; A_1, A_2, A_3) &= m_1 m_2 \sin\alpha_3 \sin\left(\alpha_3 + \frac{\delta}{2}\right) + m_1 m_3 \sin\alpha_2 \sin\left(\alpha_2 + \frac{\delta}{2}\right) \\ &\quad + m_2 m_3 \sin\alpha_1 \sin\left(\alpha_1 + \frac{\delta}{2}\right)\,, \end{aligned} \tag{202}$$

where α_k, $k = 1, 2, 3$, is the vertex gyroangle of vertex A_k of gyrotriangle $A_1A_2A_3$, be two scalars associated with the point A.

1. *The point A lies on the circumgyrocircle $C(A_1A_2A_3)$ of gyrotriangle $A_1A_2A_3$ if and only if the gyrobarycentric coordinates m_1, m_2, m_3 of A in* (201) *satisfy the circumgyrocircle condition*

$$K(A; A_1, A_2, A_3) = 0 \tag{203a}$$

or, equivalently, the gyrotrigonometric circumgyrocircle condition

$$T(A; A_1, A_2, A_3) = 0\,. \tag{203b}$$

2. *The point A lies in the interior of circumgyrocircle* $C(A_1A_2A_3)$ *if and only if*

$$K(A; A_1, A_2, A_3) > 0 \tag{204a}$$

or, equivalently,

$$T(A; A_1, A_2, A_3) > 0\,. \tag{204b}$$

3. *The point A lies in the exterior of circumgyrocircle* $C(A_1A_2A_3)$ *if and only if*

$$K(A; A_1, A_2, A_3) < 0 \tag{205a}$$

or, equivalently,

$$T(A; A_1, A_2, A_3) < 0\,. \tag{205b}$$

Moreover, the circumgyrocircle $C(A_1A_2A_3)$ *is the locus of the point A in* (201), *with parametric gyrobarycentric coordinates* m_k, $k = 1, 2, 3$, *given*

4. *by the parametric equations*

$$\begin{aligned} m_1 &= -(\gamma_{23} - 1)t \\ m_2 &= (\gamma_{12} - 1)t^2 + (\gamma_{13} - 1)t = m_3 t \\ m_3 &= (\gamma_{12} - 1)t + (\gamma_{13} - 1) \end{aligned} \tag{206}$$

with the parameter t, $t \in \mathbb{R} \cup \{-\infty, \infty\}$, *where the two parameter values* $t = -\infty$ *and* $t = \infty$ *are identified* (Fig. 15), *or, equivalently,*

5. *by the parametric equations*

$$\begin{aligned} m_1 &= -(\gamma_{23} - 1)\sin\theta \\ m_2 &= \quad (\gamma_{13} - 1)\sin\theta + (\gamma_{12} - 1)(1 - \cos\theta) \\ m_3 &= \quad (\gamma_{12} - 1)\sin\theta + (\gamma_{13} - 1)(1 + \cos\theta) \end{aligned} \tag{207}$$

with the parameter θ, $0 \le \theta \le 2\pi$, *where the two parameter values* $\theta = 0$ *and* $\theta = 2\pi$ *are identified* (Fig. 16).

6. *The circumgyrocenter, O, of gyrotriangle* $A_1A_2A_3$ *is given by* (165) *and* (166), *p. 40, and, gyrotrigonometrically, by* (39)–(41), *p. 13.*
7. *The circumgyroradius, R, of gyrotriangle* $A_1A_2A_3$ *is given by* (198).

Proof The equivalence between K and T in the theorem is proved in (183) and (184). The proof of each item of the theorem follows.

1. It follows from (194) that $K = K(A; A_1, A_2, A_3)$ vanishes if an only if $d = R$, that is, by Definition 4, if an only if the point A lies on circumgyrocircle $C(A_1A_2A_3)$.

2. It follows from (194) that $K = K(A; A_1, A_2, A_3) > 0$ if an only if $d < R$, that is, by Definition 4, if an only if the point A lies in the interior of circumgyrocircle $C(A_1A_2A_3)$.
3. It follows from (194) that $K = K(A; A_1, A_2, A_3) < 0$ if an only if $d > R$, that is, by Definition 4, if an only if the point A lies in the exterior of circumgyrocircle $C(A_1A_2A_3)$.
4. Item (4) follows from (186).
5. Item (5) follows from (189).
6. The proof of Item (6) is given in the derivation of (165) and (166), p. 40, and (39) and (41), p. 13.
7. The proof of Item (7) is given in the derivation of (198).

The proof of the theorem is thus complete.

Example 4 Let A be a point in an Einstein gyrovector space $\mathbb{R}^n_s$, $n \geq 2$, given by its gyrobarycentric representation (201) with respect to a gyrobarycentrically independent set $\{A_1, A_2, A_3\}$, with gyrobarycentric coordinates $m_1 = 0$, $m_2 = 0$, and $m_3 \neq 0$. Then, $A = A_3$, and the gyrobarycentric coordinates of A satisfy the circumgyrocircle condition (203) of Theorem 17. Hence, by Theorem 17, the point A_3 lies on the circumgyrocircle of gyrotriangle $\{A_1, A_2, A_3\}$, as obviously expected.

Example 5 Let A be a point in an Einstein gyrovector space $\mathbb{R}^n_s$, $n \geq 2$, given by its gyrobarycentric representation (201) with respect to a gyrobarycentrically independent set $\{A_1, A_2, A_3\}$, where gyrotriangle $A_1A_2A_3$ is equilateral. Then, A lies on the circumgyrocircle of gyrotriangle $A_1A_2A_3$ if and only if the gyrobarycentric coordinates m_1, m_2, m_3 of A in (201) satisfy the circumgyrocircle condition

$$m_1m_2 + m_1m_3 + m_2m_3 = 0\,. \tag{208}$$

17 Circle Barycentric Representation

Lemma 1, p. 20, enables Theorem 17 to be reduced to its Euclidean counterpart, Theorem 18. As an immediate application of Lemma 1 we, therefore, note that

$$\lim_{s\to\infty} 2\,s^2\,K(A; A_1, A_2, A_3) = m_1m_2a_{12}^2 + m_1m_3a_{13}^2 + m_2m_3a_{23}^2 \tag{209}$$

where $K = K(A; A_1, A_2, A_3)$ is given by (202), and $a_{ij} = \|-A_i + A_j\|$.

Definition 5 (Circle Interior and Exterior Points) *Let $A_1A_2A_3$ be a triangle and let $C(A_1A_2A_3)$, O and R be, respectively, the circumcircle, circumcenter, and circumradius of the triangle in a Euclidean space $\mathbb{R}^n$. Let P be a generic point in $\mathbb{A}_3^{euc}$,*

$$P = \frac{m_1A_1 + m_2A_2 + m_3A_3}{m_1 + m_2 + m_3} \in \mathbb{A}_3^{euc}\,, \tag{210}$$

where

$$\mathbb{A}_3^{euc} = A_1 + \text{Span}\{-A_1 + A_2, -A_1 + A_3\} \subset \mathbb{R}^n, \tag{211}$$

and let $d = \|-P + O\|$ be the distance of P from O. Then, P lies

1. *in the interior of $C(A_1A_2A_3)$ if $d < R$;*
2. *in the exterior of $C(A_1A_2A_3)$ if $d > R$; and*
3. *on $C(A_1A_2A_3)$ if $d = R$.*

Theorem 18 (Point to Circumcenter Distance) *Let $A_1A_2A_3$ be a triangle and let $C(A_1A_2A_3)$, O and R be, respectively, the circumcircle, circumcenter, and circumradius of the triangle in the Euclidean space $\mathbb{R}^n$, and let $A \in \mathbb{A}_3^{euc}$ be a point given by its barycentric representation*

$$A = \frac{m_1A_1 + m_2A_2 + m_3A_3}{m_1 + m_2 + m_3} \tag{212}$$

with respect to the barycentrically independent set $S = \{A_1, A_2, A_3\}$. Furthermore, let

$$d = \|-A + O\| \tag{213}$$

be the distance from A to O.

Then,

$$d = \sqrt{1 - \frac{K_{euc}}{M^2R^2}}\,R, \tag{214}$$

where

$$K_{euc} = \sum_{\substack{i,j=1\\ i<j}}^{3} m_im_ja_{ij}^2, \tag{215}$$

$a_{ij} = \|-A_i + A_j\|$, and

$$M = \sum_{k=1}^{3} m_k. \tag{216}$$

Proof Theorem 18 is the Euclidean counterpart of Theorem 16. The proof of Theorem 18 from Theorem 16 is immediate, noting (209), and noting that each equation in Theorem 18 is the Euclidean limit, $s \to \infty$, of a corresponding equation in Theorem 16.

The Euclidean counterpart of the gyrocircle gyrobarycentric representation Theorem 17 is the following elegant theorem, illustrated in Fig. 17.

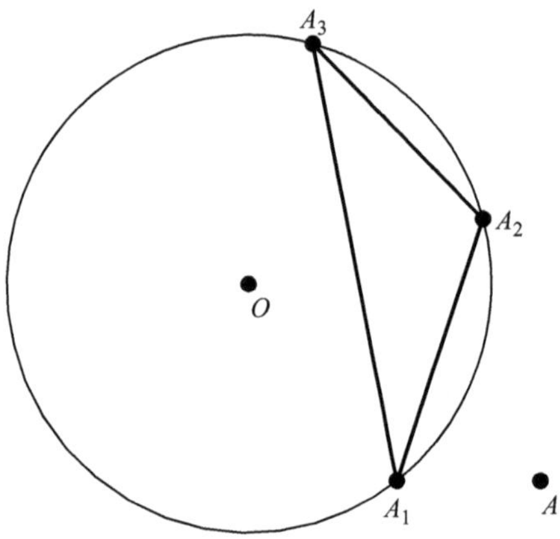

Fig. 17 Illustrating Theorems 18 and 19. The circumcircle $C(A_1A_2A_3)$ of triangle $A_1A_2A_3$ in the Euclidean plane $\mathbb{R}^2$ is the locus of the point A with barycentric representation (217), where its parametric barycentric coordinates are given by (222) or, equivalently, by (223). The radius of $C(A_1A_2A_3)$ is $R = \| -O + A_k\|$, $k = 1, 2, 3$, and the distance $d = \| -A + O\|$ from a generic point A in the triangle plane to the center O of $C(A_1A_2A_3)$ is given by (214). The latter provides an elegant condition in Theorem 19 that determines whether the point A lies on the circumcircle $C(A_1A_2A_3)$, or in its interior or exterior

Theorem 19 (Circle Barycentric Representation) *Let A, $A \in A_1 + \text{Span}\{-A_1 + A_2, -A_1 + A_3\} \subset \mathbb{R}^n$, be a point given by its barycentric representation*

$$A = \frac{m_1A_1 + m_2A_2 + m_3A_3}{m_1 + m_2 + m_3} \tag{217}$$

with respect to the barycentrically independent set $S = \{A_1, A_2, A_3\}$ in the Euclidean space $\mathbb{R}^n$, $n \geq 2$, and let

$$\begin{aligned} K_{euc}(A; A_1A_2A_3) &= m_1m_2a_{12}^2 + m_1m_3a_{13}^2 + m_2m_3a_{23}^2 \\ T_{euc}(A; A_1A_2A_3) &= m_1m_2 \sin^2\alpha_3 + m_1m_3 \sin^2\alpha_2 + m_2m_3 \sin^2\alpha_3 \end{aligned} \tag{218}$$

where α_k, $k = 1, 2, 3$, is the vertex angle of vertex A_k of triangle $A_1A_2A_3$, be two scalars associated with A.

1. *The point A lies on the circumcircle $C(A_1A_2A_3)$ of triangle $A_1A_2A_3$ if and only if the barycentric coordinates m_1, m_2, m_3 of A in (217) satisfy the circumcircle condition*

$$K_{euc}(A; A_1, A_2, A_3) = 0 \tag{219a}$$

or, equivalently, the trigonometric circumcircle condition

$$T_{euc}(A; A_1, A_2, A_3) = 0\,. \tag{219b}$$

2. *The point A lies in the interior of circumcircle* $C(A_1A_2A_3)$ *if and only if*

$$K_{euc}(A; A_1, A_2, A_3) > 0 \tag{220a}$$

or, equivalently,

$$T_{euc}(A; A_1, A_2, A_3) > 0\,. \tag{220b}$$

3. *The point A lies in the exterior of circumcircle* $C(A_1A_2A_3)$ *if and only if*

$$K_{euc}(A; A_1, A_2, A_3) < 0 \tag{221a}$$

or, equivalently,

$$T_{euc}(A; A_1, A_2, A_3) < 0\,. \tag{221b}$$

Moreover, the circumcircle $C(A_1A_2A_3)$ *is the locus of the point A in* (217), *with parametric barycentric coordinates* m_k, $k = 1, 2, 3$, *given*

4. *by the parametric equations*

$$\begin{aligned} m_1 &= -t\sin^2\alpha_1 \\ m_2 &= t\sin^2\alpha_2 + t^2\sin^2\alpha_3 = m_3 t \\ m_3 &= \sin^2\alpha_2 + t\sin^2\alpha_3 \end{aligned} \tag{222}$$

with the parameter t, $t \in \mathbb{R}\cup\{-\infty, \infty\}$, *where the two parameter values* $t = -\infty$ *and* $t = \infty$ *are identified, or, equivalently,*

5. *by the parametric equations*

$$\begin{aligned} m_1 &= -\sin^2\alpha_1\sin\theta \\ m_2 &= \sin^2\alpha_2\sin\theta + \sin^2\alpha_3(1-\cos\theta) \\ m_3 &= \sin^2\alpha_3\sin\theta + \sin^2\alpha_2(1+\cos\theta) \end{aligned} \tag{223}$$

with the parameter θ, $0 \le \theta \le 2\pi$, *where the two parameter values,* $\theta = 0$ *and* $\theta = 2\pi$ *are identified.*

6. *The circumcenter, O, of triangle* $A_1A_2A_3$ *is given trigonometrically by* (48), *p. 14.*
7. *The circumradius, R, of triangle* $A_1A_2A_3$ *is given by* (81), *p. 21.*

Proof In the Euclidean limit, $s \to \infty$, gamma factors tend to 1. Hence, in that limit, the gyrobarycentric representation of $A \in \mathbb{R}^n_s$ in (201) tends to the corresponding barycentric representations of $A \in \mathbb{R}^n$ in (217).

By Lemma 1, in the Euclidean limit, the circumgyrocircle condition (203a) (where K is given by (202)) tends to (219a) (where K_{euc} is given by (218)).

In the Euclidean limit gyrotriangle defects vanish. Hence, in that limit, the circumgyrocircle condition (203b) (where T is given by (202)) tends to (219b) (where T_{euc} is given by (218)), as desired.

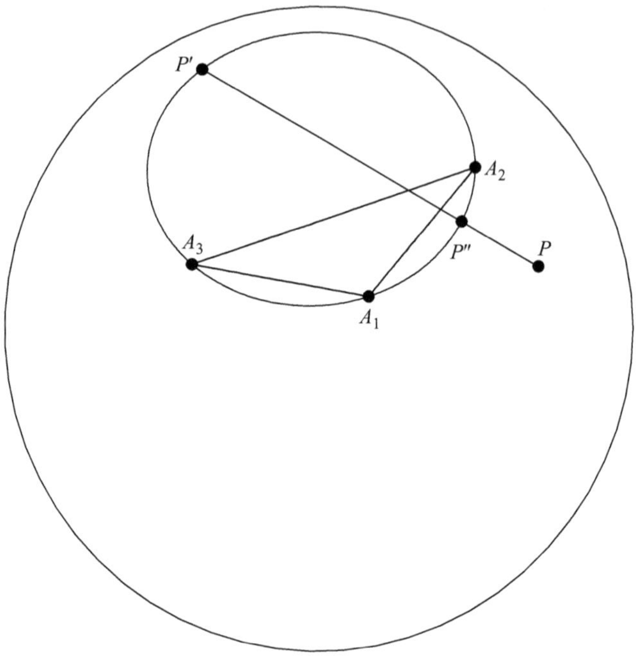

Fig. 18 The point P' lies arbitrarily on the circumgyrocircle $C(A_1A_2A_3)$ of gyrotriangle $A_1A_2A_3$ in an Einstein gyrovector plane $(\mathbb{R}_s^2, \oplus, \otimes)$, and P, $P \neq P'$, is a point in $\mathbb{R}_s^2$. Both P and P' are given by their gyrobarycentric representation with respect to the gyrobarycentrically independent set $S = \{A_1, A_2, A_3\}$ of vertices of the reference gyrotriangle $A_1A_2A_3$. The point P'', $P'' \neq P'$, that lies on the intersection of the circumgyrocircle $C(A_1A_2A_3)$ and the gyroline $L_{PP'}$ through P and P' possesses a gyrobarycentric representation with respect to the set S, with gyrobarycentric coordinates that are determined by those of P and P' according to (237) and (238)

The proof of Inequalities (220) and (221) follows, similarly, from Inequalities (204) and (205).

Similarly, the proof of (222) and (223) follows from (206) and (207) and from Euclidean limits that result immediately from Lemma 1.

The proof of Item (6) is presented in the derivation of (48), p. 21, and the proof of Item (7) is presented in the derivation of (81), p. 21.

18 Gyrocircle Gyroline Intersection

Let $P \in \mathbb{A}_3 \cap \mathbb{R}_s^n$ be a point given by its gyrobarycentric representation

$$P = \frac{m_1\gamma_{A_1} A_1 + m_2\gamma_{A_2} A_2 + m_3\gamma_{A_3} A_3}{m_1\gamma_{A_1} + m_2\gamma_{A_2} + m_3\gamma_{A_3}} \tag{224}$$

with respect to a gyrobarycentrically independent set $S = \{A_1, A_2, A_3\}$ in an Einstein gyrovector space $(\mathbb{R}_s^n, \oplus, \otimes)$. Additionally, let $P' \in \mathbb{R}_s^n$ be a point that lies arbitrarily

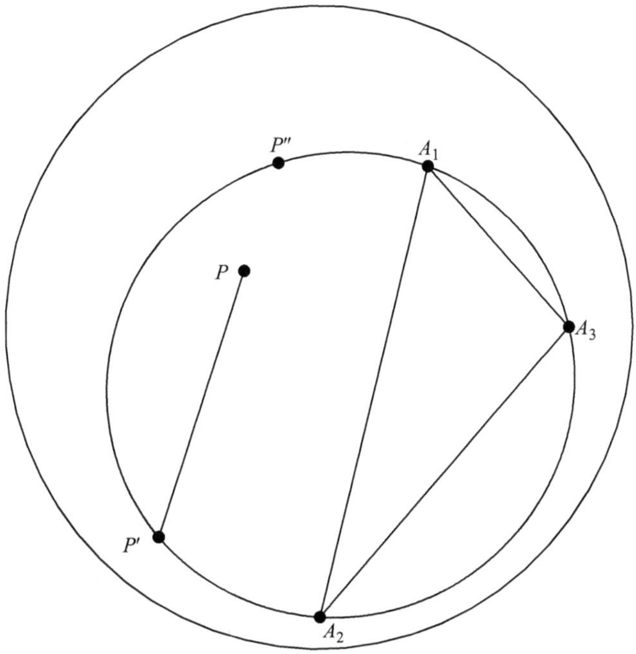

Fig. 19 In Fig. 18, the point P lies in the exterior of the circumgyrocircle $C(A_1A_2A_3)$ of gyrotriangle $A_1A_2A_3$. Here, in contrast, the point P lies in the interior of $C(A_1A_2A_3)$. As in Fig. 18, the point P'' lies on the intersection of circumgyrocircle $C(A_1A_2A_3)$ and the gyroline $L_{PP'}$ that passes through the points P and P'. The gyrobarycentric coordinates of P'' are determined by those of P and P' according to (237) and (238). Both Figs. 18 and 19 illustrate Theorem 20, p. 57

on the circumgyrocircle $C(A_1A_2A_3)$ of gyrotriangle $A_1A_2A_3$, as shown in Figs. 18 and 19, given by its parametric gyrobarycentric representation

$$P' = P'(t) = \frac{m'_1\gamma_{A_1}A_1 + m'_2\gamma_{A_2}A_2 + m'_3\gamma_{A_3}A_3}{m'_1\gamma_{A_1} + m'_2\gamma_{A_2} + m'_3\gamma_{A_3}} \tag{225}$$

with respect to the set S, where the parametric gyrobarycentric coordinates of P' are presented in (229) below.

Furthermore, let $P'' \in \mathbb{A}_3 \cap \mathbb{R}^n_s$ be a point given by its gyrobarycentric representation

$$P'' = \frac{m''_1\gamma_{A_1}A_1 + m''_2\gamma_{A_2}A_2 + m''_3\gamma_{A_3}A_3}{m''_1\gamma_{A_1} + m''_2\gamma_{A_2} + m''_3\gamma_{A_3}} \tag{226}$$

with respect to the set S, such that

1. P'' lies on the gyroline $L_{PP'}$ that passes through the points P and P'; and
2. P'' and P', $P'' \neq P'$, lie on the circumgyrocircle $C(A_1A_2A_3)$ of gyrotriangle $A_1A_2A_3$, as shown in Figs. 18 and 19.

We will determine the gyrobarycentric coordinates of P'' in terms of those of P and P' and in terms of the reference gyrotriangle $A_1A_2A_3$ in (230) and (231) below.

Following the condition that P'' lies on circumgyrocircle $C(A_1A_2A_3)$, by the gyrocircle gyrobarycentric representation Theorem 17, p. 48, the gyrobarycentric coordinates m_k'' of P'' satisfy the circumgyrocircle condition (203). That is,

$$m_1''m_2''(\gamma_{12}-1)+m_1''m_3''(\gamma_{13}-1)+m_2''m_3''(\gamma_{23}-1)=0\,. \tag{227}$$

We wish to express, in (230) below, the gyrobarycentric coordinates m_k'' of P'' in terms of the gyrobarycentric coordinates of P and P'.

Since the point P'' lies on gyroline $L_{PP'}$, its gyrobarycentric coordinates m_k'', $k=1,2,3$, are given parametrically by [26, Sect. 4.10], that is,

$$\begin{aligned}
m_1'' &= m_1(m_1'\gamma_{11}+m_2'\gamma_{12}+m_3'\gamma_{13}) - \left\{ \begin{vmatrix} m_1 & m_2 \\ m_1' & m_2' \end{vmatrix} \gamma_{12} + \begin{vmatrix} m_1 & m_3 \\ m_1' & m_3' \end{vmatrix} \gamma_{13} \right\} t_0\,, \\
m_2'' &= m_2(m_1'\gamma_{11}+m_2'\gamma_{12}+m_3'\gamma_{13}) + \left\{ \begin{vmatrix} m_1 & m_2 \\ m_1' & m_2' \end{vmatrix} \gamma_{11} - \begin{vmatrix} m_2 & m_3 \\ m_2' & m_3' \end{vmatrix} \gamma_{13} \right\} t_0\,, \\
m_3'' &= m_3(m_1'\gamma_{11}+m_2'\gamma_{12}+m_3'\gamma_{13}) + \left\{ \begin{vmatrix} m_1 & m_3 \\ m_1' & m_3' \end{vmatrix} \gamma_{11} + \begin{vmatrix} m_2 & m_3 \\ m_2' & m_3' \end{vmatrix} \gamma_{12} \right\} t_0\,,
\end{aligned} \tag{228}$$

$\gamma_{11}=1$, with the parameter $t_0\in\mathbb{R}$.

The point P' lies on circumgyrocircle $C(A_1A_2A_3)$ of gyrotriangle $A_1A_2A_3$. Hence, by (206), the gyrobarycentric coordinates m_k' of P' can be parameterized as

$$\begin{aligned}
m_1' &= -(\gamma_{23}-1)t \\
m_2' &= (\gamma_{12}-1)t^2+(\gamma_{13}-1)t = m_3't \\
m_3' &= (\gamma_{12}-1)t+(\gamma_{13}-1)\,,
\end{aligned} \tag{229}$$

with the parameter $t\in\mathbb{R}\cup\{-\infty,\infty\}$.

Inserting (229) in (228), and, successively, inserting the resulting gyrobarycentric coordinates m_k'' in the circumgyrocircle condition (227), we obtain a linear equation for the unknown t_0. The resulting solution, T_0 of t_0 is too involved and hence is not presented here. Inserting $t_0=T_0$ in (228), we obtain the desired gyrobarycentric coordinates m_k'', $k=1,2,3$, of P''. The resulting expressions of m_k'' are initially involved. Fortunately, however, these involved expressions can be factorized and, owing to the homogeneity of gyrobarycentric coordinates, nonzero common factors are irrelevant and, hence, can be omitted. The resulting gyrobarycentric coordinates m_k'' in the gyrobarycentric representation of P'' in (226) turn out to be simple and elegant. These are

$$\begin{aligned}
m_1'' &= E_1E_2 \\
m_2'' &= E_0E_1(\gamma_{13}-1)
\end{aligned} \tag{230}$$

$$m_3'' = -E_0E_2(\gamma_{12} - 1)\,,$$

where

$$\begin{aligned}
E_0 &= m_2 - m_3t \\
E_1 &= m_1(\gamma_{13} - 1) + m_2(\gamma_{23} - 1) + m_1(\gamma_{12} - 1)t \\
E_2 &= m_1(\gamma_{13} - 1) + m_1(\gamma_{12} - 1)t + m_3(\gamma_{23} - 1)t\,.
\end{aligned} \tag{231}$$

Interestingly, E_0, E_1, and E_2 are related by the equation

$$E_1 - E_2 = E_0(\gamma_{23} - 1)\,. \tag{232}$$

Formalizing, we obtain the following theorem:

Theorem 20 (The Gyroline Gyrocircle Intersection Theorem)

1. *Let $A_1A_2A_3$ be a gyrotriangle that possesses a circumgyrocircle in an Einstein gyrovector space $(\mathbb{R}^n_s, \oplus, \otimes)$, $n \geq 2$,*
2. *let $P \in \mathbb{A}_3 \cap \mathbb{R}^n_s$ be a point given by its gyrobarycentric representation*

$$P = \frac{m_1\gamma_{A_1}A_1 + m_2\gamma_{A_2}A_2 + m_3\gamma_{A_3}A_3}{m_1\gamma_{A_1} + m_2\gamma_{A_2} + m_3\gamma_{A_3}} \tag{233}$$

with respect to the gyrobarycentrically independent set $S = \{A_1, A_2, A_3\}$ of the vertices of the reference gyrotriangle $A_1A_2A_3$,

3. *let P' be a point that lies arbitrarily on the circumgyrocircle $C(A_1A_2A_3)$ of gyrotriangle $A_1A_2A_3$, $P' \neq P$, and*
4. *let P'' be the unique point that lies simultaneously on the circumgyrocircle $C(A_1A_2A_3)$ and on the gyroline $L_{PP'}$ that passes through the points P and P' where, in general, $P'' \neq P'$, as shown in* Figs. 18 *and* 19. *Finally,*
5. *let P' be parameterized by the parameter t,*

$$t \in \mathbb{R} \cup \{-\infty, \infty\}\,, \tag{234}$$

so that following the Gyrocircle Gyrobarycentric Representation Theorem (17), (206), *p. 48, P' possesses the gyrobarycentric representation*

$$P' = P'(t) = \frac{m_1'\gamma_{A_1}A_1 + m_2'\gamma_{A_2}A_2 + m_3'\gamma_{A_3}A_3}{m_1'\gamma_{A_1} + m_2'\gamma_{A_2} + m_3'\gamma_{A_3}} \tag{235}$$

with respect to the set S, where

$$\begin{aligned}
m_1' &= -(\gamma_{23} - 1)t \\
m_2' &= (\gamma_{12} - 1)t^2 + (\gamma_{13} - 1)t = m_3't \\
m_3' &= (\gamma_{12} - 1)t + (\gamma_{13} - 1)\,.
\end{aligned} \tag{236}$$

Then, the point P'' possesses the t-dependent gyrobarycentric representation

$$P'' = P''(t) = \frac{m_1''\gamma_{A_1} A_1 + m_2''\gamma_{A_2} A_2 + m_3''\gamma_{A_3} A_3}{m_1''\gamma_{A_1} + m_2''\gamma_{A_2} + m_3''\gamma_{A_3}} \tag{237}$$

with respect to the set S, where

$$\begin{aligned} m_1'' &= E_1 E_2 \\ m_2'' &= E_0 E_1(\gamma_{13} - 1) \\ m_3'' &= -E_0 E_2(\gamma_{12} - 1)\,, \end{aligned} \tag{238}$$

and where

$$\begin{aligned} E_0 &= m_2 - m_3 t \\ E_1 &= m_1(\gamma_{13} - 1) + m_2(\gamma_{23} - 1) + m_1(\gamma_{12} - 1)t \\ E_2 &= m_1(\gamma_{13} - 1) + m_1(\gamma_{12} - 1)t + m_3(\gamma_{23} - 1)t\,. \end{aligned} \tag{239}$$

19 Gyrocircle–Gyroline Tangency Points

Theorem 21 (The Gyrocircle Gyrotangents Theorem) *Let $P \in \mathbb{A}_3 \cap \mathbb{R}_s^n$ be a point given by its gyrobarycentric representation*

$$P = \frac{m_1\gamma_{A_1} A_1 + m_2\gamma_{A_2} A_2 + m_3\gamma_{A_3} A_3}{m_1\gamma_{A_1} + m_2\gamma_{A_2} + m_3\gamma_{A_3}} \tag{240}$$

with respect to a gyrobarycentrically independent set $S = \{A_1, A_2, A_3\}$ in an Einstein gyrovector plane $(\mathbb{R}_s^2, \oplus, \otimes)$, and let $C(A_1A_2A_3)$ be the circumgyrocircle of gyrotriangle $A_1A_2A_3$. Then the two tangency points $P_\pm$ of the gyrotangent gyrolines of circumgyrocircle $C(A_1A_2A_3)$ that pass through the point P, when exist as shown in Fig. 20, *are given by their gyrobarycentric representations*

$$P_\pm = \frac{m_1'\gamma_{A_1} A_1 + m_2'\gamma_{A_2} A_2 + m_3'\gamma_{A_3} A_3}{m_1'\gamma_{A_1} + m_2'\gamma_{A_2} + m_3'\gamma_{A_3}} \tag{241}$$

with respect to S. The gyrobarycentric coordinates m_k', $k = 1, 2, 3$, of $P_\pm$ are given in terms of the gyrobarycentric coordinates m_k of P and the sides of gyrotriangle $A_1A_2A_3$ by the equations

$$\begin{aligned} m_1' &= F_0 F_1(\gamma_{23} - 1) \\ m_2' &= F_1 F_2 \end{aligned} \tag{242}$$

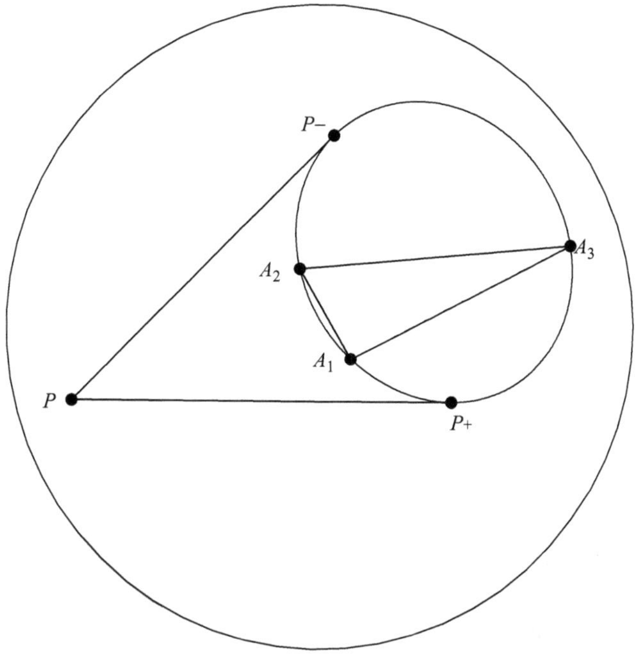

Fig. 20 Illustrating the Gyrocircle Tangents Theorem 21. The two tangency points $P_{\pm}$ of the circumgyrocircle $C(A_1A_2A_3)$ of a gyrotriangle $A_1A_2A_3$ in an Einstein gyrovector plane $(\mathbb{R}_s^2, \oplus, \otimes)$, which lies on the *tangent lines* that pass through a point P that lies in the *exterior* of the circumgyrocircle, are shown. The Euclidean counterpart of this figure is presented in Fig. 21

$$m_3' = -F_0F_2(\gamma_{12} - 1),$$

where

$$\begin{aligned} F_0 &= \quad m_1(\gamma_{12} - 1) + m_3(\gamma_{23} - 1) \\ F_1 &= \quad m_1(\gamma_{12} - 1)(\gamma_{13} - 1) \pm \sqrt{-\Delta_1\Delta_2} \\ F_2 &= -m_3(\gamma_{13} - 1)(\gamma_{23} - 1) \pm \sqrt{-\Delta_1\Delta_2}, \end{aligned} \tag{243}$$

and where

$$\begin{aligned} \Delta_1 &= m_1m_2(\gamma_{12} - 1) + m_1m_3(\gamma_{13} - 1) + m_2m_3(\gamma_{23} - 1) \\ \Delta_2 &= (\gamma_{12} - 1)(\gamma_{13} - 1)(\gamma_{23} - 1) > 0. \end{aligned} \tag{244}$$

1. *Two distinct tangency points, $P_{\pm}$, exist if and only if $\Delta_1 < 0$ or, equivalently, if and only if the point P lies in the exterior of the circumgyrocircle $C(A_1A_2A_3)$, as shown in* Fig. 20.
2. *The two distinct tangency points degenerate to a single one, $P_{\pm} = P$ if and only if $\Delta_1 = 0$ or, equivalently, if and only if the point P lies on the circumgyrocircle $C(A_1A_2A_3)$.*

3. *There are no tangency points if and only if* $\Delta_1 > 0$ *or, equivalently, if and only if the point* P *lies in the interior of the circumgyrocircle* $C(A_1A_2A_3)$.

The constant $m_{P_\pm}$ *of the gyrobarycentric representation* (241) *of* $P_\pm$ *is given by*

$$m_{P_\pm} = (\gamma_{12} - 1)(\gamma_{13} - 1)(\gamma_{23} - 1)E_1 \pm \sqrt{-\Delta_1\Delta_2}E_2\,, \tag{245}$$

where

$$\begin{aligned} E_1 &= (m_1 - m_2 + m_3)\{m_1(\gamma_{12} - 1) + m_3(\gamma_{23} - 1)\} - 2m_1m_3(\gamma_{13} - 1) \\ E_2 &= -m_1(\gamma_{12} - 1)^2 + m_3(\gamma_{23} - 1)^2 + m_1(\gamma_{12} - 1)(\gamma_{13} - 1) \\ &\quad - m_3(\gamma_{13} - 1)(\gamma_{23} - 1) + (m_1 - m_3)(\gamma_{12} - 1)(\gamma_{23} - 1). \end{aligned} \tag{246}$$

Proof The Gyroline Gyrocircle Intersection Theorem 20 enables gyrocircle–gyroline tangency points to be determined. Indeed, in the special case when the two points P' and P'' in Theorem 20, in (235) and in (237), shown in Fig. 18, are coincident, the resulting point, $P_t := P' = P''$ is a point of tangency. There are two distinct ways for the points P' and P'' to become coincident by approaching each other along the circumference of the circumgyrocircle $C(A_1A_2A_3)$ of gyrotriangle $A_1A_2A_3$. These give rise to two tangency points to the circumgyrocircle that lie on the two gyrotangent gyrolines that pass through the point P, shown in Fig. 20.

The points P' and P'' in (235) and (237) possess gyrobarycentric representations that involve the parameter $t \in \mathbb{R} \cup \{-\infty, \infty\}$. The condition $P' = P''$ that gives rise to tangency points determines the values of the parameter t that correspond to the two tangency points.

Indeed, if $P' = P''$, then it follows from the gyrobarycentric representations of P' and P'' in (235) and (237) that

$$\begin{aligned} \frac{m_2'}{m_1'} &= \frac{m_2''}{m_1''} \\ \frac{m_3'}{m_1'} &= \frac{m_3''}{m_1''}\,, \end{aligned} \tag{247}$$

thus obtaining two equations for the parameter t that gives rise to the tangency points.

Inserting (236), (238), and (239) in (247), we see that the two equations in (247) are equivalent to each other. Solving one of these equations for the unknown parameter value t, we obtain

$$t = \frac{-m_1(\gamma_{12} - 1)(\gamma_{13} - 1) \pm \sqrt{-\Delta_1\Delta_2}}{\{m_1(\gamma_{12} - 1) + m_3(\gamma_{23} - 1)\}(\gamma_{12} - 1)}\,, \tag{248}$$

where

$$\Delta_1 = m_1m_2(\gamma_{12} - 1) + m_1m_3(\gamma_{13} - 1) + m_2m_3(\gamma_{23} - 1) \tag{249}$$

$$\Delta_2 = (\gamma_{12} - 1)(\gamma_{13} - 1)(\gamma_{23} - 1) > 0\,.$$

We see from the Gyrocircle Gyrobarycentric Representation Theorem 17, p. 48, that

1. $\Delta_1 = 0$ if and only if P lies on circumgyrocircle $C(A_1A_2A_3)$;
2. $\Delta_1 > 0$ if and only if the point P in (233) lies in the interior of circumgyrocircle $C(A_1A_2A_3)$; and
3. $\Delta_1 < 0$ if and only if the point P lies in the exterior of circumgyrocircle $C(A_1A_2A_3)$.

We thus see from (248) that a value of the real parameter t that corresponds to a tangency point P_t, $P_t := P' = P''$, exists if and only if the point P lies on circumgyrocircle $C(A_1A_2A_3)$ or in its exterior. Indeed, this result is expected since it is clear from Figs. 18 and 19 that a gyrotangent gyroline to the circumgyrocircle that passes through the point P exists if and only if P lies in the exterior of the circumgyrocircle. Clearly, the two distinct tangency points $P_\pm$ of circumgyrocircle $C(A_1A_2A_3)$ degenerate to a single point, $P_\pm = P$, if and only if the point P lies on the circumgyrocircle.

Inserting the parameter value of t from (248) into the gyrobarycentric coordinates m'_k, $k = 1, 2, 3$, of P' in (236), and omitting irrelevant nonzero common factors, we obtain the gyrobarycentric coordinates in (242)–(244), as desired.

By [26, Eq. (4.27), p. 90] and the *Gyrobarycentric Representation Gyrocovariance Theorem*, [26, Theorem 4.6, pp. 90–91], the constant $m_{P_\pm}$ of the gyrobarycentric representation (241) of $P_\pm$ is given by the equation

$$m_{P_\pm}^2 = (m'_1)^2 + (m'_2)^2 + (m'_3)^2 + 2(m'_1m'_2\gamma_{12} + m'_1m'_3\gamma_{13} + m'_2m'_3\gamma_{23})\,. \quad (250)$$

The substitution of the gyrobarycentric coordinates m'_k from (242) into (250) yields (245), as desired.

20 Gyrocircle Gyrotangent Gyrolength

Let $P_\pm$ denote collectively the two tangency points, P_+ and P_-, of the gyrotangent gyrolines drawn from a point P to the circumgyrocircle $C(A_1A_2A_3)$ of a gyrotriangle $A_1A_2A_3$ in an Einstein gyrovector space $(\mathbb{R}^n_s, \oplus, \otimes)$, as shown in Fig. 20 for $n = 2$. Accordingly, let $P \in \mathrm{Span}\{\ominus A_1 \oplus A_2, \ominus A_1 \oplus A_3\} \subset \mathbb{R}^n$ be a point on the gyroplane of gyrotriangle $A_1A_2A_3$ with gyrobarycentric representation

$$P = \frac{m_1\gamma_{A_1}A_1 + m_2\gamma_{A_2}A_2 + m_3\gamma_{A_3}A_3}{m_1\gamma_{A_1} + m_2\gamma_{A_2} + m_3\gamma_{A_3}} \quad (251)$$

so that, following Theorem 21, (241), the tangency point $P_\pm$ possesses the gyrobarycentric representation

$$P_\pm = \frac{m'_1\gamma_{A_1}A_1 + m'_2\gamma_{A_2}A_2 + m'_3\gamma_{A_3}A_3}{m'_1\gamma_{A_1} + m'_2\gamma_{A_2} + m'_3\gamma_{A_3}} \quad (252)$$

with gyrobarycentric coordinates m'_k, $k = 1, 2, 3$, given by (242)–(244).
Hence, by [26, Sect. 4.9] with $N = 3$,

$$\gamma_{\ominus P \oplus P_\pm} = \frac{1}{m_P m_{P_\pm}} \left\{ (m_1 m'_2 + m'_1 m_2)\gamma_{12} + (m_1 m'_3 + m'_1 m_3)\gamma_{13} \right.$$
$$\left. + (m_2 m'_3 + m'_2 m_3)\gamma_{23} + m_1 m'_1 + m_2 m'_2 + m_3 m'_3 \right\}, \quad (253)$$

where by [26, Eq. (4.27), p. 90] and the *Gyrobarycentric Representation Gyrocovariance Theorem*, [26, Theorem 4.6, pp. 90–91], the constant m_P of the gyrobarycentric representation (251) of P is given by

$$m_P^2 = (m_1 + m_2 + m_3)^2 + 2(m_1 m_2(\gamma_{12} - 1) + m_1 m_3(\gamma_{13} - 1) + m_2 m_3(\gamma_{23} - 1)) \quad (254)$$

and where $m_{P_\pm}$ is the constant of the gyrobarycentric representation (252) of $P_\pm$, given by (245).

Inserting m'_k, $k = 1, 2, 3$, from (242)–(244) into (253), we obtain the elegant equation

$$\gamma_{\ominus P \oplus P_\pm} = \frac{m_1 + m_2 + m_3}{m_P}. \quad (255)$$

The gyrolength of the gyrotangent gyrosegment $PP_\pm$ can readily be obtained from (255). Indeed, following (255) and (4), p. 3, we have

$$\|\ominus P \oplus P_\pm\| = s \frac{\sqrt{\gamma^2_{\ominus P \oplus P_\pm} - 1}}{\gamma_{\ominus P \oplus P_\pm}}. \quad (256)$$

Unlike the gyrobarycentric coordinates m'_k and the constant $m_{P_\pm}$ in (253), the gamma factor in (255) is free of the square root in (243) and (245). This implies the gyrodistance equality

$$\|\ominus P \oplus P_+\| = \|\ominus P \oplus P_-\|. \quad (257)$$

Following (254)–(256) we have

$$\begin{aligned} \|\ominus P \oplus P_\pm\| &= s \frac{\sqrt{\gamma^2_{\ominus P \oplus P_\pm} - 1}}{\gamma_{\ominus P \oplus P_\pm}} \\ &= s \frac{\sqrt{-2(m_1 m_2(\gamma_{12} - 1) + m_1 m_3(\gamma_{13} - 1) + m_2 m_3(\gamma_{23} - 1))}}{m_1 + m_2 + m_3} \end{aligned} \quad (258)$$

Note that by Theorem 17, p. 48, the radicand on the extreme right-hand side of (258) is positive (zero) if and only if the point P lies in the exterior of of circumgyrocircle $C(A_1 A_2 A_3)$ (if and only if the point P lies on $C(A_1 A_2 A_3)$).

Formalizing results of this section, we obtain the following theorem.

Theorem 22 **(Gyrocircle Gyrotangent Gyrolength)**

1. *Let* $A_1A_2A_3$ *be a gyrotriangle that possesses a circumgyrocircle* $C(A_1A_2A_3)$ *in an Einstein gyrovector space* $(\mathbb{R}^n_s, \oplus, \otimes)$,
2. *let* $P \in \mathrm{Span}\{\ominus A_1 \oplus A_2,\ \ominus A_1 \oplus A_3\} \subset \mathbb{R}^n$ *be a point that lies in the exterior of* $C(A_1A_2A_3)$*, given by its gyrobarycentric representation*

$$P = \frac{m_1\gamma_{A_1} A_1 + m_2\gamma_{A_2} A_2 + m_3\gamma_{A_3} A_3}{m_1\gamma_{A_1} + m_2\gamma_{A_2} + m_3\gamma_{A_3}} \tag{259}$$

with respect to the set $S = \{A_1, A_2, A_3\}$*, and*
3. *let* $P_\pm$ *represent collectively the two tangency points,* P_+ *and* P_-*, of the two gyrotangent gyrolines drawn from* P *to circumgyrocircle* $C(A_1A_2A_3)$*, as shown in* Fig. 20.

Then, we have the following results:

1. *The tangency points* $P_\pm$ *possess the gyrobarycentric representations*

$$P_\pm = \frac{m'_1\gamma_{A_1} A_1 + m'_2\gamma_{A_2} A_2 + m'_3\gamma_{A_3} A_3}{m'_1\gamma_{A_1} + m'_2\gamma_{A_2} + m'_3\gamma_{A_3}} \tag{260}$$

with gyrobarycentric coordinates m'_k*,* $k = 1, 2, 3$*, given by* (242)–(244).
2. *The gyrolengths of the two gyroline gyrosegments* PP_+ *and* PP_- *are equal,*

$$\|\ominus P \oplus P_+\| = \|\ominus P \oplus P_-\| . \tag{261}$$

3. *The gamma factor of each of the two gyrosegments* PP_+ *and* PP_- *is given by the equations*

$$\gamma_{\ominus P \oplus P_\pm} = \frac{m_1 + m_2 + m_3}{m_P}, \tag{262}$$

where m_P *is the constant of the gyrobarycentric representation of* P *with respect to* S.
4. *The gyrolength of each of the two gyrotangent gyrosegments* PP_+ *and* PP_- *is given by*

$$\begin{aligned} \|\ominus P \oplus P_\pm\| &= s\frac{\sqrt{\gamma^2_{\ominus P \oplus P_\pm} - 1}}{\gamma_{\ominus P \oplus P_\pm}} \\ &= s\frac{\sqrt{-2\{m_1m_2(\gamma_{12} - 1) + m_1m_3(\gamma_{13} - 1) + m_2m_3(\gamma_{23} - 1)\}}}{m_1 + m_2 + m_3} . \end{aligned} \tag{263}$$

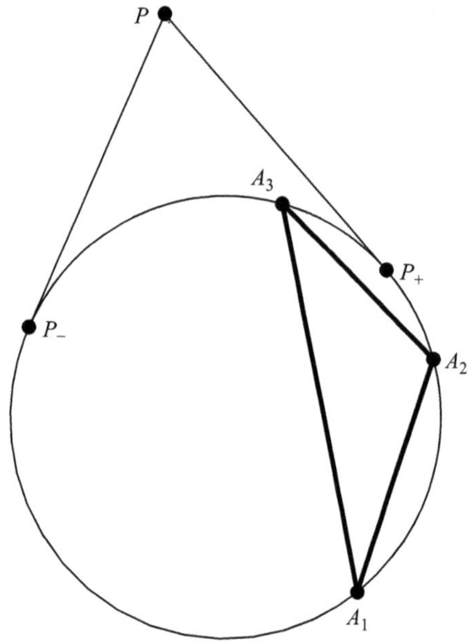

Fig. 21 Illustrating the Circle Tangents Theorem 23. The two tangency points $P_\pm$ of the circumcircle $C(A_1A_2A_3)$ of a triangle $A_1A_2A_3$ in the Euclidean plane $\mathbb{R}^2$, which lie on the tangent lines that pass through a point P that lies in the exterior of the circumcircle, are shown. This figure is the Euclidean counterpart of Fig. 20

21 Circle–Line Tangency Points

The Euclidean counterpart of the Gyrocircle Gyrotangents Theorem 21 is the following elegant theorem, illustrated in Fig. 21:

Theorem 23 (The Circle Tangents Theorem) *Let $P \in \mathbb{R}^2$ be a point given by its barycentric representation*

$$P = \frac{m_1A_1 + m_2A_2 + m_3A_3}{m_1 + m_2 + m_3} \tag{264}$$

with respect to a barycentrically independent set $S = \{A_1, A_2, A_3\}$ in the Euclidean plane $\mathbb{R}^2$, and let $C(A_1A_2A_3)$ be the circumcircle of triangle $A_1A_2A_3$. Then the two tangency points $P_\pm$ of the tangent lines of circumcircle $C(A_1A_2A_3)$ that pass through the point P, when exist, are given by their barycentric representations

$$P_\pm = \frac{m_1'A_1 + m_2'A_2 + m_3'A_3}{m_1' + m_2' + m_3'}\,, \tag{265}$$

shown in Fig. 21. *The barycentric coordinates of $P_\pm$ are given by the equations*

$$\begin{aligned} m_1' &= \quad F_0F_1\sin^2\alpha_1 \\ m_2' &= \quad F_1F_2 \\ m_3' &= -F_0F_2\sin^2\alpha_3\,, \end{aligned} \tag{266}$$

where

$$\begin{aligned} F_0 &= m_1 \sin^2 \alpha_3 + m_3 \sin^2 \alpha_1 \\ F_1 &= \{m_1 \sin^2 \alpha_2 \sin^2 \alpha_3 \pm \sin \alpha_1 \sin \alpha_2 \sin \alpha_3 \sqrt{-\Delta}\} \sin^2 \alpha_1 \\ F_2 &= \{m_3 \sin^2 \alpha_1 \sin^2 \alpha_2 \mp \sin \alpha_1 \sin \alpha_2 \sin \alpha_3 \sqrt{-\Delta}\} \sin^2 \alpha_3 , \end{aligned} \tag{267}$$

and where

$$\Delta = m_1 m_2 \sin^2 \alpha_3 + m_1 m_3 \sin^2 \alpha_2 + m_2 m_3 \sin^2 \alpha_1 . \tag{268}$$

1. *Two distinct tangency points, $P_\pm$, exist if and only if $\Delta < 0$ or, equivalently, if and only if the point P lies in the exterior of the circumcircle $C(A_1 A_2 A_3)$, as shown in* Fig. 21.
2. *The two distinct tangency points degenerate to a single one, $P_\pm = P$ if and only if $\Delta = 0$ or, equivalently, if and only if the point P lies on the circumcircle $C(A_1 A_2 A_3)$.*
3. *There are no tangency points if and only if $\Delta > 0$ or, equivalently, if and only if the point P lies in the interior of the circumcircle $C(A_1 A_2 A_3)$.*
4. *The lengths of the two tangent segments $\| - P + P_+ \|$ and $\| - P + P_+ \|$ are equal, given by*

$$\| - P + P_\pm \| = \frac{\sqrt{-\{m_1 m_2 a_{12}^2 + m_1 m_3 a_{13}^2 + m_2 m_3 a_{23}^2\}}}{m_1 + m_2 + m_3} . \tag{269}$$

Proof We will show that each result of this theorem is the Euclidean limit of a corresponding result of its hyperbolic counterpart, Theorem 21. Noting that in the Euclidean limit, $s \to \infty$, gamma factors tend to 1, the Euclidean limit of the gyrobarycentric coordinates m'_k, $k = 1, 2, 3$, in (241)–(244) is trivial. The resulting barycentric coordinates vanish, giving rise to an indeterminate barycentric representation. Thus, a straightforward application of the Euclidean limit to the gyrobarycentric coordinates m'_k in (241)–(244) in an attempt to recover the corresponding barycentric coordinates m'_k in (265)–(268) is destined to fail.

However, owing to their homogeneity, the gyrobarycentric coordinates m'_k in (241)–(244) can be written in a form that admits the Euclidean limit to a viable barycentric coordinate system. Specifically, the following equivalent form, in which each gyrobarycentric coordinate is multiplied by a common nonzero factor, is what we need, in which we replace m'_k by m''_k:

$$\begin{aligned} m''_1 &= \frac{m'_1}{(\gamma_{12} - 1)^4} = \frac{F_0}{\gamma_{12} - 1} \frac{F_1}{(\gamma_{12} - 1)^2} \frac{\gamma_{23} - 1}{\gamma_{12} - 1} \\ m''_2 &= \frac{m'_2}{(\gamma_{12} - 1)^4} = \frac{F_1}{(\gamma_{12} - 1)^2} \frac{F_2}{(\gamma_{12} - 1)^2} \\ m''_3 &= \frac{m'_3}{(\gamma_{12} - 1)^4} = -\frac{F_0}{\gamma_{12} - 1} \frac{F_1}{(\gamma_{12} - 1)^2} \frac{\gamma_{12} - 1}{\gamma_{12} - 1} , \end{aligned} \tag{270}$$

where, by (243),

$$\frac{F_0}{\gamma_{12}-1} = m_1 + m_3\frac{\gamma_{23}-1}{\gamma_{12}-1} := F_0'$$

$$\frac{F_1}{(\gamma_{12}-1)^2} = m_1\frac{\gamma_{13}-1}{\gamma_{12}-1} \pm \left\{\frac{-\Delta_1}{\gamma_{12}-1}\frac{\Delta_2}{(\gamma_{12}-1)^3}\right\}^{\frac{1}{2}} := F_1' \tag{271}$$

$$\frac{F_2}{(\gamma_{12}-1)^2} = -m_3\frac{\gamma_{13}-1}{\gamma_{12}-1}\frac{\gamma_{23}-1}{\gamma_{12}-1} \pm \left\{\frac{-\Delta_1}{\gamma_{12}-1}\frac{\Delta_2}{(\gamma_{12}-1)^3}\right\}^{\frac{1}{2}} := F_2',$$

and where

$$\frac{\Delta_1}{\gamma_{12}-1} = m_1m_2 + m_1m_3\frac{\gamma_{13}-1}{\gamma_{12}-1} + m_2m_3\frac{\gamma_{23}-1}{\gamma_{12}-1} := \Delta_1'$$

$$\frac{\Delta_2}{(\gamma_{12}-1)^3} = \frac{\gamma_{13}-1}{\gamma_{12}-1}\frac{\gamma_{23}-1}{\gamma_{12}-1} := \Delta_2'. \tag{272}$$

Hence, by (270)–(272),

$$\begin{aligned} m_1'' &= F_0'F_1'\frac{\gamma_{23}-1}{\gamma_{12}-1} \\ m_2'' &= F_1'F_2' \\ m_3'' &= -F_0'F_2'. \end{aligned} \tag{273}$$

The gyrobarycentric coordinates m_k'', $k = 1,2,3$, in (270)–(273) appear in a form that admits a nontrivial Euclidean limit, $s \to \infty$, by means of Lemma 1. Indeed, one can readily show that the Euclidean limit of the gyrobarycentric coordinate system m_k'' in (273) turns out to be the barycentric coordinate system (266)–(268), in which we rename m_k'' as m_k' ($k = 1,2,3$), and F_k' as F_k ($k = 0,1,2$).

Finally, it remains to prove (269). The length $\|-P+P_\pm\|$ of the tangent segment $PP_\pm$, shown in Fig. 21, is the Euclidean limit, $s \to \infty$, of the gyrolength $\|\ominus P \oplus P_\pm\|$ of the gyrotangent gyrosegment $PP_\pm$, shown in Fig. 20,

$$\lim_{s\to\infty} \|\ominus P \oplus P_\pm\| = \|-P+P_\pm\|. \tag{274}$$

Indeed, by (274), (263) and Lemma 1, p. 20, we have

$$\begin{aligned} \|-P+P_\pm\| &= \lim_{s\to\infty} \|\ominus P \oplus P_\pm\| \\ &= \lim_{s\to\infty} \frac{\sqrt{-2\,s^2\{m_1m_2(\gamma_{12}-1)+m_1m_3(\gamma_{13}-1)+m_2m_3(\gamma_{23}-1)\}}}{m_1+m_2+m_3} \\ &= \frac{\sqrt{-\{m_1m_2a_{12}^2+m_1m_3a_{13}^2+m_2m_3a_{23}^2\}}}{m_1+m_2+m_3}, \end{aligned} \tag{275}$$

where $a_{ij} = \|-A_i+A_j\|$, as desired.

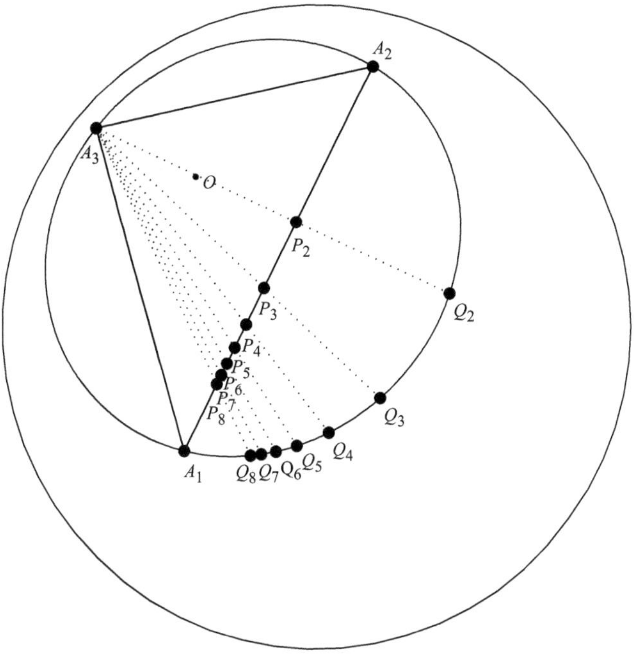

Fig. 22 Circumgyrocevians. A circumgyrocevian, $A_3Q_k, k = 2, 3, \ldots, 8$, is a gyrocevian A_3P_k of a gyrotriangle $A_1A_2A_3$ in an Einstein gyrovector space, extended to the point Q_k where it meets the gyrotriangle circumgyrocircle. The kth circumgyrocevian A_3Q_k in this figure corresponds to the parameter value $t_1 = 1/k$ of the parameter $0 < t_1 < 1$ in (276). The point A_1 corresponds to $t_1 = 0$, and the point A_2 corresponds to $t_1 = 1$

22 Circumgyrocevians

Definition 6 (Circumgyrocevians) *A circumgyrocevian of a gyrotriangle is a gyrocevian of the gyrotriangle extended to the point where it meets the gyrotriangle circumgyrocircle, as shown in* Fig. 22.

The concept of the circumgyrocevian, together with its associated Theorem 24, p. 71, is employed in Sect. 23 and, consequently, proves useful in the study of the Intersecting Gyrochords Theorem 26, p. 79.

In this section, we determine the circumgyrocevian associated with a given gyrocevian in an Einstein gyrovector space $(\mathbb{R}^n_s, \oplus, \otimes)$.

Let P be a generic point of a gyrosegment A_1A_2 in an Einstein gyrovector space $(\mathbb{R}^n_s, \oplus, \otimes)$, as shown in Fig. 23. Then, P possesses the gyrobarycentric representation

$$P = \frac{(1 - t_1)\gamma_{A_1} A_1 + t_1 \gamma_{A_2} A_2}{(1 - t_1)\gamma_{A_1} + t_1 \gamma_{A_2}} \tag{276}$$

for any value of the parameter $t_1, 0 \le t_1 \le 1$. The parameter t_1 determines the location of the point P on gyrosegment A_1A_2, as shown in Fig. 22. Thus, in particular, $t_1 = 0$ gives $P = A_1$, $t_1 = 1$ gives $P = A_2$, and $t = 1/2$ gives the gyromidpoint $M_{A_1A_2}$ of A_1 and A_2.

Applying the Gyrobarycentric Representation Gyrocovariance Theorem [26, Theorem 4.6, pp. 90–91] with $X = \ominus A_1$, we have

$$\begin{aligned} \ominus A_1 \oplus P &= \ominus A_1 \oplus \frac{(1-t_1)\gamma_{A_1} A_1 + t_1 \gamma_{A_2} A_2}{(1-t_1)\gamma_{A_1} + t_1 \gamma_{A_2}} \\ &= \frac{(1-t_1)\gamma_{\ominus A_1 \oplus A_1}(\ominus A_1 \oplus A_1) + t_1 \gamma_{\ominus A_1 \oplus A_2}(\ominus A_1 \oplus A_2)}{(1-t_1)\gamma_{\ominus A_1 \oplus A_1} + t_1 \gamma_{\ominus A_1 \oplus A_2}} \\ &= \frac{t_1 \gamma_{12} \mathbf{a}_{12}}{1 + (\gamma_{12} - 1)t_1}, \end{aligned} \tag{277}$$

where we use the gyrotriangle index notation (13).

Let A_3P be the gyroray that emanates from the point A_3 and passes through the point P in the Einstein gyrovector space, as shown in Fig. 23. Gyrolines and gyrorays in Einstein gyrovector spaces coincide with Euclidean segments, enabling methods of linear algebra to be applied for solving intersection problems for gyrolines and gyrorays.

Let us consider the gyroray

$$(\ominus A_1 \oplus A_3)(\ominus A_1 \oplus P) = \mathbf{a}_{13}(\ominus A_1 \oplus P) \tag{278}$$

that emanates from $(\ominus A_1 \oplus A_3) = \mathbf{a}_{13}$ and passes through the point $\ominus A_1 \oplus P$. This ray is the left gyrotranslation by $\ominus A_1$ of the gyroray A_3P. By methods of linear algebra, the points $A(t)$ of the gyroray (278), parameterized by t, $t \ge 0$, are given by

$$A(t) = \mathbf{a}_{13}(1-t) + (\ominus A_1 \oplus P)t\,. \tag{279}$$

Equivalently, by means of (277), (279) can be written as

$$\begin{aligned} A(t) &= \mathbf{a}_{13}(1-t) + \frac{t_1 \gamma_{12} \mathbf{a}_{12}}{1 + (\gamma_{12} - 1)t_1} t \\ &= m_2 \gamma_{12} \mathbf{a}_{12} + m_3 \gamma_{13} \mathbf{a}_{13}\,, \end{aligned} \tag{280}$$

where

$$\begin{aligned} m_2 &= \frac{t_1 t}{1 + (\gamma_{12} - 1)t_1} \\ m_3 &= \frac{1-t}{\gamma_{13}}\,. \end{aligned} \tag{281}$$

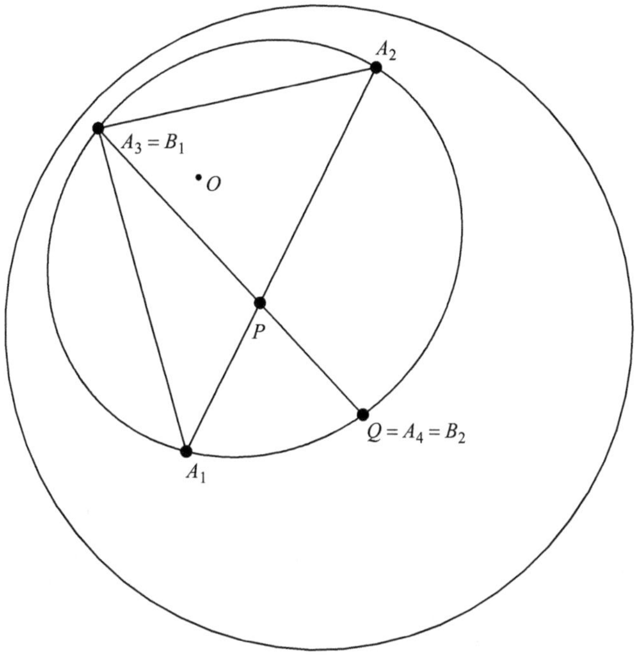

Fig. 23 Illustrating the Circumgyrocevian Theorem 24, p. 71, and the intersecting Gyrochords Theorem 26, p. 79. A circumgyrocevian, A_3Q, is shown. It is a gyrocevian A_3P of a gyrotriangle $A_1A_2A_3$ in an Einstein gyrovector space $(\mathbb{R}_s^n, \oplus, \otimes)$, $n \geq 2$, extended to the point Q where it meets the gyrotriangle circumgyrocircle. With $A_4 = Q$, this figure presents two gyrochords, A_1A_2 and $A_3A_4 = B_1B_2$, of a gyrocircle, intersecting at a point P inside the gyrocircle. Here, the point $Q \in \mathbb{R}_s^n$ is determined by its gyrobarycentric representation (291) in Theorem 24. This figure is the hyperbolic counterpart of Fig. 24

The set of points $A(t)$, $0 \leq t < \infty$, is the set of all points of the left gyrotranslated gyroray (278) by $\ominus A_1$. Hence, the set of points $A_1 \oplus A(t)$, $0 \leq t < \infty$, is the set of all points of the original gyroray

$$A_3P = \{A_1 \oplus A(t) : \ 0 \leq t < \infty\}, \tag{282}$$

which is of interest as indicated in Figs. 22, 23, and 24.

Noting that $\ominus A_1 \oplus A_1 = \mathbf{0}$ and $\gamma_{\mathbf{0}} = 1$, (280) can be written as

$$A(t) = \frac{m_1\gamma_{\ominus A_1 \oplus A_1}(\ominus A_1 \oplus A_1) + m_2\gamma_{\ominus A_1 \oplus A_2}(\ominus A_1 \oplus A_2) + m_3\gamma_{\ominus A_1 \oplus A_3}(\ominus A_1 \oplus A_3)}{m_1\gamma_{\ominus A_1 \oplus A_1} + m_2\gamma_{\ominus A_1 \oplus A_2} + m_3\gamma_{\ominus A_1 \oplus A_3}}, \tag{283}$$

where m_1 is determined by the condition

$$m_1 + m_2\gamma_{12} + m_3\gamma_{13} = 1 \tag{284}$$

that insures that (283) and (280) are identically equal.

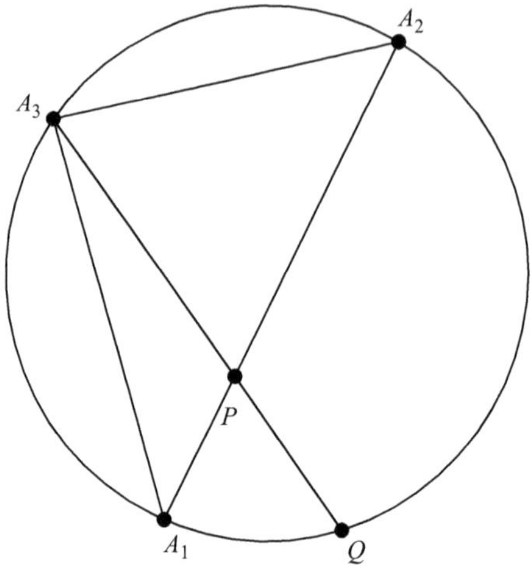

Fig. 24 Illustrating the Circumcevian Theorem 25, p. 72, and the Intersecting Chords Theorem 27, p. 80. A circular cevian, A_3Q, is shown. It is a cevian A_3P of a triangle $A_1A_2A_3$ in a Euclidean space $\mathbb{R}^n$, $n \geq 2$, extended to the point Q where it meets the triangle circumcircle. This figure presents two chords, A_1A_2 and A_3Q, of a circle, intersecting at a point P inside the circle. Here, the point $Q \in \mathbb{R}^n$ is determined by its barycentric representation (295) in Theorem 25. This figure is the Euclidean counterpart of Fig. 23

Solving (284) for m_1, where m_2 and m_3 are given by (281), we have

$$m_1 = \frac{(1-t_1)t}{1+(\gamma_{12}-1)t_1}. \tag{285}$$

By means of the Gyrobarycentric Representation Gyrocovariance Theorem [26, Theorem 4.6, pp. 90–91] with $X = \ominus A_1$, (283) can be written as

$$A(t) = \ominus A_1 \oplus \frac{m_1\gamma_{A_1}A_1 + m_2\gamma_{A_2}A_2 + m_3\gamma_{A_3}A_3}{m_1\gamma_{A_1} + m_2\gamma_{A_2} + m_3\gamma_{A_3}}, \tag{286}$$

where m_1, m_2, and m_3 are given by (285) and (281).

Following (286) we have, by a left cancellation,

$$Q(t) := A_1 \oplus A(t) = \frac{m_1\gamma_{A_1}A_1 + m_2\gamma_{A_2}A_2 + m_3\gamma_{A_3}A_3}{m_1\gamma_{A_1} + m_2\gamma_{A_2} + m_3\gamma_{A_3}}, \tag{287}$$

where the set of points $Q(t) = A_1 \oplus A(t)$, $0 \leq t < \infty$, forms the gyroray A_3P, (282), shown in Fig. 23.

We have thus obtained in (287), the gyrobarycentric representation with respect to $\{A_1, A_2, A_3\}$ of each point $Q(t)$ of the gyroray A_3P. We now wish to determine the unique point of the gyroray A_3P, other than A_3, that lies on the circumgyrocircle of gyrotriangle $A_1A_2A_3$ in Fig. 23.

By the Gyrocircle Gyrobarycentric Representation Theorem 17, p. 48, a point $Q(t)$ of the gyroray A_3P in (287) lies on the circumgyrocircle of a gyrotriangle $A_1A_2A_3$ if and only if the gyrobarycentric coordinates m_k, $k = 1, 2, 3$, of $Q(t)$ in (287) satisfy the circumgyrocircle condition (203).

Accordingly, we substitute $m_1.m_2.m_3$ from (285) and (281) into (203), obtaining a linear equation for the unknown t, the solution $t = t_0$ of which is substituted into (285) and (281), obtaining gyrobarycentric coordinates $m_1.m_2.m_3$ for the point $Q = Q(t_0)$ that lies on the circumgyrocircle of gyrotriangle $A_1A_2A_3$, shown in Figs. 22, 23, and 24. Finally, by removing an irrelevant nonzero common factor of the gyrobarycentric coordinates, we obtain the gyrobarycentric coordinates of Q,

$$\begin{aligned} m_1 &= \{\gamma_{13} - 1 + (\gamma_{23} - \gamma_{13})t_1\}(1 - t_1) \\ m_2 &= \{\gamma_{13} - 1 + (\gamma_{23} - \gamma_{13})t_1\}t_1 \\ m_3 &= -(\gamma_{12} - 1)(1 - t_1)t_1 \,, \end{aligned} \tag{288}$$

where t_1, $0 < t_1 < 1$, is a parameter that determines by (276), the location of the point P on the gyrochord A_1A_2 of the circumgyrocircle of gyrotriangle $A_1A_2A_3$ in Fig. 23.

The point Q, given by (287) with gyrobarycentric coordinates given by (288) is thus the only point of gyroray A_3P, other than A_3, that lies on the circumgyrocircle of gyrotriangle $A_1A_2A_3$ in Fig. 23. Formalizing, we thus obtain the following theorem:

Theorem 24 (The Circumgyrocevian Theorem). *Let A_3P be a gyrocevian of a gyrotriangle $A_1A_2A_3$ in an Einstein gyrovector space $(\mathbb{R}^n_s, \oplus, \otimes)$, $n \geq 2$, and let A_3Q be its corresponding circumgyrocevian, as shown in* Fig. 23. *Furthermore, let*

$$P = \frac{(1 - t_1)\gamma_{A_1} A_1 + t_1\gamma_{A_2} A_2}{(1 - t_1)\gamma_{A_1} + t_1\gamma_{A_2}} \tag{289}$$

be the gyrobarycentric representation of P with respect to the gyrobarycentrically independent set $\{A_1, A_2\}$ where the location of P on the circumgyrocircle gyrochord A_1A_2 is determined by the parameter t_1,

$$0 \leq t_1 \leq 1 \,. \tag{290}$$

Then, the point Q possesses the gyrobarycentric representation

$$Q = \frac{m_1\gamma_{A_1} A_1 + m_2\gamma_{A_2} A_2 + m_3\gamma_{A_3} A_3}{m_1\gamma_{A_1} + m_2\gamma_{A_2} + m_3\gamma_{A_3}} \tag{291}$$

with respect to the gyrobarycentrically independent set $\{A_1, A_2, A_3\}$, where the gyrobarycentric coordinates $m_1, m_2,$ and m_3 are given by

$$\begin{aligned} m_1 &= \{\gamma_{13} - 1 + (\gamma_{23} - \gamma_{13})t_1\}(1 - t_1) \\ m_2 &= \{\gamma_{13} - 1 + (\gamma_{23} - \gamma_{13})t_1\}t_1 \\ m_3 &= -(\gamma_{12} - 1)(1 - t_1)t_1 \,. \end{aligned} \tag{292}$$

Example 6 For $t_1 = 0$, the gyrobarycentric coordinates in (292) specialize to $m_1 = \gamma_{13} - 1 > 0$ and $m_2 = m_3 = 0$ implying, by (291), $Q = A_1$.

Example 7 For $t_1 = 1$, the gyrobarycentric coordinates in (292) specialize to $m_2 = \gamma_{23} - 1 > 0$ and $m_1 = m_3 = 0$ implying, by (291), $Q = A_2$.

Exploring the Euclidean limit of the Circumgyrocevian Theorem 24, we obtain the following Circumcevian Theorem:

Theorem 25 **(The Circumcevian Theorem)** *Let A_3P be a Cevian of a triangle $A_1A_2A_3$ in a Euclidean space $\mathbb{R}^n$, $n \geq 2$, and let A_3Q be its corresponding circumcevian, as shown in* Fig. 24. *Furthermore, let*

$$P = (1 - t_1)A_1 + t_1 A_2 \tag{293}$$

be the barycentric representation of P with respect to the barycentrically independent set $\{A_1, A_2\}$, so that the location of P on the circumcircle chord A_1A_2 is determined by the parameter t_1,

$$0 \leq t_1 \leq 1\,. \tag{294}$$

Then, the point Q possesses the barycentric representation

$$Q = \frac{m_1A_1 + m_2A_2 + m_3A_3}{m_1 + m_2 + m_3} \tag{295}$$

with respect to the barycentrically independent set $\{A_1, A_2, A_3\}$, where the barycentric coordinates $m_1, m_2,$ and m_3 are given by

$$\begin{aligned} m_1 &= \{\sin^2\alpha_2 + t_1 \sin(\alpha_1 - \alpha_2)\sin(\alpha_3\}(1 - t_1) \\ m_2 &= \{\sin^2\alpha_2 + t_1 \sin(\alpha_1 - \alpha_2)\sin(\alpha_3\}t_1 \\ m_3 &= -(1 - t_1)t_1 \sin^2\alpha_3\,, \end{aligned} \tag{296}$$

where α_k, $k = 1, 2, 3$, are the corresponding angles of triangle $A_1A_2A_3$.

Proof In the Euclidean limit, $s \to \infty$, the gyrobarycentric representation of $P \in \mathbb{R}^n_s$ in (289) and (290) reduces to the corresponding barycentric representations of $P \in \mathbb{R}^n$ in (293) and (294).

Obviously, in that Euclidean limit, the gyrobarycentric representation of $Q \in \mathbb{R}^n_s$ in (291) reduces to its corresponding barycentric representation of $Q \in \mathbb{R}^n$ in (295).

Hence, it remains to show that the gyrobarycentric coordinates m_k, $k = 1, 2, 3$, of $Q \in \mathbb{R}^n_s$ in (292) tend to their corresponding barycentric coordinates m_k, $k = 1, 2, 3$, of $Q \in \mathbb{R}^n$ in (296).

In the Euclidean limit, $s \to \infty$, gamma factors tend to 1. Hence, straightforward Euclidean limits of m_k, $k = 1, 2, 3$, in (292), as $s \to \infty$, give vanishing barycentric coordinates, $m_k = 0$, $k = 1, 2, 3$, resulting in an indeterminate barycentric representation of type $0/0$.

However, owing to the homogeneity of gyrobarycentric coordinates, the gyrobarycentric coordinates $m_k = 0$, $k = 1, 2, 3$, of Q in (291) and (292) can be written in the following form, which admits the Euclidean limit as $s \to \infty$,

$$\begin{aligned} m_1 &= \left\{ \frac{\gamma_{13} - 1}{\gamma_{12} - 1} + \frac{\gamma_{23} - \gamma_{13}}{\gamma_{12} - 1} t_1 \right\} (1 - t_1) \\ m_2 &= \left\{ \frac{\gamma_{13} - 1}{\gamma_{12} - 1} + \frac{\gamma_{23} - \gamma_{13}}{\gamma_{12} - 1} t_1 \right\} t_1 \\ m_3 &= -(1 - t_1) t_1 \,. \end{aligned} \tag{297}$$

Indeed, following Euclidean limits that can be derived from Lemma 1, we have the Euclidean limits:

$$\begin{aligned} \lim_{s\to\infty} \frac{\gamma_{13} - 1}{\gamma_{12} - 1} &= \frac{\sin^2 \alpha_2}{\sin^2 \alpha_3} \\ \lim_{s\to\infty} \frac{\gamma_{23} - \gamma_{13}}{\gamma_{12} - 1} &= \frac{\sin(\alpha_1 - \alpha_2)}{\sin \alpha_3} \,. \end{aligned} \tag{298}$$

Hence, in the Euclidean limit, $s \to \infty$, the gyrobarycentric coordinates m_k, $k = 1, 2, 3$, of $Q \in \mathbb{R}^n_s$ in (297) tend to the following barycentric coordinates of $Q \in \mathbb{R}^n$,

$$\begin{aligned} m_1 &= \left\{ \frac{\sin^2 \alpha_2}{\sin^2 \alpha_3} + \frac{\sin(\alpha_1 - \alpha_2)}{\sin \alpha_3} t_1 \right\} (1 - t_1) \\ m_2 &= \left\{ \frac{\sin^2 \alpha_2}{\sin^2 \alpha_3} + \frac{\sin(\alpha_1 - \alpha_2)}{\sin \alpha_3} t_1 \right\} t_1 \\ m_3 &= -(1 - t_1) t_1 \,. \end{aligned} \tag{299}$$

Owing to their homogeneity, the barycentric coordinates (299) of $Q \in \mathbb{R}^n$ can be written as (296), as desired.

23 Gyrodistances Related to the Gyrocevian

In order to prepare the stage for the Intersecting Gyrochords Theorem 26, p. 79, in this section we calculate the gyrodistance between A_k and the gyrocevian foot P, $k = 1, 2, 3$, of a gyrotriangle $A_1A_2A_3$, shown in Fig. 23.

Let $A_1A_2A_3$ be a gyrotriangle that possesses a circumgyrocircle in an Einstein gyrovector space $(\mathbb{R}^n_s, \oplus, \otimes)$, and let P, given by (289), be a generic point of the interior of chord A_1A_2 of the circumgyrocircle, as shown in Fig. 23.

Applying the Gyrobarycentric Representation Gyrocovariance Theorem [26, Theorem 4.6, pp. 90–91] to the gyrobarycentric representation (289) of P, using the gyrotriangle index notation (13), p. 7, we have

$$\ominus A_1 \oplus P = \frac{t_1 \gamma_{12} \mathbf{a}_{12}}{(1 - t_1) + t_1 \gamma_{12}}$$

$$\ominus A_2 \oplus P = \frac{(1-t_1)\gamma_{12}\mathbf{a}_{21}}{(1-t_1)\gamma_{12}+t_1}$$
$$\ominus A_3 \oplus P = \frac{(1-t_1)\gamma_{13}\mathbf{a}_{31}+t_1\gamma_{23}\mathbf{a}_{23}}{(1-t_1)\gamma_{13}+t_1\gamma_{23}} \tag{300}$$

and

$$\begin{aligned}
\gamma_{|A_1P|} &= \gamma_{\ominus A_1 \oplus P} = \frac{(1-t_1)+t_1\gamma_{12}}{m_P} \\
\gamma_{|A_2P|} &= \gamma_{\ominus A_2 \oplus P} = \frac{(1-t_1)\gamma_{12}+t_1}{m_P} \\
\gamma_{|A_3P|} &= \gamma_{\ominus A_3 \oplus P} = \frac{(1-t_1)\gamma_{13}+t_1\gamma_{23}}{m_P}\,,
\end{aligned} \tag{301}$$

where $m_P > 0$ is the constant of the gyrobarycentric representation (289) of P with respect to the set $\{A_1, A_2\}$, given by

$$m_P^2 = (1-t_1)^2 + t_1^2 + 2(1-t_1)t_1\gamma_{12} = 1 + 2(\gamma_{12}-1)(1-t_1)t_1\,. \tag{302}$$

By Identity (4), p. 3, and (301) and straightforward algebra, we have

$$\begin{aligned}
\frac{1}{s^2}\gamma_{|A_1P|}^2|A_1P|^2 &= \gamma_{|A_1P|}^2 - 1 = \frac{(\gamma_{12}^2-1)t_1^2}{1+2(\gamma_{12}-1)(1-t_1)t_1} \\
\frac{1}{s^2}\gamma_{|A_2P|}^2|A_2P|^2 &= \gamma_{|A_2P|}^2 - 1 = \frac{(\gamma_{12}^2-1)(1-t_1)^2}{1+2(\gamma_{12}-1)(1-t_1)t_1} \\
\frac{1}{s^2}\gamma_{|A_3P|}^2|A_3P|^2 &= \gamma_{|A_3P|}^2 - 1 = \frac{\{\gamma_{13}(1-t_1)+\gamma_{23}t_1\}^2}{1+2(\gamma_{12}-1)(1-t_1)t_1} - 1\,.
\end{aligned} \tag{303}$$

The equations in (303) prove useful in the proof of Theorem 26, p. 79.

Finally, by means of Identity (4), p. 3, it follows from (303) that the squared gyrodistances $|A_kP|^2 = \|\ominus A_k \oplus P\|^2$ between A_k and P, $k = 1, 2, 3$, are given by the equations

$$\begin{aligned}
\frac{1}{s^2}\|\ominus A_1 \oplus P\|^2 &= \frac{(\gamma_{12}^2-1)t_1^2}{\{(\gamma_{12}-1)t_1+1\}^2} \\
\frac{1}{s^2}\|\ominus A_2 \oplus P\|^2 &= \frac{(\gamma_{12}^2-1)(1-t_1)^2}{\{\gamma_{12}(1-t_1)+t_1\}^2} \\
\frac{1}{s^2}\|\ominus A_3 \oplus P\|^2 &= 1 - \frac{2(\gamma_{12}-1)(1-t_1)t_1+1}{\{\gamma_{13}(1-t_1)+\gamma_{23}t_1\}^2}\,.
\end{aligned} \tag{304}$$

It is clear from the first and the second equations in (304) that $|A_1P| = 0$ when $t_1 = 0$ and $|A_2P| = 0$ when $t_1 = 1$. Hence, $P = A_1$ when $t_1 = 0$ and $P = A_2$ when $t_1 = 1$, as indicated in Fig. 23.

24 A Gyrodistance Related to the Circumgyrocevian

In order to prepare the stage for the Intersecting Gyrochords Theorem 26, p. 79, in this section we calculate the gyrodistance $|PQ| = \|\ominus P \oplus Q\|$ between the gyrocevian foot P and the circumgyrocevian foot Q of a gyrotriangle $A_1A_2A_3$, shown in Fig. 23, p. 71.

By Theorem 24, the point Q, shown in Fig. 23, is given by its gyrobarycentric representation, (291),

$$Q = \frac{m_1\gamma_{A_1} A_1 + m_2\gamma_{A_2} A_2 + m_3\gamma_{A_3} A_3}{m_1\gamma_{A_1} + m_2\gamma_{A_2} + m_3\gamma_{A_3}} \tag{305a}$$

with respect to the set $\{A_1, A_2, A_3\}$, where, (292),

$$\begin{aligned} m_1 &= \{\gamma_{13} - 1 + (\gamma_{23} - \gamma_{13})t_1\}(1 - t_1) \\ m_2 &= \{\gamma_{13} - 1 + (\gamma_{23} - \gamma_{13})t_1\}t_1 \\ m_3 &= -(\gamma_{12} - 1)(1 - t_1)t_1 \end{aligned} \tag{305b}$$

and the point P is given by its gyrobarycentric representation, (289),

$$P = \frac{m_1'\gamma_{A_1} A_1 + m_2'\gamma_{A_2} A_2}{m_1'\gamma_{A_1} + m_2'\gamma_{A_2}}, \tag{306a}$$

where

$$\begin{aligned} m_1' &= 1 - t_1 \\ m_2' &= t_1 , \end{aligned} \tag{306b}$$

$0 < t_1 < 1$.

Hence, by [26, Sect. 4.9] with $N = 3$,

$$\begin{aligned} \gamma_{\ominus P \oplus Q} &= \frac{1}{m_P m_Q} \{(m_1 m_2' + m_1' m_2)\gamma_{12} + (m_1 m_3' + m_1' m_3)\gamma_{13} \\ &\quad + (m_2 m_3' + m_2' m_3)\gamma_{23} + m_1 m_1' + m_2 m_2' + m_3 m_3'\} \\ &= \frac{1}{m_P m_Q} \{(m_1 m_2' + m_1' m_2)\gamma_{12} + m_1' m_3 \gamma_{13} + m_2' m_3 \gamma_{23} + m_1 m_1' + m_2 m_2'\} \end{aligned} \tag{307}$$

noting that $m_3' = 0$ in the gyrobarycentric representation of P in (306).

As we see from the *gyrobarycentric representation constant* associated with the Gyrobarycentric Representation Gyrocovariance Theorem [26, Theorem 4.6, pp. 90–91], and by (305b) and (306b), $m_Q > 0$ and $m_P > 0$ are given by

$$m_Q^2 = m_1^2 + m_2^2 + m_3^2 + 2(m_1 m_2 \gamma_{12} + m_1 m_3 \gamma_{13} + m_2 m_3 \gamma_{23})$$

$$
\begin{aligned}
&= \{(\gamma_{13}-1)(1-t_1)+(\gamma_{23}-1)t_1-(\gamma_{12}-1)(1-t_1)t_1\}^2 \\
m_P^2 &= (m_1')^2+(m_2')^2+2m_1'm_2'\gamma_{12} \\
&= 1+2(\gamma_{12}-1)(1-t_1)t_1\,,
\end{aligned}
\tag{308}
$$

noting that $m_P^2 > 0$ for all $0 < t_1 < 1$. Also $m_Q^2 > 0$ for all $0 < t_1 < 1$ according to the Circumgyrocenter Theorem 1, p. 12, since gyrotriangle $A_1A_2A_3$ possesses a circumgyrocircle on which the point Q lies, as shown in Fig. 23.

In (308), m_Q^2 is a *perfect square* in the sense that it appears in (308) as a squared polynomial function. Since $m_Q > 0$, we have

$$
m_Q = (\gamma_{13}-1)(1-t_1)+(\gamma_{23}-1)t_1-(\gamma_{12}-1)(1-t_1)t_1 \; > \; 0\,,
\tag{309}
$$

where, indeed, $m_Q > 0$ for all $0 \le t_1 \le 1$.

Substituting (305b) and (306b) into (307) and squaring, we obtain $\gamma^2_{\ominus P\oplus Q}$ and, hence, $\gamma^2_{\ominus P\oplus Q} - 1$ expressed in terms of gamma factors and the parameter t_1, $0 < t_1 < 1$,

$$
\begin{aligned}
&\frac{1}{s^2}\gamma^2_{|PQ|}|PQ|^2 = \gamma^2_{\ominus P\oplus Q} - 1 \\
&= \frac{(\gamma_{12}-1)^2\{[\gamma_{13}(1-t_1)+\gamma_{23}t_1]^2-2(\gamma_{12}-1)(1-t_1)t_1-1\}(1-t_1)^2t_1^2}{\{2(\gamma_{12}-1)(1-t_1)t_1+1\}\{(\gamma_{12}-1)(1-t_1)t_1-\gamma_{13}(1-t_1)-\gamma_{23}t_1+1\}^2}\,,
\end{aligned}
\tag{310}
$$

noting (4), p. 3. The extreme sides of (310) prove useful in the proof of Theorem 26, p. 79.

Substituting $\gamma^2_{\ominus P\oplus Q}$ from (310) into Identity, (4), p. 3,

$$
|PQ|^2 := \|\ominus P\oplus Q\|^2 = s^2\frac{\gamma^2_{\ominus P\oplus Q}-1}{\gamma^2_{\ominus P\oplus Q}}\,,
\tag{311}
$$

we obtain the desired gyrodistance,

$$
\frac{1}{s^2}\|\ominus P\oplus Q\|^2 = \frac{(\gamma_{12}-1)^2\{[\gamma_{13}(1-t_1)+\gamma_{23}t_1]^2-2(\gamma_{12}-1)(1-t_1)t_1-1\}(1-t_1)^2t_1^2}{D^2}\,,
\tag{312a}
$$

where

$$
\begin{aligned}
D = \gamma_{13}-1+(\gamma_{12}-1)\gamma_{13}t_1-2(\gamma_{12}-1)(1-t_1)t_1-(\gamma_{12}-1)\gamma_{23}t_1^2 \\
+(\gamma_{13}-\gamma_{23})\{-\gamma_{12}+(\gamma_{12}-1)(1-t_1)^2\}t_1\,.
\end{aligned}
\tag{312b}
$$

25 Circumgyrocevian Gyrolength

In order to prepare the stage for the Intersecting Gyrochords Theorem 26, p. 79, in this section we calculate the gyrodistance $|A_3Q| = \|\ominus A_3\oplus Q\|$ between vertex A_3 and its opposing circumgyrocevian foot Q of a gyrotriangle $A_1A_2A_3$, shown in Fig. 23, p. 71.

Let A_3Q be a circumgyrocevian of a gyrotriangle $A_1A_2A_3$ in an Einstein gyrovector space $(\mathbb{R}^n_s, \oplus, \otimes)$, as shown in Fig. 23, so that the gyrobarycentric representation of Q with respect to the gyrobarycentrically independent set $\{A_1, A_2, A_3\}$ is given by (305).

Applying the Gyrobarycentric Representation Gyrocovariance Theorem [26, Theorem 4.6, pp. 90–91] with $X = \ominus A_3$ to the gyrobarycentric representation (305a) of Q, and using the gyrotriangle index notation, (13), p. 7, we have

$$\begin{aligned}\ominus A_3\oplus Q &= \frac{m_1\gamma_{\ominus A_3\oplus A_1}(\ominus A_3\oplus A_1)+m_2\gamma_{\ominus A_3\oplus A_2}(\ominus A_3\oplus A_2)+m_3\gamma_{\ominus A_3\oplus A_3}(\ominus A_3\oplus A_3)}{m_1\gamma_{\ominus A_3\oplus A_1}+m_2\gamma_{\ominus A_3\oplus A_2}+m_3\gamma_{\ominus A_3\oplus A_3}}\\ &= \frac{m_1\gamma_{13}\mathbf{a}_{31}+m_2\gamma_{23}\mathbf{a}_{32}}{m_1\gamma_{13}+m_2\gamma_{23}+m_3}\end{aligned} \tag{313}$$

and

$$\gamma_{\ominus A_3\oplus Q} = \frac{m_1\gamma_{13}+m_2\gamma_{23}+m_3}{m_Q}, \tag{314}$$

where $m_Q > 0$ is the constant of the gyrobarycentric representation (305) of Q, given by (309).

Substituting m_1, m_2, and m_3 from (305b) and m_Q from (309) into (314), we obtain the equation

$$\gamma_{\ominus A_3\oplus Q} = \frac{\gamma_{13}(\gamma_{13}-1)(1-t_1)+\gamma_{23}(\gamma_{23}-1)t_1-\{\gamma_{12}-1+(\gamma_{13}-\gamma_{23})^2\}(1-t_1)t_1}{(\gamma_{13}-1)(1-t_1)+(\gamma_{23}-1)t_1-(\gamma_{12}-1)(1-t_1)t_1}. \tag{315}$$

Equation (315) proves useful in the proof of Theorem 26 in Sect. 26.

Finally, the gyrolength $\|\ominus A_3\oplus Q\|$ of the circumgyrocevian A_3Q of gyrotriangle $A_1A_2A_3$ in Fig. 23 is obtained from (315) by means of (4), p. 3,

$$\begin{aligned}\frac{1}{s^2}\|\ominus A_3\oplus Q\|^2 &= \frac{\gamma^2_{\ominus A_3\oplus Q}-1}{\gamma^2_{\ominus A_3\oplus Q}}\\ &= \frac{\{\gamma_{13}^2t_2+\gamma_{23}^2t_1-1-[2(\gamma_{12}-1)+(\gamma_{23}-\gamma_{13})^2]t_2t_1\}\{\gamma_{13}-1+(\gamma_{23}-\gamma_{13})t_1\}^2}{\{\gamma_{13}(\gamma_{13}-1)t_2+\gamma_{23}(\gamma_{23}-1)t_1-\{\gamma_{12}-1+(\gamma_{13}-\gamma_{23})^2\}t_2t_1\}^2},\end{aligned} \tag{316}$$

where $t_2 = 1-t_1, 0 < t_1 < 1$.

26 The Intersecting Gyrochords Theorem

Following Sects. 23–25, we are now in the position to prove the Intersecting Gyrochords Theorem.

Theorem 26 (The Intersecting Gyrochords Theorem) *If two gyrochords, A_1A_2 and B_1B_2, of a gyrocircle in an Einstein gyrovector space $(\mathbb{R}^n_s, \oplus, \otimes)$ intersect at a point P, as shown in* Fig. 25, *then*

$$\frac{\gamma_{|PA_1|}|PA_1|\gamma_{|PA_2|}|PA_2|}{\gamma_{|A_1A_2|}+1} = \frac{\gamma_{|PB_1|}|PB_1|\gamma_{|PB_2|}|PB_2|}{\gamma_{|B_1B_2|}+1}. \tag{317}$$

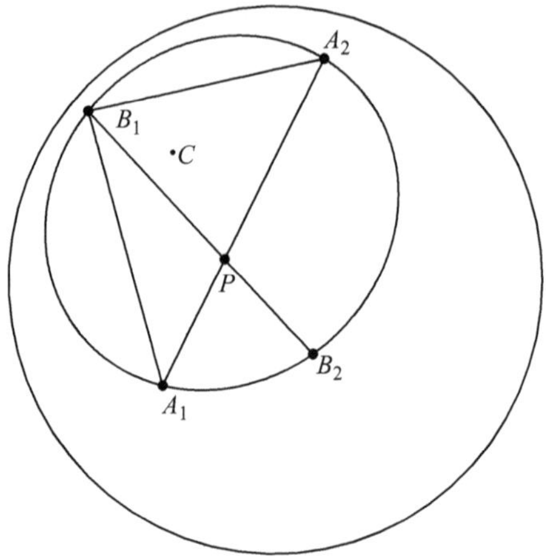

Fig. 25 The Intersecting Gyrochords Theorem 26 in the hyperbolic plane $(\mathbb{R}_s^2, \oplus, \otimes)$. Two intersecting gyrochords of a gyrocircle, A_1A_2 and B_1B_2, are shown in this hyperbolic counterpart of Fig. 26

Proof As a matter of notation we have

$$\begin{aligned}|A_1A_2| &= \|\ominus A_1 \oplus A_2\| = \gamma_{12}\\ |B_1B_2| &= |A_3Q|\,,\end{aligned} \tag{318}$$

where we use the notation in Fig. 23 in which, in particular, $B_1 = A_3$ and $B_2 = Q$.

Hence, noting that each side of (317) is positive, it can be written equivalently as

$$\left(\frac{\gamma_{|PA_1|}|PA_1|\gamma_{|PA_2|}|PA_2|}{\gamma_{12}+1}\right)^2 = \left(\frac{\gamma_{|PA_3|}|PA_3|\gamma_{|PQ|}|PQ|}{\gamma_{|A_3Q|}+1}\right)^2, \tag{319}$$

where $|PA_1| = |A_1P|$ is the gyrolength $\|\ominus P \oplus A_1\|$ of gyrosegment PA_1, *etc.*

Substituting into (319)

1. $(\gamma_{|PA_k|}|PA_k|)^2$, $k = 1, 2, 3$, from (303) in Sect. 23, and
2. $(\gamma_{|PQ|}|PQ|)^2$ from (310) in Sect. 24, and
3. $\gamma_{|A_3Q|}$ from (315) in Sect. 25,

we find that Identity (319) is valid, each side of which being equal to the right-hand side of each of the two equations in (320) below,

$$\begin{aligned}\left(\frac{\gamma_{|PA_1|}|PA_1|\gamma_{|PA_2|}|PA_2|}{\gamma_{12}+1}\right)^2 &= s^4\left(\frac{(\gamma_{12}^2-1)(1-t_1)t_1}{2(\gamma_{12}^2-1)(1-t_1)t_1+1}\right)^2\\ \left(\frac{\gamma_{|PA_3|}|PA_3|\gamma_{|PQ|}|PQ|}{\gamma_{|A_3Q|}+1}\right)^2 &= s^4\left(\frac{(\gamma_{12}^2-1)(1-t_1)t_1}{2(\gamma_{12}^2-1)(1-t_1)t_1+1}\right)^2.\end{aligned} \tag{320}$$

The two equations in (320) share the right-hand sides, so that the proof of (319) and, hence, of (317), is complete.

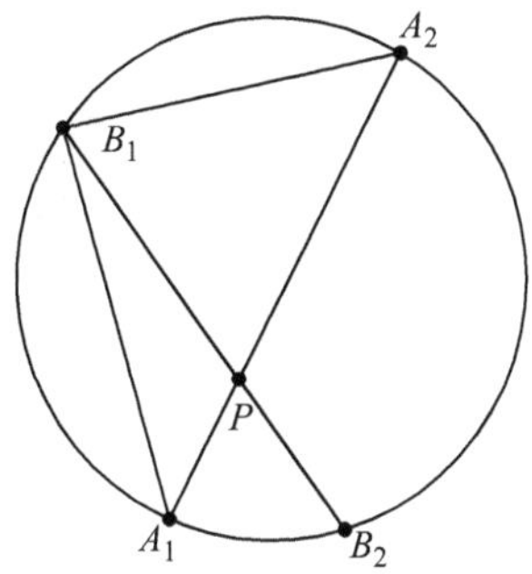

Fig. 26 The Intersecting Chords Theorem 27 in the Euclidean plane $\mathbb{R}^2$. Two intersecting chords of a circle, A_1A_2 and B_1B_2, of a circle are shown. This figure is the Euclidean counterpart of Fig. 25

In the Euclidean limit, $s \to \infty$, gyrolengths of gyrosegments tend to lengths of corresponding segments and gamma factors tend to 1. Hence, in that limit, the Intersecting Gyrochords Theorem 26 reduces to the following well-known Intersecting Chords Theorem of Euclidean geometry:

Theorem 27 (The Intersecting Chords Theorem) *If two chords, A_1A_2 and B_1B_2, of a circle in a Euclidean space $\mathbb{R}^n$ intersect at a point P, as shown in* Fig. 26, *then*

$$|PA_1||PA_2| = |PB_1||PB_2|\,. \tag{321}$$

Our ambiguous notation that emphasizes analogies between hyperbolic and Euclidean geometry should raise no confusion. Thus, in particular, one should note that:

1. in the Intersecting Gyrochords Theorem 26, $|PA_1|$ is the gyrolength,

$$|PA_1| = \|\ominus P \oplus A_1\|\,, \tag{322}$$

   of gyrosegment PA_1, etc., while, in contrast,
2. in the Intersecting Chords Theorem 27, $|PA_1|$ is the length,

$$|PA_1| = \|-P + A_1\|\,, \tag{323}$$

of segment PA_1, etc.

References

1. Crowe, M.J.: A History of Vector Analysis. Dover Publications Inc., New York (1994). (The Evolution of the Idea of a Vectorial System, Corrected reprint of the 1985 edition)
2. Einstein, A.: Zur Elektrodynamik Bewegter Körper [on the electrodynamics of moving bodies]. Ann. Physik. (Leipzig) **17**, 891–921 (1905). (We use the English translation in [3] or in [8], or in http://www.fourmilab.ch/etexts/einstein/specrel/www/)
3. Einstein, A.: Einstein's Miraculous Years: Five Papers that Changed the Face of Physics. Princeton University Press, Princeton (1998). Edited and introduced by John Stachel. Includes

bibliographical references. Einstein's dissertation on the determination of molecular dimensions, Einstein on Brownian motion, Einstein on the theory of relativity, Einstein's early work on the quantum hypothesis. A new English translation of Einstein's 1905 paper on pp. 123–160
4. Fock, V.: The Theory of Space, Time and Gravitation. The Macmillan Co., New York (1964). (Second revised edition. Translated from the Russian by Kemmer, N. A Pergamon Press Book)
5. Gray, J.: Möbius's geometrical mechanics. In: Fauvel, J., Flood, R., Wilson, R. (eds.) Möbius and His Band, Mathematics and Astronomy in Nineteenth-Century Germany, 78–103. The Clarendon Press Oxford University Press, New York (1993)
6. Kelly, P.J., Matthews, G.: The Non-Euclidean, Hyperbolic Plane. Its Structure and Consistency. Springer-Verlag, New York (1981). (Universitext)
7. Kimberling, C.: Clark Kimberling's Encyclopedia of Triangle Centers—ETC (2012). http://faculty.evansville.edu/ck6/encyclopedia/ETC.html
8. Lorentz, H.A., Einstein, A., Minkowski, H., Weyl, H.: The Principle of Relativity. Dover Publications Inc., New York (1923) (undated). A collection of original memoirs on the special and general theory of relativity, with notes by A. Sommerfeld, Translated by W. Perrett and G.B. Jeffery
9. Maor, E.: Trigonometric Delights. Princeton University Press, Princeton (1998)
10. Millman, R.S., Parker, G.D.: Geometry: A Metric Approach with Models, 2nd edn. Springer-Verlag, New York (1991)
11. Møller, C.: The Theory of Relativity. Clarendon Press, Oxford (1952)
12. Rassias, T.M.: Book review: analytic hyperbolic geometry and Albert Einstein's special theory of relativity, by Abraham A. Ungar. Nonlinear Funct. Anal. Appl. **13**(1), 167–177 (2008)
13. Rassias, T.M.: Book review: A gyrovector space approach to hyperbolic geometry, by Abraham A. Ungar. J. Geom. Symm. Phys. **18**, 93–106 (2010)
14. Sexl, R.U., Urbantke, H.K.: Relativity, Groups, Particles. Special Relativity and Relativistic Symmetry in Field and Particle Physics. Springer-Verlag, Vienna (2001). (Revised and translated from the third German (1992) edition by Urbantke)
15. Sönmez, N., Ungar, A.A.: The Einstein relativistic velocity model of hyperbolic geometry and its plane separation axiom. Adv. Appl. Clifford Alg. **23**, 209–236 (2013)
16. Ungar, A.A.: Quasidirect product groups and the Lorentz transformation group. In: Rassias, T.M. (ed.) Constantin Carathéodory: An international Tribute, vol. I, II, pp. 1378–1392. World Sci. Publishing, Teaneck (1991)
17. Ungar, A.A.: Gyrovector spaces in the service of hyperbolic geometry. In: Rassias, T.M. (ed.) Mathematical Analysis and Applications, pp. 305–360. Hadronic Press, Palm Harbor (2000)
18. Ungar, A.A.: Möbius transformations of the ball, Ahlfors' rotation and gyrovector spaces. In: Rassias, T.M. (ed.) Nonlinear Analysis in Geometry and Topology, pp. 241–287. Hadronic Press, Palm Harbor (2000)
19. Ungar, A.A.: Beyond the Einstein Addition Law and its Gyroscopic Thomas Precession: The Theory of Gyrogroups and Gyrovector Spaces, vol. 117 of Fundamental Theories of Physics. Kluwer Academic, Dordrecht (2001)
20. Ungar, A.A.: Analytic Hyperbolic Geometry: Mathematical Foundations and Applications. World Scientific Publishing Co. Pte. Ltd., Hackensack (2005)
21. Ungar, A.A.: The hyperbolic square and Möbius transformations. Banach J. Math. Anal. **1**(1), 101–116 (2007)
22. Ungar, A.A.: Analytic Hyperbolic Geometry and Albert Einstein's Special Theory of Relativity. World Scientific Publishing Co, Pte. Ltd., Hackensack (2008)
23. Ungar, A.A.: A Gyrovector Space Approach to Hyperbolic Geometry. Morgan & Claypool, San Rafael (2009)
24. Ungar, A.A.: The hyperbolic triangle incenter. Dedicated to the 30th anniversary of Themistocles M. Rassias' stability theorem. Nonlinear Funct. Anal. Appl. **14**(5), 817–841 (2009)
25. Ungar, A.A.: Barycentric Calculus in Euclidean and Hyperbolic Geometry: A Comparative Introduction. World Scientific Publishing Co, Pte. Ltd., Hackensack (2010)

26. Ungar, A.A.: Hyperbolic Triangle Centers: The Special Relativistic Approach. Springer-Verlag, New York (2010)
27. Ungar, A.A.: Gyrations: the missing link between classical mechanics with its underlying Euclidean geometry and relativistic mechanics with its underlying hyperbolic geometry. In: Pardalos, P.M., Rassias, T.M. (eds.) Essays in Mathematics and its Applications in Honor of Stephen Smale's 80th Birthday, pp. 463–504. Springer, Heidelberg (2012). arXiv 1302.5678 (math-ph)
28. Ungar, A.A.: Möbius transformation and Einstein velocity addition in the hyperbolic geometry of Bolyai and Lobachevsky. In: Pardalos, P.M., Georgiev, P.G., Srivastava, H.M. (eds.) Nonlinear Analysis, vol. 68 of Springer Optim. Appl., pp. 721–770. Springer, New York (2012). Stability, Approximation, and Inequalities. In honor of Themistocles M. Rassias on the occasion of his 60th birthday
29. Ungar, A.A.: An introduduction to hyperbolic barycentric coordinates and their applications. In: Pardalos, P., Rassias, T.M. (eds) Mathematics Without Boundaries: Surveys in Interdisciplinary Research. Springer Optim. Appl. Springer, New York (2013) (in print)
30. Walter, S.: Book review: beyond the Einstein addition law and its gyroscopic Thomas precession: the theory of gyrogroups and gyrovector spaces, by Abraham A. Ungar. Found. Phys. **32**(2), 327–330 (2002)

Multidimensional Half-Discrete Hilbert-Type Inequalities and Operator Expressions

Bicheng Yang

Mathematics Subject Classification *26D15, 31A10, 45P05, 47G10, 47A07*

Abstract In this chapter, by the use of the methods of weight functions and technique of real analysis, two kinds of multidimensional more accurate half-discrete Hilbert-type inequalities with the best possible constant factors are provided. Equivalent forms and some reverses are obtained. We also consider the related operator expressions with the norms, some particular examples and some corollaries in certain conditions.

Keywords Half-discrete Hilbert-type inequality · Weight function · Equivalent form · Reverse · Hilbert-type operator

1 Introduction

Assuming that $p > 1, \frac{1}{p} + \frac{1}{q} = 1, f(x), g(y) \geq 0, f \in L^p(\mathbf{R}_+), g \in L^q(\mathbf{R}_+)$,

$$||f||_p = \left\{ \int_0^\infty f^p(x)dx \right\}^{\frac{1}{p}} > 0,$$

$||g||_q > 0$, we have the following Hardy–Hilbert's integral inequality (cf. [1]):

$$\int_0^\infty \int_0^\infty \frac{f(x)g(y)}{x+y} dxdy < \frac{\pi}{\sin(\pi/p)} ||f||_p ||g||_q, \tag{1}$$

where the constant factor $\frac{\pi}{\sin(\pi/p)}$ is the best possible factor. If $a_m, b_n \geq 0$, $a = \{a_m\}_{m=1}^\infty \in l^p, b = \{b_n\}_{n=1}^\infty \in l^q$,

$$||a||_p = \left\{ \sum_{m=1}^\infty a_m^p \right\}^{\frac{1}{p}} > 0,$$

B. Yang (✉)
Department of Mathematics, Guangdong University of Education,
Guangdong, 510303 Guangzhou, P. R. China
e-mail: bcyang@gdei.edu.cn; bcyang818@163.com

T. M. Rassias, P. M. Pardalos (eds.), *Mathematics Without Boundaries*,
DOI 10.1007/978-1-4939-1106-6_23,

$||b||_q > 0$, then we still have the following discrete variant of the above inequality with the same best constant $\frac{\pi}{\sin(\pi/p)}$:

$$\sum_{m=1}^{\infty}\sum_{n=1}^{\infty}\frac{a_m b_n}{m+n} < \frac{\pi}{\sin(\pi/p)}||a||_p||b||_q, \tag{2}$$

which is named after Hardy–Hilbert's inequality (cf. [1]). Inequalities (1) and (2) are important in analysis and its applications (cf. [1–6]).

In 1998, by introducing an independent parameter $\lambda \in (0,1]$, Yang [7] gave an extension of (1) at $p = q = 2$. In 2009–2011, Yang [3, 4] gave some extensions of (1) and (2) as follows:

If $\lambda_1, \lambda_2, \lambda \in \mathbf{R}, \lambda_1 + \lambda_2 = \lambda, k_\lambda(x,y)$ is a non-negative homogeneous function of degree $-\lambda$, with

$$k(\lambda_1) = \int_0^\infty k_\lambda(t,1)t^{\lambda_1-1}dt \in \mathbf{R}_+,$$
$$\phi(x) = x^{p(1-\lambda_1)-1}, \psi(y) = y^{q(1-\lambda_2)-1},$$

$f(x), g(y) \geq 0$,

$$f \in L_{p,\phi}(\mathbf{R}_+) = \left\{f; ||f||_{p,\phi} := \left\{\int_0^\infty \phi(x)|f(x)|^p dx\right\}^{\frac{1}{p}} < \infty\right\},$$

$g \in L_{q,\psi}(\mathbf{R}_+), ||f||_{p,\phi}, ||g||_{q,\psi} > 0$, then we have

$$\int_0^\infty \int_0^\infty k_\lambda(x,y)f(x)g(y)dxdy < k(\lambda_1)||f||_{p,\phi}||g||_{q,\psi}, \tag{3}$$

where the constant factor $k(\lambda_1)$ is the best possible. Moreover, if $k_\lambda(x,y)$ is finite and $k_\lambda(x,y)x^{\lambda_1-1}$ $\left(k_\lambda(x,y)y^{\lambda_2-1}\right)$ is decreasing with respect to $x > 0 (y > 0)$, then for $a_m, b_n \geq 0$,

$$a \in l_{p,\phi} = \left\{a; ||a||_{p,\phi} := \left\{\sum_{n=1}^{\infty}\phi(n)|a_n|^p\right\}^{\frac{1}{p}} < \infty\right\},$$

$b = \{b_n\}_{n=1}^{\infty} \in l_{q,\psi}, ||a||_{p,\phi}, ||b||_{q,\psi} > 0$, we have

$$\sum_{m=1}^{\infty}\sum_{n=1}^{\infty} k_\lambda(m,n)a_m b_n < k(\lambda_1)||a||_{p,\phi}||b||_{q,\psi}, \tag{4}$$

where, the constant factor $k(\lambda_1)$ is still the best possible.

Clearly, for $\lambda = 1, k_1(x,y) = \frac{1}{x+y}, \lambda_1 = \frac{1}{q}, \lambda_2 = \frac{1}{p}$, (3) reduces to (1), while (4) reduces to (2). Some other results including multidimensional Hilbert-type integral inequalities are provided by [8–21].

On the topic of half-discrete Hilbert-type inequalities with the nonhomogeneous kernels, Hardy et al. provided a few results in Theorem 351 of [1]. They did not prove that the constant factors are the best possible. However, Yang [22] gave a result with the kernel $\frac{1}{(1+nx)^{\lambda}}$ by introducing a variable and proved that the constant factor is the best possible. In 2011, Yang [23] gave the following half-discrete Hardy–Hilbert's inequality with the best possible constant factor $B(\lambda_1, \lambda_2)$:

$$\int_0^{\infty} f(x) \sum_{n=1}^{\infty} \frac{a_n}{(x+n)^{\lambda}} dx < B(\lambda_1, \lambda_2) ||f||_{p,\phi} ||a||_{q,\psi}, \tag{5}$$

where, $\lambda_1 \lambda_2 > 0, 0 \leq \lambda_2 \leq 1, \lambda_1 + \lambda_2 = \lambda$,

$$B(u, v) = \int_0^{\infty} \frac{1}{(1+t)^{u+v}} t^{u-1} dt (u, v > 0)$$

is the beta function. Zhong et al. [24–30] investigated several half-discrete Hilbert-type inequalities with particular kernels.

Using the way of weight functions and the techniques of discrete and integral Hilbert-type inequalities with some additional conditions on the kernel, a half-discrete Hilbert-type inequality with a general homogeneous kernel of degree $-\lambda \in \mathbf{R}$ and a best constant factor $k(\lambda_1)$ is obtained as follows:

$$\int_0^{\infty} f(x) \sum_{n=1}^{\infty} k_{\lambda}(x, n) a_n dx < k(\lambda_1) ||f||_{p,\phi} ||a||_{q,\psi}, \tag{6}$$

which is an extension of (5) (see Yang and Chen [31]). At the same time, a half-discrete Hilbert-type inequality with a general nonhomogeneous kernel and a best constant factor is given by Yang [32] .

Remark 1

1. Many different kinds of Hilbert-type discrete, half-discrete, and integral inequalities with applications are presented in recent 20 years. Special attention is given to new results proved during 2009–2012. Included are many generalizations, extensions, and refinements of Hilbert-type discrete, half-discrete, and integral inequalities involving many special functions such as Riemann's zeta, beta, gamma, hypergeometric, trigonometric, hyperbolic, zeta, Bernoulli functions, Bernoulli numbers, and Euler constant et al.
2. In his five books, Yang [3–6, 33] presented many new results on Hilbert-type operators with general homogeneous kernels of degree of real numbers and two pairs of conjugate exponents as well as the related inequalities. These research monographs contained recent developments of discrete, half-discrete, and integral types of operators and inequalities with proofs, examples, and applications.

In this chapter, by the use of the methods of weight functions and technique of real analysis, two kinds of multidimensional more accurate half-discrete Hilbert-type inequalities with the best possible constant factors are given, which are extensions of

(6) and Yang's [32] work. Equivalent forms and some reverses with the best constant factors are obtained. We also consider the relating operator expressions with the norm, some particular examples and some corollaries in certain conditions.

2 Some Lemmas

Lemma 1 *Suppose that* $(-1)^i h^{(i)}(t) > 0 (t > 0; i = 0, 1, 2)$. *Then*

(i) *For* $b > 0, 0 < \alpha \leq 1$, *we have*

$$(-1)^i \frac{d^i}{dx^i} h\left((b + x^\alpha)^{\frac{1}{\alpha}}\right) > 0 (x > 0; i = 1, 2)$$

(ii) *For* $\int_{\frac{1}{2}}^{\infty} h(t)dt < \infty$, *we have*

$$\int_1^{\infty} h(t)dt < \sum_{n=1}^{\infty} h(n) < \int_{\frac{1}{2}}^{\infty} h(t)dt. \tag{7}$$

Proof (i) We find

$$\frac{d}{dx} h\left((b + x^\alpha)^{\frac{1}{\alpha}}\right) = h'\left((b + x^\alpha)^{\frac{1}{\alpha}}\right)(b + x^\alpha)^{\frac{1}{\alpha}-1} x^{\alpha-1} < 0,$$

$$\frac{d^2}{dx^2} h\left((b + x^\alpha)^{\frac{1}{\alpha}}\right) = \frac{d}{dx}\left[h'\left((b + x^\alpha)^{\frac{1}{\alpha}}\right)(b + x^\alpha)^{\frac{1}{\alpha}-1} x^{\alpha-1}\right]$$

$$= h''\left((b + x^\alpha)^{\frac{1}{\alpha}}\right)(b + x^\alpha)^{\frac{2}{\alpha}-2} x^{2\alpha-2}$$
$$+ \alpha\left(\frac{1}{\alpha} - 1\right) h'\left((b + x^\alpha)^{\frac{1}{\alpha}}\right)(b + x^\alpha)^{\frac{1}{\alpha}-2} x^{2\alpha-2}$$
$$+ (\alpha - 1)h'\left((b + x^\alpha)^{\frac{1}{\alpha}}\right)(b + x^\alpha)^{\frac{1}{\alpha}-1} x^{\alpha-2}$$
$$= h''\left((b + x^\alpha)^{\frac{1}{\alpha}}\right)(b + x^\alpha)^{\frac{2}{\alpha}-2} x^{2\alpha-2}$$
$$+ b(\alpha - 1)h'\left((b + x^\alpha)^{\frac{1}{\alpha}}\right)(b + x^\alpha)^{\frac{1}{\alpha}-2} x^{\alpha-2} > 0.$$

(ii) Since, $h(t)$ is a decreasing convex function, by the decreasing property and Hermite–Hadamard's inequality (cf. [36]), we have

$$\int_n^{n+1} h(t)dt < h(n) < \int_{n-\frac{1}{2}}^{n+\frac{1}{2}} h(t)dt (n \in \mathbf{N}),$$

and then

$$\int_1^{\infty} h(t)dt = \sum_{n=1}^{\infty} \int_n^{n+1} h(t)dt < \sum_{n=1}^{\infty} h(n)$$

$$< \sum_{n=1}^{\infty} \int_{n-\frac{1}{2}}^{n+\frac{1}{2}} h(t)dt = \int_{\frac{1}{2}}^{\infty} h(t)dt.$$

Hence, (7) follows. The lemma is proved.

Note If $(-1)^i h^{(i)}(t) > 0 (t > 0; i = 0, 1)$, $\int_0^\infty h(t)dt < \infty$, since

$$h(n) < \int_{n-1}^{n} h(t)dt (n \in \mathbf{N}),$$

we still have

$$\sum_{n=1}^{\infty} h(n) < \sum_{n=1}^{\infty} \int_{n-1}^{n} h(t)dt = \int_{0}^{\infty} h(t)dt.$$

If $i_0, j_0 \in \mathbf{N}$($\mathbf{N}$ is the set of positive integers), $\alpha, \beta > 0$, we put

$$||x||_\alpha := \left(\sum_{k=1}^{i_0} |x_k|^\alpha\right)^{\frac{1}{\alpha}} \left(x = (x_1, \cdots, x_{i_0}) \in \mathbf{R}^{i_0}\right),$$

$$||y||_\beta := \left(\sum_{k=1}^{j_0} |y_k|^\beta\right)^{\frac{1}{\beta}} \left(y = (y_1, \cdots, y_{j_0}) \in \mathbf{R}^{j_0}\right). \quad (8)$$

Lemma 2 *If* $s \in \mathbf{N}, \gamma, M > 0, \Psi(u)$ *is a non-negative measurable function in* $(0, 1]$, *and*

$$D_M := \left\{x \in \mathbf{R}_+^s; 0 < u = \sum_{i=1}^{s} \left(\frac{x_i}{M}\right)^\gamma \leq 1\right\},$$

then we have (cf. [6]*)*

$$\int \cdots \int_{D_M} \Psi\left(\sum_{i=1}^{s} \left(\frac{x_i}{M}\right)^\gamma\right) dx_1 \cdots dx_s$$

$$= \frac{M^s \Gamma^s\left(\frac{1}{\gamma}\right)}{\gamma^s \Gamma(\frac{s}{\gamma})} \int_0^1 \Psi(u) u^{\frac{s}{\gamma}-1} du. \quad (9)$$

In view of (9) and the condition, it follows:
(i) For

$$\mathbf{R}_+^s = \left\{x \in \mathbf{R}_+^s; 0 < u = \sum_{i=1}^{s} \left(\frac{x_i}{M}\right)^\gamma \leq 1 (M \to \infty)\right\},$$

we have

$$\int \cdots \int_{\mathbf{R}_+^s} \Psi\left(\sum_{i=1}^{s} \left(\frac{x_i}{M}\right)^\gamma\right) dx_1 \cdots dx_s$$

$$= \lim_{M \to \infty} \frac{M^s \Gamma^s(\frac{1}{\gamma})}{\gamma^s \Gamma(\frac{s}{\gamma})} \int_0^1 \Psi(u) u^{\frac{s}{\gamma}-1} du; \quad (10)$$

(ii) For

$$\{x \in \mathbf{R}_+^s; ||x||_\gamma \geq 1\}$$
$$= \left\{x \in \mathbf{R}_+^s; \frac{1}{M^\gamma} < u = \sum_{i=1}^s \left(\frac{x_i}{M}\right)^\gamma \leq 1 (M \to \infty)\right\},$$

setting $\Psi(u) = 0 (u \in (0, \frac{1}{M^\gamma}))$, we have

$$\int \cdots \int_{\{||x||_\gamma \geq 1\}} \Psi\left(\sum_{i=1}^s \left(\frac{x_i}{M}\right)^\gamma\right) dx_1 \cdots dx_s$$

$$= \lim_{M \to \infty} \frac{M^s \Gamma^s(\frac{1}{\gamma})}{\gamma^s \Gamma(\frac{s}{\gamma})} \int_{\frac{1}{M^\gamma}}^1 \Psi(u) u^{\frac{s}{\gamma}-1} du; \tag{11}$$

(iii) For

$$\{x \in \mathbf{R}_+^s; ||x||_\gamma \leq 1\}$$
$$= \left\{x \in \mathbf{R}_+^s; 0 < u = \sum_{i=1}^s \left(\frac{x_i}{M}\right)^\gamma \leq \frac{1}{M^\gamma}\right\},$$

setting $\Psi(u) = 0 (u \in (\frac{1}{M^\gamma}, \infty))$, we have

$$\int \cdots \int_{\{||x||_\gamma \leq 1\}} \Psi\left(\sum_{i=1}^s \left(\frac{x_i}{M}\right)^\gamma\right) dx_1 \cdots dx_s$$

$$= \frac{M^s \Gamma^s(\frac{1}{\gamma})}{\gamma^s \Gamma(\frac{s}{\gamma})} \int_0^{\frac{1}{M^\gamma}} \Psi(u) u^{\frac{s}{\gamma}-1} du; \tag{12}$$

(iv) For

$$\{x \in \mathbf{R}_+^s; x_i \geq 1\} = \left\{x \in \mathbf{R}_+^s; \frac{s}{M^\gamma} < u\right.$$
$$\left.= \sum_{i=1}^s \left(\frac{x_i}{M}\right)^\gamma \leq 1 (M^\gamma > s; M \to \infty)\right\},$$

setting $\Psi(u) = 0 (u \in (0, \frac{s}{M^\gamma}))$, we have

$$\int \cdots \int_{\{x_i \geq 1\}} \Psi\left(\sum_{i=1}^s \left(\frac{x_i}{M}\right)^\gamma\right) dx_1 \cdots dx_s$$

$$= \lim_{M \to \infty} \frac{M^s \Gamma^s(\frac{1}{\gamma})}{\gamma^s \Gamma(\frac{s}{\gamma})} \int_{\frac{s}{M^\gamma}}^1 \Psi(u) u^{\frac{s}{\gamma}-1} du; \tag{13}$$

(v) For

$$\{x \in \mathbf{R}_+^s; 0<x_i \leq 1\} = \{x \in \mathbf{R}_+^s; 0 < u = \sum_{i=1}^s \left(\frac{x_i}{M}\right)^\gamma \leq \frac{s}{M^\gamma}(M^\gamma < s)\right\},$$

setting $\Psi(u) = 0(u \in (\frac{s}{M^\gamma}, \infty))$, we have

$$\int \cdots \int_{\{0<x_i\leq 1\}} \Psi\left(\sum_{i=1}^s \left(\frac{x_i}{M}\right)^\gamma\right) dx_1 \cdots dx_s = \frac{M^s \Gamma^s(\frac{1}{\gamma})}{\gamma^s \Gamma(\frac{s}{\gamma})} \int_0^{\frac{s}{M^\gamma}} \Psi(u) u^{\frac{s}{\gamma}-1} du. \tag{14}$$

Lemma 3 *For $s \in \mathbf{N}, \gamma > 0, \varepsilon > 0, c = (c_1, \cdots, c_s) \in [0, \frac{1}{2}]^s$, we have*

$$\int \cdots \int_{\{x \in \mathbf{R}_+^s; x_i \geq 1\}} ||x||_\gamma^{-s-\varepsilon} dx_1 \cdots dx_s = \frac{\Gamma^s(\frac{1}{\gamma})}{\varepsilon s^{\varepsilon/\gamma} \gamma^{s-1} \Gamma(\frac{s}{\gamma})}, \tag{15}$$

$$\int \cdots \int_{\{x \in \mathbf{R}_+^s; x_i \leq 1\}} ||x||_\gamma^{-s+\varepsilon} dx_1 \cdots dx_s = \frac{s^{\varepsilon/\gamma} \Gamma^s(\frac{1}{\gamma})}{\varepsilon \gamma^{s-1} \Gamma(\frac{s}{\gamma})}, \tag{16}$$

$$\sum_m ||m - c||_\gamma^{-s-\varepsilon} = \frac{\Gamma^s\left(\frac{1}{\gamma}\right)}{\varepsilon s^{\varepsilon/\gamma} \gamma^{s-1} \Gamma(\frac{s}{\gamma})} + O(1)(\varepsilon \to 0^+). \tag{17}$$

Proof By (13), it follows

$$\int \cdots \int_{\{x \in \mathbf{R}_+^s; x_i \geq 1\}} ||x||_\gamma^{-s-\varepsilon} dx_1 \cdots dx_s$$

$$= \int \cdots \int_{\{x \in \mathbf{R}_+^s; x_i \geq 1\}} \left\{M[\sum_{i=1}^s \left(\frac{x_i}{M}\right)^\gamma]^{\frac{1}{\gamma}}\right\}^{-s-\varepsilon} dx_1 \cdots dx_s$$

$$= \frac{M^s \Gamma^s\left(\frac{1}{\gamma}\right)}{\gamma^s \Gamma(\frac{s}{\gamma})} \int_{\frac{s}{M^\gamma}}^1 \left(Mu^{1/\gamma}\right)^{-s-\varepsilon} u^{\frac{s}{\gamma}-1} du$$

$$= \frac{\Gamma^s(\frac{1}{\gamma})}{\varepsilon s^{\varepsilon/\gamma} \gamma^{s-1} \Gamma(\frac{s}{\gamma})}.$$

Hence we have (15).

By (14), it follows

$$\int \cdots \int_{\{x \in \mathbf{R}_+^s; x_i \leq 1\}} ||x||_\gamma^{-s+\varepsilon} dx_1 \cdots dx_s$$

$$= \int \cdots \int_{\{x \in \mathbf{R}_+^s; x_i \le 1\}} \left\{ M[\sum_{i=1}^{s} \left(\frac{x_i}{M}\right)^{\gamma}]^{\frac{1}{\gamma}} \right\}^{-s+\varepsilon} dx_1 \cdots dx_s$$

$$= \frac{M^s \Gamma^s(\frac{1}{\gamma})}{\gamma^s \Gamma(\frac{s}{\gamma})} \int_0^{\frac{s}{M^\gamma}} \left(M u^{1/\gamma}\right)^{-s+\varepsilon} u^{\frac{s}{\gamma}-1} du$$

$$= \frac{s^{\varepsilon/\gamma} \Gamma^s(\frac{1}{\gamma})}{\varepsilon \gamma^{s-1} \Gamma(\frac{s}{\gamma})}.$$

Hence we have (16).

By (7) and (13), it follows

$$\sum_m ||m-c||_\gamma^{-s-\varepsilon} > \int_{\{x \in \mathbf{R}_+^s; x_i \ge 1+c_i\}} ||x-c||_\gamma^{-s-\varepsilon} dx$$

$$= \int_{\{u \in \mathbf{R}_+^s; u_i \ge 1\}} ||u||_\gamma^{-s-\varepsilon} du$$

$$= \lim_{M \to \infty} \int \cdots \int_{D_M} \left\{ M \left[\sum_{i=1}^{s} \left(\frac{u_i}{M}\right)^{\gamma} \right]^{\frac{1}{\gamma}} \right\}^{-s-\varepsilon} du_1 \cdots du_s$$

$$= \lim_{M \to \infty} \frac{M^s \Gamma^s(\frac{1}{\gamma})}{\gamma^s \Gamma(\frac{s}{\gamma})} \int_{\frac{s}{M^\gamma}}^{1} \left(M v^{1/\gamma}\right)^{-s-\varepsilon} v^{\frac{s}{\gamma}-1} dv$$

$$= \frac{\Gamma^s(\frac{1}{\gamma})}{\varepsilon s^{\varepsilon/\gamma} \gamma^{s-1} \Gamma(\frac{s}{\gamma})}.$$

For $s = 1$, $0 < \sum_{m=1}^{2} ||m-c||_\gamma^{-1-\varepsilon} < \infty$; for $s \ge 2$, by (13), we have

$$0 < \sum_{\{m \in \mathbf{N}^s; \exists i_0, m_{i_0}=1,2\}} ||m-c||_\gamma^{-s-\varepsilon}$$

$$\le a + \sum_{\{m \in \mathbf{N}^{s-1}; m_i \ge 3\}} ||m-c||_\gamma^{-(s-1)-(1+\varepsilon)}$$

$$\le a + \int_{\{x \in \mathbf{R}_+^{s-1}; x_i \ge 1+c_i\}} ||x-c||_\gamma^{-(s-1)-(1+\varepsilon)} dx$$

$$= a + \int_{\{u \in \mathbf{R}_+^s; u_i \ge 1\}} ||u||_\gamma^{-(s-1)-(1+\varepsilon)} du$$

$$= a + \frac{\Gamma^{s-1}(\frac{1}{\gamma})}{(1+\varepsilon)(s-1)^{(1+\varepsilon)/\gamma} \gamma^{s-2} \Gamma(\frac{s}{\gamma})} < \infty (a \in \mathbf{R}_+).$$

Then we have

$$0 < \sum_{\{m \in \mathbf{N}^s; m_i \ge 1\}} ||m-c||_\gamma^{-s-\varepsilon}$$

$$= \sum_{\{m\in\mathbf{N}^s;\exists i_0, m_{i_0}=1,2\}} ||m-c||_\gamma^{-s-\varepsilon} + \sum_{\{m\in\mathbf{N}^s;m_i\geq 3\}} ||m-c||_\gamma^{-s-\varepsilon}$$
$$\leq \widetilde{O}(1) + \int_{\{x\in\mathbf{R}_+^s;x_i\geq 1+c_i\}} ||x-c||_\gamma^{-s-\varepsilon}dx$$
$$= \widetilde{O}(1) + \frac{\Gamma^s(\frac{1}{\gamma})}{\varepsilon s^{\varepsilon/\gamma}\gamma^{s-1}\Gamma(\frac{s}{\gamma})}(\varepsilon \to 0^+),$$

and then (17) follows. The lemma is proved.

Lemma 4 *Suppose that $h(t)$ is a non-negative measurable function in $\mathbf{R}_+, a \in \mathbf{R}$, and there exists a constant $\delta_0 > 0$, such that for any $\delta \in [0, \delta_0)$,*

$$k(a \pm \delta) := \int_0^\infty h(t)t^{(a\pm\delta)-1}dt \in \mathbf{R}.$$

Then we have

$$k(a \pm \delta) = k(a) + o(1)(\delta \to 0^+). \tag{18}$$

Proof For any $\delta \in [0, \frac{\delta_0}{2})$, it follows

$$h(t)t^{(a\pm\delta)-1} \leq g(t) := \begin{cases} h(t)t^{(a-\frac{\delta_0}{2})-1}, t \in (0,1], \\ h(t)t^{(a+\frac{\delta_0}{2})-1}, t \in (1,\infty). \end{cases}$$

Since we find

$$0 \leq \int_0^\infty g(t)dt$$
$$= \int_0^1 h(t)t^{(a-\frac{\delta_0}{2})-1}dt + \int_1^\infty h(t)t^{(a+\frac{\delta_0}{2})-1}dt$$
$$\leq \int_0^\infty h(t)t^{(a-\frac{\delta_0}{2})-1}dt + \int_0^\infty h(t)t^{(a+\frac{\delta_0}{2})-1}dt$$
$$= k\left(a - \frac{\delta_0}{2}\right) + k\left(\alpha + \frac{\delta_0}{2}\right) \in \mathbf{R},$$

then by Lebesgue control convergence theorem (cf. [34]), it follows

$$k(a \pm \delta) = \int_0^\infty h(t)t^{(a\pm\delta)-1}dt$$
$$= \int_0^\infty h(t)t^{a-1}dt + o(1)(\delta \to 0^+),$$

namely, (18) follows. The lemma is proved.

Lemma 5 *If* $\mathbf{C}$ *is the set of complex numbers and* $\mathbf{C}_\infty = \mathbf{C}\cup\{\infty\}, z_k \in \mathbf{C}\{z|\mathrm{Re}z \geq 0, \mathrm{Im}z = 0\}(k = 1, 2, \cdots, n)$ *are different points, the function* $f(z)$ *is analytic in* $\mathbf{C}_\infty$ *except for* $z_i (i = 1, 2, \cdots, n)$, *and* $z = \infty$ *is a zero point of* $f(z)$ *whose order is not less than 1, then for* $\alpha \in \mathbf{R}$, *we have*

$$\int_0^\infty f(x)x^{\alpha-1}dx = \frac{2\pi i}{1 - e^{2\pi\alpha i}} \sum_{k=1}^n \mathrm{Res}\left[f(z)z^{\alpha-1}, z_k\right], \tag{19}$$

where, $0 < \mathrm{Im}\ln z = \arg z < 2\pi$. *In particular, if* $z_k (k = 1, \cdots, n)$ *are all poles of order 1, setting* $\varphi_k(z) = (z - z_k)f(z)(\varphi_k(z_k) \neq 0)$, *then*

$$\int_0^\infty f(x)x^{\alpha-1}dx = \frac{\pi}{\sin\pi\alpha} \sum_{k=1}^n (-z_k)^{\alpha-1}\varphi_k(z_k). \tag{20}$$

Proof By [35] (P.118), we have (19). We find

$$\begin{aligned} 1 - e^{2\pi\alpha i} &= 1 - \cos 2\pi\alpha - i\sin 2\pi\alpha \\ &= -2i\sin\pi\alpha(\cos\pi\alpha + i\sin\pi\alpha) \\ &= -2ie^{i\pi\alpha}\sin\pi\alpha. \end{aligned}$$

In particular, since $f(z)z^{\alpha-1} = \frac{1}{z-z_k}\left(\varphi_k(z)z^{\alpha-1}\right)$, it is obvious that

$$Res[f(z)z^{\alpha-1}, -a_k] = z_k{}^{\alpha-1}\varphi_k(z_k) = -e^{i\pi\alpha}(-z_k)^{\alpha-1}\varphi_k(z_k).$$

Then by (19), we obtain (20). The lemma is proved.

Example 1 For $s \in \mathbf{N}$, we set

$$k_\lambda(x, y) = \frac{1}{\prod_{k=1}^s (x^{\lambda/s} + a_k y^{\lambda/s})}$$

$(0 < a_1 < \cdots < a_s, 0 < \widetilde{\lambda}_1 < \lambda)$, by (20), setting $u = t^{\lambda/s}$, we find

$$\begin{aligned} k_s(\widetilde{\lambda}_1) &= \int_0^\infty \frac{1}{\prod_{k=1}^s (t^{\lambda/s} + a_k)} t^{\widetilde{\lambda}_1 - 1}dt \\ &= \frac{s}{\lambda}\int_0^\infty \frac{1}{\prod_{k=1}^s (u + a_k)} u^{\frac{s\widetilde{\lambda}_1}{\lambda} - 1}du \end{aligned}$$

$$= \frac{\pi s}{\lambda\sin\left(\frac{\pi s\widetilde{\lambda}_1}{\lambda}\right)} \sum_{k=1}^s a_k^{\frac{s\widetilde{\lambda}_1}{\lambda} - 1} \prod_{j=1(j\neq k)}^s \frac{1}{a_j - a_k} \in \mathbf{R}_+. \tag{21}$$

In particular, (i) For $s = 1$,

$$k_\lambda(x,y) = \frac{1}{x^\lambda + a_1 y^\lambda}(a_1 > 0),$$

we have

$$k_1(\widetilde{\lambda}_1) = \left[\frac{\Gamma^{j_0}(\frac{1}{\beta})}{\beta^{j_0-1}\Gamma(\frac{j_0}{\beta})}\right]^{\frac{1}{p}} \left[\frac{\Gamma^{i_0}(\frac{1}{\alpha})}{\alpha^{i_0-1}\Gamma(\frac{i_0}{\alpha})}\right]^{\frac{1}{q}} \times \frac{\pi a_1^{\frac{\widetilde{\lambda}_1}{\lambda}-1}}{\lambda \sin(\frac{\pi\widetilde{\lambda}_1}{\lambda})};$$

(ii) For $s = 2$,

$$k_\lambda(x,y) = \frac{1}{(x^{\lambda/2} + a_1 y^{\lambda/2})(x^{\lambda/2} + a_2 y^{\lambda/2})}(0 < a_1 < a_2),$$

we have

$$k_2(\widetilde{\lambda}_1) = \left[\frac{\Gamma^{j_0}(\frac{1}{\beta})}{\beta^{j_0-1}\Gamma(\frac{j_0}{\beta})}\right]^{\frac{1}{p}} \left[\frac{\Gamma^{i_0}(\frac{1}{\alpha})}{\alpha^{i_0-1}\Gamma(\frac{i_0}{\alpha})}\right]^{\frac{1}{q}} \times \frac{2\pi}{\lambda \sin(\frac{2\pi\widetilde{\lambda}_1}{\lambda})} \frac{a_1^{\frac{2\widetilde{\lambda}_1}{\lambda}-1} - a_2^{\frac{2\widetilde{\lambda}_1}{\lambda}-1}}{a_2 - a_1};$$

(iii) For $s = 3$,

$$k_\lambda(x,y) = \frac{1}{(x^{\lambda/3} + a_1 y^{\lambda/3})(x^{\lambda/3} + a_2 y^{\lambda/3})(x^{\lambda/3} + a_3 y^{\lambda/3})}$$

$(0 < a_1 < a_2 < a_3)$, we have

$$k_3(\widetilde{\lambda}_1) = \left[\frac{\Gamma^{j_0}(\frac{1}{\beta})}{\beta^{j_0-1}\Gamma(\frac{j_0}{\beta})}\right]^{\frac{1}{p}} \left[\frac{\Gamma^{i_0}(\frac{1}{\alpha})}{\alpha^{i_0-1}\Gamma(\frac{i_0}{\alpha})}\right]^{\frac{1}{q}} \times \frac{3\pi}{\lambda \sin(\frac{3\pi\widetilde{\lambda}_1}{\lambda})} \sum_{k=1}^{3} a_k^{\frac{3\widetilde{\lambda}_1}{\lambda}-1} \prod_{j=1(j\neq k)}^{3} \frac{1}{a_j - a_k}.$$

For $\widetilde{\lambda}_1 = \frac{\lambda}{2}$, we may find

$$k_3(\frac{\lambda}{2}) = \left[\frac{\Gamma^{j_0}(\frac{1}{\beta})}{\beta^{j_0-1}\Gamma(\frac{j_0}{\beta})}\right]^{\frac{1}{p}} \left[\frac{\Gamma^{i_0}(\frac{1}{\alpha})}{\alpha^{i_0-1}\Gamma(\frac{i_0}{\alpha})}\right]^{\frac{1}{q}} \times \frac{3\pi}{\lambda(\sqrt{a_1}+\sqrt{a_2})(\sqrt{a_1}+\sqrt{a_3})(\sqrt{a_2}+\sqrt{a_3})}.$$

Example 2 We set

$$k_\lambda(x,y) = \frac{1}{x^\lambda + \sqrt{c}(xy)^{\lambda/2}\cos\gamma + \frac{c}{4}y^\lambda}$$

$(c > 0, 0 < \gamma < \frac{\pi}{2}, 0 < \widetilde{\lambda}_1 < \lambda)$, setting $z_1 = -\frac{\sqrt{c}}{2}e^{i\gamma}$, $z_2 = -\frac{\sqrt{c}}{2}e^{-i\gamma}$, by (20), it follows

$$k\left(\widetilde{\lambda}_1\right) = \int_0^\infty \frac{t^{\widetilde{\lambda}_1 - 1}}{t^\lambda + \sqrt{c}t^{\lambda/2}\cos\gamma + \frac{c}{4}}dt$$

$$= \frac{2}{\lambda}\int_0^\infty \frac{u^{\frac{2\widetilde{\lambda}_1}{\lambda}-1}}{u^2 + \sqrt{c}u\cos\gamma + \frac{c}{4}}du$$

$$= \frac{2}{\lambda}\int_0^\infty \frac{u^{\frac{2\widetilde{\lambda}_1}{\lambda}-1}}{(u - z_1)(u - z_2)}du$$

$$= \frac{2\pi}{\lambda\sin\left(\frac{2\pi\widetilde{\lambda}_1}{\lambda}\right)}\left[\left(\frac{\sqrt{c}}{2}e^{i\gamma}\right)^{\frac{2\widetilde{\lambda}_1}{\lambda}-1}\frac{\sqrt{c}}{2(e^{-i\gamma} - e^{i\gamma})}\right.$$

$$\left.+\left(\frac{\sqrt{c}}{2}e^{-i\gamma}\right)^{\frac{2\widetilde{\lambda}_1}{\lambda}-1}\frac{\sqrt{c}}{2\left(e^{i\gamma} - e^{-i\gamma}\right)}\right]$$

$$= \left(\frac{\sqrt{c}}{2}\right)^{\frac{2\widetilde{\lambda}_1}{\lambda}}\frac{2\pi\sin\gamma\left(1 - \frac{2\widetilde{\lambda}_1}{\lambda}\right)}{\lambda\sin\gamma\sin\left(\frac{2\pi\widetilde{\lambda}_1}{\lambda}\right)} \in \mathbf{R}_+.$$

3 Multidimensional More Accurate Inequalities with the General Homogeneous Kernels

In this section, we agree on that $i_0, j_0 \in \mathbf{N}$, $\alpha, \beta > 0$, $p \in \mathbf{R}\backslash\{0,1\}$, $\frac{1}{p} + \frac{1}{q} = 1$, $\lambda, \lambda_1, \lambda_2 \in \mathbf{R}$, $\lambda_1 + \lambda_2 = \lambda$, $k_\lambda(x,y)(>0)$ is a continuous homogeneous function of degree $-\lambda$ in $\mathbf{R}_+^2$, satisfying for any $u, x, y \in \mathbf{R}_+$,

$$k_\lambda(ux, uy) = u^{-\lambda}k_\lambda(x,y).$$

Definition 1 For $x = (x_1, \cdots, x_{i_0}) \in \mathbf{R}_+^{i_0}, n = (n_1, \cdots, n_{j_0}) \in \mathbf{N}^{j_0}, \tau = (\tau_1, \cdots, \tau_{j_0}) \in [0, \frac{1}{2}]^{j_0}$, define two weight coefficients $\omega_\lambda(\lambda_2, n)$ and $\varpi_\lambda(\lambda_1, x)$ as follows:

$$\omega_\lambda(\lambda_2, n) := ||n - \tau||_\beta^{\lambda_2}\int_{\mathbf{R}_+^{i_0}} k_\lambda(||x||_\alpha, ||n - \tau||_\beta)\frac{dx}{||x||_\alpha^{i_0 - \lambda_1}}, \tag{22}$$

$$\varpi_\lambda(\lambda_1, x) := ||x||_\alpha^{\lambda_1} \sum_n k_\lambda(||x||_\alpha, ||n-\tau||_\beta) \frac{1}{||n-\tau||_\beta^{j_0-\lambda_2}}, \tag{23}$$

where, $\sum_n = \sum_{n_{j_0}=1}^{\infty} \cdots \sum_{n_1=1}^{\infty}$.

Lemma 6 *As the assumptions of Definition 2, if both $\omega_\lambda(\lambda_2, n)$ and $\varpi_\lambda(\lambda_1, x)$ are finite, $f(x) = f(x_1, \cdots, x_{i_0}) \geq 0, a_n = a_{(n_1,\cdots,n_{j_0})} \geq 0$, then*

(i) For $p > 1$, we have the following inequality:

$$J_1 := \left\{ \sum_n \frac{||n-\tau||_\beta^{p\lambda_2 - j_0}}{[\omega_\lambda(\lambda_2, n)]^{p-1}} \left(\int_{\mathbf{R}_+^{i_0}} k_\lambda(||x||_\alpha, ||n-\tau||_\beta) f(x) dx \right)^p \right\}^{\frac{1}{p}}$$

$$\leq \left\{ \int_{\mathbf{R}_+^{i_0}} \varpi_\lambda(\lambda_1, x) ||x||_\alpha^{p(i_0-\lambda_1)-i_0} f^p(x) dx \right\}^{\frac{1}{p}}, \tag{24}$$

$$J_2 := \left\{ \int_{\mathbf{R}_+^{i_0}} \frac{||x||_\alpha^{q\lambda_1 - i_0}}{[\varpi_\lambda(\lambda_1, x)]^{q-1}} \left(\sum_n k_\lambda(||x||_\alpha, ||n-\tau||_\beta) a_n \right)^q dx \right\}^{\frac{1}{q}}$$

$$\leq \left\{ \sum_n \omega_\lambda(\lambda_2, n) ||n-\tau||_\beta^{q(j_0-\lambda_2)-j_0} a_n^q \right\}^{\frac{1}{q}} \tag{25}$$

(ii) For $p < 0$, or $0 < p < 1$, we have the reverses of (24) *and* (25).

Proof (i) For $p > 1$, by Hölder's inequality with weight (cf. [36]), it follows

$$\int_{\mathbf{R}_+^{i_0}} k_\lambda(||x||_\alpha, ||n-\tau||_\beta) f(x) dx$$

$$= \int_{\mathbf{R}_+^{i_0}} k_\lambda(||x||_\alpha, ||n-\tau||_\beta) \left[\frac{||x||_\alpha^{(i_0-\lambda_1)/q}}{||n-\tau||_\beta^{(j_0-\lambda_2)/p}} f(x) \right] \times \left[\frac{||n-\tau||_\beta^{(j_0-\lambda_2)/p}}{||x||_\alpha^{(i_0-\lambda_1)/q}} \right] dx$$

$$\leq \left\{ \int_{\mathbf{R}_+^{i_0}} k_\lambda(||x||_\alpha, ||n-\tau||_\beta) \frac{||x||_\alpha^{(i_0-\lambda_1)(p-1)}}{||n-\tau||_\beta^{j_0-\lambda_2}} f^p(x) dx \right\}^{\frac{1}{p}}$$

$$\times \left\{ \int_{\mathbf{R}_+^{i_0}} k_\lambda(||x||_\alpha, ||n-\tau||_\beta) \frac{||n-\tau||_\beta^{(j_0-\lambda_2)(q-1)}}{||x||_\alpha^{i_0-\lambda_1}} \right\}^{\frac{1}{q}}$$

$$= [\omega_\lambda(\lambda_2, n)]^{\frac{1}{q}} ||n-\tau||_\beta^{\frac{j_0}{p}-\lambda_2}$$

$$\times \left\{ \int_{\mathbf{R}_+^{i_0}} k_\lambda(||x||_\alpha, ||n-\tau||_\beta) \frac{||x||_\alpha^{(i_0-\lambda_1)(p-1)}}{||n-\tau||_\beta^{j_0-\lambda_2}} f^p(x) dx \right\}^{\frac{1}{p}}. \tag{26}$$

Then by Lebesgue term by term integration theorem (cf. [34]), we have

$$J_1 \leq \left\{ \sum_n \int_{\mathbf{R}_+^{i_0}} k_\lambda(||x||_\alpha, ||n-\tau||_\beta) \frac{||x||_\alpha^{(i_0-\lambda_1)(p-1)}}{||n-\tau||_\beta^{j_0-\lambda_2}} f^p(x)dx \right\}^{\frac{1}{p}}$$

$$= \left\{ \int_{\mathbf{R}_+^{i_0}} \sum_n k_\lambda(||x||_\alpha, ||n-\tau||_\beta) \frac{||x||_\alpha^{(i_0-\lambda_1)(p-1)}}{||n-\tau||_\beta^{j_0-\lambda_2}} f^p(x)dx \right\}^{\frac{1}{p}}$$

$$= \left\{ \int_{\mathbf{R}_+^{i_0}} \varpi_\lambda(\lambda_1, x) ||x||_\alpha^{p(i_0-\lambda_1)-i_0} f^p(x)dx \right\}^{\frac{1}{p}}. \tag{27}$$

Hence, (24) follows.
By the same way, we obtain

$$\sum_n k_\lambda(||x||_\alpha, ||n-\tau||_\beta) a_n \leq [\varpi_\lambda(\lambda_1, x)]^{\frac{1}{p}} ||x||_\alpha^{\frac{i_0}{q}-\lambda_1}$$

$$\times \left\{ \sum_n k_\lambda(||x||_\alpha, ||n-\tau||_\beta) \frac{||n-\tau||_\beta^{(j_0-\lambda_2)(q-1)}}{||x||_\alpha^{i_0-\lambda_1}} a_n^q \right\}^{\frac{1}{q}}.$$

Then by Lebesgue term by term integration theorem and the same way in obtaining (27), we have (25).

(ii) For $p < 0$, or $0 < p < 1$, by the reverse Hölder's inequality with weight (cf. [36]), we obtain the reverse of (26). Then by Lebesgue term by term integration theorem, we still can obtain the reverse of (25). The lemma is proved.

Lemma 7 *As the assumptions of Lemma 6, then*

(i) For $p > 1$, we have the following inequality equivalent to (24) *and* (25)*:*

$$I := \sum_n \int_{\mathbf{R}_+^{i_0}} k_\lambda(||x||_\alpha, ||n-\tau||_\beta) a_n f(x)dx$$

$$\leq \left\{ \int_{\mathbf{R}_+^{i_0}} \varpi_\lambda(\lambda_1, x) ||x||_\alpha^{p(i_0-\lambda_1)-i_0} f^p(x)dx \right\}^{\frac{1}{p}}$$

$$\times \left\{ \sum_n \omega_\lambda(\lambda_2, n) ||n-\tau||_\beta^{q(j_0-\lambda_2)-j_0} a_n^q \right\}^{\frac{1}{q}}; \tag{28}$$

(ii) For $p < 0$, or $0 < p < 1$, we have the reverse of (28) *equivalent to the reverses of* (24) *and* (25)*.*

Proof (i) For $p > 1$, by Hölder's inequality (cf. [36]), it follows

$$I = \sum_n \frac{||n-\tau||_\beta^{\frac{j_0}{q}-(j_0-\lambda_2)}}{[\omega_\lambda(\lambda_2,n)]^{\frac{1}{q}}} \left[\int_{\mathbf{R}_+^{i_0}} k_\lambda(||x||_\alpha, ||n-\tau||_\beta) f(x)dx\right]$$

$$\times \left[[\omega_\lambda(\lambda_2,n)]^{\frac{1}{q}} ||n-\tau||_\beta^{(j_0-\lambda_2)-\frac{j_0}{q}} a_n\right]$$

$$\leq J_1 \left\{\sum_n \omega_\lambda(\lambda_2,n) ||n-\tau||_\beta^{q(j_0-\lambda_2)-j_0} a_n^q\right\}^{\frac{1}{q}}.$$

Then by (24), we have (28).

On the other hand, assuming that (28) is valid, we set

$$b_n := \frac{||n-\tau||_\beta^{p\lambda_2-j_0}}{[\omega_\lambda(\lambda_2,n)]^{p-1}} \left(\int_{\mathbf{R}_+^{i_0}} k_\lambda(||x||_\alpha, ||n-\tau||_\beta) f(x)dx\right)^{p-1}, n \in \mathbf{N}^{j_0}.$$

Then it follows

$$J_1^p = \sum_n \omega_\lambda(\lambda_2,n) ||n-\tau||_\beta^{q(j_0-\lambda_2)-j_0} a_n^q.$$

If $J_1 = 0$, then (24) is trivially valid; if $J_1 = \infty$, then by (27), (24) keeps the form of equality($= \infty$). Suppose that $0 < J_1 < \infty$. By (28), we have

$$0 < \sum_n \omega_\lambda(\lambda_2,n) ||n-\tau||_\beta^{q(j_0-\lambda_2)-j_0} a_n^q = J_1^p = I$$

$$\leq \left\{\int_{\mathbf{R}_+^{i_0}} \varpi_\lambda(\lambda_1,x) ||x||_\alpha^{p(i_0-\lambda_1)-i_0} f^p(x)dx\right\}^{\frac{1}{p}}$$

$$\times \left\{\sum_n \omega_\lambda(\lambda_2,n) ||n-\tau||_\beta^{q(j_0-\lambda_2)-j_0} a_n^q\right\}^{\frac{1}{q}} < \infty.$$

It follows

$$J_1 = \left\{\sum_n \omega_\lambda(\lambda_2,n) ||n-\tau||_\beta^{q(j_0-\lambda_2)-j_0} a_n^q\right\}^{\frac{1}{p}}$$

$$\leq \left\{\int_{\mathbf{R}_+^{i_0}} \varpi_\lambda(\lambda_1,x) ||x||_\alpha^{p(i_0-\lambda_1)-i_0} f^p(x)dx\right\}^{\frac{1}{p}},$$

and then (24) follows. Hence, (24) and (28) are equivalent.

By Hölder's inequality and the same way, we can obtain

$$I \leq \left\{ \int_{\mathbf{R}_+^{i_0}} \varpi_\lambda(\lambda_1, x) ||x||_\alpha^{p(i_0-\lambda_1)-i_0} f^p(x) dx \right\}^{\frac{1}{p}} J_2.$$

Then by (25), we have (28). On the other hand, assuming that (28) is valid, we set

$$f(x) = \frac{||x||_\alpha^{q\lambda_1 - i_0}}{[\varpi_\lambda(\lambda_1, x)]^{q-1}} \left(\sum_n k_\lambda(||x||_\alpha, ||n-\tau||_\beta) a_n \right)^{q-1} \quad (x \in \mathbf{R}_+^{i_0}).$$

Then it follows

$$J_2^q = \int_{\mathbf{R}_+^{i_0}} \varpi_\lambda(\lambda_1, x) ||x||_\alpha^{p(i_0-\lambda_1)-i_0} f^p(x) dx.$$

By (28) and the same way, we can obtain

$$\begin{aligned} J_2 &= \left\{ \int_{\mathbf{R}_+^{i_0}} \varpi_\lambda(\lambda_1, x) ||x||_\alpha^{p(i_0-\lambda_1)-i_0} f^p(x) dx \right\}^{\frac{1}{q}} \\ &\leq \left\{ \sum_n \omega_\lambda(\lambda_2, n) ||n-\tau||_\beta^{q(j_0-\lambda_2)-j_0} a_n^q \right\}^{\frac{1}{q}}, \end{aligned}$$

and then (25) is equivalent to (28).

Hence (24), (25), and (28) are equivalent.

(ii) For $p < 0$, or $0 < p < 1$, by the same way, we have the reverse of (28) equivalent to the reverses of (24) and (25). The lemma is proved.

Lemma 8 *If* $\widetilde{\lambda}_1, \widetilde{\lambda}_2 \in \mathbf{R}, \widetilde{\lambda}_1 + \widetilde{\lambda}_2 = \lambda$,

$$k(\widetilde{\lambda}_1) := \int_0^\infty k_\lambda(u, 1) u^{\widetilde{\lambda}_1 - 1} du \in \mathbf{R}_+,$$

then we have

$$\varpi_\lambda\left(\widetilde{\lambda}_2, n\right) = K_2(\widetilde{\lambda}_1) := \frac{\Gamma^{i_0}\left(\frac{1}{\alpha}\right)}{\alpha^{i_0-1}\Gamma\left(\frac{i_0}{\alpha}\right)} k\left(\widetilde{\lambda}_1\right) \in \mathbf{R}_+ (n \in \mathbf{N}^{j_0}). \tag{29}$$

Moreover, if $0 < \beta \leq 1$,

$$\frac{\partial}{\partial y}\left(k_\lambda(x, y) y^{\widetilde{\lambda}_2 - j_0}\right) < 0,$$

$$\frac{\partial^2}{\partial y^2}\left(k_\lambda(x, y) y^{\widetilde{\lambda}_2 - j_0}\right) > 0,$$

there exists constants $L > 0$ and $\eta_1 > \widetilde{\lambda}_1$, satisfying

$$k_\lambda(t,1) \leq \frac{L}{t^{\eta_1}}(t \in (0,\infty)), \tag{30}$$

then we have

$$K_1(\widetilde{\lambda}_1)(1 - \theta_{\widetilde{\lambda}_1}(x)) < \omega_\lambda(\widetilde{\lambda}_1, x) < K_1(\widetilde{\lambda}_1)\left(x \in \mathbf{R}_+^{i_0}\right). \tag{31}$$

where,

$$K_1(\widetilde{\lambda}_1) := \frac{\Gamma^{j_0}(\frac{1}{\beta})}{\beta^{j_0-1}\Gamma(\frac{j_0}{\beta})} k(\widetilde{\lambda}_1) \in \mathbf{R}_+, \tag{32}$$

$$\theta_{\widetilde{\lambda}_1}(x) := \frac{1}{k(\widetilde{\lambda}_1)} \int_{\frac{||x||_\alpha}{j_0^{1/\beta}}}^{\infty} k_\lambda(t,1) t^{\widetilde{\lambda}_1 - 1} dt$$

$$= O\left(\frac{1}{||x||_\alpha^{\rho(\widetilde{\lambda}_1)}}\right) \in (0,1)(\rho(\widetilde{\lambda}_1) = \eta_1 - \widetilde{\lambda}_1 > 0). \tag{33}$$

Proof By (10), we find

$$\omega_\lambda(\widetilde{\lambda}_2, n) = ||n-\tau||_\beta^{\widetilde{\lambda}_2} \int_{\mathbf{R}_+^{i_0}} k_\lambda\left(M\left[\sum_{i=1}^{i_0}\left(\frac{x_i}{M}\right)^\alpha\right]^{\frac{1}{\alpha}}, ||n-\tau||_\beta\right)$$

$$\times \frac{1}{M^{i_0-\widetilde{\lambda}_1}[\sum_{i=1}^{i_0}(\frac{x_i}{M})^\alpha]^{\frac{1}{\alpha}(i_0-\widetilde{\lambda}_1)}} dx_1 \cdots dx_{i_0}$$

$$= \lim_{M\to\infty} \frac{M^{i_0}\Gamma^{i_0}(\frac{1}{\alpha})}{\alpha^{i_0}\Gamma(\frac{i_0}{\alpha})} ||n-\tau||_\beta^{\widetilde{\lambda}_2}$$

$$\times \int_0^1 \frac{k_\lambda\left(Mu^{\frac{1}{\alpha}}, ||n-\tau||_\beta\right)}{M^{i_0-\widetilde{\lambda}_1} u^{\frac{1}{\alpha}(i_0-\widetilde{\lambda}_1)}} u^{\frac{i_0}{\alpha}-1} du$$

$$= \frac{\Gamma^{i_0}(\frac{1}{\alpha})}{\alpha^{i_0-1}\Gamma(\frac{i_0}{\alpha})} \int_0^\infty k_\lambda(u,1) u^{\widetilde{\lambda}_1 - 1} du.$$

Hence, we have (29).

Moreover, by (7) and (10), we obtain

$$\varpi_\lambda(\widetilde{\lambda}_1, x) < ||x||_\alpha^{\widetilde{\lambda}_1} \int_{\{y \in \mathbf{R}_+^{j_0} y_i \geq \frac{1}{2}\}} k_\lambda(||x||_\alpha, ||y-\tau||_\beta) \frac{dy}{||y-\tau||_\beta^{j_0-\widetilde{\lambda}_2}}$$

$$= ||x||_\alpha^{\widetilde{\lambda}_1} \int_{\{u \in \mathbf{R}_+^{j_0}; u_i \geq \frac{1}{2} - \tau_i\}} k_\lambda(||x||_\alpha, ||u||_\beta) \frac{du}{||u||_\beta^{j_0-\widetilde{\lambda}_2}}$$

$$\leq ||x||_{\alpha}^{\widetilde{\lambda}_1} \int_{\mathbf{R}_+^{j_0}} k_\lambda(||x||_\alpha, ||u||_\beta) \frac{du}{||u||_\beta^{j_0-\widetilde{\lambda}_2}}$$

$$= \frac{\Gamma^{j_0}(\frac{1}{\beta})}{\alpha^{j_0-1}\Gamma(\frac{j_0}{\beta})} \int_0^\infty k_\lambda(v,1) v^{\widetilde{\lambda}_1-1} dv$$

$$= K_2(\widetilde{\lambda}_1).$$

By (7), (13), and (30), we find

$$\varpi_\lambda(\widetilde{\lambda}_1, x) > ||x||_\alpha^{\widetilde{\lambda}_1} \int_{\{y\in\mathbf{R}_+^{j_0} y_i \geq 1+\tau_i\}} k_\lambda(||x||_\alpha, ||y-\tau||_\beta) \frac{dy}{||y-\tau||_\beta^{j_0-\widetilde{\lambda}_2}}$$

$$= ||x||_\alpha^{\widetilde{\lambda}_1} \int_{\{u\in\mathbf{R}_+^{j_0}; u_i \geq 1\}} k_\lambda(||x||_\alpha, ||u||_\beta) \frac{du}{||u||_\beta^{j_0-\widetilde{\lambda}_2}}$$

$$= \lim_{M\to\infty} \frac{M^{j_0}\Gamma^{j_0}(\frac{1}{\beta})}{\beta^{j_0-1}\Gamma(\frac{j_0}{\beta})} \int_{\frac{j_0}{M^\beta}}^1 \frac{k_\lambda\left(||x||_\alpha, Mv^{\frac{1}{\beta}}\right)}{(Mv^{\frac{1}{\beta}})^{j_0-\widetilde{\lambda}_2}} v^{\frac{j_0}{\beta}-1} dv$$

$$= \frac{\Gamma^{j_0}(\frac{1}{\beta})}{\beta^{j_0-1}\Gamma(\frac{j_0}{\beta})} \int_0^{\frac{||x||_\alpha}{j_0^{1/\beta}}} k_\lambda(t,1) t^{\widetilde{\lambda}_1-1} dt$$

$$= \frac{\Gamma^{j_0}\left(\frac{1}{\beta}\right)}{\beta^{j_0-1}\Gamma(\frac{j_0}{\beta})} k(\widetilde{\lambda}_1)[1-\theta_{\widetilde{\lambda}_1}(x)] > 0,$$

$$0 < \theta_{\widetilde{\lambda}_1}(x) = \frac{1}{k(\widetilde{\lambda}_1)} \int_{\frac{||x||_\alpha}{j_0^{1/\beta}}}^\infty k_\lambda(t,1) t^{\widetilde{\lambda}_1-1} dt$$

$$\leq \frac{L}{k(\widetilde{\lambda}_1)} \int_{\frac{||x||_\alpha}{j_0^{1/\beta}}}^\infty t^{\widetilde{\lambda}_1-\eta_1-1} dt$$

$$= \frac{L j_0^{(\eta_1-\widetilde{\lambda}_1)/\beta}}{k(\widetilde{\lambda}_1)(\eta_1-\widetilde{\lambda}_1)} \frac{1}{||x||_\beta^{\eta_1-\widetilde{\lambda}_1}}.$$

Hence we have (31)–(33). The lemma is proved.
Setting

$$\Phi(x) := ||x||_\alpha^{p(i_0-\lambda_1)-i_0}, \Psi(n) := ||n-\tau||_\beta^{q(j_0-\lambda_2)-j_0},$$

$$\widetilde{\Phi}(x) := (1-\theta_{\lambda_1}(x))||x||_\alpha^{p(i_0-\lambda_1)-i_0} \left(\theta_{\lambda_1}(x) \in (0,1); x \in \mathbf{R}_+^{i_0}\right),$$

by Lemma 6 and Lemma 7, we have Theorem 1.

Theorem 1 *Suppose that* $\alpha > 0, 0 < \beta \leq 1, \tau = (\tau_1, \cdots, \tau_{j_0}) \in \left[0, \frac{1}{2}\right]^{j_0}$, *there exists a constant* $\delta_0 > 0$, *such that for any* $\widetilde{\lambda}_i \in (\lambda_i - \delta_0, \lambda_i + \delta_0)(i = 1, 2)$,

$$k(\widetilde{\lambda}_1) = \int_0^\infty k_\lambda(v, 1)v^{\widetilde{\lambda}_1 - 1}dv \in \mathbf{R}_+,$$

and

$$\frac{\partial}{\partial y}\left(k_\lambda(x, y)y^{\widetilde{\lambda}_2 - j_0}\right) < 0,$$

$$\frac{\partial^2}{\partial y^2}(k_\lambda(x, y)y^{\widetilde{\lambda}_2 - j_0}) > 0,$$

and there exists constants $L > 0$ *and* $\eta_1 > \widetilde{\lambda}_1$, *satisfying*

$$k_\lambda(t, 1) \leq \frac{L}{t^{\eta_1}}(t \in (0, \infty)).$$

If $p > 1,\ f(x) = f(x_1, \cdots, x_{i_0}) \geq 0, a_n = a_{(n_1, \cdots, n_{j_0})} \geq 0$,

$$0 < ||f||_{p,\Phi} = \left\{\int_{\mathbf{R}_+^{i_0}} \Phi(x)f^p(x)dx\right\}^{\frac{1}{p}} < \infty,$$

$$0 < ||a||_{q,\Psi} = \left\{\sum_n \Psi(n)a_n^q\right\}^{\frac{1}{q}} < \infty,$$

then we have the following equivalent inequalities with the best possible constant factor $K(\lambda_1)$:

$$I = \sum_n \int_{\mathbf{R}_+^{i_0}} k_\lambda(||x||_\alpha, ||n - \tau||_\beta)a_n f(x)dx < K(\lambda_1)||f||_{p,\Phi}||a||_{q,\Psi}, \tag{34}$$

$$J := \left\{\sum_n ||n - \tau||_\beta^{p\lambda_2 - j_0}\left(\int_{\mathbf{R}_+^{i_0}} k_\lambda(||x||_\alpha, ||n - \tau||_\beta)f(x)dx\right)^p\right\}^{\frac{1}{p}}$$

$$< K(\lambda_1)||f||_{p,\Phi}, \tag{35}$$

$$H := \left\{\int_{\mathbf{R}_+^{i_0}} ||x||_\alpha^{q\lambda_1 - i_0}\left(\sum_n k_\lambda(||x||_\alpha, ||n - \tau||_\beta)a_n\right)^q dx\right\}^{\frac{1}{q}}$$

$$< K(\lambda_1)||a||_{q,\Psi}, \tag{36}$$

where,

$$k(\lambda_1) = \int_0^\infty k_\lambda(u,1)u^{\lambda_1-1}du,$$

$$K(\lambda_1) = (K_1(\lambda_1))^{\frac{1}{p}}(K_2(\lambda_1))^{\frac{1}{q}}$$

$$= \left[\frac{\Gamma^{j_0}(\frac{1}{\beta})}{\beta^{j_0-1}\Gamma(\frac{j_0}{\beta})}\right]^{\frac{1}{p}} \left[\frac{\Gamma^{i_0}(\frac{1}{\alpha})}{\alpha^{i_0-1}\Gamma(\frac{i_0}{\alpha})}\right]^{\frac{1}{q}} k(\lambda_1).$$

In particular, for $i_0 = j_0 = 1, \tau \in [0, \frac{1}{2}]$,

$$\varphi(x) := x^{p(1-\lambda_1)-1}, \psi(n) := (n-\tau)^{q(1-\lambda_2)-1},$$

we have the following equivalent inequalities with the best possible constant factor $k(\lambda_1)$:

$$\sum_{n=1}^\infty \int_0^\infty k_\lambda(x, n-\tau)a_n f(x)dx < k(\lambda_1)||f||_{p,\varphi}||a||_{q,\psi}, \tag{37}$$

$$\left\{\sum_{n=1}^\infty (n-\tau)^{p\lambda_2-1}\left(\int_0^\infty k_\lambda(x,n-\tau)f(x)dx\right)^p\right\}^{\frac{1}{p}} < k(\lambda_1)||f||_{p,\varphi}, \tag{38}$$

$$\left\{\int_0^\infty x^{q\lambda_1-1}\left(\sum_{n=1}^\infty k_\lambda(x,n-\tau)a_n\right)^q dx\right\}^{\frac{1}{q}} < k(\lambda_1)||a||_{q,\psi}. \tag{39}$$

Proof By Lemma 6, Lemma 7, and Lemma 8, we have equivalent inequalities (34), (35), and (36) respectively. By Hölder's inequality, we still have

$$I \le J\left\{\sum_n ||n-\tau||_\beta^{q(j_0-\lambda_2)-j_0} a_n^q\right\}^{\frac{1}{q}}, \tag{40}$$

$$I \le \{\int_{\mathbf{R}_+^{i_0}} ||x||_\alpha^{p(i_0-\lambda_1)-i_0} f^p(x)dx\}^{\frac{1}{p}} H. \tag{41}$$

For $0 < \varepsilon < q\delta_0$, we set $\widetilde{f}(x), \widetilde{a}_n$ as follows:

$$\widetilde{f}(x) := \begin{cases} 0, 0 < x_i < 1 (i = 1, \cdots, i_0), \\ ||x||_\alpha^{\lambda_1 - \frac{\varepsilon}{p} - i_0}, x_i \ge 1 (i = 1, \cdots, i_0), \end{cases}$$

$$\widetilde{a}_n := ||n-\tau||_\beta^{\left(\lambda_2 - \frac{\varepsilon}{q}\right) - j_0}, n \in \mathbf{N}^{j_0}.$$

Then for $\widetilde{\lambda}_1 = \lambda_1 + \frac{\varepsilon}{q}, \widetilde{\lambda}_2 = \lambda_2 - \frac{\varepsilon}{q}$, in view of (15), (17), and (33), we find

$$||\widetilde{f}||_{p,\Phi}||\widetilde{a}||_{q,\Psi}$$

$$= \left\{ \int_{\{x\in \mathbf{R}_+^{i_0}; x_i \geq 1\}} ||x||_\alpha^{-i_0-\varepsilon} dx \right\}^{\frac{1}{p}} \left\{ \sum_n ||n-\tau||_\beta^{-j_0-\varepsilon} \right\}^{\frac{1}{q}}$$

$$= \frac{1}{\varepsilon} \left\{ \frac{\Gamma^{i_0}(\frac{1}{\alpha})}{i_0^{\varepsilon/\alpha} \alpha^{i_0-1} \Gamma(\frac{i_0}{\alpha})} \right\}^{\frac{1}{p}}$$

$$\times \left\{ \frac{\Gamma^{j_0}(\frac{1}{\beta})}{j_0^{\varepsilon/\beta} \beta^{j_0-1} \Gamma(\frac{j_0}{\beta})} + \varepsilon O(1) \right\}^{\frac{1}{q}},$$

$$\widetilde{I} := \int_{\mathbf{R}_+^{i_0}} \sum_n k_\lambda(||x||_\alpha, ||n-\tau||_\beta) \widetilde{a}_n \widetilde{f}(x) dx$$

$$= \int_{\{x\in \mathbf{R}_+^{i_0}; x_i \geq 1\}} ||x||_\alpha^{-i_0-\varepsilon} \varpi_\lambda(\widetilde{\lambda}_1, x) dx$$

$$\geq K_1(\widetilde{\lambda}_1) \int_{\{x\in \mathbf{R}_+^{i_0}; x_i \geq 1\}} ||x||_\alpha^{-i_0-\varepsilon} \left(1 - O\left(\frac{1}{||x||_\alpha^{\rho(\widetilde{\lambda}_1)}} \right) \right) dx$$

$$= \frac{1}{\varepsilon} K_1(\widetilde{\lambda}_1) \left[\frac{\Gamma^{i_0}\left(\frac{1}{\alpha}\right)}{i_0^{\varepsilon/\alpha} \alpha^{i_0-1} \Gamma\left(\frac{i_0}{\alpha}\right)} - \varepsilon O_{\widetilde{\lambda}_1}(1) \right].$$

If there exists a constant $K \leq K(\lambda_1)$, such that (34) is valid when replacing $K(\lambda_1)$ by K, then in particular, we have

$$K_1(\widetilde{\lambda}_1) \left[\frac{\Gamma^{i_0}(\frac{1}{\alpha})}{i_0^{\varepsilon/\alpha} \alpha^{i_0-1} \Gamma(\frac{i_0}{\alpha})} - \varepsilon O_{\widetilde{\lambda}_1}(1) \right]$$

$$\leq \varepsilon \widetilde{I} < \varepsilon K ||\widetilde{f}||_{p,\Phi} ||\widetilde{a}||_{q,\Psi}$$

$$= K \left\{ \frac{\Gamma^{i_0}(\frac{1}{\alpha})}{i_0^{\varepsilon/\alpha} \alpha^{i_0-1} \Gamma(\frac{i_0}{\alpha})} \right\}^{\frac{1}{p}}$$

$$\times \left\{ \frac{\Gamma^{j_0}(\frac{1}{\beta})}{j_0^{\varepsilon/\beta} \beta^{j_0-1} \Gamma(\frac{j_0}{\beta})} + \varepsilon O(1) \right\}^{\frac{1}{q}},$$

and then by (18), $K(\lambda_1) \leq K(\varepsilon \to 0^+)$. Hence, $K = K(\lambda_1)$ is the best possible constant factor of (34).

By the equivalency, we can prove that the constant factor $K(\lambda_1)$ in (35) and (36) is the best possible. Otherwise, we would reach a contradiction by (40) and (41) that the constant factor $K(\lambda_1)$ in (34) is not the best possible. The theorem is proved.

Theorem 2 *As the assumptions of Theorem 1, if* $p < 0$, $f(x) = f(x_1, \cdots, x_{i_0}) \geq 0$, $a_n = a_{(n_1,\cdots,n_{j_0})} \geq 0$, $0 < ||f||_{p,\Phi} < \infty, 0 < ||a||_{q,\Psi} < \infty$, *then we have the* *following equivalent inequalities with the best possible constant factor* $K(\lambda_1)$:

$$\sum_n \int_{\mathbf{R}_+^{i_0}} k_\lambda(||x||_\alpha, ||n-\tau||_\beta) a_n f(x) dx > K(\lambda_1)||f||_{p,\Phi}||a||_{q,\Psi}, \tag{42}$$

$$\left\{\sum_n ||n-\tau||_\beta^{p\lambda_2 - j_0} \left(\int_{\mathbf{R}_+^{i_0}} k_\lambda(||x||_\alpha, ||n-\tau||_\beta) f(x) dx\right)^p\right\}^{\frac{1}{p}} > K(\lambda_1)||f||_{p,\Phi}, \tag{43}$$

$$\left\{\int_{\mathbf{R}_+^{i_0}} ||x||_\alpha^{q\lambda_1 - i_0} \left(\sum_n k_\lambda(||x||_\alpha, ||n-\tau||_\beta) a_n\right)^q dx\right\}^{\frac{1}{q}} > K(\lambda_1)||a||_{q,\Psi}. \tag{44}$$

In particular, for $i_0 = j_0 = 1, \tau \in [0, \frac{1}{2}], \varphi(x)$ and $\psi(n)$ are indicated by Theorem 1, we have the following equivalent inequalities with the best possible constant factor $k(\lambda_1)$:

$$\sum_{n=1}^\infty \int_0^\infty k_\lambda(x, n-\tau) a_n f(x) dx > k(\lambda_1)||f||_{p,\varphi}||a||_{q,\psi}, \tag{45}$$

$$\left\{\sum_{n=1}^\infty (n-\tau)^{p\lambda_2 - 1} \left(\int_0^\infty k_\lambda(x, n-\tau) f(x) dx\right)^p\right\}^{\frac{1}{p}} > k(\lambda_1)||f||_{p,\varphi}, \tag{46}$$

$$\left\{\int_0^\infty x^{q\lambda_1 - 1} \left(\sum_{n=1}^\infty k_\lambda(x, n-\tau) a_n\right)^q dx\right\}^{\frac{1}{q}} > k(\lambda_1)||a||_{q,\psi}. \tag{47}$$

Proof We only prove that the constant factor $K(\lambda_1)$ in (40) is the best possible. The rest are omitted.

For $0 < \varepsilon < q\delta_0, \widetilde{\lambda}_1 = \lambda_1 + \frac{\varepsilon}{q}, \widetilde{\lambda}_2 = \lambda_2 - \frac{\varepsilon}{q}$, we set $\widetilde{f}(x), \widetilde{a}_n$ as Theorem 1. If there exists a constant $K \geq K(\lambda_1)$, such that (40) is valid when replacing $K(\lambda_1)$ by K, then in particular, we have

$$K\left\{\frac{\Gamma^{i_0}(\frac{1}{\alpha})}{i_0^{\varepsilon/\alpha}\alpha^{i_0-1}\Gamma(\frac{i_0}{\alpha})}\right\}^{\frac{1}{p}}\left\{\frac{\Gamma^{j_0}(\frac{1}{\beta})}{j_0^{\varepsilon/\beta}\beta^{j_0-1}\Gamma(\frac{j_0}{\beta})} + \varepsilon O(1)\right\}^{\frac{1}{q}}$$

$$= \varepsilon K||\widetilde{f}||_{p,\Phi}||\widetilde{a}||_{q,\Psi} < \varepsilon \widetilde{I} = \varepsilon \int_{\{x \in \mathbf{R}_+^{i_0}; x_i \geq 1\}} ||x||_\alpha^{-i_0 - \varepsilon} \varpi_\lambda(\widetilde{\lambda}_1, x) dx$$

$$< \varepsilon K_2(\widetilde{\lambda}_1) \int_{\{x \in \mathbf{R}_+^{i_0}; x_i \geq 1\}} ||x||_\alpha^{-i_0 - \varepsilon} dx$$

$$= K_2(\widetilde{\lambda}_1) \frac{\Gamma^{i_0}(\frac{1}{\alpha})}{i_0^{\varepsilon/\alpha}\alpha^{i_0-1}\Gamma(\frac{i_0}{\alpha})},$$

and by (18), $K \leq K(\lambda_1)(\varepsilon \to 0^+)$. Hence $K = K(\lambda_1)$ is the best possible constant factor of (40). The theorem is proved.

Theorem 3 *As the assumptions of Theorem 1, if* $0 < p < 1, f(x) = f(x_1, \cdots, x_{i_0}) \geq 0, a_n = a_{(n_1,\cdots,n_{j_0})} \geq 0, 0 < ||f||_{p,\widetilde{\Phi}} < \infty, 0 < ||a||_{q,\Psi} < \infty$, *then we have the following equivalent inequalities with the best possible constant factor* $K(\lambda_1)$:

$$\sum_n \int_{\mathbf{R}_+^{i_0}} k_\lambda(||x||_\alpha, ||n-\tau||_\beta) a_n f(x) dx > K(\lambda_1)||f||_{p,\widetilde{\Phi}}||a||_{q,\Psi}, \tag{48}$$

$$\left\{\sum_n ||n-\tau||_\beta^{p\lambda_2 - j_0} \left(\int_{\mathbf{R}_+^{i_0}} k_\lambda(||x||_\alpha, ||n-\tau||_\beta) f(x) dx\right)^p\right\}^{\frac{1}{p}} > K(\lambda_1)||f||_{p,\widetilde{\Phi}}, \tag{49}$$

$$\left\{\int_{\mathbf{R}_+^{i_0}} \frac{||x||_\alpha^{q\lambda_1 - i_0}}{(1-\theta_{\lambda_1}(x))^{q-1}} \left(\sum_n k_\lambda(||x||_\alpha, ||n-\tau||_\beta) a_n\right)^q dx\right\}^{\frac{1}{q}} > K(\lambda_1)||a||_{q,\Psi}. \tag{50}$$

In particular, for $i_0 = j_0 = 1, \tau \in [0, \frac{1}{2}], \varphi(x)$ and $\psi(n)$ are indicated by Theorem 1,

$$\widetilde{\varphi}(x) = (1-\theta_{\lambda_1}(x)) x^{p(1-\lambda_1)-1} (\theta_{\lambda_1}(x) \in (0,1); x \in \mathbf{R}_+),$$

we have the following equivalent inequalities with the best possible constant factor $k(\lambda_1)$:

$$\sum_{n=1}^\infty \int_0^\infty k_\lambda(x, n-\tau) a_n f(x) dx > k(\lambda_1)||f||_{p,\varphi}||a||_{q,\psi}, \tag{51}$$

$$\left\{\sum_{n=1}^\infty (n-\tau)^{p\lambda_2 - 1} \left(\int_0^\infty k_\lambda(x, n-\tau) f(x) dx\right)^p\right\}^{\frac{1}{p}} > k(\lambda_1)||f||_{p,\widetilde{\varphi}}, \tag{52}$$

$$\left\{\int_0^\infty \frac{x^{q\lambda_1 - 1}}{(1-\theta_{\lambda_1}(x))^{q-1}} \left(\sum_{n=1}^\infty k_\lambda(x, n-\tau) a_n\right)^q dx\right\}^{\frac{1}{q}} > k(\lambda_1)||a||_{q,\psi}. \tag{53}$$

Proof We only prove that the constant factor $K(\lambda_1)$ in (48) is the best possible. The rest are omitted. For $0 < \varepsilon < |q|\delta_0, \widetilde{\lambda}_1 = \lambda_1 + \frac{\varepsilon}{q}, \widetilde{\lambda}_2 = \lambda_2 - \frac{\varepsilon}{q}$, we set $\widetilde{f}(x), \widetilde{a}_n$ as Theorem 1. Then in view of (15), (17), and (33), we find

$$||\widetilde{f}||_{p,\widetilde{\Phi}}||\widetilde{a}||_{q,\Psi} = \left\{\int_{\{x \in \mathbf{R}_+^{i_0}; x_i \geq 1\}} ||x||_\alpha^{-i_0 - \varepsilon} \left(1 - O\left(\frac{1}{||x||_\alpha^{\rho(\widetilde{\lambda}_1)}}\right)\right) dx\right\}^{\frac{1}{p}}$$

$$\times \left\{ \sum_{n} ||n-\tau||_{\beta}^{-j_0-\varepsilon} \right\}^{\frac{1}{q}}$$

$$= \left\{ \frac{\Gamma^{i_0}(\frac{1}{\alpha})}{i_0^{\varepsilon/\alpha} \alpha^{i_0-1} \Gamma(\frac{i_0}{\alpha})} - \varepsilon \widetilde{O}(1) \right\}^{\frac{1}{p}}$$

$$\times \left\{ \frac{\Gamma^{j_0}(\frac{1}{\beta})}{j_0^{\varepsilon/\beta} \beta^{j_0-1} \Gamma(\frac{j_0}{\beta})} + \varepsilon O(1) \right\}^{\frac{1}{q}}.$$

If there exists a constant $K \geq K(\lambda_1)$, such that (48) is valid when replacing $K(\lambda_1)$ by K, then in particular, we have

$$K \left\{ \frac{\Gamma^{i_0}\left(\frac{1}{\alpha}\right)}{i_0^{\varepsilon/\alpha} \alpha^{i_0-1} \Gamma\left(\frac{i_0}{\alpha}\right)} - \varepsilon \widetilde{O}(1) \right\}^{\frac{1}{p}}$$

$$\times \left\{ \frac{\Gamma^{j_0}\left(\frac{1}{\beta}\right)}{j_0^{\varepsilon/\beta} \beta^{j_0-1} \Gamma\left(\frac{j_0}{\beta}\right)} + \varepsilon O(1) \right\}^{\frac{1}{q}}$$

$$= \varepsilon K ||\widetilde{f}||_{p,\widetilde{\Phi}} ||\widetilde{a}||_{q,\Psi} < \varepsilon \widetilde{I}$$

$$= \varepsilon \int_{\{x \in \mathbf{R}_+^{i_0}; x_i \geq 1\}} ||x||_{\alpha}^{-i_0-\varepsilon} \varpi_{\lambda}\left(\widetilde{\lambda}_1, x\right) dx$$

$$< \varepsilon K_2(\widetilde{\lambda}_1) \int_{\{x \in \mathbf{R}_+^{i_0}; x_i \geq 1\}} ||x||_{\alpha}^{-i_0-\varepsilon} dx$$

$$= K_2(\widetilde{\lambda}_1) \frac{\Gamma^{i_0}(\frac{1}{\alpha})}{i_0^{\varepsilon/\alpha} \alpha^{i_0-1} \Gamma(\frac{i_0}{\alpha})},$$

and by (18), $K \leq K(\lambda_1)(\varepsilon \to 0^+)$. Hence $K = K(\lambda_1)$ is the best possible constant factor of (48). The theorem is proved.

For $p > 1$, we set

$$\Phi(x) = ||x||_{\alpha}^{p(i_0-\lambda_1)-i_0} (x \in \mathbf{R}_+^{i_0}),$$

$$\Psi(n) = ||n-\tau||_{\beta}^{q(j_0-\lambda_2)-j_0} (n \in \mathbf{N}^{j_0}),$$

wherefrom

$$[\Psi(n)]^{1-p} = ||n-\tau||_{\beta}^{p\lambda_2-j_0}, [\Phi(x)]^{1-q} = ||x||_{\alpha}^{q\lambda_1-j_0}.$$

We define two real weight normal spaces $\mathbf{L}_{p,\Phi}\left(\mathbf{R}_+^{i_0}\right)$ and $\mathbf{l}_{q,\Psi}$ as follows:

$$\mathbf{L}_{p,\Phi}(\mathbf{R}_+^{i_0}) := \left\{ f; ||f||_{p,\Phi} = \left\{ \int_{\mathbf{R}_+^{i_0}} \Phi(x)|f(x)|^p dx \right\}^{\frac{1}{p}} < \infty \right\},$$

$$\mathbf{l}_{q,\Psi} := \left\{ a = \{a_n\}; ||a||_{q,\Psi} = \left\{ \sum_n \Psi(n)|a_n|^q \right\}^{\frac{1}{q}} < \infty \right\}.$$

As the assumptions of Theorem 1, in view of

$$J < K(\lambda_1)||f||_{p,\Phi}, H < K(\lambda_1)||a||_{q,\Psi},$$

we can have Definition 2.

Definition 2 Define a first kind of multidimensional half-discrete Hilbert-type operator

$$T_1 : \mathbf{L}_{p,\Phi}(\mathbf{R}_+^{i_0}) \to \mathbf{l}_{p,\Psi^{1-p}}$$

as follows: For $f \in \mathbf{L}_{p,\Phi}(\mathbf{R}_+^{i_0})$, there exists an unique representation $T_1 f \in \mathbf{l}_{p,\Psi^{1-p}}$, satisfying

$$T_1 f(n) := \int_{\mathbf{R}_+^{i_0}} k_\lambda(||x||_\alpha, ||n - \tau||_\beta) f(x) dx (n \in \mathbf{N}^{j_0}). \tag{54}$$

For $a \in \mathbf{l}_{q,\Psi}$, we define the following formal inner product of $T_1 f$ and a as follows:

$$(T_1 f, a) := \sum_n a_n \int_{\mathbf{R}_+^{i_0}} k_\lambda(||x||_\alpha, ||n - \tau||_\beta) f(x) dx. \tag{55}$$

Define a second kind of multidimensional half-discrete Hilbert-type operator

$$T_2 : \mathbf{l}_{q,\Psi} \to \mathbf{L}_{q,\Phi^{1-q}} \left(\mathbf{R}_+^{i_0}\right)$$

as follows: For $a \in \mathbf{l}_{q,\Psi}$, there exists an unique representation $T_2 a \in \mathbf{L}_{q,\Phi^{1-q}}(\mathbf{R}_+^{i_0})$, satisfying

$$T_2 a(x) := \sum_n k_\lambda(||x||_\alpha, ||n - \tau||_\beta) a_n \left(x \in \mathbf{R}_+^{i_0}\right). \tag{56}$$

For $f \in \mathbf{L}_{p,\Phi}(\mathbf{R}_+^{i_0})$, we define the following formal inner product of f and $T_2 a$ as follows:

$$(f, T_2 a) := \int_{\mathbf{R}_+^{i_0}} \sum_n k_\lambda(||x||_\alpha, ||n - \tau||_\beta) a_n f(x) dx. \tag{57}$$

Then by Theorem 1, for $0 < ||f||_{p,\Phi}, ||a||_{q,\Psi} < \infty$, we have the following equivalent inequalities:

$$(T_1 f, a) = (f, T_2 a) < K(\lambda_1)||f||_{p,\Phi}||a||_{q,\Psi}, \tag{58}$$

$$||T_1 f||_{p,\Psi^{1-p}} < K(\lambda_1)||f||_{p,\Phi}, \tag{59}$$

$$||T_2 a||_{q,\Phi^{1-q}} < K(\lambda_1)||a||_{q,\Psi}. \tag{60}$$

It follows that T_1 and T_2 are bounded with

$$||T_1|| := \sup_{f(\neq\theta)\in \mathbf{L}_{p,\Phi}(\mathbf{R}_+^{i_0})} \frac{||T_1 f||_{p,\Psi^{1-p}}}{||f||_{p,\Phi}} \leq K(\lambda_1),$$

$$||T_2|| := \sup_{a(\neq\theta)\in \mathbf{l}_{q,\Upsilon}} \frac{||T_2 a||_{q,\Phi^{1-q}}}{||a||_{q,\Psi}} \leq K(\lambda_1).$$

Since the constant factor $K(\lambda_1)$ in (59) and (60) is the best possible, we have

$$||T_1|| = ||T_2|| = K(\lambda_1)$$

$$= \left[\frac{\Gamma^{j_0}(\frac{1}{\beta})}{\beta^{j_0-1}\Gamma(\frac{j_0}{\beta})}\right]^{\frac{1}{p}} \left[\frac{\Gamma^{i_0}(\frac{1}{\alpha})}{\alpha^{i_0-1}\Gamma(\frac{i_0}{\alpha})}\right]^{\frac{1}{q}} k(\lambda_1). \tag{61}$$

Example 3 (i) We set

$$k_\lambda(x,y) = \frac{1}{x^\lambda + y^\lambda}(0 < \lambda \leq 1, \lambda_1, \lambda_2 > 0),$$

For $\delta_0 = \frac{1}{2}\min\{\lambda_1, \lambda_2\} > 0$, and $\widetilde{\lambda}_i \in (\lambda_i - \delta_0, \lambda_i + \delta_0)(i = 1, 2)$, $\widetilde{\lambda}_1 + \widetilde{\lambda}_2 = \lambda$, it follows

$$\begin{aligned} k(\widetilde{\lambda}_1) &= \int_0^\infty \frac{1}{t^\lambda + 1} t^{\widetilde{\lambda}_1 - 1} dt \\ &= \frac{1}{\lambda} \int_0^\infty \frac{1}{v+1} v^{\frac{\widetilde{\lambda}_1}{\lambda} - 1} dv \\ &= \frac{\pi}{\lambda \sin \pi(\widetilde{\lambda}_1/\lambda)} \in \mathbf{R}_+, \end{aligned}$$

and

$$\frac{\partial}{\partial y}\left(\frac{1}{x^\lambda + y^\lambda} y^{\widetilde{\lambda}_2 - j_0}\right) < 0,$$
$$\frac{\partial^2}{\partial y^2}\left(\frac{1}{x^\lambda + y^\lambda} y^{\widetilde{\lambda}_2 - j_0}\right) > 0.$$

Setting $\eta_1 \in (\lambda_1 + \delta_0, \lambda)$, then $\eta_1 > \widetilde{\lambda}_1$. Since

$$\frac{t^{\eta_1}}{t^\lambda + 1} \to 0(t \to 0^+), \frac{t^{\eta_1}}{t^\lambda + 1} \to 0(t \to \infty),$$

there exists a constant $L > 0$, such that

$$k_\lambda(t,1) = \frac{1}{t^\lambda + 1} \leq \frac{L}{t^{\eta_1}}(t \in (0, \infty)).$$

Then by Theorem 1 and (61), for $\alpha > 0, 0 < \beta \leq 1$, we have

$$||T_1|| = ||T_2|| = \left[\frac{\Gamma^{j_0}(\frac{1}{\beta})}{\beta^{j_0-1}\Gamma(\frac{j_0}{\beta})}\right]^{\frac{1}{p}}$$

$$\times \left[\frac{\Gamma^{i_0}(\frac{1}{\alpha})}{\alpha^{i_0-1}\Gamma(\frac{i_0}{\alpha})}\right]^{\frac{1}{q}} \frac{\pi}{\lambda \sin\left(\frac{\pi\lambda_1}{\lambda}\right)}. \tag{62}$$

(ii) We set

$$k_\lambda(x, y) = \frac{1}{(x+y)^\lambda}(\lambda, \lambda_1 > 0, 0 < \lambda_2 < j_0),$$

For $\delta_0 = \frac{1}{2}\min\{\lambda_1, \lambda_2, j_0 - \lambda_2\} > 0$, and $\widetilde{\lambda}_i \in (\lambda_i - \delta_0, \lambda_i + \delta_0)(i = 1, 2), \widetilde{\lambda}_1 + \widetilde{\lambda}_2 = \lambda$, it follows

$$k(\widetilde{\lambda}_1) = \int_0^\infty \frac{1}{(t+1)^\lambda} t^{\widetilde{\lambda}_1 - 1} dt = B(\widetilde{\lambda}_1, \widetilde{\lambda}_2) \in \mathbf{R}_+,$$

and

$$\frac{\partial}{\partial y}\left(\frac{1}{(x+y)^\lambda} y^{\widetilde{\lambda}_2 - j_0}\right) < 0,$$
$$\frac{\partial^2}{\partial y^2}\left(\frac{1}{(x+y)^\lambda} y^{\widetilde{\lambda}_2 - j_0}\right) > 0.$$

Setting $\eta_1 \in (\lambda_1 + \delta_0, \lambda)$, then $\eta_1 > \widetilde{\lambda}_1$. Since

$$\frac{t^{\eta_1}}{(t+1)^\lambda} \to 0(t \to 0^+), \frac{t^{\eta_1}}{(t+1)^\lambda} \to 0(t \to \infty),$$

there exists a constant $L > 0$, such that

$$k_\lambda(t, 1) = \frac{1}{(t+1)^\lambda} \leq \frac{L}{t^{\eta_1}}(t \in (0, \infty)).$$

Then by Theorem 1 and (61), for $\alpha > 0, 0 < \beta \leq 1$, we have

$$||T_1|| = ||T_2|| = \left[\frac{\Gamma^{j_0}(\frac{1}{\beta})}{\beta^{j_0-1}\Gamma(\frac{j_0}{\beta})}\right]^{\frac{1}{p}} \left[\frac{\Gamma^{i_0}(\frac{1}{\alpha})}{\alpha^{i_0-1}\Gamma(\frac{i_0}{\alpha})}\right]^{\frac{1}{q}} B(\lambda_1, \lambda_2). \tag{63}$$

(iii) We set

$$k_\lambda(x, y) = \frac{\ln(x/y)}{x^\lambda - y^\lambda}(0 < \lambda \leq 1, \lambda_1, \lambda_2 > 0).$$

For $\delta_0 = \frac{1}{2}\min\{\lambda_1, \lambda_2\} > 0$, and $\widetilde{\lambda}_i \in (\lambda_i - \delta_0, \lambda_i + \delta_0)(i = 1, 2), \widetilde{\lambda}_1 + \widetilde{\lambda}_2 = \lambda$, it follows

$$k(\widetilde{\lambda}_1) = \int_0^\infty \frac{\ln t}{t^\lambda - 1} t^{\widetilde{\lambda}_1 - 1} dt$$

$$= \frac{1}{\lambda^2}\int_0^\infty \frac{\ln v}{v-1} v^{\frac{\widetilde{\lambda}_1}{\lambda}-1} dv$$

$$= \left[\frac{\pi}{\lambda \sin \pi(\widetilde{\lambda}_1/\lambda)}\right]^2 \in \mathbf{R}_+,$$

and

$$\frac{\partial}{\partial y}\left(\frac{\ln (x/y)}{x^\lambda - y^\lambda} y^{\widetilde{\lambda}_2 - j_0}\right) < 0,$$

$$\frac{\partial^2}{\partial y^2}\left(\frac{\ln (x/y)}{x^\lambda - y^\lambda} y^{\widetilde{\lambda}_2 - j_0}\right) > 0.$$

Setting $\eta_1 \in (\lambda_1 + \delta_0, \lambda)$, then $\eta_1 > \widetilde{\lambda}_1$. Since

$$\frac{(\ln t)t^{\eta_1}}{t^\lambda - 1} \to 0(t \to 0^+), \frac{(\ln t)t^{\eta_1}}{t^\lambda - 1} \to 0(t \to \infty),$$

there exists a constant $L > 0$, such that

$$k_\lambda(t,1) = \frac{\ln t}{t^\lambda - 1} \le \frac{L}{t^{\eta_1}} (t \in (0,\infty)).$$

Then by Theorem 1 and (61), for $\alpha > 0, 0 < \beta \le 1$, we have

$$||T_1|| = ||T_2|| = \left[\frac{\Gamma^{j_0}(\frac{1}{\beta})}{\beta^{j_0-1}\Gamma(\frac{j_0}{\beta})}\right]^{\frac{1}{p}} \left[\frac{\Gamma^{i_0}(\frac{1}{\alpha})}{\alpha^{i_0-1}\Gamma(\frac{i_0}{\alpha})}\right]^{\frac{1}{q}} \times \left[\frac{\pi}{\lambda \sin \pi \left(\frac{\lambda_1}{\lambda}\right)}\right]^2. \quad (64)$$

(iv) For $s \in \mathbf{N}$, we set

$$k_\lambda(x,y) = \frac{1}{\prod_{k=1}^s (x^{\lambda/s} + a_k y^{\lambda/s})} (0 < a_1 < \cdots < a_s, \\ 0 < \lambda \le s, \lambda_1 > 0, 0 < \lambda_2 < j_0).$$

For $\delta_0 = \frac{1}{2}\min\{\lambda_1, \lambda_2, j_0 - \lambda_2\} > 0$, and $\widetilde{\lambda}_i \in (\lambda_i - \delta_0, \lambda_i + \delta_0)(i = 1,2)$, $\widetilde{\lambda}_1 + \widetilde{\lambda}_2 = \lambda$, by (21), it follows

$$k_s(\widetilde{\lambda}_1) = \int_0^\infty \frac{1}{\prod_{k=1}^s (t^{\lambda/s} + a_k)} t^{\widetilde{\lambda}_1 - 1} dt$$

$$= \frac{\pi s}{\lambda \sin (\frac{\pi s \widetilde{\lambda}_1}{\lambda})} \sum_{k=1}^s a_k^{\frac{s\widetilde{\lambda}_1}{\lambda} - 1} \prod_{j=1(j \ne k)}^s \frac{1}{a_j - a_k} \in \mathbf{R}_+,$$

and

$$\frac{\partial}{\partial y}\left(\frac{y^{\widetilde{\lambda}_2-j_0}}{\prod_{k=1}^{s}(x^{\lambda/s}+a_k y^{\lambda/s})}\right)<0,$$
$$\frac{\partial^2}{\partial y^2}\left(\frac{y^{\widetilde{\lambda}_2-j_0}}{\prod_{k=1}^{s}(x^{\lambda/s}+a_k y^{\lambda/s})}\right)>0.$$

Setting $\eta_1 \in (\lambda_1+\delta_0, \lambda)$, then $\eta_1 > \widetilde{\lambda}_1$. Since

$$\frac{t^{\eta_1}}{\prod_{k=1}^{s}(t^{\lambda/s}+a_k)} \to 0(t \to 0^+),$$
$$\frac{t^{\eta_1}}{\prod_{k=1}^{s}(t^{\lambda/s}+a_k)} \to 0(t \to \infty),$$

there exists a constant $L>0$, such that

$$k_\lambda(t,1)=\frac{1}{\prod_{k=1}^{s}(t^{\lambda/s}+a_k)} \le \frac{L}{t^{\eta_1}}(t \in (0,\infty)).$$

Then by Theorem 1 and (61), for $\alpha > 0, 0<\beta \le 1$, we have

$$||T_1|| = ||T_2|| = \left[\frac{\Gamma^{j_0}(\frac{1}{\beta})}{\beta^{j_0-1}\Gamma(\frac{j_0}{\beta})}\right]^{\frac{1}{p}}\left[\frac{\Gamma^{i_0}(\frac{1}{\alpha})}{\alpha^{i_0-1}\Gamma(\frac{i_0}{\alpha})}\right]^{\frac{1}{q}} k_s(\lambda_1). \tag{65}$$

(v) We set

$$k_\lambda(x,y)=\frac{1}{x^\lambda+\sqrt{c}(xy)^{\lambda/2}\cos\gamma+\frac{c}{4}y^\lambda}$$
$$(0<\lambda\le 1, 0<\gamma<\frac{\pi}{2}, \lambda_1,\lambda_2>0).$$

For $\delta_0 = \frac{1}{2}\min\{\lambda_1,\lambda_2\}>0$, and $\widetilde{\lambda}_i \in (\lambda_i-\delta_0, \lambda_i+\delta_0)(i=1,2)$, $\widetilde{\lambda}_1+\widetilde{\lambda}_2=\lambda$, by Example 2, it follows

$$k(\widetilde{\lambda}_1)=\int_0^\infty \frac{1}{t^\lambda+\sqrt{c}t^{\lambda/2}\cos\gamma+\frac{c}{4}} t^{\widetilde{\lambda}_1-1}dt$$
$$=\left(\frac{\sqrt{c}}{2}\right)^{\frac{2\widetilde{\lambda}_1}{\lambda}} \frac{2\pi\sin\gamma\left(1-\frac{2\widetilde{\lambda}_1}{\lambda}\right)}{\lambda\sin\gamma\sin(\frac{2\pi\widetilde{\lambda}_1}{\lambda})} \in \mathbf{R}_+,$$

and

$$\frac{\partial}{\partial y}\left(\frac{y^{\widetilde{\lambda}_2-j_0}}{x^\lambda+\sqrt{c}(xy)^{\lambda/2}\cos\gamma+\frac{c}{4}y^\lambda}\right)<0,$$

$$\frac{\partial^2}{\partial y^2}\left(\frac{y^{\widetilde{\lambda}_2-j_0}}{x^\lambda+\sqrt{c}(xy)^{\lambda/2}\cos\gamma+\frac{c}{4}y^\lambda}\right)>0.$$

Setting $\eta_1 \in (\lambda_1+\delta_0, \lambda)$, then $\eta_1 > \widetilde{\lambda}_1$. Since

$$\frac{t^{\eta_1}}{t^\lambda+\sqrt{c}t^{\lambda/2}\cos\gamma+\frac{c}{4}} \to 0(t \to 0^+),$$

$$\frac{t^{\eta_1}}{t^\lambda+\sqrt{c}t^{\lambda/2}\cos\gamma+\frac{c}{4}} \to 0(t \to \infty),$$

there exists a constant $L > 0$, such that

$$k_\lambda(t,1)=\frac{1}{t^\lambda+\sqrt{c}t^{\lambda/2}\cos\gamma+\frac{c}{4}} \le \frac{L}{t^{\eta_1}}(t \in (0,\infty)).$$

Then by Theorem 1 and (61), for $\alpha > 0, 0 < \beta \le 1$, we have

$$||T_1||=||T_2||=\left[\frac{\Gamma^{j_0}(\frac{1}{\beta})}{\beta^{j_0-1}\Gamma(\frac{j_0}{\beta})}\right]^{\frac{1}{p}}\left[\frac{\Gamma^{i_0}(\frac{1}{\alpha})}{\alpha^{i_0-1}\Gamma(\frac{i_0}{\alpha})}\right]^{\frac{1}{q}}$$

$$\times\left(\frac{\sqrt{c}}{2}\right)^{\frac{2\lambda_1}{\lambda}}\frac{2\pi\sin\gamma\left(1-\frac{2\lambda_1}{\lambda}\right)}{\lambda\sin\gamma\sin\left(\frac{2\pi\lambda_1}{\lambda}\right)}. \tag{66}$$

(vi) We set

$$k_0(x,y)=\ln\left(\frac{bx^\gamma+y^\gamma}{ax^\gamma+y^\gamma}\right)(0 \le a < b, 0 < \gamma \le 1,$$
$$0 < -\lambda_1 = \lambda_2 = \sigma < \gamma).$$

For $\delta_0 = \frac{1}{2}\min\{\sigma, \gamma-\sigma\} > 0$, and $\widetilde{\lambda}_i \in (\lambda_i-\delta_0, \lambda_i+\delta_0)(i=1,2), \widetilde{\lambda}_1+\widetilde{\lambda}_2=0$, it follows

$$k(\widetilde{\lambda}_1)=\int_0^\infty \ln\left(\frac{bt^\gamma+1}{at^\gamma+1}\right)t^{\widetilde{\lambda}_1-1}dt$$
$$=\frac{1}{\widetilde{\lambda}_1}\int_0^\infty \ln\left(\frac{bt^\gamma+1}{at^\gamma+1}\right)dt^{\widetilde{\lambda}_1}$$
$$=\frac{1}{\widetilde{\lambda}_1}\left[t^{\widetilde{\lambda}_1}\ln\left(\frac{bt^\gamma+1}{at^\gamma+1}\right)|_0^\infty\right.$$
$$\left.-\gamma\int_0^\infty\left(\frac{b}{bt^\gamma+1}-\frac{a}{at^\gamma+1}\right)t^{\widetilde{\lambda}_1+\gamma-1}dt\right]$$
$$=\frac{1}{\widetilde{\lambda}_1}\left(b^{-\frac{\widetilde{\lambda}_1}{\gamma}}-a^{-\frac{\widetilde{\lambda}_1}{\gamma}}\right)\int_0^\infty\frac{1}{u+1}u^{(1+\frac{\widetilde{\lambda}_1}{\gamma})-1}du$$

$$= \frac{-1}{\widetilde{\lambda}_1}\left(b^{-\frac{\widetilde{\lambda}_1}{\gamma}} - a^{-\frac{\widetilde{\lambda}_1}{\gamma}}\right)\frac{\pi}{\sin\pi(\frac{-\widetilde{\lambda}_1}{\gamma})} \in \mathbf{R}_+,$$

and

$$\frac{\partial}{\partial y}\left(\ln\left(\frac{bx^\gamma + y^\gamma}{ax^\gamma + y^\gamma}\right)y^{\widetilde{\lambda}_2 - j_0}\right) < 0,$$
$$\frac{\partial^2}{\partial y^2}\left(\ln\left(\frac{bx^\gamma + y^\gamma}{ax^\gamma + y^\gamma}\right)y^{\widetilde{\lambda}_2 - j_0}\right) > 0.$$

Setting $\eta_1 \in (-\sigma + \delta_0, 0)$, then $\eta_1 > \widetilde{\lambda}_1$. Since

$$t^{\eta_1}\ln\left(\frac{bt^\gamma + 1}{at^\gamma + 1}\right) \to 0(t \to 0^+),$$
$$t^{\eta_1}\ln\left(\frac{bt^\gamma + 1}{at^\gamma + 1}\right) \to 0(t \to \infty),$$

there exists a constant $L > 0$, such that

$$k_0(t,1) = \ln\left(\frac{bt^\gamma + 1}{at^\gamma + 1}\right) \le \frac{L}{t^{\eta_1}}(t \in (0,\infty)).$$

Then by Theorem 1 and (61), for $\alpha > 0, 0 < \beta \le 1$, we have

$$||T_1|| = ||T_2|| = \left[\frac{\Gamma^{j_0}(\frac{1}{\beta})}{\beta^{j_0-1}\Gamma(\frac{j_0}{\beta})}\right]^{\frac{1}{p}}\left[\frac{\Gamma^{i_0}(\frac{1}{\alpha})}{\alpha^{i_0-1}\Gamma(\frac{i_0}{\alpha})}\right]^{\frac{1}{q}} \times \frac{\left(b^{\frac{\sigma}{\gamma}} - a^{\frac{\sigma}{\gamma}}\right)\pi}{\sigma\sin\pi\left(\frac{\sigma}{\gamma}\right)}. \qquad (67)$$

(vii) We set

$$k_0(x,y) = e^{-\rho(\frac{y}{x})^\gamma}(\rho > 0, 0 < \gamma \le 1,$$
$$0 < -\lambda_1 = \lambda_2 = \sigma < \gamma).$$

For $\delta_0 = \frac{1}{2}\min\{\sigma, \gamma - \sigma\} > 0$, and $\widetilde{\lambda}_i \in (\lambda_i - \delta_0, \lambda_i + \delta_0)(i = 1,2)$, $\widetilde{\lambda}_1 + \widetilde{\lambda}_2 = 0$, it follows

$$k(\widetilde{\lambda}_1) = \int_0^\infty e^{-\frac{\rho}{t^\gamma}}t^{\widetilde{\lambda}_1 - 1}dt$$
$$= \frac{1}{\gamma}\rho^{\frac{\widetilde{\lambda}_1}{\gamma}}\int_0^\infty e^{-u}u^{-\frac{\widetilde{\lambda}_1}{\gamma}-1}du$$
$$= \frac{1}{\gamma}\rho^{\frac{\widetilde{\lambda}_1}{\gamma}}\Gamma\left(-\frac{\widetilde{\lambda}_1}{\gamma}\right) \in \mathbf{R}_+,$$

and

$$\frac{\partial}{\partial y}\left(e^{-\rho(\frac{y}{x})^{\gamma}}y^{\widetilde{\lambda}_2-j_0}\right)<0,$$

$$\frac{\partial^2}{\partial y^2}\left(e^{-\rho(\frac{y}{x})^{\gamma}}y^{\widetilde{\lambda}_2-j_0}\right)>0.$$

Setting $\eta_1\in(-\sigma+\delta_0,0)$, then $\eta_1>\widetilde{\lambda}_1$. Since

$$t^{\eta_1}e^{-\frac{\rho}{t^{\gamma}}}\to 0(t\to 0^+),t^{\eta_1}e^{-\frac{\rho}{t^{\gamma}}}\to 0(t\to\infty),$$

there exists a constant $L>0$, such that

$$k_0(t,1)=e^{-\frac{\rho}{t^{\gamma}}}\leq\frac{L}{t^{\eta_1}}(t\in(0,\infty)).$$

Then by Theorem 1 and (61), for $\alpha>0, 0<\beta\leq 1$, we have

$$||T_1||=||T_2||=\left[\frac{\Gamma^{j_0}(\frac{1}{\beta})}{\beta^{j_0-1}\Gamma(\frac{j_0}{\beta})}\right]^{\frac{1}{p}}\left[\frac{\Gamma^{i_0}(\frac{1}{\alpha})}{\alpha^{i_0-1}\Gamma(\frac{i_0}{\alpha})}\right]^{\frac{1}{q}}\times\frac{1}{\gamma\rho^{\frac{\sigma}{\gamma}}}\Gamma\left(\frac{\sigma}{\gamma}\right). \tag{68}$$

(viii) We set

$$k_0(x,y)=\arctan\rho\left(\frac{x}{y}\right)^{\gamma}$$
$$(\rho,\gamma>0,0<-\lambda_1=\lambda_2=\sigma<\min\{j_0,\gamma\}).$$

For $\delta_0=\frac{1}{2}\min\{\sigma,\gamma-\sigma,j_0-\sigma\}>0$, and $\widetilde{\lambda}_i\in(\lambda_i-\delta_0,\lambda_i+\delta_0)(i=1,2)$, $\widetilde{\lambda}_1+\widetilde{\lambda}_2=0$, it follows

$$\begin{aligned}k(\widetilde{\lambda}_1)&=\int_0^{\infty}t^{\widetilde{\lambda}_1-1}(\arctan\rho t^{\gamma})dt\\&=\frac{1}{\widetilde{\lambda}_1}\int_0^{\infty}(\arctan\rho t^{\gamma})dt^{\widetilde{\lambda}_1}\\&=\frac{1}{\widetilde{\lambda}_1}\left[(\arctan\rho t^{\gamma})t^{\widetilde{\lambda}_1}|_0^{\infty}-\int_0^{\infty}\frac{\gamma\rho t^{\widetilde{\lambda}_1+\gamma-1}}{1+(\rho t^{\gamma})^2}dt\right]\\&=\frac{-\rho^{-\frac{\widetilde{\lambda}_1}{\gamma}}}{2\widetilde{\lambda}_1}\int_0^{\infty}\frac{1}{1+u}u^{(\frac{\widetilde{\lambda}_1}{2\gamma}+\frac{1}{2})-1}du\\&=\frac{-\rho^{-\frac{\widetilde{\lambda}_1}{\gamma}}\pi}{2\widetilde{\lambda}_1\sin\pi(\frac{\widetilde{\lambda}_1}{2\gamma}+\frac{1}{2})}\end{aligned}$$

$$= \frac{-\rho^{-\frac{\widetilde{\lambda}_1}{\gamma}}\pi}{2\widetilde{\lambda}_1 \cos\pi(\frac{\widetilde{\lambda}_1}{2\gamma})} \in \mathbf{R}_+,$$

and

$$\frac{\partial}{\partial y}\left(y^{\widetilde{\lambda}_2 - j_0} \arctan \rho(\frac{x}{y})^\gamma\right) < 0,$$

$$\frac{\partial^2}{\partial y^2}\left(y^{\widetilde{\lambda}_2 - j_0} \arctan \rho\left(\frac{x}{y}\right)^\gamma\right) > 0.$$

Setting $\eta_1 \in (-\sigma + \delta_0, 0)$, then $\eta_1 > \widetilde{\lambda}_1$. Since

$$t^{\eta_1} \arctan \rho t^\gamma \to 0 (t \to 0^+),$$

$$t^{\eta_1} \arctan \rho t^\gamma \to 0 (t \to \infty),$$

there exists a constant $L > 0$, such that

$$k_0(t, 1) = \arctan \rho t^\gamma \leq \frac{L}{t^{\eta_1}} (t \in (0, \infty)).$$

Then by Theorem 1 and (61), for $\alpha > 0, 0 < \beta \leq 1$, we have

$$||T_1|| = ||T_2|| = \left[\frac{\Gamma^{j_0}(\frac{1}{\beta})}{\beta^{j_0-1}\Gamma(\frac{j_0}{\beta})}\right]^{\frac{1}{p}} \left[\frac{\Gamma^{i_0}(\frac{1}{\alpha})}{\alpha^{i_0-1}\Gamma(\frac{i_0}{\alpha})}\right]^{\frac{1}{q}} \times \frac{\rho^{\frac{\sigma}{\gamma}}\pi}{2\sigma \cos\pi\left(\frac{\sigma}{2\gamma}\right)}. \quad (69)$$

Example 4 (i) We set

$$k_0(x, y) = \operatorname{csch}\left(\frac{y}{x}\right) = \frac{2}{e^{(y/x)} - e^{-(y/x)}}$$

$$(1 < -\lambda_1 = \lambda_2 = \sigma < j_0),$$

where, $\operatorname{csch} u = \frac{2}{e^u - e^{-u}}$ is hyperbolic cosecant function (cf. [46]). For $\delta_0 = \frac{1}{2}\min\{j_0 - \sigma, \sigma - 1\} > 0$, and $\widetilde{\lambda}_i \in (\lambda_i - \delta_0, \lambda_i + \delta_0)(i = 1, 2)$, $\widetilde{\lambda}_1 + \widetilde{\lambda}_2 = 0$, it follows $-\widetilde{\lambda}_1 > -\lambda_1 - \delta_0 > 1$,

$$k(\widetilde{\lambda}_1) = \int_0^\infty t^{\widetilde{\lambda}_1 - 1} \operatorname{csch}\left(\frac{1}{t}\right) dt = \int_0^\infty \frac{2t^{\widetilde{\lambda}_1 - 1}}{e^{1/t} - e^{-1/t}} dt$$

$$= \int_0^\infty \frac{2e^{-u} u^{-\widetilde{\lambda}_1 - 1}}{1 - e^{-2u}} du = 2\int_0^\infty \sum_{k=0}^\infty e^{-(2k+1)u} u^{-\widetilde{\lambda}_1 - 1} du$$

$$= 2\sum_{k=0}^{\infty}\int_0^{\infty} e^{-(2k+1)u}u^{-\widetilde{\lambda}_1-1}du$$

$$= 2\Gamma(-\widetilde{\lambda}_1)\sum_{k=0}^{\infty}\frac{1}{(2k+1)^{-\widetilde{\lambda}_1}}.$$

For $\sigma > 1$, the Riemann's zeta function $\zeta(\cdot)$ is expressed as follows:

$$\zeta(\sigma) = \sum_{k=1}^{\infty}\frac{1}{k^{\sigma}} = \sum_{k=0}^{\infty}\frac{1}{(2k+1)^{\sigma}} + \sum_{k=1}^{\infty}\frac{1}{(2k)^{\sigma}}$$
$$= \sum_{k=0}^{\infty}\frac{1}{(2k+1)^{\sigma}} + \frac{1}{2^{\sigma}}\zeta(\sigma),$$

then we find

$$\sum_{k=0}^{\infty}\frac{1}{(2k+1)^{\sigma}} = (1-\frac{1}{2^{\sigma}})\zeta(\sigma),$$

and

$$k(\widetilde{\lambda}_1) = \Gamma\left(-\widetilde{\lambda}_1\right)\left(2-\frac{1}{2^{-\widetilde{\lambda}_1-1}}\right)\zeta\left(-\widetilde{\lambda}_1\right) \in \mathbf{R}_+.$$

We obtain

$$\frac{\partial}{\partial y}\left(\frac{2}{e^{(y/x)}-e^{-(y/x)}}\right) = \frac{-2\left[e^{(y/x)}+e^{-(y/x)}\right]}{x\left[e^{(y/x)}-e^{-(y/x)}\right]^2} < 0,$$

and

$$\frac{\partial^2}{\partial y^2}\left(\frac{2}{e^{(x/y)}-e^{-(x/y)}}\right)$$
$$= \frac{-2\left[e^{(y/x)}-e^{-(y/x)}\right]}{x^2[e^{(y/x)}-e^{-(y/x)}]^2} + \frac{4[e^{(y/x)}+e^{-(y/x)}]^2}{x^2[e^{(y/x)}-e^{-(y/x)}]^3}$$
$$= \frac{2\left[e^{2(y/x)}+e^{-2(y/x)}+6\right]}{x^2[e^{(y/x)}-e^{-(y/x)}]^3} > 0.$$

For $\widetilde{\lambda}_2 - j_0 < 0$, it follows

$$\frac{\partial}{\partial y}\left(y^{\widetilde{\lambda}_2-j_0}\mathrm{csch}\left(\frac{y}{x}\right)\right) < 0, \frac{\partial^2}{\partial y^2}\left(y^{\widetilde{\lambda}_2-j_0}\mathrm{csch}\left(\frac{y}{x}\right)\right) > 0.$$

Setting $-\sigma + \delta_0 < \eta_1 < -1$, then $\eta_1 > \widetilde{\lambda}_1$. Since

$$t^{\eta_1}\mathrm{csch}\left(\frac{1}{t}\right) \to 0(t \to 0^+), t^{\eta_1}\mathrm{csch}\left(\frac{1}{t}\right) \to 0(t \to \infty),$$

there exists a constant $L > 0$, such that

$$k_0(t,1) = \operatorname{csch}\left(\frac{1}{t}\right) \le \frac{L}{t^{\eta_1}}(t \in (0,\infty)).$$

Then by Theorem 1 and (61), for $\alpha > 0, 0 < \beta \le 1$, we have

$$||T_1|| = ||T_2|| = \left[\frac{\Gamma^{j_0}(\frac{1}{\beta})}{\beta^{j_0-1}\Gamma(\frac{j_0}{\beta})}\right]^{\frac{1}{p}} \left[\frac{\Gamma^{i_0}(\frac{1}{\alpha})}{\alpha^{i_0-1}\Gamma(\frac{i_0}{\alpha})}\right]^{\frac{1}{q}}$$

$$\times \left(2 - \frac{1}{2^{\sigma-1}}\right)\Gamma(\sigma)\zeta(\sigma). \tag{70}$$

(ii) We set

$$k_0(x,y) = \coth(\frac{y}{x}) - 1 = \frac{e^{(y/x)} + e^{-(y/x)}}{e^{(y/x)} - e^{-(y/x)}} - 1$$
$$= \frac{2e^{-(y/x)}}{e^{(y/x)} - e^{-(y/x)}}(1 < -\lambda_1 = \lambda_2 = \sigma < j_0),$$

where, $\coth u = \frac{e^u+e^{-u}}{e^u-e^{-u}}$ is hyperbolic cotangent function (cf. [46]). For $\delta_0 = \min\{j_0 - \sigma, \sigma - 1\} > 0$, and $\widetilde{\lambda}_i \in (\lambda_i - \delta_0, \lambda_i + \delta_0)(i = 1,2)$, $\widetilde{\lambda}_1 + \widetilde{\lambda}_2 = 0$, it follows

$$k(\widetilde{\lambda}_1) = \int_0^\infty t^{\widetilde{\lambda}_1-1}\left(\coth\left(\frac{1}{t}\right) - 1\right)dt = \int_0^\infty \frac{2e^{-1/t}t^{\widetilde{\lambda}_1-1}}{e^{1/t} - e^{-1/t}}dt$$
$$= \int_0^\infty \frac{2e^{-2u}u^{-\widetilde{\lambda}_1-1}}{1 - e^{-2u}}du = 2\int_0^\infty \sum_{k=0}^\infty e^{-2(k+1)u}u^{-\widetilde{\lambda}_1-1}du$$
$$= 2\sum_{k=0}^\infty \int_0^\infty e^{-2(k+1)u}u^{-\widetilde{\lambda}_1-1}du$$
$$= 2\Gamma\left(-\widetilde{\lambda}_1\right)\sum_{k=1}^\infty \frac{1}{(2k)^{-\widetilde{\lambda}_1}} = \frac{\Gamma\left(-\widetilde{\lambda}_1\right)}{2^{-\widetilde{\lambda}_1-1}}\zeta\left(-\widetilde{\lambda}_1\right) \in \mathbf{R}_+.$$

We can obtain

$$\frac{\partial}{\partial y}\left(y^{\widetilde{\lambda}_2-j_0}\left(\coth\left(\frac{y}{x}\right) - 1\right)\right) < 0,$$
$$\frac{\partial^2}{\partial y^2}\left(y^{\widetilde{\lambda}_2-j_0}\left(\coth\left(\frac{y}{x}\right) - 1\right)\right) > 0.$$

Setting $-\sigma + \delta_0 < \eta_1 < -1$, then $\eta_1 > \widetilde{\lambda}_1$. Since

$$t^{\eta_1}\left(\coth\left(\frac{1}{t}\right) - 1\right) \to 0(t \to 0^+),$$

$$t^{\eta_1}\left(\coth\left(\frac{1}{t}\right)-1\right)\to 0(t\to\infty),$$

there exists a constant $L>0$, such that

$$k_0(t,1)=\coth\left(\frac{1}{t}\right)-1\leq\frac{L}{t^{\eta_1}}(t\in(0,\infty)).$$

Then by Theorem 1 and (61), for $\alpha>0, 0<\beta\leq 1$, we have

$$||T_1||=||T_2||=\left[\frac{\Gamma^{j_0}(\frac{1}{\beta})}{\beta^{j_0-1}\Gamma(\frac{j_0}{\beta})}\right]^{\frac{1}{p}}\left[\frac{\Gamma^{i_0}(\frac{1}{\alpha})}{\alpha^{i_0-1}\Gamma(\frac{i_0}{\alpha})}\right]^{\frac{1}{q}}\frac{\Gamma(\sigma)}{2^{\sigma-1}}\zeta(\sigma). \tag{71}$$

(iii) We set

$$\begin{aligned}k_0(x,y)&=1-\tan h(\frac{y}{x})=1-\frac{e^{(y/x)}-e^{-(y/x)}}{e^{(y/x)}+e^{-(y/x)}}\\&=\frac{2e^{-(y/x)}}{e^{(y/x)}+e^{-(y/x)}}(0<-\lambda_1=\lambda_2=\sigma<j_0),\end{aligned}$$

where, $\tan hu=\frac{2}{e^u+e^{-u}}$ is hyperbolic tangent function (cf. [46]). For $\delta_0=\frac{1}{2}\min\{\lambda_2,j_0-\lambda_2\}>0$, and $\widetilde{\lambda}_i\in(\lambda_i-\delta_0,\lambda_i+\delta_0)(i=1,2)$, $\widetilde{\lambda}_1+\widetilde{\lambda}_2=\lambda,\widetilde{\lambda}_1<0$, it follows

$$\begin{aligned}k(\widetilde{\lambda}_1)&=\int_0^\infty\frac{2e^{-1/t}t^{\widetilde{\lambda}_1-1}}{e^{1/t}+e^{-1/t}}dt=\int_0^\infty\frac{2u^{-\widetilde{\lambda}_1-1}}{e^{2u}(1+e^{-2u})}du\\&=2\int_0^\infty\sum_{k=0}^\infty(-1)^k e^{-2(k+1)u}u^{-\widetilde{\lambda}_1-1}du\\&=2\sum_{k=0}^\infty(-1)^k\int_0^\infty e^{-2(k+1)u}u^{-\widetilde{\lambda}_1-1}du\\&=2\Gamma(-\widetilde{\lambda}_1)\sum_{k=1}^\infty\frac{(-1)^{k-1}}{(2k)^{-\widetilde{\lambda}_1}}\in\mathbf{R}_+.\end{aligned}$$

Since, we find

$$\frac{\partial}{\partial y}k_0(x,y)=\frac{\partial}{\partial y}\frac{2}{e^{2(y/x)}+1}=\frac{-4e^{2(y/x)}}{x(e^{2(y/x)}+1)^2}<0,$$

$$\begin{aligned}\frac{\partial^2}{\partial y^2}k_0(x,y)&=-\frac{8e^{2(y/x)}}{x^2(e^{2(y/x)}+1)^2}+\frac{16e^{4(y/x)}}{x^2(e^{2(y/x)}+1)^3}\\&=\frac{8e^{2(y/x)}(e^{2(y/x)}-1)}{x^2(e^{2(y/x)}+1)^3}>0,\end{aligned}$$

then for $\widetilde{\lambda}_2 < j_0$,we can obtain

$$\frac{\partial}{\partial y}\left(y^{\widetilde{\lambda}_2-j_0}\left(\coth\left(\frac{y}{x}\right)-1\right)\right) < 0,$$

$$\frac{\partial^2}{\partial y^2}\left(y^{\widetilde{\lambda}_2-j_0}\left(\coth\left(\frac{y}{x}\right)-1\right)\right) > 0.$$

There exists a constant $\eta_1 \in (\lambda_1 - \delta_0, 0)$, such that $0 > \eta_1 > \lambda_1 + \delta_0 > \widetilde{\lambda}_1$. In view of

$$\lim_{v\to\infty} v^{\eta_1} k_0(v,1) = \lim_{v\to\infty} \frac{2v^{\eta_1}}{e^{2/v}+1} = 0,$$

$$\lim_{v\to 0^+} v^{\eta_1} k_0(v,1) = \lim_{v\to 0^+} \frac{2v^{\eta_1}}{e^{2/v}+1}$$

$$= \lim_{v\to 0^+} \frac{-2\eta_1 v^{\eta_1+1}}{e^{2/v}} = \cdots = 0,$$

there exists a constant $L > 0$, such that

$$k_0(v,1) = \frac{2e^{-1/v}}{e^{1/v}+e^{-1/v}} \le \frac{L}{v^{\eta_1}} (v \in (0,\infty)).$$

Therefore, by Theorem 1 and (61), for $\alpha > 0, 0 < \beta \le 1$, it follows

$$||T_1|| = ||T_2|| = \left[\frac{\Gamma^{j_0}(\frac{1}{\beta})}{\beta^{j_0-1}\Gamma(\frac{j_0}{\beta})}\right]^{\frac{1}{p}} \left[\frac{\Gamma^{i_0}(\frac{1}{\alpha})}{\alpha^{i_0-1}\Gamma(\frac{i_0}{\alpha})}\right]^{\frac{1}{q}} \times \frac{\Gamma(\sigma)}{2^{\sigma-1}} \sum_{k=1}^{\infty} \frac{(-1)^{k-1}}{k^\sigma}. \tag{72}$$

Note The following references [37–45] provide an extensive theory and applications of analytic number theory related to Riemann's zeta function that will provide a source study for further research on Hilbert-type inequalities.

4 Multidimensional More Accurate Inequalities with the General Nonhomogeneous Kernels

In this section, we agree on that $i_0, j_0 \in \mathbf{N}$, $\alpha, \beta > 0, p \in \mathbf{R}\backslash\{0,1\}, \frac{1}{p} + \frac{1}{q} = 1$, $h(t)(>0)$ is a measurable function in $\mathbf{R}_+$.

Definition 3 For $x = (x_1, \cdots, x_{i_0}) \in \mathbf{R}_+^{i_0}, n = (n_1, \cdots, n_{j_0}) \in \mathbf{N}^{j_0}, \tau = (\tau_1, \cdots, \tau_{j_0}) \in [0, \frac{1}{2}]^{j_0}, \sigma \in \mathbf{R}$, define two weight coefficients $w(\sigma, n)$ and $W(\sigma, m)$ as follows:

$$w(\sigma,n) := ||n-\tau||_\beta^\sigma \int_{\mathbf{R}_+^{i_0}} h(||x||_\alpha ||n-\tau||_\beta) \frac{dx}{||x||_\alpha^{i_0-\sigma}},$$

$$W(\sigma,x) := ||x||_\alpha^\sigma \sum_n h(||x||_\alpha ||n-\tau||_\beta) \frac{1}{||n-\tau||_\beta^{j_0-\sigma}}, \tag{73}$$

where, $\sum_n = \sum_{n_{j_0}=1}^\infty \cdots \sum_{n_1=1}^\infty$.

Lemma 9 *As the assumptions of Definition 4, if both* $w(\sigma,n)$ *and* $W(\sigma,x)$ *are finite, and* $f(x) = f(x_1,\cdots,x_{i_0}) \geq 0, a_n = a_{(n_1,\cdots,n_{j_0})} \geq 0$, *then*

(i) For $p > 1$, we have the following inequality:

$$\widehat{J}_1 := \left\{ \sum_n \frac{||n-\tau||_\beta^{p\sigma-j_0}}{[w(\sigma,n)]^{p-1}} \left(\int_{\mathbf{R}_+^{i_0}} h(||x||_\alpha ||n-\tau||_\beta) f(x) dx \right)^p \right\}^{\frac{1}{p}}$$

$$\leq \left\{ \int_{\mathbf{R}_+^{i_0}} W(\sigma,x) ||x||_\alpha^{p(i_0-\sigma)-i_0} f^p(x) dx \right\}^{\frac{1}{p}}, \tag{74}$$

$$\widehat{J}_2 := \left\{ \int_{\mathbf{R}_+^{i_0}} \frac{||x||_\alpha^{q\sigma-i_0}}{[W(\sigma,x)]^{q-1}} \left(\sum_n h(||x||_\alpha ||n-\tau||_\beta) a_n \right)^q dx \right\}^{\frac{1}{q}}$$

$$\leq \left\{ \sum_n w(\sigma,n) ||n-\tau||_\beta^{q(j_0-\sigma)-j_0} a_n^q \right\}^{\frac{1}{q}}; \tag{75}$$

(ii) For $p < 0$, or $0 < p < 1$, we have the reverses of (74) and (75).

Proof (i) For $p > 1$, by Hölder's inequality with weight (cf. [36]), it follows

$$\int_{\mathbf{R}_+^{i_0}} h(||x||_\alpha ||n-\tau||_\beta) f(x) dx$$

$$= \int_{\mathbf{R}_+^{i_0}} h(||x||_\alpha ||n-\tau||_\beta) \left[\frac{||x||_\alpha^{(i_0-\sigma)/q} f(x)}{||n-\tau||_\beta^{(j_0-\sigma)/p}} \right] \left[\frac{||n-\tau||_\beta^{(j_0-\sigma)/p}}{||x||_\alpha^{(i_0-\sigma)/q}} \right] dx$$

$$\leq \left\{ \int_{\mathbf{R}_+^{i_0}} h(||x||_\alpha ||n-\tau||_\beta) \frac{||x||_\alpha^{(i_0-\sigma)(p-1)}}{||n-\tau||_\beta^{j_0-\sigma}} f^p(x) dx \right\}^{\frac{1}{p}}$$

$$\times \left\{ \int_{\mathbf{R}_+^{i_0}} h(||x||_\alpha ||n-\tau||_\beta) \frac{||n-\tau||_\beta^{(j_0-\sigma)(q-1)}}{||x||_\alpha^{i_0-\sigma}} \right\}^{\frac{1}{q}}$$

$$= [w(\sigma,n)]^{\frac{1}{q}} ||n-\tau||_\beta^{\frac{j_0}{p}-\sigma}$$

$$\times \left\{ \int_{\mathbf{R}_+^{i_0}} h(||x||_\alpha ||n-\tau||_\beta) \frac{||x||_\alpha^{(i_0-\sigma)(p-1)}}{||n-\tau||_\beta^{j_0-\sigma}} f^p(x) dx \right\}^{\frac{1}{p}}. \tag{76}$$

Then by Lebesgue term by term integration theorem (cf. [34]), we have

$$\widehat{J}_1 \leq \left\{ \sum_n \int_{\mathbf{R}_+^{i_0}} h(||x||_\alpha ||n-\tau||_\beta) \frac{||x||_\alpha^{(i_0-\sigma)(p-1)}}{||n-\tau||_\beta^{j_0-\sigma}} f^p(x) dx \right\}^{\frac{1}{p}}$$

$$= \left\{ \int_{\mathbf{R}_+^{i_0}} \sum_n h(||x||_\alpha ||n-\tau||_\beta) \frac{||x||_\alpha^{(i_0-\sigma)(p-1)}}{||n-\tau||_\beta^{j_0-\sigma}} f^p(x) dx \right\}^{\frac{1}{p}}$$

$$= \left\{ \int_{\mathbf{R}_+^{i_0}} W(\sigma, x) ||x||_\alpha^{p(i_0-\sigma)-i_0} f^p(x) dx \right\}^{\frac{1}{p}}. \tag{77}$$

Hence, (74) follows.

By the same way, we have

$$\sum_n h(||x||_\alpha ||n-\tau||_\beta) a_n \leq [W(\sigma, x)]^{\frac{1}{p}} ||x||_\alpha^{\frac{i_0}{q}-\sigma}$$

$$\times \left\{ \sum_n h\left(||x||_\alpha ||n-\tau||_\beta\right) \frac{||n-\tau||_\beta^{(j_0-\sigma)(q-1)}}{||x||_\alpha^{i_0-\sigma}} a_n^q \right\}^{\frac{1}{q}},$$

then by Lebesgue term by term integration theorem and the same way in obtaining (77), we have (75).

(ii) For $p < 0$, or $0 < p < 1$, by the reverse Hölder's inequality with weight (cf. [36]), we obtain the reverse of (76). Then by Lebesgue term by term integration theorem, we still can obtain the reverse of (75). The lemma is proved.

Lemma 10 *As the assumptions of Lemma 9, then*

(i) For $p > 1$, we have the following inequality equivalent to (74) and (75):

$$\widehat{I} := \sum_n \int_{\mathbf{R}_+^{i_0}} h(||x||_\alpha ||n-\tau||_\beta) a_n f(x) dx$$

$$\leq \left\{ \int_{\mathbf{R}_+^{i_0}} W(\sigma, x) ||x||_\alpha^{p(i_0-\sigma)-i_0} f^p(x) dx \right\}^{\frac{1}{p}}$$

$$\times \left\{ \sum_n w(\sigma, n) ||n-\tau||_\beta^{q(j_0-\sigma)-j_0} a_n^q \right\}^{\frac{1}{q}}; \tag{78}$$

(ii) For $p < 0$, or $0 < p < 1$, we have the reverse of (78) equivalent to the reverses of (74) and (75).

Proof (i) For $p > 1$, by Hölder's inequality (cf. [36]), it follows

$$\widehat{I} = \sum_n \frac{||n-\tau||_\beta^{\frac{j_0}{q}-(j_0-\sigma)}}{[w(\sigma,n)]^{\frac{1}{q}}} \left[\int_{\mathbf{R}_+^{i_0}} h(||x||_\alpha ||n-\tau||_\beta) f(x)dx\right]$$
$$\times \left[[w(\sigma,n)]^{\frac{1}{q}} ||n-\tau||_\beta^{(j_0-\sigma)-\frac{j_0}{q}} a_n\right]$$
$$\leq \widehat{J}_1 \left\{\sum_n w(\sigma,n)||n-\tau||_\beta^{q(j_0-\sigma)-j_0} a_n^q\right\}^{\frac{1}{q}}. \tag{79}$$

Then by (74), we have (78).

On the other hand, assuming that (78) is valid, we set

$$b_n := \frac{||n-\tau||_\beta^{p\sigma-j_0}}{[w(\sigma,n)]^{p-1}} \left(\int_{\mathbf{R}_+^{i_0}} h(||x||_\alpha ||n-\tau||_\beta) f(x)dx\right)^{p-1}, n \in \mathbf{N}^{j_0}.$$

Then it follows

$$\widehat{J}_1^p = \sum_n w(\sigma,n)||n-\tau||_\beta^{q(j_0-\sigma)-j_0} a_n^q.$$

If $\widehat{J}_1 = 0$, then (74) is trivially valid; if $\widehat{J}_1 = \infty$, then by (77), (74) keeps the form of equality ($= \infty$). Suppose that $0 < \widehat{J}_1 < \infty$. By (78), we have

$$0 < \sum_n w(\sigma,n)||n-\tau||_\beta^{q(j_0-\sigma)-j_0} a_n^q = \widehat{J}_1^p = \widehat{I}$$
$$\leq \left\{\int_{\mathbf{R}_+^{i_0}} W(\sigma,x)||x||_\alpha^{p(i_0-\sigma)-i_0} f^p(x)dx\right\}^{\frac{1}{p}}$$
$$\times \left\{\sum_n w(\sigma,n)||n-\tau||_\beta^{q(j_0-\sigma)-j_0} a_n^q\right\}^{\frac{1}{q}} < \infty.$$

It follows

$$\widehat{J}_1 = \left\{\sum_n w(\sigma,n)||n-\tau||_\beta^{q(j_0-\sigma)-j_0} a_n^q\right\}^{\frac{1}{p}}$$
$$\leq \left\{\int_{\mathbf{R}_+^{i_0}} W(\sigma,x)||x||_\alpha^{p(i_0-\sigma)-i_0} f^p(x)dx\right\}^{\frac{1}{p}},$$

and then (74) follows. Hence, (78) and (74) are equivalent.

By Hölder's inequality and the same way, we can obtain

$$\widehat{I} \leq \left\{ \int_{\mathbf{R}_+^{i_0}} W(\sigma, x) ||x||_\alpha^{p(i_0-\sigma)-i_0} f^p(x) dx \right\}^{\frac{1}{p}} \widehat{J}_2.$$

Then by (75), we have (78). On the other hand, assuming that (78) is valid, we set

$$f(x) = \frac{||x||_\alpha^{q\sigma - i_0}}{[W(\sigma, x)]^{q-1}} \left(\sum_n h(||x||_\alpha ||n - \tau||_\beta) a_n \right)^{q-1} \quad (x \in \mathbf{R}_+^{i_0}).$$

Then it follows

$$\widehat{J}_2^q = \int_{\mathbf{R}_+^{i_0}} W(\sigma, x) ||x||_\alpha^{p(i_0-\sigma)-i_0} f^p(x) dx.$$

By (78), we can obtain

$$\widehat{J}_2 = \left\{ \int_{\mathbf{R}_+^{i_0}} W(\sigma, x) ||x||_\alpha^{p(i_0-\sigma)-i_0} f^p(x) dx \right\}^{\frac{1}{q}}$$
$$\leq \left\{ \sum_n w(\sigma, n) ||n - \tau||_\beta^{q(j_0-\sigma)-j_0} a_n^q \right\}^{\frac{1}{q}},$$

and then (78) and (75) are equivalent.

Hence, (74), (75), and (78) are equivalent.

(ii) For $p < 0$, or $0 < p < 1$, by the same way, we have the reverse of (78) equivalent to the reverses of (74) and (75). The lemma is proved.

Lemma 11 *If* $\alpha, \beta > 0, \widetilde{\sigma} \in \mathbf{R}$,

$$\widehat{k}(\widetilde{\sigma}) := \int_0^\infty h(u) u^{\widetilde{\sigma}-1} du \in \mathbf{R}_+,$$

then we have

$$w(\widetilde{\sigma}, n) = \widehat{K}_2(\widetilde{\sigma}) := \frac{\Gamma^{i_0}(\frac{1}{\alpha})}{\alpha^{i_0-1} \Gamma(\frac{i_0}{\alpha})} \widehat{k}(\widetilde{\sigma}) \in \mathbf{R}_+ (n \in \mathbf{N}^{j_0}). \tag{80}$$

Moreover, if $0 < \beta \leq 1$,

$$\frac{d}{dt}(h(t) t^{\widetilde{\sigma}-j_0}) < 0, \frac{d^2}{dt^2}(h(t) t^{\widetilde{\sigma}-j_0}) > 0,$$

there exists constants $L > 0$, and $\eta_0 < \widetilde{\sigma}$, satisfying

$$h(t) \leq \frac{L}{t^{\eta_0}} (t \in (0, \infty)), \tag{81}$$

then we have

$$\widehat{K}_1(\widetilde{\sigma})(1-\vartheta_{\widetilde{\sigma}}(x)) < W(\widetilde{\sigma},x) < \widehat{K}_1(\widetilde{\sigma})(x \in \mathbf{R}_+^{i_0}). \tag{82}$$

where,

$$\widehat{K}_1(\widetilde{\sigma}) := \frac{\Gamma^{j_0}(\frac{1}{\beta})}{\beta^{j_0-1}\Gamma(\frac{j_0}{\beta})}\widehat{k}(\widetilde{\sigma}) \in \mathbf{R}_+, \tag{83}$$

$$\vartheta_{\widetilde{\sigma}}(x) := \frac{1}{\widehat{k}(\widetilde{\sigma})}\int_0^{j_0^{1/\beta}||x||_\alpha} h(t)t^{\widetilde{\sigma}-1}dt$$

$$= O(||x||_\alpha^{\rho(\widetilde{\sigma})}) \in (0,1)(\rho(\widetilde{\sigma}) = \widetilde{\sigma}-\eta_0 > 0). \tag{84}$$

Proof By (10), we find

$$w(\widetilde{\sigma},n) = ||n-\tau||_\beta^{\widetilde{\sigma}}\int_{\mathbf{R}_+^{i_0}} h\left(M\left[\sum_{i=1}^{i_0}\left(\frac{x_i}{M}\right)^\alpha\right]^{\frac{1}{\alpha}}||n-\tau||_\beta\right)$$

$$\times \frac{1}{M^{i_0-\widetilde{\sigma}}[\sum_{i=1}^{i_0}(\frac{x_i}{M})^\alpha]^{\frac{1}{\alpha}(i_0-\widetilde{\sigma})}}dx_1\cdots dx_{i_0}$$

$$= \lim_{M\to\infty}\left[\frac{M^{i_0}\Gamma^{i_0}(\frac{1}{\alpha})}{\alpha^{i_0}\Gamma(\frac{i_0}{\alpha})}||n-\tau||_\beta^{\widetilde{\sigma}}\right.$$

$$\left.\times\int_0^1 \frac{h\left(Mu^{\frac{1}{\alpha}}||n-\tau||_\beta\right)}{M^{i_0-\widetilde{\sigma}}u^{\frac{1}{\alpha}(i_0-\widetilde{\sigma})}}u^{\frac{i_0}{\alpha}-1}du\right]$$

$$= \frac{\Gamma^{i_0}(\frac{1}{\alpha})}{\alpha^{i_0-1}\Gamma(\frac{i_0}{\alpha})}\int_0^\infty h(t)t^{\widetilde{\sigma}-1}dt.$$

Hence, we have (80).

Moreover, by (7) and (10), we obtain

$$W(\widetilde{\sigma},x) < ||x||_\alpha^{\widetilde{\sigma}}\int_{\{y\in\mathbf{R}_+^{j_0}y_i\geq\frac{1}{2}\}} h(||x||_\alpha||y-\tau||_\beta)\frac{dy}{||y-\tau||_\beta^{j_0-\widetilde{\sigma}}}$$

$$= ||x||_\alpha^{\widetilde{\sigma}}\int_{\{u\in\mathbf{R}_+^{j_0};u_i\geq\frac{1}{2}-\tau_i\}} h(||x||_\alpha||u||_\beta)\frac{du}{||u||_\beta^{j_0-\widetilde{\sigma}}}$$

$$\leq ||x||_\alpha^{\widetilde{\sigma}}\int_{\mathbf{R}_+^{j_0}} h(||x||_\alpha||u||_\beta)\frac{du}{||u||_\beta^{j_0-\widetilde{\sigma}}}$$

$$= \frac{\Gamma^{j_0}(\frac{1}{\beta})}{\alpha^{j_0-1}\Gamma(\frac{j_0}{\beta})}\int_0^\infty h(t)t^{\widetilde{\sigma}-1}dt$$
$$= \widehat{K}_2(\widetilde{\sigma}).$$

By (7), (10), and (81), we find

$$W(\widetilde{\sigma},x) > ||x||_\alpha^{\widetilde{\sigma}}\int_{\{y\in\mathbf{R}_+^{j_0} y_i\geq 1+\tau_i\}} h(||x||_\alpha||y-\tau||_\beta)\frac{dy}{||y-\tau||_\beta^{j_0-\widetilde{\sigma}}}$$
$$= ||x||_\alpha^{\widetilde{\sigma}}\int_{\{u\in\mathbf{R}_+^{j_0};u_i\geq 1\}} h(||x||_\alpha||u||_\beta)\frac{du}{||u||_\beta^{j_0-\widetilde{\sigma}}}$$

$$= \lim_{M\to\infty}\frac{M^{j_0}\Gamma^{j_0}(\frac{1}{\beta})}{\beta^{j_0-1}\Gamma(\frac{j_0}{\beta})}\int_{\frac{j_0}{M^\beta}}^1 \frac{h\left(||x||_\alpha M v^{\frac{1}{\beta}}\right)}{(Mv^{\frac{1}{\beta}})^{j_0-\widetilde{\sigma}}} v^{\frac{j_0}{\beta}-1}dv$$
$$= \frac{\Gamma^{j_0}(\frac{1}{\beta})}{\beta^{j_0-1}\Gamma(\frac{j_0}{\beta})}\int_{j_0^{1/\beta}||x||_\alpha}^\infty h(t)t^{\widetilde{\sigma}-1}dt$$
$$= \frac{\Gamma^{j_0}(\frac{1}{\beta})}{\beta^{j_0-1}\Gamma(\frac{j_0}{\beta})}\widehat{k}(\widetilde{\sigma})[1-\vartheta_{\widetilde{\sigma}}(x)] > 0,$$

$$0 < \vartheta_{\widetilde{\sigma}}(x) = \frac{1}{\widehat{k}(\widetilde{\sigma})}\int_0^{j_0^{1/\beta}||x||_\alpha} h(t)t^{\widetilde{\sigma}-1}dt$$
$$\leq \frac{L}{\widehat{k}(\widetilde{\sigma})}\int_0^{j_0^{1/\beta}||x||_\alpha} t^{\widetilde{\sigma}-\eta_0-1}dt$$
$$= \frac{Lj_0^{(\widetilde{\sigma}-\eta_0)/\beta}}{\widehat{k}(\widetilde{\sigma})(\widetilde{\sigma}-\eta_0)}||x||_\beta^{\widetilde{\sigma}-\eta_0}.$$

Hence, we have (80), (81), and (82). The lemma is proved.

Setting

$$\Phi_\sigma(x) := ||x||_\alpha^{p(i_0-\sigma)-i_0}, \Psi_\sigma(n) := ||n-\tau||_\beta^{q(j_0-\sigma)-j_0},$$
$$\widetilde{\Phi}_\sigma(x) := (1-\vartheta_\sigma(x))||x||_\alpha^{p(i_0-\sigma)-i_0}(\vartheta_\sigma(x)\in(0,1); x\in\mathbf{R}_+^{i_0}),$$

by Lemmas 9–11, we have Theorem 4.

Theorem 4 *Suppose that* $\alpha > 0, 0 < \beta \leq 1, \tau = (\tau_1, \cdots, \tau_{j_0}) \in [0, \frac{1}{2}]^{j_0}, \sigma \in \mathbf{R}$, *there exists a constant* $\delta_0 > 0$, *such that for any* $\widetilde{\sigma} \in (\sigma-\delta_0, \sigma+\delta_0)$,

$$\widehat{k}(\widetilde{\sigma}) = \int_0^\infty h(t)t^{\widetilde{\sigma}-1}dt \in \mathbf{R}_+,$$

$$\frac{d}{dt}(h(t)t^{\widetilde{\sigma}-j_0}) < 0, \frac{d^2}{dt^2}(h(t)t^{\widetilde{\sigma}-j_0}) > 0,$$

and there exists constants $L > 0$ *and* $\eta_0 < \widetilde{\sigma}$, *satisfying*

$$h(t) \le \frac{L}{t^{\eta_0}}(t \in (0,\infty)). \tag{85}$$

If $p > 1,\ f(x) = f(x_1, \cdots, x_{i_0}) \ge 0, a_n = a_{(n_1,\cdots,n_{j_0})} \ge 0,$

$$0 < ||f||_{p,\Phi_\sigma} = \left\{ \int_{\mathbf{R}_+^{i_0}} \Phi_\sigma(x) f^p(x) dx \right\}^{\frac{1}{p}} < \infty,$$

$$0 < ||a||_{q,\Psi_\sigma} = \left\{ \sum_n \Psi_\sigma(n) a_n^q \right\}^{\frac{1}{q}} < \infty,$$

then we have the following equivalent inequalities with the best possible constant factor $\widehat{K}(\sigma)$:

$$\widehat{I} = \sum_n \int_{\mathbf{R}_+^{i_0}} h(||x||_\alpha ||n-\tau||_\beta) a_n f(x) dx < \widehat{K}(\sigma) ||f||_{p,\Phi_\sigma} ||a||_{q,\Psi_\sigma}, \tag{86}$$

$$\widehat{J} := \left\{ \sum_n ||n-\tau||_\beta^{p\sigma - j_0} \left(\int_{\mathbf{R}_+^{i_0}} h(||x||_\alpha ||n-\tau||_\beta) f(x) dx \right)^p \right\}^{\frac{1}{p}}$$

$$< \widehat{K}(\sigma) ||f||_{p,\Phi_\sigma}, \tag{87}$$

$$\widehat{H} := \left\{ \int_{\mathbf{R}_+^{i_0}} ||x||_\alpha^{q\sigma - i_0} \left(\sum_n h(||x||_\alpha ||n-\tau||_\beta) a_n \right)^q dx \right\}^{\frac{1}{q}}$$

$$< \widehat{K}(\sigma) ||a||_{q,\Psi_\sigma}, \tag{88}$$

where, $\widehat{k}(\sigma) = \int_0^\infty h(t) t^{\sigma-1} dt$, *and*

$$\widehat{K}(\sigma) = \left(\widehat{K}_1(\sigma)\right)^{\frac{1}{p}} \left(\widehat{K}_2(\sigma)\right)^{\frac{1}{q}}$$

$$= \left[\frac{\Gamma^{j_0}(\frac{1}{\beta})}{\beta^{j_0-1}\Gamma(\frac{j_0}{\beta})} \right]^{\frac{1}{p}} \left[\frac{\Gamma^{i_0}(\frac{1}{\alpha})}{\alpha^{i_0-1}\Gamma(\frac{i_0}{\alpha})} \right]^{\frac{1}{q}} \widehat{k}(\sigma). \tag{89}$$

In particular, for $i_0 = j_0 = 1$, $\tau \in \left[0, \frac{1}{2}\right]$, setting

$$\varphi_\sigma(x) := x^{p(1-\sigma)-1}, \psi_\sigma(n) := (n-\tau)^{q(1-\sigma)-1},$$

we have the following equivalent inequalities with the best possible constant factor $\widehat{k}(\sigma)$:

$$\sum_{n=1}^\infty \int_0^\infty h(x(n-\tau)) a_n f(x) dx < \widehat{k}(\sigma) ||f||_{p,\varphi_\sigma} ||a||_{q,\psi_\sigma}, \tag{90}$$

$$\left\{\sum_{n=1}^{\infty}(n-\tau)^{p\sigma-1}\left(\int_0^{\infty}h(x(n-\tau))f(x)dx\right)^p\right\}^{\frac{1}{p}}<\widehat{k}(\sigma)||f||_{p,\varphi_\sigma},\tag{91}$$

$$\left\{\int_0^{\infty}x^{q\sigma-1}\left(\sum_{n=1}^{\infty}h(x(n-\tau))a_n\right)^q dx\right\}^{\frac{1}{q}}<\widehat{k}(\sigma)||a||_{q,\psi_\sigma}.\tag{92}$$

Proof By Lemma 9, Lemma 10, and Lemma 11, we have equivalent inequalities (86), (87), and (88). By Hölder's inequality, we still have

$$\widehat{I}\leq\widehat{J}\left\{\sum_n||n-\tau||_\beta^{q(j_0-\sigma)-j_0}a_n^q\right\}^{\frac{1}{q}},\tag{93}$$

$$\widehat{I}\leq\{\int_{\mathbf{R}_+^{i_0}}||x||_\alpha^{p(i_0-\sigma)-i_0}f^p(x)dx\}^{\frac{1}{p}}\widehat{H}.\tag{94}$$

For $0<\varepsilon<q\delta_0$, we set $\widetilde{f}(x),\widetilde{a}_n$ as follows:

$$\widetilde{f}(x):=\begin{cases}||x||_\alpha^{\sigma+\frac{\varepsilon}{p}-i_0},0<x_i\leq 1(i=1,\cdots,i_0),\\ 0,x_i>1(i=1,\cdots,i_0),\end{cases}$$

$$\widetilde{a}_n:=||n-\tau||_\beta^{\left(\sigma-\frac{\varepsilon}{q}\right)-j_0},n\in\mathbf{N}^{j_0}.$$

Then for $\widetilde{\sigma}=\sigma-\frac{\varepsilon}{q}$, in view of (16), (17), and (84), we find

$$\begin{aligned}&||\widetilde{f}||_{p,\Phi_\sigma}||\widetilde{a}||_{q,\Psi_\sigma}\\&=\left\{\int_{\{x\in\mathbf{R}_+^{i_0};x_i\leq 1\}}||x||_\alpha^{-i_0+\varepsilon}dx\right\}^{\frac{1}{p}}\left\{\sum_n||n-\tau||_\beta^{-j_0-\varepsilon}\right\}^{\frac{1}{q}}\\&=\frac{1}{\varepsilon}\left\{\frac{i_0^{\varepsilon/\alpha}\Gamma^{i_0}(\frac{1}{\alpha})}{\alpha^{i_0-1}\Gamma(\frac{i_0}{\alpha})}\right\}^{\frac{1}{p}}\\&\times\left\{\frac{\Gamma^{j_0}(\frac{1}{\beta})}{j_0^{\varepsilon/\beta}\beta^{j_0-1}\Gamma(\frac{j_0}{\beta})}+\varepsilon O(1)\right\}^{\frac{1}{q}},\end{aligned}$$

$$\begin{aligned}\widetilde{I}&:=\int_{\mathbf{R}_+^{i_0}}\sum_n h(||x||_\alpha||n-\tau||_\beta)\widetilde{a}_n\widetilde{f}(x)dx\\&=\int_{\{x\in\mathbf{R}_+^{i_0};x_i\leq 1\}}||x||_\alpha^{-i_0+\varepsilon}W(\widetilde{\sigma},x)dx\\&\geq\widehat{K}_1(\widetilde{\sigma})\int_{\{x\in\mathbf{R}_+^{i_0};x_i\leq 1\}}||x||_\alpha^{-i_0+\varepsilon}(1-O\left(||x||_\alpha^{\rho(\widetilde{\sigma})}\right))\,dx\end{aligned}$$

$$= \frac{1}{\varepsilon}\widehat{K}_1(\widetilde{\sigma})\left[\frac{i_0^{\varepsilon/\alpha}\Gamma^{i_0}(\frac{1}{\alpha})}{\alpha^{i_0-1}\Gamma(\frac{i_0}{\alpha})} - \varepsilon O_{\widetilde{\sigma}}(1)\right].$$

If there exists a constant $K \leq \widehat{K}(\sigma)$, such that (84) is valid when replacing $\widehat{K}(\sigma)$ by K, then in particular, we have

$$\widehat{K}_1(\widetilde{\sigma})\left[\frac{i_0^{\varepsilon/\alpha}\Gamma^{i_0}(\frac{1}{\alpha})}{\alpha^{i_0-1}\Gamma(\frac{i_0}{\alpha})} - \varepsilon O_{\widetilde{\sigma}}(1)\right]$$
$$\leq \varepsilon \widetilde{I} < \varepsilon K ||\widetilde{f}||_{p,\Phi}||\widetilde{a}||_{q,\Psi}$$
$$= K\left\{\frac{i_0^{\varepsilon/\alpha}\Gamma^{i_0}(\frac{1}{\alpha})}{\alpha^{i_0-1}\Gamma(\frac{i_0}{\alpha})}\right\}^{\frac{1}{p}}$$
$$\times\left\{\frac{\Gamma^{j_0}(\frac{1}{\beta})}{j_0^{\varepsilon/\beta}\beta^{j_0-1}\Gamma(\frac{j_0}{\beta})} + \varepsilon O(1)\right\}^{\frac{1}{q}},$$

and then by (7), $\widehat{K}(\sigma) \leq K(\varepsilon \to 0^+)$. Hence $K = \widehat{K}(\sigma)$ is the best possible constant factor of (86).

By the equivalency, the constant factor $\widehat{K}(\sigma)$ in (87) and (88) is the best possible. Otherwise, we would reach a contradiction by (93) and (94) that the constant factor $\widehat{K}(\sigma)$ in (86) is not the best possible. The theorem is proved.

Theorem 5 *As the assumptions of Theorem 4, if* $p < 0$, $f(x) = f(x_1, \cdots, x_{i_0}) \geq 0$, $a_n = a_{(n_1,\cdots,n_{j_0})} \geq 0$, $0 < ||f||_{p,\Phi_\sigma} < \infty, 0 < ||a||_{q,\Psi_\sigma} < \infty$, *then we have the following equivalent inequalities with the best possible constant factor* $\widehat{K}(\sigma)$:

$$\sum_n \int_{\mathbf{R}_+^{i_0}} h(||x||_\alpha ||n-\tau||_\beta) a_n f(x) dx > \widehat{K}(\sigma)||f||_{p,\Phi_\sigma}||a||_{q,\Psi_\sigma}, \tag{95}$$

$$\left\{\sum_n ||n-\tau||_\beta^{p\sigma-j_0}\left(\int_{\mathbf{R}_+^{i_0}} h(||x||_\alpha ||n-\tau||_\beta) f(x) dx\right)^p\right\}^{\frac{1}{p}} > \widehat{K}(\sigma)||f||_{p,\Phi_\sigma}, \tag{96}$$

$$\left\{\int_{\mathbf{R}_+^{i_0}} ||x||_\alpha^{q\sigma-i_0}\left(\sum_n h(||x||_\alpha ||n-\tau||_\beta) a_n\right)^q dx\right\}^{\frac{1}{q}} > \widehat{K}(\sigma)||a||_{q,\Psi_\sigma}. \tag{97}$$

In particular, for $i_0 = j_0 = 1$, $\tau \in [0, \frac{1}{2}]$, setting $\varphi_\sigma(x)$ and $\psi_\sigma(n)$ as Theorem 4, we have the following equivalent inequalities with the best possible constant factor $\widehat{k}(\sigma)$:

$$\sum_{n=1}^\infty \int_0^\infty h(x(n-\tau)) a_n f(x) dx > \widehat{k}(\sigma)||f||_{p,\varphi_\sigma}||a||_{q,\psi_\sigma}, \tag{98}$$

$$\left\{\sum_{n=1}^{\infty}(n-\tau)^{p\sigma-1}\left(\int_0^{\infty}h(x(n-\tau))f(x)dx\right)^p\right\}^{\frac{1}{p}} > \widehat{k}(\sigma)||f||_{p,\varphi_\sigma}, \tag{99}$$

$$\left\{\int_0^{\infty}x^{q\sigma-1}\left(\sum_{n=1}^{\infty}h(x(n-\tau))a_n\right)^q dx\right\}^{\frac{1}{q}} > \widehat{k}(\sigma)||a||_{q,\psi_\sigma}. \tag{100}$$

Proof We only prove that the constant factor $\widehat{K}(\sigma)$ in (95) is the best possible. The rests are omitted. For $0 < \varepsilon < q\delta_0, \widetilde{\sigma} = \sigma - \frac{\varepsilon}{q}$, we set $\widetilde{f}(x), \widetilde{a}_n$ as Theorem 4. If there exists a constant $K \geq \widehat{K}(\sigma)$, such that (95) is valid as we replace $\widehat{K}(\sigma)$ by K, then in particular, by (16), (17), and (82), we have

$$\begin{aligned}
&K\left\{\frac{i_0^{\varepsilon/\alpha}\Gamma^{i_0}(\frac{1}{\alpha})}{\alpha^{i_0-1}\Gamma(\frac{i_0}{\alpha})}\right\}^{\frac{1}{p}}\left\{\frac{\Gamma^{j_0}(\frac{1}{\beta})}{j_0^{\varepsilon/\beta}\beta^{j_0-1}\Gamma(\frac{j_0}{\beta})}+\varepsilon O(1)\right\}^{\frac{1}{q}}\\
&=\varepsilon K||\widetilde{f}||_{p,\Phi_\sigma}||\widetilde{a}||_{q,\Psi_\sigma} < \varepsilon\widetilde{I}\\
&\quad=\varepsilon\int_{\{x\in\mathbf{R}_+^{i_0};x_i\leq 1\}}||x||_\alpha^{-i_0+\varepsilon}W(\widetilde{\sigma},x)dx\\
&\quad<\varepsilon\widehat{K}_2(\widetilde{\sigma})\int_{\{x\in\mathbf{R}_+^{i_0};x_i\leq 1\}}||x||_\alpha^{-i_0+\varepsilon}dx\\
&\quad=\widehat{K}_2(\widetilde{\sigma})\frac{i_0^{\varepsilon/\alpha}\Gamma^{i_0}(\frac{1}{\alpha})}{\alpha^{i_0-1}\Gamma(\frac{i_0}{\alpha})},
\end{aligned}$$

and then by (7), $K \leq \widehat{K}(\sigma)(\varepsilon \to 0^+)$. Hence $K = \widehat{K}(\sigma)$ is the best possible constant factor of (95). The theorem is proved.

Theorem 6 *As the assumptions of Theorem 4, if* $0 < p < 1$, $f(x) = f(x_1, \cdots, x_{i_0}) \geq 0$, $a_n = a_{(n_1,\cdots,n_{j_0})} \geq 0$, $0 < ||f||_{p,\widetilde{\Phi}_\sigma} < \infty, 0 < ||a||_{q,\Psi_\sigma} < \infty$, *then we have the following equivalent inequalities with the best possible constant factor* $\widehat{K}(\sigma)$:

$$\sum_n\int_{\mathbf{R}_+^{i_0}}h(||x||_\alpha||n-\tau||_\beta)a_nf(x)dx > \widehat{K}(\sigma)||f||_{p,\widetilde{\Phi}_\sigma}||a||_{q,\Psi_\sigma}, \tag{101}$$

$$\left\{\sum_n||n-\tau||_\beta^{p\sigma-j_0}\left(\int_{\mathbf{R}_+^{i_0}}h(||x||_\alpha||n-\tau||_\beta)f(x)dx\right)^p\right\}^{\frac{1}{p}} > \widehat{K}(\sigma)||f||_{p,\widetilde{\Phi}_\sigma}, \tag{102}$$

$$\left\{\int_{\mathbf{R}_{+}^{i_0}} \frac{||x||_{\alpha}^{q\sigma-i_0}}{(1-\vartheta_{\sigma}(x))^{q-1}} \left(\sum_{n} h(||x||_{\alpha}||n-\tau||_{\beta})a_n\right)^{q} dx\right\}^{\frac{1}{q}}$$

$$> \widehat{K}(\sigma)||a||_{q,\Psi_{\sigma}}. \tag{103}$$

In particular, for $i_0 = j_0 = 1$, $\tau \in [0, \frac{1}{2}]$, setting $\varphi_{\sigma}(x)$ and $\psi_{\sigma}(n)$ as Theorem 4, and

$$\widetilde{\varphi}_{\sigma}(x) := (1-\vartheta_{\sigma}(x))x^{p(1-\sigma)-1}(\vartheta_{\sigma}(x) \in (0,1); x \in \mathbf{R}_{+}),$$

we have the following equivalent inequalities with the best possible constant factor $\widehat{k}(\sigma)$:

$$\sum_{n=1}^{\infty}\int_{0}^{\infty} h(x(n-\tau))a_n f(x)dx > \widehat{k}(\sigma)||f||_{p,\widetilde{\varphi}_{\sigma}}||a||_{q,\psi_{\sigma}}, \tag{104}$$

$$\left\{\sum_{n=1}^{\infty}(n-\tau)^{p\sigma-1}\left(\int_{0}^{\infty} h(x(n-\tau))f(x)dx\right)^{p}\right\}^{\frac{1}{p}} > \widehat{k}(\sigma)||f||_{p,\widetilde{\varphi}_{\sigma}}, \tag{105}$$

$$\left\{\int_{0}^{\infty} \frac{x^{q\sigma-1}}{(1-\vartheta_{\sigma}(x))^{q-1}}\left(\sum_{n=1}^{\infty} h(x(n-\tau))a_n\right)^{q} dx\right\}^{\frac{1}{q}}$$

$$> \widehat{k}(\sigma)||a||_{q,\psi_{\sigma}}. \tag{106}$$

Proof We only prove that the constant factor $\widehat{K}(\sigma)$ in (101) is the best possible. The rest are omitted. For $0 < \varepsilon < |q|\delta_0, \widetilde{\sigma} = \sigma - \frac{\varepsilon}{q}$, we set $\widetilde{f}(x), \widetilde{a}_n$ as Theorem 4. Then by (16), (17), and (84), we obtain

$$\begin{aligned}
&||\widetilde{f}||_{p,\widetilde{\Phi}_{\sigma}}||\widetilde{a}||_{q,\Psi_{\sigma}}\\
&= \left\{\int_{\{x\in\mathbf{R}_{+}^{i_0};x_i\le 1\}} ||x||_{\alpha}^{-i_0+\varepsilon}(1-O(||x||_{\alpha}^{\rho(\sigma)}))dx\right\}^{\frac{1}{p}}\\
&\times\left\{\sum_{n}||n-\tau||_{\beta}^{-j_0-\varepsilon}\right\}^{\frac{1}{q}}\\
&= \left\{\frac{i_0^{\varepsilon/\alpha}\Gamma^{i_0}(\frac{1}{\alpha})}{\alpha^{i_0-1}\Gamma(\frac{i_0}{\alpha})} - \varepsilon\widetilde{O}(1)\right\}^{\frac{1}{p}}\\
&\times\left\{\frac{\Gamma^{j_0}(\frac{1}{\beta})}{j_0^{\varepsilon/\beta}\beta^{j_0-1}\Gamma(\frac{j_0}{\beta})} + \varepsilon O(1)\right\}^{\frac{1}{q}}.
\end{aligned}$$

If there exists a constant $K \geq \widehat{K}(\sigma)$, such that (101) is valid as we replace $\widehat{K}(\sigma)$ by K, then in particular, by (82) and (16), we have

$$K\left\{\frac{i_0^{\varepsilon/\alpha}\Gamma^{i_0}(\frac{1}{\alpha})}{\alpha^{i_0-1}\Gamma(\frac{i_0}{\alpha})}-\varepsilon\widetilde{O}(1)\right\}^{\frac{1}{p}}$$

$$\times\left\{\frac{\Gamma^{j_0}(\frac{1}{\beta})}{j_0^{\varepsilon/\beta}\beta^{j_0-1}\Gamma(\frac{j_0}{\beta})}+\varepsilon O(1)\right\}^{\frac{1}{q}}$$

$$=\varepsilon K||\widetilde{f}||_{p,\widetilde{\Phi}_\sigma}||\widetilde{a}||_{q,\Psi_\sigma}<\varepsilon\widetilde{I}$$

$$=\varepsilon\int_{\{x\in\mathbf{R}_+^{i_0};x_i\leq 1\}}||x||_\alpha^{-i_0-\varepsilon}W(\widetilde{\sigma},x)dx$$

$$<\varepsilon\widehat{K}_2(\widetilde{\sigma})\int_{\{x\in\mathbf{R}_+^{i_0};x_i\leq 1\}}||x||_\alpha^{-i_0+\varepsilon}dx$$

$$=\widehat{K}_2(\widetilde{\sigma})\frac{i_0^{\varepsilon/\alpha}\Gamma^{i_0}(\frac{1}{\alpha})}{\alpha^{i_0-1}\Gamma(\frac{i_0}{\alpha})},$$

and then by (7), $K \leq \widehat{K}(\sigma)(\varepsilon \to 0^+)$. Hence $K = \widehat{K}(\sigma)$ is the best possible constant factor of (101). The theorem is proved.

For $p > 1$, we set

$$\Phi_\sigma(x)=||x||_\alpha^{p(i_0-\sigma)-i_0}\left(x\in\mathbf{R}_+^{i_0}\right),\Psi(n)=||n-\tau||_\beta^{q(j_0-\sigma)-j_0},$$

wherefrom

$$[\Psi_\sigma(n)]^{1-p}=||n-\tau||_\beta^{p\sigma-j_0},[\Phi_\sigma(x)]^{1-q}=||x||_\alpha^{q\sigma-j_0}.$$

We define two real weight normal spaces $\mathbf{L}_{p,\Phi_\sigma}(\mathbf{R}_+^{i_0})$ and $\mathbf{l}_{q,\Psi_\sigma}$ as follows:

$$\mathbf{L}_{p,\Phi_\sigma}(\mathbf{R}_+^{i_0}):=\left\{f;||f||_{p,\Phi_\sigma}=\left\{\int_{\mathbf{R}_+^{i_0}}\Phi_\sigma(x)|f(x)|^p dx\right\}^{\frac{1}{p}}<\infty\right\},$$

$$\mathbf{l}_{q,\Psi_\sigma}:=\left\{a=\{a_n\};||a||_{q,\psi_\sigma}=\left\{\sum_n\Psi_\sigma(n)|a_n|^q\right\}^{\frac{1}{q}}<\infty\right\}.$$

As the assumptions of Theorem 4, in view of

$$\widehat{J}<\widehat{K}(\sigma)||f||_{p,\Phi_\sigma},\widehat{H}<\widehat{K}(\sigma)||a||_{q,\Psi_\sigma},$$

we can give the following definition:

Definition 4 Define a first kind of multidimensional half-discrete Hilbert-type operator

$$\widehat{T}_1 : \mathbf{L}_{p,\Phi_\sigma}(\mathbf{R}_+^{i_0}) \to \mathbf{l}_{p,\Psi_\sigma^{1-p}}$$

as follows: For $f \in \mathbf{L}_{p,\Phi_\sigma}(\mathbf{R}_+^{i_0})$, there exists an unique representation $\widehat{T}_1 f \in \mathbf{l}_{p,\Psi_\sigma^{1-p}}$, satisfying

$$\left(\widehat{T}_1 f\right)(n) := \int_{\mathbf{R}_+^{i_0}} h(||x||_\alpha ||n-\tau||_\beta) f(x) dx (n \in \mathbf{N}^{j_0}). \tag{107}$$

For $a \in \mathbf{l}_{q,\Psi_\sigma}$, we define the following formal inner product of $\widehat{T}_1 f$ and a as follows:

$$\left(\widehat{T}_1 f, a\right) := \sum_n a_n \int_{\mathbf{R}_+^{i_0}} h(||x||_\alpha ||n-\tau||_\beta) f(x) dx. \tag{108}$$

Define a second kind of multidimensional half-discrete Hilbert-type operator

$$\widehat{T}_2 : \mathbf{l}_{q,\Psi_\sigma} \to \mathbf{L}_{q,\Phi_\sigma^{1-q}}(\mathbf{R}_+^{i_0})$$

as follows: For $a \in \mathbf{l}_{q,\Psi_\sigma}$, there exists an unique representation $\widehat{T}_2 a \in \mathbf{L}_{q,\Phi_\sigma^{1-q}}(\mathbf{R}_+^{i_0})$, satisfying

$$(\widehat{T}_2 a)(x) := \sum_n h(||x||_\alpha ||n-\tau||_\beta) a_n (x \in \mathbf{R}_+^{i_0}). \tag{109}$$

For $f \in \mathbf{L}_{p,\Phi_\sigma}(\mathbf{R}_+^{i_0})$, we define the following formal inner product of f and $\widehat{T}_2 a$ as follows:

$$(f, \widehat{T}_2 a) := \int_{\mathbf{R}_+^{i_0}} \sum_n h(||x||_\alpha ||n-\tau||_\beta) a_n f(x) dx. \tag{110}$$

Then by Theorem 4, for $0 < ||f||_{p,\Phi_\sigma}, ||a||_{q,\Psi_\sigma} < \infty$, we have the following equivalent inequalities:

$$(\widehat{T}_1 f, a) = (f, \widehat{T}_2 a) < \widehat{K}(\sigma) ||f||_{p,\Phi_\sigma} ||a||_{q,\Psi_\sigma}, \tag{111}$$

$$||\widehat{T}_1 f||_{p,\Psi_\sigma^{1-p}} < \widehat{K}(\sigma) ||f||_{p,\Phi_\sigma}, \tag{112}$$

$$||\widehat{T}_2 a||_{q,\Phi_\sigma^{1-q}} < \widehat{K}(\sigma) ||a||_{q,\Psi_\sigma}. \tag{113}$$

It follows that $\widehat{T}_1$ and $\widehat{T}_2$ are bounded with

$$||\widehat{T}_1|| := \sup_{f(\neq\theta)\in \mathbf{L}_{p,\Phi_\sigma}(\mathbf{R}_+^{i_0})} \frac{||\widehat{T}_1 f||_{p,\Psi_\sigma^{1-p}}}{||f||_{p,\Phi_\sigma}} \leq \widehat{K}(\sigma),$$

$$||\widehat{T}_2|| := \sup_{a(\neq\theta)\in \mathbf{l}_{q,\Psi_\sigma}} \frac{||\widehat{T}_2 a||_{q,\Phi_\sigma^{1-q}}}{||a||_{q,\Psi_\sigma}} \leq \widehat{K}(\sigma).$$

Since the constant factor $\widehat{K}(\sigma)$ in (112) and (113) is the best possible, we have

$$||\widehat{T}_1|| = ||\widehat{T}_2|| = \widehat{K}(\sigma)$$
$$= \left[\frac{\Gamma^{j_0}(\frac{1}{\beta})}{\beta^{j_0-1}\Gamma(\frac{j_0}{\beta})}\right]^{\frac{1}{p}} \left[\frac{\Gamma^{i_0}(\frac{1}{\alpha})}{\alpha^{i_0-1}\Gamma(\frac{i_0}{\alpha})}\right]^{\frac{1}{q}} \widehat{k}(\sigma). \tag{114}$$

Example 5 (i) We set

$$h(t) = \frac{1}{t^\lambda + 1}(0 < \lambda \leq 1, 0 < \sigma < \lambda),$$

For $\delta_0 = \frac{1}{2}\min\{\sigma, \lambda - \sigma\} > 0$, and $\widetilde{\sigma} \in (\sigma - \delta_0, \sigma + \delta_0)$, it follows

$$\widehat{k}(\widetilde{\sigma}) = \int_0^\infty \frac{1}{t^\lambda + 1} t^{\widetilde{\sigma}-1} dt$$
$$= \frac{1}{\lambda}\int_0^\infty \frac{1}{v+1} v^{\frac{\widetilde{\sigma}}{\lambda}-1} dv$$
$$= \frac{\pi}{\lambda \sin \pi(\widetilde{\sigma}/\lambda)} \in \mathbf{R}_+,$$

and

$$\frac{d}{dt}\left(\frac{1}{t^\lambda + 1} t^{\widetilde{\sigma}-j_0}\right) < 0, \frac{d^2}{dt^2}\left(\frac{1}{t^\lambda + 1} t^{\widetilde{\sigma}-j_0}\right) > 0.$$

Setting $\eta_0 \in (0, \sigma - \delta_0)$, then we find $\eta_0 < \widetilde{\sigma}$ ($< \lambda$). Since

$$\frac{t^{\eta_0}}{t^\lambda + 1} \to 0(t \to 0^+), \frac{t^{\eta_0}}{t^\lambda + 1} \to 0(t \to \infty),$$

there exists a constant $L > 0$, such that

$$h(t) = \frac{1}{t^\lambda + 1} \leq \frac{L}{t^{\eta_0}}(t \in (0, \infty)).$$

Then by Theorem 4 and (114), for $\alpha > 0, 0 < \beta \leq 1$, we have

$$||\widehat{T}_1|| = ||\widehat{T}_2|| = \left[\frac{\Gamma^{j_0}(\frac{1}{\beta})}{\beta^{j_0-1}\Gamma(\frac{j_0}{\beta})}\right]^{\frac{1}{p}} \times \left[\frac{\Gamma^{i_0}(\frac{1}{\alpha})}{\alpha^{i_0-1}\Gamma(\frac{i_0}{\alpha})}\right]^{\frac{1}{q}} \frac{\pi}{\lambda \sin(\frac{\pi\sigma}{\lambda})}. \tag{115}$$

(ii) We set

$$h(t) = \frac{1}{(t+1)^\lambda}(\lambda > 0, 0 < \sigma < \min\{\lambda, j_0\}),$$

For $\delta_0 = \frac{1}{2}\min\{\sigma, \lambda - \sigma, j_0 - \sigma\} > 0$, and $\widetilde{\sigma} \in (\sigma - \delta_0, \sigma + \delta_0)$, it follows

$$\widehat{k}(\widetilde{\sigma}) = \int_0^\infty \frac{1}{(t+1)^\lambda} t^{\widetilde{\sigma}-1} dt = B(\widetilde{\sigma}, \lambda - \widetilde{\sigma}) \in \mathbf{R}_+,$$

and

$$\frac{d}{dt}\left(\frac{1}{(t+1)^\lambda} t^{\widetilde{\sigma}-j_0}\right) < 0,$$

$$\frac{d^2}{dt^2}\left(\frac{1}{(t+1)^\lambda}t^{\widetilde{\sigma}-j_0}\right) > 0.$$

Setting $\eta_0 \in (0, \sigma - \delta_0)$, then we find $\eta_0 < \widetilde{\sigma}(< \lambda)$. Since

$$\frac{t^{\eta_0}}{(t+1)^\lambda} \to 0(t \to 0^+), \frac{t^{\eta_0}}{(t+1)^\lambda} \to 0(t \to \infty),$$

there exists a constant $L > 0$, such that

$$k_\lambda(t,1) = \frac{1}{(t+1)^\lambda} \le \frac{L}{t^{\eta_0}}(t \in (0,\infty)).$$

Then by Theorem 4 and (114), for $\alpha > 0, 0 < \beta \le 1$, we have

$$||\widehat{T}_1|| = ||\widehat{T}_2|| = \left[\frac{\Gamma^{j_0}(\frac{1}{\beta})}{\beta^{j_0-1}\Gamma(\frac{j_0}{\beta})}\right]^{\frac{1}{p}} \times \left[\frac{\Gamma^{i_0}(\frac{1}{\alpha})}{\alpha^{i_0-1}\Gamma(\frac{i_0}{\alpha})}\right]^{\frac{1}{q}} B(\sigma, \lambda - \sigma). \tag{116}$$

(iii) We set

$$h(t) = \frac{\ln t}{t^\lambda - 1}(0 < \lambda \le 1, 0 < \sigma < \lambda).$$

For $\delta_0 = \frac{1}{2}\min\{\sigma, \lambda - \sigma\}$, and $\widetilde{\sigma} \in (\sigma - \delta_0, \sigma + \delta_0)$, it follows

$$\begin{aligned}\widehat{k}(\widetilde{\sigma}) &= \int_0^\infty \frac{\ln t}{t^\lambda - 1} t^{\widetilde{\sigma}-1} dt \\ &= \frac{1}{\lambda^2}\int_0^\infty \frac{\ln v}{v-1} v^{\frac{\widetilde{\sigma}}{\lambda}-1} dv \\ &= \left[\frac{\pi}{\lambda \sin \pi(\widetilde{\sigma}/\lambda)}\right]^2 \in \mathbf{R}_+,\end{aligned}$$

and

$$\frac{d}{dt}\left(\frac{\ln t}{t^\lambda - 1}t^{\widetilde{\sigma}-j_0}\right) < 0, \frac{d^2}{dt^2}\left(\frac{\ln t}{t^\lambda - 1}t^{\widetilde{\sigma}-j_0}\right) > 0.$$

Setting $\eta_0 \in (0, \sigma - \delta_0)$, then we find $\eta_0 < \widetilde{\sigma}(< \lambda)$. Since

$$\frac{(\ln t)t^{\eta_0}}{t^\lambda - 1} \to 0(t \to 0^+), \frac{(\ln t)t^{\eta_0}}{t^\lambda - 1} \to 0(t \to \infty),$$

there exists a constant $L > 0$, such that

$$h(t) = \frac{\ln t}{t^\lambda - 1} \le \frac{L}{t^{\eta_0}}(t \in (0,\infty)).$$

Then by Theorem 4 and (114), for $\alpha > 0, 0 < \beta \le 1$, we have

$$\begin{aligned}||\widehat{T}_1|| = ||\widehat{T}_2|| &= \left[\frac{\Gamma^{j_0}(\frac{1}{\beta})}{\beta^{j_0-1}\Gamma(\frac{j_0}{\beta})}\right]^{\frac{1}{p}} \left[\frac{\Gamma^{i_0}(\frac{1}{\alpha})}{\alpha^{i_0-1}\Gamma(\frac{i_0}{\alpha})}\right]^{\frac{1}{q}} \\ &\times \left[\frac{\pi}{\lambda \sin \pi(\frac{\sigma}{\lambda})}\right]^2.\end{aligned} \tag{117}$$

(iv) For $s \in \mathbf{N}$, we set

$$h(t) = \frac{1}{\prod_{k=1}^{s}(t^{\lambda/s} + a_k)}(0 < a_1 < \cdots < a_s,$$
$$0 < \lambda \leq s, 0 < \sigma < \min\{\lambda, j_0\}).$$

For $\delta_0 = \frac{1}{2}\min\{\sigma, \lambda - \sigma, j_0 - \sigma\} > 0$, and $\widetilde{\sigma} \in (\sigma - \delta_0, \sigma + \delta_0)$, it follows

$$\widehat{k}_s(\widetilde{\sigma}) = \int_0^\infty \frac{1}{\prod_{k=1}^{s}(t^{\lambda/s} + a_k)} t^{\widetilde{\sigma}-1} dt$$
$$= \frac{\pi s}{\lambda \sin(\frac{\pi s \widetilde{\sigma}}{\lambda})} \sum_{k=1}^{s} a_k^{\frac{s\widetilde{\sigma}}{\lambda}-1} \prod_{j=1(j\neq k)}^{s} \frac{1}{a_j - a_k} \in \mathbf{R}_+,$$

and

$$\frac{d}{dt}\left(\frac{t^{\widetilde{\sigma}-j_0}}{\prod_{k=1}^{s}(t^{\lambda/s} + a_k)}\right) < 0,$$
$$\frac{d^2}{dt^2}\left(\frac{t^{\widetilde{\sigma}-j_0}}{\prod_{k=1}^{s}(t^{\lambda/s} + a_k)}\right) > 0.$$

Setting $\eta_0 \in (0, \sigma - \delta_0)$, then we find $\eta_0 < \widetilde{\sigma}(< \lambda)$. Since

$$\frac{t^{\eta_0}}{\prod_{k=1}^{s}(t^{\lambda/s} + a_k)} \to 0(t \to 0^+),$$
$$\frac{t^{\eta_0}}{\prod_{k=1}^{s}(t^{\lambda/s} + a_k)} \to 0(t \to \infty),$$

there exists a constant $L > 0$, such that

$$h(t) = \frac{1}{\prod_{k=1}^{s}(t^{\lambda/s} + a_k)} \leq \frac{L}{t^{\eta_0}}(t \in (0, \infty)).$$

Then by Theorem 4 and (114), for $\alpha > 0, 0 < \beta \leq 1$, we have

$$||\widehat{T}_1|| = ||\widehat{T}_2|| = \left[\frac{\Gamma^{j_0}(\frac{1}{\beta})}{\beta^{j_0-1}\Gamma(\frac{j_0}{\beta})}\right]^{\frac{1}{p}} \left[\frac{\Gamma^{i_0}(\frac{1}{\alpha})}{\alpha^{i_0-1}\Gamma(\frac{i_0}{\alpha})}\right]^{\frac{1}{q}} \widehat{k}_s(\sigma). \tag{118}$$

(v) We set

$$h(t) = \frac{1}{t^\lambda + \sqrt{c}t^{\lambda/2}\cos\gamma + \frac{c}{4}}$$
$$(c > 0, 0 < \lambda \leq 1, 0 < \gamma < \frac{\pi}{2}, 0 < \sigma < \lambda).$$

For $\delta_0 = \frac{1}{2}\min\{\sigma, \lambda - \sigma\} > 0$, and $\widetilde{\sigma} \in (\sigma - \delta_0, \sigma + \delta_0)$, by Example 2, it follows

$$\widehat{k}(\widetilde{\sigma}) = \int_0^\infty \frac{1}{t^\lambda + \sqrt{c}t^{\lambda/2}\cos\gamma + \frac{c}{4}} t^{\widetilde{\sigma}-1} dt$$
$$= (\frac{\sqrt{c}}{2})^{\frac{2\widetilde{\sigma}}{\lambda}} \frac{2\pi \sin\gamma(1 - \frac{2\widetilde{\sigma}}{\lambda})}{\lambda \sin\gamma \sin(\frac{2\pi\widetilde{\sigma}}{\lambda})} \in \mathbf{R}_+,$$

and

$$\frac{d}{dt}\left(\frac{t^{\widetilde{\sigma}-j_0}}{t^\lambda + \sqrt{c}t^{\lambda/2}\cos\gamma + \frac{c}{4}}\right) < 0,$$
$$\frac{d^2}{dt^2}\left(\frac{t^{\widetilde{\sigma}-j_0}}{t^\lambda + \sqrt{c}t^{\lambda/2}\cos\gamma + \frac{c}{4}}\right) > 0.$$

Setting $\eta_0 \in (0, \sigma - \delta_0)$, then we find $\eta_0 < \widetilde{\sigma}(< \lambda)$. Since

$$\frac{t^{\eta_0}}{t^\lambda + \sqrt{c}t^{\lambda/2}\cos\gamma + \frac{c}{4}} \to 0(t \to 0^+),$$
$$\frac{t^{\eta_0}}{t^\lambda + \sqrt{c}t^{\lambda/2}\cos\gamma + \frac{c}{4}} \to 0(t \to \infty),$$

there exists a constant $L > 0$, such that

$$h(t) = \frac{1}{t^\lambda + \sqrt{c}t^{\lambda/2}\cos\gamma + \frac{c}{4}} \le \frac{L}{t^{\eta_0}} (t \in (0, \infty)).$$

Then by Theorem 4 and (114), for $\alpha > 0, 0 < \beta \le 1$, we have

$$||\widehat{T}_1|| = ||\widehat{T}_2|| = \left[\frac{\Gamma^{j_0}(\frac{1}{\beta})}{\beta^{j_0-1}\Gamma(\frac{j_0}{\beta})}\right]^{\frac{1}{p}} \left[\frac{\Gamma^{i_0}(\frac{1}{\alpha})}{\alpha^{i_0-1}\Gamma(\frac{i_0}{\alpha})}\right]^{\frac{1}{q}}$$
$$\times (\frac{\sqrt{c}}{2})^{\frac{2\sigma}{\lambda}} \frac{2\pi \sin\gamma \left(1 - \frac{2\sigma}{\lambda}\right)}{\lambda \sin\gamma \sin\left(\frac{2\pi\sigma}{\lambda}\right)}. \tag{119}$$

(vi) We set

$$h(t) = \ln\left(\frac{b + t^\gamma}{a + t^\gamma}\right)$$
$$(0 \le a < b, 0 < \gamma \le 1, 0 < \sigma < \gamma).$$

For $\delta_0 = \frac{1}{2}\min\{\sigma, \gamma - \sigma\} > 0$, and $\widetilde{\sigma} \in (\sigma - \delta_0, \sigma + \delta_0)$, it follows

$$\widehat{k}(\widetilde{\sigma}) = \int_0^\infty \ln\left(\frac{b + t^\gamma}{a + t^\gamma}\right) t^{\widetilde{\sigma}-1} dt$$

$$= \int_0^\infty \ln\left(\frac{by^\gamma + 1}{ay^\gamma + 1}\right) y^{-\widetilde{\sigma}-1} dy$$
$$= \frac{1}{\widetilde{\sigma}} \left(b^{\frac{\widetilde{\sigma}}{\gamma}} - a^{\frac{\widetilde{\sigma}}{\gamma}}\right) \frac{\pi}{\sin \pi(\frac{\widetilde{\sigma}}{\gamma})} \in \mathbf{R}_+,$$

and

$$\frac{d}{dt}\left(\ln\left(\frac{b+t^\gamma}{a+t^\gamma}\right) t^{\widetilde{\sigma}-j_0}\right) < 0,$$
$$\frac{d^2}{dt^2}\left(\ln\left(\frac{b+t^\gamma}{a+t^\gamma}\right) t^{\widetilde{\sigma}-j_0}\right) > 0.$$

Setting $\eta_0 \in (0, \sigma - \delta_0)$, then we find $\eta_0 < \widetilde{\sigma} < \gamma$. Since

$$t^{\eta_0} \ln\left(\frac{b+t^\gamma}{a+t^\gamma}\right) \to 0(t \to 0^+),$$
$$t^{\eta_0} \ln\left(\frac{b+t^\gamma}{a+t^\gamma}\right) \to 0(t \to \infty),$$

there exists a constant $L > 0$, such that

$$h(t) = \ln\left(\frac{b+t^\gamma}{a+t^\gamma}\right) \le \frac{L}{t^{\eta_0}} (t \in (0,\infty)).$$

Then by Theorem 4 and (114), for $\alpha > 0, 0 < \beta \le 1$, we have

$$||\widehat{T}_1|| = ||\widehat{T}_2|| = \left[\frac{\Gamma^{j_0}(\frac{1}{\beta})}{\beta^{j_0-1}\Gamma(\frac{j_0}{\beta})}\right]^{\frac{1}{p}} \left[\frac{\Gamma^{i_0}(\frac{1}{\alpha})}{\alpha^{i_0-1}\Gamma(\frac{i_0}{\alpha})}\right]^{\frac{1}{q}} \times \frac{\left(b^{\frac{\sigma}{\gamma}} - a^{\frac{\sigma}{\gamma}}\right)\pi}{\sigma \sin \pi\left(\frac{\sigma}{\gamma}\right)}. \tag{120}$$

(vii) We set

$$h(t) = e^{-\rho t^\gamma} (\rho > 0, 0 < \gamma \le 1, 0 < \sigma < \gamma).$$

For $\delta_0 = \frac{1}{2}\min\{\sigma, \gamma - \sigma\} > 0$, and $\widetilde{\sigma} \in (\sigma - \delta_0, \sigma + \delta_0)$, it follows

$$\widehat{k}(\widetilde{\sigma}) = \int_0^\infty e^{-\rho t^\gamma} t^{\widetilde{\sigma}-1} dt = \frac{1}{\gamma} \rho^{-\frac{\widetilde{\sigma}}{\gamma}} \int_0^\infty e^{-u} u^{\frac{\widetilde{\sigma}}{\gamma}-1} du$$
$$= \frac{1}{\gamma \rho^{\frac{\widetilde{\sigma}}{\gamma}}} \Gamma\left(\frac{\widetilde{\sigma}}{\gamma}\right) \in \mathbf{R}_+,$$

and

$$\frac{d}{dt}\left(e^{-\rho t^{\gamma}}t^{\widetilde{\sigma}-j_0}\right)<0,\ \frac{d^2}{dt^2}\left(e^{-\rho t^{\gamma}}t^{\widetilde{\sigma}-j_0}\right)>0.$$

Setting $\eta_0 \in (0, \sigma - \delta_0)$, then we find $\eta_0 < \widetilde{\sigma}$. Since

$$t^{\eta_0}e^{-\rho t^{\gamma}} \to 0(t \to 0^+), t^{\eta_0}e^{-\rho t^{\gamma}} \to 0(t \to \infty),$$

there exists a constant $L > 0$, such that

$$h(t) = e^{-\rho t^{\gamma}} \leq \frac{L}{t^{\eta_0}}(t \in (0, \infty)).$$

Then by Theorem 4 and (114), for $\alpha > 0, 0 < \beta \leq 1$, we have

$$\begin{aligned}||\widehat{T}_1|| = ||\widehat{T}_2|| &= \left[\frac{\Gamma^{j_0}(\frac{1}{\beta})}{\beta^{j_0-1}\Gamma(\frac{j_0}{\beta})}\right]^{\frac{1}{p}}\left[\frac{\Gamma^{i_0}(\frac{1}{\alpha})}{\alpha^{i_0-1}\Gamma(\frac{i_0}{\alpha})}\right]^{\frac{1}{q}} \\ &\times \frac{1}{\gamma\rho^{\frac{\sigma}{\gamma}}}\Gamma(\frac{\sigma}{\gamma}). \qquad (121)\end{aligned}$$

(viii) We set

$$h(t) = \arctan \rho t^{-\gamma}(\rho, \gamma > 0, 0 < \sigma < \min\{j_0, \gamma\}).$$

For $\delta_0 = \frac{1}{2}\min\{\sigma, \gamma - \sigma, j_0 - \sigma\} > 0$, and $\widetilde{\sigma} \in (\sigma - \delta_0, \sigma + \delta_0)$, it follows

$$\begin{aligned}\widehat{k}(\widetilde{\sigma}) &= \int_0^{\infty} t^{\widetilde{\sigma}-1}(\arctan \rho t^{-\gamma})dt = \frac{1}{\widetilde{\sigma}}\int_0^{\infty}(\arctan \rho t^{-\gamma})dt^{\widetilde{\sigma}} \\ &= \frac{1}{\widetilde{\sigma}}\left[(\arctan \rho t^{-\gamma})t^{\widetilde{\sigma}}|_0^{\infty} + \int_0^{\infty}\frac{\gamma\rho t^{\widetilde{\sigma}-\gamma-1}}{1+(\rho t^{-\gamma})^2}dt\right] \\ &= \frac{\rho^{\frac{\widetilde{\sigma}}{\gamma}}}{2\widetilde{\sigma}}\int_0^{\infty}\frac{1}{1+u}u^{(-\frac{\widetilde{\sigma}}{2\gamma}+\frac{1}{2})-1}du \\ &= \frac{\rho^{\frac{\widetilde{\sigma}}{\gamma}}\pi}{2\widetilde{\sigma}\sin\pi(-\frac{\widetilde{\sigma}}{2\gamma}+\frac{1}{2})} \\ &= \frac{\rho^{\frac{\widetilde{\sigma}}{\gamma}}\pi}{2\widetilde{\sigma}\cos\pi(\frac{\widetilde{\sigma}}{2\gamma})} \in \mathbf{R}_+,\end{aligned}$$

and

$$\begin{aligned}&\frac{d}{dt}\left(t^{\widetilde{\sigma}-j_0}\arctan \rho t^{-\gamma}\right) < 0, \\ &\frac{d^2}{dt^2}\left(t^{\widetilde{\sigma}-j_0}\arctan \rho t^{-\gamma}\right) > 0.\end{aligned}$$

Setting $\eta_0 \in (0, \sigma - \delta_0)$, then we find $\eta_0 < \widetilde{\sigma}$. Since

$$t^{\eta_0} \arctan \rho t^{-\gamma} \to 0(t \to 0^+),$$
$$t^{\eta_0} \arctan \rho t^{-\gamma} \to 0(t \to \infty),$$

there exists a constant $L > 0$, such that

$$h(t) = \arctan \rho t^{-\gamma} \leq \frac{L}{t^{\eta_0}}(t \in (0, \infty)).$$

Then by Theorem 4 and (114), for $\alpha > 0, 0 < \beta \leq 1$, we have

$$||\widehat{T}_1|| = ||\widehat{T}_2|| = \left[\frac{\Gamma^{j_0}(\frac{1}{\beta})}{\beta^{j_0-1}\Gamma(\frac{j_0}{\beta})}\right]^{\frac{1}{p}} \left[\frac{\Gamma^{i_0}(\frac{1}{\alpha})}{\alpha^{i_0-1}\Gamma(\frac{i_0}{\alpha})}\right]^{\frac{1}{q}} \times \frac{\rho^{\frac{\sigma}{\gamma}}\pi}{2\sigma \cos \pi \left(\frac{\sigma}{2\gamma}\right)}. \tag{122}$$

Example 6 (i) We set

$$h(t) = \text{csch} t (1 < \sigma < j_0).$$

For $\delta_0 = \frac{1}{2}\min\{j_0 - \sigma, \sigma - 1\} > 0$, and $\widetilde{\sigma} \in (\sigma - \delta_0, \sigma + \delta_0)$, it follows

$$\begin{aligned}
\widehat{k}(\widetilde{\sigma}) &= \int_0^\infty t^{\widetilde{\sigma}-1}(\text{csch} t)dt = \int_0^\infty \frac{2t^{\widetilde{\sigma}-1}}{e^t - e^{-t}}dt \\
&= \int_0^\infty \frac{2e^{-t}t^{\widetilde{\sigma}-1}}{1 - e^{-2t}}dt = 2\int_0^\infty \sum_{k=0}^\infty e^{-(2k+1)t}t^{\widetilde{\sigma}-1}dt \\
&= 2\sum_{k=0}^\infty \int_0^\infty e^{-(2k+1)t}t^{\widetilde{\sigma}-1}dt = 2\Gamma(\widetilde{\sigma})\sum_{k=0}^\infty \frac{1}{(2k+1)^{\widetilde{\sigma}}} \\
&= \Gamma(\widetilde{\sigma})\left(2 - \frac{1}{2^{\widetilde{\sigma}-1}}\right)\zeta(\widetilde{\sigma}) \in \mathbf{R}_+.
\end{aligned}$$

We obtain

$$\frac{d}{dt}(t^{\widetilde{\sigma}-j_0}\text{csch} t) < 0, \frac{d^2}{dt^2}(t^{\widetilde{\sigma}-j_0}\text{csch} t) > 0.$$

Setting $\eta_0 \in (0, \sigma - \delta_0)$, then we find $0 < \eta_0 < \widetilde{\sigma}$. Since

$$t^{\eta_0}\text{csch} t \to 0(t \to 0^+), t^{\eta_0}\text{csch} t \to 0(t \to \infty),$$

there exists a constant $L > 0$, such that

$$h(t) = \text{csch} t \leq \frac{L}{t^{\eta_0}}(t \in (0, \infty)).$$

Then by Theorem 4 and (114), for $\alpha > 0, 0 < \beta \le 1$, we have

$$||\widehat{T}_1|| = ||\widehat{T}_2|| = \left[\frac{\Gamma^{j_0}\left(\frac{1}{\beta}\right)}{\beta^{j_0-1}\Gamma\left(\frac{j_0}{\beta}\right)}\right]^{\frac{1}{p}} \left[\frac{\Gamma^{i_0}\left(\frac{1}{\alpha}\right)}{\alpha^{i_0-1}\Gamma\left(\frac{i_0}{\alpha}\right)}\right]^{\frac{1}{q}} \times \left(2 - \frac{1}{2^{\sigma-1}}\right)\Gamma(\sigma)\zeta(\sigma). \tag{123}$$

(ii) We set

$$h(t) = \coth t - 1 (1 < \sigma < j_0).$$

For $\delta_0 = \frac{1}{2}\min\{j_0 - \sigma, \sigma - 1\} > 0$, and $\widetilde{\sigma} \in (\sigma - \delta_0, \sigma + \delta_0)$, it follows

$$\begin{aligned}\widehat{k}(\widetilde{\sigma}) &= \int_0^\infty t^{\widetilde{\sigma}-1}(\coth t - 1)dt = \int_0^\infty \frac{2e^{-t}t^{\widetilde{\sigma}-1}}{e^t - e^{-t}}dt \\ &= \int_0^\infty \frac{2e^{-2t}t^{\widetilde{\sigma}-1}}{1-e^{-2t}}dt = 2\int_0^\infty \sum_{k=0}^\infty e^{-2(k+1)t}t^{\widetilde{\sigma}-1}dt \\ &= 2\sum_{k=0}^\infty \int_0^\infty e^{-2(k+1)t}t^{\widetilde{\sigma}-1}dt \\ &= 2\Gamma(\widetilde{\sigma})\sum_{k=1}^\infty \frac{1}{(2k)^{\widetilde{\sigma}}} = \frac{\Gamma(\widetilde{\sigma})}{2^{\widetilde{\sigma}-1}}\zeta(\widetilde{\sigma}) \in \mathbf{R}_+.\end{aligned}$$

We obtain

$$\frac{d}{dt}(t^{\widetilde{\sigma}-j_0}(\coth t - 1)) < 0, \frac{d^2}{dt^2}(t^{\widetilde{\sigma}-j_0}(\coth t - 1)) > 0.$$

Setting $\eta_0 \in (0, \sigma - \delta_0)$, then we find $0 < \eta_0 < \widetilde{\sigma}$. Since

$$t^{\eta_0}(\coth t - 1) \to 0 (t \to 0^+), t^{\eta_0}(\coth t - 1) \to 0 (t \to \infty),$$

there exists a constant $L > 0$, such that

$$h(t) = \coth t - 1 \le \frac{L}{t^{\eta_0}} (t \in (0, \infty)).$$

Then by Theorem 4 and (114), for $\alpha > 0, 0 < \beta \le 1$, we have

$$||\widehat{T}_1|| = ||\widehat{T}_2|| = \left[\frac{\Gamma^{j_0}(\frac{1}{\beta})}{\beta^{j_0-1}\Gamma(\frac{j_0}{\beta})}\right]^{\frac{1}{p}} \left[\frac{\Gamma^{i_0}(\frac{1}{\alpha})}{\alpha^{i_0-1}\Gamma(\frac{i_0}{\alpha})}\right]^{\frac{1}{q}} \times \frac{1}{2^{\sigma-1}}\Gamma(\sigma)\zeta(\sigma). \tag{124}$$

(iii) We set

$$h(t) = 1 - \tan h(t) = \frac{2e^{-t}}{e^t + e^{-t}}(0 < \sigma < j_0).$$

For $\delta_0 = \frac{1}{2}\min\{\sigma, j_0 - \sigma\} > 0$, and $\widetilde{\sigma} \in (\sigma - \delta_0, \sigma + \delta_0)$, it follows $0 < \widetilde{\sigma} < j_0$ and

$$\begin{aligned}\widehat{k}(\widetilde{\sigma}) &= \int_0^\infty \frac{2e^{-t}}{e^t + e^{-t}} t^{\widetilde{\sigma}-1} dt \\ &= 2\Gamma(\widetilde{\sigma}) \sum_{k=1}^\infty \frac{(-1)^{k-1}}{(2k)^{\widetilde{\sigma}}} \in \mathbf{R}_+.\end{aligned}$$

We can obtain

$$\frac{d}{dt}(t^{\widetilde{\sigma}-j_0}(1 - \tanh t)) < 0, \frac{d^2}{dt^2}\left(t^{\widetilde{\sigma}-j_0}(1 - \tanh t)\right) > 0.$$

There exists a constant $\eta_0 \in (0, \sigma - \delta_0)$, such that $0 < \eta_0 < \widetilde{\sigma}$. In view of $t^{\eta_0}h(t) \to 0(t \to 0^+)$, and $t^{\eta_0}h(t) \to 0(t \to \infty)$, there exists a constant $L > 0$, satisfying

$$h(t) \leq \frac{L}{t^{\eta_0}}(t \in (0, \infty)).$$

Therefore, by Theorem 4 and (114), for $\alpha > 0, 0 < \beta \leq 1$, it follows

$$\begin{aligned}||\widehat{T}_1|| = ||\widehat{T}_2|| &= \left[\frac{\Gamma^{j_0}(\frac{1}{\beta})}{\beta^{j_0-1}\Gamma(\frac{j_0}{\beta})}\right]^{\frac{1}{p}} \left[\frac{\Gamma^{i_0}(\frac{1}{\alpha})}{\alpha^{i_0-1}\Gamma(\frac{i_0}{\alpha})}\right]^{\frac{1}{q}} \\ &\times \frac{\Gamma(\sigma)}{2^{\sigma-1}} \sum_{k=1}^\infty \frac{(-1)^{k-1}}{k^\sigma}.\end{aligned}$$

5 Some Corollaries on Multidimensional Half-Discrete Hilbert-Type Inequalities

If in Theorems 1–3, we do not have the assumption of $\frac{\partial^2}{\partial y^2}\left(k_\lambda(x, y)y^{\widetilde{\lambda}_2-j_0}\right) > 0$ and $\beta \leq 1$, then for $\tau = 0$, setting

$$\begin{aligned}&\Phi(x) = ||x||_\alpha^{p(i_0-\lambda_1)-i_0}, \Psi^{(0)}(n) := ||n||_\beta^{q(j_0-\lambda_2)-j_0}, \\ &\widetilde{\Phi}(x) = (1 - \theta_{\lambda_1}(x))||x||_\alpha^{p(i_0-\lambda_1)-i_0} \\ &(\theta_{\lambda_1}(x) \in (0, 1); x \in \mathbf{R}_+^{i_0}),\end{aligned}$$

in view of the Note of Lemma 1 and by the same way, we have the following corollaries:

Corollary 1 *Suppose that* $\alpha, \beta > 0$, *there exists a constant* $\delta_0 > 0$, *such that for any* $\widetilde{\lambda}_i \in (\lambda_i - \delta_0, \lambda_i + \delta_0)(i = 1, 2)$,

$$k\left(\widetilde{\lambda}_1\right) = \int_0^\infty k_\lambda(v, 1) v^{\widetilde{\lambda}_1 - 1} dv \in \mathbf{R}_+,$$

$k_\lambda(x, y) y^{\widetilde{\lambda}_2 - j_0}$ *is strict decreasing with respect to* $y \in \mathbf{R}_+$, *there exists constants* $L > 0$ *and* $\eta_1 > \widetilde{\lambda}_1$, *satisfying*

$$k_\lambda(t, 1) \leq \frac{L}{t^{\eta_1}} (t \in (0, \infty)).$$

If $p > 1,\ f(x) = f(x_1, \cdots, x_{i_0}) \geq 0, a_n = a_{(n_1, \cdots, n_{j_0})} \geq 0,$

$$0 < ||f||_{p,\Phi} = \left\{ \int_{\mathbf{R}_+^{i_0}} \Phi(x) f^p(x) dx \right\}^{\frac{1}{p}} < \infty,$$

$$0 < ||a||_{q,\Psi^{(0)}} = \left\{ \sum_n \Psi^{(0)}(n) a_n^q \right\}^{\frac{1}{q}} < \infty,$$

then we have the following equivalent inequalities with the best possible constant factor $K(\lambda_1)$:

$$\sum_n \int_{\mathbf{R}_+^{i_0}} k_\lambda(||x||_\alpha, ||n||_\beta) a_n f(x) dx < K(\lambda_1) ||f||_{p,\Phi} ||a||_{q,\Psi^{(0)}}, \tag{125}$$

$$J^{(0)} := \left\{ \sum_n ||n||_\beta^{p\lambda_2 - j_0} \left(\int_{\mathbf{R}_+^{i_0}} k_\lambda(||x||_\alpha, ||n||_\beta) f(x) dx \right)^p \right\}^{\frac{1}{p}}$$

$$< K(\lambda_1) ||f||_{p,\Phi}, \tag{126}$$

$$H^{(0)} := \left\{ \int_{\mathbf{R}_+^{i_0}} ||x||_\alpha^{q\lambda_1 - i_0} \left(\sum_n k_\lambda(||x||_\alpha, ||n||_\beta) a_n \right)^q dx \right\}^{\frac{1}{q}}$$

$$< K(\lambda_1) ||a||_{q,\Psi^{(0)}}, \tag{127}$$

where,

$$k(\lambda_1) = \int_0^\infty k_\lambda(v, 1) v^{\lambda_1 - 1} dv,$$

$$K(\lambda_1) = (K_1(\lambda_1))^{\frac{1}{p}} (K_2(\lambda_1))^{\frac{1}{q}}$$
$$= \left[\frac{\Gamma^{j_0}(\frac{1}{\beta})}{\beta^{j_0-1}\Gamma(\frac{j_0}{\beta})}\right]^{\frac{1}{p}} \left[\frac{\Gamma^{i_0}(\frac{1}{\alpha})}{\alpha^{i_0-1}\Gamma(\frac{i_0}{\alpha})}\right]^{\frac{1}{q}} k(\lambda_1).$$

In particular, for $i_0 = j_0 = 1$,

$$\varphi(x) := x^{p(1-\lambda_1)-1}, \psi^{(0)}(n) := n^{q(1-\lambda_2)-1},$$

we have the following equivalent inequalities with the best possible constant factor $k(\lambda_1)$:

$$\sum_{n=1}^{\infty} \int_0^{\infty} k_\lambda(x,n) a_n f(x) dx < k(\lambda_1) ||f||_{p,\varphi} ||a||_{q,\psi^{(0)}}, \tag{128}$$

$$\left\{\sum_{n=1}^{\infty} n^{p\lambda_2-1} \left(\int_0^{\infty} k_\lambda(x,n) f(x) dx\right)^p\right\}^{\frac{1}{p}} < k(\lambda_1) ||f||_{p,\varphi}, \tag{129}$$

$$\left\{\int_0^{\infty} x^{q\lambda_1-1} \left(\sum_{n=1}^{\infty} k_\lambda(x,n) a_n\right)^q dx\right\}^{\frac{1}{q}} < k(\lambda_1) ||a||_{q,\psi^{(0)}}. \tag{130}$$

Corollary 2 *As the assumptions of Corollary 1, if* $p < 0 (0 < q < 1)$, $f(x) = f(x_1, \cdots, x_{i_0}) \geq 0$, $a_n = a_{(n_1,\cdots,n_{j_0})} \geq 0$, $0 < ||f||_{p,\Phi} < \infty, 0 < ||a||_{q,\Psi^{(0)}} < \infty$, *then we have the following equivalent inequalities with the best possible constant factor* $K(\lambda_1)$:

$$\sum_n \int_{\mathbf{R}_+^{i_0}} k_\lambda(||x||_\alpha, ||n||_\beta) a_n f(x) dx > K(\lambda_1) ||f||_{p,\Phi} ||a||_{q,\Psi^{(0)}}, \tag{131}$$

$$\left\{\sum_n ||n||_\beta^{p\lambda_2-j_0} \left(\int_{\mathbf{R}_+^{i_0}} k_\lambda(||x||_\alpha, ||n||_\beta) f(x) dx\right)^p\right\}^{\frac{1}{p}} > K(\lambda_1) ||f||_{p,\Phi}, \tag{132}$$

$$\left\{\int_{\mathbf{R}_+^{i_0}} ||x||_\alpha^{q\lambda_1-i_0} \left(\sum_n k_\lambda(||x||_\alpha, ||n||_\beta) a_n\right)^q dx\right\}^{\frac{1}{q}} > K(\lambda_1) ||a||_{q,\Psi^{(0)}}. \tag{133}$$

In particular, for $i_0 = j_0 = 1$, $\varphi(x)$ and $\psi^{(0)}(n)$ are indicated by Corollary 1, we have the following equivalent inequalities with the best possible constant factor $k(\lambda_1)$:

$$\sum_{n=1}^{\infty} \int_0^{\infty} k_\lambda(x,n) a_n f(x) dx > k(\lambda_1) ||f||_{p,\varphi} ||a||_{q,\psi^{(0)}}, \tag{134}$$

$$\left\{\sum_{n=1}^{\infty} n^{p\lambda_2-1} \left(\int_0^{\infty} k_\lambda(x,n) f(x) dx\right)^p\right\}^{\frac{1}{p}} > k(\lambda_1) ||f||_{p,\varphi}, \tag{135}$$

$$\left\{\int_0^\infty x^{q\lambda_1-1}\left(\sum_{n=1}^\infty k_\lambda(x,n)a_n\right)^q dx\right\}^{\frac{1}{q}} > k(\lambda_1)||a||_{q,\psi^{(0)}}. \tag{136}$$

Corollary 3 *As the assumptions of Corollary 1, if* $0 < p < 1(q < 0)$, $f(x) = f(x_1, \cdots, x_{i_0}) \geq 0$, $a_n = a_{(n_1,\cdots,n_{j_0})} \geq 0$, $0 < ||f||_{p,\widetilde{\Phi}} < \infty, 0 < ||a||_{q,\Psi^{(0)}} < \infty$, *then we have the following equivalent inequalities with the best possible constant factor* $K(\lambda_1)$:

$$\sum_n \int_{\mathbf{R}_+^{i_0}} k_\lambda(||x||_\alpha, ||n||_\beta)a_n f(x)dx > K(\lambda_1)||f||_{p,\widetilde{\Phi}}||a||_{q,\Psi^{(0)}}, \tag{137}$$

$$\left\{\sum_n ||n||_\beta^{p\lambda_2-j_0}\left(\int_{\mathbf{R}_+^{i_0}} k_\lambda(||x||_\alpha, ||n||_\beta)f(x)dx\right)^p\right\}^{\frac{1}{p}} > K(\lambda_1)||f||_{p,\widetilde{\Phi}}, \tag{138}$$

$$\left\{\int_{\mathbf{R}_+^{i_0}} \frac{||x||_\alpha^{q\lambda_1-i_0}}{(1-\theta_{\lambda_1}(x))^{q-1}}\left(\sum_n k_\lambda(||x||_\alpha, ||n||_\beta)a_n\right)^q dx\right\}^{\frac{1}{q}}$$
$$> K(\lambda_1)||a||_{q,\Psi^{(0)}}. \tag{139}$$

In particular, for $i_0 = j_0 = 1, \varphi(x)$ and $\psi^{(0)}(n)$ are indicated by Corollary 1,

$$\widetilde{\varphi}(x) = (1-\theta_{\lambda_1}(x))x^{p(1-\lambda_1)-1}(\theta_{\lambda_1}(x) \in (0,1); x \in \mathbf{R}_+),$$

we have the following equivalent inequalities with the best possible constant factor $k(\lambda_1)$:

$$\sum_{n=1}^\infty \int_0^\infty k_\lambda(x,n)a_n f(x)dx > k(\lambda_1)||f||_{p,\varphi}||a||_{q,\psi^{(0)}}, \tag{140}$$

$$\left\{\sum_{n=1}^\infty n^{p\lambda_2-1}\left(\int_0^\infty k_\lambda(x,n)f(x)dx\right)^p\right\}^{\frac{1}{p}} > k(\lambda_1)||f||_{p,\widetilde{\varphi}}, \tag{141}$$

$$\left\{\int_0^\infty \frac{x^{q\lambda_1-i_0}}{(1-\theta_{\lambda_1}(x))^{q-1}}\left(\sum_{n=1}^\infty k_\lambda(x,n)a_n\right)^q dx\right\}^{\frac{1}{q}}$$

$$> k(\lambda_1)||a||_{q,\psi^{(0)}}. \tag{142}$$

If in Theorems 4–6, we do not have the assumptions of $\frac{d^2}{dt^2}(h(t)t^{\widetilde{\sigma}-j_0}) > 0$, and $\beta \leq 1$, then for $\tau = 0$, setting

$$\Phi_\sigma(x) = ||x||_\alpha^{p(i_0-\sigma)-i_0}, \Psi_\sigma^{(0)}(n) := ||n||_\beta^{q(j_0-\sigma)-j_0},$$
$$\widetilde{\Phi}_\sigma(x) = (1-\vartheta_\sigma(x))||x||_\alpha^{p(i_0-\sigma)-i_0}(\vartheta_\sigma(x) \in (0,1); x \in \mathbf{R}_+^{i_0}),$$

in view of the Note of Lemma 1 and by the same way, we have the following corollaries:

Corollary 4 *Suppose that* $\alpha, \beta > 0, \sigma \in \mathbf{R}$, *there exists a constant* $\delta_0 > 0$, *such that for any* $\widetilde{\sigma} \in (\sigma - \delta_0, \sigma + \delta_0)$,

$$\widehat{k}(\widetilde{\sigma}) = \int_0^\infty h(t)t^{\widetilde{\sigma}-1}dt \in \mathbf{R}_+,$$

$h(t)t^{\widetilde{\sigma}-j_0}$ *is strict decreasing with respect to* $t \in \mathbf{R}_+$, *there exists constants* $L > 0$ *and* $\eta_0 < \widetilde{\sigma}$, *satisfying*

$$h(t) \leq \frac{L}{t^{\eta_0}}(t \in (0, \infty)).$$

If $p > 1,\ f(x) = f(x_1, \cdots, x_{i_0}) \geq 0, a_n = a_{(n_1,\cdots,n_{j_0})} \geq 0$,

$$0 < ||f||_{p,\Phi_\sigma} = \left\{\int_{\mathbf{R}_+^{i_0}} \Phi_\sigma(x) f^p(x)dx\right\}^{\frac{1}{p}} < \infty,$$

$$0 < ||a||_{q,\Psi_\sigma^{(0)}} = \left\{\sum_n \Psi_\sigma^{(0)}(n)a_n^q\right\}^{\frac{1}{q}} < \infty,$$

then we have the following equivalent inequalities with the best possible constant factor $\widehat{K}(\sigma)$:

$$\sum_n \int_{\mathbf{R}_+^{i_0}} h(||x||_\alpha ||n||_\beta)a_n f(x)dx < \widehat{K}(\sigma)||f||_{p,\Phi_\sigma} ||a||_{q,\Psi_\sigma^{(0)}}, \tag{143}$$

$$\widehat{J}^{(0)} := \left\{\sum_n ||n||_\beta^{p\sigma-j_0} \left(\int_{\mathbf{R}_+^{i_0}} h(||x||_\alpha ||n||_\beta) f(x)dx\right)^p\right\}^{\frac{1}{p}}$$
$$< \widehat{K}(\sigma)||f||_{p,\Phi_\sigma}, \tag{144}$$

$$\widehat{H}^{(0)} := \left\{\int_{\mathbf{R}_+^{i_0}} ||x||_\alpha^{q\sigma-i_0} \left(\sum_n h(||x||_\alpha ||n||_\beta)a_n\right)^q dx\right\}^{\frac{1}{q}}$$
$$< \widehat{K}(\sigma)||a||_{q,\Psi_\sigma^{(0)}}, \tag{145}$$

where, $\widehat{k}(\sigma) = \int_0^\infty h(t)t^{\sigma-1}dt$, *and*

$$\widehat{K}(\sigma) = \left(\widehat{K}_1(\sigma)\right)^{\frac{1}{p}} \left(\widehat{K}_2(\sigma)\right)^{\frac{1}{q}}$$
$$= \left[\frac{\Gamma^{j_0}\left(\frac{1}{\beta}\right)}{\beta^{j_0-1}\Gamma\left(\frac{j_0}{\beta}\right)}\right]^{\frac{1}{p}} \left[\frac{\Gamma^{i_0}\left(\frac{1}{\alpha}\right)}{\alpha^{i_0-1}\Gamma\left(\frac{i_0}{\alpha}\right)}\right]^{\frac{1}{q}} \widehat{k}(\sigma).$$

In particular, for $i_0 = j_0 = 1$, setting

$$\varphi_\sigma(x) = x^{p(1-\sigma)-1}, \psi_\sigma^{(0)}(n) := n^{q(1-\sigma)-1},$$

we have the following equivalent inequalities with the best possible constant factor $\widehat{k}(\sigma)$:

$$\sum_{n=1}^{\infty} \int_0^{\infty} h(xn) a_n f(x) dx < \widehat{k}(\sigma) ||f||_{p,\varphi_\sigma} ||a||_{q,\psi_\sigma^{(0)}}, \tag{146}$$

$$\left\{ \sum_{n=1}^{\infty} n^{p\sigma-1} \left(\int_0^{\infty} h(xn) f(x) dx \right)^p \right\}^{\frac{1}{p}} < \widehat{k}(\sigma) ||f||_{p,\varphi_\sigma}, \tag{147}$$

$$\left\{ \int_0^{\infty} x^{q\sigma-1} \left(\sum_{n=1}^{\infty} h(xn) a_n \right)^q dx \right\}^{\frac{1}{q}} < \widehat{k}(\sigma) ||a||_{q,\psi_\sigma^{(0)}}. \tag{148}$$

Corollary 5 *As the assumptions of Corollary 4, if* $p < 0 (0 < q < 1)$, $f(x) = f(x_1, \cdots, x_{i_0}) \geq 0$, $a_n = a_{(n_1,\cdots,n_{j_0})} \geq 0$, $0 < ||f||_{p,\Phi_\sigma} < \infty, 0 < ||a||_{q,\Psi_\sigma^{(0)}} < \infty$, *then we have the following equivalent inequalities with the best possible constant factor* $\widehat{K}(\sigma)$:

$$\sum_n \int_{\mathbf{R}_+^{i_0}} h(||x||_\alpha ||n||_\beta) a_n f(x) dx > \widehat{K}(\sigma) ||f||_{p,\Phi_\sigma} ||a||_{q,\Psi_\sigma^{(0)}}, \tag{149}$$

$$\left\{ \sum_n ||n||_\beta^{p\sigma-j_0} \left(\int_{\mathbf{R}_+^{i_0}} h(||x||_\alpha ||n||_\beta) f(x) dx \right)^p \right\}^{\frac{1}{p}} > \widehat{K}(\sigma) ||f||_{p,\Phi_\sigma}, \tag{150}$$

$$\left\{ \int_{\mathbf{R}_+^{i_0}} ||x||_\alpha^{q\sigma-i_0} \left(\sum_n h(||x||_\alpha ||n||_\beta) a_n \right)^q dx \right\}^{\frac{1}{q}} > \widehat{K}(\sigma) ||a||_{q,\Psi_\sigma^{(0)}}. \tag{151}$$

In particular, for $i_0 = j_0 = 1$, setting $\varphi_\sigma(x)$ and $\psi_\sigma^{(0)}(n)$ as Corollary 4, we have the following equivalent inequalities with the best possible constant factor $\widehat{k}(\sigma)$:

$$\sum_{n=1}^{\infty} \int_0^{\infty} h(xn) a_n f(x) dx > \widehat{k}(\sigma) ||f||_{p,\varphi_\sigma} ||a||_{q,\psi_\sigma^{(0)}}, \tag{152}$$

$$\left\{ \sum_{n=1}^{\infty} n^{p\sigma-1} \left(\int_0^{\infty} h(xn) f(x) dx \right)^p \right\}^{\frac{1}{p}} > \widehat{k}(\sigma) ||f||_{p,\varphi_\sigma}, \tag{153}$$

$$\left\{ \int_0^{\infty} x^{q\sigma-1} \left(\sum_{n=1}^{\infty} h(xn) a_n \right)^q dx \right\}^{\frac{1}{q}} > \widehat{k}(\sigma) ||a||_{q,\psi_\sigma^{(0)}}. \tag{154}$$

Corollary 6 *As the assumptions of Corollary 4, if* $0 < p < 1 (q < 0)$, $f(x) = f(x_1, \cdots, x_{i_0}) \geq 0$, $a_n = a_{(n_1,\cdots,n_{j_0})} \geq 0$, $0 < ||f||_{p,\widetilde{\Phi}_\sigma} < \infty, 0 < ||a||_{q,\Psi_\sigma^{(0)}} < \infty$, *then we have the following equivalent inequalities with the best possible constant factor* $\widehat{K}(\sigma)$:

$$\sum_n \int_{\mathbf{R}_+^{i_0}} h(||x||_\alpha ||n||_\beta) a_n f(x) dx > \widehat{K}(\sigma) ||f||_{p,\widetilde{\Phi}_\sigma} ||a||_{q,\Psi_\sigma^{(0)}}, \tag{155}$$

$$\left\{ \sum_n ||n||_\beta^{p\sigma - j_0} \left(\int_{\mathbf{R}_+^{i_0}} h(||x||_\alpha ||n||_\beta) f(x) dx \right)^p \right\}^{\frac{1}{p}} > \widehat{K}(\sigma) ||f||_{p,\widetilde{\Phi}_\sigma}, \tag{156}$$

$$\left\{ \int_{\mathbf{R}_+^{i_0}} \frac{||x||_\alpha^{q\sigma - i_0}}{(1 - \vartheta_\sigma(x))^{q-1}} \left(\sum_n h(||x||_\alpha ||n||_\beta) a_n \right)^q dx \right\}^{\frac{1}{q}}$$

$$> \widehat{K}(\sigma) ||a||_{q,\Psi_\sigma^{(0)}}. \tag{157}$$

In particular, for $i_0 = j_0 = 1$, setting $\varphi_\sigma(x)$ and $\psi_\sigma^{(0)}(n)$ as Corollary 4, and

$$\widetilde{\varphi}_\sigma(x) := (1 - \vartheta_\sigma(x)) x^{p(1-\sigma)-1} (\vartheta_\sigma(x) \in (0,1); x \in \mathbf{R}_+),$$

we have the following equivalent inequalities with the best possible constant factor $\widehat{k}(\sigma)$:

$$\sum_{n=1}^{\infty} \int_0^{\infty} h(xn) a_n f(x) dx > \widehat{k}(\sigma) ||f||_{p,\widetilde{\varphi}_\sigma} ||a||_{q,\psi_\sigma^{(0)}}, \tag{158}$$

$$\left\{ \sum_{n=1}^{\infty} n^{p\sigma - 1} \left(\int_0^{\infty} h(xn) f(x) dx \right)^p \right\}^{\frac{1}{p}} > \widehat{k}(\sigma) ||f||_{p,\widetilde{\varphi}_\sigma}, \tag{159}$$

$$\left\{ \int_0^{\infty} \frac{x^{q\sigma - 1}}{(1 - \vartheta_\sigma(x))^{q-1}} \left(\sum_{n=1}^{\infty} h(xn) a_n \right)^q dx \right\}^{\frac{1}{q}} > \widehat{k}(\sigma) ||a||_{q,\psi_\sigma^{(0)}}. \tag{160}$$

As the assumptions of Corollary 1, in view of

$$J^{(0)} < K(\lambda_1) ||f||_{p,\Phi}, \ H^{(0)} < K(\lambda_1) ||a||_{q,\Psi^{(0)}},$$

we can give the following definition:

Definition 5 Define a first kind of multidimensional half-discrete Hilbert-type operator

$$T_1^{(0)} : \mathbf{L}_{p,\Phi}(\mathbf{R}_+^{i_0}) \to \mathbf{l}_{p,[\Psi^{(0)}]^{1-p}}$$

as follows: For $f \in \mathbf{L}_{p,\Phi}(\mathbf{R}_+^{i_0})$, there exists an unique representation $T_1^{(0)} f \in \mathbf{l}_{p,[\Psi^{(0)}]^{1-p}}$, satisfying

$$\left(T_1^{(0)} f \right)(n) := \int_{\mathbf{R}_+^{i_0}} k_\lambda(||x||_\alpha, ||n||_\beta) f(x) dx (n \in \mathbf{N}^{j_0}). \tag{161}$$

For $a \in \mathbf{l}_{q,\Psi^{(0)}}$, we set the following formal inner product of $T_1^{(0)} f$ and a as follows:

$$\left(T_1^{(0)} f, a\right) = \sum_n a_n \int_{\mathbf{R}_+^{i_0}} k_\lambda(||x||_\alpha, ||n||_\beta) f(x) dx. \tag{162}$$

Define a second kind of multidimensional half-discrete Hilbert-type operator

$$T_2^{(0)} : \mathbf{l}_{q,\Psi^{(0)}} \to \mathbf{L}_{q,\Phi^{1-q}}(\mathbf{R}_+^{i_0})$$

as follows: For $a \in \mathbf{l}_{q,\Psi^{(0)}}$, there exists an unique representation $T_2^{(0)} a \in \mathbf{L}_{q,\Phi^{1-q}}(\mathbf{R}_+^{i_0})$, satisfying

$$\left(T_2^{(0)} a\right)(x) := \sum_n k_\lambda(||x||_\alpha, ||n||_\beta) a_n \left(x \in \mathbf{R}_+^{i_0}\right). \tag{163}$$

For $f \in \mathbf{L}_{p,\Phi}(\mathbf{R}_+^{i_0})$, we set the following formal inner product of f and $T_2^{(0)} a$ as follows:

$$(f, T_2^{(0)} a) = \int_{\mathbf{R}_+^{i_0}} \sum_n k_\lambda(||x||_\alpha, ||n||_\beta) a_n f(x) dx. \tag{164}$$

Then by Corollary 1, for $0 < ||f||_{p,\Phi}, ||a||_{q,\Psi^{(0)}} < \infty$, we have the following equivalent inequalities:

$$(T_1^{(0)} f, a) = (f, T_2^{(0)} a) < K(\lambda_1) ||f||_{p,\Phi} ||a||_{q,\Psi^{(0)}}, \tag{165}$$

$$||T_1^{(0)} f||_{p,[\Psi^{(0)}]^{1-p}} < K(\lambda_1) ||f||_{p,\Phi}, \tag{166}$$

$$||T_2^{(0)} a||_{q,\Phi^{1-q}} < K(\lambda_1) ||a||_{q,\Psi^{(0)}}. \tag{167}$$

It follows that $T_1^{(0)}$ and $T_2^{(0)}$ are bounded with

$$||T_1^{(0)}|| = \sup_{f(\neq\theta)\in \mathbf{L}_{p,\Phi}(\mathbf{R}_+^{i_0})} \frac{||T_1^{(0)} f||_{p,[\Psi^{(0)}]^{1-p}}}{||f||_{p,\Phi}} \leq K(\lambda_1)$$

$$||T_2^{(0)}|| := \sup_{a(\neq\theta)\in \mathbf{l}_{q,\Psi^{(0)}}} \frac{||T_2^{(0)} a||_{q,\Phi^{1-q}}}{||a||_{q,\Psi^{(0)}}} \leq K(\lambda_1).$$

Since the constant factor $K(\lambda_1)$ in (166) and (167) is the best possible, we have

$$||T_1^{(0)}|| = ||T_2^{(0)}|| = K(\lambda_1)$$

$$= \left[\frac{\Gamma^{j_0}\left(\frac{1}{\beta}\right)}{\beta^{j_0-1}\Gamma\left(\frac{j_0}{\beta}\right)}\right]^{\frac{1}{p}} \left[\frac{\Gamma^{i_0}\left(\frac{1}{\alpha}\right)}{\alpha^{i_0-1}\Gamma\left(\frac{i_0}{\alpha}\right)}\right]^{\frac{1}{q}} k(\lambda_1). \tag{168}$$

As the assumptions of Corollary 4, in view of

$$\widehat{J}^{(0)} < \widehat{K}(\sigma)||f||_{p,\Phi_\sigma}, \widehat{H}^{(0)} < \widehat{K}(\sigma)||a||_{q,\Psi_\sigma^{(0)}},$$

we can give the following definition:

Definition 6 Define a first kind of multidimensional half-discrete Hilbert-type operator

$$\widehat{T}_1^{(0)} : \mathbf{L}_{p,\Phi_\sigma}(\mathbf{R}_+^{i_0}) \to \mathbf{l}_{p,[\Psi_\sigma^{(0)}]^{1-p}}$$

as follows: For $f \in \mathbf{L}_{p,\Phi_\sigma}(\mathbf{R}_+^{i_0})$, there exists an unique representation $\widehat{T}_1^{(0)} f \in \mathbf{l}_{p,[\Psi_\sigma^{(0)}]^{1-p}}$, satisfying

$$\left(\widehat{T}_1^{(0)} f\right)(n) := \int_{\mathbf{R}_+^{i_0}} h(||x||_\alpha ||n||_\beta) f(x) dx (n \in \mathbf{N}^{j_0}). \tag{169}$$

For $a \in \mathbf{l}_{q,\Psi_\sigma^{(0)}}$, we set the following formal inner product of a and $\widehat{T}_1^{(0)} f$ as follows:

$$(a, \widehat{T}_1^{(0)} f) = \sum_n a_n \int_{\mathbf{R}_+^{i_0}} h\left(||x||_\alpha ||n||_\beta\right) f(x) dx. \tag{170}$$

Define a second kind of multidimensional half-discrete Hilbert-type operator

$$\widehat{T}_2^{(0)} : \mathbf{l}_{q,\Psi_\sigma} \to \mathbf{L}_{q,\Phi_\sigma^{1-q}}\left(\mathbf{R}_+^{i_0}\right)$$

as follows: For $a \in \mathbf{l}_{q,\Psi_\sigma^{(0)}}$, there exists an unique representation $\widehat{T}_2^{(0)} a \in \mathbf{L}_{q,\Phi_\sigma^{1-q}}(\mathbf{R}_+^{i_0})$, satisfying

$$\left(\widehat{T}_2^{(0)} a\right)(x) := \sum_n h\left(||x||_\alpha ||n||_\beta\right) a_n \left(x \in \mathbf{R}_+^{i_0}\right). \tag{171}$$

For $f \in \mathbf{L}_{p,\Phi_\sigma}(\mathbf{R}_+^{i_0})$, we set the following formal inner product of f and $\widehat{T}_2^{(0)} a$ as follows:

$$(f, \widehat{T}_2^{(0)} a) := \int_{\mathbf{R}_+^{i_0}} \sum_n h\left(||x||_\alpha ||n||_\beta\right) a_n f(x) dx. \tag{172}$$

Then by Corollary 4, for $0 < ||f||_{p,\Phi_\sigma}, ||a||_{q,\Psi_\sigma^{(0)}} < \infty$, we have the following equivalent inequalities:

$$(\widehat{T}_1^{(0)} f, a) = \left(f, \widehat{T}_2^{(0)} a\right) < \widehat{K}(\sigma)||f||_{p,\Phi_\sigma} ||a||_{q,\Psi_\sigma^{(0)}}, \tag{173}$$

$$||\widehat{T}_1 f||_{p,[\Psi_\sigma^{(0)}]^{1-p}} < \widehat{K}(\sigma)||f||_{p,\Phi_\sigma}, \tag{174}$$

$$||\widehat{T}_2 a||_{q,\Phi_\sigma^{1-q}} < \widehat{K}(\sigma)||a||_{q,\Psi_\sigma^{(0)}}. \tag{175}$$

It follows that $\widehat{T}_1$ and $\widehat{T}_2$ are bounded with

$$||\widehat{T}_1|| := \sup_{f(\neq\theta)\in \mathbf{L}_{p,\Phi_\sigma}(\mathbf{R}_+^{i_0})} \frac{||\widehat{T}_1 f||_{p,[\Psi_\sigma^{(0)}]^{1-p}}}{||f||_{p,\Phi_\sigma}} \le \widehat{K}(\sigma)$$

$$||\widehat{T}_2|| := \sup_{a(\neq\theta)\in \mathbf{l}_{q,\Psi_\sigma^{(0)}}} \frac{||\widehat{T}_2 a||_{q,\Phi_\sigma^{1-q}}}{||a||_{q,\Psi_\sigma^{(0)}}} \le \widehat{K}(\sigma).$$

Since the constant factor $\widehat{K}(\sigma)$ in (174) and (175) is the best possible, we have

$$||\widehat{T}_1|| = ||\widehat{T}_2|| = \widehat{K}(\sigma)$$

$$= \left[\frac{\Gamma^{j_0}(\frac{1}{\beta})}{\beta^{j_0-1}\Gamma(\frac{j_0}{\beta})}\right]^{\frac{1}{p}} \left[\frac{\Gamma^{i_0}(\frac{1}{\alpha})}{\alpha^{i_0-1}\Gamma(\frac{i_0}{\alpha})}\right]^{\frac{1}{q}} \widehat{k}(\sigma). \quad (176)$$

Example 7 (1) We set

$$k_\lambda(x,y) = \frac{(\min\{x,y\})^\gamma}{(\max\{x,y\})^{\lambda+\gamma}}$$
$$(\lambda_1 > -\gamma, -\gamma < \lambda_2 < j_0 - \gamma).$$

For $\delta_0 = \frac{1}{2}\min\{\lambda_1+\gamma, \lambda_2+\gamma, j_0-\gamma-\lambda_2\} > 0$, and $\widetilde{\lambda}_i \in (\lambda_i - \delta_0, \lambda_i + \delta_0)(i = 1,2)$, $\widetilde{\lambda}_1 + \widetilde{\lambda}_2 = \lambda$, it follows

$$k(\widetilde{\lambda}_1) = \int_0^\infty \frac{(\min\{t,1\})^\gamma}{(\max\{t,1\})^{\lambda+\gamma}} t^{\widetilde{\lambda}_1-1} dt$$
$$= \int_0^1 t^{\widetilde{\lambda}_1+\gamma-1} dt + \int_1^\infty \frac{1}{t^{\lambda+\gamma}} t^{\widetilde{\lambda}_1-1} dt$$
$$= \frac{\lambda+2\gamma}{(\widetilde{\lambda}_1+\gamma)(\widetilde{\lambda}_2+\gamma)} \in \mathbf{R}_+.$$

Since

$$k_\lambda(x,y) y^{\widetilde{\lambda}_2 - j_0} = \frac{(\min\{x,y\})^\gamma}{(\max\{x,y\})^{\lambda+\gamma}} y^{\widetilde{\lambda}_2 - j_0}$$
$$= \begin{cases} \frac{y^{\gamma+\widetilde{\lambda}_2-j_0}}{x^{\lambda+\gamma}}, 0 < y < x, \\ \frac{x^\gamma}{y^{\widetilde{\lambda}_1+\gamma+j_0}}, y \ge x \end{cases}$$

is a strict decreasing function with respect to $y \in \mathbf{R}_+$.

There exists a constant $\eta_1 \in (\max\{-\gamma, \lambda_1+\delta_0\}, \lambda+\gamma)$, such that $\eta_1 > \lambda_1 + \delta_0 > \widetilde{\lambda}_1$, $\gamma + \eta_1 > 0$ and $\lambda + \gamma - \eta_1 > 0$. In view of

$$v^{\eta_1} k_\lambda(v,1) = v^{\eta_1} \frac{(\min\{v,1\})^\gamma}{(\max\{v,1\})^{\lambda+\gamma}}$$

$$= \begin{cases} v^{\gamma+\eta_1}, 0 < v < 1, \\ \frac{1}{v^{\lambda+\gamma-\eta_1}}, v \geq 1, \end{cases}$$

we have $v^{\eta_1} k_\lambda(v,1) \to 0 (v \to 0^+)$ and $v^{\eta_1} k_\lambda(v,1) \to 0 (v \to \infty)$. Hence, there exists a constant $L > 0$, satisfying

$$k_\lambda(v,1) \leq \frac{L}{v^{\eta_1}} (v \in (0,\infty)).$$

Therefore, by Corollary 1 and (168), it follows

$$||T_1^{(0)}|| = ||T_2^{(0)}|| = \left[\frac{\Gamma^{j_0}\left(\frac{1}{\beta}\right)}{\beta^{j_0-1}\Gamma\left(\frac{j_0}{\beta}\right)}\right]^{\frac{1}{p}} \left[\frac{\Gamma^{i_0}\left(\frac{1}{\alpha}\right)}{\alpha^{i_0-1}\Gamma\left(\frac{i_0}{\alpha}\right)}\right]^{\frac{1}{q}}$$

$$\times \frac{\lambda + 2\gamma}{(\lambda_1 + \gamma)(\lambda_2 + \gamma)}. \tag{177}$$

In particular, (i) For $\gamma = 0$, we have

$$k_\lambda(x,y) = \frac{1}{(\max\{x,y\})^\lambda} (\lambda_1 > 0, 0 < \lambda_2 < j_0)$$

and

$$||T_1^{(0)}|| = ||T_2^{(0)}|| = \left[\frac{\Gamma^{j_0}\left(\frac{1}{\beta}\right)}{\beta^{j_0-1}\Gamma\left(\frac{j_0}{\beta}\right)}\right]^{\frac{1}{p}} \left[\frac{\Gamma^{i_0}\left(\frac{1}{\alpha}\right)}{\alpha^{i_0-1}\Gamma\left(\frac{i_0}{\alpha}\right)}\right]^{\frac{1}{q}} \frac{\lambda}{\lambda_1\lambda_2} \tag{178}$$

(ii) For $\gamma = -\lambda$, we have

$$k_\lambda(x,y) = \frac{1}{(\min\{x,y\})^\lambda} (\lambda < \lambda_1, \lambda < \lambda_2 < j_0 + \lambda)$$

and

$$||T_1^{(0)}|| = ||T_2^{(0)}|| = \left[\frac{\Gamma^{j_0}\left(\frac{1}{\beta}\right)}{\beta^{j_0-1}\Gamma\left(\frac{j_0}{\beta}\right)}\right]^{\frac{1}{p}} \left[\frac{\Gamma^{i_0}\left(\frac{1}{\alpha}\right)}{\alpha^{i_0-1}\Gamma\left(\frac{i_0}{\alpha}\right)}\right]^{\frac{1}{q}} \frac{(-\lambda)}{\lambda_1\lambda_2} \tag{179}$$

(iii) For $\lambda = 0, \gamma > 0$, setting $\sigma = \lambda_1 = -\lambda_2$, we have

$$k_0(x,y) = \left(\frac{\min\{x,y\}}{\max\{x,y\}}\right)^\gamma (\min\{-\gamma, \gamma - j_0\} < \sigma < \gamma)$$

and

$$||T|| = \left[\frac{\Gamma^{j_0}\left(\frac{1}{\beta}\right)}{\beta^{j_0-1}\Gamma\left(\frac{j_0}{\beta}\right)}\right]^{\frac{1}{p}} \left[\frac{\Gamma^{i_0}\left(\frac{1}{\alpha}\right)}{\alpha^{i_0-1}\Gamma\left(\frac{i_0}{\alpha}\right)}\right]^{\frac{1}{q}} \frac{2\gamma}{\gamma^2 - \sigma^2}. \tag{180}$$

(2) We set

$$h(t)=\frac{(\min\{t,1\})^{\gamma}}{(\max\{t,1\})^{\lambda+\gamma}}(-\gamma<\sigma<\lambda+\gamma<j_0-\gamma).$$

For $\delta_0=\frac{1}{2}\min\{\sigma+\gamma,\lambda+\gamma-\sigma,j_0-\sigma-\gamma\}>0$, and $\widetilde{\sigma}\in(\sigma-\delta_0,\sigma+\delta_0)$, it follows

$$\begin{aligned}\widehat{k}(\widetilde{\sigma})&=\int_0^{\infty}\frac{(\min\{t,1\})^{\gamma}}{(\max\{t,1\})^{\lambda+\gamma}}t^{\widetilde{\sigma}-1}dt\\&=\frac{\lambda+2\gamma}{(\widetilde{\sigma}+\gamma)(\lambda-\widetilde{\sigma}+\gamma)}\in\mathbf{R}_+,\end{aligned}$$

and

$$\begin{aligned}h(t)t^{\widetilde{\sigma}-j_0}&=\frac{(\min\{t,1\})^{\gamma}}{(\max\{t,1\})^{\lambda+\gamma}}t^{\widetilde{\sigma}-j_0}\\&=\begin{cases}t^{\gamma+\widetilde{\sigma}-j_0},0<t<1,\\\frac{1}{t^{\lambda+\gamma-\widetilde{\sigma}+j_0}},t\geq 1,\end{cases}\end{aligned}$$

is strict decreasing with respect to $t\in\mathbf{R}_+$.

There exists a constant $\eta_0\in(-\gamma,\min\{\sigma-\delta_0,\lambda+\gamma\})$, such that $\eta_0<\widetilde{\sigma},\gamma+\eta_0>0$ and $\lambda+\gamma-\eta_0>0$. In view of

$$\begin{aligned}t^{\eta_0}h(t)&=\frac{t^{\eta_0}(\min\{t,1\})^{\gamma}}{(\max\{t,1\})^{\lambda+\gamma}}\\&=\begin{cases}t^{\gamma+\eta_0},0<t<1,\\\frac{1}{t^{\lambda+\gamma-\eta_0}},t\geq 1,\end{cases}\end{aligned}$$

we have $t^{\eta_0}h(t)\to 0(t\to 0^+)$, and $t^{\eta_0}h(t)\to 0(t\to\infty)$. Hence, there exists a constant $L>0$, satisfying

$$h(t)\leq\frac{L}{t^{\eta_0}}(t\in(0,\infty)).$$

Therefore, by Corollary 4 and (176), it follows

$$||\widehat{T}_1^{(0)}||=||\widehat{T}_2^{(0)}||=\left[\frac{\Gamma^{j_0}\left(\frac{1}{\beta}\right)}{\beta^{j_0-1}\Gamma\left(\frac{j_0}{\beta}\right)}\right]^{\frac{1}{p}}\left[\frac{\Gamma^{i_0}\left(\frac{1}{\alpha}\right)}{\alpha^{i_0-1}\Gamma\left(\frac{i_0}{\alpha}\right)}\right]^{\frac{1}{q}}$$

$$\times\frac{\lambda+2\gamma}{(\sigma+\gamma)(\lambda-\sigma+\gamma)}. \tag{181}$$

Example 8 (i) We set

$$k_0(x,y) = \sec h(\frac{y}{x}) = \frac{2}{e^{(y/x)} + e^{-(y/x)}}$$
$$(0 < -\lambda_1 = \lambda_2 = \sigma < j_0),$$

where, $\sec h(u) = \frac{2}{e^u + e^{-u}}$ is called hyperbolic secant function (cf. [46]). For $\delta_0 = \frac{1}{2}\min\{\lambda_2, j_0 - \lambda_2\} > 0$, and $\widetilde{\lambda}_i \in (\lambda_i - \delta_0, \lambda_i + \delta_0)(i = 1, 2)$, $\widetilde{\lambda}_1 + \widetilde{\lambda}_2 = \lambda, \widetilde{\lambda}_1 < 0$, it follows

$$\begin{aligned} k(\widetilde{\lambda}_1) &= \int_0^\infty \frac{2t^{\widetilde{\lambda}_1 - 1}}{e^{1/t} + e^{-1/t}} dt = \int_0^\infty \frac{2u^{-\widetilde{\lambda}_1 - 1}}{e^u(1 + e^{-2u})} du \\ &= 2\int_0^\infty \sum_{k=0}^\infty (-1)^k e^{-(2k+1)u} u^{-\widetilde{\lambda}_1 - 1} du \\ &= 2\sum_{k=0}^\infty (-1)^k \int_0^\infty e^{-(2k+1)u} u^{-\widetilde{\lambda}_1 - 1} du \\ &= 2\Gamma\left(-\widetilde{\lambda}_1\right) \sum_{k=0}^\infty \frac{(-1)^k}{(2k+1)^{-\widetilde{\lambda}_1}} \in \mathbf{R}_+. \end{aligned}$$

Since we find

$$\begin{aligned} \frac{\partial}{\partial y} k_0(x,y) &= \frac{\partial}{\partial y} \frac{2}{e^{(y/x)} + e^{-(y/x)}} \\ &= -\frac{2\left(e^{(y/x)} - e^{-(y/x)}\right)}{x(e^{(y/x)} + e^{-(y/x)})^2} < 0, \end{aligned}$$

then for $\widetilde{\lambda}_2 < j_0$, $k_0(x,y)y^{\widetilde{\lambda}_2 - j_0}$ is a strict decreasing function with respect to $y \in \mathbf{R}_+$.

There exists a constant $\eta_1 \in (\lambda_1 - \delta_0, 0)$, such that $\eta_1 > \lambda_1 + \delta_0 > \widetilde{\lambda}_1$. In view of

$$\lim_{v\to\infty} v^{\eta_1} k_0(v,1) = \lim_{v\to\infty} \frac{2v^{\eta_1}}{e^{1/v} + e^{-1/v}} = 0,$$

$$\lim_{v\to 0^+} v^{\eta_1} k_0(v,1) = \lim_{v\to 0^+} \frac{-2\eta_1 v^{\eta_1 + 1}}{e^{1/v} - e^{-1/v}} = \cdots = 0,$$

there exists a constant $L > 0$, such that

$$k_0(v,1) = \frac{2}{e^{1/v} + e^{-1/v}} \leq \frac{L}{v^{\eta_1}} (v \in (0,\infty)).$$

Therefore, by Corollary 1 and (168), it follows

$$\begin{aligned} ||T_1^{(0)}|| = ||T_2^{(0)}|| &= \left[\frac{\Gamma^{j_0}\left(\frac{1}{\beta}\right)}{\beta^{j_0 - 1}\Gamma\left(\frac{j_0}{\beta}\right)}\right]^{\frac{1}{p}} \left[\frac{\Gamma^{i_0}\left(\frac{1}{\alpha}\right)}{\alpha^{i_0 - 1}\Gamma\left(\frac{i_0}{\alpha}\right)}\right]^{\frac{1}{q}} \\ &\quad \times 2\Gamma(\sigma) \sum_{k=0}^\infty \frac{(-1)^k}{(2k+1)^\sigma}. \end{aligned} \tag{182}$$

(ii) We set

$$h(t) = \sec h(t) = \frac{2}{e^t + e^{-t}}(0 < \sigma < j_0).$$

For $\delta_0 = \frac{1}{2}\min\{\sigma, j_0 - \sigma\} > 0$, and $\widetilde{\sigma} \in (\sigma - \delta_0, \sigma + \delta_0)$, it follows $0 < \widetilde{\sigma} < j_0$ and

$$\widehat{k}(\widetilde{\sigma}) = \int_0^\infty \frac{2}{e^t + e^{-t}} t^{\widetilde{\sigma}-1} dt$$
$$= 2\Gamma(\widetilde{\sigma}) \sum_{k=0}^\infty \frac{(-1)^k}{(2k+1)^{\widetilde{\sigma}}} \in \mathbf{R}_+,$$

and $h(t)t^{\widetilde{\sigma}-j_0}$ is strict decreasing with respect to $t \in \mathbf{R}_+$.

There exists a constant $\eta_0 \in (0, \sigma - \delta_0)$, such that $0 < \eta_0 < \widetilde{\sigma}$. In view of $t^{\eta_0}h(t) \to 0(t \to 0^+)$, and $t^{\eta_0}h(t) \to 0(t \to \infty)$, there exists a constant $L > 0$, satisfying

$$h(t) \leq \frac{L}{t^{\eta_0}}(t \in (0, \infty)).$$

Therefore, by Corollary 4 and (176), it follows

$$||\widehat{T}_1^{(0)}|| = ||\widehat{T}_2^{(0)}|| = \left[\frac{\Gamma^{j_0}\left(\frac{1}{\beta}\right)}{\beta^{j_0-1}\Gamma\left(\frac{j_0}{\beta}\right)}\right]^{\frac{1}{p}} \left[\frac{\Gamma^{i_0}\left(\frac{1}{\alpha}\right)}{\alpha^{i_0-1}\Gamma\left(\frac{i_0}{\alpha}\right)}\right]^{\frac{1}{q}}$$
$$\times 2\Gamma(\sigma) \sum_{k=0}^\infty \frac{(-1)^k}{(2k+1)^\sigma}. \tag{183}$$

Remark 2 (i) For $\tau = 0$, (34) and (86) reduces to (124) and (143). Hence, (34) and (86) are more accurate inequalities of (125) and (143) in the conditions of Theorem 1 (Theorem 4).

(ii) It is evident that (125) is a multidimensional extension of (6).

Acknowledgement This work is supported by 2012 Knowledge Construction Special Foundation Item of Guangdong Institution of Higher Learning College and University (No. 2012KJCX0079).

References

1. Hardy, G.H., Littlewood, J.E., Pólya G.: Inequalities. Cambridge University Press, Cambridge (1934)
2. Mitrinović, D.S., Pečarić, J.E., Fink, A.M.: Inequalities Involving Functions and Their Integrals and Derivatives. Kluwer Academic Publishers, Boston (1991)
3. Yang, B.C.: Hilbert-Type Integral Inequalities. Bentham Science Publishers Ltd., Sharjah (2009)

4. Yang, B.C.: Discrete Hilbert-Type Inequalities. Bentham Science Publishers Ltd., Sharjah (2011)
5. Yang, B.C.: The Norm of Operator and Hilbert-Type Inequalities. Science Press, Beijin, (2009)
6. Yang, B.C.: Hilbert-type integral operators: norms and inequalities (Chap. 42). In: Paralos P.M. et al. (eds.) Nonlinear Analysis, Stability, Approximation, and Inequalities. Springer, New York, pp. 771–859 (2012)
7. Yang, B.C.: On Hilbert's integral inequality. J. Math. Anal. Appl. **220**, 778–785 (1998)
8. Yang, B.C., Brnetić, I., Krnić, M., Pečarić, J.E.: Generalization of Hilbert and Hardy-Hilbert Integral Inequalities. Math. Inequal. Appl. **8**(2), 259–272 (2005)
9. Krnić, M., Pečarić, J.E.: Hilbert's inequalities and their reverses. Publ. Math. Debrecen. **67**(3–4), 315–331 (2005)
10. Yang, B.C., Rassias T.M.: On the way of weight coefficient and research for Hilbert-type inequalities. Math. Inequal. Appl. **6**(4), 625–658 (2003)
11. Yang, B.C., Rassias T.M.: On a Hilbert-type integral inequality in the subinterval and its operator expression. Banach J. Math. Anal. **4**(2), 100–110 (2010)
12. Azar, L.: On some extensions of Hardy-Hilbert's inequality and applications. J. Inequal. Appl. Art. no. 546829. (2009)
13. Arpad, B., Choonghong, O.: Best constant for certain multilinear integral operator. J. Inequal. Appl. Art. no. 28582. (2006)
14. Kuang, J.C., Debnath, L.: On Hilbert's type inequalities on the weighted Orlicz spaces. Pac. J. Appl. Math. **1**(1), 95–103 (2007)
15. Zhong, W.Y.: The Hilbert-type integral inequality with a homogeneous kernel of lambda-degree. J. Inequal. Appl. Art. no. 917392. (2008)
16. Hong, Y.: On Hardy-Hilbert integral inequalities with some parameters. J. Inequal. Pure Appl. Math. **6**(4), Art. 92, 1–10 (2005)
17. Zhong, W.Y., Yang, B.C.: On multiple Hardy-Hilbert's integral inequality with kernel. J. Inequal. Appl. 2007, Art.ID 27962, 17 p, (2007) doi: 10.1155/ 2007/27
18. Yang, B.C., Krnić, M.: On the norm of a multi-dimensional Hilbert-type operator. Sarajevo J. Math. **7**(20), 223–243(2011)
19. Rassias, M.T., Yang, B.C.: On half-discrete Hilbert's inequality. App. Math. Comput. in print
20. Rassias, M.T., Yang, B.C.: A multidimensional Hilbert-type integral inequality related to Riemann's zeta function. In: Daras, N. (eds.) Applications of Mathematics and Informatics in Science and Engineering. Springer, New York, in print
21. Li, Y.J., He, B.: On inequalities of Hilbert's type. Bull. Aust. Math. Soc. **76**(1), 1–13 (2007)
22. Yang, B.C.: A mixed Hilbert-type inequality with a best constant factor. Int. J. Pure Appl. Math. **20**(3), 319–328 (2005)
23. Yang, B.C.: A half-discrete Hilbert-type inequality. J. Guangdong Univ. Edu. **31**(3), 1–7 (2011)
24. Zhong, W.Y.: A mixed Hilbert-type inequality and its equivalent forms. J. Guangdong Univ. Edu. **31**(5), 18–22 (2011)
25. Zhong, W.Y.: A half discrete Hilbert-type inequality and its equivalent forms. J. Guangdong Univ. Edu. **32**(5), 8–12 (2012)
26. Zhong, J.H., Yang, B.C.: On an extension of a more accurate Hilbert-type inequality. J. Zhejiang Univ. (Science Edition) **35**(2), 121–124 (2008)
27. Zhong, J.H.: Two classes of half-discrete reverse Hilbert-type inequalities with a non-homogeneous kernel. J. Guangdong Univ. Edu. **32**(5), 11–20 (2012)
28. Zhong, W.Y., Yang, B.C.: A best extension of Hilbert inequality involving several parameters. J. Jinan Univ. (Natural Science), **28**(1), 20–23 (2007)
29. Zhong, W.Y., Yang, B.C.: A reverse Hilbert's type integral inequality with some parameters and the equivalent forms. Pure Appl. Math. **24**(2), 401–407 (2008)
30. Zhong, W.Y., Yang, B.C.: On multiple Hardy-Hilbert's integral inequality with kernel. J. Inequal. Appl. 2007, Art.ID 27962, 17 p. (2007) doi: 10.1155/ 2007/27
31. Yang, B.C., Chen, Q.: A half-discrete Hilbert-type inequality with a homogeneous kernel and an extension. J. Inequal. Appl. **124** (2011). doi:10.1186/1029-242X-2011-124

32. Yang, B.C.: A half-discrete Hilbert-type inequality with a non-homogeneous kernel and two variables. Mediterr. J. Math. (2012). doi: 10.1007/s00009- 012- 0213-50
33. Yang, B.C.: Two Types of Multiple Half-Discrete Hilbert-Type Inequalities. Lambert Academic Publishing, Germany (2012)
34. Kuang, J.C.: Introduction to Real Analysis. Hunan Education Press, Chansha (1996)
35. Pan, Y.L., Wang, H.T., Wang F.T.: On Complex Functions. Science Press, Beijing (2006)
36. Kuang, J.C.: Applied Inequalities. Shangdong Science Technic Press, Jinan (2004)
37. Alladi, K., Milovanovic, G.V., Rassias, M.T. (eds.): Analytic Number Theory, Approximation Theory and Special Functions. Springer, New York, in print.
38. Apostol, T.M.: Introduction to Analytic Number Theory. Springer, New York (1984)
39. Edwards, H.M.: Riemann's Zeta Function. Dover Publications, New York (1974)
40. Erdos, P., Suranyi, J.: Topics in the Theory of Numbers. Springer , New York (2003)
41. Hardy, G.H., Wright, E.W.: An Introduction to the Theory of Numbers, 5th edn. Clarendon Press, Oxford (1979)
42. Iwaniec, H., Kowalski, E.: Analytic Number Theory, vol. 53. American Mathematical Society, Colloquium Publications, Rhode Island (2004)
43. Landau, E.: Elementary Number Theory, 2nd edn. Chelsea, New York (1966)
44. T, R.M.: Problem–Solving and Selected Topics in Number Theory : in the Spirit of the Mathematical Olympiads (Foreword by Preda Mihailescu). Springer. New York (2011)
45. Miller, S.J., Takloo – Bighash, R.: An Invitation to Modern Number Theory. Princeton University Press, Princeton (2006)
46. Zhong, Y.Q.: On Complex Functions. Higher Education Press, Beijing (2004)

On the Extension Problems of Isometric and Nonexpansive Mappings

Xiuzhong Yang and Xiaopeng Zhao

AMS 2000 Subject Classifications: *46B04, 46B20, 46A16, 46E40.*

Abstract We present a survey of ideas and results stemming from the Tingley's problem: Let X and Y be real normed spaces with unit spheres $S_1(X)$ and $S_1(Y)$, respectively. Suppose that $V_0 : S_1(X) \to S_1(Y)$ is a surjective isometry. Is V_0 necessarily the restriction to $S_1(X)$ of a linear or affine isometry on X? Two more related isometric extension problems for nonsurjective isometry or nonexpansive mapping are also proposed and studied.A series of important results are outlined and some open problems are raised.

Keywords Isometric extension · Tingley problem · Banach space · Strictly convex · Uniformly convex · Bochner integral · F-space · β-normed space

1 Introduction

Isometries are, in the most general sense, transformations which preserve distance between elements. Such transformations are basic in the study of geometry which is concerned with rigid motions and properties preserved by them.

Definition 1.1 *if (E, d_E) and (F, d_F) are two metric spaces, then a mapping V from E to F is said to be an isometry if $d_F(Vx, Vy) = d_E(x, y)$ for all $x, y \in E$.*

The study of isometries has been a hot topic for the mathematicians in functional analysis as well as metric geometry. Most of the results on isometry have been included in books by Banach [4] and Fleming and Jamison [22, 23]. In particular,

X. Yang (✉)
College of Mathematics and Information Science, Hebei Normal University, and Hebei Key Laboratory of Computational Mathematics and Applications, 050024 Shijiazhuang, Hebei, P.R. China
e-mail: xiuzhongyang@126.com

X. Zhao
Department of Mathematics, Zhejiang University, 310027 Hangzhou, Zhejiang, P.R. China
e-mail: zhaoxiaopeng.2007@163.com

T. M. Rassias, P. M. Pardalos (eds.), *Mathematics Without Boundaries*,
DOI 10.1007/978-1-4939-1106-6_24,

a series of important papers dealt with "the Aleksandrov problem for unit distance preserving mappings", "the extension of Mazur and Ulam theorem", " the stability of isometry", etc. The interested reader can consult: Th. M. Rassias [51–56], Th. M. Rassias and S. Xiang [59–62], Th. M. Rassias and P. Šemrl [57], Th. M. Rassias and C. S. Sharma [58], C.-G. Park and Th. M.Rassias [44–50], S.-M. Jung and Th. M. Rassias [28, 29], B. Mielnik and Th. M. Rassias [43], K. Ciesielski and Th. M. Rassias [6].

In this paper our attention is focused on the study of extension of isometries defined on unit spheres. First we recall some well-known result and the history of research field with which we are concerned.

In 1932, Mazur and Ulam [42] proved that an isometry of one normed space onto another which carries θ to θ is linear. In 1972, Mankiewicz [41] generalized Mazur–Ulam theorem and showed that an isometry which maps an open connected subset of a normed space X onto an open subset of another normed space Y can be extended (uniquely) to an affine (i.e. a linear transformation composed with a translation) isometry from X to Y.

Applying the above result to the unit balls of X and Y, it follows that the whole spaces X and Y are linear isometric if and only if their unit balls are isometric, that is, in the isometry theory sense, the unit ball determines the space. In 1983, Tingley [66] showed that isometries between unit spheres of finite-dimensional Banach spaces necessarily map antipodal points to antipodal points. Furthermore, Tingley proposed the following problem:

Problem 1.2 *(Surjective Isometric Extension Problem, SIEP) Let X and Y be real normed spaces with unit spheres $S_1(X)$ and $S_1(Y)$, respectively. Suppose that $V_0 : S_1(X) \to S_1(Y)$ is a surjective isometry. Is V_0 necessarily the restriction to $S_1(X)$ of a linear or affine isometry on X?*

It is the so-called isometric extension problem, also called the Tingley problem. But D. Tingley only proved that: for finite-dimensional spaces X and Y, if V_0 is an isometric mapping from the unit sphere $S_1(X)$ onto $S_1(Y)$, then $V_0(-x) = -V_0(x)$, for all $x \in S_1(X)$. By a similar method Y. Ma [38] obtained the same conclusion for the infinite-dimensional strictly convex normed spaces and for the space l^1. In the complex spaces, for example, we take $X = Y = C$ and $V_0(x) = \overline{x}$; the answer of Tingley's problem on such a space is negative. Hence, if there is no other assumption, then throughout this paper the spaces involved are real spaces.

Tingley's problem has been extensively studied since its appearance, and many important results have been obtained. It is surprising to see that for the two concrete classic normed spaces of the same type, the answer to the Tingley problem is almost affirmative, but for the general normed spaces, it is still an open problem. Actually, without the assumption of surjectivity, we can immediately give a negative answer from the following classical counterexamples:

Example 1. 3 *Define an isometric mapping $V_0 : S_1\left(l^1_{(2)}\right) \to S_1\left(l^1_{(3)}\right)$ as follows:*

$$V_0(x_1, x_2) = \begin{cases} (0, x_1, x_2), & if\ x_1 < 0, \\ (x_1, 0, x_2), & if\ x_1 \geq 0. \end{cases}$$

for each $(x_1, x_2) \in S_1(l^1_{(2)})$.

Then V_0 *can be isometrically extended to the whole space by the natural way, however, it can not have any linear extension.*

Zhang [83] gave the following counterexample to demonstrate that there exists the nonsurjective isometry between the unit spheres which can not be extended to a linear isometry.

Example 1.4 *An isometry* V_0 *from the unit sphere of* $l^\infty_{(2)}$ *into the unit sphere of* $l^\infty_{(3)}$ *is defined by:*

$$V_0(x_1,x_2) = \begin{cases} (1, \frac{3}{4}x_2, x_2), & if\ x_1 = 1, x_2 \geq 0; \\ (-1, x_2, \frac{3}{4}x_2), & if\ x_1 = -1, x_2 \geq 0; \\ (x_1, 1 - \frac{1}{4}x_1, 1), & if\ x_1 \geq 0, x_2 = 1; \\ (x_1, 1, 1 + \frac{1}{4}x_1), & if\ x_1 < 0, x_2 = 1; \\ (x_1, x_2, x_2), & if\ x_2 < 0. \end{cases}$$

for each $(x_1, x_2) \in S_1(l^\infty_{(2)})$.

It is worthy to study the Tingley problem for nonsurjective isometric operators. In 2004, the following nonsurjective isometric extension problem appeared in [67].

Problem 1.5 *(Nonsurjective Isometric Extension Problem, NSIEP) Let* X *and* Y *be real normed spaces with unit spheres* $S_1(X)$ *and* $S_1(Y)$*, respectively. Suppose* $V_0 : S_1(X) \to S_1(Y)$ *is a nonsurjective isometry. Is* V_0 *necessarily for the restriction to* $S_1(X)$ *of a linear or affine isometry on* X*?*

Recently more and more people are interested in the study on the extension of 1-Lipschitz mapping between the metric spaces.

Definition 1.6 *Let* (E, d_E) *and* (F, d_F) *be two metric spaces. A mapping* V *from* E *to* F *is called 1-Lipschitz (or nonexpansive) if* $d_F(Vx, Vy) \leq d_E(x, y)$ *for all* $x, y \in E$. V *is called anti-1-Lipschitz if* $\leq$ *is replaced by* $\geq$.

Nonexpansive mapping is one of the key concepts in the metric fixed point theory, but what we are most interested in is the following problem:

Problem 1.7 *(Problem of isometric extension of nonexpansive mapping) Let* X *and* Y *be real normed spaces with unit spheres* $S_1(X)$ *and* $S_1(Y)$*, respectively. Suppose that* $V_0 : S_1(X) \to S_1(Y)$ *is a nonexpansive mapping. Is* V_0 *necessarily the restriction to* $S_1(X)$ *of a linear or affine isometry on* X*?*

Throughout this paper, our notation and terminology are standard. However, for the convenience of the reader we recall some notation. $S_1(E) = \{x : x \in E, \|x\| = 1\}$ denotes the unit sphere of normed spaces, *supp*$(f) = \{x : f(x) \neq 0\}$ denotes support of function f. Vector valued function spaces $E = L_p(\Omega, \Sigma, \mu), F = L_p(\Omega_1, \Sigma_1, \nu; X)$. are denoted respectively by $E = L_p(\mu)$ and $F = L_p(\nu, X)$, where $1 < p \neq 2$ and X is a normed space. $L_p(\nu, X)$ stands for the space of all (equivalent classes of) X-valued Bochner integrable functions f defined on Ω_1 with $\int_{\Omega_1} \|f\|^p d\nu < \infty$. The interested reader may also consult [5, 7–9, 32, 40, 63].

In this survey paper, we will present the results on the SIEP, the NSIEP and the problem of isometric extension of nonexpansive mapping, as well as propose some open problems.

2 On the SIEP Between Spaces of Same Type

Up to now, many mathematical workers have been studying the SIEP, and have obtained many important results. First, they were interested in the SIEP for the Banach spaces of the same type. Later, with the solution of the such a problem, they were most interested in the SIEP for the Banach spaces of different types. In the following two sections we will summarize recent studies on the SIEP and some results that they have obtained.

R. Wang was the first to solve the SIEP between spaces of the same type; he dealt with the $C_0(\Omega)$. Let Ω be a locally compact Hausdorff space and $C_0(\Omega)$ denote the space of all continuous functions, which vanish at infinity (i.e. for each $\epsilon > 0$, $\{\omega : |f(\omega)| \geq \epsilon\}$ is compact). R. Wang [70] gave the following theorem:

Theorem 2.1 *The unit spheres of $C_0(X)$ and $C_0(Y)$ are isometric if and only if X and Y are homoeomorphic; and each isometry between the unit spheres of $C_0(X)$ and $C_0(Y)$ is necessarily the restriction of an isometry between $C_0(X)$ and $C_0(Y)$ where one part is linear and the other part is conjugate-linear.*

In [71–76], R. Wang obtained a series of good results which showed that the answers to the SIEP are positive for the the following types of spaces: $C_0(\Omega, \sigma)$-type spaces [71], $C_0(\Omega, E)$-type spaces [72], $C_0^{(n)}(X)$-type spaces[74], certain l^1-sum spaces of $C_0(\Omega, E)$-type spaces [73], certain l^p-sum spaces of $C_0(\Omega, E)$-type spaces [75] and l^1-sum spaces of strictly convex normed spaces [76].

D. Zhan [81] first studied the SIEP between vector valued spaces by proving the following:

Theorem 2.2 *If (Ω, μ) is σ-finite measure space and X is a strictly convex normed space, then the isometric operator between the unit spheres of $L^1(\Omega, X)$ can be linearly extended to the whole spaces.*

In [82], he also studied the isometric extension problem on $L^p(\Omega, X)$ (where $0 < p < 1$, and X is strictly convex space) and got positive answers. Y. Xiao and R. Wang [77] discussed the isometric extension problem of the AM-spaces. First they gave the representation theorem of any isometry V_0 from the unit sphere of an AM-space onto the unit sphere of another AM-space, then as an application they show that there exists a real linear isometry V such that V_0 is a restriction of V.

G. Ding in [12–14] first derived the following representation theorem of onto isometric mappings between two unit spheres of two $l^p(\Gamma)$ $(1 \leq p \leq \infty)$-type spaces They are not trivial generalizations of the corresponding theorems given in [4, Sect. 5, Chap. 11], because these isometric mappings are only defined on the unit spheres but not defined on the whole space as a linear operator. As an application, he can show that the isometric operator between the unit spheres of such spaces can be linearly isometrically extended to the whole spaces.

Theorem 2.3 *Suppose that V_0 is a surjective isometric mapping from the unit sphere $S_1[l^p(\Gamma)]$ onto the unit sphere $S_1[l^p(\Delta)]$ for $1 \leq p \leq \infty$and $p \neq 2$, then there exists a permutation π from Δ onto Γ and a set of real numbers $\{\theta_\delta\}_{\delta\in\Delta}$ with $|\theta_\delta| = 1$ for* *all* $\delta \in \Delta$ *such that*

$$V_0(x) = \sum_{\delta \in \Delta} \theta_\delta \cdot \xi_{\pi(\delta)} d_\delta, \ \forall x \in \sum_{\gamma \in \Gamma} \xi_\gamma e_\gamma \in S_1[l^p(\Gamma)]$$

Furthermore, as an application, he can easily prove the following theorem:

Theorem 2.4 *Suppose that* V_0 *is a surjective isometric mapping from the unit sphere* $S_1[l^p(\Gamma)]$ *onto the unit sphere* $S_1[l^p(\Delta)]$ *for* $1 \leq p \leq \infty$ *and* $p \neq 2, V_0$ *can be extended to a real linear isometry from* $l^p(\Gamma)$ *onto* $l^p(\Delta)$.

Ding's results are important because he relived an important method of dealing with the SIEP between spaces of same type, that is: if you can give representation forms of onto isometric mappings between two unit spheres of two specific spaces, then there are no difficulties in solving it.

G. An [2] characterized the surjective isometries between the unit spheres of $(l_{\beta_n})(0 < \beta_n < 1, n \in N)$-type spaces which have the same standard base as the the space $l^\beta : e_n = \{\xi_i\}, \xi_i = 1, if\ i = n; = 0, if\ i \neq n.$

Definition 2.5 *Let* $0 < \beta_n < 1$ *for all integers n. The space* l^{β_n} *consists of all real sequences* $x = \{\xi_n\}$ *for which* $\sum_{n=1}^{\infty} |\xi_n|^{\beta_n}$ *is finite, with the F-norm* $\|x\| := \sum_{n=1}^{\infty} |\xi_n|^{\beta_n}$.

She proved the following representation theorem:

Theorem 2.6 *Let* $0 < \beta_n < 1$ *for all integers n. An operator* V_0 *is an isometric mapping from the unit sphere* $S_1(l^{\beta_n})$ *onto* $S_1(l^{\beta_n})$ *if and only if there exists a sequence* $\{\theta_n\}$ *of signs and a permutation* π *of the integers, such that* $\beta_n = \beta_{\pi(n)}$ *for any integer n; and* $V_0(x) = \sum_{n=1}^{\infty} \theta_n \xi_{\pi(n)} e_n$ *for any element* $x = \sum_{n=1}^{\infty} \xi_n e_n \in S_1(l^{\beta_n})$

As an application, she proved:

Theorem 2.7 *Let* $0 < \beta_n < 1$ *for all integers n. If* V_0 *is an isometric mapping from the unit sphere* $S_1(l^{\beta_n})$ *onto* $S_1(l^{\beta_n})$, *then* V_0 *can be extended to a real linear isometry from* (l^{β_n}) *onto* (l^{β_n})

Z. Hou and H. Xu [27] dealt with the case of abstract L^p-spaces, gave the representation theorem of the surjective isometric operator between the unit spheres of the L^p type spaces for $p \geq 1$, and then the SIEP holds true. In short, more than 20 years passed since Tingley proposed the isometric extension problem. Tingley's problem has almost been solved on specific classical Banach spaces. To conclude this section, I would like to propose the following open problem:

Problem 2.8 *Let* $E = L^p(\mu, X), F = L^p(\nu, Y)(1 \leq p \leq \infty)$ *be vector valued function spaces,* X, Y *be normed spaces. Does the SIEP between E and F hold true?*

3 On the SIEP Between Banach Spaces of Different Types

G. Ding was the first to consider the SIEP between differently typical spaces. In [11] he studied the problem between $S_1(E)$ and $S_1(C(\Omega))$. The main result is as follows:

Theorem 3.1 *Let E be a normed space, the subset $sm[S_1(E)]$ of the smooth points of the unit spheres of E be dense in its unit sphere $S_1(E)$. Suppose that $V_0 : S_1(E) \to S_1(C(\Omega))$ is an onto isometry (where Ω is a compact Hausdorff space). If*

$$\|V_0x - |\lambda|V_0x_0\| \le \|x - |\lambda|x_0\|, \ \ \forall\lambda \in \mathbf{R}, \tag{1}$$

for all $x \in S_1(E)$ and $x \in sm[S_1(E)]$, then V_0 can be extended to linear isometry on the whole space.

The inequality (1) is important. G. An [1] used a similar method to obtain the affirmative to the SIEP problem on the real reflexive space and its adjoint space under the same condition (1). The results ware generalized by X. Fang and J. Wang [19] and X. Yang, Z. Hou and X. Fu [80], who obtained the following result: Let E and F be real normed spaces. Suppose that $V : S_1(E) \to S_1(F)$ is onto isometry. If for each $x, x_0 \in S_1(E)$,

$$\|Vx - |\lambda|Vx_0\| \le \|x - |\lambda|x_0\|, \ \ \forall\lambda \in \mathbf{R}, \tag{2}$$

then V can be extended to linear isometry on the whole space. In fact, we [80] not only established the relationship between isometric mapping and nonexpansive mapping but also generalized the preceding results to a more general setting such as β-normed spaces (see [9]); we also got corresponding results on into isometry. Let us present our results [80] which generalized the corresponding results in [1, 11, 19, 30].

Theorem 3.2 *Let E, F be normed linear spaces, B be a bounded closed subset of E, and $\theta \in intB$. If $V : E \to F$ is nonexpansive mapping, and the restriction of V in the boundary of B $V|_{\partial B}$ is an isometric mapping, then*

(1) V must be an isometry on B.
(2) If V is positively homogeneous mapping, then V must be isometric mapping from E to F.
(3) If V is both positively homogeneous and surjective or F is strictly convex space, then V must be linear isometric operator.

Proof (1) First, it is shown that for any $y \in \text{int}B, x \in \partial B$ the equality

$$\|Vx - Vy\| = \|x - y\|.$$

holds. In fact, if there exist $x_0 \in \partial B, y_0 \in \text{int}B$ such that $\|Vx_0 - Vy_0\| < \|x_0 - y_0\|$. Let $z(t) = y + t(y - x)$, since $y \neq x$, and B bounded set, there must exist $t_1 > 0$ such that $z(t_1) \notin B$. Take $t_0 = \sup\{t > 0 : z(t) \in \text{int}\,B\}$, then $z_0 =: z(t_0)$ must be on the boundary of B . Thus

$$\|x_0 - z_0\| = \|Vx_0 - Vz_0\|$$

$$\leq \|Vx_0 - Vy_0\| + \|Vy_0 - Vz_0\|$$
$$< \|x_0 - y_0\| + \|y_0 - z_0\| = \|x_0 - z_0\|.$$

This is a contradiction.

Second, we show that if $y, z \in \operatorname{int} B$, then $\|Ty - Tz\| = \|y - z\|$.

In fact, Using a similar method, there exists a point $x \in \partial B$ such that $y \in [x, z]$. Here, $[x_1, x_2]$ denotes the linear interval which connecting x_1 and x_2 . Hence

$$\begin{aligned}\|x - y\| + \|y - z\| &= \|x - z\| \\ &= \|Vx - Vz\| \text{ (by the result above)} \\ &\leq \|Vx - Vy\| + \|Vy - Vz\| \\ &\leq \|x - y\| + \|y - z\|,\end{aligned}$$

So we get

$$\|Ty - Tz\| = \|y - z\|.$$

(2) Suppose that V is positively homogeneous. For any $x, y \in E$, considering $\theta \in \operatorname{int}B$, we can take λ such that $\lambda x, \lambda y \in B$, from which we get

$$\|V(\lambda x) - V(\lambda y)\| = \|\lambda x - \lambda y\| = \lambda\|x - y\|.$$

From the positive homogeneity of V, we get $\|Vx - Vy\| = \|x - y\|$, i.e., V is isometry.

(3) From Mazur–Ulam's theorem [42] and Baker's theorem [3] the conclusion holds.

Theorem 3.3 *Let E, F normed spaces, $V_0 : S_1(E) \to S_1(F)$ be an isometry, and for any $x, y \in S_1(E)$ we have*

$$\|V_0x - |\lambda|V_0y\| \leq \|x - |\lambda|y\|, \quad \forall\lambda \in \mathbf{R}.$$

Then V_0 can be extended to be an isometry on the whole space.

In addition, if V_0 is surjective, or F is strictly convex, then V_0 can be extended to be an linear isometry on the whole space.

Proof We define $\widetilde{V_0} : E \to F$ as follows

$$\widetilde{V_0}(x) = \begin{cases} \|x\|V_0\left(\frac{x}{\|x\|}\right), & x \neq \theta; \\ \theta, & x = \theta. \end{cases}$$

then $\|\widetilde{V_0}(x)\| = \|x\|$, and $\widetilde{V_0}$ positively homogeneous mapping. Notice that

$$\begin{aligned}&\|\widetilde{V_0}(x) - \widetilde{V_0}(y)\| \\ &= \left\|\|x\|V_0\left(\frac{x}{\|x\|}\right) - \|y\|V_0\left(\frac{y}{\|y\|}\right)\right\|\end{aligned}$$

$$
\begin{aligned}
&=\|x\|\left\|V_0\left(\frac{x}{\|x\|}\right)-\frac{\|y\|}{\|x\|}V_0\left(\frac{y}{\|y\|}\right)\right\| \\
&\leq\|x\|\left\|\frac{x}{\|x\|}-\frac{\|y\|}{\|x\|}\frac{y}{\|y\|}\right\| \\
&=\|x-y\|.
\end{aligned}
$$

Hence $\widetilde{V_0}$ is nonexpansive and positively homogeneous mapping, and the conclusion holds from Theorem 3.2.

Remark 3.4 *Our result is more general than that of* [19], *and method of proof is simpler than that of* [19]. *Our result solves the problem of extension of into or onto isometry which satisfies certain conditions, while the results in* [19] *concern the case of surjective isometry. Hence, their results are a special case of ours.*

The preceding theorem cannot be extended to general β-normed($0 < \beta < 1$), but can be extended special β- normed spaces.

In [39], In order to extend Baker's theorem to β-normed spaces, the author introduced the notion of β-strict convexity.

Definition 3.5 [39] *Let $(E, \|\cdot\|)$ be a β-normed space. E is said to be β-strictly convex, if $\|x+y\|^{\frac{1}{\beta}} = \|x\|^{\frac{1}{\beta}} + \|y\|^{\frac{1}{\beta}}$. Then there exists $\alpha > 0$ such that $x = \alpha y$.*

She proved the following results:

Lemma 3.6 [39] *$(E, \|\cdot\|)$ is a strictly convex normed space if and only if $(E, \|\cdot\|^{\beta})$ is β-strictly convex β- normed space.*

Lemma 3.7 [39] *Let E, F be two real β-normed spaces, and F be a β-strictly convex β-normed space. Then any isometric mapping from E to F is an affine.*

In β-normed spaces,we have the following results:

Theorem 3.8 *Let E, F be β-strictly convex β-normed spaces, and B be bounded closed set of E, and $\theta \in int B$. If $V : E \to F$ is nonexpansive mapping, and the restriction of V on the boundary of B denoted by $V|_{\partial B}$ is an isometry, then*

(1) V is an isometry on B.
(2) If V is of positive homogeneity, then V must be linear isometric mapping from E to F.

Theorem 3.9 *Let E, F be β-strictly convex β-normed spaces, and $V_0 : S_1(E) \to S_1(F)$ be isometric mapping such that*

$$\|V_0x - |\lambda|V_0y\| \leq \|x - |\lambda|y\|, \forall\lambda \in \mathbf{R},$$

then V_0 can be extended to be a linear mapping on the whole space.

Remark 3.10 *The study on the SIEP between spaces of different types is more difficult than that of same type, because there is no fixed method to follow. Most results hold true under certain conditions which seem too strong. The above results we have given can provide one method of research on the SIEP between spaces of different types.*

X. Fang and J. Wang [20] obtained a better result which deal with E and $C(\Omega)$:

Theorem 3.11 *Let E be a real normed space, and $C(\Omega)$ be real Banach spaces and Ω be a compact metric space. Suppose $T : S_1(E) \to S_1(C(\Omega))$ is an onto isometry. Then T can be linearly isometrically extended to the whole space E.*

In [19], Fang and Wang dealt with the case of $l^1(\Gamma)$ space:

Theorem 3.12 *Let E be a real normed space, and $l^1(\Gamma)$ be also real, where Γ is an index set. Let T be an isometric mapping from $S_1(E)$ onto $S_1(l^1(\Gamma))$. Then T can be extended to a real linear isometry from E to $l^1(\Gamma)$.*

R. Liu[34] dealt with the case for $C(\Omega)$ and real Banach spaces E:

Theorem 3.13 *Let Ω be a compact Hausdorff space, $C(\Omega)$ be the space of all real-valued continuous functions on, and E be a real Banach space. Let $V_0 : S_1(C(\Omega)) \to S_1(E)$ be a surjective isometry. Then V_0 can be extended to be a linear isometry of $C(\Omega)$ onto E.*

In the same paper, R. Liu also proved with special technique the following theorem:

Theorem 3.14 *Let E be a Banach space and Γ a nonempty index set. Assume that V is an isometric embedding from $S_1(E)$ into $S_1(l_\infty(\Gamma)$ such that for any $x \in S(E)$, $\lambda \in [-2, 2]$, and $\gamma \in \Gamma$, $Vx + \lambda e_\gamma \in V[S_1(E)]$ whenever $\|Vx + \lambda e_\gamma\| = 1$. Then there exists a positively homogeneous isometry embedding V from E into $l_\infty(\Gamma)$ such that $V|_{S(E)} = V$.*

As an application, he obtained a short proof of the main theorem which is a generalization of the main results in [16, 26, 34].

Theorem 3.15 *(See* [16, 26, 34]*) Let E be a Banach space, and let F be an $\mathcal{L}^\infty(\Gamma)$-type space for a nonempty index set Γ. Assume that V is a surjective isometry from $S(E)$ onto $S(F)$. Then there is a linear surjective isometry $\tilde{V}$ from E onto F such that $\tilde{V}|_{S(E)} = V$.*

Recently, D. Tan [65] studied SIEP between $(L_p(\mu)$ and a Banach space E and obtained a positive answer to Tingley's problem in this case:

Theorem 3.16 *Let E be a Banach space, and let $V : S_1(L_p(\mu)) \to S_1(E)$ be a surjective isometry where $1 < p < \infty, p \neq 2$. Then V is the restriction of a linear isometry from $L_p(\mu)$ onto E. In particular, E is linearly isometric to $L_p(\mu)$ if and only if their unit spheres are isometric to each other.*

We would like to propose the following open problem:

Problem 3.17 *Let $E = L_p(\nu, E)$ (where E is a any strictly convex space), F be any normed spaces. Is the answer to the SIEP is positive or negative?*

4 The NSIEP Between the Banach Spaces

In 2004, Wang Jian [67] proved that every onto isometry between the unit spheres of atomic AL_p-spaces ($0 < p < \infty, p \neq 2$) can be extended to be a linear isometry on the whole spaces as well as give the characteristic that an into isometry $V : S(E) \to S(F)$ can be extended to a linear isometry $\tilde{V} : E \to F$. Many people turned their attention to the study on the NSIEP between the Banach spaces. A series of results were obtained.

X. Yang [79] completely solved the SIEP and NSIEP between $E = L_p(\Omega, \Sigma, \mu)$ and $F = L_p(\nu, H)$ (where $1 < p \neq 2, H$ is a Hilbert space); when F is another Banach space, he obtained sufficient conditions under which an isometric mapping from unit sphere of $L_p(\mu)$ into (or onto) that of another Banach space can be extended to be a linear isometry on the whole space. Therefore, he generalized the corresponding results of [67]. We will introduce his results and method.

Throughout this section, we always assume that $1 < p \neq 2$ and orthogonal relation in the meaning of the following Definition 4.1.

Definition 4.1 *Let E be a normed space. An element x is said to be orthogonal to an element y, denoted by* $x \perp y$ *if* $\|x + y\|^p + \|x - y\|^p = 2(\|x\|^p + \|y\|^p)$.

Definition 4.2 *Let V be a mapping between two Banach spaces E and F,* $\perp$ *be an orthogonal relation in the meaning of Definition 4.1. If for any* $x, y \in E$ *with* $x \perp y$ *such that*

(1) $V(x + y) = Vx + Vy$, then V is called an orthogonally additive mapping.
(2) $V(ax + by) = aVx + bVy$, where $|a|^p + |b|^p = 1, 1 < p \neq 2$, then V is called a p-affinely orthogonal mapping.

Remark 4.3 In general, that $V : E \to F$ is an isometric mapping and that V is a p-affinely orthogonal mapping are not equivalent. It is clear that p-affinely orthogonal mapping need not be isometric mapping. For reverse implication we consider the mapping $V : \mathbf{R} \to l_2^\infty(\mathbf{R}^2$ with the max norm) given by

$$V(x) = \begin{cases} (x, x), & if \;\; x \geq 0, \\ (x, -x), & if \;\; x < 0. \end{cases}$$

Then V is an isometry, but is clearly not a p-affinely orthogonal mapping. The example implies that isometric mapping need not be p-affinely orthogonal mapping.

Lemma 4.4 [10] *Let E be a strictly convex normed linear space and F a normed linear space. If* V_0 *is a mapping from the unit sphere* $S(E)$ *of E to the unit sphere* $S(F)$ *of F such that* $-V_0\{S(E)\} \subseteq V_0\{(S(E)\}$ *and for any* $x, y \in S(E)$, $\|V_0(x) - V_0(y)\| \leq \|x - y\|$ *holds, then* V_0 *is one-to-one and* $V_0(-x) = -V_0(x)$.

Lemma 4.5 *Let* $f, g \in S(L_p(\mu))$*, then*
$f \perp g \Leftrightarrow supp(f) \cap supp(g)$ *is a null set.*
More generally, we have

Lemma 4.6 *Let* $f, g \in S(L_p(\nu, H))$*, then*
$f \perp g \Leftrightarrow supp(f) \cap supp(g)$ *is a null set.*

Proof Sufficiency is obvious. We only need to prove the necessity.
Case 1. $2 < p < +\infty$. By Hölder's inequality and the characterization of inner

product space, we have

$$\begin{aligned}\|f+g\|_p^p+\|f-g\|_p^p &= \int_{\Omega_1}\left(\|f+g\|^p+\|f-g\|^p\right)d\nu\\ &\geq 2^{1-\frac{p}{2}}\int_{\Omega_1}\left(\|f+g\|^2+\|f-g\|^2\right)^{p/2}d\nu\\ &= 2\int_{\Omega_1}\left(\|f\|^2+\|g\|^2\right)^{p/2}d\nu\\ &\geq 2\int_{\Omega_1}\left(\|f\|^p+\|g\|^p\right)d\nu,\end{aligned}$$

which implies that if the assumption holds, then the last inequality becomes equality, and $f(\omega)=0$ or $g(\omega)=0$. Therefore $supp(f)\cap supp(g)$ is a null set.

Case 2. $1<p<2$. We only need to replace each "$\geq$" by "$\leq$" in each of the above inequalities, then repeat method for Case 1 to finish the proof.

Theorem 4.7 *Let $E=L_p(\mu)$, F be any Banach space, and let $V_0: S(E)\to S(F)$ be an isometric mapping. If V_0 is a p-affinely orthogonal mapping, then there exists a linear isometric mapping V from E to F such that $V|_{S(E)}=V_0$.*

First, we give the following lemma, which can be easily proved by induction.

Lemma 4.8 *Let $E=L_p(\mu)$, F be any Banach space, and let $V_0: S(E)\to S(F)$ be a p-affinely orthogonal mapping. Assume that there exist $(f_i)_{i=1}^n\subseteq S(L_p(\mu))$ such that if $i\neq j$, then $f_i\perp f_j$. Then for any $(a_i)_{i=1}^n$ with $\sum_{i=1}^n|a_i|^p=1$, we have*

$$V_0\left(\sum_{i=1}^n a_i f_i\right)=\sum_{i=1}^n a_i V_0(f_i).$$

Lemma 4.9 *Suppose that $f=\sum_{i=1}^n a_i\chi_{E_i}\in L_p(\mu)$ is a simple function, where $\{E_i\}\in\Sigma$ are pairwise disjoint measurable sets, and $\bigcup_{i=1}^n E_i=\Omega$. Define*

$$V(f)=\begin{cases}0, & if\quad f=0,\\ \|f\|V_0\left(\frac{f}{\|f\|}\right), & if\quad f\neq 0.\end{cases}$$

Then we have

$$Vf=\sum_{i=1}^n a_i V\left(\chi_{E_i}\right). \tag{3}$$

Proof By the definition of V and Lemma 4.8, we have

$$V\left(\sum_{i=1}^{n} a_i \chi_{E_i}\right) = \left\|\sum_{i=1}^{n} a_i \chi_{E_i}\right\| V_0\left(\frac{\sum_{i=1}^{n} a_i \chi_{E_i}}{\|\sum_{i=1}^{n} a_i \chi_{E_i}\|}\right)$$
$$= \left\|\sum_{i=1}^{n} a_i \chi_{E_i}\right\| \left(\frac{\sum_{i=1}^{n} a_i \|\chi_{E_i}\|}{\|\sum_{i=1}^{n} a_i \chi_{E_i}\|}\right) V_0\left(\frac{\chi_{E_i}}{\|\chi_{E_i}\|}\right)$$
$$= \sum_{i=1}^{n} a_i V\left(\chi_{E_i}\right).$$

Hence we have completed the proof of the lemma.

The following lemma is well known and can be proved easily.

Lemma 4.10 *F is strictly convex if and only if $L_p(\nu, F)$ is strictly convex.*

Lemma 4.11 *Let $E = L_p(\mu)$, F be any Banach space. Assume that the isometric mapping $V_0 : S(E) \to S(F)$ is a p-affinely orthogonal mapping. Let V be a mapping defined by Lemma 4.9. Then for any two measurable functions $f, g \in L_p(\mu)$, we have*

$$V(f+g) = Vf + Vg. \tag{4}$$

Proof First we show (4) holds for any two simple functions $f, g \in L_p(\mu)$. Set $f = \sum_{i=1}^{n} a_i \chi_{E_i}$, $g = \sum_{j=1}^{m} b_i \chi_{F_j}$, where $\{E_i\}, \{F_j\}$ are pairwise disjoint measurable sets, respectively, $\bigcup_{i=1}^{n} E_i = \Omega, \bigcup_{j=1}^{m} F_j = \Omega$. Let $E_{i,j} = E_i \cap F_j, i, j = 1, 2, \cdots$. Then

$$V(f+g) = V\left(\sum_{i=1}^{n}\sum_{j=1}^{m} a_i \chi_{E_{ij}} + \sum_{i=1}^{n}\sum_{j=1}^{m} b_j \chi_{E_{ij}}\right)$$
$$= V\left(\sum_{i=1}^{n}\sum_{j=1}^{m} \left(a_i + b_j\right) \chi_{E_{ij}}\right)$$
$$= \sum_{i=1}^{n}\sum_{j=1}^{m} (a_i + b_j) V(\chi_{E_{ij}})$$
$$= \sum_{i=1}^{n}\sum_{j=1}^{m} a_i V\left(\chi_{E_{ij}}\right) + \sum_{i=1}^{n}\sum_{j=1}^{m} b_j V\left(\chi_{E_{ij}}\right)$$

$$= V\left(\sum_{i=1}^{n}\sum_{j=1}^{m} a_i \chi_{E_{ij}}\right) + V\left(\sum_{i=1}^{n}\sum_{j=1}^{m} b_j \chi_{E_{ij}}\right)$$
$$= V(f) + V(g).$$

Second, we prove (4) holds for any two measurable functions $f, g \in L_p(\mu)$.

For any $f, g \in L_p(\mu)$, there exist simple function sequences f_n, g_n such that $\lim_{n\to\infty} f_n = f$, $\lim_{n\to\infty} g_n = g$. We can easily obtain

$$\lim_{n\to\infty}\left\|\frac{f_n}{\|f_n\|} - \frac{f}{\|f\|}\right\|_p = 0,\ \lim_{n\to\infty}\left\|\frac{g_n}{\|g_n\|} - \frac{g}{\|g\|}\right\|_p = 0, \text{ and } \lim_{n\to\infty}\left\|\frac{f_n+g_n}{\|f_n+g_n\|} - \frac{f+g}{\|f+g\|}\right\|_p = 0.$$

Set $f_n = \sum_{i=1}^{m_n} a_{n_i}\chi_{E_{n_i}}$, $g_n = \sum_{j=1}^{k_n} b_{n_j}\chi_{F_{n_j}}$, where $\{E_{n_i}\}, \{F_{n_j}\}$ are pairwise disjoint measurable sets respectively, $\bigcup_{i=1}^{m_n} E_{n_i} = \Omega$, $\bigcup_{j=1}^{k_n} F_{n_j} = \Omega$. Then we have

$$V(f+g) = \|f+g\| V_0\left(\frac{f+g}{\|f+g\|}\right)$$
$$= \|f+g\| V_0 \lim_{n\to\infty}\left(\frac{f_n+g_n}{\|f_n+g_n\|}\right)$$
$$= \|f+g\| \lim_{n\to\infty} V_0\left(\frac{f_n+g_n}{\|f_n+g_n\|}\right)$$
$$= \|f+g\| \lim_{n\to\infty} V\left(\frac{\sum_{i=1}^{m_n} a_{n_i}\chi_{E_{n_i}} + \sum_{j=1}^{k_n} b_{n_j}\chi_{F_{n_j}}}{\|f_n+g_n\|}\right)$$
$$= \|f+g\| \left\{\lim_{n\to\infty} V\left(\frac{\sum_{i=1}^{m_n} a_{n_i}\chi_{E_{n_i}}}{\|f_n+g_n\|}\right) + \lim_{n\to\infty} V\left(\frac{\sum_{j=1}^{k_n} b_{n_j}\chi_{F_{n_j}}}{\|f_n+g_n\|}\right)\right\}$$
$$= \|f+g\| \left\{\lim_{n\to\infty} \frac{\sum_{i=1}^{m_n} a_{n_i} V(\chi_{E_{n_i}})}{\|f_n+g_n\|} + \lim_{n\to\infty} \frac{\sum_{j=1}^{k_n} b_{n_j} V(\chi_{F_{n_j}})}{\|f_n+g_n\|}\right\}$$
$$= \lim_{n\to\infty} V\left(\sum_{i=1}^{m_n} a_{n_i}\chi_{E_{n_i}}\right) + \lim_{n\to\infty} V\left(\sum_{j=1}^{k_n} b_{n_j}\chi_{F_{n_j}}\right)$$
$$= \lim_{n\to\infty} V(f_n) + \lim_{n\to\infty} V(g_n)$$
$$= \lim_{n\to\infty} \|f_n\| V_0\left(\frac{f_n}{\|f_n\|}\right) + \lim_{n\to\infty} \|g_n\| V_0\left(\frac{g_n}{\|g_n\|}\right)$$

$$= \|f\| V_0 \left(\frac{f}{\|f\|} \right) + \|g\| V_0 \left(\frac{g}{\|g\|} \right)$$
$$= V(f) + V(g).$$

Now we come to the proof of Theorem 4.7. Define mapping $V : L_p(\mu) \to F$ just as in Lemma 4.9. Then V is a homogenous mapping by definition and Lemma 4.4, and an additive mapping by Lemma 4.11, thus we complete the proof of Theorem 4.7. From Theorem 4.7, we know that problem on the extension of isometries between unit spheres of two Banach spaces is closely related to p-affinely orthogonal mapping; we give the following lemma to demonstrate under what condition can the isometry between unit spheres become a p-affinely orthogonal mapping.

Lemma 4.12 *Let $E = L_p(\mu)$, and F be a Banach space such that each isometric mapping $T : S(l_p^{(2)}) \to S(F)$ can be uniquely represented as*

$$T(ae_1 + be_2) = aT(e_1) + bT(e_2),$$

where $e_1 = (1,0), e_2 = (0,1), |a|^p + |b|^p = 1$. Then any isometric mapping $V_0 : S(E) \to S(F)$ must be a p-affinely orthogonal mapping.

Proof Let $f, g \in S(E)$ and $f \perp g$ then f and g are linear independent. Define $T_0 : span\{f, g\} \to l_p^{(2)}$ by $T_0(af + bg) = ae_1 + be_2$, then T_0 must be a surjective linear isometry. Set $U = V_0 T_0^{-1}$, then U is an isometric mapping from $S(l_p^{(2)})$ to $S(F)$. By assumption, U must have the property: $U(ae_1 + be_2) = aU(e_1) + bU(e_2)$, which implies that

$$V_0(af + bg) = aV_0(f) + bV_0(g), \quad |a|^p + |b|^p = 1.$$

Theorem 4.13 *Let $E = L_p(\mu)$, F be a Banach space such that each isometric mapping $V : S(l_p^{(2)}) \to S(F)$ can be uniquely represented as*

$$V(ae_1 + be_2) = aV(e_1) + bV(e_2),$$

where $e_1 = (1,0), e_2 = (0,1), |a|^p + |b|^p = 1$. Suppose that $V_0 : S(L_p(\mu)) \to S(F)$ be an isometric mapping. Then there exists a linear isometric mapping $\tilde{V}$ from E to F such that $\tilde{V}|_{S(E)} = V_0$.

Proof By Lemma 4.12 and Theorem 4.7, we know that the theorem holds.

Theorem 4.14 *If $F = L_p(\nu, H)$, then each isometric mapping $V_0 : S\left(l_p^{(2)}\right) \to S(F)$ can be uniquely represented as*

$$V_0(ae_1 + be_2) = aV_0(e_1) + bV_0(e_2),$$

where $e_1 = (1,0), e_2 = (0,1), |a|^p + |b|^p = 1$.

Lemma 4.15 *Let $V_0 : S(L_p(\mu)) \to S(L_p(\nu, H))$ be an isometric mapping. Then for any $f, g \in S(L_p(\mu))$, $\quad f \perp g \Leftrightarrow V_0(f) \perp V_0(g)$.*

Proof By Lemma 4.4, we get $V_0(-f) = -V_0(f)$ and

$$\|V_0(f) + V_0(g)\| = \|V_0(f) - V_0(-g)\| = \|f + g\|,$$

from which the lemma holds. Now we come to the proof of Theorem 4.14. Let $f = V_0(e_1), g = V_0(e_2)$ where $e_1 = (1,0), e_2 = (0,1)$. Since $e_1 \perp e_2$, by Lemma 4.15, we get $f \perp g$. Take measurable sets A and B such that $A \cap B = \emptyset, A \cup B = \Omega_1, supp(f) \subseteq A$ and $supp(g) \subseteq B$. Write $h = V_0(ae_1 + be_2) = h\chi_A + h\chi_B$. By Lemma 4.4 we get

$$V_0(sgn(a), 0) = sgn(a)f, \quad V_0(0, sgn(b)) = sgn(b)g.$$

$$\|sgn(a)f - h\|_p^p = \|V_0(sgn(a), 0) - V_0(a,b)\|_p^p = \|(sgn(a), 0) - (a,b)\|_p^p = (1 - |a|)^p + |b|^p.$$

Similarly $\|sgn(b)g - h\|_p^p = |a|^p + (1 - |b|)^p$. From the definition of norm we also have

$$\|sgn(a)f - h\|_p^p = \int_A \|sgn(a)f - h\|^p d\nu + \int_B \|h\|^p d\nu.$$

$$\|sgn(b)g - h\|_p^p = \int_A \|h\|^p d\nu + \int_B \|sgn(b)g - h\|^p d\nu.$$

Set $\alpha = \left(\int_A \|h\|^p d\nu\right)^{1/p}, \quad \beta = \left(\int_B \|h\|^p d\nu\right)^{1/p}$, then $\alpha^p + \beta^p = 1$ and

$$\begin{aligned}(1-\alpha)^p + \beta^p &= \left\{\left(\int_A \|sgn(a)f\|^p d\nu\right)^{1/p} - \left(\int_A \|h\|^p d\nu\right)^{1/p}\right\}^p + \int_B \|h\|^p d\nu \\ &\leq \int_A \|sgn(a)f - h\|^p d\nu + \int_B \|h\|^p d\nu \\ &= (1 - |a|)^p + |b|^p.\end{aligned}$$

Since the function $u(t) = t^p - (1-t)^p$ is strictly increasing on the interval $[0, 1]$, we have $\alpha \geq |a|$, and $\beta \geq |b|$ by the inequality above. Moreover, $\alpha^p + \beta^p = 1$ and $|a|^p + |b|^p = 1$, thus $\alpha = |a|, \beta = |b|$. Finally, we show that $h|_A = af$. Since the space $L_p(\nu, H)$ is strictly convex (Lemma 4.10), every point outside the unit sphere has a unique best approximation from the unit ball. Take an element $sgn(a)f$, which has a best approximation element af from sphere with radius $|a|$.

$$\begin{aligned}\|sgn(a)f - h|_A\|_p &= \left(\int_A \|sgn(a)f - h|_A\|^p d\nu\right)^{1/p} \\ &= \left(\int_{\Omega_1} \|sgn(a)f - h\|^p d\nu - \int_B \|h\|^p d\nu\right)^{1/p}\end{aligned}$$

$$= \left\{\|V_0(sgn(a),0) - V_0(a,b)\|^p - |b|^p\right\}^{1/p}$$
$$= 1 - |a|$$
$$= \|sgn(a)f(1 - |a|)\|_p$$
$$= \|af - sgn(a)f\|_p.$$

From the uniqueness of best approximation, we have $h|_A = af$. Similarly, we can show that $h|_B = bg$, thus the proof of the lemma is complete.

Theorem 4.16 *Let $E = L_p(\mu), F = L_p(\nu, H)$. If $V_0 : S(E) \to S(F)$ is any isometric mapping, then there exists a linear isometric mapping V from E to F such that $V|_{S(E)} = V_0$.*

Proof By Theorems 4.13 and 4.14, we complete the proof. To conclude, we propose the following problem:

Problem 4.17 *In Theorem 4.16, if we replace the image space with $F = L_p(\nu, X)$, where X is a uniformly convex Banach space or strictly convex space, does the theorem hold?*

From the proof of Theorem 4.14, under additional conditions on $L_p(\nu, X)$, we have the following theorem.

Theorem 4.18 *Let $E = L_p(\mu), F = L_p(\nu, X)$, where X is a strictly convex space, satisfying: for $f, g \in S\left(L_p(\nu, X)\right)$, then $f \perp g \Leftrightarrow supp(f) \cap supp(g)$ is a null set. If $V_0 : S(E) \to S(F)$ is any isometric mapping, then there exists a linear isometric mapping V from E to F such that $V|_{S(E)} = V_0$.*

For the NSIEP, many people devoted to study the spaces such that l^1, $L^1(\mu)$, $\mathcal{L}^\infty(\Gamma)$-type space, etc. Many important results about different spaces have been obtained by G. Ding, X. Fang, R. Liu, etc.

G. Ding [15] studied the NSIEP between spaces $l^\infty_{(2)}$ and $L^1(\mu)$, and gave some sufficient conditions such that the NSIEP hold true.

Theorem 4.19 *Let V_0 be an isometric mapping from the unit sphere $S_1(l^\infty_{(2)})$ into $S(L(T, \mu))$. Then V_0 can be extended to a linear isometry defined on the whole space $l^\infty_{(2)}$ if and only if the following conditions hold:*

$$(1)\dim span V_0\left[S_1(l^\infty_{(2)})\right] = 2;\;\; (2) - V_0\left[S_1(l^\infty_{(2)})\right] \subset V_0\left[S_1(l^\infty_{(2)})\right].$$

This result was improved by R. Wang [68] who proved the following:

Theorem 4.20 *Let V_0 be an isometric mapping from the unit sphere $S_1\left(l^\infty_{(n)}\right)$ into $S_1(E)$. Then V_0 can be extended to a linear isometry defined on the whole space $l^\infty_{(n)}$ if and only if the following conditions hold:*

$$(1)\dim span V_0\left[S_1(l^\infty_{(n)})\right] = n;\;\; (2) - V_0\left[S_1(l^\infty_{(n)})\right] \subset V_0\left[S_1(l^\infty_{(n)})\right].$$

Theorem 4.21 *Let V_0 be an isometric mapping from the unit sphere $S_1(c_0)$ into$S_1(E)$. Then V_0 can be extended to a linear isometry defined on the whole space*

c_0 if and only if the following conditions hold: (i)$\{V_0(e_i)\}_{i=1}^{\infty}$ is a Schauder base of $\overline{span V_0(S(c_0))}$; (ii) $-V_0S_1(c_0) \subset V_0S_1(c_0)$.

In [25], X. Fu dealt with the NSIEP between real normed space and $l^\infty(\Gamma)$ space:

Theorem 4.22 *Let E be a real normed space, and $l^\infty(\Gamma)$ (Γ is an index set) be also real. Let V_0 be an isometric mapping from $S_1(E)$ into $S_1(l^\infty(\Gamma))$ such that:*
(1) $\{e_\gamma\}_{\gamma\in\Gamma} \subset V_0[S_1(E)]$;
(2) For any $y_1, y_2 \in V_0[S_1(E)]$, and any $\lambda, \mu \in R$,

$$\|\lambda y_1 + \mu y_2\| = 1 \Rightarrow \lambda y_1 + \mu y_2 \in V_0[S_1(E)].$$

Then V_0 can be linearly and isometrically extended to the whole space.

In [17], G. Ding considered the general case and dealt with the NSIEP between $L^1(T,\mu)$ and Banach space, obtained the following theorem:

Theorem 4.23 *Let E be a Banach space. Let V_0 be an isometric mapping from $S_1(L^1(T,\mu))$ into $S(E)$, then V_0 can be extended to a real linear isometry on the whole $(L^1(T,\mu))$ if and only if: for every $x_1, x_2 \in S_1(L^1(T,\mu))$ and $\xi_1, \xi_2 \in R$,*

$$\|\xi_1 V_0(x_1) + \xi_2 V_0(x_2)\| = 1 \Rightarrow \xi_1 V_0(x_1) + \xi_2 V_0(x_2) \in S_1(L^1(T,\mu)).$$

Moreover, if V_0 is surjective, then by the theorem above V_0 can be extended to be a linear isometry from $(L^1(T,\mu))$ onto E.

R. Liu [35] studied the NSIEP between the spaces $c_0(\Gamma)$ and $l^\infty(\Delta)$ and gave some sufficient conditions such that the NSIEP hold true.

Theorem 4.24 *If V_0 is an isometry from the unit sphere $S_1(c_0(\Gamma))$ into $S_1(l^\infty(\Delta))$ and $(\xi_1 V_0(x_1) + \xi_2 V_0(x_2))/\|\xi_1 V_0(x_1) + \xi_2 V_0(x_2)\| \in S_1(c_0(\Gamma))$. Whenever $x_1, x_2 \in S_1(c_0(\Gamma))$ and $\xi_1, \xi_2 \in R$, such that $\xi_1 V_0(x_1) + \xi_2 V_0(x_2) \neq 0$, then V_0 can be extended to be a linear isometry on the whole space.*

Furthermore, G. Ding defined a class of sequence spaces. Given a nonempty set Γ, he considered the space of all bounded real-valued functions on Γ was equipped with the sup norm. Any closed subspace containing all $e'_\gamma s(\gamma \in \Gamma)$ is called the $\mathcal{L}^\infty(\Gamma)$-type space. For example, the spaces $l^\infty(\Gamma), c(\Gamma)$ and $c_0(\Gamma)$ (in particular, l^∞, c and c_0, etc. are the $\mathcal{L}^\infty(\Gamma)$-type space (or$\mathcal{L}^\infty(\Gamma)$, in brief).

G. Ding [16] obtained some conditions under which the answer to the NSIEP between $\mathcal{L}^\infty(\Gamma)$-type space and a Banach space E is positive. He got the following theorem:

Theorem 4.25 *Let E be a Banach space and V_0 be an isometric mapping from the unit sphere $S_1(\mathcal{L}^\infty(\Gamma))$ into the unit sphere $S_1(E)$. Then V_0 can be extended to a linear isometry on the whole space $\mathcal{L}^\infty(\Gamma)$ if and only if the following conditions hold:*

(i) $x_1, x_2 \in S_1(\mathcal{L}^\infty(\Gamma))$ and $\xi_1, \xi_2 \in R$,

$$\|\xi_1 V_0(x_1) + \xi_2 V_0(x_2)\| = 1 \Rightarrow \xi_1 V_0(x_1) + \xi_2 V_0(x_2) \in S_1(\mathcal{L}^\infty(\Gamma));$$

(ii) For every mutual disjoint subsets $\{\Gamma_1, \Gamma_2 \cdots \Gamma_n\}$ of Γ, if $V_0(x) = \sum_{k=1}^n \lambda_k V_0(\chi_{\Gamma_k})$, then $x = \sum_{k=1}^n \lambda'_k (\chi_{\Gamma_k}) + x_0$, where $supp\, x_0 \subset \left(\cup_{k=1}^n \Gamma_k\right)^c$.

In [34], R. Liu also proved the isometric extension problem for the $C(\Omega)$-space.

Theorem 4.26 *Let Ω be a compact Hausdorff space, $C(\Omega)$ be the space of all real-valued continuous functions on Ω and E a real Banach space. Let $V_0 : S_1(C(\Omega)) \to S_1(E)$ be an isometry. Then V_0 can be extended to a linear isometry of $C(\Omega)$ into E if and only if the following condition holds: $x_1, x_2 \in S_1(C(\Omega))$ and $\xi_1, \xi_2 \in R$,*

$$\|\xi_1 V_0(x_1) + \xi_2 V_0(x_2)\| = 1 \Rightarrow \xi_1 V_0(x_1) + \xi_2 V_0(x_2) \in S_1(C(\Omega))$$

Theorem 4.27 *Let E be a Banach space and Γ a nonempty set. Assume that V is an isometric embedding from $S(E)$ into $S(l^\infty(\Gamma))$ such that there is an $\mathcal{L}^\infty(\Gamma)$-type space F satisfying that $V[S(E)] = S(F)$. Then there exists a linear isometric embedding $\tilde{V}$ from E into $l^\infty(\Gamma)$ (onto F) such that $\tilde{V}|_{S(E)} = V$.*

R. Liu [33] studied the NSIEP between l^β-direct sums of strictly convex normed spaces $(0 < \beta \leq 1)$. He called $x \perp y$ if $suppx \cap suppy = \emptyset$., where the $suppx$ is the support of $x = (x_i)$which can be defined by the set $\{i; i \in \Gamma, x_i \neq 0\}$. He proved the following theorem which was the generalization of the corresponding theorem in [76].

Theorem 4.28 *Let $0 < \beta \leq 1$. $V_0 : S_1(E) \to S_1(F)$ is a nonsurjective isometry and $-V_0(S_1(E)) \subset V_0(S_1(E))$. If for all $x_\gamma \in S_1(E_\gamma)$, $y \in S_1(E)$ with $x_\gamma \perp y$, such that $\lambda_1 V_0(x_\gamma) + \lambda_2 V_0(y) \in V_0(S(E))$, where $\lambda_1, \lambda_2 \in R$ and $|\lambda_1|^p + |\lambda_2|^p = 1$, then V_0 can be linearly extended to an isometry from E into F.*

In [31] L. Li and W. Ren defined the space $\mathcal{L}^\infty(T, \Sigma, \mu)$ and obtained the following theorem which was the generalization of Theorem 4.25.

Theorem 4.29 *Let E be a normed linear space. Let V_0 be an isometric mapping from $S_1(\mathcal{L}^\infty(T, \Sigma, \mu))$ into $S(E)$. If V_0 satisfies that:*
(1) For any $x_1, x_2 \in S_1(\mathcal{L}^\infty(T, \Sigma, \mu))$ and $\lambda, \mu \in R$,

$$\|\lambda V_0(x_1) + \mu V_0(x_2)\| = 1 \Rightarrow \lambda V_0(x_1) + \mu V_0(x_2) \in V_0\left[S_1(\mathcal{L}^\infty(T, \Sigma, \mu))\right];$$

(2) For any disjoint elements $x_1$and $x_2 \in S_1(\mathcal{L}^\infty(T, \Sigma, \mu))$, we have

$$\dim(span V_0[S_1(span\{x_1, x_2\})]) = 2;$$

then V_0 can be extended to be a linear isometry on $\mathcal{L}^\infty(T, \Sigma, \mu)$.

5 On Linearly Isometric Extensions for Nonexpansive Mappings Between Unit Spheres

G. Ding [10] first discussed the isometric extension problem between Hilbert spaces without the assumption of the surjectivity, and he showed that a 1-Lipschitz mapping between the unit spheres of two Hilbert spaces can be extended to a real linear isometry of the whole space. X. Fu [24] generalized Ding's result to a local 1-Lipschitz (that is, there exists $\delta > 0$ such that $\|V_0(x_1) - V_0(x_2)\| \leq \|x_1 - x_2\|$if $\|x_1 - x_2\| \leq \delta$).

In [21], X. Fang generalized the result from Hilbert space to l^p spaces. He proved the following theorem:

Theorem 5.1 *Let T be a mapping from the unit sphere $S_1(l^p(\Gamma))$ into $S_1(l^p(\Delta))$. If T is a 1-Lipschitz mapping such that $-T[S_1(l^p(\Gamma))] \subset T[S_1(l^p(\Gamma))]$, then T can be linearly isometrically extended to the whole space for $p > 2$; if T is injective and the inverse mapping T^{-1} is a 1-Lipschitz mapping, then T can be extended to be a linear isometry from $l^p(\Gamma)$ into $l^p(\Delta)$ for $1 < p \leq 2$.*

Then G. Ding [18] generalized the result to general AL_p spaces.

Theorem 5.2 *If T is a 1-Lipschitz mapping between unit spheres of two ALp spaces with $p > 2$ and $-T[S_1(L^p(\Gamma))] \subset T[S_1(L^p(\Gamma))]$, then T can be extended to a linear isometry defined on the whole space. If $1 < p < 2$ and T is an anti $1 - Lipschitz$ mapping, then T can also be linearly and isometrically extended.*

In [64], D. Tan established the relationship between surjective nonexpansive mapping and surjective isometry for the spaces $\mathcal{L}^\infty(\Gamma)$ and $\mathcal{L}^\infty(\Delta)$, her result can be summarized as follow:

Theorem 5.3 *Let $V_0 : S_1(\mathcal{L}^\infty(\Gamma)) \to S_1(\mathcal{L}^\infty(\Delta))$ be a surjective nonexpansive mapping. Then V_0 must be an surjective isometry and there exists a family of real numbers $\{\theta_\delta\}_{\delta\in\Delta}$ with $|\theta_\delta| = 1$ and a bijection $\sigma : \Delta \to \Gamma$ such that*

$$V_0(x)(\delta) = \theta_\delta x(\sigma(\delta)), \forall x \in S_1(\mathcal{L}^\infty(\Gamma)), \forall \delta \in \Delta.$$

From the above theorem, she easily derived:

Theorem 5.4 *Let $V_0 : S_1(\mathcal{L}^\infty(\Gamma)) \to S_1(\mathcal{L}^\infty(\Delta))$ be a surjective nonexpansive mapping, then V_0 can be extended to be a linear isometry defined on the whole space $\mathcal{L}^\infty(\Gamma)$.*

It is well known that the modulus of convexity of Banach spaces X is defined by

$$\delta_X(\epsilon) = \inf\left\{1 - \frac{\|x + y\|}{2} : x, y \in S_1(X), \|x - y\| = \epsilon\right\} \ for \ \epsilon \in [0, 2].$$

X is uniformly convex, if $\delta_X(\epsilon) > 0$ for all $\epsilon \in [0, 2]$.

In [69], R. Wang introduced the concept of convex modulus set of X by using the definition of the modulus of convexity of Banach spaces as follows:

Definition 5.5 *Let X be a Banach space, the set $A \subset S_1(X)$ is said to be a convex modulus set of X, if for asny $\epsilon \in [0, 2]$ and any $\eta > 0$, there exists $x_{\epsilon,\eta}, y_{\epsilon,\eta} \in A$, such that*

$$\|x_{\epsilon,\eta} - y_{\epsilon,\eta}\| = \epsilon \ and \ 1 - \frac{\|x_{\epsilon,\eta} - y_{\epsilon,\eta}\|}{2} < \delta_X(\epsilon) + \eta.$$

He gave the following key lemma which demonstrated the relationship between convex modulus set and isometry between unit spheres:

Lemma 5.6 *Let X, Y be a Banach spaces, X being uniformly convex, and V_0 be a mapping from the unit sphere $S_1(X)$ into $S_1(Y)$. If $-V_0[S_1(X)] \subset V_0[S_1(X)]$, $V_0[S(X)]$ is a convex modulus set of X, and*

$$\|V_0(x_1) - V_0(x_2)\| \leq \|x_1 - x_2\|, \ \forall x_1, x_2 \in S_1(X),$$

then Y is also a uniformly convex Banach space and $\delta_X(\epsilon) \leq \delta_Y(\epsilon)$ for all $\epsilon \in [0, 2]$.

From the lemma, R. Wang easily got his result.

Theorem 5.7 *Let X be a Banach space, H be a Hilbert space, V_0 be a mapping from the unit sphere $S_1(H)$ into $S_1(X)$. If $-V_0[S_1(H)] \subset V_0[S_1(H)]$, $V_0[S(H)]$ is a convex modulus set of X, and*

$$\|V_0(x_1) - V_0(x_2)\| \leq \|x_1 - x_2\|, \ \forall x_1, x_2 \in S_1(H),$$

then V_0 can be extended linearly to the whole space.

In the same paper, R. Wang also introduced a new concept called semiconvex modulus set, and proved a more general result as follows:

Theorem 5.8 *Let X be a Banach space, H be a Hilbert space, V_0 be a mapping from the unit sphere $S_1(H)$ into $S_1(X)$. If $-V_0[S_1(H)] \subset V_0[S_1(H)]$, $V_0[S(H)]$ is a semiconvex modulus set of X, and*

$$\|V_0(x_1) - V_0(x_2)\| \leq \|x_1 - x_2\|, \ \forall x_1, x_2 \in S_1(H).$$

Then V_0 can be extended linearly to the whole space.

R. Liu [36] proved the following:

Theorem 5.9 *Let E and F be two Banach spaces. Suppose that $V_0 : S_1(E) \to S_1(F)$ is a 1-Lipschitz mapping: $\|V_0(x) - V_0(y)\| \leq \|x - y\|$ for any $x, y \in S_1(E)$. Assume also that V_0 is a surjective mapping such that for any $x, y \in S_1(E)$. and $r > 0$, we have*

$$\|V_0(x) - rV_0(y)\| \wedge \|V_0(x) + rV_0(-y)\| \leq \|x - ry\|$$

and $\|V_0(x) - V_0(-x)\| = 2$. Then V_0 can be extended to be a linear isometry of E onto F.

Remark 5.10 *Most of the results on the extension of the nonexpansive mapping must satisfy the condition: $-V_0[S_1(H)] \subset V_0[S_1(H)]$. This is a very strong demand. So to delete or weaken the condition is a direction of further research. There are still other topics such as X. Yang* [78] *dealt with the extension of $(\lambda, \psi, 2)$-isometries on the unit spheres; R. Liu* [37] *dealt with extension of approximate isometries between unit spheres, etc. Although results in this direction are not as rich as that of SIEP and NSIEP, we are sure more and more results will occur in the future.*

Acknowledgement The authors express their thanks to Prof. Themistocles M. Rassias for his suggestions in revising the paper. The first author was supported by NNSF of China (11371119) and NSF of Education Department of Hebei Province (Z2014031).

References

1. An, G.: On extension of isometries between the unit spheres. Acta. Math. Sin. Chin. Ser. **47** (4), 653–656 (2004)
2. An, G.: Isometries on unit spheres of (l^{β_n}). J. Math. Anal. Appl. **301**, 249–254 (2005)
3. Baker, J.A.: Isometries in normed spaces. Am. Math. Mon. **78**, 655–658 (1971)
4. Banach, S.: Theorie des operations lineares. AMS Chelsea Publishing, Warsaw (1932)
5. Benyamini, Y., Lindenstrauss, J.: Geometric Nonlinear Functional Analysis, vol. 1 (American Mathematical Society Colloquium Publications, 48). American Mathematical Society, Providence (2000)
6. Ciesielski, K., Rassias T.M.: On some properties of isometric mappings. Facta Univ. Ser. Math. Inform. **7**, 107–115 (1992)
7. Day, M.M.: Normal Linear Spaces. Springer-Verlag, Berlin (1958)
8. Diestel, J., Uhl, J.J.: Vector Measures. Mathematical Surveys and Monographs, vol. 15. American Mathematical Society (1977)
9. Ding, G.G.: The Selected Topics on Topological Linear Spaces. Guangxi Education Press, Nanning (1987)
10. Ding, G.G.: The 1-Lipschitz mapping between the unit spheres of two Hilbert spaces can be extended to a real linear linear isometry of the whole space. Sci. China Ser. A **45**(4), 479–483 (2002)
11. Ding, G.G.: On extension of isometries between unit spheres of E and $C(\Omega)$. Acta. Math. Sin. Engl. Ser. **19**(4), 793–800 (2003)
12. Ding, G.G.: The isometric extension problem in the unit spheres of $l^p(\Gamma)(p > 1)$ type spaces. Sci. China Ser. A **46**(3), 333–338 (2003)
13. Ding, G.G.: The representation theorem of onto isometric mappings between two unit spheres of $l^1(\Gamma)$ type spaces and the application to the isometric extension problem . Acta. Math. Sin. Engl. Ser. **20**(6), 1089–1094 (2004)
14. Ding, G.G.: The representation theorem of onto isometric mappings between two unit spheres of l^∞-type spaces and the application on isometric extension problem. Sci. China Ser. A **47**(5), 722–729 (2004)
15. Ding, G.G.: The isometric extension of an into mapping from the unit sphere $S(l^\infty_{(2)})$ to $S(L^1(\mu))$. Acta. Math. Sin. Engl. Ser. **22**(6), 1721–1724 (2006)
16. Ding, G.G.: The isometric extension of the into mapping from the $\mathcal{L}^\infty(\Gamma)$ -type space to some Banach space E.Illinois J. Math. **51**(2), 445–453 (2007)
17. Ding, G.G.: Extension of isometries on the unit sphere of AL-space. Sci China Ser. A **38**(5), 541–555 (2008)
18. Ding, G.G.: On linearly isometric extensions for 1-Lipschitz mappings between unit spheres of AL_p-spaces $(p > 2)$. Acta. Math. Sin. Engl. Ser. **26**(2), 331–336 (2010)
19. Fang, X.N., Wang, J.H.: On linear extension of isometries between the unit spheres. Acta Math. Sin. Chin. Ser. **48**(6), 1109–1112 (2005)
20. Fang, X.N., Wang, J.H.: On extension of isometries between the unit spheres of normed space E and $C(\Omega)$. Acta Math. Sin. Engl. Ser. **22**(6) , 1819–1824 (2006)
21. Fang, X.N., Wang, J.H.: Extension of isometries between unit spheres of normed space E and $l^1(\Gamma)$. Acta Math. Sin. Chin. Ser. **51**(1), 24–28 (2008)
22. Fleming, R.J., Jamison, J.E.: Isometries on Banach Spaces: Function Spaces. Monographs and Survey in Pure and Applied Mathematics, vol. 129. Chapman & Hall/CRC (2003)
23. Fleming, R.J., Jamison, J.E.: Isometries on Banach Spaces: Vector-valued Function Spaces. Monographs and Survey in Pure and Applied mathematics, vol. 138. Chapman & Hall/CRC (2007)
24. Fu, X.: A note on isometric extension of unit spheres in Hilbert spaces. Acta Math. Sin. Chin. Ser. **45**, 1247–1248 (2002)

25. Fu, X: Isometries on the spaces. Acta Math. Sci. **26B**(3), 502–508 (2006)
26. Fu, X.: The isometric extension of the into mapping from the unit sphere $S(E)$ to $S(l^{\infty}(\Gamma))$. Acta Math. Sin. Engl. Ser. **24**(9), 1475–1482 (2008)
27. Hou, Z., Xu, H.: The isometric representation and the extension of the mapping between the unit spheres of $L_p(T, \Sigma, \mu)$ type spaces. Acta Sci. Nat. Univ. Nankai **39**(6), 75–84 (2006)
28. Jung, S.M.: On distance-preserving mappings, R.T.M. J. Korean Math. Soc. **41**(4), 667–680 (2004)
29. Jung, S.M., Rassias, T.M.: Mappings preserving two distances. Nonlinear Funct. Anal. Appl. **10**(5), 717–723 (2005)
30. Li, L.: Extension of isometries between unit sphere. Acta Math. Sin. Chin. Ser. **48**(6), 1105–1108 (2005)
31. Li, L., Ren, W.Y.: On extension of isometries between unit spheres of L^{∞} and E. Quaest. Math. **31**, 209–218 (2008)
32. Lindenstrauss, J., Tzafriri, L.: Classical Banach spaces II: Function spaces, Ergebnisse der Mathematik und ihrer Grenzgebiete, vol. 97. Springer, Berlin (1979)
33. Liu, R.: Isometries between the unit spheres of l^{β}-sum of strictly convex normed spaces. Acta Math. Sin. Chin. Ser. **50**, 227–232 (2007)
34. Liu, R.: On extension of isometries between the unit sphere of $\mathcal{L}^{\infty}$ -type space and some Banach space E. J. Math. Anal. Appl. **333**, 959–970 (2007)
35. Liu, R.: The discussion and application of the non-surjective isometries between the unit spheres of two l^{∞}-spaces. Acta Math. Sci. **23A**(3), 385–391 (2007)
36. Liu, R: 1-Lipschitz mappings between unit spheres of Banach spaces. Acta Math. Sin. Chin. Ser. **50**(5), 1064–1070 (2007)
37. Liu, R: On extension of isometries and approximate isometries between unit spheres. J. Math. Anal. Appl. **352**(2), 749–761 (2009)
38. Ma, Y.: Isometries of the unit sphere. Acta Math. Sci. Engl. Ed. **12**(4), 366–373 (1992)
39. Ma, Y.: The Aleksandrov problem for unit distance preserving mapping. Acta Math. Sci. Ser. B Engl. Ed. **20**(3), 359–364 (2000)
40. Ma, Y.: On the Aleksandrov problem of isometric mapping. J. Math. Res. Expo. **23** (4), 623–630 (2003)
41. Mankiewicz, P.: On extension of isometries in normed linear spaces. Bull. Acad. Polon. Sci. Ser. Sci. Math. Astronomy Phys. **20**, 367–371 (1972)
42. Mazur, S., Ulam, S.: Surless transformations isometriques d'espaces vectoriels normes. Competes Rendus Acad. Sci. Paris **194**, 946–948 (1932)
43. Mielnik, B., Rassias, T.M.: On the Aleksandrov problem of conservative distances. Proc. Am. Math. Soc. **116**, 1115–1118 (1992)
44. Park, C.G., Rassias, T.M.: The N-isometric isomorphisms in linear n-normed C*-algebras. Acta Math. Sin. Engl. Ser. **22**(6) 1863–1890 (2006)
45. Park, C.G, Rassias, T.M.: Isometries on linear n-normed spaces. J. Inequalities Pure Appl. Math. **7**(5), (Article 168, 7pp) (2006)
46. Park, C.G., Rassias T.M.: Additive isometries on Banach spaces. Nonlinear Funct. Anal. Appl. **11**(5), 793–803 (2006)
47. Park, C.G., Rassias T. M.: Inequalities in additive N-isometries on linear N-normed Banach spaces. J. Inequalities Appl. Article ID 70597, Vol. 2007, 12 p (2007)
48. Park, C.G., Rassias, T.M.: Isometric additive mappings in quasi-Banach spaces. Nonlinear Funct. Anal. Appl. **12**(3), 377–385 (2007)
49. Park, C.G., Rassias T.M.: Isometric additive mappings in generalized quasi- Banach spaces. Banach J. Math. Anal. **2**(1), 59–69 (2008)
50. Park, C.G., Rassias, T.M.: d-isometric linear mappings in linear d-normed Banach modules. J. Korean Math. Soc. **45**(1), 249–271 (2008)
51. Rassias, T.M.: Is a distance one preserving mapping between metric spaces always an isometry? Am. Math. Mon. **90**, 200 (1983)

52. Rassias, T.M.: Some remarks on isometric mappings. Facta Uni. Ser. Math. Inform. **2**, 49–52 (1987)
53. Rassias, T.M.: Properties of isometric mappings. J. Math. Anal. Appl. **235**(1), 108–121 (1999)
54. Rassias, T.M.: Isometries and approximate isometries. Intl. J. Math. Math. Sci. **25**(2), 73–91 (2001)
55. Rassias T. M: On the A.D. Aleksandrov problem of conservative distances and the Mazur-Ulam theorem. Nonlinear Anal. Theory Methods Appl. **47**(4), 2597–2608 (2001)
56. Rassias, T.M.: On the Aleksandrov problem for isometric mappings. Appl. Anal. Discret. Math. **1**, 18–28 (2007)
57. Rassias, T.M., Šemrl, P.: On the Mazur-Ulam theorem and the Aleksandrov problem for unit distance preserving mappings. Proc. Am. Math. Soc. **118**, 919–925 (1993)
58. Rassias, T.M., Sharma, C.S.: Properties of isometries. J. Natural Geom. **3**, 1–38 (1993)
59. Rassias, T.M., Xiang, S.: On mappings with conservative distances and the Mazur- Ulam theorem. Publ. Fac. Electr. Eng. Univ. Belgrade Ser. Math. **11**, 1–8 (2000)
60. Rassias, T.M., Xiang, S.: On Mazur-Ulam theorem and mappings which preserve distances. Nonlinear Funct. Anal. Appl. **5**(2), 61–66 (2000)
61. Rassias, T.M., Xiang, S.: On approximate isometries in Banach spaces. Nonlinear Funct. Anal. Appl. **6**(2), 291–300 (2001)
62. Rassias, T.M., Xiang, S.: On the stability of approximate isometries. Tamsui Oxf. J. Math. Sci. **18**(1), 45–56 (2002)
63. Rolewicz, S.: Metric Linear Spaces. Polish Scientific Publishers, Warsaw (1972)
64. Tan, D.: Nonexpansive mappings on the unit spheres of some Banach spaces. Bull. Aust. Math. Soc. **80**, 139–146 (2009)
65. Tan, D.: Extension of isometries on the unit sphere of L_p spaces. Acta Math. Sin. Engl. Ser. **28**(6), 1197–1208 (2012)
66. Tingley, D.: Isometries of the unit sphere. Geom. Dedicata **22**, 371–378 (1987)
67. Wang, J.: On extension of isometries between unit spheres of AL_p-spaces ($0 < p < \infty$). Proc. Amer. Math. Soc. **132**(10), 2899–2909 (2004)
68. Wang, R.D.: A remark on extension of into isometries. Acta Math. Sin. Engl. Ser. **26**(1), 203–208 (2010)
69. Wang, R.D.: On linear extension of 1-Lipschitz mapping from Hilbert space into a normed space. Acta Math. Sci. Ser. B Engl. Ed. **30B**(1), 161–165 (2010)
70. Wang, R.S.: Isometries between the unit spheres of $C_0(\Omega)$ type spaces. Acta Math. Sci. **14**, 82–89 (1994)
71. Wang, R.S.: Isometries of $C_0(\Omega, \sigma)$ type spaces. Kobe J. Math. **12**, 31–43 (1995)
72. Wang, R.S.: Isometries on the $C_0(\Omega, E)$ type spaces. J. Math. Sci. Univ. Tokyo **2**, 117–130 (1995)
73. Wang, R.S.: Isometries on the l^1-sum of $C_0(\Omega, E)$ type spaces. J. Math Sci. Univ. Tokyo **2**, 131–154 (1995)
74. Wang, R.S.: Isometries of $C_0^{(n)}(X)$. Hokkaido Math. J. **25** (3), 465–519 (1996)
75. Wang, R.S.: Isometries on the l_p-sum of $C_0(\Omega, E)$ type spaces. J. Math. Sci. Univ. Tokyo **3**, 471–493 (1996)
76. Wang, R., Orihara, A.: Isometries between the unit spheres of l_1-sum of strictly convex normed spaces. Acta. Sci. Nat. Univ. Nankai **35**(1), 38–42 (2002)
77. Xiao, Y.H., Wang, R.S.: On extension of isometries between unit spheres of abstract M spaces. Chin. Ann. Math. Ser. A **22**(1), 97–104 (2001) (in Chinese)
78. Yang, X.Z.: Extension of isometries and $(\lambda, \psi, 2)$-isometries on the unit spheres. Acta. Math. Sin. Chin. Ser. **49**, 1397–1402 (2006)
79. Yang, X.Z.: On extension of isometries between unit spheres of $L_p(\mu)$ and $L_p(\nu; H)(1 < p \neq 2;$ *H is a Hilbert space*). J. Math. Anal. Appl. **323**, 985–992 (2006)
80. Yang, X.Z., Hou, Z.B., Fu, X.H.: On linear extension of isometries between the unit spheres of β-normed spaces. Acta. Math. Sin. Chin. Ser. **48**(6), 1199–1202 (2005)

81. Zhan, D.P.: Extensions of isometries between unit spheres. Acta. Math. Sin. Chin. Ser. **41**(2), 275–280 (1998)
82. Zhan, D.P.: Extension of isometries between some p-normed spaces. Acta Math. Sci. **19**(5), 606–612 (1999) (Chinese)
83. Zhang, L.: On the extension problem of isometric operator between unit spheres of finite dimensional l^{∞}-spaces. Acta. Sci. Natur. Univ. Nankai **39**(3), 110–112 (2006)

Advances in Opial's Type Integral Inequalities

Chang-Jian Zhao and Wing-Sum Cheung

Abstract Opial's inequality and its generalizations, extensions and discretizations play a fundamental role in the study of existence and uniqueness of initial and boundary value problems for ordinary and partial differential equations as well as difference equations. Over the years, Opial's type integral inequalities have been receiving non-diminishing attention. In this article, we establish some new Opial's type integral inequalities which in special cases yield some existing results of Rozanova, Agarwal-Pang, Pachpatte, Das and Agarwal-Sheng, and provide new and handy tools to qualitative as well as quantitative analysis of solutions to differential equations.

Keywords Opial-type inequalities

1 Introduction

In 1960, Opial [23] established the following integral inequality.

Theorem 1.1 *Suppose $f \in C^1[0,h]$ satisfies $f(0) = f(h) = 0$ and $f(x) > 0$ for all $x \in (0,h)$. Then, the integral inequality*

$$\int_0^h \left|f(x)f'(x)\right| dx \leq \frac{h}{4}\int_0^h (f'(x))^2 dx, \tag{1}$$

holds, where the constant $\frac{h}{4}$ is best possible.

Opial's inequality and its generalizations, extensions, and discretizations play a fundamental role in establishing the existence and uniqueness of initial and boundary value problems for ordinary and partial differential equations as well as difference equations [2, 4, 7, 20, 22]. The inequality (1) has received considerable attention

C.-J. Zhao (✉)
Department of Mathematics, China Jiliang University, 310018, Hangzhou, People's Republic of China
e-mail: chjzhao@163.com

W.-S. Cheung
Department of Mathematics, The University of Hong Kong, Pokfulam Road, Hong Kong, China
e-mail: wscheung@hku.hk

T. M. Rassias, P. M. Pardalos (eds.), *Mathematics Without Boundaries*,
DOI 10.1007/978-1-4939-1106-6_25, © Springer Science+Business Media, LLC 2014

and a large number of papers dealing with new proofs, extensions, generalizations, variants, and discrete analogues of Opial's inequality have appeared in the literature [9–18, 21, 24–31, 34, 37–40, 42–45]. For an extensive survey on these inequalities, see [2, 22]. For Opial type integral inequalities involving high-order partial derivatives, see [1, 3, 5, 6, 19, 36].

The main purpose of the chapter is to introduce some new Opial-type integral inequalities. The results in special cases yield some of the recent results on Opial's type inequalities and provide some new and handy tools to the study of qualitative as well as quantitative properties of solutions to differential equations.

2 Opial-Rozanova's Type Inequality

A nontrivial generalization of Opial's inequality (1) was established by Hua [18] as follows.

Theorem A *Let $x(t)$ be absolutely continuous in $[0,a]$ with $x(0)=0$. Let l be a positive integer. Then*

$$\int_0^a |x(t)x'(t)|dt \leq \frac{a^l}{l+1}\int_0^a |x'(t)|^{l+1}dt.$$

A sharper inequality was established by Godunova [17] as follows.

Theorem B *Let $f(t)$ be convex and increasing on $[0,\infty)$ with $f(0)=0$, and $x(t)$ be absolutely continuous on $[0,\tau]$ with $x(\alpha)=0$ for some $\alpha\in[0,\tau]$. Then*

$$\int_\alpha^\tau f'(|x(t)|)|x'(t)|dt \leq f\left(\int_\alpha^\tau |x'(t)|dt\right).$$

Rozanova [31] further extended the above inequality and his result is embodied in the following:

Theorem C *Let $f(t)$ and $g(t)$ be convex and increasing functions on $[0,\infty)$ with $f(0)=0$, and let $p(t)\geq 0$, $p'(t)>0, t\in[\alpha,a]$ with $p(\alpha)=0$. Let $x(t)$ be absolutely continuous on $[\alpha,a)$ with $x(\alpha)=0$. Then*

$$\int_\alpha^a p'(t)\cdot g\left(\frac{|x'(t)|}{p'(t)}\right)\cdot\left[f'\left(p(t)\cdot g\left(\frac{|x(t)|}{p(t)}\right)\right)\right]dt$$
$$\leq f\left(\int_\alpha^a p'(t)\cdot g\left(\frac{|x'(t)|}{p'(t)}\right)dt\right).$$

The aim of this Section is to establish some Opial–Rozanova type inequalities involving higher-order partial derivatives, which generalize the series of results in Theorem A, B, and C.

Theorem 2.1 [41] *Let f and g be convex and increasing functions on $[0,\infty)$ with $f(0)=0$, and let $p(s,t)\geq 0$, $D_1D_2p(s,t)>0$, $p(s,\beta)=p(\alpha,t)=0$ for all*

$s \in [\alpha, a]$, $t \in [\beta, b]$, and $D_1 D_2 p(s,t)\mid_{t=\tau} = 0$. Let $x(s,t)$ be absolutely continuous on $[\alpha, a) \times [\beta, b]$ with $x(s,\beta) = x(\alpha,t) = 0$ for all $s \in [\alpha, a], t \in [\beta, b]$. Then

$$\int_\alpha^a \int_\beta^b D_1 D_2 p(s,t) \cdot g\left(\frac{|D_1 D_2 x(s,t)|}{D_1 D_2 p(s,t)}\right) \cdot \frac{\partial}{\partial t}\left[f\left(p(s,t) \cdot g\left(\frac{|x(s,t)|}{p(s,t)}\right)\right)\right] dtds$$

$$\leq f\left(\int_\alpha^a \int_\beta^b D_1 D_2 p(s,t) \cdot g\left(\frac{|D_1 D_2 x(s,t)|}{D_1 D_2 p(s,t)}\right) dtds\right). \tag{2}$$

Proof Let $y(s,t) = \int_\alpha^s \int_\beta^t |D_1 D_2 x(\sigma,\tau)|\, d\tau d\sigma$ so that $D_1 D_2 y(s,t) = |D_1 D_2 x(s,t)|$ and $y(s,t) \geq |x(s,t)|$. Thus, from Jensen's integral inequality, we obtain

$$g\left(\frac{|x(s,t)|}{p(s,t)}\right) \leq g\left(\frac{y(s,t)}{p(s,t)}\right) \leq g\left(\frac{\int_\alpha^s \int_\beta^t D_1 D_2 p(\sigma,\tau) \frac{|D_1 D_2 x(\sigma,\tau)|}{D_1 D_2 p(\sigma,\tau)} d\tau\, d\sigma}{\int_\alpha^s \int_\beta^t D_1 D_2 p(\sigma,\tau) d\tau\, d\sigma}\right)$$
$$\leq \frac{1}{p(s,t)} \int_\alpha^s \int_\beta^t D_1 D_2 p(\sigma,\tau) g\left(\frac{|D_1 D_2 x(\sigma,\tau)|}{D_1 D_2 p(\sigma,\tau)}\right) d\tau\, d\sigma.$$

Hence,

$$\int_\alpha^a \int_\beta^b D_1 D_2 p(s,t) \cdot g\left(\frac{|D_1 D_2 x(s,t)|}{D_1 D_2 p(s,t)}\right) \cdot \frac{\partial}{\partial t}\left[f\left(p(s,t) \cdot g\left(\frac{|x(s,t)|}{p(s,t)}\right)\right)\right] dtds$$
$$\leq \int_\alpha^a \int_\beta^b D_1 D_2 p(s,t) \cdot g\left(\frac{D_1 D_2 y(s,t)}{D_1 D_2 p(s,t)}\right)$$
$$\times \frac{\partial}{\partial t}\left[f\left(\int_\alpha^s \int_\beta^t D_1 D_2 p(\sigma,\tau) \cdot g\left(\frac{D_1 D_2 y(\sigma,\tau)}{D_1 D_2 p(\sigma,\tau)}\right) d\tau\, d\sigma\right)\right] dtds.$$

On the other hand

$$\frac{\partial^2}{\partial s \partial t}\left[f\left(\int_\alpha^s \int_\beta^t D_1 D_2 p(\sigma,\tau) \cdot g\left(\frac{D_1 D_2 y(\sigma,\tau)}{D_1 D_2 p(\sigma,\tau)}\right) d\tau\, d\sigma\right)\right]$$
$$= \frac{\partial}{\partial s}\left\{\frac{\partial}{\partial t}\left[f\left(\int_\alpha^s \int_\beta^t D_1 D_2 p(\sigma,\tau) \cdot g\left(\frac{D_1 D_2 y(\sigma,\tau)}{D_1 D_2 p(\sigma,\tau)}\right) d\tau\, d\sigma\right)\right]\right.$$
$$\left.\times \int_\alpha^s p_{\sigma t}(\sigma,t) \cdot g\left(\frac{D_1 D_2 y(\sigma,\tau)}{D_1 D_2 p(\sigma,t)}\right) d\sigma\right\}$$
$$= \left\{\frac{\partial^2}{\partial s \partial t}\left[f\left(\int_\alpha^s \int_\beta^t D_1 D_2 p(\sigma,\tau) \cdot g\left(\frac{D_1 D_2 y(\sigma,\tau)}{D_1 D_2 p(\sigma,\tau)}\right) d\tau\, d\sigma\right)\right]\right\}$$
$$\times \int_\alpha^s D_1 D_2 p(\sigma,t) \cdot g\left(\frac{D_1 D_2 y(\sigma,\tau)}{p_{\sigma t}(\sigma,t)}\right) d\sigma$$
$$\times \int_\beta^t p_{s\tau}(s,\tau) \cdot g\left(\frac{D_1 D_2 y(\sigma,\tau)}{D_1 D_2 p(s,\tau)}\right) d\tau + D_1 D_2 p(s,t) \cdot g\left(\frac{D_1 D_2 y(s,t)}{D_1 D_2 p(s,t)}\right)$$

$$\times \frac{\partial}{\partial t}\left[f\left(\int_\alpha^s \int_\beta^t D_1D_2p(\sigma,\tau)\cdot g\left(\frac{D_1D_2y(\sigma,\tau)}{D_1D_2p(\sigma,\tau)}\right)d\tau\, d\sigma\right)\right]$$

$$= D_1D_2p(s,t)\cdot g\left(\frac{D_1D_2y(s,t)}{D_1D_2p(s,t)}\right)$$

$$\times \frac{\partial f}{\partial t}\left[\left(\int_\alpha^s \int_\beta^t D_1D_2p(\sigma,\tau)\cdot g\left(\frac{D_1D_2y(\sigma,\tau)}{D_1D_2p(\sigma,\tau)}\right)d\tau\, d\sigma\right)\right].$$

Hence,

$$\int_\alpha^a \int_\beta^b D_1D_2p(s,t)\cdot g\left(\frac{|D_1D_2x(s,t)|}{D_1D_2p(s,t)}\right)\cdot \frac{\partial}{\partial t}\left[f\left(p(s,t)\cdot g\left(\frac{|x(s,t)|}{p(s,t)}\right)\right)\right]dtds$$

$$\le \int_\alpha^a \int_\beta^b \frac{\partial^2}{\partial s\partial t}\left[f\left(\int_\alpha^s \int_\beta^t D_1D_2p(\sigma,\tau)\cdot g\left(\frac{D_1D_2y(\sigma,\tau)}{D_1D_2p(\sigma,\tau)}\right)d\tau\, d\sigma\right)\right]dtds$$

$$= f\left(\int_\alpha^a \int_\beta^b D_1D_2p(\sigma,\tau)\cdot g\left(\frac{D_1D_2y(\sigma,\tau)}{D_1D_2p(\sigma,\tau)}\right)d\tau\, d\sigma\right)$$

$$= f\left(\int_\alpha^a \int_\beta^b D_1D_2p(s,t)\cdot g\left(\frac{|D_1D_2x(s,t)|}{D_1D_2p(s,t)}\right)dtds\right).$$

This completes the proof.

Remark 2.1 In case, $x(s,t)$ reduces to a function of one variable $x(t)$, then with suitable modifications of the proof of Theorem 2.1, (2) becomes the inequality in Theorem C.

Remark 2.2 Taking $g(x) = x$ in (2), then the inequality becomes

$$\int_\alpha^a \int_\beta^b |D_1D_2x(s,t)|\cdot \frac{\partial}{\partial t}(f(|x(s,t)|))dtds \le f\left(\int_\alpha^a \int_\beta^b |D_1D_2x(s,t)|\, dtds\right)$$

which further reduces to the inequality in Theorem B when $x(s,t)$ reduces to $x(t)$.

Remark 2.3 For $f(t) = t^{l+1}, l \ge 0$, the inequality in Remark 2.2 reduces to

$$\int_\alpha^a \int_\beta^b |x(s,t)|^l \frac{\partial}{\partial t}(|x(s,t)|)dtds \le \frac{1}{l+1}\left(\int_\alpha^a \int_\beta^b |D_1D_2x(s,t)|\, dtds\right)^{l+1}.$$

Applying Hölder's inequality to the right hand side with indices $l+1$ and $(l+1)/l$, this yields

$$\int_\alpha^a \int_\beta^b |x(s,t)|^l \frac{\partial}{\partial t}(|x(s,t)|)dtds \le$$

$$\frac{[(a-\alpha)(b-\beta)]^l}{l+1}\int_\alpha^a \int_\beta^b |D_1D_2x(s,t)|^{l+1}\, dtds.$$

When $x(s,t)$ reduces to $x(t)$ and $\alpha = \beta = 0$, the last inequality becomes Hua's inequality in Theorem A.

Theorem 2.2 [41] *Assume that*

(i) f, g, *and* $x(s,t)$ *are as in Theorem 1.1,*
(ii) $p(s,t)$ *is increasing on* $[0,a] \times [0,b]$ *with* $p(s,\beta) = p(\alpha,t) = 0$,
(iii) h *is concave and increasing on* $[0,\infty)$,
(iv) $\phi(t)$ *is increasing on* $[0,a]$ *with* $\phi(0) = 0$,
(v) $y(s,t) = \int_0^s \int_0^t D_1 D_2 p(\sigma,\tau) g\left(\frac{|D_1 D_2 x(\sigma,\tau)|}{D_1 D_2 p(\sigma,\tau)}\right) d\tau\, d\sigma$ *satisfies*

$$D_1 D_2 f\,(y(s,t))\, D_1 D_2 y(s,t) \cdot \phi\left(\frac{1}{D_1 D_2 y(s,t)}\right) \le \frac{c_{(a,b)}}{y(a,b)} \cdot \phi'\left(\frac{t}{y(a,b)}\right).$$

Then

$$\int_0^a \int_0^b D_1 D_2 f\left(p(s,t) g\left(\frac{|x(s,t)|}{p(s,t)}\right)\right) \cdot v\left(D_1 D_2 p(s,t) g\left(\frac{|D_1 D_2 x(s,t)|}{D_1 D_2 p(s,t)}\right)\right) dt ds$$

$$\le w\left(\int_0^a \int_0^b D_1 D_2 p(s,t) g\left(\frac{|x(s,t)|}{D_1 D_2 p(s,t)}\right) dt ds\right), \tag{3}$$

where

$$v(z) = z h\left(\phi\left(\frac{1}{z}\right)\right),$$

$$w(z) = c_{(a,b)} h\left(a\phi\left(\frac{b}{z}\right)\right),$$

and

$$c_{(a,b)} = \int_0^a \int_0^b D_1 D_2 f\,(y(s,t))\, D_1 D_2 y(s,t) dt ds.$$

Proof From the hypotheses, we obtain

$$p(s,t) g\left(\frac{|x(s,t)|}{p(s,t)}\right) \le \int_0^s \int_0^t D_1 D_2 p(\sigma,\tau) g\left(\frac{|D_1 D_2 x(\sigma,\tau)|}{D_1 D_2 p(\sigma,\tau)}\right) d\tau\, d\sigma = y(s,t).$$

In view of Jensen's inequality (for concave functions), we have

$$\int_0^a \int_0^b D_1 D_2 f\left(p(s,t) g\left(\frac{|x(s,t)|}{p(s,t)}\right)\right) \cdot v\left(D_1 D_2 p(s,t) g\left(\frac{|D_1 D_2 x(s,t)|}{D_1 D_2 p(s,t)}\right)\right) dt ds$$

$$\le \int_0^a \int_0^b D_1 D_2 f\,(y(s,t)) \cdot v\,(D_1 D_2 y(s,t))\, dt ds$$

$$= \int_0^a \int_0^b D_1D_2f\,(y(s,t))\,D_1D_2y(s,t)\cdot h\left(\phi\left(\frac{1}{D_1D_2y(s,t)}\right)\right)dtds$$

$$= \frac{\int_0^a \int_0^b D_1D_2f\,(y(s,t))\,D_1D_2y(s,t)\cdot h\left(\phi\left(\frac{1}{D_1D_2y(s,t)}\right)\right)dtds}{\int_0^a \int_0^b D_1D_2f\,(y(s,t))\,D_1D_2y(s,t)dtds}$$

$$\times \int_0^a \int_0^b D_1D_2f\,(y(s,t))\,D_1D_2y(s,t)dtds$$

$$\le h\left(\frac{\int_0^a \int_0^b D_1D_2f\,(y(s,t))\,D_1D_2y(s,t)\cdot \phi\left(\frac{1}{D_1D_2y(s,t)}\right)dtds}{\int_0^a \int_0^b D_1D_2f\,(y(s,t))\,D_1D_2y(s,t)dtds}\right)\cdot c_{(a,b)}$$

$$\le h\left(\frac{\int_0^a \int_0^b \frac{c_{(a,b)}}{y(a,b)}\cdot \phi'\left(\frac{t}{y(a,b)}\right)dtds}{c_{(a,b)}}\right)\cdot c_{(a,b)}$$

$$= h\left(\frac{1}{y(a,b)}\int_0^a \left(y(a,b)\phi\left(\frac{t}{y(a,b)}\right)\Big|_{t=0}^{t=b}\right)ds\right)\cdot c_{(a,b)}$$

$$= h\left(a\phi\left(\frac{b}{y(a,b)}\right)\right)\cdot c_{(a,b)}$$

$$= w\left(\int_0^a \int_0^b D_1D_2p(s,t)g\left(\frac{|D_1D_2x(s,t)|}{D_1D_2p(s,t)}\right)dtds\right).$$

This completes the proof.

Remark 2.4 In case, $x(s,t)$ reduces to $x(t)$, then with suitable modifications of the proof of Theorem 2.2, (3) becomes

$$\int_0^a f'\left(p(t)g\left(\frac{|x(t)|}{p(t)}\right)\right)\cdot v\left(p'(t)g\left(\frac{|x'(t)|}{p'(t)}\right)\right)dt \le w\left(\int_0^a p'(t)g\left(\frac{|x'(t)|}{p'(t)}\right)dt\right),$$

which is a result of Rozanova in [31].

For $x(t) = x_1(t), x_1'(t) > 0, x_1(0) = 0, x(a) = b, g(t) = t, f(t) = \phi(t) = t^2$ and $h(t) = \sqrt{1+t}$, the inequality above further reduces to Polya's inequality (see [31]).

Remark 2.5 Taking $g(x) = x$ in (3), then the inequality becomes

$$\int_0^a \int_0^b D_1D_2f(\,|x(s,t)|\,)\cdot v(\,|D_1D_2x(s,t)|\,)dtds \le w\left(\int_0^a \int_0^b |D_1D_2x(s,t)|\,dtds\right).$$

3 Opial-Agarwal-Pang Type Inequalities

In 1995, Agarwal and Pang [1] proved the following interesting Opial's type inequalities.

Theorem A *Let $\lambda \geq 1$ be a given real number, and $p(t)$ be a nonnegative continuous function on $[0, a]$. Further, let $x(t)$ be an absolutely continuous function on $[0, a]$, with $x(0) = x(a) = 0$. Then*

$$\int_0^a p(t)|x(t)|^\lambda dtds \leq \frac{1}{2}\int_0^a [t(a-t)]^{(\lambda-1)/2} p(t)dt \int_0^a \left|x'(t)\right|^\lambda dt.$$

Theorem B *Assume that*

(i) *l, m, μ and ν are nonnegative real numbers such that $\frac{1}{\mu} + \frac{1}{\nu} = 1$, and $l\mu \geq 1$,*
(ii) *$q(t)$ is a nonnegative and continuous function on $[0, a]$,*
(iii) *$x_1(t)$ and $x_2(t)$ are absolutely continuous functions on $[0, a]$, with $x_1(0) = x_1(a) = x_2(0) = x_2(a) = 0$.*

Then

$$\begin{aligned}&\int_0^a q(t)\left[|x_1(t)|^l \left|x_2'(t)\right|^m + |x_2(t)|^l \left|x_1'(t)\right|^m\right] dt\\ &\leq \left(\frac{1}{2}\int_0^a [t(a-t)]^{(l\mu-1)/2} q^\mu(t)dt\right)^{1/\mu}\\ &\times \int_0^a \left[\frac{1}{\mu}\left(\left|x_1'(t)\right|^{l\mu} + \left|x_2'(t)\right|^{l\mu}\right) + \frac{1}{\nu}\left(\left|x_1'(t)\right|^{m\nu} + \left|x_2'(t)\right|^{m\nu}\right)\right]dt.\end{aligned}$$

The main purpose of this section is to establish some Opial–Agarwal–Pang type inequalities. Our results in special cases yield the inequalities in Theorem A and B.

Theorem 3.1 [35] *Let $\lambda \geq 1$ be a real number, and $p(s, t)$ be a nonnegative continuous function on $[0, a] \times [0, b]$. Let $x(s, t)$ be an absolutely continuous function on $[0, a] \times [0, b]$, with $x(s, 0) = x(0, t) = 0$ and $x(a, t) = x(s, b) = 0$ for all $(s, t) \in [0, a] \times [0, b]$. Then*

$$\int_0^a \int_0^b p(s,t)|x(s,t)|^\lambda dtds$$

$$\leq \frac{1}{2}\left(\int_0^a \int_0^b [st(a-s)(b-t)]^{(\lambda-1)/2} p(s,t)dtds\right)\int_0^a \int_0^b \left|\frac{\partial^2}{\partial s \partial t}x(s,t)\right|^\lambda dtds. \tag{4}$$

Proof From the hypotheses, we have

$$x(s,t) = \int_0^s \int_0^t \frac{\partial^2}{\partial s \partial t}x(s,t)dtds.$$

By Hölder's inequality with indices λ and $\lambda/(\lambda - 1)$, it follows that

$$|x(s,t)|^{\lambda/2} \leq \left[\left(\int_0^s \int_0^t \left|\frac{\partial^2}{\partial s \partial t}x(s,t)\right| dtds\right)^\lambda\right]^{1/2}$$

$$\leq (st)^{(\lambda-1)/2}\left(\int_0^s\int_0^t\left|\frac{\partial^2}{\partial s\partial t}x(s,t)\right|^\lambda dtds\right)^{1/2}.$$

Similarly

$$|x(s,t)|^{\lambda/2}\leq[(a-s)(b-t)]^{(\lambda-1)/2}\left(\int_s^a\int_t^b\left|\frac{\partial^2}{\partial s\partial t}x(s,t)\right|^\lambda dtds\right)^{1/2}.$$

A multiplication of above two inequalities, taking into account of the elementary inequality $2\sqrt{\alpha\beta}\leq\alpha+\beta$ for all $\alpha,\ \beta\geq 0$, yields

$$\begin{aligned}|x(s,t)|^\lambda &\leq [st(a-s)(b-t)]^{(\lambda-1)/2}\left(\int_0^s\int_0^t\left|\frac{\partial^2}{\partial s\partial t}x(s,t)\right|^\lambda dtds\right)^{1/2}\\ &\quad\times\left(\int_s^a\int_t^b\left|\frac{\partial^2}{\partial s\partial t}x(s,t)\right|^\lambda dtds\right)^{1/2}\\ &\leq\frac{1}{2}[st(a-s)(b-t)]^{(\lambda-1)/2}\\ &\quad\times\left(\int_0^s\int_0^t\left|\frac{\partial^2}{\partial s\partial t}x(s,t)\right|^\lambda dtds+\int_s^a\int_t^b\left|\frac{\partial^2}{\partial s\partial t}x(s,t)\right|^\lambda dtds\right)\\ &=\frac{1}{2}[st(a-s)(b-t)]^{(\lambda-1)/2}\int_0^a\int_0^b\left|\frac{\partial^2}{\partial s\partial t}x(s,t)\right|^\lambda dtds.\end{aligned}$$

Multiplying both sides of above inequality by $p(s,t)$ and integrating over t from 0 to b and then integrating the resulting inequality over s from 0 to a, we obtain

$$\begin{aligned}&\int_0^a\int_0^b p(s,t)|x(s,t)|^\lambda dtds\\ &\leq\frac{1}{2}\int_0^a\int_0^b\left([st(a-s)(b-t)]^{(\lambda-1)/2}p(s,t)\int_0^a\int_0^b\left|\frac{\partial^2}{\partial s\partial t}x(s,t)\right|^\lambda dtds\right)dtds\\ &=\frac{1}{2}\left(\int_0^a\int_0^b[st(a-s)(b-t)]^{(\lambda-1)/2}p(s,t)dtds\right)\int_0^a\int_0^b\left|\frac{\partial^2}{\partial s\partial t}x(s,t)\right|^\lambda dtds.\end{aligned}$$

Remark 3.1 In case, $x(s,t)$ reduces to $x(t)$, then with suitable modifications, (4) becomes the inequality in Theorem A.

Remark 3.2 Taking $p(s,t)=$ constant in (4), we have

$$\int_0^a\int_0^b|x(s,t)|^\lambda dtds\leq\frac{1}{2}(ab)^\lambda\left[B\left(\frac{\lambda+1}{2},\frac{\lambda+1}{2}\right)\right]^2\int_0^a\int_0^b\left|\frac{\partial^2}{\partial s\partial t}x(s,t)\right|^\lambda dtds,$$

where B is the Beta function.

Theorem 3.2 [35] *Assume that*

(i) l, m, μ, *and* ν *are nonnegative real numbers such that* $\frac{1}{\mu} + \frac{1}{\nu} = 1$, *and* $l\mu \geq 1$,
(ii) $q(s,t)$ *is a nonnegative continuous function on* $[0,a] \times [0,b]$,
(iii) $x_j(s,t)$ *is absolutely continuous on* $[0,a] \times [0,b]$, *with* $x_j(s,0) = x_j(0,t) = 0$ *and* $x_j(a,t) = x_j(s,b) = 0$ *for* $j = 1, 2$.

Then

$$\int_0^a \int_0^b q(s,t) \left[|x_1(s,t)|^l \left| \frac{\partial^2}{\partial s \partial t} x_2(s,t) \right|^m + |x_2(s,t)|^l \left| \frac{\partial^2}{\partial s \partial t} x_1(s,t) \right|^m \right] dtds$$
$$\leq \left(\frac{1}{2} \int_0^a \int_0^b [st(a-s)(b-t)]^{(l\mu-1)/2} q^\mu(s,t) dtds \right)^{1/\mu}$$
$$\times \int_0^a \int_0^b \left[\frac{1}{\mu} \left(\left| \frac{\partial^2}{\partial s \partial t} x_1(s,t) \right|^{l\mu} + \left| \frac{\partial^2}{\partial s \partial t} x_2(s,t) \right|^{l\mu} \right) \right.$$
$$\left. + \frac{1}{\nu} \left(\left| \frac{\partial^2}{\partial s \partial t} x_1(s,t) \right|^{m\nu} + \left| \frac{\partial^2}{\partial s \partial t} x_2(s,t) \right|^{m\nu} \right) \right] dtds. \tag{5}$$

Proof From (4) and in view of Hölder's inequality and Young's inequality, we have

$$\int_0^a \int_0^b q(s,t) |x_1(s,t)|^l \left| \frac{\partial^2}{\partial s \partial t} x_2(s,t) \right|^m dtds$$
$$\leq \left(\int_0^a \int_0^b q^\mu(s,t) |x_1(s,t)|^{l\mu} dtds \right)^{1/\mu} \left(\int_0^a \int_0^b \left| \frac{\partial^2}{\partial s \partial t} x_2(s,t) \right|^{m\nu} dtds \right)^{1/\nu}$$
$$\leq \left(\frac{1}{2} \int_0^a \int_0^b [st(a-s)(b-t)]^{(l\mu-1)/2} q^\mu(s,t) dtds \right)^{1/\mu}$$
$$\times \left(\int_0^a \int_0^b \left| \frac{\partial^2}{\partial s \partial t} x(s,t) \right|^{l\mu} dtds \right)^{1/\mu} \left(\int_0^a \int_0^b \left| \frac{\partial^2}{\partial s \partial t} x_2(s,t) \right|^{m\nu} dtds \right)^{1/\nu}$$
$$\leq \left(\frac{1}{2} \int_0^a \int_0^b [st(a-s)(b-t)]^{(l\mu-1)/2} q^\mu(s,t) dtds \right)^{1/\mu}$$
$$\times \int_0^a \int_0^b \left(\frac{1}{\mu} \left| \frac{\partial^2}{\partial s \partial t} x_1(s,t) \right|^{l\mu} + \frac{1}{\nu} \left| \frac{\partial^2}{\partial s \partial t} x_2(s,t) \right|^{m\nu} \right) dtds.$$

Similarly,

$$\int_0^a \int_0^b q(s,t) |x_1(s,t)|^m \left| \frac{\partial^2}{\partial s \partial t} x_2(s,t) \right|^l dtds$$
$$\leq \left(\frac{1}{2} \int_0^a \int_0^b [st(a-s)(b-t)]^{(l\mu-1)/2} q^\mu(s,t) dtds \right)^{1/\mu}$$

$$\times \int_0^a \int_0^b \left(\frac{1}{\nu} \left| \frac{\partial^2}{\partial s \partial t} x_1(s,t) \right|^{m\nu} + \frac{1}{\mu} \left| \frac{\partial^2}{\partial s \partial t} x_2(s,t) \right|^{l\mu} \right) dtds.$$

An addition of above two inequalities gives inequality (5).

Remark 3.3 In case, $x(s,t)$ reduces to $x(t)$, with suitable modifications, (5) becomes

$$\int_0^a q(t) \left[|x_1(t)|^l \left|x_2'(t)\right|^m + |x_2(t)|^l \left|x_1'(t)\right|^m \right] dt$$
$$\leq \left(\frac{1}{2} \int_0^a \int_0 [t(a-t)]^{(l\mu-1)/2} q^\mu(t) dt \right)^{1/\mu}$$
$$\times \int_0^a \left[\frac{1}{\mu} \left(\left|x_1'(t)\right|^{l\mu} + \left|x_2'(t)\right|^{l\mu} \right) + \frac{1}{\nu} \left(\left|x_1'(t)\right|^{m\nu} + \left|x_2'(t)\right|^{m\nu} \right) \right] dt$$

which is an inequality established by Agarwal and Pang in [1].

Remark 3.4 Taking $\mu = \nu = 2$ in (5), if $q(s,t) \neq$ constant, we have

$$\int_0^a \int_0^b q(s,t) \left(|x_1(s,t)|^l \left| \frac{\partial^2}{\partial s \partial t} x_2(s,t) \right|^m + |x_2(s,t)|^l \left| \frac{\partial^1}{\partial s \partial t} x_1(s,t) \right|^m \right) dtds$$
$$\leq \left(\frac{1}{8} \int_0^a \int_0^b [st(a-s)(b-t)]^{(2l-1)/2} q^2(s,t) dtds \right)^{1/2} \times$$
$$\int_0^a \int_0^b \left(\left| \tfrac{\partial^2}{\partial s \partial t} x_1(s,t) \right|^{2l} + \left| \tfrac{\partial^2}{\partial s \partial t} x_2(s,t) \right|^{2l} + \left| \tfrac{\partial^2}{\partial s \partial t} x_1(s,t) \right|^{2m} + \left| \tfrac{\partial^2}{} \partial s \partial t x_2(s,t) \right|^{2m} \right) dtds.$$

In case, $x_j(s,t)$, $j = 1, 2$ reduces to $x_j(t)$, with suitable modifications, the above inequality becomes.

$$\int_0^a q(t) \left(|x_1(t)|^l \left|x_2'(t)\right|^m + |x_2(t)|^l \left|x_1'(t)\right|^m \right) dtds$$
$$\leq \left(\tfrac{1}{8} \int_0^a [t(a-t)]^{(2l-1)/2} q^2(t) dt \right)^{1/2}$$
$$\times \int_0^a \left(\left|x_1'(t)\right|^{2l} + \left|x_2'(t)\right|^{2l} + \left|x_1'(t)\right|^{2m} + \left|x_2'(t)\right|^{2m} \right) dt,$$

which is another inequality established by Agarwal and Pang in [1]. This inequality is sharper than the following inequality (see [19]):

$$\int_0^a q(t) \left(|x_1(t)|^l \left|x_2'(t)\right|^m + |x_2(t)|^l \left|x_1'(t)\right|^m \right) dtds \leq \left(\frac{h^{2\,l-1}}{4^{l+1}} \int_0^a q^2(t) dt \right)^{1/2}$$
$$\times \int_0^a \left(\left|x_1'(t)\right|^{2l} + \left|x_2'(t)\right|^{2l} + \left|x_1'(t)\right|^{2m} + \left|x_2'(t)\right|^{2m} \right) dt.$$

The following example shows the usefulness of the preceding results to the uniqueness of solutions of initial value problems involving higher order derivatives.

Theorem 3.3 [35] *Consider the system of differential equations*

$$y_j'' = f_j(t, y_1, y_1', y_2, y_2'),\ j = 1,2 \tag{6}$$

with initial conditions

$$y_j^{(i)}(0) = y_{j,i},\ j = 1,2,\ 0 \le i \le 1\,. \tag{7}$$

Suppose $f_j : [0,\tau] \times \mathbb{R}^2 \times \mathbb{R}^2 \to \mathbb{R},\ j = 1,2$ *are continuous, satisfying the Lipschitz condition*

$$|f_j(t, y_{1,0}, y_{1,1}, y_{2,0}, y_{2,1}) - f_j(t, \bar{y}_{1,0}, \bar{y}_{1,1}, \bar{y}_{2,0}, \bar{y}_{2,1})|$$
$$\le \sum_{k=0}^{1} \big[q_{1,j,k}(t)|y_{1,k} - \bar{y}_{1,k}| + q_{2,j,k}(t)|y_{2,k} - \bar{y}_{2,k}|\big],$$

where the functions $q_{r,j,k} \ge 0, 1 \le r, j \le 2, 0 \le k \le 1$ *are continuous on* $[0,\tau]$. *Then, the problems* (6), (7) *have at most one solution on* $[0,\tau]$.

Proof If the problems $\{(6),(7)\}$ have two solutions $(y_1(t), y_2(t)), (\bar{y}_1(t), \bar{y}_2(t))$, then for the functions $x_j(t) = y_j(t) - \bar{y}_j(t), j = 1,2$, it follows that

$$|x_j''(t)|^2 \le \sum_{k=0}^{1} \Big[q_{1,j,k}(t)|x_1^{(k)}(t)||x_j''(t)| + q_{2,j,k}(t)|x_2^{(k)}(t)||x_j''(t)|\Big], \quad j = 1,2.$$

Summing up these two inequalities, and integrating from 0 to t, we obtain

$$\int_0^t \big[|x_1''(s)|^2 + |x_2''(s)|^2\big]ds \le \sum_{k=0}^{1}\int_0^t \bar{q}_k(s)\big[|x_1^{(k)}(s)||x_1''(s)| + |x_2^{(k)}(s)||x_2''(s)|\big]ds$$
$$+\sum_{k=0}^{1}\int_0^t \bar{q}_k^*(s)\big[|x_2^{(k)}(s)||x_2''(s)| + |x_1^{(k)}(s)||x_1''(s)|\big]ds,$$

where $\bar{q}_k(t) = \max(q_{1,1,k}(t), q_{2,2,k}(t))$ and $q_k^*(t) = \max(q_{2,1,k}(t), q_{1,2,k}(t))$.

For $0 \le k \le 1$, by applying inequality (5) with $x(s,t) = x(t), l = 1, \mu = 2, m = 1$ to the right side of above inequality, and with suitable modifications, it is easy to arrive at

$$\int_0^t \big[|x_1''(s)|^2 + |x_2''(s)|^2\big]ds \le K(t)\int_0^t \big[|x_1''(s)|^2 + |x_2''(s)|^2\big]ds,$$

where $K(t)$ is a continuous function with the property that $K(0) = 0$. This forces $y_1(t) = \bar{y}_1(t), y_2(t) = \bar{y}_2(t)$ for all $t \in [0,\tau]$.

4 Opial–Pachpatte Type Inequalities

The first natural extension of Opial's inequality (1) involving higher order derivatives $x^{(n)}(s)$, $n \geq 1$, is embodied in the following.

Theorem A [2] *Let $x(t) \in C^{(n)}[0,a]$ be such that $x^{(i)}(0) = 0, 0 \leq i \leq n-1 (n \geq 1)$. Then*

$$\int_0^a \left|x(t)x^{(n)}(t)\right| dt \leq \frac{1}{2}a^n \int_0^a \left|x^{(n)}(t)\right|^2 dt.$$

A sharp version of the above inequality is the following.

Theorem B [16] *Let $x(t) \in C^{(n-1)}[0,a]$ be such that $x^{(i)}(0) = 0, 0 \leq i \leq n-1 (n \geq 1)$. Suppose $x^{(n-1)}(t)$ is absolutely continuous and $\int_0^a |x^n(t)|^2 dt < \infty$. Then,*

$$\int_0^a \left|x(t)x^{(n)}(t)\right| dt \leq \frac{1}{2n!}\left(\frac{n}{2n-1}\right)^{1/2} a^n \int_0^a \left|x^{(n)}(t)\right|^2 dt.$$

A couple of generalizations of the above inequalities are following:

Theorem C [24] *Let $x(t) \in C^{(n-1)}[0,a]$ be such that $x^{(i)}(0) = 0$, $0 \leq i \leq n-1$, $n \geq 1$. Suppose $x^{(n-1)}(t)$ is absolutely continuous and $\int_0^a |x^{(n)}(t)|^2 dt < \infty$. Then*

$$\int_0^a \left|\prod_{\kappa=0}^{n} x^{(\kappa)}(t)\right| dt \leq K(n) a^{(n^2+1)/2} \left(\int_0^a \left|x^{(n)}(t)\right|^2 dt\right)^{(n+1)/2}$$

for all $0 \leq \kappa \leq n-1$, where

$$K(n) = \frac{1}{(n^2+1)(n+1)\prod_{\kappa=0}^{n-1}(n-\kappa-1)!}\left(\frac{(n^2+1)(n+1)}{\prod_{\kappa=0}^{n-1}(2n-2\kappa-1)}\right)^{1/2}.$$

Theorem D [25] *For $j = 1,2$, let $x_j(t) \in C^{n-1}[0,a]$ be such that $x_j^{(i)}(0) = 0, 0 \leq i \leq n-1$. If $x_j^{(n-1)}$ is absolutely continuous and $\int_0^a |x_j^{(n)}(t)|^2 dt < \infty$, $j = 1,2$, then*

$$\int_0^a \left\{\left|x_2(t)x_1^{(n)}(t)\right| + \left|x_1(t)x_2^{(n)}(t)\right|\right\} dt$$

$$\leq \frac{1}{2n!}\left(\frac{n}{2n-1}\right)^{1/2} a^n \left[\int_0^a \left(\left|x_1^{(n)}(t)\right|^2 + \left|x_2^{(n)}(t)\right|^2\right) dt\right].$$

The aim of this section is to establish some new Opial–Pachpatter type inequalities which further generalize the inequalities in Theorems C and D.

Theorem 4.1 [46] *Let $x(s,t) \in C^{(n-1,m-1)}([0,a] \times [0,b])$ be such that $\frac{\partial^\kappa}{\partial s^\kappa} x(s,t)|_{s=0} = 0, 0 \leq \kappa \leq n-1$ and $\frac{\partial^\lambda}{\partial t^\lambda} x(s,t)|_{t=0} = 0, 0 \leq \lambda \leq m-1$. Further,* *suppose* $\frac{\partial^n}{\partial s^n}\left(\frac{\partial^{m-1}}{\partial t^{m-1}}x(s,t)\right)$ *and* $\frac{\partial^{n-1}}{\partial s^{n-1}}\left(\frac{\partial^m}{\partial t^m}x(s,t)\right)$ *are absolutely continuous on* $[0,a]\times[0,b]$, *and*

$$\int_0^a\int_0^b\left|\frac{\partial^{n+m}}{\partial s^n\partial t^m}x(s,t)\right|^2 dtds$$

exist. Then

$$\int_0^a\int_0^b\left|\frac{\prod_{\kappa=0}^{n}\prod_{\lambda=0}^{m}\frac{\partial^{\kappa+\lambda}}{\partial s^\kappa\partial t^\lambda}x(s,t)}{\prod_{\kappa=0}^{n-1}\frac{\partial^{\kappa+m}}{\partial s^\kappa\partial t^m}x(s,t)\cdot\prod_{\lambda=0}^{m-1}\frac{\partial^{n+\lambda}}{\partial s^n\partial t^\lambda}x(s,t)}\right|dtds$$

$$\leq M(n,m)a^{(n^2+1)/2}b^{(m^2+1)/2}\left(\int_0^a\int_0^b\left|\frac{\partial^{n+m}}{\partial s^n\partial t^m}x(s,t)\right|^2 dtds\right)^{(mn+1)/2}, \quad (8)$$

where

$$M(n,m)=\frac{1}{(n^2+1)(m^2+1)(mn+1)\prod_{\kappa=0}^{n-1}\prod_{\lambda=0}^{m-1}(n-\kappa-1)!(m-\lambda-1)!}\times$$

$$\times\left(\frac{(n^2+1)(m^2+1)(mn+1)}{\prod_{\kappa=0}^{n-1}\prod_{\lambda=0}^{m-1}\left[(2n-2\kappa-1)(2m-2\lambda-1)\right]}\right)^{1/2}.$$

Proof From the hypotheses, we have

$$\left|\frac{\partial^{\kappa+\lambda}}{\partial s^\kappa\partial t^\lambda}x(s,t)\right|\leq\frac{1}{(n-\kappa-1)!(m-\lambda-1)!}$$
$$\times\int_0^s\int_0^t(s-\sigma)^{n-\kappa-1}(t-\tau)^{m-\lambda-1}\left|\frac{\partial^{n+m}}{\partial\sigma^n\partial\tau^m}x(\sigma,\tau)\right|d\tau\, d\sigma.$$

By using Cauchy–Schwarz inequality, we obtain

$$\left|\frac{\partial^{\kappa+\lambda}}{\partial s^\kappa\partial t^\lambda}x(s,t)\right|\leq\frac{s^{n-k-1/2}t^{m-\lambda-1/2}}{(n-\kappa-1)!(m-\lambda-1)!\left[(2n-2k-1)(2m-2\lambda-1)\right]^{1/2}}$$

$$\times\left(\int_0^s\int_0^t\left|\frac{\partial^{n+m}}{\partial\sigma^n\partial\tau^m}x(\sigma,\tau)\right|^2 d\tau\, d\sigma\right)^{1/2}.$$

Multiplying the above inequality by $\left|\frac{\partial^{n+m}}{\partial s^n\partial t^m}x(s,t)\right|$, we obtain

$$\left|\frac{\prod_{\kappa=0}^{n}\prod_{\lambda=0}^{m}\frac{\partial^{\kappa+\lambda}}{\partial s^\kappa\partial t^\lambda}x(s,t)}{\prod_{\kappa=0}^{n-1}\frac{\partial^{\kappa+m}}{\partial s^\kappa\partial t^m}x(s,t)\cdot\prod_{\lambda=0}^{m-1}\frac{\partial^{n+\lambda}}{\partial s^n\partial t^\lambda}x(s,t)}\right|$$

$$\leq \frac{s^{n^2/2}t^{m^2/2}}{\prod_{\kappa=0}^{n-1}\prod_{\lambda=0}^{m-1}(n-\kappa-1)!(m-\lambda-1)!\prod_{\kappa=0}^{n-1}\prod_{\lambda=0}^{m-1}\left[(2n-2k-1)(2m-2\lambda-1)\right]^{1/2}}$$

$$\times\left|\frac{\partial^{n+m}}{\partial s^n\partial t^m}x(s,t)\right|\left(\int_0^s\int_0^t\left|\frac{\partial^{n+m}}{\partial\sigma^n\partial\tau^m}x(\sigma,\tau)\right|^2 d\tau\, d\sigma\right)^{mn/2}.$$

Integrating both sides over t from 0 to b and then the resulting inequality over s from 0 to a, and by applying the Cauchy–Schwarz inequality to the right side, we get

$$\int_0^a\int_0^b\left|\frac{\prod_{\kappa=0}^{n}\prod_{\lambda=0}^{m}\frac{\partial^{\kappa+\lambda}}{\partial s^\kappa\partial t^\lambda}x(s,t)}{\prod_{\kappa=0}^{n-1}\frac{\partial^{\kappa+m}}{\partial s^\kappa\partial t^m}x(s,t)\cdot\prod_{\lambda=0}^{m-1}\frac{\partial^{n+\lambda}}{\partial s^n\partial t^\lambda}x(s,t)}\right|dtds$$

$$\leq\frac{\left(\int_0^a\int_0^b s^{n^2}t^{m^2}dtds\right)^{1/2}}{\prod_{\kappa=0}^{n-1}\prod_{\lambda=0}^{m-1}(n-\kappa-1)!(m-\lambda-1)!\prod_{\kappa=0}^{n-1}\prod_{\lambda=0}^{m-1}\left[(2n-2k-1)(2m-2\lambda-1)\right]^{1/2}}$$

$$\times\left(\int_0^a\int_0^b\left|\frac{\partial^{n+m}}{\partial s^n\partial t^m}x(s,t)\right|^2\left(\int_0^s\int_0^t\left|\frac{\partial^{n+m}}{\partial\sigma^n\partial\tau^m}x(\sigma,\tau)\right|^2 d\tau\, d\sigma\right)^{mn} dtds\right)^{1/2}.$$

On the other hand, in view of the following straight forward identities

$$\frac{\partial^2}{\partial s\partial t}\left[\left(\int_0^s\int_0^t\left|\frac{\partial^{n+m}}{\partial\sigma^n\partial\tau^m}x(\sigma,\tau)\right|^2 d\tau\, d\sigma\right)^{mn+1}\right]$$

$$=(mn+1)\left|\frac{\partial^{n+m}}{\partial s^n\partial t^m}x(s,t)\right|^2\left(\int_0^s\int_0^t\left|\frac{\partial^{n+m}}{\partial\sigma^n\partial\tau^m}x(\sigma,\tau)\right|^2 d\tau\, d\sigma\right)^{mn}$$

and

$$\left(\int_0^a\int_0^b s^{n^2}t^{m^2}dtds\right)^{1/2}=\frac{a^{(n^2+1)/2}b^{(m^2+1)/2}}{[(n^2+1)(m^2+1)]^{1/2}},$$

we easily deduce inequality (8).

Remark 4.1 In case, $x(s,t)$ reduces to $x(t)$, with suitable modifications, (8) becomes the inequality in Theorem C.

Theorem 4.2 [46] *For $i=1,2$, let $x_i(s,t)\in C^{(n-1,m-1)}([0,a]\times[0,b])$ be such that $\frac{\partial^\kappa}{\partial s^\kappa}x_i(s,t)|_{s=0}=0, 0\leq\kappa\leq n-1$, and $\frac{\partial^\lambda}{\partial t^\lambda}x_i(s,t)|_{t=0}=0, 0\leq\lambda\leq m-1$. Further, suppose $\frac{\partial^n}{\partial s^n}\left(\frac{\partial^{m-1}}{\partial t^{m-1}}x_i(s,t)\right)$ and $\frac{\partial^{n-1}}{\partial s^{n-1}}\left(\frac{\partial^m}{\partial t^m}x_i(s,t)\right)$ are absolutely continuous on $[0,a]\times[0,b]$, and $\int_0^a\int_0^b\left|\frac{\partial^{n+m}}{\partial s^n\partial t^m}x_i(s,t)\right|^2 dtds$ exists. Then,*

$$\int_0^a\int_0^b\left(\left|\frac{\partial^{\kappa+\lambda}}{\partial s^\kappa\partial t^\lambda}x_1(s,t)\frac{\partial^{n+m}}{\partial s^n\partial t^m}x_2(s,t)\right|+\left|\frac{\partial^{\kappa+\lambda}}{\partial s^\kappa\partial t^\lambda}x_2(s,t)\frac{\partial^{n+m}}{\partial s^n\partial t^m}x_1(s,t)\right|\right)dtds$$

$$\leq \sqrt{2}D(n,m)a^{n-\kappa}b^{m-\lambda}\int_0^a\int_0^b\left[\left|\frac{\partial^{n+m}}{\partial s^n\partial t^m}x_1(s,t)\right|^2+\left|\frac{\partial^{n+m}}{\partial s^n\partial t^m}x_2(s,t)\right|^2\right]dtds, \tag{9}$$

where

$$D(n,m)=\frac{1}{4(n-\kappa)!(m-\lambda)!}\left(\frac{(n-\kappa)(m-\lambda)}{(2n-2\kappa-1)(2m-2\lambda-1)}\right)^{1/2}.$$

Proof From the hypotheses, we have

$$\left|\frac{\partial^{\kappa+\lambda}}{\partial s^\kappa\partial t^\lambda}x_1(s,t)\right|\leq\frac{1}{(n-\kappa-1)!(m-\lambda-1)!}$$
$$\times\int_0^s\int_0^t(s-\sigma)^{n-\kappa-1}(t-\tau)^{m-\lambda-1}\left|\frac{\partial^{n+m}}{\partial\sigma^n\partial\tau^m}x_1(\sigma,\tau)\right|d\tau\, d\sigma.$$

Multiplying both sides by $\left|\frac{\partial^{n+m}}{\partial s^n\partial t^m}x_2(s,t)\right|$, and using Cauchy–Schwarz inequality, we obtain

$$\left|\frac{\partial^{\kappa+\lambda}}{\partial s^\kappa\partial t^\lambda}x_1(s,t)\frac{\partial^{n+m}}{\partial s^n\partial t^m}x_2(s,t)\right|$$
$$\leq\frac{s^{n-k-1/2}t^{m-\lambda-1/2}\left|\frac{\partial^{n+m}}{\partial s^n\partial t^m}x_2(s,t)\right|}{(n-\kappa-1)!(m-\lambda-1)!\left[(2n-2k-1)(2m-2\lambda-1)\right]^{1/2}}$$
$$\times\left(\int_0^s\int_0^t\left|\frac{\partial^{n+m}}{\partial\sigma^n\partial\tau^m}x_1(\sigma,\tau)\right|^2d\tau\, d\sigma\right)^{1/2}.$$

Integrating both sides over t from 0 to b and then the resulting inequality over s from 0 to a, and by applying the Cauchy-Schwarz inequality to the right side, we observe

$$\int_0^a\int_0^b\left|\frac{\partial^{\kappa+\lambda}}{\partial s^\kappa\partial t^\lambda}x_1(s,t)\frac{\partial^{n+m}}{\partial s^n\partial t^m}x_2(s,t)\right|dtds$$
$$\leq\frac{\left(\int_0^a\int_0^b s^{2n-2k-1}t^{2m-2\lambda-1}dtds\right)^{1/2}}{(n-\kappa-1)!(m-\lambda-1)!\left[(2n-2k-1)(2m-2\lambda-1)\right]^{1/2}}$$
$$\times\left(\int_0^a\int_0^b\left|\frac{\partial^{n+m}}{\partial s^n\partial t^m}x_2(s,t)\right|^2\left(\int_0^s\int_0^t\left|\frac{\partial^{n+m}}{\partial\sigma^n\partial\tau^m}x_1(\sigma,\tau)\right|^2d\tau\, d\sigma\right)dtds\right)^{1/2}.$$

Similarly

$$\int_0^a\int_0^b\left|\frac{\partial^{\kappa+\lambda}}{\partial s^\kappa\partial t^\lambda}x_2(s,t)\frac{\partial^{n+m}}{\partial s^n\partial t^m}x_1(s,t)\right|dtds$$

$$\leq \frac{\left(\int_0^a\int_0^b s^{2n-2k-1}t^{2m-2\lambda-1}dtds\right)^{1/2}}{(n-\kappa-1)!(m-\lambda-1)!\left[(2n-2k-1)(2m-2\lambda-1)\right]^{1/2}}$$

$$\times\left(\int_0^a\int_0^b\left|\frac{\partial^{n+m}}{\partial s^n\partial t^m}x_1(s,t)\right|^2\left(\int_0^s\int_0^t\left|\frac{\partial^{n+m}}{\partial\sigma^n\partial\tau^m}x_2(\sigma,\tau)\right|^2d\tau\, d\sigma\right)dtds\right)^{1/2}.$$

In view of the elementary inequality $\alpha^{1/2}+\beta^{1/2}\leq[2(\alpha+\beta)]^{1/2},\alpha,\beta\geq 0$, we have

$$\int_0^a\int_0^b\left(\left|\frac{\partial^{\kappa+\lambda}}{\partial s^\kappa\partial t^\lambda}x_1(s,t)\frac{\partial^{n+m}}{\partial s^n\partial t^m}x_2(s,t)\right|+\left|\frac{\partial^{\kappa+\lambda}}{\partial s^\kappa\partial t^\lambda}x_2(s,t)\frac{\partial^{n+m}}{\partial s^n\partial t^m}x_1(s,t)\right|\right)dtds$$

$$\leq \frac{\left(\int_0^a\int_0^b s^{2n-2k-1}t^{2m-2\lambda-1}dtds\right)^{1/2}}{(n-\kappa-1)!(m-\lambda-1)!\left[(2n-2k-1)(2m-2\lambda-1)\right]^{1/2}}$$

$$\times\left\{2\int_0^a\int_0^b\left|\frac{\partial^{n+m}}{\partial s^n\partial t^m}x_2(s,t)\right|^2\left(\int_0^s\int_0^t\left|\frac{\partial^{n+m}}{\partial\sigma^n\partial\tau^m}x_1(\sigma,\tau)\right|^2d\tau\, d\sigma\right)dtds\right.$$

$$\left.+2\int_0^a\int_0^b\left|\frac{\partial^{n+m}}{\partial s^n\partial t^m}x_1(s,t)\right|^2\left(\int_0^s\int_0^t\left|\frac{\partial^{n+m}}{\partial\sigma^n\partial\tau^m}x_2(\sigma,\tau)\right|^2d\tau\, d\sigma\right)dtds\right\}^{1/2}.$$

On the other hand, in view of the following straight forward identities

$$\frac{\partial^2}{\partial s\partial t}\left[\left(\int_0^s\int_0^t\left|\frac{\partial^{n+m}}{\partial\sigma^n\partial\tau^m}x_1(\sigma,\tau)\right|^2d\tau\, d\sigma\right)\left(\int_0^s\int_0^t\left|\frac{\partial^{n+m}}{\partial\sigma^n\partial\tau^m}x_2(\sigma,\tau)\right|^2d\tau\, d\sigma\right)\right]$$

$$=\left|\frac{\partial^{n+m}}{\partial s^n\partial t^m}x_1(s,t)\right|^2\left(\int_0^s\int_0^t\left|\frac{\partial^{n+m}}{\partial\sigma^n\partial\tau^m}x_2(\sigma,\tau)\right|^2d\tau\, d\sigma\right)+\left|\frac{\partial^{n+m}}{\partial s^n\partial t^m}x_2(s,t)\right|^2$$

$$\times\left(\int_0^s\int_0^t\left|\frac{\partial^{n+m}}{\partial\sigma^n\partial\tau^m}x_1(\sigma,\tau)\right|^2d\tau\, d\sigma\right)$$

and

$$\left(\int_0^a\int_0^b s^{2n-2k-1}t^{2m-2\lambda-1}dtds\right)^{1/2}=\frac{a^{n-\kappa}b^{m-\lambda}}{2\left[(n-\kappa)(m-\lambda)\right]^{1/2}},$$

plus the elementary inequality $(\alpha\beta)^{1/2}\leq\frac{1}{2}(\alpha+\beta),\alpha,\beta\geq 0$, inequality (9) follows immediately.

Remark 4.2 In case, $x_i(s,t)$ reduces to $x_i(t)$, $i=1,2$, with suitable modifications, (9) becomes the inequality in Theorem D.

5 Opial–Das's Type Inequalities

The main purpose of this section is to establish some new Opial–type inequalities involving higher order partial derivatives which extend Das's inequalities (see [16]) to more general situations.

Let $n, k \geq 1$. Our main results are given in the following theorems.

Theorem 5.1 [47] *Let* $x(s,t) \in C^{(n-1,m-1)}([0,a] \times [0,b])$ *be such that* $\frac{\partial^\kappa}{\partial s^\kappa} x(s,t)|_{s=0} = 0,\ 0 \leq \kappa \leq n-1;\ \frac{\partial^\lambda}{\partial t^\lambda} x(s,t)|_{t=0} = 0,\ 0 \leq \lambda \leq m-1;$ *and* $\frac{\partial^n}{\partial s^n}\left(\frac{\partial^{m-1}}{\partial t^{m-1}} x(s,t)\right)$, $\frac{\partial^{n-1}}{\partial s^{n-1}}\left(\frac{\partial^m}{\partial t^m} x(s,t)\right)$ *be absolutely continuous on* $[0,a] \times [0,b]$. *Suppose* $1/p + 1/q = 1$, $p > 1$, *and* $\int_0^a \int_0^b \left|\frac{\partial^{n+m}}{\partial s^n \partial t^m} x(s,t)\right|^q dtds$ *exists. Then,*

$$\int_0^a \int_0^b \left| \frac{\partial^{\kappa+\lambda}}{\partial s^\kappa \partial t^\lambda} x(s,t) \frac{\partial^{n+m}}{\partial s^n \partial t^m} x(s,t) \right| dtds$$

$$\leq C_{n,m,\kappa,\lambda,p,q} a^{n-\kappa-1+2/p} b^{m-\lambda-1+2/p} \left(\int_0^a \int_0^b \left| \frac{\partial^{n+m}}{\partial s^n \partial t^m} x(s,t) \right|^q dtds \right)^{(q+1)/q}, \tag{10}$$

where

$$C_{n,m,\kappa,\lambda,p,q}$$

$$= \frac{[(p(n-\kappa-1)+1)(p(n-\kappa-1)+2)(p(m-\lambda-1)+1)(p(m-\lambda-1)+2)]^{-1/p}}{2^{1/q}(n-\kappa-1)!(m-\lambda-1)!}.$$

Proof From the hypotheses, we have

$$\left| \frac{\partial^{\kappa+\lambda}}{\partial s^\kappa \partial t^\lambda} x(s,t) \right| \leq \frac{1}{(n-\kappa-1)!(m-\lambda-1)!}$$
$$\times \int_0^s \int_0^t (s-\sigma)^{n-\kappa-1}(t-\tau)^{m-\lambda-1} \left| \frac{\partial^{n+m}}{\partial \sigma^n \partial \tau^m} x(\sigma,\tau) \right| d\tau\, d\sigma.$$

Multiplying both sides by $\left|\frac{\partial^{n+m}}{\partial s^n \partial t^m} x(s,t)\right|$ and using the Hölder's inequality, we have

$$\left| \frac{\partial^{\kappa+\lambda}}{\partial s^\kappa \partial t^\lambda} x(s,t) \frac{\partial^{n+m}}{\partial s^n \partial t^m} x(s,t) \right| \leq \frac{\left|\frac{\partial^{\kappa+\lambda}}{\partial s^\kappa \partial t^\lambda} x(s,t)\right|}{(n-\kappa-1)!(m-\lambda-1)!}$$
$$\times \left(\int_0^s \int_0^t (s-\sigma)^{p(n-\kappa-1)}(t-\tau)^{p(m-\lambda-1)} d\tau\, d\sigma \right)^{\frac{1}{p}}$$
$$\times \left(\int_0^s \int_0^t \left| \frac{\partial^{n+m}}{\partial \sigma^n \partial \tau^m} x(\sigma,\tau) \right|^q d\tau\, d\sigma \right)^{\frac{1}{q}}$$

$$= \frac{s^{n-\kappa-1+1/p} t^{m-\lambda-1+1/p}}{(n-\kappa-1)!(m-\lambda-1)![(p(n-\kappa-1)+1)(p(m-\lambda-1)+1)]^{1/p}}$$

$$\times\left|\frac{\partial^{\kappa+\lambda}}{\partial s^{\kappa}\partial t^{\lambda}}x(s,t)\right|\left(\int_0^s\int_0^t\left|\frac{\partial^{n+m}}{\partial\sigma^n\partial\tau^m}x(\sigma,\tau)\right|^q d\tau\, d\sigma\right)^{\frac{1}{q}}.$$

Integrating both sides over t from 0 to b and then the resulting inequality over s from 0 to a, by applying Hölder's inequality, we obtain

$$\int_0^a\int_0^b\left|\frac{\partial^{\kappa+\lambda}}{\partial s^{\kappa}\partial t^{\lambda}}x(s,t)\frac{\partial^{n+m}}{\partial s^n\partial t^m}x(s,t)\right|dtds$$

$$\leq\frac{1}{(n-\kappa-1)!(m-\lambda-1)![(p(n-\kappa-1)+1)(p(m-\lambda-1)+1)]^{1/p}}$$

$$\times\left(\int_0^a\int_0^b s^{p(n-\kappa-1)+1}t^{p(m-\lambda-1)+1}dtds\right)^{\frac{1}{p}}$$

$$\times\left(\int_0^a\int_0^b\left|\frac{\partial^{\kappa+\lambda}}{\partial s^{\kappa}\partial t^{\lambda}}x(s,t)\right|^q\left(\int_0^s\int_0^t\left|\frac{\partial^{n+m}}{\partial\sigma^n\partial\tau^m}x(\sigma,\tau)\right|^q d\tau\, d\sigma\right)dtds\right)^{\frac{1}{q}}.$$

On the other hand, in view of the identities

$$\frac{\partial^2}{\partial s\partial t}\left[\left(\int_0^s\int_0^t\left|\frac{\partial^{n+m}}{\partial\sigma^n\partial\tau^m}x(\sigma,\tau)\right|^q d\tau\, d\sigma\right)^2\right]$$

$$=2\left|\frac{\partial^{n+m}}{\partial s^n\partial t^m}x(s,t)\right|^q\left(\int_0^s\int_0^t\left|\frac{\partial^{n+m}}{\partial\sigma^n\partial\tau^m}x(\sigma,\tau)\right|^q d\tau\, d\sigma\right)$$

and

$$\left(\int_0^a\int_0^b s^{p(n-\kappa-1)+1}t^{p(m-\lambda-1)+1}dtds\right)^{1/p}$$

$$=\frac{a^{n-\kappa-1+2/p}b^{m-\lambda-1+1/p}}{[(p(n-\kappa-1)+2)(p(m-\lambda-1)+2)]^{1/p}},$$

we have

$$\int_0^a\int_0^b\left|\frac{\partial^{\kappa+\lambda}}{\partial s^{\kappa}\partial t^{\lambda}}x(s,t)\frac{\partial^{n+m}}{\partial s^n\partial t^m}x(s,t)\right|dtds$$

$$\leq C_{n,m,\kappa,\lambda,p,q}a^{n-\kappa-1+2/p}b^{m-\lambda-1+2/p}\left(\int_0^a\int_0^b\left|\frac{\partial^{n+m}}{\partial s^n\partial t^m}x(s,t)\right|^q dtds\right)^{2/q},$$

where

$$C_{n,m,\kappa,\lambda,p,q}$$

$$=\frac{[(p(n-\kappa-1)+1)(p(n-\kappa-1)+2)(p(m-\lambda-1)+1)(p(m-\lambda-1)+2)]^{-1/p}}{2^{1/q}(n-\kappa-1)!(m-\lambda-1)!}.$$

This completes the proof.

Remark 5.1 Taking $p = q = 2$ and $\kappa = \lambda = 0$ in (10), we have

$$\int_0^a \int_0^b \left| x(s,t) \cdot \frac{\partial^{n+m}}{\partial s^n \partial t^m} x(s,t) \right| dtds \leq c_{n,m} \cdot a^n b^m \cdot \int_0^a \int_0^b \left| \frac{\partial^{n+m}}{\partial s^n \partial t^m} x(s,t) \right|^2 dtds,$$

where

$$c_{n,m} = \frac{1}{4n!m!} \left(\frac{2nm}{(2n-1)(2m-1)} \right)^{\frac{1}{2}}.$$

In case, $x(s,t)$ reduces to $x(t)$, with suitable modifications, the above inequality reduces to

$$\int_0^a |x(t)x^{(n)}(t)|dt \leq \frac{1}{2n!} \cdot \left(\frac{n}{2n-1} \right)^{\frac{1}{2}} a^n \int_0^a |x^{(n)}(t)|^2 dt,$$

which is an inequality established by Das in [16]. Obviously, for $n \geq 2$, the above inequality is sharper than the following inequality established by Willett in [32].

$$\int_0^a |x(t)x^n(t)|dt \leq \frac{1}{2} a^n \int_0^a |x^n(t)|^2 dt.$$

Remark 5.2 Taking $n = m = 1$, $\kappa = \lambda = 0$, and $p = q = 2$ in (10), we have

$$\int_0^a \int_0^b \left| x(s,t) \cdot \frac{\partial^2}{\partial s \partial t} x(s,t) \right| dtds \leq \frac{\sqrt{2}}{4} ab \int_0^a \int_0^b \left| \frac{\partial^2}{\partial s \partial t} x(s,t) \right|^2 dtds.$$

In case $x(s,t)$ reduces to $x(t)$, with suitable modifications, the above inequality reduces to

$$\int_0^a |x(t)x'(t)|dt \leq \frac{a}{2} \int_0^a |x'(t)|^2 dt$$

for all $x(t)$ which is absolutely continuous in $[0,a]$ with $x(0) = 0$. Observe that this is an inequality established by Beesack in [8].

Theorem 5.2 [47] *Let l and m be positive numbers satisfying $l+m > 1$. Further, let $x(s,t) \in C^{(n-1,m-1)}([0,a] \times [0,b])$ be such that $\frac{\partial^\kappa}{\partial s^\kappa} x(s,t)|_{s=0} = 0$, $0 \leq \kappa \leq n-1$; $\frac{\partial^\lambda}{\partial t^\lambda} x(s,t)|_{t=0} = 0$, $0 \leq \lambda \leq m-1$; and $\frac{\partial^n}{\partial s^n} \left(\frac{\partial^{m-1}}{\partial t^{m-1}} x(s,t) \right)$, $\frac{\partial^{n-1}}{\partial s^{n-1}} \left(\frac{\partial^m}{\partial t^m} x(s,t) \right)$ be absolutely continuous on $[0,a] \times [0,b]$. If $\int_0^a \int_0^b \left| \frac{\partial^{n+m}}{\partial s^n \partial t^m} x(s,t) \right|^{l+m} dtds$ exists, then*

$$\int_0^a \int_0^b \left| \frac{\partial^{\kappa+\lambda}}{\partial s^\kappa \partial t^\lambda} x(s,t) \right|^l \left| \frac{\partial^{n+m}}{\partial s^n \partial t^m} x(s,t) \right|^m dtds$$

$$\leq C^*_{n,m,\kappa,\lambda} a^{(n-\kappa)l} b^{(m-\lambda)l} \int_0^a \int_0^b \left| \frac{\partial^{n+m}}{\partial s^n \partial t^m} x(s,t) \right|^{l+m} dtds, \qquad (11)$$

where

$$C^*_{n,m,\kappa,\lambda} = \xi^{1+l\xi} m^{m\xi} \left(\frac{(n-\kappa)(m-\lambda)(1-\xi)^2}{(n-\kappa-1)(m-\lambda-1)} \right)^{l(1-\xi)} \frac{1}{[(n-\kappa)!(m-\lambda)!]^l},$$

and $\xi = \frac{1}{l+m}$.

Proof From the hypotheses, we have for $0 \le \kappa \le n-1, 0 \le \lambda \le m-1$,

$$\left|\frac{\partial^{\kappa+\lambda}}{\partial s^{\kappa}\partial t^{\lambda}}x(s,t)\right| \le \frac{1}{(n-\kappa-1)!(m-\lambda-1)!}$$

$$\times \int_0^s\int_0^t (s-\sigma)^{n-\kappa-1}(t-\tau)^{m-\lambda-1}\left|\frac{\partial^{n+m}}{\partial\sigma^n\partial\tau^m}x(\sigma,\tau)\right| d\tau\, d\sigma.$$

By Hölder's inequality with indices $l+m$ and $\frac{l+m}{l+m-1}$, we have

$$\left|\frac{\partial^{\kappa+\lambda}}{\partial s^{\kappa}\partial t^{\lambda}}x(s,t)\right| \le \frac{1}{(n-\kappa-1)!(m-\lambda-1)!}$$

$$\times\left(\int_0^s\int_0^t [(s-\sigma)^{n-\kappa-1}(t-\tau)^{m-\lambda-1}]^{\frac{l+m}{l+m-1}} d\tau\, d\sigma\right)^{\frac{l+m-1}{l+m}}$$

$$\times\left(\int_0^s\int_0^t \left|\frac{\partial^{n+m}}{\partial\sigma^n\partial\tau^m}x(\sigma,\tau)\right|^{l+m} d\tau\, d\sigma\right)^{\frac{1}{l+m}}$$

$$= Ds^{n-\kappa-\xi}t^{m-\lambda-\xi}\left(\int_0^s\int_0^t \left|\frac{\partial^{n+m}}{\partial\sigma^n\partial\tau^m}x(\sigma,\tau)\right|^{l+m} d\tau\, d\sigma\right)^{\xi},$$

where

$$D = \left(\frac{(1-\xi)^2}{(n-\kappa-\xi)(m-\lambda-\xi)}\right)^{1-\xi}\frac{1}{(n-\kappa-1)!(m-\lambda-1)!}.$$

Multiplying both sides of the above inequality by $\left|\frac{\partial^{n+m}}{\partial s^n\partial t^m}x(s,t)\right|^m$ and integrating over t from 0 to b and then the resulting inequality over s from 0 to a, we obtain

$$\int_0^a\int_0^b \left|\frac{\partial^{\kappa+\lambda}}{\partial s^{\kappa}\partial t^{\lambda}}x(s,t)\right|^l \left|\frac{\partial^{n+m}}{\partial s^n\partial t^m}x(s,t)\right|^m dtds$$

$$\le D^l\int_0^a\int_0^b s^{l(n-\kappa-\xi)}t^{l(m-\lambda-\xi)}\left|\frac{\partial^{n+m}}{\partial s^n\partial t^m}x(s,t)\right|^m$$

$$\times\left(\int_0^s\int_0^t \left|\frac{\partial^{n+m}}{\partial\sigma^n\partial\tau^m}x(\sigma,\tau)\right|^{l+m} d\tau\, d\sigma\right)^{l\xi} dtds.$$

Applying Hölder's inequality with indices $\frac{l+m}{l}$ and $\frac{l+m}{m}$ to the integral on the right hand side, we obtain

$$\int_0^a\int_0^b \left|\frac{\partial^{\kappa+\lambda}}{\partial s^{\kappa}\partial t^{\lambda}}x(s,t)\right|^l \left|\frac{\partial^{n+m}}{\partial s^n\partial t^m}x(s,t)\right|^m dtds$$

$$\leq D^l\left(\int_0^a\int_0^b s^{(n-\kappa-\xi)(l+m)}t^{(m-\lambda-\xi)(l+m)}dtds\right)^{\frac{l}{l+m}}$$

$$\times\left(\int_0^a\int_0^b\left|\frac{\partial^{\kappa+\lambda}}{\partial s^\kappa\partial t^\lambda}x(s,t)\right|^{m+l}\left(\int_0^s\int_0^t\left|\frac{\partial^{n+m}}{\partial\sigma^n\partial\tau^m}x(\sigma,\tau)\right|^{l+m}d\tau\, d\sigma\right)^{\frac{l}{m}}dtds\right)^{\frac{m}{l+m}}$$

$$= D^l\left(\int_0^a\int_0^b s^{(n-\kappa-\xi)(l+m)}t^{(m-\lambda-\xi)(l+m)}dtds\right)^{\frac{l}{l+m}}$$

$$\times\left(\frac{m}{l+m}\int_0^a\int_0^b\frac{\partial^2}{\partial s\partial t}\left\{\left(\int_0^s\int_0^t\left|\frac{\partial^{n+m}}{\partial\sigma^n\partial\tau^m}x(\sigma,\tau))\right|^{l+m}d\tau\, d\sigma\right)^{\frac{l}{m}+1}\right\}dtds\right)^{\frac{m}{l+m}}$$

$$= D^l\left(\frac{\xi^2}{(n-\kappa)(m-\lambda)}\right)^{\xi l}(m\xi)^{m\xi}a^{(n-\kappa)l}b^{(m-\lambda)l}\int_0^a\int_0^b\left|\frac{\partial^{n+m}}{\partial s^n\partial t^m}x(s,t)\right|^{l+m}dtds$$

$$= C^*_{n,m,\kappa,\lambda}a^{(n-\kappa)l}b^{(m-\lambda)l}\int_0^a\int_0^b\left|\frac{\partial^{n+m}}{\partial s^n\partial t^m}x(s,t)\right|^{l+m}dtds,$$

where

$$C^*_{n,m,\kappa,\lambda}=\xi^{1+l\xi}m^{m\xi}\left(\frac{(n-\kappa)(m-\lambda)(1-\xi)^2}{(n-\kappa-1)(m-\lambda-1)}\right)^{l(1-\xi)}\frac{1}{[(n-\kappa)!(m-\lambda)!]^l},\ \xi=\frac{1}{l+m}.$$

This completes the proof.

Remark 5.3 Taking $\kappa=\lambda=0$ in (11), we have

$$\int_0^a\int_0^b|x(s,t)|^l\left|\frac{\partial^{n+m}}{\partial s^n\partial t^m}x(s,t)\right|^m dtds\leq c^*_{n,m}a^{nl}b^{ml}\int_0^a\int_0^b\left|\frac{\partial^{n+m}}{\partial s^n\partial t^m}x(s,t)\right|^{l+m}dtds,$$

where

$$c^*_{n,m}=\xi^{l\xi+1}m^{\xi m}\left(\frac{mn(1-\xi)^2}{(n-\xi)(m-\xi)}\right)^{l(1-\xi)}\frac{1}{(n!m!)^l},\ \xi=\frac{1}{l+m}.$$

In case, $x(s,t)$ reduces to $s(t)$, with suitable modifications, the above inequality becomes

$$\int_0^a|x(t)|^l\left|x^{(n)}(t)\right|^m dt\leq\xi m^{m\xi}\left(\frac{n(1-\xi)}{n-\xi}\right)^{l(1-\xi)}(n!)^{-l}a^{nl}\int_0^a\left|x^{(n)}(t)\right|^{l+m}dt,$$

which is an inequality given by Das in [16]. In the special case where $n=1$, we have

$$\int_0^a|x(t)|^l|x'(t)|^m dt\leq\frac{m^{m/(l+m)}}{l+m}a^l\int_0^a|x'(t)|^{m+l}dt. \tag{12}$$

For $m,l\geq 1$, Yang [33] established the following inequality:

$$\int_0^a|x(t)|^l|x'(t)|^m dt\leq\frac{m}{l+m}a^l\int_0^a|x'(t)|^{m+l}dt.$$

Obviously, for $m,l\geq 1$, (12) is sharper than the above inequality.

Remark 5.4 Taking $n = m = 1$ and $\kappa = \lambda = 0$ in (11), we have

$$\int_0^a \int_0^b |x(s,t)|^l \left| \frac{\partial^2}{\partial s \partial t} x(s,t) \right|^m dtds \leq c_{1,1}^*(ab)^l \int_0^a \int_0^b \left| \frac{\partial^2}{\partial s \partial t} x(s,t) \right|^{m+l} dtds.$$

In case, $x(s,t)$ reduces to $s(t)$, with suitable modifications, the above inequality becomes

$$\int_0^a |x(t)|^l |x'(t)|^m dt \leq \xi m^{m\xi} a^l \int_0^a |x'(t)|^{m+l} dt, \quad \xi = (l+m)^{-1},$$

which is an inequality established by Yang in [33].

6 Opial-Agarwal-Sheng Type Inequalities

In the present section, we shall use an extension method of Agarwal [4] to establish some new estimates on Opial–Agarwal–Sheng type inequalities involving $2n$ independent variables.

Let $E = \prod_{i=1}^n [a_i, b_i]$, $E' = \prod_{i=1}^n [c_i, d_i]$ be bounded rectangles in R^n and let $E \times E'$ be the Cartesian product of E and E'. For $x_i, y_i \in R$, $i = 1, \ldots, n, (x, y) = (x_1, \ldots, x_n, y_1, \ldots, y_n)$ is a variable point in $E \times E'$ and $dxdy = dx_1 \cdots dx_n dy_1 \cdots dy_n$. For any continuous real-valued function $u(x, y)$ defined on $E \times E'$, we denote by $\int_E \int_{E'} u(x, y) dydx$ the $2n$-fold integral

$$\int_{a_1}^{b_1} \cdots \int_{a_n}^{b_n} \int_{c_1}^{d_1} \cdots \int_{c_n}^{d_n} u(x_1, \ldots, x_n, y_1, \ldots, y_n) dy_n \cdots dy_1 dx_n \cdots dx_1$$

and for any $(x, y) \in E \times E'$, $\int_{E(x)} \int_{E'(x)} u(s,t) dtds$ is the $2n$-fold integral

$$\int_{a_1}^{x_1} \cdots \int_{a_n}^{x_n} \int_{c_1}^{y_1} \cdots \int_{c_n}^{y_n} u(s_1, \ldots, s_n, t_1, \ldots, t_n) dt_n \cdots dt_1 ds_n \cdots ds_1.$$

We represent by $F(E \times E')$ the class of continuous functions $u(x, y) : E \times E' \to \mathbb{R}$, and by $G(E \times E')$ the subclass of functions in $F(E \times E')$ which satisfy

$$u(x, y)|_{x_i = a_i} = 0, u(x, y)|_{y_i = c_i} = 0, u(x, y)|_{x_i = b_i} = 0, u(x, y)|_{y_i = d_i} = 0$$

for all $1 \leq i \leq n$.

Theorem 6.1 [48] *Let $l, \mu, \lambda \geq 1$, be real numbers such that $\frac{1}{\mu} + \frac{1}{\lambda} = 1$, and $u(x, y) \in G(E \times E')$. Then, the following inequality holds*

$$\int_E \int_{E'} |u(x, y)|^l dydx \leq \frac{1}{2n} \left(\sum_{i=1}^n [(b_i - a_i)(c_i - d_i)]^\mu \right)^{1/\mu}$$
$$\times \left(\int_E \int_{E'} |u(x, y)|^{(l-1)\mu} dydx \right)^{1/\mu}$$

$$\times \left(\int_E \int_{E'} \|\text{grad } u(x, y)\|_\lambda^\lambda dy dx \right)^{1/\lambda}, \tag{13}$$

where

$$\|\text{grad } u(x, y)\|_\lambda = \left(\sum_{i=1}^n \left| \frac{\partial^2}{\partial x_i \partial y_i} u(x, y) \right|^\lambda \right)^{1/\lambda}.$$

Proof For each fixed i, $1 \le i \le n$, in view of

$$u(x, y)|_{x_i=a_i} = 0, u(x, y)|_{y_i=c_i} = 0, u(x, y)|_{x_i=b_i} = 0, u(x, y)|_{y_i=d_i} = 0,$$

we have

$$u^l(x, y) = u^{l-1}(x, y) \int_{a_i}^{x_i} \int_{c_i}^{y_i} \frac{\partial^2}{\partial s_i \partial t_i} u(x, y; s_i, t_i) dt_i ds_i$$

and

$$u^l(x, y) = u^{l-1}(x, y) \int_{x_i}^{b_i} \int_{y_i}^{d_i} \frac{\partial^2}{\partial s_i \partial t_i} u(x, y; s_i, t_i) dt_i ds_i$$

where

$$u(x, y; s_i, t_i) = u(x_1, \ldots, x_{i-1}, s_i, x_{i+1}, \ldots, x_n, y_1, \ldots, y_{i-1}, t_i, y_{i+1}, \ldots, y_n).$$

In view of the arithmetic-geometric means inequality and Hölder's inequality with indices μ and λ, it follows that

$$|u(x, y)|^l \le \frac{1}{2} |u(x, y)|^{l-1} \int_{a_i}^{b_i} \int_{c_i}^{d_i} \left| \frac{\partial^2}{\partial s_i \partial t_i} u(x, y; s_i, t_i) \right| dt_i ds_i$$

$$\le \frac{1}{2} |u(x, y)|^{l-1} [(b_i - a_i)(c_i - d_i)]^{1/\mu} \left(\int_{a_i}^{x_i} \int_{c_i}^{y_i} \left| \frac{\partial^2}{\partial s_i \partial t_i} u(x, y; s_i, t_i) \right|^\lambda dt_i ds_i \right)^{1/\lambda}.$$

Summing up the above inequalities for $1 \le i \le n$, integrating over $E \times E'$, and applying Hölder's inequality with indices μ and λ two times, we get

$$\int_E \int_{E'} |u(x, y)|^l dy dx \le \frac{1}{2n} \sum_{i=1}^n [(b_i - a_i)(c_i - d_i)]^{1/\mu}$$

$$\times \int_E \int_{E'} |u(x, y)|^{l-1} \left(\int_{a_i}^{b_i} \int_{c_i}^{d_i} \left| \frac{\partial^2}{\partial s_i \partial t_i} u(x, y; s_i, t_i) \right|^\lambda dt_i ds_i \right)^{1/\lambda} dy dx$$

$$\le \frac{1}{2n} \left(\int_E \int_{E'} |u(x, y)|^{(l-1)\mu} dy dx \right)^{1/\mu} \sum_{i=1}^n [(b_i - a_i)(c_i - d_i)]^{1/\mu}$$

$$\times\left(\int_E\int_{E'}\int_{a_i}^{b_i}\int_{c_i}^{d_i}\left|\frac{\partial^2}{\partial s_i\partial t_i}u(x,y;s_i,t_i)\right|^{\lambda}dt_ids_idydx\right)^{1/\lambda}$$

$$\leq\frac{1}{2n}\left(\int_E\int_{E'}|u(x,y)|^{(l-1)\mu}dydx\right)^{1/\mu}\sum_{i=1}^{n}[(b_i-a_i)(c_i-d_i)]^{1/\mu+1/\lambda}$$

$$\times\left(\int_E\int_{E'}\left|\frac{\partial^2}{\partial x_i\partial y_i}u(x,y)\right|^{\lambda}dydx\right)^{1/\lambda}$$

$$\leq\frac{1}{2n}\left(\int_E\int_{E'}|u(x,y)|^{(l-1)\mu}dydx\right)^{1/\mu}\left(\sum_{i=1}^{n}[(b_i-a_i)(c_i-d_i)]^{\mu}\right)^{1/\mu}$$

$$\times\left(\int_E\int_{E'}\|\text{grad }u(x,y)\|_{\lambda}^{\lambda}dydx\right)^{1/\lambda},$$

where

$$\|\text{grad }u(x,y)\|_{\lambda}=\left(\sum_{i=1}^{n}\left|\frac{\partial^2}{\partial x_i\partial y_i}u(x,y)\right|^{\lambda}\right)^{1/\lambda}.$$

The proof is complete.

Remark 6.1 In case, $u(x,y)$ reduces to $u(x)$ in (13), with suitable modifications, we have

$$\int_E|u(x)|^{(l)\mu}dx\leq\frac{1}{2n}\left(\int_E|u(x)|^{(l-1)\mu}dx\right)^{1/\mu}\left(\sum_{i=1}^{n}(b_i-a_i)^{\mu}\right)^{1/\mu}$$

$$\times\left(\int_E\|\text{grad }u(x)\|_{\lambda}^{\lambda}dx\right)^{1/\lambda},$$

where

$$\|\text{grad }u(x)\|_{\lambda}=\left(\sum_{i=1}^{n}\left|\frac{\partial}{\partial x_i}u(x)\right|^{\lambda}\right)^{1/\lambda}.$$

This is an important inequality established by Agarwal and Sheng in [4].

Remark 6.2 For any real numbers $l_k\geq\frac{1}{r}$, $1\leq k\leq r$, the arithmetic–geometric means inequality together with (13) gives

$$\int_E\int_{E'}\prod_{k=1}^{r}|u_k(x,y)|^{l_k}dydx\leq\frac{1}{r}\sum_{k=1}^{r}\int_E\int_{E'}|u_k(x,y)|^{rl_k}dydx$$

$$\leq\frac{1}{2nr}\left(\sum_{i=1}^{n}[(b_i-a_i)(c_i-d_i)]^{\mu}\right)^{1/\mu}\sum_{k=1}^{r}\left(\int_E\int_{E'}|u_k(x,y)|^{(rl_k-1)\mu}dydx\right)^{1/\mu}$$

$$\times\left(\int_E\int_{E'}\|\text{grad } u_k(x,y)\|_\lambda^\lambda dydx\right)^{1/\lambda}. \tag{14}$$

This is a general form of the following result which was given by Agarwal and Sheng in [4]:

$$\int_E\prod_{k=1}^r|u_k(x)|^{l_k}dx \le \frac{1}{2nr}\left(\sum_{i=1}^n(b_i-a_i)^\mu\right)^{1/\mu}\sum_{k=1}^r\left(\int_E|u_k(x)|^{(rl_k-1)\mu}dx\right)^{1/\mu}$$
$$\times\left(\int_E\|\text{grad } u_k(x)\|_\lambda^\lambda dx\right)^{1/\lambda},$$

where

$$\|\text{grad } u_k(x)\|_\lambda = \left(\sum_{i=1}^n\left|\frac{\partial}{\partial x_i}u(x)\right|^\lambda\right)^{1/\lambda}.$$

Remark 6.3 In particular, for $l_k = (p_k+2)/(2r)$, $p_k \ge 1, 1\le k\le r, \mu=\lambda=2$, inequality (14) reduces to

$$\int_E\int_{E'}\left(\prod_{k=1}^r|u_k(x,y)|^{(p_k+2)/2}\right)^{1/r}dydx$$
$$\le\frac{1}{2nr}\left(\sum_{i=1}^n[(b_i-a_i)(c_i-d_i)]^2\right)^{1/2}\sum_{k=1}^r\left(\int_E\int_{E'}|u(x,y)|^{p_k}dydx\right)^{1/2}$$
$$\times\left(\int_E\int_{E'}\|\text{grad } u_k(x,y)\|_2^2dydx\right)^{1/2},$$

which is just a general form of the following result given by Agarwal and Sheng in [4]:

$$\int_E\left(\prod_{k=1}^r|u_k(x)|^{(p_k+2)/2}\right)^{1/r}dx$$
$$\le\frac{1}{2nr}\left(\sum_{i=1}^n(b_i-a_i)^2\right)^{1/2}\sum_{k=1}^r\left(\int_E|u(x)|^{p_k}dx\right)^{1/2}\left(\int_E\|\text{grad } u_k(x)\|_2^2dx\right)^{1/2}.$$

On the other hand, the above inequality with the right-hand side multiplied by $\left(\prod_{k=1}^r((p_k+2)/2)\right)^{1/r}$ and the term $\left(\sum_{i=1}^n(b_i-a_i)^2\right)^{1/2}$ being replaced by $\sqrt{n}\beta$ has been proved by Pachpatte.

Remark 6.4 In case, $u(x,y)$ reduces to $u(x)$, then inequality (14) and its particular case $l\ge 2$, $\mu=\lambda=2$ with the right-hand side multiplied by l have also been proved by Pachpatte independently.

Theorem 6.2 [48] *Let* $\lambda \geq 1$ *and* $u(x, y) \in G(E \times E')$. *Then*

$$\int_E \int_{E'} |u(x,y)|^{2\lambda} dydx \leq \frac{\pi\lambda^2\beta^2\alpha^2}{128n} \left(\int_E \int_{E'} |u(x,y)|^{2\lambda} dydx \right)^{(\lambda-1)/\lambda}$$

$$\times \left(\int_E \int_{E'} \sum_{i=1}^{n} \left| \frac{\partial^2 u}{\partial s_i \partial t_i} + (\lambda - 1) \frac{1}{u(x,y)} \frac{\partial u}{\partial x_i} \frac{\partial u}{\partial y_i} \right|^{2\lambda} dydx \right)^{1/\lambda}, \quad (15)$$

where $\beta = \max_{1 \leq i \leq n}(b_i - a_i)$ *and* $\alpha = \max_{1 \leq i \leq n}(d_i - c_i)$.

Proof For each fixed $1 \leq i \leq n$, we have

$$u^\lambda(x,y) = \lambda \int_{a_i}^{x_i} \int_{c_i}^{y_i} \left[u^{\lambda-1}(x,y;s_i,t_i) \frac{\partial^2 u}{\partial s_i \partial t_i} + (\lambda-1) u^{\lambda-2}(x,y;s_i,t_i) \frac{\partial u}{\partial s_i} \frac{\partial u}{\partial t_i} \right] dt_i ds_i$$

and hence from the Cauchy–Schwarz inequality,

$$|u(x,y)|^\lambda \leq \lambda^2 (x_i - a_i)(y_i - c_i)$$
$$\times \int_{a_i}^{x_i} \int_{c_i}^{y_i} \left| u^{\lambda-1}(x,y;s_i,t_i) \frac{\partial^2 u}{\partial s_i \partial t_i} + (\lambda-1) u^{\lambda-2}(x,y;s_i,t_i) \frac{\partial u}{\partial s_i} \frac{\partial u}{\partial t_i} \right|^2 dt_i ds_i.$$

Similarly,

$$|u(x,y)|^\lambda \leq \lambda^2 (b_i - x_i)(d_i - y_i)$$
$$\times \int_{x_i}^{b_i} \int_{y_i}^{d_i} \left| u^{\lambda-1}(x,y;s_i,t_i) \frac{\partial^2 u}{\partial s_i \partial t_i} + (\lambda-1) u^{\lambda-2}(x,y;s_i,t_i) \frac{\partial u}{\partial s_i} \frac{\partial u}{\partial t_i} \right|^2 dt_i ds_i.$$

Multiplying the above two inequalities, summing up the resulting inequalities for $1 \leq i \leq n$, and integrating over $E \times E'$, we obtain

$$\int_E \int_{E'} |u(x,y)|^{2\lambda} dydx \leq \frac{\lambda^2}{2n} \int_E \int_{E'} \left\{ \sum_{i=1}^{n} [(x_i - a_i)(y_i - c_i)(b_i - x_i)(d_i - y_i)]^{1/2} \right.$$
$$\left. \times \int_{a_i}^{b_i} \int_{c_i}^{d_i} \left| u^{\lambda-1}(x,y;s_i,t_i) \frac{\partial^2 u}{\partial s_i \partial t_i} + (\lambda-1) u^{\lambda-2}(x,y;s_i,t_i) \frac{\partial u}{\partial s_i} \frac{\partial u}{\partial t_i} \right|^2 dt_i ds_i \right\} dydx$$
$$= \frac{\lambda^2}{2n} \sum_{i=1}^{n} \int_{a_i}^{b_i} \int_{c_i}^{d_i} [(x_i - a_i)(y_i - c_i)(b_i - x_i)(d_i - y_i)]^{1/2} dy_i dx_i$$
$$\times \int_E \int_{E'} \left| u^{\lambda-1}(x,y) \frac{\partial^2 u}{\partial s_i \partial t_i} + (\lambda-1) u^{\lambda-2}(x,y) \frac{\partial u}{\partial x_i} \frac{\partial u}{\partial y_i} \right|^2 dydx$$
$$\leq \frac{\pi\lambda^2\beta^2\alpha^2}{128n} \int_E \int_{E'} \sum_{i=1}^{n} \left| u^{\lambda-1}(x,y) \frac{\partial^2 u}{\partial s_i \partial t_i} + (\lambda-1) u^{\lambda-2}(x,y) \frac{\partial u}{\partial x_i} \frac{\partial u}{\partial y_i} \right|^2 dydx,$$

where $\beta = \max_{1\le i\le n}(b_i - a_i)$ and $\alpha = \max_{1\le i\le n}(d_i - c_i)$.

Hence, using Hölder's inequality with indices λ and $\lambda/(\lambda-1)$ on right-hand side of the above inequality, we have

$$\int_E\int_{E'} |u(x,y)|^{2\lambda}dydx \le \frac{\pi\lambda^2\beta^2\alpha^2}{128n}\left(\int_E\int_{E'} |u(x,y)|^{2\lambda}dydx\right)^{(\lambda-1)/\lambda}$$
$$\times\left(\int_E\int_{E'}\sum_{i=1}^n\left|\frac{\partial^2 u}{\partial s_i\partial t_i} + (\lambda-1)\frac{1}{u(x,y)}\frac{\partial u}{\partial x_i}\frac{\partial u}{\partial y_i}\right|^{2\lambda} dydx\right)^{1/\lambda}.$$

The proof is complete.

Remark 6.5 In case, $u(x,y)$ reduces to $u(x)$, with suitable modifications, (15) becomes the following Agarwal and Sheng inequality in [4].

$$\int_E |u(x)|^{2\lambda}dx \le \frac{\pi\lambda^2\beta^2}{16n}\left(\int_E |u(x)|^{2\lambda}dx\right)^{(\lambda-1)/\lambda}\left(\int_E \|\text{grad } u(x)\|_2^{2\lambda}dx\right)^{1/\lambda},$$

where $\beta = \max_{1\le i\le n}(b_i - a_i)$.

Theorem 6.3 [48] *Let $l \ge 0, m \ge 1$ be real numbers, and $u(x,y) \in G(E\times E')$. Then*

$$\int_E\int_{E'} |u(x,y)|^{l+m}dydx \le \frac{1}{n}\left(\frac{m+l}{2m}\right)^m\sum_{i=1}^n[(b_i-a_i)(d_i-c_i)]^m$$

$$\times\int_E\int_{E'}\left|u^{l/m}(x,y)\frac{\partial^2 u}{\partial x_i\partial y_i} + \frac{l}{m}u^{(l/m-1)}(x,y)\frac{\partial u}{\partial x_i}\frac{\partial u}{\partial y_i}\right|^m dydx. \tag{16}$$

Proof For fixed $1\le i\le n$, we obtain

$$u^{l+m}(x,y) = \frac{m+l}{m}[u(x,y)]^{(m-1)(l+m)/m}$$
$$\times\int_{a_i}^{x_i}\int_{c_i}^{y_i}\left[u^{l/m}(x,y;s_i,t_i)\frac{\partial^2 u}{\partial s_i\partial t_i} + \frac{l}{m}u^{(l/m-1)}(x,y;s_i,t_i)\frac{\partial u}{\partial s_i}\frac{\partial u}{\partial t_i}\right]dt_ids_i,$$

and hence it follows that

$$|u(x,y)|^{l+m} \le \frac{m+l}{m}|u(x,y)|^{(m-1)(l+m)/m}$$
$$\times\int_{a_i}^{x_i}\int_{c_i}^{y_i}\left|u^{l/m}(x,y;s_i,t_i)\frac{\partial^2 u}{\partial s_i\partial t_i} + \frac{l}{m}u^{(l/m-1)}(x,y;s_i,t_i)\frac{\partial u}{\partial s_i}\frac{\partial u}{\partial t_i}\right|dt_ids_i.$$

Similarly,

$$|u(x,y)|^{l+m} \le \frac{m+l}{m}|u(x,y)|^{(m-1)(l+m)/m}$$

$$\times \int_{x_i}^{b_i}\int_{y_i}^{d_i}\left|u^{l/m}(x,y;s_i,t_i)\frac{\partial^2 u}{\partial s_i \partial t_i}+\frac{l}{m}u^{(l/m-1)}(x,y;s_i,t_i)\frac{\partial u}{\partial s_i}\frac{\partial u}{\partial t_i}\right| dt_i ds_i.$$

Adding the above two inequalities and integrating the resulting inequality from c_i to d_i and then from a_i to b_i, we have

$$\int_{a_i}^{b_i}\int_{c_i}^{d_i}|u(x,y)|^{l+m}dy_i dx_i \le \frac{m+l}{2m}\left(\int_{a_i}^{b_i}\int_{c_i}^{d_i}|u(x,y)|^{(m-1)(l+m)/m}dy_i dx_i\right)$$
$$\times \int_{a_i}^{b_i}\int_{c_i}^{d_i}\left|u^{l/m}(x,y)\frac{\partial^2 u}{\partial x_i \partial y_i}+\frac{l}{m}u^{(l/m-1)}(x,y)\frac{\partial u}{\partial x_i}\frac{\partial u}{\partial y_i}\right| dy_i dx_i.$$

Next, in each integral on the right-hand side of the above inequality, we apply Hölder's inequality with indices m and $m/(m-1)$ and get

$$\int_{a_i}^{b_i}\int_{c_i}^{d_i}|u(x,y)|^{l+m}dy_i dx_i \le \frac{m+l}{2m}\left(\int_{a_i}^{b_i}\int_{c_i}^{d_i}|u(x,y)|^{l+m}dy_i dx_i\right)^{(m-1)/m}$$
$$\times[(b_i-a_i)(d_i-c_i)]^{1/m}[(b_i-a_i)(d_i-c_i)]^{(m-1)/m}$$
$$\times\left(\int_{a_i}^{b_i}\int_{c_i}^{d_i}\left|u^{l/m}(x,y)\frac{\partial^2 u}{\partial x_i \partial y_i}+\frac{l}{m}u^{(l/m-1)}(x,y)\frac{\partial u}{\partial x_i}\frac{\partial u}{\partial y_i}\right|^m dy_i dx_i\right)^{1/m},$$

which is, unless $\int_{a_i}^{b_i}\int_{c_i}^{d_i}|u(x,y)|^{l+m}dy_i dx_i = 0$ (in which case inequality (16) is obvious), the same as

$$\left(\int_{a_i}^{b_i}\int_{c_i}^{d_i}|u(x,y)|^{l+m}dy_i dx_i\right)^{1/m} \le \frac{m+l}{2m}[(b_i-a_i)(d_i-c_i)]$$
$$\times\left(\int_{a_i}^{b_i}\int_{c_i}^{d_i}\left|u^{l/m}(x,y)\frac{\partial^2 u}{\partial x_i \partial y_i}+\frac{l}{m}u^{(l/m-1)}(x,y)\frac{\partial u}{\partial x_i}\frac{\partial u}{\partial y_i}\right|^m dy_i dx_i\right)^{1/m}.$$

Finally, raising to the mth power on both sides of the above inequality, and then by induction, we arrive easily at the desired inequality (16).

Remark 6.6 In case, $u(x,y)$ reduces to $u(x)$, with suitable modifications, (16) becomes the following Agarwal and Sheng inequality in [4]:

$$\int_E |u(x)|^{l+m}dx \le \frac{1}{n}\left(\frac{m+l}{2m}\right)^m \sum_{i=1}^n (b_i-a_i)^m \int_E |u(x)|^l \left|\frac{\partial}{\partial x_i}u(x)\right|^m dx.$$

Acknowledgements C.-J. Zhao's research was supported by the National Natural Sciences Foundation of China (11371334) and the Zhejiang Provincial Natural Science Foundation of China (Y13A010019). W.S. Cheung's research was supported by a HKU Seed Grant for Basic Research.

References

1. Agarwal, R.P.: Sharp Opial-type inequalities involving r-derivatives and their applications. Tohoku Math. J. **47**(4), 567–593 (1995)
2. Agarwal, R.P., Lakshmikantham, V.: Uniqueness and Nonuniqueness Criteria for Ordinary Differential Equations. World scientific, Singapore (1993)
3. Agarwal, R.P., Pang, P.Y.H.: Opial Inequalities with Applications in Differential and Difference Equations. Kluwer, Dordrecht (1995)
4. Agarwal, R.P., Sheng, Q.: Sharp integral inequalities in n independent varibles. Nonlinear Anal. Theory Meth. Appl. **26**(2), 179–210 (1996)
5. Agarwal, R.P., Thandapani, E.: On some new integrodifferential inequalities. Anal. sti. Univ. Al. I. Cuza din Iasi. **28**, 123–126 (1982)
6. Alzer, H.: An Opial-type inequality involving higher-order derivatives of two functions. Appl. Math. Lett. **10**(4), 123–128 (1997)
7. Bainov, D., Simeonov, P.: Integral Inequalities and Applications. Kluwer, Dordrecht (1992)
8. Beesack, P.R.: On an integral inequality of Z. Opial. Trans. Am. Math. Soc. **104**, 470–475 (1997)
9. Cheung, W.S.: On Opial-type inequalities in two variables. Aequ. Math. **38**, 236–244 (1989)
10. Cheung, W.S.: Some new Opial-type inequalities. Mathematika. **37**, 136–142 (1990)
11. Cheung, W.S.: Some generalized Opial-type inequalities. J. Math. Anal. Appl. **162**, 317–321 (1991)
12. Cheung, W.S.: Opial-type inequalities with m functions in n variables. Mathematika. **39**, 319–326 (1992)
13. Cheung, W.S., Zhao, C.J.: On Opial-type integral inequalities. J. Ineq. Appl. **2007** Article ID(38347), 15 pages (2007)
14. Cheung, W.S., Zhao, C.J.: On a discrete Opial-type integral inequality. J. Ineq. Pure Appl. Math. **8**, 1–9 (2007)
15. Cheung, W.S., Zhao, D.D., Pečarić J. E.: Opial-type inequalities for differential operators. Nonlinear Anal. **66**(9), 2028–2039 (2007).
16. Das, K.M.: An inequality similar to Opial's inequality. Proc. Am. Math. Soc. **22**, 258–261 (1969)
17. Godunova, E.K., Levin, V.I.: On an inequality of Maroni. Mat. Zametki. **2**, 221–224 (1967)
18. Hua, L.K.: On an inequality of Opial. Sci. Sin. **14**, 789–790 (1965)
19. Karpuz, B., Kaymakcalan, B., Özkan, U.M.: Some multi-dimenstonal Opial-type inequalities on time scales. J. Math. Ineq. **4**(2), 207–216 (2010)
20. Li, J.D.: Opial-type integral inequalities involving several higher order derivatives. J. Math. Anal. Appl. **167**, 98–100 (1992)
21. Mitrinovič, D.S.: Analytic Inequalities. Springer-Verlag, Berlin (1970)
22. Mitrinovič, D.S., Pečarić, J.E., Fink, A.M.: Inequalities Involving Functions and Their Integrals and Derivatives. Kluwer, Dordrecht (1991)
23. Opial, Z.: Sur une inégalité. Ann. Polon. Math. **8**, 29–32 (1960)
24. Pachpatte, B.G.: On some new generalizations of Opial inequality. Demonstratio Math. **19**, 281–291 (1986)
25. Pachpatte, B.G.: On Opial-type intergral inequalities. J. Math. Anal. Appl. **120**, 547–556 (1986)
26. Pachpatte, B.G.: On integral inequalities similar to Opial's inequality. Demonstratio Math. **22**, 21–27 (1989)
27. Pachpatte, B.G.: A note on generalized Opial type inequalities. Tamkang J. Math. **24**, 229–235 (1993)
28. Pečarić, J.E.: An integral inequality In: Srivastava, H. M., Rassias, T.M. (eds.) Analysis, Geometry, and Groups: A Riemann Legacy Volume Part II, pp. 472–478, Hadronic Press, Palm Harbor (1993)

29. Pečarić, J.E., Brnetić, I.: Note on generalization of Godunova-Levin-Opial inequality. Demonstratio Math. **30**, 545–549 (1997)
30. Pečarić, J.E., Brnetić, I.: Note on the generalization of Godunova-Levin-Opial inequality in several independent variables. J. Math. Anal. Appl. **215**, 274–282 (1997)
31. Rozanova, G.I.: Integral inequalities with derivatives and with arbitrary convex functions. Moskov. Gos. Ped. Inst. Vcen. Zap. **460**, 58–65 (1972)
32. Willett, D.: The existence-uniqueness theorem for an n-th order linear ordinary differential equation. Am. Math. Month. **75**, 174–178 (1968)
33. Yang, G.S.: On a certain result of Z. Opial. Proc. Jpn. Acad. **42**, 78–83 (1966)
34. Yang, G.S.: Inequality of Opial-type in two variables. Tamkang J. Math. **13**, 255–259 (1982)
35. Zhao, C.J., Bencze, M.: On Agarwal-Pang type integral inequalities. Ukr. Math. J. **64**(2), 225–237 (2012)
36. Zhao, C.J., Cheung, W.S.: Sharp integral inequalities involving high-order partial derivatives. J. Ineq. Appl. **2008**, 10 (2008). (Article ID 571417)
37. Zhao, C.J., Cheung, W.S.: On inverse Hilbert-type inequalities. J. Ineq. Appl. **2008**, 6 (2008). (Article ID 693248,)
38. Zhao, C.J., Cheung, W.S.: On multivariate Grüss inequalities. J. Ineq. Appl. **2008**, 6 (2008). (Article ID 249438)
39. Zhao, C.J., Cheung, W.S.: On Ostrowski-type inequalities for higher-order partial derivatives. J. Ineq. Appl. **2010**, 8 (2010). Article ID 960672
40. Zhao, C.J., Cheung, W.S.: On some new Opial type inequalities. J. Ineq. Appl. **2011**, 7 (2011). doi:10.1186/1029-242X-2011-7
41. Zhao, C.J., Cheung, W.S.: On improvements of the Rozanova's inequality. J. Ineq. Appl. **2011**, 33 (2011)
42. Zhao, C.J., Cheung, W.S.: On Opial inequalities involving higher order derivatives. Bull. Korean Math. Soc. **49**(6), 1263–1274 (2012)
43. Zhao, C.J., Cheung, W.S.: On Opial-type inequalities with higher order derivatives. Appl. Math. Lett. **25**(12), 2156–2161 (2012)
44. Zhao, C.J., Cheung, W.S.: On Opial's type inequalities for an integral operator with homogeneous kernel. J. Ineq. Appl. **2012**, 123 (2012)
45. Zhao, C.J., Cheung, W.S.: On Hilbert type inequalities. J. Ineq. Appl. **2012**, 145 (2012)
46. Zhao, C.J., Cheung, W.S.: On Opial-type integral inequalities and applications. Math. Ineq. Appl. 2013, **16**(4), in press
47. Zhao, C.J., Cheung, W.S.: On Opial-Dan's type inequalities. Bull. Malaysian Math. Soc. 2013, **36**(4), in press
48. Zhao, C.J., Cheung, W.S., Bencze M.: Some sharp integral inequalities involving partial derivatives. J. Ineq. Appl. **2012**, 109 (2012)

Index

T. M. Rassias, P. M. Pardalos (eds.), *Mathematics Without Boundaries*,
DOI 10.1007/978-1-4939-1106-6,

MIX
Papier aus verantwortungsvollen Quellen
Paper from responsible sources
FSC® C105338

If you have any concerns about our products,
you can contact us on
ProductSafety@springernature.com

In case Publisher is established outside the EU,
the EU authorized representative is:
Springer Nature Customer Service Center GmbH
Europaplatz 3, 69115 Heidelberg, Germany

Printed by Libri Plureos GmbH
in Hamburg, Germany